Intermediate Algebra

Julie Miller
Molly O'Neill
Nancy Hyde

WITH CONTRIBUTIONS BY MITCHEL LEVY

 **Higher Education**

Boston   Burr Ridge, IL   Dubuque, IA   Madison, WI   New York   San Francisco   St. Louis
Bangkok   Bogotá   Caracas   Kuala Lumpur   Lisbon   London   Madrid   Mexico City
Milan   Montreal   New Delhi   Santiago   Seoul   Singapore   Sydney   Taipei   Toronto

INTERMEDIATE ALGEBRA

Published by McGraw-Hill, a business unit of The McGraw-Hill Companies, Inc., 1221 Avenue of the Americas, New York, NY 10020. Copyright © 2007 by The McGraw-Hill Companies, Inc. All rights reserved. No part of this publication may be reproduced or distributed in any form or by any means, or stored in a database or retrieval system, without the prior written consent of The McGraw-Hill Companies, Inc., including, but not limited to, in any network or other electronic storage or transmission, or broadcast for distance learning.

Some ancillaries, including electronic and print components, may not be available to customers outside the United States.

This book is printed on acid-free paper.

3 4 5 6 7 8 9 0 VNH/VNH 0 9 8 7

ISBN-13 978-0-07-302326-7—soft cover
ISBN-10 0-07-302326-4

ISBN-13 978-0-07-327410-2—hardcover
ISBN-10 0-07-327410-0

ISBN-13 978-0-07-302327-4 (Annotated Instructor's Edition)—soft cover
ISBN-10 0-07-302327-2

ISBN-13 978-0-07-327409-6 (Annotated Instructor's Edition)—hardcover
ISBN-10 0-07-327409-7

Publisher: *Elizabeth J. Haefele*
Sponsoring Editor: *David Millage*
Senior Developmental Editor: *Erin Brown*
Marketing Manager: *Barbara Owca*
Project Manager: *Jodi Rhomberg*
Senior Production Supervisor: *Sherry L. Kane*
Lead Media Project Manager: *Stacy A. Patch*
Media Technology Producer: *Amber M. Huebner*

Designer: *Laurie B. Janssen*
(USE) Cover Image: © *Frank Krahmer/zefa/Corbis*
Lead Photo Research Coordinator: *Carrie K. Burger*
Photo Research: *Pam Carley/Sound Reach*
Supplement Producer: *Melissa M. Leick*
Compositor: *TechBooks/GTS Companies, York, PA*
Typeface: *10/12 TimesTen Roman*
Printer: *Von Hoffmann Corporation*

Photo Credits

Preface: Julie Miller photo by Marc Campbell; Molly O'Neill photo by Gail Beckwith; Chapter 1 Opener: © Vol. 28/Corbis; p. 62: © PhotoDisc website; p. 67: © Corbis website; Chapter 2 Opener: © Bob Daemrich/ PhotoEdit; p. 166, top: © PhotoDisc website; p. 166, bottom: © Vol. 107/Corbis; Chapter 3 Opener: © Jeff Greenberg/PhotoEdit; p. 226: © PhotoDisc website; Chapter 4 Opener: © The McGraw-Hill Companies, Inc./Jill Braaten, photographer; p. 287: Courtesy NOAA; p. 299: © Vol. 59/Corbis; Chapter 5 Opener: © John Neubauer/PhotoEdit; p. 313: © Corbis website; p. 382: © LouAnn Wilson; p. 395: © Tom McCarthy/PhotoEdit; Chapter 6 Opener: © Dennis MacDonald/PhotoEdit; p. 447: © Corbis website; p. 450: © Douglas Peebles/ Corbis; p. 452: © Vol. 20/Corbis; Chapter 7 Opener: © Vol. 1/PhotoDisc/Getty; p. 517: © Tony Freeman/ PhotoEdit; p. 522: © Vol. 145/Corbis; p. 527: © Nathan Bemm/Corbis; Chapter 8 Opener: © Vol. 16/PhotoDisc/ Getty; p. 563: © Corbis website; p. 610: © Mary Kate Denny/PhotoEdit; Chapter 9 Opener: © Vol. 29/PhotoDisc/ Getty; p. 635: Quest/Science Photo Library/Photo Researchers, Inc.; p. 666: © The McGraw-Hill Companies, Inc./John Thoeming, photographer; p. 691: © Corbis website; Chapter 10 Opener: © Vol. 34/PhotoDisc/Getty; p. 721: © Corbis; p. 741: © PhotoDisc

Library of Congress Cataloging-in-Publication Data

Miller, Julie, 1962–
   Intermediate algebra / Julie Miller, Molly O'Neill, Nancy Hyde. — 1st ed.
      p.   cm.
   Includes index.
   ISBN 978-0-07-302326-7 — 0-07-302326-4 (acid-free paper)
   1. Algebra—Textbooks. I. O'Neill, Molly, 1953–. II. Hyde, Nancy. III. Title.

QA154.3.M554   2007
512.9—dc22   2005058431
                                                        CIP

www.mhhe.com

Contents

Preface — x

1 Review of Basic Algebraic Concepts — 1

	Preview	2
1.1	Sets of Numbers and Interval Notation	3
1.2	Operations on Real Numbers	15
1.3	Simplifying Expressions	31
1.4	Linear Equations in One Variable	39
	Midchapter Review	51
1.5	Applications of Linear Equations in One Variable	52
1.6	Literal Equations and Applications to Geometry	64
1.7	Linear Inequalities in One Variable	73
1.8	Properties of Integer Exponents and Scientific Notation	84
	Summary	94
	Review Exercises	101
	Test	105

2 Linear Equations in Two Variables — 107

	Preview	108
2.1	The Rectangular Coordinate System and Midpoint Formula	109
2.2	Linear Equations in Two Variables	118
2.3	Slope of a Line	132
	Midchapter Review	144
2.4	Equations of a Line	145
2.5	Applications of Linear Equations and Graphing	158
	Summary	172
	Review Exercises	177
	Test	180
	Cumulative Review	183

3 Systems of Linear Equations — 185

	Preview	186
3.1	Solving Systems of Linear Equations by Graphing	187
3.2	Solving Systems of Equations by Using the Substitution Method	197
3.3	Solving Systems of Equations by Using the Addition Method	203
	Midchapter Review	210
3.4	Applications of Systems of Linear Equations in Two Variables	210
3.5	Systems of Linear Equations in Three Variables and Applications	219
3.6	Solving Systems of Linear Equations by Using Matrices	228
	Summary	237
	Review Exercises	243
	Test	245
	Cumulative Review	247

4 Introduction to Relations and Functions — 249

	Preview	250
4.1	Introduction to Relations	251
4.2	Introduction to Functions	259
4.3	Graphs of Basic Functions	272
4.4	Variation	283
	Summary	292
	Review Exercises	296
	Test	299
	Cumulative Review	301

5 Polynomials — 303

	Preview	304
5.1	Addition and Subtraction of Polynomials and Polynomial Functions	305
5.2	Multiplication of Polynomials	315
5.3	Division of Polynomials	325
	Midchapter Review	335
5.4	Greatest Common Factor and Factoring by Grouping	336

5.5	Factoring Trinomials	345
5.6	Factoring Binomials	360
5.7	Solving Equations Using the Zero Product Rule	370
	Summary	384
	Review Exercises	389
	Test	393
	Cumulative Review	394

6 Rational Expressions and Rational Equations 397

	Preview	398
6.1	Rational Expressions and Rational Functions	399
6.2	Multiplication and Division of Rational Expressions	411
6.3	Addition and Subtraction of Rational Expressions	416
	Midchapter Review	426
6.4	Complex Fractions	427
6.5	Rational Equations	435
6.6	Applications of Rational Equations and Proportions	443
	Summary	453
	Review Exercises	458
	Test	460
	Cumulative Review	461

7 Radicals and Complex Numbers 465

	Preview	466
7.1	Definition of an nth-Root	467
7.2	Rational Exponents	480
7.3	Simplifying Radical Expressions	488
	Midchapter Review	495
7.4	Addition and Subtraction of Radicals	495
7.5	Multiplication of Radicals	502
7.6	Rationalization	510
7.7	Radical Equations	518
7.8	Complex Numbers	529
	Summary	540
	Review Exercises	545
	Test	549
	Cumulative Review	550

8 Quadratic Equations and Functions — 553

	Preview	554
8.1	Square Root Property and Completing the Square	555
8.2	Quadratic Formula	564
8.3	Equations in Quadratic Form	578
	Midchapter Review	585
8.4	Graphs of Quadratic Functions	586
8.5	Vertex of a Parabola and Applications	600
	Summary	613
	Review Exercises	617
	Test	620
	Cumulative Review Exercises	622

9 More Equations and Inequalities — 625

	Preview	626
9.1	Compound Inequalities	627
9.2	Polynomial and Rational Inequalities	636
	Midchapter Review	648
9.3	Absolute Value Equations	649
9.4	Absolute Value Inequalities	656
9.5	Linear Inequalities in Two Variables	668
	Summary	683
	Review Exercises	688
	Test	692
	Cumulative Review	693

10 Exponential and Logarithmic Functions — 697

	Preview	698
10.1	Algebra and Composition of Functions	699
10.2	Inverse Functions	707
10.3	Exponential Functions	718
10.4	Logarithmic Functions	729
	Midchapter Review	743
10.5	Properties of Logarithms	744
10.6	The Irrational Number, e	753
10.7	Exponential and Logarithmic Equations	767

	Summary	783
	Review Exercises	789
	Test	793
	Cumulative Review	796

11 Conic Sections — 801

	Preview	802
11.1	Distance Formula and Circles	803
11.2	More on the Parabola	812
11.3	The Ellipse and Hyperbola	821
	Midchapter Review	830
11.4	Nonlinear Systems of Equations in Two Variables	831
11.5	Nonlinear Inequalities and Systems of Inequalities	840
	Summary	849
	Review Exercises	854
	Test	857
	Cumulative Review Exercises	859

Additional Topics Appendix — A-1

A.1	Binomial Expansions	A-1
A.2	Determinants and Cramer's Rule	A-8
A.3	Sequences and Series	A-19
A.4	Arithmetic and Geometric Sequences and Series	A-28
	Student Answer Appendix	A-43

About the Authors

JULIE MILLER

Julie Miller has been on the faculty of the Mathematics Department at Daytona Beach Community College for 16 years, where she has taught developmental and upper-level courses. Prior to her work at DBCC, she worked as a software engineer for General Electric in the area of flight and radar simulation. Julie earned a bachelor of science in applied mathematics from Union College in Schenectady, New York, and a master of science in mathematics from the University of Florida. In addition to this textbook, she has authored several course supplements for college algebra, trigonometry, and precalculus, as well as several short works of fiction and nonfiction for young readers.

"My father is a medical researcher, and I got hooked on math and science when I was young and would visit his laboratory. I can remember using graph paper to plot data points for his experiments and doing simple calculations. He would then tell me what the peaks and features in the graph meant in the context of his experiment. I think that applications and hands-on experience made math come alive for me and I'd like to see math come alive for my students."

—Julie Miller

MOLLY O'NEILL

Molly O'Neill is also from Daytona Beach Community College, where she has taught for 18 years in the Mathematics Department. She has taught a variety of courses from developmental mathematics to calculus. Before she came to Florida, Molly taught as an adjunct instructor at the University of Michigan–Dearborn, Eastern Michigan University, Wayne State University, and Oakland Community College. Molly earned a bachelor of science in mathematics and a master of arts and teaching from Western Michigan University in Kalamazoo, Michigan. Besides this textbook, she has authored several course supplements for college algebra, trigonometry, and precalculus and has reviewed texts for developmental mathematics.

"I differ from many of my colleagues in that math was not always easy for me. But in seventh grade I had a teacher who taught me that if I follow the rules of mathematics, even I could solve math problems. Once I understood this, I enjoyed math to the point of choosing it for my career. I now have the greatest job because I get to do math everyday and I have the opportunity to influence my students just as I was influenced. Authoring these texts has given me another avenue to reach even more students."

—Molly O'Neill

Nancy Hyde has been a full time faculty member of the Mathematics Department at Broward Community College for 24 years. During this time she has taught the full spectrum of courses from developmental math through differential equations. She received a bachelor of science degree in math education from Florida State University and a master's degree in math education from Florida Atlantic University. She has conducted workshops and seminars for both students and teachers on the use of technology in the classroom. In addition to this textbook, she has authored a graphing calculator supplement for College Algebra.

NANCY HYDE

"I grew up in Brevard County, Florida, with my father working at Cape Canaveral. I was always excited by mathematics and physics in relation to the space program. As I studied higher levels of mathematics I became more intrigued by its abstract nature and infinite possibilities. It is enjoyable and rewarding to convey this perspective to students while helping them to understand mathematics."

—Nancy Hyde

Mitchel Levy of Broward Community College joins the team as the exercise consultant for the Miller/O'Neill/Hyde paperback series. Mitchel received his BA in mathematics in 1983 from the State University of New York at Albany and his MA in mathematical statistics from the University of Maryland, College Park in 1988. With over 17 years of teaching and extensive reviewing experience, Mitchel knows what makes exercise sets work for students. In 1987 he received the first annual "Excellence in Teaching" award for graduate teaching assistants at the University of Maryland. Mitchel was honored as the Broward Community College Professor of the year in 1994, and has co-coached the Broward math team to 3 state championships over 7 years.

MITCHEL LEVY

"I love teaching all level of mathematics from Elementary Algebra through Calculus and Statistics."

—Mitchel Levy

Preface

From the Authors

First and foremost, we would like to thank the students and colleagues who have helped us prepare this text. The content and organization are based on a wealth of resources. In addition to an accumulation of our own notes and experiences as teachers, we recognize the influence of colleagues at Daytona Beach Community College and Broward Community College as well as fellow presenters and attendees of national mathematics conferences and meetings. Perhaps our single greatest source of inspiration has been our students, who ask good, probing questions every day and challenge us to find new and better ways to convey mathematical concepts. We gratefully acknowledge the part that each has played in the writing of this book.

In designing the framework for this text, the time we have spent with our students has proven especially invaluable. Over the years we have observed that students struggle consistently with certain topics. We have also come to know the influence of forces beyond the math, particularly motivational issues. An awareness of the various pitfalls has enabled us to tailor pedagogy and techniques that directly address students' needs and promote their success. Those techniques and pedagogy are outlined here.

Active Classroom

First, we believe students retain more of what they learn when they are actively engaged in the classroom. Consequently, as we wrote each section of text, we also wrote accompanying worksheets called **Classroom Activities** to foster accountability and to encourage classroom participation. Classroom Activities resemble the examples that students encounter in the textbook. The activities can be assigned to individual students, or to pairs or groups of students. Most of the activities have been tested in the classroom with our own students. In one class in particular, the introduction of Classroom Activities transformed a group of "clock watchers" into students who literally had to be ushered out of the classroom so that the next class could come in. The activities can be found in the Instructor's Resource Manual, which is available through MathZone.

Conceptual Support

While we believe students must practice basic skills to be successful in any mathematics class, we also believe concepts are important. To this end, we have included **Concept Connections** questions and homework exercises that ask students to **"Interpret the meaning in the context of the problem."** These questions make students stop and think so they can process what they learn. In this way, students will learn the underlying concepts. They will also form an understanding of what their answers mean in the contexts of the problems they solve.

Writing Style

Many students believe that reading a mathematics text is an exercise in futility. However, students who take the time to read the text and features within the margins may cast that notion aside. In particular, the **Tips** and **Avoiding Mistakes** boxes should prove especially enlightening. They offer the types of insights and hints that are usually only revealed during classroom lecture. On the whole, students should be very comfortable with the reading level, as the language and tone are consistent with those used daily within our own developmental mathematics classes.

Real-World Applications

Another critical component of the text is the inclusion of **contemporary real-world examples and applications**. We based examples and applications on information that students encounter daily when they turn on the news, read a magazine, or surf the World Wide Web. We incorporated data for students to answer mathematical questions based on information in tables and graphs. When students encounter facts or information that is meaningful to them, they will relate better to the material and remember more of what they learn.

Study Skills

Many students in this course lack the basic study skills needed to be successful. Therefore, at the beginning of every set of homework exercises, we included a set of **Study Skills Exercises**. The exercises focus on one of nine areas: learning about the course, using the text, taking notes, completing homework assignments, test-taking, time-management, learning styles, preparing for a final exam, and defining **key terms**. Through completion of these exercises, students will be in a better position to pass the class and adopt techniques that will benefit them throughout their academic careers.

Language of Mathematics

Finally, for students to succeed in mathematics, they must be able to understand its language and notation. We place special emphasis on the skill of translating mathematical notation to English expressions and vice versa through **Translating Expressions Exercises**. These appear intermittently throughout the text. We also include key terms in the homework exercises and ask students to define these terms.

While we have made every effort to fine-tune this textbook to serve the needs of all students, we acknowledge that no textbook can satisfy every student's needs entirely. However, we do trust that the thoughtfully designed pedagogy and contents of this textbook offer any willing student the opportunity to achieve success, opening the door to a wider world of possibilities.

Listening to Students' and Instructors' Concerns

Our editorial staff has amassed the results of reviewer questionnaires, user diaries, focus groups, and symposia. We have consulted with an eight-member panel of intermediate algebra instructors and their students on the development of this book. In addition, we have read hundreds of pages of reviews from instructors across the country. At McGraw-Hill symposia, faculty from across the United States gathered to discuss issues and trends in developmental mathematics. These efforts have involved hundreds of faculty and have explored issues such as content, readability, and even the aesthetics of page layout.

What Sets This Book Apart?

While this textbook offers complete coverage of the intermediate algebra curriculum, there are several concepts that receive special emphasis.

Early Graphing and Functions

This text focuses on an algebraic approach to problem solving with graphical interpretation. The rectangular coordinate system and graphs of linear equations in two variables are presented early (Chapter 2).

Chapter 4 is devoted entirely to functions. The definitions of a function and a relation are covered, along with domain, range, function notation, and variation. These concepts are introduced through applications. Furthermore, we have carefully and deliberately emphasized the *interpretation* of functions and function notation.

After Chapter 4, functions are revisited where appropriate. Instructors have the option of continuing to use function notation with polynomials, quadratics, radicals, rational expressions, exponentials, and logarithms. However, for those instructors who do not wish to implement function notation throughout the course, function notation in the chapters following Chapter 4, can be easily skipped.

A Factoring Strategy

The specially titled review, "A Factoring Strategy," on p. 369 is designed to provide a cumulative review of the methods used to factor different types of polynomials. This review provides an opportunity for students to see a variety of polynomials and to apply the skills they have learned in the previous sections to successfully factor each polynomial.

Identifying Equations and Inequalities

A student who completes intermediate algebra should be able to recognize and solve a variety of equations and inequalities; however, the skill of distinguishing different types of equations and inequalities is often overlooked. Chapter 9—More Equations and Inequalities—is designed as a synthesis chapter in which students are exposed to a variety of equations and inequalities appropriate at this level. Note that each section of Chapter 9 is a self-contained unit and can be introduced at the instructor's discretion at another location in the text.

Calculator Usage

The use of a scientific or a graphing calculator often inspires great debate among faculty who teach developmental mathematics. Our Calculator Connections boxes offer screen shots and some keystrokes to support concepts and applications where a calculator might enhance learning. Our approach is to use a calculator as a verification tool after analytical methods have been applied. The Calculator Connections boxes are self-contained units and may be employed or easily omitted at the recommendation of the instructor.

Graphing Calculator Exercises and Scientific Calculator Exercises do appear within the section-ending Practice Exercise sets where appropriate, but they are clearly noted as such and can be assigned or omitted at the instructor's discretion.

Suggestions Welcome!

Many features of this book, and many refinements in writing, illustrations, and content, came about because of suggestions and questions from instructors and their students. We invite your comments with regard to this textbook as we work to further shape and refine its contents.

Julie Miller
millerj@dbcc.edu

Molly O'Neill
oneillm@dbcc.edu

Nancy Hyde
hyde_n@firn.edu

Acknowledgements and Reviewers

The development of this textbook would never have been possible without the creative ideas and constructive feedback offered by many reviewers. We are especially thankful to the following instructors for their valuable feedback and careful review of the manuscript.

Board of Advisors

Mary Kay Best,	*Coastal Bend College*
Lisa Buckelew,	*Oklahoma City Community College*
Gail Burkett,	*Palm Beach Community College*
Felicia Graves,	*Cuyahoga Community College*
Joe Kemble,	*Lamar University*
Pascal Roubides,	*Miami-Dade College*
Katalin Szucs,	*East Carolina University/Pitt Community College*
Paul Wozniak,	*El Camino College*

Special acknowledgment goes to the following instructors for their ideas and recommendations:

Yvonne Aucoin,	*Tidewater Community College–Norfolk*
Alina Coronel,	*Miami-Dade College–Kendall*
David French,	*Tidewater Community College–Chesapeake*
Donna Gerken,	*Miami-Dade College–Kendall*
Rebecca Hubiak,	*Tidewater Community College–Virginia Beach*
Patricia Jayne,	*Miami-Dade College–Kendall*
Mike Kirby,	*Tidewater Community College–Virginia Beach*
Charlotte Newsom,	*Tidewater Community College–Virginia Beach*
Drago Stoyanovich,	*Miami-Dade College–Kendall*

Manuscript Reviewers

Rosalie Abraham, *Florida Community College at Jacksonville*
John P. Anderson, *San Jacinto College South*
Sohrab Bakhtyari, *St. Petersburg College*
Dean Burbank, *Gulf Coast Community College*
Warren J. Burch, *Brevard Community College*
Susan Caldiero, *Cosumnes River College*
Natalie M. Creed, *Gaston College*
Benay Don, *Suffolk County Community College*
Lucy Edwards, *Las Positas College*
Jacqueline English, *Northern Oklahoma College Enid*
Thomas English, *Pennsylvania State University–Erie*
Ann Evans, *University of Massachusetts–Boston*
Carol J. Flakus, *Lower Columbia College*
Ken Harrelson, *Oklahoma City Community College*
LaVerne Harrison, *University of Wisconsin–Marshfield County*
Michelle S. Hollis, *Bowling Green Community College of Western Kentucky University*
Tina Johnson, *Midwestern State University*
Regina Keller, *Suffolk County Community College*
Richard Leedy, *Polk Community College*
Denise J. LeGrand, *University of Arkansas at Little Rock*
Sheila Leonard, *Lasell College*
Tim McBride, *Spartanburg Technical College*
Gary McCracken, *Shelton State Community College*
Donna Martin, *Florida Community College*
Danielle Morgan, *San Jacinto College South*
Dr. Christina Anne Morian, *Lincoln University*
Faith Peters, *Miami-Dade College*
Marilyn Platt, *Gaston College*
Leesa Pohl, *Donnelly College*
Nancy Ressler, *Oakton Community College*
Dr. Richard F. Riggs, *New Jersey City University*
Mansour Samimi, *Winston-Salem State University*
Nancy J. Sattler, Ph.D., *Terra Community College*
Eugenia Shipe, *Pennsylvania State University*
Laurence Stone, *Dakota County Technical College*
Donna Szott, *Community College of Allegheny County–South Campus*
Michael Tran, *Antelope Valley College*
Bernadette Turner, *Lincoln University*
David Turner, *Faulkner University*
Vince Waggener, *Trident Technical College*
Abbas Zadegan, *Florida Memorial College*

Special thanks go to Carrie Green for preparing the Instructor's Solutions Manual and the Student's Solutions Manual and for her work ensuring accuracy, and to Yolanda Davis and Patricia Jayne for their appearance in and work on the video series. Further thanks go to Kelly Jackson for preparing the Instructor Notes.

Finally, we are forever grateful to the many people behind the scenes at McGraw-Hill, our publishing family. To Erin Brown, our lifeline on this project, without you we'd be lost. To Liz Haefele, your passion for excellence has been a constant inspiration. To Michael Lange and David Dietz, thanks for your vision and input and for being there all these years. To Barb Owca and David Millage, we marvel at your creative ideas in a world that's forever changing. To Jeff Huettman and Amber Huebner for your awesome work with the technology and to Jodi Rhomberg for keeping the train on the track during production.

Most importantly, we give special thanks to all the students and instructors who use *Intermediate Algebra* in their classes.

Julie Miller   Molly O'Neill   Nancy Hyde

Dedication

To my father, Kent D. Miller, for teaching me to love science and math   —Julie Miller

To Carolyn, thanks   —Molly O'Neill

To my friend, Steve Wolfe   —Nancy Hyde

A COMMITMENT TO ACCURACY

You have a right to expect an accurate textbook, and McGraw-Hill invests considerable time and effort to make sure that we deliver one. Listed below are the many steps we take to make sure this happens.

OUR ACCURACY VERIFICATION PROCESS

First Round
Step 1: Numerous **college math instructors** review the manuscript and report on any errors that they may find, and the authors make these corrections in their final manuscript.

Second Round
Step 2: Once the manuscript has been typeset, the **authors** check their manuscript against the first page proofs to ensure that all illustrations, graphs, examples, exercises, solutions, and answers have been correctly laid out on the pages, and that all notation is correctly used.

Step 3: An outside, **professional mathematician** works through every example and exercise in the page proofs to verify the accuracy of the answers.

Step 4: A **proofreader** adds a triple layer of accuracy assurance in the first pages by hunting for errors, then a second, corrected round of page proofs is produced.

Third Round
Step 5: The **author team** reviews the second round of page proofs for two reasons: 1) to make certain that any previous corrections were properly made, and 2) to look for any errors they might have missed on the first round.

Step 6: A **second proofreader** is added to the project to examine the new round of page proofs to double check the author team's work and to lend a fresh, critical eye to the book before the third round of paging.

Fourth Round
Step 7: A **third proofreader** inspects the third round of page proofs to verify that all previous corrections have been properly made and that there are no new or remaining errors.

Step 8: Meanwhile, in partnership with **independent mathematicians,** the text accuracy is verified from a variety of fresh perspectives:
- The **test bank author** checks for consistency and accuracy as they prepare the computerized test item file.
- The **solutions manual author** works every single exercise and verifies their answers, reporting any errors to the publisher.
- A **consulting group of mathematicians,** who write material for the text's MathZone site, notifies the publisher of any errors they encounter in the page proofs.
- A video production company employing **expert math instructors** for the text's videos will alert the publisher of any errors they might find in the page proofs.

Final Round
Step 9: The **project manager,** who has overseen the book from the beginning, performs a **fourth proofread** of the textbook during the printing process, providing a final accuracy review.

⇒ What results is a mathematics textbook that is as accurate and error-free as is humanly possible, and our authors and publishing staff are confident that our many layers of quality assurance have produced textbooks that are the leaders of the industry for their integrity and correctness.

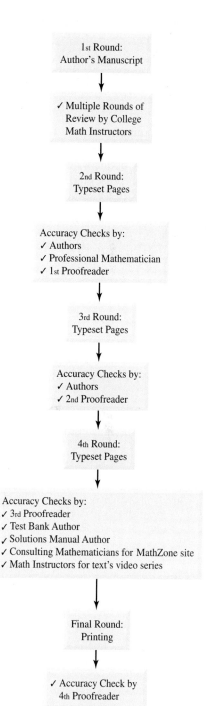

Guided Tour

Chapter Opener

Each chapter opens with an application relating to an exercise presented in the chapter. Section titles are clearly listed for easy reference.

Linear Equations in Two Variables 2

- **2.1** The Rectangular Coordinate System and Midpoint Formula
- **2.2** Linear Equations in Two Variables
- **2.3** Slope of a Line
- **2.4** Equations of a Line
- **2.5** Applications of Linear Equations and Graphing

Using graphs to interpret data (information) is an important tool in science and business. This chapter covers topics related to graphing and applications of graphing. For example, the slope of a line represents a rate of change. In Exercise 53 in Section 2.3, we see that the number of cellular phone subscriptions has increased linearly from 1998 to 2006. The slope of the line indicates that the number of cell phone subscriptions has increased by 18.75 million per year during this time period.

---

chapter 2 preview

The exercises in this chapter preview contain concepts that have not yet been presented. These exercises are provided for students who want to compare their levels of understanding before and after studying the chapter. Alternatively, you may prefer to work these exercises when the chapter is completed and before taking the exam.

Section 2.1

For Exercises 1–3, state the quadrant in which each point lies.

1. $(3, -4)$ 2. $(-2, 2)$ 3. $(-1, -4)$

4. Find the midpoint of the line segment between the points $(-4, 10)$ and $(11, -6)$.

Section 2.2

For Exercises 5–6, graph the lines.

5. $y = -2x - 3$ 6. $y = -3$

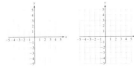

7. Given the equation $2x - 3y = -10$, find the x- and y-intercepts.

Section 2.3

For Exercises 8–10, find the slope of the line through the given points.

8. $(-5, -2)$ and $(-1, 6)$

9. $(4, -5)$ and $(-2, -5)$

10. $(8, 0)$ and $(8, -9)$

11. The slope of a line is $-\frac{4}{3}$.
    a. What is the slope of a line parallel to the given line?
    b. What is the slope of a line perpendicular to the given line?

Section 2.4

12. Given the equation $5x - 9y = 1$, determine the slope and y-intercept.

13. Write an equation for the line that passes through the points $(4, 5)$ and $(1, -1)$.

14. Write an equation for the line that passes through the point $(2, -2)$ and is perpendicular to the line $y = 4x - 1$.

Section 2.5

15. A record rainfall in Hawaii occurred in November 2000 at Kappala Ranch. The National Weather Service reported that during the storm, rain fell at an average rate of $\frac{1}{2}$ inches per hour (in./hr).
    a. Write an equation to compute the total amount of rain y (in inches) after x hr of the storm.
    b. Use the equation to compute the amount of rainfall after 15 hr.
    c. When the storm ended, a total of 36 in. of rain had fallen. How long did the storm last?

Chapter Preview

A Chapter Preview appears at the beginning of each chapter. It contains exercises, grouped by section. The exercises are based on topics not yet presented, offering students an opportunity to compare their levels of understanding before and after studying the chapter.

Objectives

A list of important learning objectives is provided at the beginning of each section. Each objective corresponds to a heading within the section and within the exercises, making it easy for students to locate topics as they study or as they work through homework exercises.

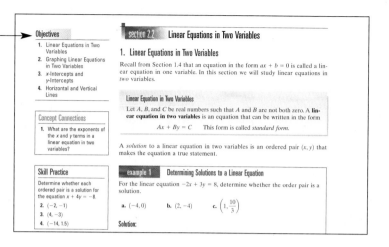

Concept Connections

Students can test their understanding of what they have read by completing the Concept Connections exercises that appear in the margins. These questions test how well students grasp concepts. Students can check their responses by referring to the answers at the bottom of the page.

Skill Practice Exercises

Every worked example is paired with a Skill Practice exercise. These exercises appear in the margin directly beside the worked examples and offer students an immediate opportunity to work problems that mirror the examples. Students can then check their work by referring to the answers at the bottom of the page.

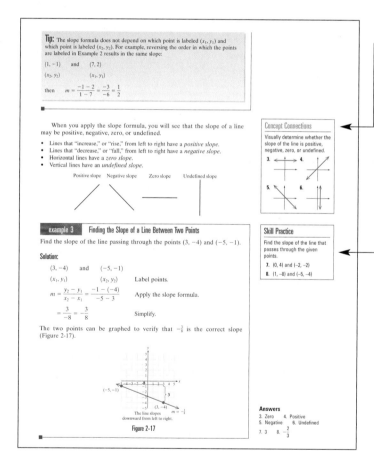

Avoiding Mistakes

Through notes labeled Avoiding Mistakes students are alerted to common errors and are shown methods to avoid them.

Tips

Tip boxes appear throughout the text and offer helpful hints and insight.

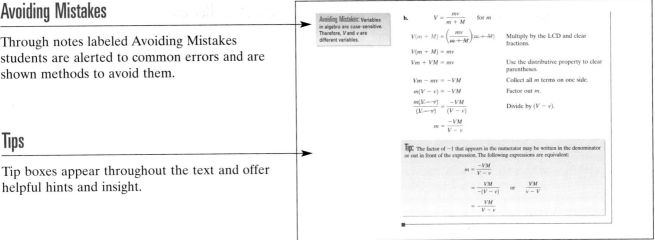

Worked Examples

Examples are set off in boxes and organized so that students can easily follow the solutions. Explanations appear beside each step and color-coding is used, where appropriate. For additional step-by-step instruction, students can run the "e-Professors" in MathZone. The e-Professors are based on worked examples from the text and use the solution methodologies presented in the text.

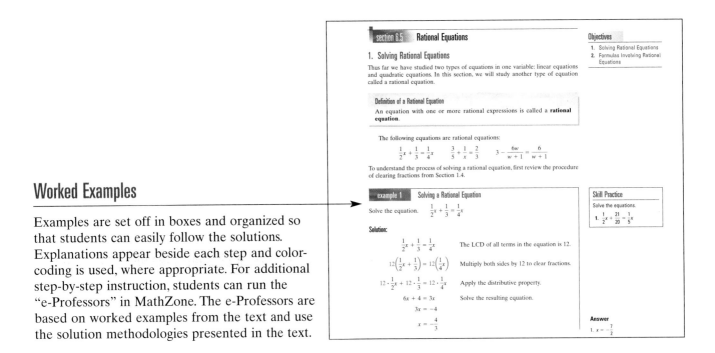

Midchapter Review

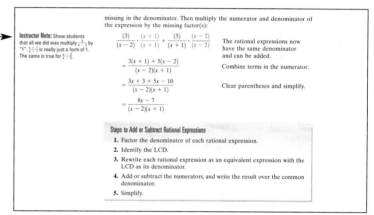

Midchapter Reviews are provided to help solidify the foundation of concepts learned in the beginning of a chapter before expanding to new ideas presented later in the chapter.

Instructor Note (AIE only)

Throughout each section of the Annotated Instructor's Edition (AIE), notes to the instructor can be found in the margins. The notes may assist with lecture preparation in that they point out items that tend to confuse students, or lead students to err.

References to Classroom Activities (AIE only)

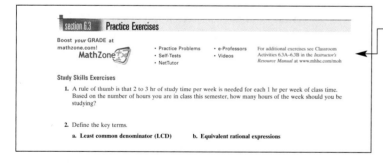

References are made to Classroom Activities at the beginning of each set of Practice Exercises in the AIE. The activities may be found in the *Instructor's Resource Manual*, which is available through MathZone, and can be used during lecture, or assigned for additional practice.

Practice Exercises

A variety of problem types appear in the section-ending Practice Exercises. Problem types are clearly labeled with either a heading or an icon for easy identification. References to MathZone are also found at the beginning of the Practice Exercises to remind students and instructors that additional help and practice problems are available. The core exercises for each section are organized by section objective. General references to examples are provided for blocks of core exercises. **Mixed Exercises** are also provided in some sections where no reference to objectives or examples is offered.

Icon Key

The following key has been prepared for easy identification of "themed" exercises appearing within the Practice Exercises.

Student Edition
Exercises Keyed to Video
Calculator Exercises

AIE only
Writing
Translating Expressions
Geometry

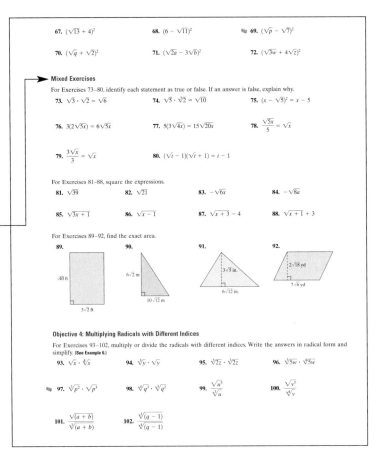

Study Skills Exercises appear at the beginning of the exercise set. They are designed to help students learn techniques to improve their study habits including exam preparation, note taking, and time management.

In the Practice Exercises, where appropriate, students are asked to define the **Key Terms** that are presented in the section. Assigning these exercises will help students to develop and expand their mathematical vocabulary.

Review Exercises also appear at the start of the Practice Exercises. The purpose of the Review Exercises is to help students retain their knowledge of concepts previously learned.

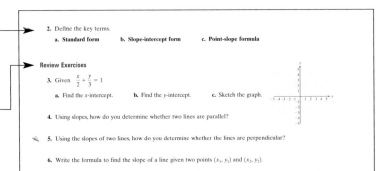

Writing Exercises offer students an opportunity to conceptualize and communicate their understanding of algebra. These, along with the **Translating Expressions Exercises** enable students to strengthen their command of mathematical language and notation and improve their reading and writing skills.

Geometry Exercises appear throughout the Practice Exercises and encourage students to review and apply geometry concepts.

---

5. Using the slopes of two lines, how do you determine whether the lines are perpendicular?

6. Write the formula to find the slope of a line given two points (x_1, y_1) and (x_2, y_2).

7. Given the two points $(-1, -2)$ and $(2, 4)$,
   a. Find the slope of the line containing the two points.
   b. Find the slope of a line parallel to the line containing the points.
   c. Find the slope of a line perpendicular to the line containing the points.

For Exercises 91–94, translate the English phrase to an algebraic expression.

91. The sum of q and the square of p
92. The product of 11 and the cube root of x
93. The quotient of 6 and the cube root of x
94. The difference of y and the principal square root of x
95. If a square has an area of 64 in.², then what are the lengths of the sides?
96. If a square has an area of 121 m², then what are the lengths of the sides?

Graphing Calculator Exercises

For Exercises 97–104, use a calculator to evaluate the expressions to 4 decimal places.

97. $\sqrt{69}$
98. $\sqrt{5798}$
99. $2 + \sqrt[3]{5}$
100. $3 - 2\sqrt[3]{10}$
101. $7\sqrt[4]{25}$
102. $-3\sqrt[5]{9}$
103. $\dfrac{3 - \sqrt{19}}{11}$
104. $\dfrac{5 + 2\sqrt{15}}{12}$

105. Graph $h(x) = \sqrt{x - 2}$ on the standard viewing window. Use the graph to confirm the domain found in Exercise 79.

106. Graph $k(x) = \sqrt{x + 1}$ on the standard viewing window. Use the graph to confirm the domain found in Exercise 80.

107. Graph $g(x) = \sqrt[3]{x - 2}$ on the standard viewing window. Use the graph to confirm the domain found in Exercise 81.

108. Graph $f(x) = \sqrt[3]{x + 1}$ on the standard viewing window. Use the graph to confirm the domain found in Exercise 82.

---

8. How does the graph of $f(x) = (x + 10)^2$ compare with the graph of $y = x^2$?

For Exercises 9–16, find the value of k to complete the square.

9. $x^2 - 8x + k$
10. $x^2 + 4x + k$
11. $y^2 + 7y + k$
12. $a^2 - a + k$
13. $b^2 + \dfrac{2}{9}b + k$
14. $m^2 - \dfrac{2}{7}m + k$
15. $t^2 - \dfrac{1}{3}t + k$
16. $p^2 + \dfrac{1}{4}p + k$

Objective 1: Writing a Quadratic Function in the Form $f(x) = a(x - h)^2 + k$

For Exercises 17–30, write the function in the form $f(x) = a(x - h)^2 + k$ by completing the square. Then identify the vertex. (See Example 1.)

17. $g(x) = x^2 - 8x + 5$
18. $h(x) = x^2 + 4x + 5$
19. $n(x) = 2x^2 + 12x + 13$
20. $f(x) = 4x^2 + 16x + 19$
21. $p(x) = -3x^2 + 6x - 5$
22. $q(x) = -2x^2 + 12x - 11$

Calculator Exercises signify situations where a calculator would provide assistance for time-consuming calculations. These exercises were carefully designed to demonstrate the types of situations in which a calculator is a handy tool rather than a "crutch." **Graphing Calculator Exercises** also appear, where appropriate, at the end of the Practice Exercise sets.

Exercises Keyed to Video are labeled with an icon to help students and instructors identify those exercises for which accompanying video instruction is available.

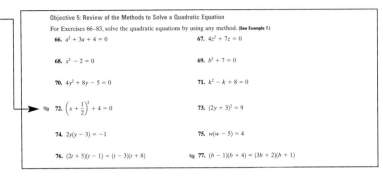

Objective 5: Review of the Methods to Solve a Quadratic Equation

For Exercises 66–83, solve the quadratic equations by using any method. (See Example 7.)

66. $a^2 + 3a + 4 = 0$
67. $4z^2 + 7z = 0$
68. $x^2 - 2 = 0$
69. $b^2 + 7 = 0$
70. $4y^2 + 8y - 5 = 0$
71. $k^2 - k + 8 = 0$
72. $\left(x + \frac{1}{2}\right)^2 + 4 = 0$
73. $(2y + 3)^2 = 9$
74. $2y(y - 3) = -1$
75. $w(w - 5) = 4$
76. $(2t + 5)(t - 1) = (t - 3)(t + 8)$
77. $(b - 1)(b + 4) = (3b + 2)(b + 1)$

Expanding Your Skills

84. An artist has been commissioned to make a stained glass window in the shape of a regular octagon. The octagon must fit inside an 18-in. square space. See the figure.

a. Let x represent the length of each side of the octagon. Verify that the legs of the small triangles formed by the corners of the square can be expressed as $\frac{18 - x}{2}$.

b. Use the Pythagorean theorem to set up an equation in terms of x that represents the relationship between the legs of the triangle and the hypotenuse.

c. Simplify the equation by clearing parentheses and clearing fractions.

d. Solve the resulting quadratic equation by using the quadratic formula. Use a calculator and round your answers to the nearest tenth of an inch.

e. There are two solutions for x. Which one is appropriate and why?

Graphing Calculator Exercises

85. Graph $Y_1 = x^3 - 27$. Compare the x-intercepts with the solutions to the equation $x^3 - 27 = 0$ found in Exercise 42.

86. Graph $Y_1 = 64x^3 + 1$. Compare the x-intercepts with the solutions to the equation $64x^3 + 1 = 0$ found in Exercise 43.

87. Graph $Y_1 = 3x^3 - 6x^2 + 6x$. Compare the x-intercepts with the solutions to the equation $3x^3 - 6x^2 + 6x = 0$ found in Exercise 44.

Expanding Your Skills, found near the end of most Practice Exercises, challenge students' knowledge of the concepts presented.

Applications based on real-world facts and figures motivate students and enable them to hone their problem-solving skills.

Optional Calculator Connections are located throughout the text where appropriate. They can be implemented at the instructor's discretion depending on the emphasis placed on the calculator in the course. The Calculator Connection display keystrokes and Graphing Calculator Exercises are found at the end of the practice exercises where appropriate.

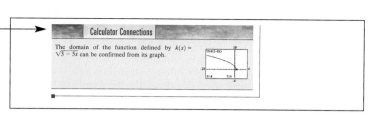

End-of-Chapter Summary and Exercises

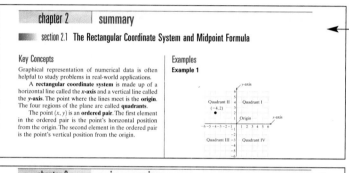

The **Summary**, located at the end of each chapter, outlines key concepts for each section and illustrates those concepts with examples.

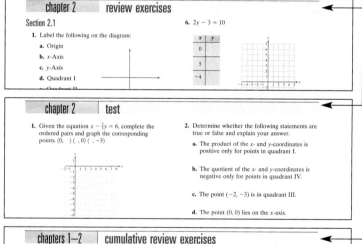

Following the Summary is a set of **Review Exercises** that are organized by section. A **Chapter Test** appears after each set of Review Exercises. Chapters 2–9 also include **Cumulative Reviews** that follow the Chapter Tests. These end-of-chapter materials provide students with ample opportunity to prepare for quizzes or exams.

SUPPLEMENTS

For the Instructor

Instructor's Resource Manual
The Instructor's Resource Manual (IRM), written by the authors, is a printable electronic supplement available through MathZone. The IRM includes discovery based classroom activities, worksheets for drill-and-practice, materials for a student portfolio, and some tips for implementing successful cooperative learning. Numerous classroom activities are available for each section of text and can be used as a complement to lecture or can be assigned for work outside of class. The activities are designed for group or individual work and take about 5–10 minutes each. With increasing demands on faculty schedules, these ready-made lessons offer a convenient means for both full-time and adjunct faculty to promote active learning in the classroom.

   www.mathzone.com

McGraw-Hill's **MathZone 2.0** is a complete **Web-based tutorial and course management system** for mathematics and statistics, designed for greater ease of use than any other system available. Free upon adoption of a McGraw-Hill textbook, the system enables instructors to **create and share courses and assignments** with colleagues, adjunct faculty members, and teaching assistants with only a few mouse clicks. All **assignments, exercises, "e-Professor" multimedia tutorials, video lectures, and NetTutor® live tutors** follow the textbook's learning objectives and problem-solving style and notation. Using MathZone's **assignment builder,** instructors can **edit questions and algorithms, import their own content**, and **create announcements and due-dates** for homework and quizzes. MathZone's **automated grading function** reports the results of easy-to-assign algorithmically-generated homework, quizzes, and tests. All student activity within MathZone is recorded and available through a **fully integrated gradebook** that can be downloaded to Microsoft Excel®. MathZone also is available on CD-ROM. (See "Supplements for the Student" for descriptions of the elements of MathZone.)

Instructor's Testing and Resource CD

This cross-platform CD-ROM provides a wealth of resources for the instructor. Among the supplements featured on the CD-ROM is a **computerized test bank** utilizing Brownstone Diploma® algorithm-based testing software to create customized exams quickly. This user-friendly program enables instructors to search for questions by topic, format, or difficulty level; to edit existing questions or to add new ones; and to scramble questions and answer keys for multiple version of a single test. Hundreds of text-specific open-ended and multiple-choice questions are included in the question bank. Sample chapter tests are also provided.

ALEKS (**A**ssessment and **LE**arning in **K**nowledge **S**paces) is an artificial intelligence-based system for mathematics learning, available over the web 24/7. Using unique adaptive questioning, ALEKS accurately assesses what topics each student knows and then determines exactly what each student is ready to learn next. ALEKS interacts with the students much as a skilled human tutor would, moving between explanation and practice as needed, correcting and analyzing errors, defining terms and changing topics on request, and helping them master the course content more quickly and easily. Moreover, the new ALEKS 3.0 now links to text-specific videos, multimedia tutorials, and textbook pages in PDF format. ALEKS also offers a robust classroom management system that allows instructors to monitor and direct student progress toward mastery of curricular goals. See www.highed.aleks.com.

Miller/O'Neill/Hyde Video Lectures on Digital Video Disk (DVD)

In the videos, qualified instructors work through selected problems from the textbook, following the solution methodology employed in the text. The video series is available on DVD or online as an assignable element of MathZone. The DVDs are closed-captioned for the hearing impaired, subtitled in Spanish, and meet the Americans with Disabilities Act Standards for Accessible Design. Instructors may use them as resources in a learning center, for online courses, and/or to provide extra help for students who require extra practice.

Annotated Instructor's Edition

In the Annotated Instructor's Edition (AIE), **answers to all exercises and tests appear adjacent to each exercise**, in a color used *only* for annotations. The AIE also contains **Instructor Notes** that appear in the margin. The notes may assist with lecture preparation. Also found in the AIE are icons within the Practice Exercises that serve to guide instructors in their preparation of homework assignments and lessons.

Instructor's Solutions Manual

The Instructor's Solutions Manual provides comprehensive, worked-out solutions to all exercises in the Chapter Previews; the Practice Exercises; the Midchapter Reviews; the end-of-chapter Review Exercises; the Chapter Tests; and the Cumulative Review Exercises.

For the Student

 www.mathzone.com

McGraw-Hill's MathZone is a powerful web-based tutorial for homework, quizzing, testing, and multimedia instruction. Also available in CD-ROM format, MathZone offers:

Practice exercises based on the text and generated in an unlimited quantity for as much practice as needed to master any objective

Video clips of classroom instructors showing how to solve exercises from the text, step-by-step

e-Professor animations that take the student through step-by-step instructions, delivered on-screen and narrated by a teacher on audio, for solving exercises from the textbook; the user controls the pace of the explanations and can review as needed

NetTutor, which offers personalized instruction by live tutors familiar with the textbook's objectives and problem-solving methods.

Every assignment, exercise, video lecture, and e-Professor is derived from the textbook.

Student's Solutions Manual

The Student's Solutions Manual provides comprehensive, worked-out solutions to the odd-numbered exercises in the Chapter Previews, the Practice Exercise sets; the Midchapter Reviews, the end-of-chapter Review Exercises, the Chapter Tests, and the Cumulative Review Exercises.

Video Lectures on Digital Video Disk (DVD)

The video series is based on exercises from the textbook. Each presenter works through selected problems, following the solution methodology employed in the text. The video series is available on DVD or online as part of MathZone. The DVDs are closed-captioned for the hearing impaired, subtitled in Spanish, and meet the Americans with Disabilities Act Standards for Accessible Design.

NetTutor

Available through MathZone, NetTutor is a revolutionary system that enables students to interact with a live tutor over the web. NetTutor's web-based, graphical chat capabilities enable students and tutors to use mathematical notation and even to draw graphs as they work through a problem together. Students can also submit questions and receive answers, browse previously-answered questions, and view previous sessions. Tutors are familiar with the textbook's objectives and problem-solving styles.

Review of Basic Algebraic Concepts

1

1.1 Sets of Numbers and Interval Notation
1.2 Operations on Real Numbers
1.3 Simplifying Expressions
1.4 Linear Equations in One Variable
1.5 Applications of Linear Equations in One Variable
1.6 Literal Equations and Applications to Geometry
1.7 Linear Inequalities in One Variable
1.8 Properties of Integer Exponents and Scientific Notation

This chapter begins with a review of the operations on positive and negative numbers. The focus then shifts to linear equations and inequalities and their applications. For example, in Exercise 96 in Section 1.2, the equation $F = \frac{9}{5}C + 32$ converts Celsius temperatures to Fahrenheit temperatures. A person traveling in Europe would use this formula to convert $-5°C$ to the familiar Fahrenheit scale to obtain $23°F$.

Chapter 1 Preview

The exercises in this chapter preview contain concepts that have not yet been presented. These exercises are provided for students who want to compare their levels of understanding before and after studying the chapter. Alternatively, you may prefer to work these exercises when the chapter is completed and before taking the exam.

Section 1.1

1. Label each number as rational or irrational.

   a. $\sqrt{8}$ b. $-\dfrac{8}{3}$ c. $1.\overline{2}$ d. π

2. Graph each set and express the set in interval notation.

   a. $\{x \mid x \geq 4\}$

   b. $\{x \mid -1 < x \leq 6\}$

Section 1.2

For Exercises 3–5, perform the indicated operations.

3. $-9 - 15$

4. $-|12| - 5(-6) + 3^2$

5. $\dfrac{5 + 5(2)}{3(1-2) + \sqrt{9}}$

Section 1.3

For Exercises 6–7, match each expression with the appropriate property.

   a. Associative property of addition
   b. Distributive property of multiplication over addition

6. $5(x + y) = 5x + 5y$

7. $(5 + x) + y = 5 + (x + y)$

8. Clear the parentheses and combine like terms.
   $4(x^2 + 2x) - 3(2x^2 - x)$

Section 1.4

For Exercises 9–12, solve the equations.

9. $5x + 3 = -17$

10. $2y - 7 = 6y - 8$

11. $\dfrac{3}{5}x - \dfrac{3}{4} = \dfrac{1}{2}x + \dfrac{7}{10}$

12. $5(4x - 1) + 6 = 2x + 9(2x - 3)$

Section 1.5

13. The sum of three consecutive odd integers is 33. Find the integers.

14. The cost of an item after 5% sales tax was added was $19.74. Find the price of the item before tax.

Section 1.6

15. The perimeter of a rectangular field is 344 feet (ft). The length is 12 ft more than 3 times the width. Find the length and width.

16. The formula for the area of a triangle is $A = \tfrac{1}{2}bh$. Solve the formula for b.

Section 1.7

For Exercises 17–18, solve the inequality and write the solution in interval notation.

17. $-5x + 2(x - 3) > 12$ 18. $-8 \leq 4c - 2 \leq 10$

Section 1.8

For 19–20, simplify the expressions. Write the answers using positive exponents.

19. $a^3 a^{-6} a$ 20. $\dfrac{p^{-5} q^7}{p^{-2} q^{-3}}$

21. Write the numbers in scientific notation.

   a. 0.0000081   b. 252,000,000

22. Write the numbers in expanded form.

   a. 4.8×10^4 b. 6.0×10^{-3}

Section 1.1 Sets of Numbers and Interval Notation

1. Set of Real Numbers

Algebra is a powerful mathematical tool that is used to solve real-world problems in science, business, and many other fields. We begin our study of algebra with a review of basic definitions and notations used to express algebraic relationships.

In mathematics, a collection of elements is called a **set**, and the symbols { } are used to enclose the elements of the set. For example, the set {a, e, i, o, u} represents the vowels in the English alphabet. The set {1, 3, 5, 7} represents the first four positive odd numbers. Another method to express a set is to *describe* the elements of the set by using **set-builder notation**. Consider the set {a, e, i, o, u} in set-builder notation.

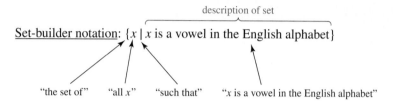

Consider the set {1, 3, 5, 7} in set-builder notation.

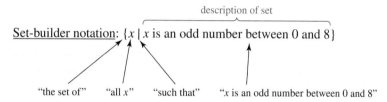

Several sets of numbers are used extensively in algebra. The numbers you are familiar with in day-to-day calculations are elements of the set of **real numbers**. These numbers can be represented graphically on a horizontal number line with a point labeled as 0. Positive real numbers are graphed to the right of 0, and negative real numbers are graphed to the left. Each point on the number line corresponds to exactly one real number, and for this reason, the line is called the **real number line** (Figure 1-1).

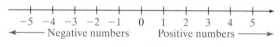

Figure 1-1

Several sets of numbers are **subsets** (or part) of the set of real numbers. These are

- The set of natural numbers
- The set of whole numbers
- The set of integers
- The set of rational numbers
- The set of irrational numbers

Objectives

1. Set of Real Numbers
2. Inequalities
3. Interval Notation
4. Union and Intersection of Sets
5. Translations Involving Inequalities

Concept Connections

Describe the following sets, using set-builder notation.

1. {January, June, July}

Answer

1. $\{x \mid x$ is the name of a month beginning with the letter J$\}$

4   Chapter 1   Review of Basic Algebraic Concepts

Concept Connections

2. List the elements in the set $\{x \mid x$ is a natural number less than 6$\}$.

Definition of the Natural Numbers, Whole Numbers, and Integers

The set of **natural numbers** is $\{1, 2, 3, \ldots\}$.
The set of **whole numbers** is $\{0, 1, 2, 3, \ldots\}$.
The set of **integers** is $\{\ldots, -3, -2, -1, 0, 1, 2, 3, \ldots\}$.

The set of rational numbers consists of all the numbers that can be defined as a ratio of two integers.

Definition of the Rational Numbers

The set of **rational numbers** is $\{\frac{p}{q} \mid p$ and q are integers and q does not equal zero$\}$.

Skill Practice

Show that the numbers are rational by writing them as a ratio of integers.

3. -9
4. 0.45
5. $0.\overline{3}$
6. 0

Tip: Any rational number can be represented by a terminating decimal or by a repeating decimal.

example 1   Identifying Rational Numbers

Show that each number is a rational number by finding two integers whose ratio equals the given number.

a. $\dfrac{-4}{7}$ b. 8 c. $0.\overline{6}$ d. 0.87

Solution:

a. $\dfrac{-4}{7}$ is a rational number because it can be expressed as the ratio of the integers -4 and 7.

b. 8 is a rational number because it can be expressed as the ratio of the integers 8 and 1 ($8 = \frac{8}{1}$). In this example we see that *an integer is also a rational number.*

c. $0.\overline{6}$ represents the repeating decimal $0.6666666\ldots$ and can be expressed as the ratio of 2 and 3 ($0.\overline{6} = \frac{2}{3}$). In this example we see that *a repeating decimal is a rational number.*

d. 0.87 is the ratio of 87 and 100 ($0.87 = \frac{87}{100}$). In this example we see that *a terminating decimal is a rational number.*

Some real numbers such as the number π (pi) cannot be represented by the ratio of two integers. In decimal form, an irrational number is a nonterminating, nonrepeating decimal. The value of π, for example, can be approximated as $\pi \approx 3.1415926535897932$. However, the decimal digits continue indefinitely with no pattern. Other examples of irrational numbers are the square roots of nonperfect squares, such as $\sqrt{3}$ and $\sqrt{11}$.

Definition of the Irrational Numbers

The set of **irrational numbers** is $\{x \mid x$ is a real number that is not rational$\}$.
Note: An irrational number cannot be written as a terminating decimal or as a repeating decimal.

Answers

2. $\{1, 2, 3, 4, 5\}$
3. $\dfrac{-9}{1}$ 4. $\dfrac{45}{100}$
5. $\dfrac{1}{3}$ 6. $\dfrac{0}{1}$

The set of real numbers consists of both the rational numbers and the irrational numbers. The relationships among the sets of numbers discussed thus far are illustrated in Figure 1-2.

Section 1.1 Sets of Numbers and Interval Notation 5

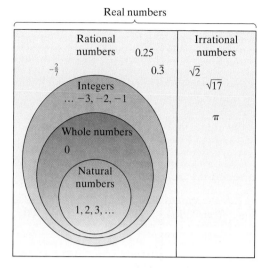

Figure 1-2

Concept Connections

7. Explain how to determine if a decimal number is rational or irrational.

example 2 — Classifying Numbers by Set

Check the set(s) to which each number belongs. The numbers may belong to more than one set.

	Natural Numbers	Whole Numbers	Integers	Rational Numbers	Irrational Numbers	Real Numbers
-6						
$\sqrt{23}$						
$-\frac{2}{7}$						
3						
2.35						

Solution:

	Natural Numbers	Whole Numbers	Integers	Rational Numbers	Irrational Numbers	Real Numbers
-6			✓	✓		✓
$\sqrt{23}$					✓	✓
$-\frac{2}{7}$				✓		✓
3	✓	✓	✓	✓		✓
2.35				✓		✓

Skill Practice

8. Check the set(s) to which each number belongs.

	1	0.47	$\sqrt{5}$	$-\frac{1}{2}$
Natural				
Whole				
Integer				
Rational				
Irrational				
Real				

Answers

7. A decimal number that terminates or repeats is rational. A decimal number that is nonterminating and nonrepeating is irrational.

8.
	1	0.47	$\sqrt{5}$	$-\frac{1}{2}$
Natural	✓			
Whole	✓			
Integer	✓			
Rational	✓	✓		✓
Irrational			✓	
Real	✓	✓	✓	✓

2. Inequalities

The relative size of two numbers can be compared by using the real number line. We say that a is less than b (written mathematically as $a < b$) if a lies to the left of b on the number line.

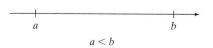

$a < b$

We say that *a* is greater than *b* (written mathematically as $a > b$) if *a* lies to the right of *b* on the number line.

$a > b$

Table 1-1 summarizes the relational operators that compare two real numbers *a* and *b*.

table 1-1

Mathematical Expression	Translation	Other Meanings
$a < b$	*a* is less than *b*	*b* exceeds *a* *b* is greater than *a*
$a > b$	*a* is greater than *b*	*a* exceeds *b* *b* is less than *a*
$a \leq b$	*a* is less than or equal to *b*	*a* is at most *b* *a* is no more than *b*
$a \geq b$	*a* is greater than or equal to *b*	*a* is no less than *b* *a* is at least *b*
$a = b$	*a* is equal to *b*	
$a \neq b$	*a* is not equal to *b*	
$a \approx b$	*a* is approximately equal to *b*	

The symbols $<$, $>$, $\leq$, $\geq$, and $\neq$ are called inequality signs, and the expressions $a < b$, $a > b$, $a \leq b$, $a \geq b$, and $a \neq b$ are called **inequalities**.

Skill Practice

Fill in the blanks with the appropriate symbol, $<$ or $>$.

9. 2 _____ -12
10. -7.2 _____ -4.6
11. $\dfrac{1}{4}$ _____ $\dfrac{2}{9}$

example 3   Ordering Real Numbers

Fill in the blank with the appropriate inequality symbol: $<$ or $>$

a. -2 _____ -5 b. $\dfrac{4}{7}$ _____ $\dfrac{3}{5}$ c. -1.3 _____ 2.8

Solution:

a. $-2 \;\;>\;\; -5$

b. To compare $\frac{4}{7}$ and $\frac{3}{5}$, write the fractions as equivalent fractions with a common denominator.

$$\frac{4}{7} \cdot \frac{5}{5} = \frac{20}{35} \quad \text{and} \quad \frac{3}{5} \cdot \frac{7}{7} = \frac{21}{35}$$

Because $\dfrac{20}{35} < \dfrac{21}{35}$, then $\dfrac{4}{7} \;\;<\;\; \dfrac{3}{5}$

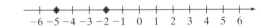

c. $-1.3 \;\;<\;\; 2.8$

Answers

9. $>$ 10. $<$ 11. $>$

3. Interval Notation

The set $\{x \mid x \geq 3\}$ represents all real numbers greater than or equal to 3. This set can be illustrated graphically on the number line.

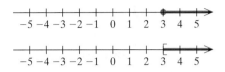

By convention, a closed circle ● or a square bracket [is used to indicate that an "endpoint" ($x = 3$) *is included* in the set. This interval is a closed interval because its endpoint is included.

The set $\{x \mid x > 3\}$ represents all real numbers strictly greater than 3. This set can be illustrated graphically on the number line.

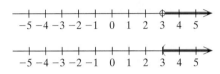

By convention, an open circle ○ or a parenthesis (is used to indicate that an "endpoint" ($x = 3$) is *not* included in the set. This interval is an open interval because its endpoint is *not* included.

Notice that the sets $\{x \mid x \geq 3\}$ and $\{x \mid x > 3\}$ consist of an infinite number of elements that cannot all be listed. Another method to represent the elements of such sets is by using **interval notation**. To understand interval notation, first consider the real number line, which extends infinitely far to the left and right. The symbol ∞ is used to represent infinity. The symbol $-\infty$ is used to represent negative infinity.

To express a set of real numbers in interval notation, sketch the graph first, using the symbols ( ) or [ ]. Then use these symbols at the endpoints to define the interval.

example 4 Expressing Sets by Using Interval Notation

Graph the sets on the number line, and express the set in interval notation.

a. $\{x \mid x \geq 3\}$ b. $\{x \mid x > 3\}$ c. $\{x \mid x \leq -\frac{3}{2}\}$

Solution:

a. **Set-Builder Notation**   **Graph**   **Interval Notation**

$\{x \mid x \geq 3\}$ [3, ∞)

```
-∞ —+—+—+—+—+—+—+—[—+—+—→ ∞
   -5 -4 -3 -2 -1  0  1  2  3  4  5
                                ↓        ↓
                               [3   ,   ∞)
```

The graph of the set $\{x \mid x \geq 3\}$ "begins" at 3 and extends infinitely far to the right. The corresponding interval notation "begins" at 3 and extends to ∞. Notice that a square bracket [is used at 3 for both the graph and the interval notation to include $x = 3$. *A parenthesis is always used at ∞ (and at $-\infty$) because there is no endpoint.*

Skill Practice

Graph and express the set, using interval notation.

12. $\{x \mid x < 0\}$

13. $\{w \mid w \geq -7\}$

14. $\{y \mid y > 3.5\}$

Answers

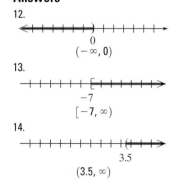

b. Set-Builder Notation   **Graph**   **Interval Notation**

$\{x \mid x > 3\}$ $(3, \infty)$

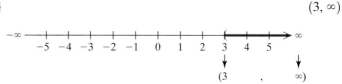

c. Set-Builder Notation   **Graph**   **Interval Notation**

$\{x \mid x \leq -\tfrac{3}{2}\}$ $(-\infty, -\tfrac{3}{2}]$

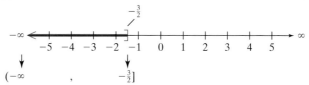

The graph of the set $\{x \mid x \leq -\tfrac{3}{2}\}$ extends infinitely far to the left. Interval notation is always written from left to right. Therefore, $-\infty$ is written first, followed by a comma, and then followed by the right-hand endpoint $-\tfrac{3}{2}$.

Using Interval Notation

- The endpoints used in interval notation are always written from left to right. That is, the smaller number is written first, followed by a comma, followed by the larger number.
- Parentheses) or (indicate that an endpoint is *excluded* from the set.
- Square brackets] or [indicate that an endpoint is *included* in the set.
- Parentheses are always used with ∞ and $-\infty$.

Table 1-2 summarizes the solution sets for four general inequalities.

Concept Connections

Graph and express in set-builder notation.

15. $[4, \infty)$

16. $\left(-\infty, \tfrac{1}{2}\right)$

table 1-2

Set-Builder Notation	Graph	Interval Notation
$\{x \mid x > a\}$	──(──→ at a	(a, ∞)
$\{x \mid x \geq a\}$	──[──→ at a	$[a, \infty)$
$\{x \mid x < a\}$	←──)── at a	$(-\infty, a)$
$\{x \mid x \leq a\}$	←──]── at a	$(-\infty, a]$

4. Union and Intersection of Sets

Two or more sets can be combined by the operations of union and intersection.

A Union B and A Intersection B

The **union** of sets A and B, denoted $A \cup B$, is the set of elements that belong to set A or to set B or to both sets A and B.

The **intersection** of two sets A and B, denoted $A \cap B$, is the set of elements common to both A and B.

Answers

15.
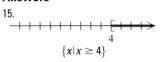
$\{x \mid x \geq 4\}$

16.
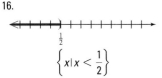
$\left\{x \mid x < \tfrac{1}{2}\right\}$

The concepts of the union and intersection of two sets are illustrated in Figures 1-3 and 1-4:

$A \cup B$
A union B
The elements in A or B or both

Figure 1-3

$A \cap B$
A intersection B
The elements in A and B

Figure 1-4

example 5 — Finding the Union and Intersection of Sets

Given the sets: $A = \{a, b, c, d, e, f\}$ $B = \{a, c, e, g, i, k\}$ $C = \{g, h, i, j, k\}$

Find: **a.** $A \cup B$ **b.** $A \cap B$ **c.** $A \cap C$

Solution:

a. $A \cup B = \{a, b, c, d, e, f, g, i, k\}$ The union of A and B includes all the elements of A along with all the elements of B. Notice that the elements a, c, and e are not listed twice.

b. $A \cap B = \{a, c, e\}$ The intersection of A and B includes only those elements that are common to both sets.

c. $A \cap C = \{\ \}$ (the **empty set**) Because A and C share no common elements, the intersection of A and C is the empty, or null, set.

Skill Practice

Given:
$A = \{r, s, t, u, v, w\}$
$B = \{s, v, w, y, z\}$
$C = \{x, y, z\}$
Find:
17. $A \cap B$
18. $A \cap C$
19. $B \cup C$

Tip: The empty set may be denoted by the symbol $\{\ \}$ or by the symbol $\emptyset$.

example 6 — Finding the Union and Intersection of Sets

Given the sets: $A = \{x \mid x < 3\}$ $B = \{x \mid x \geq -2\}$ $C = \{x \mid x \geq 5\}$

Graph the following sets. Then express each set in interval notation.

a. $A \cap B$ **b.** $A \cup C$ **c.** $A \cup B$ **d.** $A \cap C$

Solution:

It is helpful to visualize the graphs of individual sets on the number line before taking the union or intersection.

a. Graph of $A = \{x \mid x < 3\}$

Graph of $B = \{x \mid x \geq -2\}$

Graph of $A \cap B$ (the "overlap")

Interval notation: $[-2, 3)$

Note that the set $A \cap B$ represents the real numbers greater than or equal to -2 and less than 3. This relationship can be written more concisely as a compound inequality: $-2 \leq x < 3$. We can interpret this inequality as "x is between -2 and 3, including $x = -2$."

Skill Practice

Find the intersection or union by first graphing the sets on the real number line. State answers in interval notation.

20. $\{x \mid x \geq 1\} \cap \{x \mid x < 10\}$
21. $\{x \mid x \leq -2\} \cup \{x \mid x > 3\}$
22. $\{x \mid x > -2\} \cup \{x \mid x < 3\}$
23. $\{x \mid x < 1\} \cap \{x \mid x > 5\}$

Answers
17. $\{s, v, w\}$ 18. $\{\ \}$
19. $\{s, v, w, x, y, z\}$ 20. $[1, 10)$
21. $(-\infty, -2] \cup (3, \infty)$
22. $(-\infty, \infty)$ 23. $\{\ \}$

b. Graph of $A = \{x \mid x < 3\}$

Graph of $C = \{x \mid x \geq 5\}$

Graph of $A \cup C$

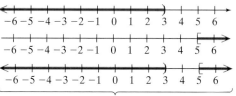

Interval notation: $(-\infty, 3) \cup [5, \infty)$

$A \cup C$ includes all elements from set A along with the elements from set C.

c. Graph of $A = \{x \mid x < 3\}$

Graph of $B = \{x \mid x \geq -2\}$

Graph of $A \cup B$

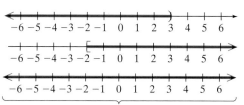

Interval notation: $(-\infty, \infty)$

$A \cup B$ includes all elements from set A along with the elements of set B. This encompasses all real numbers.

d. Graph of $A = \{x \mid x < 3\}$

Graph of $C = \{x \mid x \geq 5\}$

Graph of $A \cap C$
(the sets do not "overlap")
$A \cap C$ is the empty set $\{\ \}$.

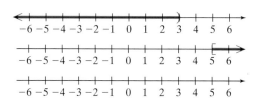

5. Translations Involving Inequalities

In Table 1-1, we learned that phrases such as *at least*, *at most*, *no more than*, *no less than*, and *between* can be translated into mathematical terms by using inequality signs.

Skill Practice

Translate the italicized phrase to a mathematical inequality.

24. The gas mileage, *m*, for a Honda Civic is *at least 30 mpg*.

25. The gas mileage, *m*, for a Harley Davidson motorcycle is *more than 45 mpg*.

example 7 — Translating Inequalities

The intensity of a hurricane is often defined according to its maximum sustained winds, for which wind speed is measured to the nearest mile per hour. Translate the italicized phrases into mathematical inequalities.

a. A tropical storm is updated to hurricane status if the sustained wind speed, *w*, *is at least 74 mph*.

b. Hurricanes are categorized according to intensity by the Saffir-Simpson scale. On a scale of 1 to 5, a category 5 hurricane is the most destructive. A category 5 hurricane has sustained winds, *w*, *exceeding 155 mph*.

c. A category 4 hurricane has sustained winds, *w*, *of at least 131 mph but no more than 155 mph*.

Solution:

a. $w \geq 74$ mph   **b.** $w > 155$ mph   **c.** 131 mph $\leq w \leq 155$ mph

Answers

24. $m \geq 30$ 25. $m > 45$

Section 1.1 Practice Exercises

Boost your GRADE at mathzone.com!

- Practice Problems
- Self-Tests
- NetTutor
- e-Professors
- Videos

Study Skills Exercises

1. In this text we will provide skills for you to enhance your learning experience. Each set of Practice Exercises will begin with an activity that focuses on one of eight areas: learning about your course, using your text, taking notes, doing homework, taking an exam (test and math anxiety), managing your time, recognizing your learning style, and studying for the final exam.

   Each activity requires only a few minutes and will help you to pass this class and become a better math student. Many of these skills can be carried over to other disciplines and help you to become a model college student.

   To begin, write down the following information.

   a. Instructor's name

   b. Days of the week that the class meets

   c. The room number in which the class meets

   d. Is there a lab requirement for this course? If so, what is the requirement and what is the location of the lab?

2. Define the key terms.

   a. Set
   b. Set-builder notation
   c. Real numbers
   d. Real number line
   e. Subsets
   f. Natural numbers
   g. Whole numbers
   h. Integers
   i. Rational numbers
   j. Irrational numbers
   k. Inequalities
   l. Interval notation
   m. Union
   n. Intersection
   o. Empty set

Objective 1: Set of Real Numbers

3. Plot the numbers on the number line.

   $\{1.7, \pi, -5, 4.\overline{2}\}$

4. Plot the numbers on the number line.

   $\{1\frac{1}{2}, 0, -3, -\frac{1}{2}, \frac{3}{4}\}$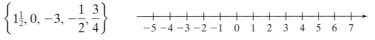

12 Chapter 1 Review of Basic Algebraic Concepts

For Exercises 5–9, show that each number is a rational number by finding two integers whose ratio equals the given number. (See Example 1.)

5. -10

6. $1\frac{1}{2}$

7. $-\frac{3}{5}$

8. -0.1

9. 0

10. Check the sets to which each number belongs. (See Example 2.)

	Real Numbers	Irrational Numbers	Rational Numbers	Integers	Whole Numbers	Natural Numbers
5						
$-\sqrt{9}$						
-1.7						
$\frac{1}{2}$						
$\sqrt{7}$						
$\frac{0}{4}$						
$0.\overline{2}$						

11. Check the sets to which each number belongs.

	Real Numbers	Irrational Numbers	Rational Numbers	Integers	Whole Numbers	Natural Numbers
$\frac{6}{8}$						
$1\frac{1}{2}$						
π						
0						
$0.\overline{8}$						
$\frac{8}{2}$						
$4.\overline{2}$						

Objective 2: Inequalities

For Exercises 12–17, fill in the blanks with the appropriate symbol: $<$ or $>$. (See Example 3.)

12. -9 ___ -1

13. 0 ___ -6

14. 0.15 ___ 0.04

15. -2.5 ___ 0.6

16. $\frac{5}{3}$ ___ $\frac{10}{7}$

17. $-\frac{21}{5}$ ___ $-\frac{17}{4}$

Objective 3: Interval Notation

For Exercises 18–25, express the set in interval notation. (See Example 4.)

18. ——(——→ 2

19. ——(——→ $\frac{5}{6}$

20. ←——]—— 0

21. ←——]—— 9

22. ——(——]—— −5 0

23. ——[——)—— −1 15

24. ——[——→ −1

25. ——[——→ 12.8

For Exercises 26–43, graph the sets and express each set in interval notation.

26. $\{x \mid x > 3\}$

27. $\{x \mid x < 3\}$

28. $\{y \mid y \leq -2\}$

29. $\{z \mid z \geq -4\}$

30. $\{w \mid w < \frac{9}{2}\}$

31. $\{p \mid p \geq -\frac{7}{3}\}$

32. $\{x \mid -2.5 < x \leq 4.5\}$

33. $\{x \mid -6 \leq x < 0\}$

34. All real numbers less than −3.

35. All real numbers greater than 2.34.

36. All real numbers greater than $\frac{5}{2}$.

37. All real numbers less than $\frac{4}{7}$.

38. All real numbers not less than 2.

39. All real numbers no more than 5.

40. All real numbers between −4 and 4.

41. All real numbers between −7 and −1.

42. All real numbers between −3 and 0, inclusive.

43. All real numbers between −1 and 6, inclusive.

For Exercises 44–51, write an expression in words that describes the set of numbers given by each interval. (Answers may vary.)

44. $(-\infty, -4)$

45. $[2, \infty)$

46. $(-2, 7]$

47. $(-3.9, 0)$

48. $[-180, 90]$

49. $(-\infty, \infty)$

50. $(3.2, \infty)$

51. $(-\infty, -1]$

Objective 4: Union and Intersection of Sets

52. Given: $M = \{-3, -1, 1, 3, 5\}$ and $N = \{-4, -3, -2, -1, 0\}$.

List the elements of the following sets:

a. $M \cap N$ **b.** $M \cup N$

53. Given: $P = \{a, b, c, d, e, f, g, h, i\}$ and $Q = \{a, e, i, o, u\}$.

List the elements of the following sets. **(See Example 5.)**

a. $P \cap Q$ **b.** $P \cup Q$

Let $A = \{x \mid x > -3\}$, $B = \{x \mid x \leq 0\}$, $C = \{x \mid -1 \leq x < 4\}$, and $D = \{x \mid 1 < x < 3\}$. For Exercises 54–61, graph the sets described here. Then express the answer in set-builder notation and in interval notation. **(See Example 6.)**

54. $A \cap B$

55. $A \cup B$

56. $B \cup C$

57. $B \cap C$

58. $C \cup D$

59. $C \cap D$

60. $B \cap D$

61. $A \cup D$

Let $X = \{x \mid x \geq -10\}$, $Y = \{x \mid x < 1\}$, $Z = \{x \mid x > -1\}$, and $W = \{x \mid x \leq -3\}$. For Exercises 62–67, find the intersection or union of the sets X, Y, Z, and W.

62. $X \cap Y$ **63.** $X \cup Y$ **64.** $Y \cup Z$

65. $Y \cap Z$ **66.** $Z \cup W$ **67.** $Z \cap W$

Objective 5: Translations Involving Inequalities

The following chart defines the ranges for normal blood pressure, high normal blood pressure, and high blood pressure (*hypertension*). All values are measured in millimeters of mercury, mm Hg. (Source: American Heart Association.)

Normal	Systolic less than 130	Diastolic less than 85
High normal	Systolic 130–139	Diastolic 85–89
Hypertension	Systolic 140 or greater	Diastolic 90 or greater

For Exercises 68–72, write an inequality using the variable p that represents each condition. **(See Example 7.)**

68. Normal systolic blood pressure

69. Diastolic pressure in hypertension

70. High normal range for systolic pressure

71. Systolic pressure in hypertension

72. Normal diastolic blood pressure

A pH scale determines whether a solution is acidic or alkaline. The pH scale runs from 0 to 14, with 0 being the most acidic and 14 being the most alkaline. A pH of 7 is neutral (distilled water has a pH of 7).

For Exercises 73–77, write the pH ranges as inequalities and label the substances as acidic or alkaline.

73. Lemon juice: 2.2 through 2.4, inclusive

74. Eggs: 7.6 through 8.0, inclusive

75. Carbonated soft drinks: 3.0 through 3.5, inclusive

76. Milk: 6.6 through 6.9, inclusive

77. Milk of magnesia: 10.0 through 11.0, inclusive

Expanding Your Skills

For Exercises 78–89, find the intersection or union of the following sets given.

78. $[1, 3) \cap (2, 7)$
79. $(-\infty, 0) \cap (-2, 5)$
80. $[-2, 4] \cap (3, \infty)$
81. $(-6, 0) \cap [-2, 9]$

82. $[-2, 7] \cup (-\infty, -1)$
83. $(-\infty, 0) \cup (-4, 1)$
84. $(2, 5) \cup (4, \infty)$
85. $[-6, -1] \cup (-2, \infty)$

86. $(-\infty, 3) \cup (-1, \infty)$
87. $(-\infty, -3) \cap (-1, \infty)$
88. $(-\infty, -8) \cap (0, \infty)$
89. $(-\infty, 8) \cup (0, \infty)$

section 1.2 Operations on Real Numbers

1. Opposite and Absolute Value

Several key definitions are associated with the set of real numbers and constitute the foundation of algebra. Two important definitions are the opposite of a real number and the absolute value of a real number.

Definition of the Opposite of a Real Number

Two numbers that are the same distance from 0 but on opposite sides of 0 on the number line are called **opposites** of each other. Symbolically, we denote the opposite of a real number a as $-a$.

The numbers -4 and 4 are opposites of each other. Similarly, the numbers $\frac{3}{2}$ and $-\frac{3}{2}$ are opposites.

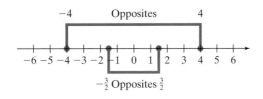

Objectives

1. Opposite and Absolute Value
2. Addition and Subtraction of Real Numbers
3. Multiplication and Division of Real Numbers
4. Exponential Expressions
5. Square Roots
6. Order of Operations
7. Evaluating Expressions

Concept Connections

1. If x is positive, then is $-x$ positive or negative?
2. If x is negative, then is $-x$ positive or negative?

Answers

1. Negative   2. Positive

The Absolute Value of a Real Number

The **absolute value** of a real number a, denoted $|a|$, is the distance between a and 0 on the number line.

Note: The absolute value of any real number is *nonnegative*.

For example: $|5| = 5$ and $|-5| = 5$

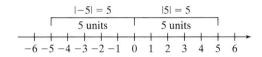

Skill Practice

Simplify.

3. $|-92|$
4. $|7.6|$
5. $-|2|$

example 1   Evaluating Absolute Value Expressions

Simplify the expressions: **a.** $|-2.5|$ **b.** $\left|\frac{5}{4}\right|$ **c.** $-|-4|$

Solution:

a. $|-2.5| = 2.5$

b. $\left|\frac{5}{4}\right| = \frac{5}{4}$

c. $-|-4| = -(4) = -4$

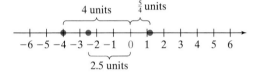

Calculator Connections

Some calculators have an absolute value function. For example,

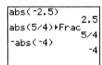

The absolute value of a number a is its distance from zero on the number line. The definition of $|a|$ may also be given algebraically depending on whether a is negative or nonnegative.

Definition of the Absolute Value of a Real Number

Let a be a real number. Then

1. If a is nonnegative (that is, $a \geq 0$), then $|a| = a$.
2. If a is negative (that is, $a < 0$), then $|a| = -a$.

This definition states that if a is a nonnegative number, then $|a|$ equals a itself. If a is a negative number, then $|a|$ equals the opposite of a. For example,

$|9| = 9$ Because 9 is positive, $|9|$ equals the number 9 itself.

$|-7| = 7$ Because -7 is negative, $|-7|$ equals the opposite of -7, which is 7.

Answers

3. 92   4. 7.6   5. -2

2. Addition and Subtraction of Real Numbers

> **Addition of Real Numbers**
> 1. To add two numbers with the *same sign*, add their absolute values and apply the common sign to the sum.
> 2. To add two numbers with *different signs*, subtract the smaller absolute value from the larger absolute value. Then apply the sign of the number having the larger absolute value.

example 2   Adding Real Numbers

Perform the indicated operations:

a. $-2 + (-6)$ **b.** $-10.3 + 13.8$ **c.** $\frac{5}{6} + \left(-1\frac{1}{4}\right)$

Skill Practice

6. $-4 + (-1)$
7. $-2.6 + 1.8$
8. $-1 + \left(-\frac{3}{7}\right)$
9. $-9 + 6 + 13 + (-5)$

Solution:

a.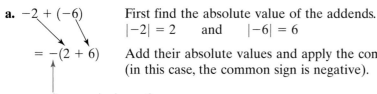

$-2 + (-6)$ First find the absolute value of the addends.
 $|-2| = 2$ and $|-6| = 6$

$= -(2 + 6)$ Add their absolute values and apply the common sign (in this case, the common sign is negative).

Common sign is negative.

$= -8$ The sum is -8.

b. $-10.3 + 13.8$ First find the absolute value of the addends.
 $|-10.3| = 10.3$ and $|13.8| = 13.8$

The absolute value of 13.8 is greater than the absolute value of -10.3. Therefore, the sum is positive.

$= +(13.8 - 10.3)$ Subtract the smaller absolute value from the larger absolute value.

Apply the sign of the number with the larger absolute value.

$= 3.5$

c. $\dfrac{5}{6} + \left(-1\dfrac{1}{4}\right)$

$= \dfrac{5}{6} + \left(-\dfrac{5}{4}\right)$ Write $-1\dfrac{1}{4}$ as an improper fraction.

$= \dfrac{5 \cdot 2}{6 \cdot 2} + \left(-\dfrac{5 \cdot 3}{4 \cdot 3}\right)$ The LCD is 12. Write each fraction with the LCD.

$= \dfrac{10}{12} + \left(-\dfrac{15}{12}\right)$ Find the absolute value of the addends.

$\left|\dfrac{10}{12}\right| = \dfrac{10}{12}$ and $\left|-\dfrac{15}{12}\right| = \dfrac{15}{12}$

The absolute value of $-\dfrac{15}{12}$ is greater than the absolute value of $\dfrac{10}{12}$. Therefore, the sum is negative.

Answers
6. -5 7. -0.8
8. $-\dfrac{10}{7}$ 9. 5

$$= -\left(\frac{15}{12} - \frac{10}{12}\right)$$ Subtract the smaller absolute value from the larger absolute value.

Apply the sign of the number with the larger absolute value.

$$= -\frac{5}{12}$$ Subtract.

Subtraction of real numbers is defined in terms of the addition process. To subtract two real numbers, add the opposite of the second number to the first number.

Subtraction of Real Numbers

If a and b are real numbers, then $\quad a - b = a + (-b)$

Skill Practice

Subtract.

10. $9 - (-8)$
11. $-5 - (-2)$
12. $-1.1 - 3$
13. $\dfrac{1}{6} - \dfrac{3}{4}$

example 3   Subtracting Real Numbers

Perform the indicated operations.

a. $-13 - 5$ b. $2.7 - (-3.8)$ c. $\dfrac{5}{2} - 4\dfrac{2}{3}$

Solution:

a. $-13 - 5$

$\quad = -13 + (-5)$ Add the opposite of the second number to the first number.

$\quad = -18$ Add.

b. $2.7 - (-3.8)$

$\quad = 2.7 + (3.8)$ Add the opposite of the second number to the first number.

$\quad = 6.5$ Add.

c. $\dfrac{5}{2} - 4\dfrac{2}{3}$

$\quad = \dfrac{5}{2} + \left(-4\dfrac{2}{3}\right)$ Add the opposite of the second number to the first number.

$\quad = \dfrac{5}{2} + \left(-\dfrac{14}{3}\right)$ Write the mixed number as a fraction.

$\quad = \dfrac{15}{6} + \left(-\dfrac{28}{6}\right)$ Get a common denominator and add.

$\quad = -\dfrac{13}{6}$ or $-2\dfrac{1}{6}$

Answers

10. 17   11. -3
12. -4.1 13. $-\dfrac{7}{12}$

3. Multiplication and Division of Real Numbers

The sign of the product of two real numbers is determined by the signs of the factors.

> **Multiplication of Real Numbers**
> 1. The product of two real numbers with the *same* sign is *positive*.
> 2. The product of two real numbers with *different* signs is *negative*.
> 3. The product of any real number and zero is *zero*.

example 4 **Multiplying Real Numbers**

Multiply the real numbers.

a. $(2)(-5.1)$ b. $-\dfrac{2}{3} \cdot \dfrac{9}{8}$ c. $\left(-3\dfrac{1}{3}\right)\left(-\dfrac{3}{10}\right)$

Solution:

a. $(2)(-5.1)$
$= -10.2$ *Different* signs. The product is negative.

b. $-\dfrac{2}{3} \cdot \dfrac{9}{8}$

$= -\dfrac{18}{24}$ *Different* signs. The product is negative.

$= -\dfrac{3}{4}$ Simplify to lowest terms.

c. $\left(-3\dfrac{1}{3}\right)\left(-\dfrac{3}{10}\right)$

$= \left(-\dfrac{10}{3}\right)\left(-\dfrac{3}{10}\right)$ Write the mixed number as a fraction.

$= \dfrac{30}{30}$ *Same* signs. The product is positive.

$= 1$ Simplify to lowest terms.

Skill Practice

Multiply.

14. $(-5)(11)$
15. $(-4)\left(\dfrac{2}{3}\right)$
16. $(-5.6)(-2.2)$

Notice from Example 4(c) that $\left(-\dfrac{10}{3}\right)\left(-\dfrac{3}{10}\right) = 1$. If the product of two numbers is 1, then the numbers are said to be **reciprocals**. That is, the reciprocal of a real number a is $\dfrac{1}{a}$. Furthermore, $a \cdot \dfrac{1}{a} = 1$.

> **Tip:** A number and its reciprocal have the same sign. For example:
> $\left(-\dfrac{10}{3}\right)\left(-\dfrac{3}{10}\right) = 1$ and $3 \cdot \dfrac{1}{3} = 1$

Answers

14. -55 15. $-\dfrac{8}{3}$
16. 12.32

Recall that subtraction of real numbers was defined in terms of addition. In a similar way, division of real numbers can be defined in terms of multiplication. *To divide two real numbers, multiply the first number by the reciprocal of the second number.* For example:

$$10 \div 5 = 2 \quad \text{or equivalently} \quad 10 \cdot \frac{1}{5} = 2$$

Because division of real numbers can be expressed in terms of multiplication, the sign rules that apply to multiplication also apply to division.

$$10 \div 2 = 10 \cdot \frac{1}{2} = 5$$
$$-10 \div (-2) = -10 \cdot \left(-\frac{1}{2}\right) = 5$$

Dividing two numbers of the same sign produces a *positive* quotient.

$$10 \div (-2) = 10\left(-\frac{1}{2}\right) = -5$$
$$-10 \div 2 = -10 \cdot \frac{1}{2} = -5$$

Dividing two numbers of opposite signs produces a *negative* quotient.

Concept Connections

17. If $x > 0$ and $y < 0$, which of the following is true about the value of xy?

 a. xy is positive.
 b. xy is negative.
 c. The sign of xy cannot be determined.

18. If $x > 0$ and $y < 0$, which of the following is true about the value of $(x + y)$?

 a. $(x + y)$ is positive.
 b. $(x + y)$ is negative.
 c. The sign of $(x + y)$ cannot be determined.

Division of Real Numbers

Assume that a and b are real numbers such that $b \neq 0$.

1. If a and b have the *same* signs, then the quotient $\frac{a}{b}$ is *positive*.
2. If a and b have *different* signs, then the quotient $\frac{a}{b}$ is *negative*.
3. $\frac{0}{b} = 0$.
4. $\frac{b}{0}$ is undefined.

The relationship between multiplication and division can be used to investigate properties 3 and 4 in the preceding box. For example,

$$\frac{0}{6} = 0 \qquad \text{Because } 6 \times 0 = 0 \checkmark$$

$$\frac{6}{0} \text{ is undefined} \qquad \text{Because there is no number that when multiplied by 0 will equal 6}$$

Note: The quotient of 0 and 0 *cannot* be determined. Evaluating an expression of the form $\frac{0}{0} = ?$ is equivalent to asking, "What number times zero will equal 0?" That is, $(0)(?) = 0$. Any real number will satisfy this requirement; however, expressions involving $\frac{0}{0}$ are usually discussed in advanced mathematics courses.

example 5 Dividing Real Numbers

Divide the real numbers. Write the answer as a fraction or whole number.

a. $\dfrac{-42}{7}$ b. $\dfrac{-96}{-144}$ c. $\dfrac{-5}{-7}$ d. $3\dfrac{1}{10} \div \left(-\dfrac{2}{5}\right)$

Answers
17. b 18. c

Solution:

a. $\dfrac{-42}{7} = -6$ *Different* signs. The quotient is negative.

> **Tip:** Recall that multiplication may be used to check a division problem. For example:
> $$\dfrac{-42}{7} = -6 \quad \Rightarrow \quad (7) \cdot (-6) = -42 \checkmark$$

b. $\dfrac{-96}{-144} = \dfrac{2}{3}$ *Same* signs. The quotient is positive. Simplify.

c. $\dfrac{-5}{-7} = \dfrac{5}{7}$ *Same* signs. The quotient is positive.

d. $3\dfrac{1}{10} \div \left(-\dfrac{2}{5}\right)$

$= \dfrac{31}{10}\left(-\dfrac{5}{2}\right)$ Write the mixed number as an improper fraction, and multiply by the reciprocal of the second number.

$= \dfrac{31}{\underset{2}{\cancel{10}}}\left(-\dfrac{\overset{1}{\cancel{5}}}{2}\right)$

$= -\dfrac{31}{4}$ *Different* signs. The quotient is negative.

> **Tip:** If the numerator and denominator of a fraction have opposite signs, then the quotient will be negative. Therefore, a fraction has the same value whether the negative sign is written in the numerator, in the denominator, or in front of a fraction.
> $$-\dfrac{31}{4} = \dfrac{-31}{4} = \dfrac{31}{-4}$$

Skill Practice

Divide.

19. $\dfrac{-28}{-4}$ 20. $\dfrac{42}{-2}$

21. $-\dfrac{2}{3} \div 4$ 22. $\dfrac{-1}{-2}$

> **Tip:** If the numerator and denominator are both negative, then the fraction is positive:
> $$\dfrac{-5}{-7} = \dfrac{5}{7}$$

---

4. Exponential Expressions

To simplify the process of repeated multiplication, exponential notation is often used. For example, the quantity $3 \cdot 3 \cdot 3 \cdot 3 \cdot 3$ can be written as 3^5 (3 to the fifth power).

> **Definition of b^n**
>
> Let b represent any real number and n represent a positive integer. Then
> $$b^n = \underbrace{b \cdot b \cdot b \cdot b \cdots b}_{n\text{-factors of } b}$$
>
> b^n is read as "b to the nth power."
> b is called the **base** and n is called the **exponent**, or **power**.
> b^2 is read as "b squared," and b^3 is read as "b cubed."

Answers

19. 7   20. -21
21. $-\dfrac{1}{6}$ 22. $\dfrac{1}{2}$

Skill Practice

Simplify.

23. 2^3
24. -10^2
25. $(-10)^2$
26. $\left(\dfrac{3}{4}\right)^3$

example 6 Evaluating Exponential Expressions

Simplify the expression.

a. 5^3
b. $(-2)^4$
c. -2^4
d. $\left(-\dfrac{1}{3}\right)^2$

Solution:

a. $5^3 = 5 \cdot 5 \cdot 5$ The base is 5, and the exponent is 3.
   $= 125$

b. $(-2)^4 = (-2)(-2)(-2)(-2)$ The base is -2, and the exponent is 4. The exponent 4 applies to the entire contents of the parentheses.
   $= 16$

c. $-2^4 = -[2 \cdot 2 \cdot 2 \cdot 2]$ The base is 2, and the exponent is 4. Because no parentheses enclose the negative sign, the exponent applies to only 2.
   $= -16$

Tip: The quantity -2^4 can also be interpreted as $-1 \cdot 2^4$.

$$-2^4 = -1 \cdot 2^4 = -1 \cdot (2 \cdot 2 \cdot 2 \cdot 2) = -16$$

d. $\left(-\dfrac{1}{3}\right)^2 = \left(-\dfrac{1}{3}\right)\left(-\dfrac{1}{3}\right)$ The base is $-\dfrac{1}{3}$, and the exponent is 2.
   $= \dfrac{1}{9}$

Calculator Connections

On many calculators, the $\boxed{x^2}$ key is used to square a number. The $\boxed{\wedge}$ key is used to raise a base to any power.

```
5^3
           125
(-2)^4
            16
-2^4
           -16
```

5. Square Roots

The inverse operation to squaring a number is to find its square roots. For example, finding a square root of 9 is equivalent to asking, "What number when squared equals 9?" One obvious answer is 3, because $(3)^2 = 9$. However, -3 is also a square root of 9 because $(-3)^2 = 9$. For now, we will focus on the **principal square root** which is always taken to be nonnegative.

The symbol $\sqrt{}$, called a **radical sign**, is used to denote the principal square root of a number. Therefore, the principal square root of 9 can be written as $\sqrt{9}$. The expression $\sqrt{64}$ represents the principal square root of 64.

Answers

23. 8
24. -100
25. 100
26. $\dfrac{27}{64}$

example 7   Evaluating Square Roots

Evaluate the expressions, if possible.

a. $\sqrt{81}$ b. $\sqrt{\dfrac{25}{64}}$ c. $\sqrt{-16}$

Solution:

a. $\sqrt{81} = 9$ because $(9)^2 = 81$

b. $\sqrt{\dfrac{25}{64}} = \dfrac{5}{8}$ because $\left(\dfrac{5}{8}\right)^2 = \dfrac{25}{64}$

c. $\sqrt{-16}$ is *not a real number* because no real number when squared will be negative.

Skill Practice

Evaluate, if possible.

27. $\sqrt{25}$ 28. $\sqrt{-4}$
29. $\sqrt{0}$ 30. $\sqrt{\dfrac{49}{100}}$

Calculator Connections

The $\boxed{\sqrt{}}$ key is used to find the square root of a non-negative real number.

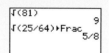

Example 7(c) illustrates that the square root of a negative number is not a real number because no real number when squared will be negative.

The Square Root of a Negative Number

Let a be a negative real number. Then $\sqrt{a}$ is not a real number.

6. Order of Operations

When algebraic expressions contain numerous operations, it is important to evaluate the operations in the proper order. Parentheses ( ), brackets [ ], and braces { } are used for grouping numbers and algebraic expressions. It is important to recognize that operations must be done first within parentheses and other grouping symbols. Other grouping symbols include absolute value bars, radical signs, and fraction bars.

Order of Operations

1. First, simplify expressions within parentheses and other grouping symbols. These include absolute value bars, fraction bars, and radicals. If embedded parentheses are present, start with the innermost parentheses.
2. Evaluate expressions involving exponents, radicals, and absolute values.
3. Perform multiplication or division in the order in which they occur from left to right.
4. Perform addition or subtraction in the order in which they occur from left to right.

Answers

27. 5   28. Not a real number
29. 0   30. $\dfrac{7}{10}$

Skill Practice

Simplify the expressions.

31. $36 \div 2^2 \cdot 3$

32. $-18 - 5 \cdot 2 + 6$

33. $\dfrac{-|5 - 7| + 11}{(-1 - 2)^2}$

example 8 Applying the Order of Operations

Simplify the following expressions.

a. $10 - 5(2 - 5)^2 + 6 \div 3 + \sqrt{16 - 7}$

b. $\dfrac{|(-3)^3 + (5^2 - 3)|}{-15 \div (-3)(2)}$

Solution:

a. $10 - 5(2 - 5)^2 + 6 \div 3 + \sqrt{16 - 7}$

$= 10 - 5(-3)^2 + 6 \div 3 + \sqrt{9}$ Simplify inside the parentheses and radical.

$= 10 - 5(9) + 6 \div 3 + 3$ Simplify exponents and radicals.

$= 10 - 45 + 2 + 3$ Do multiplication and division from left to right.

$= -35 + 2 + 3$ Do addition and subtraction from left to right.

$= -33 + 3$

$= -30$

Tip: Don't try to do too many steps at once. Taking a shortcut may result in a careless error. For each step rewrite the entire expression, changing only the operation being evaluated.

b. $\dfrac{|(-3)^3 + (5^2 - 3)|}{-15 \div (-3)(2)}$ Simplify numerator and denominator separately.

$= \dfrac{|(-3)^3 + (25 - 3)|}{5(2)}$ *Numerator:* Simplify inner parentheses.
Denominator: Do multiplication and division (left to right).

$= \dfrac{|(-3)^3 + (22)|}{10}$ *Numerator:* Simplify inner parentheses.
Denominator: Multiply.

$= \dfrac{|-27 + 22|}{10}$ Simplify exponents.

$= \dfrac{|-5|}{10}$ Add within the absolute value.

$= \dfrac{5}{10} \text{ or } \dfrac{1}{2}$ Evaluate the absolute value.

Calculator Connections

To evaluate the expression

$\dfrac{|(-3)^3 + (5^2 - 3)|}{-15 \div (-3)(2)}$

on a graphing calculator, use parentheses to enclose the absolute value expression. Likewise, it is necessary to use parentheses to enclose the entire denominator.

Answers

31. 27   32. -22 33. 1

7. Evaluating Expressions

The order of operations is followed when evaluating an algebraic expression or when evaluating a geometric formula. For a list of common geometry formulas, see the inside front cover of the text. It is important to note that some geometric formulas use Greek letters (such as π) and some formulas use variables with subscripts. A **subscript** is a number or letter written to the right of and slightly below a variable. Subscripts are used on variables to represent different quantities. For example, the area of a trapezoid is given by $A = \frac{1}{2}(b_1 + b_2)h$. The values of b_1 and b_2 (read as "b sub 1" and "b sub 2") represent the two different bases of the trapezoid (Figure 1-5). This is illustrated in Example 9.

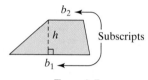

Figure 1-5

example 9 — Evaluating an Algebraic Expression

A homeowner in North Carolina wants to buy protective film for a trapezoid-shaped window. The film will adhere to shattered glass in the event that the glass breaks during a bad storm. Find the area of the window whose dimensions are given in Figure 1-6.

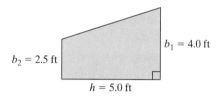

Figure 1-6

Solution:

$A = \dfrac{1}{2}(b_1 + b_2)h$

$= \dfrac{1}{2}(4.0 \text{ ft} + 2.5 \text{ ft})(5.0 \text{ ft})$ Substitute $b_1 = 4.0$ ft, $b_2 = 2.5$ ft, and $h = 5.0$ ft.

$= \dfrac{1}{2}(6.5 \text{ ft})(5.0 \text{ ft})$ Simplify inside parentheses.

$= 16.25 \text{ ft}^2$ Multiply from left to right.

The area of the window is 16.25 ft².

Skill Practice

34. Use the formula given in Example 9 to find the area of the trapezoid.

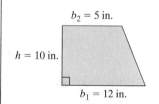

Tip: Subscripts should not be confused with *superscripts*, which are written above a variable. Superscripts are used to denote powers.

$$b_2 \neq b^2$$

Answer

34. The area is 85 in.²

Section 1.2 Practice Exercises

Boost *your* GRADE at mathzone.com!

- Practice Problems
- Self-Tests
- NetTutor
- e-Professors
- Videos

Study Skills Exercises

1. Write the following information and put this information in a place that is easily accessible.

    a. Instructor's office number   b. Instructor's telephone number

    c. Instructor's e-mail address   d. Instructor's office hours

2. Define the key terms:

    a. **Opposite**   b. **Absolute value**   c. **Reciprocal**

    d. **Base**   e. **Exponent**   f. **Power**

    g. **Principal square root**   h. **Radical sign**   i. **Order of operations**

    j. **Subscript**

Review Exercises

For Exercises 3–6, describe the set.

3. Rational numbers ∩ Integers

4. Rational numbers ∪ Irrational numbers

5. Natural numbers ∪ {0}

6. Integers ∩ Whole numbers

Objective 1: Opposite and Absolute Value

7. If the absolute value of a number can be thought of as its distance from zero, explain why an absolute value can never be negative.

8. If a number is negative, then its *opposite* will be   a. Positive   b. Negative.

9. If a number is negative, then its *reciprocal* will be   a. Positive   b. Negative.

10. If a number is negative, then its *absolute value* will be   a. Positive   b. Negative.

11. Complete the table. (See Example 1.)

Number	Opposite	Reciprocal	Absolute Value
6			
	$-\frac{1}{11}$		
		$-\frac{1}{8}$	
	$\frac{13}{10}$		
0			
		$-0.\overline{3}$	

12. Complete the table.

Number	Opposite	Reciprocal	Absolute Value
-9			
	$\frac{2}{3}$		
		14	
-1			
0			
		$2\frac{1}{9}$	

For Exercises 13–20, fill in the blank with the appropriate symbol ($<$, $>$, $=$).

13. $-|6|$ _____ $|-6|$ **14.** $-(-5)$ _____ $-|-5|$ **15.** $|-4|$ _____ $|4|$

16. $-|2|$ _____ (-2) **17.** $-|-1|$ _____ 1 **18.** -3 _____ $-|-7|$

19. $|2 + (-5)|$ _____ $|2| + |-5|$ **20.** $|4 + 3|$ _____ $|4| + |3|$

Objective 2: Addition and Subtraction of Real Numbers

For Exercises 21–36, add or subtract as indicated. (See Examples 2–3.)

21. $-8 + 4$ **22.** $3 + (-7)$ **23.** $-12 + (-7)$ **24.** $-5 + (-11)$

25. $-17 - (-10)$ **26.** $-14 - (-2)$ **27.** $5 - (-9)$ **28.** $8 - (-4)$

29. $-6 - 15$ **30.** $-21 - 4$ **31.** $1.5 - 9.6$ **32.** $4.8 - 10$

33. $\frac{2}{3} + \left(-2\frac{1}{3}\right)$ **34.** $-\frac{4}{7} + \left(1\frac{4}{7}\right)$ **35.** $-\frac{5}{9} - \frac{14}{15}$ **36.** $-6 - \frac{2}{9}$

Objective 3: Multiplication and Division of Real Numbers

For Exercises 37–50, perform the indicated operation. (See Examples 4–5.)

37. $4(-8)$
38. $-21(3)$
39. $\dfrac{2}{9} \cdot \dfrac{12}{7}$
40. $\left(-\dfrac{5}{9}\right) \cdot \left(-1\dfrac{7}{11}\right)$

41. $-\dfrac{2}{3} \div \left(-1\dfrac{5}{7}\right)$
42. $\dfrac{5}{8} \div (-5)$
43. $7 \div 0$
44. $\dfrac{1}{16} \div 0$

45. $0 \div (-3)$
46. $0 \div 11$
47. $(-1.2)(-3.1)$
48. $(4.6)(-2.25)$

49. $(5.418) \div (0.9)$
50. $(6.9) \div (7.5)$

Objective 4: Exponential Expressions

For Exercises 51–58, evaluate the expressions. (See Example 6.)

51. 4^3
52. -2^3
53. -7^2
54. -2^4

55. $(-7)^2$
56. $(-5)^2$
57. $\left(\dfrac{5}{3}\right)^3$
58. $\left(\dfrac{10}{9}\right)^2$

Objective 5: Square Roots

For Exercises 59–66, evaluate the expression, if possible. (See Example 7.)

59. $\sqrt{81}$
60. $\sqrt{1}$
61. $\sqrt{-4}$
62. $\sqrt{-36}$

63. $\sqrt{\dfrac{1}{4}}$
64. $\sqrt{\dfrac{9}{4}}$
65. $-\sqrt{49}$
66. $-\sqrt{100}$

Objective 6: Order of Operations

For Exercises 67–92, simplify by using the order of operations. (See Example 8.)

67. $-\left(\dfrac{3}{4}\right)^2$
68. $-\left(\dfrac{2}{3}\right)^2$
69. $5 + 3^3$
70. $10 - 2^4$

71. $4^3 - 1^3$
72. $(3 + 4)^2$
73. $5 \cdot 2^3$
74. $12 \div 2^2$

75. $6 + 10 \div 2 \cdot 3 - 4$
76. $12 \div 3 \cdot 4 - 18$
77. $4^2 - (5 - 2)^2 \cdot 3$
78. $5 - 3(8 \div 4)^2$

79. $2 - 5(9 - 4\sqrt{25})^2$
80. $5^2 - (\sqrt{9} + 4 \div 2)$
81. $\left(-\dfrac{3}{5}\right)^2 - \dfrac{3}{5} \cdot \dfrac{5}{9} + \dfrac{7}{10}$
82. $\dfrac{1}{2} - \left(\dfrac{2}{3} \div \dfrac{5}{9}\right) + \dfrac{5}{6}$

83. $1.75 \div 0.25 - (1.25)^2$

84. $5.4 - (0.3)^2 \div 0.09$

85. $\dfrac{\sqrt{10^2 - 8^2}}{3^2}$

86. $\dfrac{\sqrt{16 - 7} + 3^2}{\sqrt{16} - \sqrt{4}}$

87. $-|-11 + 5| + |7 - 2|$

88. $-|-8 - 3| - (-8 - 3)$

89. $\dfrac{8(-3) - 6}{-7 - (-2)}$

90. $\dfrac{6(-2) - 8}{-15 - (10)}$

91. $\left(\dfrac{1}{2}\right)^2 + \left(\dfrac{6 - 4}{5}\right)^2 + \left(\dfrac{5 + 2}{10}\right)^2$

92. $\left(\dfrac{2^3}{2^3 + 1}\right)^2 \div \left(\dfrac{8 - (-2)}{3^2}\right)^2$

For Exercises 93–94, find the average of the set of data values by adding the values and dividing by the number of values.

93. Find the average low temperature for a week in January in St. John's, Newfoundland. (Round to the nearest tenth of a degree.)

Day	Mon.	Tues.	Wed.	Thur.	Fri.	Sat.	Sun.
Low temperature	−18°C	−16°C	−20°C	−11°C	−4°C	−3°C	1°C

94. Find the average high temperature for a week in January in St. John's, Newfoundland. (Round to the nearest tenth of a degree.)

Day	Mon.	Tues.	Wed.	Thur.	Fri.	Sat.	Sun.
High temperature	−2°C	−6°C	−7°C	0°C	1°C	8°C	10°C

Objective 7: Evaluating Expressions

95. The formula $C = \frac{5}{9}(F - 32)$ converts temperatures in the Fahrenheit scale to the Celsius scale. Find the equivalent Celsius temperature for each Fahrenheit temperature.

   a. 77°F b. 212°F c. 32°F d. −40°F

96. The formula $F = \frac{9}{5}C + 32$ converts Celsius temperatures to Fahrenheit temperatures. Find the equivalent Fahrenheit temperature for each Celsius temperature.

   a. −5°C b. 0°C c. 37°C d. −40°C

Use the geometry formulas found in the inside front cover of the book to answer Exercises 97–106. **(See Example 9.)**

For Exercises 97–100, find the area.

97. Trapezoid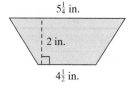

98. Parallelogram

99. Triangle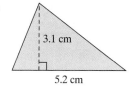

100. Rectangle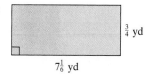

For Exercises 101–106, find the volume. (Use the π key on your calculator, and round the final answer to 1 decimal place.)

101. Sphere

$r = 1.5$ ft

102. Right circular cone

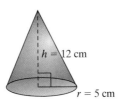

$h = 12$ cm, $r = 5$ cm

103. Right circular cone

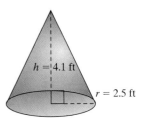

$h = 4.1$ ft, $r = 2.5$ ft

104. Sphere

$r = \frac{1}{2}$ yd

105. Cylinder

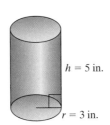

$h = 5$ in., $r = 3$ in.

106. Cylinder

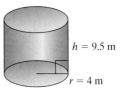

$h = 9.5$ m, $r = 4$ m

Graphing Calculator Exercises

107. Which expression when entered into a graphing calculator will yield the correct value of $\frac{12}{6-2}$?

$$12/6 - 2 \quad \text{or} \quad 12/(6 - 2)$$

108. Which expression when entered into a graphing calculator will yield the correct value of $\frac{24-6}{3}$?

$$(24 - 6)/3 \quad \text{or} \quad 24 - 6/3$$

109. Verify your solution to Exercise 85 by entering the expression into a graphing calculator:

$$(\sqrt{(10^2 - 8^2)})/3^2$$

110. Verify your solution to Exercise 86 by entering the expression into a graphing calculator:

$$(\sqrt{(16 - 7)} + 3^2)/(\sqrt{(16)} - \sqrt{(4)})$$

Section 1.3 Simplifying Expressions

Objectives
1. Recognizing Terms, Factors, and Coefficients
2. Properties of Real Numbers
3. Simplifying Expressions

1. Recognizing Terms, Factors, and Coefficients

An algebraic expression is a single term or a sum of two or more terms. A **term** is a constant or the product of a constant and one or more variables. For example, the expression

$$-6x^2 + 5xyz - 11 \quad \text{or} \quad -6x^2 + 5xyz + (-11)$$

consists of the terms $-6x^2, 5xyz,$ and -11.

The terms $-6x^2$ and $5xyz$ are **variable terms**, and the term -11 is called a **constant term**. It is important to distinguish between a term and the **factors** within a term. For example, the quantity $5xyz$ is one term, but the values $5, x, y,$ and z are factors within the term. The constant factor in a term is called the numerical coefficient or simply **coefficient** of the term. In the terms $-6x^2, 5xyz,$ and -11, the coefficients are $-6, 5,$ and -11, respectively. A term containing only variables such as xy has a coefficient of 1.

Terms are called *like* **terms** if they each have the same variables and the corresponding variables are raised to the same powers. For example:

Like Terms			*Unlike* Terms			
$-6t$	and	$4t$	$-6t$	and	$4s$	(different variables)
$1.8ab$	and	$-3ab$	$1.8xy$	and	$-3x$	(different variables)
$\frac{1}{2}c^2d^3$	and	c^2d^3	$\frac{1}{2}c^2d^3$	and	c^2d	(different powers)
4	and	6	$4p$	and	6	(different variables)

Concept Connections
1. Write two terms that are *like* terms to $7x^2$. (Answers may vary.)

example 1 — Identifying Terms, Factors, Coefficients, and *Like* Terms

a. List the terms of the expression. $\quad -4x^2 - 7x + \frac{2}{3}$
b. Identify the coefficient of the term. $\quad yz^3$
c. Identify the pair of *like* terms. $\quad 16b, 4b^2 \quad \text{or} \quad \frac{1}{2}c, -\frac{1}{6}c$

Solution:

a. The terms of the expression $-4x^2 - 7x + \frac{2}{3}$ are $-4x^2, -7x,$ and $\frac{2}{3}$.

b. The term yz^3 can be written as $1yz^3$; therefore, the coefficient is 1.

c. $\frac{1}{2}c, -\frac{1}{6}c$ are *like* terms because they have the same variable raised to the same power.

Skill Practice

Given: $-2x^2 + 5x + \frac{1}{2} - y^2$

2. List the terms of the expression.
3. Which term is the constant term?
4. Identify the coefficient of the term $-y^2$.

2. Properties of Real Numbers

Simplifying algebraic expressions requires several important properties of real numbers that are stated in Table 1-3. Assume that $a, b,$ and c represent real numbers or real-valued algebraic expressions.

Answers
1. For example: $-4x^2$ and x^2
2. $-2x^2,\ 5x,\ \frac{1}{2},\ -y^2$
3. $\frac{1}{2}$ 4. -1

table 1-3

Property Name	Algebraic Representation	Example	Description/Notes
Commutative property of addition	$a + b = b + a$	$5 + 3 = 3 + 5$	The order in which two real numbers are added or multiplied does not affect the result.
Commutative property of multiplication	$a \cdot b = b \cdot a$	$(5)(3) = (3)(5)$	
Associative property of addition	$(a + b) + c = a + (b + c)$	$(2 + 3) + 7 = 2 + (3 + 7)$	The manner in which two real numbers are grouped under addition or multiplication does not affect the result.
Associative property of multiplication	$(a \cdot b)c = a(b \cdot c)$	$(2 \cdot 3)7 = 2(3 \cdot 7)$	
Distributive property of multiplication over addition	$a(b + c) = ab + ac$	$3(5 + 2) = 3 \cdot 5 + 3 \cdot 2$	A factor outside the parentheses is multiplied by each term inside the parentheses.
Identity property of addition	0 is the identity element for addition because $a + 0 = 0 + a = a$	$5 + 0 = 0 + 5 = 5$	Any number added to the identity element 0 will remain unchanged.
Identity property of multiplication	1 is the identity element for multiplication because $a \cdot 1 = 1 \cdot a = a$	$5 \cdot 1 = 1 \cdot 5 = 5$	Any number multiplied by the identity element 1 will remain unchanged.
Inverse property of addition	a and $(-a)$ are additive inverses because $a + (-a) = 0$ and $(-a) + a = 0$	$3 + (-3) = 0$	The sum of a number and its additive inverse (opposite) is the identity element 0.
Inverse property of multiplication	a and $\frac{1}{a}$ are multiplicative inverses because $a \cdot \frac{1}{a} = 1$ and $\frac{1}{a} \cdot a = 1$ (provided $a \neq 0$)	$5 \cdot \frac{1}{5} = 1$	The product of a number and its multiplicative inverse (reciprocal) is the identity element 1.

The properties of real numbers are used to multiply algebraic expressions. To multiply a term by an algebraic expression containing more than one term, we apply the distributive property of multiplication over addition.

Skill Practice

Apply the distributive property.

5. $10(30y - 40)$

6. $-2(4x - 3y - 6)$

7. $\frac{1}{2}(4a + 7)$

8. $-(7t - 1.6s + 9.2)$

example 2 Applying the Distributive Property

Apply the distributive property.

a. $4(2x + 5)$

b. $-(-3.4q + 5.7r)$

c. $-3(a + 2b - 5c)$

d. $-\frac{2}{3}\left(-9x + \frac{3}{8}y - 5\right)$

Answers

5. $300y - 400$
6. $-8x + 6y + 12$
7. $2a + \frac{7}{2}$
8. $-7t + 1.6s - 9.2$

Solution:

a. $4(2x + 5)$

$= 4(2x) + 4(5)$ Apply the distributive property.
$= 8x + 20$ Simplify, using the associative property of multiplication.

b. $-(-3.4q + 5.7r)$ The negative sign preceding the parentheses can be interpreted as a factor of -1.

$= -1(-3.4q + 5.7r)$
$= -1(-3.4q) + (-1)(5.7r)$ Apply the distributive property.
$= 3.4q - 5.7r$

c. $-3(a + 2b - 5c)$

$= -3(a) + (-3)(2b) + (-3)(-5c)$ Apply the distributive property.
$= -3a - 6b + 15c$ Simplify.

d. $-\dfrac{2}{3}\left(-9x + \dfrac{3}{8}y - 5\right)$

$= -\dfrac{2}{3}(-9x) + \left(-\dfrac{2}{3}\right)\left(\dfrac{3}{8}y\right) + \left(-\dfrac{2}{3}\right)(-5)$ Apply the distributive property.

$= \dfrac{18}{3}x - \dfrac{6}{24}y + \dfrac{10}{3}$ Simplify.

$= 6x - \dfrac{1}{4}y + \dfrac{10}{3}$ Reduce to lowest terms.

Tip: When applying the distributive property, a negative factor preceding the parentheses will change the signs of the terms within the parentheses.

$-3(a + 2b - 5c)$
$-3a - 6b + 15c$

Notice that the parentheses are removed after the distributive property is applied. Sometimes this is referred to as clearing parentheses.

Two terms can be added or subtracted only if they are *like* terms. To add or subtract *like* terms, we use the distributive property, as shown in Example 3.

example 3   Using the Distributive Property to Add and Subtract *Like* Terms

Add and subtract as indicated.

a. $-8x + 3x$ **b.** $4.75y^2 - 9.25y^2 + y^2$

Solution:

a. $-8x + 3x$

$= x(-8 + 3)$ Apply the distributive property.
$= x(-5)$ Simplify.
$= -5x$

Skill Practice

Combine *like* terms.
9. $-4y + 7y$
10. $a^2 - 6a^2 + 3a^2$
11. $\dfrac{1}{2}s + \dfrac{2}{3} + \dfrac{3}{4} - \dfrac{5}{3}s$

Answers
9. $3y$ 10. $-2a^2$
11. $-\dfrac{7}{6}s + \dfrac{17}{12}$

b. $4.75y^2 - 9.25y^2 + y^2$

$= 4.75y^2 - 9.25y^2 + 1y^2$ Notice that y^2 is interpreted as $1y^2$.

$= y^2(4.75 - 9.25 + 1)$ Apply the distributive property.

$= y^2(-3.5)$ Simplify.

$= -3.5y^2$

Although the distributive property is used to add and subtract *like* terms, it is tedious to write each step. Observe that adding or subtracting *like* terms is a matter of combining the coefficients and leaving the variable factors unchanged. This can be shown in one step. This shortcut will be used throughout the text. For example:

$$4w + 7w = 11w \qquad 8ab^2 + 10ab^2 - 5ab^2 = 13ab^2$$

3. Simplifying Expressions

Clearing parentheses and combining *like* terms are important tools to simplifying algebraic expressions. This is demonstrated in Example 4.

Skill Practice

Simplify by clearing parentheses and combining *like* terms.

12. $7(2x - 3) + 3(x + 4)$
13. $6z - 3(4z - 2)$
14. $-4(1.5y + 2.2) - (3.5y + 1.8)$
15. $\frac{1}{2}(4x - 1) + \frac{5}{2} - 5x$

example 4 Clearing Parentheses and Combining *Like* Terms

Simplify by clearing parentheses and combining *like* terms.

a. $4 - 3(2x - 8) - 1$

b. $-(3s - 11t) - 5(2t + 8s) - 10s$

c. $2[1.5x + 4.7(x^2 - 5.2x) - 3x]$

d. $-\frac{1}{3}(3w - 6) - \left(\frac{1}{4}w + 4\right)$

Solution:

a. $4 - 3(2x - 8) - 1$

$= 4 - 6x + 24 - 1$ Apply the distributive property.

$= 4 + 24 - 1 - 6x$ Group *like* terms.

$= 27 - 6x$ Combine *like* terms.

$= 27 - 6x$

$= -6x + 27$

Tip: The expression $27 - 6x$ is equal to $-6x + 27$. However, it is customary to write the variable term first.

b. $-(3s - 11t) - 5(2t + 8s) - 10s$

$= -3s + 11t - 10t - 40s - 10s$ Apply the distributive property.

$= -3s - 40s - 10s + 11t - 10t$ Group *like* terms.

$= -53s + t$ Combine *like* terms.

Answers
12. $17x - 9$
13. $-6z + 6$
14. $-9.5y - 10.6$
15. $-3x + 2$

c. $2[1.5x + 4.7(x^2 - 5.2x) - 3x]$

$\quad = 2[1.5x + 4.7x^2 - 24.44x - 3x]$ Apply the distributive property to inner parentheses.

$\quad = 2[1.5x - 24.44x - 3x + 4.7x^2]$ Group *like* terms.

$\quad = 2[-25.94x + 4.7x^2]$ Combine *like* terms.

$\quad = -51.88x + 9.4x^2$ Apply the distributive property.

$\quad = -51.88x + 9.4x^2$

$\quad = 9.4x^2 - 51.88x$

Tip: By using the commutative property of addition, the expression $-51.88x + 9.4x^2$ can also be written as $9.4x^2 + (-51.88x)$ or simply $9.4x^2 - 51.88x$. Although the expressions are all equal, it is customary to write the terms in descending order of the powers of the variable.

d. $-\dfrac{1}{3}(3w - 6) - \left(\dfrac{1}{4}w + 4\right)$

$\quad = -\dfrac{3}{3}w + \dfrac{6}{3} - \dfrac{1}{4}w - 4$ Apply the distributive property.

$\quad = -w + 2 - \dfrac{1}{4}w - 4$ Reduce fractions.

$\quad = -\dfrac{4}{4}w - \dfrac{1}{4}w + 2 - 4$ Group *like* terms and find a common denominator.

$\quad = -\dfrac{5}{4}w - 2$ Combine *like* terms.

Concept Connections

16. State whether each polynomial is equivalent to or not equivalent to $-(x - y)$.

    a. $x + y$
    b. $-x + y$
    c. $y - x$
    d. $-x - y$

Answer
16. a. Not equivalent
      b. Equivalent
      c. Equivalent
      d. Not equivalent

section 1.3 Practice Exercises

Boost *your* GRADE at mathzone.com!

- Practice Problems
- Self-Tests
- NetTutor
- e-Professors
- Videos

Study Skills Exercises

1. Sometimes you may run into a problem with homework, or you find that you are having trouble keeping up with the pace of the class. A tutor can be a good resource. Answer the following questions.

   a. Does your college offer tutoring?   b. Is it free?   c. Where would you go to sign up for a tutor?

2. Define the key terms.

   a. Term   b. Variable term   c. Constant term

   d. Factor   e. Coefficient   f. *Like* terms

Review Exercises

3. a. Classify the number -4 as a whole number, natural number, rational number, irrational number, integer, or real number. (Choose all that apply.)

  b. What is the reciprocal of -4?

  c. What is the opposite of -4?

  d. What is the absolute value of -4?

4. a. Classify the number 0 as a whole number, natural number, rational number, irrational number, integer, or real number. (Choose all that apply.)

  b. What is the reciprocal of 0 (if it exists)?

  c. What is the opposite of 0?

  d. What is the absolute value of 0?

For Exercises 5–8, write the set in interval notation.

5. $\{x \mid x > |-3|\}$

6. $\left\{x \mid x \leq \left|-\dfrac{4}{3}\right|\right\}$

7. $\left\{w \mid -\dfrac{5}{2} < w \leq \sqrt{9}\right\}$

8. $\left\{z \mid 2 \leq z < \dfrac{11}{3}\right\}$

Objective 1: Recognizing Terms, Factors, and Coefficients

For Exercises 9–12:

  a. Determine the number of terms in the expression.

  b. Identify the constant term.

  c. List the coefficients of each term. Separate by commas. **(See Example 1.)**

9. $2x^3 - 5xy + 6$

10. $a^2 - 4ab - b^2 + 8$

11. $pq - 7 + q^2 - 4q + p$

12. $7x - 1 + 3xy$

Objective 2: Properties of Real Numbers

For Exercises 13–22, match each expression with the appropriate property.

13. $3 + \frac{1}{2} = \frac{1}{2} + 3$

14. $7.2(4 + 1) = 7.2(4) + 7.2(1)$

15. $(6 + 8) + 2 = 6 + (8 + 2)$

16. $(4 + 19) + 7 = (19 + 4) + 7$

17. $9(4 \cdot 12) = (9 \cdot 4)12$

18. $\left(\frac{1}{4} + 2\right)20 = 5 + 40$

19. $(13 \cdot 41)6 = (41 \cdot 13)6$

20. $6(x + 3) = 6x + 18$

21. $3(y + 10) = 3(10 + y)$

22. $5(3 \cdot 7) = (5 \cdot 3)7$

a. Commutative property of addition

b. Associative property of multiplication

c. Distributive property of multiplication over addition

d. Commutative property of multiplication

e. Associative property of addition

Objective 3: Simplifying Expressions

For Exercises 23–56, clear parentheses and combine *like* terms. (See Examples 2–4.)

23. $8y - 2x + y + 5y$

24. $-9a + a - b + 5a$

25. $4p^2 - 2p + 3p - 6 + 2p^2$

26. $6q - 9 + 3q^2 - q^2 + 10$

27. $2p - 7p^2 - 5p + 6p^2$

28. $5a^2 - 2a - 7a^2 + 6a + 4$

29. $m - 4n^3 + 3 + 5n^3 - 9$

30. $x + 2y^3 - 2x - 8y^3$

31. $5ab + 2ab + 8a$

32. $-6m^2n - 3mn^2 - 2m^2n$

33. $14xy^2 - 5y^2 + 2xy^2$

34. $9uv + 3u^2 + 5uv + 4u^2$

35. $8(x - 3) + 1$

36. $-4(b + 2) - 3$

37. $-2(c + 3) - 2c$

38. $4(z - 4) - 3z$

39. $-(10w - 1) + 9 + w$

40. $-(2y + 7) - 4 + 3y$

41. $-9 - 4(2 - z) + 1$

42. $3 + 3(4 - w) - 11$

43. $4(2s - 7) - (s - 2)$

44. $2(t - 3) - (t - 7)$

45. $-3(-5 + 2w) - 8w + 2(w - 1)$

46. $5 - (-4t - 7) - t - 9$

47. $8x - 4(x - 2) - 2(2x + 1) - 6$

48. $6(y - 2) - 3(2y - 5) - 3$

49. $\frac{1}{2}(4 - 2c) + 5c$

50. $\frac{2}{3}(3d + 6) - 4d$

51. $3.1(2x + 2) - 4(1.2x - 1)$

52. $4.5(5 - y) + 3(1.9y + 1)$

53. $2\left[5\left(\dfrac{1}{2}a + 3\right) - (a^2 + a) + 4\right]$

54. $-3\left[3\left(b - \dfrac{2}{3}\right) - 2(b + 4) - 6b^2\right]$

55. $[(2y - 5) - 2(y - y^2)] - 3y$

56. $[-(x + 6) + 3(x^2 + 1)] + 2x$

Expanding Your Skills

57. What is the identity element for addition? Use it in an example.

58. What is the identity element for multiplication? Use it in an example.

59. What is another name for a multiplicative inverse?

60. What is another name for an additive inverse?

61. Is the operation of subtraction commutative? Explain why or why not and give an example.

62. Is the operation of division commutative? Explain why or why not and give an example.

63. Given the rectangular regions:

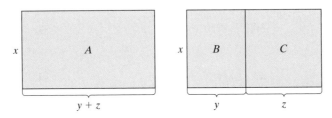

a. Write an expression for the area of region A. (Do not simplify.)

b. Write an expression for the area of region B.

c. Write an expression for the area of region C.

d. Add the expressions for the area of regions B and C.

e. Show that the area of region A is equal to the sum of the areas of regions B and C. What property of real numbers does this illustrate?

Section 1.4 Linear Equations in One Variable

1. Definition of a Linear Equation in One Variable

An **equation** is a statement that indicates that two quantities are equal. The following are equations.

$$x = -4 \qquad p + 3 = 11 \qquad -2z = -20$$

All equations have an equal sign. Furthermore, notice that the equal sign separates the equation into two parts, the left-hand side and the right-hand side. A **solution to an equation** is a value of the variable that makes the equation a true statement. Substituting a solution to an equation for the variable makes the right-hand side equal to the left-hand side.

Equation	Solution	Check	
$x = -4$	-4	$x = -4$ $\downarrow$ $-4 = -4$ ✓	Substitute -4 for x. Right-hand side equals left-hand side.
$p + 3 = 11$	8	$p + 3 = 11$ $\downarrow$ $8 + 3 = 11$ ✓	Substitute 8 for p. Right-hand side equals left-hand side.
$-2z = -20$	10	$-2z = -20$ $\downarrow$ $-2(10) = -20$ ✓	Substitute 10 for z. Right-hand side equals left-hand side.

Throughout this text we will learn to recognize and solve several different types of equations, but in this chapter we will focus on the specific type of equation called a linear equation in one variable.

> **Definition of a Linear Equation in One Variable**
>
> Let a and b be real numbers such that $a \neq 0$. A **linear equation in one variable** is an equation that can be written in the form
>
> $$ax + b = 0$$

Notice that a linear equation in one variable has only one variable. Furthermore, because the variable has an implied exponent of 1, a linear equation is sometimes called a first-degree equation.

Linear equation in one variable	Not a linear equation in one variable	
$4x - 3 = 0$	$4x^2 + 8 = 0$	(exponent for x is not 1)
$\frac{4}{5}p + \frac{3}{10} = 0$	$\frac{4}{5}p + \frac{3}{10}q = 0$	(more than one variable)

Objectives

1. Definition of a Linear Equation in One Variable
2. Solving Linear Equations
3. Clearing Fractions and Decimals
4. Conditional Equations, Contradictions, and Identities

Concept Connections

Determine if the given value is a solution to the equation.
1. $x + 2 = 8$; $x = -10$
2. $3y - 1 = 11$; $y = 4$
3. Write a linear equation whose solution is 3. (Answers may vary.)

Answers
1. No
2. Yes
3. For example: $5x - 6 = 9$

2. Solving Linear Equations

To solve a linear equation, the goal is to simplify the equation to isolate the variable. Each step used in simplifying an equation results in an equivalent equation. *Equivalent equations* have the same solution set. For example, the equations $2x + 3 = 7$ and $2x = 4$ are equivalent because $x = 2$ is the solution to both equations.

To solve an equation, we may use the addition, subtraction, multiplication, and division properties of equality. These properties state that adding, subtracting, multiplying, or dividing the same quantity on each side of an equation results in an equivalent equation.

Addition and Subtraction Properties of Equality

Let a, b, and c represent real numbers.

Addition property of equality: If $a = b$, then $a + c = b + c$.

Subtraction property of equality: If $a = b$, then $a - c = b - c$.

Multiplication and Division Properties of Equality

Let a, b, and c represent real numbers.

Multiplication property of equality: If $a = b$, then $a \cdot c = b \cdot c$.

Division property of equality: If $a = b$, then $\dfrac{a}{c} = \dfrac{b}{c}$ (provided $c \neq 0$).

Skill Practice

Solve the equations.

4. $x - 5 = -11$
5. $6y = 3$
6. $\dfrac{t}{16} = 5$
7. $-a = -2$

---

example 1 — Solving Linear Equations

Solve each equation.

a. $12 + x = 40$ **b.** $-\dfrac{1}{5}p = 2$ **c.** $4 = \dfrac{w}{2.2}$ **d.** $-x = 6$

Solution:

a.
$$12 + x = 40$$
$$12 - 12 + x = 40 - 12 \quad \text{To isolate } x, \text{ subtract 12 from both sides.}$$
$$x = 28 \quad \text{Simplify.}$$

Check: $12 + x = 40$ Check the solution in the original equation.
$$12 + (28) \stackrel{?}{=} 40$$
$$40 = 40 \checkmark \quad \text{True statement}$$

b.
$$-\dfrac{1}{5}p = 2$$
$$-5\left(-\dfrac{1}{5}p\right) = -5(2) \quad \text{To isolate } p, \text{ multiply both sides by } -5.$$
$$p = -10 \quad \text{Simplify.}$$

Answers

4. $x = -6$ 5. $y = \dfrac{1}{2}$
6. $t = 80$ 7. $a = 2$

Check: $-\dfrac{1}{5}p = 2$ Check the solution in the original equation.

$-\dfrac{1}{5}(-10) \stackrel{?}{=} 2$

$2 = 2$ ✓   True statement

c. $4 = \dfrac{w}{2.2}$

$2.2(4) = \left(\dfrac{w}{2.2}\right) \cdot 2.2$ To isolate w, multiply both sides by 2.2.

$8.8 = w$ Simplify.

Check: $4 = \dfrac{w}{2.2}$ Check the solution in the original equation.

$4 \stackrel{?}{=} \dfrac{8.8}{2.2}$

$4 = 4$ ✓   True statement

d. $-x = 6$

$-1(-x) = -1(6)$ To isolate x, multiply both sides by -1.

$x = -6$ Simplify.

Check: $-x = 6$ Check the solution in the original equation.

$-(-6) \stackrel{?}{=} 6$

$6 = 6$ ✓   True statement

For more complicated linear equations, several steps are required to isolate the variable. These steps are listed below.

Steps to Solve a Linear Equation in One Variable

1. Simplify both sides of the equation.
   - Clear parentheses.
   - Consider clearing fractions or decimals (if any are present) by multiplying both sides of the equation by a common denominator of all terms.
   - Combine *like* terms.
2. Use the addition or subtraction property of equality to collect the variable terms on one side of the equation.
3. Use the addition or subtraction property of equality to collect the constant terms on the other side of the equation.
4. Use the multiplication or division property of equality to make the coefficient of the variable term equal to 1.
5. Check your answer.

Skill Practice

Solve the equations.

8. $2x + 5 = 6x - 3$
9. $7 + 2(y - 3) = 6y + 3$
10. $4(2t + 2) - 6(t - 1) = 6 - t$
11. $3[p + 2(p - 2)] = 4(p - 3)$

example 2 — Solving Linear Equations

Solve the linear equations and check the answers.

a. $11z + 2 = 5(z - 2)$ **b.** $-3(x - 4) + 2 = 7 - (x + 1)$

c. $-4[y - 3(y - 5)] = 2(6 - 5y)$

Solution:

a.
$$11z + 2 = 5(z - 2)$$
$11z + 2 = 5z - 10$ Clear parentheses.
$11z - 5z + 2 = 5z - 5z - 10$ Subtract $5z$ from both sides.
$6z + 2 = -10$ Combine *like* terms.
$6z + 2 - 2 = -10 - 2$ Subtract 2 from both sides.
$6z = -12$
$\dfrac{6z}{6} = \dfrac{-12}{6}$ To isolate z, divide both sides of the equation by 6.
$z = -2$ Simplify.

Check: $11z + 2 = 5(z - 2)$ Check the solution in the original equation.
$11(-2) + 2 \stackrel{?}{=} 5(-2 - 2)$
$-22 + 2 \stackrel{?}{=} 5(-4)$
$-20 = -20$ ✓   True statement

b. $-3(x - 4) + 2 = 7 - (x + 1)$
$-3x + 12 + 2 = 7 - x - 1$ Clear parentheses.
$-3x + 14 = -x + 6$ Combine *like* terms.
$-3x + x + 14 = -x + x + 6$ Add x to both sides of the equation.
$-2x + 14 = 6$ Combine *like* terms.
$-2x + 14 - 14 = 6 - 14$ Subtract 14 from both sides.
$-2x = -8$
$\dfrac{-2x}{-2} = \dfrac{-8}{-2}$ To isolate x, divide both sides by -2.
$x = 4$ Simplify.

Check: $-3(x - 4) + 2 = 7 - (x + 1)$ Check the solution in the original equation.
$-3(4 - 4) + 2 \stackrel{?}{=} 7 - (4 + 1)$
$-3(0) + 2 \stackrel{?}{=} 7 - (5)$
$0 + 2 \stackrel{?}{=} 2$
$2 = 2$ ✓   True statement

Answers

8. $x = 2$ 9. $y = -\dfrac{1}{2}$
10. $t = -\dfrac{8}{3}$ 11. $p = 0$

c. $-4[y - 3(y - 5)] = 2(6 - 5y)$

$\quad -4[y - 3y + 15] = 12 - 10y$ Clear parentheses and combine *like* terms.

$\quad\quad -4[-2y + 15] = 12 - 10y$ Combine *like* terms.

$\quad\quad\quad 8y - 60 = 12 - 10y$ Clear parentheses.

$\quad 8y + 10y - 60 = 12 - 10y + 10y$ Add $10y$ to both sides of the equation.

$\quad\quad\quad 18y - 60 = 12$ Combine *like* terms.

$\quad 18y - 60 + 60 = 12 + 60$ Add 60 to both sides of the equation.

$\quad\quad\quad 18y = 72$

$\quad\quad\quad \dfrac{18y}{18} = \dfrac{72}{18}$ To isolate y, divide both sides by 18.

$\quad\quad\quad y = 4$ Simplify.

Check: $-4[y - 3(y - 5)] = 2(6 - 5y)$

$\quad\quad -4[4 - 3(4 - 5)] \stackrel{?}{=} 2(6 - 5(4))$

$\quad\quad -4[4 - 3(-1)] \stackrel{?}{=} 2(6 - 20)$

$\quad\quad -4[4 + 3] \stackrel{?}{=} 2(-14)$

$\quad\quad -4(7) \stackrel{?}{=} -28$

$\quad\quad -28 = -28 \checkmark$ True statement

3. Clearing Fractions and Decimals

When an equation contains fractions or decimals, it is sometimes helpful to clear the fractions and decimals. This is accomplished by multiplying both sides of the equation by the least common denominator (LCD) of all terms within the equation. This is demonstrated in Example 3.

example 3   Solving Linear Equations by Clearing Fractions

Solve the equation.

$$\frac{1}{4}w + \frac{1}{3}w - 1 = \frac{1}{2}(w - 4)$$

Solution:

$\quad \dfrac{1}{4}w + \dfrac{1}{3}w - 1 = \dfrac{1}{2}(w - 4)$

$\quad \dfrac{1}{4}w + \dfrac{1}{3}w - 1 = \dfrac{1}{2}w - 2$ Apply the distributive property to clear parentheses.

$\quad 12 \cdot \left(\dfrac{1}{4}w + \dfrac{1}{3}w - 1\right) = 12 \cdot \left(\dfrac{1}{2}w - 2\right)$ Multiply both sides of the equation by the LCD of all terms. In this case, the LCD is 12.

Skill Practice

Solve the equation by first clearing the fractions

12. $\dfrac{3}{4}a + \dfrac{1}{2} = \dfrac{2}{3}a + \dfrac{1}{3}$

Answer

12. $a = -2$

44   Chapter 1   Review of Basic Algebraic Concepts

$$12 \cdot \frac{1}{4}w + 12 \cdot \frac{1}{3}w + 12 \cdot (-1) = 12 \cdot \frac{1}{2}w + 12 \cdot (-2)$$ Apply the distributive property.

$$3w + 4w - 12 = 6w - 24$$

$$7w - 12 = 6w - 24$$

$$7w - 6w - 12 = 6w - 6w - 24$$ Subtract $6w$

$$w - 12 = -24$$

$$w - 12 + 12 = -24 + 12$$ Add 12 to both sides.

$$w = -12$$

Check: $$\frac{1}{4}w + \frac{1}{3}w - 1 = \frac{1}{2}(w - 4)$$

$$\frac{1}{4}(-12) + \frac{1}{3}(-12) - 1 \stackrel{?}{=} \frac{1}{2}(-12 - 4)$$

$$-3 - 4 - 1 \stackrel{?}{=} \frac{1}{2}(-16)$$

$$-8 = -8 \checkmark$$ True statement

Tip: The fractions in this equation can be eliminated by multiplying both sides of the equation by *any* common multiple of the denominators. For example, multiplying both sides of the equation by 24 produces the same solution.

$$24 \cdot \left(\frac{1}{4}w + \frac{1}{3}w - 1\right) = 24 \cdot \frac{1}{2}(w - 4)$$

$$6w + 8w - 24 = 12(w - 4)$$

$$14w - 24 = 12w - 48$$

$$2w = -24$$

$$w = -12$$

Skill Practice

Solve.

13. $\dfrac{x + 3}{4} + \dfrac{1}{8} = \dfrac{3x - 2}{2}$

---

example 4   Solving a Linear Equation with Fractions

Solve. $\dfrac{x - 2}{5} - \dfrac{x - 4}{2} = 2 + \dfrac{x + 4}{10}$

Solution:

$$\frac{x - 2}{5} - \frac{x - 4}{2} = \frac{2}{1} + \frac{x + 4}{10}$$ The LCD of all terms in the equation is 10.

$$10\left(\frac{x - 2}{5} - \frac{x - 4}{2}\right) = 10\left(\frac{2}{1} + \frac{x + 4}{10}\right)$$ Multiply both sides by 10.

Answer

13. $x = \dfrac{3}{2}$

$$\frac{\cancel{10}^{2}}{1}\cdot\left(\frac{x-2}{\cancel{5}}\right)-\frac{\cancel{10}^{5}}{1}\cdot\left(\frac{x-4}{\cancel{2}}\right)=\frac{10}{1}\cdot\left(\frac{2}{1}\right)+\frac{\cancel{10}^{1}}{1}\cdot\left(\frac{x+4}{\cancel{10}}\right)$$ Apply the distributive property.

$2(x-2) - 5(x-4) = 20 + 1(x+4)$ Clear fractions.

$2x - 4 - 5x + 20 = 20 + x + 4$ Apply the distributive property.

$-3x + 16 = x + 24$ Simplify both sides of the equation.

$-3x - x + 16 = x - x + 24$ Subtract x from both sides.

$-4x + 16 = 24$

$-4x + 16 - 16 = 24 - 16$ Subtract 16 from both sides.

$-4x = 8$

$\dfrac{-4x}{-4} = \dfrac{8}{-4}$ Divide both sides by -4.

$x = -2$ The check is left to the reader.

The same procedure used to clear fractions in an equation can be used to clear decimals.

example 5 Solving Linear Equations by Clearing Decimals

Solve the equation. $0.55x - 0.6 = 2.05x$

Solution:

Recall that any terminating decimal can be written as a fraction. Therefore, the equation $0.55x - 0.6 = 2.05x$ is equivalent to

$$\frac{55}{100}x - \frac{6}{10} = \frac{205}{100}x$$

A convenient common denominator for all terms in this equation is 100. Multiplying both sides of the equation by 100 will have the effect of "moving" the decimal point 2 places to the right.

$100(0.55x - 0.6) = 100(2.05x)$ Multiply both sides by 100 to clear decimals.

$55x - 60 = 205x$

$55x - 55x - 60 = 205x - 55x$ Subtract $55x$ from both sides.

Skill Practice

Solve the equation by first clearing the decimals.

14. $2.2x + 0.5 = 1.6x + 0.2$

Answer

14. $x = -0.5$

$$-60 = 150x$$

$$\frac{-60}{150} = \frac{150x}{150}$$ To isolate x, divide both sides by 150.

$$-\frac{60}{150} = x$$

$$x = -\frac{2}{5} = -0.4$$

Check: $\quad 0.55x - 0.6 = 2.05x$

$$0.55(-0.4) - 0.6 \stackrel{?}{=} 2.05(-0.4)$$

$$-0.22 - 0.6 \stackrel{?}{=} -0.82$$

$$-0.82 = -0.82 \checkmark \quad \text{True statement}$$

4. Conditional Equations, Contradictions, and Identities

The solution to a linear equation is the value of x that makes the equation a true statement. A linear equation has one unique solution. Some equations, however, have no solution, while others have infinitely many solutions.

I. Conditional Equations

An equation that is true for some values of the variable but false for other values is called a **conditional equation**. The equation $x + 4 = 6$ is a conditional equation because it is true on the *condition* that $x = 2$. For other values of x, the statement $x + 4 = 6$ is false.

II. Contradictions

Some equations have no solution, such as $x + 1 = x + 2$. There is no value of x that when increased by 1 will equal the same value increased by 2. If we tried to solve the equation by subtracting x from both sides, we get the contradiction $1 = 2$. This indicates that the equation has no solution. An equation that has no solution is called a **contradiction**.

$$x + 1 = x + 2$$

$$x - x + 1 = x - x + 2$$

$$1 = 2 \quad \text{(contradiction)} \quad \text{No solution}$$

III. Identities

An equation that has all real numbers as its solution set is called an **identity**. For example, consider the equation $x + 4 = x + 4$. Because the left- and right-hand sides are *identical*, any real number substituted for x will result in equal quantities on both sides. If we solve the equation, we get the identity $4 = 4$. In such a case, the solution is the set of all real numbers.

$$x + 4 = x + 4$$

$$x - x + 4 = x - x + 4$$

$$4 = 4 \quad \text{(identity)} \quad \text{The solution is all real numbers.}$$

example 6 — Identifying Conditional Equations, Contradictions, and Identities

Solve the equations. Identify each equation as a conditional equation, a contradiction, or an identity.

a. $3[x - (x + 1)] = -2$ **b.** $5(3 + c) + 2 = 2c + 3c + 17$

c. $4x - 3 = 17$

Solution:

a. $3[x - (x + 1)] = -2$

$3[x - x - 1] = -2$ Clear parentheses.

$3[-1] = -2$ Combine *like* terms.

$-3 = -2$ Contradiction

This equation is a contradiction. There is no solution.

b. $5(3 + c) + 2 = 2c + 3c + 17$

$15 + 5c + 2 = 5c + 17$ Clear parentheses and combine *like* terms.

$5c + 17 = 5c + 17$ Identity

$0 = 0$

This equation is an identity. The solution is the set of all real numbers.

c. $4x - 3 = 17$

$4x - 3 + 3 = 17 + 3$ Add 3 to both sides.

$4x = 20$

$\dfrac{4x}{4} = \dfrac{20}{4}$ To isolate x, divide both sides by 4.

$x = 5$

This equation is a conditional equation. The solution is $x = 5$.

Skill Practice

Solve the equations. Identify each equation as a conditional equation, an identity, or a contradiction.

15. $2(3x - 1) = 6(x + 1) - 8$

16. $4x + 1 - x = 6x - 2$

17. $2(-5x - 1) = 2x - 12x + 6$

Answers

15. The equation is an identity. The solution is the set of all real numbers.
16. The equation is conditional. The solution is $x = 1$.
17. The equation is a contradiction. There is no solution.

---

section 1.4 Practice Exercises

Boost *your* GRADE at mathzone.com!

- Practice Problems
- Self-Tests
- NetTutor
- e-Professors
- Videos

Study Skills Exercises

1. It is very important to attend class every day. Math is cumulative in nature, and you must master the material learned in the previous class to understand today's lesson. Because this is so important, many instructors tie attendance to the final grade. Write down the attendance policy for your class.

Chapter 1 Review of Basic Algebraic Concepts

2. Define the key terms.

    a. Equation   **b.** Solution to an equation   **c.** Linear equation in one variable

    d. Conditional equation   **e.** Contradiction   **f.** Identity

Review Exercises

For Exercises 3–6, clear parentheses and combine *like* terms.

3. $8x - 3y + 2xy - 5x + 12xy$

4. $5ab + 5a - 13 - 2a + 17$

5. $2(3z - 4) - (z + 12)$

6. $-(6w - 5) + 3(4w - 5)$

Objective 1: Definition of a Linear Equation in One Variable

For Exercises 7–12, label the equation as linear or nonlinear.

7. $2x + 1 = 5$ **8.** $10 = x + 6$ **9.** $x^2 + 7 = 9$

10. $3 + x^3 - x = 4$ **11.** $-3 = x$ **12.** $5.2 - 7x = 0$

13. Use substitution to determine which value is the solution to $2x - 1 = 5$.

    a. 2   **b.** 3   **c.** 0   **d.** -1

14. Use substitution to determine which value is the solution to $2y - 3 = -2$.

    a. 1   **b.** $\frac{1}{2}$ **c.** 0   **d.** $-\frac{1}{2}$

Objective 2: Solving Linear Equations

For Exercises 15–44, solve the equations and check your solutions. (**See Examples 1–2.**)

15. $x + 7 = 19$ **16.** $-3 + y = -28$ **17.** $64x = -2$ **18.** $\frac{t}{8} = -\frac{3}{4}$

19. $-\frac{7}{8} = -\frac{5}{6}z$ **20.** $-\frac{12}{13} = 4b$ **21.** $a + \frac{2}{5} = 2$ **22.** $-\frac{3}{8} + x = -\frac{7}{24}$

23. $2.53 = -2.3t$ **24.** $-4.8 = 6.1 + y$ **25.** $p - 2.9 = 3.8$ **26.** $-4.2a = 4.494$

27. $6q - 4 = 62$ **28.** $2w - 15 = 15$ **29.** $4y - 17 = 35$ **30.** $6z - 25 = 83$

31. $-b - 5 = 2$ **32.** $6 = -y + 1$ **33.** $3(x - 6) = 2x - 5$ **34.** $13y + 4 = 5(y - 4)$

35. $6 - (t + 2) = 5(3t - 4)$

36. $1 - 5(p + 2) = 2(p + 13)$

37. $6(a + 3) - 10 = -2(a - 4)$

38. $8(b - 2) + 3b = -9(b - 1)$

39. $-2[5 - (2z + 1)] - 4 = 2(3 - z)$

40. $3[w - (10 - w)] = 7(w + 1)$

41. $6(-y + 4) - 3(2y - 3) = -y + 5 + 5y$

42. $13 + 4w = -5(-w - 6) + 2(w + 1)$

43. $14 - 2x + 5x = -4(-2x - 5) - 6$

44. $8 - (p + 2) + 6p + 7 = p + 13$

Objective 3: Clearing Fractions and Decimals

For Exercises 45–56, solve the equations. (See Examples 3–5.)

45. $\frac{2}{3}x - \frac{1}{6} = -\frac{5}{12}x + \frac{3}{2} - \frac{1}{6}x$

46. $-\frac{1}{2}y + 4 = -\frac{9}{10}y + \frac{2}{5}$

47. $\frac{1}{5}(p - 5) = \frac{3}{5}p + \frac{1}{10}p + 1$

48. $\frac{5}{6}(q + 2) = -\frac{7}{9}q - \frac{1}{3} + 2$

49. $\frac{3x - 7}{2} + \frac{3 - 5x}{3} = \frac{3 - 6x}{5}$

50. $\frac{2y - 4}{5} = \frac{5y + 13}{4} + \frac{y}{2}$

51. $\frac{4}{3}(2q + 6) - \frac{5q - 6}{6} - \frac{q}{3} = 0$

52. $\frac{-3a + 9}{15} - \frac{2a - 5}{5} - \frac{a + 2}{10} = 0$

53. $6.3w - 1.5 = 4.8$

54. $0.2x + 53.6 = x$

55. $0.75(m - 2) + 0.25m = 0.5$

56. $0.4(n + 10) + 0.6n = 2$

Objective 4: Conditional Equations, Contradictions, and Identities

57. What is a conditional equation?

58. Explain the difference between a contradiction and an identity.

For Exercises 59–64, solve the following equations. Then label each as a conditional equation, a contradiction, or an identity. (See Example 6.)

59. $4x + 1 = 2(2x + 1) - 1$

60. $3x + 6 = 3x$

61. $-11x + 4(x - 3) = -2x - 12$

62. $5(x + 2) - 7 = 3$

63. $2x - 4 + 8x = 7x - 8 + 3x$

64. $-7x + 8 + 4x = -3(x - 3) - 1$

Mixed Exercises

For Exercises 65–96, solve the equations.

65. $-5b + 9 = -71$

66. $-3x + 18 = -66$

67. $16 = -10 + 13x$

68. $15 = -12 + 9x$

69. $10c + 3 = -3 + 12c$

70. $2w + 21 = 6w - 7$

71. $12b - 15b - 8 + 6 = 4b + 6 - 1$

72. $4z + 2 - 3z + 5 = 3 + z + 4$

73. $5(x - 2) - 2x = 3x + 7$

74. $2x + 3(x - 5) = 15$

75. $\dfrac{c}{2} - \dfrac{c}{4} + \dfrac{3c}{8} = 1$

76. $\dfrac{d}{5} - \dfrac{d}{10} + \dfrac{5d}{20} = \dfrac{7}{10}$

77. $0.75(8x - 4) = \dfrac{2}{3}(6x - 9)$

78. $-\dfrac{1}{2}(4z - 3) = -z$

79. $7(p + 2) - 4p = 3p + 14$

80. $6(z - 2) = 3z - 8 + 3z$

81. $4[3 + 5(3 - b) + 2b] = 6 - 2b$

82. $\dfrac{1}{3}(x + 3) - \dfrac{1}{6} = \dfrac{1}{6}(2x + 5)$

83. $3 - \dfrac{3}{4}x = 9$

84. $\dfrac{9}{10} - 4w = \dfrac{5}{2}$

85. $\dfrac{5}{4} + \dfrac{y - 3}{8} = \dfrac{2y + 1}{2}$

86. $\dfrac{2}{3} - \dfrac{x + 2}{6} = \dfrac{5x - 2}{2}$

87. $\dfrac{2y - 9}{10} + \dfrac{3}{2} = y$

88. $\dfrac{2}{3}x - \dfrac{5}{6}x - 3 = \dfrac{1}{2}x - 5$

89. $0.48x - 0.08x = 0.12(260 - x)$

90. $0.07w + 0.06(140 - w) = 90$

91. $0.5x + 0.25 = \dfrac{1}{3}x + \dfrac{5}{4}$

92. $0.2b + \dfrac{1}{3} = \dfrac{7}{15}$

93. $0.3b - 1.5 = 0.25(b + 2)$

94. $0.7(a - 1) = 0.25 + 0.7a$

95. $-\dfrac{7}{8}y + \dfrac{1}{4} = \dfrac{1}{2}\left(5 - \dfrac{3}{4}y\right)$

96. $5x - (8 - x) = 2[-4 - (3 + 5x) - 13]$

Expanding Your Skills

97. a. Simplify the expression. $\quad -2(y - 1) + 3(y + 2)$

b. Solve the equation. $\quad -2(y - 1) + 3(y + 2) = 0$

c. Explain the difference between simplifying an expression and solving an equation.

98. a. Simplify the expression. $\quad 4w - 8(2 + w)$

b. Solve the equation. $\quad 4w - 8(2 + w) = 0$

chapter 1 | midchapter review

1. Let
$$A = \left\{-7.1, -5\pi, -2, -\frac{1}{8}, 0, 0.\overline{3}, \sqrt{2}, \frac{7}{8}, 6, \frac{9}{2}\right\}$$

   a. List all numbers from A that are rational numbers.

   b. List all numbers from A that are natural numbers.

   c. List all numbers from A that are real numbers.

   d. List all numbers from A that are irrational numbers.

   e. List all numbers from A that are whole numbers.

2. Let $A = \{10, 20, 30, 40, 50\}$, $B = \{-10, -5, 0, 10, 20\}$, and $C = \{-5, 0, 5, 15, 25\}$. List the elements of the following sets.

   a. $A \cap B$

   b. $A \cap C$

   c. $B \cup C$

   d. $A \cup C$

For Exercises 3–12, use the order of operations to simplify the expression.

3. $-\sqrt{4} + |-5| + \sqrt{25}$

4. $|-3| + 2^3 - (-2^2)$

5. $\left(-\frac{5}{3} - \frac{1}{3}\right) \div \frac{5}{3}$

6. $-\frac{5}{6} - \left(-\frac{4}{3} - \frac{2}{5}\right)$

7. $-6 - 5(-8) + (-5)^2$

8. $-7(-3) - (-2^3)$

9. $3(-5) + 7(-5) - 4(2)$

10. $(3^2 \cdot 2 - 8) \div 5$

11. $2\sqrt{100} - 10 \div 5$

12. $-8[4 + (-2^2 \cdot 3)]$

For Exercises 13–23, identify each exercise as an expression or an equation. Then simplify the expressions and solve the equations.

13. $13 + x = -21$

14. $0.29a + 4.495 - 0.12a$

15. $5x + 4 - 6x - 8$

16. $\dfrac{b}{4} + 21 = 38$

17. $7(2 + 4n) = 11 - 4n$

18. $\dfrac{1}{8}(2p - 8) = \dfrac{1}{4}(p - 4)$

19. $7 + 8b - 12 = 3b - 8 + 5b$

20. $0.09q + 0.10(5000 + 3q)$

21. $\dfrac{1}{2}c - \dfrac{2}{5} + \dfrac{1}{10}(c + 2)$

22. $2[3(3 - y) + 2] = 6 - 2y$

23. $\dfrac{3x + 1}{4} - \dfrac{x + 5}{2} = \dfrac{7}{4}$

For Exercises 24–29, graph the sets and express each set in interval notation.

24. $\{x \mid x > -3\}$

25. $\{x \mid x \geq 6\}$

26. $\{x \mid x \leq 2\tfrac{1}{2}\}$

27. $\{x \mid x < 4.8\}$

28. $\{x \mid x < 0\} \cup \{x \mid x > 4\}$

29. $\{x \mid x \leq 13\} \cap \{x \mid x \geq 1\}$

Objectives

1. Introduction to Problem Solving
2. Applications Involving Consecutive Integers
3. Applications Involving Percents and Rates
4. Applications Involving Principal and Interest
5. Applications Involving Mixtures
6. Applications Involving Distance, Rate, and Time

section 1.5 Applications of Linear Equations in One Variable

1. Introduction to Problem Solving

One of the important uses of algebra is to develop mathematical models for understanding real-world phenomena. To solve an application problem, relevant information must be extracted from the wording of a problem and then translated into mathematical symbols. This is a skill that requires practice. The key is to stick with it and not to get discouraged.

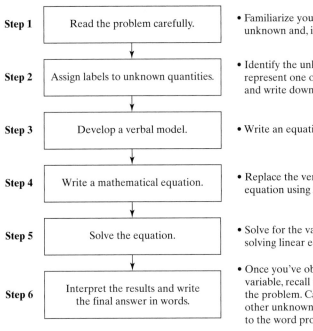

Problem-Solving Flowchart for Word Problems

- **Step 1** Read the problem carefully. — Familiarize yourself with the problem. Identify the unknown and, if possible, estimate the answer.
- **Step 2** Assign labels to unknown quantities. — Identify the unknown quantity or quantities. Let x represent one of the unknowns. Draw a picture and write down relevant formulas.
- **Step 3** Develop a verbal model. — Write an equation in *words*.
- **Step 4** Write a mathematical equation. — Replace the verbal model with a mathematical equation using x or another variable.
- **Step 5** Solve the equation. — Solve for the variable, using the steps for solving linear equations.
- **Step 6** Interpret the results and write the final answer in words. — Once you've obtained a numerical value for the variable, recall what it represents in the context of the problem. Can this value be used to determine other unknowns in the problem? Write an answer to the word problem in *words*.

Skill Practice

1. One number is 5 more than 3 times another number. The sum of the numbers is 45. Find the numbers.

example 1 Translating and Solving a Linear Equation

The sum of two numbers is 39. One number is 3 less than twice the other. What are the numbers?

Solution:

Step 1: Read the problem carefully.

Step 2: Let x represent one number.

Let $2x - 3$ represent the other number.

Step 3: (One number) + (other number) = 39

Step 4: Replace the verbal model with a mathematical equation.

(One number) + (other number) = 39

$x \quad + \quad (2x - 3) \quad = 39$

Answer

1. The numbers are 10 and 35.

Step 5: Solve for x.

$$x + (2x - 3) = 39$$
$$3x - 3 = 39$$
$$3x = 42$$
$$\frac{3x}{3} = \frac{42}{3}$$
$$x = 14$$

Step 6: Interpret your results. Refer back to step 2.

One number is x: ⟶ 14

The other number is $2x - 3$:

$2(14) - 3$ ⟶ 25

Answer: The numbers are 14 and 25.

2. Applications Involving Consecutive Integers

The word *consecutive* means "following one after the other in order." The numbers $-2, -1, 0, 1, 2$ are examples of consecutive integers. Notice that any two consecutive integers differ by 1. If x represents an integer, then $x + 1$ represents the next consecutive integer.

The numbers 2, 4, 6, 8 are consecutive even integers. The numbers 15, 17, 19, 21 are consecutive odd integers. Both consecutive odd and consecutive even integers differ by 2. If x represents an even integer, then $x + 2$ represents the next consecutive even integer. If x represents an odd integer, then $x + 2$ represents the next consecutive odd integer.

example 2 Solving a Linear Equation Involving Consecutive Integers

The sum of two consecutive odd integers is 172. Find the integers.

Solution:

Step 1: Read the problem carefully.

Step 2: Label unknowns:

Let x represent the first integer.

Let $x + 2$ represent the next odd integer.

Step 3: Write an equation in words:

(First integer) + (second integer) = 172

Step 4: Write a mathematical equation based on the verbal model.

(First integer) + (second integer) = 172

x + $(x + 2)$ = 172

Skill Practice

The sum of three consecutive integers is 66.

2. If the first integer is represented by x, write expressions for the next two integers.

3. Write a mathematical equation that describes the verbal model.

4. Solve the equation and find the three integers.

Answers

2. $x + 1$ and $x + 2$
3. $x + (x + 1) + (x + 2) = 66$
4. The integers are 21, 22, and 23.

Step 5: Solve for x.

$$x + (x + 2) = 172$$
$$2x + 2 = 172$$
$$2x = 170$$
$$x = 85$$

Step 6: Interpret your results.

One number is x: ⟶ 85

The other integer is $x + 2$:

$$85 + 2 \longrightarrow 87$$

Answer: The numbers are 85 and 87.

After completing a word problem, it is always a good idea to check that the answer is reasonable. Notice that 85 and 87 are consecutive odd integers, and the sum is equal to 172 as desired.

3. Applications Involving Percents and Rates

In many real-world applications, percents are used to represent rates.

- In 1998, the sales tax rate in the state of Tennessee was 6%.
- An ice cream machine is discounted 20%.
- A real estate sales broker receives a $4\frac{1}{2}$% commission on sales.
- A savings account earns 7% simple interest.

The following models are used to compute sales tax, commission, and simple interest. In each case the value is found by multiplying the base by the percentage.

Sales tax = (cost of merchandise)(tax rate)

Commission = (dollars in sales)(commission rate)

Simple interest = (principal)(annual interest rate)(time in years)
$$\hookrightarrow I = Prt$$

Skill Practice

5. The sales tax rate in Atlanta, Georgia, is 7%. Find the amount of sales tax paid on an automobile priced at $12,000.

example 3   Solving a Percent Application

A realtor made a 6% commission on a house that sold for $172,000. How much was her commission?

Solution:

Let x represent the commission.   Label the variables.

(Commission) = (dollars in sales)(commission rate)   Verbal model

$$x = (\$172{,}000)(0.06)$$ Mathematical model

$$x = \$10{,}320$$ Solve for x.

The realtor's commission is $10,320.   Interpret the results.

Answer

5. $840

example 4 — Solving a Percent Application

A woman invests $5000 in an account that earns $5\frac{1}{4}$% simple interest. If the money is invested for 3 years, how much money is in the account at the end of the 3-year period?

Solution:

Let x represent the total money in the account.   Label variables.

$P = \$5000$ (principal amount invested)

$r = 0.0525$ (interest rate)

$t = 3$ (time in years)

The total amount of money includes principal plus interest.

(Total money) = (principal) + (interest)	Verbal model
$x = P + Prt$	Mathematical model
$x = \$5000 + (\$5000)(0.0525)(3)$	Substitute for P, r, and t.
$x = \$5000 + \$787.50$	
$x = \$5787.50$	Solve for x.

The total amount of money in the account is $5787.50.   Interpret the results.

Skill Practice

6. A man earned $340 in 1 year on an investment that paid a 4% dividend. Find the amount of money invested.

As consumers, we often encounter situations in which merchandise has been marked up or marked down from its original cost. It is important to note that percent increase and percent decrease are based on the original cost. For example, suppose a microwave originally priced at $305 is marked down 20%.

The discount is determined by 20% of the original price: $(0.20)(\$305) = \$61.00$. The new price is $\$305.00 - \$61.00 = \$244.00$.

example 5 — Solving a Percent Increase Application

A college bookstore uses a standard markup of 22% on all books purchased wholesale from the publisher. If the bookstore sells a calculus book for $103.70, what was the original wholesale cost?

Solution:

Let x = original wholesale cost.   Label the variables.

The selling price of the book is based on the original cost of the book plus the bookstore's markup.

(Selling price) = (original price) + (markup)   Verbal model

(Selling price) = (original price) + (original price · markup rate)

$103.70 = x + (x)(0.22)$ Mathematical model

$103.70 = x + 0.22x$

Skill Practice

7. An online bookstore gives a 20% discount on paperback books. Find the original price of a book that has a selling price of $5.28 after the discount.

Answers

6. $8500   7. $6.60

$$103.70 = 1.22x \qquad \text{Combine \textit{like} terms.}$$

$$\frac{103.70}{1.22} = x$$

$$x = \$85.00 \qquad \text{Simplify.}$$

The original wholesale cost of the textbook was $85.00. Interpret the results.

4. Applications Involving Principal and Interest

Skill Practice

8. Jonathan borrowed $4000 in two loans. One loan charged 7% interest, and the other charged 1.5% interest. After 1 year, Jonathan paid $225 in interest. Find the amount borrowed in each loan account.

example 6   Solving an Investment Growth Application

Miguel had $10,000 to invest in two different mutual funds. One was a relatively safe bond fund that averaged 8% return on his investment at the end of 1 year. The other fund was a riskier stock fund that averaged 17% return in 1 year. If at the end of the year Miguel's portfolio grew to $11,475 ($1475 above his $10,000 investment), how much money did Miguel invest in each fund?

Solution:

This type of word problem is sometimes categorized as a mixture problem. Miguel is "mixing" his money between two different investments. We have to determine how the money was divided to earn $1475 in total growth.

The information in this problem can be organized in a chart. (*Note:* There are two sources of money: the amount invested and the amount earned in growth.)

	8% Bond Fund	17% Stock Fund	Total
Amount invested ($)	x	$(10{,}000 - x)$	10,000
Amount earned in growth ($)	$0.08x$	$0.17(10{,}000 - x)$	1475

Because the amount of principal is unknown for both accounts, we can let x represent the amount invested in the bond fund. If Miguel spends x dollars in the bond fund, then he has $(10{,}000 - x)$ left over to spend in the stock fund. The return for each fund is found by multiplying the principal and the percent growth rate.

To establish a mathematical model, we know that the total return ($1475) must equal the growth from the bond fund plus the growth from the stock fund:

(Growth from bond fund) + (growth from stock fund) = (total growth)

$$0.08x \quad + \quad 0.17(10{,}000 - x) \quad = \quad 1475$$

$$0.08x + 0.17(10{,}000 - x) = 1475 \qquad \text{Mathematical model}$$

$$8x + 17(10{,}000 - x) = 147{,}500 \qquad \text{Multiply by 100 to clear decimals.}$$

$$8x + 170{,}000 - 17x = 147{,}500$$

$$-9x + 170{,}000 = 147{,}500 \qquad \text{Combine \textit{like} terms.}$$

Answer

8. $3000 was borrowed at 7% interest, and $1000 was borrowed at 1.5% interest.

$$-9x = -22{,}500 \qquad \text{Subtract 170,000 from both sides.}$$

$$\frac{-9x}{-9} = \frac{-22{,}500}{-9}$$

$$x = 2500 \qquad \text{Solve for } x \text{ and interpret the results.}$$

The amount invested in the bond fund is $2500. The amount invested in the stock fund is $10,000 − x$, or $7500.

5. Applications Involving Mixtures

example 7 Solving a Mixture Application

How many liters (L) of a 60% antifreeze solution must be added to 8 L of a 10% antifreeze solution to produce a 20% antifreeze solution?

Solution:

The given information is illustrated in Figure 1-7.

Skill Practice

9. Find the number of ounces (oz) of 30% alcohol solution that must be mixed with 10 oz of a 70% solution to obtain a solution that is 40% alcohol.

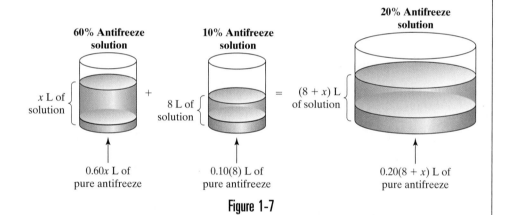

Figure 1-7

The information can be organized in a chart.

	60% Antifreeze	10% Antifreeze	Final Solution: 20% Antifreeze
Number of liters of solution	x	8	$(8 + x)$
Number of liters of pure antifreeze	$0.60x$	$0.10(8)$	$0.20(8 + x)$

Notice that an algebraic equation is derived from the second row of the table which relates the number of liters of pure antifreeze in each container.

The amount of pure antifreeze in the final solution equals the sum of the amounts of antifreeze in the first two solutions.

Answer

9. 30 oz of the 30% solution is needed.

$$\begin{pmatrix}\text{Pure antifreeze}\\\text{from solution 1}\end{pmatrix} + \begin{pmatrix}\text{pure antifreeze}\\\text{from solution 2}\end{pmatrix} = \begin{pmatrix}\text{pure antifreeze}\\\text{in the final solution}\end{pmatrix}$$

$$0.60x \quad + \quad 0.10(8) \quad = \quad 0.20(8 + x)$$

$0.60x + 0.10(8) = 0.20(8 + x)$	Mathematical model
$0.6x + 0.8 = 1.6 + 0.2x$	Apply the distributive property.
$0.6x - 0.2x + 0.8 = 1.6 + 0.2x - 0.2x$	Subtract $0.2x$ from both sides.
$0.4x + 0.8 = 1.6$	
$0.4x + 0.8 - 0.8 = 1.6 - 0.8$	Subtract 0.8 from both sides.
$0.4x = 0.8$	
$\dfrac{0.4x}{0.4} = \dfrac{0.8}{0.4}$	Divide both sides by 0.4.
$x = 2$	

Answer: 2 L of 60% antifreeze solution is necessary to make a final solution of 20% antifreeze.

6. Applications Involving Distance, Rate, and Time

The fundamental relationship among the variables distance, rate, and time is given by

$$\textbf{Distance} = \textbf{(rate)(time)} \quad \text{or} \quad \boldsymbol{d = rt}$$

For example, a motorist traveling 65 mph (miles per hour) for 3 hr (hours) will travel a distance of

$$d = (65 \text{ mph})(3 \text{ hr}) = 195 \text{ mi}$$

Skill Practice

10. Jody drove a distance of 320 mi to visit a friend. She drives part of the time at 40 mph and part at 60 mph. The trip took 6 hr. Find the amount of time she spent driving at each speed.

example 8 — Solving a Distance, Rate, Time Application

A hiker can hike $2\tfrac{1}{2}$ mph down a trail to visit Archuletta Lake. For the return trip back to her campsite (uphill), she is only able to go $1\tfrac{1}{2}$ mph. If the total time for the round trip is 4 hr 48 min (4.8 hr), find

a. The time required to walk down to the lake

b. The time required to return back to the campsite

c. The total distance the hiker traveled

Solution:

The information given in the problem can be organized in a chart.

	Distance (mi)	Rate (mph)	Time (hr)
Trip to the lake		2.5	t
Return trip		1.5	$(4.8 - t)$

Column 2: The rates of speed going to and from the lake are given in the statement of the problem.

Column 3: There are two unknown times. If we let t be the time required to go to the lake, then the time for the return trip must equal the total time minus t, or $(4.8 - t)$

Answer

10. Jody drove 2 hr at 40 mph and 4 hr at 60 mph.

Column 1: To express the distance in terms of the time t, we use the relationship $d = rt$. That is, multiply the quantities in the second and third columns.

	Distance (mi)	Rate (mph)	Time (hr)
Trip to the lake	$2.5t$	2.5	t
Return trip	$1.5(4.8 - t)$	1.5	$(4.8 - t)$

To create a mathematical model, note that the distances to and from the lake are equal. Therefore,

(Distance to lake) = (return distance)   Verbal model

$2.5t = 1.5(4.8 - t)$ Mathematical model

$2.5t = 7.2 - 1.5t$ Apply the distributive property.

$2.5t + 1.5t = 7.2 - 1.5t + 1.5t$ Add $1.5t$ to both sides.

$4.0t = 7.2$

$\dfrac{4.0t}{4.0} = \dfrac{7.2}{4.0}$

$t = 1.8$ Solve for t and interpret the results.

Answers:

a. Because t represents the time required to go down to the lake, 1.8 hr is required for the trip to the lake.

b. The time required for the return trip is $(4.8 - t)$ or $(4.8 \text{ hr} - 1.8 \text{ hr}) = 3$ hr. Therefore, the time required to return to camp is 3 hr.

c. The total distance equals the distance to the lake and back. The distance to the lake is $(2.5 \text{ mph})(1.8 \text{ hr}) = 4.5$ mi. The distance back is $(1.5 \text{ mph})(3.0 \text{ hr}) = 4.5$ mi. Therefore, the total distance the hiker walked is 9.0 mi.

section 1.5 Practice Exercises

Boost your GRADE at mathzone.com!

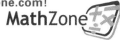

- Practice Problems
- Self-Tests
- NetTutor
- e-Professors
- Videos

Study Skills Exercises

1. In your next math class, take notes by drawing a vertical line about ¾ of the way across the paper as shown.

   On the left side, write down what your instructor puts on the board or overhead. On the right side, make your own comments about important words, procedures, or questions that you have.

2. Define the key terms.

   a. Sales tax   b. Commission   c. Simple interest

Review Exercises

For Exercises 3–11, solve the equations.

3. $7a - 2 = 11$

4. $2z + 6 = -15$

5. $4(x - 3) + 7 = 19$

6. $-3(y - 5) + 4 = 1$

7. $5(b + 4) - 3(2b + 8) = 3b$

8. $12c - 3c + 9 = 3(4 + 7c) - c$

9. $\dfrac{3}{8}p + \dfrac{3}{4} = p - \dfrac{3}{2}$

10. $\dfrac{1}{4} - 2x = 5$

11. $0.085(5)d - 0.075(4)d = 1250$

For the remaining exercises, follow the steps outlined in the Problem-Solving Flowchart found on page 52.

Objective 1: Introduction to Problem Solving

12. The larger of two numbers is 3 more than twice the smaller. The difference of the larger number and the smaller number is 8. Find the numbers.

13. One number is 3 less than another. Their sum is 15. Find the numbers. (See Example 1.)

14. The sum of 3 times a number and 2 is the same as the difference of the number and 4. Find the number.

15. Twice the sum of a number and 3 is the same as 1 subtracted from the number. Find the number.

16. The sum of two integers is 30. Ten times one integer is 5 times the other integer. Find the integers. (*Hint:* If one number is x, then the other number is $30 - x$.)

17. The sum of two integers is 10. Three times one integer is 3 less than 8 times the other integer. Find the integers. (*Hint:* If one number is x, then the other number is $10 - x$.)

Objective 2: Applications Involving Consecutive Integers

18. The sum of two consecutive page numbers in a book is 223. Find the page numbers.

19. The sum of the numbers on two consecutive raffle tickets is 808,455. Find the numbers of the tickets. (See Example 2.)

20. The sum of two consecutive odd integers is -148. Find the two integers.

21. Three times the smaller of two consecutive even integers is the same as -146 minus 4 times the larger integer. Find the integers.

22. The sum of three consecutive integers is −57. Find the integers.

23. Five times the smallest of three consecutive even integers is 10 more than twice the largest. Find the integers.

Objective 3: Applications Involving Percents and Rates

24. Leo works at a used car dealership and earns an 8% commission on sales. If he sold $39,000 in used cars, what was his commission?

25. Alysha works for a pharmaceutical company and makes 0.6% commission on all sales within her territory. If the yearly sales in her territory came to $8,200,000, what was her commission?

26. An account executive earns $600 per month plus a 3% commission on sales. The executive's goal is to earn $2400 this month. How much must she sell to achieve this goal?

27. If a salesperson in a department store sells merchandise worth over $200 in one day, she receives a 12% commission on the sales over $200. If the sales total $424 on one particular day, how much commission did she earn? **(See Example 3.)**

28. Molly had the choice of taking out a 4-year car loan at 8.5% simple interest or a 5-year car loan at 7.75% simple interest. If she borrows $15,000, which option will demand less interest?

29. Robert can take out a 3-year loan at 8% simple interest or a 2-year loan at $8\frac{1}{2}$% simple interest. If he borrows $7000, which option will demand less interest? **(See Example 4.)**

30. If Ivory Soap is $99\frac{44}{100}$% pure, then what quantity of impurities will be found in a bar of Ivory Soap that weighs 4.5 oz (ounces)?

31. In 1996 there was a Presidential election in which a third party candidate received a significant number of votes. The figure illustrates the number of votes received for Bill Clinton, Bob Dole, and Ross Perot in that election. Compute the percent of votes received by each candidate. (Round to the nearest tenth of a percent.)

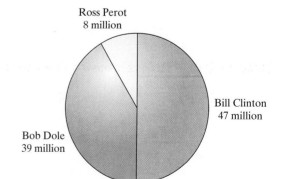

Number of votes by candidate—1996 Presidential election

Ross Perot 8 million
Bill Clinton 47 million
Bob Dole 39 million

32. The total bill (including a 6% sales tax) to have a radio installed in a car came to $265. What was the cost before tax?

33. Wayne County has a sales tax rate of 7%. How much does Mike's Honda Civic cost before tax if the total cost of the car *plus tax* is $13,888.60?

34. The price of a swimsuit after a 20% markup is $43.08. What was the price before the markup?

35. The price of a used textbook after a 35% markdown is $29.25. What was the original price? (See Example 5.)

36. In 2003, the number of people living below the poverty level in the United States was 35.2 million. This represents a 60% increase from the number in 2002. How many people lived below the poverty level in 2002?

37. In 2003, Americans spent $55.2 billion on weddings. This represents a 20% increase from the amount spent in 2001. What amount did Americans spend on weddings in 2001?

Objective 4: Applications Involving Principal and Interest

38. Darrell has a total of $12,500 in two accounts. One account pays 8% simple interest per year, and the other pays 12% simple interest. If he earned $1160 in the first year, how much did he invest in each account?

39. Lillian had $15,000 invested in two accounts, one paying 9% simple interest and one paying 10% simple interest. How much was invested in each account if the interest after 1 year is $1432? (See Example 6.)

40. Ms. Riley deposited some money in an account paying 5% simple interest and twice that amount in an account paying 6% simple interest. If the total interest from the two accounts is $765 for 1 year, how much was deposited into each account?

41. Sienna put some money in a certificate of deposit earning 4.2% simple interest. She deposited twice that amount in a money market account paying 4% simple interest. After 1 year her total interest was $488. How much did Sienna deposit in her money market account?

42. A total of $20,000 is invested between two accounts: one paying 4% simple interest and the other paying 3% simple interest. After 1 year the total interest was $720. How much was invested at each rate?

43. Mr. Hall had some money in his bank earning 4.5% simple interest. He had $5000 more deposited in a credit union earning 6% simple interest. If his total interest for 1 year was $1140, how much did he have in each account?

Objective 5: Applications Involving Mixtures

44. For a car to survive a winter in Toronto, the radiator must contain at least 75% antifreeze solution. Jacques' truck has 6 L of 50% antifreeze mixture, some of which must be drained and replaced with pure antifreeze to bring the concentration to the 75% level. How much 50% solution should be drained and replaced by pure antifreeze to have 6 L of 75% antifreeze?

45. How many ounces of water must be added to 20 oz of an 8% salt solution to make a 2% salt solution?

46. Ronald has a 12% solution of the fertilizer Super Grow. How much pure Super Grow should he add to the mixture to get 32 oz of a 17.5% concentration?

47. How many liters of an 18% alcohol solution must be added to a 10% alcohol solution to get 20 L of a 15% alcohol solution? (See Example 7.)

48. For a performance of the play *Company*, 375 tickets were sold. The price of the orchestra level seats was $25, and the balcony seats sold for $21. If the total revenue was $8875.00, how many of each type of ticket were sold?

49. Two different teas are mixed to make a blend that will be sold at a fair. Black tea sells for $2.20 per pound and orange pekoe tea sells for $3.00 per pound. How much of each should be used to obtain 4 lb of a blend selling for $2.50?

50. A nut mixture consists of almonds and cashews. Almonds are $4.98 per pound, and cashews are $6.98 per pound. How many pounds of each type of nut should be mixed to produce 16 lb selling for $5.73 per pound?

51. Two raffles are being held at a potluck dinner fund-raiser. One raffle ticket costs $2.00 per ticket for a weekend vacation. The other costs $1.00 per ticket for free passes to a movie theater. If 208 tickets were sold and a total of $320 was received, how many of each type of ticket were sold?

Objective 6: Applications Involving Distance, Rate, and Time

52. Sarah planned a trip from Daytona Beach to Detroit to visit her family. She was told that the distance from Daytona Beach to Atlanta was approximately $\frac{2}{5}$ of the distance between Daytona and Detroit. If the distance from Daytona to Detroit is approximately 1140 mi, how far is it from Daytona to Atlanta?

53. In Exercise 52, if Sarah travels at an average rate of 50 mph, how long will it take her to reach Atlanta?

54. Two families live $131\frac{1}{4}$ mi apart and want to meet for a picnic. One family has a sports car and drives at an average rate of 60 mph. The other family has an old station wagon, lots of kids, and a dog, so they average only about 45 mph. If the families leave at the same time, how long will it take them to meet?

55. Maria and Shirley hike around a lake. They start from the same point but walk in opposite directions around the 14-mi shoreline. Maria walks 0.8 mph faster than Shirley. How fast does Shirley walk if they meet in $2\frac{1}{2}$ hr? (See Example 8.)

56. Two cars leave from the same place at the same time and drive in opposite directions. One car travels an average of 10 mph faster than the other. After 2 hr the cars are 280 mi apart. Find the speed of each car.

57. Two cars are 350 km (kilometers) apart and travel toward each other on the same road. One travels 110 kph (kilometers per hour), and the other travels 90 kph. How long will it take the two cars to meet?

Objectives

1. Applications Involving Geometry
2. Literal Equations

section 1.6 Literal Equations and Applications to Geometry

1. Applications Involving Geometry

Some word problems involve the use of geometric formulas such as those listed in the inside front cover of this text.

Skill Practice

1. The length of Karen's living room is 2 ft longer than the width. The perimeter is 80 ft. Find the length and width.

example 1 Solving an Application Involving Perimeter

The length of a rectangular corral is 2 ft more than 3 times the width. The corral is situated such that one of its shorter sides is adjacent to a barn and does not require fencing. If the total amount of fencing is 774 ft, then find the dimensions of the corral.

Solution:

Read the problem and draw a sketch (Figure 1-8).

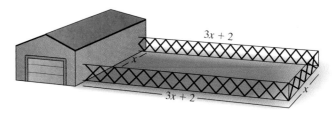

Figure 1-8

Let x represent the width. Label variables.

Let $3x + 2$ represent the length.

To create a verbal model, we might consider using the formula for the perimeter of a rectangle. However, the formula $P = 2L + 2W$ incorporates all four sides of the rectangle. The formula must be modified to include only one factor of the width.

$$\begin{pmatrix}\text{Distance around} \\ \text{three sides}\end{pmatrix} = \begin{pmatrix}2 \text{ times} \\ \text{the length}\end{pmatrix} + \begin{pmatrix}1 \text{ times} \\ \text{the width}\end{pmatrix} \quad \text{Verbal model}$$

$$774 = 2(3x + 2) + x \quad \text{Mathematical model}$$

$774 = 2(3x + 2) + x$ Solve for x.

$774 = 6x + 4 + x$ Apply the distributive property.

$774 = 7x + 4$ Combine *like* terms.

$770 = 7x$ Subtract 4 from both sides.

$110 = x$ Divide by 7 on both sides.

$x = 110$

Because x represents the width, the width of the corral is 110 ft. The length is given by

$3x + 2$ or $3(110) + 2 = 332$ Interpret the results.

The width of the corral is 110 ft, and the length is 332 ft. (To check the answer, verify that the three sides add to 774 ft.)

Answer

1. The length is 21 ft, and the width is 19 ft.

The applications involving angles utilize some of the formulas found in the front cover of this text.

example 2 Solving an Application Involving Angles

Two angles are complementary. One angle measures 10° less than 4 times the other angle. Find the measure of each angle (Figure 1-9).

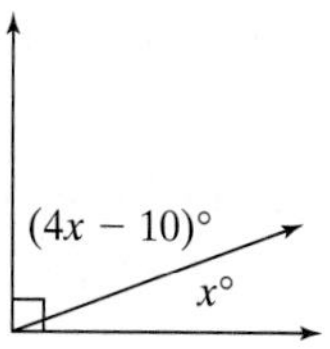

Solution:

Let x represent one angle.
Let $4x - 10$ represent the other angle.

$(4x - 10)°$
$x°$

Figure 1-9

Recall that two angles are complementary if the sum of their measures is 90°. Therefore, a verbal model is

(One angle) + (the complement of the angle) = 90°   Verbal model

$x + (4x - 10) = 90$ Mathematical equation

$5x - 10 = 90$ Solve for x.

$5x = 100$

$x = 20$

If $x = 20$, then $4x - 10 = 4(20) - 10 = 70$. The two angles are 20° and 70°.

Skill Practice

2. Two angles are supplementary, and the measure of one is 16° less than 3 times the other. Find their measures.

2. Literal Equations

Literal equations (or formulas) are equations that contain several variables. For example, the formula for the perimeter of a rectangle $P = 2L + 2W$ is an example of a literal equation. In this equation, P is expressed in terms of L and W. However, in science and other branches of applied mathematics, formulas may be more useful in alternative forms.

For example, the formula $P = 2L + 2W$ can be manipulated to solve for either L or W:

Solve for L

$P = 2L + 2W$

$P - 2W = 2L$ Subtract $2W$.

$\dfrac{P - 2W}{2} = L$ Divide by 2.

$L = \dfrac{P - 2W}{2}$

Solve for W

$P = 2L + 2W$

$P - 2L = 2W$ Subtract $2L$.

$\dfrac{P - 2L}{2} = W$ Divide by 2.

$W = \dfrac{P - 2L}{2}$

Concept Connections

Solve the equation for a.

3. $ac - bc = d$

To solve a literal equation for a specified variable, use the addition, subtraction, multiplication, and division properties of equality.

Answers

2. 49° and 131°   3. $a = \dfrac{d + bc}{c}$

Skill Practice

The formula for the area of a triangle is $A = \frac{1}{2}bh$.

4. Solve the formula for h.
5. Find the value of h when $A = 40$ in.^{2} and $b = 16$ in.

example 3   Solving a Literal Equation

The formula for the volume of a rectangular box is $V = LWH$.

a. Solve the formula $V = LWH$ for W.

b. Find the value of W if $V = 200$ in.^{3}, $L = 20$ in., and $H = 5$ in. (Figure 1-10).

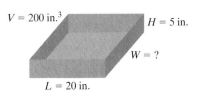

Figure 1-10

Solution:

a. $V = LWH$ The goal is to isolate the variable W.

$\dfrac{V}{LH} = \dfrac{\cancel{L}W\cancel{H}}{\cancel{L}\cancel{H}}$ Divide both sides by LH.

$\dfrac{V}{LH} = W$ Simplify.

$W = \dfrac{V}{LH}$

b. $W = \dfrac{200 \text{ in.}^3}{(20 \text{ in.})(5 \text{ in.})}$ Substitute $V = 200$ in.^{3}, $L = 20$ in., and $H = 5$ in.

$W = 2$ in.   The width is 2 in.

Skill Practice

6. The formula for the volume of a right circular cylinder is $V = \pi r^2 h$. Solve for h.

example 4   Solving a Literal Equation

The formula to find the area of a trapezoid is given by $A = \frac{1}{2}(b_1 + b_2)h$, where b_1 and b_2 are the lengths of the parallel sides and h is the height.
    Solve this formula for b_1.

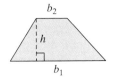

Solution:

$A = \frac{1}{2}(b_1 + b_2)h$ The goal is to isolate b_1.

$2A = 2 \cdot \frac{1}{2}(b_1 + b_2)h$ Multiply by 2 to clear fractions.

$2A = (b_1 + b_2)h$ Apply the distributive property.

$2A = b_1 h + b_2 h$

$2A - b_2 h = b_1 h$ Subtract $b_2 h$ from both sides.

$\dfrac{2A - b_2 h}{h} = \dfrac{b_1 \cancel{h}}{\cancel{h}}$ Divide by h.

$\dfrac{2A - b_2 h}{h} = b_1$

Answers

4. $h = \dfrac{2A}{b}$ 5. $h = 5$ in.

6. $h = \dfrac{V}{\pi r^2}$

Tip: When solving a literal equation for a specified variable, there is sometimes more than one way to express your final answer. This flexibility often presents difficulty for students. Students may leave their answer in one form, but the answer given in the text looks different. Yet both forms may be correct. To know if your answer is equivalent to the form given in the text you must try to manipulate it to look like the answer in the book, a process called *form fitting*.

The literal equation from Example 4 may be written in several different forms. The quantity $(2A - b_2 h)/h$ can be split into two fractions.

$$b_1 = \frac{2A - b_2 h}{h} = \frac{2A}{h} - \frac{b_2 h}{h} = \frac{2A}{h} - b_2$$

example 5 — Solving a Linear Equation in Two Variables

Given $-2x + 3y = 5$, solve for y.

Solution:

$-2x + 3y = 5$

$3y = 2x + 5$ Add $2x$ to both sides.

$\dfrac{3y}{3} = \dfrac{2x + 5}{3}$ Divide by 3 on both sides.

$y = \dfrac{2x + 5}{3}$ or $y = \dfrac{2}{3}x + \dfrac{5}{3}$

Skill Practice

Solve for y.

7. $5x + 2y = 11$

example 6 — Applying a Literal Equation

Buckingham Fountain is one of Chicago's most familiar landmarks. With 133 jets spraying a total of 14,000 gal (gallons) of water per minute, Buckingham Fountain is one of the world's largest fountains. The circumference of the fountain is approximately 880 ft.

a. The circumference of a circle is given by $C = 2\pi r$. Solve the equation for r.

b. Use the equation from part (a) to find the radius and diameter of the fountain. (Use the $\boxed{\pi}$ key on the calculator, and round the answers to 1 decimal place.)

Solution:

a. $C = 2\pi r$

$\dfrac{C}{2\pi} = \dfrac{2\pi r}{2\pi}$

$\dfrac{C}{2\pi} = r$

$r = \dfrac{C}{2\pi}$

Skill Practice

The formula to compute the surface area S of a sphere is given by $S = 4\pi r^2$.

8. Solve the equation for π.

9. A sphere has a surface area of 113 in.^{2} and a radius of 3 in. Use the formula found in part (a) to approximate π. Round to 2 decimal places.

Answers

7. $y = \dfrac{11 - 5x}{2}$ or $y = -\dfrac{5}{2}x + \dfrac{11}{2}$

8. $\pi = \dfrac{S}{4r^2}$ 9. 3.14

b. $r = \dfrac{880 \text{ ft}}{2\pi}$ Substitute $C = 880$ ft and use the π key on the calculator.

$r \approx 140.1$ ft

The radius is approximately 140.1 ft. The diameter is twice the radius ($d = 2r$); therefore the diameter is approximately 280.2 ft.

section 1.6 Practice Exercises

Boost *your* GRADE at mathzone.com!

- Practice Problems
- Self-Tests
- NetTutor
- e-Professors
- Videos

Study Skills Exercises

1. Look over the notes that you took today. Do you understand what you wrote? If there were any rules, definitions, or formulas, highlight them so that you can find them easily when studying for the test.

2. Define the key term **literal equation**.

Review Exercises

For Exercises 3–6, solve the equations.

3. $7 + 5x - (2x - 6) = 6(x + 1) + 21$

4. $\dfrac{3}{5}y - 3 + 2y = 5$

5. $3[z - (2 - 3z) - 4] = z - 7$

6. $2a - 4 + 8a = 7a - 8 + 3a$

Objective 1: Applications Involving Geometry

For Exercises 7–26, use the geometry formulas listed in the inside front cover of the text.

7. A volleyball court is twice as long as it is wide. If the perimeter is 177 ft, find the dimensions of the court. (See Example 1.)

8. Two sides of a triangle are equal in length, and the third side is 1.5 times the length of one of the other sides. If the perimeter is 14 m (meters), find the lengths of the sides.

9. The lengths of the sides of a triangle are given by three consecutive even integers. The perimeter is 24 m. What is the length of each side?

10. A triangular garden has sides that can be represented by three consecutive integers. If the perimeter of the garden is 15 ft, what are the lengths of the sides?

11. Raoul would like to build a rectangular dog run in the rear of his backyard, away from the house. The width of the yard is $11\frac{1}{2}$ yd, and Raoul wants an area of 92 yd² (square yards) for his dog.

    a. Find the dimensions of the dog run.

    b. How much fencing would Raoul need to enclose the dog run?

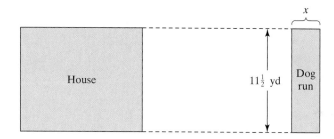

12. George built a rectangular pen for his rabbit such that the length is 7 ft less than twice the width. If the perimeter is 40 ft, what are the dimensions of the pen?

13. Antoine wants to put edging in the form of a square around a tree in his front yard. He has enough money to buy 18 ft of edging. Find the dimensions of the square that will use all the edging.

14. Joanne wants to plant a flower garden in her backyard in the shape of a trapezoid, adjacent to her house (see the figure). She also wants a front yard garden in the same shape, but with sides one-half as long. What should the dimensions be for each garden if Joanne has only a total of 60 ft of fencing?

15. The measures of two angles in a triangle are equal. The third angle measures 2 times the sum of the equal angles. Find the measures of the three angles.

16. The smallest angle in a triangle is one-half the size of the largest. The middle angle measures 25° less than the largest. Find the measures of the three angles.

17. Two angles are complementary. One angle is 5 times as large as the other angle. Find the measure of each angle. **(See Example 2.)**

18. Two angles are supplementary. One angle measures 12° less than 3 times the other. Find the measure of each angle.

In Exercises 19–26, solve for x, and then find the measure of each angle.

19.

20.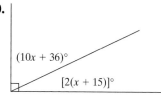

21.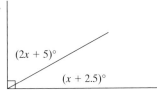

22.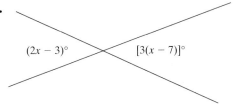

23.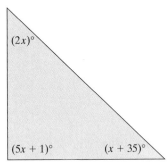

24.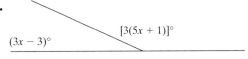

25.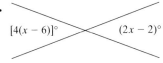

26.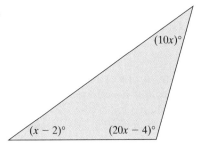

Objective 2: Literal Equations

27. Which expression(s) is (are) equivalent to $-5/(x-3)$?

   a. $-\dfrac{5}{x-3}$ **b.** $\dfrac{5}{3-x}$ **c.** $\dfrac{5}{-x+3}$

28. Which expression(s) is (are) equivalent to $(z-1)/-2$?

   a. $\dfrac{1-z}{2}$ **b.** $-\dfrac{z-1}{2}$ **c.** $\dfrac{-z+1}{2}$

29. Which expression(s) is (are) equivalent to $(-x-7)/y$?

   a. $-\dfrac{x+7}{y}$ **b.** $\dfrac{x+7}{-y}$ **c.** $\dfrac{-x-7}{-y}$

30. Which expression(s) is (are) equivalent to $-3w/(-x-y)$?

   a. $-\dfrac{3w}{-x-y}$ **b.** $\dfrac{3w}{x+y}$ **c.** $-\dfrac{-3w}{x+y}$

Section 1.6 Literal Equations and Applications to Geometry   71

For Exercises 31–48, solve for the indicated variable. (See Examples 3–4.)

31. $A = lw$ for l

32. $C_1 = \frac{5}{2}R$ for R

33. $I = Prt$ for P

34. $a + b + c = P$ for b

35. $W = K_2 - K_1$ for K_1

36. $y = mx + b$ for x

37. $F = \frac{9}{5}C + 32$ for C

38. $C = \frac{5}{9}(F - 32)$ for F

39. $K = \frac{1}{2}mv^2$ for v^2

40. $I = Prt$ for r

41. $v = v_0 + at$ for a

42. $a^2 + b^2 = c^2$ for b^2

43. $w = p(v_2 - v_1)$ for v_2

44. $A = lw$ for w

45. $ax + by = c$ for y

46. $P = 2L + 2W$ for L

47. $V = \frac{1}{3}Bh$ for B

48. $V = \frac{1}{3}\pi r^2 h$ for h

In Chapter 2 it will be necessary to change equations from the form $Ax + By = C$ to $y = mx + b$. To practice this skill, express each equation in the form $y = mx + b$ in Exercises 49–60. (See Example 5.)

49. $3x + y = 6$

50. $x + y = -4$

51. $5x - 4y = 20$

52. $-4x - 5y = 25$

53. $-6x - 2y = 13$

54. $5x - 7y = 15$

55. $3x - 3y = 6$

56. $2x - 2y = 8$

57. $9x + \frac{4}{3}y = 5$

58. $4x - \frac{1}{3}y = 5$

59. $-x + \frac{2}{3}y = 0$

60. $x - \frac{1}{4}y = 0$

For Exercises 61–62, use the relationship between distance, rate, and time given by $d = rt$. (See Example 6.)

61. a. Solve $d = rt$ for rate r.

  b. In 1998 Eddie Cheever won the Indianapolis 500 in 3 hr, 26 min, 40 sec (3.444 hr). Find his average rate of speed if the total distance is 500 mi. Round to the nearest tenth of a mile per hour.

62. a. Solve $d = rt$ for time t.

  b. In 1998 Jeff Gordon won the Daytona 500 with an average speed of 161.551 mph. Find the time it took him to complete the race if the total distance is 500 mi. Round to the nearest hundredth of an hour.

For Exercises 63–64, use the fact that the force imparted by an object is equal to its mass times acceleration, or $F = ma$.

63. a. Solve $F = ma$ for mass m.

b. The force on an object is 24.5 N (newtons), and the acceleration due to gravity is 9.8 m/sec². Find the mass of the object. (The answer will be in kilograms.)

64. a. Solve $F = ma$ for acceleration a.

b. Approximate the acceleration of a 2000-kg mass influenced by a force of 15,000 N. (The answer will be in meters per second squared, m/sec².)

In statistics the z-score formula $z = \dfrac{x - \mu}{\sigma}$ is used in studying probability. Use this formula for Exercises 65–66.

65. a. Solve $z = \dfrac{x - \mu}{\sigma}$ for x.

b. Find x when $z = 2.5$, $\mu = 100$, and $\sigma = 12$.

66. a. Solve $z = \dfrac{x - \mu}{\sigma}$ for σ.

b. Find σ when $x = 150$, $z = 2.5$, and $\mu = 110$.

Expanding Your Skills

For Exercises 67–76, solve for the indicated variable.

67. $6t - rt = 12$ for t

68. $5 = 4a + ca$ for a

69. $ax + 5 = 6x + 3$ for x

70. $cx - 4 = dx + 9$ for x

71. $A = P + Prt$ for P

72. $A = P + Prt$ for r

73. $T = mg - mf$ for m

74. $T = mg - mf$ for f

75. $ax + by = cx + z$ for x

76. $Lt + h = mt + g$ for t

Section 1.7 Linear Inequalities in One Variable

1. Solving Linear Inequalities

In Sections 1.4–1.6, we learned how to solve linear equations and their applications. In this section, we will learn the process of solving linear *inequalities*. A **linear inequality** in one variable, x, is defined as any relationship of the form: $ax + b < 0$, $ax + b \leq 0$, $ax + b > 0$, or $ax + b \geq 0$, where $a \neq 0$.

The solution to the equation $x = 3$ can be graphed as a single point on the number line.

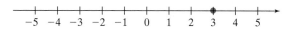

Now consider the *inequality* $x \leq 3$. The solution set to an inequality is the set of real numbers that makes the inequality a true statement. In this case, the solution set is all real numbers less than or equal to 3. Because the solution set has an infinite number of values, the values cannot be listed. Instead, we can graph the solution set or represent the set in interval notation or in set-builder notation.

Graph	Interval Notation	Set-Builder Notation
(number line from −5 to 5, shaded left of 3)	$(-\infty, 3]$	$\{x \mid x \leq 3\}$

The addition and subtraction properties of equality indicate that a value added to or subtracted from both sides of an equation results in an equivalent equation. The same is true for inequalities.

> **Addition and Subtraction Properties of Inequality**
>
> Let a, b, and c represent real numbers.
>
> *Addition property of inequality: If $a < b$, then $a + c < b + c$
>
> *Subtraction property of inequality: If $a < b$, then $a - c < b - c$
>
> *These properties may also be stated for $a \leq b$, $a > b$, and $a \geq b$.

Example 1   Solving a Linear Inequality

Solve the inequality. Graph the solution and write the solution set in interval notation.

$$3x - 7 > 2(x - 4) - 1$$

Solution:

$3x - 7 > 2(x - 4) - 1$

$3x - 7 > 2x - 8 - 1$ Apply the distributive property.

$3x - 7 > 2x - 9$

$3x - 2x - 7 > 2x - 2x - 9$ Subtract $2x$ from both sides.

Objectives

1. Solving Linear Inequalities
2. Inequalities of the Form $a < x < b$
3. Applications of Inequalities

Concept Connections

1. Graph $(-\infty, -5)$.

2. Write interval notation for the graph shown.

Skill Practice

3. Solve the inequality. Graph the solution and write the solution in interval notation.

$$4(2x - 1) > 7x + 1$$

Answers

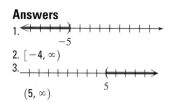

2. $[-4, \infty)$

3. (graph) $(5, \infty)$

$$x - 7 > -9$$
$$x - 7 + 7 > -9 + 7 \quad \text{Add 7 to both sides.}$$
$$x > -2$$

Graph

$\begin{array}{c}\xrightarrow{}\\-4\ -3\ -2\ -1\ \ 0\ \ 1\ \ 2\ \ 3\ \ 4\end{array}$

Interval Notation

$(-2, \infty)$

Multiplying both sides of an equation by the same quantity results in an equivalent equation. However, the same is not always true for an inequality. If you multiply or divide an inequality by a *negative* quantity, the direction of the inequality symbol must be *reversed*.

For example, consider multiplying or dividing the inequality $4 < 5$ by -1.

Multiply/divide by -1:
$$4 < 5$$
$$-4 > -5$$

The number 4 lies to the left of 5 on the number line. However, -4 lies to the right of -5. Changing the signs of two numbers changes their relative position on the number line. This is stated formally in the multiplication and division properties of inequality.

Multiplication and Division Properties of Inequality

Let a, b, and c represent real numbers.

*If c is *positive* and $a < b$, then $\quad ac < bc \quad$ and $\quad \dfrac{a}{c} < \dfrac{b}{c}$

*If c is *negative* and $a < b$, then $\quad ac > bc \quad$ and $\quad \dfrac{a}{c} > \dfrac{b}{c}$

The second statement indicates that if both sides of an inequality are multiplied or divided by a negative quantity, the inequality sign must be *reversed*.

*These properties may also be stated for $a \leq b$, $a > b$, and $a \geq b$.

Skill Practice

Solve the inequalities. Graph the solution and write the solution in interval notation.

4. $3x + 5 > 15$
5. $-4x - 12 \geq 20$
6. $5(3x + 1) < 4(5x - 5)$

example 2 Solving Linear Inequalities

Solve the inequalities. Graph the solution and write the solution set in interval notation.

a. $-2x - 5 < 2$ 	b. $-6(x - 3) \geq 2 - 2(x - 8)$

Solution:

a.
$$-2x - 5 < 2$$
$$-2x - 5 + 5 < 2 + 5 \quad \text{Add 5 to both sides.}$$
$$-2x < 7$$
$$\dfrac{-2x}{-2} > \dfrac{7}{-2} \quad \text{Divide by } -2 \text{ (reverse the inequality sign).}$$
$$x > -\dfrac{7}{2} \quad \text{or} \quad x > -3.5$$

Answers

4. $\left(\dfrac{10}{3}, \infty\right)$

5. $(-\infty, -8]$

6. $(5, \infty)$

Graph

$-\frac{7}{2}$ on number line from -5 to 5, open parenthesis at $-\frac{7}{2}$, arrow right.

Interval Notation

$\left(-\frac{7}{2}, \infty\right)$

Tip: The inequality $-2x - 5 < 2$ could have been solved by isolating x on the right-hand side of the inequality. This creates a positive coefficient on the x term and eliminates the need to divide by a negative number.

$-2x - 5 < 2$

$\quad -5 < 2x + 2$ Add $2x$ to both sides.

$\quad -7 < 2x$ Subtract 2 from both sides.

$\quad \dfrac{-7}{2} < \dfrac{2x}{2}$ Divide by 2 (because 2 is positive, do *not* reverse the inequality sign).

$\quad -\dfrac{7}{2} < x$ (Note that the inequality $-\frac{7}{2} < x$ is equivalent to $x > -\frac{7}{2}$.)

b. $-6(x - 3) \geq 2 - 2(x - 8)$

$\quad -6x + 18 \geq 2 - 2x + 16$ Apply the distributive property.

$\quad -6x + 18 \geq 18 - 2x$ Combine *like* terms.

$\quad -6x + 2x + 18 \geq 18 - 2x + 2x$ Add $2x$ to both sides.

$\quad -4x + 18 \geq 18$

$\quad -4x + 18 - 18 \geq 18 - 18$ Subtract 18 from both sides.

$\quad -4x \geq 0$

$\quad \dfrac{-4x}{-4} \leq \dfrac{0}{-4}$ Divide by -4 (*reverse* the inequality sign).

$\quad x \leq 0$

Graph

Number line from -5 to 5, closed bracket at 0, arrow left.

Interval Notation

$(-\infty, 0]$

example 3   Solving a Linear Inequality

Solve the inequality $\dfrac{-5x + 2}{-3} > x + 2$. Graph the solution and write the solution set in interval notation.

Solution:

$\dfrac{-5x + 2}{-3} > x + 2$

$-3\left(\dfrac{-5x + 2}{-3}\right) < -3(x + 2)$ Multiply by -3 to clear fractions (*reverse* the inequality sign).

$-5x + 2 < -3x - 6$

$-5x + 3x + 2 < -3x + 3x - 6$ Add $3x$ to both sides.

Skill Practice

Solve the inequality. Graph the solution and write the solution in interval notation.

7. $\dfrac{x + 1}{-3} \geq -x + 1$

Answer

7.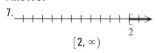

$[2, \infty)$

$$-2x + 2 < -6$$
$$-2x + 2 - 2 < -6 - 2 \quad \text{Subtract 2 from both sides.}$$
$$-2x < -8$$
$$\frac{-2x}{-2} > \frac{-8}{-2} \quad \text{Divide by } -2 \text{ (the inequality sign is reversed } again\text{).}$$
$$x > 4 \quad \text{Simplify.}$$

Graph

Interval Notation

$(4, \infty)$

In Example 3, the inequality sign was reversed twice: once for multiplying the inequality by -3 and once for dividing by -2. If you are in doubt about whether you have the inequality sign in the correct direction, you can check your final answer by using the **test point method**. That is, pick a point in the proposed solution set, and verify that it makes the original inequality true. Furthermore, any test point picked outside the solution set should make the original inequality false.

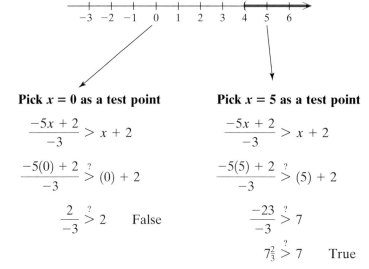

Because a test point to the right of $x = 4$ makes the inequality true, we have shaded the correct part of the number line.

2. Inequalities of the Form $a < x < b$

An inequality of the form $a < x < b$ is a type of **compound inequality**, one that defines two simultaneous conditions on the quantity x.

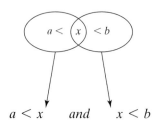

The solution to the compound inequality $a < x < b$ is the *intersection* of the inequalities $a < x$ and $x < b$. To solve a compound inequality of this form, we can actually work with the inequality as a "three-part" inequality and isolate the variable x.

Concept Connections

8. Check the solution to margin Exercise 7, using the test point method.

Answer

8. Answers may vary.

example 4 — Solving a Compound Inequality of the Form $a < x < b$

Solve the inequality $-2 \leq 3x + 1 < 5$. Graph the solution and express the solution set in interval notation.

Solution:

To solve the compound inequality $-2 \leq 3x + 1 < 5$, isolate the variable x in the "middle." The operations performed on the middle portion of the inequality must also be performed on the left-hand side and right-hand side.

$$-2 \leq 3x + 1 < 5$$

$$-2 - 1 \leq 3x + 1 - 1 < 5 - 1 \qquad \text{Subtract 1 from all three parts of the inequality.}$$

$$-3 \leq 3x < 4 \qquad \text{Simplify.}$$

$$\frac{-3}{3} \leq \frac{3x}{3} < \frac{4}{3} \qquad \text{Divide by 3 in all three parts of the inequality.}$$

$$-1 \leq x < \frac{4}{3} \qquad \text{Simplify.}$$

Graph   **Interval Notation**

$$\left[-1, \frac{4}{3}\right)$$

Skill Practice

9. Solve the compound inequality. Graph the solution and express the solution set in interval notation.

$$-8 < 5x - 3 \leq 12$$

3. Applications of Inequalities

example 5 — Solving a Compound Inequality Application

Beth received grades of 87%, 82%, 96%, and 79% on her last four algebra tests. To graduate with honors, she needs at least a B in the course.

a. What grade does she need to make on the fifth test to get a B in the course? Assume that the tests are weighted equally and that to earn a B the average of the test grades must be at least 80% but less than 90%.

b. Is it possible for Beth to earn an A in the course if an A requires an average of 90% or more?

Solution:

a. Let x represent the score on the fifth test.

The average of the five tests is given by $\dfrac{87 + 82 + 96 + 79 + x}{5}$

To earn a B, Beth requires

$80 \leq$ (average of test scores) < 90 Verbal model

$$80 \leq \frac{87 + 82 + 96 + 79 + x}{5} < 90 \qquad \text{Mathematical model}$$

Skill Practice

10. Jamie is a salesman who works on commission, so his salary varies from month to month. To qualify for an automobile loan, his salary must average $2100 for 6 months. His salaries for the past 5 months have been $1800, $2300, $1500, $2200, and $2800. What amount does he need to earn in the last month to qualify for the loan?

Answers

9.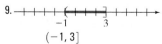
   $(-1, 3]$

10. Jamie's salary must be at least $2000.

$$5(80) \leq 5\left(\frac{87 + 82 + 96 + 79 + x}{5}\right) < 5(90) \qquad \text{Multiply by 5 to clear fractions.}$$

$$400 \leq 344 + x < 450 \qquad \text{Simplify.}$$

$$400 - 344 \leq 344 - 344 + x < 450 - 344 \qquad \text{Subtract 344 from all three parts.}$$

$$56 \leq x < 106 \qquad \text{Simplify.}$$

To earn a B in the course, Beth must score at least 56% but less than 106% on the fifth exam. Realistically, she may score between 56% and 100% because a grade over 100% is not possible.

b. To earn an A, Beth's average would have to be greater than or equal to 90%.

$$(\text{Average of test scores}) \geq 90 \qquad \text{Verbal model}$$

$$\frac{87 + 82 + 96 + 79 + x}{5} \geq 90 \qquad \text{Mathematical equation}$$

$$5\left(\frac{87 + 82 + 96 + 79 + x}{5}\right) \geq 5(90) \qquad \text{Clear fractions.}$$

$$344 + x \geq 450 \qquad \text{Simplify.}$$

$$x \geq 106 \qquad \text{Solve for } x.$$

It would be impossible for Beth to earn an A in the course because she would have to earn at least a score of 106% on the fifth test. It is impossible to earn over 100%.

Skill Practice

11. The population of Alaska has steadily increased since 1950 according to the equation $P = 10t + 117$, where t represents the number of years after 1950 and P represents the population in thousands.

For what years after 1950 was the population less than 417 thousand people?

example 6 — Solving a Linear Inequality Application

The number of registered passenger cars N in the United States has risen between 1960 and 2005 according to the equation $N = 2.5t + 64.4$, where t represents the number of years after 1960 ($t = 0$ corresponds to 1960, $t = 1$ corresponds to 1961, and so on) (Figure 1-11).

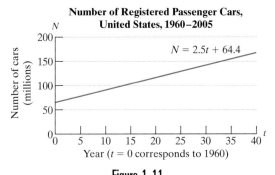

Figure 1-11

(*Source:* U.S. Department of Transportation)

a. For what years after 1960 was the number of registered passenger cars less than 89.4 million?

b. For what years was the number of registered passenger cars between 94.4 million and 101.9 million?

c. Predict the years for which the number of passenger cars will exceed 154.4 million.

Answer

11. The population was less than 417 thousand for $t < 30$. This corresponds to the years before 1980.

Solution:

a. We require $N < 89.4$ million.

$$\overbrace{N}^{} < 89.4$$

$$2.5t + 64.4 < 89.4 \qquad \text{Substitute the expression } 2.5t + 64.4 \text{ for } N.$$

$$2.5t + 64.4 - 64.4 < 89.4 - 64.4 \qquad \text{Subtract 64.4 from both sides.}$$

$$2.5t < 25$$

$$\frac{2.5t}{2.5} < \frac{25}{2.5} \qquad \text{Divide both sides by 2.5.}$$

$$t < 10 \qquad t = 10 \text{ corresponds to the year 1970.}$$

Before 1970 (but after 1960), the number of registered passenger cars was less than 89.4 million.

b. We require $94.4 < N < 101.9$. Hence

$$94.4 < 2.5t + 64.4 < 101.9 \qquad \text{Substitute the expression } 2.5t + 64.4 \text{ for } N.$$

$$94.4 - 64.4 < 2.5t + 64.4 - 64.4 < 101.9 - 64.4 \qquad \text{Subtract 64.4 from all three parts of the inequality.}$$

$$30.0 < 2.5t < 37.5$$

$$\frac{30.0}{2.5} < \frac{2.5t}{2.5} < \frac{37.5}{2.5} \qquad \text{Divide by 2.5.}$$

$$12 < t < 15 \qquad t = 12 \text{ corresponds to 1972 and } t = 15 \text{ corresponds to 1975.}$$

Between the years 1972 and 1975, the number of registered passenger cars was between 94.4 million and 101.9 million.

c. We require $N > 154.4$.

$$2.5t + 64.4 > 154.4 \qquad \text{Substitute the expression } 2.5t + 64.4 \text{ for } N.$$

$$2.5t > 90 \qquad \text{Subtract 64.4 from both sides.}$$

$$\frac{2.5t}{2.5} > \frac{90}{2.5} \qquad \text{Divide by 2.5.}$$

$$t > 36 \qquad t = 36 \text{ corresponds to the year 1996.}$$

After the year 1996, the number of registered passenger cars is predicted to exceed 154.4 million.

section 1.7 Practice Exercises

Boost *your* GRADE at mathzone.com!

- Practice Problems
- Self-Tests
- NetTutor
- e-Professors
- Videos

Study Skills Exercises

1. Some instructors allow the use of calculators. Does your instructor allow the use of a calculator? If so, what kind?

   Will you be allowed to use a calculator on tests, or just for occasional calculator problems in the text?

   Helpful Hint: If you are not permitted to use a calculator on tests, it is good to do your homework in the same way, without the calculator.

2. Define the key terms.
   a. Linear inequality
   b. Test point method
   c. Compound inequality

Review Exercises

3. Solve for v. $d = vt - 16t^2$

4. Solve for x. $4 + 5(4 - 2x) = -2(x - 1) - 4$

5. Five more than 3 times a number is 6 less than twice the number. Find the number.

6. Solve for y. $5x + 3y + 6 = 0$

7. a. The area of a triangle is given by $A = \frac{1}{2}bh$. Solve for h.

   b. If the area of a certain triangle is 10 cm^2 and the base is 3 cm, find the height.

8. Solve for t.

   $$\frac{1}{5}t - \frac{1}{2} - \frac{1}{10}t + \frac{2}{5} = \frac{3}{10}t + \frac{1}{2}$$

Objective 1: Solving Linear Inequalities

For Exercises 9–31, solve the inequalities. Graph the solution and write the solution set in interval notation. Check each answer by using the test point method. **(See Examples 1–3.)**

9. $6 \leq 4 - 2y$

10. $2x - 5 \geq 15$

11. $4z + 2 < 22$

12. $6z + 3 > 16$

13. $8w - 2 \leq 13$

14. $\frac{2}{3}t < -8$

15. $\dfrac{1}{5}p + 3 \geq -1$

16. $\dfrac{3}{4}(8y - 9) < 3$

17. $\dfrac{2}{5}(2x - 1) > 10$

18. $0.8a - 0.5 \leq 0.3a - 11$

19. $0.2w - 0.7 < 0.4 - 0.9w$

20. $-5x + 7 < 22$

21. $-3w - 6 > 9$

22. $-\dfrac{5}{6}x \leq -\dfrac{3}{4}$

23. $\dfrac{3k - 2}{-5} \leq 4$

24. $\dfrac{3p - 1}{-2} > 5$

25. $-\dfrac{3}{2}y > -\dfrac{21}{16}$

26. $0.2t + 1 > 2.4t - 10$

27. $20 \leq 8 - \dfrac{1}{3}x$

28. $3 - 4(y + 2) \leq 6 + 4(2y + 1)$

29. $1 < 3 - 4(3b - 1)$

30. $7.2k - 5.1 \geq 5.7$

31. $6h - 2.92 \leq 16.58$

Objective 2: Inequalities of the form $a < x < b$

32. Write $-3 \leq x < 2$ as two inequalities.

33. Write $5 < x \leq 7$ as two inequalities.

For Exercises 34–45, solve the compound inequalities. Graph the solution and write the solution set in interval notation. (See Example 4.)

34. $0 \leq 3a + 2 < 17$

35. $-8 < 4k - 7 < 11$

36. $5 < 4y - 3 < 21$

37. $7 \leq 3m - 5 < 10$

38. $1 \leq \dfrac{1}{5}x + 12 \leq 13$

39. $5 \leq \dfrac{1}{4}a + 1 < 9$

40. $4 > \dfrac{2x + 8}{-2} \geq -5$

41. $-5 \geq \dfrac{-y + 3}{6} \geq -8$

42. $6 \geq -2b - 3 > -6$

43. $4 > -3w - 7 \geq 2$

44. $8 > -w + 4 > -1$

45. $13 \geq -2h - 1 \geq 0$

46. Explain why $5 < x < 1$ has no solution.

47. Explain why $-1 > x > 3$ has no solution.

Objective 3: Applications of Inequalities

48. Nolvia sells copy machines, and her salary is $25,000 plus a 4% commission on sales. The equation $S = 25,000 + 0.04x$ represents her salary S in dollars in terms of her total sales x in dollars.

  a. How much money in sales does Nolvia need to earn a salary that exceeds $40,000?

  b. How much money in sales does Nolvia need to earn a salary that exceeds $80,000?

  c. Why is the money in sales required to earn a salary of $80,000 more than twice the money in sales required to earn a salary of $40,000?

49. The amount of money in a savings account A depends on the principal P, the interest rate r, and the time in years t that the money is invested. The equation $A = P + Prt$ shows the relationship among the variables for an account earning simple interest. If an investor deposits $5000 at $6\frac{1}{2}$% simple interest, the account will grow according to the formula $A = 5000 + 5000(0.065)t$. **(See Example 6.)**

  a. How many years will it take for the investment to exceed $10,000? (Round to the nearest tenth of a year.)

  b. How many years will it take for the investment to exceed $15,000? (Round to the nearest tenth of a year.)

50. The revenue R for selling x fleece jackets is given by the equation $R = 49.95x$. The cost to produce x jackets is $C = 2300 + 18.50x$. Find the number of jackets that the company needs to sell to produce a profit. (*Hint:* A profit occurs when revenue exceeds cost.)

51. The revenue R for selling x mountain bikes is $R = 249.95x$. The cost to produce x bikes is $C = 56,000 + 140x$. Find the number of bikes that the company needs to sell to produce a profit.

52. The average high and low temperatures for Vancouver, British Columbia, in January are 5.6°C and 0°C, respectively. The formula relating Celsius temperatures to Fahrenheit temperatures is given by $C = \frac{5}{9}(F - 32)$. Convert the inequality $0.0° \leq C \leq 5.6°$ to an equivalent inequality using Fahrenheit temperatures.

53. For a day in July, the temperatures in Austin, Texas ranged from 20°C to 29°C. The formula relating Celsius temperatures to Fahrenheit temperatures is given by $C = \frac{5}{9}(F - 32)$. Convert the inequality $20° \leq C \leq 29°$ to an equivalent inequality using Fahrenheit temperatures. **(See Example 5.)**

54. The poverty threshold P for four-person families between the years 1960 and 2006 can be approximated by the equation $P = 1235 + 387t$, where P is measured in dollars; and $t = 0$ corresponds to the year 1960, $t = 1$ corresponds to 1961, and so on. (Source: U.S. Bureau of the Census.)

  a. For what years after 1960 was the poverty threshold under $7040?

  b. For what years after 1960 was the poverty threshold between $4331 and $10,136?

55. Between the years 1960 and 2006, the average gas mileage (miles per gallon) for passenger cars has increased. The equation $N = 12.6 + 0.214t$ approximates the average gas mileage corresponding to the year t, where $t = 0$ represents 1960, $t = 1$ represents 1961, and so on.

   a. For what years after 1960 was the average gas mileage less than 14.1 mpg? (Round to the nearest year.)

   b. For what years was the average gas mileage between 17.1 and 18.0 mpg? (Round to the nearest year.)

Mixed Exercises

For Exercises 56–73, solve the inequalities. Graph the solution, and write the solution set in interval notation. Check each answer by using the test point method.

56. $-6p - 1 > 17$

57. $-4y + 1 \le -11$

58. $\dfrac{3}{4}x - 8 \le 1$

59. $-\dfrac{2}{5}a - 3 > 5$

60. $-1.2b - 0.4 \ge -0.4b$

61. $-0.4t + 1.2 < -2$

62. $1 < 3(2t - 4) \le 12$

63. $4 \le 2(5h - 3) < 14$

64. $-\dfrac{3}{4}c - \dfrac{5}{4} \ge 2c$

65. $-\dfrac{2}{3}q - \dfrac{1}{3} > \dfrac{1}{2}q$

66. $4 - 4(y - 2) < -5y + 6$

67. $6 - 6(k - 3) \ge -4k + 12$

68. $0 \le 2q - 1 \le 11$

69. $-10 < 7p - 1 < 1$

70. $-6(2x + 1) < 5 - (x - 4) - 6x$

71. $2(4p + 3) - p \le 5 + 3(p - 3)$

72. $6a - (9a + 1) - 3(a - 1) \ge 2$

73. $8(q + 1) - (2q + 1) + 5 > 12$

Expanding Your Skills

For Exercises 74–78, assume $a > b$. Determine which inequality sign ($>$ or $<$) should be inserted to make a true statement. Assume $a \ne 0$ and $b \ne 0$.

74. $a + c$ _____ $b + c$,   for $c > 0$

75. $a + c$ _____ $b + c$,   for $c < 0$

76. ac _____ bc,   for $c > 0$

77. ac _____ bc,   for $c < 0$

78. $\dfrac{1}{a}$ _____ $\dfrac{1}{b}$

Objectives

1. Properties of Exponents
2. Simplifying Expressions with Exponents
3. Scientific Notation

section 1.8   Properties of Integer Exponents and Scientific Notation

1. Properties of Exponents

In Section 1.2, we learned that exponents are used to represent repeated multiplication. The following properties of exponents are often used to simplify algebraic expressions.

Properties of Exponents*

Description	Property	Example	Details/Notes
Multiplication of like bases	$b^m \cdot b^n = b^{m+n}$	$b^2 \cdot b^4 = b^{2+4} = b^6$	$b^2 \cdot b^4 = (b \cdot b)(b \cdot b \cdot b \cdot b) = b^6$
Division of like bases	$\dfrac{b^m}{b^n} = b^{m-n}$	$\dfrac{b^5}{b^2} = b^{5-2} = b^3$	$\dfrac{b^5}{b^2} = \dfrac{\cancel{b} \cdot \cancel{b} \cdot b \cdot b \cdot b}{\cancel{b} \cdot \cancel{b}} = b^3$
Power rule	$(b^m)^n = b^{m \cdot n}$	$(b^4)^2 = b^{4 \cdot 2} = b^8$	$(b^4)^2 = (b \cdot b \cdot b \cdot b)(b \cdot b \cdot b \cdot b) = b^8$
Power of a product	$(ab)^m = a^m b^m$	$(ab)^3 = a^3 b^3$	$(ab)^3 = (ab)(ab)(ab) = (a \cdot a \cdot a)(b \cdot b \cdot b) = a^3 b^3$
Power of a quotient	$\left(\dfrac{a}{b}\right)^m = \dfrac{a^m}{b^m}$	$\left(\dfrac{a}{b}\right)^3 = \dfrac{a^3}{b^3}$	$\left(\dfrac{a}{b}\right)^3 = \left(\dfrac{a}{b}\right)\left(\dfrac{a}{b}\right)\left(\dfrac{a}{b}\right) = \dfrac{a \cdot a \cdot a}{b \cdot b \cdot b} = \dfrac{a^3}{b^3}$

*Assume that a and b are real numbers ($b \neq 0$) and that m and n represent positive integers.

In addition to the properties of exponents, two definitions are used to simplify algebraic expressions.

b^0 and b^{-n}

Let n be an integer, and let b be a real number such that $b \neq 0$.

1. $b^0 = 1$
2. $b^{-n} = \left(\dfrac{1}{b}\right)^n = \dfrac{1}{b^n}$

The definition of b^0 is consistent with the properties of exponents. For example, if b is a nonzero real number and n is an integer, then

$$\dfrac{b^n}{b^n} = 1$$

$$\dfrac{b^n}{b^n} = b^{n-n} = b^0$$

The expression $b^0 = 1$

Section 1.8 Properties of Integer Exponents and Scientific Notation

The definition of b^{-n} is also consistent with the properties of exponents. If b is a nonzero real number, then

$$\frac{b^3}{b^5} = \frac{\cancel{b}\cdot\cancel{b}\cdot\cancel{b}}{\cancel{b}\cdot\cancel{b}\cdot\cancel{b}\cdot b \cdot b} = \frac{1}{b^2}$$

The expression $b^{-2} = \dfrac{1}{b^2}$

$$\frac{b^3}{b^5} = b^{3-5} = b^{-2}$$

example 1 Using the Properties of Exponents

Simplify the expressions

a. $(-2)^4$ b. -2^4 c. -2^{-4} d. $(-7x)^0$ e. $-7x^0$

Solution:

a. $(-2)^4 = (-2)(-2)(-2)(-2)$
$= 16$

b. $-2^4 = -(2 \cdot 2 \cdot 2 \cdot 2)$
$= -16$

c. $-2^{-4} = \dfrac{1}{-2^4}$
$= \dfrac{1}{-(2 \cdot 2 \cdot 2 \cdot 2)}$
$= \dfrac{1}{-16}$
$= -\dfrac{1}{16}$

d. $(-7x)^0 = 1$ because $b^0 = 1$

e. $-7x^0 = -7 \cdot x^0$
$= -7 \cdot 1$
$= -7$

Skill Practice

Simplify the expressions.
1. $(-3)^2$
2. -3^2
3. -3^{-2}
4. 8^0
5. -6^0

2. Simplifying Expressions with Exponents

example 2 Simplifying Expressions with Exponents

Simplify the following expressions. Write the final answer with positive exponents only.

a. $(x^7 x^{-3})^2$ b. $\left(\dfrac{1}{5}\right)^{-3} - (2)^{-2} + 3^0$

c. $\left(\dfrac{y^3 w^{10}}{y^5 w^4}\right)^{-1}$ d. $\left(\dfrac{2a^7 b^{-4}}{8a^9 b^{-2}}\right)^{-3} (-6a^{-1}b^0)^{-2}$

Answers

1. 9   2. -9 3. $-\dfrac{1}{9}$
4. 1   5. -1

Skill Practice

Simplify the expressions. Write the final answers with positive exponents only.

6. $(a^5 b^{-3})^4$
7. $\left(\dfrac{2b^{-3}c^{-3}}{4b^{-2}c}\right)^3$
8. $(x^3 y^{-4})^2 (x^{-2} y)^{-4}$
9. $\left(\dfrac{2}{3}\right)^{-1} + 4^{-1}$

Solution:

a. $(x^7 x^{-3})^2$

$= (x^{7+(-3)})^2$ Multiply like bases by adding exponents.

$= (x^4)^2$ Apply the power rule.

$= x^8$ Multiply exponents.

b. $\left(\dfrac{1}{5}\right)^{-3} - (2)^{-2} + 3^0$

$= 5^3 - \left(\dfrac{1}{2}\right)^2 + 3^0$ Simplify negative exponents.

$= 125 - \dfrac{1}{4} + 1$ Evaluate the exponents.

$= \dfrac{500}{4} - \dfrac{1}{4} + \dfrac{4}{4}$ Write the expressions with a common denominator.

$= \dfrac{503}{4}$ Simplify.

c. $\left(\dfrac{y^3 w^{10}}{y^5 w^4}\right)^{-1}$ Work within the parentheses first.

$= (y^{3-5} w^{10-4})^{-1}$ Divide like bases by subtracting exponents.

$= (y^{-2} w^6)^{-1}$ Simplify within parentheses.

$= (y^{-2})^{-1} (w^6)^{-1}$ Apply the power rule.

$= y^2 w^{-6}$ Multiply exponents.

$= y^2 \left(\dfrac{1}{w^6}\right)$ or $\dfrac{y^2}{w^6}$ Simplify negative exponents.

d. $\left(\dfrac{2a^7 b^{-4}}{8a^9 b^{-2}}\right)^{-3} (-6a^{-1} b^0)^{-2}$ Subtract exponents within first parentheses. Simplify $\tfrac{2}{8}$ to $\tfrac{1}{4}$.

$= \left(\dfrac{a^{7-9} b^{-4-(-2)}}{4}\right)^{-3} (-6a^{-1} \cdot 1)^{-2}$ In the second parentheses, replace b^0 by 1.

$= \left(\dfrac{a^{-2} b^{-2}}{4}\right)^{-3} (-6a^{-1})^{-2}$ Simplify inside parentheses.

$= \left[\dfrac{(a^{-2})^{-3} (b^{-2})^{-3}}{4^{-3}}\right] (-6)^{-2} (a^{-1})^{-2}$ Apply the power rule.

$= \left(\dfrac{a^6 b^6}{4^{-3}}\right) (-6)^{-2} a^2$ Multiply exponents.

$= 4^3 a^6 b^6 \left[\dfrac{a^2}{(-6)^2}\right]$ Simplify negative exponents.

$= \dfrac{64 a^8 b^6}{36}$ Multiply factors in the numerator and denominator.

$= \dfrac{16 a^8 b^6}{9}$ Simplify.

Answers

6. $\dfrac{a^{20}}{b^{12}}$
7. $\dfrac{1}{8b^3 c^{12}}$
8. $\dfrac{x^{14}}{y^{12}}$
9. $\dfrac{7}{4}$

3. Scientific Notation

Scientists in a variety of fields often work with very large or very small numbers. For instance, the distance between the Earth and the Sun is approximately 93,000,000 mi. The national debt in the United States in 2004 was approximately $7,380,000,000,000. The mass of an electron is 0.000 000 000 000 000 000 000 000 000 000 911 kg.

Scientific notation was devised as a shortcut method of expressing very large and very small numbers. The principle behind scientific notation is to use a power of 10 to express the magnitude of the number. Consider the following powers of 10:

$$10^0 = 1$$

$$10^1 = 10 \qquad 10^{-1} = \frac{1}{10^1} = \frac{1}{10} = 0.1$$

$$10^2 = 100 \qquad 10^{-2} = \frac{1}{10^2} = \frac{1}{100} = 0.01$$

$$10^3 = 1000 \qquad 10^{-3} = \frac{1}{10^3} = \frac{1}{1000} = 0.001$$

$$10^4 = 10,000 \qquad 10^{-4} = \frac{1}{10^4} = \frac{1}{10,000} = 0.0001$$

Each power of 10 represents a place value in the base-10 numbering system. A number such as 50,000 may therefore be written as $5 \times 10,000$ or equivalently as 5.0×10^4. Similarly, the number 0.0035 is equal to $3.5 \times \frac{1}{1000}$ or, equivalently, 3.5×10^{-3}.

> **Definition of a Number Written in Scientific Notation**
>
> A number expressed in the form $a \times 10^n$, where $1 \leq |a| < 10$ and n is an integer, is said to be written in **scientific notation**.

Consider the following numbers in scientific notation:

The distance between the Sun and the Earth: $93{,}000{,}000 \text{ mi} = 9.3 \times 10^7 \text{ mi}$

7 places

The national debt of the United States in 2004: $\$7{,}380{,}000{,}000{,}000 = \$7.38 \times 10^{12}$

12 places

The mass of an electron: 0.000 000 000 000 000 000 000 000 000 000 911 kg

$= 9.11 \times 10^{-31}$ kg   31 places

In each case, the power of 10 corresponds to the number of place positions that the decimal point is moved. The power of 10 is sometimes called the order of magnitude (or simply the magnitude) of the number. The order of magnitude of the national debt is 10^{12} dollars (trillions). The order of magnitude of the distance between the Earth and Sun is 10^7 mi (tens of millions). The mass of an electron has an order of magnitude of 10^{-31} kg.

Concept Connections

10. What is the advantage of using scientific notation?

Answer

10. Scientific notation allows us to write very large and very small numbers in a convenient way for reading, writing, and performing operations.

Skill Practice

Write the numbers in scientific notation.

11. 2,600,000
12. 0.00088

Skill Practice

Write the numbers in standard notation.

13. -5.7×10^{-8}
14. 1.9×10^5

example 3   Writing Numbers in Scientific Notation

Fill in the table by writing the numbers in scientific notation or standard notation as indicated.

Quantity	Standard Notation	Scientific Notation
Number of NASCAR fans	75,000,000 people	
Width of an influenza virus	0.000000001 m	
Cost of hurricane Andrew		$\$2.65 \times 10^{10}$
Probability of winning the Florida state lottery		$4.35587878 \times 10^{-8}$
Approximate width of a human red blood cell	0.000007 m	
Profit of Citigroup Bank, 2003		$\$1.53 \times 10^{10}$

Solution:

Quantity	Standard Notation	Scientific Notation
Number of NASCAR fans	75,000,000 people	7.5×10^7 people
Width of an influenza virus	0.000000001 m	1.0×10^{-9} m
Cost of hurricane Andrew	$26,500,000,000	$\$2.65 \times 10^{10}$
Probability of winning the Florida state lottery	0.0000000435587878	$4.35587878 \times 10^{-8}$
Approximate width of a human red blood cell	0.000007 m	7.0×10^{-6} m
Profit of Citigroup Bank, 2003	$15,300,000,000	$\$1.53 \times 10^{10}$

Calculator Connections

Calculators use scientific notation to display very large or very small numbers. To enter scientific notation in a calculator, try using the [EE] key or the [EXP] key to express the power of 10.

```
9.3E7        93000000
7.25E-2      .0725
```

example 4   Applying Scientific Notation

a. The U.S. national debt in 2005 was approximately $7,830,000,000,000. Assuming there were approximately 290,000,000 people in the United States at that time, determine how much each individual would have to pay to pay off the debt.

b. The mean distance between the Earth and the Andromeda Galaxy is approximately 1.8×10^6 light-years. Assuming 1 light-year is 6.0×10^{12} mi, what is the distance in miles to the Andromeda Galaxy?

Answers

11. 2.6×10^6 12. 8.8×10^{-4}
13. -0.000000057 14. 190,000

Solution:

a. Divide the total U.S. national debt by the number of people:

$$\frac{7.83 \times 10^{12}}{2.9 \times 10^{8}}$$

$$= \left(\frac{7.83}{2.9}\right) \times \left(\frac{10^{12}}{10^{8}}\right) \quad \text{Divide 7.83 by 2.9 and subtract the powers of 10.}$$

$$= 2.7 \times 10^{4}$$

In standard notation, this amounts to approximately $17,692 per person.

b. Multiply the number of light-years by the number of miles per light-year.

$(1.8 \times 10^{6})(6.0 \times 10^{12})$
$= (1.8)(6.0) \times (10^{6})(10^{12})$
$= 10.8 \times 10^{18}$ Multiply 1.8 and 6.0 and add the powers of 10.

The number 10.8×10^{18} is not in "proper" scientific notation because 10.8 is not between 1 and 10.

$= (1.08 \times 10^{1}) \times 10^{18}$ Rewrite 10.8 as 1.08×10^{1}.

$= 1.08 \times (10^{1} \times 10^{18})$ Apply the associative property of multiplication.

$= 1.08 \times 10^{19}$

The distance between the Earth and the Andromeda Galaxy is 1.08×10^{19} mi.

Skill Practice

15. The thickness of a penny is 6.1×10^{-2} in. The height of the Empire State Building is 1250 ft (1.5×10^{4} in.). How many pennies would have to be stacked on top of each other to equal the height of the Empire State Building? Round to the nearest whole unit.

16. The distance from earth to the "nearby" star, Barnard's Star, is 7.6 light years (where 1 light year = 6.0×10^{12} mi). How many miles away is Barnard's Star?

Calculator Connections

Use a calculator to check the solutions to Example 4.

```
(7.83E12)/(2.9E8)
                27000
(1.8E6)*(6.0E12)
              1.08E19
```

Answers

15. Approximately 245,902 pennies
16. 4.56×10^{13} mi

section 1.8 Practice Exercises

Boost your GRADE at mathzone.com!

- Practice Problems
- Self-Tests
- NetTutor
- e-Professors
- Videos

Study Skills Exercises

1. For your next test, make a memory sheet: On a 3 × 5 card (or several 3 × 5 cards), write all the formulas and rules that you need to know. Memorize all this information. Then when your instructor hands you the test, write down all the information that you can remember, before beginning the test. Then you can take the test without worrying that you will forget something important. This process is referred to as a *memory dump*. What important definitions and concepts have you learned in this section of the text?

2. Define the key term **scientific notation**.

Review Exercises

For Exercises 3–6, solve the equation or inequality. Write the solutions to the inequalities in interval notation.

3. $\dfrac{a-2}{3} - \dfrac{3a+2}{4} = -\dfrac{1}{2}$

4. $\dfrac{2y+6}{5} + 3 = \dfrac{-y+2}{2}$

5. $6x - 2(x+3) \le 7(x+1) - 4$

6. $-5c + 3(c+2) > 6c + 8$

For Exercises 7–8, solve the equation for the indicated variable.

7. $5x - 9y = 11$ for x

8. $-2x + 3y = -8$ for y

Objective 1: Properties of Exponents

9. Explain the difference between $b^4 \cdot b^3$ and $(b^4)^3$. (*Hint*: Expand both expressions and compare.)

10. Explain the difference between ab^3 and $(ab)^3$.

For Exercises 11–16, write two examples of each property. Include examples with and without variables. (Answers may vary.)

11. $b^n \cdot b^m = b^{n+m}$

12. $(ab)^n = a^n b^n$

13. $(b^n)^m = b^{nm}$

14. $\dfrac{b^n}{b^m} = b^{n-m}$ $(b \ne 0)$

15. $\left(\dfrac{a}{b}\right)^n = \dfrac{a^n}{b^n}$ $(b \ne 0)$

16. $b^0 = 1$ $(b \ne 0)$

17. Simplify: $\left(\dfrac{2}{3}\right)^{-2}$

18. Simplify: $\left(\dfrac{1}{3}\right)^{-2}$

For Exercises 19–34, simplify. **(See Example 1.)**

19. 5^{-2}

20. 8^{-2}

21. -5^{-2}

22. -8^{-2}

23. $(-5)^{-2}$

24. $(-8)^{-2}$

25. $\left(\dfrac{1}{4}\right)^{-3}$

26. $\left(\dfrac{3}{8}\right)^{-1}$

27. $\left(-\dfrac{3}{2}\right)^{-4}$ 28. $\left(-\dfrac{1}{9}\right)^{-2}$ 29. $-\left(\dfrac{2}{5}\right)^{-3}$ 30. $-\left(\dfrac{1}{2}\right)^{-5}$

31. $(10ab)^0$ 32. $(13x)^0$ 33. $10ab^0$ 34. $13x^0$

Objective 2 : Simplifying Expressions with Exponents

For Exercises 35–84, simplify and write the answer with positive exponents only. (See Example 2.)

35. $y^3 \cdot y^5$ 36. $x^4 \cdot x^8$ 37. $\dfrac{13^8}{13^6}$ 38. $\dfrac{5^7}{5^3}$

39. $(y^2)^4$ 40. $(z^3)^4$ 41. $(3x^2)^4$ 42. $(2y^5)^3$

43. p^{-3} 44. q^{-5} 45. $7^{10} \cdot 7^{-13}$ 46. $11^{-9} \cdot 11^7$

47. $\dfrac{w^3}{w^5}$ 48. $\dfrac{t^4}{t^8}$ 49. $a^{-2}a^{-5}$ 50. $b^{-1}b^{-8}$

51. $\dfrac{r}{r^{-1}}$ 52. $\dfrac{s^{-1}}{s}$ 53. $\dfrac{z^{-6}}{z^{-2}}$ 54. $\dfrac{w^{-8}}{w^{-3}}$

55. $\dfrac{a^3}{b^{-2}}$ 56. $\dfrac{c^4}{d^{-1}}$ 57. $(6xyz^2)^0$ 58. $(-7ab^3)^0$

59. $2^4 + 2^{-2}$ 60. $3^2 + 3^{-1}$ 61. $1^{-2} + 5^{-2}$ 62. $4^{-2} + 2^{-2}$

63. $\left(\dfrac{2}{3}\right)^{-2} - \left(\dfrac{1}{2}\right)^2 + \left(\dfrac{1}{3}\right)^0$ 64. $\left(\dfrac{4}{5}\right)^{-1} + \left(\dfrac{3}{2}\right)^2 - \left(\dfrac{2}{7}\right)^0$ 65. $\dfrac{p^2 q}{p^5 q^{-1}}$

66. $\dfrac{m^{-1}n^3}{m^4 n^{-2}}$ 67. $\dfrac{-48ab^{10}}{32a^4b^3}$ 68. $\dfrac{25x^2 y^{12}}{10x^5 y^7}$ 69. $(-3x^{-4}y^5 z^2)^{-4}$

70. $(-6a^{-2}b^3 c)^{-2}$ 71. $(4m^{-2}n)(-m^6 n^{-3})$ 72. $(-6pq^{-3})(2p^4 q)$ 73. $(p^{-2}q)^3 (2pq^4)^2$

74. $(mn^3)^2 (5m^{-2}n^2)$ 75. $\left(\dfrac{x^2}{y}\right)^3 (5x^2 y)$ 76. $\left(\dfrac{a}{b^2}\right)^2 (3a^2 b^3)$ 77. $\dfrac{(-8a^2 b^2)^4}{(16a^3 b^7)^2}$

78. $\dfrac{(-3x^2 y^3)^2}{(-2xy^4)^3}$ 79. $\left(\dfrac{-2x^6 y^{-5}}{3x^{-2}y^4}\right)^{-3}$ 80. $\left(\dfrac{-6a^2 b^{-3}}{5a^{-1}b}\right)^{-2}$ 81. $\left(\dfrac{2x^{-3}y^0}{4x^6 y^{-5}}\right)^{-2}$

92   Chapter 1   Review of Basic Algebraic Concepts

82. $\left(\dfrac{a^3b^2c^0}{a^{-1}b^{-2}c^{-3}}\right)^{-2}$

83. $3xy^5\left(\dfrac{2x^4y}{6x^5y^3}\right)^{-2}$

84. $7x^{-3}y^{-4}\left(\dfrac{3x^{-1}y^5}{9x^3y^{-2}}\right)^{-3}$

Objective 3: Scientific Notation

85. Write the numbers in scientific notation. (See Example 3.)
   a. Paper is 0.0042 in. thick.
   b. One mole is 602,200,000,000,000,000,000,000 particles.
   c. The dissociation constant for nitrous acid is 0.00046.

86. Write the numbers in scientific notation.
   a. The 2000 census gives the population of the United States as approximately 281,000,000.
   b. As of 2004, the net worth of Bill Gates was $46,600,000,000.
   c. A trillion is defined as 1,000,000,000,000.

87. Write the numbers in standard notation.
   a. The number of $20 bills in circulation in 2004 was 5.2822×10^9.
   b. The dissociation constant for acetic acid is 1.8×10^{-5}.
   c. In 2004, the population of the world was approximately 6.378×10^9.

88. Write the numbers in standard notation.
   a. The amount that the U.S. government spent on education in 2001 was 4.748×10^{11}.
   b. The mass of a neutron is 1.67×10^{-24} g.
   c. The number of $2 bills in circulation in 2004 was 6.8×10^8.

For Exercises 89–94, determine which numbers are in "proper" scientific notation. If the number is not in "proper" scientific notation, correct it.

89. 35×10^4

90. 0.469×10^{-7}

91. 7.0×10^0

92. 8.12×10^1

93. 9×10^{23}

94. 6.9×10

For Exercises 95–102, perform the indicated operations and write the answer in scientific notation. (See Example 4.)

95. $(6.5 \times 10^3)(5.2 \times 10^{-8})$

96. $(3.26 \times 10^{-6})(8.2 \times 10^9)$

97. $(0.0000024)(6700000000)$

98. $(3400000000)(70000000000000)$

99. $(8.5 \times 10^{-2}) \div (2.5 \times 10^{-15})$

100. $(3 \times 10^9) \div (1.5 \times 10^{13})$

101. $(900000000) \div (360000)$

102. $(0.0000000002) \div (8000000)$

103. If one H_2O molecule contains 2 hydrogen atoms and 1 oxygen atom, and 10 H_2O molecules contain 20 hydrogen atoms and 10 oxygen atoms, how many hydrogen atoms and oxygen atoms are contained in 6.02×10^{23} H_2O molecules?

104. The star named Alpha Centauri is 4.3 light-years from the Earth. If there is approximately 6×10^9 mi in 1 light-year, how many miles away is Alpha Centauri?

105. The county of Queens, New York, has a population of approximately 2,200,000. If the area is 110 mi^2, how many people are there per square mile?

106. The county of Catawba, North Carolina, has a population of approximately 150,000. If the area is 400 mi^2, how many people are there per square mile?

Expanding Your Skills

For Exercises 107–112, simplify the expression. Assume that a and b represent positive integers and x and y are nonzero real numbers.

107. $x^{a+1}x^{a+5}$

108. $y^{a-5}y^{a+7}$

109. $\dfrac{y^{2a+1}}{y^{a-1}}$

110. $\dfrac{x^{3a-3}}{x^{a+1}}$

111. $\dfrac{x^{3b-2}y^{b+1}}{x^{2b+1}y^{2b+2}}$

112. $\dfrac{x^{2a-2}y^{a-3}}{x^{a+4}y^{a+3}}$

113. At one count per second, how many days would it take to count to 1 million? (Round to 1 decimal place.)

114. Do you know anyone who is more than 1.0×10^9 sec old? If so, who?

115. Do you know anyone who is more than 4.5×10^5 hr old? If so, who?

chapter 1 | summary

section 1.1 Sets of Numbers and Interval Notation

Key Concepts

Natural numbers: $\{1, 2, 3, \ldots\}$
Whole numbers: $\{0, 1, 2, 3, \ldots\}$
Integers: $\{\ldots, -3, -2, -1, 0, 1, 2, 3, \ldots\}$

Rational numbers: $\left\{\dfrac{p}{q} \mid p \text{ and } q \text{ are integers and } q \text{ does not equal } 0\right\}$

Irrational numbers: $\{x \mid x \text{ is a real number that is not rational}\}$

Real numbers: $\{x \mid x \text{ is rational or } x \text{ is irrational}\}$

$a < b$ "a is less than b"
$a > b$ "a is greater than b"
$a \leq b$ "a is less than or equal to b"
$a \geq b$ "a is greater than or equal to b"
$a < x < b$ "x is between a and b"

$A \cup B$ is the **union** of A and B and is the set of elements that belong to set A or set B or both sets A and B.

$A \cap B$ is the **intersection** of A and B and is the set of elements common to both A and B.

Examples

Example 1
Some rational numbers are:

$\tfrac{1}{7}, 0.5, 0.\overline{3}$

Some irrational numbers are:

$\sqrt{7}, \sqrt{2}, \pi$

Example 2

Set-Builder Notation	Interval Notation	Graph
$\{x \mid x > a\}$	(a, ∞)	
$\{x \mid x \geq a\}$	$[a, \infty)$	
$\{x \mid x < a\}$	$(-\infty, a)$	
$\{x \mid x \leq a\}$	$(-\infty, a]$	

Example 3

Union   Intersection

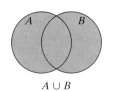
$A \cup B$

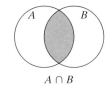
$A \cap B$

section 1.2 Operations on Real Numbers

Key Concepts

The **reciprocal** of a number $a \neq 0$ is $\frac{1}{a}$.
The **opposite** of a number a is $-a$.
The **absolute value** of a, denoted $|a|$, is its distance from zero on the number line.

Addition of Real Numbers

Same Signs: Add the absolute values of the numbers, and apply the common sign to the sum.
Unlike Signs: Subtract the smaller absolute value from the larger absolute value. Then apply the sign of the number having the larger absolute value.

Subtraction of Real Numbers

Add the opposite of the second number to the first number.

Multiplication and Division of Real Numbers

Same Signs: Product or quotient is positive.
Opposite Signs: Product or quotient is negative.

The product of any real number and 0 is 0.
The quotient of 0 and a nonzero number is 0.
The quotient of a nonzero number and 0 is undefined.

Exponents and Radicals

$b^4 = b \cdot b \cdot b \cdot b$ (b is the **base**, 4 is the **exponent**)
$\sqrt{b}$ is the **principal square root** of b (b is the radicand, $\sqrt{}$ is the radical sign).

Order of Operations

1. Simplify expressions within parentheses and other grouping symbols first.
2. Evaluate expressions involving exponents, radicals and absolute values.
3. Perform multiplication or division in order from left to right.
4. Perform addition or subtraction in order from left to right.

Examples

Example 1

Given: -5
The reciprocal is $-\frac{1}{5}$. The opposite is 5.

The absolute value is 5.

Example 2

$-3 + (-4) = -7$
$-5 + 7 = 2$

Example 3

$7 - (-5) = 7 + (5) = 12$

Example 4

$(-3)(-4) = 12 \qquad \dfrac{-15}{-3} = 5$

$(-2)(5) = -10 \qquad \dfrac{6}{-12} = -\dfrac{1}{2}$

$(-7)(0) = 0 \qquad 0 \div 9 = 0$

$-3 \div 0$ is undefined

Example 5

$6^3 = 6 \cdot 6 \cdot 6 = 216$
$\sqrt{100} = 10$

Example 6

$10 - 5(3 - 1)^2 + \sqrt{16}$
$= 10 - 5(2)^2 + \sqrt{16}$
$= 10 - 5(4) + 4$
$= 10 - 20 + 4$
$= -10 + 4$
$= -6$

section 1.3 Simplifying Expressions

Key Concepts

A **term** is a constant or the product of a constant and one or more variables.
- A **variable term** contains at least one variable.
- A **constant term** has no variable.

The **coefficient** of a term is the numerical factor of the term.

Like **terms** have the same variables, and the corresponding variables are raised to the same powers.

Distributive Property of Multiplication over Addition

$a(b + c) = ab + ac$

Two terms can be added or subtracted if they are *like* terms. Sometimes it is necessary to clear parentheses before adding or subtracting *like* terms.

Examples

Example 1

$-2x$ Variable term has coefficient -2.

x^2y Variable term has coefficient 1.

6 Constant term has coefficient 6.

Example 2

$4ab^3$ and $2ab^3$ are *like* terms.

Example 3

$$2(x + 4y) = 2x + 8y$$
$$-(a + 6b - 5c) = -a - 6b + 5c$$

Example 4

$$-4d + 12d + d$$
$$= 9d$$

Example 5

$$-2[w - 4(w - 2)] + 3$$
$$= -2[w - 4w + 8] + 3$$
$$= -2[-3w + 8] + 3$$
$$= 6w - 16 + 3$$
$$= 6w - 13$$

section 1.4 Linear Equations in One Variable

Key Concepts

A **linear equation in one variable** can be written in the form $ax + b = 0$ $(a \neq 0)$.

Steps to Solve a Linear Equation in One Variable

1. Simplify both sides of the equation.
   - Clear parentheses.
   - Consider clearing fractions or decimals (if any are present) by multiplying both sides of the equation by a common denominator of all terms.
   - Combine *like* terms.
2. Use the addition or subtraction property of equality to collect the variable terms on one side of the equation.
3. Use the addition or subtraction property of equality to collect the constant terms on the other side.
4. Use the multiplication or division property of equality to make the coefficient on the variable term equal to 1.
5. Check your answer.

An equation that has no solution is called a **contradiction**.

An equation that has all real numbers as its solutions is called an **identity**.

Examples

Example 1

$$\frac{1}{2}(x - 4) - \frac{3}{4}(x + 2) = \frac{1}{4}$$

$$\frac{1}{2}x - 2 - \frac{3}{4}x - \frac{3}{2} = \frac{1}{4}$$

$$4\left(\frac{1}{2}x - 2 - \frac{3}{4}x - \frac{3}{2}\right) = 4\left(\frac{1}{4}\right)$$

$$2x - 8 - 3x - 6 = 1$$

$$-x - 14 = 1$$

$$-x = 15$$

$$x = -15$$

Example 2

$$3x + 6 = 3(x - 5)$$

$$3x + 6 = 3x - 15$$

$$6 = -15 \quad \text{Contradiction}$$

There is no solution.

Example 3

$$-(5x + 12) - 3 = 5(-x - 3)$$

$$-5x - 12 - 3 = -5x - 15$$

$$-5x - 15 = -5x - 15$$

$$-15 = -15 \quad \text{Identity}$$

All real numbers are solutions.

section 1.5 Applications of Linear Equations in One Variable

Key Concepts

Problem-Solving Steps for Word Problems

1. Read the problem carefully.
2. Assign labels to unknown quantities.
3. Develop a verbal model.
4. Write a mathematical equation.
5. Solve the equation.
6. Interpret the results and write the final answer in words.

Sales tax: (Cost of merchandise)(tax rate)
Commission: (Dollars in sales)(rate)
Simple interest: $I = Prt$
Distance = (rate)(time) $d = rt$

Examples

Example 1

1. Estella has $8500 to invest between two accounts, one bearing 6% simple interest and the other bearing 10% simple interest. At the end of 1 year, she has earned $750 in interest. Find the amount Estella has invested in each account.

2. Let x represent the amount invested at 6%. Then $8500 - x$ is the amount invested at 10%.

	6% Account	10% Account	Total
Principal	x	$8500 - x$	8500
Interest	$0.06x$	$0.10(8500 - x)$	750

3. $\begin{pmatrix} \text{Interest from} \\ 6\% \text{ account} \end{pmatrix} + \begin{pmatrix} \text{interest from} \\ 10\% \text{ account} \end{pmatrix} = \begin{pmatrix} \text{total} \\ \text{interest} \end{pmatrix}$

4. $0.06x + 0.10(8500 - x) = 750$

5. $6x + 10(8500 - x) = 75{,}000$
   $6x + 85{,}000 - 10x = 75{,}000$
   $-4x = -10{,}000$
   $x = 2500$

6. $x = 2500$
   $8500 - x = 6000$

$2500 was invested at 6% and $6000 was invested at 10%.

section 1.6 Literal Equations and Applications to Geometry

Key Concepts

Some useful formulas for word problems:

Perimeter

Rectangle: $P = 2l + 2w$

Area

Rectangle: $A = lw$

Square: $A = x^2$

Triangle: $A = \dfrac{1}{2}bh$

Trapezoid: $A = \dfrac{1}{2}(b_1 + b_2)h$

Angles

Two angles whose measures total 90° are complementary angles.

Two angles whose measures total 180° are supplementary angles.

Vertical angles have equal measure.
$m(\angle a) = m(\angle c)$
$m(\angle b) = m(\angle d)$

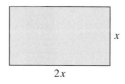

The sum of the angles of a triangle is 180°.

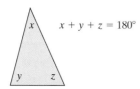

$x + y + z = 180°$

Literal equations (or formulas) are equations with several variables. To solve for a specific variable, follow the steps to solve a linear equation.

Examples

Example 1

A border of marigolds is to enclose a rectangular flower garden. If the length is twice the width and the perimeter is 25.5 ft, what are the dimensions of the garden?

$P = 2l + 2w$
$25.5 = 2(2x) + 2(x)$
$25.5 = 4x + 2x$
$25.5 = 6x$
$4.25 = x$

The width is 4.25 ft, and the length is 2(4.25) ft or 8.5 ft.

Example 2

Solve for y.

$4x - 5y = 20$

$-5y = -4x + 20$

$\dfrac{-5y}{-5} = \dfrac{-4x}{-5} + \dfrac{20}{-5}$

$y = \dfrac{4}{5}x - 4$

section 1.7 Linear Inequalities in One Variable

Key Concepts

A **linear inequality** in one variable can be written in the form

$ax + b < 0$, $\quad ax + b > 0$, $\quad ax + b \leq 0$, $\quad$ or $\quad ax + b \geq 0$

Properties of Inequalities

1. If $a < b$, then $a + c < b + c$.
2. If $a < b$, then $a - c < b - c$.
3. If c is positive and $a < b$, then $ac < bc$ and $\dfrac{a}{c} < \dfrac{b}{c}$ $(c \neq 0)$.
4. If c is negative and $a < b$, then $ac > bc$ and $\dfrac{a}{c} > \dfrac{b}{c}$ $(c \neq 0)$.

The inequality $a < x < b$ is represented by

$\xrightarrow{()}$
ab

or, in interval notation, (a, b).

Examples

Example 1
Solve.

$$\dfrac{14 - x}{-2} < -3x$$

$$-2\left(\dfrac{14 - x}{-2}\right) > -2(-3x) \quad \text{(Reverse the inequality sign.)}$$

$$14 - x > 6x$$

$$-7x > -14$$

$$\dfrac{-7x}{-7} < \dfrac{-14}{-7} \quad \text{(Reverse the inequality sign.)}$$

$$x < 2$$

$\xleftarrow{)}$
2

Interval notation: $(-\infty, 2)$

Example 2

$$-13 \leq 3x - 1 < 5$$

$$-13 + 1 \leq 3x - 1 + 1 < 5 + 1$$

$$-12 \leq 3x < 6$$

$$\dfrac{-12}{3} \leq \dfrac{3x}{3} < \dfrac{6}{3}$$

$$-4 \leq x < 2$$

$\xrightarrow{[)}$
-42

$[-4, 2)$

section 1.8 Properties of Integer Exponents and Scientific Notation

Key Concepts

Let a and b ($b \neq 0$) represent real numbers and m and n represent positive integers.

$b^m \cdot b^n = b^{m+n}$ $\quad\quad \dfrac{b^m}{b^n} = b^{m-n}$

$(b^m)^n = b^{mn}$ $\quad\quad (ab)^m = a^m b^m$

$\left(\dfrac{a}{b}\right)^m = \dfrac{a^m}{b^m}$ $\quad\quad b^0 = 1$

$b^{-n} = \left(\dfrac{1}{b}\right)^n$

A number expressed in the form $a \times 10^n$, where $1 \leq |a| < 10$ and n is an integer, is said to be written in **scientific notation**.

Examples

Example 1

$\left(\dfrac{2x^2 y}{z^{-1}}\right)^{-3} (x^{-4} y^0)$

$= \left(\dfrac{2^{-3} x^{-6} y^{-3}}{z^3}\right)(x^{-4} \cdot 1)$

$= \dfrac{2^{-3} x^{-10} y^{-3}}{z^3}$

$= \dfrac{1}{2^3 x^{10} y^3 z^3} \quad$ or $\quad \dfrac{1}{8 x^{10} y^3 z^3}$

Example 2

$0.0000002 \times 35{,}000$

$= (2.0 \times 10^{-7})(3.5 \times 10^4)$

$= 7.0 \times 10^{-3}$ or 0.007

---

chapter 1 | review exercises

Section 1.1

For Exercises 1–3, answers may vary.

1. Find a number that is a whole number but not a natural number.

2. List three rational numbers that are not integers.

3. List five integers, two of which are not whole numbers.

For Exercises 4–9, write an expression in words that describes the set of numbers given by each interval. (Answers may vary.)

4. $(7, 16)$
5. $(0, 2.6]$
6. $[-6, -3]$
7. $(8, \infty)$
8. $(-\infty, 13]$
9. $(-\infty, \infty)$

10. Explain the difference between the union and intersection of two sets. You may use the sets C and D in the following diagram to provide an example.

Let $A = \{x \mid x < 2\}$, $B = \{x \mid x \geq 0\}$, and $C = \{x \mid -1 < x < 5\}$. For Exercises 11–16, graph each set and write the set in interval notation.

11. A

12. B

13. C

14. $A \cap B$

15. $B \cap C$

16. $A \cup B$

17. True or false? $x < 3$ is equivalent to $3 > x$

18. True or false? $-2 \leq x < 5$ is equivalent to $5 > x \geq -2$

Section 1.2

For Exercises 19–20, find the opposite, reciprocal, and absolute value.

19. -8

20. $\dfrac{4}{9}$

For Exercises 21–22, simplify the exponents and the radicals.

21. $4^2, \sqrt{4}$

22. $25^2, \sqrt{25}$

For Exercises 23–32, perform the indicated operations.

23. $6 + (-8)$

24. $(-2) + (-5)$

25. $8(-2.7)$

26. $(-1.1)(7.41)$

27. $\dfrac{5}{8} \div \left(-\dfrac{13}{40}\right)$

28. $\left(-\dfrac{1}{4}\right) \div \left(-\dfrac{11}{16}\right)$

29. $\dfrac{2 - 4(3 - 7)}{-4 - 5(1 - 3)}$

30. $\dfrac{12(2) - 8}{4(-3) + 2(5)}$

31. $3^2 + 2(|-10 + 5| \div 5)$

32. $-91 + \sqrt{4}(\sqrt{25} - 13)^2$

33. Given $h = \tfrac{1}{2}gt^2 + v_0 t + h_0$, find h if $g = -32$ ft/sec^2, $v_0 = 64$ ft/sec, $h_0 = 256$ ft, and $t = 4$ sec.

34. Find the area of a parallelogram with base 42 in. and height 18 in.

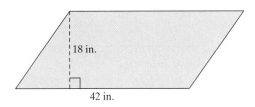

Section 1.3

For Exercises 35–38, apply the distributive property and simplify.

35. $3(x + 5y)$

36. $\dfrac{1}{2}(x + 8y - 5)$

37. $-(-4x + 10y - z)$

38. $-(13a - b - 5c)$

For Exercises 39–42, clear parentheses if necessary, and combine *like* terms.

39. $5 - 6q + 13q - 19$

40. $18p + 3 - 17p + 8p$

41. $7 - 3(y + 4) - 3y$

42. $\dfrac{3}{4}(8x - 4) + \dfrac{1}{2}(6x + 4)$

For Exercises 43–44, answers may vary.

43. Write an example of the commutative property of addition.

44. Write an example of the associative property of multiplication.

Section 1.4

45. Describe the solution set for a contradiction.

46. Describe the solution set for an identity.

For Exercises 47–56, solve the equations and identify each as a conditional equation, a contradiction, or an identity.

47. $x - 27 = -32$

48. $y + \dfrac{7}{8} = 1$

49. $7.23 + 0.6x = 0.2x$ 50. $0.1y + 1.122 = 5.2y$

51. $-(4 + 3m) = 9(3 - m)$

52. $-2(5n - 6) = 3(-n - 3)$

53. $\dfrac{x - 3}{5} - \dfrac{2x + 1}{2} = 1$

54. $3(x + 3) - 2 = 3x + 2$

55. $\dfrac{10}{8}m + 18 - \dfrac{7}{8}m = \dfrac{3}{8}m + 25$

56. $\dfrac{2}{3}m + \dfrac{1}{3}(m - 1) = -\dfrac{1}{3}m + \dfrac{1}{3}(4m - 1)$

Section 1.5

57. Explain how you would label three consecutive integers.

58. Explain how you would label two consecutive odd integers.

59. Explain what the formula $d = rt$ means.

60. Explain what the formula $I = Prt$ means.

61. To do a rope trick, a magician needs to cut a piece of rope so that one piece is one-third the length of the other piece. If she begins with a $2\frac{2}{3}$-ft rope, what lengths will the two pieces of rope be?

62. Of three consecutive even integers, the sum of the smallest two integers is equal to 6 less than the largest. Find the integers.

63. Pat averages a rate of 11 mph on his bike. One day he rode for 45 min ($\frac{3}{4}$ hr) and then got a flat tire and had to walk home. He walked the same path that he rode and it took him 2 hr. What was his average rate walking?

64. How much 10% acid solution should be mixed with a 25% acid solution to produce 3 L of a solution that is 15% acid?

65. Sharyn invests $2000 more in an account that earns 9% simple interest than she invests in an account that earns 6% simple interest. How much did she invest in each account if her total interest is $405 after 1 year?

66. In 2003, approximately 7.2 million men were in college in the United States. This represents an 8% increase over the number of men in college in 2000. Approximately how many men were in college in 2000? (Round to the nearest tenth of a million.)

67. In 2002, there were 17,430 deaths due to alcohol-related accidents in the United States. This was a 5% increase over the number of alcohol-related deaths in 1999. How many such deaths were there in 1999?

68. a. Cory made $30,403 in taxable income in 1996. If he pays 28% in federal income tax, determine the amount of tax he must pay.

    b. What is his net income (after taxes)?

Section 1.6

69. The length of a rectangle is 2 ft more than the width. Find the dimensions if the perimeter is 40 ft.

For Exercises 70–71, solve for x, and then find the measure of each angle.

70.
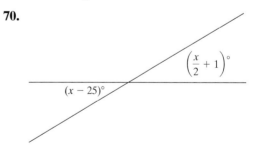
$(x - 25)°$, $\left(\dfrac{x}{2} + 1\right)°$

71.
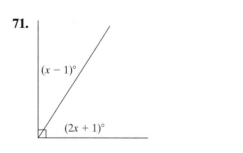
$(x - 1)°$, $(2x + 1)°$

For Exercises 72–75, solve for the indicated variable.

72. $3x - 2y = 4$ for y

73. $-6x + y = 12$ for y

74. $S = 2\pi r + \pi r^2 h$ for h

75. $A = \dfrac{1}{2}bh$ for b

76. **a.** The circumference of a circle is given by $C = 2\pi r$. Solve this equation for π.

   b. Tom measures the radius of a circle to be 6 cm and the circumference to be 37.7 cm. Use these values to approximate π. (Round to 2 decimal places.)

Section 1.7

For Exercises 77–85, solve the inequality. Graph the solution and write the solution set in interval notation.

77. $-6x - 2 > 6$

78. $-10x \leq 15$

79. $-2 \leq 3x - 9 \leq 15$

80. $5 - 7(x + 3) > 19x$

81. $4 - 3x \geq 10(-x + 5)$

82. $\dfrac{5 - 4x}{8} \geq 9$

83. $\dfrac{3 + 2x}{4} \leq 8$

84. $3 > \dfrac{4 - q}{2} \geq -\dfrac{1}{2}$

85. $-11 < -5z - 2 \leq 0$

86. One method to approximate your maximum heart rate is to subtract your age from 220. To maintain an aerobic workout, it is recommended that you sustain a heart rate of between 60% and 75% of your maximum heart rate.

   a. If the maximum heart rate h is given by the formula $h = 220 - A$, where A is a person's age, find your own maximum heart rate. (Answers will vary.)

   b. Find the interval for your own heart rate that will sustain an aerobic workout. (Answers will vary.)

Section 1.8

For Exercises 87–94, simplify the expression and write the answer with positive exponents.

87. $(3x)^3(3x)^2$

88. $(-6x^{-4})(3x^{-8})$

89. $\dfrac{24x^5y^3}{-8x^4y}$

90. $\dfrac{-18x^{-2}y^3}{-12x^{-5}y^5}$

91. $(-2a^2b^{-5})^{-3}$

92. $(-4a^{-2}b^3)^{-2}$

93. $\left(\dfrac{-4x^4y^{-2}}{5x^{-1}y^4}\right)^{-4}$

94. $\left(\dfrac{25x^2y^{-3}}{5x^4y^{-2}}\right)^{-5}$

95. Write the numbers in scientific notation.

   a. The population of Asia was 3,362,994,000 in 1998.

   b. The population of Asia is predicted to be 4,247,079,000 by 2020.

96. Write the numbers in scientific notation.

   a. A millimeter is 0.001 of a meter.

   b. A nanometer is 0.000001 of a millimeter.

97. Write the numbers in standard form.

   a. A micrometer is 1×10^{-3} of a millimeter.

   b. A nanometer is 1×10^{-9} of a meter.

98. Write the numbers in standard form.

a. The total square footage of shopping centers in the United States is approximately 5.23×10^9 ft^2.

b. The total sales of those shopping centers is $\$1.091 \times 10^{12}$. (Source: International Council of Shopping Centers.)

For Exercises 99–102, perform the indicated operations. Write the answer in scientific notation.

99. $\dfrac{2{,}500{,}000}{0.0004}$

100. $\dfrac{0.0005}{25{,}000}$

101. $(3.6 \times 10^8)(9.0 \times 10^{-2})$

102. $(7.0 \times 10^{-12})(5.2 \times 10^3)$

---

chapter 1 | test

1. a. List the integers between -5 and 2, inclusive.

b. List three rational numbers between 1 and 2. (Answers may vary.)

2. Explain the difference between the intervals $(-3, 4)$ and $[-3, 4]$.

3. Graph the sets and write each set in interval notation.

a. All real numbers less than 6

b. All real numbers at least -3

4. Given sets $A = \{x \mid x < -2\}$ and $B = \{x \mid x \geq -5\}$, graph $A \cap B$ and write the set in interval notation.

5. Write the opposite, reciprocal, and absolute value for each of the numbers.

a. $-\dfrac{1}{2}$ b. 4   c. 0

6. Simplify. $|-8| - 4(2-3)^2 \div \sqrt{4}$

7. Given $z = \dfrac{x - \mu}{\sigma/\sqrt{n}}$, find z when $n = 16$, $x = 18$, $\sigma = 1.8$, and $\mu = 17.5$. (Round the answer to 1 decimal place.)

8. True or false?

a. $(x + y) + 2 = 2 + (x + y)$ is an example of the associative property of addition.

b. $(2 \cdot 3) \cdot 5 = (3 \cdot 2) \cdot 5$ is an example of the commutative property of multiplication.

c. $(x + 3)4 = 4x + 12$ is an example of the distributive property.

d. $(10 + y) + z = 10 + (y + z)$ is an example of the associative property of addition.

9. Simplify the expressions.

a. $5b + 2 - 7b + 6 - 14$

b. $\dfrac{1}{2}(2x - 1) - \left(3x - \dfrac{3}{2}\right)$

For Exercises 10–13, solve the equations.

10. $\dfrac{x}{7} + 1 = 20$

11. $8 - 5(4 - 3z) = 2(4 - z) - 8z$

12. $0.12(x) + 0.08(60{,}000 - x) = 10{,}500$

13. $\dfrac{5 - x}{6} - \dfrac{2x - 3}{2} = \dfrac{x}{3}$

14. Label each equation as a conditional equation, an identity, or a contradiction.

   a. $(5x - 9) + 19 = 5(x + 2)$

   b. $2a - 2(1 + a) = 5$

   c. $(4w - 3) + 4 = 3(5 - w)$

15. The difference between two numbers is 72. If the larger is 5 times the smaller, find the two numbers.

16. Joëlle is determined to get some exercise and walks to the store at a brisk rate of 4.5 mph. She meets her friend Yun Ling at the store, and together they walk back at a slower rate of 3 mph. Joëlle's total walking time was 1 hr.

   a. How long did it take her to walk to the store?

   b. What is the distance to the store?

17. Shawnna banks at a credit union. Her money is distributed between two accounts: a certificate of deposit (CD) that earns 5% simple interest and a savings account that earns 3.5% simple interest. Shawnna has $100 less in her savings account than in the CD. If after 1 year her total interest is $81.50, how much did she invest in the CD?

18. A yield sign is in the shape of an equilateral triangle (all sides have equal length). Its perimeter is 81 in. Find the length of the sides.

For Exercises 19–20, solve the equations for the indicated variable.

19. $4x + 2y = 6$ for y **20.** $x = \mu + z\sigma$ for z

For Exercises 21–23, solve the inequalities. Graph the solution and write the solution set in interval notation.

21. $x + 8 > 42$ **22.** $-\dfrac{3}{2}x + 6 \geq x - 3$

23. $-2 < 3x - 1 \leq 5$

24. An elevator can accommodate a maximum weight of 2000 lb. If four passengers on the elevator have an average weight of 180 lb each, how many additional passengers of the same average weight can the elevator carry before the maximum weight capacity is exceeded?

For Exercises 25–28, simplify the expression, and write the answer with positive exponents only.

25. $\dfrac{20a^7}{4a^{-6}}$ **26.** $\dfrac{x^6 x^3}{x^{-2}}$

27. $\left(\dfrac{-3x^6}{5y^7}\right)^2$ **28.** $\dfrac{(2^{-1}xy^{-2})^{-3}(x^{-4}y)}{(x^0 y^5)^{-1}}$

29. Multiply. $(8.0 \times 10^{-6})(7.1 \times 10^5)$

30. Divide. (Write the answer in scientific notation.) $(9{,}200{,}000) \div (0.004)$

Linear Equations in Two Variables

2

- **2.1** The Rectangular Coordinate System and Midpoint Formula
- **2.2** Linear Equations in Two Variables
- **2.3** Slope of a Line
- **2.4** Equations of a Line
- **2.5** Applications of Linear Equations and Graphing

Using graphs to interpret data (information) is an important tool in science and business. This chapter covers topics related to graphing and applications of graphing. For example, the slope of a line represents a rate of change. In Exercise 53 in Section 2.3, we see that the number of cellular phone subscriptions has increased linearly from 1998 to 2006. The slope of the line indicates that the number of cell phone subscriptions has increased by 18.75 million per year during this time period.

chapter 2 preview

The exercises in this chapter preview contain concepts that have not yet been presented. These exercises are provided for students who want to compare their levels of understanding before and after studying the chapter. Alternatively, you may prefer to work these exercises when the chapter is completed and before taking the exam.

Section 2.1

For Exercises 1–3, state the quadrant in which each point lies.

1. $(3, -4)$
2. $(-2, 2)$
3. $(-1, -4)$

4. Find the midpoint of the line segment between the points $(-4, 10)$ and $(11, -6)$.

Section 2.2

For Exercises 5–6, graph the lines.

5. $y = -2x - 3$
6. $y = -3$

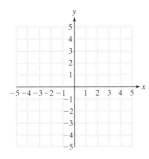
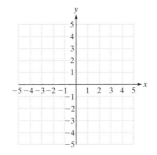

7. Given the equation $2x - 3y = -10$, find the x- and y-intercepts.

Section 2.3

For Exercises 8–10, find the slope of the line through the given points.

8. $(-5, -2)$ and $(-1, 6)$

9. $(4, -5)$ and $(-2, -5)$

10. $(8, 0)$ and $(8, -9)$

11. The slope of a line is $-\frac{4}{3}$.

    a. What is the slope of a line parallel to the given line?

    b. What is the slope of a line perpendicular to the given line?

Section 2.4

12. Given the equation $5x - 9y = 1$, determine the slope and y-intercept.

13. Write an equation for the line that passes through the points $(4, 5)$ and $(1, -1)$.

14. Write an equation for the line that passes through the point $(2, -2)$ and is perpendicular to the line $y = 4x - 1$.

Section 2.5

15. A record rainfall in Hawaii occurred in November 2000 at Kappala Ranch. The National Weather Service reported that during the storm, rain fell at an average rate of $\frac{3}{2}$ inches per hour (in./hr).

    a. Write an equation to compute the total amount of rain y (in inches) after x hr of the storm.

    b. Use the equation to compute the amount of rainfall after 15 hr.

    c. When the storm ended, a total of 36 in. of rain had fallen. How long did the storm last?

section 2.1 The Rectangular Coordinate System and Midpoint Formula

Objectives

1. The Rectangular Coordinate System
2. Plotting Points
3. The Midpoint Formula

1. The Rectangular Coordinate System

One application of algebra is the graphical representation of numerical information (or data). For example, Table 2-1 shows the percentage of individuals who participate in leisure sports activities according to the age of the individual.

table 2-1

Age (years)	Percentage of Individuals Participating in Leisure Sports Activities
20	59%
30	52%
40	44%
50	34%
60	21%
70	18%

Source: U.S. National Endowment for the Arts.

Information in table form is difficult to picture and interpret. However, when the data are presented in a graph, there appears to be a downward trend in the participation in leisure sports activities for older age groups (Figure 2-1).

Percentage of Individuals Who Participate in Leisure Sports Activities Versus Age

Figure 2-1

In this example, two variables are related: age and the percentage of individuals who participate in leisure sports activities.

To picture two variables simultaneously, we use a graph with two number lines drawn at right angles to each other (Figure 2-2). This forms a **rectangular coordinate system**. The horizontal line is called the **x-axis**, and the vertical line is called the **y-axis**. The point where the lines intersect is called the **origin**. On the x-axis, the numbers to the right of the origin are positive, and the numbers to the left are negative. On the y-axis, the numbers above the origin are positive, and the numbers below are negative. The x- and y-axes divide the graphing area into four regions called **quadrants**.

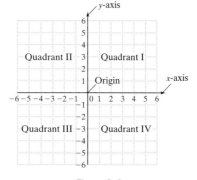

Figure 2-2

2. Plotting Points

Points graphed in a rectangular coordinate system are defined by two numbers as an **ordered pair** (x, y). The first number (called the first coordinate or abscissa) is the horizontal position from the origin. The second number (called the second coordinate or ordinate) is the vertical position from the origin. Example 1 shows how points are plotted in a rectangular coordinate system.

Skill Practice

1. Plot the point and state the quadrant or axis where it is located.

   a. (3, 5)   b. (−4, 0)
   c. (2, −1)   d. (0, 3)
   e. (−2, −2)   f. (−5, 2)

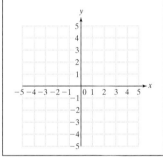

example 1 Plotting Points

Plot each point and state the quadrant or axis where it is located.

a. $(4, 1)$ b. $(-3, 4)$ c. $(4, -3)$

d. $\left(-\frac{5}{2}, -2\right)$ e. $(0, 3)$ f. $(-4, 0)$

Solution:

a. The point $(4, 1)$ is in quadrant I.

b. The point $(-3, 4)$ is in quadrant II.

c. The point $(4, -3)$ is in quadrant IV.

d. The point $\left(-\frac{5}{2}, -2\right)$ can also be written as $(-2.5, -2)$. This point is in quadrant III.

e. The point $(0, 3)$ is on the y-axis.

f. The point $(-4, 0)$ is located on the x-axis.

Tip: Notice that the points $(-3, 4)$ and $(4, -3)$ are in different quadrants. Changing the order of the coordinates changes the location of the point. That is why points are represented by *ordered* pairs (Figure 2-3).

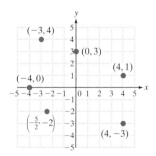

Figure 2-3

Answers

1. a. (3, 5); quadrant I
   b. (−4, 0); *x*-axis
   c. (2, −1); quadrant IV
   d. (0, 3); *y*-axis
   e. (−2, −2); quadrant III
   f. (−5, 2); quadrant II

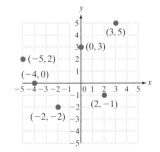

The effective use of graphs for mathematical models requires skill in identifying points and interpreting graphs.

example 2 Identifying Points

Refer to the figure to give the coordinates of each point and the quadrant or axis where it is located.

a. R
b. S
c. T
d. U
e. V

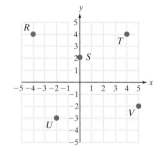

Solution:

a. *R:* The coordinates are $(-4, 4)$, and it is in quadrant II.

b. *S:* The coordinates are $(0, 2)$, and it is on the *y*-axis.

c. *T:* The coordinates are $(4, 4)$, and it is in quadrant I.

d. *U:* The coordinates are $(-2, -3)$, and it is in quadrant III.

e. *V:* The coordinates are $(5, -2)$, and it is in quadrant IV.

example 3 Interpreting a Graph

Kristine started a savings plan at the beginning of the year and plotted the amount of money she deposited in her savings account each month. The graph of her savings is shown in Figure 2-4. The values on the *x*-axis represent the first 6 months of the year, and the values on the *y*-axis represent the amount of money in dollars that she saved. Refer to Figure 2-4 to answer the questions. Let $x = 1$ represent January on the horizontal axis.

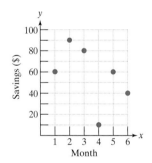

Tip: The scale on the *x*- and *y*-axes may be different. This often happens in applications. See Figure 2-4.

Figure 2-4

a. What is the *y*-coordinate when the *x*-coordinate is 6? Interpret the meaning of the ordered pair in the context of this problem.

b. In which month did she save the most? How much did she save?

c. In which month did she save the least? How much did she save?

d. How much did she save in March?

e. In which two months did she save the same amount? How much did she save in these months?

Skill Practice

2. Give the coordinates of the labeled point, and state the quadrant or axis where the point is located.

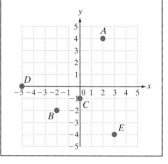

Skill Practice

The graph below shows the average price of a share of Disney stock each month for the first 6 months of 2004. The *x*-axis represents the month (January is represented by $x = 1$) and the *y*-axis represents the price per share in dollars.

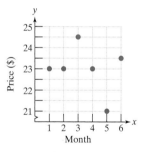

3. What was the price of a share in June?

4. What was the highest price reached by the stock and when was it?

Answers

2. $A(2, 4)$; quadrant I
   $B(-2, -2)$; quadrant III
   $C(0, -1)$; *y*-axis
   $D(-5, 0)$; *x*-axis
   $E(3, -4)$; quadrant IV
3. $23.50   4. $24.50 in March

Solution:

a. When x is 6, the y-coordinate is 40. This means that in June, Kristine saved $40.

b. The point with the greatest y-coordinate occurs when x is 2. She saved the most money, $90, in February.

c. The point with the lowest y-coordinate occurs when x is 4. She saved the least amount, $10, in April.

d. In March, the x-coordinate is 3 and the y-coordinate is 80. She saved $80 in March.

e. The two points with the same y-coordinate occur when $x = 1$ and when $x = 5$. She saved $60 in both January and May.

3. The Midpoint Formula

Consider two points in the coordinate plane and the line segment determined by the points. It is sometimes necessary to determine the point that is halfway between the endpoints of the segment. This point is called the **midpoint**. If the coordinates of the endpoints are represented by (x_1, y_1) and (x_2, y_2), then the midpoint of the segment is given by the following formula.

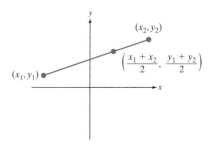

Midpoint formula: $\left(\dfrac{x_1 + x_2}{2}, \dfrac{y_1 + y_2}{2}\right)$

Tip: The midpoint of a line segment is found by taking the *average* of the x-coordinates and the *average* of the y-coordinates of the endpoints.

Skill Practice

Find the midpoint of the line segment with the given endpoints.

5. $(5, 6)$ and $(-10, 4)$
6. $(0, -6)$ and $(-4, -22)$

example 4 — Finding the Midpoint of a Segment

Find the midpoint of the line segment with the given endpoints.

a. $(-4, 6)$ and $(8, 1)$

b. $(-1.2, -3.1)$ and $(-6.6, 1.2)$

Solution:

a. $(-4, 6)$ and $(8, 1)$

$\left(\dfrac{-4 + 8}{2}, \dfrac{6 + 1}{2}\right)$ Apply the midpoint formula.

$\left(2, \dfrac{7}{2}\right)$ Simplify.

The midpoint of the segment is $(2, \tfrac{7}{2})$.

b. $(-1.2, -3.1)$ and $(-6.6, 1.2)$

$\left(\dfrac{-1.2 + -6.6}{2}, \dfrac{-3.1 + 1.2}{2}\right)$ Apply the midpoint formula.

$(-3.9, -0.95)$ Simplify.

Answers

5. $\left(-\dfrac{5}{2}, 5\right)$ 6. $(-2, -14)$

Section 2.1 The Rectangular Coordinate System and Midpoint Formula 113

example 5   Applying the Midpoint Formula

A map of a national park is created so that the ranger station is at the origin of a rectangular grid. Two hikers are located at positions (2, 3) and (−5, −2) with respect to the ranger station, where all units are in miles. The hikers would like to meet at a point halfway between them (Figure 2-5), but they are too far apart to communicate their positions to each other via radio. However, the hikers are both within radio range of the ranger station. If the ranger station relays each hiker's position to the other, at what point on the map should the hikers meet?

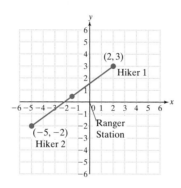

Figure 2-5

Skill Practice

7. Find the center of the circle in the figure, given that the endpoints of a diameter are (3, 2) and (7, 10).

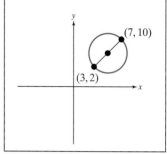

Solution:

To find the halfway point on the line segment between the two hikers, apply the midpoint formula:

$$(2, 3) \quad \text{and} \quad (-5, -2)$$
$$(x_1, y_1) \quad \text{and} \quad (x_2, y_2)$$

$$\left(\frac{x_1 + x_2}{2}, \frac{y_1 + y_2}{2}\right)$$

$$\left(\frac{2 + (-5)}{2}, \frac{3 + (-2)}{2}\right) \quad \text{Apply the midpoint formula.}$$

$$\left(\frac{-3}{2}, \frac{1}{2}\right) \quad \text{Simplify.}$$

The halfway point between the hikers is located at $\left(-\frac{3}{2}, \frac{1}{2}\right)$ or $(-1.5, 0.5)$.

Answer

7. (5, 6)

---

section 2.1   Practice Exercises

Boost your GRADE at mathzone.com!
MathZone

- Practice Problems
- Self-Tests
- NetTutor
- e-Professors
- Videos

Study Skills Exercises

1. Go to the online services that accompany this text (www.mhhe.com/moh). Name two options that this online service offers that could help you in this course.

114 Chapter 2 Linear Equations in Two Variables

2. Define the key terms.

   a. Rectangular coordinate system   **b.** *x*-Axis   **c.** *y*-Axis

   d. Origin   **e.** Quadrant   **f.** Ordered pair

   g. Midpoint

Objective 1: The Rectangular Coordinate System

3. Given the coordinates of a point, explain how to determine which quadrant the point is in.

4. What is meant by the word *ordered* in the term *ordered pair*?

Objective 2: Plotting Points

5. Plot the points on a rectangular coordinate system. **(See Example 1.)**

   a. $(-2, 1)$ **b.** $(0, 4)$ **c.** $(0, 0)$

   d. $(-3, 0)$ **e.** $\left(\dfrac{3}{2}, -\dfrac{7}{3}\right)$ **f.** $(-4.1, -2.7)$

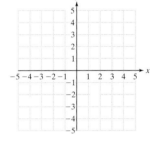

6. Plot the points on a rectangular coordinate system.

   a. $(2.1, 0)$ **b.** $\left(0, \dfrac{2}{3}\right)$ **c.** $(5, 3)$

   d. $(-1.9, 4)$ **e.** $(-2, -6)$ **f.** $\left(\dfrac{8}{5}, -\dfrac{9}{2}\right)$

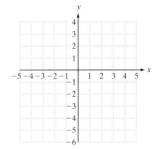

7. Plot the points on a rectangular coordinate system.

   a. $(-2, 5)$ **b.** $\left(\dfrac{5}{2}, 0\right)$ **c.** $(4, -3)$

   d. $(0, -2)$ **e.** $(2, 2)$ **f.** $(-3, -3)$

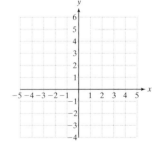

8. Plot the points on a rectangular coordinate system.

   a. $(-1, -3)$
   b. $(0, 0)$
   c. $(0, 4)$
   d. $\left(\dfrac{5}{2}, 1\right)$
   e. $(-1, 0)$
   f. $(5, -2.5)$

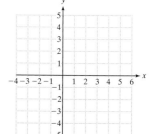

For Exercises 9–12, give the coordinates of the labeled points, and state the quadrant or axis where the point is located. (See Example 2.)

9.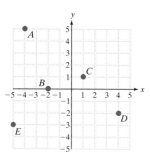

10.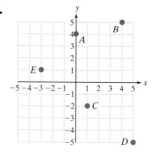

11.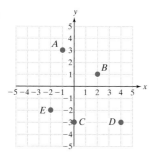

12.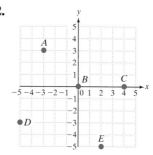

For Exercises 13–14, refer to the graphs to answer the questions. (See Example 3.)

13. The fact that obesity is increasing in both children and adults is of great concern to health care providers. One way to measure obesity is by using the body mass index. Body mass is calculated based on the height and weight of an individual. The graph shows the relationship between body mass index and weight for a person who is 5′6″ tall.

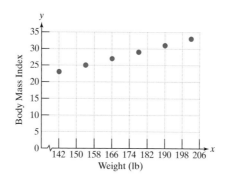

a. What is the body mass index for a 5′ 6″ person who weighs 154 lb?

b. What is the weight of a 5′ 6″ person whose body mass index is 29?

14. The graph shows the number of cases of West Nile virus reported in Colorado during the months of April through August 2004. The month of April is represented by $x = 1$ on the x-axis.

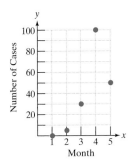

a. Which month had the greatest number of cases reported? Approximately how many cases were reported?

b. Which month had the fewest cases reported? Approximately how many cases were reported?

c. Which months had fewer than 10 cases of the virus reported?

d. Approximately how many cases of the virus were reported in August?

Objective 3: The Midpoint Formula

For Exercises 15–18, find the midpoint of the line segment. Check your answers by plotting the midpoint on the graph. **(See Example 4.)**

15.

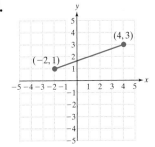

16.

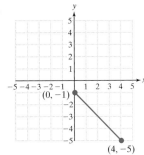

17.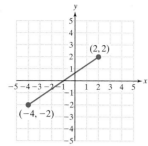

18.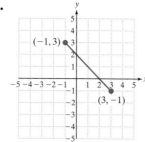

For Exercises 19–26, find the midpoint of the line segment between the two given points.

19. $(4, 0)$ and $(-6, 12)$

20. $(-7, 2)$ and $(-3, -2)$

21. $(-3, 8)$ and $(3, -2)$

22. $(0, 5)$ and $(4, -5)$

23. $(5, 2)$ and $(-6, 1)$

24. $(-9, 3)$ and $(0, -4)$

25. $(-2.4, -3.1)$ and $(1.6, 1.1)$

26. $(0.8, 5.3)$ and $(-4.2, 7.1)$

27. Two courier trucks leave the warehouse to make deliveries. One travels 20 mi north and 30 mi east. The other truck travels 5 mi south and 50 mi east. If the two drivers want to meet for lunch at a restaurant at a point halfway between them, where should they meet relative to the warehouse? (*Hint:* Label the warehouse as the origin, and find the coordinates of the restaurant. See the figure.)
(See Example 5.)

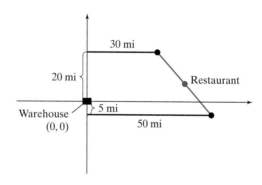

28. A map of a hiking area is drawn so that the Visitors' Center is at the origin of a rectangular grid. Two hikers are located at positions $(-1, 1)$ and $(-3, -2)$ with respect to the visitors' center where all units are in miles. A campground is located exactly halfway between the hikers. What are the coordinates of the campground?

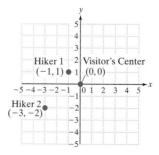

Chapter 2 Linear Equations in Two Variables

Objectives

1. Linear Equations in Two Variables
2. Graphing Linear Equations in Two Variables
3. x-Intercepts and y-Intercepts
4. Horizontal and Vertical Lines

Concept Connections

1. What are the exponents of the x and y terms in a linear equation in two variables?

Skill Practice

Determine whether each ordered pair is a solution for the equation $x + 4y = -8$.

2. $(-2, -1)$
3. $(4, -3)$
4. $(-14, 1.5)$

section 2.2 Linear Equations in Two Variables

1. Linear Equations in Two Variables

Recall from Section 1.4 that an equation in the form $ax + b = 0$ is called a linear equation in one variable. In this section we will study linear equations in *two* variables.

> **Linear Equation in Two Variables**
>
> Let A, B, and C be real numbers such that A and B are not both zero. A **linear equation in two variables** is an equation that can be written in the form
>
> $Ax + By = C$ This form is called *standard form*.

A *solution* to a linear equation in two variables is an ordered pair (x, y) that makes the equation a true statement.

example 1 Determining Solutions to a Linear Equation

For the linear equation $-2x + 3y = 8$, determine whether the order pair is a solution.

a. $(-4, 0)$ b. $(2, -4)$ c. $\left(1, \dfrac{10}{3}\right)$

Solution:

a.
$-2x + 3y = 8$

$-2(-4) + 3(0) \stackrel{?}{=} 8$

$8 + 0 = 8$ ✔ (true)

The ordered pair $(-4, 0)$ indicates that $x = -4$ and $y = 0$.

Substitute $x = -4$ and $y = 0$ into the equation.

The ordered pair $(-4, 0)$ makes the equation a true statement. The ordered pair is a solution to the equation.

b.
$-2x + 3y = 8$

$-2(2) + 3(-4) \stackrel{?}{=} 8$

$-4 + (-12) \stackrel{?}{=} 8$

$-16 \stackrel{?}{=} 8$ (false)

Test the point $(2, -4)$.

Substitute $x = 2$ and $y = -4$ into the equation.

The ordered pair $(2, -4)$ does not make the equation a true statement. The ordered pair is *not* a solution to the equation.

c.
$-2x + 3y = 8$

$-2(1) + 3\left(\dfrac{10}{3}\right) \stackrel{?}{=} 8$

$-2 + 10 = 8$ ✔ (true)

Test the point $\left(1, \dfrac{10}{3}\right)$.

Substitute $x = 1$ and $y = \dfrac{10}{3}$.

The ordered pair $\left(1, \dfrac{10}{3}\right)$ is a solution to the equation.

Answers

1. The exponents are both 1.
2. Not a solution
3. Solution
4. Solution

2. Graphing Linear Equations in Two Variables

Consider the linear equation $x + y = 3$. The solutions to the equation are ordered pairs whose x- and y-coordinates add to 3. Several solutions are given in the following list.

Solution	Check
(x, y)	$x + y = 3$
$(1, 2)$	$(1) + (2) = 3$ ✓
$(0, 3)$	$(0) + (3) = 3$ ✓
$(3, 0)$	$(3) + (0) = 3$ ✓
$(-2, 5)$	$(-2) + (5) = 3$ ✓

By graphing these ordered pairs, we see that the solution points line up (see Figure 2-6). There are actually an infinite number of solutions to the equation $x + y = 3$. The graph of all solutions to a linear equation forms a line in the xy-plane. Conversely, each ordered pair on the line is a solution to the equation.

To graph a linear equation, it is sufficient to find two solution points and draw the line between them. We will find three solution points and use the third point as a check point. This is demonstrated in Example 2.

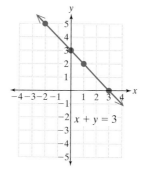

Figure 2-6

example 2 Graphing a Linear Equation in Two Variables

Graph the equation $3x + 5y = 15$.

Solution:

We will find three ordered pairs that are solutions to the equation. In the table, we have selected arbitrary values for x or y and must complete the ordered pairs.

x	y
0	
	2
5	

$\longrightarrow (0,)$
$\longrightarrow (, 2)$
$\longrightarrow (5,)$

From the first row, substitute $x = 0$.

$3x + 5y = 15$
$3(0) + 5y = 15$
$5y = 15$
$y = 3$

From the second row, substitute $y = 2$.

$3x + 5y = 15$
$3x + 5(2) = 15$
$3x + 10 = 15$
$3x = 5$
$x = \dfrac{5}{3}$

From the third row, substitute $x = 5$.

$3x + 5y = 15$
$3(5) + 5y = 15$
$15 + 5y = 15$
$5y = 0$
$y = 0$

Skill Practice

Given: $2x - y = 1$

5. Complete the table with solutions for the equation.

x	y
0	
	5
1	

6. Graph the line by plotting the points from the table.

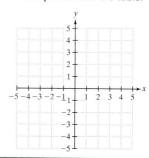

Answers

5–6.

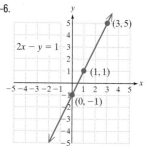

The completed list of ordered pairs is shown as follows. To graph the equation, plot the three solutions and draw the line through the points (Figure 2-7). Arrows on the ends of the line indicate that points on the line extend infinitely in both directions.

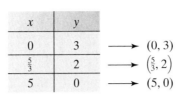
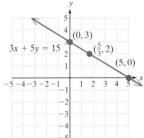

Figure 2-7

Skill Practice

7. Graph the equation $y = -\dfrac{1}{3}x + 1$.

   Hint: Select values of *x* that are multiples of 3.

example 3 Graphing a Linear Equation in Two Variables

Graph the equation $y = \dfrac{1}{2}x - 2$.

Solution:

Because the *y*-variable is isolated in the equation, it is easy to substitute a value for *x* and simplify the right-hand side to find *y*. Since any number for *x* can be used, choose numbers that are multiples of 2 that will simplify easily when multiplied by $\tfrac{1}{2}$.

x	y
0	
2	
4	

Substitute $x = 0$.

$y = \dfrac{1}{2}(0) - 2$

$y = 0 - 2$

$y = -2$

Substitute $x = 2$.

$y = \dfrac{1}{2}(2) - 2$

$y = 1 - 2$

$y = -1$

Substitute $x = 4$.

$y = \dfrac{1}{2}(4) - 2$

$y = 2 - 2$

$y = 0$

The completed list of ordered pairs is as follows. To graph the equation, plot the three solutions and draw the line through the points (Figure 2-8).

x	y	
0	-2	⟶ (0, -2)
2	-1	⟶ (2, -1)
4	0	⟶ (4, 0)

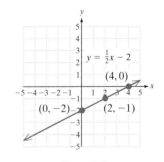

Figure 2-8

Answer

7.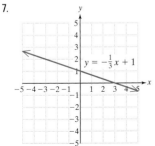

3. x-Intercepts and y-Intercepts

For many applications of graphing, it is advantageous to know the points where a graph intersects the x- or y-axis. These points are called the x- and y-intercepts.

Definition of x- and y-Intercepts

An **x-intercept*** of an equation is a point $(a, 0)$ where the graph intersects the x-axis.

A **y-intercept** of an equation is a point $(0, b)$ where the graph intersects the y-axis.

*In some applications, an x-intercept is defined as the x-coordinate of a point of intersection that a graph makes with the x-axis. For example, if an x-intercept is at the point $(3, 0)$, it is sometimes stated simply as 3 (the y-coordinate is understood to be zero). Similarly, a y-intercept is sometimes defined as the y-coordinate of a point of intersection that a graph makes with the y-axis. For example, if a y-intercept is at the point $(0, 7)$, it may be stated simply as 7 (the x-coordinate is understood to be zero).

Concept Connections

Fill in the blank in each of the statements.

8. If a point lies on the x-axis, then the _____ coordinate equals 0.

9. If a point lies on the y-axis, then the _____ coordinate equals 0.

10. Can the point (2, 3) be an x- or y-intercept? Why or why not?

In Figure 2-7, the x-intercept is $(5, 0)$. In Figure 2-8, the x-intercept is $(4, 0)$. In general, a point on the x-axis must have a y-coordinate of zero.

In Figure 2-7, the y-intercept is $(0, 3)$. In Figure 2-8, the y-intercept is $(0, -2)$. In general, a point on the y-axis must have an x-coordinate of zero.

To find the x- and y-intercepts from an equation in x and y, follow these steps:

Steps to Find the x- and y-Intercepts from an Equation

Given an equation in x and y,

1. Find the x-intercept(s) by substituting $y = 0$ into the equation and solving for x.
2. Find the y-intercept(s) by substituting $x = 0$ into the equation and solving for y.

Skill Practice

Given $y = 2x - 4$,

11. Find the x- and y-intercepts.

12. Find an additional point on the line. (Answers may vary.)

13. Use the intercepts and any additional points to graph the line.

example 4 Finding the x- and y-Intercepts of a Line

Find the x- and y-intercepts of the line $2x + 4y = 8$. Then graph the line.

Solution:

To find the x-intercept, substitute $y = 0$.

$2x + 4y = 8$
$2x + 4(0) = 8$
$2x = 8$
$x = 4$

The x-intercept is $(4, 0)$.

To find the y-intercept, substitute $x = 0$.

$2x + 4y = 8$
$2(0) + 4y = 8$
$4y = 8$
$y = 2$

The y-intercept is $(0, 2)$.

In this case, the intercepts are two distinct points and may be used to graph the line. A third point can be found to verify that the points all fall on the same line (points that lie on the same line are said to be *collinear*). Choose a different value for either x or y, such as $y = 4$.

Answers

8. y 9. x
10. No. Neither the x- nor the y-coordinate is zero.
11. x-intercept: (2, 0); y-intercept: (0, −4)
12. For example: (1, −2) or (3, 2)
13.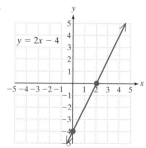

$2x + 4(4) = 8$ Substitute $y = 4$.
$2x + 16 = 8$ Solve for x.
$2x = -8$
$x = -4$ The point $(-4, 4)$ lines up with the other two points (Figure 2-9).

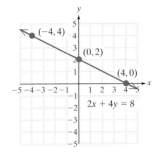

Figure 2-9

Skill Practice

Given $y = -5x$

14. Find the x- and y-intercepts.

15. Find an additional point on the line. (Answers may vary.)

16. Use the intercepts and any additional point to graph the line.

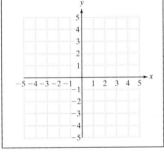

example 5   Finding the x- and y-Intercepts of a Line

Find the x- and y-intercepts of the line $y = \frac{1}{4}x$. Then graph the line.

Solution:

To find the x-intercept, substitute $y = 0$.

$y = \frac{1}{4}x$

$(0) = \frac{1}{4}x$

$0 = x$

The x-intercept is $(0, 0)$.

To find the y-intercept, substitute $x = 0$.

$y = \frac{1}{4}x$

$y = \frac{1}{4}(0)$

$y = 0$

The y-intercept is $(0, 0)$.

Notice the x- and y-intercepts are both located at the origin $(0, 0)$. In this case, the intercepts do not yield two distinct points. Therefore, another point is necessary to draw the line. We may pick any value for either x or y. However, for this equation, it would be particularly convenient to pick a value for x that is a multiple of 4 such as $x = 4$.

$y = \frac{1}{4}x$

$y = \frac{1}{4}(4)$ Substitute $x = 4$.

$y = 1$

The point $(4, 1)$ is a solution to the equation (Figure 2-10).

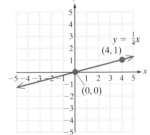

Figure 2-10

Answers

14. x-intercept: $(0, 0)$; y-intercept: $(0, 0)$
15. For example: $(1, -5)$
16.

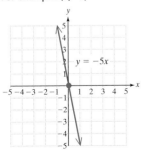

example 6   Interpreting the *x*- and *y*-Intercepts of a Line

Companies and corporations are permitted to depreciate assets that have a known useful life span. This accounting practice is called *straight-line depreciation*. In this procedure the useful life span of the asset is determined, and then the asset is depreciated by an equal amount each year until the taxable value of the asset is equal to zero.

The J. M. Gus trucking company purchases a new truck for $65,000. The truck will be depreciated at $13,000 per year. The equation that describes the depreciation line is

$$y = 65{,}000 - 13{,}000x$$

where *y* represents the value of the truck in dollars and *x* is the age of the truck in years.

a. Find the *x*- and *y*-intercepts. Plot the intercepts on a rectangular coordinate system, and draw the line that represents the straight-line depreciation.

b. What does the *x*-intercept represent in the context of this problem?

c. What does the *y*-intercept represent in the context of this problem?

Skill Practice

Acme motor company tests the engines of its trucks by running the engines in a laboratory. The engines burn 4 gal of fuel per hour. The engines begin the test with 30 gal of fuel. The equation $y = 30 - 4x$ represents the amount of fuel *y* left in the engine after *x* hr.

17. Find the *x*- and *y*-intercepts.
18. Interpret the *y*-intercept in the context of this problem.
19. Interpret the *x*-intercept in the context of this problem.

Solution:

a. To find the *x*-intercept, substitute $y = 0$.

$$0 = 65{,}000 - 13{,}000x$$
$$13{,}000x = 65{,}000$$
$$x = 5$$

The *x*-intercept is (5, 0).

To find the *y*-intercept, substitute $x = 0$.

$$y = 65{,}000 - 13{,}000(0)$$
$$y = 65{,}000$$

The *y*-intercept is (0, 65,000).

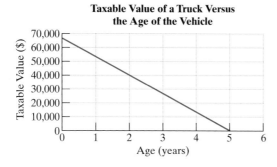

Taxable Value of a Truck Versus the Age of the Vehicle

b. The *x*-intercept (5, 0) indicates that when the truck is 5 years old, the taxable value of the truck will be $0.

c. The *y*-intercept (0, 65,000) indicates that when the truck was new (0 years old), it was worth $65,000.

4. Horizontal and Vertical Lines

Recall that a linear equation can be written in the form $Ax + By = C$, where *A* and *B* are not both zero. If either *A* or *B* is 0, then the resulting line is horizontal or vertical, respectively.

Answers

17. *x*-intercept: (7.5, 0); *y*-intercept: (0, 30)
18. The *y*-intercept (0, 30) represents the amount of fuel in the truck initially (after 0 hr). After 0 hr, the tank contains 30 gal of fuel.
19. The *x*-intercept (7.5, 0) represents the amount of fuel in the truck after 7.5 hr. After 7.5 hr the tank contains 0 gal. It is empty.

Definitions of Vertical and Horizontal Lines

1. A **vertical line** is a line that can be written in the form $x = k$, where k is a constant.
2. A **horizontal line** is a line that can be written in the form $y = k$, where k is a constant.

Skill Practice

20. Graph the line $x = -4$.

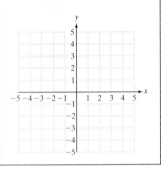

example 7 — Graphing a Vertical Line

Graph the line $x = 6$.

Solution:

Because this equation is in the form $x = k$, the line is vertical and must cross the x-axis at $x = 6$ (Figure 2-11).

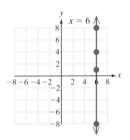

Figure 2-11

Alternative Solution:

Create a table of values for the equation $x = 6$. The choice for the x-coordinate must be 6, but y can be any real number.

x	y
6	-8
6	1
6	4
6	8

Skill Practice

21. Graph the line $-2y = 9$.

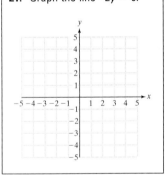

example 8 — Graphing a Horizontal Line

Graph the line $4y = -7$.

Solution:

The equation $4y = -7$ is equivalent to $y = -\frac{7}{4}$. Because the line is in the form $y = k$, the line must be horizontal and must pass through the y-axis at $y = -\frac{7}{4}$ (Figure 2-12).

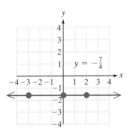

Figure 2-12

Answers

20–21.

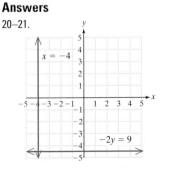

Alternative Solution:

Create a table of values for the equation $4y = -7$. The choice for the y-coordinate must be $-\frac{7}{4}$, but x can be any real number.

x	y
0	$-\frac{7}{4}$
-3	$-\frac{7}{4}$
2	$-\frac{7}{4}$

Calculator Connections

The rectangular screen where a graph is displayed is called the *viewing window*. The default settings for the display on most calculators show both the x- and y-axes from -10 to 10. This is called the *standard viewing window*.

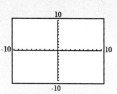

To graph an equation on a graphing calculator, it is important to isolate the y-variable in the equation. Then enter the equation into the calculator. For example:

$$x + y = 3 \Rightarrow y = -x + 3$$

Then use the *Graph* feature to graph the equation.

To graph the equation from Example 6 on a graphing calculator, the viewing window must be set to accommodate large values of y.

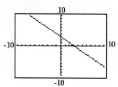

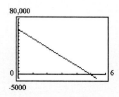

section 2.2 Practice Exercises

Boost your GRADE at mathzone.com!
MathZone

- Practice Problems
- Self-Tests
- NetTutor
- e-Professors
- Videos

Study Skills Exercises

1. Look through the pages of this chapter in your text. Write down the page number for a page that contains

   a. Avoiding Mistakes _____ b. Tip box _____ c. A key term (shown in bold) _____

2. Define the key terms.

   a. **Linear equation in two variables**   b. ***x*-Intercept**   c. ***y*-Intercept**

   d. **Vertical line**   e. **Horizontal line**

Review Exercises

3. Plot each point on a rectangular coordinate system, and identify the quadrant or axis where it is located.

    a. $A(2, -3)$ **b.** $B(-1, -1)$ **c.** $C(4, 2)$ **d.** $D(0, -4)$

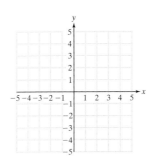

For Exercises 4–6, find the midpoint of the line segment between the given points. Check your answer by graphing the line segment and midpoint.

    4. $(-3, 1)$ and $(-15, -1)$ **5.** $(7, 8)$ and $(-4, 1)$ **6.** $(-2, 10)$ and $(-2, 0)$

Objective 1: Linear Equations in Two Variables

For Exercises 7–10, determine if the ordered pair is a solution to the linear equation. **(See Example 1.)**

7. $2x - 3y = 9$ **8.** $-5x - 2y = 6$ **9.** $x = \frac{1}{3}y + 1$ **10.** $y = -\frac{3}{2}x - 4$

    a. $(0, -3)$ **a.** $(0, 3)$ **a.** $(-1, 0)$ **a.** $(0, -4)$

    b. $(-6, 1)$ **b.** $\left(-\frac{6}{5}, 0\right)$ **b.** $(2, 3)$ **b.** $(2, -7)$

    c. $\left(1, -\frac{7}{3}\right)$ **c.** $(-2, 2)$ **c.** $(-6, 1)$ **c.** $(-4, -2)$

Objective 2: Graphing Linear Equations in Two Variables

For Exercises 11–14, complete the table. Then graph the line defined by the points. **(See Example 2.)**

11. $3x - 2y = 4$

x	y
0	
	4
-1	

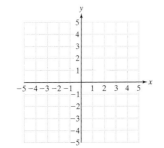

12. $4x + 3y = 6$

x	y
	2
3	
	-1

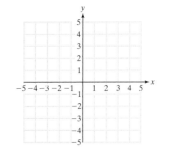

13. $y = -\dfrac{1}{5}x$

x	y
0	
5	
−5	

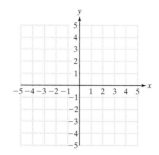

14. $y = \dfrac{1}{3}x$

x	y
0	
3	
6	

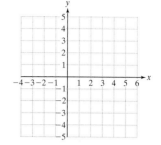

In Exercises 15–28, graph the linear equation. (**See Example 3.**)

15. $x + y = 5$

16. $x + y = -8$

17. $3x - 4y = 12$

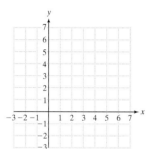
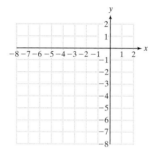
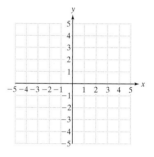

18. $5x + 3y = 15$

19. $y = -3x + 5$

20. $y = -2x + 2$

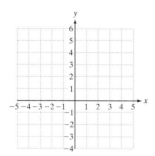
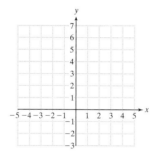
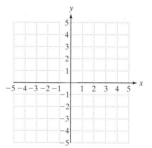

21. $y = \dfrac{2}{5}x - 1$

22. $y = \dfrac{5}{3}x + 1$

23. $x = -5y - 5$

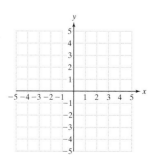
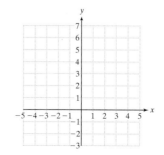
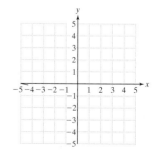

24. $x = 4y + 2$

25. $3y = 4x - 12$
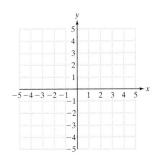

26. $2y = -3x + 2$
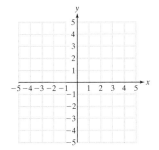

27. $x = 2y$
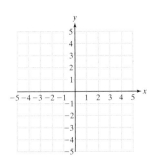

28. $x = -3y$
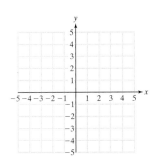

Objective 3: *x*-Intercepts and *y*-Intercepts

29. Given a linear equation, how do you find an *x*-intercept? How do you find a *y*-intercept?

30. Can the point $(4, -1)$ be an *x*- or *y*-intercept? Why or why not?

For Exercises 31–42, **a.** find the *x*-intercept, **b.** find the *y*-intercept, and **c.** graph the line. (See Examples 4–5.)

31. $2x + 3y = 18$
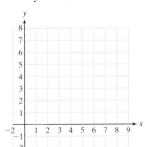

32. $2x - 5y = 10$
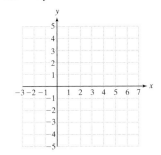

33. $x - 2y = 4$
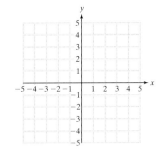

34. $x + y = 8$

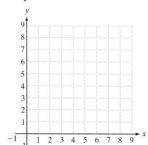

35. $5x = 3y$

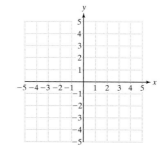

36. $3y = -5x$

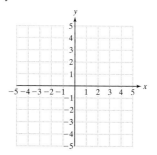

37. $y = 2x + 4$

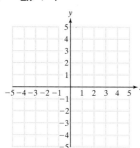

38. $y = -3x - 1$

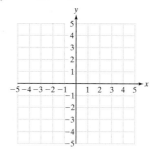

39. $y = -\dfrac{4}{3}x + 2$

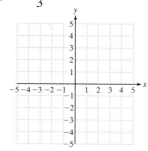

40. $y = -\dfrac{2}{5}x - 1$

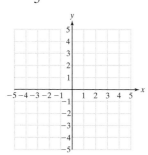

41. $x = \dfrac{1}{4}y$

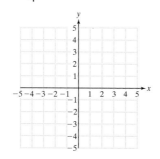

42. $x = \dfrac{2}{3}y$

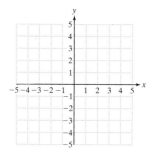

43. A salesperson makes a base salary of $10,000 a year plus a 5% commission on the total sales for the year. The yearly salary can be expressed as a linear equation as

$$y = 10{,}000 + 0.05x$$

where y represents the yearly salary and x represents the total yearly sales. **(See Example 6.)**

a. What is the salesperson's salary for a year in which his sales total $500,000?

b. What is the salesperson's salary for a year in which his sales total $300,000?

c. What does the *y*-intercept mean in the context of this problem?

d. Why is it unreasonable to use negative values for *x* in this equation?

44. A taxi company in Miami charges $2.00 for any distance up to the first mile and $1.10 for every mile thereafter. The cost of a cab ride can be modeled graphically as follows.

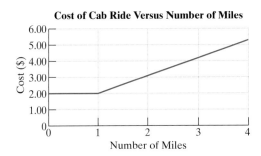

a. Explain why the first part of the model is represented by a horizontal line.

b. What does the *y*-intercept mean in the context of this problem?

c. Explain why the line representing the cost of traveling more than 1 mi is not horizontal.

d. How much would it cost to take a cab $3\frac{1}{2}$ mi?

Objective 4: Horizontal and Vertical Lines

For Exercises 45–52, identify the line as either vertical or horizontal, and graph the line. **(See Examples 7–8.)**

45. $y = -1$

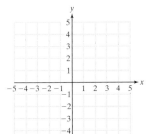

46. $y = 3$

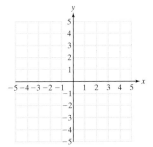

47. $x = 2$

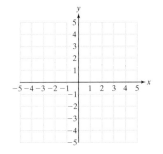

48. $x = -5$

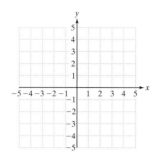

49. $2x + 6 = 5$

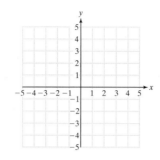

50. $-3x = 12$

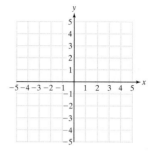

51. $-2y + 1 = 9$

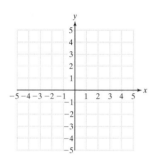

52. $-5y = -10$

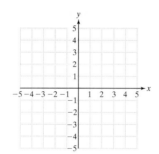

Expanding Your Skills

For Exercises 53–55, find the *x*- and *y*-intercepts.

53. $\dfrac{x}{2} + \dfrac{y}{3} = 1$

54. $\dfrac{x}{7} + \dfrac{y}{4} = 1$

55. $\dfrac{x}{a} + \dfrac{y}{b} = 1$

Graphing Calculator Exercises

For Exercises 56–61, identify which equations are linear and which are nonlinear. Then use a graphing calculator to graph the equations on the standard viewing window to support your answer.

56. $y = 2x - 3$

57. $y = |x|$

58. $y = x^2 + 1$

59. $y = \dfrac{1}{x}$

60. $y = 5$

61. $y = x^3$

For Exercises 62–65, solve the equation for *y*. Use a graphing calculator to graph the equation on the standard viewing window.

62. $2x - 3y = 7$

63. $4x + 2y = -2$

64. $3y = 9$

65. $2y + 10 = 0$

Objectives

1. Introduction to the Slope of a Line
2. The Slope Formula
3. Parallel and Perpendicular Lines
4. Applications and Interpretation of Slope

section 2.3 Slope of a Line

1. Introduction to the Slope of a Line

In Section 2.2, we learned how to graph a linear equation and to identify its *x*- and *y*-intercepts. In this section, we learn about another important feature of a line called the *slope* of a line. Geometrically, slope measures the "steepness" of a line.

Figure 2-13 shows a set of stairs with a wheelchair ramp to the side. Notice that the stairs are steeper than the ramp.

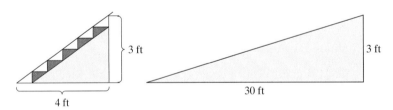

Figure 2-13

To measure the slope of a line quantitatively, consider two points on the line. The slope is the ratio of the vertical change between the two points to the horizontal change. That is, the slope is the ratio of the change in *y* to the change

Section 2.3 Slope of a Line 133

in x. As a memory device, we might think of the slope of a line as "rise over run."

$$\text{Slope} = \frac{\text{change in } y}{\text{change in } x} = \frac{\text{rise}}{\text{run}}$$

To move from point A to point B on the stairs, rise 3 ft and move to the right 4 ft (Figure 2-14).

To move from point A to point B on the wheelchair ramp, rise 3 ft and move to the right 30 ft (Figure 2-15).

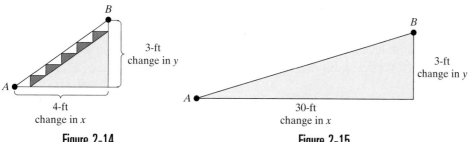

Figure 2-14

Figure 2-15

$$\text{Slope} = \frac{\text{change in } y}{\text{change in } x} = \frac{3 \text{ ft}}{4 \text{ ft}} = \frac{3}{4} \qquad \text{Slope} = \frac{\text{change in } y}{\text{change in } x} = \frac{3 \text{ ft}}{30 \text{ ft}} = \frac{1}{10}$$

The slope of the stairs is $\frac{3}{4}$ which is greater than the slope of the ramp, which is $\frac{1}{10}$.

---

example 1   Finding the Slope in an Application

Find the slope of the ladder against the wall.

Solution:

$$\text{Slope} = \frac{\text{change in } y}{\text{change in } x}$$

$$= \frac{15 \text{ ft}}{5 \text{ ft}}$$

$$= \frac{3}{1} \text{ or } 3$$

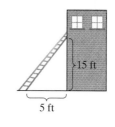

The slope is $\frac{3}{1}$ which indicates that a person climbs 3 ft for every 1 ft traveled horizontally.

Skill Practice

1. Find the slope of the roof.

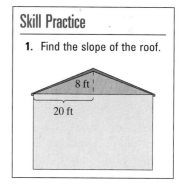

2. The Slope Formula

The slope of a line may be found by using *any* two points on the line—call these points (x_1, y_1) and (x_2, y_2). The change in y between the points can be found by

Answer

1. $\dfrac{2}{5}$

taking the difference of the y-values: $y_2 - y_1$. The change in x can be found by taking the difference of the x-values in the same order: $x_2 - x_1$.

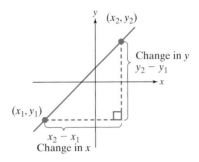

The slope of a line is often symbolized by the letter m and is given by the following formula.

Definition of the Slope of a Line

The **slope** of a line passing through the distinct points (x_1, y_1) and (x_2, y_2) is

$$m = \frac{y_2 - y_1}{x_2 - x_1} \quad \text{provided} \quad x_2 - x_1 \neq 0$$

Skill Practice

2. Find the slope of the line that passes through the points (−4, 5) and (6, 8).

example 2   Finding the Slope of a Line Through Two Points

Find the slope of the line passing through the points $(1, -1)$ and $(7, 2)$.

Solution:

To use the slope formula, first label the coordinates of each point, and then substitute their values into the slope formula.

$(1, -1)$ and $(7, 2)$
(x_1, y_1) (x_2, y_2) Label the points.

$m = \dfrac{y_2 - y_1}{x_2 - x_1} = \dfrac{2 - (-1)}{7 - 1}$ Apply the slope formula.

$= \dfrac{3}{6}$ Simplify.

$= \dfrac{1}{2}$

The slope of the line can be verified from the graph (Figure 2-16).

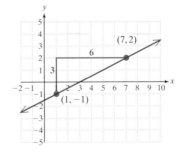

Answer

2. $\dfrac{3}{10}$

Figure 2-16

Tip: The slope formula does not depend on which point is labeled (x_1, y_1) and which point is labeled (x_2, y_2). For example, reversing the order in which the points are labeled in Example 2 results in the same slope:

$(1, -1)$ and $(7, 2)$
(x_2, y_2) $\quad\quad\quad (x_1, y_1)$

then $\quad m = \dfrac{-1 - 2}{1 - 7} = \dfrac{-3}{-6} = \dfrac{1}{2}$

When you apply the slope formula, you will see that the slope of a line may be positive, negative, zero, or undefined.

- Lines that "increase," or "rise," from left to right have a *positive slope*.
- Lines that "decrease," or "fall," from left to right have a *negative slope*.
- Horizontal lines have a *zero slope*.
- Vertical lines have an *undefined slope*.

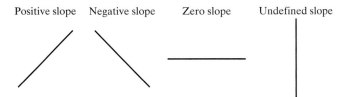

Positive slope   Negative slope   Zero slope   Undefined slope

Concept Connections

Visually determine whether the slope of the line is positive, negative, zero, or undefined.

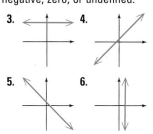

3.   4.

5.   6.

example 3 — Finding the Slope of a Line Between Two Points

Find the slope of the line passing through the points $(3, -4)$ and $(-5, -1)$.

Solution:

$(3, -4)$ and $(-5, -1)$
$(x_1, y_1) \quad\quad\quad (x_2, y_2)$ — Label points.

$m = \dfrac{y_2 - y_1}{x_2 - x_1} = \dfrac{-1 - (-4)}{-5 - 3}$ — Apply the slope formula.

$= \dfrac{3}{-8} = -\dfrac{3}{8}$ — Simplify.

The two points can be graphed to verify that $-\dfrac{3}{8}$ is the correct slope (Figure 2-17).

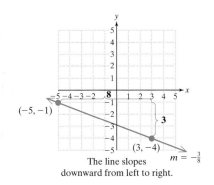

The line slopes downward from left to right.

$m = -\dfrac{3}{8}$

Figure 2-17

Skill Practice

Find the slope of the line that passes through the given points.

7. $(0, 4)$ and $(-2, -2)$
8. $(1, -8)$ and $(-5, -4)$

Answers
3. Zero 4. Positive
5. Negative 6. Undefined
7. 3   8. $-\dfrac{2}{3}$

Skill Practice

Find the slope of the line that passes through the given points.

9. $(5, -2)$ and $(5, 5)$
10. $(1, 6)$ and $(-7, 6)$

example 4   Finding the Slope of a Line Between Two Points

a. Find the slope of the line passing through the points $(-3, 4)$ and $(-3, -2)$.

b. Find the slope of the line passing through the points $(0, 2)$ and $(-4, 2)$.

Solution:

a. $(-3, 4)$ and $(-3, -2)$
(x_1, y_1) (x_2, y_2) Label points.

$$m = \frac{y_2 - y_1}{x_2 - x_1} = \frac{-2 - 4}{-3 - (-3)}$$ Apply slope formula.

$$= \frac{-6}{-3 + 3}$$

$$= \frac{-6}{0}$$ Undefined

The slope is undefined. The points form a vertical line (Figure 2-18).

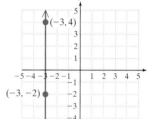

Figure 2-18

b. $(0, 2)$ and $(-4, 2)$
(x_1, y_1) and (x_2, y_2) Label the points.

$$m = \frac{y_2 - y_1}{x_2 - x_1} = \frac{2 - 2}{-4 - 0}$$ Apply the slope formula.

$$= \frac{0}{-4}$$

$$= 0$$ Simplify.

The slope is zero. The line through the two points is a horizontal line (Figure 2-19).

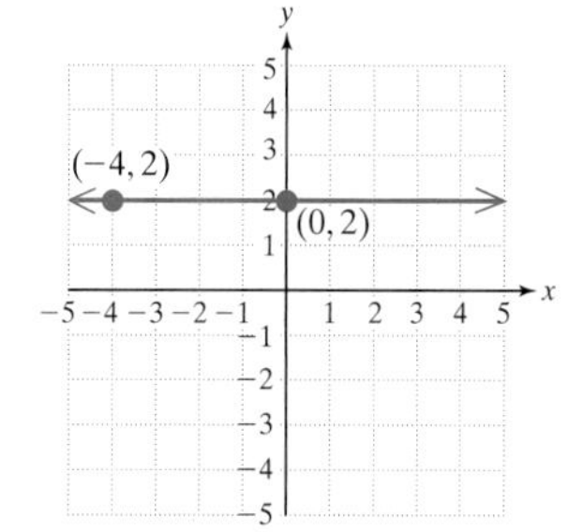

Figure 2-19

3. Parallel and Perpendicular Lines

Lines in the same plane that do not intersect are *parallel*. Nonvertical parallel lines have the same slope and different *y*-intercepts (Figure 2-20).

    Lines that intersect at a right angle are *perpendicular*. If two lines are perpendicular, then the slope of one line is the opposite of the reciprocal of the slope of the other (provided neither line is vertical) (Figure 2-21).

Slopes of Parallel Lines

If m_1 and m_2 represent the slopes of two parallel (nonvertical) lines, then

$$m_1 = m_2$$

See Figure 2-20.

Answers

9. Undefined   10. 0

Section 2.3   Slope of a Line   137

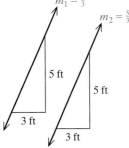

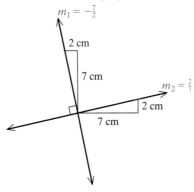

Figure 2-20   Figure 2-21

Slopes of Perpendicular Lines

If $m_1 \neq 0$ and $m_2 \neq 0$ represent the slopes of two perpendicular lines, then

$$m_1 = -\frac{1}{m_2} \quad \text{and} \quad m_2 = -\frac{1}{m_1}$$

or equivalently, $m_1 \cdot m_2 = -1$

See Figure 2-21.

---

example 5   Determining the Slope of Parallel and Perpendicular Lines

Suppose a given line has a slope of -5.

a. Find the slope of a line parallel to the given line.

b. Find the slope of a line perpendicular to the given line.

Solution:

a. The slope of a line parallel to the given line is $m = -5$ (same slope).

b. The slope of a line perpendicular to the given line is $m = \frac{1}{5}$ (the opposite of the reciprocal of -5).

---

example 6   Determining Whether Two Lines Are Parallel, Perpendicular, or Neither

Two points are given from each of two lines: L_1 and L_2. Without graphing the points, determine if the lines are parallel, perpendicular, or neither.

L_1: $(2, -3)$ and $(4, 1)$
L_2: $(5, -6)$ and $(-3, -2)$

Solution:

First determine the slope of each line. Then compare the values of the slopes to determine if the lines are parallel or perpendicular.

Skill Practice

The slope of line L_1 is $-\dfrac{4}{3}$.

11. Find the slope of a line parallel to L_1.

12. Find the slope of a line perpendicular to L_1.

Skill Practice

Two points are given for lines L_1 and L_2. Determine if the lines are parallel, perpendicular, or neither.

13. L_1: $(4, -1)$ and $(-3, 6)$
    L_2: $(-1, 3)$ and $(2, 0)$

14. L_1: $(2, 0)$ and $(-2, 3)$
    L_2: $(-6, 4)$ and $(3, 16)$

Answers

11. $-\dfrac{4}{3}$ 12. $\dfrac{3}{4}$

13. Parallel   14. Perpendicular

For line 1:   For line 2:

L_1: (2, −3) and (4, 1) L_2: (5, −6) and (−3, −2)

(x_1, y_1) (x_2, y_2) (x_1, y_1) (x_2, y_2) Label the points.

$$m = \frac{1 - (-3)}{4 - 2} \qquad\qquad m = \frac{-2 - (-6)}{-3 - (5)}$$ Apply the slope formula.

$$= \frac{4}{2} \qquad\qquad\qquad = \frac{4}{-8}$$

$$= 2 \qquad\qquad\qquad = -\frac{1}{2}$$

The slope of L_1 is 2. The slope of L_2 is $-\frac{1}{2}$. The slope of L_1 is the opposite of the reciprocal of L_2. By comparing the slopes, the lines must be perpendicular.

4. Applications and Interpretation of Slope

Skill Practice

The number of people per square mile in Alaska was 0.96 in 1990. This number increased to 1.17 in 2005.

15. Find the slope of the line that represents the population growth of Alaska. Use the points (1990, 0.96) and (2005, 1.17).

16. Interpret the meaning of the slope in the context of this problem.

example 7   Interpreting the Slope of a Line in an Application

The number of males 20 years old or older who were employed full-time in the United States varied linearly from 1970 to 2005. Approximately 43.0 million males 20 years old or older were employed full-time in 1970. By 2005, this number grew to 65.4 million (Figure 2-22).

Figure 2-22

Source: Current population survey.

a. Find the slope of the line, using the points (1970, 43.0) and (2005, 65.4). Round the answer to 2 decimal places.

b. Interpret the meaning of the slope in the context of this problem.

Solution:

a. (1970, 43.0)   and   (2005, 65.4)

     (x_1, y_1) (x_2, y_2) Label the points.

$$m = \frac{y_2 - y_1}{x_2 - x_1} = \frac{65.4 - 43.0}{2005 - 1970}$$ Apply the slope formula.

$$m = \frac{22.4}{35} \quad\text{or}\quad m = 0.64$$

Answers

15. 0.014
16. The population is increasing by 0.014 person per square mile every year.

b. The slope is approximately 0.64, meaning that the full-time workforce has increased by approximately 0.64 million men (or 640,000 men) per year between 1970 and 2005.

Section 2.3 Practice Exercises

- Practice Problems
- Self-Tests
- NetTutor
- e-Professors
- Videos

Study Skills Exercises

1. Write down the page number(s) for the Chapter Summary for this chapter. _____ Describe one way in which you can use the Summary found at the end of each chapter.

2. Define the key term **slope**.

Review Exercises

3. Find the missing coordinate so that the ordered pairs are solutions to the equation $\frac{1}{2}x + y = 4$.

   a. (0,)   **b.** (, 0)   **c.** (−4,)

For Exercises 4–7, find the *x*- and *y*-intercepts (if possible) for each equation, and sketch the graph.

4. $2x + 8 = 0$

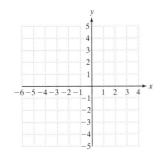

5. $4 − 2y = 0$

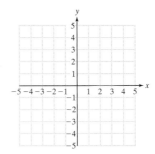

6. $2x − 2y − 6 = 0$

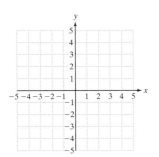

7. $x + \frac{1}{3}y = 6$

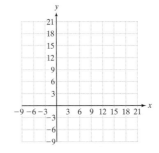

Objective 1: Introduction to the Slope of a Line

8. A 25-ft ladder is leaning against a house, as shown in the diagram. Find the slope of the ladder.

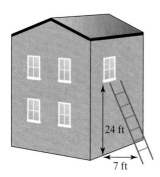

9. Find the pitch (slope) of the roof in the figure. (See Example 1.)

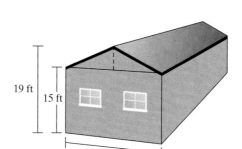

10. Find the slope of the treadmill.

11. Find the slope of the hill.

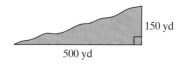

12. The road sign shown in the figure indicates the percent grade of a hill. This gives the slope of the road as the change in elevation per 100 horizontal ft. Given a 4% grade, write this as a slope in fractional form.

13. If a plane gains 1000 ft in altitude over a distance of 12,000 horizontal ft, what is the slope? Explain what this value means in the context of the problem.

Objective 2: The Slope Formula

For Exercises 14–29, use the slope formula to determine the slope of the line containing the two points. (See Examples 2–4.)

14. $(6, 0)$ and $(0, -3)$

15. $(-5, 0)$ and $(0, -4)$

16. $(-2, 3)$ and $(1, -2)$

17. $(4, 5)$ and $(-1, 0)$

18. $(-2, 5)$ and $(-7, 1)$

19. $(4, -2)$ and $(3, -1)$

20. $(0.3, -1.1)$ and $(-0.1, -0.8)$

21. $(0.4, -0.2)$ and $(0.3, -0.1)$

22. $(2, 3)$ and $(2, 7)$

23. $(-1, 5)$ and $(-1, 0)$

24. $(5, -1)$ and $(-3, -1)$

25. $(-8, 4)$ and $(1, 4)$

26. $(-4.6, 4.1)$ and $(0, 6.4)$

27. $(1.1, 4)$ and $(-3.2, -0.3)$

28. $\left(\dfrac{3}{2}, \dfrac{4}{3}\right)$ and $\left(\dfrac{7}{2}, 1\right)$

29. $\left(\dfrac{2}{3}, -\dfrac{1}{2}\right)$ and $\left(-\dfrac{1}{6}, -\dfrac{3}{2}\right)$

30. Explain how to use the graph of a line to determine whether the slope of a line is positive, negative, zero, or undefined.

31. If the slope of a line is $\dfrac{4}{3}$, how many units of change in y will be produced by 6 units of change in x?

For Exercises 32–37, estimate the slope of the line from its graph.

32.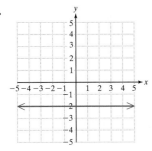

33.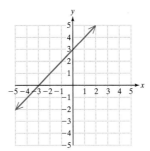

34.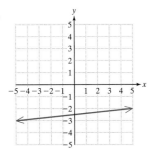

35.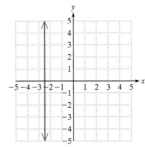

36.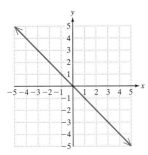

37.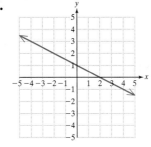

Objective 3: Parallel and Perpendicular Lines

38. Can the slopes of two perpendicular lines both be positive? Explain your answer.

142   Chapter 2   Linear Equations in Two Variables

For Exercises 39–44, the slope of a line is given.

a. Find the slope of a line parallel to the given line.

b. Find the slope of a line perpendicular to the given line. (See Example 5.)

39. $m = 5$

40. $m = 3$

41. $m = -\dfrac{4}{7}$

42. $m = -\dfrac{2}{11}$

43. $m = 0$

44. m is undefined.

In Exercises 45–52, two points are given from each of two lines L_1 and L_2. Without graphing the points, determine if the lines are perpendicular, parallel, or neither. (See Example 6.)

45. L_1: $(2, 5)$ and $(4, 9)$
L_2: $(-1, 4)$ and $(3, 2)$

46. L_1: $(-3, -5)$ and $(-1, 2)$
L_2: $(0, 4)$ and $(7, 2)$

47. L_1: $(4, -2)$ and $(3, -1)$
L_2: $(-5, -1)$ and $(-10, -16)$

48. L_1: $(0, 0)$ and $(2, 3)$
L_2: $(-2, 5)$ and $(0, -2)$

49. L_1: $(5, 3)$ and $(5, 9)$
L_2: $(4, 2)$ and $(0, 2)$

50. L_1: $(3, 5)$ and $(2, 5)$
L_2: $(2, 4)$ and $(0, 4)$

51. L_1: $(-3, -2)$ and $(2, 3)$
L_2: $(-4, 1)$ and $(0, 5)$

52. L_1: $(7, 1)$ and $(0, 0)$
L_2: $(-10, -8)$ and $(4, -6)$

Objective 4: Applications and Interpretation of Slope

53. The graph shows the number of cellular phone subscriptions (in millions) purchased in the United States for selected years. (See Example 7.)

a. Use the coordinates of the given points to find the slope of the line, and express the answer in decimal form.

b. Interpret the meaning of the slope in the context of this problem.

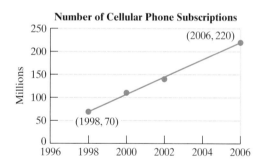

54. The number of SUVs (in millions) sold in the United States grew approximately linearly between 1994 and 2000.

a. Find the slope of the line defined by the two given points.

b. Interpret the meaning of the slope in the context of this problem.

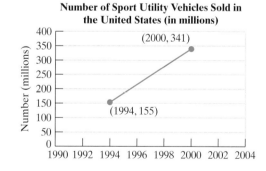

Section 2.3 Slope of a Line 143

55. The data in the graph show the average weight for boys based on age.

   a. Use the coordinates of the given points to find the slope of the line.

   b. Interpret the meaning of the slope in the context of this problem.

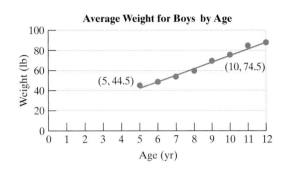

56. The data in the graph show the average weight for girls based on age.

   a. Use the coordinates of the given points to find the slope of the line, and write the answer in decimal form.

   b. Interpret the meaning of the slope in the context of this problem.

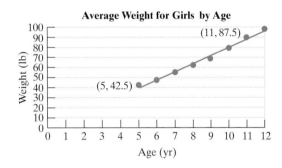

Expanding Your Skills

For Exercises 57–62, given a point P on a line and the slope m of the line, find a second point on the line (answers may vary). *Hint:* Graph the line to help you find the second point.

57. $P(0, 0)$ and $m = 2$

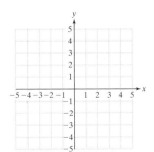

58. $P(-2, 1)$ and $m = -\dfrac{1}{3}$

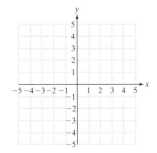

59. $P(2, -3)$ and m is undefined

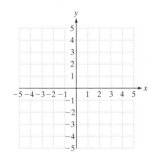

60. $P\left(0, \dfrac{4}{3}\right)$ and $m = -\dfrac{2}{3}$

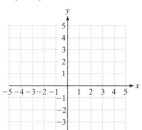

61. $P(-1, -4)$ and $m = \dfrac{4}{5}$

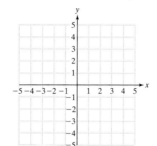

62. $P(-2, 4)$ and $m = 0$

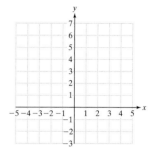

---

chapter 2 | midchapter review

1. Find the midpoint of the segment between the given points.

   a. $(-14, 20)$ and $(-6, -4)$

   b. $(1.1, 3.3)$ and $(2.3, -1.5)$

2. Determine the quadrant or axis where the point is located.

   a. $(-9, -9)$

   b. $(14, 0)$

   c. $(-2, 10)$

For Exercises 3–6, graph the lines.

3. $2x - 3y = 12$

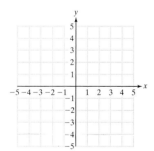

4. $y = x + 3$

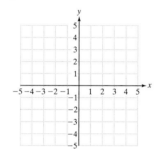

5. $x = -2$

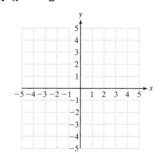

6. $y = 4$

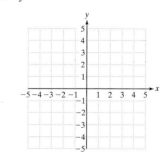

7. Find the x- and y-intercepts of the line $4x - 9y = 12$.

8. The value of a piece of farm machinery decreases over time according to the linear equation $y = -1500x + 12{,}000$, where y is the value of the machine and x is the age in years.

   a. Find the y-intercept, and interpret the meaning in the context of this problem.

   b. Find the x-intercept, and interpret the meaning in the context of this problem.

For Exercises 9–11, find the slope of the line containing the given points.

9. $(-6, 10)$ and $(-1, -2)$

10. $(7, -2)$ and $(2, 8)$

11. $(-5, 2)$ and $(-5, 4)$

12. The slope of a line is $-\dfrac{2}{3}$.

    a. Find the slope of a line parallel to the given line.

    b. Find the slope of a line perpendicular to the given line.

---

section 2.4 Equations of a Line

Objectives
1. Slope-Intercept Form
2. The Point-Slope Formula
3. Different Forms of Linear Equations

1. Slope-Intercept Form

In Section 2.2, we learned that an equation of the form $Ax + By = C$ (where A and B are not both zero) represents a line in a rectangular coordinate system. An equation of a line written in this way is said to be in **standard form**. In this section, we will learn a new form, called the **slope-intercept form**, which is useful in determining the slope and y-intercept of a line.

Let $(0, b)$ represent the y-intercept of a line. Let (x, y) represent any other point on the line. Then the slope of the line through the two points is

$m = \dfrac{y_2 - y_1}{x_2 - x_1} \quad \rightarrow \quad m = \dfrac{y - b}{x - 0}$ Apply the slope formula.

$m = \dfrac{y - b}{x}$ Simplify.

$m \cdot x = \left(\dfrac{y - b}{x}\right) x$ Clear fractions.

$mx = y - b$ Simplify.

$mx + b = y \quad$ or $\quad y = mx + b$ Solve for y: slope-intercept form.

Slope-Intercept Form of a Line

$y = mx + b$ is the slope-intercept form of a line.

m is the slope and the point $(0, b)$ is the y-intercept.

Concept Connections

Find the slope and y-intercept of the line.

1. $y = -2x - 7$

The equation $y = -4x + 7$ is written in slope-intercept form. By inspection, we can see that the slope of the line is -4 and the y-intercept is $(0, 7)$.

Answer

1. Slope: -2; y-intercept: $(0, -7)$

Skill Practice

Write the equation in slope-intercept form. Determine the slope and the y-intercept.

2. $2x - 4y = 3$

example 1 — Finding the Slope and y-Intercept of a Line

Given the line $3x + 4y = 4$, write the equation of the line in slope-intercept form, then find the slope and y-intercept.

Solution:

Write the equation in slope-intercept form, $y = mx + b$, by solving for y.

$$3x + 4y = 4$$
$$4y = -3x + 4$$
$$\frac{4y}{4} = \frac{-3x}{4} + \frac{4}{4}$$
$$y = -\frac{3}{4}x + 1 \qquad \text{The slope is } -\frac{3}{4} \text{ and the y-intercept is } (0, 1).$$

The slope-intercept form is a useful tool to graph a line. The y-intercept is a known point on the line, and the slope indicates the "direction" of the line and can be used to find a second point. Using slope-intercept form to graph a line is demonstrated in Example 2.

Skill Practice

3. Graph the line $y = \frac{1}{5}x - 2$ by using the slope and y-intercept.

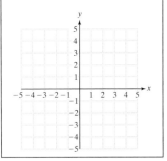

example 2 — Graphing a Line by Using the Slope and y-Intercept

Graph the line $y = -\frac{3}{4}x + 1$ by using the slope and y-intercept.

Solution:

First plot the y-intercept $(0, 1)$. The slope $m = -\frac{3}{4}$ can be written as

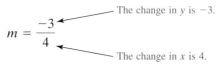

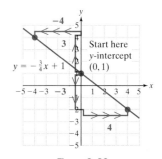

To find a second point on the line, start at the y-intercept and move *down* 3 units and to the *right* 4 units. Then draw the line through the two points (Figure 2-23).

Similarly, the slope can be written as

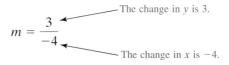

Figure 2-23

To find a second point on the line, start at the y-intercept and move *up* 3 units and to the *left* 4 units. Then draw the line through the two points (see Figure 2-23).

Answers

2. $y = \frac{1}{2}x - \frac{3}{4}$

   Slope: $\frac{1}{2}$; y-intercept: $\left(0, -\frac{3}{4}\right)$

3.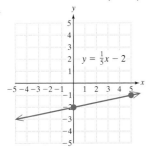

Two lines are parallel if they have the same slope and different y-intercepts. Two lines are perpendicular if the slope of one line is the opposite of the reciprocal of the slope of the other line. Otherwise, the lines are neither parallel nor perpendicular.

example 3 — Determining if Two Lines Are Parallel, Perpendicular, or Neither

Given the pair of linear equations, determine if the lines are parallel, perpendicular, or neither parallel nor perpendicular.

a. $L_1: y = -2x + 7$
$L_2: y = -2x - 1$

b. $L_1: 2y = -3x + 2$
$L_2: -4x + 6y = -12$

c. $L_1: x + y = 6$
$L_2: y = 6$

Skill Practice

Given the pair of equations, determine if the lines are parallel, perpendicular, or neither.

4. $y = -\frac{3}{4}x + 1$
$y = \frac{4}{3}x + 3$

5. $3x + y = 4$
$6x = 6 - 2y$

6. $x - y = 7$
$x = 1$

Solution:

a. The equations are written in slope-intercept form.

$L_1: y = -2x + 7$ The slope is -2 and the y-intercept is $(0, 7)$.

$L_2: y = -2x - 1$ The slope is -2 and the y-intercept is $(0, -1)$.

Because the slopes are the same and the y-intercepts are different, the lines are parallel.

b. Write each equation in slope-intercept form by solving for y.

$L_1: 2y = -3x + 2$

$\dfrac{2y}{2} = \dfrac{-3x}{2} + \dfrac{2}{2}$ Divide by 2.

$y = -\dfrac{3}{2}x + 1$

$L_2: -4x + 6y = -12$

$6y = 4x - 12$ Add $4x$ to both sides.

$\dfrac{6y}{6} = \dfrac{4}{6}x - \dfrac{12}{6}$ Divide by 6.

$y = \dfrac{2}{3}x - 2$

The slope of L_1 is $-\dfrac{3}{2}$.   The slope of L_2 is $\dfrac{2}{3}$.

The value $-\frac{3}{2}$ is the opposite of the reciprocal of $\frac{2}{3}$. Therefore, the lines are perpendicular.

c. $L_1: x + y = 6$ is equivalent to $y = -x + 6$. The slope is -1.

$L_2: y = 6$ is a horizontal line, and the slope is 0.

The slopes are not the same. Therefore, the lines are not parallel. The slope of one line is not the opposite of the reciprocal of the other slope. Therefore, the lines are not perpendicular. The lines are neither parallel nor perpendicular.

2. The Point-Slope Formula

Another useful tool to study linear equations is the point-slope formula.

Suppose a line passes through a given point (x_1, y_1) and has slope m. If (x, y) is any other point on the line, then

$m = \dfrac{y - y_1}{x - x_1}$ Slope formula

$m(x - x_1) = \dfrac{y - y_1}{x - x_1}(x - x_1)$ Clear fractions.

$m(x - x_1) = y - y_1$

or

$y - y_1 = m(x - x_1)$ Point-slope formula

Answers
4. Perpendicular   5. Parallel
6. Neither

The Point-Slope Formula

The **point-slope formula** is given by

$$y - y_1 = m(x - x_1)$$

where m is the slope of the line and (x_1, y_1) is a known point on the line.

As its name indicates, the point-slope formula is used to find an equation of a line when a point on the line is known and the slope is known.

Skill Practice

7. Use the point-slope formula to write an equation for a line passing through the point $(-2, -6)$ and with a slope of -5. Write the answer in slope-intercept form.

example 4 — Using the Point-Slope Formula to Find an Equation of a Line

Use the point-slope formula to find an equation of the line having a slope of $\frac{2}{3}$ and passing through the point $(-2, 1)$. Write the answer in slope-intercept form.

Solution:

$$m = \frac{2}{3} \quad \text{and} \quad (x_1, y_1) = (-2, 1)$$

$$y - y_1 = m(x - x_1)$$

$$y - 1 = \frac{2}{3}[x - (-2)] \quad \text{Apply the point-slope formula.}$$

$$y - 1 = \frac{2}{3}(x + 2) \quad \text{Simplify.}$$

To write the answer in slope-intercept form, clear parentheses and solve for y.

$$y - 1 = \frac{2}{3}x + \frac{4}{3} \quad \text{Clear parentheses.}$$

$$y = \frac{2}{3}x + \frac{4}{3} + 1 \quad \text{Solve for } y.$$

$$y = \frac{2}{3}x + \frac{4}{3} + \frac{3}{3} \quad \text{Find a common denominator.}$$

$$y = \frac{2}{3}x + \frac{7}{3} \quad \text{The final answer is written in slope-intercept form.}$$

Calculator Connections

We can graph the line from Example 4 to determine whether it passes through the point $(-2, 1)$. We may use a *value* function (or *eval* function) on the calculator to find the value of y when x is -2.

A *trace* feature may be used to estimate points on a line. However, when you are tracing a graph, the calculator may not return the exact coordinates of points on the graph. This is a result of the choice of scaling and the limited pixel resolution on the calculator.

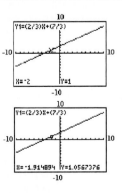

Answer

7. $y = -5x - 16$

example 5 — Finding an Equation of a Line Given Two Points

Find an equation of the line passing through the points $(5, -1)$ and $(3, 1)$. Write the answer in slope-intercept form.

Solution:

The slope formula can be used to compute the slope of the line between two points. Once the slope is known, the point-slope formula can be used to find an equation of the line.

First find the slope.

$$m = \frac{y_2 - y_1}{x_2 - x_1} = \frac{1 - (-1)}{3 - 5} = \frac{2}{-2} = -1 \qquad \text{Hence, } m = -1.$$

Next, apply the point-slope formula.

$y - y_1 = m(x - x_1)$

$y - 1 = -1(x - 3)$ Substitute $m = -1$ and use *either* point for (x_1, y_1). We will use $(3, 1)$ for (x_1, y_1).

$y - 1 = -x + 3$ Clear parentheses.

$y = -x + 3 + 1$ Solve for y.

$y = -x + 4$ The final answer is in slope-intercept form.

Skill Practice

8. Use the point-slope formula to write an equation for the line that passes through the points $(-5, 2)$ and $(-1, -1)$. Write the answer in slope-intercept form.

Tip: In Example 5, the point $(3, 1)$ was used for (x_1, y_1) in the point-slope formula. However, either point could have been used. Using the point $(5, -1)$ for (x_1, y_1) produces the same final equation:

$y - (-1) = -1(x - 5)$

$y + 1 = -x + 5$

$y = -x + 4$

example 6 — Finding an Equation of a Line Parallel to Another Line

Find an equation of the line passing through the point $(-2, -3)$ and parallel to the line $4x + y = 8$. Write the answer in slope-intercept form.

Solution:

To find an equation of a line, we must know a point on the line and the slope. The known point is $(-2, -3)$. Because the line is parallel to $4x + y = 8$, the two lines must have the same slope. Writing the equation $4x + y = 8$ in slope-intercept form, we have $y = -4x + 8$. Therefore, the slope of both lines must be -4.

We must now find an equation of the line passing through $(-2, -3)$ having a slope of -4.

$y - y_1 = m(x - x_1)$ Apply the point-slope formula.

$y - (-3) = -4[x - (-2)]$ Substitute $m = -4$ and $(-2, -3)$ for (x_1, y_1).

$y + 3 = -4(x + 2)$

$y + 3 = -4x - 8$ Clear parentheses.

$y = -4x - 11$ Write the answer in slope-intercept form.

Skill Practice

9. Find an equation of a line containing $(4, -1)$ and parallel to $2x = y - 7$. Write the answer in slope-intercept form.

Answers

8. $y = -\dfrac{3}{4}x - \dfrac{7}{4}$

9. $y = 2x - 9$

We can verify the answer to Example 6 by graphing both lines. We see that the line $y = -4x - 11$ passes through the point $(-2, -3)$ and is parallel to the line $y = -4x + 8$. See Figure 2-24.

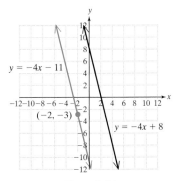

Figure 2-24

Skill Practice

10. Find an equation of the line passing through the point $(1, -6)$ and perpendicular to the line $x + 2y = 8$. Write the answer in slope-intercept form.

example 7   Finding an Equation of a Line Perpendicular to Another Line

Find an equation of the line passing through the point $(4, 3)$ and perpendicular to the line $2x + 3y = 3$. Write the answer in slope-intercept form.

Solution:

The slope of the given line can be found from its slope-intercept form.

$$2x + 3y = 3$$
$$3y = -2x + 3 \quad \text{Solve for } y.$$
$$\frac{3y}{3} = \frac{-2x}{3} + \frac{3}{3}$$
$$y = -\frac{2}{3}x + 1 \quad \text{The slope is } -\frac{2}{3}.$$

The slope of a line *perpendicular* to this line must be the opposite of the reciprocal of $-\frac{2}{3}$; hence, $m = \frac{3}{2}$. Using $m = \frac{3}{2}$ and the known point $(4, 3)$, we can apply the point-slope formula to find an equation of the line.

$$y - y_1 = m(x - x_1) \quad \text{Apply the point-slope formula.}$$
$$y - 3 = \frac{3}{2}(x - 4) \quad \text{Substitute } m = \frac{3}{2} \text{ and } (4, 3) \text{ for } (x_1, y_1).$$
$$y - 3 = \frac{3}{2}x - 6 \quad \text{Clear parentheses.}$$
$$y = \frac{3}{2}x - 6 + 3 \quad \text{Solve for } y.$$
$$y = \frac{3}{2}x - 3$$

Concept Connections

11. Explain why the point-slope formula cannot be used to find an equation for a vertical line.

Answers

10. $y = 2x - 8$
11. The slope of a vertical line is undefined. Therefore, there is no value to substitute for the slope in the point-slope formula.

Calculator Connections

From Example 7, the line $y = \frac{3}{2}x - 3$ should be perpendicular to the line $y = -\frac{2}{3}x + 1$ and should pass through the point $(4, 3)$.

Note: In this example, we are using a *square window option*, which sets the scale to display distances on the x- and y-axes as equal units of measure.

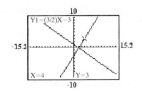

3. Different Forms of Linear Equations

A linear equation can be written in several different forms, as summarized in Table 2-2.

table 2-2

Form	Example	Comments
Standard Form $Ax + By = C$	$2x + 3y = 6$	A and B must not *both* be zero.
Horizontal Line $y = k$ (k is constant)	$y = 3$	The slope is zero, and the y-intercept is $(0, k)$.
Vertical Line $x = k$ (k is constant)	$x = -2$	The slope is undefined and the x-intercept is $(k, 0)$.
Slope-Intercept Form $y = mx + b$ Slope is m. y-Intercept is $(0, b)$.	$y = -2x + 5$ Slope $= -2$ y-Intercept is $(0, 5)$.	Solving a linear equation for y results in slope-intercept form. The coefficient of the x-term is the slope, and the constant defines the location of the y-intercept.
Point-Slope Formula $y - y_1 = m(x - x_1)$ Slope is m and (x_1, y_1) is a point on the line.	$m = -2$ $(x_1, y_1) = (3, 1)$ $y - 1 = -2(x - 3)$	This formula is typically used to build an equation of a line when a point on the line is known and the slope is known.

Although it is important to understand and apply slope-intercept form and the point-slope formula, they are not necessarily applicable to all problems. Example 8 illustrates how a little ingenuity may lead to a simple solution.

example 8   Finding an Equation of a Line

Find an equation of the line passing through the point $(-4, 1)$ and perpendicular to the x-axis.

Solution:

Any line perpendicular to the x-axis must be *vertical*. Recall that all vertical lines can be written in the form $x = k$, where k is constant. A quick sketch can help find the value of the constant (Figure 2-25).

Because the line must pass through a point whose x-coordinate is -4, the equation of the line is $x = -4$.

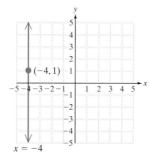

Figure 2-25

Skill Practice

12. Write an equation of the line through the point $(20, 50)$ and having a slope of 0.

Answer

12. $y = 50$

section 2.4 Practice Exercises

Boost *your* GRADE at mathzone.com!

- Practice Problems
- Self-Tests
- NetTutor
- e-Professors
- Videos

Study Skills Exercises

1. For this chapter, find the page numbers for the Chapter Review Exercises, the Chapter Test, and the Cumulative Review Exercises.

   Chapter Review Exercises, page(s)_____ Chapter Test, page(s)_____

   Cumulative Review Exercises, page(s)_____

   Compare these features and state the advantages of each.

2. Define the key terms.

   a. Standard form   b. Slope-intercept form   c. Point-slope formula

Review Exercises

3. Given $\dfrac{x}{2} + \dfrac{y}{3} = 1$

   a. Find the *x*-intercept.   b. Find the *y*-intercept.   c. Sketch the graph.

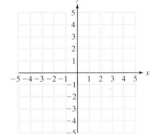

4. Using slopes, how do you determine whether two lines are parallel?

5. Using the slopes of two lines, how do you determine whether the lines are perpendicular?

6. Write the formula to find the slope of a line given two points (x_1, y_1) and (x_2, y_2).

7. Given the two points $(-1, -2)$ and $(2, 4)$,

   a. Find the slope of the line containing the two points.

   b. Find the slope of a line parallel to the line containing the points.

   c. Find the slope of a line perpendicular to the line containing the points.

Section 2.4   Equations of a Line   153

Objective 1: Slope-Intercept Form

For Exercises 8–17, determine the slope and the *y*-intercept of the line. (See Example 1.)

8. $-3x + y = 2$
9. $-7x - y = -5$
10. $17x + y = 0$
11. $x + y = 0$

12. $18 = 2y$
13. $-7 = \frac{1}{2}y$
14. $8x + 12y = 9$
15. $-9x + 10y = -4$

16. $y = 0.625x - 1.2$
17. $y = -2.5x + 1.8$

In Exercises 18–23, match the equation with the correct graph.

18. $y = \frac{3}{2}x - 2$
19. $y = -x + 3$
20. $y = \frac{13}{4}$

21. $y = x + \frac{1}{2}$
22. $x = -2$
23. $y = -\frac{1}{2}x + 2$

a.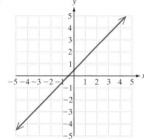
b.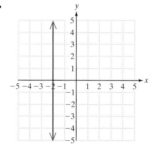
c.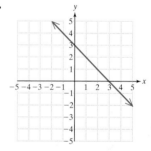

d.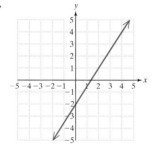
e.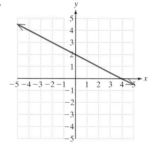
f.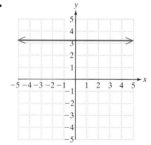

For Exercises 24–31, write the equations in slope-intercept form (if possible). Then graph each line, using the slope and *y*-intercept. (See Example 2.)

24. $y - 2 = 4x$
25. $3x = 5 - y$
26. $3x + 2y = 6$

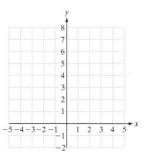
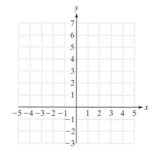
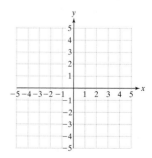

27. $x - 2y = 8$

28. $-5x = -3y - 6$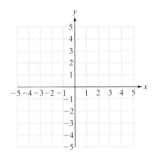

29. $-x = 8y - 2$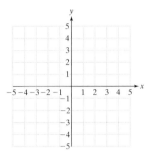

30. $2x - 5y = 0$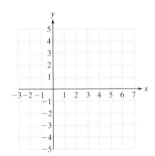

31. $3x - y = 0$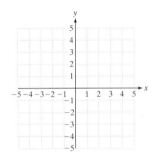

32. Given the standard form of a linear equation $Ax + By = C$, solve for y and write the equation in slope-intercept form. What is the slope of the line? What is the y-intercept?

33. Use the result of Exercise 32 to determine the slope and y-intercept of the line $3x + 7y = 9$.

For Exercises 34–39, determine if the lines are parallel, perpendicular, or neither parallel nor perpendicular. **(See Example 3.)**

34. $-3y = 5x - 1$
$6x = 10y - 12$

35. $x = 6y - 3$
$3x + \dfrac{1}{2}y = 0$

36. $3x - 4y = 12$
$\dfrac{1}{2}x - \dfrac{2}{3}y = 1$

37. $4.8x = 1.2y + 3.6$
$y - 1 = 4x$

38. $3y = 5x + 6$
$5x + 3y = 9$

39. $-y = 3x - 2$
$-6x + 2y = 6$

Objective 2: The Point-Slope Formula

For Exercises 40–67, write an equation of the line satisfying the given conditions. Write the answer in slope-intercept form or standard form. **(See Examples 4–7.)**

40. The line passes through the point $(0, -2)$ and has a slope of 3.

41. The line passes through the point $(0, 5)$ and has a slope of $-\dfrac{1}{2}$.

42. The line passes through the point $(2, 7)$ and has a slope of 2.

43. The line passes through the point (3, 10) and has a slope of −2.

44. The line passes through the point (−2, −5) and has a slope of −3.

45. The line passes through the point (−1, −6) and has a slope of 4.

46. The line passes through the point (6, −3) and has a slope of $-\frac{4}{5}$.

47. The line passes through the point (7, −2) and has a slope of $\frac{7}{2}$.

48. The line passes through (0, 4) and (3, 0).

49. The line passes through (1, 1) and (3, 7).

50. The line passes through (6, 12) and (4, 10).

51. The line passes through (−2, −1) and (3, −4).

52. The line passes through (−5, 2) and (−1, 2).

53. The line passes through (−4, −1) and (2, −1).

54. The line contains the point (3, 2) and is parallel to a line with a slope of $-\frac{3}{4}$.

55. The line contains the point (−1, 4) and is parallel to a line with a slope of $\frac{1}{2}$.

56. The line contains the point (3, 2) and is perpendicular to a line with a slope of $-\frac{3}{4}$.

57. The line contains the point (−2, 5) and is perpendicular to a line with a slope of $\frac{1}{2}$.

58. The line contains the point (2, −5) and is parallel to $y = \frac{3}{4}x + \frac{7}{4}$.

59. The line contains the point (−6, −1) and is parallel to $y = -\frac{2}{3}x - 4$.

60. The line contains the point (−8, −1) and is parallel to $x + 5y = 8$.

61. The line contains the point (4, −2) and is parallel to $3x - 4y = 8$.

62. The line contains the point (4, 0) and is parallel to the line defined by $3x = 2y$.

63. The line contains the point (−3, 0) and is parallel to the line defined by $-5x = 6y$.

64. The line is perpendicular to the line defined by $3y + 2x = 21$ and passes through the point (2, 4).

65. The line is perpendicular to $7y - x = -21$ and passes through the point (−14, 8).

66. The line is perpendicular to $\frac{1}{2}y = x$ and passes through (−3, 5).

67. The line is perpendicular to $-\frac{1}{4}y = x$ and passes through (−1, −5).

Objective 3: Different Forms of Linear Equations

For Exercises 68–75, write an equation of the line satisfying the given conditions. **(See Example 8.)**

68. The line passes through (2, −3) and has a zero slope.

69. The line contains the point $(\frac{5}{2}, 0)$ and has an undefined slope.

70. The line contains the point (2, −3) and has an undefined slope.

71. The line contains the point $(\frac{5}{2}, 0)$ and has a zero slope.

72. The line is parallel to the x-axis and passes through (4, 5).

73. The line is perpendicular to the x-axis and passes through (4, 5).

74. The line is parallel to the line $x = 4$ and passes through (5, 1).

75. The line is parallel to the line $y = -2$ and passes through (−3, 4).

Expanding Your Skills

76. Is the equation $x = -2$ in slope-intercept form? Identify the slope and y-intercept.

77. Is the equation $x = 1$ in slope-intercept form? Identify the slope and y-intercept.

78. Is the equation $y = 3$ in slope-intercept form? Identify the slope and the y-intercept.

79. Is the equation $y = -5$ in slope-intercept form? Identify the slope and the y-intercept.

Graphing Calculator Exercises

80. Use a graphing calculator to graph the lines on the same viewing window. Then explain how the lines are related.

$$y_1 = \frac{1}{2}x + 4$$

$$y_2 = \frac{1}{2}x - 2$$

81. Use a graphing calculator to graph the lines on the same viewing window. Then explain how the lines are related.

$$y_1 = -\frac{1}{3}x + 5$$

$$y_2 = -\frac{1}{3}x - 3$$

82. Use a graphing calculator to graph the lines on the same viewing window. Then explain how the lines are related.

$$y_1 = x - 2$$

$$y_2 = 2x - 2$$

$$y_3 = 3x - 2$$

83. Use a graphing calculator to graph the lines on the same viewing window. Then explain how the lines are related.

$$y_1 = -2x + 1$$

$$y_2 = -3x + 1$$

$$y_3 = -4x + 1$$

84. Use a graphing calculator to graph the lines on a square viewing window. Then explain how the lines are related.

$$y_1 = 4x - 1$$

$$y_2 = -\frac{1}{4}x - 1$$

85. Use a graphing calculator to graph the lines on a square viewing window. Then explain how the lines are related.

$$y_1 = \frac{1}{2}x - 3$$

$$y_2 = -2x - 3$$

86. Use a graphing calculator to graph the equation from Exercise 48. Use an *Eval* feature to verify that the line passes through the points $(0, 4)$ and $(3, 0)$.

87. Use a graphing calculator to graph the equation from Exercise 49. Use an *Eval* feature to verify that the line passes through the points $(1, 1)$ and $(3, 7)$.

Objectives

1. Writing a Linear Model
2. Interpreting a Linear Model
3. Finding a Linear Model from Observed Data Points

section 2.5 Applications of Linear Equations and Graphing

1. Writing a Linear Model

Algebra is a tool used to model events that occur in physical and biological sciences, sports, medicine, economics, business, and many other fields. The purpose of modeling is to represent a relationship between two or more variables with an algebraic equation.

For an equation written in slope-intercept form $y = mx + b$, the term mx is the *variable term*, and the term b is the *constant term*. The value of the term mx changes with the value of x (this is why the slope is called a rate of change). However, the term b remains constant regardless of the value of x. With these ideas in mind, a linear equation can be created if the rate of change and the constant are known.

Skill Practice

When Joe graduated from college, he had $1000 in his savings account. When he began working, he decided he would add $120 per month to his savings account.

1. Write a linear equation to compute the amount of money y in Joe's account after x months of saving.
2. Use the equation to compute the amount of money in Joe's account after 6 months.
3. Joe needs $3160 for a down payment for a car. How long will it take for Joe's account to reach this amount?

example 1 Finding a Linear Relationship

Buffalo, New York, had 2 ft (24 in.) of snow on the ground before a snowstorm. During the storm, snow fell at an average rate of $\frac{5}{8}$ in./hr.

a. Write a linear equation to compute the total snow depth y after x hr of the storm.

b. Graph the equation.

c. Use the equation to compute the depth of snow after 8 hr.

d. If the snow depth was 31.5 in. at the end of the storm, determine how long the storm lasted.

Solution:

a. The constant or base amount of snow before the storm began is 24 in. The variable amount is given by $\frac{5}{8}$ in. of snow per hour. If m is replaced by $\frac{5}{8}$ and b is replaced by 24, we have the linear equation

$$y = mx + b$$
$$y = \frac{5}{8}x + 24$$

b. The equation is in slope-intercept form, and the corresponding graph is shown in Figure 2-26.

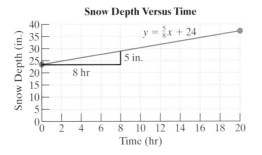

Figure 2-26

Answers

1. $y = 120x + 1000$
2. $1720 3. 18 months

c. $y = \dfrac{5}{8}x + 24$

$y = \dfrac{5}{8}(8) + 24$ Substitute $x = 8$.

$y = 5 + 24$ Solve for y.

$y = 29$ in.

The snow depth was 29 in. after 8 hr. The corresponding ordered pair is (8, 29) and can be confirmed from the graph.

d. $y = \dfrac{5}{8}x + 24$

$31.5 = \dfrac{5}{8}x + 24$ Substitute $y = 31.5$.

$8(31.5) = 8\left(\dfrac{5}{8}x + 24\right)$ Multiply by 8 to clear fractions.

$252 = 5x + 192$ Clear parentheses.

$60 = 5x$ Solve for x.

$12 = x$

The storm lasted for 12 hr. The corresponding ordered pair is (12, 31.5) and can be confirmed from the graph.

2. Interpreting a Linear Model

example 2   Interpreting a Linear Model

In 1938, President Franklin D. Roosevelt signed a bill enacting the Fair Labor Standards Act of 1938 (FLSA). In its final form, the act banned oppressive child labor and set the minimum hourly wage at 25 cents and the maximum workweek at 44 hr. Over the years, the minimum hourly wage has been increased by the government to meet the rising cost of living.

The minimum hourly wage y (in dollars per hour) in the United States between 1960 and 2005 can be approximated by the equation

$$y = 0.10x + 0.82 \qquad x \geq 0$$

where x represents the number of years since 1960 ($x = 0$ corresponds to 1960, $x = 1$ corresponds to 1961, and so on) (Figure 2-27).

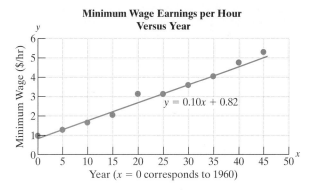

Figure 2-27

Skill Practice

The cost of long-distance service with a certain phone company is given by the equation $y = 0.12x + 6.95$, where y represents the monthly cost in dollars and x represents the number of minutes of long distance.

4. Find the slope of the line, and interpret the meaning of the slope in the context of this problem.

5. Find the y-intercept and interpret the meaning of the y-intercept in the context of this problem.

6. Use the equation to determine the cost of using 45 min of long-distance service in a month.

---

a. Find the slope of the line and interpret the meaning of the slope in the context of this problem.

b. Find the y-intercept of the line and interpret the meaning of the y-intercept in the context of this problem.

c. Use the linear equation to approximate the minimum wage in 1985.

d. Use the linear equation to predict the minimum wage in the year 2010.

Solution:

a. The equation $y = 0.10x + 0.82$ is written in slope-intercept form. The slope is 0.10 and indicates that minimum hourly wage rose an average of $0.10 per year between 1960 and 2005.

b. The y-intercept is $(0, 0.82)$. The y-intercept indicates that the minimum wage in the year 1960 ($x = 0$) was approximately $0.82 per hour. (The actual value of minimum wage in 1960 was $1.00 per hour.)

c. The year 1985 is 25 years after the year 1960. Substitute $x = 25$ into the linear equation.

$$y = 0.10x + 0.82$$
$$y = 0.10(25) + 0.82 \quad \text{Substitute } x = 25.$$
$$y = 2.50 + 0.82$$
$$y = 3.32$$

According to the linear model, the minimum wage in 1985 was approximately $3.32 per hour. (The actual minimum wage in 1985 was $3.35 per hour.)

d. The year 2010 is 50 years after the year 1960. Substitute $x = 50$ into the linear equation.

$$y = 0.10x + 0.82$$
$$y = 0.10(50) + 0.82 \quad \text{Substitute } x = 50.$$
$$y = 5.82$$

According to the linear model, minimum wage in 2010 will be approximately $5.82 per hour provided the linear trend continues. (How does this compare with the current value for minimum wage?)

Calculator Connections

A *Table* feature on a graphing calculator provides a means to evaluate the y-values of a linear equation for various values of x. For the equation $y = 0.10x + 0.82$ from Example 2, the table is set to begin at $x = 20$ and to increase x in increments of 5.

Answers

4. The slope is 0.12. In this problem, this means that the cost of the long-distance service increases by 12 cents per minute.
5. The y-intercept is $(0, 6.95)$. The meaning in this problem is that the cost of the long-distance service is $6.95 if 0 min is used.
6. $12.35

3. Finding a Linear Model from Observed Data Points

Graphing a set of data points offers a visual method to determine whether the points follow a linear pattern. If a linear trend exists, we say that there is a linear correlation between the two variables. The better the points "line up," the stronger the correlation.*

When two variables are correlated, it is often desirable to find a mathematical equation (or *model*) to describe the relationship between the variables.

example 3   Writing a Linear Model from Observed Data

Figure 2-28 represents the winning gold medal times for the women's 100-m freestyle swimming event for selected summer Olympics. Let y represent the winning time in seconds and let x represent the number of years since 1900 ($x = 0$ corresponds to 1900, $x = 1$ corresponds to 1901, and so on).

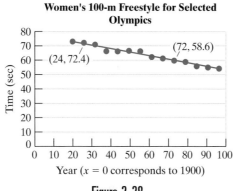

Figure 2-28

In 1924, the winning time was 72.4 sec. This corresponds to the ordered pair (24, 72.4). In 1972, the winning time was 58.6 sec, yielding the ordered pair (72, 58.6).

a. Use these ordered pairs to find a linear equation to model the winning time versus the year.

b. What is the slope of the line, and what does it mean in the context of this problem?

c. Use the linear equation to approximate the winning time for the 1964 Olympics.

d. Would it be practical to use the linear model to predict the winning time in the year 2050?

Solution:

a. The slope formula can be used to compute the slope of the line between the two points. (Round the slope to 2 decimal places.)

$$(24, 72.4) \quad \text{and} \quad (72, 58.6)$$
$$(x_1, y_1) \quad \text{and} \quad (x_2, y_2)$$

$$m = \frac{y_2 - y_1}{x_2 - x_1} = \frac{58.6 - 72.4}{72 - 24} = -0.2875 \quad \text{Hence, } m \approx -0.29.$$

Skill Practice

The figure shows data relating the cost of college textbooks in dollars to the number of pages in the book. Let y represent the cost of the book, and let x represent the number of pages.

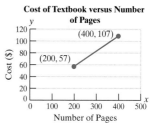

7. Use the ordered pairs indicated in the figure to write a linear equation to model the cost of textbooks versus the number of pages.

8. Use the equation to predict the cost of a textbook that has 360 pages.

Answers
7. $y = 0.25x + 7$
8. $97

*The strength of a linear correlation can be measured mathematically by using techniques often covered in statistics courses.

$$y - y_1 = m(x - x_1)$$ Apply the point-slope formula, using $m = -0.29$ and the point $(24, 72.4)$.

$$y - 72.4 = -0.29(x - 24)$$

$$y - 72.4 = -0.29x + 6.96$$ Clear parentheses.

$$y = -0.29x + 6.96 + 72.4$$ Solve for y.

$$y = -0.29x + 79.36$$ The answer is in slope-intercept form.

b. The slope is -0.29 and indicates that the winning time in the women's 100-m Olympic freestyle event has *decreased* on average by 0.29 sec/yr during this period.

c. The year 1964 is 64 years after the year 1900. Substitute $x = 64$ into the linear model.

$$y = -0.29x + 79.36$$

$$y = -0.29(64) + 79.36$$ Substitute $x = 64$.

$$y = -18.56 + 79.36$$

$$y = 60.8$$

According to the linear model, the winning time in 1964 was approximately 60.8 sec. (The actual winning time in 1964 was set by Dawn Fraser from Australia in 59.5 sec. The linear equation can only be used to *approximate* the winning time.)

d. It would not be practical to use the linear model $y = -0.29x + 79.36$ to predict the winning time in the year 2050. There is no guarantee that the linear trend will continue beyond the last observed data point in 1996. In fact, the linear trend cannot continue indefinitely; otherwise, the swimmers' times would eventually be negative. The potential for error increases for predictions made beyond the last observed data value.

section 2.5 Practice Exercises

Boost *your* GRADE at mathzone.com!

MathZone

- Practice Problems
- Self-Tests
- NetTutor
- e-Professors
- Videos

Study Skills Exercise

1. How can you utilize the margin exercises in the text?

Review Exercises

2. True or false. If an answer is false, explain why.

    a. The graph of a linear equation always has an x-intercept.

    b. The graph of a linear equation always has a y-intercept.

For Exercises 3–6, write the equation in slope-intercept form (if possible) and graph the line.

3. $2x - 2y - 4 = 0$

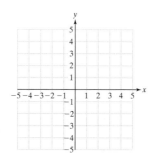

4. $2x + 3y = -12$

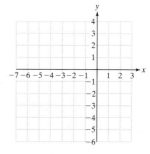

5. $2x - 4 = 6$

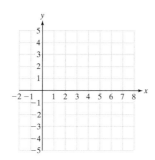

6. $3 - 2y = -6$

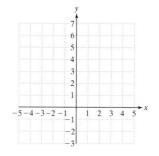

7. Given two points $(-3, -4)$ and $(3, -6)$,

   a. Find the slope of the line passing through the two points.

   b. Find an equation of the line passing through the two points. Write the answer in slope-intercept form and graph the line.

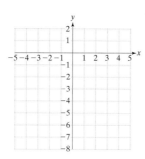

   c. Find an equation of any line parallel to the line found in part (b). (Answers may vary.)

   d. Find an equation of any line perpendicular to the line found in part (b). (Answers may vary.)

8. Given two points $(-5, -1)$ and $(-2, -4)$,

   a. Find the slope of the line passing through the two points.

   b. Find an equation of the line passing through the two points. Write the answer in slope-intercept form and graph the line.

   c. Find an equation of any line parallel to the line found in part (b). (Answers may vary.)

   d. Find an equation of any line perpendicular to the line found in part (b). (Answers may vary.)

9. Find an equation of the line parallel to the y-axis and passing through the point $(-2, -3)$. Graph the line.

10. Find an equation of the line passing through the point $(-2, -3)$ and perpendicular to the y-axis. Write the answer in slope-intercept form and graph the line.

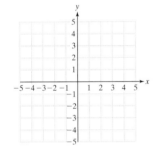
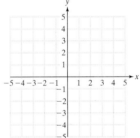

Objective 1: Writing a Linear Model

11. A car rental company charges a flat fee of $19.95 plus $0.20 per mile.

    a. Write an equation that expresses the cost y of renting a car if the car is driven for x miles. (See Example 1.)

    b. Graph the equation.

    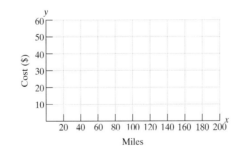

    c. What is the y-intercept and what does it mean in the context of this problem?

d. Using the equation from part (a), find the cost of driving the rental car 50, 100, and 200 mi.

e. Find the total cost of driving the rental car 100 mi if the sales tax is 6%.

f. Is it reasonable to use negative values for x in the equation? Why or why not?

12. Alex is a sales representative and earns a base salary of $1000 per month plus a 4% commission on his sales for the month.

    a. Write a linear equation that expresses Alex's monthly salary y in terms of his sales x.

    b. Graph the equation.

    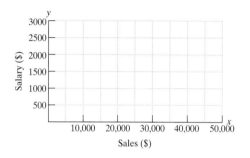

    c. What is the y-intercept and what does it represent in the context of this problem?

    d. What is the slope of the line and what does it represent in the context of this problem?

    e. How much will Alex make if his sales for a given month are $30,000?

13. Ava recently purchased a home in Crescent Beach, Florida. Her property taxes for the first year are $2742. Ava estimates that her taxes will increase at a rate of $52 per year.

    a. Write an equation to compute Ava's yearly property taxes. Let y be the amount she pays in taxes, and let x be the time in years.

    b. Graph the line.

    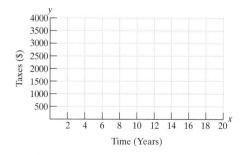

    c. What is the slope of this line? What does the slope of the line represent in the context of this problem?

d. What is the *y*-intercept? What does the *y*-intercept represent in the context of this problem?

e. What will Ava's yearly property tax be in 10 years? In 15 years?

14. Luigi Luna has started a chain of Italian restaurants called Luna Italiano. He has 19 restaurants in various locations in the northeast United States and Canada. He plans to open five new restaurants per year.

   a. Write a linear equation to express the number of stores, *y*, Luigi opens in terms of the time in years, *x*.

   b. How many stores will he have in 4 years?

   c. How many years will it take him to have 100 stores?

Objective 2: Interpreting a Linear Model

15. Sound travels at approximately one-fifth of a mile per second. Therefore, for every 5-sec difference between seeing lightning and hearing thunder, we can estimate that a storm is approximately 1 mi away. Let *y* represent the distance (in miles) that a storm is from an observer. Let *x* represent the difference in time between seeing lightning and hearing thunder. Then the distance of the storm can be approximated by the equation $y = 0.2x$, where $x \geq 0$. **(See Example 2.)**

   a. Use the linear model provided to determine how far away a storm is for the following differences in time between seeing lightning and hearing thunder: 4 sec, 12 sec, and 16 sec.

   b. If a storm is 4.2 mi away, how many seconds will pass between seeing lightning and hearing thunder?

16. The force *y* (in pounds) required to stretch a particular spring *x* inches beyond its rest (or "equilibrium") position is given by the equation $y = 2.5x$, where $0 \leq x \leq 12$.

   a. Use the equation to determine the amount of force necessary to stretch the spring 6 in. from its rest position. How much force is necessary to stretch the spring twice as far?

   b. If 45 lb of force is exerted on the spring, how far will the spring be stretched?

17. The following figure represents the median cost of new privately owned, one-family houses sold in the midwest from 1980 to 2005.

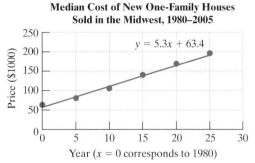

Source: U.S. Bureau of the Census and U.S. Department of Housing and Urban Development.

Let y represent the median cost of a new privately owned, one-family house sold in the midwest. Let x represent the year, where $x = 0$ corresponds to the year 1980, $x = 1$ represents 1981, and so on. Then the median cost of new privately owned, one-family houses sold in the midwest can be approximated by the equation $y = 5.3x + 63.4$, where $0 \leq x \leq 25$.

a. Use the linear equation to approximate the median cost of new privately owned, one-family houses in the midwest for the year 2005.

b. Use the linear equation to approximate the median cost for the year 1988, and compare it with the actual median cost of $101,600.

c. What is the slope of the line and what does it mean in the context of this problem?

d. What is the y-intercept and what does it mean in the context of this problem?

18. Let y represent the average number of miles driven per year for passenger cars in the United States between 1980 and 2005. Let x represent the year where $x = 0$ corresponds to 1980, $x = 1$ corresponds to 1981, and so on. The average yearly mileage for passenger cars can be approximated by the equation $y = 142x + 9060$, where $0 \leq x \leq 25$.

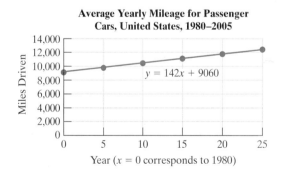

a. Use the linear equation to approximate the average yearly mileage for passenger cars in the United States in the year 2005.

b. Use the linear equation to approximate the average mileage for the year 1985, and compare it with the actual value of 9700 mi.

c. What is the slope of the line and what does it mean in the context of this problem?

d. What is the y-intercept and what does it mean in the context of this problem?

Objective 3: Finding a Linear Model from Observed Data Points

19. The figure represents the winning heights for men's pole vault in selected Olympic games. (See Example 3.)

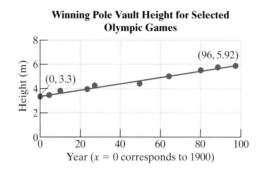

Winning Pole Vault Height for Selected Olympic Games

a. Let y represent the winning height. Let x represent the year, where $x = 0$ corresponds to the year 1900, $x = 4$ represents 1904, and so on. Use the ordered pairs given in the graph (0, 3.3) and (96, 5.92) to find a linear equation to estimate the winning pole vault height versus the year. (Round the slope to three decimal places.)

b. Use the linear equation from part (a) to approximate the winning vault for the 1920 Olympics.

c. Use the linear equation to approximate the winning vault for 1976.

d. The actual winning vault in 1920 was 4.09 m, and the actual winning vault in 1976 was 5.5 m. Are your answers from parts (b) and (c) different from these? Why?

e. What is the slope of the line? What does the slope of the line mean in the context of this problem?

20. The figure represents the winning time for the men's 100-m freestyle swimming event for selected Olympic games.

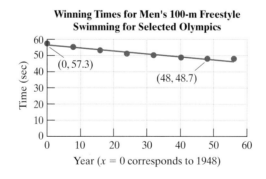

Winning Times for Men's 100-m Freestyle Swimming for Selected Olympics

a. Let y represent the winning time. Let x represent the number of years since 1948 (where $x = 0$ corresponds to the year 1948, $x = 4$ represents 1952, and so on). Use the ordered pairs given in the graph (0, 57.3) and (48, 48.7) to find a linear equation to estimate the winning time for the men's 100-m freestyle versus the year. (Round the slope to 2 decimal places.)

b. Use the linear equation from part (a) to approximate the winning 100-m time for the year 1972, and compare it with the actual winning time of 51.2 sec.

c. Use the linear equation to approximate the winning time for the year 1988.

d. What is the slope of the line and what does it mean in the context of this problem?

e. Interpret the meaning of the *x*-intercept of this line in the context of this problem. Explain why the men's swimming times will never "reach" the *x*-intercept. Do you think this linear trend will continue for the next 50 years, or will the men's swimming times begin to "level off" at some time in the future? Explain your answer.

21. At a high school football game in Miami, hot dogs were sold for $1.00 each. At the end of the night, it was determined that 650 hot dogs were sold. The following week, the price of hot dogs was raised to $1.50, and this resulted in fewer sales. Only 475 hot dogs were sold.

    a. Make a graph with the price of hot dogs on the *x*-axis and the corresponding sales on the *y*-axis. Graph the points (1.00, 650) and (1.50, 475), using suitable scaling on the *x*- and *y*-axes.

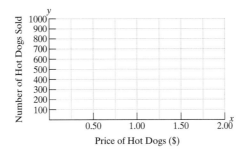

    b. Find an equation of the line through the given points. Write the equation in slope-intercept form.

    c. Use the equation from part (b) to predict the number of hot dogs that would sell if the price were changed to $1.70 per hot dog.

22. At a high school football game, soft drinks were sold for $0.50 each. At the end of the night, it was determined that 1020 drinks were sold. The following week, the price of drinks was raised to $0.75, and this resulted in fewer sales. Only 820 drinks were sold.

    a. Make a graph with the price of drinks on the *x*-axis and the corresponding sales per night on the *y*-axis. Graph the points (0.50, 1020) and (0.75, 820), using suitable scaling on the *x*- and *y*-axes.

b. Find an equation of the line through the given points. Write the equation in slope-intercept form.

c. Use the equation from part (b) to predict the number of drinks that would sell if the price were changed to $0.85 per drink.

Expanding Your Skills

23. Loraine is enrolled in an algebra class that meets 5 days per week. Her instructor gives a test every Friday. Loraine has a study plan and keeps a portfolio with notes, homework, test corrections, and vocabulary. She also records the amount of time per day that she studies and does homework. The following data represent the amount of time she studied per day and her weekly test grades.

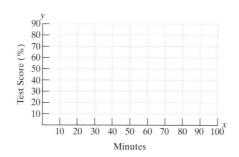

Time Studied per Day (min) x	Weekly Test Grade (percent) y
60	69
70	74
80	79
90	84
100	89

a. Graph the points on a rectangular coordinate system. Use appropriate scaling for the x- and y-axes. Do the data points appear to follow a linear trend?

b. Find a linear equation that relates Loraine's weekly test score y to the amount of time she studied per day x. (*Hint:* Pick two ordered pairs from the observed data, and find an equation of the line through the points.)

c. How many minutes should Loraine study per day in order to score at least 90% on her weekly examination? Would the equation used to determine the time Loraine needs to study to get 90% work for other students? Why or why not?

d. If Loraine is only able to spend $\frac{1}{2}$ hr/day studying her math, predict her test score for that week.

Points are *collinear* if they lie on the same line. For Exercises 24–27, use the slope formula to determine if the points are collinear.

24. $(3, -4)\ (0, -5)\ (9, -2)$ **25.** $(4, 3)\ (-4, -1)\ (2, 2)$

26. $(-2, -2)\ (0, -3)\ (-4, -1)$ **27.** $(0, 2)\ (-2, 12)\ (-1, 6)$

Graphing Calculator Exercises

28. Use a Table feature to confirm your answers to Exercise 15(a).

29. Use a Table feature to confirm your answers to Exercise 16.

30. Graph the line $y = -350x + 1000$ on the viewing window defined by $0 \leq x \leq 2$ and $0 \leq y \leq 1000$. Use the Trace key to support your answer to Exercise 21 by showing that the line passes through the points (1.00, 650) and (1.50, 475).

31. Graph the line $y = -800x + 1420$ on the viewing window defined by $0 \leq x \leq 1$ and $0 \leq y \leq 1600$. Use the Trace key to support your answer to Exercise 22 by showing that the line passes through the points (0.50, 1020) and (0.75, 820).

chapter 2 | summary

section 2.1 The Rectangular Coordinate System and Midpoint Formula

Key Concepts

Graphical representation of numerical data is often helpful to study problems in real-world applications.

A **rectangular coordinate system** is made up of a horizontal line called the ***x*-axis** and a vertical line called the ***y*-axis**. The point where the lines meet is the **origin**. The four regions of the plane are called **quadrants**.

The point (x, y) is an **ordered pair**. The first element in the ordered pair is the point's horizontal position from the origin. The second element in the ordered pair is the point's vertical position from the origin.

Examples

Example 1

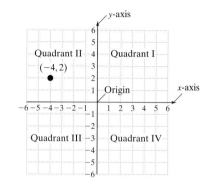

The midpoint between two points is found by using the formula

$$\left(\frac{x_1 + x_2}{2}, \frac{y_1 + y_2}{2}\right)$$

Example 2

Find the midpoint between $(-3, 1)$ and $(5, 7)$.

$$\left(\frac{-3 + 5}{2}, \frac{1 + 7}{2}\right) = (1, 4)$$

section 2.2 Linear Equations in Two Variables

Key Concepts

A **linear equation in two variables** can be written in the form $Ax + By = C$, where A, B, and C are real numbers and A and B are not both zero.

The graph of a linear equation in two variables is a line and can be represented in the rectangular coordinate system.

An **x-intercept** of an equation is a point $(a, 0)$ where the graph intersects the x-axis. To find an x-intercept, substitute 0 for y and solve for x.

A **y-intercept** of an equation is a point $(0, b)$ where the graph intersects the y-axis. To find a y-intercept, substitute 0 for x and solve for y.

A **vertical line** can be written in the form $x = k$.
A **horizontal line** can be written in the form $y = k$.

Examples

Example 1

$3x - 4y = 12$

Complete a table of ordered pairs.

x	y
0	-3
4	0
1	$-\frac{9}{4}$

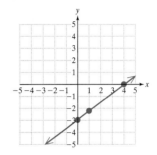

Example 2

$$2x + 3y = 8$$

x-intercept: $2x + 3(0) = 8$
$2x = 8$
$x = 4 \quad (4, 0)$

y-intercept: $2(0) + 3y = 8$
$3y = 8$
$y = \frac{8}{3} \quad \left(0, \frac{8}{3}\right)$

Example 3

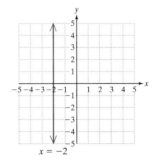

$x = -2$

Example 4

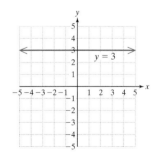

$y = 3$

section 2.3 Slope of a Line

Key Concepts

The **slope** of a line m between two distinct points (x_1, y_1) and (x_2, y_2) is given by

$$m = \frac{y_2 - y_1}{x_2 - x_1}, \quad x_2 - x_1 \neq 0$$

The slope of a line may be positive, negative, zero, or undefined.

Examples

Example 1

The slope of the line between $(1, -3)$ and $(-3, 7)$ is

$$m = \frac{7 - (-3)}{-3 - 1} = \frac{10}{-4} = -\frac{5}{2}$$

Example 2

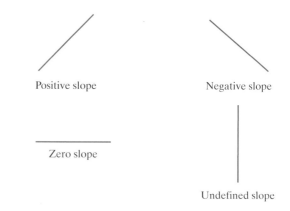

Positive slope   Negative slope

Zero slope

Undefined slope

Two parallel (nonvertical) lines have the same slope: $m_1 = m_2$.

Two lines are perpendicular if the slope of one line is the opposite of the reciprocal of the slope of the other line:

$$m_1 = \frac{-1}{m_2} \quad \text{and} \quad m_2 = \frac{-1}{m_1}$$

Or equivalently, $m_1 m_2 = -1$

Example 3

The slopes of two lines are given. Determine whether the lines are parallel, perpendicular, or neither.

a. $m_1 = -7$ and $m_2 = -7$ Parallel

b. $m_1 = -\dfrac{1}{5}$ and $m_2 = 5$ Perpendicular

c. $m_1 = -\dfrac{3}{2}$ and $m_2 = -\dfrac{2}{3}$ Neither

section 2.4 Equations of a Line

Key Concepts

Standard Form: $Ax + By = C$ (A and B are not both zero)

Horizontal line: $y = k$

Vertical line: $x = k$

Slope-intercept form: $y = mx + b$

Point-slope formula: $y - y_1 = m(x - x_1)$

Slope-intercept form is used to identify the slope and y-intercept of a line when the equation is given. Slope-intercept form can also be used to graph a line.

The point-slope formula can be used to construct an equation of a line, given a point and a slope.

Examples

Example 1

Find the slope and y-intercept. Then graph the equation.

$7x - 2y = 4$ Solve for y.

$-2y = -7x + 4$

$y = \dfrac{7}{2}x - 2$ Solve-intercept form

The slope is $\tfrac{7}{2}$; the y-intercept is $(0, -2)$.

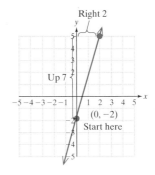

Example 2

Find an equation of the line passing through the point $(2, -3)$ and having slope $m = -4$.

Using the point-slope formula gives

$y - y_1 = m(x - x_1)$

$y - (-3) = -4(x - 2)$

$y + 3 = -4x + 8$

$y = -4x + 5$

section 2.5 Applications of Linear Equations and Graphing

Key Concepts

A linear model can be constructed to describe data for a given situation.

Examples

Example 1

The graph shows the average per capita income in the United States for 1980–2001.

The year 1980 corresponds to $x = 0$ and income is measured in $1000s.

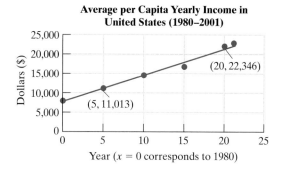

- Given two points from the data, use the point-slope formula to find an equation of the line.

Write an equation of the line, using the points $(5, 11{,}013)$ and $(20, 22{,}346)$.

Slope: $\dfrac{22{,}346 - 11{,}013}{20 - 5} = \dfrac{11{,}333}{15} \approx 756$

$$y - 11{,}013 = 756(x - 5)$$
$$y - 11{,}013 = 756x - 3780$$
$$y = 756x + 7233$$

- Interpret the meaning of the slope and y-intercept in the context of the problem.

The slope $m \approx 756$ indicates that the average income has increased by $756 per year.

The y-intercept $(0, 7233)$ means that the average income in 1980 ($x = 0$) was $7233.

- Use the equation to predict values.

Predict the average income for 2010 ($x = 30$).

$y = 756(30) + 7233$

$y = 29{,}913$

According to this model, the average income in 2010 will be approximately $29,913.

chapter 2 review exercises

Section 2.1

1. Label the following on the diagram:
   a. Origin
   b. x-Axis
   c. y-Axis
   d. Quadrant I
   e. Quadrant II
   f. Quadrant III
   g. Quadrant IV

2. Find the midpoint of the line segment between the two points $(-13, 12)$ and $(4, -18)$.

3. Find the midpoint of the line segment between the two points $(1.2, -3.7)$ and $(-4.1, -8.3)$.

4. Determine the coordinates of the points labeled in the graph.

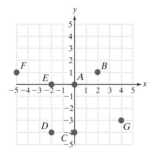

Section 2.2

For Exercises 5–7, complete the table and graph the line defined by the points.

5. $3x - 2y = -6$

x	y
0	
	0
1	

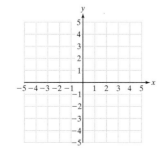

6. $2y - 3 = 10$

x	y
0	
5	
-4	

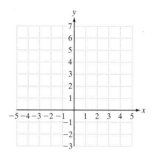

7. $6 - x = 2$

x	y
	0
	1
	-2

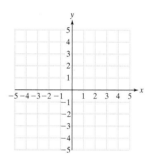

For Exercises 8–11, graph the lines. In each case find at least three points and identify the x- and y-intercepts (if possible).

8. $2x = 3y - 6$

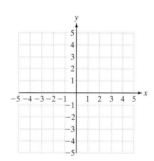

9. $5x - 2y = 0$

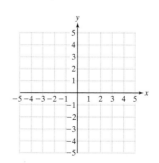

10. $2y = 6$

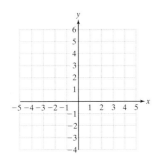

11. $-3x = 6$

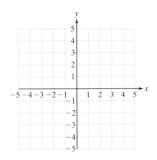

Section 2.3

12. Find the slope of the line.

a.

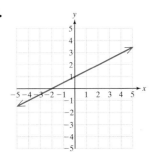

b.

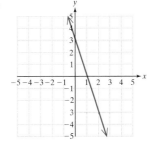

c.

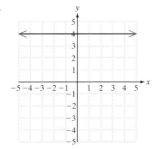

13. Draw a line with slope 2 (answers may vary).

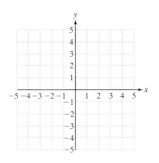

14. Draw a line with slope $-\frac{3}{4}$ (answers may vary).

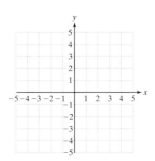

For Exercises 15–18, find the slope of the line that passes through each pair of points.

15. $(2, 6), (-1, 0)$

16. $(7, 2), (-3, -5)$

17. $(8, 2), (3, 2)$

18. $\left(-4, \dfrac{1}{2}\right), (-4, 1)$

19. Two points for each of two lines are given. Determine if the lines are parallel, perpendicular, or neither.

L_1: $(4, -6)$ and $(3, -2)$

L_2: $(3, -1)$ and $(7, 0)$

For Exercises 20–22, the slopes of two lines are given. Based on the slopes, are the lines parallel, perpendicular, or neither?

20. $m_1 = -\dfrac{1}{3}, m_2 = 3$

21. $m_1 = \dfrac{5}{4}, m_2 = \dfrac{4}{5}$

22. $m_1 = 7, m_2 = 7$

23. The graph indicates that the enrollment for a small college has been increasing linearly between 1990 and 2005.

  a. Use the two data points to find the slope of the line.

  b. Interpret the meaning of the slope in the context of this problem.

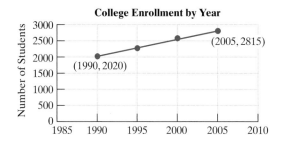

24. Approximate the slope of the stairway pictured here.

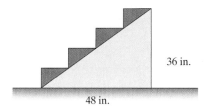

Section 2.4

25. Write a formula.

  a. Horizontal line

  b. Point-slope formula

  c. Standard form

  d. Vertical line

  e. Slope-intercept form

For Exercises 26–30, write your answer in slope-intercept form or in standard form.

26. Write an equation of the line that has slope $\frac{1}{9}$ and y-intercept $(0, 6)$.

27. Write an equation of the line that has slope $-\frac{2}{3}$ and x-intercept $(3, 0)$.

28. Write an equation of the line that passes through the points $(-8, -1)$ and $(-5, 9)$.

29. Write an equation of the line that passes through the point $(6, -2)$ and is perpendicular to the line $y = -\frac{1}{3}x + 2$.

30. Write an equation of the line that passes through the point $(0, -3)$ and is parallel to the line $4x + 3y = -1$.

31. For each of the given conditions, find an equation of the line

  a. Passing through the point $(-3, -2)$ and parallel to the x-axis.

  b. Passing through the point $(-3, -2)$ and parallel to the y-axis.

  c. Passing through the point $(-3, -2)$ and having an undefined slope.

  d. Passing through the point $(-3, -2)$ and having a zero slope.

32. Are any of the lines in Exercise 31 the same?

Section 2.5

33. Molly Kay loves the beach and decides to spend the summer selling various ice cream products on the beach. From her accounting course, she knows that her total cost is calculated as

Total cost = fixed cost + variable cost

She estimates that her fixed cost for the summer season is $20 per day. She also knows that each ice cream product costs her $0.25 from her distributor.

a. Write a relationship for the daily cost y in terms of the number of ice cream products sold per day x.

b. Graph the equation from part (a) by letting the horizontal axis represent the number of ice cream products sold per day and letting the vertical axis represent the daily cost.

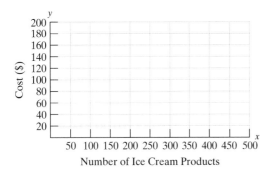

c. What does the y-intercept represent in the context of this problem?

d. What is her cost if she sells 450 ice cream products?

e. What is the slope of the line?

f. What does the slope of the line represent in the context of this problem?

34. The margin of victory for a certain college football team seems to be linearly related to the number of rushing yards gained by the star running back. The table shows the statistics.

Yards Rushed	Margin of Victory
100	20
60	10
120	24
50	7

a. Graph the data to determine if a linear trend exists. Let x represent the number of yards rushed by the star running back and y represent the points in the margin of victory.

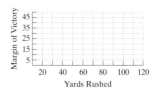

b. Find an equation for the line through the points $(50, 7)$ and $(100, 20)$.

c. Based on the equation, what would be the result of the football game if the star running back did not play?

---

chapter 2 test

1. Given the equation $x - \frac{2}{3}y = 6$, complete the ordered pairs and graph the corresponding points. $(0, \ ) \ (\ , 0) \ (\ , -3)$

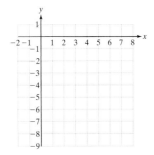

2. Determine whether the following statements are true or false and explain your answer.

   a. The product of the x- and y-coordinates is positive only for points in quadrant I.

   b. The quotient of the x- and y-coordinates is negative only for points in quadrant IV.

   c. The point $(-2, -3)$ is in quadrant III.

   d. The point $(0, 0)$ lies on the x-axis.

3. Find the midpoint of the line segment between the two points $(21, -15)$ and $(5, 32)$.

4. Explain the process for finding the x- and y-intercepts.

For Exercises 5–8, identify the x- and y-intercepts (if possible) and graph the line.

5. $6x - 8y = 24$

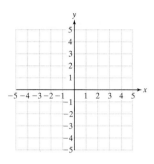

6. $x = -4$

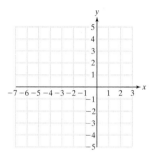

7. $3x = 5y$

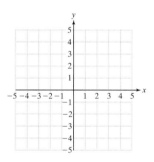

8. $2y = -6$

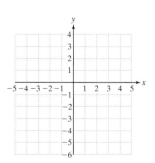

9. Find the slope of the line, given the following information:

   a. The line passes through the points $(7, -3)$ and $(-1, -8)$.

   b. The line is given by $6x - 5y = 1$.

10. Describe the relationship of the slopes of

    a. Two parallel lines

    b. Two perpendicular lines

11. The slope of a line is -7.

    a. Find the slope of a line parallel to the given line.

    b. Find the slope of a line perpendicular to the given line.

12. Two points are given for each of two lines. Determine if the lines are parallel, perpendicular, or neither.

    L_1: $(4, -4)$ and $(1, -6)$

    L_2: $(-2, 0)$ and $(0, 3)$

13. Given the equation $-3x + 4y = 4$,

    a. Write the line in slope-intercept form.

    b. Determine the slope and y-intercept.

    c. Graph the line, using the slope and y-intercept.

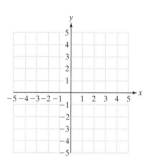

14. Determine if the lines are parallel, perpendicular, or neither.

    a. $y = -x + 4$
       $y = x - 3$

    b. $9x - 3y = 1$
       $15x - 5y = 10$

    c. $3y = 6$
       $x = 0.5$

    d. $5x - 3y = 9$
       $3x - 5y = 10$

15. Write an equation that represents a line subject to the following conditions. (Answers may vary.)

    a. A line that does not pass through the origin and has a positive slope

    b. A line with an undefined slope

    c. A line perpendicular to the y-axis. What is the slope of such a line?

    d. A slanted line that passes through the origin and has a negative slope

16. Write an equation of the line that passes through the point $(8, -\frac{1}{2})$ with slope -2. Write the answer in slope-intercept form.

17. Write an equation of the line containing the points $(2, -3)$ and $(4, 0)$.

18. Write an equation of a line containing $(4, -3)$ and parallel to $6x - 3y = 1$.

19. Write an equation of the line that passes through the point $(-10, -3)$ and is perpendicular to $3x + y = 7$. Write the answer in slope-intercept form.

20. Jack sells used cars. He is paid $800 per month plus $300 commission for each automobile he sells.

    a. Write an equation that represents Jack's monthly earnings y in terms of the number of automobiles he sells x.

    b. Graph the linear equation you found in part (a).

    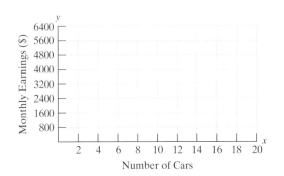

    c. What does the y-intercept mean in the context of this problem?

    d. How much will Jack earn in a month if he sells 17 automobiles?

21. The following graph represents the life expectancy for females in the United States born from 1940 through 2000.

    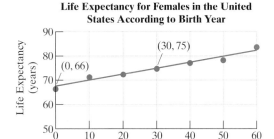

    Source: National Center for Health Statistics

    a. Approximate the y-intercept from the graph. What does the y-intercept represent in the context of this problem?

    b. Using the two points $(0, 66)$ and $(30, 75)$, determine the slope of the line. What does the slope of the line represent in the context of this problem?

    c. Use the y-intercept and the slope found in parts (a) and (b) to write an equation of the line by letting x represent the year of birth and y represent the corresponding life expectancy.

    d. Using the linear equation from part (c), approximate the life expectancy for women born in the United States in 1994. How does your answer compare with the reported life expectancy of 79 years?

chapters 1–2 | cumulative review exercises

1. Simplify the expression.

   $$\frac{5 - 2^3 \div 4 + 7}{-1 - 3(4 - 1)}$$

2. Simplify the expression $3 + \sqrt{25} - 8(\sqrt{9}) \div 6$.

3. Solve the equation for z.

   $$z - (3 + 2z) + 5 = -2z - 5$$

4. Solve the equation for b.

   $$\frac{2b - 3}{6} - \frac{b + 1}{4} = -2$$

5. A bike rider pedals 10 mph to the top of the hill and 15 mph on the way down. The total time for the round trip is 10 hr. Find the distance to the top of the hill.

6. The formula for the volume of a right circular cylinder is $V = \pi r^2 h$.

   a. Solve for h.

   b. Find h if a soda can contains 355 cm³ (which is approximately 12 oz) of soda and the diameter is 6.6 cm. Round the answer to 1 decimal place.

7. Solve the inequalities. Write your answers in interval notation.

   a. $-5x - 4 \le -2(x - 1)$ b. $-x + 4 > 1$

8. Find the slope of the line that passes through the points $(4, -5)$ and $(-6, -3)$.

9. Find the midpoint of the line segment with endpoints $(-2, -3)$ and $(0, 15)$.

For Exercises 10–11, **a.** find the x- and y-intercept, **b.** find the slope, and **c.** graph the line.

10. $3x - 5y = 10$

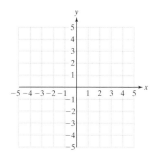

11. $2y + 4 = 10$

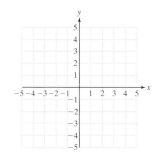

12. Find an equation for the vertical line that passes through the point $(7, 11)$.

13. Find an equation for the horizontal line that passes through the point $(19, 20)$.

14. Find an equation of the line passing through $(1, -4)$ and parallel to $2x + y = 6$. Write the answer in slope-intercept form.

15. Find an equation of the line passing through $(1, -4)$ and perpendicular to $y = \frac{1}{4}x - 2$. Write the answer in slope-intercept form.

16. At the movies, Laquita paid for drinks and popcorn for herself and her two children. She spent twice as much on popcorn as on drinks. If her total bill came to $17.94, how much did she spend on drinks and how much did she spend on popcorn?

Systems of Linear Equations

3

3.1 Solving Systems of Linear Equations by Graphing

3.2 Solving Systems of Equations by Using the Substitution Method

3.3 Solving Systems of Equations by Using the Addition Method

3.4 Applications of Systems of Linear Equations in Two Variables

3.5 Systems of Linear Equations in Three Variables and Applications

3.6 Solving Systems of Linear Equations by Using Matrices

In this chapter, we present several methods used to solve systems of linear equations in two and three variables. Systems of equations come up in a variety of applications when there are two or more unknowns. For example, in Exercise 16 from Section 3.4, a chemistry student has two different concentrations of acid solution, one of 18% acid and the other of 45% acid. A system of linear equations can be used to compute the amount of each type of solution needed to obtain 16 liters (L) of 36% acid solution.

chapter 3 preview

The exercises in this chapter preview contain concepts that have not yet been presented. These exercises are provided for students who want to compare their levels of understanding before and after studying the chapter. Alternatively, you may prefer to work these exercises when the chapter is completed and before taking the exam.

Section 3.1

1. Determine whether the ordered pair $(-5, -6)$ is a solution for the system.

$$2x - y = -4$$
$$-3x + 2y = -27$$

2. Solve the system by using the graphing method.

$$y = x - 3$$
$$2x - y = 4$$

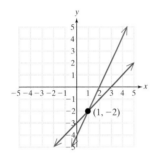

Section 3.2

For Exercises 3–4, solve by using the substitution method.

3. $3x + 5y = 3$
   $y = -2x - 5$

4. $-x + 3y = -1$
   $4x + 6y = 10$

Section 3.3

For Exercises 5–6, solve by using the addition method.

5. $3x - 2y = -7$
   $2x + 3y = 17$

6. $\frac{1}{3}x = \frac{2}{5}y - 3$
   $\frac{5}{6}x - \frac{1}{2}y = -\frac{5}{2}$

For Exercises 7–8, solve by using any method.

7. $4x - 8y = -1$
   $3x - 6y = 2$

8. $x = \frac{1}{4}y - 2$
   $y = 4x + 8$

Section 3.4

For Exercises 9–10, use a system of equations to solve each application.

9. A small plane flies 500 mi in 2 hr with the wind. The return flight against the wind takes 2.5 hr. Find the speed of the plane in still air and the speed of the wind.

10. Sally invested a total of $2500 in two accounts. One account paid 5.5% interest and the other paid 4% interest. After 1 year, Sally earned $115 in interest. Find the amount of money in each account.

Section 3.5

For Exercises 11–12, solve the system of linear equations in three variables.

11. $x - 3z = -15$
    $ 2y + 4z = 14$
    $4x - y = 3$

12. $x + 2y - 3z = 6$
    $-2x - 4y + 6z = 6$
    $3x + y - z = 1$

13. The sum of three numbers is 80. The first number is 8 less than the second number. The sum of the first and third numbers is 10 less than twice the second. Find the numbers.

Section 3.6

14. Determine the order of the matrix.

$$\begin{bmatrix} 2 & 3 & -1 \\ 4 & 5 & 6 \end{bmatrix}$$

15. Solve the system by using the Gauss-Jordan method.

$$3x - 5y = 10$$
$$-2x + 3y = -8$$

Section 3.1 Solving Systems of Linear Equations by Graphing

Objectives
1. Solutions to Systems of Linear Equations
2. Dependent and Inconsistent Systems of Linear Equations
3. Solving Systems of Linear Equations by Graphing

1. Solutions to Systems of Linear Equations

A linear equation in two variables has an infinite number of solutions that form a line in a rectangular coordinate system. Two or more linear equations form a **system of linear equations**. For example:

$$x - 3y = -5$$
$$2x + 4y = 10$$

A **solution to a system of linear equations** is an ordered pair that is a solution to *each* individual linear equation.

example 1 — Determining Solutions to a System of Linear Equations

Determine whether the ordered pairs are solutions to the system.

$$x + y = -6$$
$$3x - y = -2$$

a. $(-2, -4)$ b. $(0, -6)$

Solution:

a. Substitute the ordered pair $(-2, -4)$ into both equations:

$x + y = -6 \longrightarrow (-2) + (-4) = -6$ ✔ True

$3x - y = -2 \longrightarrow 3(-2) - (-4) = -2$ ✔ True

Because the ordered pair $(-2, -4)$ is a solution to each equation, it is a solution to the *system* of equations.

b. Substitute the ordered pair $(0, -6)$ into both equations:

$x + y = -6 \longrightarrow (0) + (-6) = -6$ ✔ True

$3x - y = -2 \longrightarrow 3(0) - (-6) \stackrel{?}{=} -2$ False

Because the ordered pair $(0, -6)$ is not a solution to the second equation, it is *not* a solution to the system of equations.

Skill Practice

Determine whether the ordered pairs are solutions to the system.

$$3x + 2y = -8$$
$$y = 2x - 18$$

1. $(-2, -1)$
2. $(8, -2)$
3. $(4, -10)$

A solution to a system of two linear equations may be interpreted graphically as a point of intersection between the two lines. Graphing the lines from Example 1, we have

$$x + y = -6$$
$$3x - y = -2$$

Notice that the lines intersect at $(-2, -4)$ (Figure 3-1).

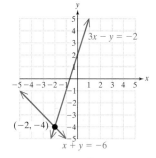

Figure 3-1

Answers
1. No   2. No   3. Yes

2. Dependent and Inconsistent Systems of Linear Equations

When two lines are drawn in a rectangular coordinate system, three geometric relationships are possible:

1. Two lines may intersect at *exactly one point*.

2. Two lines may intersect at *no point*. This occurs if the lines are parallel.

3. Two lines may intersect at *infinitely many points* along the line. This occurs if the equations represent the same line (the lines are coinciding).

If a system of linear equations has one or more solutions, the system is said to be a **consistent system**. If a linear equation has no solution, it is said to be an **inconsistent system**.

If two equations represent the same line, then all points along the line are solutions to the system of equations. In such a case, the system is characterized as a **dependent system**. An **independent system** is one in which the two equations represent different lines.

Concept Connections

For Exercises 4–5, choose the letter of the correct response.

4. Two different lines with the same slope and different *y*-intercepts intersect in how many points?
   a. 1   b. 2   c. 0
   d. Infinitely many

5. Two lines with different slopes intersect in how many points?
   a. 1   b. 2   c. 0
   d. Infinitely many

Solutions to Systems of Linear Equations in Two Variables

One unique solution	No solution	Infinitely many solutions

One point of intersection	Parallel lines	Coinciding lines
System is consistent.	System is inconsistent.	System is consistent.
System is independent.	System is independent.	System is dependent.

Skill Practice

6. Solve by using the graphing method.

   $3x + y = -5$
   $x - 2y = -4$

   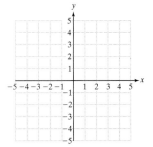

3. Solving Systems of Linear Equations by Graphing

example 2 Solving a System of Linear Equations by Graphing

Solve the system by graphing both linear equations and finding the point(s) of intersection.

$$y = -\frac{3}{2}x + \frac{1}{2}$$
$$2x + 3y = -6$$

Answers
4. c.   5. a.   6. $(-2, 1)$

Solution:

To graph each equation, write the equation in slope-intercept form $y = mx + b$.

First equation:

$y = -\dfrac{3}{2}x + \dfrac{1}{2}$ Slope: $-\dfrac{3}{2}$

Second equation:

$2x + 3y = -6$

$3y = -2x - 6$

$\dfrac{3y}{3} = \dfrac{-2x}{3} - \dfrac{6}{3}$

$y = -\dfrac{2}{3}x - 2$ Slope: $-\dfrac{2}{3}$

From their slope-intercept forms, we see that the lines have different slopes, indicating that the lines must intersect at exactly one point. We can graph the lines by using the slope and y-intercept to find the point of intersection (Figure 3-2).

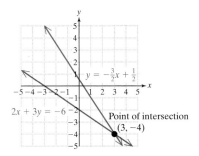

Figure 3-2

The point $(3, -4)$ appears to be the point of intersection. This can be confirmed by substituting $x = 3$ and $y = -4$ into both equations.

$y = -\dfrac{3}{2}x + \dfrac{1}{2} \longrightarrow -4 \stackrel{?}{=} -\dfrac{3}{2}(3) + \dfrac{1}{2} \longrightarrow -4 = -\dfrac{9}{2} + \dfrac{1}{2}$ ✔ True

$2x + 3y = -6 \longrightarrow 2(3) + 3(-4) \stackrel{?}{=} -6 \longrightarrow 6 - 12 = -6$ ✔ True

The solution is $(3, -4)$.

Tip: In Example 2, the lines could also have been graphed by using the x- and y-intercepts or by using a table of points. However, the advantage of writing the equations in slope-intercept form is that we can compare the slopes *and* y-intercepts of each line.

1. If the slopes differ, the lines are different and nonparallel and must cross in exactly one point.
2. If the slopes are the same and the y-intercepts are different, the lines are parallel and do not intersect.
3. If the slopes are the same and the y-intercepts are the same, the two equations represent the same line.

Skill Practice

7. Solve the system by graphing.

$2x = -4y + 6$

$-3x - y = -4$

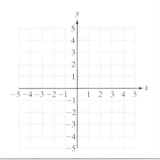

example 3 — Solving a System of Linear Equations by Graphing

Solve the system by graphing.

$$4x = 8$$
$$6y = -3x + 6$$

Solution:

The first equation $4x = 8$ can be written as $x = 2$. This is an equation of a vertical line. To graph the second equation, write the equation in slope-intercept form.

First equation: Second equation:

$4x = 8$ $6y = -3x + 6$

$x = 2$ $\dfrac{6y}{6} = \dfrac{-3x}{6} + \dfrac{6}{6}$

 $y = -\dfrac{1}{2}x + 1$

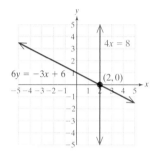

Figure 3-3

The graphs of the lines are shown in Figure 3-3. The point of intersection is $(2, 0)$. This can be confirmed by substituting $(2, 0)$ into both equations.

$4x = 8 \longrightarrow 4(2) = 8$ ✓ True

$6y = -3x + 6 \longrightarrow 6(0) = -3(2) + 6$ ✓ True

The solution is $(2, 0)$.

Skill Practice

8. Solve the system by graphing.

$y = x$

$y = x - 3$

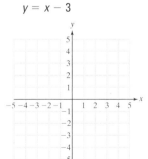

example 4 — Solving a System of Equations by Graphing

Solve the system by graphing.

$$-x + 3y = -6$$
$$6y = 2x + 6$$

Solution:

To graph the line, write each equation in slope-intercept form.

$-x + 3y = -6$ $6y = 2x + 6$

$3y = x - 6$

$\dfrac{3y}{3} = \dfrac{x}{3} - \dfrac{6}{3}$ $\dfrac{6y}{6} = \dfrac{2x}{6} + \dfrac{6}{6}$

$y = \dfrac{1}{3}x - 2$ $y = \dfrac{1}{3}x + 1$

Because the lines have the same slope but different y-intercepts, they are parallel (Figure 3-4). Two parallel lines do not intersect, which implies that the system has no solution. The system is inconsistent.

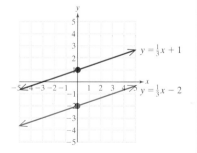

Figure 3-4

Answers

7. $(1, 1)$
8. The system is inconsistent. There is no solution.

example 5 — Solving a System of Linear Equations by Graphing

Solve the system by graphing.

$$x + 4y = 8$$
$$y = -\frac{1}{4}x + 2$$

Solution:

Write the first equation in slope-intercept form. The second equation is already in slope-intercept form.

First equation:

$x + 4y = 8$

$4y = -x + 8$

$\dfrac{4y}{4} = \dfrac{-x}{4} + \dfrac{8}{4}$

$y = -\dfrac{1}{4}x + 2$

Second equation:

$y = -\dfrac{1}{4}x + 2$

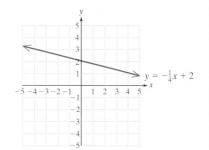

Figure 3-5

Notice that the slope-intercept forms of the two lines are identical. Therefore, the equations represent the same line (Figure 3-5). The system is dependent, and the solution to the system of equations is the set of all points on the line.

Because not all the ordered pairs in the solution set can be listed, we can write the solution in set-builder notation. Furthermore, the equations $x + 4y = 8$ and $y = -\frac{1}{4}x + 2$ represent the same line. Therefore, the solution set may be written as $\{(x, y) \mid y = -\frac{1}{4}x + 2\}$ or $\{(x, y) \mid x + 4y = 8\}$.

Skill Practice

9. Solve the system by graphing.

$$y = \frac{1}{2}x + 1$$
$$x - 2y = -2$$

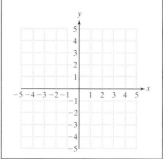

Calculator Connections

The solution to a system of equations can be found by using either a *Trace* feature or an *Intersect* feature on a graphing calculator to find the point of intersection between two curves.

For example, consider the system

$$-2x + y = 6$$
$$5x + y = -1$$

Answer

9. Dependent system;

$\{(x, y) \mid y = \dfrac{1}{2}x + 1\}$

First graph the equations together on the same viewing window. Recall that to enter the equations into the calculator, the equations must be written with the y-variable isolated.

$$\text{Isolate } y.$$
$$-2x + y = 6 \longrightarrow y = 2x + 6$$
$$5x + y = -1 \longrightarrow y = -5x - 1$$

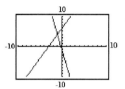

By inspection of the graph, it appears that the solution is $(-1, 4)$. The *Trace* option on the calculator may come close to $(-1, 4)$ but may not show the exact solution (Figure 3-6). However, an *Intersect* feature on a graphing calculator may provide the exact solution (Figure 3-7). See your user's manual for further details.

Using *Trace*   Using *Intersect*

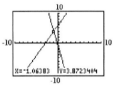
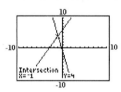

Figure 3-6   **Figure 3-7**

section 3.1 Practice Exercises

Boost *your* GRADE at mathzone.com!

- Practice Problems
- Self-Tests
- NetTutor
- e-Professors
- Videos

Study Skills Exercises

1. Instructors vary in what they emphasize on tests. For example, test material may come from the textbook, notes, handouts, homework, etc. What does your instructor emphasize?

2. Define the key terms.
   a. System of linear equations
   b. Solution to a system of linear equations
   c. Consistent system
   d. Inconsistent system
   e. Dependent system
   f. Independent system

Objective 1: Solutions to Systems of Linear Equations

For Exercises 3–8, determine which points are solutions to the given system. (See Example 1.)

3. $y = 8x - 5$
$y = 4x + 3$
$(-1, 13), (-1, 1), (2, 11)$

4. $y = -\dfrac{1}{2}x - 5$
$y = \dfrac{3}{4}x - 10$
$(4, -7), (0, -10), \left(3, -\dfrac{9}{2}\right)$

5. $2x - 7y = -30$
$y = 3x + 7$
$(0, -30), \left(\dfrac{3}{2}, 5\right), (-1, 4)$

6. $x + 2y = 4$
$y = -\dfrac{1}{2}x + 2$
$(-2, 3), (4, 0), \left(3, \dfrac{1}{2}\right)$

7. $x - y = 6$
$4x + 3y = -4$
$(4, -2), (6, 0), (2, -4)$

8. $x - 3y = 3$
$2x - 9y = 1$
$(0, 1), (4, 1), \left(8, \dfrac{5}{3}\right)$

Objective 2: Dependent and Inconsistent Systems of Linear Equations

For Exercises 9–14, the graph of a system of linear equations is given.
  a. Identify whether the system is consistent or inconsistent.
  b. Identify whether the system is dependent or independent.
  c. Identify the number of solutions to the system.

9. $y = x + 3$
$3x + y = -1$

10. $5x - 3y = 6$
$3y = 2x + 3$

11. $2x = y + 4$
$-4x + 2y = 2$

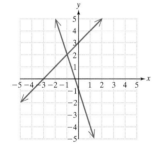
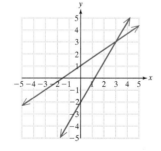
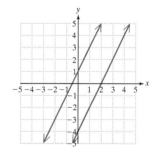

12. $y = -2x - 3$
   $-4x - 2y = 0$

13. $y = \dfrac{1}{3}x + 2$
   $-x + 3y = 6$
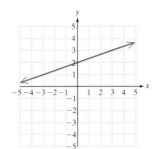

14. $y = -\dfrac{2}{3}x - 1$
   $-4x - 6y = 6$
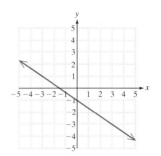

Objective 3: Solving Systems of Linear Equations by Graphing

For Exercises 15–32, solve the systems of equations by graphing. **(See Examples 2–5.)**

15. $2x + y = 4$
   $x + 2y = -1$
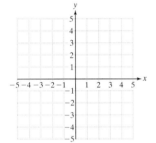

16. $4x - 3y = 12$
   $3x + 4y = -16$
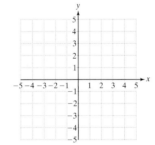

17. $y = -2x + 3$
   $y = 5x - 4$
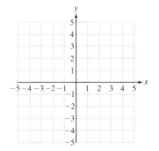

18. $y = 2x + 5$
   $y = -x + 2$
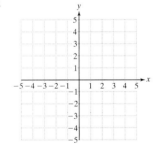

19. $y = \dfrac{1}{3}x - 5$
   $y = -\dfrac{2}{3}x - 2$
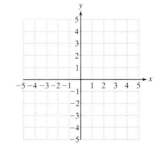

20. $y = \dfrac{1}{2}x + 2$
   $y = \dfrac{5}{2}x - 2$
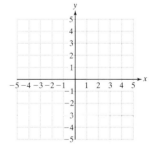

21. $x = 4$
$y = 2x - 3$

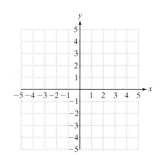

22. $3x + 2y = 6$
$y = -3$

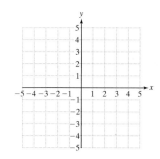

23. $y = -2x + 3$
$-2x = y + 1$

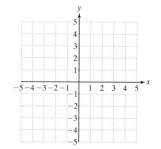

24. $y = \dfrac{1}{3}x - 2$
$x = 3y - 9$

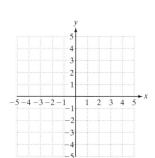

25. $y = \dfrac{2}{3}x - 1$
$2x = 3y + 3$

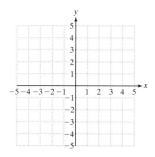

26. $4x = 16 - 8y$
$y = -\dfrac{1}{2}x + 2$

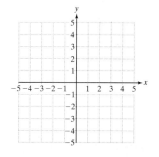

27. $2x = 4$
$\dfrac{1}{2}y = -1$

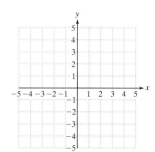

28. $y + 7 = 6$
$-5 = 2x$

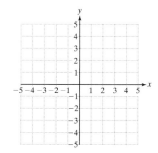

29. $-x + 3y = 6$
$6y = 2x + 12$

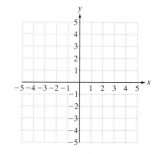

30. $3x = 2y - 4$
    $-4y = -6x - 8$

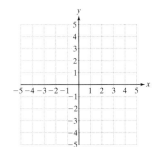

31. $2x - y = 4$
    $4x + 2 = 2y$

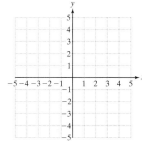

32. $x = 4y + 4$
    $-2x + 8y = -16$

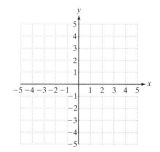

For Exercises 33–36, identify each statement as true or false.

33. A consistent system is a system that always has a unique solution.

34. A dependent system is a system that has no solution.

35. If two lines coincide, the system is dependent.

36. If two lines are parallel, the system is independent.

Graphing Calculator Exercises

For Exercises 37–42, use a graphing calculator to graph each linear equation on the same viewing window. Use a *Trace* or *Intersect* feature to find the point(s) of intersection.

37. $y = 2x - 3$
    $y = -4x + 9$

38. $y = -\frac{1}{2}x + 2$
    $y = \frac{1}{3}x - 3$

39. $x + y = 4$
    $-2x + y = -5$

40. $x - 2y = -2$
    $-3x + 2y = 6$

41. $-x + 3y = -6$
    $6y = 2x + 6$

42. $x + 4y = 8$
    $y = -\frac{1}{4}x + 2$

Section 3.2 Solving Systems of Equations by Using the Substitution Method

Objectives
1. The Substitution Method
2. Solving Inconsistent Systems and Dependent Systems

1. The Substitution Method

Graphing a system of equations is one method to find the solution of the system. In this section and Section 3.3, we will present two algebraic methods to solve a system of equations. The first is called the *substitution method*. This technique is particularly important because it can be used to solve more advanced problems including nonlinear systems of equations.

The first step in the substitution process is to isolate one of the variables from one of the equations. Consider the system

$$x + y = 16$$
$$x - y = 4$$

Solving the first equation for x yields $x = 16 - y$. Then, because x is equal to $16 - y$, the expression $16 - y$ may replace x in the second equation. This leaves the second equation in terms of y only.

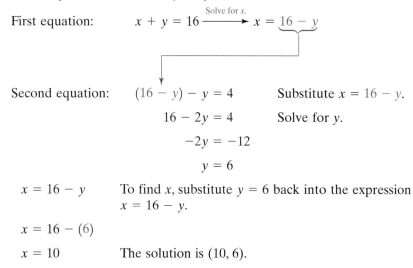

First equation:	$x + y = 16 \xrightarrow{\text{Solve for } x.} x = \underbrace{16 - y}$	
Second equation:	$(16 - y) - y = 4$	Substitute $x = 16 - y$.
	$16 - 2y = 4$	Solve for y.
	$-2y = -12$	
	$y = 6$	
$x = 16 - y$	To find x, substitute $y = 6$ back into the expression $x = 16 - y$.	
$x = 16 - (6)$		
$x = 10$	The solution is $(10, 6)$.	

Solving a System of Equations by the Substitution Method

1. Isolate one of the variables from one equation.
2. Substitute the quantity found in step 1 into the *other* equation.
3. Solve the resulting equation.
4. Substitute the value found in step 3 back into the equation in step 1 to find the value of the remaining variable.
5. Check the solution in both equations, and write the answer as an ordered pair.

Skill Practice

1. Solve by using the substitution method.
   $x = 2y + 3$
   $4x - 2y = 0$

example 1 — Using the Substitution Method to Solve a Linear Equation

Solve the system by using the substitution method.
$$-3x + 4y = 9$$
$$x = -\frac{1}{3}y + 2$$

Solution:

$$-3x + 4y = 9$$
$$x = -\frac{1}{3}y + 2$$

Step 1: In the second equation, x is already isolated.

$$-3\left(-\frac{1}{3}y + 2\right) + 4y = 9$$

Step 2: Substitute the quantity $-\frac{1}{3}y + 2$ for x in the *other* equation.

$$y - 6 + 4y = 9$$

Step 3: Solve for y.

$$5y = 15$$
$$y = 3$$

Now use the known value of y to solve for the remaining variable x.

$$x = -\frac{1}{3}y + 2$$
$$x = -\frac{1}{3}(3) + 2$$

Step 4: Substitute $y = 3$ into the equation $x = -\frac{1}{3}y + 2$.

$$x = -1 + 2$$
$$x = 1$$

Step 5: Check the ordered pair $(1, 3)$ in each original equation.

$$-3x + 4y = 9 \qquad\qquad x = -\frac{1}{3}y + 2$$
$$-3(1) + 4(3) \stackrel{?}{=} 9 \qquad\qquad 1 \stackrel{?}{=} -\frac{1}{3}(3) + 2$$
$$-3 + 12 = 9 \checkmark \text{ True} \qquad\qquad 1 = -1 + 2 \checkmark \text{ True}$$

The solution is $(1, 3)$.

Skill Practice

2. Solve by the substitution method.
   $3x + y = 8$
   $x - 2y = 12$

example 2 — Using the Substitution Method to Solve a Linear System

Solve the system by using the substitution method.
$$3x - 2y = -7$$
$$6x + y = 6$$

Solution:

The y variable in the second equation is the easiest variable to isolate because its coefficient is 1.

Answers
1. $(-1, -2)$ 2. $(4, -4)$

$$3x - 2y = -7$$
$$6x + y = 6 \longrightarrow y = \underline{-6x + 6}$$ **Step 1:** Solve the second equation for y.

$$3x - 2(-6x + 6) = -7$$ **Step 2:** Substitute the quantity $-6x + 6$ for y in the *other* equation.
$$3x + 12x - 12 = -7$$
$$15x - 12 = -7$$ **Step 3:** Solve for x.
$$15x = 5$$
$$\frac{15x}{15} = \frac{5}{15}$$
$$x = \frac{1}{3}$$

Avoiding Mistakes:
Do not substitute $y = -6x + 6$ into the same equation from which it came. This mistake will result in an identity:
$$6x + y = 6$$
$$6x + (-6x + 6) = 6$$
$$6x - 6x + 6 = 6$$
$$6 = 6$$

$$y = -6x + 6$$
$$y = -6\left(\frac{1}{3}\right) + 6$$ **Step 4:** Substitute $x = \frac{1}{3}$ into the expression $y = -6x + 6$.
$$y = -2 + 6$$
$$y = 4$$

| | | **Step 5:** Check the ordered pair $\left(\frac{1}{3}, 4\right)$ in each original equation. |

$$3x - 2y = -7 \qquad 6x + y = 6$$
$$3\left(\frac{1}{3}\right) - 2(4) \stackrel{?}{=} -7 \qquad 6\left(\frac{1}{3}\right) + 4 \stackrel{?}{=} 6$$
$$1 - 8 = -7 \checkmark \qquad 2 + 4 = 6 \checkmark$$

The solution is $\left(\frac{1}{3}, 4\right)$.

2. Solving Inconsistent Systems and Dependent Systems

example 3 Using the Substitution Method to Solve a Linear System

Solve the system by using the substitution method.
$$x = 2y - 4$$
$$-2x + 4y = 6$$

Skill Practice

3. Solve by the substitution method.
$$8x - 16y = 3$$
$$y = \frac{1}{2}x + 1$$

Solution:

$$x = 2y - 4$$ **Step 1:** The x variable is already isolated.
$$-2x + 4y = 6$$

$$-2(2y - 4) + 4y = 6$$ **Step 2:** Substitute the quantity $x = 2y - 4$ into the *other* equation.

Answer
3. Inconsistent, no solution

$$-4y + 8 + 4y = 6$$

Step 3: Solve for y.

$$8 = 6$$

There is no solution.
The system is inconsistent.

The equation reduces to a contradiction, indicating that the system has no solution. Hence the lines never intersect and must be parallel. The system is *inconsistent*.

> **Tip:** The answer to Example 3 can be verified by writing each equation in slope-intercept form and graphing the equations.
>
> **Equation 1**   **Equation 2**
>
> $x = 2y - 4$ $-2x + 4y = 6$
>
> $2y = x + 4$ $4y = 2x + 6$
>
> $\dfrac{2y}{2} = \dfrac{x}{2} + \dfrac{4}{2}$ $\dfrac{4y}{4} = \dfrac{2x}{4} + \dfrac{6}{4}$
>
> $y = \dfrac{1}{2}x + 2$ $y = \dfrac{1}{2}x + \dfrac{3}{2}$
>
>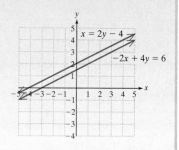
>
> Notice that the equations have the same slope, but different y-intercepts; therefore, the lines must be parallel. There is no solution to this system of equations.

Skill Practice

4. Solve the system by using substitution.

   $3x + 6y = 12$

   $2y = -x + 4$

example 4   Solving a Dependent System

Solve by using the substitution method.

$$4x - 2y = -6$$
$$y - 3 = 2x$$

Solution:

$$4x - 2y = -6$$

$$y - 3 = 2x \longrightarrow y = \underbrace{2x + 3}$$ **Step 1:** Solve for one of the variables.

$$4x - 2(2x + 3) = -6$$ **Step 2:** Substitute the quantity $2x + 3$ for y in the *other* equation.

$$4x - 4x - 6 = -6$$ **Step 3:** Solve for x. Apply the distributive property to clear the parentheses.

$$-6 = -6$$

The system reduces to the identity $-6 = -6$. Therefore, the original two equations are equivalent, and the system is dependent. The solution consists of all points on the common line. Because the equations $4x - 2y = -6$ and $y - 3 = 2x$ represent the same line, the solution may be written as

$$\{(x, y) \mid 4x - 2y = -6\} \quad \text{or} \quad \{(x, y) \mid y - 3 = 2x\}$$

Answer

4. Dependent system:
   $\{(x, y) \mid 3x + 6y = 12\}$

Tip: We can confirm the results of Example 4 by writing each equation in slope-intercept form. The slope-intercept forms are identical, indicating that the lines are the same.

$$4x - 2y = -6 \longrightarrow -2y = -4x - 6 \longrightarrow y = 2x + 3$$
$$y - 3 = 2x \longrightarrow y = 2x + 3$$

(slope-intercept form)

section 3.2 Practice Exercises

Boost your GRADE at mathzone.com!

- Practice Problems
- Self-Tests
- NetTutor
- e-Professors
- Videos

Study Skills Exercise

1. Make up a practice test for yourself. Use examples or exercises from the text. Be sure to cover each concept that was presented.

Review Exercises

For Exercises 2–5, using the slope-intercept of the lines, **a.** determine whether the system is consistent or inconsistent and **b.** determine whether the system is dependent or independent.

2. $y = 8x - 1$
   $2x - 16y = 3$

3. $4x + 6y = 1$
   $10x + 15y = \dfrac{5}{2}$

4. $2x - 4y = 0$
   $x - 2y = 9$

5. $6x + 3y = 8$
   $8x + 4y = -1$

For Exercises 6–7, solve the system by graphing.

6. $2x - 3y = 8$
   $3x + 4y = 12$

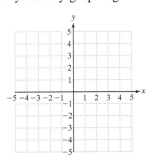

7. $y = 2x + 3$
   $6x + 3y = 9$

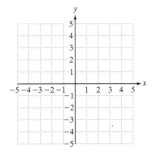

Objective 1: The Substitution Method

8. Describe the process of solving a system of linear equations by using substitution.

For Exercises 9–18, solve by using the substitution method. **(See Examples 1–2.)**

9. $4x + 12y = 4$
   $y = 5x + 11$

10. $y = -3x - 1$
    $2x - 3y = -8$

11. $x = 10y + 34$
    $-7x + y = -31$

12. $-3x + 8y = -1$
    $4x - y = 11$

13. $12x - 2y = 0$
$-7x + y = -1$

14. $3x + 12y = 24$
$x - 5y = 17$

15. $x - 3y = -4$
$2x + 3y = -5$

16. $x - y = 8$
$3x + 2y = 9$

17. $5x - 2y = 10$
$y = x - 1$

18. $2x - y = -1$
$y = -2x$

Objective 2: Solving Inconsistent Systems and Dependent Systems

For Exercises 19–26, solve the systems. (See Examples 3–4.)

19. $2x - 6y = -2$
$x = 3y - 1$

20. $-2x + 4y = 22$
$x = 2y - 11$

21. $y = \frac{1}{7}x + 3$
$x - 7y = -4$

22. $x = -\frac{3}{2}y + \frac{1}{2}$
$4x + 6y = 7$

23. $5x - y = 10$
$2y = 10x - 5$

24. $x + 4y = 8$
$3x = 3 - 12y$

25. $3x - y = 7$
$-14 + 6x = 2y$

26. $x = 4y + 1$
$-12y = -3x + 3$

27. When using the substitution method, explain how to determine whether a system of linear equations is dependent.

28. When using the substitution method, explain how to determine whether a system of linear equations is inconsistent.

Mixed Exercises

For Exercises 29–46, solve the system by using the substitution method.

29. $x = 1.3y + 1.5$
$y = 1.2x - 4.6$

30. $y = 0.8x - 1.8$
$1.1x = -y + 9.6$

31. $y = \frac{2}{3}x - \frac{1}{3}$
$x = \frac{1}{4}y + \frac{17}{4}$

32. $x = \frac{1}{6}y - \frac{5}{3}$
$y = \frac{1}{5}x + \frac{21}{5}$

33. $-2x + y = 4$
$-\frac{1}{4}x + \frac{1}{8}y = \frac{1}{4}$

34. $8x - y = 8$
$\frac{1}{3}x - \frac{1}{24}y = \frac{1}{2}$

35. $3x = 6$
$9y + 4x = -19$

36. $-8x + 11y = -25$
$3x = 30$

37. $-300x - 125y = 1350$
$y + 2 = 8$

38. $200y = 150x$
$y - 4 = 1$

39. $2x - y = 6$
$\frac{1}{6}x - \frac{1}{12}y = \frac{1}{2}$

40. $x - 4y = 8$
$\frac{1}{16}x - \frac{1}{4}y = \frac{1}{2}$

41. $y = 200x - 320$
$y = -150x + 1080$

42. $y = -54x + 300$
$y = 20x - 70$

43. $y = -2.7x - 5.1$
$y = 3.1x - 63.1$

44. $y = 6.8x + 2.3$
$y = -4.1x + 56.8$

45. $4x + 4y = 5$
$x - 4y = -\frac{5}{2}$

46. $-2x + y = -6$
$6x - 13y = -12$

Section 3.3 Solving Systems of Equations by Using the Addition Method

Objectives
1. The Addition Method
2. Solving Inconsistent Systems and Dependent Systems

1. The Addition Method

The next method we present to solve systems of linear equations is the *addition method* (sometimes called the elimination method). With the addition method, begin by writing both equations in standard form $Ax + By = C$. Then multiply one or both equations by appropriate constants to create opposite coefficients on either the *x*- or the *y*-variable. Next the equations may be added to eliminate the variable having opposite coefficients. This process is demonstrated in Example 1.

example 1 — Solving a System by the Addition Method

Solve the system by using the addition method.

$$3x - 4y = 2$$
$$4x + y = 9$$

Skill Practice
1. Solve by the addition method.
$$2x - 3y = 13$$
$$x + 2y = 3$$

Solution:

$3x - 4y = 2 \quad\quad 3x - 4y = 2$

$4x + y = 9 \xrightarrow{\text{Multiply by 4.}} 16x + 4y = 36$ — Multiply the second equation by 4. This makes the coefficients of the *y* variables *opposite*.

$$\begin{aligned} 3x - 4y &= 2 \\ 16x + 4y &= 36 \\ \hline 19x \phantom{{}+4y} &= 38 \end{aligned}$$

Now if the equations are added, the *y* variable will be eliminated.

$x = 2$ Solve for *x*.

Substitute $x = 2$ back into one of the original equations and solve for *y*.

$3x - 4y = 2$
$3(2) - 4y = 2$
$6 - 4y = 2$
$-4y = -4$
$y = 1$

Check the ordered pair (2, 1) in each original equation:

$3x - 4y = 2 \longrightarrow 3(2) - 4(1) \stackrel{?}{=} 2 \longrightarrow 6 - 4 = 2$ ✔

$4x + y = 9 \longrightarrow 4(2) + (1) \stackrel{?}{=} 9 \longrightarrow 8 + 1 = 9$ ✔

The solution is (2, 1).

Tip: Substituting $x = 2$ into the other equation, $4x + y = 9$, produces the same value for *y*.

$$4x + y = 9$$
$$4(2) + y = 9$$
$$8 + y = 9$$
$$y = 1$$

The steps to solve a system of linear equations in two variables by the addition method is outlined in the following box.

Answer
1. $(5, -1)$

> **Solving a System of Equations by the Addition Method**
> 1. Write both equations in standard form: $Ax + By = C$
> 2. Clear fractions or decimals (optional).
> 3. Multiply one or both equations by nonzero constants to create opposite coefficients for one of the variables.
> 4. Add the equations from step 3 to eliminate one variable.
> 5. Solve for the remaining variable.
> 6. Substitute the known value found in step 5 into one of the original equations to solve for the other variable.
> 7. Check the ordered pair in *both* equations.

Skill Practice

2. Solve by the addition method.
$2y = 5x - 4$
$3x - 4y = 1$

example 2   Solving a System by the Addition Method

Solve the system by using the addition method.

$$4x + 5y = 2$$
$$3x = 1 - 4y$$

Solution:

$4x + 5y = 2 \longrightarrow 4x + 5y = 2$
$3x = 1 - 4y \longrightarrow 3x + 4y = 1$

Step 1: Write both equations in standard form. There are no fractions or decimals.

We may choose to eliminate either variable. To eliminate x, change the coefficients to 12 and -12.

$4x + 5y = 2 \xrightarrow{\text{Multiply by 3.}} 12x + 15y = 6$
$3x + 4y = 1 \xrightarrow{\text{Multiply by } -4.} -12x - 16y = -4$

Step 3: Multiply the first equation by 3. Multiply the second equation by -4.

$12x + 15y = 6$
$-12x - 16y = -4$

Step 4: Add the equations.

$-y = 2$
$y = -2$

Step 5: Solve for y.

$4x + 5y = 2$
$4x + 5(-2) = 2$
$4x - 10 = 2$
$4x = 12$
$x = 3$

Step 6: Substitute $y = -2$ back into one of the original equations and solve for x.

The solution is $(3, -2)$.

Step 7: Check the ordered pair $(3, -2)$ in both original equations.

Answer

2. $\left(1, \dfrac{1}{2}\right)$

Section 3.3 Solving Systems of Equations by Using the Addition Method

Tip: To eliminate the x variable in Example 2, both equations were multiplied by appropriate constants to create $12x$ and $-12x$. We chose 12 because it is the *least common multiple* of 4 and 3.

We could have solved the system by eliminating the y-variable. To eliminate y, we would multiply the top equation by 4 and the bottom equation by -5. This would make the coefficients of the y-variable 20 and -20, respectively.

$4x + 5y = 2 \xrightarrow{\text{Multiply by 4.}} 16x + 20y = 8$

$3x + 4y = 1 \xrightarrow{\text{Multiply by }-5.} -15x - 20y = -5$

example 3   Solving a System of Equations by the Addition Method

Solve the system by using the addition method.

$$x - 2y = 6 + y$$
$$0.05y = 0.02x - 0.10$$

Skill Practice

3. Solve by the addition method.
$$0.2x + 0.3y = 1.5$$
$$5x + 3y = 20 - y$$

Solution:

$x - 2y = 6 + y \longrightarrow x - 3y = 6$
$0.05y = 0.02x - 0.10 \longrightarrow -0.02x + 0.05y = -0.10$

Step 1: Write both equations in standard form.

$x - 3y = 6$
$-0.02x + 0.05y = -0.10 \xrightarrow{\text{Multiply by 100.}} -2x + 5y = -10$

Step 2: Clear decimals.

$x - 3y = 6 \xrightarrow{\text{Multiply by 2.}} 2x - 6y = 12$
$-2x + 5y = -10 \longrightarrow \underline{-2x + 5y = -10}$

Step 3: Create opposite coefficients.

$-y = 2$

Step 4: Add the equations.

$y = -2$

Step 5: Solve for y.

$x - 2y = 6 + y$
$x - 2(-2) = 6 + (-2)$
$x + 4 = 4$
$x = 0$

Step 6: To solve for x, substitute $y = -2$ into one of the original equations.

Step 7: Check the ordered pair $(0, -2)$ in each original equation.

$x - 2y = 6 + y \longrightarrow (0) - 2(-2) \stackrel{?}{=} 6 + (-2) \longrightarrow 4 = 4$ ✔

$0.05y = 0.02x - 0.10 \longrightarrow 0.05(-2) \stackrel{?}{=} 0.02(0) - 0.10 \longrightarrow -0.10 = -0.10$ ✔

The solution is $(0, -2)$.

Answer

3. $(0, 5)$

2. Solving Inconsistent Systems and Dependent Systems

Skill Practice

4. Solve by the addition method.
$$3x + y = 4$$
$$x = -\frac{1}{3}y + \frac{4}{3}$$

example 4 — Solving a System of Equations by the Addition Method

Solve the system by using the addition method.

$$\frac{1}{5}x - \frac{1}{2}y = 1$$
$$-4x + 10y = -20$$

Solution:

$$\frac{1}{5}x - \frac{1}{2}y = 1$$
$$-4x + 10y = -20$$

Step 1: Equations are in standard form.

$$10\left(\frac{1}{5}x - \frac{1}{2}y\right) = 10 \cdot 1 \longrightarrow 2x - 5y = 10$$
$$-4x + 10y = -20$$

Step 2: Clear fractions.

$$2x - 5y = 10 \xrightarrow{\text{Multiply by 2.}} 4x - 10y = 20$$
$$-4x + 10y = -20 \longrightarrow \underline{-4x + 10y = -20}$$
$$0 = 0$$

Step 3: Multiply the first equation by 2.

Step 4: Add the equations.

Notice that both variables were eliminated. The system of equations is reduced to the identity $0 = 0$. Therefore, the two original equations are equivalent and the system is dependent. The solution set consists of an infinite number of ordered pairs (x, y) that fall on the common line of intersection $-4x + 10y = -20$, or equivalently $\frac{1}{5}x - \frac{1}{2}y = 1$. The solution set can be written in set notation as

$$\{(x, y) \mid -4x + 10y = -20\} \quad \text{or} \quad \left\{(x, y) \;\middle|\; \frac{1}{5}x - \frac{1}{2}y = 1\right\}$$

Skill Practice

5. Solve by the addition method.
$$18 + 10x = 6y$$
$$5x - 3y = 9$$

example 5 — Solving an Inconsistent System

Solve the system by using the addition method.

$$2y = -3x + 4$$
$$120x + 80y = 40$$

Solution:

$$2y = -3x + 4 \xrightarrow{\text{standard form}} 3x + 2y = 4$$
$$120x + 80y = 40 \longrightarrow 120x + 80y = 40$$

Step 1: Write the equations in standard form.

Step 2: There are no decimals or fractions.

Answers

4. Dependent system:
$\{(x, y) \mid 3x + y = 4\}$
5. Inconsistent system, no solution

$$3x + 2y = 4 \xrightarrow{\text{Multiply by } -40.} -120x - 80y = -160$$
$$120x + 80y = 40 \qquad\qquad\quad 120x + 80y = 40$$
$$\qquad\qquad\qquad\qquad\qquad\qquad\qquad\quad 0 = -120$$

Step 3: Multiply the top equation by -40.

Step 4: Add the equations.

The equations reduce to a contradiction, indicating that the system has no solution. The system is inconsistent. The two equations represent parallel lines, as shown in Figure 3-8.

There is no solution.

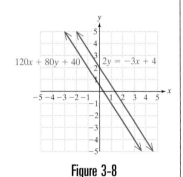

Figure 3-8

section 3.3   Practice Exercises

Boost your GRADE at mathzone.com!

- Practice Problems
- Self-Tests
- NetTutor
- e-Professors
- Videos

Study Skills Exercise

1. When taking a test, go through the test and work all the problems that you know first. Then go back and work on the problems that are more difficult. Give yourself a limit for how much time you spend on each problem (maybe 3 to 5 minutes the first time through). Circle the importance of each statement.

	Not important	Somewhat important	Very important
a. Read through the entire test first.	1	2	3
b. If time allows, go back and check each problem.	1	2	3
c. Write out all steps instead of doing the work in your head.	1	2	3

Review Exercises

For Exercises 2–4, use the slope-intercept form of the lines to determine the number of solutions for the system of equations.

2. $y = \dfrac{1}{2}x - 4$
   $y = \dfrac{1}{2}x + 1$

3. $y = 2.32x - 8.1$
   $y = 1.46x - 8.1$

4. $4x = y + 7$
   $-2y = -8x + 14$

Objective 1: The Addition Method

For Exercises 5–14, solve the system by the addition method. (See Examples 1–3.)

5. $3x - y = -1$
    $-3x + 4y = -14$

6. $5x - 2y = 15$
    $3x + 2y = -7$

7. $2x + 3y = 3$
    $-10x + 2y = -32$

8. $2x - 5y = 7$
    $3x - 10y = 13$

9. $3x + 7y = -20$
    $-5x + 3y = -84$

10. $6x - 9y = -15$
    $5x - 2y = -40$

11. $3x = 10y + 13$
    $7y = 4x - 11$

12. $-5x = 6y - 4$
    $5y = 1 - 3x$

13. $1.2x - 0.6y = 3$
    $0.8x - 1.4y = 3$

14. $1.8x + 0.8y = 1.4$
    $1.2x + 0.6y = 1.2$

Objective 2: Solving Inconsistent Systems and Dependent Systems

For Exercises 15–22, solve the systems. (See Examples 4–5.)

15. $3x - 2y = 1$
    $-6x + 4y = -2$

16. $3x - y = 4$
    $6x - 2y = 8$

17. $6y = 14 - 4x$
    $0.2x = -0.3y - 0.7$

18. $2x = 4 - y$
    $-0.1y = 0.2x - 0.2$

19. $12x - 4y = 2$
    $0.6x = 0.1 + 0.2y$

20. $10x - 15y = 5$
    $0.3y = 0.2x - 0.1$

21. $\frac{1}{2}x + y = \frac{7}{6}$
    $x + 2y = 4.5$

22. $0.2x - 0.1y = -1.2$
    $x - \frac{1}{2}y = 3$

Mixed Exercises

23. Describe a situation in which you would prefer to use the substitution method over the addition method.

24. If you used the addition method to solve the given system, would it be easier to eliminate the x or y variable? Explain.

    $3x - 5y = 4$
    $7x + 10y = 31$

For Exercises 25–44, solve by using either the addition method or the substitution method.

25. $2x - 4y = 8$
    $y = 2x + 1$

26. $8x + 6y = -8$
    $x = 6y - 10$

27. $2x + 5y = 9$
    $4x - 7y = -16$

28. $x + 5y = 7$
    $2x + 7y = 8$

29. $2x - y = 9$
    $y = -\frac{4}{3}x + 1$

30. $y = \frac{1}{2}x - 3$
    $x + 4y = 0$

31. $0.4x - 0.6y = 0.5$
    $0.2x - 0.3y = 0.7$

32. $0.3x + 0.6y = 0.7$
    $0.2x + 0.4y = 0.5$

33. $\frac{1}{4}x - \frac{1}{6}y = -2$
    $-\frac{1}{6}x + \frac{1}{5}y = 4$

34. $\frac{1}{3}x + \frac{1}{5}y = 7$
    $\frac{1}{6}x - \frac{2}{5}y = -4$

35. $\frac{1}{3}x - \frac{1}{2}y = 0$
    $x = \frac{3}{2}y$

36. $\frac{2}{5}x - \frac{2}{3}y = 0$
    $y = \frac{3}{5}x$

37. $2(x + 2y) = 20 - y$
    $-7(x - y) = 16 + 3y$

38. $-3(x + y) = -10 - 4y$
    $4(x + 2y) = 140 + 3y$

39. $-4y = 10$
    $4x + 3 = 1$

40. $-9x = 15$
    $3y + 2 = 1$

41. $5x - 3y = 18$
    $-3x + 5y = 18$

42. $6x - 3y = -3$
    $4x + 5y = -9$

43. $3x - 2 = \frac{1}{3}(11 + 5y)$
    $x + \frac{2}{3}(2y - 3) = -2$

44. $2(2y + 3) - 2x = 1 - x$
    $x + y = \frac{1}{5}(7 + y)$

Expanding Your Skills

For Exercises 45–48, use the addition method first to solve for x. Then repeat the addition method again, using the original system of equations, this time solving for y.

45. $6x - 2y = 5$
    $5x + 3y = -2$

46. $5x - 4y = 7$
    $-2x + 6y = 5$

47. $4x + 7y = 8$
    $-5x - 2y = 3$

48. $2x - 4y = 9$
    $-5x + 3y = 1$

chapter 3 | midchapter review

For Exercises 1–4, solve each system by using three different methods.

a. Use the graphing method.

b. Use the substitution method.

c. Use the addition method.

1. $4x + y = -2$
   $5x - y = -7$

2. $4x - 2y = 6$
   $x = \frac{1}{2}y + \frac{3}{2}$

3. $4x = 3y$
   $y = \frac{4}{3}x + 2$

4. $4y = 8x + 20$
   $8x = 24$

Objectives

1. Applications Involving Cost
2. Applications Involving Mixtures
3. Applications Involving Principal and Interest
4. Applications Involving Distance, Rate, and Time
5. Applications Involving Geometry

section 3.4 Applications of Systems of Linear Equations in Two Variables

1. Applications Involving Cost

In Chapter 1 we solved numerous application problems using equations that contained one variable. However, when an application has more than one unknown, sometimes it is more convenient to use multiple variables. In this section, we will solve applications containing two unknowns. When two variables are present, the goal is to set up a system of two independent equations.

Skill Practice

1. At the movie theater, Tom spent $7.75 on 3 soft drinks and 2 boxes of popcorn. Carly bought 5 soft drinks and 1 box of popcorn for total of $8.25. Use a system of equations to find the cost of a soft drink and the cost of a box of popcorn.

example 1 Solving a Cost Application

At an amusement park, five hot dogs and one drink cost $16. Two hot dogs and three drinks cost $9. Find the cost per hot dog and the cost per drink.

Solution:

Let h represent the cost per hot dog.   Label the variables.

Let d represent the cost per drink.

$\begin{pmatrix} \text{Cost of 5} \\ \text{hot dogs} \end{pmatrix} + \begin{pmatrix} \text{cost of 1} \\ \text{drink} \end{pmatrix} = \$16 \longrightarrow 5h + d = 16$ Write two equations.

$\begin{pmatrix} \text{Cost of 2} \\ \text{hot dogs} \end{pmatrix} + \begin{pmatrix} \text{cost of 3} \\ \text{drinks} \end{pmatrix} = \$9 \longrightarrow 2h + 3d = 9$

This system can be solved by either the substitution method or the addition method. We will solve by using the substitution method. The d-variable in the first equation is the easiest variable to isolate.

Answer

1. Soft drink: $1.25; popcorn: $2.00

$$5h + d = 16 \longrightarrow d = -5h + 16$$
$$2h + 3d = 9$$

Solve for d in the first equation.

Substitute the quantity $-5h + 16$ for d in the *second* equation.

$$2h + 3(-5h + 16) = 9$$
$$2h - 15h + 48 = 9$$
$$-13h + 48 = 9$$
$$-13h = -39$$
$$h = 3$$

Clear parentheses.

Solve for h.

$$d = -5(3) + 16 \longrightarrow d = 1$$

Substitute $h = 3$ in the equation $d = -5h + 16$.

Because $h = 3$, the cost per hot dog is $3.00.

Because $d = 1$, the cost per drink is $1.00.

A word problem can be checked by verifying that the solution meets the conditions specified in the problem.

5 hot dogs + 1 drink = 5($3.00) + 1($1.00) = $16.00 as expected

2 hot dogs + 3 drinks = 2($3.00) + 3($1.00) = $9.00 as expected

2. Applications Involving Mixtures

example 2   Solving an Application Involving Chemistry

One brand of cleaner used to etch concrete is 25% acid. A stronger industrial-strength cleaner is 50% acid. How many gallons of each cleaner should be mixed to produce 20 gal of a 40% acid solution?

Solution:

Let x represent the amount of 25% acid cleaner.

Let y represent the amount of 50% acid cleaner.

	25% Acid	50% Acid	40% Acid
Number of gallons of solution	x	y	20
Number of gallons of pure acid	$0.25x$	$0.50y$	0.40(20), or 8

From the first row of the table, we have

$$\begin{pmatrix} \text{Amount of} \\ 25\% \text{ solution} \end{pmatrix} + \begin{pmatrix} \text{amount of} \\ 50\% \text{ solution} \end{pmatrix} = \begin{pmatrix} \text{total amount} \\ \text{of solution} \end{pmatrix} \longrightarrow x + y = 20$$

From the second row of the table we have

$$\begin{pmatrix} \text{Amount of} \\ \text{pure acid in} \\ 25\% \text{ solution} \end{pmatrix} + \begin{pmatrix} \text{amount of} \\ \text{pure acid in} \\ 50\% \text{ solution} \end{pmatrix} = \begin{pmatrix} \text{amount of} \\ \text{pure acid in} \\ \text{resulting solution} \end{pmatrix} \longrightarrow 0.25x + 0.50y = 8$$

Skill Practice

2. A pharmacist needs 8 ounces (oz) of a solution that is 50% saline. How many ounces of 60% saline solution and 20% saline solution must be mixed to obtain the mixture needed?

Answer

2. 6 oz of 60% solution and 2 oz of 20% solution

$$x + y = 20 \longrightarrow x + y = 20$$
$$0.25x + 0.50y = 8 \longrightarrow 25x + 50y = 800 \quad \text{Multiply by 100 to clear decimals.}$$

$$x + y = 20 \xrightarrow{\text{Multiply by } -25.} -25x - 25y = -500 \quad \text{Create opposite coefficients of } x.$$
$$25x + 50y = 800 \longrightarrow \underline{25x + 50y = 800}$$
$$25y = 300 \quad \text{Add the equations to eliminate } x.$$
$$y = 12$$

$$x + y = 20 \quad \text{Substitute } y = 12 \text{ back into one of the original equations.}$$
$$x + 12 = 20$$
$$x = 8$$

Therefore, 8 gal of 25% acid solution must be added to 12 gal of 50% acid solution to create 20 gal of a 40% acid solution.

3. Applications Involving Principal and Interest

example 3 Solving a Mixture Application Involving Finance

Serena invested money in two accounts: a savings account that yields 4.5% simple interest and a certificate of deposit that yields 7% simple interest. The amount invested at 7% was twice the amount invested at 4.5%. How much did Serena invest in each account if the total interest at the end of 1 year was $1017.50?

Solution:

Let x represent the amount invested in the savings account (the 4.5% account).

Let y represent the amount invested in the certificate of deposit (the 7% account).

	4.5% Account	7% Account	Total
Principal	x	y	
Interest	$0.045x$	$0.07y$	1017.50

Because the amount invested at 7% was twice the amount invested at 4.5%, we have

$$\begin{pmatrix} \text{Amount} \\ \text{invested} \\ \text{at 7\%} \end{pmatrix} = 2 \begin{pmatrix} \text{amount} \\ \text{invested} \\ \text{at 4.5\%} \end{pmatrix} \rightarrow y = 2x$$

From the second row of the table, we have

$$\begin{pmatrix} \text{Interest} \\ \text{earned from} \\ \text{4.5\% account} \end{pmatrix} + \begin{pmatrix} \text{interest} \\ \text{earned from} \\ \text{7\% account} \end{pmatrix} = \begin{pmatrix} \text{total} \\ \text{interest} \end{pmatrix} \rightarrow 0.045x + 0.07y = 1017.50$$

Skill Practice

3. Seth invested money in two accounts, one paying 5% interest and the other paying 6% interest. The amount invested at 5% was $1000 more than the amount invested at 6%. He earned a total of $820 interest in 1 year. Use a system of equations to find the amount invested in each account.

Answer

3. $8000 invested at 5% and $7000 invested at 6%

$$y = 2x$$
$$45x + 70y = 1{,}017{,}500 \quad \text{Multiply by 1000 to clear decimals.}$$

Because the *y*-variable in the first equation is isolated, we will use the substitution method.

$$45x + 70(2x) = 1{,}017{,}500 \quad \text{Substitute the quantity } 2x \text{ into the second equation.}$$

$$45x + 140x = 1{,}017{,}500 \quad \text{Solve for } x.$$
$$185x = 1{,}017{,}500$$
$$x = \frac{1{,}017{,}500}{185}$$
$$x = 5500$$

$$y = 2x$$
$$y = 2(5500) \quad \text{Substitute } x = 5500 \text{ into the equation } y = 2x \text{ to solve for } y.$$
$$y = 11{,}000$$

Because $x = 5500$, the amount invested in the savings account is $5500.

Because $y = 11{,}000$, the amount invested in the certificate of deposit is $11,000.

Check: $11,000 is twice $5500. Furthermore,

$$\begin{pmatrix} \text{Interest} \\ \text{earned from} \\ 4.5\% \text{ account} \end{pmatrix} + \begin{pmatrix} \text{interest} \\ \text{earned from} \\ 7\% \text{ account} \end{pmatrix} = \$5500(0.045) + \$11{,}000(0.07) = 1017.50 \checkmark$$

4. Applications Involving Distance, Rate, and Time

example 4   Solving a Distance, Rate, and Time Application

A plane flies 660 mi from Atlanta to Miami in 1.2 hr when traveling with a tailwind. The return flight against the same wind takes 1.5 hr. Find the speed of the plane in still air and the speed of the wind.

Solution:

Let p represent the speed of the plane in still air.

Let w represent the speed of the wind.

The speed of the plane *with* the wind: (Plane's still airspeed) + (wind speed): $p + w$

The speed of the plane *against* the wind: (Plane's still airspeed) − (wind speed): $p - w$

Skill Practice

4. A plane flies 1200 mi from Orlando to New York in 2 hr with a tailwind. The return flight against the same wind takes 2.5 hr. Find the speed of the plane in still air and the speed of the wind.

Answer

4. Speed of plane: 540 mph; speed of wind: 60 mph

Set up a chart to organize the given information:

	Distance	Rate	Time
With a tailwind	660	$p + w$	1.2
Against a head wind	660	$p - w$	1.5

Two equations can be found by using the relationship $d = rt$ (distance = rate · time).

$$\begin{pmatrix} \text{Distance} \\ \text{with} \\ \text{wind} \end{pmatrix} = \begin{pmatrix} \text{speed} \\ \text{with} \\ \text{wind} \end{pmatrix} \begin{pmatrix} \text{time} \\ \text{with} \\ \text{wind} \end{pmatrix} \longrightarrow 660 = (p + w)(1.2)$$

$$\begin{pmatrix} \text{Distance} \\ \text{against} \\ \text{wind} \end{pmatrix} = \begin{pmatrix} \text{speed} \\ \text{against} \\ \text{wind} \end{pmatrix} \begin{pmatrix} \text{time} \\ \text{against} \\ \text{wind} \end{pmatrix} \longrightarrow 660 = (p - w)(1.5)$$

$660 = (p + w)(1.2)$ Notice that the first equation may be *divided* by 1.2 and still leave integer coefficients. Similarly, the second equation may be simplified by dividing by 1.5.

$660 = (p - w)(1.5)$

$660 = (p + w)(1.2) \xrightarrow{\text{Divide by 1.2}} \dfrac{660}{1.2} = \dfrac{(p + w)\cancel{1.2}}{\cancel{1.2}} \longrightarrow 550 = p + w$

$660 = (p - w)(1.5) \xrightarrow{\text{Divide by 1.5}} \dfrac{660}{1.5} = \dfrac{(p - w)\cancel{1.5}}{\cancel{1.5}} \longrightarrow 440 = p - w$

$550 = p + w$
$440 = p - w$
$990 = 2p$ Add the equations.
$p = 495$

$550 = (495) + w$ Substitute $p = 495$ into the equation $550 = p + w$.
$55 = w$ Solve for w.

The speed of the plane in still air is 495 mph, and the speed of the wind is 55 mph.

5. Applications Involving Geometry

example 5   Solving a Geometry Application

The sum of the two acute angles in a right triangle is 90°. The measure of one angle is 6° less than 2 times the measure of the other angle. Find the measure of each angle.

Solution:

Let x represent one acute angle.

Let y represent the other acute angle.

Skill Practice

5. Two angles are supplementary. The measure of one angle is 16° less than 3 times the measure of the other. Use a system of equations to find the measures of the angles.

Answer

5. 49° and 131°

The sum of the two acute angles is 90°: $\quad x + y = 90$

One angle is 6° less than 2 times the other angle: $\quad x = 2y - 6$

$\quad x + y = 90$
$\quad x = 2y - 6$
Because one variable is already isolated, we will use the substitution method.

$(2y - 6) + y = 90 \quad$ Substitute $x = 2y - 6$ into the first equation.
$3y - 6 = 90$
$3y = 96$
$y = 32$

$x = 2y - 6 \quad$ To find x, substitute $y = 32$ into the equation $x = 2y - 6$.
$x = 2(32) - 6$
$x = 64 - 6$
$x = 58$

The two acute angles in the triangle measure 32° and 58°.

section 3.4   Practice Exercises

Boost your GRADE at mathzone.com!

- Practice Problems
- Self-Tests
- NetTutor
- e-Professors
- Videos

Study Skills Exercise

1. Do you believe that you have math anxiety? _____ If yes, why do you think so?

Review Exercises

2. State three methods that can be used to solve a system of linear equations in two variables.

For Exercises 3–6, state which method you would prefer to use to solve the system. Then solve the system.

3. $y = 9 - 2x$
$3x - y = 16$

4. $7x - y = -25$
$2x + 5y = 14$

5. $5x + 2y = 6$
$-2x - y = 3$

6. $x = 5y - 2$
$-3x + 7y = 14$

Objective 1: Applications Involving Cost

7. The local community college theater put on a production of *Chicago*. There were 186 tickets sold, some for $16 (nonstudent price) and others for $12 (student price). If the receipts for one performance totaled $2640, how many of each type of ticket were sold? **(See Example 1.)**

8. Jack and Diane bought school supplies. Jack spent $10.65 on 4 notebooks and 5 pens. Diane spent $7.50 on 3 notebooks and 3 pens. What is the cost of 1 notebook and what is the cost of 1 pen?

9. Jacob bought lunch for his fellow office workers on Monday. He spent $7.35 on 3 hamburgers and 2 fish sandwiches. Ralph bought lunch on Tuesday and spent $7.15 for 4 hamburgers and 1 fish sandwich. What is the price of 1 hamburger, and what is the price of 1 fish sandwich?

10. A group of four golfers pays $150 to play a round of golf. Of these four, one is a member of the club and three are nonmembers. Another group of golfers consists of two members and one nonmember and pays a total of $75. What is the cost for a member to play a round of golf, and what is the cost for a nonmember?

11. Sam has a pocket full of change consisting of dimes and quarters. The total value is $3.15. There are 7 more quarters than dimes. How many of each coin are there?

12. Crystal has several dimes and quarters in her purse, totaling $2.70. There is 1 less dime than there are quarters. How many of each coin are there?

13. A coin collection consists of 50¢ pieces and $1 coins. If there are a total of 21 coins worth $15.50, how many 50¢ pieces and $1 coins are there?

14. Suzy has a piggy bank consisting of nickels and dimes. If there are 30 coins worth $1.90, how many nickels and dimes are in the bank?

Objective 2: Applications Involving Mixtures

15. A jar of one face cream contains 18% moisturizer, and another type contains 24% moisturizer. How many ounces of each should be combined to get 12 oz of a cream that is 22% moisturizer? **(See Example 2.)**

16. A chemistry student wants to mix an 18% acid solution with a 45% acid solution to get 16 L of a 36% acid solution. How many liters of the 18% solution and how many liters of the 45% solution should be mixed?

17. How much pure bleach must be combined with a solution that is 4% bleach to make 12 oz of a 12% bleach solution?

18. A fruit punch that contains 25% fruit juice is combined with a fruit drink that contains 10% fruit juice. How many ounces of each should be used to make 48 oz of a mixture that is 15% fruit juice?

Objective 3: Applications Involving Principal and Interest

19. Alex invested $27,000 in two accounts: one that pays 2% simple interest and one that pays 3% simple interest. At the end of the first year, his total return was $685. How much was invested in each account? **(See Example 3.)**

20. Sonia invested a total of $12,000 into two accounts paying 7.5% and 6% simple interest. If her total return at the end of the first year was $840, how much did she invest in each account?

21. A credit union offers 5.5% simple interest on a certificate of deposit (CD) and 3.5% simple interest on a savings account. If Mr. Roderick invested $200 more in the CD than in the savings account and the total interest after the first year was $245, how much was invested in each account?

22. Ms. Kioki divided $20,000 into two accounts paying 4% and 3% simple interest. At the end of the first year, the total interest from both accounts was $675. Find the amount invested in each account.

Objective 4: Applications Involving Distance, Rate, and Time

23. It takes a boat 2 hr to go 16 mi downstream with the current and 4 hr to return against the current. Find the speed of the boat in still water and the speed of the current. (See Example 4.)

24. The Gulf Stream is a warm ocean current that extends from the eastern side of the Gulf of Mexico up through the Florida Straits and along the southeastern coast of the United States to Cape Hatteras, North Carolina. A boat travels with the current 100 mi from Miami, Florida, to Freeport, Bahamas, in 2.5 hr. The return trip against the same current takes $3\frac{1}{3}$ hr. Find the speed of the boat in still water and the speed of the current.

25. A plane flew 720 mi in 3 hr with the wind. It would take 4 hr to travel the same distance against the wind. What is the rate of the plane in still air and the rate of wind?

26. Jeannie and Juan rollerblade in opposite directions. Juan averages 2 mph faster than Jeannie. If they began at the same place and ended up 20 mi apart after 2 hr, how fast did each of them travel?

Objective 5: Applications Involving Geometry

For Exercises 27–32, solve the applications involving geometry. If necessary, refer to the geometry formulas listed in the inside front cover of the text. (See Example 5.)

27. In a right triangle, one acute angle measures 6° more than 3 times the other. If the sum of the measures of the two acute angles must equal 90°, find the measures of the acute angles.

28. An isosceles triangle has two angles of the same measure (see figure). If the angle represented by y measures 3° less than the angle x, find the measures of all angles of the triangle. (Recall that the sum of the measures of the angles of a triangle is 180°.)

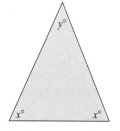

29. Two angles are supplementary. One angle measures 2° less than 3 times the other. What are the measures of the two angles?

30. The measure of one angle is 5 times the measure of another. If the two angles are supplementary, find the measures of the angles.

31. One angle measures 3° more than twice another. If the two angles are complementary, find the measures of the angles.

32. Two angles are complementary. One angle measures 15° more than 2 times the measure of the other. What are the measures of the two angles?

Expanding Your Skills

For Exercises 33–36, solve the business applications.

33. The *demand* for a certain printer cartridge is related to the price. In general, the higher the price x, the lower the demand y. The *supply* for the printer cartridges is also related to price. The higher the price, the greater the incentive for the supplier to stock the item. The supply and demand for the printer cartridges depend on the price according to the equations

    $y_d = -10x + 500$ where x is the price per cartridge in dollars and y_d is the demand measured in 1000s of cartridges

    $y_s = \dfrac{20}{3}x$ where x is the price per cartridge in dollars and y_s is the supply measured in 1000s of cartridges

    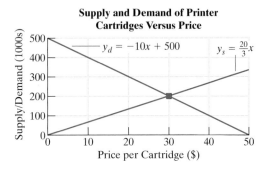

    Find the price at which the supply and demand are in equilibrium (supply = demand), and confirm your answer with the graph.

34. The supply and demand for a pack of note cards depend on the price according to the equations

    $y_d = -130x + 660$ where x is the price per pack in dollars and y_d is the demand in 1000s of note cards

    $y_s = 90x$ where x is the price per pack in dollars and y_s is the supply measured in 1000s of note cards

    Find the price at which the supply and demand are in equilibrium (supply = demand).

35. A rental car company rents a compact car for $20 a day, plus $0.25 per mile. A midsize car rents for $30 a day, plus $0.20 per mile.

    a. Write a linear equation representing the cost to rent the compact car.

    b. Write a linear equation representing the cost to rent a midsize car.

    c. Find the number of miles at which the cost to rent either car would be the same.

36. One phone company charges $0.15 per minute for long-distance calls. A second company charges only $0.10 per minute for long-distance calls, but adds a monthly fee of $4.95.

    a. Write a linear equation representing the cost for the first company.

    b. Write a linear equation representing the cost for the second company.

    c. Find the number of minutes of long-distance calling for which the total bill from either company would be the same.

Section 3.5 Systems of Linear Equations in Three Variables and Applications

Objectives

1. Solutions to Systems of Linear Equations in Three Variables
2. Solving Systems of Linear Equations in Three Variables
3. Applications of Linear Equations in Three Variables

1. Solutions to Systems of Linear Equations in Three Variables

In Sections 3.1–3.3, we solved systems of linear equations in two variables. In this section, we will expand the discussion to solving systems involving three variables.

A **linear equation in three variables** can be written in the form $Ax + By + Cz = D$, where A, B, and C are not all zero. For example, the equation $2x + 3y + z = 6$ is a linear equation in three variables. Solutions to this equation are **ordered triples** of the form (x, y, z) that satisfy the equation. Some solutions to the equation $2x + 3y + z = 6$ are

Solution: Check:

$(1, 1, 1) \longrightarrow 2(1) + 3(1) + (1) = 6$ ✔ True

$(2, 0, 2) \longrightarrow 2(2) + 3(0) + (2) = 6$ ✔ True

$(0, 1, 3) \longrightarrow 2(0) + 3(1) + (3) = 6$ ✔ True

Infinitely many ordered triples serve as solutions to the equation $2x + 3y + z = 6$.

The set of all ordered triples that are solutions to a linear equation in three variables may be represented graphically by a plane in space. Figure 3-9 shows a portion of the plane $2x + 3y + x = 6$ in a 3-dimensional coordinate system.

A solution to a system of linear equations in three variables is an ordered triple that satisfies *each* equation. Geometrically, a solution is a point of intersection of the planes represented by the equations in the system.

A system of linear equations in three variables may have *one unique solution, infinitely many solutions,* or *no solution.*

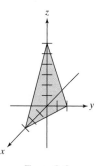

Figure 3-9

One unique solution (planes intersect at one point)
- The system is consistent.
- The system is independent.

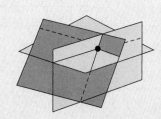

No solution (the three planes do not all intersect)
- The system is inconsistent.
- The system is independent.

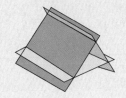 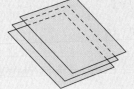 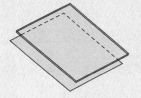

Infinitely many solutions (planes intersect at infinitely many points)
- The system is consistent.
- The system is dependent.

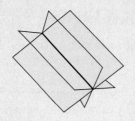 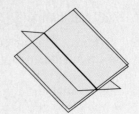 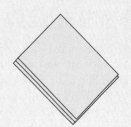

2. Solving Systems of Linear Equations in Three Variables

To solve a system involving three variables, the goal is to eliminate one variable. This reduces the system to two equations in two variables. One strategy for eliminating a variable is to pair up the original equations two at a time.

Solving a System of Three Linear Equations in Three Variables

1. Write each equation in standard form $Ax + By + Cz = D$.
2. Choose a pair of equations, and eliminate one of the variables by using the addition method.
3. Choose a different pair of equations and eliminate the *same* variable.
4. Once steps 2 and 3 are complete, you should have two equations in two variables. Solve this system by using the methods from Sections 3.2 and 3.3.
5. Substitute the values of the variables found in step 4 into any of the three original equations that contain the third variable. Solve for the third variable.
6. Check the ordered triple in each of the original equations.

Skill Practice

1. Solve the system.
$$x + 2y + z = 1$$
$$3x - y + 2z = 13$$
$$2x + 3y - z = -8$$

example 1   Solving a System of Linear Equations in Three Variables

Solve the system.
$$2x + y - 3z = -7$$
$$3x - 2y + z = 11$$
$$-2x - 3y - 2z = 3$$

Solution:

$\boxed{A} \quad 2x + y - 3z = -7$
$\boxed{B} \quad 3x - 2y + z = 11$
$\boxed{C} \quad -2x - 3y - 2z = 3$

Step 1: The equations are already in standard form.

- It is often helpful to label the equations.
- The y-variable can be easily eliminated from equations $\boxed{A}$ and $\boxed{B}$ and from equations $\boxed{A}$ and $\boxed{C}$. This is accomplished by creating opposite coefficients for the y-terms and then adding the equations.

Answer
1. $(1, -2, 4)$

Step 2: Eliminate the y-variable from equations $\boxed{A}$ and $\boxed{B}$.

$\boxed{A}$ $2x + y - 3z = -7$ $\xrightarrow{\text{Multiply by 2.}}$ $4x + 2y - 6z = -14$
$\boxed{B}$ $3x - 2y + z = 11$ $\longrightarrow$ $\underline{3x - 2y + z = 11}$
$7x - 5z = -3$ $\boxed{D}$

Step 3: Eliminate the y-variable again, this time from equations $\boxed{A}$ and $\boxed{C}$.

$\boxed{A}$ $2x + y - 3z = -7$ $\xrightarrow{\text{Multiply by 3.}}$ $6x + 3y - 9z = -21$
$\boxed{C}$ $-2x - 3y - 2z = 3$ $\longrightarrow$ $\underline{-2x - 3y - 2z = 3}$
$4x - 11z = -18$ $\boxed{E}$

Tip: It is important to note that in steps 2 and 3, the *same* variable is eliminated.

Step 4: Now equations $\boxed{D}$ and $\boxed{E}$ can be paired up to form a linear system in two variables. Solve this system.

$\boxed{D}$ $7x - 5z = -3$ $\xrightarrow{\text{Multiply by } -4.}$ $-28x + 20z = 12$
$\boxed{E}$ $4x - 11z = -18$ $\xrightarrow{\text{Multiply by 7.}}$ $\underline{28x - 77z = -126}$
$-57z = -114$
$z = 2$

Once one variable has been found, substitute this value into either equation in the two-variable system, that is, either equation $\boxed{D}$ or $\boxed{E}$.

$\boxed{D}$ $7x - 5z = -3$
$7x - 5(2) = -3$ Substitute $z = 2$ into equation $\boxed{D}$.
$7x - 10 = -3$
$7x = 7$
$x = 1$

$\boxed{A}$ $2x + y - 3z = -7$ **Step 5:** Now that two variables are known, substitute these values (x and z) into any of the original three equations to find the remaining variable y.
$2(1) + y - 3(2) = -7$
$2 + y - 6 = -7$
$y - 4 = -7$
$y = -3$ Substitute $x = 1$ and $z = 2$ into equation $\boxed{A}$.

The solution is $(1, -3, 2)$. **Step 6:** Check the ordered triple in the three original equations.

Check: $2x + y - 3z = -7 \rightarrow 2(1) + (-3) - 3(2) = -7$ ✔ True
$\phantom{\text{Check: }}3x - 2y + z = 11 \rightarrow 3(1) - 2(-3) + (2) = 11$ ✔ True
$\phantom{\text{Check: }}-2x - 3y - 2z = 3 \rightarrow -2(1) - 3(-3) - 2(2) = 3$ ✔ True

Skill Practice

2. The perimeter of a triangle is 30 in. The shortest side is 4 in. shorter than the longest side. The longest side is 6 in. less than the sum of the other two sides. Find the length of each side.

example 2 — Applying Systems of Linear Equations in Three Variables

In a triangle, the smallest angle measures 10° more than one-half the measure of the largest angle. The middle angle measures 12° more than the measure of the smallest angle. Find the measure of each angle.

Solution:

Let x represent the measure of the smallest angle.

Let y represent the measure of the middle angle.

Let z represent the measure of the largest angle.

To solve for three variables, we need to establish three independent relationships among x, y, and z.

$\boxed{A}$ $x = \dfrac{z}{2} + 10$ The smallest angle measures 10° more than one-half the measure of the largest angle.

$\boxed{B}$ $y = x + 12$ The middle angle measures 12° more than the measure of the smallest angle.

$\boxed{C}$ $x + y + z = 180$ The sum of the angles inscribed in a triangle measures 180°.

Clear fractions and write each equation in standard form.

Standard Form

$\boxed{A}$ $x = \dfrac{z}{2} + 10$ $\xrightarrow{\text{Multiply by 2.}}$ $2x = z + 20$ $\longrightarrow$ $2x - z = 20$

$\boxed{B}$ $y = x + 12$ $\longrightarrow$ $-x + y = 12$

$\boxed{C}$ $x + y + z = 180$ $\longrightarrow$ $x + y + z = 180$

Notice equation $\boxed{B}$ is missing the z-variable. Therefore, we can eliminate z again by pairing up equations $\boxed{A}$ and $\boxed{C}$.

$\boxed{A}$ $2x - z = 20$
$\boxed{C}$ $\underline{x + y + z = 180}$
       $3x + y = 200$ $\boxed{D}$

$\boxed{B}$ $-x + y = 12$ $\xrightarrow{\text{Multiply by }-1.}$ $x - y = -12$ Pair up equations $\boxed{B}$ and
$\boxed{D}$ $3x + y = 200$ $\longrightarrow$ $\underline{3x + y = 200}$ $\boxed{D}$ to form a system of two variables.
                                  $4x = 188$

                                  $x = 47$ Solve for x.

Answer

2. 8 in., 10 in., and 12 in.

From equation $\boxed{B}$ we have $-x + y = 12 \longrightarrow -47 + y = 12 \rightarrow y = 59$

From equation $\boxed{C}$ we have $x + y + z = 180 \rightarrow 47 + 59 + z = 180 \rightarrow z = 74$

The smallest angle measures 47°, the middle angle measures 59°, and the largest angle measures 74°.

---

example 3   **Solving a Dependent System of Linear Equations**

Solve the system. If there is not a unique solution, label the system as either dependent or inconsistent.

$$\boxed{A} \quad 3x + y - z = 8$$
$$\boxed{B} \quad 2x - y + 2z = 3$$
$$\boxed{C} \quad x + 2y - 3z = 5$$

Solution:

The first step is to make a decision regarding the variable to eliminate. The y-variable is particularly easy to eliminate because the coefficients of y in equations $\boxed{A}$ and $\boxed{B}$ are already opposites. The y-variable can be eliminated from equations $\boxed{B}$ and $\boxed{C}$ by multiplying equation $\boxed{B}$ by 2.

$\boxed{A} \quad 3x + y - z = 8$
$\boxed{B} \quad \underline{2x - y + 2z = 3}$
$\phantom{\boxed{A}} \quad 5x + z = 11 \quad \boxed{D}$

Pair up equations $\boxed{A}$ and $\boxed{B}$ to eliminate y.

$\boxed{B} \quad 2x - y + 2z = 3 \xrightarrow{\text{Multiply by 2.}} 4x - 2y + 4z = 6$
$\boxed{C} \quad x + 2y - 3z = 5 \longrightarrow \underline{x + 2y - 3z = 5}$
$\phantom{\boxed{B}} 5x + z = 11 \quad \boxed{E}$

Pair up equations $\boxed{B}$ and $\boxed{C}$ to eliminate y.

Because equations $\boxed{D}$ and $\boxed{E}$ are equivalent equations, it appears that this is a dependent system. By eliminating variables we obtain the identity $0 = 0$.

$\boxed{D} \quad 5x + z = 11 \xrightarrow{\text{Multiply by } -1.} -5x - z = -11$
$\boxed{E} \quad 5x + z = 11 \longrightarrow \underline{5x + z = 11}$
$\phantom{\boxed{D} xxxxxxxxxxxxxxxxxxxxxx} 0 = 0$

The result $0 = 0$ indicates that there are infinitely many solutions and that the system is dependent.

---

example 4   **Solving an Inconsistent System of Linear Equations**

Solve the system. If there is not a unique solution, identify the system as either dependent or inconsistent.

$$2x + 3y - 7z = 4$$
$$-4x - 6y + 14z = 1$$
$$5x + y - 3z = 6$$

Skill Practice

3. Solve the system. If the system does not have a unique solution, identify the system as dependent or inconsistent.

$$x + y + z = 8$$
$$2x - y + z = 6$$
$$-5x - 2y - 4z = -30$$

Skill Practice

4. Solve the system. If the system does not have a unique solution, identify the system as dependent or inconsistent.

$$x - 2y + z = 5$$
$$x - 3y + 2z = -7$$
$$-2x + 4y - 2z = 6$$

Answers

3. Dependent system
4. Inconsistent system

Solution:

We will eliminate the x-variable.

A	$2x + 3y - 7z = 4$	$\xrightarrow{\text{Multiply by 2.}}$	$4x + 6y - 14z = 8$
B	$-4x - 6y + 14z = 1$	$\longrightarrow$	$-4x - 6y + 14z = 1$
C	$5x + y - 3z = 6$		$0 = 9$ (contradiction)

The result $0 = 9$ is a contradiction, indicating that the system has no solution. The system is inconsistent.

3. Applications of Linear Equations in Three Variables

Skill Practice

5. Annette, Barb, and Carlita work in a clothing shop. One day the three had combined sales of $1480. Annette sold $120 more than Barb. Barb and Carlita combined sold $280 more than Annette. How much did each person sell?

example 5 Applying Systems of Linear Equations to Nutrition

Doctors have become increasingly concerned about the sodium intake in the U.S. diet. Recommendations by the American Medical Association indicate that most individuals should not exceed 2400 mg of sodium per day.

Liz ate 1 slice of pizza, 1 serving of ice cream, and 1 glass of soda for a total of 1030 mg of sodium. David ate 3 slices of pizza, no ice cream, and 2 glasses of soda for a total of 2420 mg of sodium. Melinda ate 2 slices of pizza, 1 serving of ice cream, and 2 glasses of soda for a total of 1910 mg of sodium. How much sodium is in one serving of each item?

Solution:

Let x represent the sodium content of 1 slice of pizza.

Let y represent the sodium content of 1 serving of ice cream.

Let z represent the sodium content of 1 glass of soda.

From Liz's meal we have: A $\;x + y + z = 1030$

From David's meal we have: B $\;3x + 2z = 2420$

From Melinda's meal we have: C $\;2x + y + 2z = 1910$

Equation B is missing the y-variable. Eliminating y from equations A and C, we have

A	$x + y + z = 1030$	$\xrightarrow{\text{Multiply by } -1.}$	$-x - y - z = -1030$
C	$2x + y + 2z = 1910$	$\longrightarrow$	$2x + y + 2z = 1910$
		D	$x + z = 880$

Solve the system formed by equations B and D.

B	$3x + 2z = 2420$	$\longrightarrow$	$3x + 2z = 2420$
D	$x + z = 880$	$\xrightarrow{\text{Multiply by } -2.}$	$-2x - 2z = -1760$
			$x = 660$

From equation D we have $x + z = 880 \longrightarrow 660 + z = 880 \longrightarrow z = 220$

From equation A we have $x + y + z = 1030 \longrightarrow 660 + y + 220 = 1030 \longrightarrow y = 150$

Therefore, 1 slice of pizza has 660 mg of sodium, 1 serving of ice cream has 150 mg of sodium, and 1 glass of soda has 220 mg of sodium.

Answer

5. Annette sold $600, Barb sold $480, and Carlita sold $400.

section 3.5 Practice Exercises

- Practice Problems
- Self-Tests
- NetTutor
- e-Professors
- Videos

Study Skills Exercises

1. Of the list below, circle the activities that you think can help someone with math anxiety.

   a. Deep breathing
   b. Reading a book about math anxiety
   c. Scheduling extra study time
   d. Keeping a positive attitude

2. Define the key terms.
   a. Linear equation in three variables
   b. Ordered triple

Review Exercises

For Exercises 3–4, solve the systems by using two methods: (**a**) the substitution method and (**b**) the addition method.

3. $3x + y = 4$
   $4x + y = 5$

4. $2x - 5y = 3$
   $-4x + 10y = 3$

5. Two cars leave Kansas City at the same time. One travels east and one travels west. After 3 hr the cars are 369 mi apart. If one car travels 7 mph slower than the other, find the speed of each car.

Objective 1: Solutions to Systems of Linear Equations in Three Variables

6. How many solutions are possible when solving a system of three equations with three variables?

7. Which of the following points are solutions to the system?

   $(2, 1, 7), (3, -10, -6), (4, 0, 2)$

   $2x - y + z = 10$
   $4x + 2y - 3z = 10$
   $x - 3y + 2z = 8$

8. Which of the following points are solutions to the system?

   $(1, 1, 3), (0, 0, 4), (4, 2, 1)$

   $-3x - 3y - 6z = -24$
   $-9x - 6y + 3z = -45$
   $9x + 3y - 9z = 33$

9. Which of the following points are solutions to the system?

   $(12, 2, -2), (4, 2, 1), (1, 1, 1)$

   $-x - y - 4z = -6$
   $x - 3y + z = -1$
   $4x + y - z = 4$

10. Which of the following points are solutions to the system?

    $(0, 4, 3), (3, 6, 10), (3, 3, 1)$

    $x + 2y - z = 5$
    $x - 3y + z = -5$
    $-2x + y - z = -4$

Objective 2: Solving Systems of Linear Equations in Three Variables

For Exercises 11–20, solve the system of equations. (See Example 1.)

11. $\begin{aligned} x + y + z &= 6 \\ -x + y - z &= -2 \\ 2x + 3y + z &= 11 \end{aligned}$

12. $\begin{aligned} x - y - z &= -11 \\ x + y - z &= 15 \\ 2x - y + z &= -9 \end{aligned}$

13. $\begin{aligned} -3x + y - z &= 8 \\ -4x + 2y + 3z &= -3 \\ 2x + 3y - 2z &= -1 \end{aligned}$

14. $\begin{aligned} 2x + 3y + 3z &= 15 \\ 3x - 6y - 6z &= -23 \\ -9x - 3y + 6z &= 8 \end{aligned}$

15. $\begin{aligned} 2x - y + z &= -1 \\ -3x + 2y - 2z &= 1 \\ 5x + 3y + 3z &= 16 \end{aligned}$

16. $\begin{aligned} 4x - 3y + 2z &= 12 \\ -3x + 2y - 3z &= -5 \\ 2x - y + 7z &= -8 \end{aligned}$

17. $\begin{aligned} 2x - 3y + 2z &= -1 \\ x + 2y &= -4 \\ x + z &= 1 \end{aligned}$

18. $\begin{aligned} x + y + z &= 2 \\ 2x - z &= 5 \\ 3y + z &= 2 \end{aligned}$

19. $\begin{aligned} 4x + 9y &= 8 \\ 8x + 6z &= -1 \\ 6y + 6z &= -1 \end{aligned}$

20. $\begin{aligned} 3x + 2z &= 11 \\ y - 7z &= 4 \\ x - 6y &= 1 \end{aligned}$

Objective 3: Applications of Linear Equations in Three Variables

21. A triangle has one angle that measures 5° more than twice the smallest angle, and the largest angle measures 11° less than 3 times the measure of the smallest angle. Find the measures of the three angles. (See Example 2.)

22. The largest angle of a triangle measures 4° less than 5 times the measure of the smallest angle. The middle angle measures twice that of the smallest angle. Find the measures of the three angles.

23. The perimeter of a triangle is 55 cm. The measure of the smallest side is 8 less than the middle side. The measure of longest side is equal to one less than the sum of the other two sides. Find the lengths of the sides.

24. The perimeter of a triangle is 5 ft. The longest side of the triangle measures 20 in. more than the shortest side. The middle side 3 times the measure of the shortest side. Find the lengths of the three sides in *inches*.

25. A movie theater charges $7 for adults, $5 for children under age 17, and $4 for seniors over age 60. For one showing of *Batman* the theater sold 222 tickets and took in $1383. If twice as many adult tickets were sold as the total of children and senior tickets, how many tickets of each kind were sold? (See Example 5.)

26. Goofie Golf has 18 holes that are par 3, par 4, or par 5. Most of the holes are par 4. In fact, there are 3 times as many par 4s as par 3s. There are 3 more par 5s than par 3s. How many of each type are there?

27. Combining peanuts, pecans, and cashews makes a party mixture of nuts. If the amount of peanuts equals the amount of pecans and cashews combined, and if there are twice as many cashews as pecans, how many ounces of each nut is used to make 48 oz of party mixture?

28. Souvenir hats, T-shirts, and jackets are sold at a rock concert. Three hats, two T-shirts, and one jacket cost $140. Two hats, two T-shirts, and two jackets cost $170. One hat, three T-shirts, and two jackets cost $180. Find the prices of the individual items.

29. In 2002, Baylor University in Waco, Texas, had twice as many students as Vanderbilt University in Nashville, Tennessee. Pace University in New York City had 2800 more students than Vanderbilt University. If the enrollment for all three schools totaled 27,200, find the enrollment for each school.

30. Annie and Maria traveled overseas for seven days and stayed in three different hotels in three different cities: Stockholm, Sweden; Oslo, Norway; and Paris, France.

    The total bill for all seven nights (not including tax) was $1040. The total tax was $106. The nightly cost (excluding tax) to stay at the hotel in Paris was $80 more than the nightly cost (excluding tax) to stay in Oslo. Find the cost per night for each hotel excluding tax.

City	Number of Nights	Cost/Night ($)	Tax Rate
Paris, France	1	x	8%
Stockholm, Sweden	4	y	11%
Oslo, Norway	2	z	10%

Mixed Exercises

For Exercises 31–40, solve the system. If there is not a unique solution, label the system as either dependent or inconsistent. (See Examples 3–4.)

31. $2x + y + 3z = 2$
    $x - y + 2z = -4$
    $x + 3y - z = 1$

32. $x + y - z = 0$
    $3x - 2y + 6z = 1$
    $7x + 3y + z = 4$

33. $6x - 2y + 2z = 2$
    $4x + 8y - 2z = 5$
    $-2x - 4y + z = -2$

34. $3x + 2y + z = 3$
    $x - 3y + z = 4$
    $-6x - 4y - 2z = 1$

35. $\frac{1}{2}x + \frac{2}{3}y = \frac{5}{2}$
    $\frac{1}{5}x - \frac{1}{2}z = -\frac{3}{10}$
    $\frac{1}{3}y - \frac{1}{4}z = \frac{3}{4}$

36. $\frac{1}{2}x + \frac{1}{4}y + z = 3$
    $\frac{1}{8}x + \frac{1}{4}y + \frac{1}{4}z = \frac{9}{8}$
    $x - y - \frac{2}{3}z = \frac{1}{3}$

37. $2x + y - 3z = -3$
    $3x - 2y + 4z = 1$
    $4x + 2y - 6z = -6$

38. $2x + y = -3$
    $2y + 16z = -10$
    $-7x - 3y + 4z = 8$

39. $-0.1y + 0.2z = 0.2$
    $0.1x + 0.1y + 0.1z = 0.2$
    $-0.1x + 0.3z = 0.2$

40. $0.1x - 0.2y = 0$
    $0.3y + 0.1z = -0.1$
    $0.4x - 0.1z = 1.2$

Expanding Your Skills

The systems in Exercises 41–44 are called homogeneous systems because each system has (0, 0, 0) as a solution. However, if a system is dependent, it will have infinitely many more solutions. For each system determine whether (0, 0, 0) is the only solution or if the system is dependent.

41. $2x - 4y + 8z = 0$
    $-x - 3y + z = 0$
    $x - 2y + 5z = 0$

42. $2x - 4y + z = 0$
    $x - 3y - z = 0$
    $3x - y + 2z = 0$

43. $4x - 2y - 3z = 0$
    $-8x - y + z = 0$
    $2x - y - \frac{3}{2}z = 0$

44. $5x + y = 0$
    $4y - z = 0$
    $5x + 5y - z = 0$

Section 3.6 Solving Systems of Linear Equations by Using Matrices

Objectives
1. Introduction to Matrices
2. Solving Systems of Linear Equations by Using the Gauss-Jordan Method

1. Introduction to Matrices

In Sections 3.2, 3.3, and 3.5, we solved systems of linear equations by using the substitution method and the addition method. We now present a third method called the Gauss-Jordan method that uses matrices to solve a linear system.

A **matrix** is a rectangular array of numbers (the plural of *matrix* is *matrices*). The rows of a matrix are read horizontally, and the columns of a matrix are read vertically. Every number or entry within a matrix is called an element of the matrix.

The **order of a matrix** is determined by the number of rows and number of columns. A matrix with m rows and n columns is an $m \times n$ (read as "m by n") matrix. Notice that with the order of a matrix, the number of rows is given first, followed by the number of columns.

Skill Practice

Determine the order of the matrix.

1. $\begin{bmatrix} -5 & 2 \\ 1 & 3 \\ 8 & 9 \end{bmatrix}$
2. $[4, -8]$
3. $\begin{bmatrix} 5 \\ 10 \\ 15 \end{bmatrix}$

example 1 Determining the Order of a Matrix

Determine the order of each matrix.

a. $\begin{bmatrix} 2 & -4 & 1 \\ 5 & \pi & \sqrt{7} \end{bmatrix}$
b. $\begin{bmatrix} 1.9 \\ 0 \\ 7.2 \\ -6.1 \end{bmatrix}$
c. $\begin{bmatrix} 1 & 0 & 0 \\ 0 & 1 & 0 \\ 0 & 0 & 1 \end{bmatrix}$
d. $\begin{bmatrix} a & b & c \end{bmatrix}$

Solution:

a. This matrix has two rows and three columns. Therefore, it is a 2×3 matrix.

b. This matrix has four rows and one column. Therefore, it is a 4×1 matrix. A matrix with one column is called a **column matrix**.

c. This matrix has three rows and three columns. Therefore, it is a 3×3 matrix. A matrix with the same number of rows and columns is called a **square matrix**.

d. This matrix has one row and three columns. Therefore, it is a 1×3 matrix. A matrix with one row is called a **row matrix**.

A matrix can be used to represent a system of linear equations written in standard form. To do so, we extract the coefficients of the variable terms and the constants within the equation. For example, consider the system

$$2x - y = 5$$
$$x + 2y = -5$$

The matrix **A** is called the **coefficient matrix**.

$$\mathbf{A} = \begin{bmatrix} 2 & -1 \\ 1 & 2 \end{bmatrix}$$

If we extract both the coefficients and the constants from the equations, we can construct the **augmented matrix** of the system:

$$\begin{bmatrix} 2 & -1 & | & 5 \\ 1 & 2 & | & -5 \end{bmatrix}$$

A vertical bar is inserted into an augmented matrix to designate the position of the equal signs.

Answers
1. 3×2 2. 1×2 3. 3×1

example 2 — Writing the Augmented Matrix of a System of Linear Equations

Write the augmented matrix for each linear system.

a. $-3x - 4y = 3$
$2x + 4y = 2$

b. $2x - 3z = 14$
$ 2y + z = 2$
$x + y = 4$

Skill Practice

4. Write the augmented matrix for the system.
$2x - y + z = 14$
$-3x + 4y = 8$
$x - y + 5z = 0$

Solution:

a. $\begin{bmatrix} -3 & -4 & | & 3 \\ 2 & 4 & | & 2 \end{bmatrix}$

b. $\begin{bmatrix} 2 & 0 & -3 & | & 14 \\ 0 & 2 & 1 & | & 2 \\ 1 & 1 & 0 & | & 4 \end{bmatrix}$

Tip: Notice that zeros are inserted to denote the coefficient of each missing term.

example 3 — Writing a Linear System from an Augmented Matrix

Write a system of linear equations represented by each augmented matrix.

a. $\begin{bmatrix} 2 & -5 & | & -8 \\ 4 & 1 & | & 6 \end{bmatrix}$

b. $\begin{bmatrix} 2 & -1 & 3 & | & 14 \\ 1 & 1 & -2 & | & -5 \\ 3 & 1 & -1 & | & 2 \end{bmatrix}$

c. $\begin{bmatrix} 1 & 0 & 0 & | & 4 \\ 0 & 1 & 0 & | & -1 \\ 0 & 0 & 1 & | & 0 \end{bmatrix}$

Skill Practice

Write a system of linear equations represented by each augmented matrix.

5. $\begin{bmatrix} 2 & 3 & | & 5 \\ -1 & 8 & | & 1 \end{bmatrix}$

6. $\begin{bmatrix} -3 & 2 & 1 & | & 4 \\ 14 & 1 & 0 & | & 20 \\ -8 & 3 & 5 & | & 6 \end{bmatrix}$

Solution:

a. $2x - 5y = -8$
$4x + y = 6$

b. $2x - y + 3z = 14$
$x + y - 2z = -5$
$3x + y - z = 2$

c. $x + 0y + 0z = 4$
$0x + y + 0z = -1$ or $x = 4$
$0x + 0y + z = 0$ $y = -1$
$ z = 0$

Concept Connections

7. A matrix has been reduced to the form shown. Give the solution to the system of linear equations represented by the matrix.

$\begin{bmatrix} 1 & 0 & 0 & | & -8 \\ 0 & 1 & 0 & | & 2 \\ 0 & 0 & 1 & | & 15 \end{bmatrix}$

2. Solving Systems of Linear Equations by Using the Gauss-Jordan Method

We know that interchanging two equations results in an equivalent system of linear equations. Interchanging two rows in an augmented matrix results in an equivalent augmented matrix. Similarly, because each row in an augmented matrix represents a linear equation, we can perform the following elementary row operations that result in an equivalent augmented matrix.

Answers

4. $\begin{bmatrix} 2 & -1 & 1 & | & 14 \\ -3 & 4 & 0 & | & 8 \\ 1 & -1 & 5 & | & 0 \end{bmatrix}$

5. $2x + 3y = 5$
$-x + 8y = 1$

6. $-3x + 2y + z = 4$
$14x + y = 20$
$-8x + 3y + 5z = 6$

7. $x = -8, y = 2, z = 15$

Elementary Row Operations

The following *elementary row operations* performed on an augmented matrix produce an equivalent augmented matrix:

1. Interchange two rows.
2. Multiply every element in a row by a nonzero real number.
3. Add a multiple of one row to another row.

When we are solving a system of linear equations by any method, the goal is to write a series of simpler but equivalent systems of equations until the solution is obvious. The *Gauss-Jordan method* uses a series of elementary row operations performed on the augmented matrix to produce a simpler augmented matrix. In particular, we want to produce an augmented matrix that has 1s along the diagonal of the matrix of coefficients and 0s for the remaining entries in the matrix of coefficients. A matrix written in this way is said to be written in **reduced row echelon form**. For example, the augmented matrix from Example 3(c) is written in reduced row echelon form.

$$\begin{bmatrix} 1 & 0 & 0 & | & 4 \\ 0 & 1 & 0 & | & -1 \\ 0 & 0 & 1 & | & 0 \end{bmatrix}$$

The solution to the corresponding system of equations is easily recognized as $x = 4$, $y = -1$, and $z = 0$.

Similarly, matrix **B** represents a solution of $x = a$ and $y = b$.

$$\mathbf{B} = \begin{bmatrix} 1 & 0 & | & a \\ 0 & 1 & | & b \end{bmatrix}$$

Skill Practice

8. Solve by using the Gauss-Jordan method.
$$x - 2y = -21$$
$$2x + y = -2$$

example 4 Solving a System of Linear Equations by Using the Gauss-Jordan Method

Solve by using the Gauss-Jordan method.

$$2x - y = 5$$
$$x + 2y = -5$$

Solution:

$$\begin{bmatrix} 2 & -1 & | & 5 \\ 1 & 2 & | & -5 \end{bmatrix}$$ Set up the augmented matrix.

$$\xrightarrow{R_1 \Leftrightarrow R_2} \begin{bmatrix} 1 & 2 & | & -5 \\ 2 & -1 & | & 5 \end{bmatrix}$$ Switch row 1 and row 2 to get a 1 in the upper left position.

$$\xrightarrow{-2R_1 + R_2 \Rightarrow R_2} \begin{bmatrix} 1 & 2 & | & -5 \\ 0 & -5 & | & 15 \end{bmatrix}$$ Multiply row 1 by -2 and add the result to row 2. This produces an entry of 0 below the upper left position.

Answer
8. $(-5, 8)$

$-\frac{1}{5}R_2 \Rightarrow R_2$ $\longrightarrow \begin{bmatrix} 1 & 2 & | & -5 \\ 0 & 1 & | & -3 \end{bmatrix}$ Multiply row 2 by $-\frac{1}{5}$ to produce a 1 along the diagonal in the second row.

$-2R_2 + R_1 \Rightarrow R_1$ $\longrightarrow \begin{bmatrix} 1 & 0 & | & 1 \\ 0 & 1 & | & -3 \end{bmatrix}$ Multiply row 2 by -2 and add the result to row 1. This produces a 0 in the first row, second column.

The matrix **C** is in reduced row echelon form. From the augmented matrix, we have $x = 1$ and $y = -3$. The solution to the system is $(1, -3)$.

$$\mathbf{C} = \begin{bmatrix} 1 & 0 & | & 1 \\ 0 & 1 & | & -3 \end{bmatrix}$$

The order in which we manipulate the elements of an augmented matrix to produce reduced row echelon form was demonstrated in Example 4. In general, the order is as follows.

- First produce a 1 in the first row, first column. Then use the first row to obtain 0s in the first column below this element.
- Next, if possible, produce a 1 in the second row, second column. Use the second row to obtain 0s above and below this element.
- Next, if possible, produce a 1 in the third row, third column. Use the third row to obtain 0s above and below this element.
- The process continues until reduced row echelon form is obtained.

example 5   Solving a System of Linear Equations by Using the Gauss-Jordan Method

Solve by using the Gauss-Jordan method.

$$x = -y + 5$$
$$-2x + 2z = y - 10$$
$$3x + 6y + 7z = 14$$

Skill Practice

Solve by using the Gauss-Jordan method.

9. $x + y + z = 2$
$x - y + z = 4$
$x + 4y + 2z = 1$

Solution:

First write each equation in the system in standard form.

$x = -y + 5 \longrightarrow x + y = 5$
$-2x + 2z = y - 10 \longrightarrow -2x - y + 2z = -10$
$3x + 6y + 7z = 14 \longrightarrow 3x + 6y + 7z = 14$

$\begin{bmatrix} 1 & 1 & 0 & | & 5 \\ -2 & -1 & 2 & | & -10 \\ 3 & 6 & 7 & | & 14 \end{bmatrix}$ Set up the augmented matrix.

$2R_1 + R_2 \Rightarrow R_2$
$-3R_1 + R_3 \Rightarrow R_3$ $\longrightarrow \begin{bmatrix} 1 & 1 & 0 & | & 5 \\ 0 & 1 & 2 & | & 0 \\ 0 & 3 & 7 & | & -1 \end{bmatrix}$ Multiply row 1 by 2 and add the result to row 2. Multiply row 1 by -3 and add the result to row 3.

Answer

9. $(1, -1, 2)$

Chapter 3 Systems of Linear Equations

$$-1R_2 + R_1 \Rightarrow R_1$$
$$-3R_2 + R_3 \Rightarrow R_3$$
$$\begin{bmatrix} 1 & 0 & -2 & | & 5 \\ 0 & 1 & 2 & | & 0 \\ 0 & 0 & 1 & | & -1 \end{bmatrix}$$

Multiply row 2 by -1 and add the result to row 1. Multiply row 2 by -3 and add the result to row 3.

$$2R_3 + R_1 \Rightarrow R_1$$
$$-2R_3 + R_2 \Rightarrow R_2$$
$$\begin{bmatrix} 1 & 0 & 0 & | & 3 \\ 0 & 1 & 0 & | & 2 \\ 0 & 0 & 1 & | & -1 \end{bmatrix}$$

Multiply row 3 by 2 and add the result to row 1. Multiply row 3 by -2 and add the result to row 2.

From the reduced row echelon form of the matrix, we have $x = 3$, $y = 2$, and $z = -1$. The solution to the system is $(3, 2, -1)$.

It is particularly easy to recognize a dependent or inconsistent system of equations from the reduced row echelon form of an augmented matrix. This is demonstrated in Examples 6 and 7.

Skill Practice

Solve by using the Gauss-Jordan method.

10. $4x - 6y = 16$
$6x - 9y = 24$

example 6 Solving a Dependent System of Equations by Using the Gauss-Jordan Method

Solve by using the Gauss-Jordan method.

$$x - 3y = 4$$
$$\frac{1}{2}x - \frac{3}{2}y = 2$$

Solution:

$$\begin{bmatrix} 1 & -3 & | & 4 \\ \frac{1}{2} & -\frac{3}{2} & | & 2 \end{bmatrix}$$ Set up the augmented matrix.

$$-\frac{1}{2}R_1 + R_2 \Rightarrow R_2 \quad \begin{bmatrix} 1 & -3 & | & 4 \\ 0 & 0 & | & 0 \end{bmatrix}$$ Multiply row 1 by $-\frac{1}{2}$ and add the result to row 2.

The second row of the augmented matrix represents the equation $0 = 0$; hence, the system is dependent. The solution is $\{(x, y) \mid x - 3y = 4\}$.

Skill Practice

11. Solve by using the Gauss-Jordan method.

$6x + 10y = 1$
$15x + 25y = 3$

example 7 Solving an Inconsistent System of Equations by Using the Gauss-Jordan Method

Solve by using the Gauss-Jordan method.

$$2x - 5y = 10$$
$$\frac{2}{5}x - y = 7$$

Answers

10. Dependent system:
    $\{(x, y) \mid 4x - 6y = 16\}$
11. Inconsistent system, no solution

Solution:

$$\begin{bmatrix} 2 & -5 & | & 10 \\ \frac{2}{5} & -1 & | & 7 \end{bmatrix}$$ Set up the augmented matrix.

$\xrightarrow{-\frac{1}{5}R_1 + R_2 \Rightarrow R_2}$ $\begin{bmatrix} 2 & -5 & | & 10 \\ 0 & 0 & | & 5 \end{bmatrix}$ Multiply row 1 by $-\frac{1}{5}$ and add the result to row 2.

The second row of the augmented matrix represents the contradiction $0 = 5$; hence, the system is inconsistent. There is no solution.

Calculator Connections

Many graphing calculators have a matrix editor in which the user defines the order of the matrix and then enters the elements of the matrix. For example, the 2×3 matrix

$$\mathbf{D} = \begin{bmatrix} 2 & -3 & | & -13 \\ 3 & 1 & | & 8 \end{bmatrix}$$

is entered as shown.

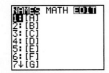

Once an augmented matrix has been entered into a graphing calculator, a *rref.* function can be used to transform the matrix into reduced row echelon form (see figure).

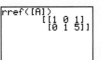

section 3.6 Practice Exercises

Boost your GRADE at mathzone.com!

- Practice Problems
- Self-Tests
- NetTutor
- e-Professors
- Videos

Study Skills Exercises

1. Prepare a one-page summary sheet with the most important information that you need for the test. On the day of the test, look at this sheet several times to refresh your memory, instead of trying to memorize new information.

2. Define the key terms.
   - **a.** Matrix
   - **b.** Order of a matrix
   - **c.** Column matrix
   - **d.** Square matrix
   - **e.** Row matrix
   - **f.** Coefficient matrix
   - **g.** Augmented matrix
   - **h.** Reduced row echelon form

Review Exercises

For Exercises 3–5, solve the system by using any method.

3. $x - 6y = 9$
$x + 2y = 13$

4. $x + y - z = 8$
$x - 2y + z = 3$
$x + 3y + 2z = 7$

5. $2x - y + z = -4$
$-x + y + 3z = -7$
$x + 3y - 4z = 22$

Objective 1: Introduction to Matrices

6. What is an augmented matrix?

7. What is a coefficient matrix?

8. How do you determine the order of a matrix?

9. What is a square matrix?

For Exercises 10–17, (a) determine the order of each matrix and (b) determine if the matrix is a row matrix, a column matrix, a square matrix, or none of these. (See Example 1.)

10. $\begin{bmatrix} 4 \\ 5 \\ -3 \\ 0 \end{bmatrix}$

11. $\begin{bmatrix} 5 \\ -1 \\ 2 \end{bmatrix}$

12. $\begin{bmatrix} -9 & 4 & 3 \\ -1 & -8 & 4 \\ 5 & 8 & 7 \end{bmatrix}$

13. $\begin{bmatrix} 3 & -9 \\ -1 & -3 \end{bmatrix}$

14. $\begin{bmatrix} 4 & -7 \end{bmatrix}$

15. $\begin{bmatrix} 0 & -8 & 11 & 5 \end{bmatrix}$

16. $\begin{bmatrix} 5 & -8.1 & 4.2 & 0 \\ 4.3 & -9 & 18 & 3 \end{bmatrix}$

17. $\begin{bmatrix} \frac{1}{3} & \frac{3}{4} & 6 \\ -2 & 1 & -\frac{7}{8} \end{bmatrix}$

For Exercises 18–21, set up the augmented matrix. (See Example 2.)

18. $x - 2y = -1$
$2x + y = -7$

19. $x - 3y = 3$
$2x - 5y = 4$

20. $x - 2y = 5 - z$
$2x + 6y + 3z = -2$
$3x - y - 2z = 1$

21. $5x - 17 = -2z$
$8x + 6z = 26 + y$
$8x + 3y - 12z = 24$

For Exercises 22–25, write a system of linear equations represented by the augmented matrix. (See Example 3.)

22. $\begin{bmatrix} 4 & 3 & | & 6 \\ 12 & 5 & | & -6 \end{bmatrix}$

23. $\begin{bmatrix} -2 & 5 & | & -15 \\ -7 & 15 & | & -45 \end{bmatrix}$

24. $\begin{bmatrix} 1 & 0 & 0 & | & 4 \\ 0 & 1 & 0 & | & -1 \\ 0 & 0 & 1 & | & 7 \end{bmatrix}$

25. $\begin{bmatrix} 1 & 0 & 0 & | & 0.5 \\ 0 & 1 & 0 & | & 6.1 \\ 0 & 0 & 1 & | & 3.9 \end{bmatrix}$

Objective 2: Solving Systems of Linear Equations by Using the Gauss-Jordan Method

26. Given the matrix **E**

$$E = \begin{bmatrix} 3 & -2 & | & 8 \\ 9 & -1 & | & 7 \end{bmatrix}$$

a. What is the element in the second row and third column?

b. What is the element in the first row and second column?

27. Given the matrix **F**

$$F = \begin{bmatrix} 1 & 8 & | & 0 \\ 12 & -13 & | & -2 \end{bmatrix}$$

a. What is the element in the second row and second column?

b. What is the element in the first row and third column?

28. Given the matrix **Z**

$$Z = \begin{bmatrix} 2 & 1 & | & 11 \\ 2 & -1 & | & 1 \end{bmatrix}$$

write the matrix obtained by multiplying the elements in the first row by $\frac{1}{2}$.

29. Given the matrix **J**

$$J = \begin{bmatrix} 1 & 1 & | & 7 \\ 0 & 3 & | & -6 \end{bmatrix}$$

write the matrix obtained by multiplying the elements in the second row by $\frac{1}{3}$.

30. Given the matrix **K**

$$K = \begin{bmatrix} 5 & 2 & | & 1 \\ 1 & -4 & | & 3 \end{bmatrix}$$

write the matrix obtained by interchanging rows 1 and 2.

31. Given the matrix **L**

$$L = \begin{bmatrix} 9 & 6 & | & 13 \\ -7 & 2 & | & 19 \end{bmatrix}$$

write the matrix obtained by interchanging rows 1 and 2.

32. Given the matrix **M**

$$M = \begin{bmatrix} 1 & 5 & | & 2 \\ -3 & -4 & | & -1 \end{bmatrix}$$

write the matrix obtained by multiplying the first row by 3 and adding the result to row 2.

33. Given the matrix **N**

$$N = \begin{bmatrix} 1 & 3 & | & -5 \\ -2 & 2 & | & 12 \end{bmatrix}$$

write the matrix obtained by multiplying the first row by 2 and adding the result to row 2.

34. Given the matrix **R**

$$R = \begin{bmatrix} 1 & 3 & 0 & | & -1 \\ 4 & 1 & -5 & | & 6 \\ -2 & 0 & -3 & | & 10 \end{bmatrix}$$

a. Write the matrix obtained by multiplying the first row by -4 and adding the result to row 2.

b. Using the matrix obtained from part (a), write the matrix obtained by multiplying the first row by 2 and adding the result to row 3.

35. Given the matrix **S**

$$S = \begin{bmatrix} 1 & 2 & 0 & | & 10 \\ 5 & 1 & -4 & | & 3 \\ -3 & 4 & 5 & | & 2 \end{bmatrix}$$

a. Write the matrix obtained by multiplying the first row by -5 and adding the result to row 2.

b. Using the matrix obtained from part (a), write the matrix obtained by multiplying the first row by 3 and adding the result to row 3.

For Exercises 36–39, use the augmented matrices **A**, **B**, and **C** to answer true or false.

$$\mathbf{A} = \begin{bmatrix} 6 & -4 & | & 2 \\ 5 & -2 & | & 7 \end{bmatrix} \quad \mathbf{B} = \begin{bmatrix} 5 & -2 & | & 7 \\ 6 & -4 & | & 2 \end{bmatrix} \quad \mathbf{C} = \begin{bmatrix} 1 & -\frac{2}{3} & | & \frac{1}{3} \\ 5 & 2 & | & 7 \end{bmatrix}$$

36. The matrix **A** is a 3×2 matrix.

37. Matrix **B** is equivalent to matrix **A**.

38. Matrix **A** is equivalent to matrix **C**.

39. Matrix **B** is equivalent to matrix **C**.

40. What does the notation $R_2 \Leftrightarrow R_1$ mean when one is performing the Gauss-Jordan method?

41. What does the notation $2R_3 \Rightarrow R_3$ mean when one is performing the Gauss-Jordan method?

42. What does the notation $-3R_1 + R_2 \Rightarrow R_2$ mean when one is performing the Gauss-Jordan method?

43. What does the notation $4R_2 + R_3 \Rightarrow R_3$ mean when one is performing the Gauss-Jordan method?

For Exercises 44–59, solve the systems by using the Gauss-Jordan method. **(See Examples 4–7.)**

44. $x - 2y = -1$
$2x + y = -7$

45. $x - 3y = 3$
$2x - 5y = 4$

46. $x + 3y = 6$
$-4x - 9y = 3$

47. $2x - 3y = -2$
$x + 2y = 13$

48. $x + 3y = 3$
$4x + 12y = 12$

49. $2x + 5y = 1$
$-4x - 10y = -2$

50. $x - y = 4$
$2x + y = 5$

51. $2x - y = 0$
$x + y = 3$

52. $x + 3y = -1$
$-3x - 6y = 12$

53. $x + y = 4$
$2x - 4y = -4$

54. $3x + y = -4$
$-6x - 2y = 3$

55. $2x + y = 4$
$6x + 3y = -1$

56. $x + y + z = 6$
$x - y + z = 2$
$x + y - z = 0$

57. $2x - 3y - 2z = 11$
$x + 3y + 8z = 1$
$3x - y + 14z = -2$

58. $x - 2y = 5 - z$
$2x + 6y + 3z = -10$
$3x - y - 2z = 5$

59. $5x - 10z = 15$
$x - y + 6z = 23$
$x + 3y - 12z = 13$

Graphing Calculator Exercises

For Exercises 60–65, use the matrix features on a graphing calculator to express each augmented matrix in reduced row echelon form. Compare your results to the solution you obtained in the indicated exercise.

60. $\begin{bmatrix} 1 & -2 & | & -1 \\ 2 & 1 & | & -7 \end{bmatrix}$

Compare with Exercise 44.

61. $\begin{bmatrix} 1 & -3 & | & 3 \\ 2 & -5 & | & 4 \end{bmatrix}$

Compare with Exercise 45.

62. $\begin{bmatrix} 1 & 3 & | & 6 \\ -4 & -9 & | & 3 \end{bmatrix}$

Compare with Exercise 46.

63. $\begin{bmatrix} 2 & -3 & | & -2 \\ 1 & 2 & | & 13 \end{bmatrix}$

Compare with Exercise 47.

64. $\begin{bmatrix} 1 & 1 & 1 & | & 6 \\ 1 & -1 & 1 & | & 2 \\ 1 & 1 & -1 & | & 0 \end{bmatrix}$

Compare with Exercise 56.

65. $\begin{bmatrix} 2 & -3 & -2 & | & 11 \\ 1 & 3 & 8 & | & 1 \\ 3 & -1 & 14 & | & -2 \end{bmatrix}$

Compare with Exercise 57.

---

chapter 3 | summary

section 3.1 Solving Systems of Equations by Graphing

Key Concepts

A **system of linear equations** in two variables can be solved by graphing.

A **solution to a system of linear equations** is an ordered pair that satisfies each equation in the system. Graphically, this represents a point of intersection of the lines.

There may be one solution, infinitely many solutions, or no solution.

One solution
Consistent
Independent

Infinitely many solutions
Consistent
Dependent

No solution
Inconsistent
Independent

A system of equations is **consistent** if there is at least one solution. A system is **inconsistent** if there is no solution.

A linear system in x and y is **dependent** if two equations represent the same line. The solution set is the set of all points on the line.

If two linear equations represent different lines, then the system of equations is **independent**.

Examples

Example 1

Solve by graphing. $\quad x + y = 3$
$\qquad\qquad\qquad\quad 2x - y = 0$

Write each equation in $y = mx + b$ form to graph.

$y = -x + 3$

$y = 2x$

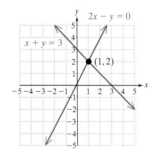

The solution is the point of intersection $(1, 2)$.

section 3.2 Solving Systems of Equations by Using the Substitution Method

Key Concepts

Substitution Method

1. Isolate one of the variables.
2. Substitute the quantity found in step 1 into the *other* equation.
3. Solve the resulting equation.
4. Substitute the value from step 3 back into the equation from step 1 to solve for the remaining variable.
5. Check the ordered pair in both equations, and write the answer as an ordered pair.

Examples

Example 1

$2y = -6x + 14$

$2x + y = 5$ $\xrightarrow{\text{Isolate a variable.}}$ $y = -2x + 5$

Substitute

$2(-2x + 5) = -6x + 14$

$-4x + 10 = -6x + 14$

$2x + 10 = 14$

$2x = 4$

$x = 2$

$y = -2x + 5$ Now solve for y.

$y = -2(2) + 5$

$y = 1$

The solution is $(2, 1)$ and checks in both equations.

section 3.3 Solving Systems of Equations by Using the Addition Method

Key Concepts

Addition Method

1. Write both equations in standard form $Ax + By = C$.
2. Clear fractions or decimals (optional).
3. Multiply one or both equations by nonzero constants to create opposite coefficients for one of the variables.
4. Add the equations from step 3 to eliminate one variable.
5. Solve for the remaining variable.
6. Substitute the known value from step 5 back into one of the original equations to solve for the other variable.
7. Check the ordered pair in both equations.

A system is consistent if there is at least one solution. A system is inconsistent if there is no solution. An inconsistent system is detected by a contradiction (such as $0 = 5$).

A system is independent if the two equations represent different lines. A system is dependent if the two equations represent the same line. This produces infinitely many solutions. A dependent system is detected by an identity (such as $0 = 0$).

Examples

Example 1

$$3x - 4y = 18 \xrightarrow{\text{Mult. by 3.}} 9x - 12y = 54$$
$$-5x - 3y = -1 \xrightarrow{\text{Mult. by } -4.} 20x + 12y = 4$$
$$29x = 58$$
$$x = 2$$

$$3(2) - 4y = 18$$
$$6 - 4y = 18$$
$$-4y = 12$$
$$y = -3$$

The solution is $(2, -3)$ and checks in both equations.

Example 2

$$2x + y = 3 \xrightarrow{\text{Mult. by 2.}} 4x + 2y = 6$$
$$-4x - 2y = 1 \xrightarrow{\phantom{\text{Mult. by 2.}}} -4x - 2y = 1$$
$$0 = 7$$

Contradiction. The system is inconsistent.

Example 3

$$x + 3y = 1 \xrightarrow{\text{Mult. by } -2.} -2x - 6y = -2$$
$$2x + 6y = 2 \xrightarrow{\phantom{\text{Mult. by } -2.}} 2x + 6y = 2$$
$$0 = 0$$

Identity. The system is dependent.

section 3.4 Applications of Systems of Linear Equations in Two Variables

Key Concepts

Solve application problems by using systems of linear equations in two variables.

- Cost applications
- Mixture applications
- Applications involving principal and interest
- Applications involving distance, rate, and time
- Geometry applications

Steps to Solve Applications:

1. Label two variables.
2. Construct two equations in words.
3. Write two equations.
4. Solve the system.
5. Write the answer.

Examples

Example 1

Mercedes invested $1500 more in a certificate of deposit that pays 6.5% simple interest than she did in a savings account that pays 4% simple interest. If her total interest at the end of 1 year is $622.50, find the amount she invested in the 6.5% account.

Let x represent the amount of money invested at 6.5%.
Let y represent the amount of money invested at 4%.

$$\begin{pmatrix} \text{Amount invested} \\ \text{at } 6.5\% \end{pmatrix} = \begin{pmatrix} \text{amount invested} \\ \text{at } 4\% \end{pmatrix} + \$1500$$

$$\begin{pmatrix} \text{Interest earned} \\ \text{from } 6.5\% \\ \text{account} \end{pmatrix} + \begin{pmatrix} \text{interest earned} \\ \text{from } 4\% \\ \text{account} \end{pmatrix} = \$622.50$$

$$x = 1500 + y$$
$$0.065x + 0.04y = 622.50$$

Using substitution gives

$$0.065(1500 + y) + 0.04y = 622.50$$
$$97.5 + 0.065y + 0.04y = 622.50$$
$$0.105y = 525$$
$$y = 5000$$

$x = 1500 + 5000 = 6500$

Mercedes invested $6500 at 6.5% and $5000 at 4%.

section 3.5 Systems of Linear Equations in Three Variables and Applications

Key Concepts

A **linear equation in three variables** can be written in the form $Ax + By + Cz = D$, where A, B, and C are not all zero. The graph of a linear equation in three variables is a plane in space.

A solution to a system of linear equations in three variables is an **ordered triple** that satisfies each equation. Graphically, a solution is a point of intersection among three planes.

A system of linear equations in three variables may have one unique solution, infinitely many solutions (dependent system), or no solution (inconsistent system).

Examples

Example 1

$\boxed{A}$ $\quad x + 2y - z = 4$
$\boxed{B}$ $\quad 3x - y + z = 5$
$\boxed{C}$ $\quad 2x + 3y + 2z = 7$

$\boxed{A}$ and $\boxed{B}$ $\quad x + 2y - z = 4$
$\phantom{\boxed{A} \text{ and } \boxed{B} \quad}\underline{3x - y + z = 5}$
$\phantom{\boxed{A} \text{ and } \boxed{B} \quad}4x + y = 9 \;\boxed{D}$

$2 \cdot \boxed{A}$ and $\boxed{C}$ $\quad 2x + 4y - 2z = 8$
$\phantom{2 \cdot \boxed{A} \text{ and } \boxed{C} \quad}\underline{2x + 3y + 2z = 7}$
$\phantom{2 \cdot \boxed{A} \text{ and } \boxed{C} \quad}4x + 7y = 15 \;\boxed{E}$

$\boxed{D}$ $\;4x + y = 9 \rightarrow -4x - y = -9$
$\boxed{E}$ $\;4x + 7y = 15 \rightarrow \underline{4x + 7y = 15}$
$\phantom{\boxed{E} \;4x + 7y = 15 \rightarrow \;}6y = 6$
$\phantom{\boxed{E} \;4x + 7y = 15 \rightarrow \;\;\;}y = 1$

Substitute $y = 1$ into either equation $\boxed{D}$ or $\boxed{E}$.

$\boxed{D}$ $\;4x + (1) = 9$
$\phantom{\boxed{D} \;\;}4x = 8$
$\phantom{\boxed{D} \;\;\;}x = 2$

Substitute $x = 2$ and $y = 1$ into equation $\boxed{A}$, $\boxed{B}$, or $\boxed{C}$.

$\boxed{A}$ $\;(2) + 2(1) - z = 4$
$\phantom{\boxed{A} \;(2) + 2(1) -\;\;}z = 0$

The solution is $(2, 1, 0)$.

section 3.6 Solving Systems of Linear Equations by Using Matrices

Key Concepts

A **matrix** is a rectangular array of numbers displayed in rows and columns. Every number or entry within a matrix is called an element of the matrix.

The **order of a matrix** is determined by the number of rows and number of columns. A matrix with m rows and n columns is an $m \times n$ matrix.

A system of equations written in standard form can be represented by an **augmented matrix** consisting of the coefficients of the terms of each equation in the system.

The Gauss-Jordan method can be used to solve a system of equations by using the following elementary row operations on an augmented matrix.

1. Interchange two rows.
2. Multiply every element in a row by a nonzero real number.
3. Add a multiple of one row to another row.

These operations are used to write the matrix in **reduced row echelon form**.

$$\begin{bmatrix} 1 & 0 & | & a \\ 0 & 1 & | & b \end{bmatrix}$$

which represents the solution, $x = a$ and $y = b$.

Examples

Example 1

$[1 \ 2 \ 5]$ is a 1×3 matrix (called a row matrix).

$\begin{bmatrix} -1 & 8 \\ 1 & 5 \end{bmatrix}$ is a 2×2 matrix (called a square matrix).

$\begin{bmatrix} 4 \\ 1 \end{bmatrix}$ is a 2×1 matrix (called a column matrix).

Example 2

The augmented matrix for

$4x + y = -12$

$x - 2y = 6$

is

$$\begin{bmatrix} 4 & 1 & | & -12 \\ 1 & -2 & | & 6 \end{bmatrix}$$

Example 3

Solve the system from Example 2 by using the Gauss-Jordan method.

$R_1 \Leftrightarrow R_2 \quad \begin{bmatrix} 1 & -2 & | & 6 \\ 4 & 1 & | & -12 \end{bmatrix}$

$-4R_1 + R_2 \Rightarrow R_2 \quad \begin{bmatrix} 1 & -2 & | & 6 \\ 0 & 9 & | & -36 \end{bmatrix}$

$\frac{1}{9}R_2 \Rightarrow R_2 \quad \begin{bmatrix} 1 & -2 & | & 6 \\ 0 & 1 & | & -4 \end{bmatrix}$

$2R_2 + R_1 \Rightarrow R_1 \quad \begin{bmatrix} 1 & 0 & | & -2 \\ 0 & 1 & | & -4 \end{bmatrix}$

Solution:

$x = -2 \quad$ and $\quad y = -4$

chapter 3 review exercises

Section 3.1

1. Determine if the ordered pair is a solution to the system.

$$-5x - 7y = 4$$
$$y = -\frac{1}{2}x - 1$$

   a. $(2, 2)$ b. $(2, -2)$

For Exercises 2–4, answer true or false.

2. An inconsistent system has one solution.

3. Parallel lines form an inconsistent system.

4. Lines with different slopes intersect in one point.

For Exercises 5–7, solve the system by graphing.

5. $y = 2x + 7$
   $y = -x - 5$

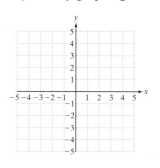

6. $6x + 2y = 4$
   $3x = -y + 2$

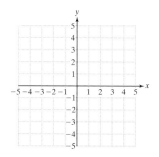

7. $y = \frac{1}{2}x - 2$
   $-4x + 8y = -8$

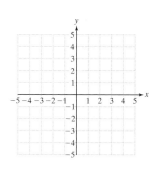

Section 3.2

For Exercises 8–11, solve the systems by using the substitution method.

8. $y = \frac{3}{4}x - 4$
   $-x + 2y = -6$

9. $3x = 11y - 9$
   $y = \frac{3}{11}x + \frac{6}{11}$

10. $4x + y = 7$
    $x + \frac{1}{4}y = \frac{7}{4}$

11. $6x + y = 5$
    $5x + y = 3$

Section 3.3

For Exercises 12–21, solve the systems by using the addition method.

12. $\frac{2}{5}x + \frac{3}{5}y = 1$
    $x - \frac{2}{3}y = \frac{1}{3}$

13. $4x + 3y = 5$
    $3x - 4y = 10$

14. $3x + 4y = 2$
    $2x + 5y = -1$

15. $3x + y = 1$
    $-x - \frac{1}{3}y = -\frac{1}{3}$

16. $2y = 3x - 8$
    $-6x = -4y + 4$

17. $x + 2y = 4$
    $3(x + y) = y + 2x + 2$

18. $-(y + 4x) = 2x - 9$
    $-2x + 2y = -10$

19. $-(4x - 35) = 3y$
    $-(x - 15) = y$

20. $-0.4x + 0.3y = 1.8$
    $0.6x - 0.2y = -1.2$

21. $0.02x - 0.01y = -0.11$
    $0.01x + 0.04y = 0.26$

Section 3.4

22. Antonio invested twice as much money in an account paying 5% simple interest as he did in an account paying 3.5% simple interest. If his total interest at the end of 1 year is $303.75, find the amount he invested in the 5% account.

23. A school carnival sold tickets to ride on a Ferris wheel. The charge was $1.50 for an adult and $1.00 for students. If 54 tickets were sold for a total of $70.50, how many of each type of ticket were sold?

24. How many liters of 20% saline solution must be mixed with 50% saline solution to produce 16 L of a 31.25% saline solution?

25. It takes a pilot $1\frac{3}{4}$ hr to travel with the wind to get from Jacksonville, Florida, to Myrtle Beach, South Carolina. Her return trip takes 2 hr flying against the wind. What is the speed of the wind and the speed of the plane in still air if the distance between Jacksonville and Myrtle Beach is 280 mi?

26. Two phone companies offer discount rates to students.

Company 1: $9.95 per month, plus $0.10 per minute for long-distance calls

Company 2: $12.95 per month, plus $0.08 per minute for long-distance calls

   a. Write a linear equation describing the total cost, y, for x min of long-distance calls from Company 1.

   b. Write a linear equation describing the total cost, y, for x min of long-distance calls from Company 2.

   c. How many minutes of long-distance calls would result in equal cost for both offers?

27. Two angles are complementary. One angle measures 6° more than 5 times the measure of the other. What are the measures of the two angles?

Section 3.5

For Exercises 28–31, solve the systems of equations. If a system does not have a unique solution, indicate whether it is dependent or inconsistent.

28. $5x + 5y + 5z = 30$
$-x + y + z = 2$
$10x + 6y - 2z = 4$

29. $5x + 3y - z = 5$
$x + 2y + z = 6$
$-x - 2y - z = 8$

30. $3x + 4z = 5$
$ 2y + 3z = 2$
$2x - 5y = 8$

31. $x + y + z = 4$
$-x - 2y - 3z = -6$
$2x + 4y + 6z = 12$

32. The perimeter of a right triangle is 30 ft. One leg is 2 ft longer than twice the shortest leg. The hypotenuse is 2 ft less than 3 times the shortest leg. Find the lengths of the sides of this triangle.

33. Three pumps are working to drain a construction site. Working together, the pumps can pump 950 gal/hr of water. The slowest pump pumps 150 gal/hr less than the fastest pump. The fastest pump pumps 150 gal/hr less than the sum of the other two pumps. How many gallons can each pump drain per hour?

Section 3.6

For Exercises 34–37, determine the order of each matrix.

34. $\begin{bmatrix} 2 & 4 & -1 \\ 5 & 0 & -3 \\ -1 & 6 & 10 \end{bmatrix}$

35. $\begin{bmatrix} -5 & 6 \\ 9 & 2 \\ 0 & -3 \end{bmatrix}$

36. $\begin{bmatrix} 0 & 13 & -4 & 16 \end{bmatrix}$

37. $\begin{bmatrix} 7 \\ 12 \\ -4 \end{bmatrix}$

For Exercises 38–39, set up the augmented matrix.

38. $x + y = 3$
$x - y = -1$

39. $x - y + z = 4$
$2x - y + 3z = 8$
$-2x + 2y - z = -9$

For Exercises 40–41, write a corresponding system of equations from the augmented matrix.

40. $\begin{bmatrix} 1 & 0 & | & 9 \\ 0 & 1 & | & -3 \end{bmatrix}$
41. $\begin{bmatrix} 1 & 0 & 0 & | & -5 \\ 0 & 1 & 0 & | & 2 \\ 0 & 0 & 1 & | & -8 \end{bmatrix}$

42. Given the matrix **C**

$$C = \begin{bmatrix} 1 & 3 & | & 1 \\ 4 & -1 & | & 6 \end{bmatrix}$$

   a. What is the element in the second row and first column?

   b. Write the matrix obtained by multiplying the first row by -4 and adding the result to row 2.

43. Given the matrix **D**

$$D = \begin{bmatrix} 1 & 2 & 0 & | & -3 \\ 4 & -1 & 1 & | & 0 \\ -3 & 2 & 2 & | & 5 \end{bmatrix}$$

   a. Write the matrix obtained by multiplying the first row by -4 and adding the result to row 2.

   b. Using the matrix obtained in part (a), write the matrix obtained by multiplying the first row by 3 and adding the result to row 3.

For Exercises 44–47, solve the system by using the Gauss-Jordan method.

44. $x + y = 3$
    $x - y = -1$

45. $4x + 3y = 6$
    $12x + 5y = -6$

46. $x - y + z = -4$
    $2x + y - 2z = 9$
    $x + 2y + z = 5$

47. $x - y + z = 4$
    $2x - y + 3z = 8$
    $-2x + 2y - z = -9$

---

chapter 3 | test

1. Determine if the ordered pair $(\frac{1}{4}, 2)$ is a solution to the system.

   $4x - 3y = -5$
   $12x + 2y = 7$

Match each figure with the appropriate description.

2.
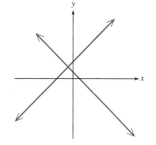

3.
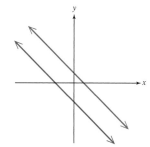

4.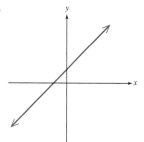

   a. The system is consistent and dependent. There are infinitely many solutions.

   b. The system is consistent and independent. There is one solution.

   c. The system is inconsistent and independent. There are no solutions.

5. Solve the system by graphing.

$4x - 2y = -4$

$3x + y = 7$

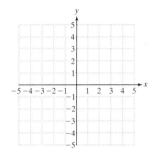

6. Solve the system by using the substitution method.

$3x + 5y = 13$

$y = x + 9$

7. Solve the system by using the addition method.

$6x + 8y = 5$

$3x - 2y = 1$

For Exercises 8–13, solve the system of equations.

8. $7y = 5x - 21$

$9y + 2x = -27$

9. $3x - 5y = -7$

$-18x + 30y = 42$

10. $\frac{1}{5}x = \frac{1}{2}y + \frac{17}{5}$

$\frac{1}{4}(x + 2) = -\frac{1}{6}y$

11. $4x = 5 - 2y$

$y = -2x + 4$

12. $-0.03y + 0.06x = 0.3$

$0.4x - 2 = -0.5y$

13. $2x + 2y + 4z = -6$

$3x + y + 2z = 29$

$x - y - z = 44$

14. How many liters of a 20% acid solution should be mixed with a 60% acid solution to produce 200 L of a 44% acid solution?

15. Two angles are complementary. Two times the measure of one angle is 60° less than the measure of the other. Find the measure of each angle.

16. Working together, Joanne, Kent, and Geoff can process 504 orders per day for their business. Kent can process 20 more orders per day than Joanne can process. Geoff can process 104 fewer orders per day than Kent and Joanne combined. Find the number of orders that each person can process per day.

17. Write an example of a 3×2 matrix.

18. Given the matrix **A**

$$\mathbf{A} = \left[\begin{array}{ccc|c} 1 & 2 & 1 & -3 \\ 4 & 0 & 1 & -2 \\ -5 & -6 & 3 & 0 \end{array} \right]$$

a. Write the matrix obtained by multiplying the first row by -4 and adding the result to row 2.

b. Using the matrix obtained in part (a), write the matrix obtained by multiplying the first row by 5 and adding the result to row 3.

For Exercises 19–20, solve by using the Gauss-Jordan method.

19. $5x - 4y = 34$

$x - 2y = 8$

20. $x + y + z = 1$

$2x + y = 0$

$-2y - z = 5$

chapters 1–3 | cumulative review exercises

For Exercises 1–2, solve the equation.

1. $-5(2x - 1) - 2(3x + 1) = 7 - 2(8x + 1)$

2. $\frac{1}{2}(a - 2) - \frac{3}{4}(2a + 1) = -\frac{1}{6}$

3. Solve the inequality. Write the answer in interval notation.
$$-3y - 2(y + 1) < 5$$

4. Identify the slope and the x- and y-intercepts of the line $5x - 2y = 15$.

For Exercises 5–6, graph the lines.

5. $y = -\frac{1}{3}x - 4$ 6. $x = -2$

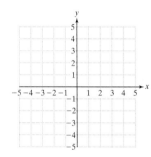 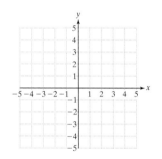

7. Find the slope of the line passing through the points $(4, -10)$ and $(6, -10)$.

8. Find an equation for the line that passes through the points $(3, -8)$ and $(2, -4)$. Write the answer in slope-intercept form.

9. Solve the system by using the addition method.
$$2x - 3y = 6$$
$$\frac{1}{2}x - \frac{3}{4}y = 1$$

10. Solve the system by using the substitution method.
$$2x + y = 4$$
$$y = 3x - 1$$

11. A child's piggy bank contains 19 coins consisting of nickels, dimes, and quarters. The total amount of money in the bank is $3.05. If the number of quarters is 1 more than twice the number of nickels, find the number of each type of coin in the bank.

12. Two video clubs rent tapes according to the following fee schedules:

    Club 1: $25 initiation fee plus $2.50 per tape
    Club 2: $10 initiation fee plus $3.00 per tape

    a. Write a linear equation describing the total cost, y, of renting x tapes from club 1.

    b. Write a linear equation describing the total cost, y, of renting x tapes from club 2.

    c. How many tapes would have to be rented to make the cost for club 1 the same as the cost for club 2?

13. Solve the system.
$$3x + 2y + 3z = 3$$
$$4x - 5y + 7z = 1$$
$$2x + 3y - 2z = 6$$

14. Determine the order of the matrix.
$$\begin{bmatrix} 4 & 5 & 1 \\ -2 & 6 & 0 \end{bmatrix}$$

15. Write an example of a 2×4 matrix.

16. List at least two different row operations.

17. Solve the system by using the Gauss-Jordan method.
$$2x - 4y = -2$$
$$4x + y = 5$$

Introduction to Relations and Functions

4

4.1 Introduction to Relations
4.2 Introduction to Functions
4.3 Graphs of Basic Functions
4.4 Variation

In this chapter, we introduce the concept of a function. In general terms, a function defines how one variable depends on one or more other variables. In Exercise 83 in Section 4.2, we learn that the score P (in percent) on a certain exam is a function of the number of hours x spent studying for the exam. Using the function

$$P(x) = \frac{100x^2}{50 + x^2}$$

we see that a student who spends 5 hr studying will receive a score of 33.3%. On the other hand, a student who spends 30 hr studying will receive a score of 94.7%.

Chapter 4 Preview

The exercises in this chapter preview contain concepts that have not yet been presented. These exercises are provided for students who want to compare their levels of understanding before and after studying the chapter. Alternatively, you may prefer to work these exercises when the chapter is completed and before taking the exam.

Section 4.1

1. The table shown relates the number of hours per week spent studying math to the final average in the math course.

Hours per Week of Study Time	Final Average in Course
8	88
2	70
0	45
6	95
8	90

   a. Write the relation as a set of ordered pairs.

   b. List the domain and range.

Section 4.2

For Exercises 2–3, determine whether each relation defines y as a function of x.

2. $\{(1, 5), (1, 6), (1, 7)\}$

3. $\{(2, -1), (5, 6), (-7, -1), (8, 8), (0, 5)\}$

4. Given $f(x) = -2x^2 - x$, find

   a. $f(0)$ b. $f(-2)$ c. $f(h)$

For Exercises 5–7, refer to the graph.

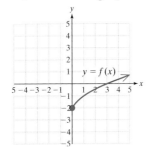

5. Find $f(1)$.

6. For what value of x is $f(x) = 0$?

7. Write the domain and range in interval notation.

8. Determine the domain of each function. Express the domain in interval notation.

   a. $f(x) = \dfrac{4x}{x - 2}$ b. $g(x) = \sqrt{x + 7}$

Section 4.3

9. Identify the functions as linear, constant, or neither.

   a. $f(x) = x^2 + 2$ b. $f(x) = -9$

   c. $f(x) = x - 4$

10. The Wiretap Phone Company charges $7.95 per month plus 8¢ per minute for long-distance service. The function describing the cost per month is $C(t) = 7.95 + 0.08t$, where t is the time in minutes.

    a. Is the function linear?

    b. What is the monthly charge for using 30 min of long-distance service?

    c. Find $C(0)$ and interpret the result in the context of this problem.

Section 4.4

11. Translate to an equivalent mathematical model. Use k as the constant of variation.

    v varies directly as the square of u.

12. The intensity I of a lightbulb varies inversely as the square of the distance d from the bulb. The intensity is 50 candlepower at 4 ft. Find the intensity at a distance of 5 ft.

section 4.1 Introduction to Relations

Objectives
1. Definition of a Relation
2. Domain and Range of a Relation
3. Applications Involving Relations

1. Definition of a Relation

In many naturally occurring phenomena, two variables may be linked by some other type of relationship. Table 4-1 shows a correspondence between the length of a woman's femur and her height. (The femur is the large bone in the thigh attached to the knee and hip.)

table 4-1

Length of Femur (cm) x	Height (in.) y		Ordered Pair
45.5	65.5	→	(45.5, 65.5)
48.2	68.0	→	(48.2, 68.0)
41.8	62.2	→	(41.8, 62.2)
46.0	66.0	→	(46.0, 66.0)
50.4	70.0	→	(50.4, 70.0)

Each data point from Table 4-1 may be represented as an ordered pair. In this case, the first value represents the length of a woman's femur and the second, the woman's height. The set of ordered pairs {(45.5, 65.5), (48.2, 68.0), (41.8, 62.2), (46.0, 66.0), (50.4, 70.0)} defines a relation between femur length and height.

2. Domain and Range of a Relation

Definition of a Relation in x and y

Any set of ordered pairs (x, y) is called a **relation in x and y**. Furthermore,

- The set of first components in the ordered pairs is called the **domain of the relation**.
- The set of second components in the ordered pairs is called the **range of the relation**.

example 1 Finding the Domain and Range of a Relation

Find the domain and range of the relation linking the length of a woman's femur to her height {(45.5, 65.5), (48.2, 68.0), (41.8, 62.2), (46.0, 66.0), (50.4, 70.0)}.

Solution:

Domain: {45.5, 48.2, 41.8, 46.0, 50.4}   Set of first coordinates

Range: {65.5, 68.0, 62.2, 66.0, 70.0}   Set of second coordinates

Skill Practice
1. Find the domain and range of the relation.
$$\left\{(0, 0), (-8, 4), \left(\frac{1}{2}, 1\right), (-3, 4), (-8, 0)\right\}$$

Answer

1. Domain $\left\{0, -8, \frac{1}{2}, -3\right\}$, range {0, 4, 1}

252   Chapter 4   Introduction to Relations and Functions

The x- and y-components that constitute the ordered pairs in a relation do not need to be numerical. For example, Table 4-2 depicts five states in the United States and the corresponding number of representatives in the House of Representatives as of July 2005.

table 4-2

State x	Number of Representatives y
Alabama	7
California	53
Colorado	7
Florida	25
Kansas	4

These data define a relation:

{(Alabama, 7), (California, 53), (Colorado, 7), (Florida, 25), (Kansas, 4)}

Skill Practice

The table depicts six types of animal and the corresponding longevity.

Animal x	Longevity (Years) y
Bear	22.5
Cat	11
Cow	20.5
Deer	12.5
Dog	11
Elephant	35

2. Write the ordered pairs indicated by the relation in the table.
3. Find the domain and range of the relation.

example 2   Finding the Domain and Range of a Relation

Find the domain and range of the relation

{(Alabama, 7), (California, 53), (Colorado, 7), (Florida, 25), (Kansas, 4)}

Solution:

Domain: {Alabama, California, Colorado, Florida, Kansas}

Range: {7, 53, 25, 4}   (*Note:* The element 7 is not listed twice.)

A relation may consist of a finite number of ordered pairs or an infinite number of ordered pairs. Furthermore, a relation may be defined by several different methods: by a list of ordered pairs, by a correspondence between the domain and range, by a graph, or by an equation.

- A relation may be defined as a set of ordered pairs.

    {(1, 2), (−3, 4), (1, −4), (3, 4)}

- A relation may be defined by a correspondence (Figure 4-1). The corresponding ordered pairs are {(1, 2), (1, −4), (−3, 4), (3, 4)}.

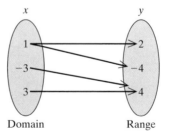

Figure 4-1

Answers

2. {(Bear, 22.5), (Cat, 11), (Cow, 20.5), (Deer, 12.5), (Dog, 11), (Elephant, 35)}
3. Domain: {Bear, Cat, Cow, Deer, Dog, Elephant}; range: {22.5, 11, 20.5, 12.5, 35}

- A relation may be defined by a graph (Figure 4-2). The corresponding ordered pairs are {(1, 2), (−3, 4), (1, −4), (3, 4)}.

- A relation may be expressed by an equation such as $x = y^2$. The solutions to this equation define an infinite set of ordered pairs of the form $\{(x, y) \mid x = y^2\}$. The solutions can also be represented by a graph in a rectangular coordinate system (Figure 4-3).

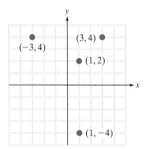

Figure 4-2

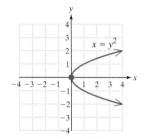

Figure 4-3

example 3 Finding the Domain and Range of a Relation

Find the domain and range of the following relations:

Solution:

a.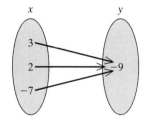

Domain: {3, 2, −7}

Range: {−9}

b.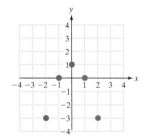

The domain elements are the *x*-coordinates of the points, and the range elements are the *y*-coordinates.

Domain: {−2, −1, 0, 1, 2}

Range: {−3, 0, 1}

c.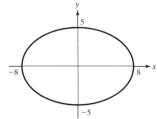

The domain consists of an infinite number of *x*-values extending from −8 to 8 (shown in red). The range consists of all *y*-values from −5 to 5 (shown in blue). Thus, the domain and range must be expressed in set-builder notation or in interval notation.

Domain: $\{x \mid x$ is a real number and $-8 \leq x \leq 8\}$ or $[-8, 8]$

Range: $\{y \mid y$ is a real number and $-5 \leq y \leq 5\}$ or $[-5, 5]$

Skill Practice

Find the domain and range of the relations.

4.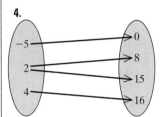

5.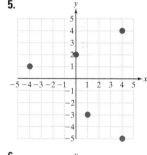

6.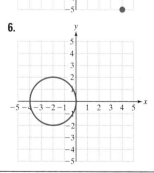

Answers
4. Domain {−5, 2, 4}, range {0, 8, 15, 16}
5. Domain {−4, 0, 1, 4}, range {−5, −3, 1, 2, 4}
6. Domain: $\{x \mid x$ is a real number and $-4 \leq x \leq 0\}$ or $[-4, 0]$
Range: $\{y \mid y$ is a real number and $-2 \leq y \leq 2\}$ or $[-2, 2]$

d. $x = y^2$

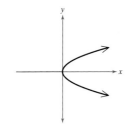

The arrows on the curve indicate that the graph extends infinitely far up and to the right and infinitely far down and to the right.

Domain: $\{x \mid x \text{ is a real number and } x \geq 0\}$ or $[0, \infty)$

Range: $\{y \mid y \text{ is any real number}\}$ or $(-\infty, \infty)$

3. Applications Involving Relations

Skill Practice

The linear equation, $y = -0.014x + 64.5$, relates the weight of a car, x, (in pounds) to its gas mileage, y, (in mpg). Use this relationship for the following exercises.

7. Find the gas mileage in miles per gallon for a car weighing 2550 lb.
8. Find the gas mileage for a car weighing 2850 lb.

example 4 Analyzing a Relation

The data in Table 4-3 depict the length of a woman's femur and her corresponding height. Based on these data, a forensics specialist or archeologist can find a linear relationship between height y and femur length x:

$$y = 0.906x + 24.3 \qquad 40 \leq x \leq 55$$

From this type of relationship, the height of a woman can be inferred based on skeletal remains.

a. Find the height of a woman whose femur is 46.0 cm.

table 4-3

Length of Femur (cm) x	Height (in.) y
45.5	65.5
48.2	68.0
41.8	62.2
46.0	66.0
50.4	70.0

b. Find the height of a woman whose femur is 51.0 cm.

c. Why is the domain restricted to $40 \leq x \leq 55$?

Solution:

a. $y = 0.906x + 24.3$
$= 0.906(46.0) + 24.3$ Substitute $x = 46.0$ cm.
$= 65.976$ The woman is approximately 66.0 in. tall.

b. $y = 0.906x + 24.3$
$= 0.906(51.0) + 24.3$ Substitute $x = 51.0$ cm.
$= 70.506$ The woman is approximately 70.5 in. tall.

c. The domain restricts femur length to values between 40 cm and 55 cm inclusive. These values are within the normal lengths for an adult female and are in the proximity of the observed data (Figure 4-4).

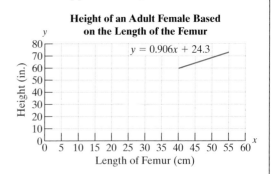

Figure 4-4

Answers

7. 28.8 mpg
8. 24.6 mpg

Section 4.1 Practice Exercises

Boost *your* GRADE at mathzone.com!

- Practice Problems
- Self-Tests
- NetTutor
- e-Professors
- Videos

Study Skills Exercises

1. After you get a test back, it is a good idea to correct the test so that you do not make the same errors again. One recommended approach is to use a clean sheet of paper, and divide the paper down the middle vertically, as shown. For each problem that you missed on the test, rework the problem correctly on the left-hand side of the paper. Then write an explanation on the right-hand side of the paper.

   Take the time this week to make corrections from your last test.

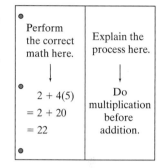

Perform the correct math here.	Explain the process here.
↓	↓
2 + 4(5) = 2 + 20 = 22	Do multiplication before addition.

2. Define the key terms.

   a. **Relation in x and y**

   b. **Domain of a relation**

   c. **Range of a relation**

Objective 1: Definition of a Relation

For Exercises 3–6, write each relation as a set of ordered pairs.

3.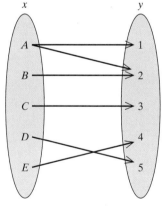

4.

State x	Year of Statehood y
Connecticut	1788
Colorado	1876
Maryland	1788
Illinois	1818
Missouri	1821

5. Reference daily intake (RDI) for proteins

Group x	RDI (g) y
Pregnant women	60
Nursing mothers	65
Infants younger than 1 year old	14
Children from 1 to 4 years old	16
Adults	50

6.

x	y
0	3
−2	$\frac{1}{2}$
5	10
−7	1
−2	8
5	1

256   Chapter 4   Introduction to Relations and Functions

Objective 2: Domain and Range of a Relation

7. List the domain and range of Exercise 3. (See Examples 1–2.)

8. List the domain and range of Exercise 4.

9. List the domain and range of Exercise 5.

10. List the domain and range of Exercise 6.

For Exercises 11–24, find the domain and range of the relations. Use interval notation where appropriate. (See Example 3.)

11.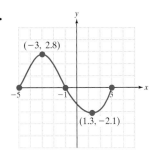

12.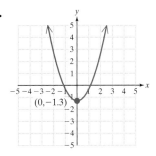

13.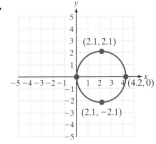

14.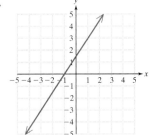

15.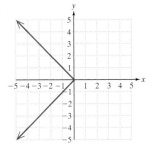

16.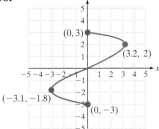

17. *Hint:* The open circle indicates that the end point is not included in the relation.

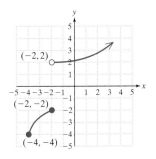

18.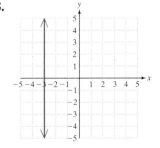

19.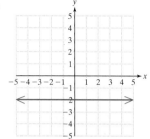

20.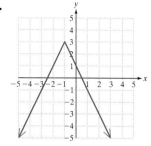

21.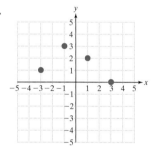

22.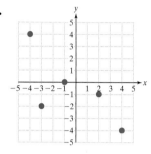

23.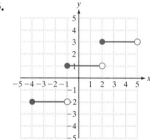

24.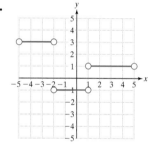

Objective 3: Applications Involving Relations

25. The following table gives a relation between the month of the year and the average precipitation for that month for Miami, Florida. **(See Example 4.)**

Month x	Precipitation (in.) y	Month x	Precipitation (in.) y
Jan.	2.01	July	5.70
Feb.	2.08	Aug.	7.58
Mar.	2.39	Sept.	7.63
Apr.	2.85	Oct.	5.64
May	6.21	Nov.	2.66
June	9.33	Dec.	1.83

Source: U.S. National Oceanic and Atmospheric Administration

a. What is the range element corresponding to April?

b. What is the range element corresponding to June?

c. Which element in the domain corresponds to the least value in the range?

d. Complete the ordered pair: (, 2.66)

e. Complete the ordered pair: (Sept.,)

f. What is the domain of this relation?

26. The table gives a relation between a person's age and the person's maximum recommended heart rate.

    a. What is the domain?

    b. What is the range?

    c. The range element 200 corresponds to what element in the domain?

    d. Complete the ordered pair: (50,   )

    e. Complete the ordered pair: (, 190)

Age (years) x	Maximum Recommended Heart Rate (Beats per Minute) y
20	200
30	190
40	180
50	170
60	160

27. The population of Canada, y, (in millions) can be approximated by the relation $y = 0.146x + 31$, where x represents the number of years since 2000.

    a. Approximate the population of Canada in the year 2006.

    b. In what year will the population of Canada reach approximately 32,752,000?

28. As of April 2004, the world record times for selected women's track and field events are shown in the table.

    The women's world record time y (in seconds) required to run x meters can be approximated by the relation $y = -10.78 + 0.159x$.

    a. Predict the time required for a 500-m race.

    b. Use this model to predict the time for a 1000-m race. Is this value exactly the same as the data value given in the table? Explain.

Distance (m)	Time (sec)	Winner's Name and Country
100	10.49	Florence Griffith Joyner (United States)
200	21.34	Florence Griffith Joyner (United States)
400	47.60	Marita Koch (East Germany)
800	113.28	Jarmila Kratochvilova (Czechoslovakia)
1000	148.98	Svetlana Masterkova (Russia)
1500	230.46	Qu Yunxia (China)

Expanding Your Skills

29. a. Define a relation with four ordered pairs such that the first element of the ordered pair is the name of a friend and the second element is your friend's place of birth.

    b. State the domain and range of this relation.

30. a. Define a relation with four ordered pairs such that the first element is a state and the second element is its capital.

    b. State the domain and range of this relation.

31. Use a mathematical equation to define a relation whose second component y is 1 less than 2 times the first component x.

32. Use a mathematical equation to define a relation whose second component y is 3 more than the first component x.

---

section 4.2 Introduction to Functions

Objectives
1. Definition of a Function
2. Vertical Line Test
3. Function Notation
4. Finding Function Values from a Graph
5. Domain of a Function

1. Definition of a Function

In this section we introduce a special type of relation called a function.

> **Definition of a Function**
>
> Given a relation in x and y, we say "y is a **function** of x" if for every element x in the domain, there corresponds exactly one element y in the range.

To understand the difference between a relation that is a function and a relation that is not a function, consider Example 1.

example 1 Determining Whether a Relation Is a Function

Determine which of the relations define y as a function of x.

a.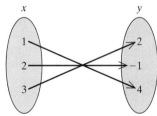

b.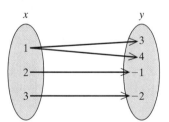

c.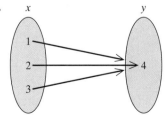

Skill Practice

Determine if the relations define y as a function of x.

1.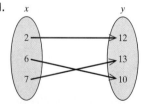

2. $\{(4,2), (-5,4), (0,0), (8,4)\}$

3. $\{(-1,6), (8,9), (-1,4), (-3,10)\}$

Answers
1. Yes 2. Yes 3. No

Solution:

a. This relation is defined by the set of ordered pairs $\{(1, 4), (2, -1), (3, 2)\}$.

Notice that for each x in the domain there is only one corresponding y in the range. Therefore, this relation is a function.

When $x = 1$, there is only one possibility for y: $y = 4$

When $x = 2$, there is only one possibility for y: $y = -1$

When $x = 3$, there is only one possibility for y: $y = 2$

b. This relation is defined by the set of ordered pairs

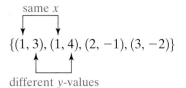

When $x = 1$, there are *two* possible range elements: $y = 3$ and $y = 4$. Therefore, this relation is *not* a function.

c. This relation is defined by the set of ordered pairs $\{(1, 4), (2, 4), (3, 4)\}$.

When $x = 1$, there is only one possibility for y: $y = 4$

When $x = 2$, there is only one possibility for y: $y = 4$

When $x = 3$, there is only one possibility for y: $y = 4$

Because each value of x in the domain has only one corresponding y value, this relation is a function.

2. Vertical Line Test

A relation that is not a function has at least one domain element x paired with more than one range value y. For example, the ordered pairs $(4, 2)$ and $(4, -2)$ do not constitute a function because two different y-values correspond to the same x. These two points are aligned vertically in the xy-plane, and a vertical line drawn through one point also intersects the other point. Thus if a vertical line drawn through a graph of a relation intersects the graph in more than one point, the relation cannot be a function. This idea is stated formally as the **vertical line test**.

> **The Vertical Line Test**
>
> Consider a relation defined by a set of points (x, y) in a rectangular coordinate system. The graph defines y as a function of x if no vertical line intersects the graph in more than one point.

The vertical line test also implies that if any vertical line drawn through the graph of a relation intersects the relation in more than one point, then the relation does *not define y as a function of x*.

The vertical line test can be demonstrated by graphing the ordered pairs from the relations in Example 1.

a. $\{(1, 4), (2, -1), (3, 2)\}$ **b.** $\{(1, 3), (1, 4), (2, -1), (3, -2)\}$

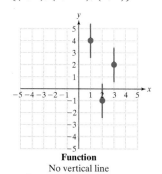
Function
No vertical line intersects more than once.

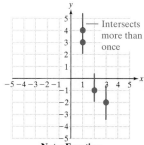
Intersects more than once
Not a Function
A vertical line intersects in more than one point.

example 2   Using the Vertical Line Test

Use the vertical line test to determine whether the following relations define y as a function of x.

a. 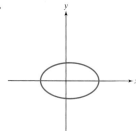   b.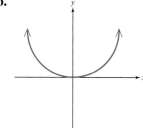

Skill Practice

Use the vertical line test to determine whether the relations define y as a function of x.

4.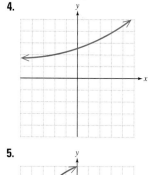

5.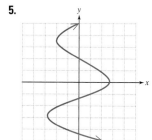

Solution:

a.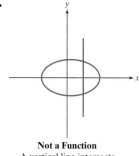
Not a Function
A vertical line intersects in more than one point.

b.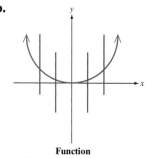
Function
No vertical line intersects in more than one point.

3. Function Notation

A function is defined as a relation with the added restriction that each value in the domain must have only one corresponding y-value in the range. In mathematics, functions are often given by rules or equations to define the relationship between two or more variables. For example, the equation $y = 2x$ defines the set of ordered pairs such that the y-value is twice the x-value.

When a function is defined by an equation, we often use **function notation**. For example, the equation $y = 2x$ may be written in function notation as

Answers
4. Yes   5. No

$$f(x) = 2x$$ where f is the name of the function, x is an input value from the domain of the function, and $f(x)$ is the function value (or y-value) corresponding to x

The notation $f(x)$ is read as "f of x" or "the value of the function f at x."

A function may be evaluated at different values of x by substituting x-values from the domain into the function. For example, to evaluate the function defined by $f(x) = 2x$ at $x = 5$, substitute $x = 5$ into the function.

$$f(x) = 2x$$
$$f(5) = 2(5)$$
$$f(5) = 10$$

Tip: The function value $f(5) = 10$ can be written as the ordered pair $(5, 10)$.

Thus, when $x = 5$, the corresponding function value is 10. We say "f of 5 is 10" or "f at 5 is 10."

The names of functions are often given by either lowercase or uppercase letters, such as f, g, h, p, K, and M.

Skill Practice

Given the function defined by $f(x) = -2x - 3$, find the function values.

6. $f(1)$ 7. $f(0)$
8. $f(-3)$ 9. $f\left(\dfrac{1}{2}\right)$

example 3 Evaluating a Function

Given the function defined by $g(x) = \frac{1}{2}x - 1$, find the function values.

a. $g(0)$ **b.** $g(2)$ **c.** $g(4)$ **d.** $g(-2)$

Solution:

a. $g(x) = \dfrac{1}{2}x - 1$

$g(0) = \dfrac{1}{2}(0) - 1$

$= 0 - 1$

$= -1$ We say, "g of 0 is -1." This is equivalent to the ordered pair $(0, -1)$.

b. $g(x) = \dfrac{1}{2}x - 1$

$g(2) = \dfrac{1}{2}(2) - 1$

$= 1 - 1$

$= 0$ We say "g of 2 is 0." This is equivalent to the ordered pair $(2, 0)$.

c. $g(x) = \dfrac{1}{2}x - 1$

$g(4) = \dfrac{1}{2}(4) - 1$

$= 2 - 1$

$= 1$ We say "g of 4 is 1." This is equivalent to the ordered pair $(4, 1)$.

Answers
6. -5 7. -3 8. 3
9. -4

d. $g(x) = \dfrac{1}{2}x - 1$

$g(-2) = \dfrac{1}{2}(-2) - 1$

$ = -1 - 1$

$ = -2 \qquad$ We say "g of −2 is −2." This is equivalent to the ordered pair $(-2, -2)$.

Notice that $g(0)$, $g(2)$, $g(4)$, and $g(-2)$ correspond to the ordered pairs $(0, -1)$, $(2, 0)$, $(4, 1)$, and $(-2, -2)$. In the graph, these points "line up." The graph of *all* ordered pairs defined by this function is a line with a slope of $\tfrac{1}{2}$ and y-intercept of $(0, -1)$ (Figure 4-5). This should not be surprising because the function defined by $g(x) = \tfrac{1}{2}x - 1$ is equivalent to $y = \tfrac{1}{2}x - 1$.

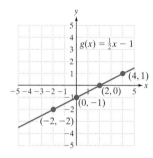

Figure 4-5

Calculator Connections

The values of $g(x)$ in Example 3 can be found using a *Table* feature.

$Y_1 = \tfrac{1}{2}x - 1$

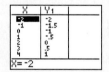

Function values can also be evaluated by using a *Value* (or *Eval*) feature. The value of $g(4)$ is shown here.

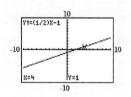

A function may be evaluated at numerical values or at algebraic expressions as shown in Example 4.

example 4   Evaluating Functions

Given the functions defined by $f(x) = x^2 - 2x$ and $g(x) = 3x + 5$, find the function values.

a. $f(t)$ **b.** $g(w + 4)$ **c.** $f(-t)$

Solution:

a. $f(x) = x^2 - 2x$

$f(t) = (t)^2 - 2(t) \qquad$ Substitute $x = t$ for all values of x in the function.

$ = t^2 - 2t \qquad$ Simplify.

Skill Practice

Given the function defined by $g(x) = 4x - 3$, find the function values.

10. $g(a)$ **11.** $g(x + h)$

12. $g(-x)$

Answers

10. $4a - 3$ **11.** $4x + 4h - 3$
12. $-4x - 3$

b. $g(x) = 3x + 5$

$g(w + 4) = 3(w + 4) + 5$ Substitute $x = w + 4$ for all values of x in the function.

$\qquad\qquad = 3w + 12 + 5$

$\qquad\qquad = 3w + 17$ Simplify.

c. $f(x) = x^2 - 2x$ Substitute $-t$ for x.

$f(-t) = (-t)^2 - 2(-t)$

$\qquad\quad = t^2 + 2t$ Simplify.

4. Finding Function Values from a Graph

We can find function values by looking at a graph of the function. The value of $f(a)$ refers to the y-coordinate of a point with x-coordinate a.

Skill Practice

Refer to the function graphed below.

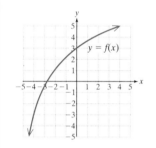

13. Find $f(0)$.
14. Find $f(-4)$.
15. For what value of x is $f(x) = 5$?
16. For what value of x is $f(x) = 1$?

example 5 Finding Function Values from a Graph

Consider the function pictured in Figure 4-6.

a. Find $h(-1)$.

b. Find $h(2)$.

c. For what value of x is $h(x) = 3$?

d. For what values of x is $h(x) = 0$?

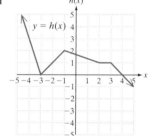

Figure 4-6

Solution:

a. $h(-1) = 2$ This corresponds to the ordered pair $(-1, 2)$.

b. $h(2) = 1$ This corresponds to the ordered pair $(2, 1)$.

c. $h(x) = 3$ for $x = -4$ This corresponds to the ordered pair $(-4, 3)$.

d. $h(x) = 0$ for $x = -3$ and $x = 4$ These are the ordered pairs $(-3, 0)$ and $(4, 0)$.

5. Domain of a Function

A function is a relation, and it is often necessary to determine its domain and range. Consider a function defined by the equation $y = f(x)$. The **domain** of f is the set of all x-values that when substituted into the function, produce a real number. The **range** of f is the set of all y-values corresponding to the values of x in the domain.

To find the domain of a function defined by $y = f(x)$, keep these guidelines in mind.

Exclude values of x that make the denominator of a fraction zero.
Exclude values of x that make the expression within a square root negative.

Answers

13. 3   14. -3
15. 4   16. -2

example 6 Finding the Domain of a Function

Find the domain of the functions. Write the answers in interval notation.

a. $f(x) = \dfrac{x+7}{2x-1}$ **b.** $h(x) = \dfrac{x-4}{x^2+9}$

c. $k(t) = \sqrt{t+4}$ **d.** $g(t) = t^2 - 3t$

Skill Practice

Find the domain of the functions. Write the answers in interval notation.

17. $f(x) = \dfrac{2x+1}{x-9}$

18. $g(x) = \sqrt{x-2}$

19. $h(x) = x + 6$

Solution:

a. The function will not be a real number when the denominator is zero, that is, when

$$2x - 1 = 0$$
$$2x = 1$$
$$x = \dfrac{1}{2}$$ The value $x = \tfrac{1}{2}$ must be *excluded* from the domain.

Interval notation: $\left(-\infty, \dfrac{1}{2}\right) \cup \left(\dfrac{1}{2}, \infty\right)$

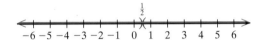

b. The quantity x^2 is greater than or equal to 0 for all real numbers x, and the number 9 is positive. Therefore, the sum $x^2 + 9$ must be *positive* for all real numbers x. The denominator of $h(x) = (x-4)/(x^2+9)$ will never be zero; the domain is therefore the set of all real numbers.

Interval notation: $(-\infty, \infty)$

c. The function defined by $k(t) = \sqrt{t+4}$ will not be a real number when the radicand is negative; hence the domain is the set of all t-values that make the radicand *greater than or equal to zero:*

$$t + 4 \geq 0$$
$$t \geq -4$$

Interval notation: $[-4, \infty)$

d. The function defined by $g(t) = t^2 - 3t$ has no restrictions on its domain because any real number substituted for t will produce a real number. The domain is the set of all real numbers.

Interval notation: $(-\infty, \infty)$

Answers

17. $(-\infty, 9) \cup (9, \infty)$ 18. $[2, \infty)$
19. $(-\infty, \infty)$

Section 4.2 Practice Exercises

Boost your GRADE at mathzone.com!

- Practice Problems
- Self-Tests
- NetTutor
- e-Professors
- Videos

Study Skills Exercises

1. Meet some of the other students in your class. They can be good resources for asking questions and discussing the material that was covered in class. Write the names of two fellow students.

2. Define the key terms.
   a. Function
   b. Function notation
   c. Domain
   d. Range
   e. Vertical line test

Review Exercises

For Exercises 3–4, **a.** write the relation as a set of ordered pairs, **b.** identify the domain, and **c.** identify the range.

3.
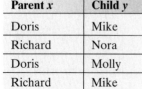

Parent x	Child y
Doris	Mike
Richard	Nora
Doris	Molly
Richard	Mike

4.
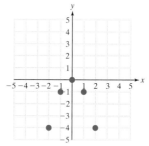

Objective 1: Definition of a Function

For Exercises 5–10, determine if the relation defines y as a function of x. (See Example 1.)

5.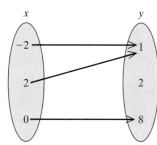

6.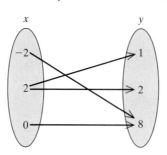

7.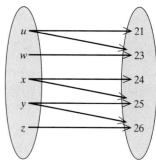

8.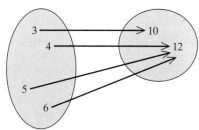

9. $\{(1, 2), (3, 4), (5, 4), (-9, 3)\}$

10. $\left\{(0, -1.1), \left(\dfrac{1}{2}, 8\right), (1.1, 8), \left(4, \dfrac{1}{2}\right)\right\}$

Objective 2: Vertical Line Test

For Exercises 11–16, use the vertical line test to determine whether the relation defines y as a function of x. (See Example 2.)

11.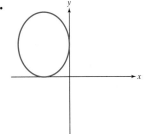

12.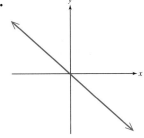

13.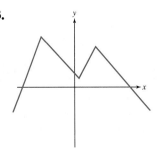

14.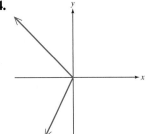

15.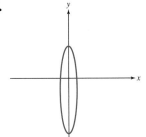

16.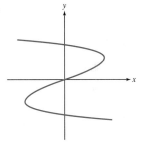

Objective 3: Function Notation

Consider the functions defined by $f(x) = 6x - 2$, $g(x) = -x^2 - 4x + 1$, $h(x) = 7$, and $k(x) = |x - 2|$. For Exercises 17–48, find the following. (See Examples 3–4.)

17. $g(2)$

18. $k(2)$

19. $g(0)$

20. $h(0)$

21. $k(0)$

22. $f(0)$

23. $f(t)$

24. $g(a)$

25. $h(u)$

26. $k(v)$

27. $g(-3)$

28. $h(-5)$

29. $k(-2)$

30. $f(-6)$

31. $f(x + 1)$

32. $h(x + 1)$

33. $g(x - 2)$

34. $k(x - 3)$

35. $k(x + h)$

36. $g(x + h)$

37. $h(a + b)$

38. $f(x + h)$

39. $f(-a)$

40. $g(-b)$

41. $k(-c)$

42. $h(-x)$

43. $f\left(\dfrac{1}{2}\right)$

44. $g\left(\dfrac{1}{4}\right)$

45. $h\left(\dfrac{1}{7}\right)$

46. $k\left(\dfrac{3}{2}\right)$

47. $f(-2.8)$

48. $k(-5.4)$

Consider the functions $p = \{(\frac{1}{2}, 6), (2, -7), (1, 0), (3, 2\pi)\}$ and $q = \{(6, 4), (2, -5), (\frac{3}{4}, \frac{1}{5}), (0, 9)\}$. For Exercises 49–56, find the function values.

49. $p(2)$ **50.** $p(1)$ **51.** $p(3)$ **52.** $p\left(\dfrac{1}{2}\right)$

53. $q(2)$ **54.** $q\left(\dfrac{3}{4}\right)$ **55.** $q(6)$ **56.** $q(0)$

Objective 4: Finding Function Values from a Graph

57. The graph of $y = f(x)$ is given. **(See Example 5.)**

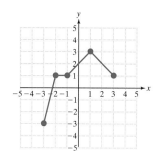

  a. Find $f(0)$.

  b. Find $f(3)$.

  c. Find $f(-2)$.

  d. For what value(s) of x is $f(x) = -3$?

  e. For what value(s) of x is $f(x) = 3$?

  f. Write the domain of f.

  g. Write the range of f.

58. The graph of $y = g(x)$ is given.

  a. Find $g(-1)$.

  b. Find $g(1)$.

  c. Find $g(4)$.

  d. For what value(s) of x is $g(x) = 3$?

  e. For what value(s) of x is $g(x) = 0$?

  f. Write the domain of g.

  g. Write the range of g.

59. The graph of $y = H(x)$ is given.

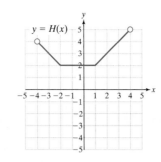

a. Find $H(-3)$.

b. Find $H(4)$.

c. Find $H(3)$.

d. For what value(s) of x is $H(x) = 3$?

e. For what value(s) of x is $H(x) = 2$?

f. Write the domain of H.

g. Write the range of H.

60. The graph of $y = K(x)$ is given.

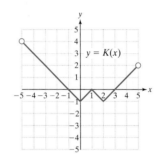

a. Find $K(0)$.

b. Find $K(-5)$.

c. Find $K(1)$.

d. For what value(s) of x is $K(x) = 0$?

e. For what value(s) of x is $K(x) = 3$?

f. Write the domain of K.

g. Write the range of K.

Objective 5: Domain of a Function

61. Explain how to determine the domain of the function defined by

$$f(x) = \frac{x + 6}{x - 2}$$

62. Explain how to determine the domain of the function defined by $g(x) = \sqrt{x - 3}$.

For Exercises 63–78, find the domain. Write the answers in interval notation. (See Example 6.)

63. $k(x) = \dfrac{x - 3}{x + 6}$

64. $m(x) = \dfrac{x - 1}{x - 4}$

65. $f(t) = \dfrac{5}{t}$

66. $g(t) = \dfrac{t - 7}{t}$

67. $h(p) = \dfrac{p - 4}{p^2 + 1}$

68. $n(p) = \dfrac{p + 8}{p^2 + 2}$

69. $h(t) = \sqrt{t + 7}$

70. $k(t) = \sqrt{t - 5}$

71. $f(a) = \sqrt{a - 3}$

72. $g(a) = \sqrt{a + 2}$

73. $m(x) = \sqrt{1 - 2x}$

74. $n(x) = \sqrt{12 - 6x}$

75. $p(t) = 2t^2 + t - 1$

76. $q(t) = t^3 + t - 1$

77. $f(x) = x + 6$

78. $g(x) = 8x - \pi$

Mixed Exercises

79. The height (in feet) of a ball that is dropped from an 80-ft building is given by $h(t) = -16t^2 + 80$, where t is time in seconds after the ball is dropped.

   a. Find $h(1)$ and $h(1.5)$

   b. Interpret the meaning of the function values found in part (a).

80. A ball is dropped from a 50-m building. The height (in meters) after t sec is given by $h(t) = -4.9t^2 + 50$.

   a. Find $h(1)$ and $h(1.5)$.

   b. Interpret the meaning of the function values found in part (a).

81. If Alicia rides a bike at an average of 11.5 mph, the distance that she rides can be represented by $d(t) = 11.5t$, where t is the time in hours.

   a. Find $d(1)$ and $d(1.5)$.

   b. Interpret the meaning of the function values found in part (a).

82. If Miguel walks at an average of 5.9 km/hr, the distance that he walks can be represented by $d(t) = 5.9t$, where t is the time in hours.

   a. Find $d(1)$ and $d(2)$.

   b. Interpret the meaning of the function values found in part (a).

83. Brian's score on an exam is a function of the number of hours he spends studying. The function defined by $P(x) = \dfrac{100x^2}{50 + x^2}$ $(x \geq 0)$ indicates that he will achieve a score of $P\%$ if he studies for x hr.

   a. Evaluate $P(0)$, $P(5)$, $P(10)$, $P(15)$, $P(20)$, and $P(25)$. (Round to 1 decimal place.) Interpret $P(25)$ in the context of this problem.

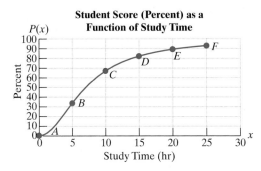

Student Score (Percent) as a Function of Study Time

   b. Match the function values found in part (a) with the points A, B, C, D, E, and F on the graph.

Expanding Your Skills

For Exercises 84–85, find the domain. Write the answers in interval notation.

84. $q(x) = \dfrac{2}{\sqrt{x + 2}}$

85. $p(x) = \dfrac{8}{\sqrt{x - 4}}$

For Exercises 86–95, refer to the functions $y = f(x)$ and $y = g(x)$, defined as follows:

$$f = \{(-3, 5), (-7, -3), (-\tfrac{3}{2}, 4), (1.2, 5)\}$$
$$g = \{(0, 6), (2, 6), (6, 0), (1, 0)\}$$

86. Identify the domain of f.

87. Identify the range of f.

88. Identify the range of g.

89. Identify the domain of g.

90. For what value(s) of x is $f(x) = 5$?

91. For what value(s) of x is $f(x) = -3$?

92. For what value(s) of x is $g(x) = 0$?

93. For what value(s) of x is $g(x) = 6$?

94. Find $f(-7)$.

95. Find $g(0)$.

Graphing Calculator Exercises

96. Graph $k(t) = \sqrt{t - 5}$. Use the graph to support your answer to Exercise 70.

97. Graph $h(t) = \sqrt{t + 7}$. Use the graph to support your answer to Exercise 69.

98. a. Graph $h(t) = -4.9t^2 + 50$ on a viewing window defined by $0 \leq t \leq 3$ and $0 \leq y \leq 60$.

b. Use the graph to approximate the function at $t = 1$. Use these values to support your answer to Exercise 80.

99. a. Graph $h(t) = -16t^2 + 80$ on a viewing window defined by $0 \leq t \leq 2$ and $0 \leq y \leq 100$.

b. Use the graph to approximate the function at $t = 1$. Use these values to support your answer to Exercise 79.

section 4.3 Graphs of Basic Functions

Objectives

1. Linear and Constant Functions
2. Graphs of Basic Functions
3. Definition of a Quadratic Function
4. Finding the x- and y-Intercepts of a Function Defined by $y = f(x)$

1. Linear and Constant Functions

A function may be expressed as a mathematical equation that relates two or more variables. In this section, we will look at several elementary functions.

We know from Section 2.2 that an equation in the form $y = k$ where k is a constant is a horizontal line. In function notation, this can be written as $f(x) = k$. For example, the function defined by $f(x) = 3$ is a horizontal line as shown in Figure 4-7.

We say that a function defined by $f(x) = k$ is a constant function because for any value of x, the function value is constant.

An equation of the form $y = mx + b$ is represented graphically by a line with slope m and y-intercept $(0, b)$. In function notation, this may be written as $f(x) = mx + b$. A function in this form is called a linear function. For example, the function defined by $f(x) = 2x - 3$ is a linear function with slope $m = 2$ and y-intercept $(0, -3)$ (Figure 4-8).

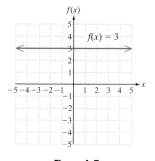

Figure 4-7

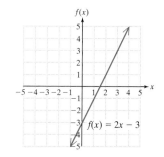

Figure 4-8

Definition of a Linear Function and a Constant Function

Let m and b represent real numbers such that $m \neq 0$. Then

A function that can be written in the form $f(x) = mx + b$ is a **linear function**.
A function that can be written in the form $f(x) = b$ is a **constant function**.

Note: The graphs of linear and constant functions are lines.

2. Graphs of Basic Functions

At this point, we are able to recognize the equations and graphs of linear and constant functions. In addition to linear and constant functions, the following equations define six basic functions that will be encountered in the study of algebra:

Equation		Function Notation
$y = x$		$f(x) = x$
$y = x^2$		$f(x) = x^2$
$y = x^3$	equivalent function notation $\longrightarrow$	$f(x) = x^3$
$y = |x|$		$f(x) = |x|$
$y = \sqrt{x}$		$f(x) = \sqrt{x}$
$y = \dfrac{1}{x}$		$f(x) = \dfrac{1}{x}$

The graph of the function defined by $f(x) = x$ is linear, with slope $m = 1$ and y-intercept $(0, 0)$ (Figure 4-9).

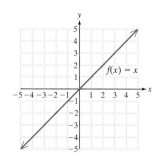

Figure 4-9

To determine the shapes of the other basic functions, we can plot several points to establish the pattern of the graph. Analyzing the equation itself may also provide insight to the domain, range, and shape of the function. To demonstrate this, we will graph $f(x) = x^2$ and $g(x) = \frac{1}{x}$.

example 1 — Graphing Basic Functions

Graph the functions defined by

a. $f(x) = x^2$ **b.** $g(x) = \dfrac{1}{x}$

Solution:

a. The domain of the function given by $f(x) = x^2$ (or equivalently $y = x^2$) is all real numbers.

To graph the function, choose arbitrary values of x within the domain of the function. Be sure to choose values of x that are positive and values that are negative to determine the behavior of the function to the right and left of the origin (Table 4-4). The graph of $f(x) = x^2$ is shown in Figure 4-10.

Skill Practice

1. Graph $f(x) = -x^2$ by first making a table of points.

Answer

1.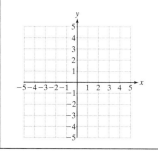

Skill Practice

2. Graph $h(x) = |x| - 1$ by first making a table of points.

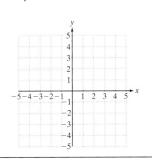

The function values are equated to the square of x, so $f(x)$ will always be greater than or equal to zero. Hence, the y-coordinates on the graph will never be negative. The range of the function is $\{y \mid y$ is a real number and $y \geq 0\}$. The arrows on each branch of the graph imply that the pattern continues indefinitely.

table 4-4

x	$f(x) = x^2$
0	0
1	1
2	4
3	9
-1	1
-2	4
-3	9

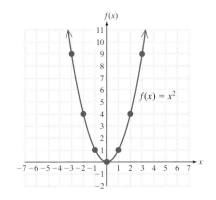

Figure 4-10

b. $g(x) = \dfrac{1}{x}$ Notice that $x = 0$ is not in the domain of the function. From the equation $y = \frac{1}{x}$, the y-values will be the reciprocal of the x-values. The graph defined by $g(x) = \frac{1}{x}$ is shown in Figure 4-11.

x	$g(x) = \dfrac{1}{x}$
1	1
2	$\frac{1}{2}$
3	$\frac{1}{3}$
-1	-1
-2	$-\frac{1}{2}$
-3	$-\frac{1}{3}$

x	$g(x) = \dfrac{1}{x}$
$\frac{1}{2}$	2
$\frac{1}{3}$	3
$\frac{1}{4}$	4
$-\frac{1}{2}$	-2
$-\frac{1}{3}$	-3
$-\frac{1}{4}$	-4

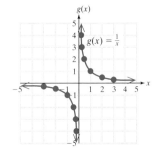

Figure 4-11

Notice that as x approaches ∞ and $-\infty$, the y-values approach zero, and the graph approaches the x-axis. In this case, the x-axis is called a *horizontal asymptote*. Similarly, the graph of the function approaches the y-axis as x gets close to zero. In this case, the y-axis is called a *vertical asymptote*.

Calculator Connections

The graphs of the functions defined by $f(x) = x^2$ and $g(x) = \frac{1}{x}$ are shown in the following calculator displays.

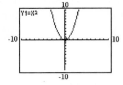 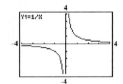

Answer

2.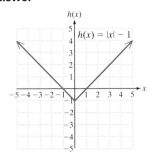

For your reference, we have provided the graphs of six basic functions in the following table.

Summary of Six Basic Functions and Their Graphs

Function	Graph	Domain and Range
1. $f(x) = x$		Domain $(-\infty, \infty)$ Range $(-\infty, \infty)$
2. $f(x) = x^2$		Domain $(-\infty, \infty)$ Range $[0, \infty)$
3. $f(x) = x^3$		Domain $(-\infty, \infty)$ Range $(-\infty, \infty)$
4. $f(x) = |x|$		Domain $(-\infty, \infty)$ Range $[0, \infty)$
5. $f(x) = \sqrt{x}$		Domain $[0, \infty)$ Range $[0, \infty)$
6. $f(x) = \dfrac{1}{x}$		Domain $(-\infty, 0) \cup (0, \infty)$ Range $(-\infty, 0) \cup (0, \infty)$

The shapes of these six graphs will be developed in the homework exercises. These functions are used often in the study of algebra. Therefore, we recommend that you associate an equation with its graph and commit each to memory.

3. Definition of a Quadratic Function

In Example 1 we graphed the function defined by $f(x) = x^2$ by plotting points. This function belongs to a special category called **quadratic functions.** A quadratic function can be written in the form $f(x) = ax^2 + bx + c$, where a, b, and c

are real numbers and $a \neq 0$. The graph of a quadratic function is in the shape of a **parabola**. The leading coefficient, a, determines the direction of the parabola.

If $a > 0$, then the parabola opens upward, for example, $f(x) = x^2$. The minimum point on a parabola opening upward is called the vertex (Figure 4-12).

If $a < 0$, then the parabola opens downward, for example, $f(x) = -x^2$. The maximum point on a parabola opening downward is called the vertex (Figure 4-13).

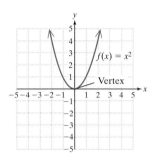

Figure 4-12

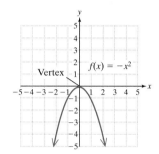

Figure 4-13

Skill Practice

Identify whether the function is constant, linear, quadratic, or none of these.

3. $m(x) = -2x^2 - 3x + 7$
4. $n(x) = -6$
5. $W(x) = \dfrac{4}{3}x - \dfrac{1}{2}$
6. $R(x) = \dfrac{4}{3x} - \dfrac{1}{2}$

example 2   Identifying Linear and Constant Functions

Identify each function as linear, constant, quadratic, or none of these.

a. $f(x) = -4$
b. $f(x) = x^2 + 3x + 2$
c. $f(x) = 7 - 2x$
d. $f(x) = \dfrac{4x + 8}{8}$

Solution:

a. $f(x) = -4$ is a constant function. It is in the form $f(x) = b$, where $b = -4$.

b. $f(x) = x^2 + 3x + 2$ is a quadratic function. It is in the form $ax^2 + bx + c = 0$, where $a \neq 0$.

c. $f(x) = 7 - 2x$ is linear. Writing it in the form $f(x) = mx + b$, we get $f(x) = -2x + 7$, where $m = -2$ and $b = 7$.

d. $f(x) = \dfrac{4x + 8}{8}$ is linear. Writing it in the form $f(x) = mx + b$, we get

$f(x) = \dfrac{4x}{8} + \dfrac{8}{8}$

$= \dfrac{1}{2}x + 1$, where $m = \dfrac{1}{2}$ and $b = 1$.

4. Finding the *x*- and *y*-Intercepts of a Function Defined by $y = f(x)$

In Section 2.2, we learned that to find the *x*-intercept, we substitute $y = 0$ and solve the equation for *x*. Using function notation, this is equivalent to finding the real solutions of the equation $f(x) = 0$. To find the *y*-intercept, substitute $x = 0$ and solve the equation for *y*. In function notation, this is equivalent to finding $f(0)$.

Answers
3. Quadratic
4. Constant
5. Linear
6. None of these

Finding the x- and y-Intercepts of a Function

Given a function defined by $y = f(x)$,

1. The x-intercepts are the real solutions to the equation $f(x) = 0$.
2. The y-intercept is given by $f(0)$.

example 3 Finding the x- and y-Intercepts of a Function

Given the function defined by $f(x) = 2x - 4$:

a. Find the x-intercept(s).
b. Find the y-intercept.
c. Graph the function.

Skill Practice

Consider $f(x) = -5x + 1$.

7. Find the x-intercept.
8. Find the y-intercept.
9. Graph the function.

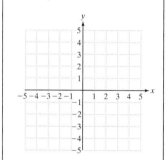

Solution:

a. To find the x-intercept(s), find the real solutions to the equation $f(x) = 0$.

$f(x) = 2x - 4$
$0 = 2x - 4$ Substitute $f(x) = 0$.
$4 = 2x$
$2 = x$ The x-intercept is $(2, 0)$.

b. To find the y-intercept, evaluate $f(0)$.

$f(0) = 2(0) - 4$ Substitute $x = 0$.
$f(0) = -4$ The y-intercept is $(0, -4)$.

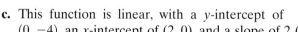

Figure 4-14

c. This function is linear, with a y-intercept of $(0, -4)$, an x-intercept of $(2, 0)$, and a slope of 2 (Figure 4-14).

example 4 Finding the x- and y-Intercepts of a Function

For the function pictured in Figure 4-15, estimate

a. The real values of x for which $f(x) = 0$.
b. The value of $f(0)$.

Solution:

a. The real values of x for which $f(x) = 0$ are the x-intercepts of the function. For this graph, the x-intercepts are located at $x = -2$, $x = 2$, and $x = 3$.

b. The value of $f(0)$ is the value of y at $x = 0$. That is, $f(0)$ is the y-intercept, $f(0) = 6$.

Figure 4-15

Answers

7. $(\frac{1}{5}, 0)$ 8. $(0, 1)$
9.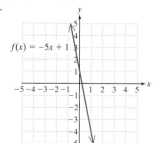

Skill Practice

Use the function shown below.

10. Estimate the real value(s) of x for which $f(x) = 0$

11. Estimate the value of $f(0)$.

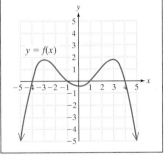

example 5 Finding the x- and y-Intercepts of a Function

For the function pictured in Figure 4-16, estimate

a. The real values of x for which $f(x) = 0$.

b. The value of $f(0)$.

Solution:

a. There are no x-intercepts for this graph; therefore, there are no real values of x for which $f(x) = 0$.

b. From the graph, the value of $f(0)$ is 3.

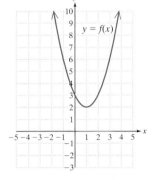

Figure 4-16

Answers
10. $x = -4, x = -1, x = 1, x = 4$
11. $f(0) = -\frac{1}{2}$

---

section 4.3 Practice Exercises

Boost your GRADE at mathzone.com!

- Practice Problems
- Self-Tests
- NetTutor
- e-Professors
- Videos

Study Skills Exercises

1. In a study group, check which activities you might try to help you learn and understand the material.
   - ☐ Quiz each other by asking each other questions.
   - ☐ Practice teaching each other.
   - ☐ Share and comparing class notes.
   - ☐ Support and encouraging each other.
   - ☐ Work together on exercises and sample problems.

2. Define the key terms.

   a. Linear function **b.** Constant function **c.** Quadratic function **d.** Parabola

Review Exercises

3. Given: $g = \{(6, 1), (5, 2), (4, 3), (3, 4)\}$

   a. Is this relation a function?

   b. List the elements in the domain.

   c. List the elements in the range.

4. Given: $f = \{(7, 3), (2, 3), (-5, 3)\}$

   a. Is this relation a function?

   b. List the elements in the domain.

   c. List the elements in the range.

5. Given: $f(x) = \sqrt{x + 4}$

   a. Evaluate $f(0), f(-3), f(-4)$, and $f(-5)$, if possible.

   b. Write the domain of this function in interval notation.

6. Given: $g(x) = \dfrac{2}{x - 3}$

   a. Evaluate $g(2), g(4), g(5)$, and $g(3)$, if possible.

   b. Write the domain of this function in interval notation.

7. The force (measured in pounds) to stretch a certain spring x inches is given by $f(x) = 3x$. Evaluate $f(3)$ and $f(10)$, and interpret the results in the context of this problem.

8. The velocity in feet per second of a falling object is given by $V(t) = 32t$, where t is the time in seconds after the object was released. Evaluate $V(2)$ and $V(5)$, and interpret the results in the context of this problem.

Objective 1: Linear and Constant Functions

9. Fill in the blank with the word *vertical* or *horizontal*. The graph of a constant function is a _____ line.

10. For the linear function $f(x) = mx + b$, identify the slope and y-intercept.

11. Graph the constant function $f(x) = 2$. Then use the graph to identify the domain and range of f.

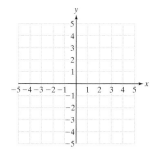

12. Graph the linear function $g(x) = -2x + 1$. Then use the graph to identify the domain and range of g.

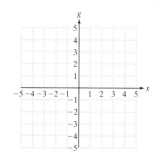

Objective 2: Graphs of Basic Functions

For Exercises 13–18, sketch a graph by completing the table and plotting the points. (See Example 1.)

13. $g(x) = |x|$

x	g(x)
−2	
−1	
0	
1	
2	

14. $f(x) = \dfrac{1}{x}$

x	f(x)
−2	
−1	
$-\frac{1}{2}$	
$-\frac{1}{4}$	

x	f(x)
	$\frac{1}{4}$
	$\frac{1}{2}$
	1
	2

15. $h(x) = x^3$

x	h(x)
−2	
−1	
0	
1	
2	

16. $k(x) = x$

x	k(x)
−2	
−1	
0	
1	
2	

17. $p(x) = \sqrt{x}$

x	p(x)
0	
1	
4	
9	
16	

18. $q(x) = x^2$

x	q(x)
−2	
−1	
0	
1	
2	

Objective 3: Definition of a Quadratic Function

For Exercises 19–30, determine if the function is constant, linear, quadratic, or none of these. (See Example 2.)

19. $f(x) = 2x^2 + 3x + 1$

20. $g(x) = -x^2 + 4x + 12$

21. $k(x) = -3x - 7$

22. $h(x) = -x - 3$

23. $m(x) = \dfrac{4}{3}$

24. $n(x) = 0.8$

25. $p(x) = \dfrac{2}{3x} + \dfrac{1}{4}$

26. $Q(x) = \dfrac{1}{5x} - 3$

27. $t(x) = \dfrac{2}{3}x + \dfrac{1}{4}$

28. $r(x) = \dfrac{1}{5}x - 3$

29. $w(x) = \sqrt{4-x}$

30. $T(x) = -|x + 10|$

Objective 4: Finding the x- and y-Intercepts of a Function Defined by $y = f(x)$

For Exercises 31–38, find the x- and y-intercepts, and graph the function. **(See Example 3.)**

31. $f(x) = 5x - 10$

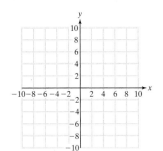

32. $f(x) = -3x + 12$

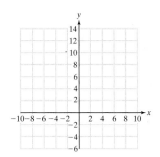

33. $g(x) = -6x + 5$

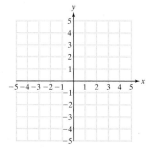

34. $h(x) = 2x + 9$

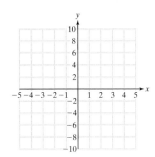

35. $f(x) = 18$

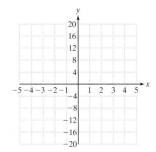

36. $g(x) = -7$

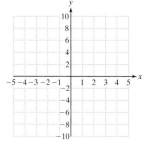

37. $g(x) = \dfrac{2}{3}x + \dfrac{1}{4}$

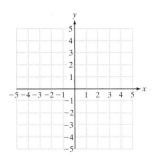

38. $h(x) = -\dfrac{5}{6}x + \dfrac{1}{2}$

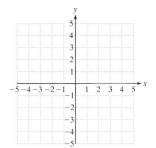

For Exercises 39–44, use the function pictured to estimate

a. The real values of x for which $f(x) = 0$.   **b.** The value of $f(0)$. **(See Examples 4–5).**

39.

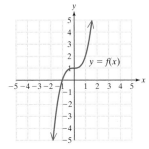

40.

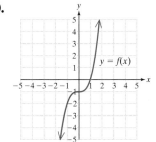

41.

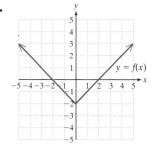

42.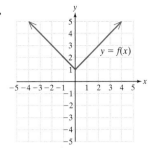

43.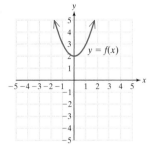

44.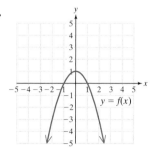

For Exercises 45–54,

a. Identify the domain of the function.

b. Identify the y-intercept of the function.

c. Match the function with its graph by recognizing the basic shape of the function and using the results from parts (a) and (b). Plot additional points if necessary.

45. $q(x) = 2x^2$

46. $p(x) = -2x^2 + 1$

47. $h(x) = x^3 + 1$

48. $k(x) = x^3 - 2$

49. $r(x) = \sqrt{x + 1}$

50. $s(x) = \sqrt{x} + 4$

51. $f(x) = \dfrac{1}{x - 3}$

52. $g(x) = \dfrac{1}{x + 1}$

53. $k(x) = |x + 2|$

54. $h(x) = |x - 1| + 2$

i.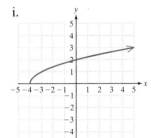

ii.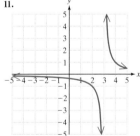

iii.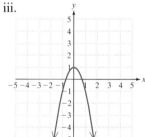

iv.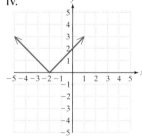

v.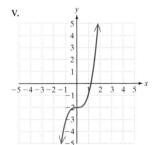

vi.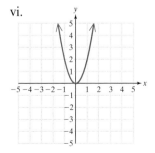

vii.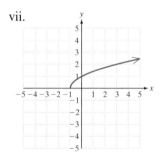

viii.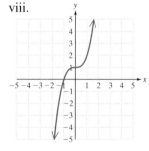

ix. 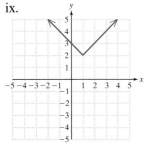   x.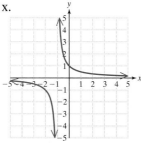

Graphing Calculator Exercises

For Exercises 55–60, use a graphing calculator to graph the basic functions. Verify your answers from the table on page 275.

55. $f(x) = x$ **56.** $f(x) = x^2$ **57.** $f(x) = x^3$

58. $f(x) = |x|$ **59.** $f(x) = \sqrt{x}$ **60.** $f(x) = \dfrac{1}{x}$

---

section 4.4 Variation

1. Definition of Direct and Inverse Variation

In this section, we introduce the concept of variation. Direct and inverse variation models can show how one quantity varies in proportion to another.

Objectives
1. Definition of Direct and Inverse Variation
2. Translations Involving Variation
3. Applications of Variation

Definition of Direct and Inverse Variation

Let k be a nonzero constant real number. Then the following statements are equivalent:

1. y varies **directly** as x.
   y is directly proportional to x. $\quad\}\quad y = kx$

2. y varies **inversely** as x.
   y is inversely proportional to x. $\quad\}\quad y = \dfrac{k}{x}$

Note: The value of k is called the constant of variation.

For a car traveling at 30 mph, the equation $d = 30t$ indicates that the distance traveled is *directly proportional* to the time of travel. For positive values of k, when two variables are directly related, as one variable increases, the other variable will also increase. Likewise, if one variable decreases, the other will decrease. In the equation $d = 30t$, the longer the time of the trip, the greater the distance traveled. The shorter the time of the trip, the shorter the distance traveled.

For positive values of k, when two variables are *inversely related*, as one variable increases, the other will decrease, and vice versa. Consider a car traveling between Toronto and Montreal, a distance of 500 km. The time required to make the trip is inversely proportional to the speed of travel: $t = 500/r$. As the rate of speed r increases, the quotient $500/r$ will decrease. Hence the time will decrease. Similarly, as the rate of speed decreases, the trip will take longer.

2. Translations Involving Variation

The first step in using a variation model is to translate an English phrase into an equivalent mathematical equation.

Skill Practice

Translate to a variation model.

1. The time t it takes to drive a particular distance is inversely proportional to the speed s.
2. The amount of your paycheck P varies directly with the number of hours h that your work.
3. q varies inversely as the square of t.

example 1 Translating to a Variation Model

Translate each expression into an equivalent mathematical model.

a. The circumference of a circle varies directly as the radius.

b. At a constant temperature, the volume of a gas varies inversely as the pressure.

c. The length of time of a meeting is directly proportional to the *square* of the number of people present.

Solution:

a. Let C represent circumference and r represent radius. The variables are directly related, so use the model $C = kr$.

b. Let V represent volume and P represent pressure. Because the variables are inversely related, use the model $V = \dfrac{k}{P}$.

c. Let t represent time and let N be the number of people present at a meeting. Because t is directly related to N^2, use the model $t = kN^2$.

Sometimes a variable varies directly as the product of two or more other variables. In this case, we have joint variation.

Definition of Joint Variation

Let k be a nonzero constant real number. Then the following statements are equivalent:

$\left.\begin{array}{l} y \text{ varies } \textbf{jointly} \text{ as } w \text{ and } z. \\ y \text{ is jointly proportional to } w \text{ and } z. \end{array}\right\} \quad y = kwz$

Answers

1. $t = \dfrac{k}{s}$
2. $P = kh$
3. $q = \dfrac{k}{t^2}$

example 2 — Translating to a Variation Model

Translate each expression into an equivalent mathematical model.

a. y varies jointly as u and the square root of v.

b. The gravitational force of attraction between two planets varies jointly as the product of their masses and inversely as the square of the distance between them.

Solution:

a. $y = ku\sqrt{v}$

b. Let m_1 and m_2 represent the masses of the two planets. Let F represent the gravitational force of attraction and d represent the distance between the planets. The variation model is $F = \dfrac{km_1 m_2}{d^2}$.

Skill Practice

Translate to a variation model.

4. a varies jointly as b and c.
5. x varies directly as the square root of y and inversely as z.

---

3. Applications of Variation

Consider the variation models $y = kx$ and $y = k/x$. In either case, if values for x and y are known, we can solve for k. Once k is known, we can use the variation equation to find y if x is known, or to find x if y is known. This concept is the basis for solving many problems involving variation.

> **Steps to Find a Variation Model**
>
> 1. Write a general variation model that relates the variables given in the problem. Let k represent the constant of variation.
> 2. Solve for k by substituting known values of the variables into the model from step 1.
> 3. Substitute the value of k into the original variation model from step 1.

---

example 3 — Solving an Application Involving Direct Variation

The variable z varies directly as w. When w is 16, z is 56.

a. Write a variation model for this situation. Use k as the constant of variation.

b. Solve for the constant of variation.

c. Find the value of z when w is 84.

Solution:

a. $z = kw$

b. $z = kw$

$56 = k(16)$ Substitute known values for z and w. Then solve for the unknown value of k.

$\dfrac{56}{16} = \dfrac{k(16)}{16}$ To isolate k, divide both sides by 16.

$\dfrac{7}{2} = k$ Simplify $\dfrac{56}{16}$ to $\dfrac{7}{2}$.

Skill Practice

The variable q varies directly as the square of v. When v is 2, q is 40.

6. Write a variation model for this relationship.
7. Solve for the constant of variation.
8. Find q when $v = 7$.

Answers

4. $a = kbc$ 5. $x = \dfrac{k\sqrt{y}}{z}$
6. $q = kv^2$ 7. $k = 10$
8. $q = 490$

c. With the value of k known, the variation model can now be written as $z = \frac{7}{2}w$.

$$z = \frac{7}{2}(84)$$ To find z when $w = 84$, substitute $w = 84$ into the equation.
$$z = 294$$

Skill Practice

The amount of water needed by a mountain hiker varies directly as the time spent hiking. The hiker needs 2.4 L for a 3-hr hike.

9. Write a model that relates the amount of water needed to the time of the hike.

10. How much water will be needed for a 5-hr hike?

example 4 — Solving an Application Involving Direct Variation

The speed of a racing canoe in still water varies directly as the square root of the length of the canoe.

a. If a 16-ft canoe can travel 6.2 mph in still water, find a variation model that relates the speed of a canoe to its length.

b. Find the speed of a 25-ft canoe.

Solution:

a. Let s represent the speed of the canoe and L represent the length. The general variation model is $s = k\sqrt{L}$. To solve for k, substitute the known values for s and L.

$$s = k\sqrt{L}$$
$$6.2 = k\sqrt{16}$$ Substitute $s = 6.2$ mph and $L = 16$ ft.
$$6.2 = k \cdot 4$$
$$\frac{6.2}{4} = \frac{4k}{4}$$ Solve for k.
$$k = 1.55$$
$$s = 1.55\sqrt{L}$$ Substitute $k = 1.55$ into the model $s = k\sqrt{L}$.

b. $s = 1.55\sqrt{L}$
$ = 1.55\sqrt{25}$ Find the speed when $L = 25$ ft.
$ = 7.75$ mph

Skill Practice

11. The yield on a bond varies inversely as the price. The yield on a particular bond is 4% when the price is $100. Find the yield when the price is $80.

example 5 — Solving an Application Involving Inverse Variation

The loudness of sound measured in decibels (dB) varies inversely as the square of the distance between the listener and the source of the sound. If the loudness of sound is 17.92 dB at a distance of 10 ft from a stereo speaker, what is the decibel level 20 ft from the speaker?

Solution:

Let L represent the loudness of sound in decibels and d represent the distance in feet. The inverse relationship between decibel level and the square of the distance is modeled by

$$L = \frac{k}{d^2}$$

$$17.92 = \frac{k}{(10)^2}$$ Substitute $L = 17.92$ dB and $d = 10$ ft.

Answers

9. $w = 0.8\,t$ 10. 4 L 11. 5%

$$17.92 = \frac{k}{100}$$

$$(17.92)100 = \frac{k}{\cancel{100}} \cdot \cancel{100} \quad \text{Solve for } k \text{ (clear fractions).}$$

$$k = 1792$$

$$L = \frac{1792}{d^2} \quad \text{Substitute } k = 1792 \text{ into the original model } L = \frac{k}{d^2}.$$

With the value of k known, we can find L for any value of d.

$$L = \frac{1792}{(20)^2} \quad \text{Find the loudness when } d = 20 \text{ ft.}$$

$$= 4.48 \text{ dB}$$

Notice that the loudness of sound is 17.92 dB at a distance 10 ft from the speaker. When the distance from the speaker is increased to 20 ft, the decibel level decreases to 4.48 dB. This is consistent with an inverse relationship. For $k > 0$, as one variable is increased, the other is decreased. It also seems reasonable that the farther one moves away from the source of a sound, the softer the sound becomes.

example 6   Solving an Application Involving Joint Variation

From August 23 to August 28, 1992, Hurricane Andrew carved a path of destruction from the Caribbean to south Florida and from the Louisiana coast to the North Carolina mountains. The destructive power of a hurricane or other severe storm is related to the *square* of the wind speed. During Hurricane Andrew, the National Weather Service reported wind gusts over 180 mph. These winds were strong enough to send a piece of plywood through a tree.

The kinetic energy of an object varies jointly as the weight of the object at sea level and as the square of its velocity. During a hurricane, a $\frac{1}{2}$-lb stone traveling at 60 mph has 81 joules (J) of kinetic energy. Suppose the wind speed doubles to 120 mph. Find the kinetic energy.

Solution:

Let E represent the kinetic energy, let w represent the weight, and let v represent the velocity of the stone. The variation model is

$$E = kwv^2$$

$$81 = k(0.5)(60)^2 \quad \text{Substitute } E = 81 \text{ J}, w = 0.5 \text{ lb, and } v = 60 \text{ mph.}$$

Skill Practice

12. The amount of simple interest earned in an account varies jointly as the interest rate and time of the investment. An account earns $40 in 2 years at 4% interest. How much interest would be earned in 3 years at a rate of 5%?

Answer

12. $75

$$81 = k(0.5)(3600) \quad \text{Simplify exponents.}$$
$$81 = k(1800)$$
$$\frac{81}{1800} = \frac{k(\cancel{1800})}{\cancel{1800}} \quad \text{Divide by 1800.}$$
$$0.045 = k \quad \text{Solve for } k.$$

With the value of k known, the model $E = kwv^2$ can be written as $E = 0.045wv^2$. We now find the kinetic energy of a $\frac{1}{2}$-lb stone traveling at 120 mph.

$$E = 0.045(0.5)(120)^2$$
$$= 324$$

The kinetic energy of a $\frac{1}{2}$-lb stone traveling at 120 mph is 324 J.

In Example 6, when the velocity increased by 2 times, the kinetic energy increased by 4 times (note that 324 J = 4·81 J). This factor of 4 occurs because the kinetic energy is proportional to the *square* of the velocity. When the velocity increased by 2 times, the kinetic energy increased by 2^2 times.

section 4.4 Practice Exercises

Boost your GRADE at mathzone.com!

MathZone

- Practice Problems
- Self-Tests
- NetTutor
- e-Professors
- Videos

Study Skills Exercises

1. Budgeting enough time to do homework and to study for a class is one of the most important steps to success in a class. Use the weekly calendar below to help you plan your time for your studies this week. Also write other obligations such as the time required for your job, for your family, for sleeping, and for eating. Be realistic when you estimate the time for each activity.

Time	Mon.	Tues.	Wed.	Thurs.	Fri.	Sat.	Sun.
7–8 A.M.							
8–9							
9–10							
10–11							
11–12							
12–1 P.M.							
1–2							
2–3							
3–4							
4–5							
5–6							
6–7							
7–8							
8–9							
9–10							

2. Define the key terms.
   a. Direct variation
   b. Inverse variation
   c. Joint variation

Review Exercises

For Exercises 3–8, refer to the graph.

3. Find $f(-3)$.

4. Find $f(0)$.

5. Find the value(s) of x for which $f(x) = 1$.

6. Find the value(s) of x for which $f(x) = 2$.

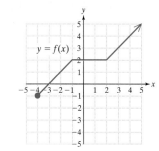

7. Write the domain of f.

8. Write the range of f.

Objective 1: Definition of Direct and Inverse Variation

9. Suppose y varies directly as x, and $k > 0$.
   a. If x increases, then will y increase or decrease?
   b. If x decreases, will y increase or decrease?

10. Suppose y varies inversely as x, and $k > 0$.
    a. If x increases, then will y increase or decrease?
    b. If x decreases, then will y increase or decrease?

Objective 2: Translations Involving Variation

For Exercises 11–18, write a variation model. Use k as the constant of variation. (See Examples 1–2.)

11. T varies directly as q.

12. P varies inversely as r.

13. W varies inversely as the square of p.

14. Y varies directly as the square root of z.

15. Q is directly proportional to x and inversely proportional to the cube of y.

16. M is directly proportional to the square of p and inversely proportional to the cube of n.

17. L varies jointly as w and the square root of v.

18. X varies jointly as the square of y and w.

Objective 3: Applications of Variation

For Exercises 19–24, find the constant of variation k. (See Example 3.)

19. y varies directly as x, and when x is 4, y is 18.

20. m varies directly as x and when x is 8, m is 22.

21. p is inversely proportional to q and when q is 16, p is 32.

22. T is inversely proportional to x and when x is 40, T is 200.

23. y varies jointly as w and v. When w is 50 and v is 0.1, y is 8.75.

24. N varies jointly as t and p. When t is 1 and p is 7.5, N is 330.

Solve Exercises 25–30 by using the steps found on page 285. **(See Example 3.)**

25. Z varies directly as the square of w, and $Z = 14$ when $w = 4$. Find Z when $w = 8$.

26. Q varies inversely as the square of p, and $Q = 4$ when $p = 3$. Find Q when $p = 2$.

27. L varies jointly as a and the square root of b, and $L = 72$ when $a = 8$ and $b = 9$. Find L when $a = \frac{1}{2}$ and $b = 36$.

28. Y varies jointly as the cube of x and the square root of w, and $Y = 128$ when $x = 2$ and $w = 16$. Find Y when $x = \frac{1}{2}$ and $w = 64$.

29. B varies directly as m and inversely as n, and $B = 20$ when $m = 10$ and $n = 3$. Find B when $m = 15$ and $n = 12$.

30. R varies directly as s and inversely as t, and $R = 14$ when $s = 2$ and $t = 9$. Find R when $s = 4$ and $t = 3$.

For Exercises 31–42, use a variation model to solve for the unknown value. **(See Examples 4–6.)**

31. The amount of pollution entering the atmosphere varies directly as the number of people living in an area. If 80,000 people cause 56,800 tons of pollutants, how many tons enter the atmosphere in a city with a population of 500,000?

32. The area of a picture projected on a wall varies directly as the square of the distance from the projector to the wall. If a 10-ft distance produces a 16-ft^2 picture, what is the area of a picture produced when the projection unit is moved to a distance 20 ft from the wall?

33. The stopping distance of a car is directly proportional to the square of the speed of the car. If a car traveling at 40 mph has a stopping distance of 109 ft, find the stopping distance of a car that is traveling at 25 mph. (Round your answer to 1 decimal place.)

34. The intensity of a light source varies inversely as the square of the distance from the source. If the intensity is 48 lumens (lm) at a distance of 5 ft, what is the intensity when the distance is 8 ft?

35. The current in a wire varies directly as the voltage and inversely as the resistance. If the current is 9 amperes (A) when the voltage is 90 volts (V) and the resistance is 10 ohms (Ω), find the current when the voltage is 185 V and the resistance is 10 Ω.

36. The power in an electric circuit varies jointly as the current and the square of the resistance. If the power is 144 watts (W) when the current is 4 A and the resistance is 6 Ω, find the power when the current is 3 A and the resistance is 10 Ω.

37. The resistance of a wire varies directly as its length and inversely as the square of its diameter. A 40-ft wire with 0.1-in. diameter has a resistance of 4 Ω. What is the resistance of a 50-ft wire with a diameter of 0.20 in.?

38. The frequency of a vibrating string is inversely proportional to its length. A 24-in. piano string vibrates at 252 cycles/sec. What would be the frequency of an 18-in. string?

39. The weight of a medicine ball varies directly as the cube of its radius. A ball with a radius of 3 in. weighs 4.32 lb. How much would a medicine ball weigh if its radius were 5 in.?

40. The surface area of a cube varies directly as the square of the length of an edge. The surface area is 24 ft² when the length of an edge is 2 ft. Find the surface area of a cube with an edge that is 5 ft.

41. The strength of a wooden beam varies jointly as the width of the beam and the square of the thickness of the beam and inversely as the length of the beam. A beam that is 48 in. long, 6 in. wide, and 2 in. thick can support a load of 417 lb. Find the maximum load that can be safely supported by a board that is 12 in. wide, 72 in. long, and 4 in. thick.

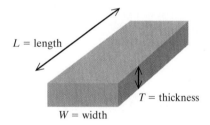

42. The period of a pendulum is the length of time required to complete one swing back and forth. The period varies directly as the square root of the length of the pendulum. If it takes 1.8 sec for a 0.81-m pendulum to complete one period, what is the period of a 1-m pendulum?

Expanding Your Skills

43. The area A of a square varies directly as the square of the length l of its sides.

   a. Write a general variation model with k as the constant of variation.

   b. If the length of the sides is doubled, what effect will that have on the area?

   c. If the length of a side is tripled, what effect will that have on the area?

44. In a physics laboratory, a spring is fixed to the ceiling. With no weight attached to the end of the spring, the spring is said to be in its equilibrium position. As weights are applied to the end of the spring, the force stretches the spring a distance d from its equilibrium position. A student in the laboratory collects the following data:

Force F (lb)	2	4	6	8	10
Distance d (cm)	2.5	5.0	7.5	10.0	12.5

a. Based on the data, do you suspect a direct relationship between force and distance or an inverse relationship?

b. Find a variation model that describes the relationship between force and distance.

chapter 4 summary

section 4.1 Introduction to Relations

Key Concepts

Any set of ordered pairs (x, y) is called a **relation in x and y**.

The **domain** of a relation is the set of first components in the ordered pairs in the relation. The **range** of a relation is the set of second components in the ordered pairs.

Examples

Example 1

Let $A = \{(0, 0), (1, 1), (2, 4), (3, 9), (-1, 1), (-2, 4), (-3, 9)\}$.

Domain of A: $\{0, 1, 2, 3, -1, -2, -3\}$

Range of A: $\{0, 1, 4, 9\}$

Example 2

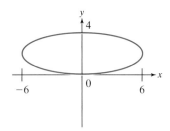

Domain: $[-6, 6]$
Range: $[0, 4]$

section 4.2 Introduction to Functions

Key Concepts

Given a relation in x and y, we say "**y is a function of x**" if for every element x in the domain, there corresponds exactly one element y in the range.

The Vertical Line Test for Functions

Consider a relation defined by a set of points (x, y) in a rectangular coordinate system. Then the graph defines y as a function of x if no vertical line intersects the graph in more than one point.

Function Notation

$f(x)$ is the value of the function f at x.

The domain of a function defined by $y = f(x)$ is the set of x-values that when substituted into the function produces a real number. In particular,

- Exclude values of x that make the denominator of a fraction zero.
- Exclude values of x that make the expression within a square root negative.

Examples

Example 1

Function $\{(1, 3), (2, 5), (6, 3)\}$
Nonfunction $\{(1, 3), (2, 5), (1, 4)\}$

Example 2

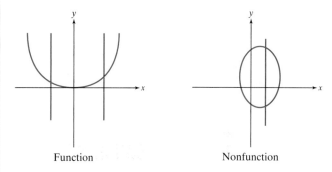

Function   Nonfunction

Example 3

Given $f(x) = -3x^2 + 5x$, find $f(-2)$.

$f(-2) = -3(-2)^2 + 5(-2)$
$= -12 - 10$
$= -22$

Example 4

Find the domain.

1. $f(x) = \dfrac{x + 4}{x - 5}; (-\infty, 5) \cup (5, \infty)$
2. $f(x) = \sqrt{x - 3}; [3, \infty)$
3. $f(x) = 3x^2 - 5; (-\infty, \infty)$

section 4.3 Graphs of Basic Functions

Key Concepts

A function of the form $f(x) = mx + b\ (m \neq 0)$ is a **linear function**. Its graph is a line with slope m and y-intercept $(0, b)$.

A function of the form $f(x) = k$ is a **constant function**. Its graph is a horizontal line.

A function of the form $f(x) = ax^2 + bx + c\ (a \neq 0)$ is a **quadratic function**. Its graph is a **parabola**.

Graphs of basic functions:

$f(x) = x$ 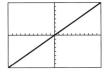 $f(x) = x^2$ 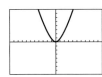 $f(x) = x^3$

$f(x) = |x|$ 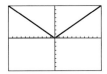 $f(x) = \sqrt{x}$ 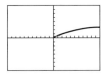 $f(x) = \dfrac{1}{x}$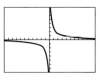

The x-intercepts of a function are determined by finding the real solutions to the equation $f(x) = 0$.

The y-intercept of a function is at $f(0)$.

Examples

Example 1

$f(x) = 2x - 3$

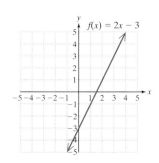

Example 2

$f(x) = 3$

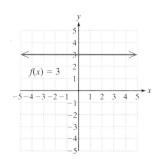

Example 3

$f(x) = x^2 - 2x - 1$

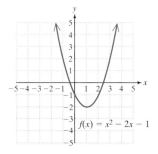

Example 4

Find the x- and y-intercepts for the function pictured.

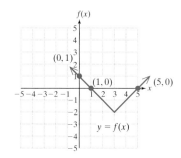

$f(x) = 0$, when $x = 1$ and $x = 5$.

The x-intercepts are $(1, 0)$ and $(5, 0)$.

$f(0) = 1$. The y-intercept is $(0, 1)$.

section 4.4 Variation

Key Concepts

Direct Variation

y varies directly as x.
y is directly proportional to x. $\Big\}$ $y = kx$

Inverse Variation

y varies inversely as x.
y is inversely proportional to x. $\Big\}$ $y = \dfrac{k}{x}$

Joint Variation

y varies jointly as w and z.
y is jointly proportional to w and z. $\Big\}$ $y = kwz$

Steps to Find a Variation Model

1. Write a general variation model that relates the variables given in the problem. Let k represent the constant of variation.

2. Solve for k by substituting known values of the variables into the model from step 1.

3. Substitute the value of k into the original variation model from step 1.

Examples

Example 1

t varies directly as the square root of x.

$t = k\sqrt{x}$

Example 2

W is inversely proportional to the cube of x.

$W = \dfrac{k}{x^3}$

Example 3

y is jointly proportional to x and to the square of z.

$y = kxz^2$

Example 4

C varies directly as the square root of d and inversely as t. If $C = 12$ when d is 9 and t is 6, find C if d is 16 and t is 12.

Step 1: $C = \dfrac{k\sqrt{d}}{t}$

Step 2: $12 = \dfrac{k\sqrt{9}}{6} \Rightarrow 12 = \dfrac{k \cdot 3}{6} \Rightarrow k = 24$

Step 3: $C = \dfrac{24\sqrt{d}}{t} \Rightarrow C = \dfrac{24\sqrt{16}}{12} \Rightarrow C = 8$

chapter 4 review exercises

Section 4.1

1. Write a relation with four ordered pairs for which the first element is the name of a parent and the second element is the name of the parent's child.

For Exercises 2–5, find the domain and range.

2. $\left\{\left(\dfrac{1}{3}, 10\right), \left(6, -\dfrac{1}{2}\right), \left(\dfrac{1}{4}, 4\right), \left(7, \dfrac{2}{5}\right)\right\}$

3.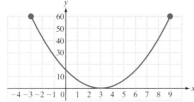

4.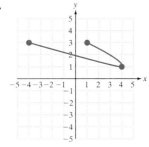

5.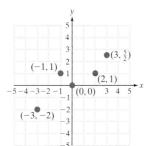

Section 4.2

6. Sketch a relation that is *not* a function. (Answers may vary.)

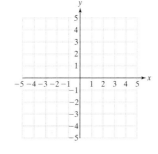

7. Sketch a relation that *is* a function. (Answers may vary.)

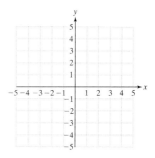

For Exercises 8–13:

a. Determine whether the relation defines y as a function of x.

b. Find the domain.

c. Find the range.

8.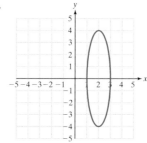

9.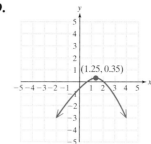

10. $\{(1, 3), (2, 3), (3, 3), (4, 3)\}$

11. $\{(0, 2), (0, 3), (4, 4), (0, 5)\}$

12.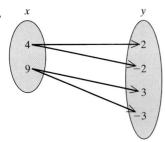

13.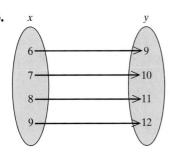

For Exercises 14–21, find the function values given $f(x) = 6x^2 - 4$.

14. $f(0)$ **15.** $f(1)$

16. $f(-1)$ **17.** $f(t)$

18. $f(b)$ **19.** $f(\pi)$

20. $f(\square)$ **21.** $f(-2)$

For Exercises 22–25, write the domain of each function in interval notation.

22. $g(x) = 7x^3 + 1$ **23.** $h(x) = \dfrac{x + 10}{x - 11}$

24. $k(x) = \sqrt{x - 8}$ **25.** $w(x) = \sqrt{x + 2}$

26. Anita is a waitress and makes $6 per hour plus tips. Her tips average $5 per table. In one 8-hr shift, Anita's pay can be described by $p(x) = 48 + 5x$, where x represents the number of tables she waits on. Find out how much Anita will earn if she waits on

     a. 10 tables   **b.** 15 tables   **c.** 20 tables

Section 4.3

For Exercises 27–32, sketch the functions from memory.

27. $h(x) = x$

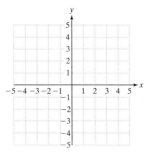

28. $f(x) = x^2$

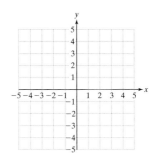

29. $g(x) = x^3$

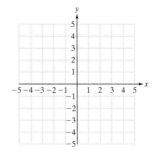

30. $w(x) = |x|$

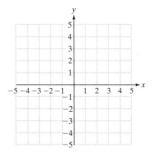

31. $s(x) = \sqrt{x}$

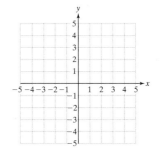

32. $r(x) = \dfrac{1}{x}$

For Exercises 33–34, sketch the functions.

33. $q(x) = 3$

34. $k(x) = 2x + 1$

For Exercises 35–36, find the x- and y-intercepts.

35. $p(x) = 4x - 7$

36. $q(x) = -2x + 9$

37. The function defined by $b(t) = 0.7t + 4.5$ represents the per capita consumption of bottled water in the United States between 1985 and 2005. The values of $b(t)$ are measured in gallons, and $t = 0$ corresponds to the year 1985. (Source: U.S. Department of Agriculture.)

  a. Evaluate $b(0)$ and $b(7)$ and interpret the results in the context of this problem.

  b. What is the slope of this function? Interpret the slope in the context of this problem.

For Exercises 38–43, refer to the graph.

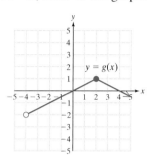

38. Find $g(-2)$.

39. Find $g(4)$.

40. For what value(s) of x is $g(x) = 0$?

41. For what value(s) of x is $g(x) = -4$?

42. Write the domain of g.

43. Write the range of g.

44. Given: $s(x) = (x - 2)^2$

  a. Find $s(4), s(3), s(2), s(1),$ and $s(0)$.

  b. What is the domain of s?

45. Given: $r(x) = 2\sqrt{x - 4}$

  a. Find $r(4), r(5),$ and $r(8)$.

  b. What is the domain of r?

46. Given: $h(x) = \dfrac{3}{x - 3}$

  a. Find $h(-3), h(-1), h(0), h(2), h(4), h(5),$ and $h(7)$.

  b. What is the domain of h?

47. Given: $k(x) = -|x + 3|$

  a. Find $k(-5), k(-4), k(-3), k(-2),$ and $k(-1)$.

  b. What is the domain of k?

Section 4.4

48. The force applied to a spring varies directly with the distance that the spring is stretched. When 6 lb of force is applied, the spring stretches 2 ft.

  a. Write a variation model using k as the constant of variation.

  b. Find k.

  c. How many feet will the spring stretch when 5 lb of pressure is applied?

49. Suppose y varies directly with the cube of x and $y = 32$ when $x = 2$. Find y when $x = 4$.

50. Suppose y varies jointly with x and the square root of z, and $y = 3$ when $x = 3$ and $z = 4$. Find y when $x = 8$ and $z = 9$.

51. The distance d that one can see to the horizon varies directly as the square root of the height above sea level. If a person 25 m above sea level can see 30 km, how far can a person see if she is 64 m above sea level?

---

chapter 4 | test

For Exercises 1–3, **a.** determine if the relation defines y as a function of x, **b.** identify the domain, and **c.** identify the range.

1.
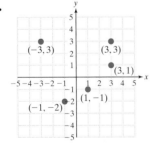

2.
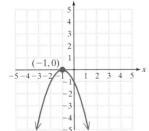

3. Explain how to find the x- and y-intercepts of a function defined by $y = f(x)$.

Graph the functions.

4. $f(x) = -3x - 1$
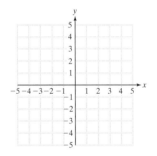

5. $k(x) = -2$
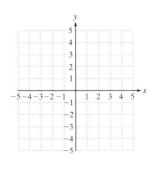

6. $p(x) = x^2$

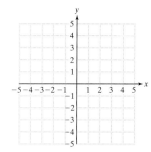

7. $w(x) = |x|$

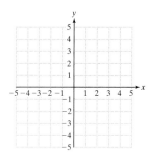

For Exercises 8–10, write the domain in interval notation.

8. $f(x) = \dfrac{x - 5}{x + 7}$

9. $f(x) = \sqrt{x + 7}$

10. $h(x) = (x + 7)(x - 5)$

11. Given: $r(x) = x^2 - 2x + 1$

   a. Find $r(-2), r(-1), r(0), r(2),$ and $r(3)$.

   b. What is the domain of r?

12. The function defined by $s(t) = 1.6t + 36$ approximates the per capita consumption of soft drinks in the United States between 1985 and 2005. The values of $s(t)$ are measured in gallons, and $t = 0$ corresponds to the year 1985. (Source: U.S. Department of Agriculture.)

   a. Evaluate $s(0)$ and $s(7)$ and interpret the results in the context of this problem.

   b. What is the slope of the function? Interpret the slope in the context of this problem.

For Exercises 13–20, refer to the graph.

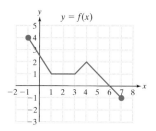

13. Find $f(1)$.

14. Find $f(4)$.

15. Write the domain of f.

16. Write the range of f.

17. True or false? The value $y = 5$ is in the range of f.

18. Find the x-intercept of the function.

19. For what value(s) of x is $f(x) = 0$?

20. For what value(s) of x is $f(x) = 1$?

For Exercises 21–24, determine if the function is constant, linear, quadratic, or none of these.

21. $f(x) = -3x^2$

22. $g(x) = -3x$

23. $h(x) = -3$

24. $k(x) = -\dfrac{3}{x}$

25. Find the x- and y-intercepts for $f(x) = \dfrac{3}{4}x + 9$.

26. Write a variation model using k as the constant of variation. The variable x varies directly as y and inversely as the square of t.

27. The period of a pendulum varies directly as the square root of the length of the pendulum. If the period of the pendulum is 2.2 sec when the length is 4 ft, find the period when the length is 9 ft.

chapters 1–4 | cumulative review exercises

1. Solve the equation.

   $$\frac{1}{3}t + \frac{1}{5} = \frac{1}{10}(t - 2)$$

2. Simplify. $5 - 3(2 - \sqrt{25}) + 2 - 10 \div 5$

3. Write the inequalities in interval notation.

   a. x is greater than or equal to 6.

   b. x is less than 17.

   c. x is between -2 and 3, inclusive.

4. Solve the inequality. Write the solution set in interval notation.

   $$4 \leq -6y + 5$$

5. Determine the volume of the cone pictured here. Round your answer to the nearest whole unit.

   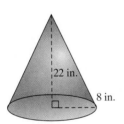

6. Find an equation of the line passing through the origin and perpendicular to $3x - 4y = 1$. Write your final answer in slope-intercept form.

7. Find the pitch (slope) of the roof.

   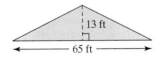

8. a. Explain how to find the x- and y-intercepts of a function $y = f(x)$.

   b. Find the y-intercept of the function defined by $f(x) = 3x + 2$.

   c. Find the x-intercept(s) of the function defined by $f(x) = 3x + 2$.

9. Is the ordered triple $(2, 1, 0)$ a solution to the following system of equations? Why or why not?

   $$x + 2y - z = 4$$
   $$2x - 3y - z = 1$$
   $$-3x + 2y + 2z = 8$$

10. Solve the system by using the substitution method.

    $$-y - 2x = -10$$
    $$4x - 20 = -5y$$

11. Solve the system by using the addition method.

    $$5x + 7y = -9$$
    $$-3x - 2y = -10$$

12. Solve the system.

    $$-\frac{1}{4}x + \frac{1}{3}y = -1$$
    $$\frac{1}{2}x - \frac{3}{10}y = 2$$

13. One positive number is two-thirds of another positive number. The larger number is 12 more than the smaller. Find the numbers.

14. State the domain and range of the relation. Is the relation a function? $\{(3, -1), (4, -5), (3, -8)\}$

15. The linear function defined by $N(x) = 420x + 5260$ provides a model for the number of full-time-equivalent (FTE) students attending a community college from 1988 to 2006. Assume that $x = 0$ corresponds to the year 1988.

    a. Use this model to find the number of FTE students who attended the college in 1996.

    b. If this linear trend continues, predict the year in which the number of FTE students will reach 14,920.

16. State the domain and range of the relation. Is the relation a function?

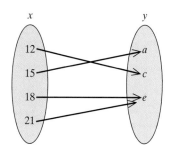

17. Given: $f(x) = \frac{1}{2}x - 1$ and $g(x) = 3x^2 - 2x$

  a. Find $f(4)$. **b.** Find $g(-3)$.

For Exercises 18–19, write the domain of the functions in interval notation.

18. $f(x) = \dfrac{1}{x - 15}$

19. $g(x) = \sqrt{x - 6}$

20. Simple interest varies jointly as the interest rate and as the time the money is invested. If an investment yields $1120 interest at 8% for 2 years, how much interest will the investment yield at 10% for 5 years?

Polynomials

5

5.1 Addition and Subtraction of Polynomials and Polynomial Functions

5.2 Multiplication of Polynomials

5.3 Division of Polynomials

5.4 Greatest Common Factor and Factoring by Grouping

5.5 Factoring Trinomials

5.6 Factoring Binomials

5.7 Solving Equations by Using the Zero Product Rule

In this chapter, we study addition, subtraction, multiplication, and division of polynomials, along with an important operation called factoring. A polynomial function is a function defined by a sum of terms of the form ax^n, where a is a real number and n is a whole number. In Exercise 68 in Section 5.1, the function $P(t) = 0.022t^2 + 2.012t + 102$ is used to estimate the population P (in millions) of Mexico, t years after 2000.

chapter 5 preview

The exercises in this chapter preview contain concepts that have not yet been presented. These exercises are provided for students who want to compare their levels of understanding before and after studying the chapter. Alternatively, you may prefer to work these exercises when the chapter is completed and before taking the exam.

Section 5.1

1. Given the polynomial: $-7x^5 - 2x^3 + 4x^6 - 2$

   a. Write the polynomial in descending order.

   b. Determine the degree of the polynomial.

   c. Determine the leading coefficient.

For Exercises 2–3, perform the indicated operations.

2. $(4x^2y - 3xy + 2xy^2) + (-2x^2y - 3xy - 6xy^2)$

3. $(-5t^2 + 2t - 3) - (2t^2 - 4t - 1)$

4. Given the polynomial function $P(x) = -3x^2 + 2x - 1$, find $P(-2)$.

Section 5.2

For Exercises 5–9, multiply the polynomials.

5. $3x^2y^3(4x^2y^2 + 2xy^3 - 10x^3y)$

6. $(6x - 5)(2x + 1)$

7. $(3z - 2)(4z^2 - 2z - 5)$

8. $(3t + 2)^2$

9. $(4d - c)(4d + c)$

Section 5.3

For Exercises 10–11, divide the polynomials.

10. $(-16p^4 + 24p^3 - 8p^2) \div (8p^2)$

11. $\dfrac{6x^2 - 5x + 6}{2x + 1}$

12. Use synthetic division to divide the polynomials.
$$\dfrac{5y^2 + 6y - 3}{y + 2}$$

Section 5.4

13. Factor out the greatest common factor.
$14x^3y^4 - 20x^2y^3 + 18xy$

14. Factor by grouping. $ab + ac - 2b - 2c$

Section 5.5

For Exercises 15–18, factor the polynomials completely.

15. $a^2 + a - 42$

16. $4y^2 + 16y + 7$

17. $4x^3 + 14x^2 + 10x$

18. $9c^2 + 12cd + 4d^2$

Section 5.6

For Exercises 19–20, factor completely.

19. $25y^2 - 49$

20. $z^3 + 8$

Section 5.7

For Exercises 21–23, solve the equations.

21. $x^2 - 7x = 18$

22. $2(a + 6)(2a + 5)(a - 10) = 0$

23. $2c(c + 5) + 8 = 3(3c + 4) - c^2$

24. The length of a rectangle is 1 ft more than twice the width. The area is 36 ft². Find the length and width.

Section 5.1 Addition and Subtraction of Polynomials and Polynomial Functions

1. Polynomials: Basic Definitions

One commonly used algebraic expression is called a polynomial. A **polynomial** in x is defined as a finite sum of terms of the form ax^n, where a is a real number and the exponent n is a whole number. For each term, a is called the **coefficient**, and n is called the **degree of the term**. For example:

Term (Expressed in the Form ax^n)	Coefficient	Degree
$3x^5$	3	5
x^{14} → rewrite as $1x^{14}$	1	14
7 → rewrite as $7x^0$	7	0
$\frac{1}{2}p$ → rewrite as $\frac{1}{2}p^1$	$\frac{1}{2}$	1

Objectives
1. Polynomials: Basic Definitions
2. Addition of Polynomials
3. Subtraction of Polynomials
4. Polynomial Functions

Concept Connections
Give the coefficient and the degree of each term.

1. $-5y$ 2. $\frac{3}{4}c^8$

If a polynomial has exactly one term, it is categorized as a **monomial**. A two-term polynomial is called a **binomial**, and a three-term polynomial is called a **trinomial**. Usually the terms of a polynomial are written in descending order according to degree. In descending order, the highest-degree term is written first and is called the **leading term**. Its coefficient is called the **leading coefficient**. The **degree of a polynomial** is the largest degree of all its terms. Thus, the leading term determines the degree of the polynomial.

	Expression	Descending Order	Leading Coefficient	Degree of Polynomial
Monomials	$2x^9$	$2x^9$	2	9
	-49	-49	-49	0
Binomials	$10y - 7y^2$	$-7y^2 + 10y$	-7	2
	$6 - \frac{2}{3}b$	$-\frac{2}{3}b + 6$	$-\frac{2}{3}$	1
Trinomials	$w + 2w^3 + 9w^6$	$9w^6 + 2w^3 + w$	9	6
	$2.5a^4 - a^8 + 1.3a^3$	$-a^8 + 2.5a^4 + 1.3a^3$	-1	8

Concept Connections
Write the polynomial in descending order. Then give the degree and the leading coefficient.

3. $12 + 6t^3 - t^4$
4. $6x^3y^4$
5. Write a trinomial whose degree is 5. (Answers may vary.)

Polynomials may have more than one variable. In such a case, the degree of a term is the sum of the exponents of the variables contained in the term. For example, the term $2x^3y^4z$ has degree 8 because the exponents applied to x, y, and z are 3, 4, and 1, respectively.

The following polynomial has a degree of 12 because the highest degree of its terms is 12.

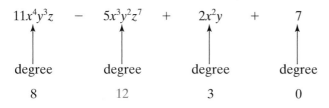

Answers
1. Degree: 1; coefficient: -5
2. Degree: 8; coefficient: $\frac{3}{4}$
3. $-t^4 + 6t^3 + 12$
   Degree: 4; leading coefficient: -1
4. $6x^3y^4$
   Degree: 7; coefficient: 6
5. For example: $2x^5 - 4x^2 + 1$

2. Addition of Polynomials

To add or subtract two polynomials, we combine *like* terms. Recall that two terms are *like* **terms** if they each have the same variables and the corresponding variables are raised to the same powers.

Skill Practice

Add the polynomials.

6. $(2x^2 + 5x - 2)$
   $+ (6x^2 - 8x - 8)$

7. $(-5a^2b - 6ab^2)$
   $+ (2a^2b + ab^2)$

---

example 1 Adding Polynomials

Add the polynomials.

a. $(3t^3 + 2t^2 - 5t) + (t^3 - 6t)$ b. $\left(\dfrac{2}{3}w^2 - w + \dfrac{1}{8}\right) + \left(\dfrac{4}{3}w^2 + 8w - \dfrac{1}{4}\right)$

c. $(a^2b + 7ab + 6) + (5a^2b - 2ab - 7)$

Solution:

a. $(3t^3 + 2t^2 - 5t) + (t^3 - 6t)$

$\quad = 3t^3 + t^3 + 2t^2 + (-5t) + (-6t)$ Group *like* terms.

$\quad = 4t^3 + 2t^2 - 11t$ Add *like* terms.

b. $\left(\dfrac{2}{3}w^2 - w + \dfrac{1}{8}\right) + \left(\dfrac{4}{3}w^2 + 8w - \dfrac{1}{4}\right)$

$\quad = \dfrac{2}{3}w^2 + \dfrac{4}{3}w^2 + (-w) + 8w + \dfrac{1}{8} + \left(-\dfrac{1}{4}\right)$ Group *like* terms.

$\quad = \dfrac{6}{3}w^2 + 7w + \left(\dfrac{1}{8} - \dfrac{2}{8}\right)$ Add fractions with common denominators.

$\quad = 2w^2 + 7w - \dfrac{1}{8}$ Simplify.

c. $(a^2b + 7ab + 6) + (5a^2b - 2ab - 7)$

$\quad = a^2b + 5a^2b + 7ab + (-2ab) + 6 + (-7)$ Group *like* terms.

$\quad = 6a^2b + 5ab - 1$ Add *like* terms.

Tip: Addition of polynomials can be performed vertically by vertically aligning *like* terms.

$$(a^2b + 7ab + 6) + (5a^2b - 2ab - 7) \longrightarrow \begin{array}{r} a^2b + 7ab + 6 \\ + \ 5a^2b - 2ab - 7 \\ \hline 6a^2b + 5ab - 1 \end{array}$$

---

3. Subtraction of Polynomials

Subtraction of two polynomials is similar to subtracting real numbers. Add the opposite of the second polynomial to the first polynomial.

The opposite (or additive inverse) of a real number a is $-a$. Similarly, if A is a polynomial, then $-A$ is its opposite.

Answers

6. $8x^2 - 3x - 10$
7. $-3a^2b - 5ab^2$

Section 5.1 Addition and Subtraction of Polynomials and Polynomial Functions

example 2 — Finding the Opposite of a Polynomial

Find the opposite of the polynomials.

a. $4x$ **b.** $5a - 2b - c$ **c.** $5.5y^4 - 2.4y^3 + 1.1y - 3$

Solution:

a. The opposite of $4x$ is $-(4x)$, or $-4x$.

b. The opposite of $5a - 2b - c$ is $-(5a - 2b - c)$ or equivalently $-5a + 2b + c$.

c. The opposite of $5.5y^4 - 2.4y^3 + 1.1y - 3$ is $-(5.5y^4 - 2.4y^3 + 1.1y - 3)$ or equivalently $-5.5y^4 + 2.4y^3 - 1.1y + 3$.

Skill Practice

Find the opposite of the polynomials.

8. $-7z$
9. $2p - 3q + r + 1$

Tip: Notice that the sign of each term is changed when finding the opposite of a polynomial.

Definition of Subtraction of Polynomials

If A and B are polynomials, then $A - B = A + (-B)$.

---

example 3 — Subtracting Polynomials

Subtract the polynomials.

a. $(3x^2 + 2x - 5) - (4x^2 - 7x + 2)$ **b.** $(6x^2y - 2xy + 5) - (x^2y - 3)$

Solution:

a. $(3x^2 + 2x - 5) - (4x^2 - 7x + 2)$
$= (3x^2 + 2x - 5) + (-4x^2 + 7x - 2)$ Add the opposite of the second polynomial.
$= 3x^2 + (-4x^2) + 2x + 7x + (-5) + (-2)$ Group *like* terms.
$= -x^2 + 9x - 7$ Combine *like* terms.

b. $(6x^2y - 2xy + 5) - (x^2y - 3)$
$= (6x^2y - 2xy + 5) + (-x^2y + 3)$ Add the opposite of the second polynomial.
$= 6x^2y + (-x^2y) + (-2xy) + 5 + 3$ Group *like* terms.
$= 5x^2y - 2xy + 8$ Combine *like* terms.

Skill Practice

Subtract the polynomials.

10. $(6a^2b - 2ab)$
    $- (-3a^2b + 2ab + 3)$

11. $\left(\dfrac{1}{3}p^3 + \dfrac{3}{4}p^2 - p\right)$
    $- \left(\dfrac{1}{2}p^3 + \dfrac{1}{3}p^2 + \dfrac{1}{2}p\right)$

Tip: Subtraction of polynomials can be performed vertically by vertically aligning *like* terms. Then add the opposite of the second polynomial. "Placeholders" (shown in bold) may be used to help line up *like* terms.

$(6x^2y - 2xy + 5) - (x^2y - 3) \longrightarrow \quad 6x^2y - 2xy + 5 \quad \xrightarrow{\text{Add the opposite.}} \quad 6x^2y - 2xy + 5$
$\qquad\qquad\qquad\qquad\qquad\qquad\qquad\qquad -(x^2y + \mathbf{0xy} - 3) \qquad\qquad\qquad + -x^2y - \mathbf{0xy} + 3$
$\qquad\qquad\qquad\qquad\qquad\qquad\qquad\qquad\qquad\qquad\qquad\qquad\qquad\qquad\qquad\qquad 5x^2y - 2xy + 8$

Answers

8. $7z$ 9. $-2p + 3q - r - 1$
10. $9a^2b - 4ab - 3$
11. $-\dfrac{1}{6}p^3 + \dfrac{5}{12}p^2 - \dfrac{3}{2}p$

Skill Practice

12. Subtract $(8t^2 - 4t - 3)$ from $(-6t^2 + t + 2)$.

---

example 4   Subtracting Polynomials

Subtract $\quad \dfrac{1}{2}x^4 - \dfrac{3}{4}x^2 + \dfrac{1}{5} \quad$ from $\quad \dfrac{3}{2}x^4 + \dfrac{1}{2}x^2 - 4x$

Solution:

In general, to subtract a from b, we write $b - a$. Therefore, to subtract

$$\dfrac{1}{2}x^4 - \dfrac{3}{4}x^2 + \dfrac{1}{5} \quad \text{from} \quad \dfrac{3}{2}x^4 + \dfrac{1}{2}x^2 - 4x$$

we have

$$\left(\dfrac{3}{2}x^4 + \dfrac{1}{2}x^2 - 4x\right) - \left(\dfrac{1}{2}x^4 - \dfrac{3}{4}x^2 + \dfrac{1}{5}\right)$$

$= \dfrac{3}{2}x^4 + \dfrac{1}{2}x^2 - 4x - \dfrac{1}{2}x^4 + \dfrac{3}{4}x^2 - \dfrac{1}{5}$ Subtract the polynomials.

$= \dfrac{3}{2}x^4 - \dfrac{1}{2}x^4 + \dfrac{1}{2}x^2 + \dfrac{3}{4}x^2 - 4x - \dfrac{1}{5}$ Group *like* terms.

$= \dfrac{3}{2}x^4 - \dfrac{1}{2}x^4 + \dfrac{2}{4}x^2 + \dfrac{3}{4}x^2 - 4x - \dfrac{1}{5}$ Write *like* terms with a common denominator.

$= \dfrac{2}{2}x^4 + \dfrac{5}{4}x^2 - 4x - \dfrac{1}{5}$ Combine *like* terms.

$= x^4 + \dfrac{5}{4}x^2 - 4x - \dfrac{1}{5}$ Simplify.

---

4. Polynomial Functions

A **polynomial function** is a function defined by a finite sum of terms of the form ax^n, where a is a real number and n is a whole number. For example, the functions defined here are polynomial functions:

$f(x) = 3x - 8$

$g(x) = 4x^5 - 2x^3 + 5x - 3$

$h(x) = -\dfrac{1}{2}x^4 + \dfrac{3}{5}x^3 - 4x^2 + \dfrac{5}{9}x - 1$

$k(x) = 7 \quad$ ($7 = 7x^0$ which is of the form ax^n, where $n = 0$ is a whole number)

The following functions are *not* polynomial functions:

$m(x) = \dfrac{1}{x} - 8 \quad \left(\dfrac{1}{x} = x^{-1}\text{, the exponent } -1 \text{ is not a whole number}\right)$

$q(x) = |x| \quad\quad (|x| \text{ is not of the form } ax^n)$

Answer

12. $-14t^2 + 5t + 5$

example 5 — Evaluating a Polynomial Function

Given $P(x) = x^3 + 2x^2 - x - 2$, find the function values.

a. $P(-3)$ **b.** $P(-1)$ **c.** $P(0)$ **d.** $P(2)$

Skill Practice

Given: $P(x) = -2x^3 - 4x + 6$

13. Find $P(0)$.
14. Find $P(-2)$.

Solution:

a. $P(x) = x^3 + 2x^2 - x - 2$

$P(-3) = (-3)^3 + 2(-3)^2 - (-3) - 2$

$= -27 + 2(9) + 3 - 2$

$= -27 + 18 + 3 - 2$

$= -8$

b. $P(-1) = (-1)^3 + 2(-1)^2 - (-1) - 2$

$= -1 + 2(1) + 1 - 2$

$= -1 + 2 + 1 - 2$

$= 0$

c. $P(0) = (0)^3 + 2(0)^2 - (0) - 2$

$= -2$

d. $P(2) = (2)^3 + 2(2)^2 - (2) - 2$

$= 8 + 2(4) - 2 - 2$

$= 8 + 8 - 2 - 2$

$= 12$

The function values can be confirmed from the graph of $y = P(x)$ (Figure 5-1).

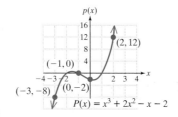

Figure 5-1

---

example 6 — Using Polynomial Functions in Applications

The length of a rectangle is 4 m less than 3 times the width. Let x represent the width. Write a polynomial function P that represents the perimeter of the rectangle and simplify the result.

Solution:

Let x represent the width. Then $3x - 4$ is the length. The perimeter of a rectangle is given by $P = 2L + 2W$. Thus

$P(x) = 2(3x - 4) + 2(x)$

$= 6x - 8 + 2x$

$= 8x - 8$

Skill Practice

15. The longest side of a triangle is 2 ft less than 4 times the shortest side. The middle side is 3 ft more than twice the shortest side. Let x represent the shortest side. Find a polynomial function P that represents the perimeter of the triangle, and simplify the result.

Answers

13. $P(0) = 6$ 14. $P(-2) = 30$
15. $P(x) = x + (4x - 2) + (2x + 3)$
    $= 7x + 1$

Skill Practice

The yearly cost of tuition at public two-year colleges from 1992 to 2006 can be approximated by

$T(x) = -0.08x^2 + 61x + 1135$

for $0 \leq x \leq 14$, where x represents the number of years since 1992.

16. Find $T(13)$ and interpret the result.

17. Use the function T to approximate the cost of tuition in the year 1997.

example 7 — Applying a Polynomial Function

The percent of females between the ages of 18 and 24 who engaged in cigarette smoking in the United States can be approximated by $F(x) = -0.125x^2 + 0.165x + 29.1$, where x is the number of years since 1997 and $F(x)$ is measured as a percent (Figure 5-2).

a. Evaluate $F(2)$ to 1 decimal place, and interpret the meaning in the context of this problem.

b. What percent of females between the ages of 18 and 24 smoked in the year 2005? Round to the nearest tenth of a percent.

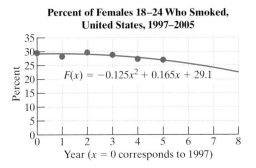

Source: Center for Disease Control.

Figure 5-2

Solution:

a. $F(2) = -0.125(2)^2 + 0.165(2) + 29.1$ Substitute $x = 2$ into the function.

   ≈ 28.9

In the year 1999 ($x = 2$ years since 1997), approximately 28.9% of females between the ages of 18 and 24 smoked.

b. The year 2005 is 8 years since 1997. Substitute $x = 8$ into the function.

   $F(8) = -0.125(8)^2 + 0.165(8) + 29.1$ Substitute $x = 8$ into the function.

   $\approx 22.4\%$

Approximately 22.4% of females in the 18–24 age group smoked in 2005.

Answers

16. $T(13) \approx 1914$. In the year 2005, tuition for public two-year colleges averaged approximately $1914.

17. $1438

---

section 5.1 — Practice Exercises

Boost your GRADE at mathzone.com!

- Practice Problems
- Self-Tests
- NetTutor
- e-Professors
- Videos

Study Skills Exercises

1. After you do a section of homework, check the answers to the odd-numbered exercises in the back of the text. Choose a method to identify the exercises that you got wrong or had trouble with (i.e., circle the number or put a star by the number). List some reasons why it is important to label these problems.

2. Define the key terms.

   a. Polynomial   b. Coefficient   c. Degree of the term   d. Monomial

   e. Binomial   f. Trinomial   g. Leading term   h. Leading coefficient

   i. Degree of a polynomial   j. *Like* terms   k. Polynomial function

Objective 1: Polynomials: Basic Definitions

For Exercises 3–8, write the polynomial in descending order. Then identify the leading coefficient and the degree.

3. $a^2 - 6a^3 - a$
4. $2b - b^4 + 5b^2$
5. $6x^2 - x + 3x^4 - 1$
6. $8 - 4y + y^5 - y^2$
7. $100 - t^2$
8. $-51 + s^2$

For Exercises 9–14, write a polynomial in one variable that is described by the following. (Answers may vary.)

9. A monomial of degree 5
10. A monomial of degree 4
11. A trinomial of degree 2
12. A trinomial of degree 3
13. A binomial of degree 4
14. A binomial of degree 2

Objective 2: Addition of Polynomials

For Exercises 15–22, add the polynomials and simplify. (See Example 1.)

15. $(-4m^2 + 4m) + (5m^2 + 6m)$
16. $(3n^3 + 5n) + (2n^3 - 2n)$
17. $(3x^4 - x^3 - x^2) + (3x^3 - 7x^2 + 2x)$
18. $(6x^3 - 2x^2 - 12) + (x^2 + 3x + 9)$
19. $\left(\dfrac{1}{2}w^3 + \dfrac{2}{9}w^2 - 1.8w\right) + \left(\dfrac{3}{2}w^3 - \dfrac{1}{9}w^2 + 2.7w\right)$
20. $\left(2.9t^4 - \dfrac{7}{8}t + \dfrac{5}{3}\right) + \left(-8.1t^4 - \dfrac{1}{8}t - \dfrac{1}{3}\right)$
21. Add $(9x^2 - 5x + 1)$ to $(8x^2 + x - 15)$.
22. Add $(-x^3 + 5x)$ to $(10x^3 + x^2 - 10)$.

Objective 3: Subtraction of Polynomials

For Exercises 23–28, write the opposite of the given polynomial. (See Example 2.)

23. $-30y^3$
24. $-2x^2$
25. $4p^3 + 2p - 12$
26. $8t^2 - 4t - 3$
27. $-11ab^2 + a^2b$
28. $-23rs - 4r + 9s$

For Exercises 29–38, subtract the polynomials and simplify. (See Examples 3–4.)

29. $(13z^5 - z^2) - (7z^5 + 5z^2)$
30. $(8w^4 + 3w^2) - (12w^4 - w^2)$
31. $(-3x^3 + 3x^2 - x + 6) - (-x^3 - x^2 - x + 1)$
32. $(-8x^3 + 6x + 7) - (-5x^3 - 2x - 4)$
33. $\left(\dfrac{1}{5}a^2 - \dfrac{1}{2}ab + \dfrac{1}{10}b^2 + 3\right) - \left(-\dfrac{3}{10}a^2 + \dfrac{2}{5}ab - \dfrac{1}{2}b^2 - 5\right)$

34. $\left(\dfrac{4}{7}a^2 - \dfrac{1}{7}ab + \dfrac{1}{14}b^2 - 7\right) - \left(\dfrac{1}{2}a^2 - \dfrac{2}{7}ab - \dfrac{9}{14}b^2 + 1\right)$

35. Subtract $(9x^2 - 5x + 1)$ from $(8x^2 + x - 15)$.

36. Subtract $(-x^3 + 5x)$ from $(10x^3 + x^2 - 10)$.

37. Find the difference of $(3x^5 - 2x^3 + 4)$ and $(x^4 + 2x^3 - 7)$.

38. Find the difference of $(7x^{10} - 2x^4 - 3x)$ and $(-4x^3 - 5x^4 + x + 5)$.

Mixed Exercises

For Exercises 39–50, add or subtract as indicated. Write the answers in descending order.

39. $(8y^2 - 4y^3) - (3y^2 - 8y^3)$ **40.** $(-9y^2 - 8) - (4y^2 + 3)$ **41.** $(-2r - 6r^4) + (-r^4 - 9r)$

42. $(-8s^9 + 7s^2) + (7s^9 - s^2)$ **43.** $(5xy + 13x^2 + 3y) - (4x^2 - 8y)$ **44.** $(6p^2q - 2q) - (-2p^2q + 13)$

45. $(11ab - 23b^2) + (7ab - 19b^2)$ **46.** $(-4x^2y + 9) + (8x^2y - 12)$ **47.** $[2p - (3p + 5)] + (4p - 6) + 2$

48. $-(q - 2) - [4 - (2q - 3) + 5]$ **49.** $5 - [2m^2 - (4m^2 + 1)]$ **50.** $[4n^3 - (n^3 + 4)] + 3n^3$

Objective 4: Polynomial Functions

For Exercises 51–58, determine whether the given function is a polynomial function. If it is a polynomial function, state the degree. If not, state the reason why.

51. $g(x) = -7$ **52.** $g(x) = 4x$ **53.** $h(x) = \dfrac{2}{3}x^2 - 5$

54. $k(x) = -7x^4 - 0.3x + x^3$ **55.** $p(x) = 8x^3 + 2x^2 - \dfrac{3}{x}$ **56.** $q(x) = x^2 - 4x^{-3}$

57. $M(x) = |x| + 5x$ **58.** $N(x) = x^2 + |x|$

59. Given $P(x) = -x^4 + 2x - 5$, find the function values. **(See Example 5.)**
   a. $P(2)$ **b.** $P(-1)$ **c.** $P(0)$ **d.** $P(1)$

60. Given $N(x) = -x^2 + 5x$, find the function values.
   a. $N(1)$ **b.** $N(-1)$ **c.** $N(2)$ **d.** $N(0)$

61. Given $H(x) = \frac{1}{2}x^3 - x + \frac{1}{4}$, find the function values.
   a. $H(0)$ **b.** $H(2)$ **c.** $H(-2)$ **d.** $H(-1)$

62. Given $K(x) = \frac{2}{3}x^2 + \frac{1}{9}$, find the function values.
   a. $K(0)$ **b.** $K(3)$ **c.** $K(-3)$ **d.** $K(-1)$

63. A rectangular garden is designed to be 3 ft longer than it is wide. Let x represent the width of the garden. Find a function P that represents the perimeter in terms of x. **(See Example 6.)**

64. A flowerbed is in the shape of a triangle with the larger side 3 times the middle side and the smallest side 2 ft shorter than the middle side. Let x represent the length of the middle side. Find a function P that represents the perimeter in terms of x.

65. The cost in dollars of producing x toy cars is $C(x) = 2.2x + 1$. The revenue received is $R(x) = 5.98x$. To calculate profit, subtract the cost from the revenue.

   a. Write and simplify a function P that represents profit in terms of x.

   b. Find the profit of producing 50 toy cars.

66. The cost in dollars of producing x lawn chairs is $C(x) = 2.5x + 10.1$. The revenue for selling x chairs is $R(x) = 6.99x$. To calculate profit, subtract the cost from the revenue.

   a. Write and simplify a function P that represents profit in terms of x.

   b. Find the profit of producing 100 lawn chairs.

67. The function defined by $D(x) = 8x^2 + 86x + 2766$ approximates the yearly dormitory charges for private four-year colleges between the years 1995 and 2005. $D(x)$ is measured in dollars, and $x = 0$ corresponds to the year 1995. Find the function values and interpret their meaning in the context of this problem. **(See Example 7.)**

   a. $D(0)$ **b.** $D(2)$

   c. $D(4)$ **d.** $D(6)$

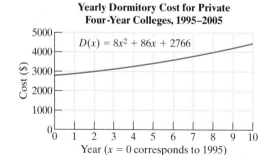

Yearly Dormitory Cost for Private Four-Year Colleges, 1995–2005

Source: U.S. National Center for Education Statistics.

68. The population of Mexico can be modeled by $P(t) = 0.022t^2 + 2.012t + 102$, where t is the number of years since 2000 and P is the number of people in millions.

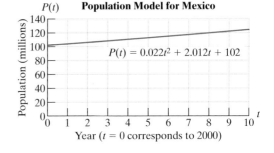

Population Model for Mexico

  a. Evaluate $P(0)$ and $P(6)$, and interpret their meaning in the context of this problem. Round to 1 decimal place if necessary.

  b. If this trend continues, what will the population of Mexico be in the year 2010? Round to 1 decimal place if necessary.

69. The number of women, w, who were due child support in the United States between 1995 and 2005 can be approximated by
$$W(t) = 143t + 5865$$
where t is the number of years after 1995 and $W(t)$ is measured in thousands. (Source: U.S. Bureau of the Census.)

  a. Evaluate $W(0)$, $W(5)$, and $W(10)$.

  b. Interpret the meaning of the function value of $W(10)$.

70. The total amount of child support due (in billions of dollars) in the United States between 1995 and 2005 can be approximated by
$$D(t) = 0.925t + 22.333$$
where t is the number of years after 1995 and $D(t)$ is the amount due (in billions of dollars).

  a. Evaluate $D(0)$, $D(4)$, and $D(8)$.

  b. Interpret the meaning of the function value of $D(8)$.

Expanding Your Skills

71. A toy rocket is shot from ground level at an angle of 60° from the horizontal. See the figure. The x- and y-positions of the rocket (measured in feet) vary with time t according to
$$x(t) = 25t$$
$$y(t) = -16t^2 + 43.3t$$

  a. Evaluate $x(0)$ and $y(0)$, and write the values as an ordered pair. Interpret the meaning of these function values in the context of this problem. Match the ordered pair with a point on the graph.

  b. Evaluate $x(1)$ and $y(1)$ and write the values as an ordered pair. Interpret the meaning of these function values in the context of this problem. Match the ordered pair with a point on the graph.

  c. Evaluate $x(2)$ and $y(2)$, and write the values as an ordered pair. Match the ordered pair with a point on the graph.

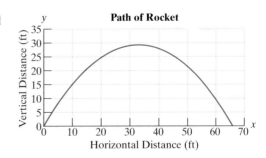

Path of Rocket

Section 5.2 Multiplication of Polynomials

Objectives
1. Multiplying Polynomials
2. Special Case Products: Difference of Squares and Perfect Square Trinomials
3. Translations Involving Polynomials
4. Applications Involving a Product of Polynomials

1. Multiplying Polynomials

The properties of exponents covered in Section 1.8 can be used to simplify many algebraic expressions including the multiplication of monomials. To multiply monomials, first use the associative and commutative properties of multiplication to group coefficients and *like* bases. Then simplify the result by using the properties of exponents.

Example 1 — Multiplying Monomials

Multiply the monomials.

a. $(3x^2y^7)(5x^3y)$ b. $(-3x^4y^3)(-2x^6yz^8)$

Solution:

a. $(3x^2y^7)(5x^3y)$
$= (3 \cdot 5)(x^2 \cdot x^3)(y^7 \cdot y)$ Group coefficients and like bases.
$= 15x^5y^8$ Add exponents and simplify.

b. $(-3x^4y^3)(-2x^6yz^8)$
$= [(-3)(-2)](x^4 \cdot x^6)(y^3 \cdot y)(z^8)$ Group coefficients and like bases.
$= 6x^{10}y^4z^8$ Add exponents and simplify.

Skill Practice

Multiply the polynomials.
1. $(-8r^3s)(-4r^4s^4)$

The distributive property is used to multiply polynomials: $a(b + c) = ab + ac$.

Example 2 — Multiplying a Polynomial by a Monomial

Multiply the polynomials.

a. $5y^3(2y^2 - 7y + 6)$ b. $-4a^3b^7c\left(2ab^2c^4 - \frac{1}{2}a^5b\right)$

Solution:

a. $5y^3(2y^2 - 7y + 6)$
$= (5y^3)(2y^2) + (5y^3)(-7y) + (5y^3)(6)$ Apply the distributive property.
$= 10y^5 - 35y^4 + 30y^3$ Simplify each term.

b. $-4a^3b^7c\left(2ab^2c^4 - \frac{1}{2}a^5b\right)$
$= (-4a^3b^7c)(2ab^2c^4) + (-4a^3b^7c)\left(-\frac{1}{2}a^5b\right)$ Apply the distributive property.
$= -8a^4b^9c^5 + 2a^8b^8c$ Simplify each term.

Skill Practice

Multiply the polynomials.
2. $-6b^2(2b^2 + 3b - 8)$
3. $8t^3\left(\frac{1}{2}t^3 - \frac{1}{4}t^2\right)$

Answers
1. $32r^7s^5$
2. $-12b^4 - 18b^3 + 48b^2$
3. $4t^6 - 2t^5$

Thus far, we have illustrated polynomial multiplication involving monomials. Next, the distributive property will be used to multiply polynomials with more than one term. For example:

$(x + 3)(x + 5) = (x + 3)x + (x + 3)5$ Apply the distributive property.

$= (x + 3)x + (x + 3)5$ Apply the distributive property again.

$= x \cdot x + 3 \cdot x + x \cdot 5 + 3 \cdot 5$

$= x^2 + 3x + 5x + 15$

$= x^2 + 8x + 15$ Combine *like* terms.

Note: Using the distributive property results in multiplying each term of the first polynomial by each term of the second polynomial:

$(x + 3)(x + 5) = x \cdot x + x \cdot 5 + 3 \cdot x + 3 \cdot 5$

$= x^2 + 5x + 3x + 15$

$= x^2 + 8x + 15$

Skill Practice

Multiply the polynomials.

4. $(2y - 1)(3y^2 - 2y - 1)$
5. $(4t + 5)(2t + 3)$
6. $(3a^4 + 2a^3 + a) \cdot (2a^2 - a - 5)$

example 3 Multiplying Polynomials

Multiply the polynomials.

a. $(2x^2 + 4)(3x^2 - x + 5)$ b. $(3y + 2)(7y - 6)$

Solution:

a. $(2x^2 + 4)(3x^2 - x + 5)$ Multiply each term in the first polynomial by each term in the second.

$= (2x^2)(3x^2) + (2x^2)(-x) + (2x^2)(5)$
$ + (4)(3x^2) + (4)(-x) + (4)(5)$ Apply the distributive property.

$= 6x^4 - 2x^3 + 10x^2 + 12x^2 - 4x + 20$ Simplify each term.

$= 6x^4 - 2x^3 + 22x^2 - 4x + 20$ Combine *like* terms.

Tip: Multiplication of polynomials can be performed vertically by a process similar to column multiplication of real numbers.

$(2x^2 + 4)(3x^2 - x + 5) \longrightarrow$
$$\begin{array}{r} 3x^2 - x + 5 \\ \times\ 2x^2 + 4 \\ \hline 12x^2 - 4x + 20 \\ 6x^4 - 2x^3 + 10x^2 \\ \hline 6x^4 - 2x^3 + 22x^2 - 4x + 20 \end{array}$$

Note: When multiplying by the column method, it is important to align *like* terms vertically before adding terms.

Answers

4. $6y^3 - 7y^2 + 1$
5. $8t^2 + 22t + 15$
6. $6a^6 + a^5 - 17a^4 - 8a^3 - a^2 - 5a$

b. $(3y + 2)(7y - 6)$ Multiply each term in the first polynomial by each term in the second.

$= (3y)(7y) + (3y)(-6) + (2)(7y) + (2)(-6)$ Apply the distributive property.

$= 21y^2 - 18y + 14y - 12$ Simplify each term.

$= 21y^2 - 4y - 12$ Combine *like* terms.

Tip: The acronym, FOIL (first outer inner last) can be used as a memory device to multiply the two binomials.

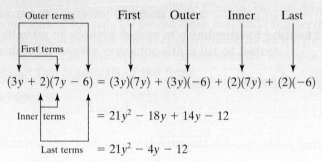

Note: It is important to realize that the acronym FOIL may only be used when finding the product of two *binomials*.

2. Special Case Products: Difference of Squares and Perfect Square Trinomials

In some cases the product of two binomials takes on a special pattern.

I. The first special case occurs when multiplying the sum and difference of the same two terms. For example:

$(2x + 3)(2x - 3)$

$= 4x^2 - 6x + 6x - 9$

$= 4x^2 - 9$

Notice that the "middle terms" are opposites. This leaves only the difference between the square of the first term and the square of the second term. For this reason, the product is called a *difference of squares*.

Note: The sum and difference of the same two terms are called **conjugates**. For example, we call $2x + 3$ the conjugate of $2x - 3$ and vice versa. In general, $a + b$ and $a - b$ are conjugates of each other.

II. The second special case involves the square of a binomial. For example:

$(3x + 7)^2$

$= (3x + 7)(3x + 7)$

$= 9x^2 + 21x + 21x + 49$

$= 9x^2 + 42x + 49$

$= (3x)^2 + 2(3x)(7) + (7)^2$

When squaring a binomial, the product will be a trinomial called a *perfect square trinomial*. The first and third terms are formed by squaring the terms of the binomial. The middle term is twice the product of the terms in the binomial.

Note: The expression $(3x - 7)^2$ also results in a perfect square trinomial, but the middle term is negative.

$$(3x - 7)(3x - 7) = 9x^2 - 21x - 21x + 49 = 9x^2 - 42x + 49$$

The following table summarizes these special case products.

Special Case Product Formulas

1. $(a + b)(a - b) = a^2 - b^2$ The product is called a **difference of squares**.
2. $(a + b)^2 = a^2 + 2ab + b^2$
   $(a - b)^2 = a^2 - 2ab + b^2$ The product is called a **perfect square trinomial**.

It is advantageous for you to become familiar with these special case products because they will be presented again when we factor polynomials.

Skill Practice

Multiply the polynomials.
7. $(c - 3)^2$
8. $(5x - 4)(5x + 4)$
9. $(7s^2 + 2t)^2$

example 4   Finding Special Products

Use the special product formulas to multiply the polynomials.

a. $(5x - 2)^2$ **b.** $(6c - 7d)(6c + 7d)$ **c.** $(4x^3 + 3y^2)^2$

Solution:

a. $(5x - 2)^2$ $a = 5x, b = 2$

     $= (5x)^2 - 2(5x)(2) + (2)^2$ Apply the formula $a^2 - 2ab + b^2$.

     $= 25x^2 - 20x + 4$ Simplify each term.

b. $(6c - 7d)(6c + 7d)$ $a = 6c, b = 7d$

     $= (6c)^2 - (7d)^2$ Apply the formula $a^2 - b^2$.

     $= 36c^2 - 49d^2$ Simplify each term.

c. $(4x^3 + 3y^2)^2$ $a = 4x^3, b = 3y^2$

     $= (4x^3)^2 + 2(4x^3)(3y^2) + (3y^2)^2$ Apply the formula $a^2 + 2ab + b^2$.

     $= 16x^6 + 24x^3y^2 + 9y^4$ Simplify each term.

The special case products can be used to simplify more complicated algebraic expressions.

example 5   Using Special Products

Multiply the following expressions.

a. $(x + y)^3$ **b.** $[x + (y + z)][x - (y + z)]$

Answers

7. $c^2 - 6c + 9$ 8. $25x^2 - 16$
9. $49s^4 + 28s^2t + 4t^2$

Solution:

a. $(x + y)^3$

$= (x + y)^2(x + y)$ Rewrite as the square of a binomial and another factor.

$= (x^2 + 2xy + y^2)(x + y)$ Expand $(x + y)^2$ by using the special case product formula.

$= (x^2)(x) + (x^2)(y) + (2xy)(x) + (2xy)(y) + (y^2)(x) + (y^2)(y)$ Apply the distributive property.

$= x^3 + x^2y + 2x^2y + 2xy^2 + xy^2 + y^3$ Simplify each term.

$= x^3 + 3x^2y + 3xy^2 + y^3$ Combine *like* terms.

b. $[x + (y + z)][x - (y + z)]$ This product is in the form $(a + b)(a - b)$, where $a = x$ and $b = (y + z)$.

$= (x)^2 - (y + z)^2$ Apply the formula $a^2 - b^2$.

$= (x)^2 - (y^2 + 2yz + z^2)$ Expand $(y + z)^2$ by using the special case product formula.

$= x^2 - y^2 - 2yz - z^2$ Apply the distributive property.

Skill Practice

Multiply the polynomials.

10. $(b + 2)^3$
11. $[a + (b + 3)][a - (b + 3)]$

3. Translations Involving Polynomials

example 6   Translating Between English Form and Algebraic Form

Complete the table.

English Form	Algebraic Form
The square of the sum of x and y	
	$x^2 + y^2$
The square of the product of 3 and x	

Skill Practice

Translate to algebraic form:

12. The square of the difference of a and b
13. The difference of the square of a and the square of b
14. Translate to English form: $a - b^2$.

Solution:

English Form	Algebraic Form	Notes
The square of the sum of x and y	$(x + y)^2$	The *sum* is squared, not the individual terms.
The sum of the squares of x and y	$x^2 + y^2$	The individual terms x and y are squared first. Then the sum is taken.
The square of the product of 3 and x	$(3x)^2$	The product of 3 and x is taken. Then the result is squared.

Answers

10. $b^3 + 6b^2 + 12b + 8$
11. $a^2 - b^2 - 6b - 9$
12. $(a - b)^2$ 13. $a^2 - b^2$
14. The difference of a and the square of b

4. Applications Involving a Product of Polynomials

Skill Practice

A rectangular photograph is mounted on a square piece of cardboard whose sides have length x. The border that surrounds the photo is 3 in. on each side and 4 in. on both top and bottom.

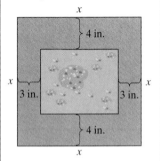

15. Write an expression for the area of the photograph and multiply.

16. Determine the area of the photograph if x is 12.

example 7 — Applying a Product of Polynomials

A box is created from a sheet of cardboard 20 in. on a side by cutting a square from each corner and folding up the sides (Figures 5-3 and 5-4). Let x represent the length of the sides of the squares removed from each corner.

a. Find an expression for the volume of the box in terms of x.

b. Find the volume if a 4-in. square is removed.

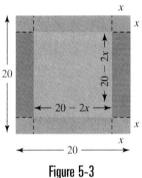

Figure 5-3

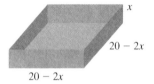

Figure 5-4

Solution:

a. The volume of a rectangular box is given by the formula $V = lwh$. The length and width can both be expressed as $20 - 2x$. The height of the box is x. Hence the volume is given by

$$V = l \cdot w \cdot h$$
$$= (20 - 2x)(20 - 2x)x$$
$$= (20 - 2x)^2 x$$
$$= (400 - 80x + 4x^2)x$$
$$= 400x - 80x^2 + 4x^3$$
$$= 4x^3 - 80x^2 + 400x$$

b. If a 4-in. square is removed from the corners of the box, we have $x = 4$ in. Thus the volume is

$$V = 4(4)^3 - 80(4)^2 + 400(4)$$
$$= 4(64) - 80(16) + 400(4)$$
$$= 256 - 1280 + 1600$$
$$= 576$$

The volume is 576 in.^{3}

Answers

15. $A = (x - 8)(x - 6)$;
    $A = x^2 - 14x + 48$
16. 24 in.^{2}

Section 5.2 Multiplication of Polynomials 321

section 5.2 Practice Exercises

Boost *your* GRADE at mathzone.com!

- Practice Problems
- Self-Tests
- NetTutor
- e-Professors
- Videos

Study Skills Exercises

1. Do you need complete silence, or do you listen to music while you do your homework? Try something different today so that you can compare and choose the best situation for you.

2. Define the key terms.
   a. Difference of squares
   b. Conjugates
   c. Perfect square trinomial

Review Exercises

3. Given $f(x) = 4x^3 - 5$, find the function values.
   a. $f(3)$
   b. $f(0)$
   c. $f(-2)$

4. Given $g(x) = x^4 - x^2 - 3$, find the function values.
   a. $g(-1)$
   b. $g(2)$
   c. $g(0)$

For Exercises 5–6, perform the indicated operations.

5. $(3x^2 - 7x - 2) + (-x^2 + 3x - 5)$

6. $(3x^2 - 7x - 2) - (-x^2 + 3x - 5)$

7. Can the terms $2x^3$ and $3x^2$ be combined by addition? Can they be combined by multiplication? Explain.

8. Write the distributive property of multiplication over addition. Give an example of the distributive property. (Answers may vary.)

Objective 1: Multiplying Polynomials

For Exercises 9–38, multiply the polynomials by using the distributive property and the special product formulas. (See Examples 1–3.)

9. $3ab(a + b)$
10. $2a(3 - a)$
11. $\frac{1}{5}(2a - 3)$

12. $\frac{1}{3}(6b + 4)$
13. $2m^3n^2(m^2n^3 - 3mn^2 + 4n)$
14. $3p^2q(p^3q^3 - pq^2 - 4p)$

15. $(x + y)(x - 2y)$
16. $(3a + 5)(a - 2)$
17. $(6x - 1)(5 + 2x)$

18. $(7 + 3x)(x - 8)$
19. $(4a - 9)(2a - 1)$
20. $(3b + 5)(b - 5)$

21. $(y^2 - 12)(2y^2 + 3)$

22. $(4p^2 - 1)(2p^2 + 5)$

23. $(5s + 3t)(5s - 2t)$

24. $(4a + 3b)(4a - b)$

25. $(n^2 + 10)(5n + 3)$

26. $(m^2 + 8)(3m + 7)$

27. $(1.3a - 4b)(2.5a + 7b)$

28. $(2.1x - 3.5y)(4.7x + 2y)$

29. $(2x + y)(3x^2 + 2xy + y^2)$

30. $(h - 5k)(h^2 - 2hk + 3k^2)$

31. $(x - 7)(x^2 + 7x + 49)$

32. $(x + 3)(x^2 - 3x + 9)$

33. $(4a - b)(a^3 - 4a^2b + ab^2 - b^3)$

34. $(3m + 2n)(m^3 + 2m^2n - mn^2 + 2n^3)$

35. $\left(\dfrac{1}{2}a - 2b + c\right)(a + 6b - c)$

36. $(x + y - 2z)(5x - y + z)$

37. $(-x^2 + 2x + 1)(3x - 5)$

38. $\left(\dfrac{1}{2}a^2 - 2ab + b^2\right)(2a + b)$

Objective 2: Special Case Products: Difference of Squares and Perfect Square Trinomials

For Exercises 39–54, multiply by using the special case products. (See Example 4.)

39. $(a - 8)(a + 8)$

40. $(b + 2)(b - 2)$

41. $(3p + 1)(3p - 1)$

42. $(5q - 3)(5q + 3)$

43. $\left(x - \dfrac{1}{3}\right)\left(x + \dfrac{1}{3}\right)$

44. $\left(\dfrac{1}{2}x + \dfrac{1}{3}\right)\left(\dfrac{1}{2}x - \dfrac{1}{3}\right)$

45. $(3h - k)(3h + k)$

46. $(x - 7y)(x + 7y)$

47. $(3h - k)^2$

48. $(x - 7y)^2$

49. $(t - 7)^2$

50. $(w + 9)^2$

51. $(u + 3v)^2$

52. $(a - 4b)^2$

53. $\left(h + \dfrac{1}{6}k\right)^2$

54. $\left(\dfrac{2}{5}x + 1\right)^2$

55. Multiply the expressions. Explain their similarities.
   a. $(A - B)(A + B)$
   b. $[(x + y) - B][(x + y) + B]$

56. Multiply the expressions. Explain their similarities.
   a. $(A + B)(A - B)$
   b. $[A + (3h + k)][A - (3h + k)]$

For Exercises 57–62, multiply the expressions. (See Example 5.)

57. $[(w + v) - 2][(w + v) + 2]$

58. $[(x + y) - 6][(x + y) + 6]$

59. $[2 - (x + y)][2 + (x + y)]$

60. $[a - (b + 1)][a + (b + 1)]$

61. $[(3a - 4) + b][(3a - 4) - b]$

62. $[(5p - 7) - q][(5p - 7) + q]$

63. Explain how to multiply $(x + y)^3$.

64. Explain how to multiply $(a - b)^3$.

For Exercises 65–68, multiply the expressions. (See Example 5.)

65. $(2x + y)^3$ **66.** $(x - 5y)^3$ **67.** $(4a - b)^3$ **68.** $(3a + 4b)^3$

69. Explain how you would multiply the binomials

$(x - 2)(x + 6)(2x + 1)$

70. Explain how you would multiply the binomials

$(a + b)(a - b)(2a + b)(2a - b)$

For Exercises 71–74, multiply the expressions containing more than two factors. (*Hint:* First multiply two of the factors, then multiply that product by the third factor.)

71. $2a^2(a + 5)(3a + 1)$ **72.** $-5y(2y - 3)(y + 3)$ **73.** $(x + 3)(x - 3)(x + 5)$ **74.** $(t + 2)(t - 3)(t + 1)$

Objective 3: Translations Involving Polynomials

For Exercises 75–78, translate from English form to algebraic form. (See Example 6.)

75. The square of a plus the cube of b

76. The square of the sum of r and t

77. The difference of x squared and y cubed

78. The square of the product of 3 and a

For Exercises 79–82, translate from algebraic form to English form. (See Example 6.)

79. $p^3 + q^2$ **80.** $a^3 - b^3$ **81.** xy^2 **82.** $(c + d)^3$

Objective 4: Applications Involving a Product of Polynomials

83. A rectangular garden has a walk around it of width x. The garden is 20 ft by 15 ft. Find a function f in terms of x that represents the combined area of the garden and walk. Simplify the result. (See Example 7.)

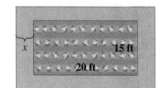

84. An 8-in. by 10-in. photograph is in a frame of width x. Find a function g in terms of x that represents the area of the frame. Simplify the result.

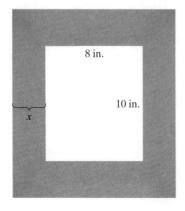

For Exercises 85–90, write an expression for the area and simplify your answer.

85. Square

86. Square

87. Rectangle

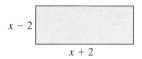

88. Rectangle

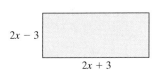

89. Triangle

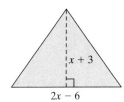

90. Triangle

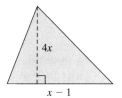

For Exercises 91–94, write an expression for the volume and simplify your answer.

91.

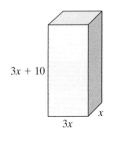

92.

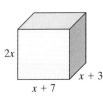

93. Cube

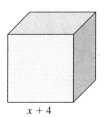

94. Cube

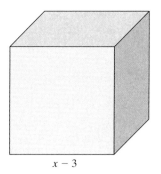

Expanding Your Skills

95. Explain how to multiply $(x + 2)^4$.

96. Explain how to multiply $(y - 3)^4$.

97. $(2x - 3)$ multiplied by what binomial will result in the trinomial $10x^2 - 27x + 18$? Check your answer by multiplying the binomials.

98. $(4x + 1)$ multiplied by what binomial will result in the trinomial $12x^2 - 5x - 2$? Check your answer by multiplying the binomials.

99. $(4y + 3)$ multiplied by what binomial will result in the trinomial $8y^2 + 2y - 3$? Check your answer by multiplying the binomials.

100. $(3y - 2)$ multiplied by what binomial will result in the trinomial $3y^2 - 17y + 10$? Check your answer by multiplying the binomials.

section 5.3 Division of Polynomials

1. Division by a Monomial

Division of polynomials is presented in this section as two separate cases. The first case illustrates division by a monomial divisor. The second case illustrates division by a polynomial with two or more terms.

To divide a polynomial by a monomial, divide each individual term in the polynomial by the divisor and simplify the result.

Objectives
1. Division by a Monomial
2. Long Division
3. Synthetic Division

To Divide a Polynomial by a Monomial

If a, b, and c are polynomials such that $c \neq 0$, then

$$\frac{a+b}{c} = \frac{a}{c} + \frac{b}{c} \qquad \text{Similarly,} \qquad \frac{a-b}{c} = \frac{a}{c} - \frac{b}{c}$$

---

example 1   Dividing a Polynomial by a Monomial

Divide the polynomials.

a. $\dfrac{3x^4 - 6x^3 + 9x}{3x}$

b. $(10c^3d - 15c^2d^2 + 2cd^3) \div (5c^2d^2)$

Solution:

a. $\dfrac{3x^4 - 6x^3 + 9x}{3x}$

$= \dfrac{3x^4}{3x} - \dfrac{6x^3}{3x} + \dfrac{9x}{3x}$ Divide each term in the numerator by $3x$.

$= x^3 - 2x^2 + 3$ Simplify each term, using the properties of exponents.

b. $(10c^3d - 15c^2d^2 + 2cd^3) \div (5c^2d^2)$

$= \dfrac{10c^3d - 15c^2d^2 + 2cd^3}{5c^2d^2}$

$= \dfrac{10c^3d}{5c^2d^2} - \dfrac{15c^2d^2}{5c^2d^2} + \dfrac{2cd^3}{5c^2d^2}$ Divide each term in the numerator by $5c^2d^2$.

$= \dfrac{2c}{d} - 3 + \dfrac{2d}{5c}$ Simplify each term.

Skill Practice

Divide.

1. $\dfrac{18y^3 - 6y^2 - 12y}{6y}$

2. $(-24a^3b^2 - 16a^2b^3 + 8ab) \div (-8ab)$

---

2. Long Division

If the divisor has two or more terms, a long division process similar to the division of real numbers is used.

Answers
1. $3y^2 - y - 2$
2. $3a^2b + 2ab^2 - 1$

Skill Practice

Divide.

3. $(4x^2 + 6x - 8) \div (x + 3)$
4. $\dfrac{12p^3 + 2p^2 - 5p - 3}{3p - 1}$

Concept Connections

5. When is long division required for dividing polynomials?

example 2   Using Long Division to Divide Polynomials

Divide the polynomials by using long division.

$$(3x^2 - 14x - 10) \div (x - 2)$$

Solution:

$x - 2 \overline{)3x^2 - 14x - 10}$ Divide the leading term in the dividend by the leading term in the divisor.

$(3x^2)/x = 3x$. This is the first term in the quotient.

$$\begin{array}{r} 3x \phantom{{}-14x-10} \\ x-2\overline{)3x^2 - 14x - 10} \\ 3x^2 - 6x \end{array}$$

Multiply $3x$ by the divisor and record the result: $3x(x - 2) = 3x^2 - 6x$.

$$\begin{array}{r} 3x \phantom{{}-14x-10} \\ x-2\overline{)3x^2 - 14x - 10} \\ \underline{-3x^2 + 6x} \\ -8x \end{array}$$

Next, subtract the quantity $3x^2 - 6x$. To do this, add its opposite.

$$\begin{array}{r} 3x - 8 \\ x-2\overline{)3x^2 - 14x - 10} \\ \underline{-3x^2 + 6x} \downarrow \\ -8x - 10 \\ -8x + 16 \end{array}$$

Bring down next column and repeat the process.

Divide the leading term by x: $-8x/x = -8$

Multiply the divisor by -8 and record the result: $-8(x - 2) = -8x + 16$.

$$\begin{array}{r} 3x - 8 \\ x-2\overline{)3x^2 - 14x - 10} \\ \underline{-3x^2 + 6x} \\ -8x - 10 \\ \underline{+8x - 16} \\ -26 \end{array}$$

Subtract the quantity $(-8x + 16)$ by adding its opposite.

The remainder is -26. We do not continue because the degree of the remainder is less than the degree of the divisor.

Summary:

The quotient is	$3x - 8$
The remainder is	-26
The divisor is	$x - 2$
The dividend is	$3x^2 - 14x - 10$

The solution to a long division problem is often written in the form: Quotient + remainder/divisor. Hence

$$(3x^2 - 14x - 10) \div (x - 2) = \mathbf{3x - 8} + \dfrac{\mathbf{-26}}{\mathbf{x - 2}}$$

This answer can also be written as

$$3x - 8 - \dfrac{26}{x - 2}$$

Answers

3. $4x - 6 + \dfrac{10}{x + 3}$
4. $4p^2 + 2p - 1 + \dfrac{-4}{3p - 1}$
5. Long division is required when the divisor has two or more terms.

The division of polynomials can be checked in the same fashion as the division of real numbers. To check, we know that

Dividend = (divisor)(quotient) + remainder

$$3x^2 - 14x - 10 \stackrel{?}{=} (x - 2)(3x - 8) + (-26)$$
$$\stackrel{?}{=} 3x^2 - 8x - 6x + 16 + (-26)$$
$$= 3x^2 - 14x - 10 \checkmark$$

example 3 — Using Long Division to Divide Polynomials

Divide the polynomials by using long division: $(-2x^3 - 10x^2 + 56) \div (2x - 4)$

Solution:

First note that the dividend has a missing power of x and can be written as $-2x^3 - 10x^2 + 0x + 56$. The term $0x$ is a placeholder for the missing term. It is helpful to use the placeholder to keep the powers of x lined up.

$$\begin{array}{r} -x^2 \\ 2x-4 \overline{\smash{\big)}-2x^3 - 10x^2 + 0x + 56} \\ \underline{-2x^3 + 4x^2} \end{array}$$

Leave space for the missing power of x.
Divide $-2x^3/2x = -x^2$ to get the first term of the quotient.

$$\begin{array}{r} -x^2 - 7x \\ 2x-4 \overline{\smash{\big)}-2x^3 - 10x^2 + 0x + 56} \\ \underline{2x^3 - 4x^2} \\ -14x^2 + 0x \\ -14x^2 + 28x \end{array}$$

Subtract by adding the opposite.
Bring down the next column.

Divide $(-14x^2)/(2x) = -7x$ to get the next term in the quotient.

$$\begin{array}{r} -x^2 - 7x - 14 \\ 2x-4 \overline{\smash{\big)}-2x^3 - 10x^2 + 0x + 56} \\ \underline{2x^3 - 4x^2} \\ -14x^2 + 0x \\ \underline{14x^2 - 28x} \\ -28x + 56 \\ -28x + 56 \end{array}$$

Subtract by adding the opposite.
Bring down the next column.

Divide $(-28x)/(2x) = -14$ to get the next term in the quotient.

$$\begin{array}{r} -x^2 - 7x - 14 \\ 2x-4 \overline{\smash{\big)}-2x^3 - 10x^2 + 0x + 56} \\ \underline{2x^3 - 4x^2} \\ -14x^2 + 0x \\ \underline{14x^2 - 28x} \\ -28x + 56 \\ \underline{28x - 56} \\ 0 \end{array}$$

Subtract by adding the opposite.

← The remainder is 0.

The quotient is $-x^2 - 7x - 14$ and the remainder is 0.

Skill Practice

Divide.

6. $\dfrac{4y^3 - 2y + 7}{2y + 2}$

Tip: Both the divisor and dividend must be written in descending order before you do polynomial division.

Answer

6. $2y^2 - 2y + 1 + \dfrac{5}{2y + 2}$

Because the remainder is zero, $2x - 4$ divides *evenly* into $-2x^3 - 10x^2 + 56$. For this reason, the divisor and quotient are *factors* of $-2x^3 - 10x^2 + 56$. To check, we have

$$\text{Dividend} = (\text{divisor})(\text{quotient}) + \text{remainder}$$

$$-2x^3 - 10x^2 + 56 \stackrel{?}{=} (2x - 4)(-x^2 - 7x - 14) + 0$$
$$\stackrel{?}{=} -2x^3 - 14x^2 - 28x + 4x^2 + 28x + 56$$
$$= -2x^3 - 10x^2 + 56 \checkmark$$

Skill Practice

Divide.

7. $(x^3 + 2x^2 + 1) \div (x^2 + 1)$

example 4   Using Long Division to Divide Polynomials

Divide.

$$(6x^4 + 15x^3 - 5x^2 - 4) \div (3x^2 - 4)$$

Solution:

The dividend has a missing power of x and can be written as $6x^4 + 15x^3 - 5x^2 + 0x - 4$.

The divisor has a missing power of x and can be written as $3x^2 + 0x - 4$.

$$\begin{array}{r} 2x^2 \\ 3x^2 + 0x - 4 \overline{\smash{)}6x^4 + 15x^3 - 5x^2 + 0x - 4} \\ \underline{6x^4 + 0x^3 - 8x^2 } \end{array}$$

Leave space for missing powers of x.

Divide $6x^4/3x^2 = 2x^2$ to get the first term of the quotient.

$$\begin{array}{r} 2x^2 + 5x \\ 3x^2 + 0x - 4 \overline{\smash{)}6x^4 + 15x^3 - 5x^2 + 0x - 4} \\ \underline{-6x^4 - 0x^3 + 8x^2 } \\ 15x^3 + 3x^2 + 0x \\ 15x^3 + 0x^2 - 20x \end{array}$$

Subtract by adding the opposite.

Bring down the next column.

$$\begin{array}{r} 2x^2 + 5x + 1 \\ 3x^2 + 0x - 4 \overline{\smash{)}6x^4 + 15x^3 - 5x^2 + 0x - 4} \\ \underline{-6x^4 - 0x^3 + 8x^2 } \\ 15x^3 + 3x^2 + 0x \\ \underline{-15x^3 - 0x^2 + 20x } \\ 3x^2 + 20x - 4 \\ 3x^2 + 0x - 4 \end{array}$$

Subtract by adding the opposite.

Bring down the next column.

$$\begin{array}{r} 2x^2 + 5x + 1 \\ 3x^2 + 0x - 4 \overline{\smash{)}6x^4 + 15x^3 - 5x^2 + 0x - 4} \\ \underline{-6x^4 - 0x^3 + 8x^2 } \\ 15x^3 + 3x^2 + 0x \\ \underline{-15x^3 - 0x^2 + 20x } \\ 3x^2 + 20x - 4 \\ \underline{-3x^2 - 0x + 4} \\ 20x \end{array}$$

Subtract by adding the opposite.

The remainder is $20x$. The degree of $20x$ is less than the degree of $3x^2 - 4$.

Answer

7. $x + 2 + \dfrac{-x - 1}{x^2 + 1}$

Therefore,

$$(6x^4 + 15x^3 - 5x^2 - 4) \div (3x^2 - 4) = 2x^2 + 5x + 1 + \frac{20x}{3x^2 - 4}$$

3. Synthetic Division

In this section we introduced the process of long division to divide two polynomials. Next, we will learn another technique, called **synthetic division**, to divide two polynomials. Synthetic division may be used when dividing a polynomial by a first-degree divisor of the form $x - r$, where r is a constant. Synthetic division is considered a "shortcut" because it uses the coefficients of the divisor and dividend without writing the variables.

Consider dividing the polynomials $(3x^2 - 14x - 10) \div (x - 2)$.

$$\begin{array}{r} 3x - 8 \\ x - 2 \overline{\smash{)}3x^2 - 14x - 10} \\ \underline{-(3x^2 - 6x)} \\ -8x - 10 \\ \underline{-(-8x + 16)} \\ -26 \end{array}$$

Concept Connections

Determine whether synthetic division can be used to divide the polynomials. Explain your answer.

8. $(x^3 + 6x^2 + 2x - 1)$
 $\div (2x + 3)$
9. $(y^2 - 5y + 8) \div (y^2 + 1)$
10. $(a^3 + 5a^2 + a - 2)$
 $\div (a - 5)$

First note that the divisor $x - 2$ is in the form $x - r$, where $r = 2$. Hence synthetic division can also be used to find the quotient and remainder.

Step 1: Write the value of r in a box. $\longrightarrow$ $\underline{2}\,|\,3 \;\; -14 \;\; -10$ $\longleftarrow$ **Step 2:** Write the coefficients of the dividend to the right of the box.

$$ 3$$

Step 3: Skip a line and draw a horizontal line below the list of coefficients. **Step 4:** Bring down the leading coefficient from the dividend and write it below the line.

$$\underline{2}\,|\,3 \;\; -14 \;\; -10$$
$$ 6 $$
$$ 3 \;\; -8 $$

Step 5: Multiply the value of r by the number below the line ($2 \times 3 = 6$). Write the result in the next column above the line. **Step 6:** Add the numbers in the column above the line ($-14 + 6$), and write the result below the line.

Repeat steps 5 and 6 until all columns have been completed.

Step 7: To get the final result, we use the numbers below the line. The number in the last column is the remainder. The other numbers are the coefficients of the quotient.

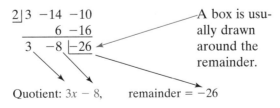

Quotient: $3x - 8$, remainder $= -26$

A box is usually drawn around the remainder.

Answers

8. No. The leading coefficient of the divisor is not equal to 1.
9. No. The divisor is not first-degree.
10. Yes. The divisor is in the form $(x - r)$, where $r = 5$.

The degree of the quotient will always be 1 less than that of the dividend. Because the dividend is a second-degree polynomial, the quotient will be a first-degree polynomial. In this case, the quotient is $3x - 8$ and the remainder is -26.

Skill Practice

Divide the polynomials by using synthetic division. Identify the quotient and the remainder.

11. $(5y^2 - 4y + 2y^3 - 5) \div (y + 3)$

---

example 5   Using Synthetic Division to Divide Polynomials

Divide the polynomials $(5x + 4x^3 - 6 + x^4) \div (x + 3)$ by using synthetic division.

Solution:

As with long division, the terms of the dividend and divisor should be written in descending order. Furthermore, missing powers must be accounted for by using placeholders (shown here in bold).

Hence,
$$5x + 4x^3 - 6 + x^4$$
$$= x^4 + 4x^3 + \mathbf{0}x^2 + 5x - 6$$

To use synthetic division, the divisor must be in the form $(x - r)$. The divisor $x + 3$ can be written as $x - (-3)$. Hence, $r = -3$.

Step 1: Write the value of r in a box.

Step 2: Write the coefficients of the dividend to the right of the box.

$$\underline{-3|}\ 1\ \ 4\ \ 0\ \ 5\ \ -6$$
$$1$$

Step 3: Skip a line and draw a horizontal line below the list of coefficients.

Step 4: Bring down the leading coefficient from the dividend and write it below the line.

Step 5: Multiply the value of r by the number below the line $(-3 \times 1 = -3)$. Write the result in the next column above the line.

$$\underline{-3|}\ 1\ \ 4\ \ 0\ \ 5\ \ -6$$
$$-3$$
$$1\ \ 1$$

Step 6: Add the numbers in the column above the line: $4 + (-3) = 1$.

Repeat steps 5 and 6:

$$\underline{-3|}\ 1\ \ \ \ 4\ \ \ \ 0\ \ \ \ 5\ \ \ -6$$
$$-3\ -3\ \ \ 9\ \ -42$$
$$1\ \ \ 1\ -3\ \ 14\ \underline{|-48}$$

The quotient is $x^3 + x^2 - 3x + 14$.

The remainder is -48.

— remainder
— constant
— x-term coefficient
— x^2-term coefficient
— x^3-term coefficient

Answer

11. Quotient: $2y^2 - y - 1$; remainder: -2

Tip: It is interesting to compare the long division process to the synthetic division process. For Example 5, long division is shown on the left, and synthetic division is shown on the right. Notice that the same pattern of coefficients used in long division appears in the synthetic division process.

$$\begin{array}{r} x^3 + x^2 - 3x + 14 \\ x+3\overline{\smash{)}x^4 + 4x^3 + 0x^2 + 5x - 6} \\ \underline{-(x^4 + 3x^3)} \\ x^3 + 0x^2 \\ \underline{-(x^3 + 3x^2)} \\ -3x^2 + 5x \\ \underline{-(-3x^2 - 9x)} \\ 14x - 6 \\ \underline{-(14x + 42)} \\ -48 \end{array}$$

$$\begin{array}{r|rrrrr} -3 & 1 & 4 & 0 & 5 & -6 \\ & & -3 & -3 & 9 & -42 \\ \hline & 1 & 1 & -3 & 14 & \boxed{-48} \end{array}$$

$x^3x^2x\text{constant}\text{remainder}$

Quotient: $x^3 + x^2 - 3x + 14$
Remainder: -48

---

example 6 — Using Synthetic Division to Divide Polynomials

Divide the polynomials by using synthetic division. Identify the quotient and remainder.

a. $(2m^7 - 3m^5 + 4m^4 - m + 8) \div (m + 2)$ **b.** $(p^4 - 81) \div (p - 3)$

Solution:

a. Insert placeholders (bold) for missing powers of m.

$(2m^7 - 3m^5 + 4m^4 - m + 8) \div (m + 2)$

$(2m^7 + \mathbf{0m^6} - 3m^5 + 4m^4 + \mathbf{0m^3} + \mathbf{0m^2} - m + 8) \div (m + 2)$

Because $m + 2$ can be written as $m - (-2)$, $r = -2$.

$$\begin{array}{r|rrrrrrrr} -2 & 2 & 0 & -3 & 4 & 0 & 0 & -1 & 8 \\ & & -4 & 8 & -10 & 12 & -24 & 48 & -94 \\ \hline & 2 & -4 & 5 & -6 & 12 & -24 & 47 & \boxed{-86} \end{array}$$

Quotient: $2m^6 - 4m^5 + 5m^4 - 6m^3 + 12m^2 - 24m + 47$
Remainder: -86

The quotient is 1 degree less than dividend.

b. $(p^4 - 81) \div (p - 3)$

$(p^4 + \mathbf{0p^3} + \mathbf{0p^2} + \mathbf{0p} - 81) \div (p - 3)$ Insert placeholders (bold) for missing powers of p.

$$\begin{array}{r|rrrrr} 3 & 1 & 0 & 0 & 0 & -81 \\ & & 3 & 9 & 27 & 81 \\ \hline & 1 & 3 & 9 & 27 & \boxed{0} \end{array}$$

Quotient: $p^3 + 3p^2 + 9p + 27$
Remainder: 0

Skill Practice

Divide the polynomials by using synthetic division. Identify the quotient and the remainder.

12. $(4c^4 - 3c^2 - 6c - 3) \div (c - 2)$

13. $(x^3 + 1) \div (x + 1)$

Answers

12. Quotient: $4c^3 + 8c^2 + 13c + 20$; remainder: 37
13. Quotient: $x^2 - x + 1$; remainder: 0

section 5.3 Practice Exercises

Boost your GRADE at mathzone.com!

- Practice Problems
- Self-Tests
- NetTutor
- e-Professors
- Videos

Study Skills Exercises

1. **a.** How often do you do your math homework? Every day? Three times a week? Once a week?

   b. Math should be studied every day. List the days and times when you plan to work on math this week.
   Mon.   Tue.   Wed.   Thurs.   Fri.   Sat.   Sun.

2. Define the key term **synthetic division**.

Review Exercises

3. **a.** Add $(3x + 1) + (2x - 5)$.

   b. Multiply $(3x + 1)(2x - 5)$.

4. **a.** Subtract $(a - 10b) - (5a + b)$.

   b. Multiply $(a - 10b)(5a + b)$.

5. **a.** Subtract $(2y^2 + 1) - (y^2 - 5y + 1)$.

   b. Multiply $(2y^2 + 1)(y^2 - 5y + 1)$.

6. **a.** Add $(x^2 - x) + (6x^2 + x + 2)$.

   b. Multiply $(x^2 - x)(6x^2 + x + 2)$.

For Exercises 7–11, answers may vary.

7. Write an example of a product of a monomial and a binomial and simplify.

8. Write an example of a product of two binomials and simplify.

9. Write an example of a sum of a binomial and a trinomial and simplify.

10. Write an example of the square of a binomial and simplify.

11. Write an example of the product of conjugates and simplify.

Objective 1: Division by a Monomial

For Exercises 12–25, divide the polynomials. Check your answer by multiplication. **(See Example 1.)**

12. $(36y + 24y^2 + 6y^3) \div (3y)$

13. $(6p^2 - 18p^4 + 30p^5) \div (6p)$

14. $(4x^3y + 12x^2y^2 - 4xy^3) \div (4xy)$

15. $(25m^5n - 10m^4n + m^3n) \div (5m^3n)$

16. $(-8y^4 - 12y^3 + 32y^2) \div (-4y^2)$

17. $(12y^5 - 8y^6 + 16y^4 - 10y^3) \div (2y^3)$

18. $(3p^4 - 6p^3 + 2p^2 - p) \div (-6p)$

19. $(-4q^3 + 8q^2 - q) \div (-12q)$

20. $(a^3 + 5a^2 + a - 5) \div (a)$

21. $(2m^5 - 3m^4 + m^3 - m^2 + 9m) \div (m^2)$

22. $(6s^3t^5 - 8s^2t^4 + 10st^2) \div (-2st^4)$

23. $(-8r^4w^2 - 4r^3w + 2w^3) \div (-4r^3w)$

24. $(8p^4q^7 - 9p^5q^6 - 11p^3q - 4) \div (p^2q)$

25. $(20a^5b^5 - 20a^3b^2 + 5a^2b + 6) \div (a^2b)$

Objective 2: Long Division

26. **a.** Divide $(2x^3 - 7x^2 + 5x - 1) \div (x - 2)$, and identify the divisor, quotient, and remainder.

    b. Explain how to check by using multiplication.

27. **a.** Divide $(x^3 + 4x^2 + 7x - 3) \div (x + 3)$, and identify the divisor, quotient, and remainder.

    b. Explain how to check by using multiplication.

For Exercises 28–43, divide the polynomials by using long division. Check your answer by multiplication.
(See Examples 2–4.)

28. $(x^2 + 11x + 19) \div (x + 4)$

29. $(x^3 - 7x^2 + 13x + 3) \div (x - 2)$

30. $(3y^3 - 7y^2 - 4y + 3) \div (y - 3)$

31. $(z^3 - 2z^2 + 2z - 5) \div (z - 4)$

32. $(-12a^2 + 77a - 121) \div (3a - 11)$

33. $(28x^2 - 29x + 6) \div (4x - 3)$

34. $(18y^2 + 9y - 20) \div (3y + 4)$

35. $(-3y^2 + 2y + 1) \div (-y + 1)$

36. $(8a^3 + 1) \div (2a + 1)$

37. $(81x^4 - 1) \div (3x + 1)$

38. $(x^4 - x^3 - x^2 + 4x - 2) \div (x^2 + x - 1)$

39. $(2a^5 - 7a^4 + 11a^3 - 22a^2 + 29a - 10) \div (2a^2 - 5a + 2)$

40. $(x^4 - 3x^2 + 10) \div (x^2 - 2)$

41. $(3y^4 - 25y^2 - 18) \div (y^2 - 3)$

42. $(n^4 - 16) \div (n - 2)$

43. $(m^3 + 27) \div (m + 3)$

Objective 3: Synthetic Division

44. Explain the conditions under which you may use synthetic division to divide polynomials.

45. Can synthetic division be used to divide $(4x^4 + 3x^3 - 7x + 9)$ by $(2x + 5)$? Explain why or why not.

46. Can synthetic division be used to divide $(6y^5 - 3y^2 + 2y - 14)$ by $(y^2 - 3)$? Explain why or why not.

47. Can synthetic division be used to divide $(3y^4 - y + 1)$ by $(y - 5)$? Explain why or why not.

48. The following table represents the result of a synthetic division.

$$\underline{5|}\ \ 1\ \ -2\ \ -4\ \ \ \ 3$$
$$\phantom{\underline{5|}\ \ 1\ \ }\ \ \ \ 5\ \ \ 15\ \ \ 55$$
$$\phantom{\underline{5|}\ }\ \ 1\ \ \ \ 3\ \ \ 11\ \ \underline{|58}$$

Use x as the variable.

a. Identify the divisor.

b. Identify the quotient.

c. Identify the remainder.

49. The following table represents the result of a synthetic division.

$$\underline{-2|}\ \ 2\ \ \ \ 3\ \ \ \ 0\ \ -1\ \ \ \ 6$$
$$\phantom{\underline{-2|}\ \ 2\ \ }\ -4\ \ \ 2\ \ -4\ \ 10$$
$$\phantom{\underline{-2|}\ }\ \ 2\ \ -1\ \ \ \ 2\ \ -5\ \ \underline{|16}$$

Use x as the variable.

a. Identify the divisor.

b. Identify the quotient.

c. Identify the remainder.

For Exercises 50–61, divide by using synthetic division. Check your answer by multiplication. **(See Examples 5–6.)**

50. $(x^2 - 2x - 48) \div (x - 8)$

51. $(x^2 - 4x - 12) \div (x - 6)$

52. $(t^2 - 3t - 4) \div (t + 1)$

53. $(h^2 + 7h + 12) \div (h + 3)$

54. $(5y^2 + 5y + 1) \div (y - 1)$

55. $(3w^2 + w - 5) \div (w + 2)$

56. $(3y^3 + 7y^2 - 4y + 3) \div (y + 3)$

57. $(z^3 - 2z^2 + 2z - 5) \div (z + 3)$

58. $(x^3 - 3x^2 + 4) \div (x - 2)$

59. $(3y^4 - 25y^2 - 18) \div (y - 3)$

60. $(4w^4 - w^2 + 6w - 3) \div \left(w - \dfrac{1}{2}\right)$

61. $(-12y^4 - 5y^3 - y^2 + y + 3) \div \left(y + \dfrac{3}{4}\right)$

Mixed Exercises

For Exercises 62–73, divide the polynomials by using an appropriate method.

62. $(-x^3 - 8x^2 - 3x - 2) \div (x + 4)$

63. $(8xy^2 - 9x^2y + 6x^2y^2) \div (x^2y^2)$

64. $(22x^2 - 11x + 33) \div (11x)$

65. $(2m^3 - 4m^2 + 5m - 33) \div (m - 3)$

66. $(12y^3 - 17y^2 + 30y - 10) \div (3y^2 - 2y + 5)$

67. $(90h^{12} - 63h^9 + 45h^8 - 36h^7) \div (9h^9)$

68. $(4x^4 + 6x^3 + 3x - 1) \div (2x^2 + 1)$

69. $(y^4 - 3y^3 - 5y^2 - 2y + 5) \div (y + 2)$

70. $(16k^{11} - 32k^{10} + 8k^8 - 40k^4) \div (8k^8)$

71. $(4m^3 - 18m^2 + 22m - 10) \div (2m^2 - 4m + 3)$

72. $(5x^3 + 9x^2 + 10x) \div (5x^2)$

73. $(15k^4 + 3k^3 + 4k^2 + 4) \div (3k^2 - 1)$

chapter 5 midchapter review

1. Give the leading coefficient and the degree of the polynomial $-5x^6 + 2x^5 - 7x^3 + 5x$.

2. Given $P(x) = -2x^3 + 3x^2 + x$, find $P(-2)$.

For Exercises 3–14, perform the indicated operations.

3. $(5t^2 - 6t + 2) - (3t^2 - 7t + 3)$

4. $-5x^2(3x^2 + x - 2)$

5. $(3x + 1)^2$

6. $\dfrac{24a^3 - 8a^2 + 16a}{8a}$

7. $(6z + 5)(6z - 5)$

8. $(6y^3 + 2y^2 + y - 2) + (3y^3 - 4y + 3)$

9. $\dfrac{4x^2 + 6x + 1}{2x - 1}$

10. $(5a + 2)(2a^2 + 3a + 1)$

11. $(t^3 - 4t^2 + t - 9) + (t + 12) - (2t^2 - 6t)$

12. $(2b^3 - 3b - 10) \div (b - 2)$

13. $(p - 5)(p + 5) - (2p^2 + 3)$

14. $(k + 4)^2 + (-4k + 9)$

Chapter 5 Polynomials

Objectives

1. Factoring Out the Greatest Common Factor
2. Factoring Out a Negative Factor
3. Factoring Out a Binomial Factor
4. Factoring by Grouping

section 5.4 Greatest Common Factor and Factoring by Grouping

1. Factoring Out the Greatest Common Factor

The next three sections of this chapter are devoted to a mathematical operation called factoring. To factor an integer means to write the integer as a product of two or more integers. To factor a polynomial means to express the polynomial as a product of two or more polynomials.

In the product $5 \cdot 7 = 35$, for example, 5 and 7 are factors of 35.

In the product $(2x + 1)(x - 6) = 2x^2 - 11x - 6$, the quantities $(2x + 1)$ and $(x - 6)$ are factors of $2x^2 - 11x - 6$.

The **greatest common factor (GCF)** of a polynomial is the greatest factor that divides each term of the polynomial evenly. For example, the greatest common factor of $9x^4 + 18x^3 - 6x^2$ is $3x^2$. In other words, $3x^2$ is the greatest factor that divides evenly into each term. To factor out the greatest common factor from a polynomial, follow these steps:

Steps to Remove the Greatest Common Factor

1. Identify the greatest common factor of all terms of the polynomial.
2. Write each term as the product of the GCF and another factor.
3. Use the distributive property to factor out the greatest common factor.

Note: To check the factorization, multiply the polynomials.

Skill Practice

Factor out the greatest common factor.

1. $45y^5 - 15y^2 + 30y$
2. $16a^2b^5 + 12a^3b^3 + 4a^3b^2$

example 1 Factoring Out the Greatest Common Factor

Factor out the greatest common factor.

a. $12x^3 + 30x^2$ **b.** $12c^2d^3 - 30c^3d^2 - 3cd$

Solution:

a. $12x^3 + 30x^2$ The GCF is $6x^2$.

$ = 6x^2(2x) + 6x^2(5)$ Write each term as the product of the GCF and another factor.

$ = 6x^2(2x + 5)$ Factor out $6x^2$ by using the distributive property.

Tip: Any factoring problem can be checked by multiplying the factors:

Check: $6x^2(2x + 5) = 12x^3 + 30x^2$ ✔

b. $12c^2d^3 - 30c^3d^2 - 3cd$ The GCF is $3cd$.

$ = 3cd(4cd^2) - 3cd(10c^2d) - 3cd(1)$ Write each term as the product of the GCF and another factor.

$ = 3cd(4cd^2 - 10c^2d - 1)$ Factor out $3cd$ by using the distributive property.

Check: $3cd(4cd^2 - 10c^2d - 1) = 12c^2d^3 - 30c^3d^2 - 3cd$ ✔

Avoiding Mistakes: In Example 1(b), the GCF of $3cd$ is equal to one of the terms of the polynomial. In such a case, you must leave a 1 in place of that term after the GCF is factored out.

$3cd(4cd^2 - 10c^2d - 1)$

Answers

1. $15y(3y^4 - y + 2)$
2. $4a^2b^2(4b^3 + 3ab + a)$

2. Factoring Out a Negative Factor

Sometimes it is advantageous to factor out the *opposite* of the GCF, particularly when the leading coefficient of the polynomial is negative. This is demonstrated in Example 2. Notice that this *changes the signs* of the remaining terms inside the parentheses.

example 2   Factoring Out a Negative Factor

Factor out the quantity $-5a^2b$ from the polynomial $-5a^4b - 10a^3b^2 + 15a^2b^3$.

Solution:

$-5a^4b - 10a^3b^2 + 15a^2b^3$ The GCF is $5a^2b$. However, in this case we will factor out the opposite of the GCF, $-5a^2b$.

$= -5a^2b(a^2) + -5a^2b(2ab) + -5a^2b(-3b^2)$ Write each term as the product of $-5a^2b$ and another factor.

$= -5a^2b(a^2 + 2ab - 3b^2)$ Factor out $-5a^2b$ by using the distributive property.

Check: $-5a^2b(a^2 + 2ab - 3b^2) = -5a^4b - 10a^3b^2 + 15a^2b^3$ ✔

Skill Practice

3. Factor out the quantity $-6xy$ from the polynomial $24x^4y^3 - 12x^2y + 18xy^2$.

3. Factoring Out a Binomial Factor

The distributive property may also be used to factor out a common factor that consists of more than one term. This is shown in Example 3.

example 3   Factoring Out a Binomial Factor

Factor out the greatest common factor.

$$x^3(x + 2) - x(x + 2) - 9(x + 2)$$

Solution:

$x^3(x + 2) - x(x + 2) - 9(x + 2)$ The GCF is the quantity $(x + 2)$.

$= (x + 2)(x^3) - (x + 2)(x) - (x + 2)(9)$ Write each term as the product of $(x + 2)$ and another factor.

$= (x + 2)(x^3 - x - 9)$ Factor out $(x + 2)$ by using the distributive property.

Skill Practice

4. Factor out the greatest common factor.
$a^2(b + 2) + 5(b + 2)$

Answers

3. $-6xy(-4x^3y^2 + 2x - 3y)$
4. $(b + 2)(a^2 + 5)$

4. Factoring by Grouping

When two binomials are multiplied, the product before simplifying contains four terms. For example:

$$(3a + 2)(2b - 7) = (3a + 2)(2b) + (3a + 2)(-7)$$
$$= (3a + 2)(2b) + (3a + 2)(-7)$$
$$= 6ab + 4b - 21a - 14$$

In Example 4, we learn how to reverse this process. That is, given a four-term polynomial, we will factor it as a product of two binomials. The process is called **factoring by grouping**.

Steps to Factor by Grouping

To factor a four-term polynomial by grouping:

1. Identify and factor out the GCF from all four terms.
2. Factor out the GCF from the first pair of terms. Factor out the GCF from the second pair of terms. (Sometimes it is necessary to factor out the *opposite* of the GCF.)
3. If the two terms share a common binomial factor, factor out the binomial factor.

Skill Practice

Factor by grouping.

5. $7c^2 + cd + 14c + 2d$

Avoiding Mistakes: In step 2, the expression $3a(2b - 7) + 2(2b - 7)$ is not yet factored because it is a *sum*, not a product. To factor the expression, you must carry it one step further.

$$3a(2b - 7) + 2(2b - 7)$$
$$= (2b - 7)(3a + 2)$$

The factored form must be represented as a product.

example 4   Factoring by Grouping

Factor by grouping.

$$6ab - 21a + 4b - 14$$

Solution:

$6ab - 21a + 4b - 14$ **Step 1:** Identify and factor out the GCF from all four terms. In this case the GCF is 1.

$= 6ab - 21a \;\vdots\; + 4b - 14$ Group the first pair of terms and the second pair of terms.

$= 3a(2b - 7) + 2(2b - 7)$ **Step 2:** Factor out the GCF from each pair of terms.

Note: The two terms now share a common binomial factor of $(2b - 7)$.

$= (2b - 7)(3a + 2)$ **Step 3:** Factor out the common binomial factor.

Check: $(2b - 7)(3a + 2) = 2b(3a) + 2b(2) - 7(3a) - 7(2)$
$= 6ab + 4b - 21a - 14$ ✔

Answer

5. $(7c + d)(c + 2)$

Section 5.4 Greatest Common Factor and Factoring by Grouping

example 5 — Factoring by Grouping

Factor by grouping.
$$x^3 + 3x^2 - 3x - 9$$

Skill Practice

Factor by grouping.
6. $a^3 - 4a^2 - 3a + 12$

Solution:

$x^3 + 3x^2 - 3x - 9$ **Step 1:** Identify and factor out the GCF from all four terms. In this case the GCF is 1.

$= x^3 + 3x^2 \mid - 3x - 9$ Group the first pair of terms and the second pair of terms.

$= x^2(x + 3) - 3(x + 3)$ **Step 2:** Factor out x^2 from the first pair of terms.

Factor out -3 from the second pair of terms (this causes the signs to change in the second parentheses). The terms now contain a common binomial factor.

$= (x + 3)(x^2 - 3)$ **Step 3:** Factor out the common binomial $(x + 3)$.

Check: $(x + 3)(x^2 - 3) = x(x^2) + x(-3) + 3(x^2) + 3(-3)$
$= x^3 - 3x + 3x^2 - 9$ ✓

Tip: One frequent question is, can the order be switched between factors? The answer is yes. Because multiplication is commutative, the order in which two or more factors are written does not matter. Thus, the following factorizations are equivalent:

$$(x + 3)(x^2 - 3) = (x^2 - 3)(x + 3)$$

example 6 — Factoring by Grouping

Factor by grouping.
$$24p^2q^2 - 18p^2q + 60pq^2 - 45pq$$

Skill Practice

Factor the polynomial.
7. $24x^2y - 12x^2 + 20xy - 10x$

Solution:

$24p^2q^2 - 18p^2q + 60pq^2 - 45pq$

$= 3pq(8pq - 6p + 20q - 15)$ **Step 1:** Remove the GCF $3pq$ from all four terms.

$= 3pq(8pq - 6p \mid + 20q - 15)$ Group the first pair of terms and the second pair of terms.

Answers
6. $(a^2 - 3)(a - 4)$
7. $2x(6x + 5)(2y - 1)$

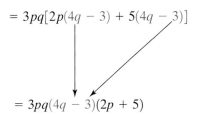

$= 3pq[2p(4q - 3) + 5(4q - 3)]$	**Step 2:** Factor out the GCF from each pair of terms. The terms share the binomial factor $(4q - 3)$.
$= 3pq(4q - 3)(2p + 5)$	**Step 3:** Factor out the common binomial $(4q - 3)$.

Check: $3pq[(4q - 3)(2p + 5)] = 3pq[4q(2p) + 4q(5) - 3(2p) - 3(5)]$
$= 3pq[8pq + 20q - 6p - 15]$
$= 24p^2q^2 + 60pq^2 - 18p^2q - 45pq$ ✓

Notice that in step 3 of factoring by grouping, a common binomial is factored from the two terms. These binomials must be *exactly* the same in each term. If the two binomial factors differ, try rearranging the original four terms.

Skill Practice

Factor the polynomial.

8. $3ry + 2s + sy + 6r$

---

example 7   Factoring by Grouping Where Rearranging Terms Is Necessary

Factor the polynomial.

$$4x + 6pa - 8a - 3px$$

Solution:

$4x + 6pa - 8a - 3px$ **Step 1:** Identify and factor out the GCF from all four terms. In this case the GCF is 1.

$= 4x + 6pa \ \vdots \ - 8a - 3px$

$= 2(2x + 3pa) - 1(8a + 3px)$ **Step 2:** The binomial factors in each term are different.

$= 4x - 8a \ \vdots \ - 3px + 6pa$ *Try rearranging the original four terms* in such a way that the first pair of coefficients is in the same ratio as the second pair of coefficients. Notice that the ratio 4 to 8 is the same as the ratio 3 to 6.

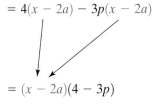

$= 4(x - 2a) - 3p(x - 2a)$ **Step 2:** Factor out 4 from the first pair of terms.

Factor out $-3p$ from the second pair of terms.

$= (x - 2a)(4 - 3p)$ **Step 3:** Factor out the common binomial factor.

Check: $(x - 2a)(4 - 3p) = x(4) + x(-3p) - 2a(4) - 2a(-3p)$
$= 4x - 3xp - 8a + 6ap$ ✓

Answer

8. $(3r + s)(2 + y)$

section 5.4 Practice Exercises

Boost *your* GRADE at mathzone.com!

- Practice Problems
- Self-Tests
- NetTutor
- e-Professors
- Videos

Study Skills Exercises

1. Does your school have a learning resource center or a tutoring center? If so, do you remember the location and hours of operation? Write them here.

   Location of Learning Resource Center or Tutoring Center:

   Hours of operation:

2. Define the key terms.

   a. Greatest common factor (GCF)   b. Factoring by grouping

Review Exercises

For Exercises 3–8, perform the indicated operation.

3. $(7t^4 + 5t^3 - 9t) - (-2t^4 + 6t^2 - 3t)$

4. $(5x^3 - 9x + 5) + (4x^3 + 3x^2 - 2x + 1) - (6x^3 - 3x^2 + x + 1)$

5. $(5y^2 - 3)(y^2 + y + 2)$

6. $(a + 6b)^2$

7. $\dfrac{6v^3 - 12v^2 + 2v}{-2v}$

8. $\dfrac{3x^3 + 2x^2 - 4}{x + 2}$

Objective 1: Factoring Out the Greatest Common Factor

9. What is meant by a common factor in a polynomial? What is meant by the greatest common factor?

10. Explain how to find the greatest common factor of a polynomial.

For Exercises 11–26, factor out the greatest common factor. (See Example 1.)

11. $3x + 12$
12. $15x - 10$
13. $6z^2 + 4z$
14. $49y^3 - 35y^2$

15. $4p^6 - 4p$
16. $5q^2 - 5q$
17. $12x^4 - 36x^2$
18. $51w^4 - 34w^3$

19. $9st^2 + 27t$
20. $8a^2b^3 + 12a^2b$
21. $9a^2 + 27a + 18$
22. $3x^2 - 15x + 9$

23. $10x^2y + 15xy^2 - 35xy$

24. $12c^3d - 15c^2d + 3cd$

25. $13b^2 - 11a^2b - 12ab$

26. $6a^3 - 2a^2b + 5a^2$

Objective 2: Factoring Out a Negative Factor

For Exercises 27–32, factor out the indicated quantity. (See Example 2.)

27. $-x^2 - 10x + 7$: Factor out the quantity -1.

28. $-5y^2 + 10y + 3$: Factor out the quantity -1.

29. $12x^3y - 6x^2y - 3xy$: Factor out the quantity $-3xy$.

30. $32a^4b^2 + 24a^3b + 16a^2b$: Factor out the quantity $-8a^2b$.

31. $-2t^3 + 11t^2 - 3t$: Factor out the quantity $-t$.

32. $-7y^2z - 5yz - z$: Factor out the quantity $-z$.

Objective 3: Factoring Out a Binomial Factor

For Exercises 33–40, factor out the GCF. (See Example 3.)

33. $2a(3z - 2b) - 5(3z - 2b)$

34. $5x(3x + 4) + 2(3x + 4)$

35. $2x^2(2x - 3) + (2x - 3)$

36. $z(w - 9) + (w - 9)$

37. $y(2x + 1)^2 - 3(2x + 1)^2$

38. $a(b - 7)^2 + 5(b - 7)^2$

39. $3y(x - 2)^2 + 6(x - 2)^2$

40. $10z(z + 3)^2 - 2(z + 3)^2$

41. Solve the equation $U = Av + Acw$ for A by first factoring out A.

42. Solve the equation $S = rt + wt$ for t by first factoring out t.

43. Solve the equation $ay + bx = cy$ for y.

44. Solve the equation $cd + 2x = ac$ for c.

45. Construct a polynomial that has a greatest common factor of $3x^2$. (Answers may vary.)

46. Construct two different trinomials that have a greatest common factor of $5x^2y^3$. (Answers may vary.)

47. Construct a binomial that has a greatest common factor of $(c + d)$. (Answers may vary.)

Objective 4: Factoring by Grouping

48. If a polynomial has four terms, what technique would you use to factor it?

49. Factor the polynomials by grouping.

    a. $2ax - ay + 6bx - 3by$

    b. $10w^2 - 5w - 6bw + 3b$

    c. Explain why you factored out $3b$ from the second pair of terms in part (a) but factored out the quantity $-3b$ from the second pair of terms in part (b).

50. Factor the polynomials by grouping.

    a. $3xy + 2bx + 6by + 4b^2$

    b. $15ac + 10ab - 6bc - 4b^2$

    c. Explain why you factored out $2b$ from the second pair of terms in part (a) but factored out the quantity $-2b$ from the second pair of terms in part (b).

For Exercises 51–70, factor each polynomial by grouping (if possible). **(See Examples 4–7.)**

51. $y^3 + 4y^2 + 3y + 12$

52. $ab + b + 2a + 2$

53. $6p - 42 + pq - 7q$

54. $2t - 8 + st - 4s$

55. $2mx + 2nx + 3my + 3ny$

56. $4x^2 + 6xy - 2xy - 3y^2$

57. $10ax - 15ay - 8bx + 12by$

58. $35a^2 - 15a + 14a - 6$

59. $x^3 - x^2 - 3x + 3$

60. $2rs + 4s - r - 2$

61. $6p^2q + 18pq - 30p^2 - 90p$

62. $5s^2t + 20st - 15s^2 - 60s$

63. $100x^3 - 300x^2 + 200x - 600$

64. $2x^5 - 10x^4 + 6x^3 - 30x^2$

65. $6ax - by + 2bx - 3ay$

66. $5pq - 12 - 4q + 15p$

67. $4a - 3b - ab + 12$

68. $x^2y + 6x - 3x^3 - 2y$

69. $7y^3 - 21y^2 + 5y - 10$

70. $5ax + 10bx - 2ac + 4bc$

71. Explain why the grouping method failed for Exercise 69.

72. Explain why the grouping method failed for Exercise 70.

73. The area of a rectangle of width w is given by $A = 2w^2 + w$. Factor the right-hand side of the equation to find an expression for the length of the rectangle.

74. The amount in a savings account bearing simple interest at an interest rate r for t years is given by $A = P + Prt$ where P is the principal amount invested.

    a. Solve the equation for P.

    b. Compute the amount of principal originally invested if the account is worth $12,705 after 3 years at a 7% interest rate.

Expanding Your Skills

For Exercises 75–82, factor out the greatest common factor and simplify.

75. $(a + 3)^4 + 6(a + 3)^5$

76. $(4 - b)^4 - 2(4 - b)^3$

77. $24(3x + 5)^3 - 30(3x + 5)^2$

78. $10(2y + 3)^2 + 15(2y + 3)^3$

79. $(t + 4)^2 - (t + 4)$

80. $(p + 6)^2 - (p + 6)$

81. $15w^2(2w - 1)^3 + 5w^3(2w - 1)^2$

82. $8z^4(3z - 2)^2 + 12z^3(3z - 2)^3$

section 5.5 Factoring Trinomials

1. Factoring Trinomials: Grouping Method

In Section 5.4, we learned how to factor out the greatest common factor from a polynomial and how to factor a four-term polynomial by grouping. In this section we present two methods to factor trinomials. The first method is called the grouping method (sometimes referred to as the "*ac*" method). The second method is called the trial-and-error method.

The product of two binomials results in a four-term expression that can sometimes be simplified to a trinomial. To factor the trinomial, we want to reverse the process.

Multiply: $(2x + 3)(x + 2)$ $\xrightarrow{\text{Multiply the binomials.}}$ $2x^2 + 4x + 3x + 6$

$\xrightarrow{\text{Add the middle terms.}}$ $2x^2 + 7x + 6$

Factor: $2x^2 + 7x + 6$ $\xrightarrow{\text{Rewrite the middle term as a sum or difference of terms.}}$ $2x^2 + 4x + 3x + 6$

$\xrightarrow{\text{Factor by grouping.}}$ $(2x + 3)(x + 2)$

To factor a trinomial $ax^2 + bx + c$ by the grouping method, we rewrite the middle term bx as a sum or difference of terms. The goal is to produce a four-term polynomial that can be factored by grouping. The process is outlined as follows.

The Grouping Method to Factor $ax^2 + bx + c$ ($a \neq 0$)

1. Multiply the coefficients of the first and last terms, ac.
2. Find two integers whose product is ac and whose sum is b. (If no pair of integers can be found, then the trinomial cannot be factored further and is called a **prime polynomial**.)
3. Rewrite the middle term bx as the sum of two terms whose coefficients are the integers found in step 2.
4. Factor by grouping.

The grouping method for factoring trinomials is illustrated in Example 1. Before we begin, however, keep these two important guidelines in mind.

- For any factoring problem you encounter, always factor out the GCF from all terms first.
- To factor a trinomial, write the trinomial in the form $ax^2 + bx + c$.

example 1 Factoring a Trinomial by the Grouping Method

Factor the trinomial by the grouping method.

$$6x^2 + 19x + 10$$

Objectives

1. Factoring Trinomials: Grouping Method
2. Factoring Trinomials: Trial-and-Error Method
3. Factoring Trinomials with a Leading Coefficient of 1
4. Factoring Perfect Square Trinomials
5. Factoring by Using Substitution
6. Summary of Factoring Trinomials

Concept Connections

Given: the trinomial $3x^2 + 8x + 4$

1. Identify a, b, c.
2. What is the value of ac?
3. Find the two factors of ac whose sum is equal to b.

Skill Practice

4. Use the results from margin Exercise 3 to factor the polynomial by grouping.
   $3x^2 + 8x + 4$
5. Factor by grouping
   $5y^2 - 3y - 2$

Answers

1. $a = 3$; $b = 8$; $c = 4$ 2. 12
3. 6 and 2   4. $(3x + 2)(x + 2)$
5. $(5y + 2)(y - 1)$

Solution:

$6x^2 + 19x + 10$ — Factor out the GCF from all terms. In this case, the GCF is 1.

$6x^2 + 19x + 10$
$a = 6 \quad b = 19 \quad c = 10$

Step 1: The trinomial is written in the form $ax^2 + bx + c$. Find the product of a and c: $ac = (6)(10) = 60$.

Factors of 60	Factors of 60
$1 \cdot 60$	$(-1)(-60)$
$2 \cdot 30$	$(-2)(-30)$
$3 \cdot 20$	$(-3)(-20)$
$4 \cdot 15$	$(-4)(-15)$
$5 \cdot 12$	$(-5)(-12)$
$6 \cdot 10$	$(-6)(-10)$

Step 2: List all the factors of ac, and search for the pair whose sum equals the value of b.

That is, list the factors of 60 and find the pair whose *sum* equals 19.

The numbers 4 and 15 satisfy both conditions: $4 \cdot 15 = 60$ and $4 + 15 = 19$.

$6x^2 + 19x + 10$

$= 6x^2 + 4x + 15x + 10$

Step 3: Write the middle term of the trinomial as the sum of two terms whose coefficients are the selected pair of numbers 4 and 15.

$= 6x^2 + 4x \ | \ + 15x + 10$

Step 4: Factor by grouping.

$= 2x(3x + 2) + 5(3x + 2)$

$= (3x + 2)(2x + 5)$

Check: $(3x + 2)(2x + 5) = 6x^2 + 15x + 4x + 10$
$= 6x^2 + 19x + 10$ ✓

Tip: One frequently asked question is whether the order matters when we rewrite the middle term of the trinomial as two terms (step 3). The answer is no. From Example 1, the two middle terms in step 3 could have been reversed to obtain the same result.

$6x^2 + 19x + 10 = 6x^2 + 15x + 4x + 10$
$= 3x(2x + 5) + 2(2x + 5)$
$= (2x + 5)(3x + 2)$

This example also points out that the order in which two factors are written does not matter. The expression $(3x + 2)(2x + 5)$ is equivalent to $(2x + 5)(3x + 2)$ because multiplication is a commutative operation.

---

example 2 Factoring a Trinomial by the Grouping Method

Factor.

$$f(x) = 12x^2 - 5x - 2$$

Skill Practice

6. Factor $f(x) = 10x^2 + x - 3$.

Answer

6. $f(x) = (5x + 3)(2x - 1)$

Solution:

$f(x) = 12x^2 - 5x - 2$ The GCF is 1.

$a = 12 \quad b = -5 \quad c = -2$

Step 1: The function is written in the form $f(x) = ax^2 + bx + c$. Find the product $ac = 12(-2) = -24$.

Factors of –24	Factors of –24
(1)(−24)	(−1)(24)
(2)(−12)	(−2)(12)
(3)(−8)	(−3)(8)
(4)(−6)	(−4)(6)

Step 2: List all the factors of −24, and find the pair whose sum equals −5.

The numbers 3 and −8 produce a product of −24 and a sum of −5.

$f(x) = 12x^2 - 5x - 2$

$= 12x^2 + 3x - 8x - 2$

Step 3: Write the middle term of the trinomial as two terms whose coefficients are the selected numbers 3 and −8.

$= 12x^2 + 3x \mid -8x - 2$

Step 4: Factor by grouping.

$= 3x(4x + 1) - 2(4x + 1)$

$= (4x + 1)(3x - 2)$

The check is left for the reader.

---

example 3 Factoring a Trinomial by the Grouping Method

Factor the trinomial by the grouping method.

$$-5p^3 + 50p^2 - 125p$$

Solution:

$-5p^3 + 50p^2 - 125p$

$= -5p(p^2 - 10p + 25)$ Factor out $-5p$.

Step 1: Find the product $a \cdot c = (1)(25) = 25$.

Factors of 25	Factors of 25
1 · 25	(−1)(−25)
5 · 5	(−5)(−5)

Step 2: The numbers −5 and −5 form a product of 25 and a sum of −10.

$-5p(p^2 - 10p + 25)$

Step 3: Write the middle term of the trinomial as two terms whose coefficients are −5 and −5.

$= -5p[p^2 - 5p \mid -5p + 25]$

Step 4: Factor by grouping.

$= -5p[p(p - 5) - 5(p - 5)]$

$= -5p[(p - 5)(p - 5)]$

$= -5p(p - 5)^2$

Check: $-5p(p - 5)^2 = -5p(p^2 - 10p + 25)$

$= -5p^3 + 50p^2 - 125p$ ✔

Skill Practice

Factor by grouping.

7. $-4z^3 - 2z^2 + 20z$
8. $8c^2 + 14cd + 5d^2$

Answers

7. $-2z(2z + 5)(z - 2)$
8. $(2c + d)(4c + 5d)$

Tip: Notice when the GCF is removed from the original trinomial, the new trinomial has smaller coefficients. This makes the factoring process simpler because the product ac is smaller. It is much easier to list the factors of 25 than the factors of 625.

Original trinomial	With the GCF factored out
$-5p^3 + 50p^2 - 125p$	$-5p(p^2 - 10p + 25)$
$ac = (-5)(-125) = 625$	$ac = (1)(25) = 25$

2. Factoring Trinomials: Trial-and-Error Method

Another method that is widely used to factor trinomials of the form $ax^2 + bx + c$ is the trial-and-error method. To understand how the trial-and-error method works, first consider the multiplication of two binomials:

$$(2x + 3)(1x + 2) = 2x^2 + \underbrace{4x + 3x}_{\substack{\text{sum of products} \\ \text{of inner terms} \\ \text{and outer terms}}} + 6 = 2x^2 + 7x + 6$$

Product of $2 \cdot 1$; Product of $3 \cdot 2$

To factor the trinomial $2x^2 + 7x + 6$, this operation is reversed. Hence

$$2x^2 + 7x + 6 = (\Box x \quad \Box)(\Box x \quad \Box)$$

with factors of 2 and factors of 6.

We need to fill in the blanks so that the product of the first terms in the binomials is $2x^2$ and the product of the last terms in the binomials is 6. Furthermore, the factors of $2x^2$ and 6 must be chosen so that the sum of the products of the inner terms and outer terms equals $7x$.

To produce the product $2x^2$, we might try the factors $2x$ and x within the binomials.

$$(2x \quad \Box)(x \quad \Box)$$

To produce a product of 6, the remaining terms in the binomials must either both be positive or both be negative. To produce a positive middle term, we will try positive factors of 6 in the remaining blanks until the correct product is found. The possibilities are $1 \cdot 6, 2 \cdot 3, 3 \cdot 2,$ and $6 \cdot 1$.

$(2x + 1)(x + 6) = 2x^2 + 12x + 1x + 6 = 2x^2 + 13x + 6$	Wrong middle term
$(2x + 2)(x + 3) = 2x^2 + 6x + 2x + 6 = 2x^2 + 8x + 6$	Wrong middle term
$\mathbf{(2x + 3)(x + 2) = 2x^2 + 4x + 3x + 6 = 2x^2 + 7x + 6}$	Correct!
$(2x + 6)(x + 1) = 2x^2 + 2x + 6x + 6 = 2x^2 + 8x + 6$	Wrong middle term

The correct factorization of $2x^2 + 7x + 6$ is $(2x + 3)(x + 2)$. ✔

As this example shows, we factor a trinomial of the form $ax^2 + bx + c$ by shuffling the factors of a and c within the binomials until the correct product

---

Concept Connections

Two binomials are multiplied, and the result is shown.

$(3x + 5)(2x + 3)$ Product.
$6x^2 + 9x + 10x + 15$ Multiply.
$6x^2 + 19x + 15$ Result.

9. Use arrows to show which two terms of the binomials were multiplied to obtain the first term of the result.

   $(3x + 5)(2x + 3)$

10. Use arrows to show which two terms of the binomials were multiplied to obtain the last term of the result.

    $(3x + 5)(2x + 3)$

11. Use arrows to show which pairs of terms were multiplied and then added to obtain the middle term of the result.

    $(3x + 5)(2x + 3)$

Answers

9. $(3x + 5)(2x + 3)$
10. $(3x + 5)(2x + 3)$
11. $(3x + 5)(2x + 3)$

is obtained. However, sometimes it is not necessary to test all the possible combinations of factors. In Example 3, the GCF of the original trinomial is 1. Therefore, any binomial factor that shares a common factor *greater than 1* does not need to be considered. In this case the possibilities $(2x + 2)(x + 3)$ and $(2x + 6)(x + 1)$ cannot work.

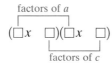

The steps to factor a trinomial by the trial-and-error method are outlined as follows.

The Trial-and-Error Method to Factor $ax^2 + bx + c$

1. Factor out the greatest common factor.
2. List all pairs of positive factors of a and pairs of positive factors of c. Consider the reverse order for either list of factors.
3. Construct two binomials of the form

$$(\Box x \quad \Box)(\Box x \quad \Box)$$

with factors of a on top and factors of c on bottom.

Test each combination of factors and signs until the correct product is found. If no combination of factors produces the correct product, the trinomial cannot be factored further and is a **prime polynomial**.

example 4 — Factoring a Trinomial by the Trial-and-Error Method

Factor the trinomial by the trial-and-error method.

$$10x^2 - 9x - 1$$

Solution:

$10x^2 - 9x - 1$ **Step 1:** Factor out the GCF from all terms. The GCF is 1. The trinomial is written in the form $ax^2 + bx + c$.

To factor $10x^2 - 9x - 1$, two binomials must be constructed in the form

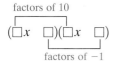

Step 2: To produce the product $10x^2$, we might try $5x$ and $2x$ or $10x$ and $1x$. To produce a product of -1, we will try the factors $1(-1)$ and $-1(1)$.

Step 3: Construct all possible binomial factors, using different combinations of the factors of $10x^2$ and -1.

$(5x + 1)(2x - 1) = 10x^2 - 5x + 2x - 1 = 10x^2 - 3x - 1$ Wrong middle term

$(5x - 1)(2x + 1) = 10x^2 + 5x - 2x - 1 = 10x^2 + 3x - 1$ Wrong middle term

Skill Practice

Factor by trial and error.
12. $5y^2 - 9y + 4$

Answer
12. $(5y - 4)(y - 1)$

The numbers 1 and -1 did not produce the correct trinomial when coupled with $5x$ and $2x$, so we try $10x$ and $1x$.

$(10x + 1)(1x - 1)$ $= 10x^2 - 10x + 1x - 1 = \mathbf{10x^2 - 9x - 1}$ Correct!

$(10x - 1)(1x + 1) = 10x^2 + 10x - 1x - 1 = 10x^2 + 9x - 1$ Wrong middle term

Hence $10x^2 - 9x - 1 = (10x + 1)(x - 1)$

In Example 4, the factors of -1 must have opposite signs to produce a negative product. Therefore, one binomial factor is a sum and one is a difference. Determining the correct signs is an important aspect of factoring trinomials. We suggest the following guidelines:

Tip: Given the trinomial $ax^2 + bx + c$ ($a > 0$), the signs can be determined as follows:

1. If c is *positive*, then the signs in the binomials must be the same (either both positive or both negative). The correct choice is determined by the middle term. If the middle term is positive, then both signs must be positive. If the middle term is negative, then both signs must be negative.

   c is positive.

   Example: $20x^2 + 43x + 21$
   $(4x + 3)(5x + 7)$
   same signs

   c is positive.

   Example: $20x^2 - 43x + 21$
   $(4x - 3)(5x - 7)$
   same signs

2. If c is *negative*, then the signs in the binomials must be different. The middle term in the trinomial determines which factor gets the positive sign and which factor gets the negative sign.

   c is negative.

   Example: $x^2 + 3x - 28$
   $(x + 7)(x - 4)$
   different signs

   c is negative.

   Example: $x^2 - 3x - 28$
   $(x - 7)(x + 4)$
   different signs

Skill Practice

Factor by trial-and-error.

13. $3a^2 - 7a - 6$
14. $4t^2 + 5t - 6$

example 5   Factoring a Trinomial

Factor the trinomial by the trial-and-error method.

$$8y^2 + 13y - 6$$

Solution:

$8y^2 + 13y - 6$ **Step 1:** The GCF is 1.

$(\Box y \ \Box)(\Box y \ \Box)$

Factors of 8 **Factors of 6** **Step 2:** List the positive factors of 8 and positive factors of 6. Consider the reverse order in one list of factors.

$1 \cdot 8$ $1 \cdot 6$
$2 \cdot 4$ $2 \cdot 3$
          $3 \cdot 2$ ⎱
          $6 \cdot 1$ ⎰ (reverse order)

Answers

13. $(3a + 2)(a - 3)$
14. $(4t - 3)(t + 2)$

(2y   1)(4y   6)
(2y   2)(4y   3)
(2y   3)(4y   2)
(2y   6)(4y   1)
(1y   1)(8y   6)
(1y   3)(8y   2)

Step 3: Construct all possible binomial factors by using different combinations of the factors of 8 and 6.

Without regard to signs, these factorizations cannot work because the terms in the binomial share a common factor greater than 1.

Test the remaining factorizations. Keep in mind that to produce a product of -6, the signs within the parentheses must be opposite (one positive and one negative). Also, the sum of the products of the inner terms and outer terms must be combined to form $13y$.

(1y   6)(8y   1)   *Incorrect.*   Wrong middle term.

Regardless of signs, the product of inner terms $48y$ and the product of outer terms $1y$ cannot be combined to form the middle term $13y$.

(1y   2)(8y   3)   *Correct.*   The terms $16y$ and $3y$ can be combined to form the middle term $13x$, provided the signs are applied correctly. We require $+16y$ and $-3y$.

Hence, the correct factorization of $8y^2 + 13y - 6$ is $(y + 2)(8y - 3)$.

---

example 6   Factoring a Trinomial by the Trial-and-Error Method

Factor the trinomial by the trial-and-error method.

$$-80x^3y + 208x^2y^2 - 20xy^3$$

Solution:

$-80x^3y + 208x^2y^2 - 20xy^3$

$= -4xy(20x^2 - 52xy + 5y^2)$ **Step 1:** Factor out $-4xy$.

$= -4xy(\Box x \ \Box y)(\Box x \ \Box y)$

Factors of 20 **Factors of 5**   **Step 2:** List the positive factors of 20 and positive factors of 5. Consider the reverse order in one list of factors.
$1 \cdot 20$ $1 \cdot 5$
$2 \cdot 10$ $5 \cdot 1$
$4 \cdot 5$

Skill Practice

Factor by trial and error.
15. $-4z^3 - 22z^2 - 30z$
16. $40x^2y + 30xy^2 - 100y^3$

Answers
15. $-2z(2z + 5)(z + 3)$
16. $10y(4x - 5y)(x + 2y)$

Step 3: Construct all possible binomial factors by using different combinations of the factors of 20 and factors of 5. The signs in the parentheses must both be negative.

$-4xy(1x - 1y)(20x - 5y)$
$-4xy(2x - 1y)(10x - 5y)$ *Incorrect.* Each of these binomials contains a common factor.
$-4xy(4x - 1y)(5x - 5y)$

$-4xy(1x - 5y)(20x - 1y)$ *Incorrect.* Wrong middle term.
$-4xy(x - 5y)(20x - 1y)$
$= -4xy(20x^2 - 101xy + 5y^2)$

$-4xy(2x - 5y)(10x - 1y)$ **Correct.** $-4xy(2x - 5y)(10x - 1y)$
$= \mathbf{-4xy(20x^2 - 52xy + 5y^2)}$
$= -80x^3y + 208x^2y^2 - 20xy^3$

$-4xy(4x - 5y)(5x - 1y)$ *Incorrect.* Wrong middle term.
$-4xy(4x - 5y)(5x - 1y)$
$= -4xy(20x^2 - 29x + 5y^2)$

The correct factorization of $-80x^3y + 208x^2y^2 - 20xy^3$ is $-4xy(2x - 5y)(10x - y)$.

3. Factoring Trinomials with a Leading Coefficient of 1

If a trinomial has a leading coefficient of 1, the factoring process simplifies significantly. Consider the trinomial $x^2 + bx + c$. To produce a leading term of x^2, we can construct binomials of the form $(x + \square)(x + \square)$. The remaining terms may be satisfied by two numbers p and q whose product is c and whose sum is b:

$$(x + \overbrace{p)(x + q}^{\text{factors of } c}) = x^2 + qx + px + pq = x^2 + \underbrace{(p + q)}_{\text{sum} = b}x + \underbrace{pq}_{\text{product} = c}$$

This process is demonstrated in Example 7.

Skill Practice

Factor.

17. $c^2 + 6c - 27$
18. $3x^2 - 3x - 60$

example 7   Factoring a Trinomial with a Leading Coefficient of 1

Factor the trinomial.

$$x^2 - 10x + 16$$

Solution:

$x^2 - 10x + 16$ Factor out the GCF from all terms. In this case, the GCF is 1.

$= (x \ \square)(x \ \square)$ The trinomial is written in the form $x^2 + bx + c$. To form the product x^2, use the factors x and x.

Answers
17. $(c + 9)(c - 3)$
18. $3(x - 5)(x + 4)$

Next, look for two numbers whose product is 16 and whose sum is -10. Because the middle term is negative, we will consider only the negative factors of 16.

Factors of 16	Sum
$-1(-16)$	$-1 + (-16) = -17$
$-2(-8)$	**$-2 + (-8) = -10$**
$-4(-4)$	$-4 + (-4) = -8$

The numbers are -2 and -8.

Hence $x^2 - 10x + 16 = (x - 2)(x - 8)$

Check: $(x - 2)(x - 8) = x^2 - 8x - 2x + 16$
$= x^2 - 10x + 16$ ✓

4. Factoring Perfect Square Trinomials

Recall from Section 5.2 that the square of a binomial always results in a **perfect square trinomial**.

$$(a + b)^2 = (a + b)(a + b) = a^2 + ab + ab + b^2 = a^2 + 2ab + b^2$$
$$(a - b)^2 = (a - b)(a - b) = a^2 - ab - ab + b^2 = a^2 - 2ab + b^2$$

For example, $(2x + 7)^2 = (2x)^2 + 2(2x)(7) + (7)^2 = 4x^2 + 28x + 49$

$a = 2x \quad b = 7 \qquad a^2 + 2ab + b^2$

To factor the trinomial $4x^2 + 28x + 49$, the grouping method or the trial-and-error method can be used. However, recognizing that the trinomial is a perfect square trinomial, we can use one of the following patterns to reach a quick solution.

Factored Form of a Perfect Square Trinomial

$$a^2 + 2ab + b^2 = (a + b)^2$$
$$a^2 - 2ab + b^2 = (a - b)^2$$

Concept Connections

Determine whether or not the polynomials are perfect square trinomials.

19. $b^2 + 10b + 25$
20. $4x^2 + 2x + 1$

Tip: To determine if a trinomial is a perfect square trinomial, follow these steps:
1. Check if the first and third terms are both perfect squares with positive coefficients.
2. If this is the case, identify a and b, and determine if the middle term equals $2ab$.

example 8 Factoring Perfect Square Trinomials

Factor the trinomials completely.

a. $x^2 + 12x + 36$ b. $4x^2 - 36x + 81$

Skill Practice

Factor completely.

21. $x^2 + 2x + 1$
22. $9y^2 - 12y + 4$

Answers
19. Yes 20. No 21. $(x + 1)^2$
22. $(3y - 2)^2$

Solution:

a. $x^2 + 12x + 36$ The GCF is 1.

- The first and third terms are positive.
- The first term is a perfect square: $x^2 = (x)^2$
- The third term is a perfect square: $36 = (6)^2$
- The middle term is twice the product of x and 6:

$$12x = 2(x)(6)$$

$= x^2 + 12x + 36$ (perfect squares)

$= (x)^2 + 2(x)(6) + (6)^2$ Hence the trinomial is in the form $a^2 + 2ab + b^2$, where $a = x$ and $b = 6$.

$= (x + 6)^2$ Factor as $(a + b)^2$.

b. $4x^2 - 36x + 81$ The GCF is 1.

- The first and third terms are positive.
- The first term is a perfect square: $4x^2 = (2x)^2$.
- The third term is a perfect square: $81 = (9)^2$.
- The middle term:

$$-36x = -2(2x)(9)$$

$= 4x^2 - 36x + 81$ (perfect squares)

$= (2x)^2 - 2(2x)(9) + (9)^2$ The trinomial is in the form $a^2 - 2ab + b^2$, where $a = 2x$ and $b = 9$.

$= (2x - 9)^2$ Factor as $(a - b)^2$.

5. Factoring by Using Substitution

Sometimes it is convenient to use substitution to convert a polynomial to a simpler form before factoring.

Skill Practice

Factor by using substitution.

23. $2x^4 + 7x^2 + 3$

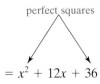 **example 9**   Using Substitution to Factor a Polynomial

Factor completely. $6y^6 - 5y^3 - 4$

Solution:

$6y^6 - 5y^3 - 4$ Let $u = y^3$.

$= 6u^2 - 5u - 4$ Substitute u for y^3 in the trinomial.

$= (2u + 1)(3u - 4)$ Factor the trinomial.

$= (2y^3 + 1)(3y^3 - 4)$ Reverse substitute. Replace u with y^3.

The factored form of $6y^6 - 5y^3 - 4$ is $(2y^3 + 1)(3y^3 - 4)$.

Answer

23. $(x^2 + 3)(2x^2 + 1)$

6. Summary of Factoring Trinomials

Summary: Factoring Trinomials of the Form $ax^2 + bx + c$ ($a \neq 0$)

When factoring trinomials, the following guidelines should be considered:

1. Factor out the greatest common factor.
2. Check to see if the trinomial is a perfect square trinomial. If so, factor it as either $(a + b)^2$ or $(a - b)^2$. (With a perfect square trinomial, you do not need to use the grouping method or trial-and-error method.)
3. If the trinomial is not a perfect square, use either the grouping method or the trial-and-error method to factor.
4. Check the factorization by multiplication.

example 10   Factoring Trinomials

Factor the trinomials completely.

a. $80s^3t + 80s^2t^2 + 20st^3$ **b.** $5w^2 + 50w + 45$ **c.** $2p^2 + 9p + 14$

Skill Practice

Factor completely.

24. $6v^2 - 12v - 18$
25. $-x^2 + 6x - 9$
26. $3r^2 - 13rs + 10s^2$

Solution:

a. $80s^3t + 80s^2t^2 + 20st^3$

$= 20st(4s^2 + 4st + t^2)$ The GCF is $20st$.

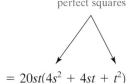
perfect squares

- The first and third terms are positive.
- The first and third terms are perfect squares: $4s^2 = (2s)^2$ and $t^2 = (t)^2$
- Because $4st = 2(2s)(t)$, the trinomial is in the form $a^2 + 2ab + b^2$, where $a = 2s$ and $b = t$.

$= 20st(4s^2 + 4st + t^2)$

$= 20st(2s + t)^2$ Factor as $(a + b)^2$.

b. $5w^2 + 50w + 45$

$= 5(w^2 + 10w + 9)$ The GCF is 5.

perfect squares

The first and third terms are perfect squares: $w^2 = (w)^2$ and $9 = (3)^2$.

$= 5(w^2 + 10w + 9)$ However, the middle term $10w \neq 2(w)(3)$. Therefore, this is *not* a perfect square trinomial.

$= 5(w + 9)(w + 1)$ To factor, use either the grouping method or the trial-and-error method.

c. $2p^2 + 9p + 14$ The GCF is 1. The trinomial is not a perfect square trinomial because neither 2 nor 14 is a perfect square. Therefore, try factoring by either the grouping method or the trial-and-error method. We use the trial-and-error method here.

Answers

24. $6(v + 1)(v - 3)$
25. $-(x - 3)^2$
26. $(3r - 10s)(r - s)$

Factors of 2	Factors of 14	After constructing all factors of 2 and 14, we see that no combination of factors will produce the correct result.
$2 \cdot 1$	$1 \cdot 14$	
	$14 \cdot 1$	
	$2 \cdot 7$	
	$7 \cdot 2$	

$(2p + 14)(p + 1)$ *Incorrect*: $(2p + 14)$ contains a common factor of 2.

$(2p + 2)(p + 7)$ *Incorrect*: $(2p + 2)$ contains a common factor of 2.

$(2p + 1)(p + 14) = 2p^2 + 28p + p + 14 \longrightarrow 2p^2 + 29p + 14$ *Incorrect.*
(wrong middle term)

$(2p + 7)(p + 2) = 2p^2 + 4p + 7p + 14 \longrightarrow 2p^2 + 11p + 14$ *Incorrect*
(wrong middle term)

Because none of the combinations of factors results in the correct product, we say that the trinomial $2p^2 + 9p + 14$ is prime. This polynomial cannot be factored by the techniques presented here.

section 5.5   Practice Exercises

Boost your GRADE at mathzone.com!

MathZone

- Practice Problems
- Self-Tests
- NetTutor
- e-Professors
- Videos

Study Skills Exercises

1. When you run into a homework exercise that is difficult for you, how long do you work on that problem before leaving it and continuing with the rest of the assignment? 10 minutes? 30 minutes? 1 hour? 2 hours? Of the times listed, which do you think is best? Explain.

2. Define the key terms:

    a. prime polynomial   **b. perfect square trinomial**

Review Exercises

For Exercises 3–8, factor the polynomial completely.

3. $36c^2d^7e^{11} + 12c^3d^5e^{15} - 6c^2d^4e^7$

4. $5x^3y^3 + 15x^4y^2 - 35x^2y^4$

5. $2x(3a - b) - (3a - b)$

6. $6(v - 8) - 3u(v - 8)$

7. $wz^2 + 2wz - 33az - 66a$

8. $3a^2x + 9ab - abx - 3b^2$

Objectives 1–3: Factoring Trinomials

In Exercises 9–46, factor the trinomial completely by using any method. Remember to look for a common factor first. **(See Examples 1–7.)**

9. $b^2 - 12b + 32$

10. $a^2 - 12a + 27$

11. $y^2 + 10y - 24$

12. $w^2 + 3w - 54$

13. $x^2 + 13x + 30$

14. $t^2 + 9t + 8$

15. $c^2 - 6c - 16$

16. $z^2 - 3z - 28$

17. $2x^2 - 7x - 15$

18. $2y^2 - 13y + 15$

19. $a + 6a^2 - 5$

20. $10b^2 - 3 - 29b$

21. $s^2 + st - 6t^2$

22. $p^2 - pq - 20q^2$

23. $3x^2 - 60x + 108$

24. $4c^2 + 12c - 72$

25. $2c^2 - 2c - 24$

26. $3x^2 + 12x - 15$

27. $2x^2 + 8xy - 10y^2$

28. $20z^2 + 26zw - 28w^2$

29. $33t^2 - 18t + 2$

30. $5p^2 - 10p + 7$

31. $3x^2 + 14xy + 15y^2$

32. $2a^2 + 15ab - 27b^2$

33. $5u^3v - 30u^2v^2 + 45uv^3$

34. $3a^3 + 30a^2b + 75ab^2$

35. $x^3 - 5x^2 - 14x$

36. $p^3 + 2p^2 - 24p$

37. $-23z - 5 + 10z^2$

38. $3 + 16y^2 + 14y$

39. $b^2 + 2b + 15$

40. $x^2 - x - 1$

41. $-2t^2 + 12t + 80$

42. $-3c^2 + 33c - 72$

43. $14a^2 + 13a - 12$

44. $12x^2 - 16x + 5$

45. $6a^2b + 22ab + 12b$

46. $6cd^2 + 9cd - 42c$

Objective 4: Factoring Perfect Square Trinomials

47. a. Multiply the binomials $(x + 5)(x + 5)$.

  b. How do you factor $x^2 + 10x + 25$?

48. a. Multiply the binomials $(2w - 5)(2w - 5)$.

  b. How do you factor $4w^2 - 20w + 25$?

49. a. Multiply the binomials $(3x - 2y)^2$.

  b. How do you factor $9x^2 - 12xy + 4y^2$?

50. a. Multiply the binomials $(x + 7y)^2$.

  b. How do you factor $x^2 + 14xy + 49y^2$?

For Exercises 51–56, fill in the blank to make the trinomial a perfect square trinomial.

51. $9x^2 + (\underline{}) + 25$

52. $16x^4 - (\underline{}) + 1$

53. $b^2 - 12b + (\underline{})$

54. $4w^2 + 28w + (\underline{})$

55. $(\underline{})z^2 + 16z + 1$

56. $(\underline{})x^2 - 42x + 49$

For Exercises 57–64, factor out the greatest common factor. Then determine if the polynomial is a perfect square trinomial. If it is, factor it. **(See Example 8.)**

57. $y^2 - 8y + 16$

58. $x^2 + 10x + 25$

59. $w^2 - 5w + 9$

60. $2a^2 + 14a + 98$

61. $9a^2 - 30ab + 25b^2$

62. $16x^4 - 48x^2y + 9y^2$

63. $16t^2 - 80tv + 20v^2$

64. $12x^2 - 12xy + 3y^2$

Objective 5: Factoring by Using Substitution

For Exercises 65–68, factor the polynomial in part (a). Then use substitution to help factor the polynomials in parts (b) and (c). **(See Example 9.)**

65. a. $u^2 - 10u + 25$

  b. $x^4 - 10x^2 + 25$

  c. $(a + 1)^2 - 10(a + 1) + 25$

66. a. $u^2 + 12u + 36$

  b. $y^4 + 12y^2 + 36$

  c. $(b - 2)^2 + 12(b - 2) + 36$

67. a. $u^2 + 11u - 26$

  b. $w^6 + 11w^3 - 26$

  c. $(y - 4)^2 + 11(y - 4) - 26$

68. a. $u^2 + 17u + 30$

  b. $z^6 + 17z^3 + 30$

  c. $(x + 3)^2 + 17(x + 3) + 30$

For Exercises 69–78, factor out by using substitution. (See Example 9.)

69. $3y^6 + 11y^3 + 6$

70. $3x^4 - 5x^2 - 12$

71. $4p^4 + 5p^2 + 1$

72. $t^4 + 3t^2 + 2$

73. $x^4 + 15x^2 + 36$

74. $t^6 - 16t^3 + 63$

75. $(3x - 1)^2 - (3x - 1) - 6$

76. $(2x + 5)^2 - (2x + 5) - 12$

77. $2(x - 5)^2 + 9(x - 5) + 4$

78. $4(x - 3)^2 + 7(x - 3) + 3$

Objective 6: Summary of Factoring Trinomials

For Exercises 79–88, factor completely by using an appropriate method. (See Example 10.)

79. $3x^3 - 9x^2 + 5x - 15$

80. $ay + ax - 5cy - 5cx$

81. $a^2 + 12a + 36$

82. $9 - 6b + b^2$

83. $81w^2 + 90w + 25$

84. $49a^2 - 28ab + 4b^2$

85. $3x(a + b) - 6(a + b)$

86. $4p(t - 8) + 2(t - 8)$

87. $12a^2bc^2 + 4ab^2c^2 - 6abc^3$

88. $18x^2z - 6xyz + 30xz^2$

For Exercises 89–96, factor the expressions that define each function. (See Example 2.)

89. $f(x) = 2x^2 + 13x - 7$

90. $g(x) = 3x^2 + 14x + 8$

91. $m(t) = t^2 - 22t + 121$

92. $n(t) = t^2 + 20t + 100$

93. $P(x) = x^3 + 4x^2 + 3x$

94. $Q(x) = x^4 + 6x^3 + 8x^2$

95. $h(a) = a^3 + 5a^2 - 6a - 30$

96. $k(a) = a^3 - 4a^2 + 2a - 8$

Expanding Your Skills

97. A student factored $4y^2 - 10y + 4$ as $(2y - 1)(2y - 4)$ on her factoring test. Why did her professor deduct several points, even though $(2y - 1)(2y - 4)$ does multiply out to $4y^2 - 10y + 4$?

98. A student factored $9w^2 + 36w + 36$ as $(3w + 6)^2$ on his factoring test. Why did his instructor deduct several points, even though $(3w + 6)^2$ does multiply out to $9w^2 + 36w + 36$?

Objectives

1. Difference of Squares
2. Using a Difference of Squares in Grouping
3. Sum and Difference of Cubes
4. Summary of Factoring Binomials
5. Factoring Binomials of the Form $x^6 - y^6$

Concept Connections

Determine whether or not each polynomial is a difference of squares.

1. $a^2 + 25$
2. $b^2 - 6$
3. $x^2 - 100$

Skill Practice

Factor completely.

4. $4z^2 - 1$
5. $7y^3 - 63y$
6. $b^4 - 16$

Answers

1. No   2. No   3. Yes
4. $(2z - 1)(2z + 1)$
5. $7y(y + 3)(y - 3)$
6. $(b^2 + 4)(b - 2)(b + 2)$

---

section 5.6   Factoring Binomials

1. Difference of Squares

Up to this point we have learned to

- Factor out the greatest common factor from a polynomial.
- Factor a four-term polynomial by grouping.
- Recognize and factor perfect square trinomials.
- Factor trinomials by the grouping method and by the trial-and-error method.

Next, we will learn how to factor binomials that fit the pattern of a difference of squares. Recall from Section 5.2 that the product of two conjugates results in a **difference of squares**

$$(a + b)(a - b) = a^2 - b^2$$

Therefore, to factor a difference of squares, the process is reversed. Identify a and b and construct the conjugate factors.

Factored Form of a Difference of Squares

$$a^2 - b^2 = (a + b)(a - b)$$

---

example 1   Factoring the Difference of Squares

Factor the binomials completely.

a. $16x^2 - 9$ b. $98c^2d - 50d^3$ c. $z^4 - 81$

Solution:

a. $16x^2 - 9$ The GCF is 1. The binomial is a difference of squares.

    $= (4x)^2 - (3)^2$ Write in the form $a^2 - b^2$, where $a = 4x$ and $b = 3$.

    $= (4x + 3)(4x - 3)$ Factor as $(a + b)(a - b)$.

b. $98c^2d - 50d^3$

    $= 2d(49c^2 - 25d^2)$ The GCF is $2d$. The resulting binomial is a difference of squares.

    $= 2d[(7c)^2 - (5d)^2]$ Write in the form $a^2 - b^2$, where $a = 7c$ and $b = 5d$.

    $= 2d(7c + 5d)(7c - 5d)$ Factor as $(a + b)(a - b)$.

c. $z^4 - 81$ The GCF is 1. The binomial is a difference of squares.

    $= (z^2)^2 - (9)^2$ Write in the form $a^2 - b^2$, where $a = z^2$ and $b = 9$.

    $= (z^2 + 9)(z^2 - 9)$ Factor as $(a + b)(a - b)$.

     $z^2 - 9$ is also a difference of squares.

    $= (z^2 + 9)(z + 3)(z - 3)$

Suppose a and b share no common factors. Then the difference of squares $a^2 - b^2$ can be factored as $(a + b)(a - b)$. However, the sum of squares $a^2 + b^2$ cannot be factored over the real numbers. To see why, consider the expression $a^2 + b^2$. The factored form would require two binomials of the form

$$(a \quad b)(a \quad b) \stackrel{?}{=} a^2 + b^2$$

If all possible combinations of signs are considered, none produces the correct product.

$(a + b)(a - b) = a^2 - b^2$ Wrong sign

$(a + b)(a + b) = a^2 + 2ab + b^2$ Wrong middle term

$(a - b)(a - b) = a^2 - 2ab + b^2$ Wrong middle term

After exhausting all possibilities, we see that if a and b share no common factors, then the sum of squares $a^2 + b^2$ is a prime polynomial.

2. Using a Difference of Squares in Grouping

Sometimes a difference of squares can be used along with other factoring techniques.

example 2   Using a Difference of Squares in Grouping

Factor completely.

a. $y^3 - 6y^2 - 4y + 24$ **b.** $25w^2 + 90w + 81 - p^2$

Solution:

a. $y^3 - 6y^2 - 4y + 24$ The GCF is 1.

$= y^3 - 6y^2 \mid -4y + 24$ The polynomial has four terms. Factor by grouping.

$= y^2(y - 6) - 4(y - 6)$

$= (y - 6)(y^2 - 4)$ $y^2 - 4$ is a difference of squares.

$= (y - 6)(y + 2)(y - 2)$

b. This polynomial is not factorable with "2 by 2" grouping as in part (a). Try grouping three terms. Notice the first three terms constitute a perfect square trinomial. Hence, we will use "3 by 1" grouping.

$\underline{25w^2 + 90w + 81} \mid -p^2$ Group "3 by 1." Factor $25w^2 + 90w + 81 = (5w + 9)^2$.

$= (5w + 9)^2 - p^2$ $(5w + 9)^2 - p^2$ is a difference of squares $a^2 - b^2$, where $a = 5w + 9$ and $b = p$.

$= [(5w + 9) + p][(5w + 9) - p]$ Factor as $a^2 - b^2 = (a + b)(a - b)$.

$= (5w + 9 + p)(5w + 9 - p)$ Simplify.

Skill Practice

Factor completely.

7. $25x^2 + 100$

8. $a^3 + 5a^2 - 9a - 45$

9. $y^2 + 14y + 49 - z^2$

Answers

7. $25(x^2 + 4)$
8. $(a + 5)(a - 3)(a + 3)$
9. $(y + 7 + z)(y + 7 - z)$

Tip: The expression $(5w + 9)^2 - p^2$ can also be factored by using substitution. Let $u = 5w + 9$. Then

$$(5w + 9)^2 - p^2$$
$$= u^2 - p^2 \qquad \text{Substitute } u = 5w + 9.$$
$$= [u + p][u - p] \qquad \text{Factor as a difference of squares.}$$
$$= [(5w + 9) + p][(5w + 9) - p] \qquad \text{Substitute back.}$$
$$= (5w + 9 + p)(5w + 9 - p)$$

3. Sum and Difference of Cubes

For binomials that represent the sum or difference of cubes, factor by using the following formulas.

Factoring a Sum and Difference of Cubes

Sum of cubes: $\quad a^3 + b^3 = (a + b)(a^2 - ab + b^2)$

Difference of cubes: $\quad a^3 - b^3 = (a - b)(a^2 + ab + b^2)$

Multiplication can be used to confirm the formulas for factoring a sum or difference of cubes.

$$(a + b)(a^2 - ab + b^2) = a^3 - a^2b + ab^2 + a^2b - ab^2 + b^3 = a^3 + b^3 \checkmark$$
$$(a - b)(a^2 + ab + b^2) = a^3 + a^2b + ab^2 - a^2b - ab^2 - b^3 = a^3 - b^3 \checkmark$$

Tip: To help remember the placement of the signs in factoring the sum or difference of cubes, remember SOAP: Same sign, Opposite signs, Always Positive.

To help you remember the formulas for factoring a sum or difference of cubes, keep the following guidelines in mind.

- The factored form is the product of a binomial and a trinomial.
- The first and third terms in the trinomial are the squares of the terms within the binomial factor.
- Without regard to sign, the middle term in the trinomial is the product of terms in the binomial factor.

$$x^3 + 8 = (x)^3 + (2)^3 = (x + 2)[(x)^2 - (x)(2) + (2)^2]$$

Square the first term of the binomial. Square the last term of the binomial. Product of terms in the binomial.

- The sign within the binomial factor is the same as the sign of the original binomial.
- The first and third terms in the trinomial are always positive.
- The sign of the middle term in the trinomial is opposite the sign within the binomial.

$$x^3 + 8 = (x)^3 + (2)^3 = (x + 2)[(x)^2 - (x)(2) + (2)^2]$$

same sign, opposite signs, positive

example 3 — Factoring a Difference of Cubes

Factor. $8x^3 - 27$

Solution:

$8x^3 - 27$ $8x^3$ and 27 are perfect cubes.

$= (2x)^3 - (3)^3$ Write as $a^3 - b^3$, where $a = 2x$ and $b = 3$.

$a^3 - b^3 = (a - b)(a^2 + ab + b^2)$

$(2x)^3 - (3)^3 = (2x - 3)[(2x)^2 + (2x)(3) + (3)^2]$ Apply the difference of cubes formula.

$= (2x - 3)(4x^2 + 6x + 9)$ Simplify.

Check: $(2x - 3)(4x^2 + 6x + 9) = 8x^3 + 12x^2 + 18x - 12x^2 - 18x - 27$

$= 8x^3 - 27$ ✔

Skill Practice

Factor completely.

10. $125p^3 - 8$

example 4 — Factoring the Sum of Cubes

Factor. $125t^3 + 64z^6$

Solution:

$125t^3 + 64z^6$ $125t^3$ and $64z^6$ are perfect cubes.

$= (5t)^3 + (4z^2)^3$ Write as $a^3 + b^3$, where $a = 5t$ and $b = 4z^2$.

$a^3 + b^3 = (a + b)(a^2 - ab + b^2)$ Apply the sum of cubes formula.

$(5t)^3 + (4z^2)^3 = [(5t) + (4z^2)][(5t)^2 - (5t)(4z^2) + (4z^2)^2]$

$= (5t + 4z^2)(25t^2 - 20tz^2 + 16z^4)$ Simplify.

Skill Practice

Factor completely.

11. $x^3 + 1000$

4. Summary of Factoring Binomials

After factoring out the greatest common factor, the next step in any factoring problem is to recognize what type of pattern it follows. Exponents that are divisible by 2 are perfect squares, and those divisible by 3 are perfect cubes. The formulas for factoring binomials are summarized here.

Factoring Binomials

1. Difference of squares: $a^2 - b^2 = (a + b)(a - b)$
2. Difference of cubes: $a^3 - b^3 = (a - b)(a^2 + ab + b^2)$
3. Sum of cubes: $a^3 + b^3 = (a + b)(a^2 - ab + b^2)$

Answers

10. $(5p - 2)(25p^2 + 10p + 4)$
11. $(x + 10)(x^2 - 10x + 100)$

Skill Practice

Factor the binomials.

12. $x^2 - \dfrac{1}{25}$
13. $24a^4 - 3a$
14. $16y^3 + 4y$
15. $18p^4 - 50t^2$

---

example 5   **Review of Factoring Binomials**

Factor the binomials.

a. $m^3 - \dfrac{1}{8}$ b. $9k^2 + 24m^2$ c. $128y^6 + 54x^3$ d. $50y^6 - 8x^2$

Solution:

a. $m^3 - \dfrac{1}{8}$ m^3 is a perfect cube: $m^3 = (m)^3$.
$\dfrac{1}{8}$ is a perfect cube: $\dfrac{1}{8} = \left(\dfrac{1}{2}\right)^3$.

$= (m)^3 - \left(\dfrac{1}{2}\right)^3$ This is a difference of cubes, where $a = m$ and $b = \dfrac{1}{2}$:
$a^3 - b^3 = (a - b)(a^2 + ab + b^2)$.

$= \left(m - \dfrac{1}{2}\right)\left(m^2 + \dfrac{1}{2}m + \dfrac{1}{4}\right)$ Factor.

b. $9k^2 + 24m^2$ Factor out the GCF.

$= 3(3k^2 + 8m^2)$ The resulting binomial is not a difference of squares or a sum or difference of cubes. It cannot be factored further over the real numbers.

c. $128y^6 + 54x^3$ Factor out the GCF.

$= 2(64y^6 + 27x^3)$ Both 64 and 27 are perfect cubes, and the exponents of both x and y are multiples of 3. This is a sum of cubes, where $a = 4y^2$ and $b = 3x$.

$= 2[(4y^2)^3 + (3x)^3]$ $a^3 + b^3 = (a + b)(a^2 - ab + b^2)$.

$= 2(4y^2 + 3x)(16y^4 - 12xy^2 + 9x^2)$ Factor.

d. $50y^6 - 8x^2$ Factor out the GCF.

$= 2(25y^6 - 4x^2)$ Both 25 and 4 are perfect squares. The exponents of both x and y are multiples of 2. This is a difference of squares, where $a = 5y^3$ and $b = 2x$.

$= 2[(5y^3)^2 - (2x)^2]$ $a^2 - b^2 = (a + b)(a - b)$.

$= 2(5y^3 + 2x)(5y^3 - 2x)$

Answers

12. $\left(x + \dfrac{1}{5}\right)\left(x - \dfrac{1}{5}\right)$
13. $3a(2a - 1)(4a^2 + 2a + 1)$
14. $4y(4y^2 + 1)$
15. $2(3p^2 + 5t)(3p^2 - 5t)$

5. Factoring Binomials of the Form $x^6 - y^6$

example 6 Factoring Binomials

Factor the binomial $x^6 - y^6$ as

a. A difference of cubes

b. A difference of squares

Skill Practice

Factor completely.

16. $a^6 - 64$

Solution:

Notice that the expressions x^6 and y^6 are both perfect squares and perfect cubes because the exponents are both multiples of 2 and of 3. Consequently, $x^6 - y^6$ can be factored initially as either a difference of cubes or a difference of squares.

a. $x^6 - y^6$ *difference of cubes*

$= (x^2)^3 - (y^2)^3$ Write as $a^3 - b^3$, where $a = x^2$ and $b = y^2$.

$= (x^2 - y^2)[(x^2)^2 + (x^2)(y^2) + (y^2)^2]$ Apply the formula $a^3 - b^3 = (a - b)(a^2 + ab + b^2)$.

$= (x^2 - y^2)(x^4 + x^2y^2 + y^4)$ Factor $x^2 - y^2$ as a difference of squares.

$= (x + y)(x - y)(x^4 + x^2y^2 + y^4)$ The expression $x^4 + x^2y^2 + y^4$ cannot be factored by using the skills learned thus far.

b. $x^6 - y^6$ *difference of squares*

$= (x^3)^2 - (y^3)^2$ Write as $a^2 - b^2$, where $a = x^3$ and $b = y^3$.

$= (x^3 + y^3)(x^3 - y^3)$ Apply the formula $a^2 - b^2 = (a + b)(a - b)$.

    sum of cubes   *difference of cubes*

Factor $x^3 + y^3$ as a sum of cubes. Factor $x^3 - y^3$ as a difference of cubes.

$= (x + y)(x^2 - xy + y^2)(x - y)(x^2 + xy + y^2)$

If given a choice between factoring a binomial as a difference of squares or as a difference of cubes, it is recommended that you factor initially as a difference of squares. As Example 6 illustrates, factoring as a difference of squares leads to a more complete factorization. Hence,

$$a^6 - b^6 = (a - b)(a^2 + ab + b^2)(a + b)(a^2 - ab + b^2)$$

Answer

16. $(a - 2)(a + 2)(a^2 + 2a + 4)(a^2 - 2a + 4)$

section 5.6 Practice Exercises

Boost your GRADE at mathzone.com!

- Practice Problems
- Self-Tests
- NetTutor
- e-Professors
- Videos

Study Skills Exercises

1. Helping other students by showing them how to work a math problem will help you as well. It will increase your understanding and make you aware if there is a concept that you do not fully understand. List some concepts from this chapter that you would feel confident teaching to someone else.

2. Define the key terms.
   a. Difference of squares
   b. Sum of cubes
   c. Difference of cubes

Review Exercises

For Exercises 3–10, factor completely.

3. $4x^2 - 20x + 25$
4. $9t^2 - 42t + 49$
5. $10x + 6xy + 5 + 3y$

6. $21a + 7ab - 3b - b^2$
7. $32p^2 - 28p - 4$
8. $6q^2 + 37q - 35$

9. $45a^2 - 9ac$
10. $11xy^2 - 55y^3$

Objective 1: Difference of Squares

11. Explain how to identify and factor a difference of squares.

12. Can you factor $25x^2 + 4$?

For Exercises 13–22, factor the binomials. Identify the binomials that are prime. (See Example 1.)

13. $x^2 - 9$
14. $y^2 - 25$
15. $16 - w^2$
16. $81 - b^2$

17. $8a^2 - 162b^2$
18. $50c^2 - 72d^2$
19. $25u^2 + 1$
20. $w^2 + 4$

21. $(2m - 5)^2 - 36$
22. $(3n + 1)^2 - 49$

Objective 2: Using the Difference of Squares in Grouping

For Exercises 23–32, use the difference of squares along with factoring by grouping. (See Example 2.)

23. $x^3 - x^2 - 16x + 16$
24. $x^3 + 5x^2 - x - 5$
25. $4x^3 + 12x^2 - x - 3$

26. $5x^3 - x^2 - 45x + 9$ **27.** $x^2 + 12x + 36 - a^2$ **28.** $a^2 + 10a + 25 - b^2$

29. $p^2 + 2pq + q^2 - 81$ **30.** $m^2 - 2mn + n^2 - 9$ **31.** $p^2 - (y^2 - 6y + 9)$

32. $b^2 - (x^2 + 4x + 4)$

Objective 3: Sum and Difference of Cubes

33. Explain how to identify and factor a sum of cubes.

34. Explain how to identify and factor a difference of cubes.

For Exercises 35–42, factor the sum or difference of cubes. **(See Examples 3–4.)**

35. $8x^3 - 1$ (Check by multiplying.) **36.** $y^3 + 64$ (Check by multiplying.)

37. $125c^3 + 27$ **38.** $216u^3 - v^3$ **39.** $x^3 - 1000$ **40.** $8y^3 - 27$

41. $64t^3 + 1$ **42.** $125r^3 + 1$

Objective 4: Summary of Factoring Binomials

For Exercises 43–66, factor completely. **(See Example 5.)**

43. $36y^2 - \dfrac{1}{25}$ **44.** $16p^2 - \dfrac{1}{9}$ **45.** $18d^{12} - 32$ **46.** $3z^8 - 12$

47. $242v^2 + 32$ **48.** $8p^2 + 200$ **49.** $4x^2 - 16$ **50.** $9m^2 - 81n^2$

51. $25 - 49q^2$ **52.** $1 - 25p^2$ **53.** $(t + 2s)^2 - 36$ **54.** $(5x + 4)^2 - y^2$

55. $27 - t^3$ **56.** $8 + y^3$ **57.** $27a^3 + \dfrac{1}{8}$ **58.** $b^3 + \dfrac{27}{125}$

59. $2m^3 + 16$ **60.** $3x^3 - 375$ **61.** $x^4 - y^4$ **62.** $81u^4 - 16v^4$

63. $a^9 + b^9$ **64.** $27m^9 - 8n^9$ **65.** $(p - 3)^3 - 64$ **66.** $(q - 1)^3 + 125$

Objective 5: Factoring Binomials of the Form $x^6 - y^6$

For Exercises 67–74, factor completely. (See Example 6.)

67. $a^6 - b^6$ (*Hint:* First factor as a difference of squares.)

68. $64x^6 - y^6$

69. $64 - y^6$

70. $1 - p^6$

71. $h^6 + k^6$ (*Hint:* Factor as a sum of cubes.)

72. $27q^6 + 125p^6$

73. $8x^6 + 125$

74. $t^6 + 1$

Mixed Exercises

75. Find a difference of squares that has $(2x + 3)$ as one of its factors.

76. Find a difference of squares that has $(4 - p)$ as one of its factors.

77. Find a difference of cubes that has $(4a^2 + 6a + 9)$ as its trinomial factor.

78. Find a sum of cubes that has $(25c^2 - 10cd + 4d^2)$ as its trinomial factor.

79. Find a sum of cubes that has $(4x^2 + y)$ as its binomial factor.

80. Find a difference of cubes that has $(3t - r^2)$ as its binomial factor.

81. Consider the shaded region:

  a. Find an expression that represents the area of the shaded region.

  b. Factor the expression found in part (a).

  c. Find the area of the shaded region if $x = 6$ in. and $y = 4$ in.

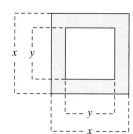

82. A manufacturer needs to know the area of a metal washer. The outer radius of the washer is R and the inner radius is r.

  a. Find an expression that represents the area of the washer.

  b. Factor the expression found in part (a).

  c. Find the area of the washer if $R = \frac{1}{2}$ in. and $r = \frac{1}{4}$ in. (Round to the nearest 0.01 in.^{2})

Expanding Your Skills

For Exercises 83–86, factor the polynomials by using the difference of squares, sum of cubes, or difference of cubes with grouping.

83. $x^2 - y^2 + x + y$

84. $25c^2 - 9d^2 + 5c - 3d$

85. $5wx^3 + 5wy^3 - 2zx^3 - 2zy^3$

86. $3xu^3 - 3xv^3 - 5yu^3 + 5yv^3$

A Factoring Strategy

Factoring Strategy
1. Factor out the greatest common factor (Section 5.4).
2. Identify whether the polynomial has two terms, three terms, or more than three terms.
3. If the polynomial has more than three terms, try factoring by grouping (Section 5.4 and Section 5.6).
4. If the polynomial has three terms, check first for a perfect square trinomial. Otherwise, factor the trinomial with the grouping method or the trial-and-error method (Section 5.5).
5. If the polynomial has two terms, determine if it fits the pattern for a difference of squares, difference of cubes, or sum of cubes. Remember, a sum of squares is not factorable over the real numbers (Section 5.6).

1. What is meant by a prime factor?

2. What is the first step in factoring any polynomial?

3. When factoring a binomial, what patterns do you look for?

4. When factoring a trinomial, what pattern do you look for first, before using the grouping method or trial-and-error method?

5. What do you look for when factoring a perfect square trinomial?

6. What do you look for when factoring a four-term polynomial?

For Exercises 7–26,
   a. After factoring out the GCF, identify by inspection what type of factoring pattern would be most appropriate:

     Difference of squares   Perfect square trinomial
     Sum of cubes   Trinomial (grouping or trial-and-error)
     Difference of cubes   Grouping

   b. Factor the polynomial completely.

7. $6x^2 - 21x - 45$

8. $8m^3 - 10m^2 - 3m$

9. $8a^2 - 50$

10. $ab + ay - b^2 - by$

11. $14u^2 - 11uv + 2v^2$

12. $9p^2 - 12pq + 4q^2$

13. $16x^3 - 2$

14. $9m^2 + 16n^2$

15. $27y^3 + 125$

16. $3x^2 - 16$

17. $128p^6 + 54q^3$

18. $5b^2 - 30b + 45$

19. $16a^4 - 1$

20. $81u^2 - 90uv + 25v^2$

21. $p^2 - 12p + 36 - c^2$

22. $4x^2 + 16$

23. $12ax - 6ay + 4bx - 2by$

24. $125y^3 - 8$

25. $5y^2 + 14y - 3$

26. $2m^4 - 128$

Objectives

1. Solving Equations by Using the Zero Product Rule
2. Applications of Quadratic Equations
3. Definition of a Quadratic Function
4. Applications of Quadratic Functions

section 5.7 Solving Equations by Using the Zero Product Rule

1. Solving Equations by Using the Zero Product Rule

In Section 1.4 we defined a linear equation in one variable as an equation of the form $ax + b = 0$ $(a \neq 0)$. A linear equation in one variable is sometimes called a first-degree polynomial equation because the highest degree of all its terms is 1. A second-degree polynomial equation is called a quadratic equation.

> **Definition of a Quadratic Equation in One Variable**
>
> If a, b, and c are real numbers such that $a \neq 0$, then a **quadratic equation** is an equation that can be written in the form
>
> $$ax^2 + bx + c = 0$$

The following equations are quadratic because they can each be written in the form $ax^2 + bx + c = 0 \, (a \neq 0)$.

$$-4x^2 + 4x = 1 \qquad x(x - 2) = 3 \qquad (x - 4)(x + 4) = 9$$
$$-4x^2 + 4x - 1 = 0 \qquad x^2 - 2x = 3 \qquad x^2 - 16 = 9$$
$$x^2 - 2x - 3 = 0 \qquad x^2 - 25 = 0$$
$$x^2 + 0x - 25 = 0$$

One method to solve a quadratic equation is to factor the equation and apply the zero product rule. The **zero product rule** states that if the product of two factors is zero, then one or both of its factors is equal to zero.

> **The Zero Product Rule**
>
> If $ab = 0$, then $a = 0$ or $b = 0$.

Concept Connections

Determine whether the equations are quadratic or linear.

1. $5x^2 = 2x + 1$
2. $a(a + 2) - 3 = 0$
3. $4(b - 3) + 2 = 0$

For example, the quadratic equation $x^2 - x - 12 = 0$ can be written in factored form as $(x - 4)(x + 3) = 0$. By the zero product rule, one or both factors must be zero. Hence, either $x - 4 = 0$ or $x + 3 = 0$. Therefore, to solve the quadratic equation, set each factor to zero and solve for x.

$(x - 4)(x + 3) = 0$ \qquad Apply the zero product rule.

$x - 4 = 0$ \quad or \quad $x + 3 = 0$ \qquad Set each factor to zero.

$x = 4$ \quad or \quad $x = -3$ \qquad Solve each equation for x.

Quadratic equations, like linear equations, arise in many applications of mathematics, science, and business. The following steps summarize the factoring method to solve a quadratic equation.

Answers

1. Quadratic
2. Quadratic
3. Linear

Steps to Solve a Quadratic Equation by Factoring

1. Write the equation in the form $ax^2 + bx + c = 0$.
2. Factor the equation completely.
3. Apply the zero product rule. That is, set each factor equal to zero and solve the resulting equations.*

*The solution(s) found in step 3 may be checked by substitution in the original equation.

example 1 — Solving Quadratic Equations

Solve.

a. $2x^2 - 5x = 12$
b. $\frac{1}{2}x^2 + \frac{2}{3}x = 0$
c. $9x(4x + 2) - 10x = 8x + 25$
d. $2x(x + 5) + 3 = 2x^2 - 5x + 1$

Solution:

a.
$$2x^2 - 5x = 12$$
$$2x^2 - 5x - 12 = 0 \quad \text{Write the equation in the form } ax^2 + bx + c = 0.$$
$$(2x + 3)(x - 4) = 0 \quad \text{Factor the polynomial completely.}$$

$2x + 3 = 0$ or $x - 4 = 0$ Set each factor equal to zero.
$2x = -3$ or $x = 4$ Solve each equation.
$x = -\frac{3}{2}$ or $x = 4$

Check: $x = -\frac{3}{2}$ Check: $x = 4$

$2x^2 - 5x = 12$ $2x^2 - 5x = 12$

$2\left(-\frac{3}{2}\right)^2 - 5\left(-\frac{3}{2}\right) \stackrel{?}{=} 12$ $2(4)^2 - 5(4) \stackrel{?}{=} 12$

$2\left(\frac{9}{4}\right) + \frac{15}{2} \stackrel{?}{=} 12$ $2(16) - 20 \stackrel{?}{=} 12$

$\frac{18}{4} + \frac{30}{4} \stackrel{?}{=} 12$ $32 - 20 = 12$ ✓

$\frac{48}{4} = 12$ ✓

b.
$$\frac{1}{2}x^2 + \frac{2}{3}x = 0 \quad \text{The equation is already in the form } ax^2 + bx + c = 0. \text{ (Note: } c = 0.\text{)}$$
$$6\left(\frac{1}{2}x^2 + \frac{2}{3}x\right) = 6(0) \quad \text{Clear fractions.}$$
$$3x^2 + 4x = 0$$

Skill Practice

Determine whether each equation is quadratic or linear. Then solve by using the appropriate method.

4. $y^2 - 2y = 35$
5. $3x^2 = 7x$
6. $t^2 - 3t + 1 = t^2 + 2t + 11$
7. $5a(2a - 3) + 4(a + 1) = 3a(3a - 2)$

Answers

4. Quadratic; $y = 7$ or $y = -5$
5. Quadratic; $x = 0$ or $x = \frac{7}{3}$
6. Linear; $t = -2$
7. Quadratic; $a = 4$ or $a = 1$

$$x(3x + 4) = 0 \qquad \text{Factor completely.}$$

$$x = 0 \quad \text{or} \quad 3x + 4 = 0 \qquad \text{Set each factor equal to zero.}$$

$$x = 0 \quad \text{or} \quad x = -\frac{4}{3} \qquad \text{Solve each equation for } x.$$

Check: $x = 0$ 　　　　　　　　Check: $x = -\frac{4}{3}$

$$\frac{1}{2}x^2 + \frac{2}{3}x = 0 \qquad\qquad \frac{1}{2}x^2 + \frac{2}{3}x = 0$$

$$\frac{1}{2}(0)^2 + \frac{2}{3}(0) \stackrel{?}{=} 0 \qquad \frac{1}{2}\left(-\frac{4}{3}\right)^2 + \frac{2}{3}\left(-\frac{4}{3}\right) \stackrel{?}{=} 0$$

$$0 = 0 \checkmark \qquad\qquad \frac{1}{2}\left(\frac{16}{9}\right) - \frac{8}{9} \stackrel{?}{=} 0$$

$$\frac{8}{9} - \frac{8}{9} = 0 \checkmark$$

c. $9x(4x + 2) - 10x = 8x + 25$

$36x^2 + 18x - 10x = 8x + 25$ 　　Clear parentheses.

$36x^2 + 8x = 8x + 25$ 　　Combine *like* terms.

$36x^2 - 25 = 0$ 　　Make one side of the equation equal to zero. The equation is in the form $ax^2 + bx + c = 0$. (*Note:* $b = 0$.)

$(6x - 5)(6x + 5) = 0$ 　　Factor completely.

$6x - 5 = 0 \quad \text{or} \quad 6x + 5 = 0$ 　　Set each factor equal to zero.

$6x = 5 \quad \text{or} \quad 6x = -5$ 　　Solve each equation.

$\dfrac{6x}{6} = \dfrac{5}{6} \quad \text{or} \quad \dfrac{6x}{6} = \dfrac{-5}{6}$

$x = \dfrac{5}{6} \quad \text{or} \quad x = -\dfrac{5}{6}$ 　　The check is left to the reader.

d. $2x(x + 5) + 3 = 2x^2 - 5x + 1$

$2x^2 + 10x + 3 = 2x^2 - 5x + 1$ 　　Clear parentheses.

$15x + 2 = 0$ 　　Make one side of the equation equal to zero. The equation is not quadratic. It is in the form $ax + b = 0$, which is linear. Solve by using the method for linear equations.

$15x = -2$

$x = -\dfrac{2}{15}$

The check is left to the reader.

The zero product rule can be used to solve higher-degree polynomial equations provided one side of the equation is zero and the other is written in factored form.

example 2 — Solving Higher-Degree Polynomial Equations

Solve the equations.

a. $-2(y + 7)(y - 1)(10y + 3) = 0$ **b.** $z^3 + 3z^2 - 4z - 12 = 0$

Skill Practice

Solve the equations

8. $3(w + 2)(2w + 1)(w - 8) = 0$

9. $x^3 + x^2 - 9x - 9 = 0$

Solution:

a. $-2(y + 7)(y - 1)(10y + 3) = 0$

One side of the equation is zero, and the other side is already factored.

$-2 = 0$ or $y + 7 = 0$ or $y - 1 = 0$ or $10y + 3 = 0$ Set each factor equal to zero.

No solution $y = -7$ or $y = 1$ or $y = -\dfrac{3}{10}$ Solve each equation for y.

Notice that when the constant factor is set to zero, the result is the contradiction $-2 = 0$. The constant factor does not produce a solution to the equation. Therefore, the only solutions are $y = -7$, $y = 1$, and $y = -\dfrac{3}{10}$. Each solution can be checked in the original equation.

b.
$z^3 + 3z^2 - 4z - 12 = 0$ This is a higher-degree polynomial equation.

$z^3 + 3z^2 \;|\; - 4z - 12 = 0$ One side of the equation is zero. Now factor. Because there are four terms, try factoring by grouping.

$z^2(z + 3) - 4(z + 3) = 0$

$(z + 3)(z^2 - 4) = 0$ $z^2 - 4$ can be factored further as a difference of squares.

$(z + 3)(z - 2)(z + 2) = 0$

$z + 3 = 0$ or $z - 2 = 0$ or $z + 2 = 0$ Set each factor equal to zero.

$z = -3$ or $z = 2$ or $z = -2$ Solve each equation.

2. Applications of Quadratic Equations

example 3 — Application of a Quadratic Equation

The product of two consecutive odd integers is 20 more than the smaller integer. Find the integers.

Answers

8. $w = -2$ or $w = -\dfrac{1}{2}$ or $w = 8$
9. $x = -1$ or $x = 3$ or $x = -3$

Skill Practice

10. The product of two consecutive even integers is 40 more than 5 times the smaller integer. Find the integers.

Solution:

Let x represent the smaller odd integer and $x + 2$ represent the next consecutive odd integer. The equation representing their product is

$$x(x + 2) = x + 20$$
$$x^2 + 2x = x + 20 \quad \text{Clear parentheses.}$$
$$x^2 + x - 20 = 0 \quad \text{Make the equation equal to zero.}$$
$$(x + 5)(x - 4) = 0 \quad \text{Factor.}$$
$$x + 5 = 0 \quad \text{or} \quad x - 4 = 0 \quad \text{Set each factor equal to zero.}$$
$$x = -5 \quad \text{or} \quad x = 4 \quad \text{Solve each equation.}$$

Since we are looking for consecutive odd integers, $x = 4$ is not a solution. Since $x = -5$ and $x + 2 = -3$, the integers are -5 and -3.

Skill Practice

11. The width of a rectangle is 5 in. less than 3 times the length. The area is 2 in.² Find the length and width.

example 4   Application Problem

The length of a basketball court is 6 ft less than 2 times the width. If the total area is 4700 ft², find the dimensions of the court.

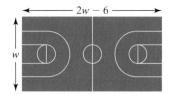

Figure 5-5

Solution:

If the width of the court is represented by w, then the length can be represented by $2w - 6$ (Figure 5-5).

$$A = (\text{length})(\text{width}) \quad \text{Area of a rectangle}$$
$$4700 = (2w - 6)w \quad \text{Mathematical equation}$$
$$4700 = 2w^2 - 6w$$
$$2w^2 - 6w - 4700 = 0 \quad \text{Set the equation equal to zero and factor.}$$
$$2(w^2 - 3w - 2350) = 0 \quad \text{Factor out the GCF.}$$
$$2(w - 50)(w + 47) = 0 \quad \text{Factor the trinomial.}$$
$$2 \neq 0 \quad \text{or} \quad w - 50 = 0 \quad \text{or} \quad w + 47 = 0 \quad \text{Set each factor equal to zero.}$$
contradiction
$$\qquad\qquad\qquad w = 50 \quad \text{or} \quad w \neq -47 \quad \text{A negative width is not possible.}$$

The width is 50 ft.

The length is $2w - 6 = 2(50) - 6 = 94$ ft.

Answers

10. 8 and 10
11. Width: 1 in.; length: 2 in.

example 5   An Application of a Quadratic Equation

A region of coastline off Biscayne Bay is approximately in the shape of a right angle. The corresponding triangular area has sandbars and is marked off on navigational charts as being shallow water. If one leg of the triangle is 0.5 mi shorter than the other leg, and the hypotenuse is 2.5 mi, find the lengths of the legs of the triangle (Figure 5-6).

Skill Practice

12. The longer leg of a right triangle measures 7 ft more than the shorter leg. The hypotenuse is 8 ft longer than the shorter leg. Find the lengths of the sides of the triangle.

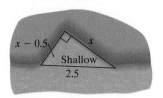

Figure 5-6

Solution:

Let x represent the longer leg.

Then $x - 0.5$ represents the shorter leg.

$$a^2 + b^2 = c^2 \qquad \text{Pythagorean theorem}$$

$$x^2 + (x - 0.5)^2 = (2.5)^2$$

Tip: Recall that the square of a binomial results in a perfect square trinomial.

$$(a - b)^2 = a^2 - 2ab + b^2$$
$$(x - 0.5)^2 = (x)^2 - 2(x)(0.5) + (0.5)^2$$
$$= x^2 - x + 0.25$$

$$x^2 + (x)^2 - 2(x)(0.5) + (0.5)^2 = 6.25$$

$$x^2 + x^2 - x + 0.25 = 6.25$$

$$2x^2 - x - 6 = 0 \qquad \text{Write the equation in the form } ax^2 + bx + c = 0.$$

$$(2x + 3)(x - 2) = 0 \qquad \text{Factor.}$$

$2x + 3 = 0 \quad \text{or} \quad x - 2 = 0 \qquad$ Set both factors to zero.

$x = -\dfrac{3}{2} \quad \text{or} \quad x = 2 \qquad$ Solve both equations for x.

The side of a triangle can not be negative, so we reject the solution $x = -\dfrac{3}{2}$.

Therefore, one leg of the triangle is 2 mi.

The other leg is $x - 0.5 = 2 - 0.5 = 1.5$ mi.

3. Definition of a Quadratic Function

In Section 4.3, we graphed several basic functions by plotting points, including $f(x) = x^2$. This function is called a quadratic function, and its graph is in the shape of a **parabola**. In general, any second-degree polynomial function is a quadratic function.

Answer

12. The sides are 5, 12, and 13 ft.

Concept Connections

Determine whether the functions are quadratic.

13. $f(x) = 2 - x^2$
14. $f(t) = 2t^2 + 5t + 3$
15. $f(x) = 2x - 9$

Definition of a Quadratic Function

Let a, b, and c represent real numbers such that $a \neq 0$. Then a function in the form $f(x) = ax^2 + bx + c$ is called a **quadratic function**.

The graph of a quadratic function is a parabola that opens up or down. The leading coefficient a determines the direction of the parabola. For the quadratic function defined by $f(x) = ax^2 + bx + c$:

If $a > 0$, the parabola opens up. For example, $f(x) = x^2$

If $a < 0$, the parabola opens down. For example, $g(x) = -x^2$

Recall from Section 4.3 that the x-intercepts of a function $y = f(x)$ are the real solutions to the equation $f(x) = 0$. The y-intercept is found by evaluating $f(0)$.

Skill Practice

16. Find the x- and y-intercepts of the function defined by $f(x) = x^2 + 8x + 12$.

example 6 Finding the x- and y-Intercepts of a Quadratic Function

Find the x- and y-intercepts.

$$f(x) = x^2 - x - 12$$

Solution:

To find the x-intercept, substitute $f(x) = 0$.

$f(x) = x^2 - x - 12$

$0 = x^2 - x - 12$ Substitute 0 for $f(x)$. The result is a quadratic equation.

$0 = (x - 4)(x + 3)$ Factor.

$x - 4 = 0$ or $x + 3 = 0$ Set each factor equal to zero.

$x = 4$ or $x = -3$ Solve each equation.

The x-intercepts are $(4, 0)$ and $(-3, 0)$.

To find the y-intercept, find $f(0)$.

$f(x) = x^2 - x - 12$

$f(0) = (0)^2 - (0) - 12$ Substitute $x = 0$.

$= -12$

The y-intercept is $(0, -12)$.

Answers
13. Quadratic
14. Quadratic
15. Not quadratic
16. x-intercepts: $(-6, 0)$ and $(-2, 0)$; y-intercept: $(0, 12)$

Calculator Connections

The graph of $f(x) = x^2 - x - 12$ supports the solution to Example 6. The graph appears to cross the x-axis at -3 and 4. The y-intercept is given as $(0, -12)$.

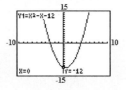

4. Applications of Quadratic Functions

example 7 Application of a Quadratic Function

A rocket is shot vertically upward with an initial velocity of 288 ft/sec. The function given by $h(t) = -16t^2 + 288t$ relates the rocket's height h (in feet) to the time t after launch (in seconds).

a. Find $h(0)$, $h(5)$, $h(10)$, and $h(15)$, and interpret the meaning of these function values in the context of the rocket's height and time after launch.

b. Find the t-intercepts of the function, and interpret their meaning in the context of the rocket's height and time after launch.

c. Find the time(s) at which the rocket is at a height of 1152 ft.

Solution:

a. $h(t) = -16t^2 + 288t$

$h(0) = -16(0)^2 + 288(0) = 0$

$h(5) = -16(5)^2 + 288(5) = 1040$

$h(10) = -16(10)^2 + 288(10) = 1280$

$h(15) = -16(15)^2 + 288(15) = 720$

$h(0) = 0$ indicates that at $t = 0$ sec, the height of the rocket is 0 ft.

$h(5) = 1040$ indicates that 5 sec after launch, the height of the rocket is 1040 ft.

$h(10) = 1280$ indicates that 10 sec after launch, the height of the rocket is 1280 ft.

$h(15) = 720$ indicates that 15 sec after launch, the height of the rocket is 720 ft.

b. The t-intercepts of the function are represented by the real solutions of the equation $h(t) = 0$.

$-16t^2 + 288t = 0$	Set $h(t) = 0$.
$-16t(t - 18) = 0$	Factor.
$-16t = 0$ or $t - 18 = 0$	Apply the zero product rule.
$t = 0$ or $t = 18$	

The rocket is at ground level initially (at $t = 0$ sec) and then again after 18 sec when it hits the ground.

Skill Practice

An object is dropped from the top of a building that is 144 ft high. The function given by $h(t) = -16t^2 + 144$ relates the height h of the object (in feet) to the time t in seconds after it is dropped.

17. Find $h(0)$ and interpret the meaning of the function value in the context of this problem.

18. Find the t-intercept(s) and interpret the meaning in the context of this problem.

Answers

17. $h(0) = 144$, which is the initial height of the object (after 0 sec).

18. The t-intercept is $(3, 0)$ which means the object is at ground level (0 ft high) after 3 sec. The intercept $(-3, 0)$ does not make sense for this problem since time cannot be negative.

c. Set $h(t) = 1152$ and solve for t.

$$h(t) = -16t^2 + 288t$$
$$1152 = -16t^2 + 288t \quad \text{Substitute } 1152 \text{ for } h(t).$$
$$16t^2 - 288t + 1152 = 0 \quad \text{Set the equation equal to zero.}$$
$$16(t^2 - 18t + 72) = 0 \quad \text{Factor out the GCF.}$$
$$16(t - 6)(t - 12) = 0 \quad \text{Factor.}$$
$$t = 6 \quad \text{or} \quad t = 12$$

The rocket will reach a height of 1152 ft after 6 sec (on the way up) and after 12 sec (on the way down). (See Figure 5-7.)

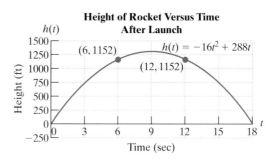

Figure 5-7

section 5.7 Practice Exercises

Boost your GRADE at mathzone.com!
MathZone

- Practice Problems
- Self-Tests
- NetTutor
- e-Professors
- Videos

Study Skills Exercises

1. There's a saying, "Leave no stone unturned." In math, this means "leave no homework problem undone." Did you do all the assigned homework in Section 5.6? Do you understand the concepts well enough to move on to the homework in this section?

2. Define the key terms.
   a. Quadratic equation
   b. Zero product rule
   c. Quadratic function
   d. Parabola

Review Exercises

For Exercises 3–8, factor completely.

3. $10x^2 + 3x$

4. $7x^2 - 28$

5. $2p^2 - 9p - 5$

6. $3q^2 - 4q - 4$

7. $t^3 - 1$

8. $z^2 - 11z + 30$

Objective 1: Solving Equations by Using the Zero Product Rule

9. What conditions are necessary to solve an equation by using the zero product rule?

10. State the zero product rule.

For Exercises 11–16, determine which of the equations are written in the correct form to apply the zero product rule directly. If an equation is not in the correct form, explain what is wrong.

11. $2x(x - 3) = 0$
12. $(u + 1)(u - 3) = 10$
13. $3p^2 - 7p + 4 = 0$
14. $t^2 - t - 12 = 0$
15. $a(a + 3)^2 = 5$
16. $\left(\frac{2}{3}x - 5\right)\left(x + \frac{1}{2}\right) = 0$

For Exercises 17–50, solve the equation. (See Examples 1–2.)

17. $(x + 3)(x + 5) = 0$
18. $(x + 7)(x - 4) = 0$
19. $(2w + 9)(5w - 1) = 0$
20. $(3a + 1)(4a - 5) = 0$
21. $x(x + 4)(10x - 3) = 0$
22. $t(t - 6)(3t - 11) = 0$
23. $0 = 5(y - 0.4)(y + 2.1)$
24. $0 = -4(z - 7.5)(z - 9.3)$
25. $x^2 + 6x - 27 = 0$
26. $2x^2 + x - 15 = 0$
27. $2x^2 + 5x = 3$
28. $-11x = 3x^2 - 4$
29. $10x^2 = 15x$
30. $5x^2 = 7x$
31. $6(y - 2) - 3(y + 1) = 8$
32. $4x + 3(x - 9) = 6x + 1$
33. $-9 = y(y + 6)$
34. $-62 = t(t - 16) + 2$
35. $9p^2 - 15p - 6 = 0$
36. $6y^2 + 2y = 48$
37. $(x + 1)(2x - 1)(x - 3) = 0$
38. $2x(x - 4)^2(4x + 3) = 0$
39. $(y - 3)(y + 4) = 8$
40. $(t + 10)(t + 5) = 6$
41. $(2a - 1)(a - 1) = 6$
42. $w(6w + 1) = 2$
43. $p^2 + (p + 7)^2 = 169$
44. $x^2 + (x + 2)^2 = 100$
45. $3t(t + 5) - t^2 = 2t^2 + 4t - 1$
46. $a^2 - 4a - 2 = (a + 3)(a - 5)$
47. $2x^3 - 8x^2 - 24x = 0$
48. $2p^3 + 20p^2 + 42p = 0$
49. $w^3 = 16w$
50. $12x^3 = 27x$

Objective 2: Applications of Quadratic Equations

51. If 5 is added to the square of a number, the result is 30. Find all such numbers.

52. Four less than the square of a number is 77. Find all such numbers.

53. The square of a number is equal to 12 more than the number. Find all such numbers.

54. The square of a number is equal to 20 more than the number. Find all such numbers.

55. The product of two consecutive integers is 42. Find the integers. **(See Example 3.)**

56. The product of two consecutive integers is 110. Find the integers.

57. The product of two consecutive odd integers is 63. Find the integers.

58. The product of two consecutive even integers is 120. Find the integers.

59. A rectangular pen is to contain 35 ft^2 of area. If the width is 2 ft less than the length, find the dimensions of the pen. **(See Example 4.)**

60. The length of a rectangular photograph is 7 in. more than the width. If the area is 78 in.^{2}, what are the dimensions of the photograph?

61. The length of a rectangular room is 5 yd more than the width. If the area is 300 yd^2, find the length and the width of the room.

62. The top of a dining room table it twice as long as it is wide. Find the dimensions of the table if the area is 18 ft^2.

63. The height of a triangle is 1 in. more than the base. If the height is increased by 2 in. while the base remains the same, the new area becomes 20 in.^{2}
    a. Find the base and height of the original triangle.

    b. Find the area of the original triangle.

64. The base of a triangle is 2 cm more than the height. If the base is increased by 4 cm while the height remains the same, the new area is 56 cm^2.
    a. Find the base and height of the original triangle.

    b. Find the area of the original triangle.

65. The area of a triangular garden is 25 ft². The base is twice the height. Find the base and the height of the triangle.

66. The height of a triangle is 1 in. more than twice the base. If the area is 18 in.², find the base and height of the triangle.

67. The sum of the squares of two consecutive positive integers is 41. Find the integers.

68. The sum of the squares of two consecutive, positive even integers is 164. Find the integers.

69. Justin must travel from Summersville to Clayton. He can drive 10 mi through the mountains at 40 mph. Or he can drive east and then north on superhighways at 60 mph. The alternative route forms a right angle as shown in the diagram. The eastern leg is 2 mi less than the northern leg.

   a. Find the total distance Justin would travel in going the alternative route. (See Example 5.)

   b. If Justin wants to minimize the time of the trip, which route should he take?

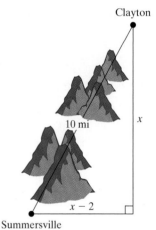

70. A 17-ft ladder is standing up against a wall. The distance between the base of the ladder and the wall is 7 ft less than the distance between the top of the ladder and the base of the wall. Find the distance between the base of the ladder and the wall.

71. A right triangle has side lengths represented by three consecutive even integers. Find the lengths of the three sides, measured in meters.

72. The hypotenuse of a right triangle is 3 m more than twice the short leg. The longer leg is 2 m more than twice the shorter leg. Find the lengths of the sides.

Objective 3: Definition of a Quadratic Function

For Exercises 73–76,

   a. Find the values of x for which $f(x) = 0$.   b. Find $f(0)$. (See Example 6.)

73. $f(x) = x^2 - 3x$ 74. $f(x) = 4x^2 + 2x$ 75. $f(x) = 5(x - 7)$ 76. $f(x) = 4(x + 5)$

For Exercises 77–80, find the x- and y-intercepts for the functions defined by $y = f(x)$.

77. $f(x) = \frac{1}{2}(x - 2)(x + 1)(2x)$ 78. $f(x) = (x + 1)(x - 2)(x + 3)^2$

79. $f(x) = x^2 - 2x + 1$ 80. $f(x) = x^2 + 4x + 4$

For Exercises 81–84, find the x-intercepts of each function and use that information to match the function with its graph.

81. $g(x) = (x + 3)(x - 3)$

82. $h(x) = x(x - 2)(x + 4)$

83. $f(x) = 4(x + 1)$

84. $k(x) = (x + 1)(x + 3)(x - 2)(x - 1)$

a.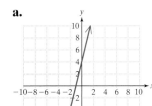

b.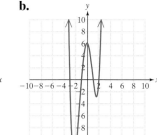

c.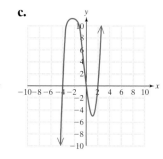

d.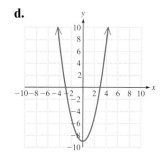

Objective 4: Applications of Quadratic Functions

85. A rocket is fired upward from ground level with an initial velocity of 490 m/sec. The height of the rocket $s(t)$ in meters is a function of the time t in seconds after launch. (See Example 7.)

$$s(t) = -4.9t^2 + 490t$$

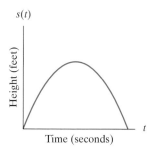

a. What characteristics of s indicate that it is a quadratic function?

b. Find the t-intercepts of the function.

c. What do the t-intercepts mean in the context of this problem?

d. At what times is the rocket at a height of 485.1 m?

86. A certain company makes water purification systems. The factory can produce x water systems per year. The profit $P(x)$ the company makes is a function of the number of systems x it produces.

$$P(x) = -2x^2 + 1000x$$

a. Is this function linear or quadratic?

b. Find the number of water systems x that would produce a zero profit.

c. What points on the graph do the answers in part (b) represent?

d. Find the number of systems for which the profit is $80,000.

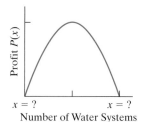

For Exercises 87–90, factor the functions represented by $f(x)$. Explain how the factored form relates to the graph of the function. Can the graph of the function help you determine the factors of the function?

87. $f(x) = x^2 - 7x + 10$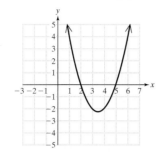

88. $f(x) = x^2 - 2x - 3$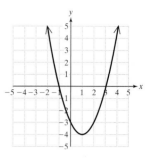

89. $f(x) = x^2 + 2x + 1$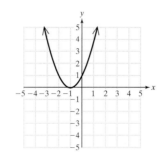

90. $f(x) = x^2 - 8x + 16$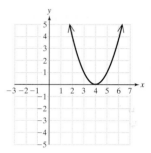

Expanding Your Skills

For Exercises 91–94, find an equation that has the given solutions. For example, 2 and -1 are solutions to $(x - 2)(x + 1) = 0$ or $x^2 - x - 2 = 0$. In general, x_1 and x_2 are solutions to the equation $a(x - x_1)(x - x_2) = 0$, where a can be any nonzero real number. For each problem, there is more than one correct answer depending on your choice of a.

91. $x = -3$ and $x = 1$

92. $x = 2$ and $x = -2$

93. $x = 0$ and $x = -5$

94. $x = 0$ and $x = -3$

Graphing Calculator Exercises

For Exercises 95–98, graph Y_1. Use the *Zoom* and *Trace* features to approximate the x-intercepts. Then solve $Y_1 = 0$ and compare the solutions to the x-intercepts.

95. $Y_1 = -x^2 + x + 2$

96. $Y_1 = -x^2 - x + 20$

97. $Y_1 = x^2 - 6x + 9$

98. $Y_1 = x^2 + 4x + 4$

chapter 5 | summary

section 5.1 Addition and Subtraction of Polynomials and Polynomial Functions

Key Concepts

A **polynomial** in x is defined by a finite sum of terms of the form ax^n, where a is a real number and n is a whole number.

- a is the **coefficient** of the term.
- n is the **degree of the term**.

The **degree of a polynomial** is the largest degree of its terms.

The term of a polynomial with the largest degree is the **leading term**. Its coefficient is the **leading coefficient**.
A one-term polynomial is a **monomial**.
A two-term polynomial is a **binomial**.
A three-term polynomial is a **trinomial**.

To add or subtract polynomials, add or subtract *like* terms.

Examples

Example 1

$7y^4 - 2y^2 + 3y + 8$

is a polynomial with leading coefficient 7 and degree 4.

Example 2

$f(x) = 4x^3 - 6x - 11$

f is a polynomial function with leading term $4x^3$ and leading coefficient 4. The degree of f is 3.

Example 3

For $f(x) = 4x^3 - 6x - 11$, find $f(-1)$.
$$f(-1) = 4(-1)^3 - 6(-1) - 11$$
$$= -9$$

Example 4

$(-4x^3y + 3x^2y^2) - (7x^3y - 5x^2y^2)$
$= -4x^3y + 3x^2y^2 - 7x^3y + 5x^2y^2$
$= -11x^3y + 8x^2y^2$

section 5.2 Multiplication of Polynomials

Key Concepts

To multiply polynomials, multiply each term in the first polynomial by each term in the second polynomial.

Special Products

1. Multiplication of **conjugates**

   $(x + y)(x - y) = x^2 - y^2$

   The product is called a **difference of squares**.

2. Square of a binomial

   $(x + y)^2 = x^2 + 2xy + y^2$
   and $(x - y)^2 = x^2 - 2xy + y^2$

   The product is called a **perfect square trinomial**.

Examples

Example 1

$(x - 2)(3x^2 - 4x + 11)$
$= 3x^3 - 4x^2 + 11x - 6x^2 + 8x - 22$
$= 3x^3 - 10x^2 + 19x - 22$

Example 2

$(3x + 5)(3x - 5)$
$= (3x)^2 - (5)^2$
$= 9x^2 - 25$

Example 3

$(4y + 3)^2$
$= (4y)^2 + (2)(4y)(3) + (3)^2$
$= 16y^2 + 24y + 9$

section 5.3 Division of Polynomials

Key Concepts

Division of polynomials:

1. For division by a monomial, use the properties

$$\frac{a+b}{c} = \frac{a}{c} + \frac{b}{c} \quad \text{and} \quad \frac{a-b}{c} = \frac{a}{c} - \frac{b}{c}$$

for $c \neq 0$.

2. If the divisor has more than one term, use long division.

3. **Synthetic division** may be used to divide a polynomial by a binomial in the form $x - r$, where r is a constant.

Examples

Example 1

$$\frac{-12a^2 - 6a + 9}{-3a}$$

$$= \frac{-12a^2}{-3a} - \frac{6a}{-3a} + \frac{9}{-3a}$$

$$= 4a + 2 - \frac{3}{a}$$

Example 2

$(3x^2 - 5x + 1) \div (x + 2)$

$$\begin{array}{r}
3x - 11 \\
x + 2 \overline{\smash{\big)}\, 3x^2 - 5x + 1} \\
\underline{-(3x^2 + 6x)} \\
-11x + 1 \\
\underline{-(-11x - 22)} \\
23
\end{array}$$

Answer: $3x - 11 + \dfrac{23}{x + 2}$

Example 3

$(3x^2 - 5x + 1) \div (x + 2)$

$$\begin{array}{r|rrr}
-2 & 3 & -5 & 1 \\
 & & -6 & 22 \\
\hline
 & 3 & -11 & \underline{|23}
\end{array}$$

Answer: $3x - 11 + \dfrac{23}{x + 2}$

section 5.4 Greatest Common Factor and Factoring by Grouping

Key Concepts

The **greatest common factor (GCF)** is the largest factor common to all terms of a polynomial. To factor out the GCF from a polynomial, use the distributive property.

A four-term polynomial may be **factored by grouping**.

Steps to Factor by Grouping

1. Identify and factor out the GCF from all four terms.
2. Factor out the GCF from the first pair of terms. Factor out the GCF from the second pair of terms. (Sometimes it is necessary to factor out the *opposite* of the GCF.)
3. If the two pairs of terms share a common binomial factor, factor out the binomial factor.

Examples

Example 1

$3x^2(a + b) - 6x(a + b)$

$= 3x(a + b)x - 3x(a + b)(2)$

$= 3x(a + b)(x - 2)$

Example 2

$60xa - 30xb - 80ya + 40yb$

$= 10[6xa - 3xb - 8ya + 4yb]$

$= 10[3x(2a - b) - 4y(2a - b)]$

$= 10(2a - b)(3x - 4y)$

section 5.5 Factoring Trinomials

Key Concepts

Grouping Method

To factor trinomials of the form $ax^2 + bx + c$:

1. Factor out the GCF.
2. Find the product ac.
3. Find two integers whose product is ac and whose sum is b. (If no pair of numbers can be found, then the trinomial is prime.)
4. Rewrite the middle term bx as the sum of two terms whose coefficients are the numbers found in step 3.
5. Factor the polynomial by grouping.

Trial-and-Error Method

To factor trinomials in the form $ax^2 + bx + c$:

1. Factor out the GCF.
2. List the pairs of factors of a and the pairs of factors of c. Consider the reverse order in either list.
3. Construct two binomials of the form

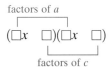

4. Test each combination of factors until the product of the outer terms and the product of inner terms add to the middle term.
5. If no combination of factors works, the polynomial is prime.

The factored form of a **perfect square trinomial** is the square of a binomial:

$a^2 + 2ab + b^2 = (a + b)^2$
$a^2 - 2ab + b^2 = (a - b)^2$

Examples

Example 1

$10y^2 + 35y - 20 = 5(2y^2 + 7y - 4)$

$$ac = (2)(-4) = -8$$

Find two integers whose product is -8 and whose sum is 7. The numbers are 8 and -1.

$5[2y^2 + 8y - 1y - 4]$
$= 5[2y(y + 4) - 1(y + 4)]$
$= 5(y + 4)(2y - 1)$

Example 2

$10y^2 + 35y - 20 = 5(2y^2 + 7y - 4)$

The pairs of factors of 2 are $2 \cdot 1$.
The pairs of factors of -4 are

$$-1 \cdot 4 \qquad 1 \cdot (-4)$$
$$-2 \cdot 2 \qquad 2 \cdot (-2)$$
$$-4 \cdot 1 \qquad 4 \cdot (-1)$$

$(2y - 2)(y + 2) = 2y^2 + 2y - 4$ No
$(2y - 4)(y + 1) = 2y^2 - 2y - 4$ No
$(2y + 1)(y - 4) = 2y^2 - 7y - 4$ No
$(2y + 2)(y - 2) = 2y^2 - 2y - 4$ No
$(2y + 4)(y - 1) = 2y^2 + 2y - 4$ No
$(2y - 1)(y + 4) = 2y^2 + 7y - 4$ Yes

Therefore, $10y^2 + 35y - 20$ factors as $5(2y - 1)(y + 4)$.

Example 3

$9w^2 - 30wz + 25z^2$
$= (3w)^2 - 2(3w)(5z) + (5z)^2$
$= (3w - 5z)^2$

section 5.6 Factoring Binomials

Key Concepts

Factoring Binomials: Summary

Difference of squares:
$a^2 - b^2 = (a + b)(a - b)$

Difference of cubes:
$a^3 - b^3 = (a - b)(a^2 + ab + b^2)$

Sum of cubes:
$a^3 + b^3 = (a + b)(a^2 - ab + b^2)$

Examples

Example 1
$25u^2 - 9v^4 = (5u + 3v^2)(5u - 3v^2)$

Example 2
$8c^3 - d^6 = (2c - d^2)(4c^2 + 2cd^2 + d^4)$

Example 3
$27w^9 + 64x^3$
$= (3w^3 + 4x)(9w^6 - 12w^3x + 16x^2)$

section 5.7 Solving Equations by Using the Zero Product Rule

Key Concepts

An equation of the form $ax^2 + bx + c = 0$, where $a \neq 0$, is a **quadratic equation**.

The **zero product rule** states that if $a \cdot b = 0$, then $a = 0$ or $b = 0$. The zero product rule can be used to solve a quadratic equation or higher-degree polynomial equation that is factored and equal to zero.

$f(x) = ax^2 + bx + c\ (a \neq 0)$ defines a **quadratic function**. The x-intercepts of a function defined by $y = f(x)$ are determined by finding the real solutions to the equation $f(x) = 0$. The y-intercept of a function $y = f(x)$ is at $f(0)$.

Examples

Example 1
$0 = x(2x - 3)(x + 4)$
$x = 0$ or $2x - 3 = 0$ or $x + 4 = 0$
$x = \dfrac{3}{2}$ or $x = -4$

Example 2
Find the x-intercepts.
$f(x) = 3x^2 - 8x + 5$
$0 = 3x^2 - 8x + 5$
$0 = (3x - 5)(x - 1)$
$3x - 5 = 0$ or $x - 1 = 0$
$x = \dfrac{5}{3}$ or $x = 1$

The x-intercepts are $\left(\dfrac{5}{3}, 0\right)$ and $(1, 0)$.

Find the y-intercept.
$f(x) = 3x^2 - 8x + 5$
$f(0) = 3(0)^2 - 8(0) + 5$
$f(0) = 5$
The y-intercept is $(0, 5)$.

chapter 5 review exercises

Section 5.1

For Exercises 1–2, identify the polynomial as a monomial, binomial, or trinomial; then give the degree of the polynomial.

1. $6x^4 + 10x - 1$
2. 18

3. Given the polynomial function defined by $g(x) = 4x - 7$, find the function values.
   a. $g(0)$
   b. $g(-4)$
   c. $g(3)$

4. Given the polynomial function defined by $p(x) = -x^4 - x + 12$, find the function values.
   a. $p(0)$
   b. $p(1)$
   c. $p(-2)$

5. The number of new sites established by Starbucks in the years from 1990 to 2006 can be approximated by the function $S(x) = 4.567x^2 + 40.43x - 40.13$, where $x = 0$ represents the year 1990.
   a. Evaluate $S(5)$ and $S(13)$ to the nearest whole unit. Match the function values with points on the graph (see the figure).
   b. Interpret the meaning of the function value for $S(13)$.

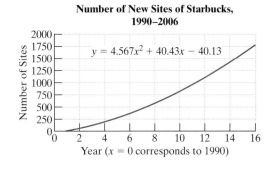

Number of New Sites of Starbucks, 1990–2006
$y = 4.567x^2 + 40.43x - 40.13$

For Exercises 6–13, add or subtract the polynomials as indicated.

6. $(x^2 - 2x - 3xy - 7) + (-3x^2 - x + 2xy + 6)$

7. $(7xy - 3xz + 5yz) + (13xy - 15xz - 8yz)$

8. $(8a^2 - 4a^3 - 3a) - (3a^2 - 9a - 7a^3)$

9. $(3a^2 - 2a - a^3) - (5a^2 - a^3 - 8a)$

10. $\left(\dfrac{5}{8}x^4 - \dfrac{1}{4}x^2 - \dfrac{1}{2}\right) - \left(-\dfrac{3}{8}x^4 + \dfrac{3}{4}x^2 + \dfrac{1}{2}\right)$

11. $\left(\dfrac{5}{6}x^4 + \dfrac{1}{2}x^2 - \dfrac{1}{3}\right) - \left(-\dfrac{1}{6}x^4 - \dfrac{1}{4}x^2 - \dfrac{1}{3}\right)$

12. $(7x - y) - [-(2x + y) - (-3x - 6y)]$

13. $-(4x - 4y) - [(4x + 2y) - (3x + 7y)]$

14. Add $-4x + 6$ to $-7x - 5$.

15. Add $2x^2 - 4x$ to $2x^2 - 7x$.

16. Subtract $-4x + 6$ from $-7x - 5$.

17. Subtract $2x^2 - 4x$ from $2x^2 - 7x$.

Section 5.2

For Exercises 18–35, multiply the polynomials.

18. $2x(x^2 - 7x - 4)$
19. $-3x(6x^2 - 5x + 4)$
20. $(x + 6)(x - 7)$
21. $(x - 2)(x - 9)$
22. $\left(\dfrac{1}{2}x + 1\right)\left(\dfrac{1}{2}x - 5\right)$
23. $\left(-\dfrac{1}{5} + 2y\right)\left(\dfrac{1}{5} + y\right)$
24. $(3x + 5)(9x^2 - 15x + 25)$
25. $(x - y)(x^2 + xy + y^2)$
26. $(2x - 5)^2$
27. $\left(\dfrac{1}{2}x + 4\right)^2$
28. $(3y - 11)(3y + 11)$
29. $(6w - 1)(6w + 1)$
30. $\left(\dfrac{2}{3}t + 4\right)\left(\dfrac{2}{3}t - 4\right)$
31. $\left(z + \dfrac{1}{4}\right)\left(z - \dfrac{1}{4}\right)$
32. $[(x + 2) - b][(x + 2) + b]$

33. $[c-(w+3)][c+(w+3)]$

34. $(2x+1)^3$ 35. $(y^2-3)^3$

36. A square garden is surrounded by a walkway of uniform width x. If the sides of the garden are given by the expression $2x+3$, find and simplify a polynomial that represents

   a. The area of the garden.

   b. The area of the walkway and garden.

   c. The area of the walkway only.

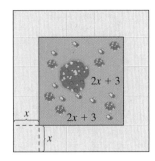

37. The length of a rectangle is 2 ft more than 3 times the width. Let x represent the width of the rectangle.

   a. Write a function P that represents the perimeter of the rectangle.

   b. Write a function A that represents the area of the rectangle.

38. In parts (a) and (b), one of the statements is true and the other is false. Identify the true statement and explain why the false statement is incorrect.

   a. $2x^2 + 5x = 7x^3$ $(2x^2)(5x) = 10x^3$

   b. $4x - 7x = -3x$ $4x - 7x = -3$

Section 5.3

For Exercises 39–40, divide the polynomials.

39. $(6x^3 + 12x^2 - 9x) \div (3x)$

40. $(10x^4 + 15x^3 - 20x^2) \div (-5x^2)$

41. a. Divide $(9y^4 + 14y^2 - 8) \div (3y+2)$.

   b. Identify the quotient and the remainder.

   c. Explain how you can check your answer.

For Exercises 42–45, divide the polynomials by using long division.

42. $(x^2 + 7x + 10) \div (x+5)$

43. $(x^2 + 8x - 16) \div (x+4)$

44. $(2x^5 - 4x^4 + 2x^3 - 4) \div (x^2 - 3x)$

45. $(2x^5 + 3x^3 + x^2 - 4) \div (x^2 + x)$

46. Explain the conditions under which you may use synthetic division.

47. The following table is the result of a synthetic division.

3⌋	2	5	−2	6	1
		6	33	93	297
	2	11	31	99	⌊298

   Use x as the variable.

   a. Identify the divisor.

   b. Identify the quotient.

   c. Identify the remainder.

For Exercises 48–52, divide the polynomials by using synthetic division.

48. $(t^3 - 3t^2 + 8t - 12) \div (t-2)$

49. $(x^2 + 7x + 14) \div (x+5)$

50. $(x^2 + 8x + 20) \div (x+4)$

51. $(w^3 - 6w^2 + 8) \div (w-3)$

52. $(p^4 - 16) \div (p-2)$

Section 5.4

For Exercises 53–57, factor by removing the greatest common factor.

53. $-x^3 - 4x^2 + 11x$

54. $21w^3 - 7w + 14$

55. $5x(x - 7) - 2(x - 7)$

56. $3t(t + 4) + 5(t + 4)$

57. $2x^2 - 26x$

For Exercises 58–61, factor by grouping (remember to take out the GCF first).

58. $m^3 - 8m^2 + m - 8$

59. $24x^3 - 36x^2 + 72x - 108$

60. $4ax^2 + 2bx^2 - 6ax - 3xb$

61. $y^3 - 6y^2 + y - 6$

Section 5.5

62. What characteristics determine a perfect square trinomial?

For Exercises 63–72, factor the polynomials by using any method.

63. $18x^2 + 27xy + 10y^2$
64. $2 + 7k + 6k^2$

65. $60a^2 + 65a^3 - 20a^4$
66. $8b^2 - 40b + 50$

67. $n^2 + 10n + 25$
68. $2x^2 + 5x + 12$

69. $y^3 - y(10 - 3y)$
70. $m + 18 - m(m - 2)$

71. $9x^2 - 12x + 4$
72. $25q^2 + 30q + 9$

Section 5.6

For Exercises 73–79, factor the binomials.

73. $25 - y^2$
74. $x^3 - \dfrac{1}{27}$

75. $b^2 + 64$
76. $a^3 + 64$

77. $h^3 + 9h$
78. $k^4 - 16$

79. $9y^3 - 4y$

For Exercises 80–81, factor by grouping and by using the difference of squares.

80. $x^2 - 8xy + 16y^2 - 9$ (*Hint*: Group three terms that constitute a perfect square trinomial, then factor as a difference of squares.)

81. $a^2 + 12a + 36 - b^2$

Section 5.7

82. How do you determine if an equation is quadratic?

83. What shape is the graph of a quadratic function?

For Exercises 84–87, label the equation as quadratic or linear.

84. $x^2 + 6x = 7$
85. $(x - 3)(x + 4) = 9$

86. $2x - 5 = 3$
87. $x + 3 = 5x^2$

For Exercises 88–91, use the zero product rule to solve the equations.

88. $x^2 - 2x - 15 = 0$

89. $8x^2 = 59x - 21$

90. $2t(t + 5) + 1 = 3t - 3 - t^2$

91. $3(x - 1)(x + 5)(2x - 9) = 0$

For Exercises 92–95, find the *x*- and *y*-intercepts of the function. Then match the function with its graph.

92. $f(x) = -4x^2 + 4$

93. $g(x) = 2x^2 - 2$

94. $h(x) = 5x^3 - 10x^2 - 20x + 40$

95. $k(x) = -\frac{1}{8}x^2 + \frac{1}{2}$

a.
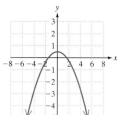

b.
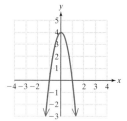

c.
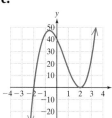

d.
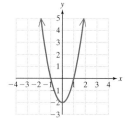

96. A moving van has the capacity to hold 1200 ft³ in volume. If the van is 10 ft high and the length is 1 ft less than twice the width, find the dimensions of the van.

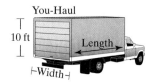

97. A missile is shot upward from a submarine 1280 ft below sea level. The initial velocity of the missile is 672 ft/sec. A function that approximates the height of the missile (relative to sea level) is given by

$$h(t) = -16t^2 + 672t - 1280$$

where $h(t)$ is the height in feet and t is the time in seconds.

a. Complete the table to determine the height of the missile for the given values of t.

Time t (sec)	Height $h(t)$ (ft)
0	
1	
3	
10	
20	
30	
40	
42	

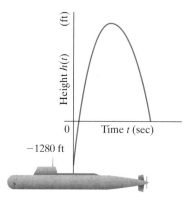

b. Interpret the meaning of a negative value of $h(t)$.

c. Factor the function to find the time required for the missile to emerge from the water and the time required for the missile to reenter the water. (*Hint*: The height of the missile will be zero at sea level.)

chapter 5 | test

1. For the function defined by $F(x) = 5x^3 - 2x^2 + 8$, find the function values $F(-1)$, $F(2)$, and $F(0)$.

2. The number of serious violent crimes in the United States for the years 1990–2003 can be approximated by the function $C(x) = -0.0145x^2 + 3.8744$, where $x = 0$ corresponds to the year 1990 and $C(x)$ is in millions.

   a. Evaluate $C(2)$, $C(6)$, and $C(12)$. Match the function values with points on the graph (see the figure).

   b. Interpret the meaning of the function value for $C(12)$.

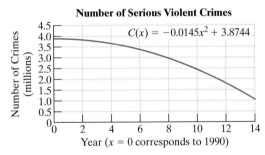

Source: Bureau of Justice Statistics.

3. Perform the indicated operations. Write the answer in descending order.
$$(5x^2 - 7x + 3) - (x^2 + 5x - 25) + (4x^2 + 4x - 20)$$

For Exercises 4–6, multiply the polynomials. Write the answer in descending order.

4. $(2a - 5)(a^2 - 4a - 9)$

5. $\left(\dfrac{1}{3}x - \dfrac{3}{2}\right)(6x + 4)$

6. $(5x - 4y^2)(5x + 4y^2)$

7. Explain why $(5x + 7)^2 \neq 25x^2 + 49$.

8. Write and simplify an expression that describes the area of the square.

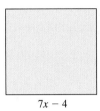

$7x - 4$

9. Divide the polynomials.
$$(2x^3y^4 + 5x^2y^2 - 6xy^3 - xy) \div (2xy)$$

10. Divide the polynomials.
$$(10p^3 + 13p^2 - p + 3) \div (2p + 3)$$

11. Divide the polynomials by using synthetic division. $(y^4 - 2y + 5) \div (y - 2)$

12. Explain the strategy for factoring a polynomial expression.

13. Explain the process to solve a polynomial equation by the zero product rule.

14. Solve $4x - 64x^3 = 0$.

15. Factor $3y^2 + 23y - 8$.

16. Factor $a^3 - 6a^2 - a + 6$.

17. Solve $x^2 + \dfrac{1}{2}x + \dfrac{1}{16} = 0$.

18. Factor $3x^3 + 24$.

19. Factor $16x^4 - 16$.

20. Factor $49x^2 - 70xy + 25y^2$.

For Exercises 21–24, find the x- and y-intercepts of the function. Then match the function with its graph.

21. $f(x) = x^2 - 6x + 8$

22. $k(x) = x^3 + 4x^2 - 9x - 36$

23. $p(x) = -2x^2 - 8x - 6$

24. $q(x) = x^3 - x^2 - 12x$

a.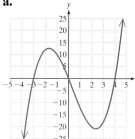

b.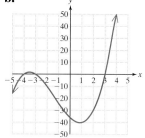

c.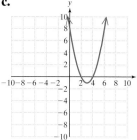

d.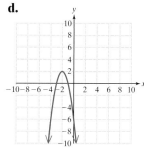

25. A child launches a toy rocket from the ground. The height of the rocket h can be determined by its horizontal distance from the launch pad x by

$$h(x) = -\frac{x^2}{256} + x$$

where x and h are in feet and $x \geq 0$ and $h \geq 0$.

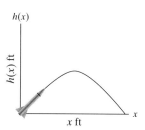

How many feet from the launch pad will the rocket hit the ground?

26. The recent population, P (in millions) of Japan can be approximated by:

$$P(t) = -0.01t^2 - 0.062t + 127.7,$$

where $t = 0$ represents the year 2000.

a. Approximate the number of people in Japan in the year 2006.

b. If the trend continues, predict the population of Japan in the year 2015.

---

chapters 1–5 | cumulative review exercises

1. Graph the inequality and express the set in interval notation: All real numbers at least 5, but not more than 12

2. Simplify the expression $3x^2 - 5x + 2 - 4(x^2 + 3)$.

3. Graph from memory.

a. $y = x^2$

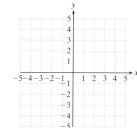

b. $y = |x|$

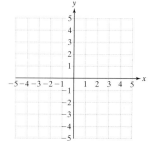

4. Simplify the expression $\left(\frac{1}{3}\right)^{-2} - \left(\frac{1}{2}\right)^{3}$.

5. In 1998, the population of Mexico was approximately 9.85×10^{7}. At the current growth rate of 1.7%, this number is expected to double after 42 years. How many people does this represent? Express your answer in scientific notation.

6. In the 1990 Orange Bowl football championship, Florida State scored 5 points more than Notre Dame. The total number of points scored was 57. Find the number of points scored by each team.

7. Find the value of each angle in the triangle.

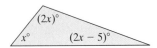

8. Divide $(x^{3} + 64) \div (x + 4)$.

9. Determine the slope and y-intercept of the line $4x - 3y = -9$, and graph the line.

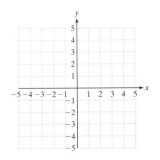

10. If y varies directly with x and inversely with z, and $y = 6$ when $x = 9$ and $z = \frac{1}{2}$, find y when $x = 3$ and $z = 4$.

11. Simplify the expression.

$$\left(\frac{36a^{-2}b^{4}}{18b^{-6}}\right)^{-3}$$

12. Solve the system.

$$2x - y + 2z = 1$$
$$-3x + 5y - 2z = 11$$
$$x + y - 2z = -1$$

13. Determine whether the relation is a function.

a. $\{(2, 1), (3, 1), (-8, 1), (5, 1)\}$

b.
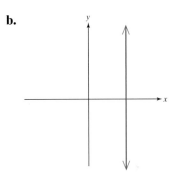

14. A telephone pole is leaning after a storm (see figure). What is the slope of the pole?

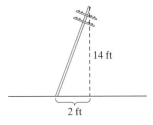

15. Given $P(x) = \frac{1}{6}x^{2} + x - 5$, find the function value $P(6)$.

16. Solve for x: $\frac{1}{3}x - \frac{1}{6} = \frac{1}{2}(x - 3)$.

17. Given $3x - 2y = 5$, solve for y.

18. A student scores 76, 85, and 92 on her first three algebra tests.

   a. Is it possible for her to score high enough on for the fourth test to bring her test average up to 90? Assume that each test is weighted equally and that the maximum score on a test is 100 points.

   b. What is the range of values required for the fourth test so that the student's test average will be between 80 and 89, inclusive?

19. How many liters of a 40% acid solution and how many liters of a 15% acid solution must be mixed to obtain 25 L of a 30% acid solution?

20. Multiply the polynomials $(4b - 3)(2b^2 + 1)$.

21. Add the polynomials.
$$(5a^2 + 3a - 1) + (3a^3 - 5a + 6)$$

22. Divide the polynomials $(6w^3 - 5w^2 - 2w) \div (2w^2)$

For Exercises 23–25, solve the equations.

23. $y^2 - 5y = 14$ 24. $25x^2 = 36$

25. $a^3 + 9a^2 + 20a = 0$

Rational Expressions and Rational Equations

6

- **6.1** Rational Expressions and Rational Functions
- **6.2** Multiplication and Division of Rational Expressions
- **6.3** Addition and Subtraction of Rational Expressions
- **6.4** Complex Fractions
- **6.5** Rational Equations
- **6.6** Applications of Rational Equations and Proportions

In this chapter, we define a rational expression as a ratio of two polynomials. First we focus on adding, subtracting, multiplying, and dividing rational expressions. The chapter concludes with solving rational equations and showing how they are used in applications. In Section 6.6, we solve problems involving rates. For example, in Exercise 51 of Section 6.6, two carpenters working at different rates are scheduled to build a kitchen. The construction foreman can set up a rational equation to determine how long it will take to get the job done with both carpenters working on the job.

chapter 6 preview

The exercises in this chapter preview contain concepts that have not yet been presented. These exercises are provided for students who want to compare their levels of understanding before and after studying the chapter. Alternatively, you may prefer to work these exercises when the chapter is completed and before taking the exam.

Section 6.1

1. Evaluate the rational expression for $x = -2$:
$$\frac{3x - 3}{x + 5}$$

2. Simplify the rational expression to lowest terms:
$$\frac{2y^2 + y - 3}{y^2 - 1}$$

3. Determine the domain of the rational expression $\frac{a}{a^2 + a - 6}$. Write the answer in set-builder notation.

4. Determine the domain of the rational function $g(x) = \frac{x + 2}{6x}$. Write the answer in interval notation.

Section 6.2

For Exercises 5–7, multiply or divide as indicated.

5. $\dfrac{6a^3b}{c^2} \cdot \dfrac{3c^3}{2ab^3}$

6. $\dfrac{t^2 + 12t + 32}{t^2 - 16} \div \dfrac{t^2 + 6t - 16}{t^2 - 2t - 8}$

7. $\dfrac{12 - 4x}{2x + 4} \cdot \dfrac{2x^2 + 9x + 10}{x^2 - 2x - 3}$

Section 6.3

For Exercises 8–12, add or subtract as indicated.

8. $\dfrac{7a}{a + 4} + \dfrac{2a}{a + 4}$

9. $\dfrac{3t - 4}{10t} - \dfrac{t - 8}{10t}$

10. $\dfrac{1}{6x^2y} - \dfrac{2}{9xy^3}$

11. $\dfrac{4}{z^2 - 3z} + \dfrac{3}{z^2 - 7z + 12}$

12. $\dfrac{q}{q^2 + 2q - 24} - \dfrac{2q}{2q^2 - 7q - 4}$

Section 6.4

For Exercises 13–14, simplify the expressions.

13. $\dfrac{\frac{1}{c} + \frac{1}{d}}{\frac{2}{c} - \frac{3}{d}}$

14. $\dfrac{6 + \frac{1}{x + 1}}{\frac{3}{x + 1} - 2}$

Section 6.5

For Exercises 15–17, solve the equations.

15. $\dfrac{5}{t} + \dfrac{1}{3} = \dfrac{5}{2}$

16. $\dfrac{12}{x^2 - 9} - \dfrac{3}{x + 3} = \dfrac{2}{x - 3}$

17. $\dfrac{a}{2} - \dfrac{3}{a} = \dfrac{1}{2}$

18. Solve the equation for y. $\dfrac{a}{x} + \dfrac{b}{y} = 1$

Section 6.6

19. Solve the proportion $\dfrac{7}{4} = \dfrac{105}{x}$.

20. Karen can wax her SUV in 2 hr. Clarann can wax the same SUV in 3 hr. If they work together, how long will it take them to wax the SUV?

21. Kathy can run 3 mi to the beach in the same time Dennis can ride his bike 7 mi to work. Kathy runs 8 mph slower than Dennis rides his bike. Find their speeds.

section 6.1 Rational Expressions and Rational Functions

1. Evaluating Rational Expressions

In Chapter 5, we introduced polynomials and polynomial functions. In this chapter, polynomials will be used to define rational expressions and rational functions.

Objectives
1. Evaluating Rational Expressions
2. Domain of a Rational Expression
3. Simplifying Rational Expressions to Lowest Terms
4. Simplifying Ratios of -1
5. Rational Functions

> **Definition of a Rational Expression**
>
> An expression is a **rational expression** if it can be written in the form $\frac{p}{q}$, where p and q are polynomials and $q \neq 0$.

A rational expression is the quotient of two polynomials. The following expressions are examples of rational expressions.

$$\frac{5}{x}, \quad \frac{2y+1}{y^2-4y+3}, \quad \frac{c^3+5}{c-4}, \quad \frac{a}{a+10}$$

example 1   Evaluating a Rational Expression

Evaluate the expression for the given values of x.

$$\frac{x-2}{x+3}$$

a. 1   b. 2   c. 0   d. -1 e. -3

Solution:

a. $\dfrac{(1)-2}{(1)+3} = \dfrac{-1}{4}$

b. $\dfrac{(2)-2}{(2)+3} = \dfrac{0}{5} = 0$

c. $\dfrac{(0)-2}{(0)+3} = \dfrac{-2}{3}$

d. $\dfrac{(-1)-2}{(-1)+3} = \dfrac{-3}{2}$

e. $\dfrac{(-3)-2}{(-3)+3} = \dfrac{-5}{0}$ (undefined)

Skill Practice

Evaluate the expression for the given values of y.

$$\frac{3y}{y-4}$$

1. $y = -1$ 2. $y = 10$
3. $y = 0$ 4. $y = 4$

2. Domain of a Rational Expression

The domain of an expression is the set of values that when substituted into the expression produces a real number. Therefore, the domain of a rational expression must exclude the values that make the denominator zero. For example:

Given $\frac{2}{x-3}$, the domain is the set of all real numbers, x, excluding $x = 3$. This can be written in set-builder notation as:

$$\{x \mid x \text{ is a real number and } x \neq 3\}$$

> **Steps to Find the Domain of a Rational Expression**
>
> 1. Set the denominator equal to zero and solve the resulting equation.
> 2. The domain is the set of all real numbers *excluding* the values found in step 1.

Answers

1. $\dfrac{3}{5}$ 2. 5
3. 0   4. Undefined

Skill Practice

Find the domains of the expressions.

5. $\dfrac{-6}{v-9}$

6. $\dfrac{y+7}{2y}$

7. $\dfrac{x+10}{2x^2-x-1}$

8. $\dfrac{t}{t^2+1}$

example 2   Finding the Domain of a Rational Expression

Write the domain in set-builder notation.

a. $\dfrac{5}{4t-1}$ **b.** $\dfrac{x+4}{2x^2-11x+5}$ **c.** $\dfrac{x}{x^2+4}$

Solution:

a. $\dfrac{5}{4t-1}$ The expression is undefined when the denominator is zero.

$4t - 1 = 0$ Set the denominator equal to zero and solve for t.

$4t = 1$

$t = \dfrac{1}{4}$ The value $t = \dfrac{1}{4}$ must be excluded from the domain.

Domain: $\{t \mid t \text{ is a real number and } t \neq \tfrac{1}{4}\}$

b. $\dfrac{x+4}{2x^2-11x+5}$

$2x^2 - 11x + 5 = 0$ Set the denominator equal to zero and solve for x.

$(2x - 1)(x - 5) = 0$ This is a factorable quadratic equation.

$2x - 1 = 0$ or $x - 5 = 0$

$x = \dfrac{1}{2}$ or $x = 5$

Domain: $\{x \mid x \text{ is a real number and } x \neq \tfrac{1}{2}, x \neq 5\}$

Tip: The domain of a rational expression excludes values for which the denominator is zero. The values for which the numerator is zero *do not affect* the domain of the expression.

c. $\dfrac{x}{x^2+4}$

Because the quantity x^2 is nonnegative for any real number x, the denominator $x^2 + 4$ cannot equal zero; therefore, no real numbers are excluded from the domain.

Domain: $\{x \mid x \text{ is a real number}\}$

3. Simplifying Rational Expressions to Lowest Terms

A rational expression is an expression in the form $\tfrac{p}{q}$, where p and q are polynomials and $q \neq 0$. As with fractions, it is often advantageous to simplify rational expressions to lowest terms.

The method for simplifying rational expressions to lowest terms mirrors the process to simplify fractions. In each case, factor the numerator and denominator. Common factors in the numerator and denominator form a ratio of 1 and can be simplified.

Answers

5. $\{v \mid v \text{ is a real number and } v \neq 9\}$
6. $\{y \mid y \text{ is a real number and } y \neq 0\}$
7. $\{x \mid x \text{ is a real number and } x \neq -\tfrac{1}{2}, x \neq 1\}$
8. All real numbers

Section 6.1   Rational Expressions and Rational Functions   401

Simplifying a fraction: $\dfrac{15}{35} \xrightarrow{\text{factor}} \dfrac{3 \cdot \cancel{5}}{7 \cdot \cancel{5}} = \dfrac{3}{7}(1) = \dfrac{3}{7}$

Simplifying a rational expression: $\dfrac{x^2 - x - 12}{x^2 - 16} \xrightarrow{\text{factor}} \dfrac{(x+3)(\cancel{x-4})}{(x+4)(\cancel{x-4})} = \dfrac{(x+3)}{(x+4)}(1) = \dfrac{x+3}{x+4}$

This process is stated formally as the fundamental principle of rational expressions.

Fundamental Principle of Rational Expressions

Let p, q, and r represent polynomials. Then

$$\dfrac{pr}{qr} = \dfrac{p}{q} \quad \text{for } q \neq 0 \text{ and } r \neq 0$$

example 3   Simplifying Rational Expressions to Lowest Terms

$$\dfrac{2x^3 + 12x^2 + 16x}{6x + 24}$$

a. Factor the numerator and denominator.
b. Determine the domain and write the domain in set-builder notation.
c. Simplify the expression.

Solution:

a. $\dfrac{2x^3 + 12x^2 + 16x}{6x + 24}$

$= \dfrac{2x(x^2 + 6x + 8)}{6(x + 4)}$ Factor the numerator and denominator.

$= \dfrac{2x(x + 2)(x + 4)}{6(x + 4)}$

b. To determine the domain, set the denominator equal to zero and solve for x.

$6(x + 4) = 0$

$x = -4$

The domain of the function is $\{x \mid x \text{ is a real number and } x \neq -4\}$.

c. $\dfrac{\cancel{2}x(\cancel{x+4})(x+2)}{\cancel{2} \cdot 3(\cancel{x+4})}$ Simplify the ratio of common factors to 1.

$= \dfrac{x(x + 2)}{3}$ provided $x \neq -4$

Skill Practice

Factor the numerator and the denominator to determine the domain. Then simplify the rational expression.

9. $\dfrac{a^2 + 3a - 28}{2a + 14}$

Avoiding Mistakes: The domain of a rational expression is always determined *before* simplifying the expression.

Answer

9. $\{a \mid a \text{ is a real number and } a \neq -7\}; \dfrac{a - 4}{2}$

It is important to note that the expressions

$$\frac{2x^3 + 12x^2 + 16x}{6x + 24} \quad \text{and} \quad \frac{x(x+2)}{3}$$

are equal for all values of x that make each expression a real number. Therefore,

$$\frac{2x^3 + 12x^2 + 16x}{6x + 24} = \frac{x(x+2)}{3}$$

for all values of x except $x = -4$. (At $x = -4$, the original expression is undefined.) This is why the *domain of a rational expression is always determined before the expression is simplified.*

The objective to simplifying a rational expression is to create an equivalent expression that is simpler to evaluate. Consider the expression from Example 3 in its original form and in its simplified form. If we substitute an arbitrary value of x into the expression (such as $x = 3$), we see that the simplified form is easier to evaluate.

Original Expression

$$\frac{2x^3 + 12x^2 + 16x}{6x + 24}$$

Substitute $x = 3$

$$\frac{2(3)^3 + 12(3)^2 + 16(3)}{6(3) + 24}$$

$$= \frac{2(27) + 12(9) + 48}{18 + 24}$$

$$= \frac{54 + 108 + 48}{42}$$

$$= \frac{210}{42}$$

$$= 5$$

Simplified Form

$$\frac{x(x+2)}{3}$$

$$\frac{(3)[(3) + 2]}{3}$$

$$= \frac{3(5)}{3}$$

$$= 5$$

Skill Practice

Simplify to lowest terms.

10. $\dfrac{8y^2 - 14y + 3}{2y^2 - y - 3}$

example 4   Simplifying a Rational Expression to Lowest Terms

Simplify to lowest terms. $\quad \dfrac{t^3 + 8}{t^2 + 6t + 8}$

Solution:

$$\frac{t^3 + 8}{t^2 + 6t + 8}$$

Tip: $t^3 + 8$ is a sum of cubes.

Recall: $\quad a^3 + b^3 = (a + b)(a^2 - ab + b^2)$

$t^3 + 8 = (t + 2)(t^2 - 2t + 4)$

$$= \frac{(t+2)(t^2 - 2t + 4)}{(t+2)(t+4)}$$

The restrictions on the domain are $t \neq -2$ and $t \neq -4$.

$$= \frac{\cancel{(t+2)}(t^2 - 2t + 4)}{\cancel{(t+2)}(t+4)}$$

Reduce common factors whose ratio is 1.

$$= \frac{t^2 - 2t + 4}{t + 4} \quad \text{provided } t \neq -2, t \neq -4$$

Answer

10. $\dfrac{4y - 1}{y + 1}$ provided $y \neq \frac{3}{2}, y \neq -1$

Avoiding Mistakes: The fundamental principle of rational expressions indicates that common *factors* in the numerator and denominator may be simplified.

$$\frac{pr}{qr} = \frac{p}{q} \cdot \frac{r}{r} = \frac{p}{q} \cdot (1) = \frac{p}{q}$$

Because this property is based on the identity property of multiplication, reducing applies only to factors (remember that factors are multiplied). Therefore, terms that are added or subtracted cannot be reduced. For example:

$$\frac{3x}{3y} = \frac{\overset{1}{\cancel{3}}x}{\underset{1}{\cancel{3}}y} = (1) \cdot \frac{x}{y} = \frac{x}{y} \qquad \text{However,} \quad \frac{x+3}{y+3} \text{ cannot be simplified.}$$

↑ Reduce common factor.　　　　　　　　　↑ cannot reduce common terms

---

example 5 Simplifying a Rational Expression to Lowest Terms

Simplify the rational expression. $\dfrac{2x^2y^5}{8x^4y^3}$

Skill Practice

Simplify the rational expression.

11. $\dfrac{9a^5b^3}{18a^8b}$

Solution:

$\dfrac{2x^2y^5}{8x^4y^3}$ 　　This expression has the restriction that $x \neq 0$ and $y \neq 0$.

$= \dfrac{2x^2y^5}{2^3x^4y^3}$ 　　Factor the denominator.

$= 2^{1-3}x^{2-4}y^{5-3}$ 　　This expression can be simplified by using the properties of exponents.

$= 2^{-2}x^{-2}y^2$

$= \dfrac{y^2}{4x^2}$ 　provided $x \neq 0$ and $y \neq 0$ 　　Remove negative exponents. Include the restrictions on x and y.

---

4. Simplifying Ratios of −1

When two factors are identical in the numerator and denominator, they form a ratio of 1 and can be simplified. Sometimes we encounter two factors that are *opposites* and form a ratio of −1. For example:

Simplified Form	Details/Notes
$\dfrac{-5}{5} = -1$	The ratio of a number and its opposite is −1.
$\dfrac{100}{-100} = -1$	The ratio of a number and its opposite is −1.

Answer

11. $\dfrac{b^2}{2a^3}$; $a \neq 0$ and $b \neq 0$

Concept Connections

12. Which of the following expressions simplifies to -1?

a. $\dfrac{x+y}{y+x}$ b. $\dfrac{x-y}{y-x}$

c. $\dfrac{x-y}{x+y}$

$$\dfrac{x+7}{-x-7} = -1 \qquad \dfrac{x+7}{-x-7} = \dfrac{x+7}{-1(x+7)} = \dfrac{\overset{1}{\cancel{x+7}}}{-1(\cancel{x+7})} = \dfrac{1}{-1} = -1$$

Factor out -1.

$$\dfrac{2-x}{x-2} = -1 \qquad \dfrac{2-x}{x-2} = \dfrac{-1(-2+x)}{x-2} = \dfrac{-1(\cancel{x-2})}{\cancel{x-2}} = \dfrac{-1}{1} = -1$$

Recognizing factors that are opposites is useful when simplifying rational expressions. For example, $a-b$ and $b-a$ are opposites because the opposite of $a-b$ can be written $-(a-b) = -a+b = b-a$. Therefore, in general, $\dfrac{a-b}{b-a} = -1$.

Skill Practice

Simplify the expressions.

13. $\dfrac{20-5x}{x^2-x-12}$

14. $\dfrac{b^2-a^2}{a^2+ab-2b^2}$

example 6 Simplifying Rational Expressions

Simplify the rational expressions to lowest terms.

a. $\dfrac{x-5}{25-x^2}$ b. $\dfrac{2x-6}{15-5x}$

Solution:

a. $\dfrac{x-5}{25-x^2}$

$= \dfrac{x-5}{(5-x)(5+x)}$ Factor. The restrictions on x are $x \ne 5$ and $x \ne -5$.

Notice that $x-5$ and $5-x$ are opposites and form a ratio of -1.

$= \dfrac{\overset{-1}{\cancel{x-5}}}{(\cancel{5-x})(5+x)}$ In general, $\dfrac{a-b}{b-a} = -1$.

$= (-1)\left(\dfrac{1}{5+x}\right)$

$= -\dfrac{1}{5+x}$ or $-\dfrac{1}{x+5}$ provided $x \ne 5$, $x \ne -5$

Tip: The factor of -1 may be applied in front of the rational expression, or it may be applied to the numerator or to the denominator. Therefore, the final answer may be written in several forms.

$$-\dfrac{1}{x+5} \quad \text{or} \quad \dfrac{-1}{x+5} \quad \text{or} \quad \dfrac{1}{-(x+5)}$$

Answers

12. b **13.** $\dfrac{-5}{x+3}$

14. $\dfrac{-(a+b)}{a+2b}$ or $-\dfrac{a+b}{a+2b}$

b. $\dfrac{2x-6}{15-5x}$

$= \dfrac{2(x-3)}{5(3-x)}$ Factor. The restriction on x is $x \neq 3$.

$= \dfrac{2\overset{-1}{\cancel{(x-3)}}}{5\cancel{(3-x)}}$ Recall: $\dfrac{a-b}{b-a} = -1$.

$= -\dfrac{2}{5}$ provided $x \neq 3$

5. Rational Functions

Thus far in the text, we have introduced several types of functions including constant, linear, and quadratic functions. Now we will introduce another category of functions called rational functions.

> **Definition of a Rational Function**
>
> A function f is a **rational function** if it can be written in the form $f(x) = \dfrac{p(x)}{q(x)}$, where p and q are polynomial functions and $q(x) \neq 0$.

For example, the functions f, g, h, and k are rational functions.

$$f(x) = \dfrac{1}{x}, \quad g(x) = \dfrac{2}{x-3}, \quad h(a) = \dfrac{a+6}{a^2-5}, \quad k(x) = \dfrac{x+4}{2x^2-11x+5}$$

In Section 4.3, we introduced the rational function defined by $f(x) = \dfrac{1}{x}$. Recall that $f(x) = \dfrac{1}{x}$ has a restriction on its domain that $x \neq 0$ and the graph of $f(x) = \dfrac{1}{x}$ has a vertical asymptote at $x = 0$ (Figure 6-1).

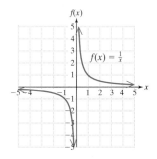

Figure 6-1

To evaluate, simplify, and find the domain of rational functions, we follow procedures similar to those shown for rational expressions.

Skill Practice

Given $h(x) = \frac{x+2}{x+6}$, find the function values.

15. $h(0)$ 16. $h(2)$
17. $h(-2)$ 18. $h(-6)$
19. Write the domain of h in interval notation.

example 7 — Evaluating a Rational Function

Given $g(x) = \dfrac{2}{x-3}$

a. Find the function values (if they exist) $g(0)$, $g(1)$, $g(2)$, $g(2.5)$, $g(2.9)$, $g(3)$, $g(3.1)$, $g(3.5)$, $g(4)$, and $g(5)$.

b. Write the domain of the function in interval notation.

Solution:

a.
$g(0) = \dfrac{2}{(0)-3} = -\dfrac{2}{3}$ 　　 $g(3) = \dfrac{2}{(3)-3} = \dfrac{2}{0}$ undefined

$g(1) = \dfrac{2}{(1)-3} = -1$ 　　 $g(3.1) = \dfrac{2}{(3.1)-3} = \dfrac{2}{0.1} = 20$

$g(2) = \dfrac{2}{(2)-3} = -2$ 　　 $g(3.5) = \dfrac{2}{(3.5)-3} = \dfrac{2}{0.5} = 4$

$g(2.5) = \dfrac{2}{(2.5)-3} = \dfrac{2}{-0.5} = -4$ 　　 $g(4) = \dfrac{2}{(4)-3} = 2$

$g(2.9) = \dfrac{2}{(2.9)-3} = \dfrac{2}{-0.1} = -20$ 　　 $g(5) = \dfrac{2}{(5)-3} = 1$

b. $g(x) = \dfrac{2}{x-3}$ 　　 The value of the function is undefined when the denominator equals zero.

$x - 3 = 0$ 　　 Set the denominator equal to zero and solve for x.

$x = 3$ 　　 The value $x = 3$ must be excluded from the domain.

Domain: $(-\infty, 3) \cup (3, \infty)$

The graph of $g(x) = \dfrac{2}{x-3}$ is shown in Figure 6-2. Notice that the function has a vertical asymptote at $x = 3$ where the function is undefined (Figure 6-2).

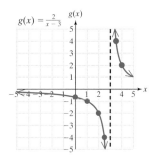

Figure 6-2

Answers

15. $\dfrac{1}{3}$ 16. $\dfrac{1}{2}$
17. 0   18. Undefined
19. $(-\infty, -6) \cup (-6, \infty)$

Calculator Connections

A *Table* feature can be used to check the function values in Example 7, $g(x) = \frac{2}{x-3}$

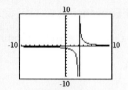

The graph of a rational function may be misleading on a graphing calculator. For example, $g(x) = 2/(x - 3)$ has a vertical asymptote at $x = 3$, but the vertical asymptote is *not* part of the graph (Figure 6-3). A graphing calculator may try to "connect the dots" between two consecutive points, one to the left of $x = 3$ and one to the right of $x = 3$. This creates a line that is nearly vertical and appears to be part of the graph.

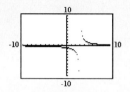

Figure 6-3

To show that this line is not part of the graph, we can graph the function in a *Dot Mode* (see the owner's manual for your calculator). The graph of $g(x) = 2/(x - 3)$ in dot mode indicates that the line $x = 3$ is not part of the function (Figure 6-4).

Figure 6-4

section 6.1 Practice Exercises

Boost *your* GRADE at mathzone.com!
MathZone

- Practice Problems
- Self-Tests
- NetTutor
- e-Professors
- Videos

Study Skills Exercises

1. A test is a *grading* tool for your instructor. How can you turn it into a *learning* tool for you?

2. Define the key terms.
   a. Rational expression
   b. Rational function

Objective 1: Evaluating Rational Expressions

3. Evaluate the expression $\frac{5}{x+1}$ for $x = 0$, $x = 2$, $x = -1$, $x = -6$. **(See Example 1.)**

4. Evaluate the expression $\frac{y-3}{y-2}$ for $y = 0$, $y = 2$, $y = 3$, $y = -4$.

5. Evaluate the expression $\dfrac{3a + 1}{a^2 + 1}$ for $a = 1$, $a = 0$, $a = -\dfrac{1}{3}$, $a = -1$.

6. Evaluate the expression $\dfrac{2z - 8}{z^2 + 9}$ for $z = 4$, $z = -4$, $z = 3$, $z = -3$.

Objective 2: Domain of a Rational Expression

For Exercises 7–16, determine the domain of the rational expression. Write the answer in set-builder notation. (See Example 2.)

7. $\dfrac{9}{y}$ **8.** $\dfrac{-10}{a}$ **9.** $\dfrac{v + 1}{v - 8}$ **10.** $\dfrac{t + 9}{t + 3}$

11. $\dfrac{3x - 1}{2x - 5}$ **12.** $\dfrac{6t + 5}{3t + 8}$ **13.** $\dfrac{q + 1}{q^2 + 6q - 27}$ **14.** $\dfrac{b^2}{2b^2 + 3b - 5}$

15. $\dfrac{c}{c^2 + 25}$ **16.** $\dfrac{d}{d^2 + 16}$

Objective 3: Simplifying Rational Expressions to Lowest Terms

17. Given: $\dfrac{x^2 + 6x + 8}{x^2 + 3x - 4}$, (see Example 3.)

    a. Factor the numerator and denominator.

    b. Determine the domain and write the domain in set-builder notation.

    c. Simplify the expression.

18. Given: $f(x) = \dfrac{x^2 - 6x}{2x^2 - 11x - 6}$

    a. Factor the numerator and denominator.

    b. Determine the domain and write the domain in set-builder notation.

    c. Simplify the expression.

19. Given: $p(x) = \dfrac{x^2 - 18x + 81}{x^2 - 81}$

   a. Factor the numerator and denominator.

   b. Determine the domain and write the domain in set-builder notation.

   c. Simplify the expression.

20. Given: $q(x) = \dfrac{x^2 + 14x + 49}{x^2 - 49}$

   a. Factor the numerator and denominator.

   b. Determine the domain and write the domain in set-builder notation.

   c. Simplify the expression.

For Exercises 21–42, simplify the rational expressions. (See Examples 4–5.)

21. $\dfrac{100x^3y^5}{36xy^8}$

22. $\dfrac{48ab^3c^2}{6a^7bc^0}$

23. $\dfrac{7w^{11}z^6}{14w^3z^3}$

24. $\dfrac{12r^9s^3}{24r^8s^4}$

25. $\dfrac{-3m^4n}{12m^6n^4}$

26. $\dfrac{-5x^3y^2}{20x^4y^2}$

27. $\dfrac{6a + 18}{9a + 27}$

28. $\dfrac{5y - 15}{3y - 9}$

29. $\dfrac{x - 5}{x^2 - 25}$

30. $\dfrac{3z - 6}{3z^2 - 12}$

31. $\dfrac{-7c}{21c^2 - 35c}$

32. $\dfrac{2p + 3}{2p^2 + 7p + 6}$

33. $\dfrac{2t^2 + 7t - 4}{-2t^2 - 5t + 3}$

34. $\dfrac{y^2 + 8y - 9}{y^2 - 5y + 4}$

35. $\dfrac{(p + 1)(2p - 1)^4}{(p + 1)^2(2p - 1)^2}$

36. $\dfrac{r(r - 3)^5}{r^3(r - 3)^2}$

37. $\dfrac{9 - z^2}{2z^2 + z - 15}$

38. $\dfrac{2c^2 + 2c - 12}{8 - 2c - c^2}$

39. $\dfrac{2z^3 + 16}{10 + 3z - z^2}$

40. $\dfrac{5p^2 - p - 4}{p^3 - 1}$

41. $\dfrac{10x^3 - 25x^2 + 4x - 10}{-4 - 10x^2}$

42. $\dfrac{8x^3 - 12x^2 + 6x - 9}{16x^4 - 9}$

Objective 4: Simplifying Ratios of −1

For Exercises 43–56, simplify the rational expressions. (See Example 6.)

43. $\dfrac{r + 6}{6 + r}$

44. $\dfrac{a + 2}{2 + a}$

45. $\dfrac{b + 8}{-b - 8}$

46. $\dfrac{7 + w}{-7 - w}$

47. $\dfrac{10-x}{x-10}$

48. $\dfrac{y-14}{14-y}$

49. $\dfrac{2t-2}{1-t}$

50. $\dfrac{5p-10}{2-p}$

51. $\dfrac{c+4}{c-4}$

52. $\dfrac{b+2}{b-2}$

53. $\dfrac{y-x}{12x^2-12y^2}$

54. $\dfrac{4w^2-49z^2}{14z-4w}$

55. $\dfrac{4x-2x^2}{5x-10}$

56. $\dfrac{2y-6}{3y^2-y^3}$

Objective 5: Rational Functions

57. Let $h(x) = \dfrac{-3}{x-1}$. Find the function values $h(0), h(1), h(-3), h(-1), h(\tfrac{1}{2})$. **(See Example 7.)**

58. Let $k(x) = \dfrac{2}{x+1}$. Find the function values $k(0), k(1), k(-3), k(-1), k(\tfrac{1}{2})$.

For Exercises 59–62, find the domain of each function and use that information to match the function with its graph.

59. $m(x) = \dfrac{1}{x+4}$

60. $n(x) = \dfrac{1}{x+1}$

61. $q(x) = \dfrac{1}{x-4}$

62. $p(x) = \dfrac{1}{x-1}$

a.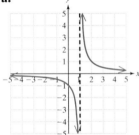

b.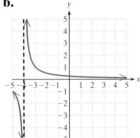

c.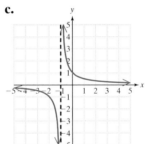

d.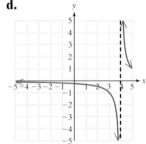

For Exercises 63–72, write the domain of the function in interval notation.

63. $f(x) = \dfrac{1}{3x-1}$

64. $r(t) = \dfrac{1}{2t-1}$

65. $q(t) = \dfrac{t+2}{8}$

66. $p(x) = \dfrac{x-5}{3}$

67. $w(x) = \dfrac{x-2}{3x-6}$

68. $s(t) = \dfrac{-2}{4t+6}$

69. $m(x) = \dfrac{3}{6x^2-7x-10}$

70. $n(x) = \dfrac{1}{2x^2+11x+12}$

71. $r(x) = \dfrac{x+1}{6x^3-x^2-15x}$

72. $s(t) = \dfrac{t+5}{t^3-8t^2+16t}$

Section 6.2 Multiplication and Division of Rational Expressions

Objectives
1. Multiplication of Rational Expressions
2. Division of Rational Expressions

1. Multiplication of Rational Expressions

Recall that to multiply fractions, we multiply the numerators and multiply the denominators. The same is true for multiplying rational expressions.

> **Multiplication of Rational Expressions**
>
> Let p, q, r, and s represent polynomials, such that $q \neq 0$ and $s \neq 0$. Then
>
> $$\frac{p}{q} \cdot \frac{r}{s} = \frac{pr}{qs}$$

For example:

Multiply the fractions
$$\frac{2}{3} \cdot \frac{5}{7} = \frac{10}{21}$$

Multiply the rational expressions
$$\frac{2x}{3y} \cdot \frac{5z}{7} = \frac{10xz}{21y}$$

Sometimes it is possible to simplify a ratio of common factors to 1 *before* multiplying. To do so, we must first factor the numerators and denominators of each fraction.

$$\frac{7}{10} \cdot \frac{15}{21} \quad \xrightarrow{\text{Factor.}} \quad \frac{\cancel{7}}{2 \cdot \cancel{5}} \cdot \frac{\cancel{5} \cdot \cancel{3}}{\cancel{3} \cdot \cancel{7}} = \frac{1}{2}$$

The same process is also used to multiply rational expressions.

> **Multiplying Rational Expressions**
>
> 1. Factor the numerators and denominators of all rational expressions.
> 2. Simplify the ratios of common factors to 1.
> 3. Multiply the remaining factors in the numerator, and multiply the remaining factors in the denominator.

example 1 Multiplying Rational Expressions

Multiply.

a. $\dfrac{5a - 5b}{10} \cdot \dfrac{2}{a^2 - b^2}$

b. $\dfrac{4w - 20p}{2w^2 - 50p^2} \cdot \dfrac{2w^2 + 7wp - 15p^2}{3w + 9p}$

Solution:

a. $\dfrac{5a - 5b}{10} \cdot \dfrac{2}{a^2 - b^2}$

$= \dfrac{5(a - b)}{5 \cdot 2} \cdot \dfrac{2}{(a - b)(a + b)}$ Factor numerator and denominator.

$= \dfrac{\cancel{5}\cancel{(a - b)}}{\cancel{5} \cdot \cancel{2}} \cdot \dfrac{\cancel{2}}{\cancel{(a - b)}(a + b)}$ Simplify.

$= \dfrac{1}{a + b}$

Avoiding Mistakes: If all factors in the numerator simplify to 1, do not forget to write the factor of 1 in the numerator.

Skill Practice

Multiply.

1. $\dfrac{3y - 6}{6y} \cdot \dfrac{y^2 + 3y + 2}{y^2 - 4}$

2. $\dfrac{p^2 + 8p + 16}{10p + 10} \cdot \dfrac{2p + 6}{p^2 + 7p + 12}$

Answers

1. $\dfrac{y + 1}{2y}$ 2. $\dfrac{p + 4}{5(p + 1)}$

b. $\dfrac{4w - 20p}{2w^2 - 50p^2} \cdot \dfrac{2w^2 + 7wp - 15p^2}{3w + 9p}$

$= \dfrac{4(w - 5p)}{2(w^2 - 25p^2)} \cdot \dfrac{(2w - 3p)(w + 5p)}{3(w + 3p)}$ Factor numerator and denominator.

$= \dfrac{2 \cdot 2(w - 5p)}{2(w - 5p)(w + 5p)} \cdot \dfrac{(2w - 3p)(w + 5p)}{3(w + 3p)}$ Factor further.

$= \dfrac{2 \cdot 2\cancel{(w - 5p)}}{\cancel{2}\cancel{(w - 5p)}\cancel{(w + 5p)}} \cdot \dfrac{(2w - 3p)\cancel{(w + 5p)}}{3(w + 3p)}$ Simplify common factors.

$= \dfrac{2(2w - 3p)}{3(w + 3p)}$

Notice that the expression is left in factored form to show that it has been simplified to lowest terms.

2. Division of Rational Expressions

Recall that to divide fractions, multiply the first fraction by the reciprocal of the second fraction.

Divide: $\dfrac{15}{14} \div \dfrac{10}{49}$ Multiply by the reciprocal of the second fraction. $\dfrac{15}{14} \cdot \dfrac{49}{10} = \dfrac{3 \cdot \cancel{5}}{2 \cdot \cancel{7}} \cdot \dfrac{\cancel{7} \cdot 7}{2 \cdot \cancel{5}} = \dfrac{21}{4}$

The same process is used for dividing rational expressions.

Division of Rational Expressions

Let p, q, r, and s represent polynomials, such that $q \neq 0$, $r \neq 0$, $s \neq 0$. Then

$$\dfrac{p}{q} \div \dfrac{r}{s} = \dfrac{p}{q} \cdot \dfrac{s}{r} = \dfrac{ps}{qr}$$

example 2 Dividing Rational Expressions

Divide. **a.** $\dfrac{8t^3 + 27}{9 - 4t^2} \div \dfrac{4t^2 - 6t + 9}{2t^2 - t - 3}$ **b.** $\dfrac{\dfrac{5c}{6d}}{\dfrac{10}{d^2}}$

Solution:

a. $\dfrac{8t^3 + 27}{9 - 4t^2} \div \dfrac{4t^2 - 6t + 9}{2t^2 - t - 3}$

$= \dfrac{8t^3 + 27}{9 - 4t^2} \cdot \dfrac{2t^2 - t - 3}{4t^2 - 6t + 9}$ Multiply the first fraction by the reciprocal of the second.

$= \dfrac{(2t + 3)(4t^2 - 6t + 9)}{(3 - 2t)(3 + 2t)} \cdot \dfrac{(2t - 3)(t + 1)}{4t^2 - 6t + 9}$ Factor numerator and denominator. Notice $8t^3 + 27$ is a sum of cubes. Furthermore, $4t^2 - 6t + 9$ does not factor over the real numbers.

Skill Practice

Divide the rational expressions.

3. $\dfrac{x^2 + x}{5x^3 - x^2} \div \dfrac{10x^2 + 12x + 2}{25x^2 - 1}$

4. $\dfrac{\dfrac{4rs}{5t}}{\dfrac{2r}{t}}$

Answers

3. $\dfrac{1}{2x}$ **4.** $\dfrac{2s}{5}$

$$= \frac{(2t+3)(4t^2-6t+9)}{(3-2t)(3+2t)} \cdot \frac{(2t-3)(t+1)}{4t^2-6t+9} \quad \text{Simplify to lowest terms.}$$

$$= (-1)\frac{(t+1)}{1}$$

$$= -(t+1) \quad \text{or} \quad -t-1$$

b. $\dfrac{\dfrac{5c}{6d}}{\dfrac{10}{d^2}}$ ⟵ This fraction bar denotes division ($\div$).

$$= \frac{5c}{6d} \div \frac{10}{d^2}$$

$$= \frac{5c}{6d} \cdot \frac{d^2}{10} \quad \text{Multiply the first fraction by the reciprocal of the second.}$$

$$= \frac{5c}{6d} \cdot \frac{d \cdot d}{2 \cdot 5}$$

$$= \frac{cd}{12}$$

Tip: In Example 2(a), the factors $(2t-3)$ and $(3-2t)$ are opposites and form a ratio of -1.

$$\frac{2t-3}{3-2t} = -1$$

The factors $(2t+3)$ and $(3+2t)$ are equal and form a ratio of 1.

$$\frac{2t+3}{3+2t} = 1$$

section 6.2 Practice Exercises

Boost your GRADE at mathzone.com!

- Practice Problems
- Self-Tests
- NetTutor
- e-Professors
- Videos

Study Skills Exercise

1. Do you remember your instructor's name, office hours, office location, and office phone? Write them here.

   Instructor's name: Instructor's office hours:

   Instructor's office location: Instructor's office phone:

Review Exercises

2. Write a rational expression whose domain is $(-\infty, 4) \cup (4, \infty)$. (Answers may vary.)

3. Write a rational expression whose domain is $(-\infty, 3) \cup (3, \infty)$. (Answers may vary.)

4. Write a rational function whose domain is $(-\infty, -5) \cup (-5, \infty)$. (Answers may vary.)

5. Write a rational function whose domain is $(-\infty, -6) \cup (-6, \infty)$. (Answers may vary.)

For Exercises 6–9, simplify the rational expressions to lowest terms.

6. $\dfrac{5x^2yz^3}{20xyz}$

7. $\dfrac{7x + 14}{7x^2 - 7x - 42}$

8. $\dfrac{25 - x^2}{x^2 - 10x + 25}$

9. $\dfrac{a^3b^2c^5}{2a^3bc^2}$

Objective 1: Multiplication of Rational Expressions

For Exercises 10–21, multiply the rational expressions. (See Example 1.)

10. $\dfrac{8w^2}{9} \cdot \dfrac{3}{2w^4}$

11. $\dfrac{16}{z^7} \cdot \dfrac{z^4}{8}$

12. $\dfrac{5p^2q^4}{12pq^3} \cdot \dfrac{6p^2}{20q^2}$

13. $\dfrac{27r^5}{7s} \cdot \dfrac{28rs^3}{9r^3s^2}$

14. $\dfrac{3z + 12}{8z^3} \cdot \dfrac{16z^3}{9z + 36}$

15. $\dfrac{x^2y}{x^2 - 4x - 5} \cdot \dfrac{2x^2 - 13x + 15}{xy^3}$

16. $\dfrac{3y^2 + 18y + 15}{6y + 6} \cdot \dfrac{y - 5}{y^2 - 25}$

17. $\dfrac{10w - 8}{w + 2} \cdot \dfrac{3w^2 - w - 14}{25w^2 - 16}$

18. $\dfrac{x - 5y}{x^2 + xy} \cdot \dfrac{y^2 - x^2}{10y - 2x}$

19. $\dfrac{3x - 15}{4x^2 - 2x} \cdot \dfrac{10x - 20x^2}{5 - x}$

20. $x(x + 5)^2 \cdot \dfrac{2}{x^2 - 25}$

21. $y(y^2 - 4) \cdot \dfrac{y}{y + 2}$

Objective 2: Division of Rational Expressions

For Exercises 22–33, divide the rational expressions. (See Example 2.)

22. $\dfrac{6x^2y^2}{(x - 2)} \div \dfrac{3xy^2}{(x - 2)^2}$

23. $\dfrac{(r + 3)^2}{4r^3s} \div \dfrac{r + 3}{rs}$

24. $\dfrac{t^2 + 5t}{t + 1} \div (t + 5)$

25. $\dfrac{6p + 7}{p + 2} \div (36p^2 - 49)$

26. $\dfrac{a}{a - 10} \div \dfrac{a^3 + 6a^2 - 40a}{a^2 - 100}$

27. $\dfrac{b^2 - 6b + 9}{b^2 - b - 6} \div \dfrac{b^2 - 9}{4}$

28. $\dfrac{2x^2 + 5xy + 2y^2}{4x^2 - y^2} \div \dfrac{x^2 + xy - 2y^2}{2x^2 + xy - y^2}$

29. $\dfrac{6s^2 + st - 2t^2}{6s^2 - 5st + t^2} \div \dfrac{3s^2 + 17st + 10t^2}{6s^2 + 13st - 5t^2}$

30. $\dfrac{x^4 - x^3 + x^2 - x}{2x^3 + 2x^2 + x + 1} \div \dfrac{x^3 - 4x^2 + x - 4}{2x^3 - 8x^2 + x - 4}$

31. $\dfrac{a^3 + a + a^2 + 1}{a^3 + a^2 + ab^2 + b^2} \div \dfrac{a^3 + a + a^2b + b}{2a^2 + 2ab + ab^2 + b^3}$

32. $\dfrac{3y - y^2}{y^3 - 27} \div \dfrac{y}{y^2 + 3y + 9}$

33. $\dfrac{8x - 4x^2}{xy - 2y + 3x - 6} \div \dfrac{3x + 6}{y + 3}$

Mixed Exercises

For Exercises 34–45, perform the indicated operations.

34. $\dfrac{8a^4b^3}{3c} \div \dfrac{a^7b^2}{9c}$

35. $\dfrac{3x^5}{2x^2y^7} \div \dfrac{4x^3y}{6y^6}$

36. $\dfrac{2}{25x^2} \cdot \dfrac{5x}{12} \div \dfrac{2}{15x}$

37. $\dfrac{4y}{7} \div \dfrac{y^2}{14} \cdot \dfrac{3}{y}$

38. $\dfrac{10x^2 - 13xy - 3y^2}{8x^2 - 10xy - 3y^2} \cdot \dfrac{2y + 8x}{2x^2 + 2y^2}$

39. $\dfrac{6a^2 + ab - b^2}{10a^2 + 5ab} \cdot \dfrac{2a^3 + 4a^2b}{3a^2 + 5ab - 2b^2}$

40. $\dfrac{(a+b)^2}{a-b} \cdot \dfrac{a^3 - b^3}{a^2 - b^2} \div \dfrac{a^2 + ab + b^2}{(a-b)^2}$

41. $\dfrac{m^2 - n^2}{(m-n)^2} \div \dfrac{m^2 - 2mn + n^2}{m^2 - mn + n^2} \cdot \dfrac{(m-n)^4}{m^3 + n^3}$

42. $\dfrac{x^2 - 4y^2}{x + 2y} \div (x + 2y) \cdot \dfrac{2y}{x - 2y}$

43. $\dfrac{x^2 - 6xy + 9y^2}{x^2 - 4y^2} \cdot \dfrac{x^2 - 5xy + 6y^2}{3y - x} \div \dfrac{x^2 - 9y^2}{x + 2y}$

44. $\dfrac{8x^3 - 27y^3}{4x^2 - 9y^2} \div \dfrac{8x^2 + 12xy + 18y^2}{2x + 3y}$

45. $\dfrac{25m^2 - 1}{125m^3 - 1} \div \dfrac{5m + 1}{25m^2 + 5m + 1}$

Expanding Your Skills

For Exercises 46–49, write an expression for the area of the figure and simplify.

46.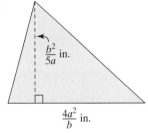

Height: $\dfrac{b^2}{5a}$ in.; Base: $\dfrac{4a^2}{b}$ in.

47.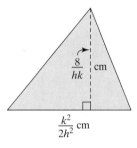

Height: $\dfrac{8}{hk}$ cm; Base: $\dfrac{k^2}{2h^2}$ cm

48.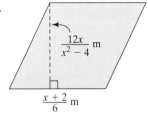

Height: $\dfrac{12x}{x^2 - 4}$ m; Base: $\dfrac{x + 2}{6}$ m

49.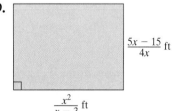

$\dfrac{5x - 15}{4x}$ ft; $\dfrac{x^2}{x - 3}$ ft

Objectives

1. Addition and Subtraction of Rational Expressions with Like Denominators
2. Identifying the Least Common Denominator
3. Writing Equivalent Rational Expressions
4. Addition and Subtraction of Rational Expressions with Unlike Denominators

section 6.3 Addition and Subtraction of Rational Expressions

1. Addition and Subtraction of Rational Expressions with Like Denominators

To add or subtract rational expressions, the expressions must have the same denominator. As with fractions, we add or subtract rational expressions with the same denominator by combining the terms in the numerator and then writing the result over the common denominator. Then, if possible, we simplify the expression to lowest terms.

Addition and Subtraction of Rational Expressions

Let p, q, and r represent polynomials where $q \neq 0$. Then

1. $\dfrac{p}{q} + \dfrac{r}{q} = \dfrac{p+r}{q}$
2. $\dfrac{p}{q} - \dfrac{r}{q} = \dfrac{p-r}{q}$

Skill Practice

Add or subtract as indicated.

1. $\dfrac{5}{12} - \dfrac{1}{12}$
2. $\dfrac{b^2+1}{b-2} + \dfrac{b-7}{b-2}$
3. $\dfrac{4t+1}{t+7} - \dfrac{t+3}{t+7}$

example 1 Adding and Subtracting Rational Expressions with Like Denominators

Add or subtract as indicated.

a. $\dfrac{1}{8} + \dfrac{3}{8}$
b. $\dfrac{5x}{2x-1} + \dfrac{3}{2x-1}$
c. $\dfrac{x^2}{x-4} - \dfrac{x+12}{x-4}$

Solution:

a. $\dfrac{1}{8} + \dfrac{3}{8} = \dfrac{1+3}{8}$ Add the terms in the numerator.

$= \dfrac{4}{8}$

$= \dfrac{1}{2}$ Simplify the fraction.

b. $\dfrac{5x}{2x-1} + \dfrac{3}{2x-1} = \dfrac{5x+3}{2x-1}$ Add the terms in the numerator. The answer is already in lowest terms.

c. $\dfrac{x^2}{x-4} - \dfrac{x+12}{x-4}$

$= \dfrac{x^2 - (x+12)}{x-4}$ Combine the terms in the numerator. Use parentheses to group the terms in the numerator that follow the subtraction sign. This will help you remember to apply the distributive property.

$= \dfrac{x^2 - x - 12}{x-4}$ Apply the distributive property.

$= \dfrac{(x-4)(x+3)}{(x-4)}$ Factor the numerator and denominator.

Answers

1. $\dfrac{1}{3}$
2. $b+3$
3. $\dfrac{3t-2}{t+7}$

$$= \frac{\overset{1}{(x-4)}(x+3)}{(x-4)}$$

Simplify the rational expression.

$$= x + 3$$

2. Identifying the Least Common Denominator

If two rational expressions have different denominators, each expression must be rewritten with a common denominator before adding or subtracting the expressions. The **least common denominator (LCD)** of two or more rational expressions is defined as the least common multiple of the denominators.

For example, consider the fractions $\frac{1}{20}$ and $\frac{1}{8}$. By inspection, you can probably see that the least common denominator is 40. To understand why, find the prime factorization of both denominators.

$$20 = 2^2 \cdot 5 \quad \text{and} \quad 8 = 2^3$$

A common multiple of 20 and 8 must be a multiple of 5, a multiple of 2^2, and a multiple of 2^3. However, any number that is a multiple of $2^3 = 8$ is automatically a multiple of $2^2 = 4$. Therefore, it is sufficient to construct the least common denominator as the product of unique prime factors, where each factor is raised to its highest power.

The LCD of $\dfrac{1}{2^2 \cdot 5}$ and $\dfrac{1}{2^3}$ is $2^3 \cdot 5 = 40$.

Steps to Find the LCD of Two or More Rational Expressions

1. Factor all denominators completely.
2. The LCD is the product of unique prime factors from the denominators, where each factor is raised to the highest power to which it appears in any denominator.

---

example 2   Finding the LCD of Rational Expressions

Find the LCD of the following rational expressions.

a. $\dfrac{1}{12}, \dfrac{5}{18}, \dfrac{7}{30}$

b. $\dfrac{1}{2x^3y}, \dfrac{5}{16xy^2z}$

c. $\dfrac{x+2}{2x-10}, \dfrac{x^2+3}{x^2-25}, \dfrac{6}{x^2+8x+16}$

d. $\dfrac{x+4}{x-3}, \dfrac{1}{3-x}$

Solution:

a. $\dfrac{1}{12}, \dfrac{5}{18}, \dfrac{7}{30}$

$\dfrac{1}{2^2 \cdot 3}, \dfrac{5}{2 \cdot 3^2}, \dfrac{7}{2 \cdot 3 \cdot 5}$ Factor the denominators completely.

LCD $= 2^2 \cdot 3^2 \cdot 5 = 180$ The LCD is the product of the factors 2, 3, and 5. Each factor is raised to its highest power.

Skill Practice

Find the LCD of the rational expressions.

4. $\dfrac{7}{40}, \dfrac{1}{15}, \dfrac{5}{6}$

5. $\dfrac{1}{9a^3b^2}, \dfrac{5}{18a^4b}$

6. $\dfrac{5}{x^2-4}, \dfrac{1}{2x+6}, \dfrac{6}{x^2+5x+6}$

7. $\dfrac{6}{z-7}, \dfrac{1}{7-z}$

Answers

4. 120   5. $18a^4b^2$
6. $2(x+2)(x-2)(x+3)$
7. $z-7$ or $7-z$

b. $\dfrac{1}{2x^3y}, \dfrac{5}{16xy^2z}$

$\dfrac{1}{2x^3y}, \dfrac{5}{2^4xy^2z}$ Factor the denominators completely.

LCD $= 2^4x^3y^2z$ The LCD is the product of the factors 2, x, y, and z. Each factor is raised to its highest power.

c. $\dfrac{x+2}{2x-10}, \dfrac{x^2+3}{x^2-25}, \dfrac{6}{x^2+8x+16}$

$\dfrac{x+2}{2(x-5)}, \dfrac{x^2+3}{(x-5)(x+5)}, \dfrac{6}{(x+4)^2}$ Factor the denominators completely.

LCD $= 2(x-5)(x+5)(x+4)^2$ The LCD is the product of the factors 2, $(x-5)$, $(x+5)$, and $(x+4)$. Each factor is raised to its highest power.

d. $\dfrac{x+4}{x-3}, \dfrac{1}{3-x}$ The denominators are already factored.

Notice that $x - 3$ and $3 - x$ are opposite factors. If -1 is factored from either expression, the binomial factors will be the same.

$\dfrac{x+4}{x-3}, \dfrac{1}{-1(-3+x)}$ $\dfrac{x+4}{-1(-x+3)}, \dfrac{1}{3-x}$

      Factor out -1.   Factor out -1.

  same binomial factors   same binomial factors

LCD $= (x-3)(-1)$ LCD $= (-1)(3-x)$

$= -x + 3$ $= -3 + x$

$= 3 - x$ $= x - 3$

The LCD can be taken as either $(3-x)$ or $(x-3)$.

3. Writing Equivalent Rational Expressions

Rational expressions can be added if they have common denominators. Once the LCD has been determined, each rational expression must be converted to an **equivalent rational expression** with the indicated denominator.

Using the identity property of multiplication, we know that for $q \neq 0$ and $r \neq 0$,

$$\dfrac{p}{q} = \dfrac{p}{q} \cdot 1 = \dfrac{p}{q} \cdot \dfrac{r}{r} = \dfrac{pr}{qr}$$

This principle is used to convert a rational expression to an equivalent expression with a different denominator. For example, $\frac{1}{2}$ can be converted to an equivalent expression with a denominator of 12 as follows:

$$\dfrac{1}{2} = \dfrac{1}{2} \cdot \dfrac{6}{6} = \dfrac{1 \cdot 6}{2 \cdot 6} = \dfrac{6}{12}$$

In this example, we multiplied $\frac{1}{2}$ by a convenient form of 1. The ratio $\frac{6}{6}$ was chosen so that the product produced a new denominator of 12. Notice that multiplying $\frac{1}{2}$ by $\frac{6}{6}$ is equivalent to multiplying the numerator and denominator of the original expression by 6. In general, if the numerator and denominator of a rational expression are both multiplied by the same nonzero quantity, the value of the expression remains unchanged.

example 3 — Creating Equivalent Rational Expressions

Convert each expression to an equivalent rational expression with the indicated denominator.

a. $\dfrac{7}{5p^2} = \dfrac{}{20p^6}$ **b.** $\dfrac{w}{w+5} = \dfrac{}{w^2 + 3w - 10}$

Solution:

a. $\dfrac{7}{5p^2} = \dfrac{}{20p^6}$ Multiply the numerator and denominator of the fraction by the missing factor of $4p^4$.

$\dfrac{7 \cdot 4p^4}{5p^2 \cdot 4p^4} = \dfrac{28p^4}{20p^6}$

b. $\dfrac{w}{w+5} = \dfrac{}{w^2 + 3w - 10}$

$\dfrac{w}{w+5} = \dfrac{}{(w+5)(w-2)}$ Factor the denominator.

$\dfrac{w}{w+5} = \dfrac{w \cdot (w-2)}{(w+5) \cdot (w-2)}$ Multiply numerator and denominator by the missing factor of $(w-2)$.

$= \dfrac{w^2 - 2w}{(w+5)(w-2)}$

Skill Practice

Fill in the blank to make an equivalent fraction with the given denominator.

8. $\dfrac{1}{8xy} = \dfrac{}{16x^3y^2}$

9. $\dfrac{5b}{b-3} = \dfrac{}{b^2 - 9}$

Tip: Notice that in Example 3(b) we multiplied the polynomials in the numerator but left the denominator in factored form. This convention is followed because when we add and subtract rational expressions, the terms in the numerators must be combined.

4. Addition and Subtraction of Rational Expressions with Unlike Denominators

To add or subtract rational expressions with unlike denominators, we must convert each expression to an equivalent expression with the same denominator. For example, consider adding the expressions $\dfrac{3}{x-2} + \dfrac{5}{x+1}$. The LCD is $(x-2)(x+1)$. For each expression, identify the factors from the LCD that are

Answers

8. $2x^2y$ **9.** $5b^2 + 15b$

missing in the denominator. Then multiply the numerator and denominator of the expression by the missing factor(s):

$$\frac{(3)}{(x-2)} \cdot \frac{(x+1)}{(x+1)} + \frac{(5)}{(x+1)} \cdot \frac{(x-2)}{(x-2)}$$ The rational expressions now have the same denominator and can be added.

$$= \frac{3(x+1) + 5(x-2)}{(x-2)(x+1)}$$ Combine terms in the numerator.

$$= \frac{3x + 3 + 5x - 10}{(x-2)(x+1)}$$ Clear parentheses and simplify.

$$= \frac{8x - 7}{(x-2)(x+1)}$$

Steps to Add or Subtract Rational Expressions

1. Factor the denominator of each rational expression.
2. Identify the LCD.
3. Rewrite each rational expression as an equivalent expression with the LCD as its denominator.
4. Add or subtract the numerators, and write the result over the common denominator.
5. Simplify.

Skill Practice

Add or subtract as indicated.

10. $\dfrac{4}{5y} + \dfrac{1}{3y^3}$

11. $\dfrac{2x+3}{x^2+x-2} - \dfrac{5}{3x-3}$

12. $\dfrac{a^2+a+24}{a^2-9} + \dfrac{5}{a+3}$

example 4 — Adding and Subtracting Rational Expressions with Unlike Denominators

Add or subtract as indicated.

a. $\dfrac{3}{7b} + \dfrac{4}{b^2}$ b. $\dfrac{3t-2}{t^2+4t-12} - \dfrac{5}{2t+12}$ c. $\dfrac{2}{x} + \dfrac{x}{x+3} - \dfrac{3x+18}{x^2+3x}$

Solution:

a. $\dfrac{3}{7b} + \dfrac{4}{b^2}$ Step 1: The denominators are already factored.

Step 2: The LCD is $7b^2$.

$= \dfrac{3}{7b} \cdot \dfrac{b}{b} + \dfrac{4}{b^2} \cdot \dfrac{7}{7}$ Step 3: Write each expression with the LCD.

$= \dfrac{3b}{7b^2} + \dfrac{28}{7b^2}$ Step 4: Add the numerators, and write the result over the LCD.

$= \dfrac{3b + 28}{7b^2}$ Step 5: Simplify.

b. $\dfrac{3t-2}{t^2+4t-12} - \dfrac{5}{2t+12}$

$= \dfrac{3t-2}{(t+6)(t-2)} - \dfrac{5}{2(t+6)}$ Step 1: Factor the denominators.

Answers

10. $\dfrac{12y^2+5}{15y^3}$

11. $\dfrac{1}{3(x+2)}$

12. $\dfrac{a+3}{a-3}$

$$= \frac{(2)}{(2)} \cdot \frac{(3t-2)}{(t+6)(t-2)} - \frac{5}{2(t+6)} \cdot \frac{(t-2)}{(t-2)}$$

Step 2: The LCD is $2(t+6)(t-2)$.

Step 3: Write each expression with the LCD.

$$= \frac{2(3t-2) - 5(t-2)}{2(t+6)(t-2)}$$

Step 4: Add the numerators and write the result over the LCD.

$$= \frac{6t - 4 - 5t + 10}{2(t+6)(t-2)}$$

Step 5: Simplify.

$$= \frac{t+6}{2(t+6)(t-2)}$$

Combine *like* terms.

$$= \frac{\overset{1}{\cancel{t+6}}}{2\cancel{(t+6)}(t-2)}$$

Simplify.

$$= \frac{1}{2(t-2)}$$

c. $\dfrac{2}{x} + \dfrac{x}{x+3} - \dfrac{3x+18}{x^2+3x}$

$$= \frac{2}{x} + \frac{x}{x+3} - \frac{3x+18}{x(x+3)}$$

Step 1: Factor the denominators.

Step 2: The LCD is $x(x+3)$.

$$= \frac{2}{x} \cdot \frac{(x+3)}{(x+3)} + \frac{x}{(x+3)} \cdot \frac{x}{x} - \frac{3x+18}{x(x+3)}$$

Step 3: Write each expression with the LCD.

$$= \frac{2(x+3) + x^2 - (3x+18)}{x(x+3)}$$

Step 4: Add the numerators, and write the result over the LCD.

$$= \frac{2x + 6 + x^2 - 3x - 18}{x(x+3)}$$

Step 5: Simplify.

$$= \frac{x^2 - x - 12}{x(x+3)}$$

Combine *like* terms.

$$= \frac{(x-4)(x+3)}{x(x+3)}$$

Factor the numerator.

$$= \frac{(x-4)\overset{1}{\cancel{(x+3)}}}{x\underset{1}{\cancel{(x+3)}}}$$

Simplify to lowest terms.

$$= \frac{x-4}{x}$$

Skill Practice

Add or subtract as indicated.

13. $\dfrac{3}{-y} - \dfrac{5}{y}$

14. $\dfrac{3a}{a-5} + \dfrac{15}{5-a}$

example 5 — Adding and Subtracting Rational Expressions with Unlike Denominators

Add or subtract as indicated.

a. $\dfrac{6}{w} + \dfrac{4}{-w}$ b. $\dfrac{x^2}{x-y} + \dfrac{y^2}{y-x}$

Solution:

a. $\dfrac{6}{w} + \dfrac{4}{-w}$
   — **Step 1:** The denominators are already factored.
   — **Step 2:** The denominators are opposites and differ by a factor of -1. The LCD can either be taken as w or $-w$. We will use an LCD of w.

$= \dfrac{6}{w} + \dfrac{4}{-w} \cdot \dfrac{(-1)}{(-1)}$
   — **Step 3:** Write each expression with the LCD. Note that $(-w)(-1) = w$.

$= \dfrac{6}{w} + \dfrac{-4}{w}$

$= \dfrac{6 + (-4)}{w}$
   — **Step 4:** Add the numerators, and write the result over the LCD.

$= \dfrac{2}{w}$
   — **Step 5:** Simplify.

b. $\dfrac{x^2}{x-y} + \dfrac{y^2}{y-x}$
   — **Step 1:** The denominators are already factored.
   — **Step 2:** The denominators are opposites and differ by a factor of -1. The LCD can be taken as either $(x-y)$ or $(y-x)$. We will use an LCD of $(x-y)$.

$= \dfrac{x^2}{(x-y)} + \dfrac{y^2}{(y-x)} \cdot \dfrac{(-1)}{(-1)}$
   — **Step 3:** Write each expression with the LCD. Note that $(y-x)(-1) = -y + x = x - y$.

$= \dfrac{x^2}{x-y} + \dfrac{-y^2}{x-y}$

$= \dfrac{x^2 - y^2}{x-y}$
   — **Step 4:** Combine the numerators, and write the result over the LCD.

$= \dfrac{(x+y)\cancel{(x-y)}}{\cancel{x-y}}$
   — **Step 5:** Factor and simplify to lowest terms.

$= x + y$

Answers

13. $\dfrac{-8}{y}$ or $\dfrac{8}{-y}$

14. 3

Section 6.3 Practice Exercises

Boost your GRADE at mathzone.com!

- Practice Problems
- Self-Tests
- NetTutor
- e-Professors
- Videos

Study Skills Exercises

1. A rule of thumb is that 2 to 3 hr of study time per week is needed for each 1 hr per week of class time. Based on the number of hours you are in class this semester, how many hours of the week should you be studying?

2. Define the key terms.
   a. Least common denominator (LCD)
   b. Equivalent rational expressions

Review Exercises

For Exercises 3–6, perform the indicated operation.

3. $\dfrac{x}{x-y} \div \dfrac{x^2}{y-x}$

4. $\dfrac{9b+9}{4b+8} \cdot \dfrac{2b+4}{3b-3}$

5. $\dfrac{(5-a)^2}{10a-2} \cdot \dfrac{25a^2-1}{a^2-10a+25}$

6. $\dfrac{x^2-z^2}{14x^2z^4} \div \dfrac{x^2+2xz+z^2}{3xz^3}$

Objective 1: Addition and Subtraction of Rational Expressions with Like Denominators

For Exercises 7–16, add or subtract as indicated and simplify if possible. (See Example 1.)

7. $\dfrac{3}{5x} + \dfrac{7}{5x}$

8. $\dfrac{1}{2x^2} - \dfrac{5}{2x^2}$

9. $\dfrac{x}{x^2-2x-3} - \dfrac{3}{x^2-2x-3}$

10. $\dfrac{x}{x^2+4x-12} + \dfrac{6}{x^2+4x-12}$

11. $\dfrac{5x-1}{(2x+9)(x-6)} - \dfrac{3x-6}{(2x+9)(x-6)}$

12. $\dfrac{4-x}{8x+1} - \dfrac{5x-6}{8x+1}$

13. $\dfrac{6}{x-5} + \dfrac{3}{5-x}$

14. $\dfrac{8}{2-x} + \dfrac{7}{x-2}$

15. $\dfrac{x-2}{x-6} - \dfrac{x+2}{6-x}$

16. $\dfrac{x-10}{x-8} - \dfrac{x+10}{8-x}$

Objective 2: Identifying the Least Common Denominator

For Exercises 17–28, find the least common denominator (LCD). (See Example 2.)

17. $\dfrac{5}{8}, \dfrac{3}{20x}$

18. $\dfrac{y}{15a}, \dfrac{y^2}{35}$

19. $\dfrac{-7}{24x}, \dfrac{5}{75x^2}$

20. $\dfrac{2}{7y^3}, \dfrac{-13}{5y^2}$

21. $\dfrac{-5}{6m^4}, \dfrac{1}{15mn^7}$

22. $\dfrac{13}{12cd^5}, \dfrac{9}{8c^3}$

23. $\dfrac{6}{(x-4)(x+2)}, \dfrac{-8}{(x-4)(x-6)}$

24. $\dfrac{x}{(2x-1)(x-7)}, \dfrac{2}{(2x-1)(x+1)}$

25. $\dfrac{3}{x(x-1)(x+7)^2}, \dfrac{-1}{x^2(x+7)}$

26. $\dfrac{14}{(x-2)^2(x+9)}, \dfrac{41}{x(x-2)(x+9)}$

27. $\dfrac{5}{x-6}, \dfrac{x-5}{x^2-8x+12}$

28. $\dfrac{7a}{a+4}, \dfrac{a+12}{a^2-16}$

Objective 3: Writing Equivalent Rational Expressions

For Exercises 29–34, fill in the blank to make an equivalent fraction with the given denominator. (See Example 3.)

29. $\dfrac{5}{3x} = \dfrac{}{9x^2y}$

30. $\dfrac{-5}{xy} = \dfrac{}{4x^2y^3}$

31. $\dfrac{2x}{x-1} = \dfrac{}{x(x-1)(x+2)}$

32. $\dfrac{5x}{2x-5} = \dfrac{}{(2x-5)(x+8)}$

33. $\dfrac{y}{y+6} = \dfrac{}{y^2+5y-6}$

34. $\dfrac{t^2}{t-8} = \dfrac{}{t^2-6t-16}$

Objective 4: Addition and Subtraction of Rational Expressions with Unlike Denominators

For Exercises 35–66, add or subtract as indicated. (See Examples 4–5.)

35. $\dfrac{4}{3p} - \dfrac{5}{2p^2}$

36. $\dfrac{6}{5a^2b} - \dfrac{1}{10ab}$

37. $\dfrac{x+3}{x^2} + \dfrac{x+5}{2x}$

38. $\dfrac{x+2}{5x^2} + \dfrac{x+4}{15x}$

39. $\dfrac{s-1}{s} - \dfrac{t+1}{t}$

40. $\dfrac{x+2}{x} - \dfrac{y-2}{y}$

41. $\dfrac{4a-2}{3a+12} - \dfrac{a-2}{a+4}$

42. $\dfrac{10}{b(b+5)} + \dfrac{2}{b}$

43. $\dfrac{6}{w(w-2)} + \dfrac{3}{w}$

44. $\dfrac{6y+5}{5y-25} - \dfrac{y+2}{y-5}$

45. $w+2+\dfrac{1}{w-2}$

46. $h-3+\dfrac{1}{h+3}$

47. $\dfrac{6b}{b-4} - \dfrac{1}{b+1}$

48. $\dfrac{a}{a-3} - \dfrac{5}{a+6}$

49. $\dfrac{t+5}{t-5} - \dfrac{10t-5}{t^2-25}$

50. $\dfrac{s+8}{s-8} - \dfrac{16s+64}{s^2-64}$

51. $\dfrac{x+2}{x^2-36} - \dfrac{x}{x^2+9x+18}$

52. $\dfrac{7}{x^2-x-2} + \dfrac{x}{x^2+4x+3}$

53. $\dfrac{9}{x^2-2x+1} - \dfrac{x-3}{x^2-x}$

54. $\dfrac{2}{4z^2-12z+9} - \dfrac{z+1}{2z^2-3z}$

55. $\dfrac{w}{6} + \dfrac{w+4}{-6}$

56. $\dfrac{4y}{3} + \dfrac{5}{-3}$

57. $\dfrac{n}{5-n} + \dfrac{2n-5}{n-5}$

58. $\dfrac{c}{7-c} + \dfrac{2c-7}{c-7}$

59. $\dfrac{2}{3x-15} + \dfrac{x}{25-x^2}$

60. $\dfrac{5}{9-x^2} - \dfrac{4}{x^2+4x+3}$

61. $\dfrac{k}{k+7} - \dfrac{2}{k^2+6k-7} + \dfrac{4k}{k-1}$

62. $\dfrac{t}{t-2} - \dfrac{1}{t^2-t-2} + \dfrac{3t}{t+1}$

63. $\dfrac{2x}{x^2-y^2} - \dfrac{1}{x-y} + \dfrac{1}{y-x}$

64. $\dfrac{3w-1}{2w^2+w-3} - \dfrac{2-w}{w-1} - \dfrac{w}{1-w}$

65. $\dfrac{3}{y} + \dfrac{2}{y-6}$

66. $\dfrac{-8}{p} + \dfrac{p}{p+5}$

Expanding Your Skills

For Exercises 67–70, write an expression that represents the perimeter of the figure and simplify.

67.

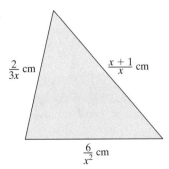

68.

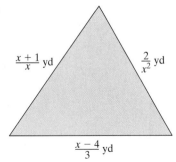

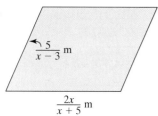

69. parallelogram with side $\dfrac{5}{x-3}$ m and base $\dfrac{2x}{x+5}$ m

70. rectangle with side $\dfrac{3}{x+2}$ ft and side $\dfrac{x}{x+1}$ ft

chapter 6 | midchapter review

1. Explain the process to simplify a rational expression to lowest terms.

2. Explain the process to multiply two rational expressions.

3. Explain the process to divide two rational expressions.

4. Explain the process to add and subtract rational expressions.

For Exercises 5–14, perform the indicated operations.

5. $\dfrac{2}{2y-3} - \dfrac{3}{2y} + 1$

6. $(x+5) + \left(\dfrac{7}{x-4}\right)$

7. $\dfrac{5x^2 - 6x + 1}{x^2 - 1} \div \dfrac{16x^2 - 9}{4x^2 + 7x + 3}$

8. $\dfrac{a^2 - 25}{3a^2 + 3ab} \cdot \dfrac{a^2 + 4a + ab + 4b}{a^2 + 9a + 20}$

9. $\dfrac{4}{y+1} + \dfrac{y+2}{y^2-1} - \dfrac{3}{y-1}$

10. $\dfrac{8w^2}{w^3 - 16w} - \dfrac{4w}{w^2 - 4w}$

11. $\dfrac{a^2 - 16}{2x + 6} \cdot \dfrac{x+3}{a-4}$

12. $\dfrac{t^2 - 9}{t} \div \dfrac{t+3}{t+2}$

13. $\dfrac{6xy}{x^2 - y^2} + \dfrac{x+y}{y-x}$

14. $(x^2 - 6x + 8)\left(\dfrac{3}{x-2}\right)$

For Exercises 15–16, simplify the expressions.

15. $\dfrac{x - 3 - bx + 3b}{bx + 3b - x - 3}$

16. $\dfrac{m^2 - n^2}{n - m}$

17. For $f(x) = \dfrac{4x + 4}{x^2 - 1}$

  a. Factor the numerator and denominator completely.

  b. Write the domain in set-builder notation.

  c. Simplify the expression to lowest terms.

18. For $g(x) = \dfrac{3x + 6}{x^2 - 3x - 10}$

  a. Factor the numerator and denominator completely.

  b. Write the domain in set-builder notation.

  c. Simplify the expression to lowest terms.

Section 6.4 Complex Fractions

1. Simplifying Complex Fractions by Method I

A **complex fraction** is a fraction whose numerator or denominator contains one or more fractions. For example:

$$\frac{\frac{5x^2}{y}}{\frac{10x}{y^2}} \quad \text{and} \quad \frac{2 + \frac{1}{2} - \frac{1}{3}}{\frac{3}{4} + \frac{1}{6}}$$

are complex fractions.

Two methods will be presented to simplify complex fractions. The first method (Method I) follows the order of operations to simplify the numerator and denominator separately before dividing. The process is summarized as follows.

Objectives
1. Simplifying Complex Fractions by Method I
2. Simplifying Complex Fractions by Method II
3. Using Complex Fractions in Applications

Steps to Simplify a Complex Fraction—Method I
1. Add or subtract expressions in the numerator to form a single fraction. Add or subtract expressions in the denominator to form a single fraction.
2. Divide the rational expressions from step 1 by multiplying the numerator of the complex fraction by the reciprocal of the denominator of the complex fraction.
3. Simplify to lowest terms, if possible.

example 1 Simplifying a Complex Fraction by Method I

Simplify the expression. $\dfrac{\frac{5x^2}{y}}{\frac{10x}{y^2}}$

Solution:

$\dfrac{\frac{5x^2}{y}}{\frac{10x}{y^2}}$ **Step 1:** The numerator and denominator of the complex fraction are already single fractions.

$= \dfrac{5x^2}{y} \cdot \dfrac{y^2}{10x}$ **Step 2:** Multiply the numerator of the complex fraction by the reciprocal of the denominator.

$= \dfrac{5 \cdot x \cdot x}{y} \cdot \dfrac{y \cdot y}{2 \cdot 5 \cdot x}$ Factor the numerators and denominators.

$= \dfrac{\cancel{5} \cdot \cancel{x} \cdot x}{\cancel{y}} \cdot \dfrac{\cancel{y} \cdot y}{2 \cdot \cancel{5} \cdot \cancel{x}}$ **Step 3:** Simplify.

$= \dfrac{xy}{2}$

Skill Practice

Simplify the expression.

1. $\dfrac{\frac{18a^3}{b^2}}{\frac{6a^2}{b}}$

Answer

1. $\dfrac{3a}{b}$

428 Chapter 6 Rational Expressions and Rational Equations

Sometimes it is necessary to simplify the numerator and denominator of a complex fraction before the division is performed. This is illustrated in Example 2.

Skill Practice

Simplify the expression.

2. $\dfrac{\frac{1}{4} - \frac{5}{6}}{\frac{1}{2} + \frac{1}{3} + 1}$

example 2   Simplifying a Complex Fraction by Method I

Simplify the expression. $\qquad \dfrac{2 + \frac{1}{2} - \frac{1}{3}}{\frac{3}{4} + \frac{1}{6}}$

Solution:

$\dfrac{2 + \frac{1}{2} - \frac{1}{3}}{\frac{3}{4} + \frac{1}{6}}$ **Step 1:** Combine fractions in numerator and denominator separately.

$= \dfrac{\frac{12}{6} + \frac{3}{6} - \frac{2}{6}}{\frac{9}{12} + \frac{2}{12}}$ The LCD in the numerator is 6. The LCD in the denominator is 12.

$= \dfrac{\frac{13}{6}}{\frac{11}{12}}$ Form single fractions in the numerator and denominator.

$= \dfrac{13}{6} \cdot \dfrac{12}{11}$ **Step 2:** Multiply by the reciprocal of $\frac{11}{12}$, which is $\frac{12}{11}$.

$= \dfrac{13}{\cancel{6}} \cdot \dfrac{\cancel{12}^{2}}{11}$

$= \dfrac{26}{11}$ **Step 3:** Simplify.

Skill Practice

Simplify the expression.

3. $\dfrac{x^{-1} + y^{-1}}{y^{-1} + xy^{-2}}$

example 3   Simplifying Complex Fractions by Method I

Simplify the expression. $\qquad \dfrac{a - a^{-1}b^2}{a^{-1} - b^{-1}}$

Solution:

$\dfrac{a - a^{-1}b^2}{a^{-1} - b^{-1}}$

$= \dfrac{a - \dfrac{b^2}{a}}{\dfrac{1}{a} - \dfrac{1}{b}}$ Rewrite the expression with positive exponents.

Answers

2. $-\dfrac{7}{22}$ 3. $\dfrac{y}{x}$

$$= \dfrac{\dfrac{a}{1} \cdot \dfrac{a}{a} - \dfrac{b^2}{a}}{\dfrac{1}{a} \cdot \dfrac{b}{b} - \dfrac{1}{b} \cdot \dfrac{a}{a}}$$

Step 1: Simplify numerator and denominator separately. Numerator LCD $= a$. Denominator LCD $= ab$.

$$= \dfrac{\dfrac{a^2}{a} - \dfrac{b^2}{a}}{\dfrac{b}{ab} - \dfrac{a}{ab}}$$

$$= \dfrac{\dfrac{a^2 - b^2}{a}}{\dfrac{b - a}{ab}}$$

Form single fractions in the numerator and denominator.

$$= \dfrac{a^2 - b^2}{a} \cdot \dfrac{ab}{b - a}$$

Step 2: Multiply the numerator of the complex fraction by the reciprocal of the denominator.

$$= \dfrac{\overset{-1}{\cancel{(a - b)}}(a + b)}{\cancel{a}} \cdot \dfrac{\overset{1}{\cancel{ab}}}{\cancel{b - a}}$$

Step 3: Factor and simplify. Recall $\dfrac{(a - b)}{(b - a)} = -1$.

$$= \dfrac{(-1)(a + b)}{1} \cdot \dfrac{b}{1}$$

$$= -b(a + b)$$

Simplify.

2. Simplifying Complex Fractions by Method II

We will now use a second method to simplify complex fractions—Method II. Recall that multiplying the numerator and denominator of a rational expression by the same quantity does not change the value of the expression. This is the basis for Method II.

Steps to Simplify a Complex Fraction—Method II

1. Multiply the numerator and denominator of the complex fraction by the LCD of *all* individual fractions within the expression.
2. Apply the distributive property, and simplify the numerator and denominator.
3. Simplify to lowest terms, if possible.

example 4   Simplifying Complex Fractions by Method II

Simplify by using Method II. $\dfrac{4 - \dfrac{6}{x}}{2 - \dfrac{3}{x}}$

Skill Practice

Simplify the expression.

4. $\dfrac{y - \dfrac{1}{y}}{1 - \dfrac{1}{y^2}}$

Answer

4. y

Solution:

$$\frac{4 - \dfrac{6}{x}}{2 - \dfrac{3}{x}}$$

The LCD of all individual terms is x.

$$= \frac{x \cdot \left(4 - \dfrac{6}{x}\right)}{x \cdot \left(2 - \dfrac{3}{x}\right)}$$

Step 1: Multiply the number and denominator of the complex fraction by the LCD, which is x.

$$= \frac{x \cdot (4) - x \cdot \left(\dfrac{6}{x}\right)}{x \cdot (2) - x \cdot \left(\dfrac{3}{x}\right)}$$

Step 2: Apply the distributive property.

$$= \frac{4x - 6}{2x - 3}$$

Step 3: Simplify numerator and denominator.

$$= \frac{2(2x - 3)}{2x - 3}$$

Factor and simplify to lowest terms.

$$= \frac{2(2x - 3)}{2x - 3}$$

$$= 2$$

Skill Practice

Simplify the expression.

5. $\dfrac{c^{-1}b - b^{-1}c}{b^{-1} + c^{-1}}$

---

example 5   Simplifying Complex Fractions by Method II

Simplify by using Method II. $\dfrac{x^{-1} - x^{-2}}{1 + 2x^{-1} - 3x^{-2}}$

Solution:

$$\frac{x^{-1} - x^{-2}}{1 + 2x^{-1} - 3x^{-2}}$$

$$= \frac{\dfrac{1}{x} - \dfrac{1}{x^2}}{1 + \dfrac{2}{x} - \dfrac{3}{x^2}}$$

Rewrite the expression with positive exponents. The LCD of all individual terms is x^2.

$$= \frac{x^2 \cdot \left(\dfrac{1}{x} - \dfrac{1}{x^2}\right)}{x^2 \cdot \left(1 + \dfrac{2}{x} - \dfrac{3}{x^2}\right)}$$

Step 1: Multiply the numerator and denominator of the complex fraction by the LCD x^2.

Answer

5. $b - c$

$$= \frac{x^2\left(\frac{1}{x}\right) - x^2\left(\frac{1}{x^2}\right)}{x^2(1) + x^2\left(\frac{2}{x}\right) - x^2\left(\frac{3}{x^2}\right)}$$ **Step 2:** Apply the distributive property.

$$= \frac{x - 1}{x^2 + 2x - 3}$$

$$= \frac{x - 1}{(x + 3)(x - 1)}$$ **Step 3:** Factor and simplify to lowest terms.

$$= \frac{\overset{1}{\cancel{x - 1}}}{(x + 3)\underset{1}{\cancel{(x - 1)}}}$$

$$= \frac{1}{x + 3}$$

example 6 — Simplifying Complex Fractions by Method II

Simplify the expression by Method II. $\dfrac{\dfrac{1}{w + 3} - \dfrac{1}{w - 3}}{1 + \dfrac{9}{w^2 - 9}}$

Skill Practice

Simplify the expression.

6. $\dfrac{\dfrac{2}{x + 1} - \dfrac{1}{x - 1}}{\dfrac{x}{x - 1} - \dfrac{1}{x + 1}}$

Solution:

$$\frac{\dfrac{1}{w + 3} - \dfrac{1}{w - 3}}{1 + \dfrac{9}{w^2 - 9}}$$

$$= \frac{\dfrac{1}{w + 3} - \dfrac{1}{w - 3}}{1 + \dfrac{9}{(w + 3)(w - 3)}}$$ Factor all denominators to find the LCD.

The LCD of $\dfrac{1}{1}, \dfrac{1}{w + 3}, \dfrac{1}{w - 3}$, and $\dfrac{9}{(w + 3)(w - 3)}$ is $(w + 3)(w - 3)$.

$$= \frac{(w + 3)(w - 3)\left(\dfrac{1}{w + 3} - \dfrac{1}{w - 3}\right)}{(w + 3)(w - 3)\left[1 + \dfrac{9}{(w + 3)(w - 3)}\right]}$$ **Step 1:** Multiply numerator and denominator of the complex fraction by $(w + 3)(w - 3)$.

$$= \frac{\cancel{(w + 3)}(w - 3)\left(\dfrac{1}{\cancel{w + 3}}\right) - (w + 3)\cancel{(w - 3)}\left(\dfrac{1}{\cancel{w - 3}}\right)}{(w + 3)(w - 3)1 + \cancel{(w + 3)}\cancel{(w - 3)}\left[\dfrac{9}{\cancel{(w + 3)}\cancel{(w - 3)}}\right]}$$ **Step 2:** Distributive property.

Answer

6. $\dfrac{x - 3}{x^2 + 1}$

$$= \frac{(w-3)-(w+3)}{(w+3)(w-3)+9} \quad \textbf{Step 3:} \quad \text{Simplify.}$$

$$= \frac{w-3-w-3}{w^2-9+9} \quad \text{Apply the distributive property.}$$

$$= \frac{-6}{w^2}$$

$$= -\frac{6}{w^2}$$

3. Using Complex Fractions in Applications

example 7 Finding the Slope of a Line

Find the slope of the line that passes through the given points.

$$\left(\frac{1}{2}, \frac{5}{6}\right) \quad \text{and} \quad \left(\frac{9}{8}, \frac{5}{4}\right)$$

Solution:

Let $(x_1, y_1) = \left(\frac{1}{2}, \frac{5}{6}\right)$ and $(x_2, y_2) = \left(\frac{9}{8}, \frac{5}{4}\right)$. Label the points.

$$\text{Slope} = \frac{y_2 - y_1}{x_2 - x_1}$$

$$= \frac{\dfrac{5}{4} - \dfrac{5}{6}}{\dfrac{9}{8} - \dfrac{1}{2}} \quad \text{Apply the slope formula.}$$

Using Method II to simplify the complex fractions, we have

$$m = \frac{24\left(\dfrac{5}{4} - \dfrac{5}{6}\right)}{24\left(\dfrac{9}{8} - \dfrac{1}{2}\right)} \quad \text{Multiply numerator and denominator by the LCD, 24.}$$

$$= \frac{\overset{6}{24}\left(\dfrac{5}{4}\right) - \overset{4}{24}\left(\dfrac{5}{6}\right)}{\overset{3}{24}\left(\dfrac{9}{8}\right) - \overset{12}{24}\left(\dfrac{1}{2}\right)} \quad \text{Apply the distributive property.}$$

$$= \frac{30 - 20}{27 - 12} \quad \text{Simplify.}$$

$$= \frac{10}{15} = \frac{2}{3}$$

The slope is $\dfrac{2}{3}$.

Skill Practice

7. Find the slope of the line that contains the points $\left(-\dfrac{3}{5}, \dfrac{1}{4}\right)$ and $\left(\dfrac{7}{10}, -\dfrac{5}{2}\right)$.

Answer

7. $m = -\dfrac{55}{26}$

Section 6.4 Practice Exercises

Boost your GRADE at mathzone.com! MathZone

- Practice Problems
- Self-Tests
- NetTutor
- e-Professors
- Videos

Study Skills Exercises

1. Careless mistakes are usually caused by losing focus on what you are doing. What are some of the distractions that you encounter when doing homework?

   List some ways you can avoid distractions while doing your homework.

2. Define the key term **complex fraction**.

Review Exercises

For Exercises 3–4, simplify to lowest terms.

3. $\dfrac{x^3 + y^3}{5x + 5y}$

4. $f(t) = \dfrac{6t^2 - 27t + 30}{12t - 30}$

For Exercises 5–8, perform the indicated operations.

5. $\dfrac{5}{x^2} + \dfrac{3}{2x}$

6. $\dfrac{2y - 4}{y + 1} \cdot \dfrac{y^2 + 3y + 2}{y^2 - 4}$

7. $\dfrac{3}{a - 5} - \dfrac{1}{a + 1}$

8. $\dfrac{7}{12 - 6b} \div \dfrac{14b}{b^2 + b - 6}$

Objective 1: Simplifying Complex Fractions by Method I

For Exercises 9–16, simplify the complex fractions by using Method I. **(See Examples 1–3.)**

9. $\dfrac{\frac{5x^2}{9y^2}}{\frac{3x}{y^2 x}}$

10. $\dfrac{\frac{3w^2}{4rs}}{\frac{15wr}{s^2}}$

11. $\dfrac{\frac{x - 6}{3x}}{\frac{3x - 18}{9}}$

12. $\dfrac{\frac{a + 4}{6}}{\frac{16 - a^2}{3}}$

13. $\dfrac{\frac{2}{3} + \frac{1}{6}}{\frac{1}{2} - \frac{1}{4}}$

14. $\dfrac{\frac{7}{8} + \frac{3}{4}}{\frac{1}{3} - \frac{5}{6}}$

15. $\dfrac{2 - \frac{1}{y}}{4 + \frac{1}{y}}$

16. $\dfrac{\frac{1}{x} - 3}{\frac{1}{x} + 3}$

Objective 2: Simplifying Complex Fractions by Method II

For Exercises 17–44, simplify the complex fractions by using Method II. **(See Examples 4–6.)**

17. $\dfrac{\frac{7y}{y + 3}}{\frac{1}{4y + 12}}$

18. $\dfrac{\frac{6x}{x - 5}}{\frac{1}{4x - 20}}$

19. $\dfrac{1 + \frac{1}{3}}{\frac{5}{6} - 1}$

20. $\dfrac{2 + \frac{4}{5}}{-1 + \frac{3}{10}}$

21. $\dfrac{\dfrac{3q}{p} - q}{q - \dfrac{q}{p}}$
22. $\dfrac{\dfrac{b}{a} + 3b}{b + \dfrac{2b}{a}}$
23. $\dfrac{\dfrac{2}{a} + \dfrac{3}{a^2}}{\dfrac{4}{a^2} - \dfrac{9}{a}}$
24. $\dfrac{\dfrac{2}{y^2} + \dfrac{1}{y}}{\dfrac{4}{y^2} - \dfrac{1}{y}}$

25. $\dfrac{t^{-1} - 1}{1 - t^{-2}}$
26. $\dfrac{d^{-2} - c^{-2}}{c^{-1} - d^{-1}}$
27. $\dfrac{-8}{\dfrac{6w}{w-1} - 4}$
28. $\dfrac{6}{2z - \dfrac{10}{z-4}}$

29. $\dfrac{\dfrac{y}{y+3}}{\dfrac{y}{y+3} + y}$
30. $\dfrac{\dfrac{4}{w-4}}{\dfrac{4}{w-4} - 1}$
31. $\dfrac{1 - \dfrac{1}{x} - \dfrac{6}{x^2}}{1 - \dfrac{4}{x} + \dfrac{3}{x^2}}$
32. $\dfrac{1 + \dfrac{1}{x} - \dfrac{12}{x^2}}{\dfrac{9}{x^2} + \dfrac{3}{x} - 2}$

33. $\dfrac{2 - \dfrac{2}{t+1}}{2 + \dfrac{2}{t}}$
34. $\dfrac{3 + \dfrac{3}{p-1}}{3 - \dfrac{3}{p}}$
35. $\dfrac{\dfrac{2}{a} - \dfrac{3}{a+1}}{\dfrac{2}{a+1} - \dfrac{3}{a}}$
36. $\dfrac{\dfrac{5}{b} + \dfrac{4}{b+1}}{\dfrac{4}{b} - \dfrac{5}{b+1}}$

37. $\dfrac{\dfrac{1}{y+2} + \dfrac{4}{y-3}}{\dfrac{2}{y-3} - \dfrac{7}{y+2}}$
38. $\dfrac{\dfrac{1}{t-4} + \dfrac{1}{t+5}}{\dfrac{6}{t+5} + \dfrac{2}{t-4}}$
39. $\dfrac{\dfrac{2}{x+h} - \dfrac{2}{x}}{h}$
40. $\dfrac{\dfrac{1}{2x+2h} - \dfrac{1}{2x}}{h}$

41. $\dfrac{x^{-2}}{x + 3x^{-1}}$
42. $\dfrac{x^{-1} + x^{-2}}{5x^{-2}}$
43. $\dfrac{2a^{-1} + 3b^{-2}}{a^{-1} - b^{-1}}$
44. $\dfrac{2m^{-1} + n^{-1}}{m^{-2} - 4n^{-1}}$

Objective 3: Using Complex Fractions in Applications

45. The slope formula is used to find the slope of the line passing through the points (x_1, y_1) and (x_2, y_2). Write the slope formula from memory.

For Exercises 46–49, find the slope of the line that passes through the given points. **(See Example 7.)**

46. $\left(1\dfrac{1}{2}, \dfrac{2}{5}\right), \left(\dfrac{1}{4}, -2\right)$
47. $\left(-\dfrac{3}{7}, \dfrac{3}{5}\right), (-1, -3)$
48. $\left(\dfrac{5}{8}, \dfrac{9}{10}\right), \left(-\dfrac{1}{16}, -\dfrac{1}{5}\right)$
49. $\left(\dfrac{1}{4}, \dfrac{1}{3}\right), \left(\dfrac{1}{8}, \dfrac{1}{6}\right)$

Expanding Your Skills

50. Show that $(x + x^{-1})^{-1} = \dfrac{x}{x^2 + 1}$ by writing the expression on the left without negative exponents and simplifying.

51. Show that $(x^{-1} + y^{-1})^{-1} = \dfrac{xy}{x+y}$ by writing the expression on the left without negative exponents and simplifying.

52. Simplify. $\dfrac{x}{1 - \left(1 + \dfrac{1}{x}\right)^{-1}}$

53. Simplify. $\dfrac{x}{1 - \left(1 - \dfrac{1}{x}\right)^{-1}}$

section 6.5 Rational Equations

Objectives
1. Solving Rational Equations
2. Formulas Involving Rational Equations

1. Solving Rational Equations

Thus far we have studied two types of equations in one variable: linear equations and quadratic equations. In this section, we will study another type of equation called a rational equation.

> **Definition of a Rational Equation**
>
> An equation with one or more rational expressions is called a **rational equation**.

The following equations are rational equations:

$$\frac{1}{2}x + \frac{1}{3} = \frac{1}{4}x \qquad \frac{3}{5} + \frac{1}{x} = \frac{2}{3} \qquad 3 - \frac{6w}{w+1} = \frac{6}{w+1}$$

To understand the process of solving a rational equation, first review the procedure of clearing fractions from Section 1.4.

example 1 Solving a Rational Equation

Solve the equation. $\dfrac{1}{2}x + \dfrac{1}{3} = \dfrac{1}{4}x$

Solution:

$\dfrac{1}{2}x + \dfrac{1}{3} = \dfrac{1}{4}x$ The LCD of all terms in the equation is 12.

$12\left(\dfrac{1}{2}x + \dfrac{1}{3}\right) = 12\left(\dfrac{1}{4}x\right)$ Multiply both sides by 12 to clear fractions.

$12 \cdot \dfrac{1}{2}x + 12 \cdot \dfrac{1}{3} = 12 \cdot \dfrac{1}{4}x$ Apply the distributive property.

$6x + 4 = 3x$ Solve the resulting equation.

$3x = -4$

$x = -\dfrac{4}{3}$

Skill Practice

Solve the equation.

1. $\dfrac{1}{2}x + \dfrac{21}{20} = \dfrac{1}{5}x$

Answer

1. $x = -\dfrac{7}{2}$

Check:
$$\frac{1}{2}x + \frac{1}{3} = \frac{1}{4}x$$

$$\frac{1}{2}\left(-\frac{4}{3}\right) + \frac{1}{3} \stackrel{?}{=} \frac{1}{4}\left(-\frac{4}{3}\right)$$

$$-\frac{2}{3} + \frac{1}{3} \stackrel{?}{=} -\frac{1}{3}$$

$$-\frac{1}{3} = -\frac{1}{3} \checkmark$$

The same process of clearing fractions is used to solve rational equations when variables are present in the denominator.

Skill Practice

2. Solve. $\dfrac{3}{y} + \dfrac{4}{3} = -1$

example 2 Solving a Rational Equation

Solve the equation. $\quad \dfrac{3}{5} + \dfrac{1}{x} = \dfrac{2}{3}$

Solution:

$$\frac{3}{5} + \frac{1}{x} = \frac{2}{3}$$ The LCD of all terms in the equation is $15x$.

$$15x\left(\frac{3}{5} + \frac{1}{x}\right) = 15x\left(\frac{2}{3}\right)$$ Multiply by $15x$ to clear fractions.

$$15x \cdot \frac{3}{5} + 15x \cdot \frac{1}{x} = 15x \cdot \frac{2}{3}$$ Apply the distributive property.

$$9x + 15 = 10x$$ Solve the resulting equation.

$$15 = x$$

Check: $x = 15$
$$\frac{3}{5} + \frac{1}{x} = \frac{2}{3}$$

$$\frac{3}{5} + \frac{1}{(15)} \stackrel{?}{=} \frac{2}{3}$$

$$\frac{9}{15} + \frac{1}{15} \stackrel{?}{=} \frac{2}{3}$$

$$\frac{10}{15} = \frac{2}{3} \checkmark$$

example 3 Solving a Rational Equation

Solve the equation. $\quad 3 - \dfrac{6w}{w+1} = \dfrac{6}{w+1}$

Answer

2. $y = -\dfrac{9}{7}$

Solution:

$$3 - \frac{6w}{w+1} = \frac{6}{w+1}$$

The LCD of all terms in the equation is $w + 1$.

$$(w+1)(3) - (w+1)\left(\frac{6w}{w+1}\right) = (w+1)\left(\frac{6}{w+1}\right)$$

Multiply by $(w+1)$ on both sides to clear fractions.

$$(w+1)(3) - (\cancel{w+1})\left(\frac{6w}{\cancel{w+1}}\right) = (\cancel{w+1})\left(\frac{6}{\cancel{w+1}}\right)$$

Apply the distributive property.

$$3w + 3 - 6w = 6$$

Solve the resulting equation.

$$-3w = 3$$

$$w = -1$$

Check:
$$3 - \frac{6w}{w+1} = \frac{6}{w+1}$$

$$3 - \frac{6(-1)}{(-1)+1} \stackrel{?}{=} \frac{6}{(-1)+1}$$

The denominator is 0 for the value of $w = -1$.

Because the value $w = -1$ makes the denominator zero in one (or more) of the rational expressions within the equation, the equation is *undefined* for $w = -1$. No other potential solutions exist, so the equation $3 - \dfrac{6w}{w+1} = \dfrac{6}{w+1}$ has no solution.

> **Skill Practice**
>
> Solve the equations.
>
> 3. $\dfrac{4}{b+2} + \dfrac{2}{b-2} = \dfrac{8}{b^2-4}$
>
> 4. $\dfrac{3}{z-5} - 4 = \dfrac{2z}{z-5}$

Examples 1–3 show that the steps for solving a rational equation mirror the process of clearing fractions from Section 1.4. However, there is one significant difference. The solutions of a rational equation must be defined in each rational expression in the equation. When $w = -1$ is substituted into the expression $\dfrac{6w}{w+1}$ or $\dfrac{6}{w+1}$, the denominator is zero and the expression is undefined. Hence $w = -1$ cannot be a solution to the equation

$$3 - \frac{6w}{w+1} = \frac{6}{w+1}$$

The steps for solving a rational equation are summarized as follows.

Steps for Solving a Rational Equation

1. Factor the denominators of all rational expressions. Identify any values of the variable for which any expression is undefined.
2. Identify the LCD of all terms in the equation.
3. Multiply both sides of the equation by the LCD.
4. Solve the resulting equation.
5. Check the potential solutions in the original equation. Note that any value from step 1 for which the equation is undefined cannot be a solution to the equation.

Answers

3. No solution ($b = 2$ does not check.)
4. $z = \dfrac{23}{6}$

Chapter 6 Rational Expressions and Rational Equations

Skill Practice

Solve the equations.

5. $\dfrac{y}{4} - \dfrac{1}{2} = \dfrac{2}{y}$

6. $\dfrac{6}{x+2} - \dfrac{20x}{x^2 - x - 6} = \dfrac{x}{x+2}$

example 4 Solving Rational Equations

Solve the equations.

a. $1 + \dfrac{3}{x} = \dfrac{28}{x^2}$ b. $\dfrac{36}{p^2 - 9} = \dfrac{2p}{p+3} - 1$

Solution:

a. $1 + \dfrac{3}{x} = \dfrac{28}{x^2}$ The LCD of all terms in the equation is x^2. Expressions will be undefined for $x = 0$.

$x^2\left(1 + \dfrac{3}{x}\right) = x^2\left(\dfrac{28}{x^2}\right)$ Multiply both sides by x^2 to clear fractions.

$x^2 \cdot 1 + x^2 \cdot \dfrac{3}{x} = x^2 \cdot \dfrac{28}{x^2}$ Apply the distributive property.

$x^2 + 3x = 28$ The resulting equation is quadratic.

$x^2 + 3x - 28 = 0$ Set the equation equal to zero and factor.

$(x + 7)(x - 4) = 0$

$x = -7$ or $x = 4$

Check: $x = -7$ Check: $x = 4$

$1 + \dfrac{3}{x} = \dfrac{28}{x^2}$ $1 + \dfrac{3}{x} = \dfrac{28}{x^2}$

$1 + \dfrac{3}{-7} \stackrel{?}{=} \dfrac{28}{(-7)^2}$ $1 + \dfrac{3}{4} \stackrel{?}{=} \dfrac{28}{(4)^2}$

$\dfrac{49}{49} - \dfrac{21}{49} \stackrel{?}{=} \dfrac{28}{49}$ $\dfrac{16}{16} + \dfrac{12}{16} \stackrel{?}{=} \dfrac{28}{16}$

$\dfrac{28}{49} = \dfrac{28}{49}$ ✓ $\dfrac{28}{16} = \dfrac{28}{16}$ ✓   Both solutions check.

b. $\dfrac{36}{p^2 - 9} = \dfrac{2p}{p+3} - 1$

$\dfrac{36}{(p+3)(p-3)} = \dfrac{2p}{p+3} - 1$ The LCD is $(p+3)(p-3)$. Expressions will be undefined for $p = 3$ and $p = -3$.

Multiply both sides by the LCD to clear fractions.

$(p+3)(p-3)\left[\dfrac{36}{(p+3)(p-3)}\right] = (p+3)(p-3)\left(\dfrac{2p}{p+3}\right) - (p+3)(p-3)1$

$\cancel{(p+3)(p-3)}\left[\dfrac{36}{\cancel{(p+3)(p-3)}}\right] = \cancel{(p+3)}(p-3)\left(\dfrac{2p}{\cancel{p+3}}\right) - (p+3)(p-3)1$

$36 = 2p(p-3) - (p+3)(p-3)$ Solve the resulting equation.

$36 = 2p^2 - 6p - (p^2 - 9)$ The equation is quadratic.

Answers

5. $y = 4;\ y = -2$
6. $x = -9$ ($x = -2$ does not check.)

$36 = 2p^2 - 6p - p^2 + 9$

$36 = p^2 - 6p + 9$

$0 = p^2 - 6p - 27$ Set the equation equal to zero and factor.

$0 = (p - 9)(p + 3)$

$p = 9$ or $p = -3$

Check: $p = 9$

$$\frac{36}{p^2 - 9} = \frac{2p}{p + 3} - 1$$

$$\frac{36}{(9)^2 - 9} \stackrel{?}{=} \frac{2(9)}{(9) + 3} - 1$$

$$\frac{36}{72} \stackrel{?}{=} \frac{18}{12} - 1$$

$$\frac{1}{2} \stackrel{?}{=} \frac{3}{2} - 1$$

$$\frac{1}{2} = \frac{1}{2} \checkmark$$

Check: $p = -3$

$$\frac{36}{p^2 - 9} = \frac{2p}{p + 3} - 1$$

$$\frac{36}{(-3)^2 - 9} \stackrel{?}{=} \frac{2(-3)}{(-3) + 3} - 1$$

Denominator is zero.

Here $p = -3$ is *not* a solution to the original equation because it is undefined in the original equation. However, $p = 9$ checks in the original equation.

The solution is $p = 9$.

2. Formulas Involving Rational Equations

example 5 Solving Literal Equations Involving Rational Expressions

Solve for the indicated variable.

a. $P = \dfrac{A}{1 + rt}$ for r **b.** $V = \dfrac{mv}{m + M}$ for m

Solution:

a. $P = \dfrac{A}{1 + rt}$ for r

$(1 + rt)P = (1 + rt)\left(\dfrac{A}{1 + rt}\right)$ Multiply both sides by the LCD $= (1 + rt)$.

$(1 + rt)P = \cancel{(1 + rt)}\left(\dfrac{A}{\cancel{1 + rt}}\right)$ Clear fractions.

$P + Prt = A$ Apply the distributive property to clear parentheses.

$Prt = A - P$ Isolate the r-term on one side.

$\dfrac{Prt}{Pt} = \dfrac{A - P}{Pt}$

$r = \dfrac{A - P}{Pt}$

Skill Practice

7. Solve the equation for q.

$$\frac{1}{q} + \frac{1}{r} = \frac{1}{s}$$

8. Solve the equation for x.

$$y = \frac{ax + b}{x + d}$$

Answers

7. $q = \dfrac{rs}{r - s}$ or $q = \dfrac{-rs}{s - r}$

8. $x = \dfrac{b - yd}{y - a}$ or $x = \dfrac{yd - b}{a - y}$

Avoiding Mistakes: Variables in algebra are case-sensitive. Therefore, V and v are different variables.

b. $\quad V = \dfrac{mv}{m+M} \quad$ for m

$V(m+M) = \left(\dfrac{mv}{\cancel{m+M}}\right)(\cancel{m+M})\qquad$ Multiply by the LCD and clear fractions.

$V(m+M) = mv$

$Vm + VM = mv \qquad$ Use the distributive property to clear parentheses.

$Vm - mv = -VM \qquad$ Collect all m terms on one side.

$m(V-v) = -VM \qquad$ Factor out m.

$\dfrac{m\cancel{(V-v)}}{\cancel{(V-v)}} = \dfrac{-VM}{(V-v)} \qquad$ Divide by $(V-v)$.

$m = \dfrac{-VM}{V-v}$

Tip: The factor of -1 that appears in the numerator may be written in the denominator or out in front of the expression. The following expressions are equivalent:

$$m = \dfrac{-VM}{V-v}$$

$$= \dfrac{VM}{-(V-v)} \quad \text{or} \quad \dfrac{VM}{v-V}$$

$$= -\dfrac{VM}{V-v}$$

---

section 6.5 Practice Exercises

Boost your GRADE at mathzone.com!
MathZone

- Practice Problems
- Self-Tests
- NetTutor
- e-Professors
- Videos

Study Skills Exercises

1. Organization is an important ingredient to success. A calendar or pocket planner is a valuable resource for keeping track of assignments and test dates. What is the date of the next test in this class?

2. Define the key term **rational equation**.

Review Exercises

For Exercises 3–8, perform the indicated operations.

3. $\dfrac{1}{x^2 - 16} + \dfrac{1}{x^2 + 8x + 16}$

4. $\dfrac{3}{y^2 - 1} - \dfrac{2}{y^2 - 2y + 1}$

5. $\dfrac{m^2 - 9}{m^2 - 3m} \div (m^2 - m - 12)$

6. $\dfrac{2t^2 + 7t + 3}{4t^2 - 1} \div (t + 3)$

7. $\dfrac{1 + x^{-1}}{1 - x^{-2}}$

8. $\dfrac{x + y}{x^{-1} + y^{-1}}$

Objective 1: Solving Rational Equations

9. **a.** Identify the LCD of all terms in the equation.

$$\dfrac{x + 2}{3} - \dfrac{x - 4}{4} = \dfrac{1}{2}$$

  b. Solve the equation.

  c. Check the answer in the original equation.

10. **a.** Identify the LCD of all terms in the equation.

$$\dfrac{x + 6}{3} - \dfrac{x + 8}{5} = 0$$

  b. Solve the equation.

  c. Check the answer in the original equation.

11. **a.** Identify the LCD of all terms in the equation.

$$\dfrac{x}{x - 5} + \dfrac{1}{5} = \dfrac{5}{x - 5}$$

  b. Solve the equation.

12. **a.** Identify the LCD of all terms in the equation.

$$\dfrac{x}{x - 2} + \dfrac{2}{3} = \dfrac{2}{x - 2}$$

  b. Solve the equation.

For Exercises 13–40, solve the rational equations. (See Examples 1–4.)

13. $\dfrac{3y}{4} - 6 = \dfrac{y}{4}$

14. $\dfrac{2w}{5} - 8 = \dfrac{4w}{5}$

15. $\dfrac{5}{4p} - \dfrac{7}{6} + 3 = 0$

16. $\dfrac{7}{15w} - \dfrac{3}{10} - 2 = 0$

17. $\dfrac{1}{2} - \dfrac{3}{2x} = \dfrac{4}{x} - \dfrac{5}{12}$

18. $\dfrac{2}{3x} + \dfrac{1}{4} = \dfrac{11}{6x} - \dfrac{1}{3}$

19. $\dfrac{3}{x - 4} + 2 = \dfrac{5}{x - 4}$

20. $\dfrac{5}{x + 3} - 2 = \dfrac{7}{x + 3}$

21. $\dfrac{1}{3} + \dfrac{2}{w - 3} = 1$

22. $\dfrac{3}{5} + \dfrac{7}{p + 2} = 2$

23. $\dfrac{12}{x} - \dfrac{12}{x - 5} = \dfrac{2}{x}$

24. $\dfrac{25}{y} - \dfrac{25}{y - 2} = \dfrac{2}{y}$

25. $\dfrac{1}{4}a - 4a^{-1} = 0$

26. $\dfrac{1}{3}t - 12t^{-1} = 0$

27. $3a^{-2} - 4a^{-1} = -1$

28. $-3w^{-1} = 2 + w^{-1}$

29. $8t^{-1} + 2 = 3t^{-1}$

30. $6z^{-2} - 5z^{-1} = 0$

31. $\dfrac{4}{t - 2} - \dfrac{8}{t^2 - 2t} = -2$

32. $\dfrac{x}{x + 6} = \dfrac{72}{x^2 - 36} + 4$

33. $\dfrac{6}{5y + 10} - \dfrac{1}{y - 5} = \dfrac{4}{y^2 - 3y - 10}$

34. $\dfrac{-3}{x^2 - 7x + 12} - \dfrac{2}{x^2 + x - 12} = \dfrac{10}{x^2 - 16}$

35. $\dfrac{6}{x^2 - 4x + 3} - \dfrac{1}{x - 3} = \dfrac{1}{4x - 4}$

36. $\dfrac{1}{4x^2 - 36} - \dfrac{5}{x + 3} + \dfrac{2}{x - 3} = 0$

37. $\dfrac{3}{k - 2} - \dfrac{5}{k + 2} + \dfrac{6k}{4 - k^2} = 0$

38. $\dfrac{h}{2} - \dfrac{h}{h - 4} = -\dfrac{4}{h - 4}$

39. $\dfrac{5}{x^2 - 7x + 12} = \dfrac{2}{x - 3} + \dfrac{5}{x - 4}$

40. $\dfrac{9}{x^2 + 7x + 10} = \dfrac{5}{x + 2} - \dfrac{3}{x + 5}$

Objective 2: Formulas Involving Rational Equations

For Exercises 41–58, solve the formula for the indicated variable. (See Example 5.)

41. $K = \dfrac{ma}{F}$ for m

42. $K = \dfrac{ma}{F}$ for a

43. $K = \dfrac{IR}{E}$ for E

44. $K = \dfrac{IR}{E}$ for R

45. $I = \dfrac{E}{R + r}$ for R

46. $I = \dfrac{E}{R + r}$ for r

47. $h = \dfrac{2A}{B + b}$ for B

48. $\dfrac{V}{\pi h} = r^2$ for h

49. $x = \dfrac{at + b}{t}$ for t

50. $\dfrac{T + mf}{m} = g$ for m

51. $\dfrac{x - y}{xy} = z$ for x

52. $\dfrac{w - n}{wn} = P$ for w

53. $a + b = \dfrac{2A}{h}$ for h

54. $1 + rt = \dfrac{A}{P}$ for P

55. $\dfrac{1}{R} = \dfrac{1}{R_1} + \dfrac{1}{R_2}$ for R

56. $\dfrac{b + a}{ab} = \dfrac{1}{f}$ for b

57. $v = \dfrac{s_2 - s_1}{t_2 - t_1}$ for t_2

58. $a = \dfrac{v_2 - v_1}{t_2 - t_1}$ for v_1

Mixed Review

59. a. Simplify. $\dfrac{3}{w - 5} + \dfrac{10}{w^2 - 25} - \dfrac{1}{w + 5}$

   b. Solve. $\dfrac{3}{w - 5} + \dfrac{10}{w^2 - 25} - \dfrac{1}{w + 5} = 0$

   c. What is the difference in the type of problem given in parts (a) and (b)?

60. a. Simplify. $\dfrac{x}{2x + 4} + \dfrac{2}{3x + 6} - 1$

   b. Solve. $\dfrac{x}{2x + 4} + \dfrac{2}{3x + 6} = 1$

   c. What is the difference in the type of problem given in parts (a) and (b)?

For Exercises 61–70, perform the indicated operation and simplify, or solve the equation for the variable.

61. $\dfrac{2}{a^2 + 4a + 3} + \dfrac{1}{a + 3}$

62. $\dfrac{1}{c + 6} + \dfrac{4}{c^2 + 8c + 12}$

63. $\dfrac{7}{y^2 - y - 2} + \dfrac{1}{y + 1} - \dfrac{3}{y - 2} = 0$

64. $\dfrac{3}{b + 2} - \dfrac{1}{b - 1} - \dfrac{5}{b^2 + b - 2} = 0$

65. $\dfrac{x}{x - 1} - \dfrac{12}{x^2 - x}$

66. $\dfrac{3}{5t - 20} + \dfrac{4}{t - 4}$

67. $\dfrac{3}{w} - 5 = \dfrac{7}{w} - 1$

68. $\dfrac{-3}{y^2} - \dfrac{1}{y} = -2$

69. $\dfrac{4p + 1}{8p - 12} + \dfrac{p - 3}{2p - 3}$

70. $\dfrac{x + 1}{2x + 4} - \dfrac{x^2}{x + 2}$

Expanding Your Skills

71. If 5 is added to the reciprocal of a number, the result is $\frac{16}{3}$. Find the number.

72. If $\frac{2}{3}$ is added to the reciprocal of a number, the result is $\frac{17}{3}$. Find the number.

73. If 7 is decreased by the reciprocal of a number, the result is $\frac{9}{2}$. Find the number.

74. If a number is added to its reciprocal, the result is $\frac{13}{6}$. Find the number.

---

section 6.6 Applications of Rational Equations and Proportions

1. Solving Proportions

A proportion is a rational equation that equates two ratios.

Objectives
1. Solving Proportions
2. Applications of Proportions
3. Similar Triangles
4. Applications of Rational Equations

> **Definition of Ratio and Proportion**
>
> 1. The **ratio** of a to b is $\dfrac{a}{b}$ ($b \neq 0$) and can also be expressed as $a:b$ or $a \div b$.
> 2. An equation that equates two ratios or rates is called a **proportion**. Therefore, if $b \neq 0$ and $d \neq 0$, then $\dfrac{a}{b} = \dfrac{c}{d}$ is a proportion.

The process for solving rational equations can be used to solve proportions.

Skill Practice

Solve the proportion.

1. $\dfrac{8}{5} = \dfrac{12}{x}$

example 1 Solving a Proportion

Solve the proportion. $\dfrac{5}{19} = \dfrac{95}{y}$

Solution:

$$\dfrac{5}{19} = \dfrac{95}{y}$$ The LCD is $19y$. Note that $y \neq 0$.

$$19y\left(\dfrac{5}{19}\right) = 19y\left(\dfrac{95}{y}\right)$$ Multiply both sides by the LCD.

$$\cancel{19}y\left(\dfrac{5}{\cancel{19}}\right) = 19\cancel{y}\left(\dfrac{95}{\cancel{y}}\right)$$ Clear fractions.

$$5y = 1805$$ Solve the resulting equation.

$$\dfrac{5y}{5} = \dfrac{1805}{5}$$

$$y = 361$$ The solution checks in the original equation.

Tip: For any proportion

$$\dfrac{a}{b} = \dfrac{c}{d} \qquad b \neq 0, d \neq 0$$

the cross products of terms are equal. Hence, $ad = bc$. Finding the cross product is a quick way to clear fractions in a proportion.* Consider Example 1:

$$\dfrac{5}{19} \nwarrow\!\!\!\nearrow \dfrac{95}{y}$$

$$5y = (19)(95) \qquad \text{Equate the cross products.}$$

$$5y = 1805$$

$$y = 361$$

*It is important to realize that this method is only valid for proportions.

2. Applications of Proportions

Skill Practice

2. The ratio of cats to dogs at an animal rescue facility is 8 to 5. How many dogs are in the facility if there are 400 cats?

example 2 Solving a Proportion

In the U.S. Senate in 2005, the ratio of Democrats to Republicans was 4 to 5. If there were 44 Democrats, how many Republicans were there?

Solution:

One method of solving this problem is to set up a proportion. Write two equivalent ratios depicting the number of Democrats to the number of Republicans. Let x represent the unknown number of Republicans in the Senate.

Given ratio → $\dfrac{4}{5} = \dfrac{44}{x}$ ← Number of Democrats / Number of Republicans

Answers

1. $x = \dfrac{15}{2}$ or 7.5   2. 250 dogs

$$5x\left(\frac{4}{5}\right) = 5x\left(\frac{44}{x}\right) \quad \text{Multiply both sides by the LCD } 5x.$$

$$4x = 5(44) \quad \text{Clear fractions.}$$

$$4x = 220$$

$$x = 55$$

There were 55 Republicans in the U.S. Senate in 2005.

example 3 — Solving a Proportion

The ratio of male to female police officers in a certain town is 11:3. If the total number of officers is 112, how many are men and how many are women?

Solution:

Let x represent the number of male police officers.

Then $112 - x$ represents the number of female police officers.

$$\frac{\text{Male}}{\text{Female}} \to \frac{11}{3} = \frac{x}{112-x} \leftarrow \frac{\text{Number of males}}{\text{Number of females}}$$

$$3(112 - x)\left(\frac{11}{3}\right) = 3(112-x)\left(\frac{x}{112-x}\right) \quad \text{Multiply both sides by } 3(112-x).$$

$$11(112 - x) = 3x \quad \text{The resulting equation is linear.}$$

$$1232 - 11x = 3x$$

$$1232 = 14x$$

$$\frac{1232}{14} = \frac{14x}{14}$$

$$x = 88$$

The number of male police officers is $x = 88$.

The number of female officers is $112 - x = 112 - 88 = 24$.

Skill Practice

3. Professor Wolfe has a ratio of passing students to failing students of 5 to 4. One semester he had a total of 207 students. How many students passed and how many failed?

3. Similar Triangles

Proportions are used in geometry with **similar triangles**. Two triangles are said to be similar if their corresponding angles are equal. In such a case, the lengths of the corresponding sides are proportional. The triangles in Figure 6-5 are similar. Therefore, the following ratios are equivalent.

$$\frac{a}{x} = \frac{b}{y} = \frac{c}{z}$$

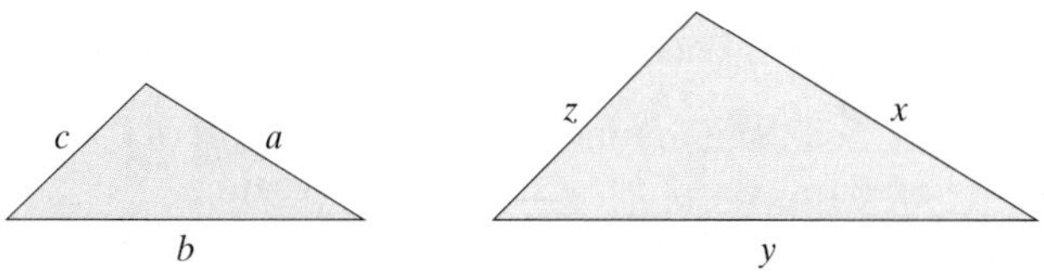

Figure 6-5

Answer

3. 115 passed and 92 failed.

Skill Practice

4. Solve for *x*, given that the triangles are similar.

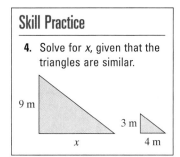

example 4 — Using Similar Triangles in an Application

The shadow cast by a yardstick is 2 ft long. The shadow cast by a tree is 11 ft long. Find the height of the tree.

Solution:

Let *x* represent the height of the tree.   Label the variables.

We will assume that the measurements were taken at the same time of day. Therefore, the angle of the sun is the same on both objects, and we can set up similar triangles (Figure 6-6).

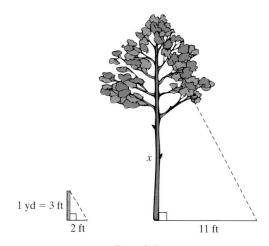

Figure 6-6

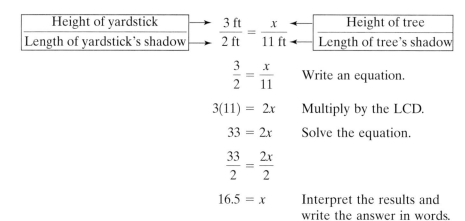

$$\frac{3}{2} = \frac{x}{11} \quad \text{Write an equation.}$$

$$3(11) = 2x \quad \text{Multiply by the LCD.}$$

$$33 = 2x \quad \text{Solve the equation.}$$

$$\frac{33}{2} = \frac{2x}{2}$$

$$16.5 = x \quad \text{Interpret the results and write the answer in words.}$$

The tree is 16.5 ft high.

4. Applications of Rational Equations

example 5 — Solving a Business Application of Rational Expressions

To produce mountain bikes, the manufacturer has a fixed cost of $56,000 plus a variable cost of $140 per bike. The cost in dollars to produce *x* bikes is given by

Answer

4. $x = 12$ m

$C(x) = 56,000 + 140x$. The average cost per bike $\overline{C}(x)$ is defined as $\overline{C}(x) = \dfrac{C(x)}{x}$ or

$$\overline{C}(x) = \dfrac{56,000 + 140x}{x}$$

Find the number of bikes the manufacturer must produce so that the average cost per bike is $180.

Solution:

Given

$$\overline{C}(x) = \dfrac{56,000 + 140x}{x}$$

$180 = \dfrac{56,000 + 140x}{x}$ Replace $\overline{C}(x)$ by 180 and solve for x.

$x(180) = x \cdot \dfrac{56,000 + 140x}{x}$ Multiply by the LCD x.

$180x = 56,000 + 140x$ The resulting equation is linear.

$40x = 56,000$ Solve for x.

$x = \dfrac{56,000}{40}$

$= 1400$

The manufacturer must produce 1400 bikes for the average cost to be $180.

Skill Practice

5. To manufacture ski equipment, a company has a fixed cost of $48,000 plus a variable cost of $200 per pair of skis. The cost to produce x pairs of skis is $C(x) = 48,000 + 200x$. The average cost per pair of skis is

$$\overline{C}(x) = \dfrac{48,000 + 200x}{x}$$

Find the number of sets of skis the manufacturer must produce so that the average cost per set is $250.

---

example 6   Solving an Application Involving Distance, Rate, and Time

An athlete's average speed on her bike is 14 mph faster than her average speed running. She can bike 31.5 mi in the same time that it takes her to run 10.5 mi. Find her speed running and her speed biking.

Solution:

Because the speed biking is given in terms of the speed running, let x represent the running speed.

Let x represent the speed running.

Then $x + 14$ represents the speed biking.

Organize the given information in a chart.

	Distance	Rate	Time
Running	10.5	x	$\dfrac{10.5}{x}$
Biking	31.5	$x + 14$	$\dfrac{31.5}{x + 14}$

Because $d = rt$, then $t = \dfrac{d}{r}$

Skill Practice

6. Devon can cross-country ski 5 km/hr faster than his sister, Shanelle. Devon skis 45 km in the same amount of time that Shanelle skis 30 km. Find their speeds.

Answers

5. 960 pairs of skis
6. Shanelle skis 10 km/hr and Devon skis 15 km/hr.

448   Chapter 6   Rational Expressions and Rational Equations

The time required to run 10.5 mi is the same as the time required to bike 31.5 mi, so we can equate the two expressions for time:

$$\frac{10.5}{x} = \frac{31.5}{x+14}$$ The LCD is $x(x+14)$.

$$x(x+14)\left(\frac{10.5}{x}\right) = x(x+14)\left(\frac{31.5}{x+14}\right)$$ Multiply by $x(x+14)$ to clear fractions.

$$10.5(x+14) = 31.5x$$ The resulting equation is linear.

$$10.5x + 147 = 31.5x$$ Solve for x.

$$-21x = -147$$

$$x = 7$$

The athlete's speed running is 7 mph.

The speed biking is $x + 14$ or $7 + 14 = 21$ mph.

Skill Practice

7. Antonio can install a new roof in 4 days. Carlos can install the same size roof in 6 days. How long will it take them to install a roof if they work together?

example 7   Solving an Application Involving "Work"

JoAn can wallpaper a bathroom in 3 hr. Bonnie can wallpaper the same bathroom in 5 hr. How long would it take them if they worked together?

Solution:

Let x represent the time required for both people working together to complete the job.

One method to approach this problem is to determine the portion of the job that each person can complete in 1 hr and extend that rate to the portion of the job completed in x hr.

- JoAn can perform the job in 3 hr. Therefore, she completes $\frac{1}{3}$ of the job in 1 hr and $\frac{1}{3}x$ jobs in x hr.
- Bonnie can perform the job in 5 hr. Therefore, she completes $\frac{1}{5}$ of the job in 1 hr and $\frac{1}{5}x$ jobs in x hr.

	Work Rate	Time	Portion of Job Completed
JoAn	$\frac{1}{3}$ job/hr	x hr	$\frac{1}{3}x$ jobs
Bonnie	$\frac{1}{5}$ job/hr	x hr	$\frac{1}{5}x$ jobs

The sum of the portions of the job completed by each person must equal one whole job:

$$\begin{pmatrix}\text{Portion of job}\\\text{completed}\\\text{by JoAn}\end{pmatrix} + \begin{pmatrix}\text{portion of job}\\\text{completed}\\\text{by Bonnie}\end{pmatrix} = \begin{pmatrix}1\\\text{whole}\\\text{job}\end{pmatrix}$$

$$\frac{1}{3}x + \frac{1}{5}x = 1$$ The LCD is 15.

Answer

7. $\frac{12}{5}$ or $2\frac{2}{5}$ days

$$15\left(\frac{1}{3}x + \frac{1}{5}x\right) = 15(1) \qquad \text{Multiply by the LCD.}$$

$$15 \cdot \frac{1}{3}x + 15 \cdot \frac{1}{5}x = 15 \cdot 1 \qquad \text{Apply the distributive property.}$$

$$5x + 3x = 15 \qquad \text{Solve the resulting linear equation.}$$

$$8x = 15$$

$$x = \frac{15}{8} \quad \text{or} \quad x = 1\frac{7}{8}$$

Together JoAn and Bonnie can wallpaper the bathroom in $1\frac{7}{8}$ hr.

section 6.6 Practice Exercises

Boost your GRADE at mathzone.com!

- Practice Problems
- Self-Tests
- NetTutor
- e-Professors
- Videos

Study Skills Exercises

1. Setting goals is an important prerequisite to success. This includes goals for completing course requirements. List several goals that you wish to complete this week, this month, and this semester.

   a. Week   **b.** Month   **c.** Semester

2. Define the key terms.

   a. Ratio   **b.** Proportion   **c.** Similar triangles

Review Exercises

For Exercises 3–10, perform the indicated operation and simplify, or solve the equation for the variable.

3. $3 - \dfrac{6}{x} = x + 8$

4. $2 + \dfrac{6}{x} = x + 7$

5. $\dfrac{5}{3x - 6} - \dfrac{3}{4x - 8}$

6. $\dfrac{4}{5t - 1} + \dfrac{1}{10t - 2}$

7. $\dfrac{2}{y - 1} - \dfrac{5}{4} = \dfrac{-1}{y + 1}$

8. $\dfrac{5}{w - 2} = 7 - \dfrac{10}{w + 2}$

9. $\dfrac{5}{x^2 - y^2} + \dfrac{3x}{x^3 + x^2 y}$

10. $\dfrac{7a}{a^2 + 2ab + b^2} + \dfrac{4}{a^2 + ab}$

Objective 1: Solving Proportions

For Exercises 11–24, solve the proportions. (See Example 1.)

11. $\dfrac{y}{6} = \dfrac{20}{15}$

12. $\dfrac{12}{18} = \dfrac{14}{x}$

13. $\dfrac{9}{75} = \dfrac{m}{50}$

14. $\dfrac{n}{15} = \dfrac{12}{45}$

15. $\dfrac{p-1}{4} = \dfrac{p+3}{3}$

16. $\dfrac{q-5}{2} = \dfrac{q+2}{3}$

17. $\dfrac{x+1}{5} = \dfrac{4}{15}$

18. $\dfrac{t-1}{7} = \dfrac{2}{21}$

19. $\dfrac{5-2x}{x} = \dfrac{1}{4}$

20. $\dfrac{2y+3}{y} = \dfrac{3}{2}$

21. $\dfrac{2}{y-1} = \dfrac{y-3}{4}$

22. $\dfrac{1}{x-5} = \dfrac{x-3}{3}$

23. $\dfrac{1}{49w} = \dfrac{w}{9}$

24. $\dfrac{1}{4z} = \dfrac{z}{25}$

Objective 2: Applications of Proportions

25. A preschool advertises that it has a 3-to-1 ratio of children to adults. If 18 children are enrolled, how many adults must be on the staff? (See Example 2.)

26. An after-school care facility tries to maintain a 4-to-1 ratio of children to adults. If the facility hired five adults, what is the maximum number of children that can enroll?

27. A 3.5-oz box of candy has a total of 21.0 g of fat. How many grams of fat would a 14-oz box of candy contain?

28. A 6-oz box of candy has 350 calories. How many calories would a 10-oz box contain?

29. A fisherman in the North Atlantic catches eight swordfish for a total of 1840 lb. Approximately how many swordfish were caught if a commercial fishing boat arrives in port with 230,000 lb of swordfish?

30. If a 64-oz bottle of laundry detergent costs $4.00, how much would an 80-oz bottle cost?

31. Pam drives her Toyota Prius 243 mi in city driving on 4.5 gal of gas. At this rate how many gallons of gas are required to drive 621 mi?

32. On a map, the distance from Sacramento, California, to San Francisco, California, is 8 cm. The legend gives the actual distance as 96 mi. On the same map, Fatima measured 7 cm from Sacramento to Modesto, California. What is the actual distance?

33. Yellowstone National Park in Wyoming has the largest population of free-roaming bison. To approximate the number of bison, 200 are captured and tagged and then left free to roam. Later, a sample of 120 bison is observed and 6 have tags. Approximate the population of bison in the park.

34. Laws have been instituted in Florida to help save the manatee. To establish the number of manatees in Florida, 150 manatees were tagged. A new sample was taken later, and among the 40 manatees in the sample, 3 were marked. Approximate the number of manatees in Florida.

35. The ratio of men to women accountants in a large accounting firm is 2 to 1. If the total number of accountants is 81, how many are men and how many are women? (See Example 3.)

36. The ratio of Hank's income spent on rent to his income spent on car payments is 3 to 1. If he spends a total of $1640 per month on the rent and car payment, how much does he spend on each item?

37. The ratio of single men in their 20s to single women in their 20s is 119 to 100 (Source: U.S. Census). In a random group of 1095 single college students in their 20s, how many are men and how many are women?

38. A chemist mixes water and alcohol in a 7 to 8 ratio. If she makes a 450-L solution, how much is water and how much is alcohol?

Objective 3: Similar Triangles

For Exercises 39–42, the triangles shown here are similar. Find the lengths of the missing sides. (See Example 4.)

39.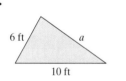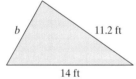

40.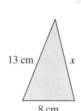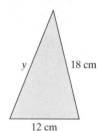

41.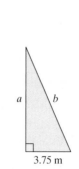

42.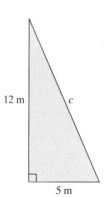

Objective 4: Applications of Rational Equations

43. The profit in dollars for selling x mountain bikes is given by $P(x) = 360x - 56{,}000$. The average profit per bike is defined as

$$\overline{P}(x) = \frac{P(x)}{x} \quad \text{or} \quad \overline{P}(x) = \frac{360x - 56{,}000}{x}$$

Find the number of bikes the manufacturer must produce so that the average profit per bike is $110. **(See Example 5.)**

44. The profit in dollars when x ballpoint pens are sold is given by $P(x) = 0.47x - 100$. The average profit per unit is defined as

$$\overline{P}(x) = \frac{P(x)}{x} \quad \text{or} \quad \overline{P}(x) = \frac{0.47x - 100}{x}$$

Find the number of pens that must be produced so that the average profit per pen is $0.27.

45. A motorist travels 80 mi while driving in a bad rainstorm. In sunny weather, the motorist drives 20 mph faster and covers 120 mi in the same amount of time. Find the speed of the motorist in the rainstorm and the speed in sunny weather. **(See Example 6.)**

46. Brooke walks 2 km/hr slower than her older sister Adrianna. If Brooke can walk 12 km in the same amount of time that Adrianna can walk 18 km, find their speeds.

47. The current in a stream is 2 mph. Find the speed of a boat in still water if it goes 26 mi downstream (with the current) in the same amount of time it takes to go 18 mi upstream (against the current).

48. A bus leaves a terminal at 9:00. A car leaves 3 hr later and averages a speed 21 mph faster than that of the bus. If the car overtakes the bus after 196 mi, find the average speed of the bus and the average speed of the car.

49. A bicyclist rides 24 mi against a wind and returns 24 mi with the wind. His average speed for the return trip is 8 mph faster. If x represents the bicyclist's speed going out against the wind, then the total time required for the round trip is given by

$$t(x) = \frac{24}{x} + \frac{24}{x + 8}, \quad \text{where } x > 0$$

How fast did the cyclist ride against the wind if the total time of the trip was 3.2 hr?

50. A boat travels 45 mi to an island and 45 mi back again. Changes in the wind and tide made the average speed on the return trip 3 mph slower than the speed on the way out. If the total time of the trip took 6 hr 45 min (6.75 hr), find the speed going to the island and the speed of the return trip.

51. One carpenter can complete a kitchen in 8 days. With the help of another carpenter, they can do the job together in 4 days. How long would it take the second carpenter if he worked alone? **(See Example 7.)**

52. One painter can paint a room in 6 hr. Another painter can paint the same room in 8 hr. How long would it take them working together?

53. Gus works twice as fast as Sid. Together they can dig a garden in 4 hr. How long would it take each person working alone?

54. It takes a child 3 times longer to vacuum a house than an adult. If it takes 1 hr for one adult and one child working together to vacuum a house, how long would it take each person working alone?

Expanding Your Skills

55. Find the value of y so that the slope of the line between the points $(3, 1)$ and $(11, y)$ is $\frac{1}{2}$.

56. Find the value of x so that the slope of the line between the points $(-2, -5)$ and $(x, 10)$ is 3.

57. Find the value of x so that the slope of the line between the points $(4, -2)$ and $(x, 2)$ is 4.

58. Find the value of y so that the slope of the line between the points $(3, 2)$ and $(-1, y)$ is $-\frac{3}{4}$.

chapter 6 summary

section 6.1 Rational Expressions and Rational Functions

Key Concepts

A **rational expression** is in the form $\frac{p}{q}$ where p and q are polynomials and $q \neq 0$.

A **rational function** is a function of the form $\frac{p(x)}{q(x)}$, where $p(x)$ and $q(x)$ are polynomial functions and $q(x) \neq 0$.

The **domain** of a rational expression or a rational function excludes the values for which the denominator is zero.

To simplify a rational expression to lowest terms, factor the numerator and denominator completely. Then simplify factors whose ratio is 1 or -1. A rational expression written in lowest terms will still have the same restrictions on the domain as the original expression.

Examples

Example 1

Find the domain of the function

$$f(x) = \frac{x - 3}{(x + 4)(2x - 1)}$$

Domain: $\left\{ x \mid x \text{ is a real number and } x \neq -4, x \neq \frac{1}{2} \right\}$

Example 2

Simplify to lowest terms.

$$\frac{t^2 - 6t - 16}{5t + 10}$$

$$\frac{(t - 8)(t + 2)}{5(t + 2)} = \frac{t - 8}{5} \quad \text{provided } t \neq -2$$

section 6.2 Multiplication and Division of Rational Expressions

Key Concepts

To multiply rational expressions, factor the numerators and denominators completely. Then simplify factors whose ratio is 1 or −1.

To divide rational expressions, multiply by the reciprocal of the divisor.

Examples

Example 1

$$\frac{b^2 - a^2}{a^2 - 2ab + b^2} \cdot \frac{a^2 - 3ab + 2b^2}{2a + 2b}$$

$$= \frac{\overset{-1}{(b-a)}\overset{1}{(b+a)}}{(a-b)^2} \cdot \frac{(a - 2b)\overset{1}{(a-b)}}{2(a+b)} \qquad \text{Factor.}$$

$$= -\frac{a - 2b}{2} \quad \text{or} \quad \frac{2b - a}{2} \qquad \text{Simplify.}$$

Example 2

$$\frac{2c^2 d^5}{15e^4} \div \frac{6c^4 d^3}{20e}$$

$$= \frac{2c^2 d^5}{15e^4} \cdot \frac{20e}{6c^4 d^3} \qquad \text{Multiply by the reciprocal.}$$

$$= \frac{2c^2 d^5}{3 \cdot 5e^4} \cdot \frac{2 \cdot 2 \cdot 5e}{2 \cdot 3c^4 d^3} \qquad \text{Factor.}$$

$$= \frac{4d^2}{9c^2 e^3} \qquad \text{Simplify.}$$

section 6.3 Addition and Subtraction of Rational Expressions

Key Concepts

To add or subtract rational expressions, the expressions must have the same denominator.

The **least common denominator (LCD)** is the product of unique factors from the denominators, in which each factor is raised to its highest power.

<u>Steps to Add or Subtract Rational Expressions</u>
1. Factor the denominator of each rational expression.
2. Identify the LCD.
3. Rewrite each rational expression as an equivalent expression with the LCD as its denominator. [This is accomplished by multiplying the numerator and denominator of each rational expression by the missing factor(s) from the LCD.]
4. Add or subtract the numerators, and write the result over the common denominator.
5. Simplify.

Examples

Example 1

For $\dfrac{1}{3(x-1)^3(x+2)}$ and $\dfrac{-5}{6(x-1)(x+7)^2}$

LCD $= 6(x-1)^3(x+2)(x+7)^2$

Example 2

$\dfrac{c}{c^2 - c - 12} - \dfrac{1}{2c - 8}$

$= \dfrac{c}{(c-4)(c+3)} - \dfrac{1}{2(c-4)}$ Factor the denominators.

The LCD is $2(c-4)(c+3)$

$\dfrac{2}{2} \cdot \dfrac{c}{(c-4)(c+3)} - \dfrac{1}{2(c-4)} \cdot \dfrac{(c+3)}{(c+3)}$ Write equivalent fractions with LCD.

$= \dfrac{2c - (c+3)}{2(c-4)(c+3)}$ Subtract.

$= \dfrac{2c - c - 3}{2(c-4)(c+3)}$ Simplify.

$= \dfrac{c - 3}{2(c-4)(c+3)}$

section 6.4 Complex Fractions

Key Concepts

Complex fractions can be simplified by using Method I or Method II.

Method I uses the order of operations to simplify the numerator and denominator separately before multiplying by the reciprocal of the denominator of the complex fraction.

To use Method II, multiply the numerator and denominator of the complex fraction by the LCD of all the individual fractions. Then simplify the result.

Examples

Example 1

Simplify by using Method I.

$$\frac{-\dfrac{5}{x}}{\dfrac{10}{x^2}}$$

$$= -\frac{5}{x} \cdot \frac{x^2}{10}$$

$$= -\frac{x}{2}$$

Example 2

Simplify by using Method II.

$$\frac{1 - \dfrac{4}{w^2}}{1 - \dfrac{1}{w} - \dfrac{6}{w^2}} \quad \text{The LCD is } w^2.$$

$$= \frac{w^2\left(1 - \dfrac{4}{w^2}\right)}{w^2\left(1 - \dfrac{1}{w} - \dfrac{6}{w^2}\right)}$$

$$= \frac{w^2 - 4}{w^2 - w - 6}$$

$$= \frac{(w-2)(\cancel{w+2})}{(w-3)(\cancel{w+2})}$$

$$= \frac{w-2}{w-3}$$

section 6.5 Rational Equations

Key Concepts

Steps to Solve a Rational Equation
1. Factor the denominators of all rational expressions. Identify any restrictions on the variable.
2. Identify the LCD of all expressions in the equation.
3. Multiply both sides of the equation by the LCD.
4. Solve the resulting equation.
5. Check each potential solution.

Examples

Example 1

$$\frac{1}{w} - \frac{1}{2w-1} = \frac{-2w}{2w-1}$$

The LCD is $w(2w-1)$.

$$w(2w-1)\frac{1}{w} - w(2w-1) \cdot \frac{1}{2w-1}$$

$$= w(2w-1) \cdot \frac{-2w}{2w-1}$$

$$(2w-1)1 - w(1) = w(-2w)$$

$$2w - 1 - w = -2w^2 \quad \text{(quadratic equation)}$$

$$2w^2 + w - 1 = 0$$

$$(2w-1)(w+1) = 0$$

$$w \neq \frac{1}{2} \quad \text{or} \quad w = -1$$

$$w = -1 \left(w = \frac{1}{2} \text{ does not check}\right)$$

section 6.6 Applications of Rational Equations and Proportions

Key Concepts

An equation that equates two **ratios** is called a **proportion**:

$$\frac{a}{b} = \frac{c}{d} \quad \text{provided } b \neq 0, d \neq 0$$

Examples

Example 1

A sample of 85 g of a particular ice cream contains 17 g of fat. How much fat does 324 g of the same ice cream contain?

$$\frac{17 \text{ g fat}}{85 \text{ g ice cream}} = \frac{x \text{ g fat}}{324 \text{ g ice cream}}$$

$$\frac{17}{85} = \frac{x}{324} \quad \text{Multiply by the LCD.}$$

$$(85 \cdot 324) \cdot \frac{17}{85} = (85 \cdot 324) \cdot \frac{x}{324}$$

$$5508 = 85x$$

$$x = 64.8 \text{ g}$$

chapter 6 review exercises

Section 6.1

1. For the rational expression $\dfrac{t-2}{t+9}$

   a. Evaluate the expression (if possible) for $t = 0, 1, 2, -3, -9$

   b. Write the domain of the expression in set-builder notation.

2. For the rational expression $\dfrac{k+1}{k-5}$

   a. Evaluate the expression (if possible) for $k = 0, 1, 5, -1, -2$

   b. Write the domain of the expression in set-builder notation.

3. Let $k(y) = \dfrac{y}{y^2 - 1}$.

   a. Find the function values (if they exist): $k(2)$, $k(0), k(1), k(-1), k\left(\dfrac{1}{2}\right)$.

   b. Identify the domain for k. Write the answer in interval notation.

4. Let $h(x) = \dfrac{x}{x^2 + 1}$.

   a. Find the function values (if they exist): $h(1)$, $h(0), h(-1), h(-3), h\left(\dfrac{1}{2}\right)$.

   b. Identify the domain for h. Write the answer in interval notation.

For Exercises 5–12, simplify the rational expressions.

5. $\dfrac{28a^3b^3}{14a^2b^3}$

6. $\dfrac{25x^2yz^3}{125xyz}$

7. $\dfrac{x^2 - 4x + 3}{x - 3}$

8. $\dfrac{k^2 + 3k - 10}{k^2 - 5k + 6}$

9. $\dfrac{x^3 - 27}{9 - x^2}$

10. $\dfrac{a^4 - 81}{3 - a}$

11. $\dfrac{2t^2 + 3t - 5}{7 - 6t - t^2}$

12. $\dfrac{y^3 - 4y}{y^2 - 5y + 6}$

For Exercises 13–16, write the domain of each function in set-builder notation. Use that information to match the function with its graph.

13. $f(x) = \dfrac{1}{x - 3}$

14. $m(x) = \dfrac{1}{x + 2}$

15. $k(x) = \dfrac{6}{x^2 - 3x}$

16. $p(x) = \dfrac{-2}{x^2 + 4}$

a.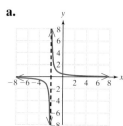

b.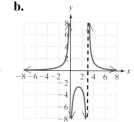

c.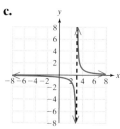

d.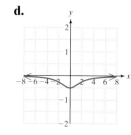

Section 6.2

For Exercises 17–28, multiply or divide as indicated.

17. $\dfrac{3a + 9}{a^2} \cdot \dfrac{a^3}{6a + 18}$

18. $\dfrac{4 - y}{5} \div \dfrac{2y - 8}{15}$

19. $\dfrac{x - 4y}{x^2 + xy} \div \dfrac{20y - 5x}{x^2 - y^2}$

20. $(x^2 + 5x - 24)\left(\dfrac{x + 8}{x - 3}\right)$

21. $\dfrac{7k + 28}{2k + 4} \cdot \dfrac{k^2 - 2k - 8}{k^2 + 2k - 8}$

22. $\dfrac{ab + 2a + b + 2}{ab - 3b + 2a - 6} \cdot \dfrac{ab - 3b + 4a - 12}{ab - b + 4a - 4}$

23. $\dfrac{x^2 + 8x - 20}{x^2 + 6x - 16} \div \dfrac{x^2 + 6x - 40}{x^2 + 3x - 40}$

24. $\dfrac{2b - b^2}{b^3 - 8} \cdot \dfrac{b^2 + 2b + 4}{b^2}$

25. $\dfrac{2w}{21} \div \dfrac{3w^2}{7} \cdot \dfrac{4}{w}$

26. $\dfrac{5y^2 - 20}{y^3 + 2y^2 + y + 2} \div \dfrac{7y}{y^3 + y}$

27. $\dfrac{x^2 + x - 20}{x^2 - 4x + 4} \cdot \dfrac{x^2 + x - 6}{12 + x - x^2} \div \dfrac{2x + 10}{10 - 5x}$

28. $(9k^2 - 25) \cdot \left(\dfrac{k + 5}{3k - 5}\right)$

Section 6.3

For Exercises 29–40, add or subtract as indicated.

29. $\dfrac{1}{x} + \dfrac{1}{x^2} - \dfrac{1}{x^3}$

30. $\dfrac{1}{x + 2} + \dfrac{5}{x - 2}$

31. $\dfrac{y}{2y - 1} + \dfrac{3}{1 - 2y}$

32. $\dfrac{a + 2}{2a + 6} - \dfrac{3}{a + 3}$

33. $\dfrac{4k}{k^2 + 2k + 1} + \dfrac{3}{k^2 - 1}$

34. $4x + 3 - \dfrac{2x + 1}{x + 4}$

35. $\dfrac{2}{a + 3} + \dfrac{2a^2 - 2a}{a^2 - 2a - 15}$

36. $\dfrac{6}{x^2 + 4x + 3} + \dfrac{7}{x^2 + 5x + 6}$

37. $\dfrac{2}{3x - 5} - 8$

38. $\dfrac{7}{4k^2 - k - 3} + \dfrac{1}{4k^2 - 7k + 3}$

39. $\dfrac{6a}{3a^2 - 7a + 2} - \dfrac{2}{3a - 1} + \dfrac{3a}{a - 2}$

40. $\dfrac{y}{y - 3} - \dfrac{2y - 5}{y + 2} - 4$

Section 6.4

For Exercises 41–48, simplify the complex fraction.

41. $\dfrac{\dfrac{2x}{3x^2 - 3}}{\dfrac{4x}{6x - 6}}$

42. $\dfrac{\dfrac{k + 2}{3}}{\dfrac{5}{k - 2}}$

43. $\dfrac{\dfrac{2}{x} + \dfrac{1}{xy}}{\dfrac{4}{x^2}}$

44. $\dfrac{\dfrac{4}{y} - 1}{\dfrac{1}{y} - \dfrac{4}{y^2}}$

45. $\dfrac{\dfrac{1}{a - 1} + 1}{\dfrac{1}{a + 1} - 1}$

46. $\dfrac{\dfrac{3}{x - 1} - \dfrac{1}{1 - x}}{\dfrac{2}{x - 1} - \dfrac{2}{x}}$

47. $\dfrac{1 + xy^{-1}}{x^2 y^{-2} - 1}$

48. $\dfrac{5a^{-1} + (ab)^{-1}}{3a^{-2}}$

For Exercises 49–50, find the slope of the line containing the two points.

49. $\left(\dfrac{2}{3}, -\dfrac{7}{4}\right)$ and $\left(\dfrac{13}{6}, -\dfrac{5}{3}\right)$

50. $\left(\dfrac{8}{15}, -\dfrac{1}{3}\right)$ and $\left(\dfrac{13}{10}, \dfrac{9}{5}\right)$

Section 6.5

For Exercises 51–56, solve the equation.

51. $\dfrac{x + 3}{x^2 - x} - \dfrac{8}{x^2 - 1} = 0$

52. $\dfrac{y}{y+3} + \dfrac{3}{y-3} = \dfrac{18}{y^2-9}$

53. $x - 9 = \dfrac{72}{x-8}$

54. $\dfrac{3x+1}{x+5} = \dfrac{x-1}{x+1} + 2$

55. $5y^{-2} + 1 = 6y^{-1}$

56. $1 + \dfrac{7}{6}m^{-1} = \dfrac{13}{6}m^{-1}$

57. Solve for x. $\quad c = \dfrac{ax+b}{x}$

58. Solve for P. $\quad \dfrac{A}{rt} = P + \dfrac{P}{rt}$

Section 6.6

For Exercises 59–62, solve the proportions.

59. $\dfrac{5}{4} = \dfrac{x}{6}$

60. $\dfrac{x}{36} = \dfrac{6}{7}$

61. $\dfrac{x+2}{3} = \dfrac{5(x+1)}{4}$

62. $\dfrac{x}{x+2} = \dfrac{-3}{5}$

63. In one game Peyton Manning completed 34 passes for 357 yd. At this rate how many yards would be gained for 22 passes?

64. Erik bought $136 Canadian with $100 American. At this rate, how many Canadian dollars can he buy with $235 American?

65. The average cost to produce x units is given by

$$\overline{C}(x) = \dfrac{1.5x + 4200}{x}$$

How many units need to be produced to have an average cost per unit of $54?

66. Stephen drove his car 45 mi. He ran out of gas and had to walk 3 mi to a gas station. His speed driving is 15 times his speed walking. If the total time for the drive and walk was $1\tfrac{1}{2}$ hr, what was his speed driving?

67. Two pipes can fill a tank in 6 hr. The larger pipe works twice as fast as the smaller pipe. How long would it take each pipe to fill the tank if they worked separately?

---

chapter 6 | test

1. For the expression $\dfrac{2x+6}{x^2-x-12}$

   a. Write the domain in set-builder notation.

   b. Simplify the expression to lowest terms.

2. For the function $h(x) = \dfrac{2x-14}{x^2-49}$

   a. Evaluate $h(0), h(5), h(7),$ and $h(-7)$, if possible.

   b. Write the domain of h in set-builder notation.

3. Write the domain of $k(x) = \dfrac{5x-3}{7}$ in interval notation.

For Exercises 4–5, simplify to lowest terms.

4. $\dfrac{12m^3n^7}{18mn^8}$

5. $\dfrac{9x^2-9}{3x^2+2x-5}$

6. Find the slope of the line containing the points $\left(\dfrac{1}{12}, -\dfrac{3}{4}\right)$ and $\left(\dfrac{5}{6}, -\dfrac{8}{3}\right)$.

For Exercises 7–13, simplify.

7. $\dfrac{2x-5}{25-4x^2} \cdot (2x^2 - x - 15)$

8. $\dfrac{x^2}{x-4} - \dfrac{8x-16}{x-4}$

9. $\dfrac{4x}{x+1} + x + \dfrac{2}{x+1}$

10. $\dfrac{3 + \dfrac{3}{k}}{4 + \dfrac{4}{k}}$

11. $\dfrac{2u^{-1} + 2v^{-1}}{4u^{-3} + 4v^{-3}}$

12. $\dfrac{ax + bx + 2a + 2b}{ax - 3a + bx - 3b} \cdot \dfrac{x - 3}{x - 5} \div \dfrac{x + 2}{ax - 5a}$

13. $\dfrac{3}{x^2 + 8x + 15} - \dfrac{1}{x^2 + 7x + 12} - \dfrac{1}{x^2 + 9x + 20}$

For Exercises 14–16, solve the equation.

14. $\dfrac{7}{z+1} - \dfrac{z-5}{z^2-1} = \dfrac{6}{z}$

15. $\dfrac{3}{y^2 - 9} + \dfrac{4}{y+3} = 1$

16. $\dfrac{4x}{x-4} = 3 + \dfrac{16}{x-4}$

17. Solve for T. $\quad \dfrac{1 + Tv}{T} = p$

18. Solve for m_1. $\quad F = \dfrac{Gm_1m_2}{r^2}$

19. If the reciprocal of a number is added to 3 times the number, the result is $\tfrac{13}{2}$. Find the number.

20. The following triangles are similar. Find the lengths of the missing sides.

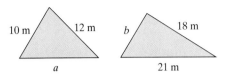

21. On a certain map, the distance between New York and Los Angeles is 8.2 in., and the actual distance is 2820 mi. What is the distance between two cities that are 5.7 in. apart on the same map? Round to the nearest mile.

22. The average profit per unit for a manufacturer to sell x units is given by

$$\overline{P}(x) = \dfrac{1.75x - 1500}{x}$$

How many units must be sold to obtain an average profit of $0.25 per unit?

23. Lance can ride 48 mi on his bike against the wind. With the wind at his back, he rides 4 mph faster and can ride 60 mi in the same amount of time. Find his speed riding against the wind and his speed riding with the wind.

24. Gail can type a chapter in a book in 4 hr. Jack can type a chapter in a book in 10 hr. How long would it take them to type a chapter if they worked together?

---

chapters 1–6 | cumulative review exercises

1. Check the sets to which each number belongs.

Set \ Number	−22	π	6	$-\sqrt{2}$
Real numbers				
Irrational numbers				
Rational numbers				
Integers				
Whole numbers				
Natural numbers				

2. At the age of 30, tennis player Steffi Graf announced her retirement from tennis. After 17 years on the tour, her total prize money amounted to 2.1839777×10^7.

   a. Write the amount of prize money in expanded form.

   b. What was Steffi Graf's average winnings per year? Round to the nearest dollar.

3. Perform the indicated operations.
$(2x - 3)(x - 4) - (x - 5)^2$

4. The area of a trapezoid is given by $A = \frac{1}{2}h(b_1 + b_2)$.

   a. Solve for b_1.

   b. Find b_1 when $h = 4$ cm, $b_2 = 6$ cm, and $A = 32$ cm^2.

5. The dimensions of a rectangular swimming pool are such that the length is 10 m less than twice the width. If the total perimeter is 160 m, find the length and width.

6. Solve the system of equations.
$$x - 3y + z = 1$$
$$2x - y - 2z = 2$$
$$x + 2y - 3z = -1$$

7. Find an equation of the line through $(-3, 5)$ that is perpendicular to the line $y = 3x$. Write the answer in slope-intercept form.

8. The value $V(x)$ of a car (in dollars) decreases with age according to $V(x) = 15{,}000 - 1250x$, $0 \leq x \leq 12$, where x represents the age of the car in years.

   a. Is this function constant, linear, quadratic, or other?

   b. Sketch the function over its domain.

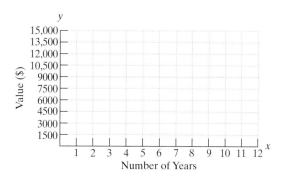

   c. What is the y-intercept? What does the y-intercept mean in the context of this problem?

   d. What is the x-intercept? What does the x-intercept mean in the context of this problem?

   e. What is the slope? Interpret the slope in the context of this problem.

   f. What is the value of $V(5)$? Interpret the value of $V(5)$ in the context of the problem.

   g. After how many years will the car be worth $5625?

9. The speed of a car varies inversely as the time to travel a fixed distance. A car traveling the speed limit of 60 mph travels between two points in 10 sec. How fast is a car moving if it takes only 8 sec to cover the same distance?

10. Find the x-intercepts. $f(x) = -12x^3 + 17x^2 - 6x$

11. Factor $64y^3 - 8z^6$ completely over the real numbers.

12. Write the domain of the functions f and g.

   a. $f(x) = \dfrac{x+7}{2x-3}$ **b.** $g(x) = \dfrac{x+3}{x^2 - x - 12}$

13. Perform the indicated operations.
$$\frac{2x^2 + 11x - 21}{4x^2 - 10x + 6} \div \frac{2x^2 - 98}{x^2 - x + xa - a}$$

14. Reduce to lowest terms.
$$\frac{x^2 - 6x + 8}{20 - 5x}$$

15. Perform the indicated operations.
$$\frac{x}{x^2 + 5x - 50} - \frac{1}{x^2 - 7x + 10} + \frac{1}{x^2 + 8x - 20}$$

16. Simplify the complex fraction.
$$\frac{1 - \dfrac{49}{c^2}}{\dfrac{7}{c} + 1}$$

17. Solve the equation.
$$\frac{4y}{y+2} - \frac{y}{y-1} = \frac{9}{y^2 + y - 2}$$

18. Max knows that the distance between Roanoke, Virginia, and Washington, D.C., is 195 mi. On a certain map, the distance between the two cities is 6.5 in. On the same map, the distance between Roanoke and Cincinnati, Ohio, is 9.25 in. Find the distance in miles between Roanoke and Cincinnati. Round to the nearest mile.

19. Determine whether the equation represents a horizontal or vertical line. Then identify the slope of the line.

  a. $x = -5$ **b.** $2y = 8$

20. Simplify. $\left(\dfrac{2x^{-4}y}{3z^2}\right)^{-1}(4xy^7)$

Radicals and Complex Numbers

7

7.1 Definition of an *n*th Root
7.2 Rational Exponents
7.3 Simplifying Radical Expressions
7.4 Addition and Subtraction of Radicals
7.5 Multiplication of Radicals
7.6 Rationalization
7.7 Radical Equations
7.8 Complex Numbers

In this chapter we study radical expressions. This includes operations on square roots, cube roots, fourth roots, and so on. We revisit the Pythagorean theorem, and then later in the chapter we solve applications of radical equations. In Exercise 74 in Section 7.7, the cost for an airline to fly x thousand passengers from New York to Atlanta is given by $C(x) = \sqrt{0.3x + 1}$, where C is the cost in millions of dollars. If the airline has $4 million in costs for this route, then the solution to the equation $4 = \sqrt{0.3x + 1}$ represents the number of passengers (in thousands) who made the flight from New York to Atlanta.

chapter 7 preview

The exercises in this chapter preview contain concepts that have not yet been presented. These exercises are provided for students who want to compare their levels of understanding before and after studying the chapter. Alternatively, you may prefer to work these exercises when the chapter is completed and before taking the exam.

Section 7.1

For Exercises 1–2, simplify, if possible.

1. $\sqrt{-9}$

2. $\sqrt{\dfrac{100}{49}}$

3. Determine the domain of $g(x) = \sqrt{x + 10}$. Write the answer in interval notation.

4. Simplify. $\sqrt{a^2}$

Section 7.2

5. Convert to radical notation. $c^{2/3}$

For Exercises 6–7, evaluate the expression.

6. $8^{1/3}$

7. $27^{-2/3}$

8. Simplify the expression. Write the answer using positive exponents. $\dfrac{x}{x^{1/3}}$

Section 7.3

For Exercises 9–11, simplify the radicals. Assume all variables represent positive real numbers.

9. $\sqrt[3]{80}$

10. $\sqrt{40x^3y^6}$

11. $\dfrac{\sqrt[4]{64}}{\sqrt[4]{4}}$

Section 7.4

For Exercises 12–13, add or subtract as indicated. Assume all variables represent positive real numbers.

12. $7\sqrt{6} - 4\sqrt{6}$

13. $7\sqrt{3x^2} + 3x\sqrt{3}$

Section 7.5

For Exercises 14–17, multiply the radical expressions. Assume all variables represent positive real numbers.

14. $4\sqrt{x}(3\sqrt{x} + 5)$

15. $(2\sqrt{3} - 4\sqrt{2})(\sqrt{3} + 3\sqrt{2})$

16. $(\sqrt{y} + 5)^2$

17. $(\sqrt{5} + 2)(\sqrt{5} - 2)$

Section 7.6

For Exercises 18–20, simplify by rationalizing the denominators.

18. $\dfrac{10}{\sqrt{2}}$

19. $\dfrac{x}{\sqrt[3]{y}}$

20. $\dfrac{8}{\sqrt{5} + 1}$

Section 7.7

For Exercises 21–23, solve the equations.

21. $\sqrt{5x + 1} = 6$

22. $\sqrt{8y - 4} + 8 = 2$

23. $\sqrt{3t + 1} + 1 = \sqrt{3t + 6}$

Section 7.8

For Exercises 24–25, simplify the expression in terms of i.

24. $\sqrt{-49}$

25. i^7

For Exercises 26–28, simplify the expressions. Write the answers in $a + bi$ form.

26. $(-6 - i) - (4 + i)$

27. $(3 + 4i)(5 + 2i)$

28. $\dfrac{6 + 2i}{1 + 3i}$

section 7.1 Definition of an *n*th Root

1. Definition of a Square Root

The inverse operation to squaring a number is to find its square roots. For example, finding a square root of 36 is equivalent to asking, "what number when squared equals 36?"

One obvious answer to this question is 6 because $(6)^2 = 36$, but -6 will also work, because $(-6)^2 = 36$.

Objectives

1. Definition of a Square Root
2. Definition of an *n*th Root
3. Roots of Variable Expressions
4. Pythagorean Theorem
5. Radical Functions

Definition of a Square Root

b is a **square root** of a if $b^2 = a$.

---

example 1   Identifying Square Roots

Identify the square roots of the real numbers.

  a. 25   **b.** 49   **c.** 0   **d.** -9

Solution:

a. 5 is a square root of 25 because $(5)^2 = 25$.

    -5 is a square root of 25 because $(-5)^2 = 25$.

b. 7 is a square root of 49 because $(7)^2 = 49$.

    -7 is a square root of 49 because $(-7)^2 = 49$.

c. 0 is a square root of 0 because $(0)^2 = 0$.

d. There are no real numbers that when squared will equal a negative number; therefore, there are no real-valued square roots of -9.

Skill Practice

Identify the square roots of the real numbers.

1. 64   2. 16
3. 1   4. -100

Tip: All positive real numbers have two real-valued square roots: one positive and one negative. Zero has only one square root, which is 0 itself. Finally, for any negative real number, there are no real-valued square roots.

Recall from Section 1.2 that the positive square root of a real number can be denoted with a **radical sign** $\sqrt{}$.

Notation for Positive and Negative Square Roots

Let a represent a positive real number. Then

1. $\sqrt{a}$ is the *positive* square root of a. The positive square root is also called the **principal square root**.
2. $-\sqrt{a}$ is the *negative* square root of a.
3. $\sqrt{0} = 0$

Answers

1. -8 and 8
2. -4 and 4
3. -1 and 1
4. No real-valued square roots

Skill Practice

Simplify the square roots.

5. $\sqrt{81}$
6. $\sqrt{\dfrac{36}{49}}$
7. $\sqrt{0.09}$

example 2 — Simplifying a Square Root

Simplify the square roots.

a. $\sqrt{36}$ b. $\sqrt{\dfrac{4}{9}}$ c. $\sqrt{0.04}$

Solution:

a. $\sqrt{36}$ denotes the positive square root of 36.

$\sqrt{36} = 6$

b. $\sqrt{\dfrac{4}{9}}$ denotes the positive square root of $\dfrac{4}{9}$.

$\sqrt{\dfrac{4}{9}} = \dfrac{2}{3}$

c. $\sqrt{0.04}$ denotes the positive square root of 0.04.

$\sqrt{0.04} = 0.2$

The numbers 36, $\dfrac{4}{9}$, and 0.04 are **perfect squares** because their square roots are rational numbers. Radicals that cannot be simplified to rational numbers are irrational numbers. Recall that an irrational number cannot be written as a terminating or repeating decimal. For example, the symbol $\sqrt{13}$ is used to represent the *exact* value of the square root of 13. The symbol $\sqrt{42}$ is used to represent the *exact* value of the square root of 42. These values can be approximated by a rational number by using a calculator.

$$\sqrt{13} \approx 3.605551275 \qquad \sqrt{42} \approx 6.480740698$$

Tip: Before using a calculator to evaluate a square root, try estimating the value first.

$\sqrt{13}$ must be a number between 3 and 4 because $\sqrt{9} < \sqrt{13} < \sqrt{16}$.

$\sqrt{42}$ must be a number between 6 and 7 because $\sqrt{36} < \sqrt{42} < \sqrt{49}$.

Calculator Connections

Use a calculator to approximate the values of $\sqrt{13}$ and $\sqrt{42}$.

```
√(13)
        3.605551275
√(42)
        6.480740698
```

A negative number cannot have a real number as a square root because no real number when squared is negative. For example, $\sqrt{-25}$ is *not* a real number because there is no real number b for which $(b)^2 = -25$.

Answers

5. 9   6. $\dfrac{6}{7}$ 7. 0.3

example 3 Evaluating Square Roots

Simplify the square roots, if possible.

a. $\sqrt{-144}$ b. $-\sqrt{144}$ c. $\sqrt{-0.01}$ d. $-\sqrt{\dfrac{1}{9}}$

Solution:

a. $\sqrt{-144}$ is *not* a real number.

b. $-\sqrt{144}$
$= -1 \cdot \sqrt{144}$
$= -1 \cdot 12$
$= -12$

Tip: For the expression $-\sqrt{144}$, the factor of -1 is *outside* the radical.

c. $\sqrt{-0.01}$ is *not* a real number.

d. $-\sqrt{\dfrac{1}{9}}$
$= -1 \cdot \sqrt{\dfrac{1}{9}}$
$= -1 \cdot \dfrac{1}{3}$
$= -\dfrac{1}{3}$

Skill Practice

Simplify the square roots, if possible.

8. $-\sqrt{64}$
9. $\sqrt{-81}$
10. $\sqrt{-\dfrac{1}{4}}$
11. $-\sqrt{0.25}$

2. Definition of an *n*th Root

Finding a square root of a number is the inverse process of squaring a number. This concept can be extended to finding a third root (called a cube root), a fourth root, and in general an **nth root**.

> **Definition of an *n*th Root**
>
> b is an *n*th root of a if $b^n = a$.

The radical sign $\sqrt{}$ is used to denote the principal square root of a number. The symbol $\sqrt[n]{}$ is used to denote the principal *n*th root of a number. In the expression $\sqrt[n]{a}$, *n* is called the **index** of the radical, and *a* is called the **radicand**. For a square root, the index is 2, but it is usually not written ($\sqrt[2]{a}$ is denoted simply as $\sqrt{a}$). A radical with an index of 3 is called a **cube root**, denoted by $\sqrt[3]{a}$.

> **Definition of $\sqrt[n]{a}$**
>
> 1. If *n* is a positive *even* integer and $a > 0$, then $\sqrt[n]{a}$ is the principal (positive) *n*th root of *a*.
> 2. If $n > 1$ is an *odd* integer, then $\sqrt[n]{a}$ is the *n*th root of *a*.
> 3. If $n > 1$ is an integer, then $\sqrt[n]{0} = 0$.

Concept Connections

12. Every positive number has how many real-valued square roots?
13. Every negative number has how many real-valued square roots?

Answers
8. -8 9. Not a real number
10. Not a real number 11. -0.5
12. Two 13. Zero

For the purpose of simplifying radicals, it is helpful to know the following powers:

Perfect Cubes	Perfect Fourth Powers	Perfect Fifth Powers
$1^3 = 1$	$1^4 = 1$	$1^5 = 1$
$2^3 = 8$	$2^4 = 16$	$2^5 = 32$
$3^3 = 27$	$3^4 = 81$	$3^5 = 243$
$4^3 = 64$	$4^4 = 256$	$4^5 = 1024$
$5^3 = 125$	$5^4 = 625$	$5^5 = 3125$

Skill Practice

Simplify, if possible.
14. $\sqrt[4]{16}$
15. $\sqrt{-4}$
16. $\sqrt[5]{32}$
17. $\sqrt[4]{-81}$

example 4 Identifying the Principal nth Root of a Real Number

Simplify the expressions, if possible.

a. $\sqrt{4}$ b. $\sqrt[3]{64}$ c. $\sqrt[5]{-32}$ d. $\sqrt[4]{81}$
e. $\sqrt[6]{1,000,000}$ f. $\sqrt{-100}$ g. $\sqrt[4]{-16}$

Solution:

a. $\sqrt{4} = 2$ because $(2)^2 = 4$
b. $\sqrt[3]{64} = 4$ because $(4)^3 = 64$
c. $\sqrt[5]{-32} = -2$ because $(-2)^5 = -32$
d. $\sqrt[4]{81} = 3$ because $(3)^4 = 81$
e. $\sqrt[6]{1,000,000} = 10$ because $(10)^6 = 1,000,000$
f. $\sqrt{-100}$ is not a real number. No real number when squared equals -100.
g. $\sqrt[4]{-16}$ is not a real number. No real number when raised to the fourth power equals -16.

Examples 4(f) and 4(g) illustrate that an nth root of a negative quantity is not a real number if the index is even. This is because no real number raised to an even power is negative.

Calculator Connections

A calculator can be used to approximate nth roots by using the $\sqrt[x]{}$ function. On most calculators, the index is entered first.

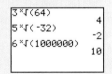

3. Roots of Variable Expressions

Finding an nth root of a variable expression is similar to finding an nth root of a numerical expression. For roots with an even index, however, particular care must be taken to obtain a nonnegative result.

Answers
14. 2 15. Not a real number
16. 2 17. Not a real number

Definition of $\sqrt[n]{a^n}$

1. If n is a positive *odd* integer, then $\sqrt[n]{a^n} = a$.
2. If n is a positive *even* integer, then $\sqrt[n]{a^n} = |a|$.

The absolute value bars are necessary for roots with an even index because the variable a may represent a positive quantity or a negative quantity. By using absolute value bars, $\sqrt[n]{a^n} = |a|$ is nonnegative and represents the principal nth root of a.

example 5 — Simplifying Expressions of the Form $\sqrt[n]{a^n}$

Simplify the expressions.

a. $\sqrt[4]{(-3)^4}$ b. $\sqrt[5]{(-3)^5}$ c. $\sqrt{(x+2)^2}$ d. $\sqrt[3]{(a+b)^3}$ e. $\sqrt{y^4}$

Solution:

a. $\sqrt[4]{(-3)^4} = |-3| = 3$ Because this is an *even*-indexed root, absolute value bars are necessary to make the answer positive.

b. $\sqrt[5]{(-3)^5} = -3$ This is an *odd*-indexed root, so absolute value bars are not necessary.

c. $\sqrt{(x+2)^2} = |x+2|$ Because this is an *even*-indexed root, absolute value bars are necessary. The sign of the quantity $x+2$ is unknown; however, $|x+2| \geq 0$ regardless of the value of x.

d. $\sqrt[3]{(a+b)^3} = a+b$ This is an *odd*-indexed root, so absolute value bars are not necessary.

e. $\sqrt{y^4} = \sqrt{(y^2)^2}$

   $= |y^2|$ Because this is an even-indexed root, use absolute value bars.

   $= y^2$ However, because y^2 is nonnegative, the absolute value bars are not necessary.

Skill Practice

Simplify the expressions.

18. $\sqrt{(-4)^2}$
19. $\sqrt[3]{(-4)^3}$
20. $\sqrt{y^2}$
21. $\sqrt[4]{(a+1)^4}$
22. $\sqrt[4]{v^8}$

If n is an even integer, then $\sqrt[n]{a^n} = |a|$; however, if the variable a is assumed to be *nonnegative*, then the absolute value bars may be dropped. That is, $\sqrt[n]{a^n} = a$ provided $a \geq 0$. In many examples and exercises, we will make the assumption that the variables within a radical expression are positive real numbers. In such a case, the absolute value bars are not needed to evaluate $\sqrt[n]{a^n}$.

It is helpful to become familiar with the patterns associated with perfect squares and perfect cubes involving variable expressions.

The following powers of x are perfect squares:

Perfect Squares

$(x^1)^2 = x^2$
$(x^2)^2 = x^4$
$(x^3)^2 = x^6$
$(x^4)^2 = x^8$

Tip: In general, any expression raised to an even power (a multiple of 2) is a perfect square.

Answers

18. 4   19. −4   20. $|y|$
21. $|a+1|$ 22. v^2

The following powers of x are perfect cubes:

Perfect Cubes

$(x^1)^3 = x^3$
$(x^2)^3 = x^6$
$(x^3)^3 = x^9$
$(x^4)^3 = x^{12}$

Tip: In general, any expression raised to a power that is a multiple of 3 is a perfect cube.

These patterns may be extended to higher powers.

Skill Practice

Simplify the expressions. Assume all variables represent positive real numbers.

23. $\sqrt{t^6}$
24. $\sqrt[3]{y^{12}}$
25. $\sqrt[4]{x^{12}y^4}$
26. $\sqrt[5]{\dfrac{32}{b^{10}}}$

example 6   Simplifying nth Roots

Simplify the expressions. Assume that all variables are positive real numbers.

a. $\sqrt{y^8}$ b. $\sqrt[3]{27a^3}$ c. $\sqrt[5]{\dfrac{a^5}{b^5}}$ d. $-\sqrt[4]{\dfrac{81x^4y^8}{16}}$

Solution:

a. $\sqrt{y^8} = \sqrt{(y^4)^2} = y^4$

b. $\sqrt[3]{27a^3} = \sqrt[3]{(3a)^3} = 3a$

c. $\sqrt[5]{\dfrac{a^5}{b^5}} = \sqrt[5]{\left(\dfrac{a}{b}\right)^5} = \dfrac{a}{b}$

d. $-\sqrt[4]{\dfrac{81x^4y^8}{16}} = -\sqrt[4]{\left(\dfrac{3xy^2}{2}\right)^4} = -\dfrac{3xy^2}{2}$

4. Pythagorean Theorem

In Section 5.7, we used the Pythagorean theorem in several applications. For the triangle shown in Figure 7-1, the **Pythagorean theorem** may be stated as $a^2 + b^2 = c^2$. In this formula, a and b are the legs of the right triangle, and c is the hypotenuse. Notice that the hypotenuse is the longest side of the right triangle and is opposite the 90° angle.

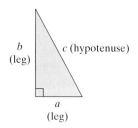

Figure 7-1

Skill Practice

27. Use the Pythagorean theorem and the definition of the principal square root to find the length of the unknown side of the right triangle.

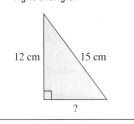

example 7   Applying the Pythagorean Theorem

Use the Pythagorean theorem and the definition of the principal square root of a positive real number to find the length of the unknown side.

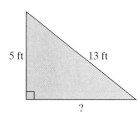

Answers

23. t^3 24. y^4 25. x^3y
26. $\dfrac{2}{b^2}$ 27. 9 cm

Solution:

$$a^2 + b^2 = c^2$$
$$(5)^2 + b^2 = (13)^2$$

Label the sides of the triangle.

Apply the Pythagorean theorem.

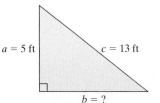

$$25 + b^2 = 169$$ Simplify.

$$b^2 = 169 - 25$$ Isolate b^2.

$$b^2 = 144$$

$$b = 12$$

By definition, b must be one of the square roots of 144. Because b represents the length of a side of a triangle, choose the positive square root of 144.

The third side is 12 ft long.

---

example 8   Applying the Pythagorean Theorem

Two boats leave a dock at 12:00 noon. One travels due north at 6 mph, and the other travels due east at 8 mph (Figure 7-2). How far apart are the two boats after 2 hr?

Solution:

The boat traveling north travels a distance of (6 mph)(2 hr) = 12 mi. The boat traveling east travels a distance of (8 mph)(2 hr) = 16 mi. The course of the boats forms a right triangle where the hypotenuse represents the distance between them.

$$a^2 + b^2 = c^2$$
$$(12)^2 + (16)^2 = c^2$$ Apply the Pythagorean theorem.
$$144 + 256 = c^2$$ Simplify.
$$400 = c^2$$
$$\sqrt{400} = c$$

By definition, c must be one of the square roots of 400. Choose the positive square root of 400 to represent distance between the two boats.

$$20 = c$$

The boats are 20 mi apart.

Figure 7-2

Skill Practice

28. Two cars leave from the same place at the same time. One travels west at 40 mph, and the other travels north at 30 mph. How far apart are they after 2 hr?

---

5. Radical Functions

If n is an integer greater than 1, then a function written in the form $f(x) = \sqrt[n]{x}$ is called a **radical function**. Note that if n is an even integer, then the function will be a real number only if the radicand is nonnegative. Therefore, the domain is restricted to nonnegative real numbers, or equivalently, $[0, \infty)$. If n is an odd integer, then the domain is all real numbers.

Answer

28. 100 mi

Skill Practice

For each function, write the domain in interval notation.

29. $f(x) = \sqrt{x + 5}$
30. $g(t) = \sqrt[3]{t - 9}$
31. $h(a) = \sqrt{1 - 2a}$

example 9 Determining the Domain of Radical Functions

For each function, write the domain in interval notation.

a. $g(t) = \sqrt[4]{t - 2}$ **b.** $h(a) = \sqrt[3]{a - 4}$ **c.** $k(x) = \sqrt{3 - 5x}$

Solution:

a. $g(t) = \sqrt[4]{t - 2}$ The index is even. The radicand must be nonnegative.

$t - 2 \geq 0$ Set the radicand greater than or equal to zero.

$t \geq 2$ Solve for t.

The domain is $[2, \infty)$.

Calculator Connections

The domain of $g(t) = \sqrt[4]{t - 2}$ can be confirmed from its graph.

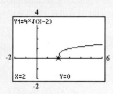

b. $h(a) = \sqrt[3]{a - 4}$ The index is odd; therefore, the domain is all real numbers.

The domain is $(-\infty, \infty)$.

Calculator Connections

The domain of $h(a) = \sqrt[3]{a - 4}$ can be confirmed from its graph.

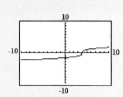

c. $k(x) = \sqrt{3 - 5x}$ The index is even; therefore, the radicand must be nonnegative.

$3 - 5x \geq 0$ Set the radicand greater than or equal to zero.

$-5x \geq -3$ Solve for x.

$\dfrac{-5x}{-5} \leq \dfrac{3}{-5}$ Reverse the inequality sign.

$x \leq \dfrac{3}{5}$

The domain is $\left(-\infty, \dfrac{3}{5}\right]$.

Answers

29. $[-5, \infty)$ 30. $(-\infty, \infty)$
31. $\left(-\infty, \dfrac{1}{2}\right]$

Calculator Connections

The domain of the function defined by $k(x) = \sqrt{3 - 5x}$ can be confirmed from its graph.

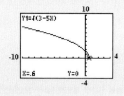

section 7.1 Practice Exercises

Boost your GRADE at mathzone.com!

- Practice Problems
- Self-Tests
- NetTutor
- e-Professors
- Videos

Study Skills Exercises

1. Sometimes test anxiety can be eliminated by adequate preparation and practice. List some places where you can find extra problems for practice.

2. Define the key terms.

   a. Square root b. Radical sign c. Principal square root d. Perfect square

   e. nth root f. Index g. Radicand h. Cube root

   i. Pythagorean theorem j. Radical function

Objective 1: Definition of a Square Root

3. a. Find the square roots of 64. (See Example 1.)

   b. Find $\sqrt{64}$.

   c. Explain the difference between the answers in part (a) and part (b).

4. a. Find the square roots of 121.

   b. Find $\sqrt{121}$.

   c. Explain the difference between the answers in part (a) and part (b).

5. a. What is the principal square root of 81?

   b. What is the negative square root of 81?

6. a. What is the principal square root of 100?

   b. What is the negative square root of 100?

7. Using the definition of a square root, explain why $\sqrt{-36}$ is not a real number.

476 Chapter 7 Radicals and Complex Numbers

For Exercises 8–19, evaluate the roots without using a calculator. Identify those that are not real numbers. (See Examples 2–3.)

8. $\sqrt{25}$
9. $\sqrt{49}$
10. $-\sqrt{25}$
11. $-\sqrt{49}$

12. $\sqrt{-25}$
13. $\sqrt{-49}$
14. $\sqrt{\dfrac{100}{121}}$
15. $\sqrt{\dfrac{64}{9}}$

16. $\sqrt{0.64}$
17. $\sqrt{0.81}$
18. $-\sqrt{0.0144}$
19. $-\sqrt{0.16}$

Objective 2: Definition of an *n*th Root

20. Using the definition of an *n*th root, explain why $\sqrt[4]{-16}$ is not a real number.

For Exercises 21–22, evaluate the roots without using a calculator. Identify those that are not real numbers.

21. a. $\sqrt{64}$
    b. $\sqrt[3]{64}$
    c. $-\sqrt{64}$

    d. $-\sqrt[3]{64}$
    e. $\sqrt{-64}$
    f. $\sqrt[3]{-64}$

22. a. $\sqrt{16}$
    b. $\sqrt[4]{16}$
    c. $-\sqrt{16}$

    d. $-\sqrt[4]{16}$
    e. $\sqrt{-16}$
    f. $\sqrt[4]{-16}$

For Exercises 23–36, evaluate the roots without using a calculator. Identify those that are not real numbers. (See Example 4.)

23. $\sqrt[3]{-27}$
24. $\sqrt[3]{-125}$
25. $\sqrt[3]{\dfrac{1}{8}}$
26. $\sqrt[5]{\dfrac{1}{32}}$

27. $\sqrt[5]{32}$
28. $\sqrt[4]{1}$
29. $\sqrt[3]{-\dfrac{125}{64}}$
30. $\sqrt[3]{-\dfrac{8}{27}}$

31. $\sqrt[4]{-1}$
32. $\sqrt[6]{-1}$
33. $\sqrt[6]{1{,}000{,}000}$
34. $\sqrt[4]{10{,}000}$

35. $-\sqrt[3]{0.008}$
36. $-\sqrt[4]{0.0016}$

Objective 3: Roots of Variable Expressions

For Exercises 37–54, simplify the radical expressions. (See Example 5.)

37. $\sqrt{a^2}$
38. $\sqrt[4]{a^4}$
39. $\sqrt[3]{a^3}$
40. $\sqrt[5]{a^5}$

41. $\sqrt[6]{a^6}$
42. $\sqrt[7]{a^7}$
43. $\sqrt{x^4}$
44. $\sqrt[3]{y^{12}}$

45. $\sqrt{x^2y^4}$

46. $\sqrt[3]{(u+v)^3}$

47. $-\sqrt[3]{\dfrac{x^3}{y^3}}$, $y \neq 0$

48. $\sqrt[4]{\dfrac{a^4}{b^8}}$, $b \neq 0$

49. $\dfrac{2}{\sqrt[4]{x^4}}$, $x \neq 0$

50. $\sqrt{(-5)^2}$

51. $\sqrt[3]{(-9)^3}$

52. $\sqrt[6]{(50)^6}$

53. $\sqrt[10]{(-2)^{10}}$

54. $\sqrt[5]{(-2)^5}$

For Exercises 55–70, simplify the expressions. Assume all variables are positive real numbers. **(See Example 6.)**

55. $\sqrt{x^2y^4}$

56. $\sqrt{16p^2}$

57. $\sqrt{\dfrac{a^6}{b^2}}$

58. $\sqrt{\dfrac{w^2}{z^4}}$

59. $-\sqrt{\dfrac{25}{q^2}}$

60. $-\sqrt{\dfrac{p^6}{81}}$

61. $\sqrt{9x^2y^4z^2}$

62. $\sqrt{4a^4b^2c^6}$

63. $\sqrt{\dfrac{h^2k^4}{16}}$

64. $\sqrt{\dfrac{4x^2}{y^8}}$

65. $-\sqrt[3]{\dfrac{t^3}{27}}$

66. $\sqrt[4]{\dfrac{16}{w^4}}$

67. $\sqrt[5]{32y^{10}}$

68. $\sqrt[3]{64x^6y^3}$

69. $\sqrt[6]{64p^{12}q^{18}}$

70. $\sqrt[4]{16r^{12}s^8}$

Objective 4: Pythagorean Theorem

For Exercises 71–74, find the length of the third side of each triangle by using the Pythagorean theorem. **(See Example 7.)**

71.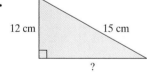

72.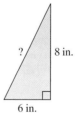

73.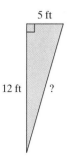

74.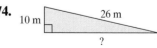

For Exercises 75–78, use the Pythagorean theorem. (See Example 8.)

75. Roberto and Sherona began running from the same place at the same time. They ran along two different paths that formed right angles with each other. Roberto ran 4 mi and stopped, while Sherona ran 3 mi and stopped. How far apart were they when they stopped?

76. Leine and Laura began hiking from their campground. Laura headed south while Leine headed east. Laura walked 12 mi and Leine walked 5 mi. How far apart were they when they stopped walking?

77. Two bikers take off from the same place at the same time. One travels north at 4 mph, and the other travels east at 3 mph. How far apart are they after 5 hr?

78. Professor Ortiz leaves campus on her bike, heading west at 6 mph. Professor Wilson leaves campus at the same time and walks south at 2.5 mph. How far apart are they after 4 hr?

Objective 5: Radical Functions

For Exercises 79–82, evaluate the function for the given values of x. Then write the domain of the function in interval notation.

79. $h(x) = \sqrt{x-2}$
   a. $h(0)$
   b. $h(1)$
   c. $h(2)$
   d. $h(3)$
   e. $h(6)$

80. $k(x) = \sqrt{x+1}$
   a. $k(-3)$
   b. $k(-2)$
   c. $k(-1)$
   d. $k(0)$
   e. $k(3)$

81. $g(x) = \sqrt[3]{x-2}$
   a. $g(-6)$
   b. $g(1)$
   c. $g(2)$
   d. $g(3)$

82. $f(x) = \sqrt[3]{x+1}$
   a. $f(-9)$
   b. $f(-2)$
   c. $f(0)$
   d. $f(7)$

For each function defined in Exercises 83–86, write the domain in interval notation. **(See Example 9.)**

83. $q(x) = \sqrt{x+5}$
84. $p(x) = \sqrt{1-x}$
85. $R(x) = \sqrt[3]{x+1}$
86. $T(x) = \sqrt{x-10}$

For Exercises 87–90, match the function with the graph. Use the domain information from Exercises 83–86.

87. $p(x) = \sqrt{1-x}$
88. $q(x) = \sqrt{x+5}$
89. $T(x) = \sqrt{x-10}$
90. $R(x) = \sqrt[3]{x+1}$

a.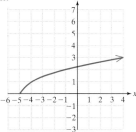

b.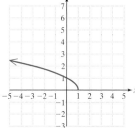

c.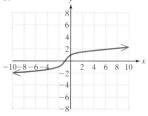

d.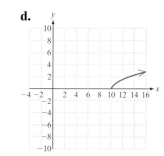

Mixed Exercises

For Exercises 91–94, translate the English phrase to an algebraic expression.

91. The sum of q and the square of p

92. The product of 11 and the cube root of x

93. The quotient of 6 and the cube root of x

94. The difference of y and the principal square root of x

95. If a square has an area of 64 in.^{2}, then what are the lengths of the sides?

96. If a square has an area of 121 m^2, then what are the lengths of the sides?

Graphing Calculator Exercises

For Exercises 97–104, use a calculator to evaluate the expressions to 4 decimal places.

97. $\sqrt{69}$

98. $\sqrt{5798}$

99. $2 + \sqrt[3]{5}$

100. $3 - 2\sqrt[4]{10}$

101. $7\sqrt[4]{25}$

102. $-3\sqrt[3]{9}$

103. $\dfrac{3 - \sqrt{19}}{11}$

104. $\dfrac{5 + 2\sqrt{15}}{12}$

105. Graph $h(x) = \sqrt{x - 2}$ on the standard viewing window. Use the graph to confirm the domain found in Exercise 79.

106. Graph $k(x) = \sqrt{x + 1}$ on the standard viewing window. Use the graph to confirm the domain found in Exercise 80.

107. Graph $g(x) = \sqrt[3]{x - 2}$ on the standard viewing window. Use the graph to confirm the domain found in Exercise 81.

108. Graph $f(x) = \sqrt[3]{x + 1}$ on the standard viewing window. Use the graph to confirm the domain found in Exercise 82.

Chapter 7 Radicals and Complex Numbers

Objectives

1. Definition of $a^{1/n}$ and $a^{m/n}$
2. Converting Between Rational Exponents and Radical Notation
3. Properties of Rational Exponents
4. Applications Involving Rational Exponents

section 7.2 Rational Exponents

1. Definition of $a^{1/n}$ and $a^{m/n}$

In Section 1.8, the properties for simplifying expressions with integer exponents were presented. In this section, the properties are expanded to include expressions with rational exponents. We begin by defining expressions of the form $a^{1/n}$.

Definition of $a^{1/n}$

Let a be a real number, and let n be an integer such that $n > 1$. If $\sqrt[n]{a}$ is a real number, then

$$a^{1/n} = \sqrt[n]{a}$$

Skill Practice

Convert the expression to radical form and simplify, if possible.

1. $16^{1/4}$
2. $(-8)^{1/3}$
3. $36^{-1/2}$
4. $(-49)^{1/2}$
5. $-49^{1/2}$

---

example 1 — Evaluating Expressions of the Form $a^{1/n}$

Convert the expression to radical form and simplify, if possible.

a. $(-8)^{1/3}$ **b.** $81^{1/4}$ **c.** $-100^{1/2}$ **d.** $(-100)^{1/2}$ **e.** $16^{-1/2}$

Solution:

a. $(-8)^{1/3} = \sqrt[3]{-8} = -2$

b. $81^{1/4} = \sqrt[4]{81} = 3$

c. $-100^{1/2} = -1 \cdot 100^{1/2}$ The exponent applies only to the base of 100.
$\phantom{-100^{1/2}} = -1\sqrt{100}$
$\phantom{-100^{1/2}} = -10$

d. $(-100)^{1/2}$ is not a real number because $\sqrt{-100}$ is not a real number.

e. $16^{-1/2} = \dfrac{1}{16^{1/2}}$ Write the expression with a positive exponent. Recall that $b^{-n} = \dfrac{1}{b^n}$.

$\phantom{16^{-1/2}} = \dfrac{1}{\sqrt{16}}$

$\phantom{16^{-1/2}} = \dfrac{1}{4}$

If $\sqrt[n]{a}$ is a real number, then we can define an expression of the form $a^{m/n}$ in such a way that the multiplication property of exponents still holds true. For example,

$16^{3/4}$
$(16^{1/4})^3 = (\sqrt[4]{16})^3 = (2)^3 = 8$
$(16^3)^{1/4} = \sqrt[4]{16^3} = \sqrt[4]{4096} = 8$

Answers

1. 2   2. -2 3. $\dfrac{1}{6}$

4. Not a real number   5. -7

Section 7.2 Rational Exponents 481

Definition of $a^{m/n}$

Let a be a real number, and let m and n be positive integers such that m and n share no common factors and $n > 1$. If $\sqrt[n]{a}$ is a real number, then

$$a^{m/n} = (a^{1/n})^m = (\sqrt[n]{a})^m \quad \text{and} \quad a^{m/n} = (a^m)^{1/n} = \sqrt[n]{a^m}$$

The rational exponent in the expression $a^{m/n}$ is essentially performing two operations. The numerator of the exponent raises the base to the mth power. The denominator takes the nth root.

example 2 Evaluating Expressions of the Form $a^{m/n}$

Convert each expression to radical form and simplify.

a. $8^{2/3}$ b. $100^{5/2}$ c. $\left(\dfrac{1}{25}\right)^{3/2}$ d. $4^{-3/2}$

Solution:

a. $8^{2/3} = (\sqrt[3]{8})^2$ Take the cube root of 8 and square the result.
   $= (2)^2$ Simplify.
   $= 4$

b. $100^{5/2} = (\sqrt{100})^5$ Take the square root of 100 and raise the result to the fifth power.
   $= (10)^5$ Simplify.
   $= 100{,}000$

c. $\left(\dfrac{1}{25}\right)^{3/2} = \left(\sqrt{\dfrac{1}{25}}\right)^3$ Take the square root of $\tfrac{1}{25}$ and cube the result.
   $= \left(\dfrac{1}{5}\right)^3$ Simplify.
   $= \dfrac{1}{125}$

d. $4^{-3/2} = \dfrac{1}{4^{3/2}}$ Write the expression with positive exponents.
   $= \dfrac{1}{(\sqrt{4})^3}$ Take the square root of 4 and cube the result.
   $= \dfrac{1}{2^3}$ Simplify.
   $= \dfrac{1}{8}$

Skill Practice

Convert each expression to radical form and simplify.

6. $9^{3/2}$
7. $8^{5/3}$
8. $32^{-4/5}$
9. $\left(\dfrac{1}{27}\right)^{4/3}$

Answers

6. 27   7. 32
8. $\dfrac{1}{16}$ 9. $\dfrac{1}{81}$

Calculator Connections

A calculator can be used to confirm the results of Example 2(a)–2(c).

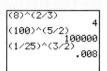

2. Converting Between Rational Exponents and Radical Notation

Skill Practice

Convert each expression to radical notation. Assume all variables represent positive real numbers.

10. $t^{4/5}$
11. $(2y^3)^{1/4}$
12. $10p^{1/2}$
13. $q^{-2/3}$

example 3 — Using Radical Notation and Rational Exponents

Convert each expression to radical notation. Assume all variables represent positive real numbers.

a. $a^{3/5}$ **b.** $(5x^2)^{1/3}$ **c.** $3y^{1/4}$ **d.** $z^{-3/4}$

Solution:

a. $a^{3/5} = \sqrt[5]{a^3}$ or $\left(\sqrt[5]{a}\right)^3$

b. $(5x^2)^{1/3} = \sqrt[3]{5x^2}$

c. $3y^{1/4} = 3\sqrt[4]{y}$ Note that the coefficient 3 is not raised to the $\frac{1}{4}$ power.

d. $z^{-3/4} = \dfrac{1}{z^{3/4}} = \dfrac{1}{\sqrt[4]{z^3}}$

Skill Practice

Convert to an equivalent expression using rational exponents. Assume all variables represent positive real numbers.

14. $\sqrt[3]{x^2}$
15. $\sqrt{5y}$
16. $5\sqrt{y}$

example 4 — Using Radical Notation and Rational Exponents

Convert each expression to an equivalent expression by using rational exponents. Assume that all variables represent positive real numbers.

a. $\sqrt[4]{b^3}$ **b.** $\sqrt{7a}$ **c.** $7\sqrt{a}$

Solution:

a. $\sqrt[4]{b^3} = b^{3/4}$

b. $\sqrt{7a} = (7a)^{1/2}$

c. $7\sqrt{a} = 7a^{1/2}$

3. Properties of Rational Exponents

In Section 1.8, several properties and definitions were introduced to simplify expressions with integer exponents. These properties also apply to rational exponents.

Answers

10. $\sqrt[5]{t^4}$ 11. $\sqrt[4]{2y^3}$
12. $10\sqrt{p}$ 13. $\dfrac{1}{\sqrt[3]{q^2}}$
14. $x^{2/3}$ 15. $(5y)^{1/2}$
16. $5y^{1/2}$

Properties of Exponents and Definitions

Let a and b be nonzero real numbers. Let m and n be rational numbers such that a^m, a^n, and b^n are real numbers.

Description	Property	Example
1. Multiplying like bases	$a^m a^n = a^{m+n}$	$x^{1/3} x^{4/3} = x^{5/3}$
2. Dividing like bases	$\dfrac{a^m}{a^n} = a^{m-n}$	$\dfrac{x^{3/5}}{x^{1/5}} = x^{2/5}$
3. The power rule	$(a^m)^n = a^{mn}$	$(2^{1/3})^{1/2} = 2^{1/6}$
4. Power of a product	$(ab)^m = a^m b^m$	$(xy)^{1/2} = x^{1/2} y^{1/2}$
5. Power of a quotient	$\left(\dfrac{a}{b}\right)^m = \dfrac{a^m}{b^m}$	$\left(\dfrac{4}{25}\right)^{1/2} = \dfrac{4^{1/2}}{25^{1/2}} = \dfrac{2}{5}$

Description	Definition	Example
1. Negative exponents	$a^{-m} = \left(\dfrac{1}{a}\right)^m = \dfrac{1}{a^m}$	$(8)^{-1/3} = \left(\dfrac{1}{8}\right)^{1/3} = \dfrac{1}{2}$
2. Zero exponent	$a^0 = 1$	$5^0 = 1$

example 5 — Simplifying Expressions with Rational Exponents

Use the properties of exponents to simplify the expressions. Assume all variables represent positive real numbers.

a. $y^{2/5} y^{3/5}$ **b.** $(s^4 t^8)^{1/4}$ **c.** $\left(\dfrac{81 c d^{-2}}{3 c^{-2} d^4}\right)^{1/3}$

Solution:

a. $y^{2/5} y^{3/5} = y^{(2/5)+(3/5)}$ Multiply like bases by adding exponents.

$\qquad = y^{5/5}$ Simplify.

$\qquad = y$

b. $(s^4 t^8)^{1/4} = s^{4/4} t^{8/4}$ Apply the power rule. Multiply exponents.

$\qquad = st^2$ Simplify.

c. $\left(\dfrac{81 c d^{-2}}{3 c^{-2} d^4}\right)^{1/3} = (27 c^{1-(-2)} d^{-2-4})^{1/3}$ Simplify inside parentheses. Subtract exponents.

$\qquad = (27 c^3 d^{-6})^{1/3}$

$\qquad = \left(\dfrac{27 c^3}{d^6}\right)^{1/3}$ Simplify the negative exponent.

$\qquad = \dfrac{27^{1/3} c^{3/3}}{d^{6/3}}$ Apply the power rule. Multiply exponents.

$\qquad = \dfrac{3c}{d^2}$ Simplify.

Skill Practice

Use the properties of exponents to simplify the expressions. Assume all variables represent positive real numbers.

17. $x^{1/2} \cdot x^{3/4}$

18. $(a^3 b^2)^{1/3}$

19. $\left(\dfrac{32 y^2 z^{-3}}{2 y^{-2} z^5}\right)^{1/4}$

Answers

17. $x^{5/4}$
18. $ab^{2/3}$
19. $\dfrac{2y}{z^2}$

4. Applications Involving Rational Exponents

Skill Practice

20. The formula for the radius of a sphere is
$$r = \left(\frac{3V}{4\pi}\right)^{1/3}$$
where V is the volume. Find the radius of a sphere whose volume is 113.04 in.^{3} (Use 3.14 for π.)

example 6 — Applying Rational Exponents

Suppose P dollars in principal is invested in an account that earns interest annually. If after t years the investment grows to A dollars, then the annual rate of return r on the investment is given by

$$r = \left(\frac{A}{P}\right)^{1/t} - 1$$

Find the annual rate of return on $5000 which grew to $12,500 after 6 years.

Solution:

$$r = \left(\frac{A}{P}\right)^{1/t} - 1$$

$$= \left(\frac{12{,}500}{5000}\right)^{1/6} - 1 \qquad \text{where } A = \$12{,}500,\ P = \$5000,\ \text{and}\ t = 6$$

$$= (2.5)^{1/6} - 1$$

$$\approx 1.165 - 1$$

$$\approx 0.165\ \text{or}\ 16.5\%$$

The annual rate of return is 16.5%.

Calculator Connections

The expression
$$r = \left(\frac{12{,}500}{5000}\right)^{1/6} - 1$$
is easily evaluated on a graphing calculator.

```
(12500/5000)^(1/
6)-1
         .1649930508
```

Answer
20. 3 in.

---

section 7.2 — Practice Exercises

Boost your GRADE at mathzone.com!

- Practice Problems
- Self-Tests
- NetTutor
- e-Professors
- Videos

Study Skills Exercises

1. Don't be afraid to mark up a book page. Use a highlighter to emphasize key points, definitions, rules, formulas, or processes. What would you highlight in this section?

2. Define the key terms.

   a. $a^{1/n}$ b. $a^{m/n}$

Review Exercises

For the exercises in this set, assume that all variables represent positive real numbers unless otherwise stated.

3. Given: $\sqrt[3]{27}$
   a. Identify the index.
   b. Identify the radicand.

4. Given: $\sqrt{18}$
   a. Identify the index.
   b. Identify the radicand.

For Exercises 5–8, evaluate the radicals (if possible).

5. $\sqrt{25}$
6. $\sqrt[3]{8}$
7. $\sqrt[4]{81}$
8. $(\sqrt[4]{16})^3$

Objective 1: Definition of $a^{1/n}$ and $a^{m/n}$

For Exercises 9–18, convert the expressions to radical form and simplify. (See Example 1.)

9. $144^{1/2}$
10. $16^{1/4}$
11. $-144^{1/2}$
12. $-16^{1/4}$

13. $(-144)^{1/2}$
14. $(-16)^{1/4}$
15. $(-64)^{1/3}$
16. $(-32)^{1/5}$

17. $(25)^{-1/2}$
18. $(27)^{-1/3}$

19. Explain how to interpret the expression $a^{m/n}$ as a radical.

20. Explain why $(\sqrt[3]{8})^4$ is easier to evaluate than $\sqrt[3]{8^4}$.

For Exercises 21–24, simplify the expression, if possible. (See Example 2.)

21. a. $16^{3/4}$ b. $-16^{3/4}$ c. $(-16)^{3/4}$ d. $16^{-3/4}$ e. $-16^{-3/4}$ f. $(-16)^{-3/4}$

22. a. $81^{3/4}$ b. $-81^{3/4}$ c. $(-81)^{3/4}$ d. $81^{-3/4}$ e. $-81^{-3/4}$ f. $(-81)^{-3/4}$

23. a. $25^{3/2}$ b. $-25^{3/2}$ c. $(-25)^{3/2}$ d. $25^{-3/2}$ e. $-25^{-3/2}$ f. $(-25)^{-3/2}$

24. a. $4^{3/2}$ b. $-4^{3/2}$ c. $(-4)^{3/2}$ d. $4^{-3/2}$ e. $-4^{-3/2}$ f. $(-4)^{-3/2}$

For Exercises 25–50, simplify the expression. (See Example 2.)

25. $64^{-3/2}$
26. $81^{-3/2}$
27. $243^{3/5}$
28. $1^{5/3}$

29. $-27^{-4/3}$
30. $-16^{-5/4}$
31. $\left(\dfrac{100}{9}\right)^{-3/2}$
32. $\left(\dfrac{49}{100}\right)^{-1/2}$

33. $(-4)^{-3/2}$
34. $(-49)^{-3/2}$
35. $(-8)^{1/3}$
36. $(-9)^{1/2}$

37. $-8^{1/3}$ **38.** $-9^{1/2}$ **39.** $27^{-2/3}$ **40.** $125^{-1/3}$

41. $\dfrac{1}{36^{-1/2}}$ **42.** $\dfrac{1}{16^{-1/2}}$ **43.** $\dfrac{1}{1000^{-1/3}}$ **44.** $\dfrac{1}{81^{-3/4}}$

45. $\left(\dfrac{1}{8}\right)^{2/3} + \left(\dfrac{1}{4}\right)^{1/2}$ **46.** $\left(\dfrac{1}{8}\right)^{-2/3} + \left(\dfrac{1}{4}\right)^{-1/2}$ **47.** $\left(\dfrac{1}{16}\right)^{-3/4} - \left(\dfrac{1}{49}\right)^{-1/2}$ **48.** $\left(\dfrac{1}{16}\right)^{1/4} - \left(\dfrac{1}{49}\right)^{1/2}$

49. $\left(\dfrac{1}{4}\right)^{1/2} + \left(\dfrac{1}{64}\right)^{-1/3}$ **50.** $\left(\dfrac{1}{36}\right)^{1/2} + \left(\dfrac{1}{64}\right)^{-5/6}$

Objective 2: Converting Between Rational Exponents and Radical Notation

For Exercises 51–58, convert each expression to radical notation. **(See Example 3.)**

51. $q^{2/3}$ **52.** $t^{3/5}$ **53.** $6y^{3/4}$ **54.** $8b^{4/9}$

55. $(x^2y)^{1/3}$ **56.** $(c^2d)^{1/6}$ **57.** $(qr)^{-1/5}$ **58.** $(7x)^{-1/4}$

For Exercises 59–66, write each expression by using rational exponents rather than radical notation. **(See Example 4.)**

59. $\sqrt[3]{x}$ **60.** $\sqrt[4]{a}$ **61.** $10\sqrt{b}$ **62.** $-2\sqrt[3]{t}$

63. $\sqrt[3]{y^2}$ **64.** $\sqrt[6]{z^5}$ **65.** $\sqrt[4]{a^2b^3}$ **66.** $\sqrt{abc}$

Objective 3: Properties of Rational Exponents

For Exercises 67–90, simplify the expressions by using the properties of rational exponents. Write the final answer using positive exponents only. **(See Example 5.)**

67. $x^{1/4}x^{-5/4}$ **68.** $2^{2/3}2^{-5/3}$ **69.** $\dfrac{p^{5/3}}{p^{2/3}}$ **70.** $\dfrac{q^{5/4}}{q^{1/4}}$

71. $(y^{1/5})^{10}$ **72.** $(x^{1/2})^8$ **73.** $6^{-1/5}6^{3/5}$ **74.** $a^{-1/3}a^{2/3}$

75. $\dfrac{4t^{-1/3}}{t^{4/3}}$ **76.** $\dfrac{5s^{-1/3}}{s^{5/3}}$ **77.** $(a^{1/3}a^{1/4})^{12}$ **78.** $(x^{2/3}x^{1/2})^6$

79. $(5a^2c^{-1/2}d^{1/2})^2$ **80.** $(2x^{-1/3}y^2z^{5/3})^3$ **81.** $\left(\dfrac{x^{-2/3}}{y^{-3/4}}\right)^{12}$ **82.** $\left(\dfrac{m^{-1/4}}{n^{-1/2}}\right)^{-4}$

83. $\left(\dfrac{16w^{-2}z}{2wz^{-8}}\right)^{1/3}$ **84.** $\left(\dfrac{50p^{-1}q}{2pq^{-3}}\right)^{1/2}$ **85.** $(25x^2y^4z^6)^{1/2}$ **86.** $(8a^6b^3c^9)^{2/3}$

87. $(x^2y^{-1/3})^6(x^{1/2}yz^{2/3})^2$ **88.** $(a^{-1/3}b^{1/2})^4(a^{-1/2}b^{3/5})^{10}$ **89.** $\left(\dfrac{x^{3m}y^{2m}}{z^{5m}}\right)^{1/m}$ **90.** $\left(\dfrac{a^{4n}b^{3n}}{c^n}\right)^{1/n}$

Objective 4: Applications Involving Rational Exponents

91. If the area A of a square is known, then the length of its sides, s, can be computed by the formula $s = A^{1/2}$.

    a. Compute the length of the sides of a square having an area of 100 in.² (See Example 6.)

    b. Compute the length of the sides of a square having an area of 72 in.² Round your answer to the nearest 0.1 in.

92. The radius r of a sphere of volume V is given by $r = \left(\dfrac{3V}{4\pi}\right)^{1/3}$. Find the radius of a sphere having a volume of 85 in.³ Round your answer to the nearest 0.1 in.

93. If P dollars in principal grows to A dollars after t years with annual interest, then the interest rate is given by $r = \left(\dfrac{A}{P}\right)^{1/t} - 1$.

    a. In one account, $10,000 grows to $16,802 after 5 years. Compute the interest rate. Round your answer to a tenth of a percent.

    b. In another account $10,000 grows to $18,000 after 7 years. Compute the interest rate. Round your answer to a tenth of a percent.

    c. Which account produced a higher average yearly return?

94. Is $(a + b)^{1/2}$ the same as $a^{1/2} + b^{1/2}$? Why or why not?

Expanding Your Skills

For Exercises 95–100, write the expression as a single radical.

95. $\sqrt{\sqrt[3]{x}}$ **96.** $\sqrt[3]{\sqrt{x}}$ **97.** $\sqrt[4]{\sqrt{y}}$ **98.** $\sqrt{\sqrt[4]{y}}$

99. $\sqrt[5]{\sqrt[3]{w}}$ **100.** $\sqrt[3]{\sqrt[4]{w}}$

For Exercises 101–108, use a calculator to approximate the expressions and round to 4 decimal places, if necessary.

101. $9^{1/2}$ **102.** $125^{-1/3}$ **103.** $50^{-1/4}$ **104.** $(172)^{3/5}$

105. $\sqrt[3]{5^2}$ **106.** $\sqrt[4]{6^3}$ **107.** $\sqrt{10^3}$ **108.** $\sqrt[3]{16}$

Objectives

1. Multiplication and Division Properties of Radicals
2. Simplifying Radicals by Using the Multiplication Property of Radicals
3. Simplifying Radicals by Using the Division Property of Radicals

section 7.3 Simplifying Radical Expressions

1. Multiplication and Division Properties of Radicals

You may have already noticed certain properties of radicals involving a product or quotient.

Multiplication and Division Properties of Radicals

Let a and b represent real numbers such that $\sqrt[n]{a}$ and $\sqrt[n]{b}$ are both real. Then

1. $\sqrt[n]{ab} = \sqrt[n]{a} \cdot \sqrt[n]{b}$ *Multiplication property of radicals*

2. $\sqrt[n]{\dfrac{a}{b}} = \dfrac{\sqrt[n]{a}}{\sqrt[n]{b}}$ $b \neq 0$ *Division property of radicals*

Properties 1 and 2 follow from the properties of rational exponents.

$$\sqrt[n]{ab} = (ab)^{1/n} = a^{1/n}b^{1/n} = \sqrt[n]{a} \cdot \sqrt[n]{b}$$

$$\sqrt[n]{\dfrac{a}{b}} = \left(\dfrac{a}{b}\right)^{1/n} = \dfrac{a^{1/n}}{b^{1/n}} = \dfrac{\sqrt[n]{a}}{\sqrt[n]{b}}$$

The multiplication and division properties of radicals indicate that a product or quotient within a radicand can be written as a product or quotient of radicals, provided the roots are real numbers. For example:

$$\sqrt{144} = \sqrt{16} \cdot \sqrt{9}$$

$$\sqrt{\dfrac{25}{36}} = \dfrac{\sqrt{25}}{\sqrt{36}}$$

The reverse process is also true. A product or quotient of radicals can be written as a single radical provided the roots are real numbers and they have the same indices.

$$\sqrt{3} \cdot \sqrt{12} = \sqrt{36}$$

$$\dfrac{\sqrt[3]{8}}{\sqrt[3]{125}} = \sqrt[3]{\dfrac{8}{125}}$$

In algebra it is customary to simplify radical expressions as much as possible.

Concept Connections

Multiply or divide the radicals.

1. $\sqrt{5} \cdot \sqrt{6}$
2. $\dfrac{\sqrt{10}}{\sqrt{5}}$

Simplified Form of a Radical

Consider any radical expression where the radicand is written as a product of prime factors. The expression is in *simplified form* if all the following conditions are met:

1. The radicand has no factor raised to a power greater than or equal to the index.
2. The radicand does not contain a fraction.
3. There are no radicals in the denominator of a fraction.

Answers

1. $\sqrt{30}$ 2. $\sqrt{2}$

2. Simplifying Radicals by Using the Multiplication Property of Radicals

The expression $\sqrt{x^2}$ is not simplified because it fails condition 1. Because x^2 is a perfect square, $\sqrt{x^2}$ is easily simplified:

$$\sqrt{x^2} = x \quad \text{for } x \geq 0$$

However, how is an expression such as $\sqrt{x^9}$ simplified? This and many other radical expressions are simplified by using the multiplication property of radicals. The following examples illustrate how nth powers can be removed from the radicands of nth roots.

example 1 — Using the Multiplication Property to Simplify a Radical Expression

Use the multiplication property of radicals to simplify the expression $\sqrt{x^9}$. Assume $x \geq 0$.

Solution:

The expression $\sqrt{x^9}$ is equivalent to $\sqrt{x^8 \cdot x}$. Applying the multiplication property of radicals, we have

$\sqrt{x^9} = \sqrt{x^8 \cdot x}$

$\quad = \sqrt{x^8} \cdot \sqrt{x}$ Apply the multiplication property of radicals.

Note that x^8 is a perfect square because $x^8 = (x^4)^2$.

$\quad = x^4\sqrt{x}$ Simplify.

Skill Practice

Simplify the expressions. Assume all variables represent positive real numbers.

3. $\sqrt{a^3}$
4. $\sqrt{y^{31}}$

In Example 1, the expression x^9 is not a perfect square. Therefore, to simplify $\sqrt{x^9}$, it was necessary to write the expression as the product of the largest perfect square and a remaining or "left-over" factor: $\sqrt{x^9} = \sqrt{x^8 \cdot x}$. This process also applies to simplifying nth roots, as shown in Example 2.

example 2 — Using the Multiplication Property to Simplify a Radical Expression

Use the multiplication property of radicals to simplify each expression. Assume all variables represent positive real numbers.

a. $\sqrt[4]{b^7}$ b. $\sqrt[3]{w^7 z^9}$

Solution:

The goal is to rewrite each radicand as the product of the largest perfect square (perfect cube, perfect fourth power, and so on) and a left-over factor.

a. $\sqrt[4]{b^7} = \sqrt[4]{b^4 \cdot b^3}$ b^4 is the largest perfect fourth power in the radicand.

$\quad = \sqrt[4]{b^4} \cdot \sqrt[4]{b^3}$ Apply the multiplication property of radicals.

$\quad = b\sqrt[4]{b^3}$ Simplify.

Skill Practice

Simplify the expressions. Assume all variables represent positive real numbers.

5. $\sqrt[4]{v^{25}}$
6. $\sqrt[3]{p^8 q^{12}}$

Answers

3. $a\sqrt{a}$ 4. $y^{15}\sqrt{y}$
5. $v^6\sqrt[4]{v}$ 6. $p^2 q^4 \sqrt[3]{p^2}$

b. $\sqrt[3]{w^7z^9} = \sqrt[3]{(w^6z^9) \cdot (w)}$ w^6z^9 is the largest perfect cube in the radicand.

$= \sqrt[3]{w^6z^9} \cdot \sqrt[3]{w}$ Apply the multiplication property of radicals.

$= w^2z^3\sqrt[3]{w}$ Simplify.

Each expression in Example 2 involves a radicand that is a product of variable factors. If a numerical factor is present, sometimes it is necessary to factor the coefficient before simplifying the radical.

Skill Practice

Simplify the radicals. Assume all variables represent positive real numbers.

7. $\sqrt{24}$
8. $5\sqrt{18}$
9. $\sqrt[4]{32a^{10}b^{19}}$

example 3 — Using the Multiplication Property to Simplify Radicals

Use the multiplication property of radicals to simplify the expressions. Assume all variables represent positive real numbers.

a. $\sqrt{56}$ **b.** $6\sqrt{50}$ **c.** $\sqrt[3]{40x^3y^5z^7}$

Solution:

a. $\sqrt{56} = \sqrt{2^3 \cdot 7}$ Factor the radicand.

$= \sqrt{(2^2) \cdot (2 \cdot 7)}$ 2^2 is the largest perfect square in the radicand.

$= \sqrt{2^2} \cdot \sqrt{2 \cdot 7}$ Apply the multiplication property of radicals.

$= 2\sqrt{14}$ Simplify.

$\begin{array}{r} 2|56 \\ 2|28 \\ 2|14 \\ 7 \end{array}$

Avoiding Mistakes:

The multiplication property of radicals allows us to simplify a product of *factors* within a radical. For example:

$\sqrt{x^2y^2} = \sqrt{x^2} \cdot \sqrt{y^2} = xy$

However, this rule does not apply to *terms* that are added or subtracted within the radical. For example:

$\sqrt{x^2 + y^2}$ and $\sqrt{x^2 - y^2}$

cannot be simplified.

Calculator Connections

A calculator can be used to support the solution to Example 3(a). The decimal approximation for $\sqrt{56}$ and $2\sqrt{14}$ agree for the first 10 digits. This in itself does not make $\sqrt{56} = 2\sqrt{14}$. It is the multiplication of property of radicals that guarantees that the expressions are equal.

```
√(56)
       7.483314774
2*√(14)
       7.483314774
```

b. $6\sqrt{50} = 6\sqrt{2 \cdot 5^2}$ Factor the radicand.

$= 6 \cdot \sqrt{5^2} \cdot \sqrt{2}$ Apply the multiplication property of radicals.

$= 6 \cdot 5 \cdot \sqrt{2}$ Simplify.

$= 30\sqrt{2}$ Simplify.

$\begin{array}{r} 2|50 \\ 5|25 \\ 5 \end{array}$

c. $\sqrt[3]{40x^3y^5z^7}$

$= \sqrt[3]{2^3 5x^3y^5z^7}$ Factor the radicand.

$= \sqrt[3]{(2^3x^3y^3z^6) \cdot (5y^2z)}$ $2^3x^3y^3z^6$ is the largest perfect cube.

$= \sqrt[3]{2^3x^3y^3z^6} \cdot \sqrt[3]{5y^2z}$ Apply the multiplication property of radicals.

$= 2xyz^2\sqrt[3]{5y^2z}$ Simplify.

$\begin{array}{r} 2|40 \\ 2|20 \\ 2|10 \\ 5 \end{array}$

Answers

7. $2\sqrt{6}$ 8. $15\sqrt{2}$
9. $2a^2b^4\sqrt[4]{2a^2b^3}$

3. Simplifying Radicals by Using the Division Property of Radicals

The division property of radicals indicates that a radical of a quotient can be written as the quotient of the radicals and vice versa, provided all roots are real numbers.

example 4 — Using the Division Property to Simplify Radicals

Use the division property of radicals to simplify the expressions. Assume all variables represent positive real numbers.

a. $\sqrt{\dfrac{a^7}{a^3}}$ b. $\dfrac{\sqrt[3]{3}}{\sqrt[3]{81}}$ c. $\dfrac{7\sqrt{50}}{15}$ d. $\sqrt[4]{\dfrac{2c^5}{32cd^8}}$

Solution:

a. $\sqrt{\dfrac{a^7}{a^3}}$ The radicand contains a fraction. However, the fraction can be reduced to lowest terms.

$= \sqrt{a^4}$

$= a^2$ Simplify the radical.

b. $\dfrac{\sqrt[3]{3}}{\sqrt[3]{81}}$ The expression has a radical in the denominator.

$= \sqrt[3]{\dfrac{3}{81}}$ Because the radicands have a common factor, write the expression as a single radical (division property of radicals).

$= \sqrt[3]{\dfrac{1}{27}}$ Reduce to lowest terms.

$= \dfrac{1}{3}$ Simplify.

c. $\dfrac{7\sqrt{50}}{15}$ Simplify $\sqrt{50}$.

$= \dfrac{7\sqrt{5^2 \cdot 2}}{15}$ 5^2 is the largest perfect square in the radicand.

$= \dfrac{7\sqrt{5^2} \cdot \sqrt{2}}{15}$ Multiplication property of radicals

$= \dfrac{7 \cdot 5\sqrt{2}}{15}$ Simplify the radicals.

$= \dfrac{7 \cdot \overset{1}{5}\sqrt{2}}{\underset{3}{15}}$ Reduce to lowest terms.

$= \dfrac{7\sqrt{2}}{3}$

Skill Practice

Use the division property of radicals to simplify the expressions. Assume all variables represent positive real numbers.

10. $\sqrt{\dfrac{v^{21}}{v^5}}$ 11. $\dfrac{\sqrt[5]{64}}{\sqrt[5]{2}}$

12. $\dfrac{2\sqrt{300}}{30}$ 13. $\sqrt[3]{\dfrac{54x^{17}y}{2x^2y^{19}}}$

Answers

10. v^8 11. 2

12. $\dfrac{2\sqrt{3}}{3}$ 13. $\dfrac{3x^5}{y^6}$

d. $\sqrt[4]{\dfrac{2c^5}{32cd^8}}$ The radicand contains a fraction.

$= \sqrt[4]{\dfrac{c^4}{16d^8}}$ Simplify the factors in the radicand.

$= \dfrac{\sqrt[4]{c^4}}{\sqrt[4]{16d^8}}$ Apply the division property of radicals.

$= \dfrac{c}{2d^2}$ Simplify.

section 7.3 Practice Exercises

Boost your GRADE at mathzone.com!
MathZone

- Practice Problems
- Self-Tests
- NetTutor
- e-Professors
- Videos

For the exercises in this set, assume that all variables represent positive real numbers unless otherwise stated.

Study Skills Exercise

1. When writing a radical expression, be sure that you understand the difference between an exponent on a coefficient and an index to a radical. Write an algebraic expression for each of the following.

   a. x cubed times the square root of y

   b. x times the cube root of y

Review Exercises

For Exercises 2–4, simplify the expression. Write the answer with positive exponents only.

2. $(a^2b^{-4})^{1/2}\left(\dfrac{a}{b^{-3}}\right)$

3. $\left(\dfrac{p^4}{q^{-6}}\right)^{-1/2}(p^3q^{-2})$

4. $(x^{1/3}y^{5/6})^{-6}$

5. Write $x^{4/7}$ in radical notation.

6. Write $y^{2/5}$ in radical notation.

7. Write $\sqrt{y^9}$ by using rational exponents.

8. Write $\sqrt[3]{x^2}$ by using rational exponents.

Objective 2: Simplifying Radicals by Using the Multiplication Property of Radicals

For Exercises 9–30, simplify the radicals. (See Examples 1–3.)

9. $\sqrt{x^{11}}$

10. $\sqrt{p^{15}}$

11. $\sqrt[3]{q^7}$

12. $\sqrt[3]{r^{17}}$

13. $\sqrt{a^5b^4}$

14. $\sqrt{c^9d^6}$

15. $-\sqrt[4]{x^8y^{13}}$

16. $-\sqrt[4]{p^{16}q^{17}}$

17. $\sqrt{28}$
18. $\sqrt{63}$
19. $\sqrt{80}$
20. $\sqrt{108}$

21. $\sqrt[3]{54}$
22. $\sqrt[3]{250}$
23. $\sqrt{25ab^3}$
24. $\sqrt{64m^5n^{20}}$

25. $\sqrt{18a^6b^3}$
26. $\sqrt{72m^5n^2}$
27. $\sqrt[3]{-16x^6yz^3}$
28. $\sqrt[3]{-192a^6bc^2}$

29. $\sqrt[4]{80w^4z^7}$
30. $\sqrt[4]{32p^8qr^5}$

Objective 3: Simplifying Radicals by Using the Division Property of Radicals

For Exercises 31–42, simplify the radicals. **(See Example 4.)**

31. $\sqrt{\dfrac{x^3}{x}}$
32. $\sqrt{\dfrac{y^5}{y}}$
33. $\dfrac{\sqrt{p^7}}{\sqrt{p^3}}$
34. $\dfrac{\sqrt{q^{11}}}{\sqrt{q^5}}$

35. $\sqrt{\dfrac{50}{2}}$
36. $\sqrt{\dfrac{98}{2}}$
37. $\dfrac{\sqrt[3]{3}}{\sqrt[3]{24}}$
38. $\dfrac{\sqrt[3]{3}}{\sqrt[3]{81}}$

39. $\dfrac{5\sqrt[3]{16}}{6}$
40. $\dfrac{7\sqrt{18}}{9}$
41. $\dfrac{5\sqrt[3]{72}}{12}$
42. $\dfrac{3\sqrt[3]{250}}{10}$

Mixed Exercises

For Exercises 43–58, simplify the radicals.

43. $5\sqrt{18}$
44. $2\sqrt{24}$
45. $-6\sqrt{75}$
46. $-8\sqrt{8}$

47. $\sqrt{25x^4y^3}$
48. $\sqrt{125p^3q^2}$
49. $\sqrt[3]{27x^2y^3z^4}$
50. $\sqrt[3]{108a^3bc^2}$

51. $\sqrt[3]{\dfrac{16a^2b}{2a^2b^4}}$
52. $\sqrt[4]{\dfrac{3s^2t^4}{10,000}}$
53. $\sqrt[5]{\dfrac{32x}{y^{10}}}$
54. $\sqrt[3]{\dfrac{-16j^3}{k^3}}$

55. $\dfrac{\sqrt{50x^3y}}{\sqrt{9y^4}}$
56. $\dfrac{\sqrt[3]{-27a^4}}{\sqrt[3]{8a}}$
57. $\sqrt{2^3a^{14}b^8c^{31}d^{22}}$
58. $\sqrt{7^5u^{12}v^{20}w^{65}x^{80}}$

For Exercises 59–62, write a mathematical expression for the English phrase and simplify.

59. The quotient of 1 and the cube root of w^6

60. The principal square root of the quotient of h and 49

61. The principal square root of the quantity k raised to the third power

62. The cube root of $2x^4$

For Exercises 63–66, find the third side of the right triangle. Write your answer as a radical and simplify.

63.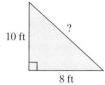

64.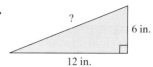

65.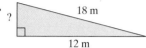

66.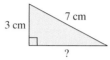

67. On a baseball diamond, the bases are 90 ft apart. Find the exact distance from home plate to second base. Then round to the nearest tenth of a foot.

68. Linda is at the beach flying a kite. The kite is directly over a sand castle 60 ft away from Linda. If 100 ft of kite string is out (ignoring any sag in the string), how high is the kite? (Assume that Linda is 5 ft tall.) See figure.

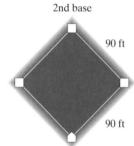

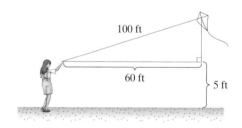

Expanding Your Skills

69. Tom has to travel from town A to town C across a small mountain range. He can travel one of two routes. He can travel on a four-lane highway from A to B and then from B to C at an average speed of 55 mph. Or he can travel on a two-lane road directly from town A to town C, but his average speed will be only 35 mph. If Tom is in a hurry, which route will take him to town C faster?

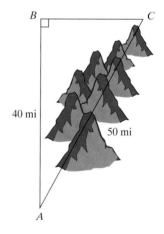

70. One side of a rectangular pasture is 80 ft in length. The diagonal distance is 110 yd. If fencing costs $3.29 per foot, how much will it cost to fence the pasture?

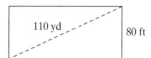

chapter 7 | midchapter review

In the following exercises, assume all variables represent positive real numbers.

For Exercises 1–10, simplify the expressions.

1. $\sqrt[3]{64}$

2. $\sqrt{81}$

3. $\sqrt[4]{\dfrac{1}{16}}$

4. $\sqrt[3]{\dfrac{27}{125}}$

5. $\sqrt[3]{54x^2w^4}$

6. $\sqrt{20t^2u^3v^7}$

7. $\sqrt{\dfrac{50x^3y^5}{x}}$

8. $\sqrt[3]{\dfrac{16w^5z^{10}}{2w^2}}$

9. $\sqrt[3]{(4r-3)^3}$

10. $\sqrt{\dfrac{a^2b^2}{(a+b)^2}}$

11. Explain how to simplify $\sqrt[5]{x^{51}}$.

For Exercises 12–19, simplify the exponential expressions. Leave no negative exponents in your final answer.

12. $8^{2/3}$

13. $16^{3/4}$

14. $\left(\dfrac{1}{25}\right)^{-1/2}$

15. $\left(\dfrac{4}{25}\right)^{-1/2}$

16. $\left(\dfrac{s^{-2}t^4}{r^6}\right)^{1/2}$

17. $\left(\dfrac{p^2q^{-3}}{r^{-5}}\right)^{1/2}$

18. $(u^2v^3)^{-1/6}(u^3v^6)^{1/3}$

19. $(x^{1/2}y^{1/4})(x^{1/2}y^{1/3})$

20. Explain how to simplify $125^{2/3}$.

21. Rewrite the expressions, using rational exponents.

   a. $\sqrt[3]{7x^2}$

   b. $\sqrt{21a}$

22. Rewrite the expressions in radical notation.

   a. $x^{2/3}$

   b. $(c^{1/4})^3$

---

section 7.4 | Addition and Subtraction of Radicals

1. Definition of *Like* Radicals

Objectives
1. Definition of *Like* Radicals
2. Addition and Subtraction of Radicals

Definition of *Like* Radicals

Two radical terms are said to be *like* radicals if they have the same index and the same radicand.

The following are pairs of *like* radicals:

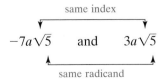

$-7a\sqrt{5}$ and $3a\sqrt{5}$ — same index, same radicand

Indices and radicands are the same.

Concept Connections

1. Give an example of a pair of *like* radicals. (Answers may vary.)
2. Give an example of a pair of radicals that are not *like*. (Answers may vary.)

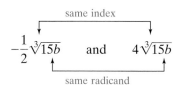

Indices and radicands are the same.

These pairs are not *like* radicals:

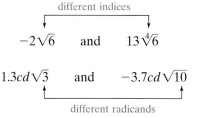

Radicals have different indices.

Radicals have different radicands.

2. Addition and Subtraction of Radicals

To add or subtract *like* radicals, use the distributive property. For example:

$$2\sqrt{5} + 6\sqrt{5} = (2 + 6)\sqrt{5}$$
$$= 8\sqrt{5}$$

$$9\sqrt[3]{2y} - 4\sqrt[3]{2y} = (9 - 4)\sqrt[3]{2y}$$
$$= 5\sqrt[3]{2y}$$

Skill Practice

Add or subtract as indicated.

3. $5\sqrt{6} - 8\sqrt{6}$
4. $\sqrt{10} + \sqrt{10}$
5. $5\sqrt[3]{xy} - 3\sqrt[3]{xy} + 7\sqrt[3]{xy}$
6. $\dfrac{5}{6}y\sqrt{2} + \dfrac{1}{4}y\sqrt{2}$

example 1 Adding and Subtracting Radicals

Add or subtract as indicated.

a. $6\sqrt{11} + 2\sqrt{11}$
b. $\sqrt{3} + \sqrt{3}$
c. $-2\sqrt[3]{ab} + 7\sqrt[3]{ab} - \sqrt[3]{ab}$
d. $\dfrac{1}{4}x\sqrt{3y} - \dfrac{3}{2}x\sqrt{3y}$

Solution:

a. $6\sqrt{11} + 2\sqrt{11}$
 $= (6 + 2)\sqrt{11}$ Apply the distributive property.
 $= 8\sqrt{11}$ Simplify.

b. $\sqrt{3} + \sqrt{3}$
 $= 1\sqrt{3} + 1\sqrt{3}$ Note that $\sqrt{3} = 1\sqrt{3}$.
 $= (1 + 1)\sqrt{3}$ Apply the distributive property.
 $= 2\sqrt{3}$ Simplify.

c. $-2\sqrt[3]{ab} + 7\sqrt[3]{ab} - \sqrt[3]{ab}$
 $= (-2 + 7 - 1)\sqrt[3]{ab}$ Apply the distributive property.
 $= 4\sqrt[3]{ab}$ Simplify.

Answers

1. For example: $2\sqrt{3x}$ and $4\sqrt{3x}$
2. For example: $6\sqrt{3}$ and $9\sqrt{7}$
3. $-3\sqrt{6}$ 4. $2\sqrt{10}$
5. $9\sqrt[3]{xy}$ 6. $\dfrac{13}{12}y\sqrt{2}$

d. $\frac{1}{4}x\sqrt{3y} - \frac{3}{2}x\sqrt{3y}$

$= \left(\frac{1}{4} - \frac{3}{2}\right)x\sqrt{3y}$ Apply the distributive property.

$= \left(\frac{1}{4} - \frac{6}{4}\right)x\sqrt{3y}$ Get a common denominator.

$= -\frac{5}{4}x\sqrt{3y}$ Simplify.

Avoiding Mistakes: The process of adding *like* radicals with the distributive property is similar to adding *like* terms. The end result is that the numerical coefficients are added and the radical factor is unchanged.

$$\sqrt{5} + \sqrt{5} = 1\sqrt{5} + 1\sqrt{5} = 2\sqrt{5}$$

Be careful: $\sqrt{5} + \sqrt{5} \neq \sqrt{10}$

In general: $\sqrt{x} + \sqrt{y} \neq \sqrt{x+y}$

Sometimes it is necessary to simplify radicals before adding or subtracting.

example 2 Adding and Subtracting Radicals

Simplify the radicals and add or subtract as indicated. Assume all variables represent positive real numbers.

a. $3\sqrt{8} + \sqrt{2}$
b. $8\sqrt{x^3y^2} - 3y\sqrt{x^3}$
c. $\sqrt{50x^2y^5} - 13y\sqrt{2x^2y^3} + xy\sqrt{98y^3}$

Solution:

a. $3\sqrt{8} + \sqrt{2}$ The radicands are different. Try simplifying the radicals first.

$= 3 \cdot 2\sqrt{2} + \sqrt{2}$ Simplify: $\sqrt{8} = 2\sqrt{2}$

$= 6\sqrt{2} + \sqrt{2}$

$= (6+1)\sqrt{2}$ Apply the distributive property.

$= 7\sqrt{2}$ Simplify.

Skill Practice

Simplify the radicals and add or subtract as indicated. Assume all variables represent positive real numbers.

7. $\sqrt{75} + 2\sqrt{3}$
8. $4\sqrt{a^2b} - 6a\sqrt{b}$
9. $-3\sqrt{2y^3} + 5y\sqrt{18y} - 2\sqrt{50y^3}$

Calculator Connections

To check the solution to Example 2(a), use a calculator to evaluate the expressions $3\sqrt{8} + \sqrt{2}$ and $7\sqrt{2}$. The decimal approximations agree to 10 digits.

```
3√(8)+√(2)
       9.899494937
7√(2)
       9.899494937
```

Answers
7. $7\sqrt{3}$
8. $-2a\sqrt{b}$
9. $2y\sqrt{2y}$

b. $8\sqrt{x^3y^2} - 3y\sqrt{x^3}$ The radicands are different. Simplify the radicals first.

$= 8xy\sqrt{x} - 3xy\sqrt{x}$ Simplify $\sqrt{x^3y^2} = xy\sqrt{x}$ and $\sqrt{x^3} = x\sqrt{x}$.

$= (8xy - 3xy)\sqrt{x}$ Apply the distributive property.

$= 5xy\sqrt{x}$ Simplify.

c. $\sqrt{50x^2y^5} - 13y\sqrt{2x^2y^3} + xy\sqrt{98y^3}$

$= 5xy^2\sqrt{2y} - 13xy^2\sqrt{2y} + 7xy^2\sqrt{2y}$

Simplify each radical.
$$\begin{cases} \sqrt{50x^2y^5} = \sqrt{5^2 2x^2y^5} \\ \qquad\qquad = 5xy^2\sqrt{2y} \\ -13y\sqrt{2x^2y^3} = -13xy^2\sqrt{2y} \\ xy\sqrt{98y^3} = xy\sqrt{7^2 2y^3} \\ \qquad\qquad = 7xy^2\sqrt{2y} \end{cases}$$

$= (5xy^2 - 13xy^2 + 7x^2y)\sqrt{2y}$ Apply the distributive property.

$= -xy^2\sqrt{2y}$

---

section 7.4 Practice Exercises

Boost your GRADE at mathzone.com!

- Practice Problems
- Self-Tests
- NetTutor
- e-Professors
- Videos

For the exercises in this set, assume that all variables represent positive real numbers, unless otherwise stated.

Study Skills Exercises

1. In addition to studying the material for a test, here are some activities that people use when preparing for a test. Circle the importance of each statement.

	not important	somewhat important	very important
a. Get a good night's sleep the night before the test.	1	2	3
b. Eat a good breakfast on the day of the test.	1	2	3
c. Wear comfortable clothes on the day of the test.	1	2	3
d. Arrive early to class on the day of the test.	1	2	3

2. Define the key term *like* **radicals**.

Review Exercises

For Exercises 3–6, simplify the radicals.

3. $\sqrt[3]{-16s^4 t^9}$

4. $-\sqrt[4]{x^7 y^4}$

5. $\sqrt{36a^2 b^3}$

6. $\dfrac{\sqrt[3]{7b^8}}{\sqrt[3]{56b^2}}$

7. Write the expression $(4x^2)^{3/2}$ as a radical and simplify.

8. Convert to rational exponents and simplify. $\sqrt[5]{3^5 x^{15} y^{10}}$

For Exercises 9–10, simplify the expressions. Write the answer with positive exponents only.

9. $y^{2/3} y^{1/4}$

10. $(x^{1/2} y^{-3/4})^{-4}$

Objective 1: Definition of *Like* Radicals

For Exercises 11–12, determine if the radical terms are *like*.

11. a. $\sqrt{2}$ and $\sqrt[3]{2}$

    b. $\sqrt{2}$ and $3\sqrt{2}$

    c. $\sqrt{2}$ and $\sqrt{5}$

12. a. $7\sqrt[3]{x}$ and $\sqrt[3]{x}$

    b. $\sqrt[3]{x}$ and $\sqrt[4]{x}$

    c. $2\sqrt[4]{x}$ and $x\sqrt[4]{2}$

13. Explain the similarities between the following pairs of expressions.

    a. $7\sqrt{5} + 4\sqrt{5}$ and $7x + 4x$

    b. $-2\sqrt{6} - 9\sqrt{3}$ and $-2x - 9y$

14. Explain the similarities between the following pairs of expressions.

    a. $-4\sqrt{3} + 5\sqrt{3}$ and $-4z + 5z$

    b. $13\sqrt{7} - 18$ and $13a - 18$

Objective 2: Addition and Subtraction of Radicals

For Exercises 15–32, add or subtract the radical expressions, if possible. (See Example 1.)

15. $3\sqrt{5} + 6\sqrt{5}$

16. $5\sqrt{a} + 3\sqrt{a}$

17. $3\sqrt[3]{t} - 2\sqrt[3]{t}$

18. $6\sqrt[3]{7} - 2\sqrt[3]{7}$

19. $6\sqrt{10} - \sqrt{10}$

20. $13\sqrt{11} - \sqrt{11}$

21. $\sqrt[4]{3} + 7\sqrt[4]{3} - \sqrt[4]{14}$

22. $2\sqrt{11} + 3\sqrt{13} + 5\sqrt{11}$

23. $8\sqrt{x} + 2\sqrt{y} - 6\sqrt{x}$

24. $10\sqrt{10} - 8\sqrt{10} + \sqrt{2}$

25. $\sqrt[3]{ab} + a\sqrt[3]{b}$

26. $x\sqrt[4]{y} - y\sqrt[4]{x}$

27. $\sqrt{2t} + \sqrt[3]{2t}$

28. $\sqrt[4]{5c} + \sqrt[3]{5c}$

29. $\dfrac{5}{6}z\sqrt[3]{6} + \dfrac{7}{9}z\sqrt[3]{6}$

30. $\dfrac{3}{4}a\sqrt[4]{b} + \dfrac{1}{6}a\sqrt[4]{b}$

31. $0.81x\sqrt{y} - 0.11x\sqrt{y}$

32. $7.5\sqrt{pq} - 6.3\sqrt{pq}$

33. Explain the process for adding the two radicals. Then find the sum. $3\sqrt{2} + 7\sqrt{50}$

34. Explain the process for adding the two radicals. Then find the sum. $\sqrt{8} + \sqrt{32}$

For Exercises 35–60, add or subtract the radical expressions as indicated. (See Example 2.)

35. $\sqrt{36} + \sqrt{81}$

36. $3\sqrt{80} - 5\sqrt{45}$

37. $2\sqrt{12} + \sqrt{48}$

38. $5\sqrt{32} + 2\sqrt{50}$

39. $4\sqrt{7} + \sqrt{63} - 2\sqrt{28}$

40. $8\sqrt{3} - 2\sqrt{27} + \sqrt{75}$

41. $5\sqrt{18} + \sqrt{32} - 4\sqrt{50}$

42. $7\sqrt{72} - \sqrt{8} + 4\sqrt{50}$

43. $\sqrt[3]{81} - \sqrt[3]{24}$

44. $17\sqrt[3]{81} - 2\sqrt[3]{24}$

45. $3\sqrt{2a} - \sqrt{8a} - \sqrt{72a}$

46. $\sqrt{12t} - \sqrt{27t} + 5\sqrt{3t}$

47. $2s^2\sqrt[3]{s^2t^6} + 3t^2\sqrt[3]{8s^8}$

48. $4\sqrt[3]{x^4} - 2x\sqrt[3]{x}$

49. $7\sqrt[3]{x^4} - x\sqrt[3]{x}$

50. $6\sqrt[3]{y^{10}} - 3y^2\sqrt[3]{y^4}$

51. $5p\sqrt{20p^2} + p^2\sqrt{80}$

52. $2q\sqrt{48q^2} - \sqrt{27q^4}$

53. $\sqrt[3]{a^2b} - \sqrt[3]{8a^2b}$

54. $w\sqrt{80} - 3\sqrt{125w^2}$

55. $11\sqrt[3]{54cd^3} - 2\sqrt[3]{2cd^3} + d\sqrt[3]{16c}$

56. $x\sqrt[3]{64x^5y^2} - x^2\sqrt[3]{x^2y^2} + 5\sqrt[3]{x^8y^2}$

57. $\dfrac{3}{2}ab\sqrt{24a^3} + \dfrac{4}{3}\sqrt{54a^5b^2} - a^2b\sqrt{150a}$

58. $mn\sqrt{72n} + \dfrac{2}{3}n\sqrt{8m^2n} - \dfrac{5}{6}\sqrt{50m^2n^3}$

59. $x\sqrt[3]{16} - 2\sqrt[3]{27x} + \sqrt[3]{54x^3}$

60. $5\sqrt[4]{y^5} - 2y\sqrt[4]{y} + \sqrt[4]{16y^7}$

Mixed Exercises

For Exercises 61–66, answer true or false. If an answer is false, explain why.

61. $\sqrt{x} + \sqrt{y} = \sqrt{x+y}$

62. $\sqrt{x} + \sqrt{x} = 2\sqrt{x}$

63. $5\sqrt[3]{x} + 2\sqrt[3]{x} = 7\sqrt[3]{x}$

64. $6\sqrt{x} + 5\sqrt[3]{x} = 11\sqrt{x}$

65. $\sqrt{y} + \sqrt{y} = \sqrt{2y}$

66. $\sqrt{c^2 + d^2} = c + d$

For Exercises 67–70, translate the English phrase to an algebraic expression. Simplify each expression, if possible.

67. The sum of the principal square root of 48 and the principal square root of 12

68. The sum of the cube root of 16 and the cube root of 2

69. The difference of 5 times the cube root of x^6 and the square of x

70. The sum of the cube of y and the principal fourth root of y^{12}

For Exercises 71–74, write an English phrase that translates the mathematical expression. (Answers may vary.)

71. $\sqrt{18} - 5^2$ **72.** $4^3 - \sqrt[3]{4}$ **73.** $\sqrt[4]{x} + y^3$ **74.** $a^4 + \sqrt{a}$

For Exercises 75–76, find the exact value of the perimeter, and then approximate the value to 1 decimal place.

75.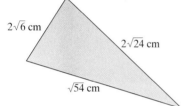

76.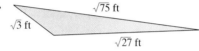

Expanding Your Skills

77. a. An irregularly shaped garden is shown in the figure. All distances are expressed in yards. Find the perimeter. *Hint:* Use the Pythagorean theorem to find the length of each side. Write the final answer in radical form.

b. Approximate your answer to 2 decimal places.

c. If edging costs $1.49 per foot and sales tax is 6%, find the total cost of edging the garden.

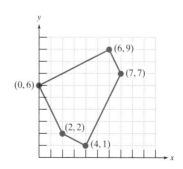

Objectives

1. Multiplication Property of Radicals
2. Expressions of the Form $(\sqrt[n]{a})^n$
3. Special Case Products
4. Multiplying Radicals with Different Indices

section 7.5 Multiplication of Radicals

1. Multiplication Property of Radicals

In this section we will learn how to multiply radicals by using the multiplication property of radicals first introduced in Section 7.3.

The Multiplication Property of Radicals

Let a and b represent real numbers such that $\sqrt[n]{a}$ and $\sqrt[n]{b}$ are both real. Then

$$\sqrt[n]{ab} = \sqrt[n]{a} \cdot \sqrt[n]{b}$$

To multiply two radical expressions, we use the multiplication property of radicals along with the commutative and associative properties of multiplication.

Skill Practice

Multiply the expressions and simplify the results. Assume all variables represent positive real numbers.

1. $\sqrt{5} \cdot \sqrt{3}$
2. $(4\sqrt{6})(-2\sqrt{6})$
3. $(3ab\sqrt{b})(-2\sqrt{ab})$
4. $(2\sqrt[3]{4ab})(5\sqrt[3]{2a^2b})$

Concept Connections

5. Explain why the multiplication property of radicals cannot be used to multiply
$\sqrt{-2} \cdot \sqrt{-8}$

example 1 Multiplying Radical Expressions

Multiply the expressions and simplify the result. Assume all variables represent positive real numbers.

a. $(3\sqrt{2})(5\sqrt{6})$
b. $(2x\sqrt{y})(-7\sqrt{xy})$
c. $(15c\sqrt[3]{cd})\left(\dfrac{1}{3}\sqrt[3]{cd^2}\right)$

Solution:

a. $(3\sqrt{2})(5\sqrt{6})$

$= (3 \cdot 5)(\sqrt{2} \cdot \sqrt{6})$ Commutative and associative properties of multiplication

$= 15\sqrt{12}$ Multiplication property of radicals

$= 15\sqrt{2^2 \cdot 3}$

$= 15 \cdot 2\sqrt{3}$ Simplify the radical.

$= 30\sqrt{3}$

b. $(2x\sqrt{y})(-7\sqrt{xy})$

$= (2x)(-7)(\sqrt{y} \cdot \sqrt{xy})$ Commutative and associative properties of multiplication

$= -14x\sqrt{xy^2}$ Multiplication property of radicals

$= -14xy\sqrt{x}$ Simplify the radical.

c. $(15c\sqrt[3]{cd})\left(\dfrac{1}{3}\sqrt[3]{cd^2}\right)$

$= \left(15c \cdot \dfrac{1}{3}\right)(\sqrt[3]{cd} \cdot \sqrt[3]{cd^2})$ Commutative and associative properties of multiplication

$= 5c\sqrt[3]{c^2d^3}$ Multiplication property of radicals

$= 5cd\sqrt[3]{c^2}$ Simplify the radical.

Answers

1. $\sqrt{15}$ 2. -48
3. $-6ab^2\sqrt{a}$ 4. $20a\sqrt[3]{b^2}$
5. The multiplication property of radicals cannot be used because the radicals $\sqrt{-2}$ and $\sqrt{-8}$ are not real numbers.

When multiplying radical expressions with more than one term, we use the distributive property.

example 2 — Multiplying Radical Expressions

Multiply the radical expressions. Assume all variables represent positive real numbers.

a. $3\sqrt{11}(2 + \sqrt{11})$
b. $(\sqrt{5} + 3\sqrt{2})(2\sqrt{5} - \sqrt{2})$
c. $(2\sqrt{14} + \sqrt{7})(6 - \sqrt{14} + 8\sqrt{7})$
d. $(-10a\sqrt{b} + 7b)(a\sqrt{b} + 2b)$

Skill Practice

Multiply the radical expressions. Assume all variables represent positive real numbers.

6. $5\sqrt{5}(2\sqrt{5} - 2)$
7. $(2\sqrt{3} - 3\sqrt{10}) \cdot (\sqrt{3} + 2\sqrt{10})$
8. $(\sqrt{6} + 3\sqrt{2}) \cdot (-4\sqrt{6} + 2\sqrt{2})$
9. $(x\sqrt{y} + y)(3x\sqrt{y} - 2y)$

Solution:

a. $3\sqrt{11}(2 + \sqrt{11})$

$= 3\sqrt{11} \cdot (2) + 3\sqrt{11} \cdot \sqrt{11}$ Apply the distributive property.

$= 6\sqrt{11} + 3\sqrt{11^2}$ Multiplication property of radicals

$= 6\sqrt{11} + 3 \cdot 11$ Simplify the radical.

$= 6\sqrt{11} + 33$

b. $(\sqrt{5} + 3\sqrt{2})(2\sqrt{5} - \sqrt{2})$

$= 2\sqrt{5^2} - \sqrt{10} + 6\sqrt{10} - 3\sqrt{2^2}$ Apply the distributive property.

$= 2 \cdot 5 + 5\sqrt{10} - 3 \cdot 2$ Simplify radicals and combine *like* radicals.

$= 10 + 5\sqrt{10} - 6$

$= 4 + 5\sqrt{10}$ Combine *like* terms.

c. $(2\sqrt{14} + \sqrt{7})(6 - \sqrt{14} + 8\sqrt{7})$

$= 12\sqrt{14} - 2\sqrt{14^2} + 16\sqrt{98} + 6\sqrt{7} - \sqrt{98} + 8\sqrt{7^2}$ Apply the distributive property.

$= 12\sqrt{14} - 2 \cdot 14 + 16 \cdot 7\sqrt{2} + 6\sqrt{7} - 7\sqrt{2} + 8 \cdot 7$ Simplify the radicals.

(*Note:* $\sqrt{98} = \sqrt{7^2 \cdot 2} = 7\sqrt{2}$)

$= 12\sqrt{14} - 28 + 112\sqrt{2} + 6\sqrt{7} - 7\sqrt{2} + 56$ Simplify.

$= 12\sqrt{14} + 105\sqrt{2} + 6\sqrt{7} + 28$ Combine *like* terms.

d. $(-10a\sqrt{b} + 7b)(a\sqrt{b} + 2b)$

$= -10a^2\sqrt{b^2} - 20ab\sqrt{b} + 7ab\sqrt{b} + 14b^2$ Apply the distributive property.

$= -10a^2b - 13ab\sqrt{b} + 14b^2$ Simplify and combine *like* terms.

Answers

6. $50 - 10\sqrt{5}$ 7. $-54 + \sqrt{30}$
8. $-12 - 20\sqrt{3}$
9. $3x^2y + xy\sqrt{y} - 2y^2$

2. Expressions of the Form $(\sqrt[n]{a})^n$

The multiplication property of radicals can be used to simplify an expression of the form $(\sqrt{a})^2$, where $a \geq 0$.

$$(\sqrt{a})^2 = \sqrt{a} \cdot \sqrt{a} = \sqrt{a^2} = a \quad \text{where } a \geq 0$$

This logic can be applied to nth roots. If $\sqrt[n]{a}$ is a real number, then $(\sqrt[n]{a})^n = a$.

Skill Practice

Simplify.
10. $(\sqrt{14})^2$
11. $(\sqrt[7]{q})^7$
12. $(\sqrt[5]{3z})^5$

---

example 3   Simplifying Radical Expressions

Simplify the expressions. Assume all variables represent positive real numbers.

  a. $(\sqrt{11})^2$ **b.** $(\sqrt[5]{z})^5$ **c.** $(\sqrt[3]{pq})^3$

Solution:

  a. $(\sqrt{11})^2 = 11$ **b.** $(\sqrt[5]{z})^5 = z$ **c.** $(\sqrt[3]{pq})^3 = pq$

---

3. Special Case Products

From Example 2, you may have noticed a similarity between multiplying radical expressions and multiplying polynomials.

In Section 5.2 we learned that the square of a binomial results in a perfect square trinomial:

$$(a + b)^2 = a^2 + 2ab + b^2$$
$$(a - b)^2 = a^2 - 2ab + b^2$$

The same patterns occur when squaring a radical expression with two terms.

Skill Practice

Square the radical expressions. Assume all variables represent positive real numbers.
13. $(\sqrt{a} - 5)^2$
14. $(4\sqrt{x} + 3)^2$

---

example 4   Squaring a Two-Term Radical Expression

Square the radical expressions. Assume all variables represent positive real numbers.

  a. $(\sqrt{d} + 3)^2$ **b.** $(5\sqrt{y} - \sqrt{2})^2$

Solution:

  a. $(\sqrt{d} + 3)^2$ This expression is in the form $(a + b)^2$, where $a = \sqrt{d}$ and $b = 3$.

$$= (\sqrt{d})^2 + 2(\sqrt{d})(3) + (3)^2 \quad \text{Apply the formula } (a+b)^2 = a^2 + 2ab + b^2.$$

$$= d + 6\sqrt{d} + 9 \quad \text{Simplify.}$$

Answers
10. 14   11. q
12. $3z$ 13. $a - 10\sqrt{a} + 25$
14. $16x + 24\sqrt{x} + 9$

Tip: The product $(\sqrt{d} + 3)^2$ can also be found by using the distributive property:

$$(\sqrt{d} + 3)^2 = (\sqrt{d} + 3)(\sqrt{d} + 3) = \sqrt{d} \cdot \sqrt{d} + \sqrt{d} \cdot 3 + 3 \cdot \sqrt{d} + 3 \cdot 3$$
$$= \sqrt{d^2} + 3\sqrt{d} + 3\sqrt{d} + 9$$
$$= d + 6\sqrt{d} + 9$$

b. $(5\sqrt{y} - \sqrt{2})^2$ This expression is in the form $(a - b)^2$, where $a = 5\sqrt{y}$ and $b = \sqrt{2}$.

$$= (5\sqrt{y})^2 - 2(5\sqrt{y})(\sqrt{2}) + (\sqrt{2})^2 \quad \text{Apply the formula } (a - b)^2 = a^2 - 2ab + b^2.$$

$$= 25y - 10\sqrt{2y} + 2 \quad \text{Simplify.}$$

Recall from Section 5.2 that the product of two conjugate binomials results in a difference of squares.

$$(a + b)(a - b) = a^2 - b^2$$

The same pattern occurs when multiplying two conjugate radical expressions.

Concept Connections

15. Give the conjugate of $(\sqrt{a} + \sqrt{b})$.

example 5 Multiplying Conjugate Radical Expressions

Multiply the radical expressions. Assume all variables represent positive real numbers.

a. $(\sqrt{3} + 2)(\sqrt{3} - 2)$ **b.** $\left(\dfrac{1}{3}\sqrt{s} - \dfrac{3}{4}\sqrt{t}\right)\left(\dfrac{1}{3}\sqrt{s} + \dfrac{3}{4}\sqrt{t}\right)$

Skill Practice

Multiply the conjugates. Assume all variables represent positive real numbers.

16. $(\sqrt{5} + 3)(\sqrt{5} - 3)$

17. $\left(\dfrac{1}{2}\sqrt{a} + \dfrac{2}{5}\sqrt{b}\right) \cdot \left(\dfrac{1}{2}\sqrt{a} - \dfrac{2}{5}\sqrt{b}\right)$

Solution:

a. $(\sqrt{3} + 2)(\sqrt{3} - 2)$ The expression is in the form $(a + b)(a - b)$, where $a = \sqrt{3}$ and $b = 2$.

$$= (\sqrt{3})^2 - (2)^2 \quad \text{Apply the formula } (a + b)(a - b) = a^2 - b^2.$$
$$= 3 - 4 \quad \text{Simplify.}$$
$$= -1$$

Answers

15. $(\sqrt{a} - \sqrt{b})$
16. -4
17. $\dfrac{1}{4}a - \dfrac{4}{25}b$

Tip: The product $(\sqrt{3} + 2)(\sqrt{3} - 2)$ can also be found by using the distributive property.

$$(\sqrt{3} + 2)(\sqrt{3} - 2) = \sqrt{3} \cdot \sqrt{3} + \sqrt{3} \cdot (-2) + 2 \cdot \sqrt{3} + 2 \cdot (-2)$$
$$= 3 - 2\sqrt{3} + 2\sqrt{3} - 4$$
$$= 3 - 4$$
$$= -1$$

b. $\left(\dfrac{1}{3}\sqrt{s} - \dfrac{3}{4}\sqrt{t}\right)\left(\dfrac{1}{3}\sqrt{s} + \dfrac{3}{4}\sqrt{t}\right)$
   This expression is in the form $(a - b)(a + b)$, where $a = \dfrac{1}{3}\sqrt{s}$ and $b = \dfrac{3}{4}\sqrt{t}$.

$$= \left(\dfrac{1}{3}\sqrt{s}\right)^2 - \left(\dfrac{3}{4}\sqrt{t}\right)^2$$ Apply the formula $(a + b)(a - b) = a^2 - b^2$.

$$= \dfrac{1}{9}s - \dfrac{9}{16}t$$ Simplify.

4. Multiplying Radicals with Different Indices

The product of two radicals can be simplified provided the radicals have the same index. If the radicals have different indices, then we can use the properties of rational exponents to obtain a common index.

example 6 — Multiplying Radicals with Different Indices

Multiply the expressions. Write the answers in radical form.

a. $\sqrt[3]{5} \cdot \sqrt[4]{5}$ **b.** $\sqrt[3]{7} \cdot \sqrt{2}$

Solution:

a. $\sqrt[3]{5} \cdot \sqrt[4]{5}$

$= 5^{1/3} 5^{1/4}$ Rewrite each expression with rational exponents.

$= 5^{(1/3) + (1/4)}$ Because the bases are equal, we can add the exponents.

$= 5^{(4/12) + (3/12)}$ Write the fractions with a common denominator.

$= 5^{7/12}$ Simplify the exponent.

$= \sqrt[12]{5^7}$ Rewrite the expression as a radical.

b. $\sqrt[3]{7} \cdot \sqrt{2}$

$= 7^{1/3} 2^{1/2}$ Rewrite each expression with rational exponents.

$= 7^{2/6} 2^{3/6}$ Write the rational exponents with a common denominator.

$= (7^2 2^3)^{1/6}$ Apply the power rule of exponents.

$= \sqrt[6]{7^2 2^3}$ Rewrite the expression as a single radical.

$= \sqrt[6]{392}$ Simplify.

Skill Practice

Multiply the expressions. Write the answers in radical form. Assume all variables represent positive real numbers.

18. $\sqrt{x} \cdot \sqrt[3]{x}$
19. $\sqrt[4]{a^3} \cdot \sqrt[3]{b}$

Answers

18. $\sqrt[6]{x^5}$ 19. $\sqrt[12]{a^9 b^4}$

Section 7.5 Practice Exercises

Boost your GRADE at mathzone.com!
MathZone

- Practice Problems
- Self-Tests
- NetTutor
- e-Professors
- Videos

For the exercises in this set, assume that all variables represent positive real numbers, unless otherwise stated.

Study Skills Exercise

1. You should try not to cram for tests. Instead, math is a subject that should be studied every day. This text gives you opportunities to review and practice as you work through the book.

   a. Find the page number for the midchapter review for this chapter.

   b. Find the page number for the review exercises for this chapter.

Review Exercises

2. Given $f(x) = \sqrt{-3x + 1}$, evaluate

   a. $f(-1)$ b. $f(-5)$

3. Given $g(x) = \sqrt{5 - x}$, evaluate

   a. $g(-20)$ b. $g(-11)$

For Exercises 4–7, simplify the radicals.

4. $\sqrt[3]{(x-y)^3}$

5. $\sqrt[5]{(2+h)^5}$

6. $\sqrt[3]{-16x^5 y^6 z^7}$

7. $-\sqrt{20a^2 b^3 c}$

For Exercises 8–12, simplify the expressions. Write the answer with positive exponents only.

8. $\dfrac{1}{9^{-1/2}}$

9. $\left(\dfrac{1}{8}\right)^{-1/3}$

10. $x^{1/3} y^{1/4} x^{-1/6} y^{1/3}$

11. $p^{1/8} q^{1/2} p^{-1/4} q^{3/2}$

12. $\dfrac{b^{1/4}}{b^{3/2}}$

For Exercises 13–14, add or subtract as indicated.

13. $-2\sqrt[3]{7} + 4\sqrt[3]{7}$

14. $4\sqrt{8x^3} - x\sqrt{50x}$

Objective 1: Multiplication Property of Radicals

For Exercises 15–50, multiply the radical expressions. (See Examples 1–2.)

15. $\sqrt[3]{7} \cdot \sqrt[3]{3}$

16. $\sqrt[4]{6} \cdot \sqrt[4]{2}$

17. $\sqrt{2} \cdot \sqrt{10}$

18. $\sqrt[3]{4} \cdot \sqrt[3]{12}$

19. $\sqrt[4]{16} \cdot \sqrt[4]{64}$

20. $\sqrt{5x^3} \cdot \sqrt{10x^4}$

21. $(4\sqrt[3]{4})(2\sqrt[3]{5})$

22. $(2\sqrt{5})(3\sqrt{7})$

23. $(8a\sqrt{b})(-3\sqrt{ab})$

24. $(p\sqrt[4]{q^3})(\sqrt[4]{pq})$

25. $\sqrt{30} \cdot \sqrt{12}$

26. $\sqrt{20} \cdot \sqrt{54}$

27. $\sqrt{6x}\sqrt{12x}$

28. $(\sqrt{3ab^2})(\sqrt{21a^2b})$

29. $\sqrt{5a^3b^2}\sqrt{20a^3b^3}$

30. $\sqrt[3]{m^2n^2} \cdot \sqrt[3]{48m^4n^2}$

31. $(4\sqrt{3xy^3})(-2\sqrt{6x^3y^2})$

32. $(2\sqrt[4]{3x})(4\sqrt[4]{27x^6})$

33. $(\sqrt[3]{4a^2b})(\sqrt[3]{2ab^3})(\sqrt[3]{54a^2b})$

34. $(\sqrt[3]{9x^3y})(\sqrt[3]{6xy})(\sqrt[3]{8x^2y^5})$

35. $\sqrt{3}(4\sqrt{3} - 6)$

36. $3\sqrt{5}(2\sqrt{5} + 4)$

37. $\sqrt{2}(\sqrt{6} - \sqrt{3})$

38. $\sqrt{5}(\sqrt{3} + \sqrt{7})$

39. $-3\sqrt{x}(\sqrt{x} + 7)$

40. $-2\sqrt{y}(8 - \sqrt{y})$

41. $(\sqrt{3} + 2\sqrt{10})(4\sqrt{3} - \sqrt{10})$

42. $(8\sqrt{7} - \sqrt{5})(\sqrt{7} + 3\sqrt{5})$

43. $(\sqrt{x} + 4)(\sqrt{x} - 9)$

44. $(\sqrt{w} - 2)(\sqrt{w} - 9)$

45. $(\sqrt[3]{y} + 2)(\sqrt[3]{y} - 3)$

46. $(4 + \sqrt[5]{p})(5 + \sqrt[5]{p})$

47. $(\sqrt{a} - 3\sqrt{b})(9\sqrt{a} - \sqrt{b})$

48. $(11\sqrt{m} + 4\sqrt{n})(\sqrt{m} + \sqrt{n})$

49. $(\sqrt{p} + 2\sqrt{q})(8 + 3\sqrt{p} - \sqrt{q})$

50. $(5\sqrt{s} - \sqrt{t})(\sqrt{s} + 5 + 6\sqrt{t})$

Objective 2: Expressions of the Form $(\sqrt[n]{a})^n$

For Exercises 51–58, simplify the expressions. Assume all variables represent positive real numbers. **(See Example 3.)**

51. $(\sqrt{15})^2$

52. $(\sqrt{58})^2$

53. $(\sqrt{3y})^2$

54. $(\sqrt{19yz})^2$

55. $(\sqrt[3]{6})^3$

56. $(\sqrt[5]{24})^5$

57. $\sqrt{709} \cdot \sqrt{709}$

58. $\sqrt{401} \cdot \sqrt{401}$

Objective 3: Special Case Products

59. **a.** Write the formula for the product of two conjugates. $(x + y)(x - y) =$

    b. Multiply $(x + 5)(x - 5)$.

60. **a.** Write the formula for squaring a binomial. $(x + y)^2 =$

    b. Multiply $(x + 5)^2$.

For Exercises 61–72, multiply the special products. **(See Examples 4–5.)**

61. $(\sqrt{3} + x)(\sqrt{3} - x)$

62. $(y + \sqrt{6})(y - \sqrt{6})$

63. $(\sqrt{6} + \sqrt{2})(\sqrt{6} - \sqrt{2})$

64. $(\sqrt{15} + \sqrt{5})(\sqrt{15} - \sqrt{5})$

65. $(8\sqrt{x} + 2\sqrt{y})(8\sqrt{x} - 2\sqrt{y})$

66. $(4\sqrt{s} + 11\sqrt{t})(4\sqrt{s} - 11\sqrt{t})$

Section 7.5 Multiplication of Radicals 509

67. $(\sqrt{13} + 4)^2$ **68.** $(6 - \sqrt{11})^2$ **69.** $(\sqrt{p} - \sqrt{7})^2$

70. $(\sqrt{q} + \sqrt{2})^2$ **71.** $(\sqrt{2a} - 3\sqrt{b})^2$ **72.** $(\sqrt{3w} + 4\sqrt{z})^2$

Mixed Exercises

For Exercises 73–80, identify each statement as true or false. If an answer is false, explain why.

73. $\sqrt{3} \cdot \sqrt{2} = \sqrt{6}$ **74.** $\sqrt{5} \cdot \sqrt[3]{2} = \sqrt{10}$ **75.** $(x - \sqrt{5})^2 = x - 5$

76. $3(2\sqrt{5x}) = 6\sqrt{5x}$ **77.** $5(3\sqrt{4x}) = 15\sqrt{20x}$ **78.** $\dfrac{\sqrt{5x}}{5} = \sqrt{x}$

79. $\dfrac{3\sqrt{x}}{3} = \sqrt{x}$ **80.** $(\sqrt{t} - 1)(\sqrt{t} + 1) = t - 1$

For Exercises 81–88, square the expressions.

81. $\sqrt{39}$ **82.** $\sqrt{21}$ **83.** $-\sqrt{6x}$ **84.** $-\sqrt{8a}$

85. $\sqrt{3x + 1}$ **86.** $\sqrt{x - 1}$ **87.** $\sqrt{x + 3} - 4$ **88.** $\sqrt{x + 1} + 3$

For Exercises 89–92, find the exact area.

89. 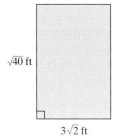 **90.** 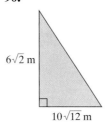 **91.** 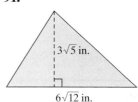 **92.**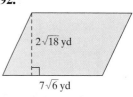

Objective 4: Multiplying Radicals with Different Indices

For Exercises 93–102, multiply or divide the radicals with different indices. Write the answers in radical form and simplify. (See Example 6.)

93. $\sqrt{x} \cdot \sqrt[4]{x}$ **94.** $\sqrt[3]{y} \cdot \sqrt{y}$ **95.** $\sqrt[5]{2z} \cdot \sqrt[3]{2z}$ **96.** $\sqrt[3]{5w} \cdot \sqrt[4]{5w}$

97. $\sqrt[3]{p^2} \cdot \sqrt{p^3}$ **98.** $\sqrt[4]{q^3} \cdot \sqrt[3]{q^2}$ **99.** $\dfrac{\sqrt{u^3}}{\sqrt[3]{u}}$ **100.** $\dfrac{\sqrt{v^5}}{\sqrt[4]{v}}$

101. $\dfrac{\sqrt{(a + b)}}{\sqrt[3]{(a + b)}}$ **102.** $\dfrac{\sqrt[3]{(q - 1)}}{\sqrt[4]{(q - 1)}}$

For Exercises 103–106, multiply the radicals with different indices.

103. $\sqrt[3]{x} \cdot \sqrt[6]{y}$
104. $\sqrt{a} \cdot \sqrt[6]{b}$
105. $\sqrt[4]{8} \cdot \sqrt{3}$
106. $\sqrt{11} \cdot \sqrt[6]{2}$

Expanding Your Skills

107. Multiply $(\sqrt[3]{a} + \sqrt[3]{b})(\sqrt[3]{a^2} - \sqrt[3]{ab} + \sqrt[3]{b^2})$.

108. Multiply $(\sqrt[3]{a} - \sqrt[3]{b})(\sqrt[3]{a^2} + \sqrt[3]{ab} + \sqrt[3]{b^2})$.

section 7.6 Rationalization

Objectives

1. Simplified Form of a Radical
2. Rationalizing the Denominator—One Term
3. Rationalizing the Denominator—Two Terms

1. Simplified Form of a Radical

Recall that for a radical expression to be in simplified form the following three conditions must be met.

Simplified Form of a Radical

Consider any radical expression in which the radicand is written as a product of prime factors. The expression is in simplified form if all the following conditions are met:

1. The radicand has no factor raised to a power greater than or equal to the index.
2. The radicand does not contain a fraction.
3. No radicals are in the denominator of a fraction.

The third condition restricts radicals from the denominator of an expression. The process to remove a radical from the denominator is called **rationalizing the denominator**. In many cases, rationalizing the denominator creates an expression that is computationally simpler. In Examples 3 and 4 we will show that

$$\frac{6}{\sqrt{3}} = 2\sqrt{3} \quad \text{and} \quad \frac{-2}{2 + \sqrt{6}} = 2 - \sqrt{6}$$

We will demonstrate the process to rationalize the denominator as two separate cases:

- Rationalizing the denominator (one term)
- Rationalizing the denominator (two terms involving square roots)

2. Rationalizing the Denominator—One Term

To begin the first case, recall that the nth root of a perfect nth power simplifies completely.

$$\sqrt{x^2} = x \quad x \geq 0$$
$$\sqrt[3]{x^3} = x$$
$$\sqrt[4]{x^4} = x \quad x \geq 0$$
$$\sqrt[5]{x^5} = x$$

. . .

Concept Connections

Determine whether each expression is in simplified form. If it is not, explain why.

1. $\sqrt[5]{a^7}$
2. $\dfrac{\sqrt{3}}{2}$
3. $\sqrt{\dfrac{1}{3}}$
4. $\dfrac{2}{\sqrt{3}}$

Answers

1. Not simplified. The radicand has a factor raised to a power greater than the index.
2. Simplified
3. Not simplified. The radicand contains a fraction.
4. Not simplified. There is a radical in the denominator.

Therefore, to rationalize a radical expression, use the multiplication property of radicals to create an nth root of an nth power.

example 1 — Rationalizing Radical Expressions

Fill in the missing radicand to rationalize the radical expressions. Assume all variables represent positive real numbers.

a. $\sqrt{a} \cdot \sqrt{?} = \sqrt{a^2} = a$ b. $\sqrt[3]{y} \cdot \sqrt[3]{?} = \sqrt[3]{y^3} = y$

c. $\sqrt[4]{2z^3} \cdot \sqrt[4]{?} = \sqrt[4]{2^4 z^4} = 2z$

Skill Practice

Fill in the missing radicand to rationalize the radical expression.

5. $\sqrt{7} \cdot \sqrt{?}$
6. $\sqrt[5]{t^2} \cdot \sqrt[5]{?}$
7. $\sqrt[3]{5x^2} \cdot \sqrt[3]{?}$

Solution:

a. $\sqrt{a} \cdot \sqrt{?} = \sqrt{a^2} = a$ What multiplied by $\sqrt{a}$ will equal $\sqrt{a^2}$?
   $\sqrt{a} \cdot \sqrt{a} = \sqrt{a^2} = a$

b. $\sqrt[3]{y} \cdot \sqrt[3]{?} = \sqrt[3]{y^3} = y$ What multiplied by $\sqrt[3]{y}$ will equal $\sqrt[3]{y^3}$?
   $\sqrt[3]{y} \cdot \sqrt[3]{y^2} = \sqrt[3]{y^3} = y$

c. $\sqrt[4]{2z^3} \cdot \sqrt[4]{?} = \sqrt[4]{2^4 z^4} = 2z$ What multiplied by $\sqrt[4]{2z^3}$ will equal $\sqrt[4]{2^4 z^4}$?
   $\sqrt[4]{2z^3} \cdot \sqrt[4]{2^3 z} = \sqrt[4]{2^4 z^4} = 2z$

To rationalize the denominator of a radical expression, multiply the numerator and denominator by an appropriate expression to create an nth root of an nth power in the denominator.

example 2 — Rationalizing the Denominator—One Term

Simplify the expression

$$\frac{5}{\sqrt[3]{a}} \qquad a \neq 0$$

Skill Practice

Simplify the expression. Assume $y > 0$.

8. $\dfrac{2}{\sqrt[4]{y}}$

Solution:

To remove the radical from the denominator, a cube root of a perfect cube is needed in the denominator. Multiply numerator and denominator by $\sqrt[3]{a^2}$ because $\sqrt[3]{a} \cdot \sqrt[3]{a^2} = \sqrt[3]{a^3} = a$.

$$\frac{5}{\sqrt[3]{a}} = \frac{5}{\sqrt[3]{a}} \cdot \frac{\sqrt[3]{a^2}}{\sqrt[3]{a^2}}$$

$$= \frac{5\sqrt[3]{a^2}}{\sqrt[3]{a^3}} \qquad \text{Multiply the radicals.}$$

$$= \frac{5\sqrt[3]{a^2}}{a} \qquad \text{Simplify.}$$

Answers

5. 7
6. t^3
7. $5^2 x$
8. $\dfrac{2\sqrt[4]{y^3}}{y}$

Note that for $a \neq 0$, the expression $\dfrac{\sqrt[3]{a^2}}{\sqrt[3]{a^2}} = 1$. In Example 2, multiplying the expression $\dfrac{5}{\sqrt[3]{a}}$ by this ratio does not change its value.

Skill Practice

Simplify the expressions.

9. $\dfrac{12}{\sqrt{2}}$
10. $\sqrt{\dfrac{8}{3}}$
11. $\dfrac{18}{\sqrt[3]{3y^2}}$
12. $\dfrac{\sqrt[3]{16x^4}}{\sqrt[3]{2x}}$

example 3 — Rationalizing the Denominator—One Term

Simplify the expressions. Assume all variables represent positive real numbers.

a. $\dfrac{6}{\sqrt{3}}$ **b.** $\sqrt{\dfrac{y^5}{7}}$ **c.** $\dfrac{15}{\sqrt[3]{25s}}$ **d.** $\dfrac{\sqrt{125p^3}}{\sqrt{5p}}$

Solution:

a. To rationalize the denominator, a square root of a perfect square is needed. Multiply numerator and denominator by $\sqrt{3}$ because $\sqrt{3} \cdot \sqrt{3} = \sqrt{3^2} = 3$.

$\dfrac{6}{\sqrt{3}} = \dfrac{6}{\sqrt{3}} \cdot \dfrac{\sqrt{3}}{\sqrt{3}}$ Rationalize the denominator.

$= \dfrac{6\sqrt{3}}{\sqrt{3^2}}$ Multiply the radicals.

$= \dfrac{6\sqrt{3}}{3}$ Simplify.

$= 2\sqrt{3}$ Reduce to lowest terms.

Calculator Connections

A calculator can be used to support the solution to a simplified radical. The calculator approximations of the expressions $6/\sqrt{3}$ and $2\sqrt{3}$ agree to 10 decimal digits.

```
6/√(3)
        3.464101615
2√(3)
        3.464101615
```

b. $\sqrt{\dfrac{y^5}{7}}$ The radical contains an irreducible fraction.

$= \dfrac{\sqrt{y^5}}{\sqrt{7}}$ Apply the division property of radicals.

$= \dfrac{y^2\sqrt{y}}{\sqrt{7}}$ Remove factors from the radical in the numerator.

$= \dfrac{y^2\sqrt{y}}{\sqrt{7}} \cdot \dfrac{\sqrt{7}}{\sqrt{7}}$ Rationalize the denominator.
Note: $\sqrt{7} \cdot \sqrt{7} = \sqrt{7^2} = 7$.

$= \dfrac{y^2\sqrt{7y}}{\sqrt{7^2}}$

$= \dfrac{y^2\sqrt{7y}}{7}$ Simplify.

Avoiding Mistakes: A factor within a radicand cannot be simplified with a factor outside the radicand. For example, $\dfrac{\sqrt{7y}}{7}$ cannot be simplified.

Answers

9. $6\sqrt{2}$
10. $\dfrac{2\sqrt{6}}{3}$
11. $\dfrac{6\sqrt[3]{9y}}{y}$
12. $2x$

c. $\dfrac{15}{\sqrt[3]{25s}}$

$= \dfrac{15}{\sqrt[3]{5^2 s}} \cdot \dfrac{\sqrt[3]{5s^2}}{\sqrt[3]{5s^2}}$ Because $25 = 5^2$, one additional factor of 5 is needed to form a perfect cube. Two additional factors of s are needed to make a perfect cube. Multiply numerator and denominator by $\sqrt[3]{5s^2}$.

$= \dfrac{15\sqrt[3]{5s^2}}{\sqrt[3]{5^3 s^3}}$

$= \dfrac{15\sqrt[3]{5s^2}}{5s}$ Simplify the perfect cube.

$= \dfrac{\overset{3}{\cancel{15}}\sqrt[3]{5s^2}}{\underset{1}{\cancel{5}s}}$ Reduce to lowest terms.

$= \dfrac{3\sqrt[3]{5s^2}}{s}$

Tip: In the expression $\dfrac{15\sqrt[3]{5s^2}}{5s}$, the factor of 15 and the factor of 5 may be reduced because both are outside the radical.

$$\dfrac{15\sqrt[3]{5s^2}}{5s} = \dfrac{15}{5} \cdot \dfrac{\sqrt[3]{5s^2}}{s} = \dfrac{3\sqrt[3]{5s^2}}{s}$$

d. $\dfrac{\sqrt{125p^3}}{\sqrt{5p}}$ Notice that the radicands in the numerator and denominator share common factors.

$= \sqrt{\dfrac{125p^3}{5p}}$ Rewrite the expression by using the division property of radicals.

$= \sqrt{25p^2}$ Simplify the fraction within the radicand.

$= 5p$ Simplify the radical.

3. Rationalizing the Denominator—Two Terms

Example 4 demonstrates how to rationalize a two-term denominator involving square roots.

First recall from the multiplication of polynomials that the product of two conjugates results in a difference of squares.

$$(a + b)(a - b) = a^2 - b^2$$

If either a or b has a square root factor, the expression will simplify without a radical. That is, the expression is *rationalized*. For example:

$$(2 + \sqrt{6})(2 - \sqrt{6}) = (2)^2 - (\sqrt{6})^2$$
$$= 4 - 6$$
$$= -2$$

Concept Connections

13. Fill in the blank to rationalize the expression.
$(5 - \sqrt{x})(?)$

Answer

13. $(5 + \sqrt{x})$

514   Chapter 7   Radicals and Complex Numbers

Skill Practice

Simplify by rationalizing the denominator.

14. $\dfrac{4}{3 + \sqrt{5}}$

---

example 4   Rationalizing the Denominator—Two Terms

Simplify the expression by rationalizing the denominator.

$$\dfrac{-2}{2 + \sqrt{6}}$$

Solution:

$\dfrac{-2}{2 + \sqrt{6}}$

$= \dfrac{(-2)}{(2 + \sqrt{6})} \cdot \dfrac{(2 - \sqrt{6})}{(2 - \sqrt{6})}$ Multiply the numerator and denominator by the conjugate of the denominator.

conjugates

$= \dfrac{-2(2 - \sqrt{6})}{(2)^2 - (\sqrt{6})^2}$ In the denominator, apply the formula $(a + b)(a - b) = a^2 - b^2$.

$= \dfrac{-2(2 - \sqrt{6})}{4 - 6}$ Simplify.

$= \dfrac{-2(2 - \sqrt{6})}{-2}$

$= \dfrac{-2(2 - \sqrt{6})}{-2}$

$= 2 - \sqrt{6}$

Skill Practice

Simplify by rationalizing the denominator. Assume that $y \geq 0$.

15. $\dfrac{\sqrt{3} - \sqrt{y}}{\sqrt{3} + \sqrt{y}}$

---

example 5   Rationalizing the Denominator—Two Terms

Rationalize the denominator of the expression. Assume all variables represent positive real numbers and $c \neq d$.

$$\dfrac{\sqrt{c} + \sqrt{d}}{\sqrt{c} - \sqrt{d}}$$

Solution:

$\dfrac{\sqrt{c} + \sqrt{d}}{\sqrt{c} - \sqrt{d}}$

$= \dfrac{(\sqrt{c} + \sqrt{d})}{(\sqrt{c} - \sqrt{d})} \cdot \dfrac{(\sqrt{c} + \sqrt{d})}{(\sqrt{c} + \sqrt{d})}$ Multiply numerator and denominator by the conjugate of the denominator.

conjugates

$= \dfrac{(\sqrt{c} + \sqrt{d})^2}{(\sqrt{c})^2 - (\sqrt{d})^2}$ In the denominator apply the formula $(a + b)(a - b) = a^2 - b^2$.

$= \dfrac{(\sqrt{c} + \sqrt{d})^2}{c - d}$ Simplify.

$= \dfrac{(\sqrt{c})^2 + 2\sqrt{c}\sqrt{d} + (\sqrt{d})^2}{c - d}$ In the numerator apply the formula $(a + b)^2 = a^2 + 2ab + b^2$.

$= \dfrac{c + 2\sqrt{cd} + d}{c - d}$

Answers

14. $3 - \sqrt{5}$

15. $\dfrac{3 - 2\sqrt{3y} + y}{3 - y}$

section 7.6 Practice Exercises

Boost *your* GRADE at mathzone.com!

MathZone

- Practice Problems
- Self-Tests
- NetTutor
- e-Professors
- Videos

For the exercises in this set, assume that all variables represent positive real numbers unless otherwise stated.

Study Skills Exercises

1. Do not wait until the day before a test to get help. Seek help from your instructor as soon as you realize that there is a concept that you do not understand. List the concepts in this chapter for which you could use some help.

2. Define the key term **rationalizing the denominator**.

Review Exercises

For Exercises 3–10, perform the indicated operations.

3. $2y\sqrt{45} + 3\sqrt{20y^2}$
4. $3x\sqrt{72x} - 9\sqrt{50x^3}$
5. $(-6\sqrt{y} + 3)(3\sqrt{y} + 1)$
6. $(\sqrt{w} + 12)(2\sqrt{w} - 4)$

7. $(8 - \sqrt{t})^2$
8. $(\sqrt{p} + 4)^2$
9. $(\sqrt{2} + \sqrt{7})(\sqrt{2} - \sqrt{7})$
10. $(\sqrt{3} + 5)(\sqrt{3} - 5)$

Objective 2: Rationalizing the Denominator—One Term

The radical expressions in Exercises 11–18 have radicals in the denominator. Fill in the missing radicands to rationalize the radical expressions in the denominators. **(See Example 1.)**

11. $\dfrac{x}{\sqrt{5}} = \dfrac{x}{\sqrt{5}} \cdot \dfrac{\sqrt{?}}{\sqrt{?}}$
12. $\dfrac{2}{\sqrt{x}} = \dfrac{2}{\sqrt{x}} \cdot \dfrac{\sqrt{?}}{\sqrt{?}}$

13. $\dfrac{7}{\sqrt[3]{x}} = \dfrac{7}{\sqrt[3]{x}} \cdot \dfrac{\sqrt[3]{?}}{\sqrt[3]{?}}$
14. $\dfrac{5}{\sqrt[4]{y}} = \dfrac{5}{\sqrt[4]{y}} \cdot \dfrac{\sqrt[4]{?}}{\sqrt[4]{?}}$

15. $\dfrac{8}{\sqrt{3z}} = \dfrac{8}{\sqrt{3z}} \cdot \dfrac{\sqrt{??}}{\sqrt{??}}$
16. $\dfrac{10}{\sqrt{7w}} = \dfrac{10}{\sqrt{7w}} \cdot \dfrac{\sqrt{??}}{\sqrt{??}}$

17. $\dfrac{1}{\sqrt[4]{2a^2}} = \dfrac{1}{\sqrt[4]{2a^2}} \cdot \dfrac{\sqrt[4]{??}}{\sqrt[4]{??}}$
18. $\dfrac{1}{\sqrt[3]{6b^2}} = \dfrac{1}{\sqrt[3]{6b^2}} \cdot \dfrac{\sqrt[3]{??}}{\sqrt[3]{??}}$

For Exercises 19–50, rationalize the denominator. **(See Examples 2–3.)**

19. $\dfrac{1}{\sqrt{3}}$
20. $\dfrac{1}{\sqrt{7}}$
21. $\sqrt{\dfrac{1}{x}}$
22. $\sqrt{\dfrac{1}{z}}$

23. $\dfrac{6}{\sqrt{2y}}$ 24. $\dfrac{9}{\sqrt{3t}}$ 25. $\dfrac{-2a}{\sqrt{a}}$ 26. $\dfrac{-7b}{\sqrt{b}}$

27. $\dfrac{6}{\sqrt{8}}$ 28. $\dfrac{2}{\sqrt{48}}$ 29. $\dfrac{3}{\sqrt[3]{2}}$ 30. $\dfrac{2}{\sqrt[3]{7}}$

31. $\dfrac{-6}{\sqrt[4]{x}}$ 32. $\dfrac{-2}{\sqrt[5]{y}}$ 33. $\dfrac{7}{\sqrt[3]{4}}$ 34. $\dfrac{1}{\sqrt[3]{9}}$

35. $\sqrt[3]{\dfrac{4}{w^2}}$ 36. $\sqrt[3]{\dfrac{5}{z^2}}$ 37. $\sqrt[4]{\dfrac{16}{3}}$ 38. $\sqrt[4]{\dfrac{81}{8}}$

39. $\dfrac{2}{\sqrt[3]{4x^2}}$ 40. $\dfrac{6}{\sqrt[3]{3y^2}}$ 41. $\sqrt[3]{\dfrac{16x^3}{y}}$ 42. $\sqrt{\dfrac{5}{9x}}$

43. $\dfrac{\sqrt{x^4 y^5}}{\sqrt{10x}}$ 44. $\sqrt[4]{\dfrac{10x^2}{15xy^3}}$ 45. $\dfrac{8}{7\sqrt{24}}$ 46. $\dfrac{5}{3\sqrt{50}}$

47. $\dfrac{1}{\sqrt{x^7}}$ 48. $\dfrac{1}{\sqrt{y^5}}$ 49. $\dfrac{2}{\sqrt{8x^5}}$ 50. $\dfrac{6}{\sqrt{27t^7}}$

Objective 3: Rationalizing the Denominator—Two Terms

51. What is the conjugate of $\sqrt{2} - \sqrt{6}$? 52. What is the conjugate of $\sqrt{11} + \sqrt{5}$?

53. What is the conjugate of $\sqrt{x} + 23$? 54. What is the conjugate of $17 - \sqrt{y}$?

For Exercises 55–74, rationalize the denominators. (See Examples 4–5.)

55. $\dfrac{4}{\sqrt{2} + 3}$ 56. $\dfrac{6}{4 - \sqrt{3}}$ 57. $\dfrac{8}{\sqrt{6} - 2}$ 58. $\dfrac{-12}{\sqrt{5} - 3}$

59. $\dfrac{\sqrt{7}}{\sqrt{3} + 2}$ 60. $\dfrac{\sqrt{8}}{\sqrt{3} + 1}$ 61. $\dfrac{-1}{\sqrt{p} + \sqrt{q}}$ 62. $\dfrac{6}{\sqrt{a} - \sqrt{b}}$

63. $\dfrac{x - 5}{\sqrt{x} + \sqrt{5}}$ 64. $\dfrac{y - 2}{\sqrt{y} - \sqrt{2}}$ 65. $\dfrac{-7}{2\sqrt{a} - 5\sqrt{b}}$ 66. $\dfrac{-4}{3\sqrt{w} - 2\sqrt{z}}$

67. $\dfrac{3\sqrt{5} - \sqrt{3}}{\sqrt{3} + \sqrt{5}}$ 68. $\dfrac{2\sqrt{5} + \sqrt{6}}{\sqrt{6} - \sqrt{5}}$ 69. $\dfrac{3\sqrt{10}}{2 + \sqrt{10}}$ 70. $\dfrac{4\sqrt{7}}{3 + \sqrt{7}}$

71. $\dfrac{2\sqrt{3}+\sqrt{7}}{3\sqrt{3}-\sqrt{7}}$ **72.** $\dfrac{5\sqrt{2}-\sqrt{5}}{5\sqrt{2}+\sqrt{5}}$ **73.** $\dfrac{\sqrt{5}+4}{2-\sqrt{5}}$ **74.** $\dfrac{3+\sqrt{2}}{\sqrt{2}-5}$

For Exercises 75–78, translate the English phrase to an algebraic expression. Then simplify the expression.

75. 16 divided by the cube root of 4

76. 21 divided by the principal fourth root of 27

77. 4 divided by the difference of x and the principal square root of 2

78. 8 divided by the sum of y and the principal square root of 3

79. The time T (in seconds) for a pendulum to make one complete swing back and forth is given by

$$T(x) = 2\pi\sqrt{\dfrac{x}{32}}$$

where x is the length of the pendulum in feet.

Determine the exact time required for one swing for a pendulum that is 1 ft long. Then approximate the time to the nearest hundredth of a second.

80. An object is dropped off a building x meters tall. The time T (in seconds) required for the object to hit the ground is given by

$$T(x) = \sqrt{\dfrac{10x}{49}}$$

Find the exact time required for the object to hit the ground if it is dropped off the First National Plaza Building in Chicago, a height of 230 m. Then round the time to the nearest hundredth of a second.

Expanding Your Skills

For Exercises 81–86, simplify each term of the expression. Then add or subtract as indicated.

81. $\dfrac{\sqrt{6}}{2} + \dfrac{1}{\sqrt{6}}$ **82.** $\dfrac{1}{\sqrt{7}} + \sqrt{7}$ **83.** $\sqrt{15} - \sqrt{\dfrac{3}{5}} + \sqrt{\dfrac{5}{3}}$ **84.** $\sqrt{\dfrac{6}{2}} - \sqrt{12} + \sqrt{\dfrac{2}{6}}$

85. $\sqrt[3]{25} + \dfrac{3}{\sqrt[3]{5}}$ **86.** $\dfrac{1}{\sqrt[3]{4}} + \sqrt[3]{54}$

For Exercises 87–90, rationalize the numerator by multiplying both numerator and denominator by the conjugate of the numerator.

87. $\dfrac{\sqrt{3}+6}{2}$ **88.** $\dfrac{\sqrt{7}-2}{5}$ **89.** $\dfrac{\sqrt{a}-\sqrt{b}}{\sqrt{a}+\sqrt{b}}$ **90.** $\dfrac{\sqrt{p}+\sqrt{q}}{\sqrt{p}-\sqrt{q}}$

Objectives

1. Solutions to Radical Equations
2. Solving Radical Equations Involving One Radical
3. Solving Radical Equations Involving More than One Radical
4. Applications of Radical Equations and Functions

Section 7.7 Radical Equations

1. Solutions to Radical Equations

An equation with one or more radicals containing a variable is called a **radical equation**. For example, $\sqrt[3]{x} = 5$ is a radical equation. Recall that $(\sqrt[n]{a})^n = a$, provided $\sqrt[n]{a}$ is a real number. The basis of solving a radical equation is to eliminate the radical by raising both sides of the equation to a power equal to the index of the radical.

To solve the equation $\sqrt[3]{x} = 5$, cube both sides of the equation.

$$\sqrt[3]{x} = 5$$
$$(\sqrt[3]{x})^3 = (5)^3$$
$$x = 125$$

By raising each side of a radical equation to a power equal to the index of the radical, a new equation is produced. Note, however, that some of or all the solutions to the new equation may *not* be solutions to the original radical equation. For this reason, it is necessary to *check all potential solutions* in the original equation. For example, consider the equation $x = 4$. By squaring both sides we produce a quadratic equation.

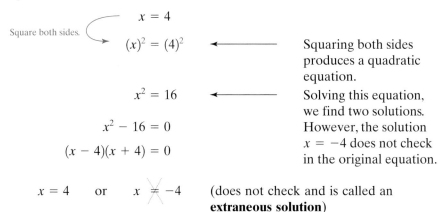

Square both sides.
$$x = 4$$
$$(x)^2 = (4)^2 \quad \longleftarrow \text{Squaring both sides produces a quadratic equation.}$$
$$x^2 = 16 \quad \longleftarrow \text{Solving this equation, we find two solutions. However, the solution } x = -4 \text{ does not check in the original equation.}$$
$$x^2 - 16 = 0$$
$$(x - 4)(x + 4) = 0$$

$x = 4$ or $x \ne -4$ (does not check and is called an **extraneous solution**)

Concept Connections

1. The equation shown must be raised to what power to clear the radical?
   $\sqrt[4]{a} = 2$

Steps to Solve a Radical Equation

1. Isolate the radical. If an equation has more than one radical, choose one of the radicals to isolate.
2. Raise each side of the equation to a power equal to the index of the radical.
3. Solve the resulting equation. If the equation still has a radical, repeat steps 1 and 2.
*4. Check the potential solutions in the original equation.

*Extraneous solutions can arise only when both sides of the equation are raised to an *even power*. Therefore, an equation with odd-index roots will not have an extraneous solution. However, it is still recommended that you check *all* potential solutions regardless of the type of root.

Answer

1. Fourth power

2. Solving Radical Equations Involving One Radical

example 1 Solving Equations Containing One Radical

Solve the equations.

a. $\sqrt{3x - 2} + 4 = 5$
b. $(w - 1)^{1/3} - 2 = 2$
c. $7 = \sqrt[4]{x + 3} + 9$
d. $y + \sqrt{y - 2} = 8$

Solution:

a. $\sqrt{3x - 2} + 4 = 5$

$\sqrt{3x - 2} = 1$ Isolate the radical.
$(\sqrt{3x - 2})^2 = (1)^2$ Because the index is 2, square both sides.
$3x - 2 = 1$ Simplify.
$3x = 3$ Solve the resulting equation.
$x = 1$

Check: $x = 1$ Check $x = 1$ as a potential solution.
$\sqrt{3x - 2} + 4 = 5$
$\sqrt{3(1) - 2} + 4 \stackrel{?}{=} 5$
$\sqrt{1} + 4 \stackrel{?}{=} 5$
$5 = 5$ ✔   True, $x = 1$ is a solution to the original equation.

b. $(w - 1)^{1/3} - 2 = 2$ Note that $(w - 1)^{1/3} = \sqrt[3]{w - 1}$.
$\sqrt[3]{w - 1} - 2 = 2$
$\sqrt[3]{w - 1} = 4$ Isolate the radical.
$(\sqrt[3]{w - 1})^3 = (4)^3$ Because the index is 3, cube both sides.
$w - 1 = 64$ Simplify.
$w = 65$

Check: $w = 65$
$(w - 1)^{1/3} - 2 = 2$ Check $w = 65$ as a potential solution.
$\sqrt[3]{65 - 1} - 2 \stackrel{?}{=} 2$
$\sqrt[3]{64} - 2 \stackrel{?}{=} 2$
$4 - 2 \stackrel{?}{=} 2$
$2 = 2$ ✔   True, $w = 65$ is a solution to the original equation.

c. $7 = \sqrt[4]{x + 3} + 9$
$-2 = \sqrt[4]{x + 3}$ Isolate the radical.
$(-2)^4 = (\sqrt[4]{x + 3})^4$ Because the index is 4, raise both sides to the fourth power.
$16 = x + 3$
$x = 13$ Solve for x.

Skill Practice

Solve the equations.

2. $\sqrt{5y + 1} - 2 = 4$
3. $(t + 2)^{1/3} + 5 = 3$
4. $\sqrt[4]{b - 1} + 6 = 3$
5. $\sqrt{x + 1} + 5 = x$

Tip: After isolating the radical in Example 1(c), the equation shows a fourth root equated to a negative number:

$-2 = \sqrt[4]{x + 3}$

By definition, a principal fourth root of any real number must be nonnegative. Therefore, there can be no real solution to this equation.

Answers
2. $y = 7$ 3. $t = -10$
4. No solution
5. $x = 8$ ($x = 3$ does not check.)

Check: $x = 13$

$7 = \sqrt[4]{x + 3} + 9$

$7 \stackrel{?}{=} \sqrt[4]{(13) + 3} + 9$

$7 \stackrel{?}{=} \sqrt[4]{16} + 9$

$7 \neq 2 + 9$ $\quad$ $x = 13$ is *not* a solution to the original equation.

The equation $7 = \sqrt[4]{x + 3} + 9$ has no solution.

d. $y + \sqrt{y - 2} = 8$

$\sqrt{y - 2} = 8 - y$ $\quad$ Isolate the radical.

$(\sqrt{y - 2})^2 = (8 - y)^2$ $\quad$ Because the index is 2, square both sides.

Note that $(8 - y)^2 = (8 - y)(8 - y) = 64 - 16y + y^2$.

$y - 2 = 64 - 16y + y^2$ $\quad$ Simplify.

$0 = y^2 - 17y + 66$ $\quad$ The equation is quadratic. Set one side equal to zero. Write the other side in descending order.

$0 = (y - 11)(y - 6)$ $\quad$ Factor.

$y - 11 = 0$ $\quad$ or $\quad$ $y - 6 = 0$ $\quad$ Set each factor equal to zero.

$y = 11$ $\quad$ or $\quad$ $y = 6$ $\quad$ Solve.

Check: $y = 11$ $\qquad\qquad\qquad$ Check: $y = 6$

$y + \sqrt{y - 2} = 8$ $\qquad\qquad$ $y + \sqrt{y - 2} = 8$

$11 + \sqrt{11 - 2} \stackrel{?}{=} 8$ $\qquad$ $6 + \sqrt{6 - 2} \stackrel{?}{=} 8$

$11 + \sqrt{9} \stackrel{?}{=} 8$ $\qquad\qquad$ $6 + \sqrt{4} \stackrel{?}{=} 8$

$11 + 3 \stackrel{?}{=} 8$ $\qquad\qquad\quad$ $6 + 2 \stackrel{?}{=} 8$

$14 \neq 8$ $\qquad\qquad\qquad\quad$ $8 = 8$ ✔

$y = 11$ is not a solution to the original equation. $\qquad$ True, $y = 6$ is a solution to the original equation.

Avoiding Mistakes: Be sure to square both sides of the equation, not the individual terms.

3. Solving Radical Equations Involving More than One Radical

Skill Practice

Solve the equation.

6. $\sqrt[5]{2y - 1} = \sqrt[5]{10y + 3}$

example 2 $\quad$ Solving Equations with Two Radicals

Solve the radical equation.

$$\sqrt[3]{2x - 4} = \sqrt[3]{1 - 8x}$$

Answer

6. $y = -\dfrac{1}{2}$

Solution:

$$\sqrt[3]{2x - 4} = \sqrt[3]{1 - 8x}$$

$(\sqrt[3]{2x - 4})^3 = (\sqrt[3]{1 - 8x})^3$ Because the index is 3, cube both sides.

$2x - 4 = 1 - 8x$ Simplify.

$10x - 4 = 1$ Solve the resulting equation. Add $8x$ to both sides.

$10x = 5$ Combine *like* terms and add 4 to both sides.

$x = \dfrac{1}{2}$ Solve for x.

Check: $x = \dfrac{1}{2}$

$$\sqrt[3]{2x - 4} = \sqrt[3]{1 - 8x}$$

$$\sqrt[3]{2\left(\dfrac{1}{2}\right) - 4} \stackrel{?}{=} \sqrt[3]{1 - 8\left(\dfrac{1}{2}\right)}$$

$$\sqrt[3]{1 - 4} \stackrel{?}{=} \sqrt[3]{1 - 4}$$

$\sqrt[3]{-3} = \sqrt[3]{-3}$ ✔   Therefore, $x = \dfrac{1}{2}$ is a solution to the original equation.

Calculator Connections

The expressions on the right- and left-hand sides of the equation $\sqrt[3]{2x - 4} = \sqrt[3]{1 - 8x}$ are each functions of x. Consider the graphs of the functions:

$$Y_1 = \sqrt[3]{2x - 4} \quad \text{and} \quad Y_2 = \sqrt[3]{1 - 8x}$$

The x-coordinate of the point of intersection of the two functions is the solution to the equation $\sqrt[3]{2x - 4} = \sqrt[3]{1 - 8x}$. The point of intersection can be approximated by using *Zoom* and *Trace* or by using an *Intersect* function.

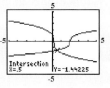

example 3   Solving Equations with Two Radicals

Solve the equation. $\sqrt{3m + 1} - \sqrt{m + 4} = 1$

Solution:

$$\sqrt{3m + 1} - \sqrt{m + 4} = 1$$

$\sqrt{3m + 1} = \sqrt{m + 4} + 1$ Isolate one of the radicals.

$(\sqrt{3m + 1}) = (\sqrt{m + 4} + 1)^2$ Square both sides.

Skill Practice

Solve the equation.

7. $\sqrt{3c + 1} - \sqrt{c - 1} = 2$

Answer

7. $c = 1; c = 5$

$$3m + 1 = m + 4 + 2\sqrt{m+4} + 1$$

Note: $(\sqrt{m+4} + 1)^2$
$= (\sqrt{m+4})^2 + 2(1)\sqrt{m+4} + (1)^2$
$= m + 4 + 2\sqrt{m+4} + 1$

$3m + 1 = m + 5 + 2\sqrt{m+4}$	Combine *like* terms.
$2m - 4 = 2\sqrt{m+4}$	Isolate the radical again.
$m - 2 = \sqrt{m+4}$	Divide both sides by 2.
$(m - 2)^2 = (\sqrt{m+4})^2$	Square both sides again.
$m^2 - 4m + 4 = m + 4$	The resulting equation is quadratic.
$m^2 - 5m = 0$	Set the quadratic equation equal to zero.
$m(m - 5) = 0$	Factor.

$m = 0$ or $m = 5$

Check: $m = 0$

$\sqrt{3(0) + 1} - \sqrt{(0) + 4} \stackrel{?}{=} 1$

$\sqrt{1} - \sqrt{4} \stackrel{?}{=} 1$

$1 - 2 \neq 1$

Check: $m = 5$

$\sqrt{3(5) + 1} - \sqrt{(5) + 4} \stackrel{?}{=} 1$

$\sqrt{16} - \sqrt{9} \stackrel{?}{=} 1$

$4 - 3 = 1$ ✔

The solution is $m = 5$ (the value $m = 0$ does not check).

4. Applications of Radical Equations and Functions

example 4 — Applying a Radical Equation in Geometry

For a pyramid with a square base, the length of a side of the base b is given by

$$b = \sqrt{\frac{3V}{h}}$$

where V is the volume and h is the height.

The Pyramid of the Pharoah Khufu (known as the Great Pyramid) at Giza has a square base (Figure 7-3). If the distance around the bottom of the pyramid is 921.6 m and the height is 146.6 m, what is the volume of the pyramid?

Skill Practice

8. The length of the legs, s, of an isosceles right triangle is $s = \sqrt{2A}$, where A is the area.

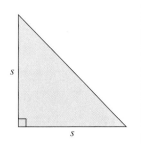

If the legs of the triangle are 9 in., find the area.

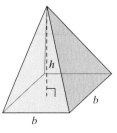

Figure 7-3

Answer

8. $\dfrac{81}{2}$ in.²

Solution:

The length of a side b (in meters) is given by $\dfrac{921.6}{4} = 230.4$ m.

$$b = \sqrt{\dfrac{3V}{h}}$$

$$230.4 = \sqrt{\dfrac{3V}{146.6}} \qquad \text{Substitute } b = 230.4 \text{ and } h = 146.6.$$

$$(230.4)^2 = \left(\sqrt{\dfrac{3V}{146.6}}\right)^2 \qquad \text{Because the index is 2, square both sides.}$$

$$53{,}084.16 = \dfrac{3V}{146.6} \qquad \text{Simplify.}$$

$$(53{,}084.16)(146.6) = \dfrac{3V}{\cancel{146.6}}(\cancel{146.6}) \qquad \text{Multiply both sides by 146.6.}$$

$$(53{,}084.16)(146.6) = 3V$$

$$\dfrac{(53{,}084.16)(146.6)}{3} = \dfrac{\cancel{3}V}{\cancel{3}} \qquad \text{Divide both sides by 3.}$$

$$2{,}594{,}046 \approx V$$

The volume of the Great Pyramid at Giza is approximately $2{,}594{,}046$ m^3.

---

example 5 Applying a Radical Function

On a certain surface, the speed $s(x)$ (in miles per hour) of a car before the brakes were applied can be approximated from the length of its skid marks x (in feet) by

$$s(x) = 3.8\sqrt{x} \qquad x \geq 0$$

See Figure 7-4.

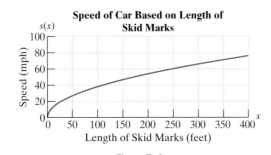

Figure 7-4

a. Find the speed of a car before the brakes were applied if its skid marks are 361 ft long.

b. How long would you expect the skid marks to be if the car had been traveling the speed limit of 50 mph? (Round to the nearest foot.)

Skill Practice

When an object is dropped from a height of 64 ft, the time t in seconds it takes to reach a height h in feet is given by

$$t = \dfrac{1}{4}\sqrt{64 - h}$$

9. Find the time to reach a height of 28 ft from the ground.

10. What is the height after 1 sec?

Answers

9. $\dfrac{3}{2}$ sec 10. 48 ft

Solution:

a. $s(x) = 3.8\sqrt{x}$

   $s(361) = 3.8\sqrt{361}$ Substitute $x = 361$.

   $= 3.8(19)$

   $= 72.2$

   If the skid marks are 361 ft, the car was traveling approximately 72.2 mph before the brakes were applied.

b. $s(x) = 3.8\sqrt{x}$

   $50 = 3.8\sqrt{x}$ Substitute $s(x) = 50$ and solve for x.

   $\dfrac{50}{3.8} = \sqrt{x}$ Isolate the radical.

   $\left(\dfrac{50}{3.8}\right)^2 = x$

   $x \approx 173$

   If the car had been going the speed limit (50 mph), then the length of the skid marks would have been approximately 173 ft.

section 7.7 Practice Exercises

Boost your GRADE at mathzone.com!
MathZone

- Practice Problems
- Self-Tests
- NetTutor
- e-Professors
- Videos

Study Skills Exercises

1. When you take a test, which of the following should you do? Check all that apply.

   ☐ Answer each question.

   ☐ Show all work.

   ☐ Write neatly.

   ☐ Include your scratch paper with your test.

   ☐ Check your work if time allows.

   ☐ Correct your test after you get it back so that you do not make the same errors again.

2. Define the key terms.

   a. Radical equation   **b. Extraneous solution**

Section 7.7 Radical Equations 525

Review Exercises

For Exercises 3–10, simplify the radical expressions, if possible. Assume all variables represent positive real numbers.

3. $\sqrt{48}$

4. $\sqrt{18w^4}$

5. $\sqrt{\dfrac{9w^3}{16}}$

6. $\sqrt{\dfrac{a^2}{3}}$

7. $\sqrt{-25}$

8. $\sqrt[3]{54c^4}$

9. $\sqrt{\dfrac{49}{5t^3}}$

10. $\sqrt[3]{-\dfrac{8}{27}}$

For Exercises 11–16, simplify each expression. Assume all radicands represent positive real numbers.

11. $(\sqrt{4x-6})^2$

12. $(\sqrt{5y+2})^2$

13. $(\sqrt[3]{9p+7})^3$

14. $(\sqrt[3]{4t+13})^3$

15. $(-\sqrt{2x})^2$

16. $(-\sqrt{5y})^2$

Objective 2: Solving Radical Equations Involving One Radical

For Exercises 17–32, solve the equations. (See Example 1.)

17. $\sqrt{4x} = 6$

18. $\sqrt{2x} = 8$

19. $\sqrt{5y+1} = 4$

20. $\sqrt{9z-5} = 11$

21. $(2z-3)^{1/2} = 9$

22. $(8+3a)^{1/2} = 5$

23. $\sqrt[3]{x-2} - 3 = 0$

24. $\sqrt[3]{2x-5} - 2 = 0$

25. $(15-w)^{1/3} = -5$

26. $(k+18)^{1/3} = -2$

27. $3 + \sqrt{x-16} = 0$

28. $12 + \sqrt{2x+1} = 0$

29. $2\sqrt{6a+7} - 2a = 0$

30. $2\sqrt{3-w} - w = 0$

31. $\sqrt[4]{2x-5} = -1$

32. $\sqrt[4]{x+16} = -4$

For Exercises 33–36, assume all variables represent positive real numbers.

33. Solve for V: $r = \sqrt[3]{\dfrac{3V}{4\pi}}$

34. Solve for V: $r = \sqrt{\dfrac{V}{h\pi}}$

35. Solve for h^2: $r = \pi\sqrt{r^2 + h^2}$

36. Solve for d: $s = 1.3\sqrt{d}$

For Exercises 37–42, square the expression as indicated.

37. $(a+5)^2$

38. $(b+7)^2$

39. $(\sqrt{5a} - 3)^2$

40. $(2 + \sqrt{b})^2$

41. $(\sqrt{r-3} + 5)^2$

42. $(2 - \sqrt{2t-4})^2$

For Exercises 43–48, solve the radical equations, if possible. (See Example 1.)

43. $\sqrt{a^2 + 2a + 1} = a + 5$

44. $\sqrt{b^2 - 5b - 8} = b + 7$

45. $\sqrt{25w^2 - 2w - 3} = 5w - 4$

46. $\sqrt{4p^2 - 2p + 1} = 2p - 3$

47. $\sqrt{5y+1} + 2 = y + 3$

48. $\sqrt{2x-2} + 3 = x + 2$

Objective 3: Solving Radical Equations Involving More than One Radical

For Exercises 49–70, solve the radical equations, if possible. (See Examples 2–3.)

49. $\sqrt[4]{h+4} = \sqrt[4]{2h-5}$

50. $\sqrt[4]{3b+6} = \sqrt[4]{7b-6}$

51. $\sqrt[3]{5a+3} - \sqrt[3]{a-13} = 0$

52. $\sqrt[3]{k-8} - \sqrt[3]{4k+1} = 0$

53. $\sqrt{5a-9} = \sqrt{5a} - 3$

54. $\sqrt{8+b} = 2 + \sqrt{b}$

55. $\sqrt{2h+5} - \sqrt{2h} = 1$

56. $\sqrt{3k-5} - \sqrt{3k} = -1$

57. $\sqrt{t-9} - \sqrt{t} = 3$

58. $\sqrt{y-16} - \sqrt{y} = 4$

59. $6 = \sqrt{x^2+3} - x$

60. $2 = \sqrt{y^2+5} - y$

61. $\sqrt{3t-7} = 2 - \sqrt{3t+1}$

62. $\sqrt{p-6} = \sqrt{p+2} - 4$

63. $\sqrt{z+1} + \sqrt{2z+3} = 1$

64. $\sqrt{2y+6} = \sqrt{7-2y} + 1$

65. $\sqrt{6m+7} - \sqrt{3m+3} = 1$

66. $\sqrt{5w+1} - \sqrt{3w} = 1$

67. $2 + 2\sqrt{2t+3} + 2\sqrt{3t-5} = 0$

68. $6 + 3\sqrt{3x+1} + 3\sqrt{x-1} = 0$

69. $4\sqrt{y} + 6 = 13$

70. $\sqrt{5x-8} = 2\sqrt{x-1}$

Objective 4: Applications of Radical Equations and Functions

71. If an object is dropped from an initial height h, its velocity at impact with the ground is given by

$$v = \sqrt{2gh}$$

where g is the acceleration due to gravity and h is the initial height.

a. Find the initial height (in feet) of an object if its velocity at impact is 44 ft/sec. (Assume that the acceleration due to gravity is $g = 32$ ft/sec^2.) (See Example 4.)

b. Find the initial height (in meters) of an object if its velocity at impact is 26 m/sec. (Assume that the acceleration due to gravity is $g = 9.8$ m/sec^2.) Round to the nearest tenth of a meter.

72. The time T (in seconds) required for a pendulum to make one complete swing back and forth is approximated by

$$T = 2\pi\sqrt{\frac{L}{g}}$$

where g is the acceleration due to gravity and L is the length of the pendulum (in feet).

a. Find the length of a pendulum that requires 1.36 sec to make one complete swing back and forth. (Assume that the acceleration due to gravity is $g = 32$ ft/sec^2.) Round to the nearest tenth of a foot.

b. Find the time required for a pendulum to complete one swing back and forth if the length of the pendulum is 4 ft. (Assume that the acceleration due to gravity is $g = 32$ ft/sec^2.) Round to the nearest tenth of a second.

73. The time $t(d)$ in seconds it takes an object to drop d meters is given by

$$t(d) = \sqrt{\frac{d}{4.9}}$$

  a. Approximate the height of the Texas Commerce Tower in Houston if it takes an object 7.89 sec to drop from the top. Round to the nearest meter. (See Example 5.)

  b. Approximate the height of the Shanghai World Financial Center if it takes an object 9.69 sec to drop from the top. Round to the nearest meter.

74. The airline cost for x thousand passengers to travel round trip from New York to Atlanta is given by

$$C(x) = \sqrt{0.3x + 1}$$

where $C(x)$ is measured in millions of dollars and $x \geq 0$.

  a. Find the airline's cost for 10,000 passengers ($x = 10$) to travel from New York to Atlanta.

  b. If the airline charges $320 per passenger, find the profit made by the airline for flying 10,000 passengers from New York to Atlanta.

  c. Approximate the number of passengers who traveled from New York to Atlanta if the total cost for the airline was $4 million.

Expanding Your Skills

75. The number of hours needed to cook a turkey that weighs x lb can be approximated by

$$t(x) = 0.90\sqrt[5]{x^3}$$

where $t(x)$ is the time in hours and x is the weight of the turkey in pounds.

  a. Find the weight of a turkey that cooked for 4 hr. Round to the nearest pound.

  b. Find $t(18)$ and interpret the result. Round to the nearest tenth of an hour.

For Exercises 76–79, use the Pythagorean theorem to find a, b, or c.

$$a^2 + b^2 = c^2$$

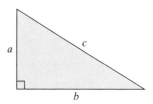

76. Find b when $a = 2$ and $c = y$.

77. Find b when $a = h$ and $c = 5$.

78. Find a when $b = x$ and $c = 8$.

79. Find a when $b = 14$ and $c = k$.

Graphing Calculator Exercises

80. Refer to Exercise 17. Graph Y_1 and Y_2 on a viewing window defined by $-10 \leq x \leq 20$ and $-5 \leq y \leq 10$.

$$Y_1 = \sqrt{4x} \quad \text{and} \quad Y_2 = 6$$

Use an *Intersect* feature to approximate the x-coordinate of the point of intersection of the two graphs to support your solution to Exercise 17.

81. Refer to Exercise 18. Graph Y_1 and Y_2 on a viewing window defined by $-10 \leq x \leq 40$ and $-5 \leq y \leq 10$.

$$Y_1 = \sqrt{2x} \quad \text{and} \quad Y_2 = 8$$

Use an *Intersect* feature to approximate the x-coordinate of the point of intersection of the two graphs to support your solution to Exercise 18.

82. Refer to Exercise 49. Graph Y_1 and Y_2 on a viewing window defined by $-5 \leq x \leq 20$ and $-1 \leq y \leq 4$.

$$Y_1 = \sqrt[4]{x + 4} \quad \text{and} \quad Y_2 = \sqrt[4]{2x - 5}$$

Use an *Intersect* feature to approximate the x-coordinate of the point of intersection of the two graphs to support your solution to Exercise 49.

83. Refer to Exercise 50. Graph Y_1 and Y_2 on a viewing window defined by $-5 \leq x \leq 20$ and $-1 \leq y \leq 4$.

$$Y_1 = \sqrt[4]{3x + 6} \quad \text{and} \quad Y_2 = \sqrt[4]{7x - 6}$$

Use an *Intersect* feature to approximate the x-coordinate of the point of intersection of the two graphs to support your solution to Exercise 50.

section 7.8 Complex Numbers

1. Definition of i

In Section 7.1, we learned that there are no real-valued square roots of a negative number. For example, $\sqrt{-9}$ is not a real number because no real number when squared equals -9. However, the square roots of a negative number are defined over another set of numbers called the **imaginary numbers**. The foundation of the set of imaginary numbers is the definition of the imaginary number i as $i = \sqrt{-1}$.

Definition of i

$$i = \sqrt{-1}$$

Note: From the definition of i, it follows that $i^2 = -1$.

Using the imaginary number i, we can define the square root of any negative real number.

Definition of $\sqrt{-b}$ for $b > 0$

Let b be a real number such that $b > 0$. Then $\sqrt{-b} = i\sqrt{b}$.

Objectives

1. Definition of i
2. Powers of i
3. Definition of a Complex Number
4. Addition, Subtraction, and Multiplication of Complex Numbers
5. Division and Simplification of Complex Numbers

Concept Connections

1. $i = ?$
2. $i^2 = ?$

example 1 Simplifying Expressions in Terms of i

Simplify the expressions in terms of i.

a. $\sqrt{-64}$ b. $\sqrt{-50}$ c. $-\sqrt{-4}$ d. $\sqrt{-29}$

Solution:

a. $\sqrt{-64} = i\sqrt{64}$
$= 8i$

b. $\sqrt{-50} = i\sqrt{50}$
$= i\sqrt{5^2 \cdot 2}$
$= 5i\sqrt{2}$

c. $-\sqrt{-4} = -1 \cdot \sqrt{-4}$
$= -1 \cdot 2i$
$= -2i$

d. $\sqrt{-29} = i\sqrt{29}$

Skill Practice

Simplify the expressions in terms of i.

3. $\sqrt{-81}$ 4. $\sqrt{-20}$
5. $\sqrt{-7}$ 6. $-\sqrt{-36}$

Avoiding Mistakes: In an expression such as $i\sqrt{29}$, the i is usually written in front of the square root. The expression $\sqrt{29}\,i$ is also correct, but may be misinterpreted as $\sqrt{29i}$ (with i incorrectly placed under the radical).

The multiplication and division properties of radicals were presented in Sections 7.3 and 7.5 as follows:

If a and b represent real numbers such that $\sqrt[n]{a}$ and $\sqrt[n]{b}$ are both real, then

$$\sqrt[n]{ab} = \sqrt[n]{a} \cdot \sqrt[n]{b} \quad \text{and} \quad \sqrt[n]{\frac{a}{b}} = \frac{\sqrt[n]{a}}{\sqrt[n]{b}} \quad b \neq 0$$

Answers
1. $\sqrt{-1}$ 2. -1 3. $9i$
4. $2i\sqrt{5}$ 5. $i\sqrt{7}$ 6. $-6i$

The conditions that $\sqrt[n]{a}$ and $\sqrt[n]{b}$ must both be real numbers prevent us from applying the multiplication and division properties of radicals for square roots with a negative radicand. Therefore, to multiply or divide radicals with a negative radicand, write the radical in terms of the imaginary number i first. This is demonstrated in Example 2.

Skill Practice

Simplify the expressions.

7. $\dfrac{\sqrt{-36}}{\sqrt{-9}}$

8. $\sqrt{-16} \cdot \sqrt{-49}$

9. $\sqrt{-2} \cdot \sqrt{-2}$

---

example 2   Simplifying a Product or Quotient in Terms of i

Simplify the expressions.

a. $\dfrac{\sqrt{-100}}{\sqrt{-25}}$ **b.** $\sqrt{-25} \cdot \sqrt{-9}$ **c.** $\sqrt{-5} \cdot \sqrt{-5}$

Solution:

a. $\dfrac{\sqrt{-100}}{\sqrt{-25}}$

$= \dfrac{10i}{5i}$ Simplify each radical in terms of i before dividing.

$= 2$

b. $\sqrt{-25} \cdot \sqrt{-9}$

$= 5i \cdot 3i$ Simplify each radical in terms of i first *before* multiplying.

$= 15i^2$ Multiply.

$= 15(-1)$ Recall that $i^2 = -1$.

$= -15$ Simplify.

c. $\sqrt{-5} \cdot \sqrt{-5}$

$= i\sqrt{5} \cdot i\sqrt{5}$

$= i^2 \cdot (\sqrt{5})^2$

$= -1 \cdot 5$

$= -5$

Avoiding Mistakes: In Example 2, we wrote the radical expressions in terms of i first, before multiplying or dividing. If we had mistakenly applied the multiplication or division property first, we would have obtained an incorrect answer.

Correct: $\sqrt{-25} \cdot \sqrt{-9}$
$= (5i)(3i) = 15i^2$
$= 15(-1) = -15$ ↑ correct

Be careful: $\sqrt{-25} \cdot \sqrt{-9}$
$\neq \sqrt{225} = 15$
↑ (incorrect answer)

Answers

7. 2   8. -28 9. -2

2. Powers of i

From the definition of $i = \sqrt{-1}$, it follows that

$i = i$

$i^2 = -1$

$i^3 = -i$ because $i^3 = i^2 \cdot i = (-1)i = -i$

$i^4 = 1$ because $i^4 = i^2 \cdot i^2 = (-1)(-1) = 1$

$i^5 = i$ because $i^5 = i^4 \cdot i = (1)i = i$

$i^6 = -1$ because $i^6 = i^4 \cdot i^2 = (1)(-1) = -1$

This pattern of values $i, -1, -i, 1, i, -1, -i, 1, \ldots$ continues for all subsequent powers of i. Here is a list of several powers of i.

Powers of i

$i^1 = i$	$i^5 = i$	$i^9 = i$
$i^2 = -1$	$i^6 = -1$	$i^{10} = -1$
$i^3 = -i$	$i^7 = -i$	$i^{11} = -i$
$i^4 = 1$	$i^8 = 1$	$i^{12} = 1$

To simplify higher powers of i, we can decompose the expression into multiples of i^4 ($i^4 = 1$) and write the remaining factors as $i, i^2,$ or i^3.

example 3   Simplifying Powers of i

Simplify the powers of i.

a. i^{13} b. i^{18} c. i^{107} d. i^{32}

Solution:

a. $i^{13} = (i^{12}) \cdot (i)$

$\phantom{i^{13}} = (i^4)^3 \cdot (i)$

$\phantom{i^{13}} = (1)^3(i)$ Recall that $i^4 = 1$.

$\phantom{i^{13}} = i$ Simplify.

b. $i^{18} = (i^{16}) \cdot (i^2)$

$\phantom{i^{18}} = (i^4)^4 \cdot (i^2)$

$\phantom{i^{18}} = (1)^4 \cdot (-1)$ $i^4 = 1$ and $i^2 = -1$

$\phantom{i^{18}} = -1$ Simplify.

c. $i^{107} = (i^{104}) \cdot (i^3)$

$\phantom{i^{107}} = (i^4)^{26}(i^3)$

$\phantom{i^{107}} = (1)^{26}(-i)$ $i^4 = 1$ and $i^3 = -i$

$\phantom{i^{107}} = -i$ Simplify.

Skill Practice

Simplify the powers of i.

10. i^8

11. i^{22}

12. i^{45}

13. i^{31}

Answers

10. 1   11. -1

12. i 13. $-i$

d. $i^{32} = (i^4)^8$

$\phantom{i^{32}} = (1)^8 \quad i^4 = 1$

$\phantom{i^{32}} = 1 \quad$ Simplify.

3. Definition of a Complex Number

We have already learned the definitions of the integers, rational numbers, irrational numbers, and real numbers. In this section, we define the complex numbers.

Definition of a Complex Number

A **complex number** is a number of the form $a + bi$, where a and b are real numbers and $i = \sqrt{-1}$.

Notes:

- If $b = 0$, then the complex number $a + bi$ is a real number.
- If $b \neq 0$, then we say that $a + bi$ is an imaginary number.
- The complex number $a + bi$ is said to be written in standard form. The quantities a and b are called the real and imaginary parts (respectively) of the complex number.
- The complex numbers $a - bi$ and $a + bi$ are called **conjugates**.

Concept Connections

14. Give an example of a complex number whose imaginary part is zero. (Answers may vary.)
15. Give an example of complex number whose real part is zero. (Answers may vary.)
16. Give the conjugate of $-7 + 2i$.

From the definition of a complex number, it follows that all real numbers are complex numbers and all imaginary numbers are complex numbers. Figure 7-5 illustrates the relationship among the sets of numbers we have learned so far.

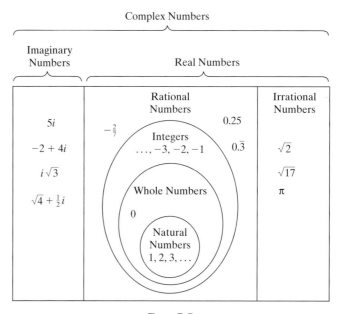

Figure 7-5

Answers

14. For example: -4
15. For example: $2i$
16. $-7 - 2i$

example 4 Identifying the Real and Imaginary Parts of a Complex Number

Identify the real and imaginary parts of the complex numbers.

a. $-8 + 2i$ b. $\dfrac{3}{2}$ c. $-1.75i$

Solution:

a. $-8 + 2i$ -8 is the real part, and 2 is the imaginary part.

b. $\dfrac{3}{2} = \dfrac{3}{2} + 0i$ Rewrite $\tfrac{3}{2}$ in the form $a + bi$.
$\tfrac{3}{2}$ is the real part, and 0 is the imaginary part.

c. $-1.75i$
$= 0 + -1.75i$ Rewrite $-1.75i$ in the form $a + bi$.
0 is the real part, and -1.75 is the imaginary part.

Skill Practice

Identify the real and imaginary parts of the complex numbers.

17. $22 - 14i$
18. -50
19. $15i$

Tip: Example 4(b) illustrates that a real number is also a complex number.

$$\dfrac{3}{2} = \dfrac{3}{2} + 0i$$

Example 4(c) illustrates that an imaginary number is also a complex number.

$$-1.75i = 0 + -1.75i$$

4. Addition, Subtraction, and Multiplication of Complex Numbers

The operations of addition, subtraction, and multiplication of real numbers also apply to imaginary numbers. To add or subtract complex numbers, combine the real parts and combine the imaginary parts. The commutative, associative, and distributive properties that apply to real numbers also apply to complex numbers.

example 5 Adding, Subtracting, and Multiplying Complex Numbers

a. Add: $(1 - 5i) + (-3 + 7i)$
b. Subtract: $\left(-\tfrac{1}{4} + \tfrac{3}{5}i\right) - \left(\tfrac{1}{2} - \tfrac{1}{10}i\right)$
c. Multiply: $(10 - 5i)(2 + 3i)$
d. Multiply: $(1.2 + 0.5i)(1.2 - 0.5i)$

Solution:

a. $(1 - 5i) + (-3 + 7i) = (1 + -3) + (-5 + 7)i$ Add real parts. Add imaginary parts.

$\qquad\qquad\qquad\qquad\quad = -2 + 2i$ Simplify.

Skill Practice

Perform the indicated operations.

20. $\left(\dfrac{1}{2} - \dfrac{1}{4}i\right) + \left(\dfrac{3}{5} + \dfrac{2}{3}i\right)$
21. $(-6 + 11i) - (-9 - 12i)$
22. $(4 - 6i)(2 - 3i)$
23. $(1.5 + 0.8i)(1.5 - 0.8i)$

Answers

17. real: 22; imaginary: -14
18. real: -50; imaginary: 0
19. real: 0; imaginary: 15
20. $\dfrac{11}{10} + \dfrac{5}{12}i$ 21. $3 + 23i$
22. $-10 - 24i$ 23. 2.89

b. $\left(-\dfrac{1}{4} + \dfrac{3}{5}i\right) - \left(\dfrac{1}{2} - \dfrac{1}{10}i\right) = -\dfrac{1}{4} + \dfrac{3}{5}i - \dfrac{1}{2} + \dfrac{1}{10}i$ Apply the distributive property.

$$= \left(-\dfrac{1}{4} - \dfrac{1}{2}\right) + \left(\dfrac{3}{5} + \dfrac{1}{10}\right)i$$ Add real parts. Add imaginary parts.

$$= \left(-\dfrac{1}{4} - \dfrac{2}{4}\right) + \left(\dfrac{6}{10} + \dfrac{1}{10}\right)i$$ Get common denominators.

$$= -\dfrac{3}{4} + \dfrac{7}{10}i$$ Simplify.

c. $(10 - 5i)(2 + 3i)$

$$= (10)(2) + (10)(3i) + (-5i)(2) + (-5i)(3i)$$ Apply the distributive property.

$$= 20 + 30i - 10i - 15i^2$$

$$= 20 + 20i - (15)(-1)$$ Recall $i^2 = -1$.

$$= 20 + 20i + 15$$

$$= 35 + 20i$$ Write in the form $a + bi$.

d. $(1.2 + 0.5i)(1.2 - 0.5i)$

The expressions $(1.2 + 0.5i)$ and $(1.2 - 0.5i)$ are conjugates. The product is a difference of squares.

$$(a + b)(a - b) = a^2 - b^2$$

$(1.2 + 0.5i)(1.2 - 0.5i) = (1.2)^2 - (0.5i)^2$ Apply the formula, where $a = 1.2$ and $b = 0.5i$.

$$= 1.44 - 0.25i^2$$

$$= 1.44 - 0.25(-1)$$ Recall $i^2 = -1$.

$$= 1.44 + 0.25$$

$$= 1.69$$

Tip: The complex numbers $(1.2 + 0.5i)$ and $(1.2 - 0.5i)$ can also be multiplied by using the distributive property:

$(1.2 + 0.5i)(1.2 - 0.5i) = 1.44 - 0.6i + 0.6i - 0.25i^2$

$$= 1.44 - 0.25(-1)$$

$$= 1.69$$

5. Division and Simplification of Complex Numbers

The product of a complex number and its conjugate is a real number. For example:

$$(5 + 3i)(5 - 3i) = 25 - 9i^2$$
$$= 25 - 9(-1)$$
$$= 25 + 9$$
$$= 34$$

To divide by a complex number, multiply the numerator and denominator by the conjugate of the denominator. This produces a real number in the denominator so that the resulting expression can be written in the form $a + bi$.

example 6 Dividing by a Complex Number

Divide the complex numbers and write the answer in the form $a + bi$.

$$\frac{4 - 3i}{5 + 2i}$$

Solution:

$\dfrac{4 - 3i}{5 + 2i}$ Multiply the numerator and denominator by the conjugate of the denominator:

$$\frac{(4 - 3i)}{(5 + 2i)} \cdot \frac{(5 - 2i)}{(5 - 2i)} = \frac{(4)(5) + (4)(-2i) + (-3i)(5) + (-3i)(-2i)}{(5)^2 - (2i)^2}$$

$$= \frac{20 - 8i - 15i + 6i^2}{25 - 4i^2} \quad \text{Simplify numerator and denominator.}$$

$$= \frac{20 - 23i + 6(-1)}{25 - 4(-1)} \quad \text{Recall } i^2 = -1.$$

$$= \frac{20 - 23i - 6}{25 + 4}$$

$$= \frac{14 - 23i}{29} \quad \text{Simplify.}$$

$$= \frac{14}{29} - \frac{23i}{29} \quad \text{Write in the form } a + bi.$$

Skill Practice

Divide the complex numbers. Write the answer in the form $a + bi$.

24. $\dfrac{2 + i}{3 - 2i}$

example 7 Simplifying Complex Numbers

Simplify the complex numbers.

a. $\dfrac{6 + \sqrt{-18}}{9}$ **b.** $\dfrac{4 - \sqrt{-36}}{2}$

Answers

24. $\dfrac{4}{13} + \dfrac{7}{13}i$

Skill Practice

Simplify the complex numbers.

25. $\dfrac{8 - \sqrt{-24}}{6}$

26. $\dfrac{-12 + \sqrt{-64}}{4}$

Solution:

a. $\dfrac{6 + \sqrt{-18}}{9} = \dfrac{6 + i\sqrt{18}}{9}$ Write the radical in terms of i.

$= \dfrac{6 + 3i\sqrt{2}}{9}$ Simplify $\sqrt{18} = 3\sqrt{2}$.

$= \dfrac{3(2 + i\sqrt{2})}{9}$ Factor the numerator.

$= \dfrac{\overset{1}{3}(2 + i\sqrt{2})}{\underset{3}{9}}$ Simplify.

$= \dfrac{2 + i\sqrt{2}}{3}$ or $\dfrac{2}{3} + i\dfrac{\sqrt{2}}{3}$

b. $\dfrac{4 - \sqrt{-36}}{2} = \dfrac{4 - i\sqrt{36}}{2}$ Write the radical in terms of i.

$= \dfrac{4 - 6i}{2}$ Simplify $\sqrt{36} = 6$.

$= \dfrac{2(2 - 3i)}{2}$ Factor the numerator.

$= \dfrac{\overset{1}{2}(2 - 3i)}{\underset{1}{2}}$ Simplify.

$= 2 - 3i$

Answers

25. $\dfrac{4 - i\sqrt{6}}{3}$ or $\dfrac{4}{3} - i\dfrac{\sqrt{6}}{3}$

26. $-3 + 2i$

---

section 7.8 Practice Exercises

Boost your GRADE at mathzone.com!

- Practice Problems
- Self-Tests
- NetTutor
- e-Professors
- Videos

Study Skills Exercises

1. Test yourself.

   Yes _____ No _____ Did you have sufficient time to study for the test on this chapter? If not, what could you have done to create more time for studying?

   Yes _____ No _____ Did you work all the assigned homework problems in this chapter?

   Yes _____ No _____ If you encountered difficulty in this chapter, did you see your instructor or tutor for help?

   Yes _____ No _____ Have you taken advantage of the textbook supplements such as the *Student Solutions Manual* and the Online Learning Center?

2. Define the key terms.

   a. Imaginary numbers

   b. i

   c. Complex number

   d. Conjugate

Review Exercises

For Exercises 3–6, perform the indicated operations.

3. $-2\sqrt{5} - 3\sqrt{50} + \sqrt{125}$

4. $\sqrt[3]{2x}(\sqrt[3]{2x} - \sqrt[3]{4x^2})$

5. $(3 - \sqrt{x})(3 + \sqrt{x})$

6. $(\sqrt{5} + \sqrt{2})^2$

For Exercises 7–10, solve the equations.

7. $\sqrt[3]{3p + 7} - \sqrt[3]{2p - 1} = 0$

8. $\sqrt[3]{t - 5} - \sqrt[3]{2t + 1} = 0$

9. $\sqrt{36c + 15} = 6\sqrt{c} + 1$

10. $\sqrt{4a + 29} = 2\sqrt{a} + 5$

Objective 1: Definition of i

11. Simplify the expressions $\sqrt{-1}$ and $-\sqrt{1}$.

12. Simplify i^2.

For Exercises 13–34, simplify the expressions. (See Examples 1–2.)

13. $\sqrt{-144}$

14. $\sqrt{-81}$

15. $\sqrt{-3}$

16. $\sqrt{-17}$

17. $\sqrt{-20}$

18. $\sqrt{-75}$

19. $2\sqrt{-25} \cdot 3\sqrt{-4}$

20. $(-4\sqrt{-9})(-3\sqrt{-1})$

21. $3\sqrt{-18} + 5\sqrt{-32}$

22. $5\sqrt{-45} + 3\sqrt{-80}$

23. $7\sqrt{-63} - 4\sqrt{-28}$

24. $7\sqrt{-3} - 4\sqrt{-27}$

25. $\sqrt{-7} \cdot \sqrt{-7}$

26. $\sqrt{-11} \cdot \sqrt{-11}$

27. $\sqrt{-9} \cdot \sqrt{-16}$

28. $\sqrt{-25} \cdot \sqrt{-36}$

29. $\sqrt{-15} \cdot \sqrt{-6}$

30. $\sqrt{-12} \cdot \sqrt{-50}$

31. $\dfrac{\sqrt{-50}}{\sqrt{25}}$

32. $\dfrac{\sqrt{-27}}{\sqrt{9}}$

33. $\dfrac{\sqrt{-90}}{\sqrt{10}}$

34. $\dfrac{\sqrt{-125}}{\sqrt{45}}$

Objective 2: Powers of i

For Exercises 35–46, simplify the powers of i. (See Example 3.)

35. i^7

36. i^{38}

37. i^{64}

38. i^{75}

39. i^{41} 40. i^{25} 41. i^{52} 42. i^0

43. i^{23} 44. i^{103} 45. i^6 46. i^{82}

Objective 3: Definition of a Complex Number

47. What is the conjugate of a complex number $a + bi$?

48. True or false?
    a. Every real number is a complex number.
    b. Every complex number is a real number.

For Exercises 49–56, identify the real and imaginary parts of the complex number. (See Example 4.)

49. $-5 + 12i$ 50. $22 - 16i$ 51. $-6i$ 52. $10i$

53. 35 54. -1 55. $\dfrac{3}{5} + i$ 56. $-\dfrac{1}{2} - \dfrac{1}{4}i$

Objective 4: Addition, Subtraction, and Multiplication of Complex Numbers

For Exercises 57–80, perform the indicated operations. Write the answer in the form $a + bi$. (See Example 5.)

57. $(2 - i) + (5 + 7i)$ 58. $(5 - 2i) + (3 + 4i)$ 59. $\left(\dfrac{1}{2} + \dfrac{2}{3}i\right) - \left(\dfrac{1}{5} - \dfrac{5}{6}i\right)$

60. $\left(\dfrac{11}{10} - \dfrac{7}{5}i\right) - \left(-\dfrac{2}{5} + \dfrac{3}{5}i\right)$ 61. $(1 + 3i) + (4 - 3i)$ 62. $(-2 + i) + (1 - i)$

63. $(2 + 3i) - (1 - 4i) + (-2 + 3i)$ 64. $(2 + 5i) - (7 - 2i) + (-3 + 4i)$

65. $(8i)(3i)$ 66. $(2i)(4i)$ 67. $6i(1 - 3i)$ 68. $-i(3 + 4i)$

69. $(2 - 10i)(3 + 2i)$ 70. $(4 + 7i)(1 - i)$ 71. $(-5 + 2i)(5 + 2i)$ 72. $(4 - 11i)(-4 - 11i)$

73. $(4 + 5i)^2$ 74. $(3 - 2i)^2$ 75. $(2 + i)(3 - 2i)(4 + 3i)$ 76. $(3 - i)(3 + i)(4 - i)$

77. $(-4 - 6i)^2$ 78. $(-3 - 5i)^2$ 79. $\left(-\dfrac{1}{2} - \dfrac{3}{4}i\right)\left(-\dfrac{1}{2} + \dfrac{3}{4}i\right)$ 80. $\left(-\dfrac{2}{3} + \dfrac{1}{6}i\right)\left(-\dfrac{2}{3} - \dfrac{1}{6}i\right)$

Objective 5: Division and Simplification of Complex Numbers

For Exercises 81–90, divide the complex numbers. Write the answer in the form $a + bi$. **(See Example 6.)**

81. $\dfrac{2}{1 + 3i}$

82. $\dfrac{-2}{3 + i}$

83. $\dfrac{-i}{4 - 3i}$

84. $\dfrac{3 - 3i}{1 - i}$

85. $\dfrac{5 + 2i}{5 - 2i}$

86. $\dfrac{7 + 3i}{4 - 2i}$

87. $\dfrac{3 + 7i}{-2 - 4i}$

88. $\dfrac{-2 + 9i}{-1 - 4i}$

89. $\dfrac{13i}{-5 - i}$

90. $\dfrac{15i}{-2 - i}$

For Exercises 91–96, simplify the complex numbers. **(See Example 7.)**

91. $\dfrac{2 + \sqrt{-16}}{8}$

92. $\dfrac{6 - \sqrt{-4}}{4}$

93. $\dfrac{-6 + \sqrt{-72}}{6}$

94. $\dfrac{-20 + \sqrt{-500}}{10}$

95. $\dfrac{-8 - \sqrt{-48}}{4}$

96. $\dfrac{-18 - \sqrt{-72}}{3}$

chapter 7 | summary

section 7.1 Definition of an nth root

Key Concepts

b is an **nth root** of a if $b^n = a$.

The expression $\sqrt{a}$ represents the principal square root of a.

The expression $\sqrt[n]{a}$ represents the principal nth root of a.

$\sqrt[n]{a^n} = |a|$ if n is even.

$\sqrt[n]{a^n} = a$ if n is odd.

$\sqrt[n]{a}$ is not a real number if $a < 0$ and n is even.

$f(x) = \sqrt[n]{x}$ defines a **radical function**.

The Pythagorean Theorem
$a^2 + b^2 = c^2$

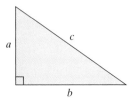

Examples

Example 1

2 is a square root of 4.

-2 is a square root of 4.

-3 is a cube root of -27.

Example 2

$\sqrt{36} = 6 \qquad \sqrt[3]{-64} = -4$

Example 3

$\sqrt[4]{(x+3)^4} = |x+3| \qquad \sqrt[5]{(x+3)^5} = x+3$

Example 4

$\sqrt[4]{-16}$ is not a real number.

Example 5

For $g(x) = \sqrt{x}$ the domain is $[0, \infty)$.

For $h(x) = \sqrt[3]{x}$ the domain is $(-\infty, \infty)$.

Example 6

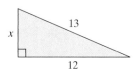

$x^2 + 12^2 = 13^2$

$x^2 + 144 = 169$

$\qquad x^2 = 25$

$\qquad x = \sqrt{25}$

$\qquad x = 5$

section 7.2 Rational Exponents

Key Concepts

Let a be a real number and n be an integer such that $n > 1$. If $\sqrt[n]{a}$ exists, then

$$a^{1/n} = \sqrt[n]{a}$$

$$a^{m/n} = \left(\sqrt[n]{a}\right)^m = \sqrt[n]{a^m}$$

Examples

Example 1

$121^{1/2} = \sqrt{121} = 11$

Example 2

$27^{2/3} = \left(\sqrt[3]{27}\right)^2 = (3)^2 = 9$

section 7.3 Simplifying Radical Expressions

Key Concepts

Let a and b represent real numbers such that $\sqrt[n]{a}$ and $\sqrt[n]{b}$ are both real. Then

$\sqrt[n]{ab} = \sqrt[n]{a} \cdot \sqrt[n]{b}$ Multiplication property

$\sqrt[n]{\dfrac{a}{b}} = \dfrac{\sqrt[n]{a}}{\sqrt[n]{b}}$ Division property

A radical expression whose radicand is written as a product of prime factors is in simplified form if all the following conditions are met:
1. The radicand has no factor raised to a power greater than or equal to the index.
2. The radicand does not contain a fraction.
3. No radicals are in the denominator of a fraction.

Examples

Example 1

$\sqrt{3} \cdot \sqrt{5} = \sqrt{15}$

Example 2

$\sqrt{\dfrac{x}{9}} = \dfrac{\sqrt{x}}{\sqrt{9}} = \dfrac{\sqrt{x}}{3}$

Example 3

$\sqrt[3]{16x^5y^7}$

$= \sqrt[3]{2^4 x^5 y^7}$

$= \sqrt[3]{2^3 x^3 y^6 \cdot 2x^2 y}$

$= \sqrt[3]{2^3 x^3 y^6} \cdot \sqrt[3]{2x^2 y}$

$= 2xy^2 \sqrt[3]{2x^2 y}$

section 7.4 Addition and Subtraction of Radicals

Key Concepts

***Like* radicals** have radical factors with the same index and the same radicand.

Use the distributive property to add and subtract *like* radicals.

Examples

Example 1

$3x\sqrt{7} - 5x\sqrt{7} + x\sqrt{7}$
$= (3 - 5 + 1) \cdot x\sqrt{7}$
$= -x\sqrt{7}$

Example 2

$x\sqrt[4]{16x} - 3\sqrt[4]{x^5}$
$= 2x\sqrt[4]{x} - 3x\sqrt[4]{x}$
$= (2x - 3x)\sqrt[4]{x}$
$= -x\sqrt[4]{x}$

section 7.5 Multiplication of Radicals

Key Concepts

The Multiplication Property of Radicals

If $\sqrt[n]{a}$ and $\sqrt[n]{b}$ are real numbers, then

$\sqrt[n]{ab} = \sqrt[n]{a} \cdot \sqrt[n]{b}$

To multiply or divide radicals with different indices, convert to rational exponents and use the properties of exponents.

Examples

Example 1

$3\sqrt{2}(\sqrt{2} + 5\sqrt{7} - \sqrt{6})$
$= 3\sqrt{4} + 15\sqrt{14} - 3\sqrt{12}$
$= 3 \cdot 2 + 15\sqrt{14} - 3 \cdot 2\sqrt{3}$
$= 6 + 15\sqrt{14} - 6\sqrt{3}$

Example 2

$\sqrt{p} \cdot \sqrt[5]{p^2}$
$= p^{1/2} \cdot p^{2/5}$
$= p^{5/10} \cdot p^{4/10}$
$= p^{9/10}$
$= \sqrt[10]{p^9}$

section 7.6 Rationalization

Key Concepts

The process of removing a radical from the denominator of an expression is called **rationalizing the denominator**.

Rationalizing a denominator with one term

Rationalizing a denominator with two terms

Examples

Example 1
Rationalize:

$$\frac{4}{\sqrt[4]{2y^3}}$$

$$= \frac{4}{\sqrt[4]{2y^3}} \cdot \frac{\sqrt[4]{2^3 y}}{\sqrt[4]{2^3 y}}$$

$$= \frac{4\sqrt[4]{8y}}{\sqrt[4]{2^4 y^4}}$$

$$= \frac{4\sqrt[4]{8y}}{2y}$$

$$= \frac{2\sqrt[4]{8y}}{y}$$

Example 2
Rationalize the denominator:

$$\frac{\sqrt{2}}{\sqrt{x} - \sqrt{3}}$$

$$= \frac{\sqrt{2}}{(\sqrt{x} - \sqrt{3})} \cdot \frac{(\sqrt{x} + \sqrt{3})}{(\sqrt{x} + \sqrt{3})}$$

$$= \frac{\sqrt{2x} + \sqrt{6}}{x - 3}$$

section 7.7 Radical Equations

Key Concepts

Steps to Solve a Radical Equation

1. Isolate the radical. If an equation has more than one radical, choose one of the radicals to isolate.
2. Raise each side of the equation to a power equal to the index of the radical.
3. Solve the resulting equation. If the equation still has a radical, repeat steps 1 and 2.
4. Check the potential solutions in the original equation.

Examples

Example 1

Solve:

$$\sqrt{b-5} - \sqrt{b+3} = 2$$
$$\sqrt{b-5} = \sqrt{b+3} + 2$$
$$(\sqrt{b-5})^2 = (\sqrt{b+3} + 2)^2$$
$$b - 5 = b + 3 + 4\sqrt{b+3} + 4$$
$$b - 5 = b + 7 + 4\sqrt{b+3}$$
$$-12 = 4\sqrt{b+3}$$
$$-3 = \sqrt{b+3}$$
$$(-3)^2 = (\sqrt{b+3})^2$$
$$9 = b + 3$$
$$6 = b$$

Check:

$$\sqrt{6-5} - \sqrt{6+3} \stackrel{?}{=} 2$$
$$\sqrt{1} - \sqrt{9} \stackrel{?}{=} 2$$
$$1 - 3 \neq 2 \quad \text{Does not check.}$$

No solution.

section 7.8 Complex Numbers

Key Concepts

$i = \sqrt{-1}$ and $i^2 = -1$

For a real number $b > 0$, $\sqrt{-b} = i\sqrt{b}$

A **complex number** is in the form $a + bi$, where a and b are real numbers. The a is called the real part, and the b is called the imaginary part.

To add or subtract complex numbers, combine the real parts and combine the imaginary parts.

Multiply complex numbers by using the distributive property.

Divide complex numbers by multiplying the numerator and denominator by the **conjugate** of the denominator.

Examples

Example 1

$\sqrt{-4} \cdot \sqrt{-9}$
$= (2i)(3i)$
$= 6i^2$
$= -6$

Example 2

$(3 - 5i) - (2 + i) + (3 - 2i)$
$= 3 - 5i - 2 - i + 3 - 2i$
$= 4 - 8i$

Example 3

$(1 + 6i)(2 + 4i)$
$= 2 + 4i + 12i + 24i^2$
$= 2 + 16i + 24(-1)$
$= -22 + 16i$

Example 4

$\dfrac{3}{2 - 5i}$

$= \dfrac{3}{2 - 5i} \cdot \dfrac{(2 + 5i)}{(2 + 5i)} = \dfrac{6 + 15i}{4 - 25i^2}$

$= \dfrac{6 + 15i}{29}$ or $\dfrac{6}{29} + \dfrac{15}{29}i$

---

chapter 7 | review exercises

For the exercises in this set, assume that all variables represent positive real numbers unless otherwise stated.

Section 7.1

1. True or false?

   a. The principal nth root of an even-indexed root is always positive.

   b. The principal nth root of an odd-indexed root is always positive.

2. Explain why $\sqrt{(-3)^2} \neq -3$.

3. Are the following statements true or false?

   a. $\sqrt{a^2 + b^2} = a + b$

   b. $\sqrt{(a + b)^2} = a + b$

For Exercises 4–6, simplify the radicals.

4. $\sqrt{\dfrac{50}{32}}$

5. $\sqrt[4]{625}$

6. $\sqrt{(-6)^2}$

7. Evaluate the function values for $f(x) = \sqrt{x-1}$.
   a. $f(10)$ b. $f(1)$ c. $f(8)$

   d. Write the domain of f in interval notation.

8. Evaluate the function values for $g(t) = \sqrt{5+t}$.
   a. $g(-5)$ b. $g(-4)$ c. $g(4)$

   d. Write the domain of g in interval notation.

9. Translate the English expression to an algebraic expression: Four more than the quotient of the cube root of $2x$ and the principal fourth root of $2x$.

For Exercises 10–11, simplify the expression. Assume that x and y represent *any* real number.

10. a. $\sqrt{x^2}$ b. $\sqrt[3]{x^3}$

    c. $\sqrt[4]{x^4}$ d. $\sqrt[5]{(x+1)^5}$

11. a. $\sqrt{4y^2}$ b. $\sqrt[3]{27y^3}$

    c. $\sqrt[100]{y^{100}}$ d. $\sqrt[101]{y^{101}}$

12. Use the Pythagorean theorem to find the length of the third side of the triangle.

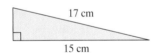

Section 7.2

13. Are the properties of exponents the same for rational exponents and integer exponents? Give an example. (Answers may vary.)

14. In the expression $x^{m/n}$ what does n represent?

15. Explain the process of eliminating a negative exponent from an algebraic expression.

For Exercises 16–20, simplify the expressions. Write the answer with positive exponents only.

16. $(-125)^{1/3}$ 17. $16^{-1/4}$

18. $\left(\dfrac{1}{16}\right)^{-3/4} - \left(\dfrac{1}{8}\right)^{-2/3}$ 19. $(b^{1/2} \cdot b^{1/3})^{12}$

20. $\left(\dfrac{x^{-1/4}y^{-1/3}z^{3/4}}{2^{1/3}x^{-1/3}y^{2/3}}\right)^{-12}$

For Exercises 21–22, rewrite the expressions by using rational exponents.

21. $\sqrt[4]{x^3}$ 22. $\sqrt[3]{2y^2}$

For Exercises 23–25, use a calculator to approximate the expressions to 4 decimal places.

23. $10^{1/3}$ 24. $17.8^{2/3}$ 25. $147^{4/5}$

26. An initial investment of P dollars is made in an account in which the return is compounded quarterly. The amount in the account can be determined by

$$A = P\left(1 + \frac{r}{4}\right)^{1/3}$$

where r is the annual rate of return and t is the time in months.

When she is 20 years old, Jenna invests $5000 in a mutual fund that grows by an average of 11% per year compounded quarterly. How much money does she have

a. After 6 months   b. After 1 year

c. At age 40   d. At age 50

e. At age 65

Section 7.3

27. List the criteria for a rational expression to be simplified.

For Exercises 28–31, simplify the radicals.

28. $\sqrt{108}$ 29. $\sqrt[4]{x^5yz^4}$

30. $\sqrt{5x} \cdot \sqrt{20x}$ 31. $\sqrt[3]{\dfrac{-16x^7y^6}{z^9}}$

32. Write an English phrase that describes the following mathematical expressions: (Answers may vary.)

a. $\sqrt{\dfrac{2}{x}}$ **b.** $(x + 1)^3$

33. An engineering firm made a mistake when building a $\frac{1}{4}$-mi bridge in the Florida Keys. The bridge was made without adequate expansion joints to prevent buckling during the heat of summer. During mid-June, the bridge expanded 1.5 ft, causing a vertical bulge in the middle. Calculate the height of the bulge h in feet. (*Note:* 1 mi = 5280 ft.) Round to the nearest foot.

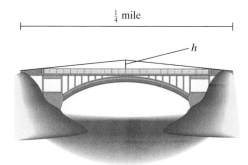

Section 7.4

34. Complete the following statement: Radicals may be added or subtracted if . . .

For Exercises 35–38, determine whether the radicals may be combined, and explain your answer.

35. $\sqrt[3]{2x} - 2\sqrt{2x}$ **36.** $2 + \sqrt{x}$

37. $\sqrt[4]{3xy} + 2\sqrt[4]{3xy}$ **38.** $-4\sqrt{32} + 7\sqrt{50}$

For Exercises 39–42, add or subtract as indicated.

39. $4\sqrt{7} - 2\sqrt{7} + 3\sqrt{7}$

40. $2\sqrt[3]{64} + 3\sqrt[3]{54} - 16$

41. $\sqrt{50} + 7\sqrt{2} - \sqrt{8}$

42. $x\sqrt[3]{16x^2} - 4\sqrt[3]{2x^5} + 5x\sqrt[3]{54x^2}$

For Exercises 43–44, answer true or false. If an answer is false, explain why. Assume all variables represent positive real numbers.

43. $5 + 3\sqrt{x} = 8\sqrt{x}$

44. $\sqrt{y} + \sqrt{y} = \sqrt{2y}$

Section 7.5

For Exercises 45–56, multiply the radicals and simplify the answer.

45. $\sqrt{3} \cdot \sqrt{12}$ **46.** $\sqrt[4]{4} \cdot \sqrt[4]{8}$

47. $-2\sqrt{3}(\sqrt{3} - 3\sqrt{3})$ **48.** $-3\sqrt{5}(2\sqrt{3} - \sqrt{5})$

49. $(2\sqrt{x} - 3)(2\sqrt{x} + 3)$ **50.** $(\sqrt{y} + 4)(\sqrt{y} - 4)$

51. $(\sqrt{7y} - \sqrt{3x})^2$ **52.** $(2\sqrt{3w} + 5)^2$

53. $(-\sqrt{z} - \sqrt{6})(2\sqrt{z} + 7\sqrt{6})$

54. $(3\sqrt{a} - \sqrt{5})(\sqrt{a} + 2\sqrt{5})$

55. $\sqrt[3]{u} \cdot \sqrt{u^5}$

56. $\sqrt{2} \cdot \sqrt[4]{w^3}$

Section 7.6

For Exercises 57–64, rationalize the denominator.

57. $\sqrt{\dfrac{7}{2y}}$ **58.** $\sqrt{\dfrac{5}{3w}}$

59. $\dfrac{4}{\sqrt[3]{9p^2}}$ **60.** $\dfrac{-2}{\sqrt[3]{2x}}$

61. $\dfrac{-5}{\sqrt{15} - \sqrt{10}}$ **62.** $\dfrac{-6}{\sqrt{7} - \sqrt{5}}$

63. $\dfrac{t - 3}{\sqrt{t} - \sqrt{3}}$ **64.** $\dfrac{w - 7}{\sqrt{w} - \sqrt{7}}$

65. Translate the mathematical expression to an English phrase. (Answers may vary.)

$$\frac{\sqrt{2}}{x^2}$$

Section 7.7

Solve the radical equations in Exercises 66–73, if possible.

66. $\sqrt{2y} = 7$

67. $\sqrt{a-6} - 5 = 0$

68. $\sqrt[3]{2w-3} + 5 = 2$

69. $\sqrt[4]{p+12} - \sqrt[4]{5p-16} = 0$

70. $\sqrt{t} + \sqrt{t-5} = 5$

71. $\sqrt{8x+1} = -\sqrt{x-13}$

72. $\sqrt{2m^2+4} - \sqrt{9m} = 0$

73. $\sqrt{x+2} = 1 - \sqrt{2x+5}$

74. A tower is supported by stabilizing wires. Find the exact length of each wire, and then round to the nearest tenth of a meter.

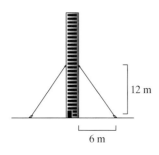

75. The velocity, $v(d)$, of an ocean wave depends on the water depth d as the wave approaches land.

$$v(d) = \sqrt{32d}$$

where $v(d)$ is in feet per second and d is in feet.

a. Find $v(20)$ and interpret its value. Round to 1 decimal place.

b. Find the depth of the water at a point where a wave is traveling at 16 ft/sec.

Section 7.8

76. Define a complex number.

77. Define an imaginary number.

78. Consider the following expressions.

$$\frac{3}{4+6i} \quad \text{and} \quad \frac{3}{\sqrt{4}+\sqrt{6}}$$

Compare the process of dividing by a complex number to the process of rationalizing the denominator.

For Exercises 79–82, rewrite the expressions in terms of i.

79. $\sqrt{-16}$

80. $-\sqrt{-5}$

81. $\sqrt{-75} \cdot \sqrt{-3}$

82. $\dfrac{-\sqrt{-24}}{\sqrt{6}}$

For Exercises 83–86, simplify the powers of i.

83. i^{38}

84. i^{101}

85. i^{19}

86. $i^{1000} + i^{1002}$

For Exercises 87–90, perform the indicated operations. Write the final answer in the form $a + bi$.

87. $(-3+i) - (2-4i)$

88. $(2i+4)(2i-4)$

89. $(4-3i)(4+3i)$

90. $(5-i)^2$

For Exercises 91–92, write the expressions in the form $a + bi$, and determine the real and imaginary parts.

91. $\dfrac{17-4i}{-4}$

92. $\dfrac{-16-8i}{8}$

For Exercises 93–94, divide and simplify. Write the final answer in the form $a + bi$.

93. $\dfrac{2-i}{3+2i}$

94. $\dfrac{10+5i}{2-i}$

For Exercises 95–96, simplify the expression.

95. $\dfrac{-8+\sqrt{-40}}{12}$

96. $\dfrac{6-\sqrt{-144}}{3}$

chapter 7 | test

1. a. What is the principal square root of 36?

   b. What is the negative square root of 36?

2. Which of the following are real numbers?
   a. $-\sqrt{100}$ **b.** $\sqrt{-100}$
   c. $-\sqrt[3]{1000}$ **d.** $\sqrt[3]{-1000}$

3. Simplify.
   a. $\sqrt[3]{y^3}$ **b.** $\sqrt[4]{y^4}$

For Exercises 4–11, simplify the radicals. Assume that all variables represent positive numbers.

4. $\sqrt[4]{81}$

5. $\sqrt{\dfrac{16}{9}}$

6. $\sqrt[3]{32}$

7. $\sqrt{a^4 b^3 c^5}$

8. $\sqrt{3x} \cdot \sqrt{6x^3}$

9. $\sqrt{\dfrac{32w^6}{3w}}$

10. $\sqrt[6]{7} \cdot \sqrt{y}$

11. $\dfrac{\sqrt[3]{10}}{\sqrt[4]{10}}$

12. a. Evaluate the function values $f(-8), f(-6), f(-4)$, and $f(-2)$ for $f(x) = \sqrt{-2x - 4}$.

   b. Write the domain of f in interval notation.

13. Use a calculator to evaluate $\dfrac{-3 - \sqrt{5}}{17}$ to 4 decimal places.

For Exercises 14–15, simplify the expressions. Assume that all variables represent positive numbers.

14. $-27^{1/3}$

15. $\dfrac{t^{-1} \cdot t^{1/2}}{t^{1/4}}$

16. Add or subtract as indicated
$$3\sqrt{5} + 4\sqrt{5} - 2\sqrt{20}$$

17. Multiply the radicals.
   a. $3\sqrt{x}(\sqrt{2} - \sqrt{5})$
   b. $(2\sqrt{5} - 3\sqrt{x})(4\sqrt{5} + \sqrt{x})$

18. Rationalize the denominator. Assume $x > 0$.
   a. $\dfrac{-2}{\sqrt[3]{x}}$ **b.** $\dfrac{\sqrt{x} + 2}{3 - \sqrt{x}}$

19. Rewrite the expressions in terms of i.
   a. $\sqrt{-8}$ **b.** $2\sqrt{-16}$ **c.** $\dfrac{2 + \sqrt{-8}}{4}$

For Exercises 20–26, perform the indicated operation and simplify completely. Write the final answer in the form $a + bi$.

20. $(3 - 5i) - (2 + 6i)$

21. $(4 + i)(8 + 2i)$

22. $\sqrt{-16} \cdot \sqrt{-49}$

23. $(4 - 7i)^2$

24. $(2 - 10i)(2 + 10i)$

25. $\dfrac{3 - 2i}{3 - 4i}$

26. $(10 + 3i)[(-5i + 8) - (5 - 3i)]$

27. If the volume V of a sphere is known, the radius of the sphere can be computed by
$$r(V) = \sqrt[3]{\dfrac{3v}{4\pi}}.$$
Find $r(10)$ to 2 decimal places. Interpret the meaning in the context of the problem.

28. A patio 20 ft wide has a slanted roof, as shown in the picture. Find the length of the roof if there is an 8-in. overhang. Round the answer to the nearest foot.

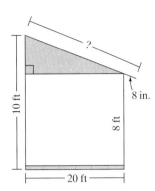

For Exercises 29–31, solve the radical equation.

29. $\sqrt[3]{2x + 5} = -3$

30. $\sqrt{5x + 8} = \sqrt{5x - 1} + 1$

31. $\sqrt{t + 7} - \sqrt{2t - 3} = 2$

---

chapters 1–7 | cumulative review exercises

1. Simplify the expression.
$6^2 - 2[5 - 8(3 - 1) + 4 \div 2]$

2. Simplify the expression.
$3x - 3(-2x + 5) - 4y + 2(3x + 5) - y$

3. Solve the equation: $\quad 9(2y + 8) = 20 - (y + 5)$

4. Solve the inequality. Write the answer in interval notation.
$2a - 4 < -14$

5. Write an equation of the line that is parallel to the line $2x + y = 9$ and passes through the point $(3, -1)$. Write the answer in slope-intercept form.

6. On the same coordinate system, graph the line $2x + y = 9$ and the line that you derived in Exercise 5. Verify that these two lines are indeed parallel.

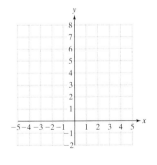

7. Solve the system of equations by using the addition method.
$$2x - 3y = 0$$
$$-4x + 3y = -1$$

8. Determine if $(2, -2, \frac{1}{2})$ is a solution to the system.
$$2x + y - 4z = 0$$
$$x - y + 2z = 5$$
$$3x + 2y + 2z = 4$$

9. Write a system of linear equations from the augmented matrix. Use x, y, and z for the variables.

$$\begin{bmatrix} 1 & 0 & 0 & | & 6 \\ 0 & 1 & 0 & | & 3 \\ 0 & 0 & 1 & | & 8 \end{bmatrix}$$

10. Given the function defined by $f(x) = 4x - 2$.

   a. Find $f(-2), f(0), f(4),$ and $f(\frac{1}{2})$.

   b. Write the ordered pairs that correspond to the function values in part (a).

   c. Graph $y = f(x)$.

   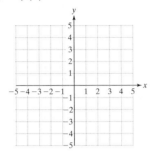

11. Determine if the graph defines y as a function of x.

   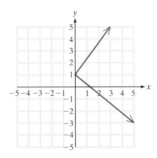

12. Simplify the expression. Write the final answer with positive exponents only.

   $$\left(\frac{a^3 b^{-1} c^3}{ab^{-5} c^2}\right)^2$$

13. Simplify the expression. Write the final answer with positive exponents only.

   $$\left(\frac{a^{3/2} b^{-1/4} c^{1/3}}{ab^{-5/4} c^0}\right)^{12}$$

14. Multiply or divide as indicated, and write the answer in scientific notation.

   a. $(3.5 \times 10^7)(4 \times 10^{-12})$

   b. $\dfrac{6.28 \times 10^5}{2.0 \times 10^{-4}}$

15. Multiply the polynomials $(2x + 5)(x - 3)$. What is the degree of the product?

16. Perform the indicated operations and simplify.
   $\sqrt{3}(\sqrt{5} + \sqrt{6} + \sqrt{3})$

17. Divide $(x^2 - x - 12) \div (x + 3)$.

18. Simplify and subtract: $\sqrt[4]{\dfrac{1}{16}} - \sqrt[3]{\dfrac{8}{27}}$

19. Simplify: $\sqrt[3]{\dfrac{54c^4}{cd^3}}$

20. Add: $4\sqrt{45b^3} + 5b\sqrt{80b}$

21. Divide: $\dfrac{13i}{3 + 2i}$ Write the answer in the form $a + bi$.

22. Solve the equation.

   $$\dfrac{5}{y - 2} - \dfrac{3}{y - 4} = \dfrac{6}{y^2 - 6y + 8}$$

23. Add: $\dfrac{3}{x^2 + 5x} + \dfrac{-2}{x^2 - 25}$

24. Divide: $\dfrac{a + 10}{2a^2 - 11a - 6} \div \dfrac{a^2 + 12a + 20}{6 - a}$

Quadratic Equations and Functions

8

8.1 Completing the Square and the Square Root Property
8.2 Quadratic Formula
8.3 Equations in Quadratic Form
8.4 Graphs of Quadratic Functions
8.5 Vertex of a Parabola and Applications

In Chapter 5 we defined a quadratic equation as $ax^2 + bx + c = 0$, where a, b, and c are real numbers and $a \neq 0$. We solved quadratic equations by factoring and applying the zero product rule. In this chapter, we present two additional techniques to solve quadratic equations. However, these techniques can be used to solve quadratic equations even if the equations are not factorable.

Quadratic equations and quadratic functions have many applications. In Exercise 53 in Section 8.2, for example, the braking distance (in feet) necessary to bring a car to a complete stop is dependent on the speed of the car at the time the brakes are applied. This is given by the function $d(v) = \frac{v^2}{20} + v$, where v is the speed in miles per hour. Suppose a motorist sees a deer in the road 100 ft ahead. To find the maximum speed, v, that the motorist could be traveling and still brake for the deer, we would solve the equation $100 = \frac{v^2}{20} + v$.

chapter 8 preview

The exercises in this chapter preview contain concepts that have not yet been presented. These exercises are provided for students who want to compare their levels of understanding before and after studying the chapter. Alternatively, you may prefer to work these exercises when the chapter is completed and before taking the exam.

Section 8.1

1. Find the value of k to complete the square. Then factor the trinomial. $y^2 - 18y + k$

For Exercises 2–3, solve by using the square root property.

2. $y^2 + 5 = 0$
3. $(a + 2)^2 = 12$

4. Solve by completing the square. $2x^2 + 12x + 2 = 0$

5. Solve for r. $V = \pi r^2 h \quad (r > 0)$

Section 8.2

For Exercises 6–7, solve by using the quadratic formula.

6. $y^2 + 6y - 1 = 0$
7. $2x^2 - 2x + 3 = 0$

8. Given: $6t^2 - 4t + 1 = 0$

   a. Find the value of the discriminant.

   b. Use the discriminant to give the number and type of solutions.

9. One leg of a right triangle is 3 times as long as the other. The length of the hypotenuse is 10 in. Find the length of each leg.

10. Solve. $y(y + 7) = 3$

Section 8.3

For Exercises 11–13, solve the equations.

11. $a = \sqrt{a + 3} - 1$
12. $y^4 - 1 = 0$

13. $z^{2/3} + 2z^{1/3} - 3 = 0$

Section 8.4

For Exercises 14–15, graph the functions.

14. $f(x) = -x^2 - 1$
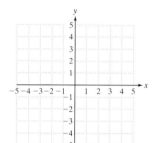

15. $g(x) = (x - 1)^2 + 2$
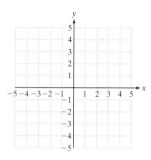

Section 8.5

16. Given: $f(x) = x^2 + 4x - 12$

    a. Write the equation for the function in the form $f(x) = a(x + h)^2 + k$.

    b. Determine the vertex.

    c. Find the x- and y-intercepts.

    d. Determine the maximum or minimum value.

    e. Determine the axis of symmetry.

17. An object is launched into the air from the top of a 240-ft building with an initial velocity of 32 ft/sec. The height of the object after t sec is given by $h(t) = -16t^2 + 32t + 240$.

    a. Find the height of the object after 3 sec.

    b. Find the maximum height reached by the object.

    c. Determine the time when the object hits the ground. (*Hint:* The height is 0.)

section 8.1 Completing the Square and the Square Root Property

1. Solving Quadratic Equations by Using the Square Root Property

In Section 5.7, we learned to solve quadratic equations by factoring and applying the zero product rule; however, the zero product rule can be used only if the equation is factorable. In this section and Section 8.2, we will learn two techniques to solve *all* quadratic equations, factorable and nonfactorable.

The first technique will use the **square root property**.

Objectives
1. Solving Quadratic Equations by Using the Square Root Property
2. Solving Quadratic Equations by Completing the Square
3. Literal Equations

The Square Root Property

For any real number, k, if $x^2 = k$, then $x = \sqrt{k}$ or $x = -\sqrt{k}$.

Note: The solution may also be written as $x = \pm\sqrt{k}$, read x equals "plus or minus the square root of k."

---

example 1   Solving Quadratic Equations by Using the Square Root Property

Use the square root property to solve the equations.

a. $x^2 = 81$ **b.** $3x^2 + 75 = 0$ **c.** $(w + 3)^2 = 20$

Solution:

a. $x^2 = 81$ The equation is in the form $x^2 = k$.

$x = \pm\sqrt{81}$ Apply the square root property.

$x = \pm 9$

The solutions are $x = 9$ and $x = -9$.

b. $3x^2 + 75 = 0$ Rewrite the equation to fit the form $x^2 = k$.

$3x^2 = -75$

$x^2 = -25$ The equation is now in the form $x^2 = k$.

$x = \pm\sqrt{-25}$ Apply the square root property.

$= \pm 5i$

Skill Practice

Use the square root property to solve the equations.

1. $a^2 = 100$
2. $b^2 + 7 = 0$
3. $8x^2 + 72 = 0$
4. $(t - 5)^2 = 18$

Tip: The equation $x^2 = 81$ can also be solved by using the zero product rule.

$x^2 = 81$
$x^2 - 81 = 0$
$(x - 9)(x + 9) = 0$
$x = 9$ or $x = -9$

Avoiding Mistakes: A common mistake is to forget the $\pm$ symbol when solving the equation $x^2 = k$:

$$x = \pm\sqrt{k}$$

The solutions are $x = 5i$ and $x = -5i$.

Check: $x = 5i$
$3x^2 + 75 = 0$
$3(5i)^2 + 75 \stackrel{?}{=} 0$
$3(25i^2) + 75 \stackrel{?}{=} 0$
$3(-25) + 75 \stackrel{?}{=} 0$
$-75 + 75 = 0$ ✓

Check: $x = -5i$
$3x^2 + 75 = 0$
$3(-5i)^2 + 75 \stackrel{?}{=} 0$
$3(25i^2) + 75 \stackrel{?}{=} 0$
$3(-25) + 75 \stackrel{?}{=} 0$
$-75 + 75 = 0$ ✓

Answers
1. $a = \pm 10$
2. $b = \pm i\sqrt{7}$
3. $x = \pm 3i$
4. $t = 5 \pm 3\sqrt{2}$

c. $(w + 3)^2 = 20$ The equation is in the form $x^2 = k$, where $x = (w + 3)$.

$w + 3 = \pm\sqrt{20}$ Apply the square root property.

$w + 3 = \pm\sqrt{2^2 \cdot 5}$ Simplify the radical.

$w + 3 = \pm 2\sqrt{5}$

$w = -3 \pm 2\sqrt{5}$ Solve for w.

The solutions are $w = -3 + 2\sqrt{5}$ and $w = -3 - 2\sqrt{5}$.

Check: $w = -3 + 2\sqrt{5}$
$(w + 3)^2 = 20$
$(-3 + 2\sqrt{5} + 3)^2 \stackrel{?}{=} 20$
$(2\sqrt{5})^2 \stackrel{?}{=} 20$
$4 \cdot 5 \stackrel{?}{=} 20$
$20 = 20$ ✓

Check: $w = -3 - 2\sqrt{5}$
$(w + 3)^2 = 20$
$(-3 - 2\sqrt{5} + 3)^2 \stackrel{?}{=} 20$
$(-2\sqrt{5})^2 \stackrel{?}{=} 20$
$4 \cdot 5 \stackrel{?}{=} 20$
$20 = 20$ ✓

2. Solving Quadratic Equations by Completing the Square

In Example 1(c), we used the square root property to solve an equation where the square of a binomial was equal to a constant.

$$(w + 3)^2 = 20$$
$$w + 3 = \pm\sqrt{20}$$
$$w = -3 \pm 2\sqrt{5}$$

In general, an equation of the form $(x + h)^2 = k$ can be solved by using the square root property. Furthermore, any equation $ax^2 + bx + c = 0$ $(a \neq 0)$ can be rewritten in the form $(x + h)^2 = k$ by using a process called **completing the square**.

We begin our discussion of completing the square with some vocabulary. For a trinomial $ax^2 + bx + c$ $(a \neq 0)$ the term ax^2 is called the quadratic term. The term bx is called the linear term, and the term c is called the constant term.

Next, notice that the factored form of a perfect square trinomial is the square of a binomial.

Perfect Square Trinomial	Factored Form
$x^2 + 10x + 25$	$(x + 5)^2$
$t^2 - 6t + 9$	$(t - 3)^2$
$p^2 - 14p + 49$	$(p - 7)^2$

Furthermore, for a perfect square trinomial with a leading coefficient of 1, the constant term is the square of one-half the coefficient of the linear term. For example:

$x^2 + 10x + 25$: $\left[\frac{1}{2}(10)\right]^2 = (5)^2 = 25$

$t^2 - 6t + 9$: $\left[\frac{1}{2}(-6)\right]^2 = (-3)^2 = 9$

$p^2 - 14p + 49$: $\left[\frac{1}{2}(-14)\right]^2 = (-7)^2 = 49$

In general, an expression of the form $x^2 + bx$ will be a perfect square trinomial if the square of one-half the linear term coefficient $(\frac{1}{2}b)^2$ is added to the expression.

Concept Connections

5. Which of the equations are in a form such that the square root property can be used directly to solve the equation?
   a. $t^2 = -10$
   b. $b^2 + 10b + 9 = 0$
   c. $(2x + 1)^2 = 8$

Answer
5. a and c

example 2 — Completing the Square

Complete the square for each expression. Then factor the expression as the square of a binomial.

a. $x^2 + 12x$ **b.** $x^2 - 26x$ **c.** $x^2 + 11x$ **d.** $x^2 - \dfrac{4}{7}x$

Solution:

The expressions are in the form $x^2 + bx$. Add the square of one-half the linear term coefficient $(\tfrac{1}{2}b)^2$.

a. $x^2 + 12x$

$x^2 + 12x + 36$ \qquad Add $\tfrac{1}{2}$ of 12, squared: $[\tfrac{1}{2}(12)]^2 = (6)^2 = 36$.

$(x + 6)^2$ \qquad Factored form

b. $x^2 - 26x$

$x^2 - 26x + 169$ \qquad Add $\tfrac{1}{2}$ of -26, squared: $[\tfrac{1}{2}(-26)]^2 = (-13)^2 = 169$.

$(x - 13)^2$ \qquad Factored form

c. $x^2 + 11x$

$x^2 + 11x + \dfrac{121}{4}$ \qquad Add $\tfrac{1}{2}$ of 11, squared: $[\tfrac{1}{2}(11)]^2 = (\tfrac{11}{2})^2 = \tfrac{121}{4}$.

$\left(x + \dfrac{11}{2}\right)^2$ \qquad Factored form

d. $x^2 - \dfrac{4}{7}x$

$x^2 - \dfrac{4}{7}x + \dfrac{4}{49}$ \qquad Add $\tfrac{1}{2}$ of $-\tfrac{4}{7}$, squared. $[\tfrac{1}{2}(-\tfrac{4}{7})]^2 = (-\tfrac{2}{7})^2 = \tfrac{4}{49}$

$\left(x - \dfrac{2}{7}\right)^2$ \qquad Factored form

Skill Practice

Complete the square for each expression and then factor as the square of a binomial.

6. $x^2 + 20x + \underline{}$
7. $y^2 - 16y + \underline{}$
8. $a^2 - 5a + \underline{}$
9. $w^2 + \dfrac{7}{3}w + \underline{}$

The process of completing the square can be used to write a quadratic equation $ax^2 + bx + c = 0$ $(a \neq 0)$ in the form $(x - h)^2 = k$. Then the square root property can be used to solve the equation. The following steps outline the procedure.

Solving a Quadratic Equation in the Form $ax^2 + bx + c = 0$ $(a \neq 0)$ by Completing the Square and Applying the Square Root Property

1. Divide both sides by a to make the leading coefficient 1.
2. Isolate the variable terms on one side of the equation.
3. Complete the square. (Add the square of one-half the linear term coefficient to both sides of the equation. Then factor the resulting perfect square trinomial.)
4. Apply the square root property and solve for x.

Answers

6. 100; $(x + 10)^2$
7. 64; $(y - 8)^2$
8. $\dfrac{25}{4}$; $\left(a - \dfrac{5}{2}\right)^2$
9. $\dfrac{49}{36}$; $\left(w + \dfrac{7}{6}\right)^2$

Skill Practice

Solve by completing the square and applying the square root property.

10. $z^2 - 4z - 2 = 0$
11. $4x^2 - 16x + 20 = 0$
12. $2y(y - 1) = 3 - y$

example 3 — Solving Quadratic Equations by Completing the Square and Applying the Square Root Property

Solve the quadratic equations by completing the square and applying the square root property.

a. $x^2 - 6x + 13 = 0$ **b.** $2x(2x - 10) = -30 + 6x$

Solution:

a. $x^2 - 6x + 13 = 0$

$x^2 - 6x + 13 = 0$ **Step 1:** Since the leading coefficient a is equal to 1, we do not have to divide by a. We can proceed to step 2.

$x^2 - 6x = -13$ **Step 2:** Isolate the variable terms on one side.

$x^2 - 6x + 9 = -13 + 9$ **Step 3:** To complete the square, add $[\frac{1}{2}(-6)]^2 = 9$ to both sides of the equation.

$(x - 3)^2 = -4$ Factor the perfect square trinomial.

$x - 3 = \pm\sqrt{-4}$ **Step 4:** Apply the square root property.

$x - 3 = \pm 2i$ Simplify the radical.

$x = 3 \pm 2i$ Solve for x.

The solutions are imaginary numbers and can be written as $x = 3 + 2i$ and $x = 3 - 2i$.

b. $2x(2x - 10) = -30 + 6x$

$4x^2 - 20x = -30 + 6x$ Clear parentheses.

$4x^2 - 26x + 30 = 0$ Write the equation in the form $ax^2 + bx + c = 0$.

$\dfrac{4x^2}{4} - \dfrac{26x}{4} + \dfrac{30}{4} = \dfrac{0}{4}$ **Step 1:** Divide both sides by the leading coefficient 4.

$x^2 - \dfrac{13}{2}x + \dfrac{15}{2} = 0$

$x^2 - \dfrac{13}{2}x = -\dfrac{15}{2}$ **Step 2:** Isolate the variable terms on one side.

$x^2 - \dfrac{13}{2}x + \dfrac{169}{16} = -\dfrac{15}{2} + \dfrac{169}{16}$ **Step 3:** Add $[\frac{1}{2}(-\frac{13}{2})]^2 = (-\frac{13}{4})^2 = \frac{169}{16}$ to both sides.

$\left(x - \dfrac{13}{4}\right)^2 = -\dfrac{120}{16} + \dfrac{169}{16}$ Factor the perfect square trinomial. Rewrite the right-hand side with a common denominator.

$\left(x - \dfrac{13}{4}\right)^2 = \dfrac{49}{16}$

Answers

10. $z = 2 \pm \sqrt{6}$
11. $x = 2 \pm i$
12. $y = \dfrac{3}{2}$ and $y = -1$

$$x - \frac{13}{4} = \pm\sqrt{\frac{49}{16}}$$ **Step 4:** Apply the square root property.

$$x - \frac{13}{4} = \pm\frac{7}{4}$$ Simplify the radical.

$$x = \frac{13}{4} \pm \frac{7}{4}$$

$$x = \frac{13}{4} + \frac{7}{4} = \frac{20}{4} = 5$$

$$x = \frac{13}{4} - \frac{7}{4} = \frac{6}{4} = \frac{3}{2}$$

The solutions are rational numbers: $x = \frac{3}{2}$ and $x = 5$. The check is left to the reader.

Tip: In general, if the solutions to a quadratic equation are rational numbers, the equation can be solved by factoring and using the zero product rule. Consider the equation from Example 3(b).

$$2x(2x - 10) = -30 + 6x$$
$$4x^2 - 20x = -30 + 6x$$
$$4x^2 - 26x + 30 = 0$$
$$2(2x^2 - 13x + 15) = 0$$
$$2(x - 5)(2x - 3) = 0$$
$$x = 5 \quad \text{or} \quad x = \frac{3}{2}$$

3. Literal Equations

example 4 Solving a Literal Equation

Ignoring air resistance, the distance d (in meters) that an object falls in t sec is given by the equation

$$d = 4.9t^2 \quad \text{where } t \geq 0$$

a. Solve the equation for t. Do not rationalize the denominator.

b. Using the equation from part (a), determine the amount of time required for an object to fall 500 m. Round to the nearest second.

Solution:

a. $d = 4.9t^2$

$\dfrac{d}{4.9} = t^2$ Isolate the quadratic term. The equation is in the form $t^2 = k$.

$t = \pm\sqrt{\dfrac{d}{4.9}}$ Apply the square root property.

$= \sqrt{\dfrac{d}{4.9}}$ Because $t \geq 0$, reject the negative solution.

Skill Practice

The formula for the area of a circle is $A = \pi r^2$, where r is the radius.

13. Solve for r. (Do not rationalize the denominator.)

14. Use the equation in Exercise 13 to find the radius when the area is 15.7 cm². (Use 3.14 for π and round to 2 decimal places.)

Answers

13. $r = \sqrt{\dfrac{A}{\pi}}$

14. The radius is $\sqrt{5} \approx 2.24$ cm.

b. $t = \sqrt{\dfrac{d}{4.9}}$

$= \sqrt{\dfrac{500}{4.9}}$ Substitute $d = 500$.

$t \approx 10.1$

The object will require approximately 10.1 sec to fall 500 m.

section 8.1 Practice Exercises

Boost your GRADE at mathzone.com!
- Practice Problems
- Self-Tests
- NetTutor
- e-Professors
- Videos

Study Skills Exercises

1. Write down the number of hours devoted to study, work, classes, and entertainment for one week. Compare the amount of time devoted to studying for school to the time spent on entertainment, such as watching TV and movies.

	Work	Classes	Studying	Entertainment
Monday				
Tuesday				
Wednesday				
Thursday				
Friday				
Saturday				
Sunday				

2. Define the key terms.

   a. Square root property   **b. Completing the square**

Objective 1: Solving Quadratic Equations by Using the Square Root Property

For Exercises 3–18, solve the equations by using the square root property. **(See Example 1.)**

3. $x^2 = 100$
4. $y^2 = 4$
5. $a^2 = 5$
6. $k^2 - 7 = 0$

7. $3v^2 + 33 = 0$
8. $-2m^2 = 50$
9. $(p - 5)^2 = 9$
10. $(q + 3)^2 = 4$

11. $(x - 2)^2 - 5 = 0$
12. $(y + 3)^2 - 7 = 0$
13. $(h - 4)^2 = -8$
14. $(t + 5)^2 = -18$

15. $6p^2 - 3 = 2$

16. $15 = 4 + 3w^2$

17. $\left(x - \dfrac{3}{2}\right)^2 + \dfrac{7}{4} = 0$

18. $\left(m + \dfrac{4}{5}\right)^2 - \dfrac{3}{25} = 0$

19. State two methods that can be used to solve the equation $x^2 - 36 = 0$. Then solve the equation by using both methods.

20. State two methods that can be used to solve the equation $x^2 - 9 = 0$. Then solve the equation by using both methods.

Objective 2: Solving Quadratic Equations by Completing the Square

For Exercises 21–30, find the value of k so that the expression is a perfect square trinomial. Then factor the trinomial. **(See Example 2.)**

21. $x^2 - 6x + k$

22. $x^2 + 12x + k$

23. $t^2 + 8t + k$

24. $v^2 - 18v + k$

25. $c^2 - c + k$

26. $x^2 + 9x + k$

27. $y^2 + 5y + k$

28. $a^2 - 7a + k$

29. $b^2 + \dfrac{2}{5}b + k$

30. $m^2 - \dfrac{2}{7}m + k$

31. Summarize the steps used in solving a quadratic equation by completing the square and applying the square root property.

32. What types of quadratic equations can be solved by completing the square and applying the square root property?

For Exercises 33–52, solve the quadratic equation by completing the square and applying the square root property. **(See Example 3.)**

33. $t^2 + 8t + 15 = 0$

34. $m^2 + 6m + 8 = 0$

35. $x^2 + 6x = 16$

36. $x^2 - 4x = -3$

37. $p^2 + 4p + 6 = 0$

38. $q^2 + 2q + 2 = 0$

39. $y^2 - 3y - 10 = 0$

40. $-24 = -2y^2 + 2y$

41. $2a^2 + 4a + 5 = 0$

42. $3a^2 + 6a - 7 = 0$

43. $9x^2 - 36x + 40 = 0$

44. $9y^2 - 12y + 5 = 0$

45. $p^2 - \dfrac{2}{5}p = \dfrac{2}{25}$

46. $n^2 - \dfrac{2}{3}n = \dfrac{1}{9}$

47. $(2w + 5)(w - 1) = 2$

48. $(3p - 5)(p + 1) = -3$

49. $n(n - 4) = 7$

50. $m(m + 10) = 2$

51. $2x(x + 6) = 14$

52. $3x(x - 2) = 24$

Objective 3: Literal Equations

53. The distance (in feet) that an object falls in t sec is given by the equation $d = 16t^2$, where $t \geq 0$. (See Example 4.)

   a. Solve the equation for t.

   b. Using the equation from part (a), determine the amount of time required for an object to fall 1024 ft. Round to the nearest second.

54. The volume of a can that is 4 in. tall is given by the equation $V = 4\pi r^2$, where r is the radius of the can, measured in inches.

   a. Solve the equation for r. Do not rationalize the denominator.

   b. Using the equation from part (a), determine the radius of a can with volume of 12.56 in.^{3}. Use 3.14 for π.

For Exercises 55–60, solve for the indicated variable.

55. $A = \pi r^2$ for r $(r > 0)$

56. $E = mc^2$ for c $(c > 0)$

57. $a^2 + b^2 + c^2 = d^2$ for a $(a > 0)$

58. $a^2 + b^2 = c^2$ for b $(b > 0)$

59. $V = \dfrac{1}{3}\pi r^2 h$ for r $(r > 0)$

60. $V = \dfrac{1}{3}s^2 h$ for s

61. A corner shelf is to be made from a triangular piece of plywood, as shown in the diagram. Find the distance x that the shelf will extend along the walls. Assume that the walls are at right angles. Round the answer to a tenth of a foot.

62. A square has an area of 50 in.^{2} What are the lengths of the sides? (Round to 1 decimal place.)

63. The amount of money A in an account with an interest rate r compounded annually is given by

$$A = P(1 + r)^t$$

where P is the initial principal and t is the number of years the money is invested.

a. If a $10,000 investment grows to $11,664 after 2 years, find the interest rate.

b. If a $6000 investment grows to $7392.60 after 2 years, find the interest rate.

c. Jamal wants to invest $5000. He wants the money to grow to at least $6500 in 2 years to cover the cost of his son's first year at college. What interest rate does Jamal need for his investment to grow to $6500 in 2 years? Round to the nearest percent.

64. The volume of a box with a square bottom and a height of 4 in. is given by $V(x) = 4x^2$, where x is the length (in inches) of the sides of the bottom of the box.

    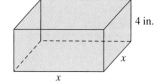

    a. If the volume of the box is 289 in.^{3}, find the dimensions of the box.

    b. Are there two possible answers to part (a)? Why or why not?

65. A textbook company has discovered that the profit for selling its books is given by

    $$P(x) = -\frac{1}{8}x^2 + 5x$$

    where x is the number of textbooks produced (in thousands) and $P(x)$ is the corresponding profit (in thousands of dollars). The graph of the function is shown at right.

    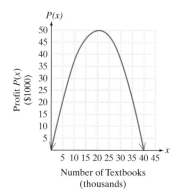

    a. Approximate the number of books required to make a profit of $20,000. [*Hint:* Let $P(x) = 20$. Then complete the square to solve for x.] Round to 1 decimal place.

    b. Why are there two answers to part (a)?

66. If we ignore air resistance, the distance (in feet) that an object travels in free fall can be approximated by $d(t) = 16t^2$, where t is the time in seconds after the object was dropped.

    a. If the CN Tower in Toronto is 1815 ft high, how long will it take an object to fall from the top of the building? Round to 1 decimal place.

    b. If the Renaissance Tower in Dallas is 886 ft high, how long will it take an object to fall from the top of the building? Round to 1 decimal place.

Objectives

1. Derivation of the Quadratic Formula
2. Solving Quadratic Equations by Using the Quadratic Formula
3. Using the Quadratic Formula in Applications
4. Discriminant
5. Review of the Methods to Solve a Quadratic Equation

section 8.2 Quadratic Formula

1. Derivation of the Quadratic Formula

If we solve a general quadratic equation $ax^2 + bx + c = 0$ $(a \neq 0)$ by completing the square and using the square root property, the result is a formula that gives the solutions for x in terms of a, b, and c.

$$ax^2 + bx + c = 0$$ Begin with a quadratic equation in standard form.

$$\frac{ax^2}{a} + \frac{bx}{a} + \frac{c}{a} = \frac{0}{a}$$ Divide by the leading coefficient.

$$x^2 + \frac{b}{a}x + \frac{c}{a} = 0$$

$$x^2 + \frac{b}{a}x = -\frac{c}{a}$$ Isolate the terms containing x.

$$x^2 + \frac{b}{a}x + \left(\frac{1}{2} \cdot \frac{b}{a}\right)^2 = \left(\frac{1}{2} \cdot \frac{b}{a}\right)^2 - \frac{c}{a}$$ Add the square of $\frac{1}{2}$ the linear term coefficient to both sides of the equation.

$$\left(x + \frac{b}{2a}\right)^2 = \frac{b^2}{4a^2} - \frac{c}{a}$$ Factor the left side as a perfect square.

$$\left(x + \frac{b}{2a}\right)^2 = \frac{b^2 - 4ac}{4a^2}$$ Combine fractions on the right side by getting a common denominator.

$$x + \frac{b}{2a} = \pm\sqrt{\frac{b^2 - 4ac}{4a^2}}$$ Apply the square root property.

$$x + \frac{b}{2a} = \frac{\pm\sqrt{b^2 - 4ac}}{2a}$$ Simplify the denominator.

$$x = -\frac{b}{2a} \pm \frac{\sqrt{b^2 - 4ac}}{2a}$$ Subtract $\frac{b}{2a}$ from both sides.

$$= \frac{-b \pm \sqrt{b^2 - 4ac}}{2a}$$ Combine fractions.

The solution to the equation $ax^2 + bx + c = 0$ for x in terms of the coefficients a, b, and c is given by the **quadratic formula**.

Concept Connections

1. Given the quadratic equation $6x^2 - 2x + 7 = 0$, identify the values of a, b, and c.

The Quadratic Formula

For any quadratic equation of the form $ax^2 + bx + c = 0$ $(a \neq 0)$ the solutions are

$$x = \frac{-b \pm \sqrt{b^2 - 4ac}}{2a}$$

Answer

1. $a = 6$; $b = -2$; $c = 7$

2. Solving Quadratic Equations by Using the Quadratic Formula

example 1 Solving a Quadratic Equation by Using the Quadratic Formula

Solve the quadratic equation by using the quadratic formula.

$$3x^2 + 8x = -5$$

Solution:

$$3x^2 + 8x = -5$$

$$3x^2 + 8x + 5 = 0 \quad \text{Write the equation in the form } ax^2 + bx + c = 0.$$

$$a = 3 \quad b = 8 \quad c = 5 \quad \text{Identify } a, b, \text{ and } c.$$

$$x = \frac{-(8) \pm \sqrt{(8)^2 - 4(3)(5)}}{2(3)} \quad \text{Apply the quadratic formula.}$$

$$= \frac{-8 \pm \sqrt{64 - 60}}{6} \quad \text{Simplify.}$$

$$= \frac{-8 \pm \sqrt{4}}{6}$$

$$= \frac{-8 \pm 2}{6}$$

There are two rational solutions.

$$x = \frac{-8 + 2}{6} = \frac{-6}{6} = -1$$

$$x = \frac{-8 - 2}{6} = \frac{-10}{6} = -\frac{5}{3}$$

Both solutions check in the original equation.

Skill Practice

Solve the equation by using the quadratic formula.

2. $6x^2 - 5x = 4$

---

example 2 Solving a Quadratic Equation by Using the Quadratic Formula

Solve the quadratic equation by using the quadratic formula.

$$x(x + 6) - 1 = 0$$

Solution:

$$x(x + 6) - 1 = 0$$

$$x^2 + 6x - 1 = 0 \quad \text{Write the equation in the form } ax^2 + bx + c = 0.$$

$$a = 1 \quad b = 6 \quad c = -1 \quad \text{Identify } a, b, \text{ and } c.$$

$$x = \frac{-(6) \pm \sqrt{(6)^2 - 4(1)(-1)}}{2(1)} \quad \text{Apply the quadratic formula.}$$

$$= \frac{-6 \pm \sqrt{36 + 4}}{2} \quad \text{Simplify.}$$

Skill Practice

Solve the equation by using the quadratic formula.

3. $y(y + 3) = 2$

Answers

2. $x = -\frac{1}{2}, x = \frac{4}{3}$

3. $y = \frac{-3 \pm \sqrt{17}}{2}$

$$= \frac{-6 \pm \sqrt{40}}{2}$$ The radical can be simplified.

$$= \frac{-6 \pm \sqrt{4 \cdot 10}}{2}$$

$$= \frac{-6 \pm 2\sqrt{10}}{2}$$

$$= \frac{2(-3 \pm \sqrt{10})}{2}$$ Simplify to lowest terms.

$$= -3 \pm \sqrt{10}$$ The solutions are irrational numbers.

The solutions can be written as

$$x = -3 + \sqrt{10} \approx -0.162 \quad \text{and} \quad x = -3 - \sqrt{10} \approx -6.162$$

Calculator Connections

Consider the equation $x(x + 6) - 1 = 0$ from Example 2. In standard form, this equation is written as $x^2 + 6x - 1 = 0$. Using the quadratic formula, we have

$$x = \frac{-(6) \pm \sqrt{(6)^2 - 4(1)(-1)}}{2(1)}$$

A calculator can be used to apply the quadratic formula directly; however, each solution must be entered separately. The solution can be checked on the calculator by using the *Ans* variable. This contains the result of the calculator's most recent computation.

$$x = \frac{-(6) + \sqrt{(6)^2 - 4(1)(-1)}}{2(1)} \approx 0.1622776602$$

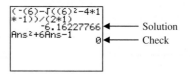
Solution
Check

$$x = \frac{-(6) - \sqrt{(6)^2 - 4(1)(-1)}}{2(1)} \approx -6.16227766$$

Solution
Check

3. Using the Quadratic Formula in Applications

example 3   Using the Quadratic Formula in an Application

A delivery truck travels south from Hartselle, Alabama, to Birmingham, Alabama, along Interstate 65. The truck then heads east to Atlanta, Georgia, along Interstate 20. The distance from Birmingham to Atlanta is 8 mi less than twice the distance from Hartselle to Birmingham. If the straight-line distance from Hartselle to Atlanta is 165 mi, find the distance from Hartselle to Birmingham and from Birmingham to Atlanta. (Round the answers to the nearest mile.)

Solution:

The motorist travels due south and then due east. Therefore, the three cities form the vertices of a right triangle (Figure 8-1).

Let x represent the distance between Hartselle and Birmingham.

Then $2x - 8$ represents the distance between Birmingham and Atlanta.

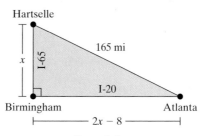

Figure 8-1

Use the Pythagorean theorem to establish a relationship among the three sides of the triangle.

$$(x)^2 + (2x - 8)^2 = (165)^2$$
$$x^2 + 4x^2 - 32x + 64 = 27{,}225$$
$$5x^2 - 32x - 27{,}161 = 0 \quad \text{Write the equation in the form } ax^2 + bx + c = 0.$$

$a = 5 \quad b = -32 \quad c = -27{,}161$ Identify a, b, and c.

$$x = \frac{-(-32) \pm \sqrt{(-32)^2 - 4(5)(-27{,}161)}}{2(5)} \quad \text{Apply the quadratic formula.}$$

$$x = \frac{32 \pm \sqrt{1024 + 543{,}220}}{10} \quad \text{Simplify.}$$

$$x = \frac{32 \pm \sqrt{544{,}244}}{10}$$

$$x = \frac{32 + \sqrt{544{,}244}}{10} \approx 76.97 \text{ mi} \quad \text{or}$$

$$x = \frac{32 - \sqrt{544{,}244}}{10} \approx -70.57 \text{ mi} \quad \text{We reject the negative distance.}$$

Recall that x represents the distance from Hartselle to Birmingham; therefore, to the nearest mile, the distance between Hartselle and Birmingham is 77 mi.

The distance between Birmingham and Atlanta is $2x - 8 = 2(77) - 8 = 146$ mi.

Skill Practice

4. Steve and Tammy leave a campground, hiking on two different trails. Steve heads south and Tammy heads east. By lunchtime they are 9 mi apart. Steve walked 3 mi more than twice as many miles as Tammy. Find the distance each person hiked. (Round to the nearest tenth of a mile.)

Answer

4. Tammy hiked 2.8 mi, and Steve hiked 8.6 mi.

Skill Practice

5. An object is launched from the top of a 96-ft building with an initial velocity of 64 ft/sec. The height $h(t)$ of the rocket is given by

$$h(t) = -16t^2 + 64t + 96$$

Find the time it takes for the object to hit the ground. [*Hint:* $h(t) = 0$ when the object hits the ground.]

example 4 — Analyzing a Quadratic Function

A model rocket is launched straight up from the side of a 144-ft cliff (Figure 8-2). The initial velocity is 112 ft/sec. The height of the rocket $h(t)$ is given by

$$h(t) = -16t^2 + 112t + 144$$

where $h(t)$ is measured in feet and t is the time in seconds. Find the time(s) at which the rocket is 208 ft above the ground.

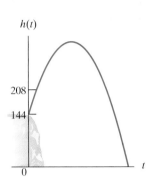

Figure 8-2

Solution:

$$h(t) = -16t^2 + 112t + 144$$

$$208 = -16t^2 + 112t + 144 \quad \text{Substitute 208 for } h(t).$$

$$16t^2 - 112t + 64 = 0 \quad \text{Write the equation in the form } at^2 + bt + c = 0.$$

$$\frac{16t^2}{16} - \frac{112t}{16} + \frac{64}{16} = \frac{0}{16} \quad \text{Divide by 16. This makes the coefficients smaller, and it is less cumbersome to apply the quadratic formula.}$$

$$t^2 - 7t + 4 = 0 \quad \text{The equation is not factorable. Apply the quadratic formula.}$$

$$t = \frac{-(-7) \pm \sqrt{(-7)^2 - 4(1)(4)}}{2(1)} \quad \text{Let } a = 1, b = -7, \text{ and } c = 4.$$

$$= \frac{7 \pm \sqrt{33}}{2}$$

$$t = \frac{7 + \sqrt{33}}{2} \approx 6.37$$

$$t = \frac{7 - \sqrt{33}}{2} \approx 0.63$$

The rocket will reach a height of 208 ft after approximately 0.63 sec (on the way up) and after 6.37 sec (on the way down).

4. Discriminant

The solutions to a quadratic equation may be rational, irrational, or imaginary numbers. The *number* and the *type* of solutions can be determined by noting the value of the square root term in the quadratic formula. The radicand of the square root $b^2 - 4ac$ is called the discriminant.

Answer

5. $2 + \sqrt{10} \approx 5.16$ sec

Using the Discriminant to Determine the Number and Type of Solutions to a Quadratic Equation

Consider the equation $ax^2 + bx + c = 0$, where a, b, and c are rational numbers and $a \neq 0$. The expression $b^2 - 4ac$ is called the **discriminant**. Furthermore,

1. If $b^2 - 4ac > 0$, then there will be two real solutions.
   a. If $b^2 - 4ac$ is a perfect square, the solutions will be rational numbers.
   b. If $b^2 - 4ac$ is not a perfect square, the solutions will be irrational numbers.
2. If $b^2 - 4ac < 0$, then there will be two imaginary solutions.
3. If $b^2 - 4ac = 0$, then there will be one rational solution.

example 5   Using the Discriminant

Use the discriminant to determine the type and number of solutions for each equation.

a. $2x^2 - 5x + 9 = 0$ b. $3x^2 = -x + 2$
c. $-2x(2x - 3) = -1$ d. $3.6x^2 = -1.2x - 0.1$

Solution:

For each equation, first write the equation in standard form $ax^2 + bx + c = 0$. Then determine the discriminant.

Equation	Discriminant	Solution Type and Number
a. $2x^2 - 5x + 9 = 0$	$b^2 - 4ac$ $= (-5)^2 - 4(2)(9)$ $= 25 - 72$ $= -47$	Because $-47 < 0$, there will be two imaginary solutions.
b. $3x^2 = -x + 2$ $3x^2 + x - 2 = 0$	$b^2 - 4ac$ $= (1)^2 - 4(3)(-2)$ $= 1 - (-24)$ $= 25$	Because $25 > 0$ and 25 is a perfect square, there will be two rational solutions.
c. $-2x(2x - 3) = -1$ $-4x^2 + 6x = -1$ $-4x^2 + 6x + 1 = 0$	$b^2 - 4ac$ $= (6)^2 - 4(-4)(1)$ $= 36 - (-16)$ $= 52$	Because $52 > 0$, but 52 is *not* a perfect square, there will be two irrational solutions.

Skill Practice

For Exercises 6–9,
Determine the discriminant. Use the discriminant to determine the type and number of solutions for the equation.

6. $3y^2 + y + 3 = 0$
7. $4t^2 = 6t - 2$
8. $3t(t + 1) = 9$
9. $\frac{2}{3}x^2 - \frac{2}{3}x + \frac{1}{6} = 0$

Answers
6. -35; two imaginary solutions
7. 4; two rational solutions
8. 117; two irrational solutions
9. 0; one rational solution

d.
$$3.6x^2 = -1.2x - 0.1$$
$$3.6x^2 + 1.2x + 0.1 = 0$$

$b^2 - 4ac$
$= (1.2)^2 - 4(3.6)(0.1)$
$= 1.44 - 1.44$
$= 0$

Because the discriminant equals 0, there will be only one rational solution.

■

With the discriminant we can determine the number of real-valued solutions to the equation $ax^2 + bx + c = 0$, and thus the number of x-intercepts to the function $f(x) = ax^2 + bx + c$. The following illustrations show the graphical interpretation of the three cases of the discriminant.

$f(x) = x^2 - 4x + 3$

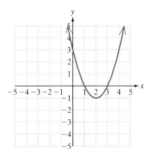

Use $x^2 - 4x + 3 = 0$ to find the value of the discriminant.

$$b^2 - 4ac = (-4)^2 - 4(1)(3)$$
$$= 4$$

Since $b^2 - 4ac = 4$,
$b^2 - 4ac > 0$
There are two real solutions and two x-intercepts, $(1, 0)$ and $(3, 0)$.

$f(x) = x^2 - x + 1$

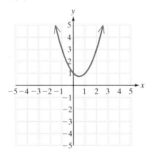

Use $x^2 - x + 1 = 0$ to find the value of the discriminant.

$$b^2 - 4ac = (-1)^2 - 4(1)(1)$$
$$= -3$$

Since $b^2 - 4ac = -3$,
$b^2 - 4ac < 0$
There are no real solutions and no x-intercepts.

$f(x) = x^2 - 2x + 1$

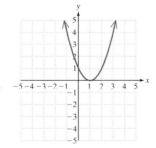

Use $x^2 - 2x + 1 = 0$ to find the value of the discriminant.

$$b^2 - 4ac = (-2)^2 - 4(1)(1)$$
$$= 0$$

$b^2 - 4ac = 0$
There is one real solution and one x-intercept $(1, 0)$.

example 6 Finding the x- and y-Intercepts of a Quadratic Function

Given: $f(x) = \dfrac{1}{4}x^2 + \dfrac{1}{4}x + \dfrac{1}{2}$

a. Find the y-intercept. **b.** Find the x-intercept(s).

Skill Practice

10. Find the x- and y-intercepts of the function given by

$$f(x) = \dfrac{1}{2}x^2 - x - \dfrac{5}{2}$$

Solution:

a. The y-intercept is the value of $f(0) = \dfrac{1}{4}(0)^2 + \dfrac{1}{4}(0) + \dfrac{1}{2}$

$$= \dfrac{1}{2}$$

The y-intercept is $(0, \tfrac{1}{2})$.

b. The x-intercepts are the real solutions to the equation $f(x) = 0$. In this case, we have

$$f(x) = \dfrac{1}{4}x^2 + \dfrac{1}{4}x + \dfrac{1}{2} = 0$$

$$4\left(\dfrac{1}{4}x^2 + \dfrac{1}{4}x + \dfrac{1}{2}\right) = 4(0) \quad \text{Multiply by 4 to clear fractions.}$$

$$x^2 + x + 2 = 0 \quad \text{The equation is in the form } ax^2 + bx + c = 0, \text{ where } a = 1, b = 1, \text{ and } c = 2.$$

$$x = \dfrac{-(1) \pm \sqrt{(1)^2 - 4(1)(2)}}{2(1)} \quad \text{Apply the quadratic formula.}$$

$$= \dfrac{-1 \pm \sqrt{-7}}{2} \quad \text{Simplify.}$$

$$= -\dfrac{1}{2} \pm \dfrac{\sqrt{7}}{2} i$$

These solutions are *imaginary numbers*. Because there are no real solutions to the equation $f(x) = 0$, the function has no x-intercepts. The function is graphed in Figure 8-3.

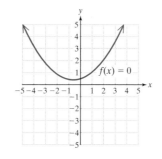

Figure 8-3

5. Review of the Methods to Solve a Quadratic Equation

Three methods have been presented to solve quadratic equations.

Methods to Solve a Quadratic Equation
- Factor and use the zero product rule (Section 5.7).
- Use the square root property. Complete the square, if necessary (Section 8.1).
- Use the quadratic formula (Section 8.2).

Using the zero product rule is often the simplest method, but it works only if you can factor the equation. The square root property and the quadratic formula can be used to solve any quadratic equation. Before solving a quadratic equation, take a minute to analyze it first. Each problem must be evaluated individually before choosing the most efficient method to find its solutions.

Answer

10. x-intercepts:
$(1 + \sqrt{6}, 0) \approx (3.45, 0)$,
$(1 - \sqrt{6}, 0) \approx (-1.45, 0)$;
y-intercept: $\left(0, -\dfrac{5}{2}\right)$

Skill Practice

Solve the quadratic equations by using any method.

11. $2t(t-1) + t^2 = 5$

12. $0.2x^2 + 1.1x = -0.3$

13. $(2y-6)^2 = -12$

example 7 — Solving Quadratic Equations by Using Any Method

Solve the quadratic equations by using any method.

a. $(x+3)^2 + x^2 - 9x = 8$ **b.** $\dfrac{x^2}{2} + \dfrac{5}{2} = -x$ **c.** $(p-2)^2 - 11 = 0$

Solution:

a.
$$(x+3)^2 + x^2 - 9x = 8$$
$$x^2 + 6x + 9 + x^2 - 9x - 8 = 0 \quad \text{Clear parentheses and write the equation in the form } ax^2 + bx + c = 0.$$
$$2x^2 - 3x + 1 = 0 \quad \text{This equation is factorable.}$$
$$(2x-1)(x-1) = 0 \quad \text{Factor.}$$
$$2x - 1 = 0 \quad \text{or} \quad x - 1 = 0 \quad \text{Apply the zero product rule.}$$
$$x = \tfrac{1}{2} \quad \text{or} \quad x = 1 \quad \text{Solve for } x.$$

This equation could have been solved by using any of the three methods, but factoring was the most efficient method.

b.
$$\dfrac{x^2}{2} + \dfrac{5}{2} = -x \quad \text{Clear fractions and write the equation in the form } ax^2 + bx + c = 0.$$
$$x^2 + 5 = -2x$$
$$x^2 + 2x + 5 = 0 \quad \text{This equation does not factor. Use the quadratic formula.}$$
$$a = 1 \quad b = 2 \quad c = 5 \quad \text{Identify } a, b, \text{ and } c.$$
$$x = \dfrac{-(2) \pm \sqrt{(2)^2 - 4(1)(5)}}{2(1)} \quad \text{Apply the quadratic formula.}$$
$$= \dfrac{-2 \pm \sqrt{-16}}{2} \quad \text{Simplify.}$$
$$= \dfrac{-2 \pm 4i}{2} \quad \text{Simplify the radical.}$$
$$= -1 \pm 2i \quad \text{Reduce to lowest terms.}$$
$$\dfrac{-2 \pm 4i}{2} = \dfrac{\cancel{2}(-1 \pm 2i)}{\cancel{2}}$$

This equation could also have been solved by completing the square and applying the square root property.

c.
$$(p-2)^2 - 11 = 0$$
$$(p-2)^2 = 11 \quad \text{The equation is in the form } x^2 = k, \text{ where } x = (p-2).$$
$$p - 2 = \pm\sqrt{11} \quad \text{Apply the square root property.}$$
$$p = 2 \pm \sqrt{11} \quad \text{Solve for } p.$$

This problem could have been solved by the quadratic formula, but that would have involved clearing parentheses and collecting *like* terms first.

Answers

11. $t = -1;\ t = \dfrac{5}{3}$

12. $x = \dfrac{-11 \pm \sqrt{97}}{4}$

13. $y = 3 \pm i\sqrt{3}$

section 8.2 Practice Exercises

Boost *your* GRADE at mathzone.com!

- Practice Problems
- Self-Tests
- NetTutor
- e-Professors
- Videos

Study Skills Exercises

1. Set goals for studying. Before you begin a homework assignment, approximate the time that it will take you to complete the assignment. This type of goal can make you more efficient in your work. Write down the time you expect it will take to finish the homework for this section.

2. Define the key terms.
   a. Quadratic formula
   b. Discriminant

Review Exercises

For Exercises 3–8, solve by completing the square and using the square root property.

3. $(x + 5)^2 = 49$
4. $16 = (2x - 3)^2$
5. $2x^2 + 12x = 16$
6. $3x^2 - 30x = 3$
7. $x^2 - 2x + 10 = 0$
8. $x^2 - 4x + 15 = 0$

For Exercises 9–12, simplify the expressions.

9. $\dfrac{16 - \sqrt{640}}{4}$
10. $\dfrac{18 + \sqrt{180}}{3}$
11. $\dfrac{14 - \sqrt{-147}}{7}$
12. $\dfrac{10 - \sqrt{-175}}{5}$

Objective 2: Solving Quadratic Equations by Using the Quadratic Formula

13. What form should a quadratic equation be in so that the quadratic formula can be applied?

14. Describe the circumstances in which the square root property can be used as a method for solving a quadratic equation.

15. Write the quadratic formula from memory.

For Exercises 16–41, solve the equation by using the quadratic formula. **(See Examples 1–2.)**

16. $a^2 + 11a - 12 = 0$
17. $5b^2 - 14b - 3 = 0$
18. $9y^2 - 2y + 5 = 0$
19. $2t^2 + 3t - 7 = 0$
20. $12p^2 - 4p + 5 = 0$
21. $5n^2 - 4n + 6 = 0$
22. $z^2 = 2z + 35$
23. $12x^2 - 5x = 2$
24. $a^2 + 3a = 8$

25. $k^2 + 4 = 6k$

26. $25x^2 - 20x + 4 = 0$

27. $9y^2 = -12y - 4$

28. $w^2 - 6w + 14 = 0$

29. $2m^2 + 3m = -2$

30. $(x + 2)(x - 3) = 1$

31. $3y(y + 1) - 7y(y + 2) = 6$

32. $-4a^2 - 2a + 3 = 0$

33. $-2m^2 - 5m + 3 = 0$

34. $\frac{1}{2}y^2 + \frac{2}{3} = -\frac{2}{3}y$

35. $\frac{2}{3}p^2 - \frac{1}{6}p + \frac{1}{2} = 0$

36. $\frac{1}{5}h^2 + h + \frac{3}{5} = 0$

(*Hint:* Clear the fractions first.)

37. $\frac{1}{4}w^2 + \frac{7}{4}w + 1 = 0$

38. $0.01x^2 + 0.06x + 0.08 = 0$

(*Hint:* Clear the decimals first.)

39. $0.5y^2 - 0.7y + 0.2 = 0$

40. $0.3t^2 + 0.7t - 0.5 = 0$

41. $0.01x^2 + 0.04x - 0.07 = 0$

42. a. Factor $x^3 - 27$.

b. Use the zero product rule and the quadratic formula to solve $x^3 - 27 = 0$. There should be three solutions (one real and two imaginary).

43. a. Factor $64x^3 + 1$.

b. Use the zero product rule and the quadratic formula to solve $64x^3 + 1 = 0$. There should be three solutions (one real and two imaginary).

44. a. Factor $3x^3 - 6x^2 + 6x$.

b. Use the zero product rule and the quadratic formula to solve $3x^3 - 6x^2 + 6x = 0$. There should be three complex solutions.

45. a. Factor $5x^3 + 5x^2 + 10x$.

b. Use the zero product rule and the quadratic formula to solve $5x^3 + 5x^2 + 10x = 0$. There should be three complex solutions.

46. The volume of a cube is 27 ft³. Find the lengths of the sides.

47. The volume of a rectangular box is 64 ft³. If the width is 3 times longer than the height, and the length is 9 times longer than the height, find the dimensions of the box.

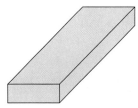

Objective 3: Using the Quadratic Formula in Applications

48. The hypotenuse of a right triangle is 10.2 m long. One leg is 2.1 m shorter than the other leg. Find the lengths of the legs. Round to 1 decimal place.

49. The hypotenuse of a right triangle is 17 ft long. One leg is 3.4 ft longer than the other leg. Find the lengths of the legs. (See Example 3.)

50. The number of lawyers N in the United States from the year 1951 through 1989 can be approximated by $N(t) = 1060t^2 - 7976t + 202{,}209$, where t represents the number of years after 1951 (Source: Datapedia of the United States).

  a. Approximate the number of lawyers in the United States in the year 1978. Round to the nearest thousand.

  b. In what year after 1951 did the number of lawyers in the United States hit 400,000? (Round to the nearest year.)

  c. In what year after 1951 did the number of lawyers hit a half-million? (Round to the nearest year.)

  d. If this trend continues, predict the number of lawyers in the United States in the year 2010. Round to the nearest thousand.

51. A ball player throws a baseball in the air from a cliff that is 48 ft in the air. The initial velocity is 48 ft/sec. The height (in feet) of the object after t sec is given by $h(t) = -16t^2 + 48t + 48$. Find the time at which the height of the object is 64 ft. (See Example 4.)

52. An astronaut on the moon throws a rock into the air from the deck of a spacecraft that is 8 m high. The initial velocity of the rock is 2.4 m/sec. The height (in meters) of the rock after t sec is given by $h(t) = -0.8t^2 + 2.4t + 8$. Find the time at which the height of the rock is 6 m.

53. The braking distance (in feet) of a car going v mph is given by

$$d(v) = \frac{v^2}{20} + v \qquad v \geq 0$$

  a. How fast would a car be traveling if its braking distance were 150 ft? Round to the nearest mile per hour.

  b. How fast would a car be traveling if its braking distance were 100 ft? Round to the nearest mile per hour.

Objective 4: Discriminant

For Exercises 54–61:

    a. Write the equation in the form $ax^2 + bx + c = 0, a > 0.$

    b. Find the value of the discriminant.

    c. Use the discriminant to determine the number and type of solutions. (See Example 5.)

54. $x^2 + 2x = -1$ **55.** $12y - 9 = 4y^2$ **56.** $19m^2 = 8m$

57. $2n - 5n^2 = 0$ **58.** $5p^2 - 21 = 0$ **59.** $3k^2 = 7$

60. $4n(n - 2) - 5n(n - 1) = 4$ **61.** $(2x + 1)(x - 3) = -9$

For Exercises 62–65, find the x- and y-intercepts of the quadratic function. (See Example 6.)

62. $f(x) = x^2 - 5x + 3$ **63.** $g(x) = -x^2 + x - 1$

64. $f(x) = 2x^2 + x + 5$ **65.** $h(x) = -3x^2 + 2x + 2$

Objective 5: Review of the Methods to Solve a Quadratic Equation

For Exercises 66–83, solve the quadratic equations by using any method. (See Example 7.)

66. $a^2 + 3a + 4 = 0$ **67.** $4z^2 + 7z = 0$

68. $x^2 - 2 = 0$ **69.** $b^2 + 7 = 0$

70. $4y^2 + 8y - 5 = 0$ **71.** $k^2 - k + 8 = 0$

72. $\left(x + \dfrac{1}{2}\right)^2 + 4 = 0$ **73.** $(2y + 3)^2 = 9$

74. $2y(y - 3) = -1$ **75.** $w(w - 5) = 4$

76. $(2t + 5)(t - 1) = (t - 3)(t + 8)$ **77.** $(b - 1)(b + 4) = (3b + 2)(b + 1)$

78. $a^2 - 2a = 3$ **79.** $x^2 - x = 30$

80. $32z^2 - 20z - 3 = 0$

81. $8k^2 - 14k + 3 = 0$

82. $3p^2 - 27 = 0$

83. $5h^2 - 125 = 0$

Expanding Your Skills

84. An artist has been commissioned to make a stained glass window in the shape of a regular octagon. The octagon must fit inside an 18-in. square space. See the figure.

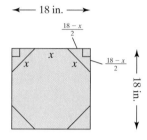

a. Let x represent the length of each side of the octagon. Verify that the legs of the small triangles formed by the corners of the square can be expressed as $\dfrac{18 - x}{2}$.

b. Use the Pythagorean theorem to set up an equation in terms of x that represents the relationship between the legs of the triangle and the hypotenuse.

c. Simplify the equation by clearing parentheses and clearing fractions.

d. Solve the resulting quadratic equation by using the quadratic formula. Use a calculator and round your answers to the nearest tenth of an inch.

e. There are two solutions for x. Which one is appropriate and why?

Graphing Calculator Exercises

85. Graph $Y_1 = x^3 - 27$. Compare the x-intercepts with the solutions to the equation $x^3 - 27 = 0$ found in Exercise 42.

86. Graph $Y_1 = 64x^3 + 1$. Compare the x-intercepts with the solutions to the equation $64x^3 + 1 = 0$ found in Exercise 43.

87. Graph $Y_1 = 3x^3 - 6x^2 + 6x$. Compare the x-intercepts with the solutions to the equation $3x^3 - 6x^2 + 6x = 0$ found in Exercise 44.

88. Graph $Y_1 = 5x^3 + 5x^2 + 10x$. Compare the x-intercepts with the solutions to the equation $5x^3 + 5x^2 + 10x = 0$ found in Exercise 45.

89. The recent population (in thousands) of Ecuador can be approximated by $P(t) = 1.12t^2 + 204.4t + 6697$, where $t = 0$ corresponds to the year 1974.

   a. Approximate the number of people in Ecuador in the year 2000. (Round to the nearest thousand.)

   b. If this trend continues, approximate the number of people in Ecuador in the year 2010. (Round to the nearest thousand.)

   c. In what year after 1974 did the population of Ecuador reach 10 million? Round to the nearest year. (*Hint:* 10 million equals 10,000 thousands.)

   d. Use a graphing calculator to graph the function P on the window $0 \le x \le 20$, $4000 \le y \le 12{,}000$. Use a *Trace* feature to approximate the year in which the population in Ecuador was 10 million (10,000 thousands).

90. The recent population (in thousands) of New Zealand can be approximated by $P(t) = 0.089t^2 + 25.7t + 3601$, where $t = 0$ corresponds to the year 1995.

   a. Approximate the number of people in New Zealand in the year 2005. (Round to the nearest thousand.)

   b. If this trend continues, approximate the number of people in New Zealand in the year 2010. (Round to the nearest thousand.)

   c. In what year after 1995 did the population of New Zealand reach 3.8 million? Round to the nearest year. (*Hint:* 3.8 million equals 3800 thousands.)

   d. Use a graphing calculator to graph the function $P(t)$ on the window $0 \le x \le 50$, $3000 \le y \le 5000$. Use the *Trace* feature to determine the year in which the population in New Zealand was 3.8 million.

---

section 8.3 Equations in Quadratic Form

Objectives

1. Equations Reducible to a Quadratic
2. Solving Equations by Using Substitution

1. Equations Reducible to a Quadratic

We have learned to solve a variety of different types of equations, including linear, radical, rational, and polynomial equations. Sometimes, however, it is necessary to use a quadratic equation as a tool to solve other types of equations. For instance, the equation in Example 1 is a radical equation that reduces to a quadratic equation after both sides are squared.

example 1 Solving an Equation in Quadratic Form

Solve the equation.
$$x - \sqrt{x} - 12 = 0$$

Solution:

$x - \sqrt{x} - 12 = 0$	This is a radical equation.
$x - 12 = \sqrt{x}$	Isolate the radical.
$(x - 12)^2 = (\sqrt{x})^2$	Square both sides.
$x^2 - 24x + 144 = x$	The resulting equation is quadratic.
$x^2 - 25x + 144 = 0$	Write the equation in the form $ax^2 + bx + c = 0$.
$(x - 9)(x - 16) = 0$	The equation is factorable.
$x = 9$ or $x = 16$	Apply the zero product rule.

Check: $x = 9$ Check: $x = 16$

$x - \sqrt{x} - 12 = 0$ $x - \sqrt{x} - 12 = 0$

$(9) - \sqrt{9} - 12 \stackrel{?}{=} 0$ $(16) - \sqrt{16} - 12 \stackrel{?}{=} 0$

$9 - 3 - 12 \neq 0$ $16 - 4 - 12 = 0$ ✓

So $x = 16$ is the only solution ($x = 9$ does not check).

Skill Practice

Solve the equation.

1. $\sqrt{z} + 2 = z$

Avoiding Mistakes: Recall that if we raise both sides of a radical equation to an even power, the potential solutions must be checked in the original equation. In Example 1, the value $x = 9$ is an extraneous solution.

example 2 Solving an Equation Quadratic in Form

Solve the equation. $w^4 - 81 = 0$

Solution:

This equation is a higher-order polynomial equation.

$$w^4 - 81 = 0$$
$$(w^2 - 9)(w^2 + 9) = 0 \qquad \text{The equation is factorable.}$$
$$(w + 3)(w - 3)(w^2 + 9) = 0$$

$w + 3 = 0$ or $w - 3 = 0$ or $w^2 + 9 = 0$ Apply the zero product rule.

$w = -3$ or $w = 3$ or $w^2 = -9$ Solve for w.

$$w = \pm\sqrt{-9}$$
$$w = \pm 3i$$

The solutions are $w = 3$, $w = -3$, $w = 3i$, and $w = -3i$.

Skill Practice

Solve the equation.

2. $a^4 - 1 = 0$

2. Solving Equations by Using Substitution

In this section, we will see that some equations that are not quadratic can be manipulated to appear as **equations in quadratic form** by using substitution.

Answers

1. $z = 4$ ($z = 1$ does not check.)
2. $a = \pm i$, $a = \pm 1$

Chapter 8 Quadratic Equations and Functions

Skill Practice

Solve the equation.

3. $(3t^2 - 10)^2 + 5(3t^2 - 10) - 14 = 0$

example 3 — Solving an Equation Quadratic in Form

Solve the equation.

$$(2x^2 - 5)^2 - 16(2x^2 - 5) + 39 = 0$$

Solution:

$(2x^2 - 5)^2 - 16(2x^2 - 5) + 39 = 0$ — If the substitution $u = (2x^2 - 5)$ is made, the equation becomes quadratic in the variable u.

Substitute $u = (2x^2 - 5)$.

$u^2 - 16u + 39 = 0$ — The equation is in the form $au^2 + bu + c = 0$.

$(u - 13)(u - 3) = 0$ — The equation is factorable.

$u = 13$ or $u = 3$ — Apply the zero product rule.

Reverse substitute.

$2x^2 - 5 = 13$ or $2x^2 - 5 = 3$

$2x^2 = 18$ or $2x^2 = 8$

$x^2 = 9$ or $x^2 = 4$ — Write the equations in the form $x^2 = k$.

$x = \pm\sqrt{9}$ or $x = \pm\sqrt{4}$ — Apply the square root property.

$= \pm 3$ or $= \pm 2$

The solutions are $x = 3$, $x = -3$, $x = 2$, $x = -2$. — Substituting these values in the original equation verifies that these are all valid solutions.

Avoiding Mistakes: When using substitution, it is critical to reverse substitute to solve the equation in terms of the original variable.

Skill Practice

Solve the equation.

4. $y^{2/3} - y^{1/3} = 12$

example 4 — Solving an Equation Quadratic in Form

Solve the equation.

$$p^{2/3} - 2p^{1/3} = 8$$

Solution:

$p^{2/3} - 2p^{1/3} = 8$

$p^{2/3} - 2p^{1/3} - 8 = 0$ — Set the equation equal to zero.

$(p^{1/3})^2 - 2(p^{1/3})^1 - 8 = 0$ — Make the substitution $u = p^{1/3}$.

Substitute $u = p^{1/3}$.

$u^2 - 2u - 8 = 0$ — Then the equation is in the form $au^2 + bu + c = 0$.

Answers

3. $t = 1$; $t = -1$; $t = 2$; $t = -2$
4. $y = 64$; $y = -27$

$$(u-4)(u+2) = 0 \quad \text{The equation is factorable.}$$

$$u = 4 \quad \text{or} \quad u = -2 \quad \text{Apply the zero product rule.}$$

Reverse substitute.

$$p^{1/3} = 4 \quad \text{or} \quad p^{1/3} = -2$$

$$\sqrt[3]{p} = 4 \quad \text{or} \quad \sqrt[3]{p} = -2 \quad \text{The equations are radical equations.}$$

$$(\sqrt[3]{p})^3 = (4)^3 \quad \text{or} \quad (\sqrt[3]{p})^3 = (-2)^3 \quad \text{Cube both sides.}$$

$$p = 64 \quad \text{or} \quad p = -8$$

Check: $p = 64$ Check: $p = -8$

$$p^{2/3} - 2p^{1/3} = 8 \quad\quad p^{2/3} - 2p^{1/3} = 8$$

$$(64)^{2/3} - 2(64)^{1/3} \stackrel{?}{=} 8 \quad\quad (-8)^{2/3} - 2(-8)^{1/3} \stackrel{?}{=} 8$$

$$16 - 2(4) \stackrel{?}{=} 8 \quad\quad 4 - 2(-2) \stackrel{?}{=} 8$$

$$8 = 8 \checkmark \quad\quad 4 + 4 = 8 \checkmark$$

The solutions are $p = 64$ and $p = -8$.

---

example 5   Solving a Quadratic Equation by Using Substitution

Solve the equation $(t - 5)^2 - 4(t - 5) + 13 = 0$ by using the substitution $u = t - 5$.

Solution:

$$(t - 5)^2 - 4(t - 5) + 13 = 0$$

This equation is quadratic; however, we can make it a simpler quadratic equation by letting $u = (t - 5)$.

Substitute $u = t - 5$.

$$u^2 - 4u + 13 = 0 \quad \text{This equation does not factor.}$$

$$u = \frac{-(-4) \pm \sqrt{(-4)^2 - 4(1)(13)}}{2(1)} \quad \text{Apply the quadratic formula: } a = 1, b = -4, c = 13.$$

$$= \frac{4 \pm \sqrt{16 - 52}}{2}$$

$$= \frac{4 \pm \sqrt{-36}}{2}$$

$$u = \frac{4 \pm 6i}{2}$$

$$u = \frac{4 + 6i}{2} = 2 + 3i$$

$$u = \frac{4 - 6i}{2} = 2 - 3i$$

Skill Practice

Solve the equation.

5. $(a + 1)^2 + 6(a + 1) + 25 = 0$

Answer

5. $a = -4 \pm 4i$

$$u = 2 + 3i \quad \text{or} \quad u = 2 - 3i$$

Reverse substitute.

$$t - 5 = 2 + 3i \quad \text{or} \quad t - 5 = 2 - 3i$$

$$t = 7 + 3i \quad \text{or} \quad t = 7 - 3i \quad \text{Both values check in the original equation.}$$

The solutions are $t = 7 + 3i$ and $t = 7 - 3i$.

section 8.3 Practice Exercises

Boost your GRADE at mathzone.com!

- Practice Problems
- Self-Tests
- NetTutor
- e-Professors
- Videos

Study Skills Exercises

1. Label the statements below as true or false.

   a. To do math, you must be born with a special skill.

   b. There is only one way to solve a math problem.

   c. Two answers that look different can both be correct.

2. Define the key term **equations in quadratic form**.

Review Exercises

For Exercises 3–8, solve the quadratic equations.

3. $16 = (2x - 3)^2$

4. $\left(x - \dfrac{3}{2}\right)^2 = \dfrac{7}{4}$

5. $n(n - 6) = -13$

6. $x(x + 8) = -16$

7. $6k^2 + 7k = 6$

8. $2x^2 - 8x - 44 = 0$

Objective 1: Equations Reducible to a Quadratic

For Exercises 9–26, solve the equation. For the equations involving radicals, be sure that you check all solutions in the original equation. (See Examples 1–2.)

9. $y + 6\sqrt{y} = 16$

10. $p - 8\sqrt{p} = -15$

11. $2x + 3\sqrt{x} - 2 = 0$

12. $3t + 5\sqrt{t} - 2 = 0$

13. $\sqrt{4b + 1} - \sqrt{b - 2} = 3$

14. $\sqrt{6a + 7} - \sqrt{3a + 3} = 1$

15. $\sqrt{5k + 6} + \sqrt{3k + 4} = 2$

16. $\sqrt{z + 15} - \sqrt{2z + 7} = 1$

17. $\sqrt{4t + 1} = t + 1$

18. $\sqrt{x^2 + 20} = 3\sqrt{x}$ **19.** $x^4 - 16 = 0$ **20.** $t^4 - 625 = 0$

21. $n^4 - 256 = 0$ **22.** $m^4 - 81 = 0$ **23.** $a^3 + 8 = 0$

24. $b^3 - 1 = 0$ **25.** $5p^3 - 5 = 0$ **26.** $2t^3 + 54 = 0$

Objective 2: Solving Equations by Using Substitution

27. Refer to Example 1. Use the method of substitution with $u = \sqrt{x}$ to solve the equation in Example 1.

28. Refer to Exercise 9. Use the method of substitution with $u = \sqrt{y}$ to solve the equation in Exercise 9.

29. a. Solve the quadratic equation by factoring. $u^2 + 10u + 24 = 0$

  b. Solve the equation by using substitution. (See Example 3.) $(y^2 + 5y)^2 + 10(y^2 + 5y) + 24 = 0$

30. a. Solve the quadratic equation by factoring. $u^2 - 2u - 35 = 0$

  b. Solve the equation by using substitution. $(w^2 - 6w)^2 - 2(w^2 - 6w) - 35 = 0$

31. a. Solve the quadratic equation by factoring. $u^2 - 2u - 24 = 0$

  b. Solve the equation by using substitution. $(x^2 - 5x)^2 - 2(x^2 - 5x) - 24 = 0$

32. a. Solve the quadratic equation by factoring. $u^2 - 4u + 3 = 0$

  b. Solve the equation by using substitution. $(2p^2 + p)^2 - 4(2p^2 + p) + 3 = 0$

For Exercises 33–48, solve using substitution. (See Examples 3–5.)

33. $(4x + 5)^2 + 3(4x + 5) + 2 = 0$ **34.** $2(5x + 3)^2 - (5x + 3) - 28 = 0$

35. $16\left(\dfrac{x + 6}{4}\right)^2 + 8\left(\dfrac{x + 6}{4}\right) + 1 = 0$ **36.** $9\left(\dfrac{x + 3}{2}\right)^2 - 6\left(\dfrac{x + 3}{2}\right) + 1 = 0$

37. $(x^2 - 2x)^2 + 2(x^2 - 2x) = 3$ **38.** $(x^2 + x)^2 - 8(x^2 + x) = -12$

39. $x^4 - 13x^2 + 36 = 0$ **40.** $y^4 - 5y^2 + 4 = 0$

41. $4m^4 - 9m^2 + 2 = 0$

42. $x^4 - 7x^2 + 12 = 0$

43. $x^6 - 9x^3 + 8 = 0$

44. $x^6 - 26x^3 - 27 = 0$

45. $m^{2/3} - m^{1/3} - 6 = 0$

46. $2n^{2/3} + 7n^{1/3} - 15 = 0$

47. $2t^{2/5} + 7t^{1/5} + 3 = 0$

48. $p^{2/5} + p^{1/5} - 2 = 0$

Expanding Your Skills

49. Solve $x^2 - 4 = 0$ in three ways:
   a. By factoring
   b. By using the square root property
   c. By using the quadratic formula

50. Solve $9x^2 - 16 = 0$ in three ways:
   a. By factoring
   b. By using the square root property
   c. By using the quadratic formula

For Exercises 51–54, solve the equation. *Hint:* Factor by grouping first.

51. $a^3 + 16a - a^2 - 16 = 0$

52. $b^3 + 9b - b^2 - 9 = 0$

53. $x^3 + 5x - 4x^2 - 20 = 0$

54. $y^3 + 8y - 3y^2 - 24 = 0$

Graphing Calculator Exercises

55. a. Solve the equation $x^4 + 4x^2 + 4 = 0$.

   b. How many solutions are real and how many solutions are imaginary?

   c. How many x-intercepts do you anticipate for the function defined by $y = x^4 + 4x^2 + 4$?

   d. Graph $Y_1 = x^4 + 4x^2 + 4$ on a standard viewing window.

56. a. Solve the equation $x^4 - 2x^2 + 1 = 0$.

   b. How many solutions are real and how many solutions are imaginary?

   c. How many x-intercepts do you anticipate for the function defined by $y = x^4 - 2x^2 + 1$?

   d. Graph $Y_1 = x^4 - 2x^2 + 1$ on a standard viewing window.

57. a. Solve the equation $x^4 - x^3 - 6x^2 = 0$.

b. How many solutions are real and how many solutions are imaginary?

c. How many x-intercepts do you anticipate for the function defined by $y = x^4 - x^3 - 6x^2$?

d. Graph $Y_1 = x^4 - x^3 - 6x^2$ on a standard viewing window.

58. a. Solve the equation $x^4 - 10x^2 + 9 = 0$.

b. How many solutions are real and how many solutions are imaginary?

c. How many x-intercepts do you anticipate for the function defined by $y = x^4 - 10x^2 + 9$?

d. Graph $Y_1 = x^4 - 10x^2 + 9$ on a standard viewing window.

chapter 8 | midchapter review

For Exercises 1–2, solve by using the square root property.

1. $a^2 + 5 = 0$
2. $(y + 4)^2 = 12$

For Exercises 3–4, find the value of k to complete the square. Then factor the trinomial.

3. $c^2 - 20c + k$
4. $v^2 + 11v + k$

For Exercises 5–6, solve by completing the square.

5. $x^2 - 10x + 5 = 0$
6. $3t^2 + 12t + 18 = 0$

7. Given the quadratic equation $ax^2 + bx + c = 0$, state the quadratic formula.

For Exercises 8–9, solve by using the quadratic formula.

8. $6x^2 + x + 2 = 0$
9. $4y^2 - 2y = 6 - 5y$

10. Given: $3x^2 + 4x - 4 = 0$

a. Find the value of the discriminant.

b. Use the discriminant to determine the number and type of solutions.

For Exercises 11–12, solve the equations.

11. $a^4 - 81 = 0$

12. $x^{2/3} + 3x^{1/3} - 10 = 0$

Objectives

1. Quadratic Functions of the Form $f(x) = x^2 + k$
2. Quadratic Functions of the Form $f(x) = (x - h)^2$
3. Quadratic Functions of the Form $f(x) = ax^2$
4. Quadratic Functions of the Form $f(x) = a(x - h)^2 + k$

section 8.4 Graphs of Quadratic Functions

In Section 5.7, we defined a quadratic function as a function of the form $f(x) = ax^2 + bx + c$ $(a \neq 0)$. We also learned that the graph of a quadratic function is a **parabola**. In this section we will learn how to graph parabolas.

A parabola opens up if $a > 0$ (Figure 8-4) and opens down if $a < 0$ (Figure 8-5). If a parabola opens up, the **vertex** is the lowest point on the graph. If a parabola opens down, the **vertex** is the highest point on the graph. The **axis of symmetry** is the vertical line that passes through the vertex.

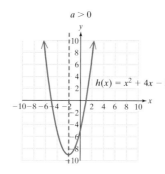

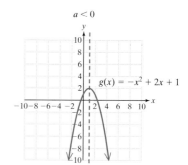

Figure 8-4   Figure 8-5

1. Quadratic Functions of the Form $f(x) = x^2 + k$

One technique for graphing a function is to plot a sufficient number of points on the function until the general shape and defining characteristics can be determined. Then sketch a curve through the points.

Skill Practice

1. Refer to the graph of $f(x) = x^2 + k$ to determine the value of k.

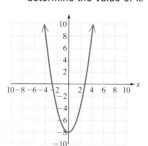

example 1 Graphing Quadratic Functions in the Form $f(x) = x^2 + k$

Graph the functions f, g, and h on the same coordinate system.

$$f(x) = x^2 \qquad g(x) = x^2 + 1 \qquad h(x) = x^2 - 2$$

Solution:

Several function values for f, g, and h are shown in Table 8-1 for selected values of x. The corresponding graphs are pictured in Figure 8-6.

Table 8-1

x	$f(x) = x^2$	$g(x) = x^2 + 1$	$h(x) = x^2 - 2$
-3	9	10	7
-2	4	5	2
-1	1	2	-1
0	0	1	-2
1	1	2	-1
2	4	5	2
3	9	10	7

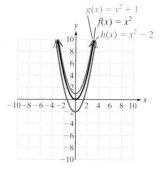

Figure 8-6

Answer

1. $k = -8$

Notice that the graphs of $g(x) = x^2 + 1$ and $h(x) = x^2 - 2$ take on the same shape as $f(x) = x^2$. However, the y-values of g are 1 greater than the y-values of f. Hence the graph of $g(x) = x^2 + 1$ is the same as the graph of $f(x) = x^2$ shifted *up* 1 unit. Likewise the y-values of h are 2 less than those of f. The graph of $h(x) = x^2 - 2$ is the same as the graph of $f(x) = x^2$ shifted *down* 2 units.

The functions in Example 1 illustrate the following properties of quadratic functions of the form $f(x) = x^2 + k$.

Graphs of $f(x) = x^2 + k$

If $k > 0$, then the graph of $f(x) = x^2 + k$ is the same as the graph of $y = x^2$ shifted *up* k units.

If $k < 0$, then the graph of $f(x) = x^2 + k$ is the same as the graph of $y = x^2$ shifted *down* $|k|$ units.

Calculator Connections

Try experimenting with a graphing calculator by graphing functions of the form $y = x^2 + k$ for several values of k.

 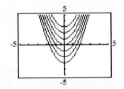

example 2   Graphing Quadratic Functions of the Form $f(x) = x^2 + k$

Sketch the functions defined by

a. $m(x) = x^2 - 4$ **b.** $n(x) = x^2 + \dfrac{7}{2}$

Solution:

a. $m(x) = x^2 - 4$

$m(x) = x^2 + (-4)$

Because $k = -4$, the graph is obtained by shifting the graph of $y = x^2$ down $|-4|$ units (Figure 8-7).

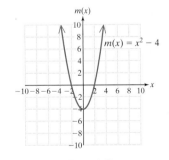

Figure 8-7

Skill Practice

2. Graph the functions f, g, and h on the same coordinate system.

$f(x) = x^2$

$g(x) = x^2 + 3$

$h(x) = x^2 - 5$

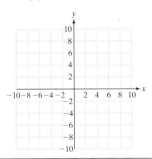

Answer

2.

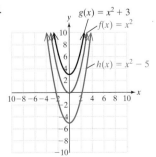

b. $n(x) = x^2 + \dfrac{7}{2}$

Because $k = \dfrac{7}{2}$, the graph is obtained by shifting the graph of $y = x^2$ up $\dfrac{7}{2}$ units (Figure 8-8).

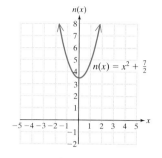

Figure 8-8

2. Quadratic Functions of the Form $f(x) = (x - h)^2$

The graph of $f(x) = x^2 + k$ represents a vertical shift (up or down) of the function $y = x^2$. Example 3 shows that functions of the form $f(x) = (x - h)^2$ represent a horizontal shift (left or right) of the function $y = x^2$.

Skill Practice

3. Refer to the graph of $f(x) = (x - h)^2$ to determine the value of h.

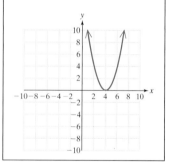

example 3   Graphing Quadratic Functions of the Form $f(x) = (x - h)^2$

Graph the functions f, g, and h on the same coordinate system.

$$f(x) = x^2 \qquad g(x) = (x + 1)^2 \qquad h(x) = (x - 2)^2$$

Solution:

Several function values for f, g, and h are shown in Table 8-2 for selected values of x. The corresponding graphs are pictured in Figure 8-9.

Table 8-2

x	$f(x) = x^2$	$g(x) = (x + 1)^2$	$h(x) = (x - 2)^2$
-4	16	9	36
-3	9	4	25
-2	4	1	16
-1	1	0	9
0	0	1	4
1	1	4	1
2	4	9	0
3	9	16	1
4	16	25	4
5	25	36	9

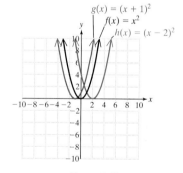

Figure 8-9

Example 3 illustrates the following properties of quadratic functions of the form $f(x) = (x - h)^2$.

Answer

3. $h = 4$

Graphs of $f(x) = (x - h)^2$

If $h > 0$, then the graph of $f(x) = (x - h)^2$ is the same as the graph of $y = x^2$ shifted h units to the *right*.

If $h < 0$, then the graph of $f(x) = (x - h)^2$ is the same as the graph of $y = x^2$ shifted $|h|$ units to the *left*.

From Example 3 we have

$$h(x) = (x - 2)^2 \quad \text{and} \quad g(x) = (x + 1)^2$$
$$g(x) = [x - (-1)]^2$$

$y = x^2$ shifted 2 units to the right

$y = x^2$ shifted $|-1|$ unit to the left

example 4   Graphing Functions of the Form $f(x) = (x - h)^2$

Sketch the functions p and q.

a. $p(x) = (x - 7)^2$ **b.** $q(x) = (x + 1.6)^2$

Solution:

a. $p(x) = (x - 7)^2$

Because $h = 7 > 0$, shift the graph of $y = x^2$ to the *right* 7 units (Figure 8-10).

b. $q(x) = (x + 1.6)^2$

$q(x) = [x - (-1.6)]^2$

Because $h = -1.6 < 0$, shift the graph of $y = x^2$ to the *left* 1.6 units (Figure 8-11).

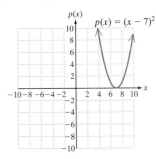

Figure 8-10

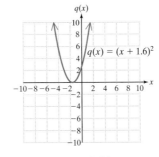

Figure 8-11

Skill Practice

4. Graph the functions f, g, and h on the same coordinate system.

$f(x) = x^2$

$g(x) = (x + 3)^2$

$h(x) = (x - 6)^2$

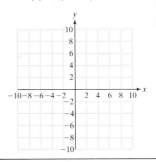

Answer

4.

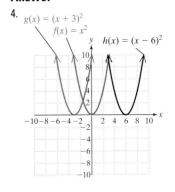

3. Quadratic Functions of the Form $f(x) = ax^2$

Examples 5 and 6 investigate functions of the form $f(x) = ax^2$ $(a \neq 0)$.

Skill Practice

5. Graph the functions f, g, and h on the same coordinate system.

$$f(x) = x^2$$
$$g(x) = 3x^2$$
$$h(x) = \frac{1}{3}x^2$$

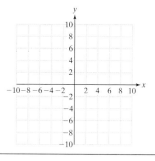

example 5 Graphing Functions of the Form $f(x) = ax^2$

Graph the functions f, g, and h on the same coordinate system.

$$f(x) = x^2 \qquad g(x) = 2x^2 \qquad h(x) = \frac{1}{2}x^2$$

Solution:

Several function values for f, g, and h are shown in Table 8-3 for selected values of x. The corresponding graphs are pictured in Figure 8-12.

Table 8-3

x	$f(x) = x^2$	$g(x) = 2x^2$	$h(x) = \frac{1}{2}x^2$
-3	9	18	$\frac{9}{2}$
-2	4	8	2
-1	1	2	$\frac{1}{2}$
0	0	0	0
1	1	2	$\frac{1}{2}$
2	4	8	2
3	9	18	$\frac{9}{2}$

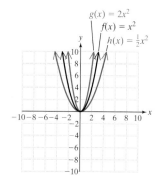

Figure 8-12

In Example 5, the function values defined by $g(x) = 2x^2$ are twice those of $f(x) = x^2$. The graph of $g(x) = 2x^2$ is the same as the graph of $f(x) = x^2$ *stretched vertically* by a factor of 2 (the graph appears narrower than $f(x) = x^2$).

In Example 5, the function values defined by $h(x) = \frac{1}{2}x^2$ are one-half those of $f(x) = x^2$. The graph of $h(x) = \frac{1}{2}x^2$ is the same as the graph of $f(x) = x^2$ *shrunk vertically* by a factor of $\frac{1}{2}$ (the graph appears wider than $f(x) = x^2$).

example 6 Graphing Functions of the Form $f(x) = ax^2$

Graph the functions f, g, and h on the same coordinate system.

$$f(x) = -x^2 \qquad g(x) = -3x^2 \qquad h(x) = -\frac{1}{3}x^2$$

Solution:

Several function values for f, g, and h are shown in Table 8-4 for selected values of x. The corresponding graphs are pictured in Figure 8-13.

Answer

5.

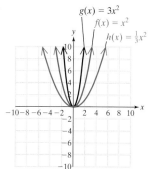

x	$f(x) = -x^2$	$g(x) = -3x^2$	$h(x) = -\frac{1}{3}x^2$
-3	-9	-27	-3
-2	-4	-12	$-\frac{4}{3}$
-1	-1	-3	$-\frac{1}{3}$
0	0	0	0
1	-1	-3	$-\frac{1}{3}$
2	-4	-12	$-\frac{4}{3}$
3	-9	-27	-3

Table 8-4

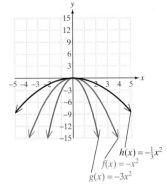

Figure 8-13

Skill Practice

6. Graph the functions f, g, and h on the same coordinate system.

$f(x) = -x^2$

$g(x) = -2x^2$

$h(x) = -\dfrac{1}{2}x^2$

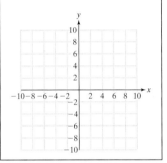

Example 6 illustrates that if the coefficient of the square term is negative, the parabola opens down. The graph of $g(x) = -3x^2$ is the same as the graph of $f(x) = -x^2$ with a *vertical stretch* by a factor of $|-3|$. The graph of $h(x) = -\frac{1}{3}x^2$ is the same as the graph of $f(x) = -x^2$ with a *vertical shrink* by a factor of $|-\frac{1}{3}|$.

Graphs of $f(x) = ax^2$

1. If $a > 0$, the parabola opens up. Furthermore,
   - If $0 < a < 1$, then the graph of $f(x) = ax^2$ is the same as the graph of $y = x^2$ with a *vertical shrink* by a factor of a.
   - If $a > 1$, then the graph of $f(x) = ax^2$ is the same as the graph of $y = x^2$ with a *vertical stretch* by a factor of a.

2. If $a < 0$, the parabola opens down. Furthermore,
   - If $0 < |a| < 1$, then the graph of $f(x) = ax^2$ is the same as the graph of $y = -x^2$ with a *vertical shrink* by a factor of $|a|$.
   - If $|a| > 1$, then the graph of $f(x) = ax^2$ is the same as the graph of $y = -x^2$ with a *vertical stretch* by a factor of $|a|$.

4. Quadratic Functions of the Form $f(x) = a(x - h)^2 + k$

We can summarize our findings from Examples 1–6 by graphing functions of the form $f(x) = a(x - h)^2 + k \quad (a \neq 0)$.

The graph of $y = x^2$ has its vertex at the origin $(0, 0)$. The graph of $f(x) = a(x - h)^2 + k$ is the same as the graph of $y = x^2$ shifted to the right or left h units and shifted up or down k units. Therefore, the vertex is shifted from $(0, 0)$ to (h, k). The axis of symmetry is the vertical line through the vertex. Hence the axis of symmetry must be the line $x = h$.

Answer

6.
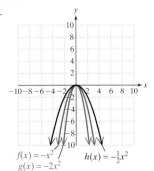

Concept Connections

7. Refer to the graph of $f(x) = ax^2 + k$ to choose which is true about the value of a.

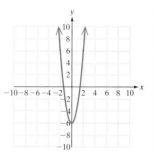

a. $a > 0$ b. $a < 0$
c. $a = 0$ d. Cannot be determined.

Graphs of $f(x) = a(x - h)^2 + k$

1. The vertex is located at (h, k).
2. The axis of symmetry is the line $x = h$.
3. If $a > 0$, the parabola opens up, and k is the **minimum value** of the function.
4. If $a < 0$, the parabola opens down, and k is the **maximum value** of the function.

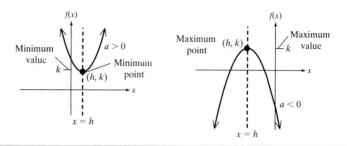

Skill Practice

8. Given the function defined by $g(x) = 3(x + 1)^2 - 3$
   a. Identify the vertex.
   b. Sketch the graph.
   c. Identify the axis of symmetry.
   d. Identify the maximum or minimum value of the function.

example 7 — Graphing a Function of the Form $f(x) = a(x - h)^2 + k$

Given the function defined by
$$f(x) = 2(x - 3)^2 + 4$$

a. Identify the vertex.
b. Sketch the function.
c. Identify the axis of symmetry.
d. Identify the maximum or minimum value of the function.

Solution:

a. The function is in the form $f(x) = a(x - h)^2 + k$, where $a = 2$, $h = 3$, and $k = 4$. Therefore, the vertex is at $(3, 4)$.

b. The graph of f is the same as the graph of $y = x^2$ shifted to the right 3 units, shifted up 4 units, and stretched vertically by a factor of 2 (Figure 8-14).

c. The axis of symmetry is the line $x = 3$.

d. Because $a > 0$, the function opens up. Therefore, the minimum function value is 4. Notice that the minimum value is the minimum y-value on the graph.

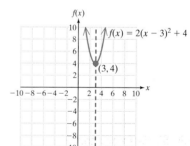

Figure 8-14

Answers

7. a
8. a. Vertex: $(-1, -3)$
   b.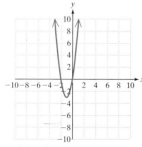
   c. Axis of symmetry: $x = -1$
   d. Minimum value: -3

example 8   Graphing a Function of the Form $f(x) = a(x - h)^2 + k$

Given the function defined by

$$g(x) = -(x + 2)^2 - \frac{7}{4}$$

a. Identify the vertex.
b. Sketch the function.
c. Identify the axis of symmetry.
d. Identify the maximum or minimum value of the function.

Solution:

a. $g(x) = -(x + 2)^2 - \dfrac{7}{4}$

$ = -1[x - (-2)]^2 + \left(-\dfrac{7}{4}\right)$

The function is in the form $g(x) = a(x - h)^2 + k$, where $a = -1$, $h = -2$, and $k = -\frac{7}{4}$. Therefore, the vertex is at $(-2, -\frac{7}{4})$.

b. The graph of g is the same as the graph of $y = x^2$ shifted to the left 2 units, shifted down $\frac{7}{4}$ units, and opening down (Figure 8-15).

c. The axis of symmetry is the line $x = -2$.

d. The parabola opens down, so the maximum function value is $-\frac{7}{4}$.

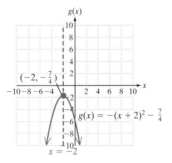

Figure 8-15

Skill Practice

9. Given the function defined by $h(x) = -\dfrac{1}{2}(x - 4)^2 + 2$
   a. Identify the vertex.
   b. Sketch the graph.
   c. Identify the axis of symmetry.
   d. Identify the maximum or minimum value of the function.

Answers

9. a. Vertex: (4, 2)
   b.
   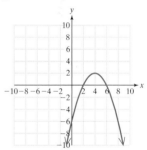
   c. Axis of symmetry: $x = 4$
   d. Maximum value: 2

---

section 8.4   Practice Exercises

Boost your GRADE at mathzone.com!   MathZone

- Practice Problems
- Self-Tests
- NetTutor
- e-Professors
- Videos

Study Skills Exercises

1. To help you stay motivated for this class, list three reasons why you are taking the class.

2. Define the key terms.
   a. Parabola   b. Vertex   c. Axis of symmetry
   d. Maximum value   e. Minimum value

Review Exercises

For Exercises 3–8, solve the equations.

3. $x^2 + x - 5 = 0$
4. $(y - 3)^2 = -4$
5. $\sqrt{2a + 7} = a + 1$
6. $5t(t - 2) = -3$

7. $2z^2 - 3z - 9 = 0$
8. $x^{2/3} + 5x^{1/3} + 6 = 0$

Objective 1: Quadratic Functions of the Form $f(x) = x^2 + k$

9. Describe how the value of k affects the graphs of functions of the form $f(x) = x^2 + k$.

For Exercises 10–19, graph the functions. (See Examples 1–2.)

10. $g(x) = x^2 + 1$
11. $f(x) = x^2 + 2$
12. $p(x) = x^2 - 3$

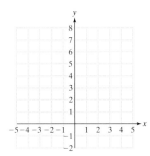
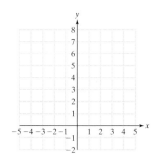
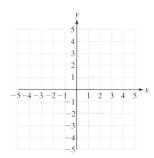

13. $q(x) = x^2 - 4$
14. $T(x) = x^2 + \dfrac{3}{4}$
15. $S(x) = x^2 + \dfrac{3}{2}$

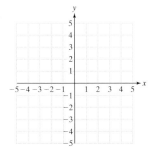
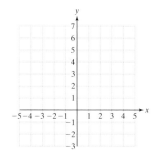
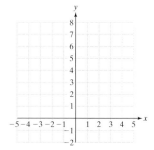

16. $M(x) = x^2 - \dfrac{5}{4}$
17. $n(x) = x^2 - \dfrac{1}{3}$
18. $P(x) = x^2 + \dfrac{1}{2}$

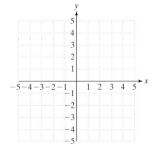
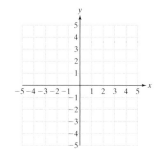
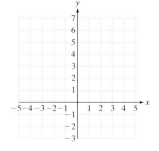

19. $Q(x) = x^2 + \dfrac{1}{4}$

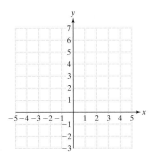

Objective 2: Quadratic Functions of the Form $f(x) = (x - h)^2$

20. Describe how the value of h affects the graphs of functions of the form $f(x) = (x - h)^2$.

For Exercises 21–30, graph the functions. (See Examples 3–4.)

21. $r(x) = (x + 1)^2$ **22.** $h(x) = (x + 2)^2$ **23.** $k(x) = (x - 3)^2$ **24.** $L(x) = (x - 4)^2$

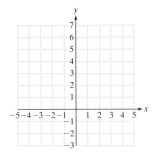 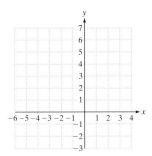 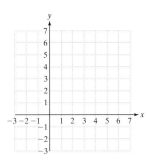 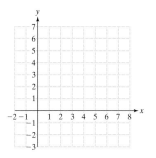

25. $A(x) = \left(x + \dfrac{3}{4}\right)^2$ **26.** $r(x) = \left(x + \dfrac{3}{2}\right)^2$ **27.** $W(x) = \left(x - \dfrac{5}{4}\right)^2$ **28.** $V(x) = \left(x - \dfrac{1}{3}\right)^2$

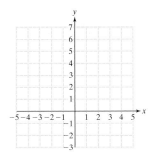 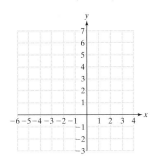 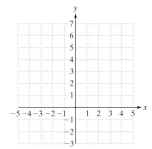 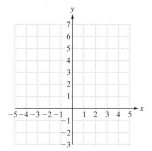

29. $M(x) = \left(x + \dfrac{1}{2}\right)^2$ **30.** $N(x) = \left(x + \dfrac{1}{4}\right)^2$

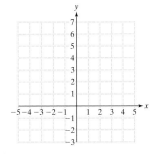 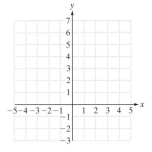

Objective 3: Quadratic Functions of the Form $f(x) = ax^2$

31. Describe how the value of a affects the graph of a function of the form $f(x) = ax^2$, where $a \neq 0$.

32. How do you determine whether the graph of a function defined by $h(x) = ax^2 + bx + c$ $(a \neq 0)$ opens up or down?

For Exercises 33–40, graph the functions. (See Examples 5–6.)

33. $f(x) = 2x^2$ **34.** $g(x) = 3x^2$ **35.** $h(x) = \dfrac{1}{2}x^2$ **36.** $f(x) = \dfrac{1}{3}x^2$

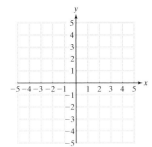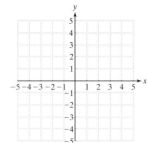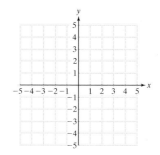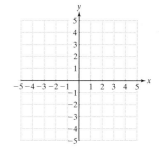

37. $c(x) = -x^2$ **38.** $g(x) = -2x^2$ **39.** $v(x) = -\dfrac{1}{3}x^2$ **40.** $f(x) = -\dfrac{1}{2}x^2$

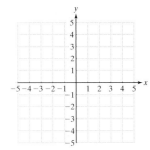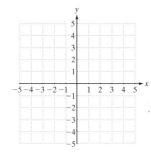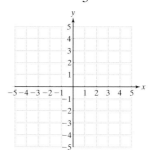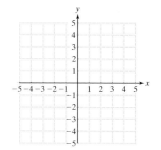

Objective 4: Quadratic Functions of the Form $f(x) = a(x - h)^2 + k$

For Exercises 41–48, match the function with its graph.

41. $f(x) = -\dfrac{1}{4}x^2$ **42.** $g(x) = (x + 3)^2$ **43.** $k(x) = (x - 3)^2$ **44.** $h(x) = \dfrac{1}{4}x^2$

45. $t(x) = x^2 + 2$ **46.** $m(x) = x^2 - 4$ **47.** $p(x) = (x + 1)^2 - 3$ **48.** $n(x) = -(x - 2)^2 + 3$

a. 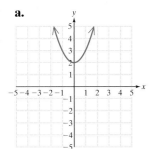   b. 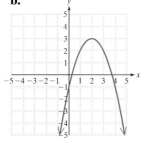   c. 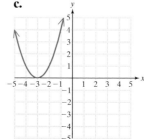   d.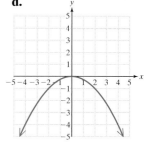

e. f. g. h.

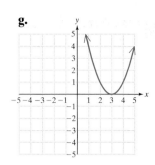
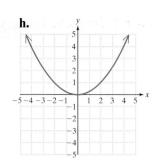

For Exercises 49–66, graph the parabola and the axis of symmetry. Label the coordinates of the vertex, and write the equation of the axis of symmetry. **(See Examples 7–8.)**

49. $y = (x - 3)^2 + 2$ **50.** $y = (x - 2)^2 + 3$ **51.** $y = (x + 1)^2 - 3$ **52.** $y = (x + 3)^2 - 1$

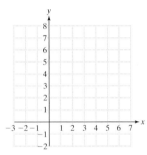
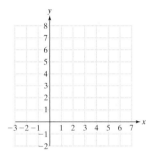
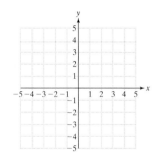
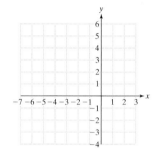

53. $y = -(x - 4)^2 - 2$ **54.** $y = -(x - 2)^2 - 4$ **55.** $y = -(x + 3)^2 + 3$ **56.** $y = -(x + 2)^2 + 2$

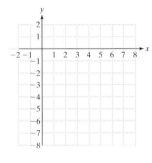
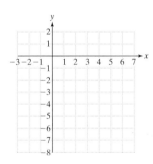
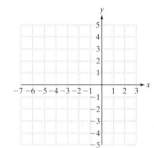
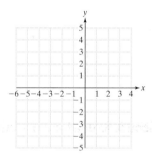

57. $y = (x + 1)^2 + 1$ **58.** $y = (x - 4)^2 - 4$ **59.** $y = 3(x - 1)^2$ **60.** $y = -3(x - 1)^2$

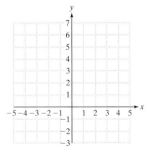
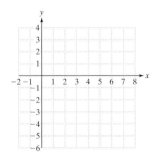
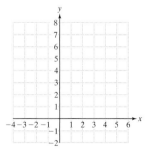
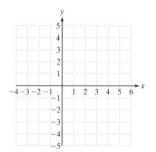

61. $y = -4x^2 + 3$

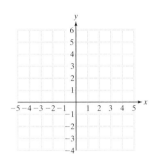

62. $y = 4x^2 + 3$

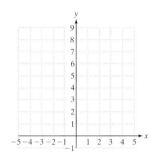

63. $y = 2(x + 3)^2 - 1$

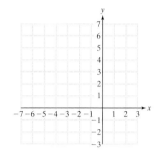

64. $y = -2(x + 3)^2 - 1$

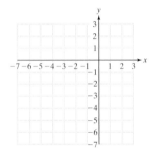

65. $y = -\dfrac{1}{4}(x - 1)^2 + 2$

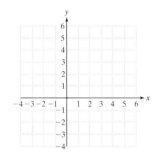

66. $y = \dfrac{1}{4}(x - 1)^2 + 2$

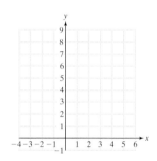

For Exercises 67–78, write the coordinates of the vertex and determine if the vertex is a maximum point or a minimum point. Then write the maximum or minimum value.

67. $f(x) = 4(x - 6)^2 - 9$

68. $g(x) = 3(x - 4)^2 - 7$

69. $p(x) = -\dfrac{2}{5}(x - 2)^2 + 5$

70. $h(x) = -\dfrac{3}{7}(x - 5)^2 + 10$

71. $k(x) = \dfrac{1}{2}(x + 8)^2$

72. $m(x) = \dfrac{2}{9}(x + 11)^2$

73. $n(x) = -6x^2 + \dfrac{21}{4}$

74. $q(x) = -4x^2 + \dfrac{1}{6}$

75. $A(x) = 2(x - 7)^2 - \dfrac{3}{2}$

76. $B(x) = 5(x - 3)^2 - \dfrac{1}{4}$

77. $F(x) = 7x^2$

78. $G(x) = -7x^2$

79. True or false: The function defined by $g(x) = -5x^2$ has a maximum value but no minimum value.

80. True or false: The function defined by $f(x) = 2(x - 5)^2$ has a maximum value but no minimum value.

81. True or false: If the vertex $(-2, 8)$ represents a minimum point, then the minimum value is -2.

82. True or false: If the vertex $(-2, 8)$ represents a maximum point, then the maximum value is 8.

83. A suspension bridge is 120 ft long. Its supporting cable hangs in a shape that resembles a parabola. The function defined by $H(x) = \frac{1}{90}(x - 60)^2 + 30$ (where $0 \le x \le 120$) approximates the height of the supporting cable a distance of x ft from the end of the bridge (see figure).

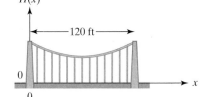

a. What is the location of the vertex of the parabolic cable?

b. What is the minimum height of the cable?

c. How high are the towers at either end of the supporting cable?

84. A 50-m bridge over a crevasse is supported by a parabolic arch. The function defined by $f(x) = -0.16(x - 25)^2 + 100$ (where $0 \le x \le 50$) approximates the height of the supporting arch x meters from the end of the bridge (see figure).

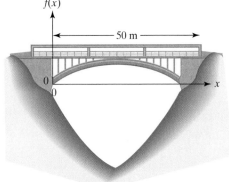

a. What is the location of the vertex of the arch?

b. What is the maximum height of the arch (relative to the origin)?

Graphing Calculator Exercises

For Exercises 85–88, verify the maximum and minimum points found in Exercises 67–70, by graphing each function on the calculator.

85. $Y_1 = 4(x - 6)^2 - 9$ (Exercise 67)

86. $Y_1 = 3(x - 4)^2 - 7$ (Exercise 68)

87. $Y_1 = -\frac{2}{5}(x - 2)^2 + 5$ (Exercise 69)

88. $Y_1 = -\frac{3}{7}(x - 5)^2 + 10$ (Exercise 70)

Objectives

1. Writing a Quadratic Function in the Form $f(x) = a(x - h)^2 + k$
2. Vertex Formula
3. Determining the Vertex and Intercepts of a Quadratic Function
4. Vertex of a Parabola: Applications

Section 8.5 Vertex of a Parabola and Applications

1. Writing a Quadratic Function in the Form $f(x) = a(x - h)^2 + k$

The graph of a quadratic function is a parabola, and if the function is written in the form $f(x) = a(x - h)^2 + k$ $(a \neq 0)$, then the vertex is at (h, k). A quadratic function can be written in the form $f(x) = a(x - h)^2 + k$ $(a \neq 0)$ by completing the square. The process is similar to the steps outlined in Section 8.1 except that all algebraic manipulations are performed on the right-hand side of the equation.

Example 1   Writing a Quadratic Function in the Form $f(x) = a(x - h)^2 + k$ $(a \neq 0)$

Given: $f(x) = x^2 + 8x + 13$

a. Write the function in the form $f(x) = a(x - h)^2 + k$.
b. Identify the vertex, axis of symmetry, and minimum function value.

Skill Practice

1. Given: $f(x) = x^2 + 8x - 1$
   a. Write the equation in the form $f(x) = a(x - h)^2 + k$.
   b. Identify the vertex, axis of symmetry, and minimum value of the function.

Avoiding Mistakes: Do not factor out the leading coefficient from the constant term.

Solution:

a. $f(x) = x^2 + 8x + 13$ Rather than dividing by the leading coefficient on both sides, we will factor out the leading coefficient from the variable terms on the right-hand side.

$= 1(x^2 + 8x) + 13$

$= 1(x^2 + 8x \quad) + 13$ Next, complete the square on the expression within the parentheses: $[\frac{1}{2}(8)]^2 = 16$.

$= 1(x^2 + 8x + 16 - 16) + 13$ Rather than add 16 to both sides of the function, we *add and subtract 16* within the parentheses on the right-hand side. This has the effect of adding 0 to the right-hand side.

$= 1(x^2 + 8x + 16) - 16 + 13$ Use the associative property of addition to regroup terms and isolate the perfect square trinomial within the parentheses.

$= (x + 4)^2 - 3$ Factor and simplify.

b. $f(x) = (x + 4)^2 - 3$

The vertex is at $(-4, -3)$.

The axis of symmetry is $x = -4$.

Because $a > 0$, the parabola opens up.

The minimum value is -3 (Figure 8-16).

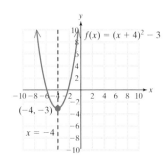

Figure 8-16

Answers

1. a. $f(x) = (x + 4)^2 - 17$
   b. Vertex: $(-4, -17)$; axis of symmetry: $x = -4$; minimum value: -17

example 2 Analyzing a Quadratic Function

Given: $f(x) = -2x^2 + 12x - 16$

a. Write the function in the form $f(x) = a(x - h)^2 + k$.
b. Find the vertex, axis of symmetry, and maximum function value.
c. Find the x- and y-intercepts.
d. Sketch the graph of the function.

Solution:

a. $f(x) = -2x^2 + 12x - 16$ To find the vertex, write the function in the form $f(x) = (x - h)^2 + k$.

$= -2(x^2 - 6x \quad\quad) - 16$ If the leading coefficient is not 1, factor the coefficient from the variable terms.

$= -2(x^2 - 6x + 9 - 9) - 16$ Add and subtract the quantity $[\frac{1}{2}(-6)]^2 = 9$ within the parentheses.

$= -2(x^2 - 6x + 9) + (-2)(-9) - 16$ To remove the term -9 from the parentheses, we must first apply the distributive property. When -9 is removed from the parentheses, it carries with it a factor of -2.

$= -2(x - 3)^2 + 18 - 16$ Factor and simplify.
$= -2(x - 3)^2 + 2$

b. $f(x) = -2(x - 3)^2 + 2$
The vertex is $(3, 2)$. The line of symmetry is $x = 3$. Because $a < 0$, the parabola opens down and the maximum value is 2.

c. The y-intercept is given by $f(0) = -2(0)^2 + 12(0) - 16 = -16$.
The y-intercept is $(0, -16)$.

To find the x-intercept(s), find the real solutions to the equation $f(x) = 0$.

$f(x) = -2x^2 + 12x - 16$
$0 = -2x^2 + 12x - 16$ Substitute $f(x) = 0$.
$0 = -2(x^2 - 6x + 8)$ Factor.
$0 = -2(x - 4)(x - 2)$
$x = 4 \quad\text{or}\quad x = 2$

The x-intercepts are $(4, 0)$ and $(2, 0)$.

Skill Practice

2. Given:
$g(x) = -x^2 + 6x - 5$

a. Write the equation in the form $g(x) = a(x - h)^2 + k$.
b. Identify the vertex, axis of symmetry, and maximum value of the function.
c. Determine the x- and y-intercepts.
d. Sketch the graph.

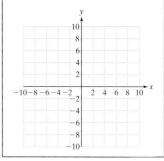

Answers

2. a. $g(x) = -(x - 3)^2 + 4$
b. Vertex: $(3, 4)$; axis of symmetry: $x = 3$; maximum value: 4
c. x-intercepts: $(5, 0)$ and $(1, 0)$; y-intercept: $(0, -5)$
d.

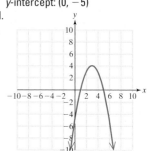

d. Using the information from parts (a)–(c), sketch the graph (Figure 8-17).

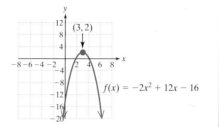

Figure 8-17

2. Vertex Formula

Completing the square and writing a quadratic function in the form $f(x) = a(x - h)^2 + k$ $(a \neq 0)$ is one method to find the vertex of a parabola. Another method is to use the vertex formula. The **vertex formula** can be derived by completing the square on the function defined by $f(x) = ax^2 + bx + c$ $(a \neq 0)$.

$f(x) = ax^2 + bx + c \quad (a \neq 0)$

$= a\left(x^2 + \dfrac{b}{a}x \phantom{+ \dfrac{b^2}{4a^2}}\right) + c$ \quad Factor a from the variable terms.

$= a\left(x^2 + \dfrac{b}{a}x + \dfrac{b^2}{4a^2} - \dfrac{b^2}{4a^2}\right) + c$ \quad Add and subtract $[\tfrac{1}{2}(b/a)]^2 = b^2/(4a^2)$ within the parentheses.

$= a\left(x^2 + \dfrac{b}{a}x + \dfrac{b^2}{4a^2}\right) + (a)\left(-\dfrac{b^2}{4a^2}\right) + c$ \quad Apply the distributive property and remove the term $-b^2/(4a^2)$ from the parentheses.

$= a\left(x + \dfrac{b}{2a}\right)^2 - \dfrac{b^2}{4a} + c$ \quad Factor the trinomial and simplify.

$= a\left(x + \dfrac{b}{2a}\right)^2 + c - \dfrac{b^2}{4a}$ \quad Apply the commutative property of addition to reverse the last two terms.

$= a\left(x + \dfrac{b}{2a}\right)^2 + \dfrac{4ac}{4a} - \dfrac{b^2}{4a}$ \quad Obtain a common denominator.

$= a\left(x + \dfrac{b}{2a}\right)^2 + \dfrac{4ac - b^2}{4a}$

$= a\left[x - \left(-\dfrac{b}{2a}\right)\right]^2 + \dfrac{4ac - b^2}{4a}$

$ \qquad \downarrow \qquad\qquad\qquad \downarrow$

$f(x) = a(x \quad - \quad h)^2 \quad + \quad k$

The function is in the form $f(x) = a(x - h)^2 + k$, where

$$h = \dfrac{-b}{2a} \quad \text{and} \quad k = \dfrac{4ac - b^2}{4a}$$

Hence, the vertex is at

$$\left(\dfrac{-b}{2a}, \dfrac{4ac - b^2}{4a}\right)$$

Although the *y*-coordinate of the vertex is given as $(4ac - b^2)/(4a)$, it is usually easier to determine the *x*-coordinate of the vertex first and then find *y* by evaluating the function at $x = -b/(2a)$.

The Vertex Formula

For $f(x) = ax^2 + bx + c$ $(a \neq 0)$, the vertex is given by

$$\left(\frac{-b}{2a}, \frac{4ac - b^2}{4a}\right) \quad \text{or} \quad \left(\frac{-b}{2a}, f\left(-\frac{b}{2a}\right)\right)$$

3. Determining the Vertex and Intercepts of a Quadratic Function

example 3   Determining the Vertex and Intercepts of a Quadratic Function

Given: $h(x) = x^2 - 2x + 5$

a. Use the vertex formula to find the vertex.

b. Find the *x*- and *y*-intercepts.

c. Sketch the function.

Solution:

a. $h(x) = x^2 - 2x + 5$

$a = 1 \quad b = -2 \quad c = 5 \qquad$ Identify *a*, *b*, and *c*.

The *x*-coordinate of the vertex is $\dfrac{-b}{2a} = \dfrac{-(-2)}{2(1)} = 1$.

The *y*-coordinate of the vertex is $h(1) = (1)^2 - 2(1) + 5 = 4$.

The vertex is $(1, 4)$.

b. The *y*-intercept is given by $h(0) = (0)^2 - 2(0) + 5 = 5$.

The *y*-intercept is $(0, 5)$.

To find the *x*-intercept(s), find the real solutions to the equation $h(x) = 0$.

$h(x) = x^2 - 2x + 5$

$0 = x^2 - 2x + 5 \qquad$ This quadratic equation is not factorable. Apply the quadratic formula: $a = 1, b = -2, c = 5$.

$$x = \frac{-(-2) \pm \sqrt{(-2)^2 - 4(1)(5)}}{2(1)}$$

$$= \frac{2 \pm \sqrt{4 - 20}}{2(1)}$$

$$= \frac{2 \pm \sqrt{-16}}{2}$$

$$= \frac{2 \pm 4i}{2}$$

$$= 1 \pm 2i$$

Skill Practice

3. Given: $f(x) = x^2 + 4x + 6$

a. Use the vertex formula to find the vertex of the parabola.

b. Determine the *x*- and *y*-intercepts.

c. Sketch the graph.

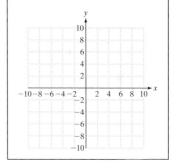

Answers

3. a. Vertex: $(-2, 2)$
b. *x*-intercepts: none;
y-intercepts: $(0, 6)$
c.

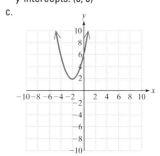

The solutions to the equation $h(x) = 0$ are not real numbers. Therefore, there are no x-intercepts.

c.

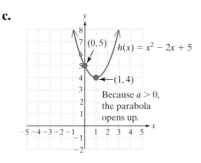

Figure 8-18

Tip: The location of the vertex and the direction that the parabola opens can be used to determine whether the function has any x-intercepts.

Given $h(x) = x^2 - 2x + 5$, the vertex $(1, 4)$ is *above* the x-axis. Furthermore, because $a > 0$, the parabola opens upward. Therefore, it is not possible for the function h to cross the x-axis (Figure 8-18).

4. Vertex of a Parabola: Applications

Skill Practice

4. An object is launched into the air with an initial velocity of 48 ft/sec from the top of a building 288 ft high. The height $h(t)$ of the object after t seconds is given by

   $h(t) = -16t^2 + 48t + 288$

   a. Find the time it takes for the object to reach its maximum height.
   b. Find the maximum height.

example 4 Applying a Quadratic Function

The crew from Extravaganza Entertainment launches fireworks at an angle of 60° from the horizontal. The height of one particular type of display can be approximated by the following function:

$$h(t) = -16t^2 + 128\sqrt{3}\,t$$

where $h(t)$ is measured in feet and t is measured in seconds.

a. How long will it take the fireworks to reach their maximum height? Round to the nearest second.

b. Find the maximum height. Round to the nearest foot.

Solution:

$h(t) = -16t^2 + 128\sqrt{3}\,t$ This parabola opens downward; therefore, the maximum height of the fireworks will occur at the vertex of the parabola.

$a = -16 \quad b = 128\sqrt{3} \quad c = 0$ Identify a, b, and c, and apply the vertex formula.

The x-coordinate of the vertex is

$$\frac{-b}{2a} = \frac{-128\sqrt{3}}{2(-16)} = \frac{-128\sqrt{3}}{-32} \approx 6.9$$

The y-coordinate of the vertex is approximately

$$h(6.9) = -16(6.9)^2 + 128\sqrt{3}(6.9) \approx 768$$

The vertex is $(6.9, 768)$.

a. The fireworks will reach their maximum height in 6.9 sec.

b. The maximum height is 768 ft.

Answers

4. a. 1.5 sec   b. 324 ft

Calculator Connections

Some graphing calculators have *Minimum* and *Maximum* features that enable the user to approximate the minimum and maximum values of a function. Otherwise, *Zoom* and *Trace* can be used.

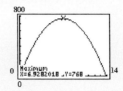

example 5 Applying a Quadratic Function

A group of students start a small company that sells used CDs on the Internet. The weekly profit (in dollars) of the company is given by the function

$$P(x) = -2x^2 + 80x - 600$$

where x represents the number of CDs produced.

a. Find the x-intercepts of the profit function, and interpret the meaning of the x-intercepts in the context of this problem.

b. Find the y-intercept of the profit function, and interpret its meaning in the context of this problem.

c. Find the vertex of the profit function, and interpret its meaning in the context of this problem.

d. Sketch the profit function.

Solution:

a. $P(x) = -2x^2 + 80x - 600$ The x-intercepts are the real solutions of the equation $P(x) = 0$.

$0 = -2x^2 + 80x - 600$

$0 = -2(x^2 - 40x + 300)$

$0 = -2(x - 10)(x - 30)$ Solve by factoring.

$x = 10$ or $x = 30$

The x-intercepts are $(10, 0)$ and $(30, 0)$. The x-intercepts represent points where the profit is zero. These are called break-even points. The break-even points occur when 10 CDs are produced and also when 30 CDs are produced.

b. The y-intercept is $P(0) = -2(0)^2 + 80(0) - 600 = -600$. The y-intercept is $(0, -600)$. The y-intercept indicates that if no CDs are produced, the company has a $600 loss.

c. The x-coordinate of the vertex is

$$\frac{-b}{2a} = \frac{-80}{2(-2)} = 20$$

The y-coordinate is $P(20) = -2(20)^2 + 80(20) - 600 = 200$.

The vertex is $(20, 200)$.
A maximum weekly profit of $200 is obtained when 20 CDs are produced.

Skill Practice

5. The weekly profit function for a tutoring service is given by $P(x) = -4x^2 + 120x - 500$, where x is the number of hours of tutoring and $P(x)$ represents profit in dollars.

   a. Find the x-intercepts of the profit function, and interpret their meaning in the context of this problem.

   b. Find the y-intercept of the profit function, and interpret its meaning in the context of this problem.

   c. Find the vertex of the profit function, and interpret its meaning in the context of this problem.

   d. Sketch the profit function.

Answers

5. **a.** x-intercepts: $(5, 0)$ and $(25, 0)$. The x-intercepts represent the break-even points, where the profit is 0.

   b. y-intercept: $(0, -500)$. If the service does no tutoring, it will have a $500 loss.

   c. Vertex: $(15, 400)$. A maximum weekly profit of $400 is obtained when 15 hr of tutoring is done.

   d.
   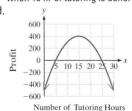
   Number of Tutoring Hours

d. Using the information from parts (a)–(c), sketch the profit function (Figure 8-19).

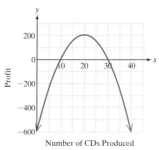

Figure 8-19

Skill Practice

6. In a recent year, an Internet company started up, became successful, and then quickly went bankrupt. The price of a share of the company's stock is given by the function

   $S(x) = -2.25x^2 + 40.5x + 42.75$

   where x is the number of months that the stock was on the market. Find the number of months at which the stock price was a maximum.

example 6   Applying a Quadratic Function

The average number of visits to office-based physicians is a function of the age of the patient.

$$N(x) = 0.0014x^2 - 0.0658x + 2.65$$

where x is a patient's age in years and $N(x)$ is the average number of doctor visits per year (Figure 8-20). Find the age for which the number of visits to office-based physicians is a minimum.

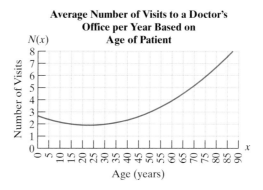

Figure 8-20

Source: U.S. National Center for Health Statistics.

Solution:

$N(x) = 0.0014x^2 - 0.0658x + 2.65$

$\dfrac{-b}{2a} = \dfrac{-(-0.0658)}{2(0.0014)} = 23.5$ Find the x-coordinate of the vertex.

The average number of visits to office-based physicians is lowest for people approximately 23.5 years old.

Answer

6. The maximum price for the stock occurred after 9 months.

section 8.5 Practice Exercises

Boost your GRADE at mathzone.com!

- Practice Problems
- Self-Tests
- NetTutor
- e-Professors
- Videos

Study Skills Exercises

1. Make a list of all the section titles in the chapter that you are studying. Write each section title on a separate sheet of paper or index card. Go back and fill in the list of objectives under each section title. When you are studying for the test, try to make up an exercise that corresponds to each objective and then work the exercise. To get started, write a problem for the objective Solving Quadratic Equations by Using the Square Root Property.

2. Define the key term **vertex formula**.

Review Exercises

3. How does the graph of $f(x) = -2x^2$ compare with the graph of $y = x^2$?

4. How does the graph of $p(x) = \frac{1}{4}x^2$ compare with the graph of $y = x^2$?

5. How does the graph of $Q(x) = x^2 - \frac{8}{3}$ compare with the graph of $y = x^2$?

6. How does the graph of $r(x) = x^2 + 7$ compare with the graph of $y = x^2$?

7. How does the graph of $s(x) = (x - 4)^2$ compare with the graph of $y = x^2$?

8. How does the graph of $t(x) = (x + 10)^2$ compare with the graph of $y = x^2$?

For Exercises 9–16, find the value of k to complete the square.

9. $x^2 - 8x + k$
10. $x^2 + 4x + k$
11. $y^2 + 7y + k$
12. $a^2 - a + k$

13. $b^2 + \frac{2}{9}b + k$
14. $m^2 - \frac{2}{7}m + k$
15. $t^2 - \frac{1}{3}t + k$
16. $p^2 + \frac{1}{4}p + k$

Objective 1: Writing a Quadratic Function in the Form $f(x) = a(x - h)^2 + k$

For Exercises 17–30, write the function in the form $f(x) = a(x - h)^2 + k$ by completing the square. Then identify the vertex. (See Example 1.)

17. $g(x) = x^2 - 8x + 5$
18. $h(x) = x^2 + 4x + 5$
19. $n(x) = 2x^2 + 12x + 13$

20. $f(x) = 4x^2 + 16x + 19$
21. $p(x) = -3x^2 + 6x - 5$
22. $q(x) = -2x^2 + 12x - 11$

23. $k(x) = x^2 + 7x - 10$

24. $m(x) = x^2 - x - 8$

25. $f(x) = x^2 + 8x + 1$

26. $g(x) = x^2 + 5x - 2$

27. $F(x) = 5x^2 + 10x + 1$

28. $G(x) = 4x^2 + 4x - 7$

29. $P(x) = -2x^2 + x$

30. $Q(x) = 3x^2 - 12x$

Objective 2: Vertex Formula

For Exercises 31–44, find the vertex by using the vertex formula. (See Example 3.)

31. $Q(x) = x^2 - 4x + 7$

32. $T(x) = x^2 - 8x + 17$

33. $r(x) = -3x^2 - 6x - 5$

34. $s(x) = -2x^2 - 12x - 19$

35. $N(x) = x^2 + 8x + 1$

36. $M(x) = x^2 + 6x - 5$

37. $m(x) = \frac{1}{2}x^2 + x + \frac{5}{2}$

38. $n(x) = \frac{1}{2}x^2 + 2x + 3$

39. $k(x) = -x^2 + 2x + 2$

40. $h(x) = -x^2 + 4x - 3$

41. $f(x) = 2x^2 + 4x + 6$

42. $g(x) = 3x^2 + 12x + 9$

43. $A(x) = -\frac{1}{3}x^2 + x$

44. $B(x) = -\frac{2}{3}x^2 - 2x$

Objective 3: Determining the Vertex and Intercepts of a Quadratic Function

For Exercises 45–50

    a. Find the vertex.

    b. Find the y-intercept.

    c. Find the x-intercept(s), if they exist.

    d. Use this information to graph the function. (See Examples 2–3.)

45. $y = x^2 + 9x + 8$

46. $y = x^2 + 7x + 10$

47. $y = 2x^2 - 2x + 4$

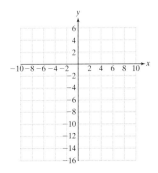
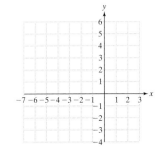
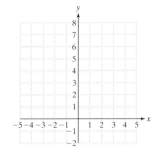

48. $y = 2x^2 - 12x + 19$

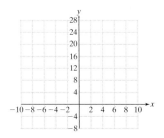

49. $y = -x^2 + 3x - \dfrac{9}{4}$

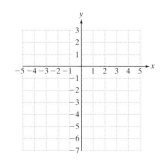

50. $y = -x^2 - \dfrac{3}{2}x - \dfrac{9}{16}$

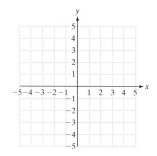

Objective 4: Vertex of a Parabola: Applications

51. Mia sells MP3 players. The average cost to produce MP3 players is given by the equation $C(x) = 2x^2 - 40x + 2200$, where x is the number of MP3 players she produces in a week. How many players must she produce to minimize her average cost? **(See Example 4.)**

52. Ben sells iPods. The average cost to produce iPods is given by the equation $C(x) = 3x^2 - 120x - 1300$, where x is the number of iPods produced per month. Determine the number of iPods that Ben needs to produce to minimize the average cost.

53. The pressure x in an automobile tire can affect its wear. Both overinflated and underinflated tires can lead to poor performance and poor mileage. For one particular tire, the function P represents the number of miles that a tire lasts (in thousands) for a given pressure x.

$$P(x) = -0.857x^2 + 56.1x - 880$$

where x is the tire pressure in pounds per square inch (psi).

  a. Find the tire pressure that will yield the maximum mileage. Round to the nearest pound per square inch.

  b. What is the maximum number of miles that a tire can last? Round to the nearest thousand.

54. A baseball player throws a ball, and the height of the ball (in feet) can be approximated by

$$y(x) = -0.011x^2 + 0.577x + 5$$

where x is the horizontal position of the ball measured in feet from the origin.

  a. For what value of x will the ball reach its highest point? Round to the nearest foot.

  b. What is the maximum height of the ball?

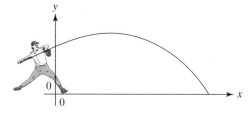

55. For a fund-raising activity, a charitable organization produces cookbooks to sell in the community. The profit (in dollars) depends on the number of cookbooks produced, x, according to

$$P(x) = -\frac{1}{50}x^2 + 12x - 550, \text{ where } x \geq 0.$$ (See Example 5.)

a. How much profit is made when 100 cookbooks are produced?

b. Find the y-intercept of the profit function, and interpret its meaning in the context of this problem.

c. How many cookbooks must be produced for the organization to break even? (*Hint:* Find the x-intercepts.)

d. Find the vertex.

e. Sketch the function.

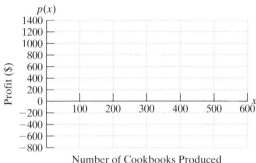
Number of Cookbooks Produced

f. How many cookbooks must be produced to maximize profit? What is the maximum profit?

56. A jewelry maker sells bracelets at art shows. The profit (in dollars) depends on the number of bracelets produced, x, according to

$$P(x) = -\frac{1}{10}x^2 + 42x - 1260 \text{ where } x \geq 0.$$

a. How much profit does the jeweler make when 10 bracelets are produced?

b. Find the y-intercept of the profit function, and interpret its meaning in the context of this problem.

c. How many bracelets must be produced for the jeweler to break even? Round to the nearest whole unit.

d. Find the vertex.

e. Sketch the function.

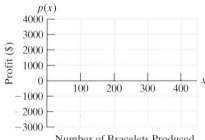
Number of Bracelets Produced

f. How many bracelets must be produced to maximize profit? What is the maximum profit?

57. Gas mileage depends in part on the speed of the car. The gas mileage of a subcompact car is given by the function $m(x) = -0.04x^2 + 3.6x - 49$, where x represents the speed in miles per hour and $m(x)$ is given in miles per gallon. At what speed will the car get the maximum gas mileage? (See Example 6.)

58. Gas mileage depends in part on the speed of the car. The gas mileage of a luxury car is given by the function $L(x) = -0.015x^2 + 1.44x - 21$, where x represents the speed in miles per hour and $L(x)$ is given in miles per gallon. At what speed will the car get the maximum gas mileage?

59. Tetanus bacillus bacterium is cultured to produce tetanus toxin used in an inactive form for the tetanus vaccine. The amount of toxin produced per batch increases with time and then becomes unstable. The amount of toxin (in grams) as a function of time t (in hours) can be approximated by the following function.

$$b(t) = -\frac{1}{1152}t^2 + \frac{1}{12}t$$

How many hours will it take to produce the maximum yield?

60. The bacterium *Pseudomonas aeruginosa* is cultured with an initial population of 10^4 active organisms. The population of active bacteria increases up to a point, and then due to a limited food supply and an increase of waste products, the population of living organisms decreases. Over the first 48 hr, the population can be approximated by the following function.

$$P(t) = -1718.75t^2 + 82{,}500t + 10{,}000 \quad \text{where } 0 \leq t \leq 48$$

Find the time required for the population to reach its maximum value. Round to the nearest hour.

Expanding Your Skills

61. A farmer wants to fence a rectangular corral adjacent to the side of a barn; however, she has only 200 ft of fencing and wants to enclose the largest possible area. See the figure.

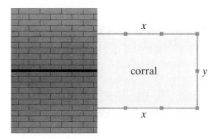

a. If x represents the length of the corral and y represents the width, explain why the dimensions of the corral are subject to the constraint $2x + y = 200$.

b. The area of the corral is given by $A = xy$. Use the constraint equation from part (a) to express A as a function of x, where $0 < x < 100$.

c. Use the function from part (b) to find the dimensions of the corral that will yield the maximum area. [*Hint:* Find the vertex of the function from part (b).]

62. A veterinarian wants to construct two equal-sized pens of maximum area out of 240 ft of fencing. See the figure.

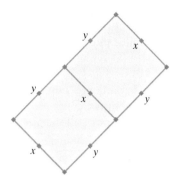

a. If x represents the length of each pen and y represents the width of each pen, explain why the dimensions of the pens are subject to the constraint $3x + 4y = 240$.

b. The area of each individual pen is given by $A = xy$. Use the constraint equation from part (a) to express A as a function of x, where $0 < x < 80$.

c. Use the function from part (b) to find the dimensions of an individual pen that will yield the maximum area. [*Hint:* Find the vertex of the function from part (b).]

Graphing Calculator Exercises

For Exercises 63–68, graph the functions in Exercises 45–50 on a graphing calculator. Use the *Max* or *Min* feature or *Zoom* and *Trace* to approximate the vertex.

63. $Y_1 = x^2 + 9x + 8$ (Exercise 45)

64. $Y_1 = x^2 + 7x + 10$ (Exercise 46)

65. $Y_1 = 2x^2 - 2x + 4$ (Exercise 47)

66. $Y_1 = 2x^2 - 12x + 19$ (Exercise 48)

67. $Y_1 = -x^2 + 3x - \dfrac{9}{4}$ (Exercise 49)

68. $Y_1 = -x^2 - \dfrac{3}{2}x - \dfrac{9}{16}$ (Exercise 50)

chapter 8 | summary

section 8.1 Completing the Square and the Square Root Property

Key Concepts

The **square root property** states that

If $x^2 = k$ then $x = \pm\sqrt{k}$

Follow these steps to solve a quadratic equation in the form $ax^2 + bx + c = 0\ (a \neq 0)$ by completing the square and applying the square root property:

1. Divide both sides by a to make the leading coefficient 1.
2. Isolate the variable terms on one side of the equation.
3. Complete the square: Add the square of one-half the linear term coefficient to both sides of the equation. Then factor the resulting perfect square trinomial.
4. Apply the square root property and solve for x.

Examples

Example 1

$(x - 5)^2 = -13$

$x - 5 = \pm\sqrt{-13}$ (square root property)

$x = 5 \pm i\sqrt{13}$

Example 2

$2x^2 - 6x - 5 = 0$

$\dfrac{2x^2}{2} - \dfrac{6x}{2} - \dfrac{5}{2} = \dfrac{0}{2}$

$x^2 - 3x = \dfrac{5}{2}$

Note: $\left[\dfrac{1}{2} \cdot (-3)\right]^2 = \dfrac{9}{4}$

$x^2 - 3x + \dfrac{9}{4} = \dfrac{5}{2} + \dfrac{9}{4}$

$\left(x - \dfrac{3}{2}\right)^2 = \dfrac{19}{4}$

$x - \dfrac{3}{2} = \pm\sqrt{\dfrac{19}{4}}$

$x = \dfrac{3}{2} \pm \dfrac{\sqrt{19}}{2}$ or $x = \dfrac{3 \pm \sqrt{19}}{2}$

Chapter 8 Quadratic Equations and Functions

section 8.2 Quadratic Formula

Key Concepts

The solutions to a quadratic equation $ax^2 + bx + c = 0$ $(a \neq 0)$ are given by the **quadratic formula**

$$x = \frac{-b \pm \sqrt{b^2 - 4ac}}{2a}$$

The **discriminant** of a quadratic equation $ax^2 + bx + c = 0$ is $b^2 - 4ac$. If a, b, and c are rational numbers, then

1. If $b^2 - 4ac > 0$, then there will be two real solutions. Moreover,
   a. If $b^2 - 4ac$ is a perfect square, the solutions will be rational numbers.
   b. If $b^2 - 4ac$ is not a perfect square, the solutions will be irrational numbers.
2. If $b^2 - 4ac < 0$, then there will be two imaginary solutions.
3. If $b^2 - 4ac = 0$, then there will be one rational solution.

Three methods to solve a quadratic equation are

1. Factoring and applying the zero product rule.
2. Completing the square and applying the square root property
3. Using the quadratic formula

Example

Example 1

$0.03x^2 - 0.02x + 0.04 = 0$

$3x^2 - 2x + 4 = 0$ (multiply by 100)

$a = 3 \quad b = -2 \quad c = 4$

$x = \dfrac{-(-2) \pm \sqrt{(-2)^2 - 4(3)(4)}}{2(3)}$

$= \dfrac{2 \pm \sqrt{4 - 48}}{6}$

$= \dfrac{2 \pm \sqrt{-44}}{6}$ The discriminant is -44. Therefore, we will have two imaginary solutions.

$= \dfrac{2 \pm 2i\sqrt{11}}{6}$

$= \dfrac{1 \pm i\sqrt{11}}{3}$

section 8.3 Equations in Quadratic Form

Key Concepts

Equations may be written in quadratic form.

Substitution may be used to solve equations that are in quadratic form.

Examples

Example 1

$$16y^4 - 1 = 0$$
$$(4y^2)^2 - 1^2 = 0$$
$$(4y^2 - 1)(4y^2 + 1) = 0 \quad \text{Factor.}$$

$4y^2 - 1 = 0$ or $4y^2 + 1 = 0$ Apply the zero product rule.

$4y^2 = 1$ or $4y^2 = -1$

$y^2 = \dfrac{1}{4}$ or $y^2 = -\dfrac{1}{4}$

$y = \pm\sqrt{\dfrac{1}{4}}$ or $y = \pm\sqrt{-\dfrac{1}{4}}$ Apply the square root property.

$y = \pm\dfrac{1}{2}$ or $y = \pm\dfrac{1}{2}i$

Example 2

$x^{2/3} - x^{1/3} - 12 = 0$

Let $u = x^{1/3}$. Therefore, $u^2 = (x^{1/3})^2 = x^{2/3}$

$$u^2 - u - 12 = 0$$
$$(u - 4)(u + 3) = 0$$

$u = 4$ or $u = -3$

$x^{1/3} = 4$ or $x^{1/3} = -3$

$x = 64$ or $x = -27$ Cube both sides.

Chapter 8 Quadratic Equations and Functions

section 8.4 Graphs of Quadratic Functions

Key Concepts

A quadratic function of the form $f(x) = x^2 + k$ shifts the graph of $y = x^2$ up k units if $k > 0$ and down $|k|$ units if $k < 0$.

Examples

Example 1

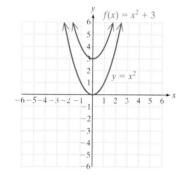

A quadratic function of the form $f(x) = (x - h)^2$ shifts the graph of $y = x^2$ right h units if $h > 0$ and left $|h|$ units if $h < 0$.

Example 2

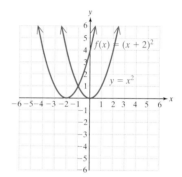

The graph of a quadratic function of the form $f(x) = ax^2$ is a parabola that opens up when $a > 0$ and opens down when $a < 0$. If $|a| > 1$, the graph of $y = x^2$ is stretched vertically by a factor of $|a|$. If $0 < |a| < 1$, the graph of $y = x^2$ is shrunk vertically by a factor of $|a|$.

Example 3

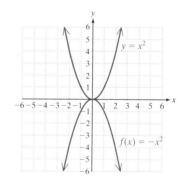

A quadratic function of the form $f(x) = a(x - h)^2 + k$ has vertex (h, k). If $a > 0$, the vertex represents the minimum point. If $a < 0$, the vertex represents the maximum point.

Example 4

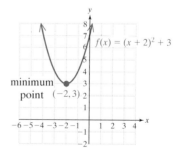

section 8.5 Vertex of a Parabola and Applications

Key Concepts

Completing the square is a technique used to write a quadratic function $f(x) = ax^2 + bx + c$ $(a \neq 0)$ in the form $f(x) = a(x - h)^2 + k$ for the purpose of identifying the vertex (h, k).

The **vertex formula** finds the vertex of a quadratic function $f(x) = ax^2 + bx + c$ $(a \neq 0)$.

The vertex is

$$\left(\frac{-b}{2a}, \frac{4ac - b^2}{4a}\right) \quad \text{or} \quad \left(\frac{-b}{2a}, f\left(\frac{-b}{2a}\right)\right)$$

Examples

Example 1

$f(x) = 3x^2 + 6x + 11$

$= 3(x^2 + 2x) + 11$

$= 3(x^2 + 2x + 1 - 1) + 11$

$= 3(x^2 + 2x + 1) - 3 + 11$

$= 3(x + 1)^2 + 8$

$= 3[x - (-1)]^2 + 8$

The vertex is $(-1, 8)$. Because $a = 3 > 0$, the vertex $(-1, 8)$ is a minimum point.

Example 2

$f(x) = -5x^2 + 4x - 1$

$a = -5 \quad b = 4 \quad c = -1$

$x = \dfrac{-4}{2(-5)} = \dfrac{2}{5}$

$f\left(\dfrac{2}{5}\right) = -5\left(\dfrac{2}{5}\right)^2 + 4\left(\dfrac{2}{5}\right) - 1 = -\dfrac{1}{5}$

The vertex is $(\frac{2}{5}, -\frac{1}{5})$. Because $a = -5 < 0$, the vertex $(\frac{2}{5}, -\frac{1}{5})$ is a maximum point.

---

chapter 8 | review exercises

Section 8.1

For Exercises 1–8, solve the equations by using the square root property.

1. $x^2 = 5$
2. $2y^2 = -8$
3. $a^2 = 81$
4. $3b^2 = -19$
5. $(x - 2)^2 = 72$
6. $(2x - 5)^2 = -9$
7. $(3y - 1)^2 = 3$
8. $3(m - 4)^2 = 15$

9. The length of each side of an equilateral triangle is 10 in. Find the height of the triangle. Round the answer to the nearest tenth of an inch.

10. Use the square root property to find the length of the sides of a square whose area is 81 in.²

11. Use the square root property to find the length of the sides of a square whose area is 150 in.² Round the answer to the nearest tenth of an inch.

For Exercises 12–15, find the value of k so that the expression is a perfect square trinomial. Then factor the trinomial.

12. $x^2 + 16x + k$

13. $x^2 - 9x + k$

14. $y^2 + \frac{1}{2}y + k$

15. $z^2 - \frac{2}{5}z + k$

For Exercises 16–21, solve the equation by completing the square and applying the square root property.

16. $w^2 + 4w + 13 = 0$

17. $4y^2 - 12y + 13 = 0$

18. $3x^2 + 2x = 1$

19. $b^2 + \frac{7}{2}b = 2$

20. $2x^2 = 12x + 6$

21. $-t^2 + 8t - 25 = 0$

Section 8.2

22. Explain how the discriminant can determine the type and number of solutions to a quadratic equation with rational coefficients.

For Exercises 23–28, determine the type (rational, irrational, or imaginary) and number of solutions for the equations by using the discriminant.

23. $x^2 - 5x = -6$

24. $2y^2 = -3y$

25. $z^2 + 23 = 17z$

26. $a^2 + a + 1 = 0$

27. $10b + 1 = -25b^2$

28. $3x^2 + 15 = 0$

For Exercises 29–36, solve the equations by using the quadratic formula.

29. $y^2 - 4y + 1 = 0$

30. $m^2 - 5m + 25 = 0$

31. $6a(a - 1) = 10 + a$

32. $3x(x - 3) = x - 8$

33. $b^2 - \frac{4}{25} = \frac{3}{5}b$

34. $k^2 + 0.4k = 0.05$

35. $32 + 4x - x^2 = 0$

36. $8y - y^2 = 0$

37. The landing distance that a certain plane will travel on a runway is determined by the initial landing speed at the instant the plane touches down. The function D relates landing distance in feet to initial landing speed s:

$$D(s) = \frac{1}{10}s^2 - 3s + 22$$

where s is in feet per second.

a. Find the landing distance for a plane traveling 150 ft/sec at touchdown.

b. If the landing speed is too fast, the pilot may run out of runway. If the speed is too slow, the plane may stall. Find the maximum initial landing speed of a plane for a runway that is 1000 ft long. Round to 1 decimal place.

38. The recent population (in thousands) of Kenya can be approximated by $P(t) = 4.62t^2 + 564.6t + 13{,}128$, where t is the number of years since 1974.

a. If this trend continues, approximate the number of people in Kenya in the year 2025.

b. In what year after 1974 will the population of Kenya reach 50 million? (*Hint:* 50 million equals 50,000 thousand.)

Section 8.3

For Exercises 39–48, solve the equations by using substitution, if necessary.

39. $x - 4\sqrt{x} - 21 = 0$

40. $n - 6\sqrt{n} + 8 = 0$

41. $y^4 - 11y^2 + 18 = 0$

42. $2m^4 - m^2 - 3 = 0$

43. $t^{2/5} + t^{1/5} - 6 = 0$

44. $p^{2/5} - 3p^{1/5} + 2 = 0$

45. $\sqrt{4a-3} - \sqrt{8a+1} = -2$

46. $\sqrt{2b-5} - \sqrt{b-2} = 2$

47. $(x^2+5)^2 + 2(x^2+5) - 8 = 0$

48. $(x^2-3)^2 - 5(x^2-3) + 4 = 0$

Section 8.4

For Exercises 49–56, graph the functions.

49. $g(x) = x^2 - 5$

50. $f(x) = x^2 + 3$

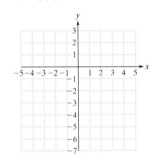
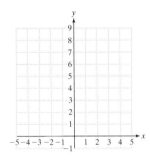

51. $h(x) = (x-5)^2$

52. $k(x) = (x+3)^2$

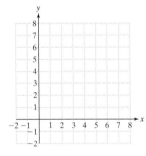
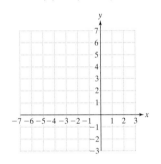

53. $m(x) = -2x^2$

54. $n(x) = -4x^2$

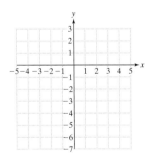
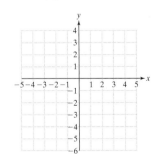

55. $p(x) = -2(x-5)^2 - 5$

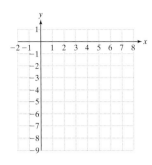

56. $q(x) = -4(x+3)^2 + 3$

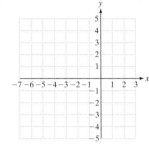

For Exercises 57–58, write the coordinates of the vertex of the parabola and determine if the vertex is a maximum point or a minimum point. Then write the maximum or the minimum value.

57. $t(x) = \dfrac{1}{3}(x-4)^2 + \dfrac{5}{3}$

58. $s(x) = -\dfrac{5}{7}(x-1)^2 - \dfrac{1}{7}$

For Exercises 59–60, write the equation of the axis of symmetry of the parabola.

59. $a(x) = -\dfrac{3}{2}\left(x + \dfrac{2}{11}\right)^2 - \dfrac{4}{13}$

60. $w(x) = -\dfrac{4}{3}\left(x - \dfrac{3}{16}\right)^2 + \dfrac{2}{9}$

Section 8.5

For Exercises 61–64, write the function in the form $f(x) = a(x-h)^2 + k$ by completing the square. Then write the coordinates of the vertex.

61. $z(x) = x^2 - 6x + 7$

62. $b(x) = x^2 - 4x - 44$

63. $p(x) = -5x^2 - 10x - 13$

64. $q(x) = -3x^2 - 24x - 54$

For Exercises 65–68, find the coordinates of the vertex of each function by using the vertex formula.

65. $f(x) = -2x^2 + 4x - 17$

66. $g(x) = -4x^2 - 8x + 3$

67. $m(x) = 3x^2 - 3x + 11$

68. $n(x) = 3x^2 + 2x - 7$

69. For the quadratic equation $y = -(x + 2)^2 + 4$

    a. Write the coordinates of the vertex.

    b. Find the x- and y-intercepts.

    c. Use this information to sketch a graph of the parabola.

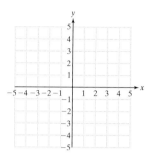

70. The height of a projectile fired vertically into the air from the ground is given by the equation $h(t) = -16t^2 + 96t$, where t represents the number of seconds that the projectile has been in the air. How long will it take the projectile to reach its maximum height?

chapter 8 test

For Exercises 1–3, solve the equation by using the square root property.

1. $(x + 3)^2 = 25$

2. $(p - 2)^2 = 12$

3. $(m + 1)^2 = -1$

4. Find the value of k so that the expression is a perfect square trinomial. Then factor the trinomial $d^2 + 7d + k$.

For Exercises 5–6, solve the equation by completing the square and applying the square root property.

5. $2x^2 + 12x - 36 = 0$

6. $2x^2 = 3x - 7$

For Exercises 7–8

    a. Write the equation in standard form $ax^2 + bx + c = 0$.

    b. Identify a, b, and c.

    c. Find the discriminant.

    d. Determine the number and type (rational, irrational, or imaginary) of solutions.

7. $x^2 - 3x = -12$

8. $y(y - 2) = -1$

For Exercises 9–10, solve the equation by using the quadratic formula.

9. $3x^2 - 4x + 1 = 0$

10. $x(x + 6) = -11 - x$

11. The base of a triangle is 3 ft less than twice the height. The area of the triangle is 14 ft². Find the base and the height. Round the answers to the nearest tenth of a foot.

12. A circular garden has an area of approximately 450 ft². Find the radius. Round the answer to the nearest tenth of a foot.

For Exercises 13–15, solve the equation by using substitution, if necessary.

13. $x - \sqrt{x} - 6 = 0$

14. $y^{2/3} + 2y^{1/3} = 8$

15. $(3y - 8)^2 - 13(3y - 8) + 30 = 0$

For Exercises 16–19, find the x- and y-intercepts of the function. Then match the function with its graph.

16. $f(x) = x^2 - 6x + 8$

17. $k(x) = x^3 + 4x^2 - 9x - 36$

18. $p(x) = -2x^2 - 8x - 6$

19. $q(x) = x^3 - x^2 - 12x$

a.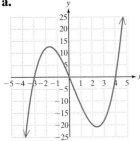

b.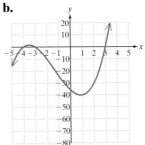

c.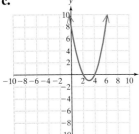

d.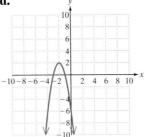

20. A child launches a toy rocket from the ground. The height of the rocket can be determined by its horizontal distance from the launch pad x by
$$h(x) = -\frac{x^2}{256} + x$$
where x and $h(x)$ are in feet.

How many feet from the launch pad will the rocket hit the ground?

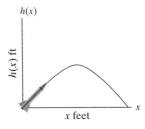

21. The recent population (in millions) of India can be approximated by $P(t) = 0.135t^2 + 12.6t + 600$, where $t = 0$ corresponds to the year 1974.

a. If this trend continues, approximate the number of people in India in the year 2014.

b. Approximate the year in which the population of India reached 1 billion (1000 million). (Round to the nearest year.)

22. Explain the relationship between the graphs of $y = x^2$ and $y = x^2 - 2$.

23. Explain the relationship between the graphs of $y = x^2$ and $y = (x + 3)^2$.

24. Explain the relationship between the graphs of $y = 4x^2$ and $y = -4x^2$.

25. Given the function defined by
$$f(x) = -(x - 4)^2 + 2$$

a. Identify the vertex of the parabola.

b. Does this parabola open up or down?

c. Does the vertex represent the maximum or minimum point of the function?

d. What is the maximum or minimum value of the function f?

e. What is the axis of symmetry for this parabola?

26. For the function defined by $g(x) = 2x^2 - 20x + 51$, find the vertex by using two methods.

a. Complete the square to write g in the form $g(x) = a(x - h)^2 + k$. Identify the vertex.

b. Use the vertex formula to find the vertex.

27. A farmer has 400 ft of fencing with which to enclose a rectangular field. The field is situated such that one of its sides is adjacent to a river and requires no fencing. The area of the field (in square feet) can be modeled by

$$A(x) = -\frac{x^2}{2} + 200x$$

where x is the length of the side parallel to the river (measured in feet).

Use the function to determine the maximum area that can be enclosed.

chapters 1–8 | cumulative review exercises

1. Given: $A = \{2, 4, 6, 8, 10\}$ and $B = \{2, 8, 12, 16\}$
   a. Find $A \cup B$.   **b.** Find $A \cap B$.

2. Perform the indicated operations and simplify.
$$(2x^2 - 5) - (x + 3)(5x - 2)$$

3. Simplify completely. $4^0 - \left(\dfrac{1}{2}\right)^{-3} - 81^{1/2}$

4. Perform the indicated operations. Write the answer in scientific notation:
$$(3.0 \times 10^{12})(6.0 \times 10^{-3})$$

5. a. Factor completely. $x^3 + 2x^2 - 9x - 18$

   b. Divide by using long division. Identify the quotient and remainder.
$$(x^3 + 2x^2 - 9x - 18) \div (x - 3)$$

6. Multiply. $(\sqrt{x} - \sqrt{2})(\sqrt{x} + \sqrt{2})$

7. Simplify. $\dfrac{4}{\sqrt{2x}}$

8. Jacques invests a total of $10,000 in two mutual funds. After 1 year, one fund produced 12% growth, and the other lost 3%. Find the amount invested in each fund if the total investment grew by $900.

9. Solve the system of equations.
$$\frac{1}{9}x - \frac{1}{3}y = -\frac{13}{9}$$
$$x - \frac{1}{2}y = \frac{9}{2}$$

10. An object is fired straight up into the air from an initial height of 384 ft with an initial velocity of 160 ft/sec. The height in feet is given by
$$h(t) = -16t^2 + 160t + 384$$
where t is the time in seconds after launch.

   a. Find the height of the object after 3 sec.

   b. Find the height of the object after 7 sec.

   c. Find the time required for the object to hit the ground.

11. Solve the equation. $(x - 3)^2 + 16 = 0$

12. Solve the equation. $2x^2 + 5x - 1 = 0$

13. What number would have to be added to the quantity $x^2 + 10x$ to make it a perfect square trinomial?

14. Factor completely. $2x^3 + 250$

15. Graph the line.
    $3x - 5y = 10$

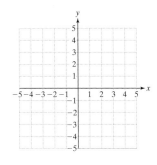

16. a. Find the x-intercepts of the function defined by $g(x) = 2x^2 - 9x + 10$.

    b. What is the y-intercept of $y = g(x)$?

17. The poverty threshold for four-person families in 1960 was $3020. The poverty threshold for four-person families in 1990 was $13,370. Let y represent the poverty threshold, and let x represent the year, where $x = 0$ corresponds to 1960. (Source: U.S. Bureau of the Census.)

    a. Plot the ordered pairs $(0, 3020)$ and $(30, 13,370)$.

    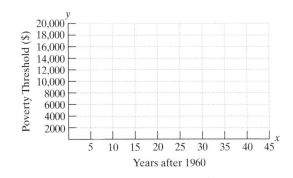

    b. Find a linear model that represents the poverty threshold y as a function of the year x.

18. Michael Jordan was the NBA leading scorer for 10 of 12 seasons between 1987 and 1998. In his 1998 season, he scored a total of 2357 points consisting of 1-point free throws, 2-point field goals, and 3-point field goals. He scored 286 more 2-point shots than he did free throws. The number of 3-point shots was 821 less than the number of 2-point shots. Determine the number of free throws, 2-point shots, and 3-point shots scored by Michael Jordan during his 1998 season.

19. Explain why this relation is *not* a function.

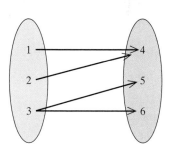

20. Graph the function defined by $f(x) = \dfrac{1}{x}$.

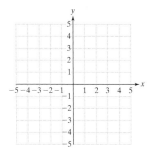

21. The quantity y varies directly as x and inversely as z. If $y = 15$ when $x = 50$ and $z = 10$, find y when $x = 65$ and $z = 5$.

22. The total number of flights (including passenger flights and cargo flights) at a large airport can be approximated by $F(x) = 300{,}000 + 0.008x$, where x is the number of passengers.

  a. Is this function linear, quadratic, constant, or other?

  b. Find the y-intercept and interpret its meaning in the context of this problem.

  c. What is the slope of the function and what does the slope mean in the context of this problem?

23. Given the function defined by $g(x) = \sqrt{2 - x}$, find the function values (if they exist) over the set of real numbers.

  a. $g(-7)$ **b.** $g(0)$ **c.** $g(3)$

24. Let $m(x) = \sqrt{x + 4}$ and $n(x) = x^2 + 2$. Find

  a. The domain of m **b.** The domain of n

25. Consider the function $y = f(x)$ graphed here. Find

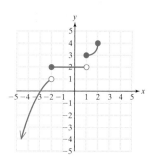

  a. The domain   **b.** The range

  c. $f(-2)$ **d.** $f(1)$ **e.** $f(0)$

  f. For what value(s) of x is $f(x) = 0$?

26. Solve. $\sqrt{8x + 5} = \sqrt{2x} + 2$

27. Solve for f. $\dfrac{1}{p} + \dfrac{1}{q} = \dfrac{1}{f}$

28. Solve. $\dfrac{15}{t^2 - 2t - 8} = \dfrac{1}{t - 4} + \dfrac{2}{t + 2}$

29. Simplify. $\dfrac{y - \dfrac{4}{y - 3}}{y - 4}$

30. Given: the function defined by $f(x) = 2(x - 3)^2 + 1$

  a. Write the coordinates of the vertex.

  b. Does the graph of the function open upward or downward?

  c. Write the coordinates of the y-intercept.

  d. Find the x-intercepts, if possible.

  e. Sketch the function.

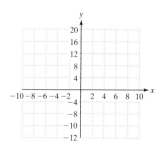

31. Use the method of completing the square to solve the equation.

$$x^2 - 16x + 2 = 0$$

32. Use the method of completing the square to find the vertex of the parabola. Check your answer by using the vertex formula.

$$f(x) = x^2 - 16x + 2$$

More Equations and Inequalities

9

9.1 Compound Inequalities
9.2 Polynomial and Rational Inequalities
9.3 Absolute Value Equations
9.4 Absolute Value Inequalities
9.5 Linear Inequalities in Two Variables

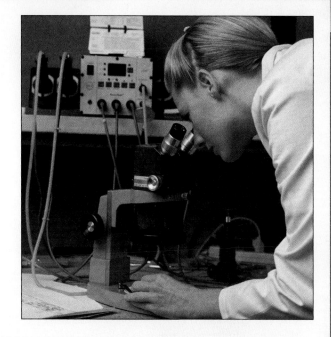

After studying this chapter, you will be able to recognize and solve a variety of equations and inequalities. Applications of inequalities are used often in science and medicine. For example, in Exercises 67–70 in Section 9.1, a doctor uses compound inequalities to express the normal ranges for white and red blood cells in human blood. In Exercises 57–61 in Section 9.4, we use absolute value inequalities to express measurement error.

chapter 9 preview

The exercises in this chapter preview contain concepts that have not yet been presented. These exercises are provided for students who want to compare their levels of understanding before and after studying the chapter. Alternatively, you may prefer to work these exercises when the chapter is completed and before taking the exam.

Section 9.1

For Exercises 1–2, solve the compound inequalities and write the solution in interval notation.

1. $-1 < 5y - 6 \leq 2$

2. $\frac{1}{4}x + 5 < 1$ or $-2x + 3 < 7$

3. A cholesterol level between 200 and 239 is considered borderline high. Let x represent the level of cholesterol, and write an inequality representing borderline high cholesterol.

Section 9.2

4. Use the graph of $f(x) = -x^2 + x + 6$ to solve the inequality $-x^2 + x + 6 \geq 0$. Write the answer in interval notation.

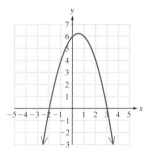

For Exercises 5–8, solve the inequalities. Write the solutions in interval notation.

5. $2a^2 - 5a - 3 > 0$

6. $p^2(p - 5)(p + 4) \leq 0$

7. $t^2 + 6t + 9 < 0$

8. $\dfrac{x + 3}{x - 6} \leq 0$

Section 9.3

For Exercises 9–11, solve the absolute value equations.

9. $|2c - 8| + 5 = 7$

10. $|4y - 1| + 12 = 6$

11. $|6x - 1| = |4x - 3|$

Section 9.4

For Exercises 12–14, solve the absolute value inequalities. Write the solutions in interval notation.

12. $|b + 6| > 8$

13. $|3x - 5| \leq 4$

14. $|6y + 8| > -2$

Section 9.5

For Exercises 15–18, graph the solution to the inequality or compound inequality.

15. $y > -3x - 1$

16. $2x + 3y \leq -6$

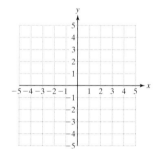
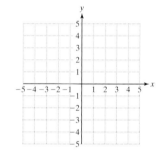

17. $x + y > 2$ and $y < \dfrac{1}{2}x + 3$

18. $x \geq -2$ or $2x - y \leq -1$

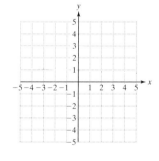
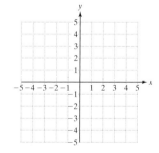

Section 9.1 Compound Inequalities

1. Union and Intersection

In Chapter 1 we graphed simple inequalities and expressed the solution set in interval notation and in set-builder notation. In this chapter, we will solve **compound inequalities** that involve the union or intersection of two or more inequalities.

A Union B and A Intersection B

The **union** of sets A and B, denoted $A \cup B$, is the set of elements that belong to set A or to set B or to both sets A and B.

The **intersection** of two sets A and B, denoted $A \cap B$, is the set of elements common to both A and B.

The concepts of the union and intersection of two sets are illustrated in Figures 9-1 and 9-2.

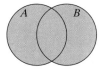

$A \cup B$
A union B
The elements in A or B or both

Figure 9-1

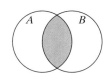

$A \cap B$
A intersection B
The elements in A and B

Figure 9-2

Objectives
1. Union and Intersection
2. Solving Compound Inequalities: And
3. Solving Compound Inequalities: Or
4. Applications of Compound Inequalities

example 1 — Finding the Union and Intersection of Two Intervals

Find the union or intersection as indicated.

a. $\left(-\infty, \frac{1}{2}\right) \cap [-3, 4)$ **b.** $(-\infty, -2) \cup [-4, 3)$

Solution:

a. $\left(-\infty, \frac{1}{2}\right) \cap [-3, 4)$ To find the intersection, graph each interval separately. Then find the real numbers common to both intervals.

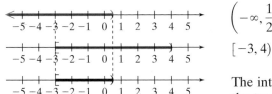

$\left(-\infty, \frac{1}{2}\right)$

$[-3, 4)$

The intersection is the "overlap" of the two intervals: $[-3, \frac{1}{2})$

The intersection is $[-3, \frac{1}{2})$.

b. $(-\infty, -2) \cup [-4, 3)$ To find the union, graph each interval separately. The union is the collection of real numbers that lie in the first interval, the second interval, or both intervals.

Skill Practice

1. Find the intersection
$(-\infty, 2) \cap [-5, \infty)$
Write the answer in interval notation.

2. Find the union
$(-\infty, -3) \cup (-\infty, 0)$
Write the answer in interval notation.

Answers
1. $[-5, 2)$ 2. $(-\infty, 0)$

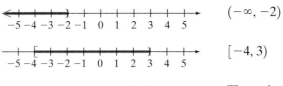

$(-\infty, -2)$

$[-4, 3)$

The union consists of all real numbers in the red interval along with the real numbers in the blue interval: $(-\infty, 3)$

The union is $(-\infty, 3)$.

2. Solving Compound Inequalities: And

The solution to two inequalities joined by the word *and* is the intersection of their solution sets. The solution to two inequalities joined by the word *or* is the union of their solution sets.

Steps to Solve a Compound Inequality

1. Solve and graph each inequality separately.
2. • If the inequalities are joined by the word *and*, find the intersection of the two solution sets.
   • If the inequalities are joined by the word *or*, find the union of the two solution sets.
3. Express the solution set in interval notation or in set-builder notation.

As you work through the examples in this section, remember that multiplying or dividing an inequality by a negative factor reverses the direction of the inequality sign.

Skill Practice

Solve the compound inequalities.

3. $5x + 2 \geq -8$ and $-4x > -24$

4. $-\dfrac{1}{3}y - \dfrac{1}{2} > 2$ and $6y + 2 > 2$

5. $-2.1x > 4.2$ and $3.5x < -10.5$

example 2 Solving Compound Inequalities: And

Solve the compound inequalities.

a. $-2x < 6$ and $x + 5 \leq 7$

b. $4.4a + 3.1 < -12.3$ and $-2.8a + 9.1 < -6.3$

c. $-\dfrac{2}{3}x \leq 6$ and $-\dfrac{1}{2}x < 1$

Solution:

a. $-2x < 6$ and $x + 5 \leq 7$ Solve each inequality separately.

$\dfrac{-2x}{-2} > \dfrac{6}{-2}$ and $x \leq 2$ Reverse the first inequality sign.

$x > -3$ and $x \leq 2$

Answers
3. $\{x \mid -2 \leq x < 6\}; [-2, 6)$
4. No solution
5. $\{x \mid x < -3\}; (-\infty, -3)$

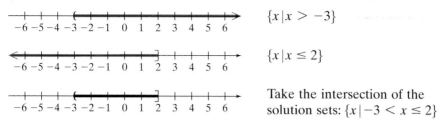

The solution is $\{x \mid -3 < x \leq 2\}$ or equivalently, in interval notation, $(-3, 2]$.

b. $4.4a + 3.1 < -12.3$ and $-2.8a + 9.1 < -6.3$

$\qquad 4.4a < -15.4$ and $-2.8a < -15.4$ Solve each inequality separately.

$\qquad \dfrac{4.4a}{4.4} < \dfrac{-15.4}{4.4}$ and $\dfrac{-2.8a}{-2.8} > \dfrac{-15.4}{-2.8}$ Reverse the second inequality sign.

$\qquad a < -3.5$ and $a > 5.5$

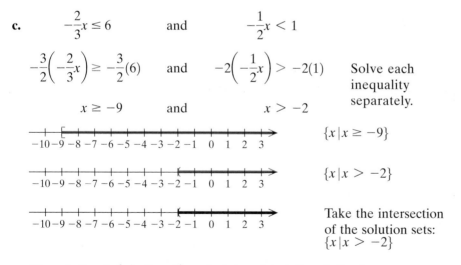

There are no real numbers that are simultaneously less than -3.5 and greater than 5.5. Hence, there is no solution.

c. $-\dfrac{2}{3}x \leq 6$ and $-\dfrac{1}{2}x < 1$

$\qquad -\dfrac{3}{2}\left(-\dfrac{2}{3}x\right) \geq -\dfrac{3}{2}(6)$ and $-2\left(-\dfrac{1}{2}x\right) > -2(1)$ Solve each inequality separately.

$\qquad x \geq -9$ and $x > -2$

[number line] $\{x \mid x \geq -9\}$

[number line] $\{x \mid x > -2\}$

[number line] Take the intersection of the solution sets: $\{x \mid x > -2\}$

The solution is $\{x \mid x > -2\}$ or, in interval notation, $(-2, \infty)$.

In Section 1.7, we learned that the inequality $a < x < b$ is the intersection of two simultaneous conditions implied on x.

$\qquad a < x < b$ is equivalent to $a < x$ and $x < b$

Skill Practice

Solve the inequality.

6. $-6 < 5 - 2x \leq 1$

example 3 Solving Compound Inequalities: And

Solve the inequality $-2 \leq -3x + 1 < 5$.

Solution:

$$-2 \leq -3x + 1 < 5$$

$-2 \leq -3x + 1$	and	$-3x + 1 < 5$	Set up the intersection of two inequalities.
$-3 \leq -3x$	and	$-3x < 4$	Solve each inequality.
$\dfrac{-3}{-3} \geq \dfrac{-3x}{-3}$	and	$\dfrac{-3x}{-3} > \dfrac{4}{-3}$	Reverse the direction of the inequality signs.
$1 \geq x$	and	$x > -\dfrac{4}{3}$	
$x \leq 1$	and	$x > -\dfrac{4}{3}$	Rewrite the inequalities.

$$-\dfrac{4}{3} < x \leq 1$$ Take the intersection of the solution sets.

The solution is $\{x \mid -\tfrac{4}{3} < x \leq 1\}$ or, equivalently in interval notation, $(-\tfrac{4}{3}, 1]$.

Tip: As an alternative approach to Example 3, we can isolate the variable x in the "middle" portion of the inequality. Recall that the operations performed on the middle part of the inequality must also be performed on the left- and right-hand sides.

$$-2 \leq -3x + 1 < 5$$

$-2 - 1 \leq -3x + 1 - 1 < 5 - 1$ Subtract 1 from all three parts of the inequality.

$-3 \leq -3x < 4$ Simplify.

$\dfrac{-3}{-3} \geq \dfrac{-3x}{-3} > \dfrac{4}{-3}$ Divide by -3 in all three parts of the inequality. (Remember to reverse inequality signs.)

$1 \geq x > -\dfrac{4}{3}$ Simplify.

$-\dfrac{4}{3} < x \leq 1$ Rewrite the inequality.

The solution is $\{x \mid -\tfrac{4}{3} < x \leq 1\}$ or, equivalently in interval notation, $(-\tfrac{4}{3}, 1]$.

Answer

3. Solving Compound Inequalities: Or

example 4   Solving Compound Inequalities: Or

Solve the compound inequalities.

a. $-3y - 5 > 4$ or $4 - y \leq 6$

b. $4x + 3 < 16$ or $-2x < 3$

c. $\dfrac{1}{3}x < 2$ or $-\dfrac{1}{2}x + 1 > 0$

Skill Practice

Solve the compound inequalities.

7. $-10t - 8 \geq 12$ or $3t - 6 > 3$

8. $x - 7 > -2$ or $-6x > -48$

9. $2.5x > -10$ or $-0.75x < 3$

Solution:

a. $-3y - 5 > 4$ or $4 - y \leq 6$

$\qquad -3y > 9$ or $-y \leq 2$ Solve each inequality separately.

$\qquad \dfrac{-3y}{-3} < \dfrac{9}{-3}$ or $\dfrac{-y}{-1} \geq \dfrac{2}{-1}$ Reverse the inequality signs.

$\qquad y < -3$ or $y \geq -2$

$\{y \mid y < -3\}$

$\{y \mid y \geq -2\}$

Take the union of the solution sets $\{y \mid y < -3$ or $y \geq -2\}$.

The solution is $\{y \mid y < -3 \text{ or } y \geq -2\}$ or, equivalently in interval notation, $(-\infty, -3) \cup [-2, \infty)$.

b. $4x + 3 < 16$ or $-2x < 3$

$\qquad 4x < 13$ or $x > -\dfrac{3}{2}$ Solve each inequality separately.

$\qquad x < \dfrac{13}{4}$ or $x > -\dfrac{3}{2}$

$\{x \mid x < \frac{13}{4}\}$

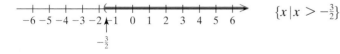
$\{x \mid x > -\frac{3}{2}\}$

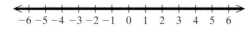
Take the union of the solution sets.

The union of the solution sets is $\{x \mid x \text{ is any real number}\}$ or equivalently $(-\infty, \infty)$.

Answers

7. $\{t \mid t \leq -2 \text{ or } t > 3\}$; $(-\infty, -2] \cup (3, \infty)$
8. All real numbers; $(-\infty, \infty)$
9. $\{x \mid x > -4\}$; $(-4, \infty)$

c. $\frac{1}{3}x < 2$ or $-\frac{1}{2}x + 1 > 0$

$3\left(\frac{1}{3}x\right) < 3(2)$ or $-\frac{1}{2}x > -1$ Solve each inequality separately.

$x < 6$ or $-2\left(-\frac{1}{2}x\right) < -2(-1)$

$x < 6$ or $x < 2$

$\{x | x < 6\}$

$\{x | x < 2\}$

Take the union of the solution sets: $\{x | x < 6\}$.

The union of the solution sets is $\{x | x < 6\}$ or, in interval notation, $(-\infty, 6)$.

4. Applications of Compound Inequalities

Compound inequalities are used in many applications, as shown in Examples 5 and 6.

Skill Practice

The length of a normal human pregnancy is from 37 to 41 weeks, inclusive.

10. Write an inequality representing the normal length of a pregnancy.

11. Write a compound inequality representing an abnormal length for a pregnancy.

example 5 Translating Compound Inequalities

The normal level of thyroid-stimulating hormone (TSH) for adults ranges from 0.4 to 4.8 microunits per milliliter (μU/mL). Let x represent the amount of TSH measured in microunits per milliliter.

a. Write an inequality representing the normal range of TSH.

b. Write a compound inequality representing abnormal TSH levels.

Solution:

a. $0.4 \leq x \leq 4.8$ b. $x < 0.4$ or $x > 4.8$

Skill Practice

12. The sum of twice a number and 11 is between 21 and 31. Find all such numbers.

example 6 Translating and Solving a Compound Inequality

The sum of a number and 4 is between -5 and 12. Find all such numbers.

Solution:

Let x represent a number.

$-5 < x + 4 < 12$ Translate the inequality.

$-5 - 4 < x + 4 - 4 < 12 - 4$ Subtract 4 from all three parts of the inequality.

$-9 < x < 8$

The number may be any real number between -9 and 8: $\{x | -9 < x < 8\}$.

Answers

10. $37 \leq w \leq 41$
11. $w < 37$ or $w > 41$
12. Any real number between 5 and 10: $\{n | 5 < n < 10\}$

section 9.1 Practice Exercises

Boost your GRADE at mathzone.com!

- Practice Problems
- Self-Tests
- NetTutor
- e-Professors
- Videos

Study Skills Exercises

1. List three benefits of successfully completing this course.

2. Define the key terms.
   a. Compound inequality   b. Intersection   c. Union

Review Exercises

For Exercises 3–8, review solving linear inequalities from Section 1.7. Write the answers in interval notation.

3. $6u + 5 > 2$

4. $-2 + 3z \leq 4$

5. $-\dfrac{3}{4}p \leq 12$

6. $-6q > -\dfrac{1}{3}$

7. $-1.5 < 0.1x - 8.1$

8. $4 \geq 2.6 + 7t$

Objective 1: Union and Intersection

For Exercises 9–14, find the intersection and union of sets as indicated. Write the answer in set-builder notation. (See Example 1.)

9. a. $(-2, 5) \cap [-1, \infty)$
   b. $(-2, 5) \cup [-1, \infty)$

10. a. $(-\infty, 4) \cap [-1, 5)$
    b. $(-\infty, 4) \cup [-1, 5)$

11. a. $\left(-\dfrac{5}{2}, 3\right) \cap \left(-1, \dfrac{9}{2}\right)$
    b. $\left(-\dfrac{5}{2}, 3\right) \cup \left(-1, \dfrac{9}{2}\right)$

12. a. $(-3.4, 1.6) \cap (-2.2, 4.1)$
    b. $(-3.4, 1.6) \cup (-2.2, 4.1)$

13. a. $(-4, 5] \cap (0, 2]$
    b. $(-4, 5] \cup (0, 2]$

14. a. $[-1, 5) \cap (0, 3)$
    b. $[-1, 5) \cup (0, 3)$

Objective 2: Solving Compound Inequalities: And

For Exercises 15–24, solve the inequality and graph the solution. Write the answer in interval notation. (See Example 2.)

15. $y - 7 \geq -9$ and $y + 2 \leq 5$

16. $a + 6 > -2$ and $5a < 30$

17. $2t + 7 < 19$ and $5t + 13 > 28$

18. $5p + 2p \geq -21$ and $-9p + 3p \geq -24$

19. $21k - 11 \leq 6k + 19$ and $3k - 11 < -k + 7$

20. $6w - 1 > 3w - 11$ and $-3w + 7 \leq 8w - 13$

21. $\dfrac{2}{3}(2p - 1) \geq 10$ and $\dfrac{4}{5}(3p + 4) \geq 20$

22. $5(a + 3) + 9 < 2$ and $3(a - 2) + 6 < 10$

23. $-2 < -x - 12$ and $-14 < 5(x - 3) + 6x$

24. $-8 \geq -3y - 2$ and $3(y - 7) + 16 > 4y$

25. Write $-4 \leq t < \dfrac{3}{4}$ as two separate inequalities.

26. Write $-2.8 < y \leq 15$ as two separate inequalities.

27. Explain why $6 < x < 2$ has no solution.

28. Explain why $4 < t < 1$ has no solution.

29. Explain why $-5 > y > -2$ has no solution.

30. Explain why $-3 > w > -1$ has no solution.

For Exercises 31–40, solve the inequality and graph the solution set. Write the answer in interval notation. **(See Example 3.)**

31. $0 \leq 2b - 5 < 9$

32. $-6 < 3k - 9 \leq 0$

33. $-1 < \dfrac{a}{6} \leq 1$

34. $-3 \leq \dfrac{1}{2}x < 0$

35. $-\dfrac{2}{3} < \dfrac{y - 4}{-6} < \dfrac{1}{3}$

36. $\dfrac{1}{3} > \dfrac{t - 4}{-3} > -2$

37. $5 \leq -3x - 2 \leq 8$

38. $-1 < -2x + 4 \leq 5$

39. $12 > 6x + 3 \geq 0$

40. $-4 \geq 2x - 5 > -7$

Objective 3: Solving Compound Inequalities: Or

For Exercises 41–54, solve the inequality and graph the solution set. Write the answer in interval notation. **(See Example 4.)**

41. $h + 4 < 0$ or $6h > -12$

42. $5y > 12$ or $y - 3 < -2$

43. $2y - 1 \geq 3$ or $y < -2$

44. $x < 0$ or $3x + 1 \geq 7$

45. $1.2 > 7.2z - 9.6$ or $3.1 \leq 6.3 - 1.6z$

46. $9.5 > 3.1 + 0.8z$ or $-2.8 > 6.1 + 0.89z$

47. $5(x - 1) \geq -5$ or $5 - x \leq 11$

48. $-p + 7 \geq 10$ or $3(p - 1) \leq 12$

49. $\dfrac{5}{3}v \leq 5$ or $-v - 6 < 1$

50. $\dfrac{3}{8}u + 1 > 0$ or $-2u \geq -4$

51. $\dfrac{3t-1}{10} > \dfrac{1}{2}$ or $\dfrac{3t-1}{10} < -\dfrac{1}{2}$

52. $\dfrac{6-x}{12} > \dfrac{1}{4}$ or $\dfrac{6-x}{12} < -\dfrac{1}{6}$

53. $0.5w + 5 < 2.5w - 4$ or $0.3w \le -0.1w - 1.6$

54. $1.25a + 3 \le 0.5a - 6$ or $2.5a - 1 \ge 9 - 1.5a$

Mixed Exercises

For Exercises 55–66, solve the inequality. Write the answer in interval notation.

55. a. $3x - 5 < 19$ and $-2x + 3 < 23$

b. $3x - 5 < 19$ or $-2x + 3 < 23$

56. a. $0.5(6x + 8) > 0.8x - 7$ and $4(x + 1) < 7.2$

b. $0.5(6x + 8) > 0.8x - 7$ or $4(x + 1) < 7.2$

57. a. $8x - 4 \ge 6.4$ or $0.3(x + 6) \le -0.6$

b. $8x - 4 \ge 6.4$ and $0.3(x + 6) \le -0.6$

58. a. $-2r + 4 \le -8$ or $3r + 5 \le 8$

b. $-2r + 4 \le -8$ and $3r + 5 \le 8$

59. $-4 \le \dfrac{2 - 4x}{3} < 8$

60. $-1 < \dfrac{3 - x}{2} \le 0$

61. $5 \ge -4(t - 3) + 3t$ or $6 < 12t + 8(4 - t)$

62. $3 > -(w - 3) + 4w$ or $-5 \ge -3(w - 5) + 6w$

63. $-7(3 - x) < 9[x - 3(x + 1)]$ and $6x - (4 - x) + 3 < -3[8 - 2(x + 1)]$

64. $-5[x + 3(2 - x)] \le 4(3 - 5x) + 7$ and $4 - [-4 - (2x - 3)] \ge 9 - (10 - 6x)$

65. $\dfrac{-x + 3}{2} > \dfrac{4 + x}{5}$ or $\dfrac{1 - x}{4} > \dfrac{2 - x}{3}$

66. $\dfrac{y - 7}{-3} < \dfrac{1}{4}$ or $\dfrac{y + 1}{-2} > -\dfrac{1}{3}$

Objective 4: Applications of Compound Inequalities

67. The normal number of white blood cells for human blood is between 4800 and 10,800 cells per cubic millimeter, inclusive. Let x represent the number of white blood cells per cubic millimeter.

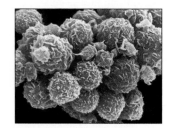

a. Write an inequality representing the normal range of white blood cells per cubic millimeter. **(See Example 5.)**

b. Write a compound inequality representing abnormal levels of white blood cells per cubic millimeter.

68. The normal number of platelets in human blood is between 2.0×10^5 and 3.5×10^5 platelets per cubic millimeter, inclusive. Let x represent the number of platelets per cubic millimeter.

   a. Write an inequality representing a normal platelet count per cubic millimeter.

   b. Write a compound inequality representing abnormal platelet counts per cubic millimeter.

69. Normal hemoglobin levels in human blood for adult males are between 13 and 16 grams per deciliter (g/dL), inclusive. Let x represent the level of hemoglobin measured in grams per deciliter.

   a. Write an inequality representing normal hemoglobin levels for adult males.

   b. Write a compound inequality representing abnormal levels of hemoglobin for adult males.

70. Normal hemoglobin levels in human blood for adult females are between 12 and 15 g/dL, inclusive. Let x represent the level of hemoglobin measured in grams per deciliter.

   a. Write an inequality representing normal hemoglobin levels for adult females.

   b. Write a compound inequality representing abnormal levels of hemoglobin for adult females.

71. Twice a number is between -3 and 12. Find all such numbers. (See Example 6.)

72. The difference of a number and 6 is between 0 and 8. Find all such numbers.

73. One plus twice a number is either greater than 5 or less than -1. Find all such numbers.

74. One-third of a number is either less than -2 or greater than 5. Find all such numbers.

Objectives

1. Solving Inequalities Graphically
2. Test Point Method
3. Solving Polynomial Inequalities by Using the Test Point Method
4. Solving Rational Inequalities by Using the Test Point Method
5. Inequalities with "Special Case" Solution Sets

section 9.2 Polynomial and Rational Inequalities

1. Solving Inequalities Graphically

In Sections 1.7 and 9.1, we solved simple and compound linear inequalities. In this section we will solve polynomial and rational inequalities. We begin by defining a quadratic inequality.

Quadratic inequalities are inequalities that can be written in any of the following forms:

$$ax^2 + bx + c \geq 0 \qquad ax^2 + bx + c \leq 0$$
$$ax^2 + bx + c > 0 \qquad ax^2 + bx + c < 0 \qquad \text{where } a \neq 0$$

Recall from Section 8.4 that a quadratic function defined by $f(x) = ax^2 + bx + c$ ($a \neq 0$) is a parabola that opens up or down. The quadratic inequality $f(x) > 0$ or equivalently $ax^2 + bx + c > 0$ is asking the question, "For what values of x is the function positive (above the x-axis)?" The inequality $f(x) < 0$ or equivalently $ax^2 + bx + c < 0$ is asking, "For what values of x is the function negative (below the x-axis)?" The graph of a quadratic function can be used to answer these questions.

example 1 — Using a Graph to Solve a Quadratic Inequality

Use the graph of $f(x) = x^2 - 6x + 8$ in Figure 9-3 to solve the inequalities.

a. $x^2 - 6x + 8 < 0$ **b.** $x^2 - 6x + 8 > 0$

Solution:

From Figure 9-3, we see that the graph of $f(x) = x^2 - 6x + 8$ is a parabola opening upward. The function factors as $f(x) = (x - 2)(x - 4)$. The x-intercepts are at $x = 2$ and $x = 4$, and the y-intercept is $(0, 8)$.

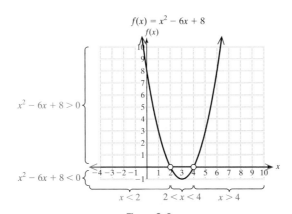

Figure 9-3

a. The solution to $x^2 - 6x + 8 < 0$ is the set of all real numbers x for which $f(x) < 0$. Graphically, this is the set of all x-values corresponding to the points where the parabola is below the x-axis (shown in red). Hence

$x^2 - 6x + 8 < 0$ for $\{x \mid 2 < x < 4\}$ or equivalently, $(2, 4)$

b. The solution to $x^2 - 6x + 8 > 0$ is the set of x-values for which $f(x) > 0$. This is the set of x-values where the parabola is above the x-axis (shown in blue). Hence

$x^2 - 6x + 8 > 0$ for $\{x \mid x < 2 \text{ or } x > 4\}$ or $(-\infty, 2) \cup (4, \infty)$

Notice that the points $x = 2$ and $x = 4$ define the boundaries of the solution sets to the inequalities in Example 1. These values are the solutions to the related equation $x^2 - 6x + 8 = 0$.

Skill Practice

Refer to the graph of $f(x) = x^2 + 3x - 4$ to solve the inequalities.

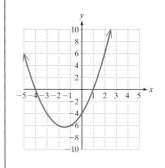

1. $x^2 + 3x - 4 > 0$
2. $x^2 + 3x - 4 < 0$

Answers
1. $\{x \mid x < -4 \text{ or } x > 1\}$; $(-\infty, -4) \cup (1, \infty)$
2. $\{x \mid -4 < x < 1\}$; $(-4, 1)$

> **Tip:** The inequalities in Example 1 are strict inequalities. Therefore, the values $x = 2$ and $x = 4$ (where $f(x) = 0$) are not included in the solution set. However, the corresponding inequalities using the symbols $\leq$ and $\geq$ do include the points where $f(x) = 0$. Hence,
>
> The solution to $\quad x^2 - 6x + 8 \leq 0 \quad$ is $\quad \{x \mid 2 \leq x \leq 4\}$
>
> The solution to $\quad x^2 - 6x + 8 \geq 0 \quad$ is $\quad \{x \mid x \leq 2 \text{ or } x \geq 4\}$

Skill Practice

Refer to the graph of $g(x) = \dfrac{1}{x-2}$ to solve the inequalities.

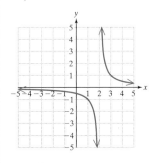

3. $\dfrac{1}{x-2} > 0$

4. $\dfrac{1}{x-2} < 0$

example 2 Using a Graph to Solve a Rational Inequality

Use the graph of $g(x) = \dfrac{1}{x+1}$ in Figure 9-4 to solve the inequalities.

a. $\dfrac{1}{x+1} < 0$
b. $\dfrac{1}{x+1} > 0$

Solution:

a. The graph of $g(x) = \dfrac{1}{x+1}$ shown in Figure 9-4 indicates that $g(x)$ is below the x-axis for $x < -1$ (shown in red). Therefore, the solution to $\dfrac{1}{x+1} < 0$ is $\{x \mid x < -1\}$ or, equivalently, $(-\infty, -1)$.

b. The graph of $g(x) = \dfrac{1}{x+1}$ shown in Figure 9-4 indicates that $g(x)$ is above the x-axis for $x > -1$ (shown in blue). Therefore, the solution to $\dfrac{1}{x+1} > 0$ is $\{x \mid x > -1\}$ or, equivalently, $(-1, \infty)$.

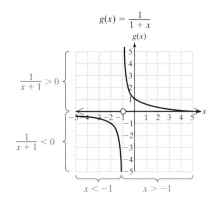

Figure 9-4

Notice that the point $x = -1$ defines the boundary of the solution sets to the inequalities in Example 2. The point $x = -1$ is a point where the inequality is undefined.

2. Test Point Method

The **boundary points** of an inequality consist of the real solutions to the related equation and the points where the inequality is undefined. Examples 1 and 2 demonstrate that the boundary points of an inequality provide the boundaries of the solution set. This is the basis of the **test point method** to solve inequalities.

Answers

3. $\{x \mid x > 2\}$ or equivalently $(2, \infty)$
4. $\{x \mid x < 2\}$ or equivalently $(-\infty, 2)$

Solving Inequalities by Using the Test Point Method

1. Find the boundary points of the inequality.
2. Plot the boundary points on the number line. This divides the number line into regions.
3. Select a test point from each region and substitute it into the original inequality.
   - If a test point makes the original inequality true, then that region is part of the solution set.
4. Test the boundary points in the original inequality.
   - If a boundary point makes the original inequality true, then that point is part of the solution set.

Concept Connections

5. Determine the boundary points of the inequality.
   $$a^2 - 8a + 12 > 0$$

3. Solving Polynomial Inequalities by Using the Test Point Method

example 3 Solving Polynomial Inequalities by Using the Test Point Method

Solve the inequalities by using the test point method.

a. $2x^2 + 5x < 12$ b. $x(x-2)(x+4)^2(x-4) > 0$

Skill Practice

Solve the inequalities by using the test point method. Write the answers in interval notation.

6. $x^2 + x > 6$
7. $t(t-5)(t+2)^2 > 0$

Solution:

a. $\quad 2x^2 + 5x < 12$ **Step 1:** Find the boundary points. Because polynomials are defined for all values of x, the only boundary points are the real solutions to the related equation.

$2x^2 + 5x = 12$ Solve the related equation.

$2x^2 + 5x - 12 = 0$

$(2x - 3)(x + 4) = 0$

$x = \frac{3}{2} \quad x = -4$ The boundary points are $\frac{3}{2}$ and -4.

Step 2: Plot the boundary points.

Step 3: Select a test point from each region.

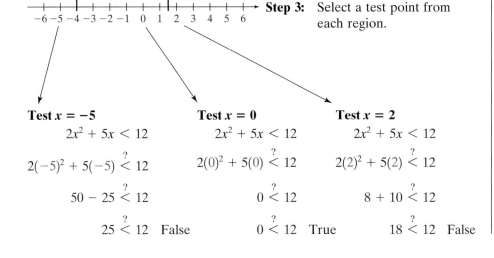

Answers
5. $a = 6$; $a = 2$
6. $(-\infty, -3) \cup (2, \infty)$
7. $(-\infty, -2) \cup (-2, 0) \cup (5, \infty)$

Test $x = -4$

$2x^2 + 5x < 12$

$2(-4)^2 + 5(-4) \stackrel{?}{<} 12$

$32 - 20 \stackrel{?}{<} 12$

$12 \stackrel{?}{<} 12$ False

Test $x = \frac{3}{2}$

$2x^2 + 5x < 12$

$2\left(\frac{3}{2}\right)^2 + 5\left(\frac{3}{2}\right) \stackrel{?}{<} 12$

$2\left(\frac{9}{4}\right) + \frac{15}{2} \stackrel{?}{<} 12$

$\frac{9}{2} + \frac{15}{2} \stackrel{?}{<} 12$

$\frac{24}{2} \stackrel{?}{<} 12$ False

Step 4: Test the boundary points.

Tip: The strict inequality, $<$, excludes values of x for which $2x^2 + 5x = 12$. This implies that the boundary points are not included in the solution set.

Neither boundary point makes the inequality true. Therefore, the boundary points are not included in the solution set.

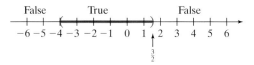

The solution is $\{x \mid -4 < x < \frac{3}{2}\}$ or equivalently in interval notation $(-4, \frac{3}{2})$.

Calculator Connections

Graph $Y_1 = 2x^2 + 5x$ and $Y_2 = 12$.
Notice that $Y_1 < Y_2$ for $\{x \mid -4 < x < \frac{3}{2}\}$.

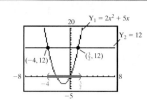

b. $x(x - 2)(x + 4)^2(x - 4) > 0$

$x(x - 2)(x + 4)^2(x - 4) = 0$

$x = 0 \quad x = 2 \quad x = -4 \quad x = 4$

Step 1: Find the boundary points.

Step 2: Plot the boundary points.

Step 3: Select a test point from each region.

Test $x = -5$: $-5(-5 - 2)(-5 + 4)^2(-5 - 4) \stackrel{?}{>} 0 \qquad -315 \stackrel{?}{>} 0$ False

Test $x = -1$: $-1(-1 - 2)(-1 + 4)^2(-1 - 4) \stackrel{?}{>} 0 \qquad -135 \stackrel{?}{>} 0$ False

Test $x = 1$: $1(1 - 2)(1 + 4)^2(1 - 4) \stackrel{?}{>} 0 \qquad 75 \stackrel{?}{>} 0$ True

Test $x = 3$: $3(3 - 2)(3 + 4)^2(3 - 4) \stackrel{?}{>} 0 \qquad -147 \stackrel{?}{>} 0$ False

Test $x = 5$: $5(5 - 2)(5 + 4)^2(5 - 4) \stackrel{?}{>} 0 \qquad 1215 \stackrel{?}{>} 0$ True

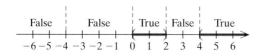

Step 4: The boundary points are not included because the inequality, >, is strict.

The solution is $\{x \mid 0 < x < 2 \text{ or } x > 4\}$ or, equivalently in interval notation, $(0, 2) \cup (4, \infty)$.

Calculator Connections

Graph $Y_1 = x(x - 2)(x + 4)^2(x - 4)$. Y_1 is positive (above the x-axis) for $\{x \mid 0 < x < 2 \text{ or } x > 4\}$ or equivalently $(0, 2) \cup (4, \infty)$.

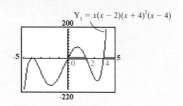

4. Solving Rational Inequalities by Using the Test Point Method

The test point method can be used to solve rational inequalities. A **rational inequality** is an inequality in which one or more terms is a rational expression. The solution set to a rational inequality must exclude all values of the variable that make the inequality undefined. That is, exclude all values that make the denominator equal to zero for any rational expression in the inequality.

example 4   Solving a Rational Inequality by Using the Test Point Method

Solve the inequality by using the test point method. $\quad \dfrac{x + 2}{x - 4} \leq 3$

Solution:

$$\dfrac{x + 2}{x - 4} \leq 3$$

Step 1: Find the boundary points. Note that the inequality is undefined for $x = 4$. Hence $x = 4$ is automatically a boundary point. To find any other boundary points, solve the related equation.

$$\dfrac{x + 2}{x - 4} = 3$$

$$(x - 4)\left(\dfrac{x + 2}{x - 4}\right) = (x - 4)(3) \qquad \text{Clear fractions.}$$

$$x + 2 = 3(x - 4) \qquad \text{Solve for } x.$$

$$x + 2 = 3x - 12$$

$$-2x = -14$$

$$x = 7$$

Skill Practice

Solve the inequality by using the test point method. Write the answer in interval notation.

8. $\dfrac{x - 5}{x + 4} \leq -1$

Answer

8. $\left(-4, \dfrac{1}{2}\right]$

The solution to the related equation is $x = 7$, and the inequality is undefined for $x = 4$. Therefore, the boundary points are $x = 4$ and $x = 7$.

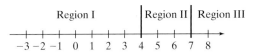

Step 2: Plot boundary points.

Step 3: Select test points.

Test $x = 0$

$$\frac{x+2}{x-4} \leq 3$$

$$\frac{0+2}{0-4} \stackrel{?}{\leq} 3$$

$$-\frac{1}{2} \stackrel{?}{\leq} 3 \quad \text{True}$$

Test $x = 5$

$$\frac{x+2}{x-4} \leq 3$$

$$\frac{5+2}{5-4} \stackrel{?}{\leq} 3$$

$$\frac{7}{1} \stackrel{?}{\leq} 3 \quad \text{False}$$

Test $x = 8$

$$\frac{x+2}{x-4} \leq 3$$

$$\frac{8+2}{8-4} \stackrel{?}{\leq} 3$$

$$\frac{10}{4} \stackrel{?}{\leq} 3$$

$$\frac{5}{2} \leq 3 \quad \text{True}$$

Step 4: Test the boundary points.

Test $x = 4$:

$$\frac{x+2}{x-4} \leq 3$$

$$\frac{4+2}{4-4} \stackrel{?}{\leq} 3$$

$$\frac{6}{0} \stackrel{?}{\leq} 3 \quad \text{Undefined}$$

Test $x = 7$:

$$\frac{x+2}{x-4} \leq 3$$

$$\frac{7+2}{7-4} \stackrel{?}{\leq} 3$$

$$\frac{9}{3} \stackrel{?}{\leq} 3 \quad \text{True}$$

The boundary point $x = 4$ cannot be included in the solution set, because it is undefined in the inequality. The boundary point $x = 7$ makes the original inequality true and must be included in the solution set.

The solution is $\{x \mid x < 4 \text{ or } x \geq 7\}$ or, equivalently in interval notation, $(\infty, 4) \cup [7, \infty)$.

Calculator Connections

Graph $Y_1 = \dfrac{x+2}{x-4}$ and $Y_2 = 3$.

Y_1 has a vertical asymptote at $x = 4$. Furthermore, $Y_1 = Y_2$ at $x = 7$. $Y_1 \leq Y_2$ (that is, Y_1 is below Y_2) for $x < 4$ and for $x \geq 7$.

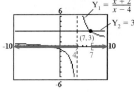

5. Inequalities with "Special Case" Solution Sets

The solution to an inequality is often one or more regions on the real number line. Sometimes, however, the solution to an inequality may be a single point on the number line, the empty set, or the set of all real numbers.

Section 9.2 Polynomial and Rational Inequalities

example 5 Solving Inequalities

Solve the inequalities.

a. $-\dfrac{16}{x^2 + 2} < 0$ b. $-\dfrac{16}{x^2 + 2} > 0$

Skill Practice

Solve the inequalities.

9. $\dfrac{1}{y^2 + 9} > 0$

10. $\dfrac{1}{y^2 + 9} < 0$

Solution:

a. Since the expressions 16 and $x^2 + 2$ are greater than zero for all real numbers x, their ratio is positive. The opposite of their ratio, $-\dfrac{16}{x^2 + 2}$, will be negative for all values of x. That is, $-\dfrac{16}{x^2 + 2} < 0$.

The solution is all real numbers, $(-\infty, \infty)$.

b. The expression $-\dfrac{16}{x^2 + 2} < 0$ for all real numbers x.

Therefore, the inequality $-\dfrac{16}{x^2 + 2} > 0$ has no solution.

Calculator Connections

The graph of $Y_1 = -\dfrac{16}{x^2 + 2}$ is below the x-axis for all x-values on the viewing window. Therefore $-\dfrac{16}{x^2 + 2} < 0$ for all x on the display window. Furthermore, there are no values of x for which $-\dfrac{16}{x^2 + 2} \geq 0$.

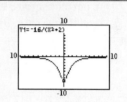

example 6 Solving Inequalities

Solve the inequalities.

a. $x^2 + 6x + 9 \geq 0$ b. $x^2 + 6x + 9 > 0$
c. $x^2 + 6x + 9 \leq 0$ d. $x^2 + 6x + 9 < 0$

Skill Practice

Solve the inequalities.

11. $x^2 - 4x + 4 \geq 0$
12. $x^2 - 4x + 4 > 0$
13. $x^2 - 4x + 4 \leq 0$
14. $x^2 - 4x + 4 < 0$

Solution:

a. $x^2 + 6x + 9 \geq 0$ Notice that $x^2 + 6x + 9$ is a perfect square trinomial.
   $(x + 3)^2 \geq 0$ Factor $x^2 + 6x + 9 = (x + 3)^2$.

The quantity $(x + 3)^2$ is a perfect square and is greater than or equal to zero for all real numbers, x. The solution is all real numbers, $(-\infty, \infty)$.

True
True ↓ True
← | | | ● | | | | | | →
−5 −4 −3 −2 −1 0 1 2 3 4 5

Answers

9. All real numbers; $(-\infty, \infty)$
10. No solution
11. All real numbers; $(-\infty, \infty)$
12. $(-\infty, 2) \cup (2, \infty)$
13. $x = 2$
14. No solution

b. $x^2 + 6x + 9 > 0$

$(x+3)^2 > 0$

This is the same inequality as in part (a) with the exception that the inequality is strict. The solution set does not include the point where $x^2 + 6x + 9 = 0$. Therefore, the boundary point $x = -3$ is *not* included in the solution set.

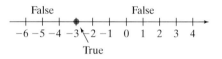

The solution set is $\{x \mid x < -3 \text{ or } x > -3\}$ or equivalently $(-\infty, -3) \cup (-3, \infty)$.

c. $x^2 + 6x + 9 \leq 0$

$(x+3)^2 \leq 0$

A perfect square cannot be less than zero. However, $(x+3)^2$ is equal to zero at $x = -3$. Therefore, the solution set is $\{-3\}$.

Tip: The graph of $f(x) = x^2 + 6x + 9$, or equivalently $f(x) = (x+3)^2$, is equal to zero at $x = 3$ and positive (above the x-axis) for all other values of x on its domain.

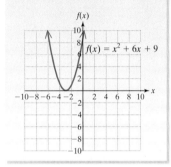

d. $x^2 + 6x + 9 < 0$

$(x+3)^2 < 0$

A perfect square cannot be negative; therefore, there are no real numbers x such that $(x+3)^2 < 0$. There is no solution.

section 9.2 Practice Exercises

Boost your GRADE at mathzone.com!

- Practice Problems
- Self-Tests
- NetTutor
- e-Professors
- Videos

Study Skills Exercises

1. Taking 12 credit-hours is the equivalent of a full-time job. Often students try to work too many hours while taking classes at school. Write down how many hours you work per week and the number of credit-hours you are taking this term.

    Number of hours worked per week _____

    Number of credit-hours taken this term _____

2. Define the key terms.

    a. Quadratic inequality   **b. Boundary points**   **c. Test point method**

    d. Rational inequality

Review Exercises

For Exercises 3–8, solve the compound inequalities. Write the solutions in interval notation.

3. $6x - 10 > 8$ or $8x + 2 < 5$

4. $3(a - 1) + 2 > 0$ or $2a > 5a + 12$

5. $5(k - 2) > -25$ and $7(1 - k) > 7$

6. $2y + 4 \geq 10$ and $5y - 3 \leq 13$

7. $6 \geq 4 - 2x \geq -2$

8. $-4 > 5 - x > -6$

Objective 1: Solving Inequalities Graphically

For Exercises 9–12, estimate from the graph the intervals for which the inequality is true. (See Examples 1–2.)

9.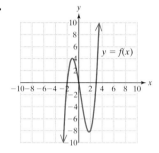

   a. $f(x) > 0$ b. $f(x) < 0$

   c. $f(x) \leq 0$ d. $f(x) \geq 0$

10.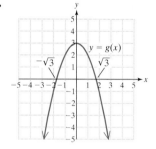

    a. $g(x) < 0$ b. $g(x) > 0$

    c. $g(x) \geq 0$ d. $g(x) \leq 0$

11.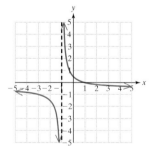

    a. $h(x) \geq 0$ b. $h(x) \leq 0$

    c. $h(x) < 0$ d. $h(x) > 0$

12.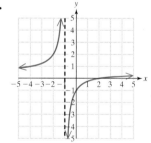

    a. $k(x) \leq 0$ b. $k(x) \geq 0$

    c. $k(x) > 0$ d. $k(x) < 0$

Objective 3: Solving Polynomial Inequalities by Using the Test Point Method

For Exercises 13–18, solve the equation and related inequalities.

13. a. $3(4 - x)(2x + 1) = 0$

    b. $3(4 - x)(2x + 1) < 0$

    c. $3(4 - x)(2x + 1) > 0$

14. a. $5(y + 6)(3 - 5y) = 0$

    b. $5(y + 6)(3 - 5y) < 0$

    c. $5(y + 6)(3 - 5y) > 0$

15. a. $x^2 + 7x = 30$

b. $x^2 + 7x < 30$

c. $x^2 + 7x > 30$

16. a. $q^2 - 4q = 5$

b. $q^2 - 4q \leq 5$

c. $q^2 - 4q \geq 5$

17. a. $2p(p - 2) = p + 3$

b. $2p(p - 2) \leq p + 3$

c. $2p(p - 2) \geq p + 3$

18. a. $3w(w + 4) = 10 - w$

b. $3w(w + 4) < 10 - w$

c. $3w(w + 4) > 10 - w$

For Exercises 19–34, solve the polynomial inequality. Write the answer in interval notation. **(See Example 3.)**

19. $(t - 7)(t + 1) < 0$

20. $(p - 4)(p + 2) > 0$

21. $-6(4 + 2x)(5 - x) > 0$

22. $-8(2t + 5)(6 - t) < 0$

23. $m(m + 1)^2(m + 5) \leq 0$

24. $w^2(3 - w)(w + 2) \geq 0$

25. $a^2 - 12a \leq -32$

26. $w^2 + 20w \geq -64$

27. $b^2 - 121 < 0$

28. $c^2 - 25 < 0$

29. $3p(p - 2) - 3 \geq 2p$

30. $2t(t + 3) - t \leq 12$

31. $x^3 - x^2 \leq 12x$

32. $x^3 + 36 > 4x^2 + 9x$

33. $w^3 + w^2 > 4w + 4$

34. $2p^3 - 5p^2 \leq 3p$

Objective 4: Solving Rational Inequalities by Using the Test Point Method

For Exercises 35–38, solve the equation and related inequalities.

35. a. $\dfrac{10}{x - 5} = 5$

b. $\dfrac{10}{x - 5} < 5$

c. $\dfrac{10}{x - 5} > 5$

36. a. $\dfrac{8}{a + 1} = 4$

b. $\dfrac{8}{a + 1} > 4$

c. $\dfrac{8}{a + 1} < 4$

37. a. $\dfrac{z + 2}{z - 6} = -3$

b. $\dfrac{z + 2}{z - 6} \leq -3$

c. $\dfrac{z + 2}{z - 6} \geq -3$

38. a. $\dfrac{w - 8}{w + 6} = 2$

b. $\dfrac{w - 8}{w + 6} \leq 2$

c. $\dfrac{w - 8}{w + 6} \geq 2$

For Exercises 39–50, solve the rational inequalities. Write the answer in interval notation. (See Example 4.)

39. $\dfrac{2}{x-1} \geq 0$ **40.** $\dfrac{-3}{x+2} \leq 0$ **41.** $\dfrac{b+4}{b-4} > 0$ **42.** $\dfrac{a+1}{a-3} < 0$

43. $\dfrac{3}{2x-7} < -1$ **44.** $\dfrac{8}{4x+9} > 1$ **45.** $\dfrac{x+1}{x-5} \geq 4$ **46.** $\dfrac{x-2}{x+6} \leq 5$

47. $\dfrac{1}{x} \leq 2$ **48.** $\dfrac{1}{x} \geq 3$ **49.** $\dfrac{(x+2)^2}{x} > 0$ **50.** $\dfrac{(x-3)^2}{x} < 0$

Objective 5: Inequalities with "Special Case" Solution Sets

For Exercises 51–62, solve the inequalities. (See Examples 5–6.)

51. $x^2 + 10x + 25 \geq 0$ **52.** $x^2 + 6x + 9 < 0$ **53.** $x^2 + 2x + 1 < 0$ **54.** $x^2 + 8x + 16 \geq 0$

55. $\dfrac{x^2}{x^2+4} < 0$ **56.** $\dfrac{x^2}{x^2+4} \geq 0$ **57.** $x^4 + 3x^2 \leq 0$ **58.** $x^4 + 2x^2 \leq 0$

59. $x^2 + 24x + 144 > 0$ **60.** $x^2 + 12x + 36 < 0$ **61.** $x^2 + 24x + 144 \leq 0$ **62.** $x^2 + 12x + 36 \geq 0$

Mixed Exercises

For Exercises 63–80, identify the inequality as one of the following types: linear, quadratic, rational, or polynomial (degree > 2). Then solve the inequality and write the answer in interval notation.

63. $2y^2 - 8 \leq 24$ **64.** $8p^2 - 18 > 0$ **65.** $(5x+2)^2 > -4$

66. $(3-7x)^2 < -1$ **67.** $4(x-2) < 6x - 3$ **68.** $-7(3-y) > 4 + 2y$

69. $\dfrac{2x+3}{x+1} \leq 2$ **70.** $\dfrac{5x-1}{x+3} \geq 5$ **71.** $4x^3 - 40x^2 + 100x > 0$

72. $2y^3 - 12y^2 + 18y < 0$ **73.** $2p^3 > 4p^2$ **74.** $w^3 \leq 5w^2$

75. $\dfrac{1}{x^2+3} < -4$ **76.** $\dfrac{1}{4t^2+5} > -2$ **77.** $x^2 - 2 < 0$

78. $y^2 - 3 > 0$ **79.** $x^2 + 5x - 2 \geq 0$ **80.** $t^2 + 7t + 3 \leq 0$

Graphing Calculator Exercises

81. To solve the inequality $\dfrac{x}{x-2} > 0$

enter Y_1 as $x/(x-2)$ and determine where the graph is above the x-axis. Write the solution in interval notation.

82. To solve the inequality $\dfrac{x}{x-2} < 0$

enter Y_1 as $x/(x-2)$ and determine where the graph is below the x-axis. Write the solution in interval notation.

83. To solve the inequality $x^2 - 1 < 0$, enter Y_1 as $x^2 - 1$ and determine where the graph is below the x-axis. Write the solution in interval notation.

84. To solve the inequality $x^2 - 1 > 0$, enter Y_1 as $x^2 - 1$ and determine where the graph is above the x-axis. Write the solution in interval notation.

For Exercises 85–88, determine the solution by graphing the inequalities.

85. $x^2 + 10x + 25 \leq 0$ **86.** $-x^2 + 10x - 25 \geq 0$ **87.** $\dfrac{8}{x^2 + 2} < 0$ **88.** $\dfrac{-6}{x^2 + 3} > 0$

---

chapter 9 | midchapter review

Solve the inequalities. Write the answers in interval notation.

1. $\dfrac{1}{3}x + 5 < 2$ or $\dfrac{2}{5}x \geq 4$

2. $3a^2 - 4a > 7$

3. $2 \geq t - 3$ and $-5t + 8 < t$

4. $\dfrac{4}{y^2} > 0$ **5.** $y^2 - y - 12 < 0$

6. $10k - 3 > 7$ and $-k + 6 > 8$

7. $w^3 + 2w^2 - w - 2 \geq 0$

8. $\dfrac{t+1}{t-2} \leq 0$

9. $\dfrac{y}{y-5} > \dfrac{1}{5}$

10. $p + 1.8 > 7.9$ and $5p - 0.4 \geq 0.6$

11. $n^2 + 2n + 1 > 0$ **12.** $\dfrac{x+1}{2x+3} > \dfrac{2}{3}$

13. $x^2 + 22x < -121$ **14.** $(m-3)^2 < 4m$

15. $0.25k^2 + 0.09 < 0.3k$

16. $\dfrac{1}{2}p - \dfrac{2}{3} < \dfrac{1}{6}p - 4$ **17.** $\dfrac{a+2}{a-5} \geq 0$

18. $(x+3)^2(x-1) \leq 0$

Section 9.3 Absolute Value Equations

1. Solving Absolute Value Equations

An equation of the form $|x| = a$ is called an **absolute value equation**. The solution includes all real numbers whose absolute value equals a. For example, the solutions to the equation $|x| = 4$ are $x = 4$ as well as $x = -4$, because $|4| = 4$ and $|-4| = 4$. In Chapter 1, we introduced a geometric interpretation of $|x|$. The absolute value of a number is its distance from zero on the number line (Figure 9-5). Therefore, the solutions to the equation $|x| = 4$ are the values of x that are 4 units away from zero.

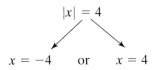
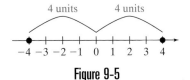

Figure 9-5

Objectives

1. Solving Absolute Value Equations
2. Solving Equations Having Two Absolute Values

Concept Connections

1. Why does the equation $|x| = -2$ have no solutions?

Solutions to Absolute Value Equations of the Form $|x| = a$

If a is a real number, then

1. If $a \geq 0$, the solutions to the equation $|x| = a$ are $x = a$ and $x = -a$.
2. If $a < 0$, there is no solution to the equation $|x| = a$.

To solve an absolute value equation of the form $|x| = a$ ($a \geq 0$), rewrite the equation as $x = a$ or $x = -a$.

---

example 1   Solving Absolute Value Equations

Solve the absolute value equations.

   a. $|x| = 5$ **b.** $|w| - 2 = 12$ **c.** $|p| = 0$ **d.** $|x| = -6$

Solution:

 a.   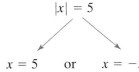   The equation is in the form $|x| = a$, where $a = 5$.

     $x = 5$ or $x = -5$ Rewrite the equation as $x = a$ or $x = -a$.

b. $|w| - 2 = 12$ Isolate the absolute value to write the equation in the form $|w| = a$.

     $|w| = 14$

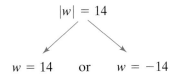

     $w = 14$ or $w = -14$ Rewrite the equation as $w = a$ or $w = -a$.

Skill Practice

Solve the absolute value equations.

2. $|y| = 7$
3. $|v| + 6 = 10$
4. $|w| = 0$
5. $|z| = -12$

Answers

1. The absolute value represents the distance from zero. This cannot be negative.
2. $y = 7$ or $y = -7$
3. $v = 4$ or $v = -4$
4. $w = 0$ 5. No solution

c. $|p| = 0$

$p = 0$ or $p = -0$ Rewrite as two equations. Notice that the second equation $p = -0$ is the same as the first equation. Intuitively, $p = 0$ is the only number whose absolute value equals 0.

d. $|x| = -6$

No solution   This equation is of the form $|x| = a$, but a is negative. There is no number whose absolute value is negative.

We have solved absolute value equations of the form $|x| = a$. Notice that x can represent any algebraic quantity. For example, to solve the equation $|2w - 3| = 5$, we still rewrite the absolute value equation as two equations. In this case, we set the quantity $2w - 3$ equal to 5 and to -5, respectively.

$$|2w - 3| = 5$$

$2w - 3 = 5$ or $2w - 3 = -5$

Steps to Solve an Absolute Value Equation

1. Isolate the absolute value. That is, write the equation in the form $|x| = a$, where a is a constant real number.
2. If $a < 0$, there is no solution.
3. Otherwise, if $a \geq 0$, rewrite the absolute value equation as $x = a$ or $x = -a$.
4. Solve the individual equations from step 3.
5. Check the answers in the original absolute value equation.

Skill Practice

Solve the absolute value equations.

6. $|4x + 1| = 9$

7. $|3z + 10| + 3 = 1$

---

example 2   Solving Absolute Value Equations

Solve the absolute value equations.

a. $|2w - 3| = 5$ **b.** $|2c - 5| + 6 = 2$

Solution:

a. $|2w - 3| = 5$ The equation is already in the form $|x| = a$, where $x = 2w - 3$.

$2w - 3 = 5$ or $2w - 3 = -5$ Rewrite as two equations.

$2w = 8$ or $2w = -2$ Solve each equation.

$w = 4$ or $w = -1$

Answers

6. $x = 2$ or $x = -\dfrac{5}{2}$
7. No solution

Check: $x = 4$ Check: $x = -1$ Check the solutions in the original equation.

$|2w - 3| = 5$ $|2w - 3| = 5$

$|2(4) - 3| \stackrel{?}{=} 5$ $|2(-1) - 3| \stackrel{?}{=} 5$

$|8 - 3| \stackrel{?}{=} 5$ $|-2 - 3| \stackrel{?}{=} 5$

$|5| \stackrel{?}{=} 5$ ✓ $|-5| \stackrel{?}{=} 5$ ✓

Calculator Connections

To confirm the answers to Example 2(a), graph $Y_1 = \text{abs}(2x - 3)$ and $Y_2 = 5$. The solutions to the equation $|2w - 3| = 5$ are the x-coordinates of the points of intersection $(4, 5)$ and $(-1, 5)$.

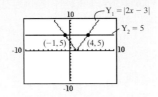

b. $|2c - 5| + 6 = 2$

$|2c - 5| = -4$ Isolate the absolute value. The equation is in the form $|x| = a$, where $x = 2c - 5$ and $a = -4$. Because $a < 0$, there is no solution.

No solution   There are no numbers x that will make an absolute value equal to a negative number.

Avoiding Mistakes: Always isolate the absolute value first. Otherwise you will get answers that do not check.

Calculator Connections

The graphs of $Y_1 = \text{abs}(2x - 5) + 6$ and $Y_2 = 2$ do not intersect.
   Therefore, there is no solution to the equation $|2x - 5| + 6 = 2$.

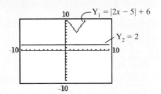

---

example 3   Solving Absolute Value Equations

Solve the absolute value equations.

a. $-2\left|\dfrac{2}{5}p + 3\right| - 7 = -19$ **b.** $|4.1 - p| + 6.9 = 6.9$

Skill Practice

Solve the absolute value equations.

8. $3\left|\dfrac{3}{2}a + 1\right| + 2 = 14$

9. $|1.2 + x| - 3.5 = -3.5$

Solution:

a. $-2\left|\dfrac{2}{5}p + 3\right| - 7 = -19$

$-2\left|\dfrac{2}{5}p + 3\right| = -12$ Isolate the absolute value.

Answers

8. $a = 2$ or $a = -\dfrac{10}{3}$

9. $x = -1.2$

$$\frac{-2\left|\frac{2}{5}p + 3\right|}{-2} = \frac{-12}{-2}$$

$$\left|\frac{2}{5}p + 3\right| = 6$$

$\frac{2}{5}p + 3 = 6$ or $\frac{2}{5}p + 3 = -6$ Rewrite as two equations.

$2p + 15 = 30$ or $2p + 15 = -30$ Multiply by 5 to clear fractions.

$2p = 15$ or $2p = -45$

$p = \frac{15}{2}$ or $p = -\frac{45}{2}$ Both solutions check in the original equation.

b. $|4.1 - p| + 6.9 = 6.9$

$|4.1 - p| = 0$ Isolate the absolute value.

$4.1 - p = 0$ or $4.1 - p = -0$ Rewrite as two equations. Notice that the equations are the same.

$-p = -4.1$ Subtract 4.1 from both sides.

$p = 4.1$ Check: $p = 4.1$

$|4.1 - p| + 6.9 = 6.9$

$|4.1 - 4.1| + 6.9 \stackrel{?}{=} 6.9$

$|0| + 6.9 \stackrel{?}{=} 6.9$

The solution is 4.1.

$6.9 = 6.9$ ✓

2. Solving Equations Having Two Absolute Values

Some equations have two absolute values. The solutions to the equation $|x| = |y|$ are $x = y$ or $x = -y$. That is, if two quantities have the same absolute value, then the quantities are equal or the quantities are opposites.

Equality of Absolute Values

$|x| = |y|$ implies that $x = y$ or $x = -y$.

example 4 Solving an Equation Having Two Absolute Values

Solve the equations.

a. $|2w - 3| = |5w + 1|$ **b.** $|x - 4| = |x + 8|$

Solution:

a.
$$|2w - 3| = |5w + 1|$$

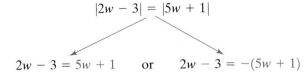

$2w - 3 = 5w + 1$ or $2w - 3 = -(5w + 1)$ Rewrite as two equations, $x = y$ or $x = -y$.

$2w - 3 = 5w + 1$ or $2w - 3 = -5w - 1$ Solve for w.

$-3w - 3 = 1$ or $7w - 3 = -1$

$-3w = 4$ or $7w = 2$

$w = -\dfrac{4}{3}$ or $w = \dfrac{2}{7}$

The solutions are $-\dfrac{4}{3}$ and $\dfrac{2}{7}$.   Both values check in the original equation.

Avoiding Mistakes: To take the opposite of the quantity $5w + 1$, use parentheses and apply the distributive property.

Skill Practice

Solve the equations.

10. $|3 - 2x| = |3x - 1|$

11. $|4t + 3| = |4t - 5|$

b. $|x - 4| = |x + 8|$

$x - 4 = x + 8$ or $x - 4 = -(x + 8)$ Rewrite as two equations, $x = y$ or $x = -y$.

$-4 = 8$ or $x - 4 = -x - 8$ Solve for x.
↑
contradiction $2x - 4 = -8$

$2x = -4$

$x = -2$

The only solution is -2. $x = -2$ checks in the original equation.

Calculator Connections

Graph $Y_1 = \text{abs}(x - 4)$ and $Y_2 = \text{abs}(x + 8)$. There is one point of intersection at $(-2, 6)$. Therefore, the solution to $|x - 4| = |x + 8|$ is $x = -2$.

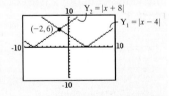

Answers

10. $x = \dfrac{4}{5}$ or $x = -2$

11. $t = \dfrac{1}{4}$

section 9.3 Practice Exercises

Boost your GRADE at mathzone.com!
MathZone

- Practice Problems
- Self-Tests
- NetTutor
- e-Professors
- Videos

Study Skills Exercises

1. It is always helpful to read the material in a section and make some notes before it is presented in class. Writing notes ahead of time will free you to listen more in class and pay special attention to the concepts that need clarification. Refer to your class syllabus, and list the next two sections that will be covered in class and a time that you can read them beforehand.

2. Define the key term **absolute value equation**.

Review Exercises

For Exercises 3–10, solve the inequalities. Write the answers in interval notation.

3. $3(a + 2) - 6 > 2$ and $-2(a - 3) + 14 > -3$
4. $3x - 5 \geq 7x + 3$ or $2x - 1 \leq 4x - 5$

5. $\dfrac{4}{y - 4} \geq 3$
6. $\dfrac{3}{t + 1} \leq 2$

7. The sum of a number and 6 is between -1 and 13. Find all such numbers.
8. The difference of 12 and a number is between -3 and -1. Find all such numbers.

9. $3(x - 2)(x + 4)(2x - 1) < 0$
10. $x^3 - 7x^2 - 8x > 0$

Objective 1: Solving Absolute Value Equations

For Exercises 11–42, solve the absolute value equations. (See Examples 1–3.)

11. $|p| = 7$
12. $|q| = 10$
13. $|x| + 5 = 11$
14. $|x| - 3 = 20$

15. $|y| = \sqrt{2}$
16. $|y| = \dfrac{5}{8}$
17. $|w| - 3 = -5$
18. $|w| + 4 = -8$

19. $|3q| = 0$
20. $|4p| = 0$
21. $\left|3x - \dfrac{1}{2}\right| = \dfrac{1}{2}$
22. $|4x + 1| = 6$

23. $\left|\dfrac{7z}{3} - \dfrac{1}{3}\right| + 3 = 6$
24. $\left|\dfrac{w}{2} + \dfrac{3}{2}\right| - 2 = 7$
25. $\left|\dfrac{5y + 2}{2}\right| = 6$
26. $\left|\dfrac{2t - 1}{3}\right| = 5$

27. $|0.2x - 3.5| = -5.6$
28. $|1.81 + 2x| = -2.2$
29. $1 = -4 + \left|2 - \dfrac{1}{4}w\right|$
30. $-12 + |6 - 2x| = -6$

31. $10 = 4 + |2y + 1|$

32. $-1 = -|5x + 7|$

33. $-2|3b - 7| - 9 = -9$

34. $-3|5x + 1| + 4 = 4$

35. $-2|x + 3| = 5$

36. $-3|x - 5| = 7$

37. $0 = |6x - 9|$

38. $7 = |4k - 6| + 7$

39. $\left|-\dfrac{1}{5} - \dfrac{1}{2}k\right| = \dfrac{9}{5}$

40. $\left|-\dfrac{1}{6} - \dfrac{2}{9}h\right| = \dfrac{1}{2}$

41. $-3|2 - 6x| + 5 = -10$

42. $5|1 - 2x| - 7 = 3$

Objective 2: Solving Equations Having Two Absolute Values

For Exercises 43–56, solve the absolute value equations. (See Example 4.)

43. $|4x - 2| = |-8|$

44. $|3x + 5| = |-5|$

45. $|4w + 3| = |2w - 5|$

46. $|3y + 1| = |2y - 7|$

47. $|2y + 5| = |7 - 2y|$

48. $|9a + 5| = |9a - 1|$

49. $\left|\dfrac{4w - 1}{6}\right| = \left|\dfrac{2w}{3} + \dfrac{1}{4}\right|$

50. $\left|\dfrac{3p + 2}{4}\right| = \left|\dfrac{1}{2}p - 2\right|$

51. $|2h - 6| = |2h + 5|$

52. $|6n - 7| = |4 - 6n|$

53. $|3.5m - 1.2| = |8.5m + 6|$

54. $|11.2n + 9| = |7.2n - 2.1|$

55. $|4x - 3| = -|2x - 1|$

56. $-|3 - 6y| = |8 - 2y|$

57. Write an absolute value equation whose solution is the set of real numbers 6 units from zero on the number line.

58. Write an absolute value equation whose solution is the set of real numbers $\dfrac{7}{2}$ units from zero on the number line.

59. Write an absolute value equation whose solution is the set of real numbers $\dfrac{4}{3}$ units from zero on the number line.

60. Write an absolute value equation whose solution is the set of real numbers 9 units from zero on the number line.

Expanding Your Skills

For Exercises 61–66, solve the absolute value equations.

61. $|5y - 3| + \sqrt{5} = 1 + \sqrt{5}$

62. $|2x - \sqrt{3}| + 4 = 4 + \sqrt{3}$

63. $|\sqrt{3} + x| = 7$

64. $\sqrt{2} + |w - 8| = 3 + 4\sqrt{2}$

65. $|w - \sqrt{6}| = |3w + \sqrt{6}|$

66. $\left|\dfrac{\sqrt{5}}{2}x - 4\right| = 6$

Graphing Calculator Exercises

For Exercises 67–74, enter the left side of the equation as Y_1 and enter the right side of the equation as Y_2. Then use the *Intersect* feature or *Zoom* and *Trace* to approximate the x-values where the two graphs intersect (if they intersect).

67. $|4x - 3| = 5$

68. $|x - 4| = 3$

69. $|8x + 1| + 8 = 1$

70. $|3x - 2| + 4 = 2$

71. $|x - 3| = |x + 2|$

72. $|x + 4| = |x - 2|$

73. $|2x - 1| = |-x + 3|$

74. $|3x| = |2x - 5|$

Objectives

1. Solving Absolute Value Inequalities by Method I
2. Solving Absolute Value Inequalities by Using Method II, the Test Point Method
3. Translating to an Absolute Value Expression

section 9.4 Absolute Value Inequalities

1. Solving Absolute Value Inequalities by Method I

In Section 9.3, we studied absolute value equations in the form $|x| = a$. In this section we will solve absolute value *inequalities*. An inequality in any of the forms $|x| < a$, $|x| \le a$, $|x| > a$, or $|x| \ge a$ is called an **absolute value inequality**.

Recall that an absolute value represents distance from zero on the real number line. Consider the following absolute value equation and inequalities.

1. $|x| = 3$

$x = 3$ or $x = -3$

Solution:

The set of all points 3 units from zero on the number line

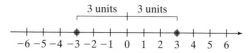

Section 9.4 Absolute Value Inequalities 657

2. $|x| > 3$

   $x < -3$ or $x > 3$

   Solution:

   The set of all points more than 3 units from zero

   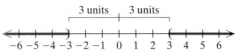

3. $|x| < 3$

   $-3 < x < 3$

   Solution:

   The set of all points less than 3 units from zero

   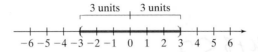

Solutions to Absolute Value Equations and Inequalities

Let a be a real number such that $a > 0$. Then

Equation/Inequality	Solution (Equivalent Form)	Graph
$|x| = a$	$x = -a$ or $x = a$	
$|x| > a$	$x < -a$ or $x > a$	
$|x| < a$	$-a < x < a$	

Concept Connections

Write an absolute value inequality to describe the set shown.

1.
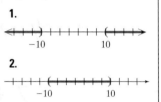

2.

To solve an absolute value inequality, first isolate the absolute value and then rewrite the absolute value inequality in its equivalent form.

example 1 Solving Absolute Value Inequalities

Solve the inequalities.

a. $|3w + 1| - 4 < 7$ **b.** $3 \leq 1 + \left|\frac{1}{2}t - 5\right|$

Skill Practice

Solve the inequalities. Write the solutions in interval notation.

3. $|2t + 5| + 2 \leq 11$

4. $\left|\frac{1}{3}c - 1\right| + 1 > 5$

Solution:

a. $|3w + 1| - 4 < 7$

$|3w + 1| < 11$ ← Isolate the absolute value first.

The inequality is in the form $|x| < a$, where $x = 3w + 1$.

$-11 < 3w + 1 < 11$ Rewrite in the equivalent form $-a < x < a$.

$-12 < 3w < 10$ Solve for w.

$-4 < w < \frac{10}{3}$

The solution is $\{w | -4 < w < \frac{10}{3}\}$ or, equivalently in interval notation, $(-4, \frac{10}{3})$.

Answers

1. $|x| > 10$ 2. $|x| < 10$
3. $[-7, 2]$
4. $(-\infty, -9) \cup (15, \infty)$

Calculator Connections

Graph $Y_1 = \text{abs}(3x + 1) - 4$ and $Y_2 = 7$. On the given display window, $Y_1 < Y_2$ (Y_1 is below Y_2) for $-4 < x < \frac{10}{3}$.

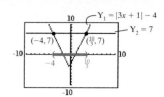

b. $3 \leq 1 + \left|\frac{1}{2}t - 5\right|$

$1 + \left|\frac{1}{2}t - 5\right| \geq 3$ Write the inequality with the absolute value on the left.

$\left|\frac{1}{2}t - 5\right| \geq 2$ Isolate the absolute value.

The inequality is in the form $|x| \geq a$, where $x = \frac{1}{2}t - 5$.

$\frac{1}{2}t - 5 \leq -2$ or $\frac{1}{2}t - 5 \geq 2$ Rewrite in the equivalent form $x \leq -a$ or $x \geq a$.

$\frac{1}{2}t \leq 3$ or $\frac{1}{2}t \geq 7$ Solve the compound inequality.

$2\left(\frac{1}{2}t\right) \leq 2(3)$ or $2\left(\frac{1}{2}t\right) \geq 2(7)$ Clear fractions.

$t \leq 6$ or $t \geq 14$ 

The solution is $\{t \mid t \leq 6 \text{ or } t \geq 14\}$ or, equivalently in interval notation, $(-\infty, 6] \cup [14, \infty)$.

Calculator Connections

Graph $Y_1 = \text{abs}((1/2)x - 5) + 1$ and $Y_2 = 3$. On the given display window, $Y_1 \geq Y_2$ for $x \leq 6$ or $x \geq 14$.

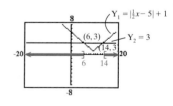

By definition, the absolute value of a real number will always be nonnegative. Therefore, the absolute value of any expression will always be greater than a negative number. Similarly, an absolute value can never be less than a negative number. Let a represent a positive real number. Then

- The solution to the inequality $|x| > -a$ is all real numbers, $(-\infty, \infty)$.
- There is no solution to the inequality $|x| < -a$.

example 2 Solving Absolute Value Inequalities

Solve the inequalities.

a. $|3d - 5| + 7 < 4$ b. $|3d - 5| + 7 > 4$

Solution:

a. $|3d - 5| + 7 < 4$

$|3d - 5| < -3$

No solution

Isolate the absolute value. An absolute value expression cannot be less than a negative number. Therefore, there is no solution.

b. $|3d - 5| + 7 > 4$

$|3d - 5| > -3$

All real numbers, $(-\infty, \infty)$

Isolate the absolute value. The inequality is in the form $|x| > a$, where a is negative. An absolute value of any real number is greater than a negative number. Therefore, the solution is all real numbers.

Skill Practice

Solve the inequalities.

5. $|4p + 2| + 6 < 2$
6. $|4p + 2| + 6 > 2$

Calculator Connections

By graphing $Y_1 = \text{abs}(3x - 5) + 7$ and $Y_2 = 4$, we see that $Y_1 > Y_2$ (Y_1 is above Y_2) for all real numbers x on the given display window.

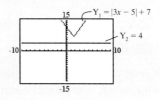

example 3 Solving Absolute Value Inequalities

Solve the inequalities.

a. $|4x + 2| \geq 0$ b. $|4x + 2| > 0$

Solution:

a. $|4x + 2| \geq 0$ ← The absolute value is already isolated.

The absolute value of any real number is nonnegative. Therefore, the solution is all real numbers, $(-\infty, \infty)$.

b. $|4x + 2| > 0$

An absolute value will be greater than zero at all points *except where it is equal to zero*. That is, the point(s) for which $|4x + 2| = 0$ must be excluded from the solution set.

$|4x + 2| = 0$

$4x + 2 = 0$ or $4x + 2 = -0$ The second equation is the same as the first.

Skill Practice

Solve the inequalities.

7. $|3x - 1| \geq 0$
8. $|3x - 1| > 0$

Answers

5. No solution
6. All real numbers; $(-\infty, \infty)$
7. $(-\infty, \infty)$
8. $\left\{x \mid x \neq \dfrac{1}{3}\right\}$ or $\left(-\infty, \dfrac{1}{3}\right) \cup \left(\dfrac{1}{3}, \infty\right)$

$$4x = -2$$

$$x = -\frac{1}{2}$$

Therefore, exclude $x = -\frac{1}{2}$ from the solution.

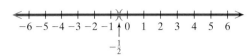

The solution is $\{x \mid x \neq -\frac{1}{2}\}$ or equivalently in interval notation, $(-\infty, -\frac{1}{2}) \cup (-\frac{1}{2}, \infty)$.

Calculator Connections

Graph $Y_1 = \text{abs}(4x + 2)$. From the graph, $Y_1 = 0$ at $x = -\frac{1}{2}$ (the x-intercept).
On the given display window, $Y_1 > 0$ for $x < -\frac{1}{2}$ or $x > -\frac{1}{2}$.

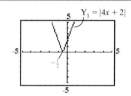

2. Solving Absolute Value Inequalities by Using Method II, the Test Point Method

For each problem in Example 1, the absolute value inequality was converted to an equivalent compound inequality. However, sometimes students have difficulty setting up the appropriate compound inequality. To avoid this problem, you may want to use the test point method to solve absolute value inequalities.

Solving Inequalities by Using the Test Point Method

1. Find the boundary points of the inequality. (Boundary points are the real solutions to the related equation and points where the inequality is undefined.)
2. Plot the boundary points on the number line. This divides the number line into regions.
3. Select a test point from each region and substitute it into the original inequality.
   - If a test point makes the original inequality true, then that region is part of the solution set.
4. Test the boundary points in the original inequality.
   - If a boundary point makes the original inequality true, then that point is part of the solution set.

To demonstrate the use of the test point method, we will repeat the absolute value inequalities from Example 1. Notice that regardless of the method used, the absolute value is always isolated *first* before any further action is taken.

example 4 Solving Absolute Value Inequalities by the Test Point Method

Solve the inequalities by using the test point method.

a. $|3w + 1| - 4 < 7$ **b.** $3 \leq 1 + \left|\frac{1}{2}t - 5\right|$

Skill Practice

Solve the inequalities by using the test point method.

9. $|2t + 5| + 2 \leq 11$
10. $\left|\frac{1}{2}c + 4\right| + 1 > 6$

Solution:

a. $|3w + 1| - 4 < 7$

$|3w + 1| < 11$ ← Isolate the absolute value.

$|3w + 1| = 11$ **Step 1:** Solve the related equation.

$3w + 1 = 11$ or $3w + 1 = -11$ Write as an equivalent system of two equations.

$3w = 10$ or $3w = -12$

$w = \frac{10}{3}$ or $w = -4$ These are the only boundary points.

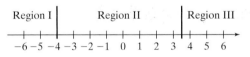

Step 2: Plot the boundary points.

Step 3: Select a test point from each region.

Test $w = -5$:

$|3(-5) + 1| - 4 \stackrel{?}{<} 7$

$|-14| - 4 \stackrel{?}{<} 7$

$14 - 4 \stackrel{?}{<} 7$

$10 \stackrel{?}{<} 7$ False

Test $w = 0$:

$|3(0) + 1| - 4 \stackrel{?}{<} 7$

$|1| - 4 \stackrel{?}{<} 7$

$-3 \stackrel{?}{<} 7$ True

Test $w = 4$:

$|3(4) + 1| - 4 \stackrel{?}{<} 7$

$|13| - 4 \stackrel{?}{<} 7$

$13 - 4 \stackrel{?}{<} 7$

$9 \stackrel{?}{<} 7$ False

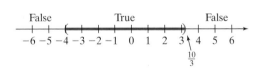

Step 4: The last step in the test point method is to determine whether the boundary points are part of the solution set. Because the original inequality is a strict inequality, the boundary points (where equality occurs) are not included.

The solution is $\{w | -4 < w < \frac{10}{3}\}$ or, equivalently in interval notation, $(-4, \frac{10}{3})$.

Answers

9. $[-7, 2]$
10. $(-\infty, -18) \cup (2, \infty)$

b. $3 \leq 1 + \left|\dfrac{1}{2}t - 5\right|$

$1 + \left|\dfrac{1}{2}t - 5\right| \geq 3$ Write the inequality with the absolute value on the left.

$\left|\dfrac{1}{2}t - 5\right| \geq 2$ Isolate the absolute value.

$\left|\dfrac{1}{2}t - 5\right| = 2$ **Step 1:** Solve the related equation.

$\dfrac{1}{2}t - 5 = 2$ or $\dfrac{1}{2}t - 5 = -2$ Write the union of two equations.

$\dfrac{1}{2}t = 7$ or $\dfrac{1}{2}t = 3$

$t = 14$ or $t = 6$ These are the boundary points.

Region I | Region II | Region III **Step 2:** Plot the boundary points.
⟵————6————14————⟶

Step 3: Select a test point from each region.

Test $t = 0$:

$3 \stackrel{?}{\leq} 1 + \left|\dfrac{1}{2}(0) - 5\right|$

$3 \stackrel{?}{\leq} 1 + |0 - 5|$

$3 \stackrel{?}{\leq} 1 + |-5|$

$3 \leq 1 + 5$ True

Test $t = 10$:

$3 \stackrel{?}{\leq} 1 + \left|\dfrac{1}{2}(10) - 5\right|$

$3 \stackrel{?}{\leq} 1 + |5 - 5|$

$3 \stackrel{?}{\leq} 1 + |0|$

$3 \stackrel{?}{\leq} 1$ False

Test $t = 16$:

$3 \stackrel{?}{\leq} 1 + \left|\dfrac{1}{2}(16) - 5\right|$

$3 \stackrel{?}{\leq} 1 + |8 - 5|$

$3 \stackrel{?}{\leq} 1 + |3|$

$3 \leq 4$ True

Step 4: The original inequality uses the sign $\geq$. Therefore, the boundary points (where equality occurs) must be part of the solution set.

True   False   True
⟵———]————————[———⟶
     6   14

The solution is $\{x \mid x \leq 6 \text{ or } x \geq 14\}$ or, equivalently in interval notation, $(-\infty, 6] \cup [14, \infty)$.

Skill Practice

Solve the inequalities.

11. $\left|\dfrac{3}{5}x + 3\right| < 0$

12. $\left|\dfrac{3}{5}x + 3\right| \leq 0$

example 5 Solving Absolute Value Inequalities

Solve the inequalities.

a. $\left|\dfrac{1}{3}x + 4\right| < 0$ **b.** $\left|\dfrac{1}{3}x + 4\right| \leq 0$

Answers

11. No solution 12. $\{-5\}$

Solution:

a. $\left|\dfrac{1}{3}x + 4\right| < 0$ ←—— The absolute value is already isolated.

No solution — Because the absolute value of any real number is nonnegative, an absolute value cannot be strictly less than zero. Therefore, there is no solution to this inequality.

b. $\left|\dfrac{1}{3}x + 4\right| \le 0$ ←—— The absolute value is already isolated.

An absolute value will never be less than zero. However, an absolute value may be equal to zero. Therefore, the only solutions to this inequality are the solutions to the related equation.

$\left|\dfrac{1}{3}x + 4\right| = 0$ Set up the related equation.

$\dfrac{1}{3}x + 4 = 0$

$\dfrac{1}{3}x = -4$

$x = -12$ This is the only boundary point.

The solution set is $\{-12\}$.

Calculator Connections

Graph $Y_1 = \text{abs}((1/3)x + 4)$. Notice that on the given viewing window the graph of Y_1 does not extend below the x-axis. Therefore, there is no solution to the inequality $Y_1 < 0$.
  Because $Y_1 = 0$ at $x = -12$, the inequality $Y_1 \le 0$ has a solution at $x = -12$.

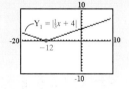

3. Translating to an Absolute Value Expression

Absolute value expressions can be used to describe distances. The distance between c and d is given by $|c - d|$. For example, the distance between -2 and 3 on the number line is $|(-2) - 3| = 5$ as expected.

example 6 Expressing Distances with Absolute Value

Write an absolute value inequality to represent the following phrases.

a. All real numbers x, whose distance from zero is greater than 5 units

b. All real numbers x, whose distance from -7 is less than 3 units

---

Skill Practice

Write an absolute value inequality to represent the following phrases.

13. All real numbers whose distance from zero is greater than 10 units

14. All real numbers whose distance from 4 is greater than 6

Answers
13. $|x| > 10$
14. $|x - 4| > 6$

Solution:

a. All real numbers x, whose distance from zero is greater than 5 units

$|x - 0| > 5$ or simply $|x| > 5$

b. All real numbers x, whose distance from -7 is less than 3 units

$|x - (-7)| < 3$ or simply $|x + 7| < 3$

Absolute value expressions can also be used to describe boundaries for measurement error.

Skill Practice

15. Vonzell molded a piece of metal in her machine shop. She measured the thickness at 12 mm. Her machine is accurate to the nearest 0.05 mm. Write an absolute value inequality to express an interval for the true measurement of the thickness, t, of the metal.

example 7 — Expressing Measurement Error with Absolute Value

Latoya measured a certain compound on a scale in the chemistry lab at school. She measured 8 g of the compound, but the scale is only accurate to the nearest tenth of a gram. Write an absolute value inequality to express an interval for the true mass, x, of the compound she measured.

Solution:

Because the scale is only accurate to the nearest tenth of a gram, the true mass, x, of the compound may deviate by as much as 0.1 g above or below 8 g. This may be expressed as an absolute value inequality:

$|x - 8.0| \leq 0.1$ or equivalently $7.9 \leq x \leq 8.1$

Answer

15. $|t - 12| \leq 0.05$ or $11.95 \leq t \leq 12.05$

---

section 9.4 — Practice Exercises

Boost your GRADE at mathzone.com!

- Practice Problems
- Self-Tests
- NetTutor
- e-Professors
- Videos

Study Skills Exercises

1. Most people cannot concentrate on studying for more than an hour without taking a break. To make the most of your time, write down a schedule for your next study session. Include breaks where you can eat a meal, walk the dog, or perform other simple tasks that need to be completed.

2. Define the key term **absolute value inequality**.

Review Exercises

For Exercises 3–6, solve the equations.

3. $|10x - 6| = -5$ 4. $2 = |5 - 7x| + 1$ 5. $|6x| = |9x + 5|$ 6. $|3y - 1| = |3y + 4|$

Section 9.4 Absolute Value Inequalities 665

For Exercises 7–10, solve the inequalities and graph the solution set. Write the solution in interval notation.

7. $-15 < 3w - 6 \leq -9$

8. $5 - 2y \leq 1$ and $3y + 2 \geq 14$

9. $m - 7 \leq -5$ or $m - 7 \geq -10$

10. $3b - 2 < 7$ or $b - 2 > 4$

Objectives 1 and 2: Solving Absolute Value Inequalities

For Exercises 11–18, solve the equations and inequalities. For each inequality, graph the solution set and express the solution in interval notation.

11. a. $|x| = 5$

b. $|x| > 5$

c. $|x| < 5$

12. a. $|a| = 4$

b. $|a| > 4$

c. $|a| < 4$

13. a. $|x - 3| = 7$

b. $|x - 3| > 7$

c. $|x - 3| < 7$

14. a. $|w + 2| = 6$

b. $|w + 2| > 6$

c. $|w + 2| < 6$

15. a. $|p| = -2$

b. $|p| > -2$

c. $|p| < -2$

16. a. $|x| = -14$

b. $|x| > -14$

c. $|x| < -14$

17. a. $|y + 1| = -6$

b. $|y + 1| > -6$

c. $|y + 1| < -6$

18. a. $|z - 4| = -3$

b. $|z - 4| > -3$

c. $|z - 4| < -3$

For Exercises 19–22, match the graph with the inequality:

19.

20.

21.

22.

a. $|x - 2| < 4$ **b.** $|x - 1| > 4$ **c.** $|x - 3| < 2$ **d.** $|x - 5| > 1$

For Exercises 23–52, solve the absolute value inequalities by using either method I **(see Examples 1–3)** or the test point method, method II **(see Examples 4–5)**. Graph the solution set and write the solution in interval notation.

23. $|x| > 6$

24. $|x| \leq 6$

25. $|t| \leq 3$

26. $|p| > 3$

27. $|y + 2| \geq 0$

28. $0 \leq |7n + 2|$

29. $5 \leq |2x - 1|$

30. $|x - 2| \geq 7$

31. $|k - 7| < -3$

32. $|h + 2| < -9$

33. $\left|\dfrac{w - 2}{3}\right| - 3 \leq 1$

34. $\left|\dfrac{x + 3}{2}\right| - 2 \geq 4$

35. $14 \leq |9 - 4y|$

36. $1 > |2m - 7|$

37. $\left|\dfrac{2x + 1}{4}\right| < 5$

38. $\left|\dfrac{x - 4}{5}\right| \leq 7$

39. $8 < |4 - 3x| + 12$

40. $-16 < |5x - 1| - 1$

41. $5 - |2m + 1| > 5$

42. $3 - |5x + 3| > 3$

43. $|p + 5| \leq 0$

44. $|y + 1| - 4 \leq -4$

45. $|z - 6| + 5 > 5$

46. $|2c - 1| - 4 > -4$

47. $5|2y - 6| + 3 \geq 13$

48. $7|y + 1| - 3 \geq 11$

49. $-3|6 - t| + 1 > -5$

50. $-4|8 - x| + 2 > -14$

51. $|0.02x + 0.06| - 0.1 < 0.05$

52. $|0.05x - 0.04| - 0.01 < 0.11$

Objective 3: Translating to an Absolute Value Expression

For Exercises 53–56, write an absolute value inequality equivalent to the expression given. (See Example 6.)

53. All real numbers whose distance from 0 is greater than 7

54. All real numbers whose distance from −3 is less than 4

55. All real numbers whose distance from 2 is at most 13

56. All real numbers whose distance from 0 is at least 6

57. A 32-oz jug of orange juice may not contain exactly 32 oz of juice. The possibility of measurement error exists when the jug is filled in the factory. If the maximum measurement error is ±0.05 oz, write an absolute value inequality representing the range of volumes, x, in which the orange juice jug may be filled. (See Example 7.)

58. The length of a board is measured to be 32.3 in. The maximum measurement error is ±0.2 in. Write an absolute value inequality that represents the range for the length of the board, x.

59. A bag of potato chips states that its weight is $6\frac{3}{4}$ oz. The maximum measurement error is $\pm\frac{1}{8}$ oz. Write an absolute value inequality that represents the range for the weight, x, of the bag of chips.

60. A $\frac{7}{8}$-in. bolt varies in length by at most $\pm\frac{1}{16}$ in. Write an absolute value inequality that represents the range for the length, x, of the bolt.

61. The width, w, of a bolt is supposed to be 2 cm but may have a 0.01-cm margin of error. Solve $|w - 2| \leq 0.01$, and interpret the solution to the inequality in the context of this problem.

62. In the 2004 election, Senator Barak Obama was projected to receive 70% of the votes with a margin of error of 3%. Solve $|p - 0.70| \leq 0.03$, and interpret the solution to the inequality in the context of this problem.

Graphing Calculator Exercises

To solve an absolute value inequality by using a graphing calculator, let Y_1 equal the left side of the inequality and let Y_2 equal the right side of the inequality. Graph both Y_1 and Y_2 on a standard viewing window and use an *Intersect* feature or *Zoom* and *Trace* to approximate the intersection of the graphs. To solve $Y_1 > Y_2$, determine all x-values where the graph of Y_1 is above the graph of Y_2. To solve $Y_1 < Y_2$, determine all x-values where the graph of Y_1 is below the graph of Y_2.

For Exercises 63–72, solve the inequalities using a graphing calculator.

63. $|x + 2| > 4$

64. $|3 - x| > 6$

65. $\left|\dfrac{x + 1}{3}\right| < 2$

66. $\left|\dfrac{x - 1}{4}\right| < 1$

67. $|x - 5| < -3$

68. $|x + 2| < -2$

69. $|2x + 5| > -4$

70. $|1 - 2x| > -4$

71. $|6x + 1| \leq 0$

72. $|3x - 4| \leq 0$

Objectives

1. Introduction to Linear Inequalities in Two Variables
2. Graphing Linear Inequalities in Two Variables
3. Compound Linear Inequalities in Two Variables
4. Graphing a Feasible Region

Concept Connections

1. Determine whether each ordered pair is a solution to the inequality $x + y > 6$
   a. (5, 3)
   b. (1, 2)
   c. (4, 2)
   d. (−6, 20)

Skill Practice

Graph the solution set of the inequality.

2. $x - y > 4$

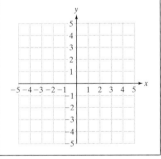

Answers

1. a. Yes b. No c. No
   d. Yes
2.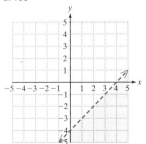

section 9.5 Linear Inequalities in Two Variables

1. Introduction to Linear Inequalities in Two Variables

A **linear inequality in two variables** x and y is an inequality that can be written in one of the following forms: $ax + by < c$, $ax + by > c$, $ax + by \leq c$, or $ax + by \geq c$, provided a and b are not both zero.

A solution to a linear inequality in two variables is an ordered pair that makes the inequality true. For example, solutions to the inequality $x + y < 6$ are ordered pairs (x, y) such that the sum of the x- and y-coordinates is less than 6. This inequality has an infinite number of solutions, and therefore it is convenient to express the solution set as a graph.

2. Graphing Linear Inequalities in Two Variables

To graph a linear inequality in two variables, we will use the test point method. The first step is to graph the related equation. This will be a line that separates the xy-plane into three regions: (1) the region below the line, (2) the region above the line, and (3) the line itself. Then, by selecting a test point (ordered pair) from each region and substituting it into the inequality, we can determine which region or regions represent the solution set.

Test Point Method—Summary

1. Graph the related equation. The equation will be a boundary line in the xy-plane.
   - If the original inequality is the strict inequality $<$ or $>$, then the line is not part of the solution set. Graph the boundary as a *dashed line*.
   - If the original inequality uses $\leq$ or $\geq$, then the line is part of the solution set. Graph the boundary as a *solid line*.
2. From each region above and below the line, select an ordered pair as a test point and substitute it into the original inequality.
   - If a test point makes the inequality true, then that region is part of the solution set.

example 1 Graphing a Linear Inequality in Two Variables

Graph the solution set of the inequality $x + y < 6$.

Solution:

Step 1: Graph the related equation $x + y = 6$ by using a dashed line (Figure 9-6).

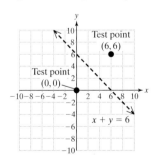

Figure 9-6

Section 9.5 Linear Inequalities in Two Variables 669

Tip: To graph the related equation, you may either create a table of points, or use the slope-intercept form of the line.

Table of points

x	y
0	6
6	0
4	2

Slope-intercept form

$x + y = 6 \longrightarrow y = -x + 6$

Step 2: Choose test points (ordered pairs) above and below the line, and substitute the points into the original inequality.

Test Point Above: (6, 6)

$x + y < 6$

$(6) + (6) \stackrel{?}{<} 6$

$12 \stackrel{?}{<} 6$ False

The test point (6, 6) *is not* a solution to the original inequality.

Test Point Below: (0, 0)

$x + y < 6$

$(0) + (0) \stackrel{?}{<} 6$

$0 \stackrel{?}{<} 6$ True

The test point (0, 0) *is* a solution to the original inequality. Shade the region below the boundary. See Figure 9-7.

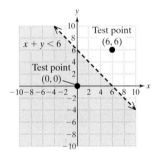

Figure 9-7

example 2 Graphing a Linear Inequality in Two Variables

Graph the solution set of the inequality $3x - 5y \leq 10$.

Solution:

$3x - 5y \leq 10$

$3x - 5y = 10$

$y = \dfrac{3}{5}x - 2$

Step 1: Graph the related equation to form the boundary of the solution set. (Here we use the slope-intercept form to graph the line.)

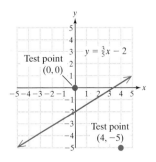

Skill Practice

Graph the solution set of the inequality.

3. $2x + 3y \leq 9$

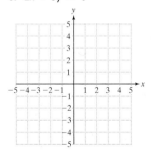

Answer

3.

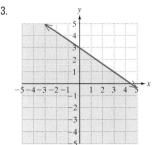

Step 2: Choose test points above and below the line.

Test Point Above: (0, 0)	**Test Point Below: (4, −5)**
$3x - 5y \leq 10$	$3x - 5y \leq 10$
$3(0) - 5(0) \stackrel{?}{\leq} 10$	$3(4) - 5(-5) \stackrel{?}{\leq} 10$
$0 \stackrel{?}{\leq} 10$ True	$12 + 25 \stackrel{?}{\leq} 10$
	$37 \stackrel{?}{\leq} 10$ False

Because the test point $(0, 0)$ above the boundary is true in the original inequality, shade the region above the line (Figure 9-8).

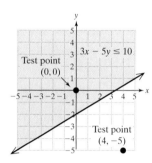

Figure 9-8

Tip: An inequality can also be graphed by first solving the inequality for y. Then

- Shade below the line if the inequality is of the form $y < mx + b$ or $y \leq mx + b$.
- Shade above the line if the inequality is of the form $y > mx + b$ or $y \geq mx + b$.

From Example 2, we have

$3x - 5y \leq 10$

$-5y \leq -3x + 10$

$\dfrac{-5y}{-5} \geq \dfrac{-3x}{-5} + \dfrac{10}{-5}$ Reverse the inequality sign.

$y \geq \dfrac{3}{5}x - 2$ Shade *above* the line.

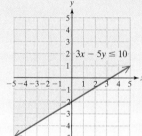

3. Compound Linear Inequalities in Two Variables

Some applications require us to find the union or intersection of two or more linear inequalities.

Section 9.5 Linear Inequalities in Two Variables

example 3   Graphing a Compound Linear Inequality in Two Variables

Graph the solution set of the compound inequality $2x + y < 1$ and $2y > x - 4$.

Solution:

First Inequality		Second Inequality	
$2x + y < 1$		$2y > x - 4$	
$2x + y = 1$	Related equation	$2y = x - 4$	Related equation
$y = -2x + 1$	Slope-intercept form	$y = \frac{1}{2}x - 2$	Slope-intercept form

For each inequality, draw the boundary line. Then pick test points above and below the line to determine the appropriate region to shade.

$2x + y < 1$ **Test Point Above: (1, 1)**   **Test Point Below: (0, 0)**

$$2(1) + (1) \stackrel{?}{<} 1 \qquad 2(0) + (0) \stackrel{?}{<} 1$$

$$3 \stackrel{?}{<} 1 \quad \text{False} \qquad 0 \stackrel{?}{<} 1 \quad \text{True}$$
(Figure 9-9)

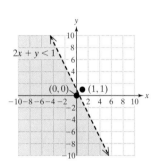

Figure 9-9

$2y > x - 4$ **Test Point Above: (0, 0)**   **Test Point Below: (5, −1)**

$$2(0) \stackrel{?}{>} (0) - 4 \qquad 2(-1) \stackrel{?}{>} (5) - 4$$

$$0 \stackrel{?}{>} -4 \quad \text{True} \qquad -2 \stackrel{?}{>} 1 \quad \text{False}$$
(Figure 9-10)

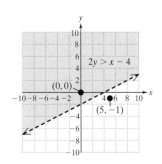

Figure 9-10

Skill Practice

4. Graph the solution set of the system of inequalities
$x - 3y > 4$ and $y > 2x + 3$

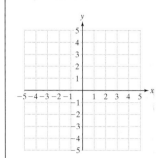

Answer

4.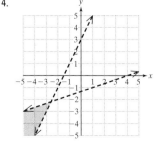

The solution to the compound inequality $2x + y < 1$ and $2y > x - 4$ is the intersection of the two individual solution sets. Therefore, the solution is the region of the plane below the line $y = -2x + 1$ and above the line $y = \frac{1}{2}x - 2$. This is the region shown in purple in Figure 9-11. The solution is graphed in Figure 9-12.

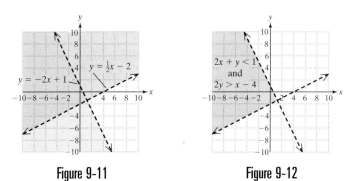

Figure 9-11 Figure 9-12

Skill Practice

5. Graph the solution set of the compound inequality
$x \geq 2$ or $3x + 2y \leq 6$

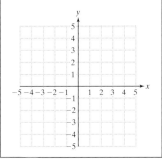

example 4 — Graphing a Compound Linear Inequality in Two Variables

Graph the solution set of the inequality $3y \geq 6$ or $y - x \leq 0$.

Solution:

First Inequality	Second Inequality
$3y \geq 6$	$y - x \leq 0$
$3y = 6$ Related equation	$y - x = 0$ Related equation
$y = 2$	$y = x$

For each inequality, draw the boundary line. Then, pick test points above and below the line to determine the appropriate region to shade:

$3y \geq 6$ **Test Point Above: (0, 3)** **Test Point Below: (0, 1)**

$3(3) \stackrel{?}{\geq} 6$ $\qquad$ $3(1) \stackrel{?}{\geq} 6$

$9 \stackrel{?}{\geq} 6$ True (Figure 9-13) $\qquad$ $3 \stackrel{?}{\geq} 6$ False

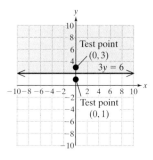

Figure 9-13

Answer

5.

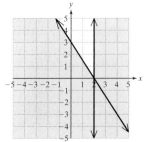

$y - x < 0$	**Test Point Above: (1, 2)**	**Test Point Below: (1, −2)**
	$(2) - (1) \overset{?}{\leq} 0$	$(-2) - (1) \overset{?}{\leq} 0$
	$1 \overset{?}{\leq} 0$ False	$-3 \overset{?}{\leq} 0$ True
		(Figure 9-14)

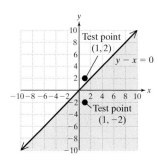

Figure 9-14

The solution to the compound inequality $3y \geq 6$ or $y - x \leq 0$ is the union of the two solution sets (Figure 9-15).

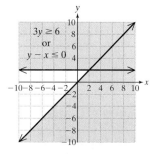

Figure 9-15

example 5 Graphing Compound Linear Inequalities

Describe the region of the plane defined by the following systems of inequalities.

a. $x > -2$ and $y > 1$ **b.** $x \leq 0$ and $y \geq 0$

Solution:

a. $x > -2$ represents the points to the *right* of the vertical line $x = -2$.

$y > 1$ represents the points *above* the horizontal line $y = 1$.

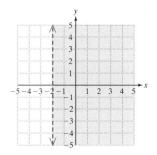
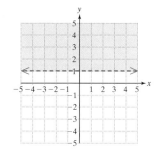

Skill Practice

Graph the region defined by the system of inequalities.

6. $x \leq 0$ and $y \leq 0$

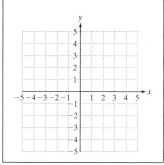

Answer

6.

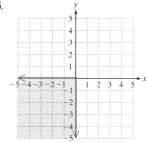

The region is quadrant-III, including the boundaries.

The graph of the compound inequality $x > -2$ and $y > 1$ is the intersection of the points to the right of $x = -2$ and above $y = 1$.

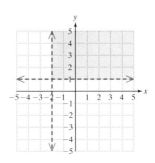

b. $x \leq 0$ $x \leq 0$ in the second and third quadrants.

$y \geq 0$ $y \geq 0$ in the first and second quadrants.

The intersection of these regions is the set of points in the second quadrant (with the boundary included).

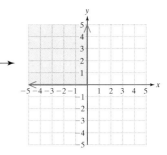

Skill Practice

A local pet rescue group has a total of 30 cages that can be used to hold cats and dogs. Let x represent the number of cages used for cats, and let y represent the number used for dogs.

7. Write a set of inequalities to express the fact that the number of cat and dog cages cannot be negative.

8. Write an inequality to describe the constraint on the total number of cages for cats and dogs.

9. Graph the system of inequalities to find the feasible region describing the available cages.

4. Graphing a Feasible Region

When two variables are related under certain constraints, a system of linear inequalities can be used to show a region of feasible values for the variables.

example 6   Graphing a Feasible Region

Susan has two tests on Friday: one in chemistry and one in psychology. Because the two classes meet in consecutive hours, she has no study time between tests. Susan estimates that she has a maximum of 12 hr of study time before the tests, and she must divide her time between chemistry and psychology.

Let x represent the number of hours Susan spends studying chemistry.

Let y represent the number of hours Susan spends studying psychology.

a. Find a set of inequalities to describe the constraints on Susan's study time.

b. Graph the constraints to find the feasible region defining Susan's study time.

Solution:

a. Because Susan cannot study chemistry or psychology for a negative period of time, we have $x \geq 0$ and $y \geq 0$.

Furthermore, her total time studying cannot exceed 12 hr: $x + y \leq 12$.

A system of inequalities that defines the constraints on Susan's study time is

$$x \geq 0$$
$$y \geq 0$$
$$x + y \leq 12$$

Answers

7. $x \geq 0$ and $y \geq 0$
8. $x + y \leq 30$
9.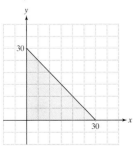

b. The first two conditions $x \geq 0$ and $y \geq 0$ represent the set of points in the first quadrant. The third condition $x + y \leq 12$ represents the set of points below and including the line $x + y = 12$ (Figure 9-16).

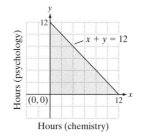

Figure 9-16

Discussion:

1. Refer to the feasible region drawn in Example 6(b). Is the ordered pair (8, 5) part of the feasible region?

    No. The ordered pair (8, 5) indicates that Susan spent 8 hr studying chemistry and 5 hr studying psychology. This is a total of 13 hr, which exceeds the constraint that Susan only had 12 hr to study. The point (8, 5) lies outside the feasible region, above the line $x + y = 12$ (Figure 9-17).

2. Is the ordered pair (7, 3) part of the feasible region?

    Yes. The ordered pair (7, 3) indicates that Susan spent 7 hr studying chemistry and 3 hr studying psychology.

    This point lies within the feasible region and satisfies all three constraints.

    $x \geq 0 \longrightarrow \quad 7 \geq 0 \quad$ True

    $y \geq 0 \longrightarrow \quad 3 \geq 0 \quad$ True

    $x + y \leq 12 \longrightarrow (7) + (3) \leq 12 \quad$ True

    Notice that the ordered pair (7, 3) corresponds to a point where Susan is not making full use of the 12 hr of study time.

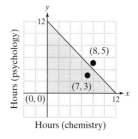

Figure 9-17

3. Suppose there was one additional constraint imposed on Susan's study time. She knows she needs to spend at least twice as much time studying chemistry as she does studying psychology. Graph the feasible region with this additional constraint.

    Because the time studying chemistry must be at least twice the time studying psychology, we have $x \geq 2y$.

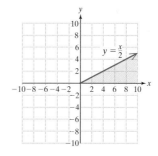

    This inequality may also be written as $y \leq \dfrac{x}{2}$.

    Figure 9-18 on page 676 shows the feasible region with the additional constraint $y \leq \dfrac{x}{2}$.

4. At what point in the feasible region is Susan making the most efficient use of her time for both classes?

    First and foremost, Susan must make use of *all* 12 hr. This occurs for points along the line $x + y = 12$. Susan will also want to study for both classes with approximately twice as much time devoted to chemistry. Therefore,

Susan will be deriving the maximum benefit at the point of intersection of the line $x + y = 12$ and the line $y = \dfrac{x}{2}$.

Using the substitution method, replace $y = \dfrac{x}{2}$ into the equation $x + y = 12$.

$x + \dfrac{x}{2} = 12$

$2x + x = 24$ Clear fractions.

$3x = 24$

$x = 8$ Solve for x.

$y = \dfrac{(8)}{2}$ To solve for y, substitute $x = 8$ into the equation $y = \dfrac{x}{2}$.

$y = 4$

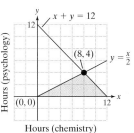

Figure 9-18

Therefore Susan should spend 8 hr studying chemistry and 4 hr studying psychology.

section 9.5 Practice Exercises

Boost *your* GRADE at mathzone.com!

MathZone

- Practice Problems
- Self-Tests
- NetTutor
- e-Professors
- Videos

Study Skills Exercises

1. A good way to determine what will be on a test is to look at both your notes and the exercises assigned by your instructor. List five kinds of problems that you think will be on the test for this chapter.

2. Define the key term **linear inequality in two variables**.

Review Exercises

For Exercises 3–8, solve the inequalities.

3. $-3 < 2k - 5 < 3$

4. $\dfrac{1}{2} < \dfrac{3}{4}y < \dfrac{3}{5}$

5. $|6a - 1| - 4 \leq 2$

6. $|3b + 5| - 8 \leq 5$

7. $|2t + 1| + 4 \geq 7$

8. $|2h - 6| - 1 \geq 3$

Objective 1: Introduction to Linear Inequalities in Two Variables

For Exercises 9–12, decide if the following points are solutions to the inequality.

9. $2x - y > 8$

   a. $(3, -5)$ **c.** $(4, -2)$

   b. $(-1, -10)$ **d.** $(0, 0)$

10. $3y + x < 5$

   a. $(-1, 7)$ **c.** $(0, 0)$

   b. $(5, 0)$ **d.** $(2, -3)$

Section 9.5 Linear Inequalities in Two Variables   677

11. $y \leq -2$

   a. $(5, -3)$ **c.** $(0, 0)$

   b. $(-4, -2)$ **d.** $(3, 2)$

12. $x \geq 5$

   a. $(4, 5)$ **c.** $(8, 8)$

   b. $(5, -1)$ **d.** $(0, 0)$

Objective 2: Graphing Linear Inequalities in Two Variables

For Exercises 13–18, decide which inequality symbol should be used $(<, >, \geq, \leq)$ by looking at the graph.

13.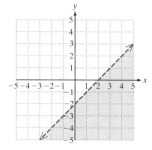

$x - y \;\underline{\quad}\; 2$

14.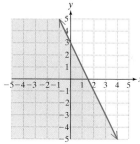

$y \;\underline{\quad}\; -2x + 3$

15.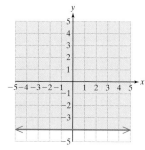

$y \;\underline{\quad}\; -4$

16.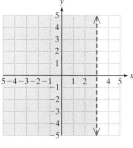

$x \;\underline{\quad}\; 3$

17.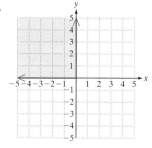

$x \;\underline{\quad}\; 0$ and $y \;\underline{\quad}\; 0$

18.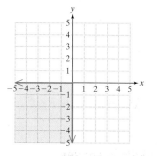

$x \;\underline{\quad}\; 0$ and $y \;\underline{\quad}\; 0$

For Exercises 19–40, graph the solution set. **(See Examples 1–2.)**

19. $x - 2y > 4$

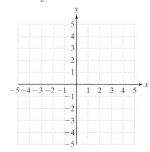

20. $x - 3y > 6$

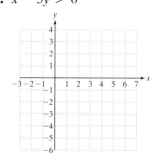

21. $5x - 2y < 10$

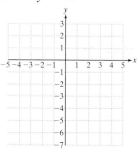

22. $x - 3y < 8$

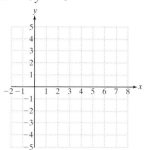

23. $2x \leq -6y + 12$

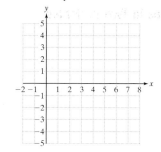

24. $4x < 3y + 12$

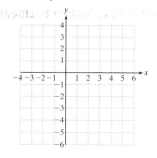

25. $y \geq -2$

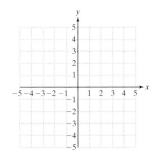

26. $y \geq 5$

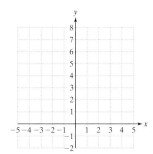

27. $4x < 5$

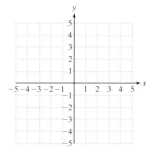

28. $x + 6 < 7$

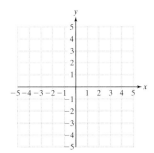

29. $y \geq \dfrac{2}{5}x - 4$

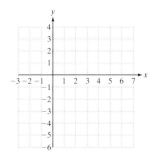

30. $y \geq -\dfrac{5}{2}x - 4$

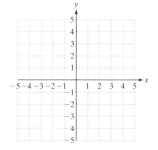

31. $y \leq \dfrac{1}{3}x + 6$

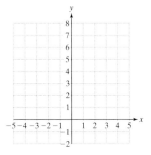

32. $y \leq -\dfrac{1}{4}x + 2$

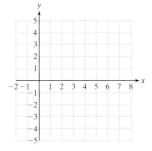

33. $y - 5x > 0$

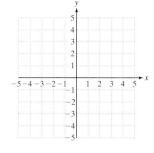

34. $y - \dfrac{1}{2}x > 0$

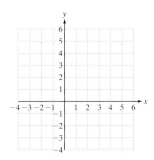

35. $\dfrac{x}{5} + \dfrac{y}{4} < 1$

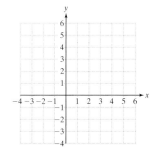

36. $x + \dfrac{y}{2} \geq 2$

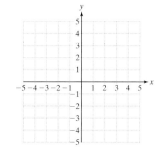

37. $0.1x + 0.2y \leq 0.6$

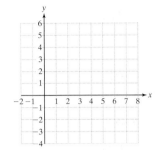

38. $0.3x - 0.2y < 0.6$

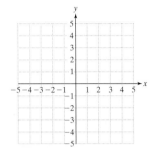

39. $x \leq -\dfrac{2}{3}y$

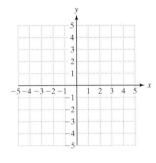

40. $x \geq -\dfrac{5}{4}y$

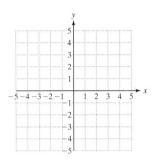

Objective 3: Compound Linear Inequalities in Two Variables

For Exercises 41–56, graph the solution set of each compound inequality. (See Examples 3–5.)

41. $y < 4$ and $y > -x + 2$

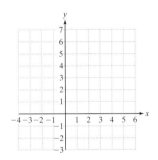

42. $y < 3$ and $x + 2y < 6$

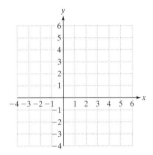

43. $2x + y \leq 5$ or $x \geq 3$

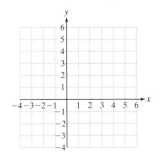

44. $x + 3y \geq 7$ or $x \leq -2$

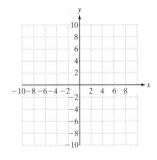

45. $x + y < 3$ and $4x + y < 6$

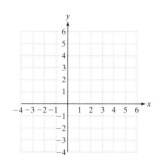

46. $x + y < 4$ and $3x + y < 9$

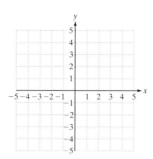

47. $2x - y \leq 2$ or $2x + 3y \geq 6$

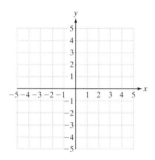

48. $3x + 2y \geq 4$ or $x - y \leq 3$

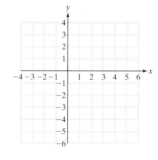

49. $x > 4$ and $y < 2$

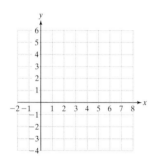

50. $x < 3$ and $y > 4$

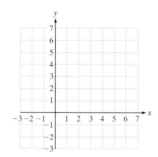

51. $x \leq -2$ or $y \leq 0$

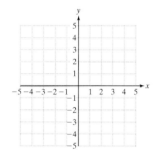

52. $x \geq 0$ or $y \geq -3$

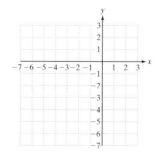

53. $x > 0$ and $x + y < 6$

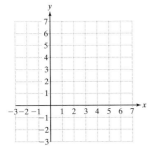

54. $x < 0$ and $x + y < 2$

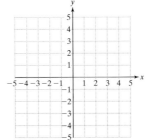

55. $y \leq 0$ or $x - y \leq -4$

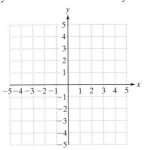

56. $y \geq 0$ or $x - y \geq -3$

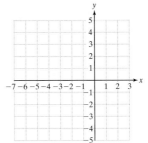

Objective 4: Graphing a Feasible Region

For Exercises 57–62, graph the feasible regions. **(See Example 6.)**

57. $x + y \leq 3$ and
$x \geq 0, y \geq 0$

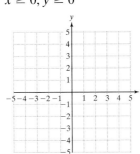

58. $x - y \leq 2$ and
$x \geq 0, y \geq 0$

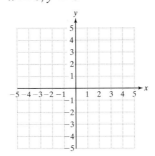

59. $y < \frac{1}{2}x - 3$ and
$x \leq 0, y \geq -5$

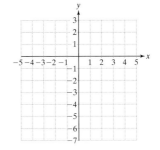

60. $y > \frac{1}{2}x - 3$ and
$x \geq -2, y \leq 0$

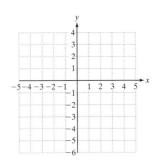

61. $x \geq 0, y \geq 0$
$x + y \leq 8$ and
$3x + 5y \leq 30$

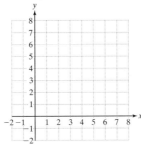

62. $x \geq 0, y \geq 0$
$x + y \leq 5$ and
$x + 2y \leq 6$

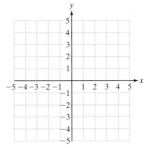

63. In scheduling two drivers for delivering pizza, James needs to have at least 65 hr scheduled this week. His two drivers, Karen and Todd, are not allowed to get overtime, so each one can work at most 40 hr. Let x represent the number of hours that Karen can be scheduled, and let y represent the number of hours Todd can be scheduled. **(See Example 6.)**

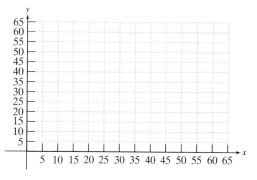

a. Write two inequalities that express the fact that Karen and Todd cannot work a negative number of hours.

b. Write two inequalities that express the fact that neither Karen nor Todd is allowed overtime (i.e., each driver can have at most 40 hr).

c. Write an inequality that expresses the fact that the total number of hours from both Karen and Todd needs to be at least 65 hr.

d. Graph the feasible region formed by graphing the inequalities.

e. Is the point (35, 40) in the feasible region? What does the point (35, 40) represent in the context of this problem?

f. Is the point (20, 40) in the feasible region? What does the point (20, 40) represent in the context of this problem?

64. A manufacturer produces two models of desks. Model A requires $1\frac{1}{2}$ hr to stain and finish and $1\frac{1}{4}$ hr to assemble. Model B requires 2 hr to stain and finish and $\frac{3}{4}$ hr to assemble. The total amount of time available for staining and finishing is 12 hr and for assembling is 6 hr. Let x represent the number of Model A desks, and let y represent the number of Model B desks.

a. Write two inequalities that express the fact that the number of desks to be produced cannot be negative.

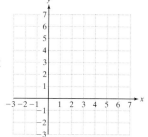

b. Write an inequality in terms of the number of Model A and Model B desks that can be produced if the total time for staining and finishing is at most 12 hr.

c. Write an inequality in terms of the number of Model A and Model B desks that can be produced if the total time for assembly is no more than 6 hr.

d. Identify the feasible region formed by graphing the preceding inequalities.

e. Is the point (4, 1) in the feasible region? What does the point (4, 1) represent in the context of this problem?

f. Is the point (5, 4) in the feasible region? What does the point (5, 4) represent in the context of this problem?

chapter 9 | summary

section 9.1 Compound Inequalities

Key Concepts

Solve two or more inequalities joined by *and* by finding the intersection of their solution sets. Solve two or more inequalities joined by *or* by finding the union of the solution sets.

Examples

Example 1

$-7x + 3 \geq -11$ and $1 - x < 4.5$
$-7x \geq -14$ and $-x < 3.5$
$x \leq 2$ and $x > -3.5$

$x \leq 2$

$x > -3.5$

The solution is $\{x \mid -3.5 < x \leq 2\}$ or equivalently $(-3.5, 2]$.

Example 2

$$-6 < \frac{3}{4}(x - 1) < 6$$

$$\frac{4}{3} \cdot -6 < \frac{4}{3} \cdot \frac{3}{4}(x - 1) < \frac{4}{3} \cdot 6$$

$$-8 < x - 1 < 8$$

The solution is $\{x \mid -7 < x < 9\}$ or equivalently $(-7, 9)$.

Example 3

$5y + 1 \geq 6$ or $2y - 5 \leq -11$
$5y \geq 5$ or $2y \leq -6$
$y \geq 1$ or $y \leq -3$

$y \geq 1$

$y \leq -3$

The solution is $\{y \mid y \geq 1 \text{ or } y \leq -3\}$ or equivalently $(-\infty, -3] \cup [1, \infty)$.

section 9.2 Polynomial and Rational Inequalities

Key Concepts

The Test Point Method to Solve Polynomial and Rational Inequalities

1. Find the boundary points of the inequality. (Boundary points are the real solutions to the related equation and points where the inequality is undefined.)
2. Plot the boundary points on the number line. This divides the number line into regions.
3. Select a test point from each region and substitute it into the original inequality.
   - If a test point makes the original inequality true, then that region is part of the solution set.
4. Test the boundary points in the original inequality.
   - If a boundary point makes the original inequality true, then that point is part of the solution set.

Examples

Example 1

$$\frac{28}{2x - 3} \leq 4$$

The inequality is undefined for $x = \frac{3}{2}$. Find other possible boundary points by solving the related equation.

$$\frac{28}{2x - 3} = 4 \quad \text{Related equation}$$

$$(2x - 3) \cdot \frac{28}{2x - 3} = (2x - 3) \cdot 4$$

$$28 = 8x - 12$$

$$40 = 8x$$

$$x = 5$$

The boundaries are $x = \frac{3}{2}$ and $x = 5$

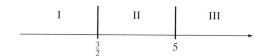

Region I: Test $x = 1$: $\quad \dfrac{28}{2(1) - 3} \stackrel{?}{\leq} 4 \quad$ True

Region II: Test $x = 2$: $\quad \dfrac{28}{2(2) - 3} \stackrel{?}{\leq} 4 \quad$ False

Region III: Test $x = 6$: $\quad \dfrac{28}{2(6) - 3} \stackrel{?}{\leq} 4 \quad$ True

The boundary point $x = \frac{3}{2}$ is not included because $\frac{28}{2x - 3}$ is undefined there. The boundary $x = 5$ does check in the original inequality.

The solution is $(-\infty, \frac{3}{2}) \cup [5, \infty)$.

section 9.3 Absolute Value Equations

Key Concepts

The equation $|x| = a$ is an absolute value equation. For $a \geq 0$, the solution to the equation $|x| = a$ is $x = a$ or $x = -a$.

Steps to Solve an Absolute Value Equation
1. Isolate the absolute value to write the equation in the form $|x| = a$.
2. If $a < 0$, there is no solution.
3. Otherwise, rewrite the equation $|x| = a$ as $x = a$ or $x = -a$.
4. Solve the equations from step 3.
5. Check answers in the original equation.

The solution to the equation $|x| = |y|$ is $x = y$ or $x = -y$.

Examples

Example 1

$|2x - 3| + 5 = 10$

$\quad |2x - 3| = 5 \quad$ Isolate the absolute value.

$\quad 2x - 3 = 5 \quad$ or $\quad 2x - 3 = -5$

$\quad 2x = 8 \quad$ or $\quad 2x = -2$

$\quad x = 4 \quad$ or $\quad x = -1$

Example 2

$|x + 2| + 5 = 1$

$\quad |x + 2| = -4 \quad$ No solution

Example 3

$|2x - 1| = |x + 4|$

$2x - 1 = x + 4 \quad$ or $\quad 2x - 1 = -(x + 4)$

$\quad\quad x = 5 \quad$ or $\quad 2x - 1 = -x - 4$

$\quad\quad\quad\quad\quad$ or $\quad\quad 3x = -3$

$\quad\quad\quad\quad\quad$ or $\quad\quad\quad x = -1$

section 9.4 Absolute Value Inequalities

Key Concepts

Solutions to Absolute Value Inequalities

$|x| > a \Rightarrow x < -a$ or $x > a$

$|x| < a \Rightarrow -a < x < a$

Test Point Method to Solve Inequalities
1. Find the boundary points of the inequality. (Boundary points are the real solutions to the related equation and points where the inequality is undefined.)
2. Plot the boundary points on the number line. This divides the number line into regions.
3. Select a test point from each region and substitute it into the original inequality.
   - If a test point makes the original inequality true, then that region is part of the solution set.
4. Test the boundary points in the original inequality.
   - If a boundary point makes the original inequality true, then that point is part of the solution set.

If *a* is negative (*a* < 0), then
1. $|x| < a$ has no solution.
2. $|x| > a$ is true for all real numbers.

Examples

Example 1

$|5x - 2| < 12$

$-12 < 5x - 2 < 12$

$-10 < 5x < 14$

$-2 < x < \dfrac{14}{5}$

The solution is $\left(-2, \dfrac{14}{5}\right)$.

Example 2

$|x - 3| + 2 \geq 7$

$|x - 3| \geq 5$ Isolate the absolute value.

$|x - 3| = 5$ Related equation

$x - 3 = 5$ or $x - 3 = -5$

$x = 8$ or $x = -2$ Boundary points

```
      I        |        II       |       III
   ───────────────────────────────────────────▶
               -2                8
```

Region I:

Test $x = -3$: $|(-3) - 3| + 2 \overset{?}{\geq} 7$ True

Region II:

Test $x = 0$: $|(0) - 3| + 2 \overset{?}{\geq} 7$ False

Region III:

Test $x = 9$: $|(9) - 3| + 2 \overset{?}{\geq} 7$ True

```
       True       False       True
   ◀──────────]           [──────────▶
              -2           8
```

The solution is $(-\infty, -2] \cup [8, \infty)$.

Example 3

$|x + 5| > -2$

The solution is all real numbers because an absolute value will always be greater than a negative number.

$(-\infty, \infty)$

section 9.5 Linear Inequalities in Two Variables

Key Concepts

A **linear inequality in two variables** is an inequality of the form $ax + by < c$, $ax + by > c$, $ax + by \leq c$, or $ax + by \geq c$.

Use the test point method to solve a linear inequality in two variables. That is, graph the related equation and shade above or below the line.

If an inequality is strict ($<$, $>$), then a dashed line is used for the boundary. If the inequality contains $\leq$ or $\geq$, then a solid line is drawn.

The union or intersection of two or more linear inequalities is the union or intersection of the solution sets.

Examples

Example 1

Graph the solution to the inequality $2x - y < 4$.

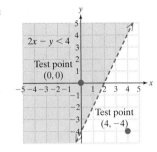

Test Points: $(0, 0)$ and $(4, -4)$

$2(0) - (0) \stackrel{?}{<} 4$ True; shade above.

$2(4) - (-4) \stackrel{?}{<} 4$ False; do not shade below.

Example 2

Graph.

$x < 0$ and $y > 2$

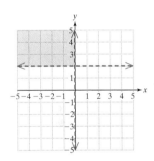

Example 3

Graph.

$x \leq 0$ or $y \geq 2$

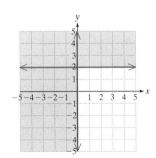

chapter 9 review exercises

Section 9.1

For Exercises 1–10, solve the compound inequalities. Write the solutions in interval notation.

1. $4m > -11$ and $4m - 3 \leq 13$

2. $4n - 7 < 1$ and $7 + 3n \geq -8$

3. $-3y + 1 \geq 10$ and $-2y - 5 \leq -15$

4. $\dfrac{1}{2} - \dfrac{h}{12} \leq \dfrac{-7}{12}$ and $\dfrac{1}{2} - \dfrac{h}{10} > -\dfrac{1}{5}$

5. $\dfrac{2}{3}t - 3 \leq 1$ or $\dfrac{3}{4}t - 2 > 7$

6. $2(3x + 1) < -10$ or $3(2x - 4) \geq 0$

7. $-7 < -7(2w + 3)$ or $-2 < -4(3w - 1)$

8. $5(p + 3) + 4 > p - 1$ or $4(p - 1) + 2 > p + 8$

9. $2 \geq -(b - 2) - 5b \geq -6$

10. $-4 \leq \dfrac{1}{2}(x - 1) < -\dfrac{3}{2}$

11. The product of $\tfrac{1}{3}$ and the sum of a number and 3 is between -1 and 5. Find all such numbers.

12. Normal levels of total cholesterol vary according to age. For adults between 25 and 40 years old, the normal range is generally accepted to be between 140 and 225 mg/dL (milligrams per deciliter), inclusive.

    a. Write an inequality representing the normal range for total cholesterol for adults between 25 and 40 years old.

    b. Write a compound inequality representing abnormal ranges for total cholesterol for adults between 25 and 40 years old.

13. Normal levels of total cholesterol vary according to age. For adults younger than 25 years old, the normal range is generally accepted to be between 125 and 200 mg/dL, inclusive.

    a. Write an inequality representing the normal range for total cholesterol for adults younger than 25 years old.

    b. Write a compound inequality representing abnormal ranges for total cholesterol for adults younger than 25 years old.

14. In certain applications in statistics, a data value that is more than 3 standard deviations from the mean is said to be an *outlier* (a value unusually far from the average). If μ represents the mean of population and σ represents the population standard deviation, then the inequality $|x - \mu| > 3\sigma$ can be used to test whether a data value, x, is an outlier.

    The mean height, μ, of adult men is 69.0 in. (5′9″) and the standard deviation, σ, of the height of adult men is 3.0. Determine whether the heights of the following men are outliers.

    a. Shaquille O'Neal, 7′1″ = 85 in.

    b. Charlie Ward, 6′3″ = 75 in.

    c. Elmer Fudd, 4′5″ = 53 in.

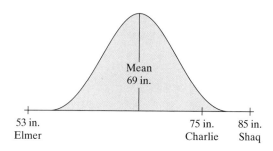

15. Explain the difference between the solution sets of the following compound inequalities.

    a. $x \leq 5$ and $x \geq -2$

    b. $x \leq 5$ or $x \geq -2$

Section 9.2

16. Solve the equation and inequalities. How do your answers to parts (a), (b), and (c) relate to the graph of $g(x) = x^2 - 4$?

   a. $x^2 - 4 = 0$

   b. $x^2 - 4 < 0$

   c. $x^2 - 4 > 0$

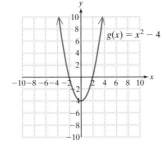

17. Solve the equation and inequalities. How do your answers to parts (a), (b), (c), and (d) relate to the graph of $k(x) = \dfrac{4x}{(x-2)}$?

   a. $\dfrac{4x}{x-2} = 0$

   b. For which values is $k(x)$ undefined?

   c. $\dfrac{4x}{x-2} \geq 0$

   d. $\dfrac{4x}{x-2} \leq 0$

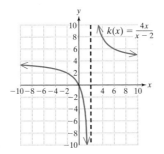

For Exercises 18–29, solve the inequalities. Write the answers in interval notation.

18. $w^2 - 4w - 12 < 0$

19. $t^2 + 6t + 9 \geq 0$

20. $\dfrac{12}{x+2} \leq 6$

21. $\dfrac{8}{p-1} \geq -4$

22. $3y(y-5)(y+2) > 0$

23. $-3c(c+2)(2c-5) < 0$

24. $-x^2 - 4x \geq 4$

25. $y^2 + 4y > 5$

26. $\dfrac{w+1}{w-3} > 1$

27. $\dfrac{2a}{a+3} \leq 2$

28. $t^2 + 10t + 25 \leq 0$

29. $-x^2 - 4x < 4$

Section 9.3

For Exercises 30–41, solve the absolute value equations.

30. $|x| = 10$

31. $|x| = 17$

32. $|8.7 - 2x| = 6.1$

33. $|5.25 - 5x| = 7.45$

34. $16 = |x+2| + 9$

35. $5 = |x-2| + 4$

36. $|4x - 1| + 6 = 4$

37. $|3x - 1| + 7 = 3$

38. $\left|\dfrac{7x-3}{5}\right| + 4 = 4$

39. $\left|\dfrac{4x+5}{-2}\right| - 3 = -3$

40. $|3x - 5| = |2x + 1|$

41. $|8x + 9| = |8x - 1|$

42. Which absolute value expression represents the distance between 3 and -2 on the number line?

   $|3 - (-2)|$ $|-2 - 3|$

Section 9.4

43. Write the compound inequality $x < -5$ or $x > 5$ as an absolute value inequality.

44. Write the compound inequality $-4 < x < 4$ as an absolute value inequality.

For Exercises 45–46, write an absolute value inequality that represents the solution sets graphed here.

45. [number line graph: open interval from -6 to 6]

46. [number line graph: regions outside $-\tfrac{2}{3}$ and $\tfrac{2}{3}$]

For Exercises 47–60, solve the absolute value inequalities. Graph the solution set and write the solution in interval notation.

47. $|x + 6| \geq 8$

48. $|x + 8| \leq 3$

49. $2|7x - 1| + 4 > 4$

50. $4|5x + 1| - 3 > -3$

51. $|3x + 4| - 6 \leq -4$

52. $|5x - 3| + 3 \leq 6$

53. $\left|\dfrac{x}{2} - 6\right| < 5$

54. $\left|\dfrac{x}{3} + 2\right| < 2$

55. $|4 - 2x| + 8 \geq 8$

56. $|9 + 3x| + 1 \geq 1$

57. $-2|5.2x - 7.8| < 13$

58. $-|2.5x + 15| < 7$

59. $|3x - 8| < -1$

60. $|x + 5| < -4$

61. State one possible situation in which an absolute value inequality will have no solution.

62. State one possible situation in which an absolute value inequality will have a solution of all real numbers.

63. The Neilsen ratings estimated that the percent, p, of the television viewing audience watching *American Idol* was 20% with a 3% margin of error. Solve the inequality $|p - 0.20| \leq 0.03$ and interpret the answer in the context of this problem.

64. The length, L, of a screw is supposed to be $3\tfrac{3}{8}$ in. Due to variation in the production equipment, there is a $\tfrac{1}{4}$-in. margin of error. Solve the inequality $|L - 3\tfrac{3}{8}| \leq \tfrac{1}{4}$ and interpret the answer in the context of this problem.

Section 9.5

For Exercises 65–72, solve the inequalities by graphing.

65. $2x > -y + 5$

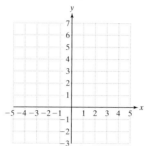

66. $2x \leq -8 - 3y$

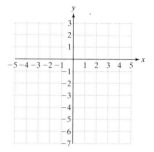

67. $y \geq -\dfrac{2}{3}x + 3$

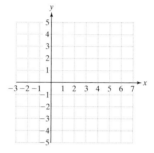

68. $y > \dfrac{3}{4}x - 2$

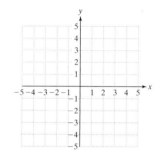

69. $x > -3$

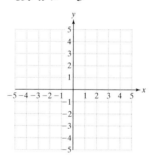

70. $x \leq 2$

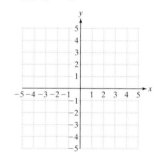

71. $x \geq \dfrac{1}{2}y$

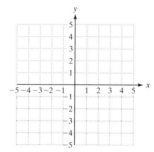

72. $x < \dfrac{2}{5}y$

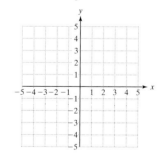

For Exercises 73–76, graph the system of inequalities.

73. $2x - y > -2$ and $2x - y \leq 2$

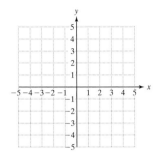

74. $3x + y \geq 6$ or $3x + y < -6$

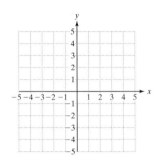

75. $x \geq 0$ $\quad y \geq 0$ $\quad$ and $\quad y \geq -\dfrac{3}{2}x + 4$

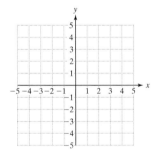

76. $x \geq 0$ $\quad y \geq 0$ $\quad$ and $\quad y \leq -\dfrac{2}{3}x + 4$

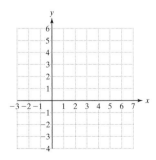

77. A pirate's treasure is buried on a small, uninhabited island in the eastern Caribbean. A shipwrecked sailor finds a treasure map at the base of a coconut palm tree. The treasure is buried within the intersection of three linear inequalities. The palm tree is located at the origin, and the positive y-axis is oriented due north. The scaling on the map is in 1-yd increments. Find the region where the sailor should dig for the treasure.

$$-2x + y \leq 4$$
$$y \leq -x + 6$$
$$y \geq 0$$

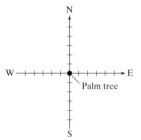 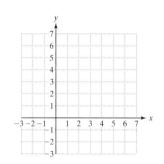

78. Suppose a farmer has 100 acres of land on which to grow oranges and grapefruit. Furthermore, because of demand from his customers, he wants to plant at least 4 times as many acres of orange trees as grapefruit trees.

Let x represent the number of acres of orange trees.

Let y represent the number of acres of grapefruit trees.

a. Write two inequalities that express the fact that the farmer cannot use a negative number of acres to plant orange and grapefruit trees.

b. Write an inequality that expresses the fact that the total number of acres used for growing orange and grapefruit trees is at most 100.

c. Write an inequality that expresses the fact that the farmer wants to plant at least 4 times as many orange trees as grapefruit trees.

d. Sketch the inequalities in parts (a)–(c) to find the feasible region for the farmer's distribution of orange and grapefruit trees.

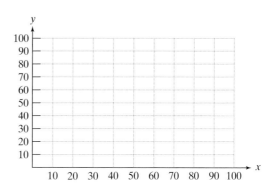

chapter 9 test

1. Solve the compound inequalities. Write the answers in interval notation.

   a. $-2 \leq 3x - 1 \leq 5$

   b. $-\dfrac{3}{5}x - 1 \leq 8$ or $-\dfrac{2}{3}x \geq 16$

   c. $-2x - 3 > -3$ and $x + 3 \geq 0$

2. **a.** $5x + 1 \leq 6$ or $2x + 4 > -6$

   b. $2x - 3 > 1$ and $x + 4 < -1$

3. The normal range in humans of the enzyme adenosine deaminase (ADA), is between 9 and 33 IU (international units), inclusive. Let x represent the ADA level in international units.

   a. Write an inequality representing the normal range for ADA.

   b. Write a compound inequality representing abnormal ranges for ADA.

For Exercises 4–9, solve the polynomial and rational inequalities.

4. $\dfrac{2x - 1}{x - 6} \leq 0$

5. $50 - 2a^2 > 0$

6. $y^3 + 3y^2 - 4y - 12 < 0$

7. $\dfrac{3}{w + 3} > 2$

8. $\dfrac{p^2}{2 + p^2} < 0$

9. $t^2 + 22t + 121 \leq 0$

10. Solve the absolute value equations.

    a. $\left|\dfrac{1}{2}x + 3\right| - 4 = 4$

    b. $|3x + 4| = |x - 12|$

11. Solve the following equation and inequalities. How do your answers to parts (a)–(c) relate to the graph of $f(x) = |x - 3| - 4$?

    a. $|x - 3| - 4 = 0$

    b. $|x - 3| - 4 < 0$

    c. $|x - 3| - 4 > 0$

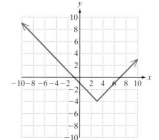

For Exercises 12–15, solve the absolute value inequalities. Write the answers in interval notation.

12. $|3 - 2x| + 6 < 2$

13. $|3x - 8| > 9$

14. $|0.4x + 0.3| - 0.2 < 7$

15. $|7 - 3x| + 1 > -3$

16. The mass of a small piece of metal is measured to be 15.41 g. If the measurement error is at most ±0.01 g, write an absolute value inequality that represents the possible mass, x, of the piece of metal.

17. Graph the solution to the inequality $2x - 5y \geq 10$.

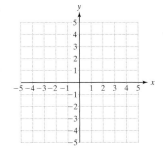

18. Solve the system of inequalities by graphing. $x + y < 3$ and $3x - 2y > -6$

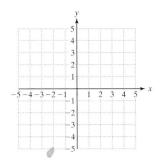

19. After menopause, women are at higher risk for hip fractures as a result of low calcium. As early as their teen years, women need at least 1200 mg of calcium per day (the USDA recommended daily allowance). One 8-oz glass of skim milk contains 300 mg of calcium, and one Antacid (regular strength) contains 400 mg of calcium. Let x represent the number of 8-oz glasses of milk that a woman drinks per day. Let y represent the number of Antacid tablets (regular strength) that a woman takes per day.

   a. Write two inequalities that express the fact that the number of glasses of milk and the number of Antacid taken each day cannot be negative.

   b. Write a linear inequality in terms of x and y for which the daily calcium intake is at least 1200 mg.

   c. Graph the inequalities.

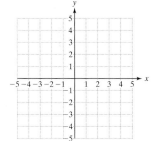

---

chapters 1–9 | cumulative review exercises

1. Perform the indicated operations.
   $$(2x - 3)(x - 4) - (x - 5)^2$$

2. Solve the equation. $\quad -9m + 3 = 2m(m - 4)$

For Exercises 3–4, solve the equation and inequalities. Write the solution to the inequalities in interval notation.

3. a. $2|3 - p| - 4 = 2$

   b. $2|3 - p| - 4 < 2$

   c. $2|3 - p| - 4 > 2$

4. a. $\left|\dfrac{y - 2}{4}\right| - 6 = -3$

   b. $\left|\dfrac{y - 2}{4}\right| - 6 < -3$

   c. $\left|\dfrac{y - 2}{4}\right| - 6 > -3$

5. Graph the inequality. $\quad 4x - y > 12$

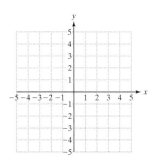

6. The time in minutes required for a rat to run through a maze depends on the number of trials, n, that the rat has practiced.

$$t(n) = \frac{3n + 15}{n + 1} \qquad n \geq 1$$

   a. Find $t(1)$, $t(50)$, and $t(500)$, and interpret the results in the context of this problem. Round to 2 decimal places, if necessary.

   b. Does there appear to be a limiting time in which the rat can complete the maze?

   c. How many trials are required so that the rat is able to finish the maze in under 5 min?

7. a. Solve the inequality. $\quad 2x^2 + x - 10 \geq 0$

   b. How does the answer in part (a) relate to the graph of the function $f(x) = 2x^2 + x - 10$?

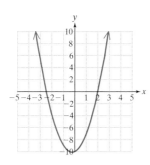

8. Shade the region defined by the compound inequality.

$$3x + y \leq -2 \quad \text{or} \quad y \geq 1$$

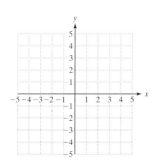

9. Simplify the expression.

$$2 - 3(x - 5) + 2[4 - (2x + 6)]$$

10. McDonald's corporation is the world's largest food service retailer. At the end of 1996, McDonald's operated 2.1×10^4 restaurants in over 100 countries. Worldwide sales in 1996 were nearly $\$3.18 \times 10^{10}$. Find the average sales per restaurant in 1996. Write the answer in scientific notation.

11. a. Divide the polynomials.

$$\frac{2x^4 - x^3 + 5x - 7}{x^2 + 2x - 1}$$

   Identify the quotient and remainder.

   b. Check your answer by multiplying.

12. The area of a trapezoid is given by $A = \frac{1}{2}h(b_1 + b_2)$.

   a. Solve for b_1.

   b. Find b_1 when $h = 4$ cm, $b_2 = 6$ cm, and $A = 32$ cm^2.

13. The speed of a car varies inversely as the time to travel a fixed distance. A car traveling the speed limit of 60 mph travels between two points in 10 sec. How fast is a car moving if it takes only 8 sec to cover the same distance?

14. Two angles are supplementary. One angle measures 9° more than twice the other angle. Find the measures of the angles.

15. Chemique invests $3000 less in an account earning 5% simple interest than she does in an account bearing 6.5% simple interest. At the end of one year, she earns a total $770 in interest. Find the amount invested in each account.

16. Determine algebraically whether the lines are parallel, perpendicular, or neither.

$$4x - 2y = 5$$
$$-3x + 6y = 10$$

17. Find the x- and y-intercepts and slope (if they exist) of the lines. Then graph the lines.

   a. $3x + 5 = 8$ **b.** $\frac{1}{2}x + y = 4$

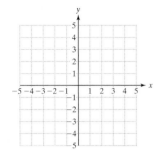
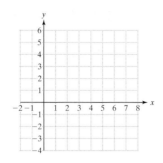

18. Find an equation of the line with slope $-\frac{2}{3}$ passing through the point $(4, -7)$. Write the final answer in slope-intercept form.

19. Solve the system of equations.
$$3x + y = z + 2$$
$$y = 1 - 2x$$
$$3z = -2y$$

20. Identify the order of the matrices.

   a. $\begin{bmatrix} 2 & 4 & 5 \\ -1 & 0 & 1 \\ 9 & 2 & 3 \\ 3 & 0 & 1 \end{bmatrix}$ **b.** $\begin{bmatrix} 5 & 6 & 3 \\ 6 & 0 & -1 \\ 0 & 1 & -2 \end{bmatrix}$

21. Against a headwind, a plane can travel 6240 mi in 13 hr. On the return trip flying with the same wind, the plane can fly 6240 mi in 12 hr. Find the wind speed and the speed of the plane in still air.

22. The profit that a company makes manufacturing computer desks is given by
$$P(x) = -\tfrac{1}{5}(x - 20)(x - 650) \qquad x \geq 0$$
where x is the number of desks produced and $P(x)$ is the profit in dollars.

   a. Is this function constant, linear, or quadratic?

   b. Find $P(0)$ and interpret the result in the context of this problem.

   c. Find the values of x where $P(x) = 0$. Interpret the results in the context of this problem.

23. Given $h(x) = \sqrt{50 - x}$, find the domain of h.

24. Simplify completely. $\dfrac{x^{-1} - y^{-1}}{y^{-2} - x^{-2}}$

25. Divide. $\dfrac{a^3 + 64}{16 - a^2} \div \dfrac{a^3 - 4a^2 + 16a}{a^2 - 3a - 4}$

26. Perform the indicated operations.
$$\frac{1}{x^2 - 7x + 10} + \frac{1}{x^2 + 8x - 20}$$

Exponential and Logarithmic Functions

10

10.1 Algebra and Composition of Functions
10.2 Inverse Functions
10.3 Exponential Functions
10.4 Logarithmic Functions
10.5 Properties of Logarithms
10.6 The Irrational Number e
10.7 Exponential and Logarithmic Equations

In this chapter we introduce exponential and logarithmic functions. Exponential functions are used to model a variety of naturally occurring phenomena, such as population growth, growth of an investment with compound interest, and decay of radioactive matter. For example, plutonium 238, written as ^{238}Pu, is used to power spacecraft. In Exercise 85 in Section 10.7, we see that 2 kg of ^{238}Pu decays according to the function $P(t) = 2e^{-0.0079t}$. $P(t)$ represents the number of kilograms of ^{238}Pu remaining after t years. This function is necessary for scientists to determine the amount of power available many years after the spacecraft has left the Earth.

chapter 10 preview

The exercises in this chapter preview contain concepts that have not yet been presented. These exercises are provided for students who want to compare their levels of understanding before and after studying the chapter. Alternatively, you may prefer to work these exercises when the chapter is completed and before taking the exam.

Section 10.1

For Exercises 1–4, use the given functions.

$f(x) = 2x^2 + 2x \qquad g(x) = 3x + 1 \qquad h(x) = \sqrt{x + 3}$

1. Find $(g - f)(x)$.
2. Find $(g \cdot h)(x)$.
3. Find $\left(\dfrac{g}{f}\right)(x)$.
4. Find $(g \circ f)(x)$.

Section 10.2

5. Given $f(x) = -4x + 2$, find f^{-1}.

6. Fill in the blank. If f and g are inverse functions, then $(f \circ g)(x) = $ _____.

Section 10.3

For Exercises 7–8, graph the function.

7. $f(x) = 4^x$
8. $g(x) = \left(\dfrac{1}{3}\right)^x$

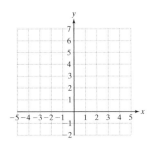
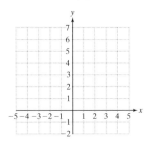

Section 10.4

For Exercises 9–10, evaluate the expression.

9. $\log_{10} 100$
10. $\log_2 \left(\dfrac{1}{8}\right)$

11. Identify the domain of the function given by $f(x) = \log(x + 5)$.

12. Graph the function $y = \log_3 x$.

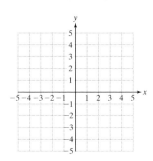

Section 10.5

For Exercises 13–14, use the properties of logarithms to combine the logarithmic expressions, and simplify as much as possible.

13. $\log_4 x + \log_4 y$
14. $\log 60 - \log 4$

15. Expand into an expression involving sums and differences of logarithms of x, y, and z.

$\log_3 \left(\dfrac{xy^2}{z}\right)$

Section 10.6

16. Evaluate $\ln e$.

17. Change to a base-10 logarithmic expression. $\log_7 82$

Section 10.7

For Exercises 18–21, solve the equations.

18. $\log_5 x + \log_5 (x + 4) = 1$
19. $\ln(y + 2) + \ln y = \ln(18 - y)$
20. $3^{2x+4} = 27$
21. $5^t = 30$

Section 10.1 Algebra and Composition of Functions

Algebra and Composition of Functions

1. Algebra of Functions

Addition, subtraction, multiplication, and division can be used to create a new function from two or more functions. The domain of the new function will be the intersection of the domains of the original functions.

Objectives
1. Algebra of Functions
2. Composition of Functions
3. Multiple Operations on Functions

Sum, Difference, Product, and Quotient of Functions

Given two functions f and g, the functions $f + g$, $f - g$, $f \cdot g$, and $\frac{f}{g}$ are defined as

$$(f + g)(x) = f(x) + g(x)$$
$$(f - g)(x) = f(x) - g(x)$$
$$(f \cdot g)(x) = f(x) \cdot g(x)$$
$$\left(\frac{f}{g}\right)(x) = \frac{f(x)}{g(x)} \quad \text{provided } g(x) \neq 0$$

For example, suppose $f(x) = |x|$ and $g(x) = 3$. Taking the sum of the functions produces a new function denoted by $(f + g)$. In this case, $(f + g)(x) = |x| + 3$. Graphically, the y-values of the function $(f + g)$ are given by the sum of the corresponding y-values of f and g. This is depicted in Figure 10-1. The function $(f + g)$ appears in red. In particular, notice that $(f + g)(2) = f(2) + g(2) = 2 + 3 = 5$.

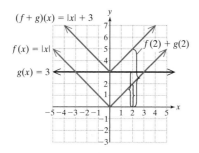

Figure 10-1

---

example 1 — Adding, Subtracting, and Multiplying Functions

Given: $g(x) = 4x \quad h(x) = x^2 - 3x \quad k(x) = \sqrt{x - 2}$

a. Find $(g + h)(x)$ and write the domain of $(g + h)$ in interval notation.
b. Find $(h - g)(x)$ and write the domain of $(h - g)$ in interval notation.
c. Find $(g \cdot k)(x)$ and write the domain of $(g \cdot k)$ in interval notation.

Solution:

a. $(g + h)(x) = g(x) + h(x)$
$= (4x) + (x^2 - 3x)$
$= 4x + x^2 - 3x$
$= x^2 + x \qquad$ The domain is all real numbers $(-\infty, \infty)$.

Skill Practice

Given:
$f(x) = x - 1$
$g(x) = 5x^2 + x$

Perform the indicated operations. Write the domain of the resulting function in interval notation.

1. $(f + g)(x)$
2. $(g - f)(x)$

Answers
1. $5x^2 + 2x - 1$; domain: $(-\infty, \infty)$
2. $5x^2 + 1$; domain: $(-\infty, \infty)$

b. $(h - g)(x) = h(x) - g(x)$
$= (x^2 - 3x) - (4x)$
$= x^2 - 3x - 4x$
$= x^2 - 7x$ The domain is all real numbers $(-\infty, \infty)$.

c. $(g \cdot k)(x) = g(x) \cdot k(x)$
$= (4x)(\sqrt{x - 2})$
$= 4x\sqrt{x - 2}$ The domain is $[2, \infty)$ because $x - 2 \geq 0$ for $x \geq 2$.

Skill Practice

Given:
$f(x) = x - 1$
$g(x) = 5x^2 + x$
$h(x) = \sqrt{5 - x}$

Perform the indicated operations. Write the domain of the resulting function in interval notation.

3. $(f \cdot h)(x)$

4. $\left(\dfrac{f}{g}\right)(x)$

example 2 Dividing Functions

Given the functions defined by $h(x) = x^2 - 3x$ and $k(x) = \sqrt{x - 2}$, find $\left(\dfrac{k}{h}\right)(x)$ and write the domain of $\left(\dfrac{k}{h}\right)$ in interval notation.

Solution:

$\left(\dfrac{k}{h}\right)(x) = \dfrac{\sqrt{x - 2}}{x^2 - 3x}$

To find the domain, we must consider the restrictions on x imposed by the square root and by the fraction.

- From the numerator we have $x - 2 \geq 0$ or, equivalently, $x \geq 2$.
- From the denominator we have $x^2 - 3x \neq 0$ or, equivalently, $x(x - 3) \neq 0$. Hence, $x \neq 3$ and $x \neq 0$.

Thus, the domain of $\dfrac{k}{h}$ is the set of real numbers greater than or equal to 2, but not equal to 3 or 0. This is shown graphically in Figure 10-2.

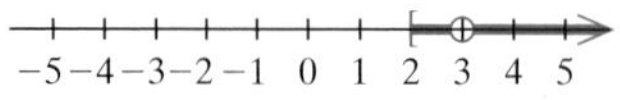

Figure 10-2

The domain is $[2, 3) \cup (3, \infty)$.

2. Composition of Functions

Composition of Functions

The **composition** of f and g, denoted $f \circ g$, is defined by the rule

$(f \circ g)(x) = f(g(x))$ provided that $g(x)$ is in the domain of f†

The composition of g and f, denoted $g \circ f$, is defined by the rule

$(g \circ f)(x) = g(f(x))$ provided that $f(x)$ is in the domain of g‡

†$f \circ g$ is also read as "f compose g."
‡$g \circ f$ is also read as "g compose f."

Answers
3. $(x - 1)\sqrt{5 - x}$; domain: $(-\infty, 5]$
4. $\dfrac{x - 1}{5x^2 + x}$; domain: $\left(-\infty, -\dfrac{1}{5}\right) \cup \left(-\dfrac{1}{5}, 0\right) \cup (0, \infty)$

For example, given $f(x) = 2x - 3$ and $g(x) = x + 5$, we have

$(f \circ g)(x) = f(g(x))$
$= f(x + 5)$ Substitute $g(x) = x + 5$ into the function f.
$= 2(x + 5) - 3$
$= 2x + 10 - 3$
$= 2x + 7$

In this composition, the function g is the innermost operation and acts on x first. Then the output value of function g becomes the domain element of the function f, as shown in the figure.

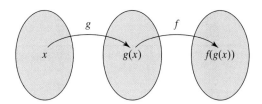

example 3 Composing Functions

Given: $f(x) = x - 5$, $g(x) = x^2$, and $n(x) = \sqrt{x + 2}$

a. Find $(f \circ g)(x)$ and write the domain of $(f \circ g)$ in interval notation.

b. Find $(g \circ f)(x)$ and write the domain of $(g \circ f)$ in interval notation.

c. Find $(n \circ f)(x)$ and write the domain of $(n \circ f)$ in interval notation.

Solution:

a. $(f \circ g)(x) = f(g(x))$
   $= f(x^2)$ Evaluate the function f at x^2.
   $= (x^2) - 5$ Replace x by x^2 in the function f.
   $= x^2 - 5$ The domain is all real numbers $(-\infty, \infty)$.

b. $(g \circ f)(x) = g(f(x))$
   $= g(x - 5)$ Evaluate the function g at $(x - 5)$.
   $= (x - 5)^2$ Replace x by the quantity $(x - 5)$ in function g.
   $= x^2 - 10x + 25$ The domain is all real numbers $(-\infty, \infty)$.

Tip: Example 3(a) and (b) illustrates that the order in which two functions are composed may result in different functions. That is, $f \circ g$ does not necessarily equal $g \circ f$.

c. $(n \circ f)(x) = n(f(x))$
   $= n(x - 5)$ Evaluate the function n at $x - 5$.
   $= \sqrt{(x - 5) + 2}$ Replace x by the quantity $(x - 5)$ in function n.
   $= \sqrt{x - 3}$ The domain is $[3, \infty)$.

Skill Practice

Given:
$f(x) = 2x^2$
$g(x) = x + 3$
$h(x) = \sqrt{x - 1}$

5. Find $(f \circ g)(x)$. Write the domain of $(f \circ g)$ in interval notation.

6. Find $(g \circ f)(x)$. Write the domain of $(g \circ f)$ in interval notation.

7. Find $(h \circ g)(x)$. Write the domain of $(h \circ g)$ in interval notation.

Answers
5. $2x^2 + 12x + 18$; domain:$(-\infty, \infty)$
6. $2x^2 + 3$; domain: $(-\infty, \infty)$
7. $\sqrt{x + 2}$; domain: $[-2, \infty)$

3. Multiple Operations on Functions

Skill Practice

Given:
$h(x) = x + 4$
$k(x) = x^2 - 3$

8. Find $(h \cdot k)(-2)$.
9. Find $\left(\dfrac{h}{k}\right)(4)$.
10. Find $(k \circ h)(1)$.

example 4 — Combining Functions

Given the functions defined by $f(x) = x - 7$ and $h(x) = 2x^3$, find the function values, if possible.

a. $(f \cdot h)(3)$ **b.** $\left(\dfrac{h}{f}\right)(7)$ **c.** $(h \circ f)(2)$

Solution:

a. $(f \cdot h)(3) = f(3) \cdot h(3)$
$= (3 - 7) \cdot 2(3)^3$
$= (-4) \cdot 2(27)$
$= -216$

b. The function $\dfrac{h}{f}$ has restrictions on its domain.

$$\left(\dfrac{h}{f}\right)(x) = \dfrac{h(x)}{f(x)} = \dfrac{2x^3}{x - 7}$$

Therefore, $x = 7$ is not in the domain, and $\left(\dfrac{h}{f}\right)(7) = \dfrac{h(7)}{f(7)}$ is undefined.

> **Avoiding Mistakes:** If you had tried evaluating the function $\dfrac{h}{f}$ at $x = 7$, the denominator would be zero and the function undefined.
>
> $$\dfrac{h(7)}{f(7)} = \dfrac{2(7)^3}{7 - 7}$$

c. $(h \circ f)(2) = h(f(2))$ Evaluate $f(2)$ first. $f(2) = 2 - 7 = -5$.
$= h(-5)$ Substitute the result into function h.
$= 2(-5)^3$
$= 2(-125)$
$= -250$

Skill Practice

Find the values from the graph.

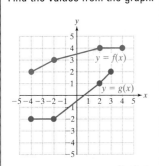

11. $(f + g)(-4)$
12. $\left(\dfrac{f}{g}\right)(2)$
13. $(g \circ f)(-2)$

example 5 — Finding Function Values from a Graph

For the functions f and g pictured, find the function values if possible.

a. $(f - g)(-3)$
b. $\left(\dfrac{g}{f}\right)(5)$
c. $(f \circ g)(4)$

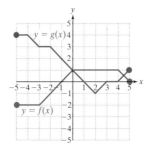

Answers

8. 2   9. $\dfrac{8}{13}$ 10. 22
11. 0   12. 4   13. 2

Solution:

a. $(f - g)(-3) = f(-3) - g(-3)$ Evaluate the difference of $f(-3)$ and $g(-3)$.

$= -2 - (3)$ Estimate function values from the graph.

$= -5$

b. $\left(\dfrac{g}{f}\right)(5) = \dfrac{g(5)}{f(5)}$ Evaluate the quotient of $g(5)$ and $f(5)$.

$= \dfrac{1}{0}$ (undefined)

The function $\dfrac{g}{f}$ is undefined at 5 because the denominator is zero.

c. $(f \circ g)(4) = f(g(4))$ From the red graph, find the value of $g(4)$ first.

$= f(0)$ From the blue graph, find the value of f at $x = 0$.

$= 1$

section 10.1 Practice Exercises

Boost your GRADE at mathzone.com!

- Practice Problems
- Self-Tests
- NetTutor
- e-Professors
- Videos

Study Skills Exercises

1. Your old tests and quizzes provide good material to study for the final exam. Use your old tests to make a list of the chapters on which you need to concentrate. Ask your professor for help if there are still concepts that you do not understand.

2. Define the key term **composition**.

Objective 1: Algebra of Functions

For Exercises 3–14, refer to the functions defined below.

$f(x) = x + 4$ $g(x) = 2x^2 + 4x$

$h(x) = \sqrt{x - 1}$ $k(x) = \dfrac{1}{x}$

Find the indicated functions. Write the domain in interval notation. **(See Examples 1–2.)**

3. $(f + g)(x)$ **4.** $(f - g)(x)$ **5.** $(g - f)(x)$

6. $(f + h)(x)$ **7.** $(f \cdot h)(x)$ **8.** $(h \cdot k)(x)$

9. $(g \cdot f)(x)$

10. $(f \cdot k)(x)$

11. $\left(\dfrac{h}{f}\right)(x)$

12. $\left(\dfrac{g}{f}\right)(x)$

13. $\left(\dfrac{f}{g}\right)(x)$

14. $\left(\dfrac{f}{h}\right)(x)$

Objective 2: Composition of Functions

For Exercises 15–22, find the indicated functions and their domains. Use f, g, h, and k as defined in Exercises 3–14. (See Example 3.)

15. $(f \circ g)(x)$

16. $(g \circ f)(x)$

17. $(f \circ k)(x)$

18. $(k \circ f)(x)$

19. $(k \circ h)(x)$

20. $(h \circ k)(x)$

21. $(k \circ g)(x)$

22. $(g \circ k)(x)$

23. Based on your answers to Exercises 15 and 16 is it true in general that $(f \circ g)(x) = (g \circ f)(x)$?

24. Based on your answers to Exercises 17 and 18, is it true in general that $(f \circ k)(x) = (k \circ f)(x)$?

For Exercises 25–30, find $(f \circ g)(x)$ and $(g \circ f)(x)$.

25. $f(x) = x^2 - 3x + 1$, $g(x) = 5x$

26. $f(x) = 3x^2 + 8$, $g(x) = 2x - 4$

27. $f(x) = |x|$, $g(x) = x^3 - 1$

28. $f(x) = \dfrac{1}{x + 2}$, $g(x) = |x + 2|$

29. For $h(x) = 5x - 4$, find $(h \circ h)(x)$.

30. For $k(x) = -x^2 + 1$, find $(k \circ k)(x)$.

Objective 3: Multiple Operations on Functions

For Exercises 31–44, refer to the functions defined below.

$$m(x) = x^3 \qquad n(x) = x - 3$$
$$r(x) = \sqrt{x + 4} \qquad p(x) = \dfrac{1}{x + 2}$$

Find the function values if possible. (See Example 4.)

31. $(m \cdot r)(0)$

32. $(n \cdot p)(0)$

33. $(m + r)(-4)$

34. $(n - m)(4)$

35. $(r \circ n)(3)$

36. $(n \circ r)(5)$

37. $(p \circ m)(-1)$

38. $(m \circ n)(5)$

39. $(m \circ p)(2)$

40. $(r \circ m)(2)$

41. $(r + p)(-3)$

42. $(n + p)(-2)$

43. $(m \circ p)(-2)$

44. $(r \circ m)(-2)$

For Exercises 45–58, approximate the function values from the graph, if possible. (See Example 5.)

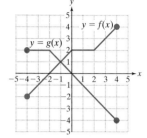

45. $(f + g)(2)$ **46.** $(g - f)(3)$ **47.** $(f \cdot g)(-1)$

48. $(g \cdot f)(-4)$ **49.** $\left(\dfrac{g}{f}\right)(0)$ **50.** $\left(\dfrac{f}{g}\right)(-2)$

51. $\left(\dfrac{f}{g}\right)(0)$ **52.** $\left(\dfrac{g}{f}\right)(-2)$ **53.** $(g \circ f)(-1)$

54. $(f \circ g)(0)$ **55.** $(f \circ g)(-4)$ **56.** $(g \circ f)(-4)$

57. $(g \circ g)(2)$ **58.** $(f \circ f)(-2)$

For Exercises 59–70, approximate the function values from the graph, if possible.

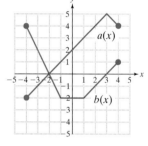

59. $(a - b)(-1)$ **60.** $(a + b)(0)$ **61.** $(b \cdot a)(1)$

62. $(a \cdot b)(2)$ **63.** $(b \circ a)(0)$ **64.** $(a \circ b)(-2)$

65. $(a \circ b)(-4)$ **66.** $(b \circ a)(-3)$ **67.** $\left(\dfrac{b}{a}\right)(3)$

68. $\left(\dfrac{a}{b}\right)(4)$ **69.** $(a \circ a)(-2)$ **70.** $(b \circ b)(1)$

71. The cost in dollars of producing x toy cars is $C(x) = 2.2x + 1$. The revenue received is $R(x) = 5.98x$. To calculate profit, subtract the cost from the revenue.

   a. Write and simplify a function P that represents profit in terms of x.

   b. Find the profit of producing 50 toy cars.

72. The cost in dollars of producing x lawn chairs is $C(x) = 2.5x + 10.1$. The revenue for selling x chairs is $R(x) = 6.99x$. To calculate profit, subtract the cost from the revenue.

   a. Write and simplify a function P that represents profit in terms of x.

   b. Find the profit in producing 100 lawn chairs.

73. The functions defined by $D(t) = 0.925t + 26.958$ and $R(t) = 0.725t + 20.558$ approximate the amount of child support (in billions of dollars) that was due and the amount of child support actually received in the United States between 2000 and 2006. In each case, $t = 0$ corresponds to 2000.

a. Find the function F defined by $F(t) = D(t) - R(t)$. What does F represent in the context of this problem?

b. Find $F(0)$, $F(2)$, and $F(4)$. What do these function values represent in the context of this problem?

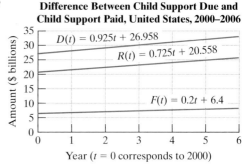

Difference Between Child Support Due and Child Support Paid, United States, 2000–2006

Source: U.S. Bureau of the Census.

74. If t represents the number of years after 1900, then the rural and urban populations in the South (United States) between 1900 and 1970 can be approximated by

$$r(t) = -3.497t^2 + 266.2t + 20{,}220$$

where $t = 0$ corresponds to 1900 and $r(t)$ represents the rural population in thousands.

$$u(t) = 0.0566t^3 + 0.952t^2 + 177.8t + 4593$$

where $t = 0$ corresponds to 1900 and $u(t)$ represents the urban population in thousands.

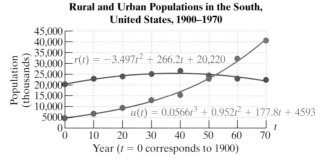

Rural and Urban Populations in the South, United States, 1900–1970

Source: Historical Abstract of the United States.

a. Find the function T defined by $T(t) = r(t) + u(t)$. What does the function T represent in the context of this problem?

b. Use the function T to approximate the total population in the South for the year 1940.

75. Joe rides a bicycle and his wheels revolve at 80 revolutions per minute (rpm). Therefore, the total number of revolutions, r, is given by $r(t) = 80t$, where t is the time in minutes. For each revolution of the wheels of the bike, he travels approximately 7 ft. Therefore, the total distance he travels, D, depends on the total number of revolutions, r, according to the function $D(r) = 7r$.

a. Find $(D \circ r)(t)$ and interpret its meaning in the context of this problem.

b. Find Joe's total distance in feet after 10 min.

76. The area, A, of a square is given by the function $a(x) = x^2$, where x is the length of the side of the square. If carpeting costs $9.95 per square yard, then the cost, C, to carpet a square room is given by $C(a) = 9.95a$, where a is the area of the room in square yards.

a. Find $(C \circ a)(x)$ and interpret its meaning in the context of this problem.

b. Find the cost to carpet a square room if its floor dimensions are 15 yd by 15 yd.

Section 10.2 Inverse Functions

Objectives
1. Introduction to Inverse Functions
2. Definition of a One-to-One Function
3. Finding an Equation of the Inverse of a Function
4. Definition of the Inverse of a Function

1. Introduction to Inverse Functions

In Section 4.2, we defined a function as a set of ordered pairs (x, y) such that for every element x in the domain, there corresponds exactly one element y in the range. For example, the function f relates the weight of a package of deli meat, x, to its cost, y.

$$f = \{(1, 2.99), (1.5, 4.49), (4, 11.96)\}$$

That is, 1 lb of meat sells for \$2.99, 1.5 lb sells for \$4.49, and 4 lb sells for \$11.96. Now suppose we create a new function in which the values of x and y are interchanged. The new function, called the **inverse of** f, denoted f^{-1}, relates the price of meat, x, to its weight, y.

$$f^{-1} = \{(2.99, 1), (4.49, 1.5), (11.96, 4)\}$$

Notice that interchanging the x- and y-values has the following outcome. The domain of f is the same as the range of f^{-1}, and the range of f is the domain of f^{-1}.

Avoiding Mistakes: f^{-1} denotes the inverse of a function. The -1 does not represent an exponent.

2. Definition of a One-to-One Function

A necessary condition for a function f to have an inverse function is that no two ordered pairs in f have different x-coordinates and the same y-coordinate. A function that satisfies this condition is called a **one-to-one function**. The function relating the weight of a package of meat to its price is a one-to-one function. However, consider the function g defined by

$$g = \{(1, 4), (2, 3), (-2, 4)\}$$
same y
different x

This function is not one-to-one because the range element 4 has two different x-coordinates, 1 and -2. Interchanging the x- and y-values produces a relation that violates the definition of a function.

$\{(4, 1), (3, 2), (4, -2)\}$ This relation is not a function because for $x = 4$ there are two different y-values, $y = 1$ and $y = -2$.

same x
different y

In Section 4.2, you learned the vertical line test to determine visually if a graph represents a function. Similarly, we use a **horizontal line test** to determine whether a function is one-to-one.

Concept Connections

Determine whether each of the relations is a one-to-one function. If it is not, state why.
1. $\{(4, 2), (1, 2), (3, 0), (-4, 4)\}$
2. $\{(0, 0), (5, -1), (2, 3), (5, 6)\}$
3. $\{(3, 4), (2, 2), (0, 8), (-1, 7)\}$

Horizontal Line Test

Consider a function defined by a set of points (x, y) in a rectangular coordinate system. The graph of the ordered pairs defines y as a *one-to-one function* of x if no horizontal line intersects the graph in more than one point.

Answers
1. Not one-to-one. The range element 2 has two different x-coordinates.
2. Not a function. The domain element 5 is paired with two different y-coordinates.
3. One-to-one function

To understand the horizontal line test, consider the functions f and g.

$f = \{(1, 2.99), (1.5, 4.49), (4, 11.96)\}$ $g = \{(1, 4), (2, 3), (-2, 4)\}$

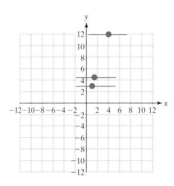 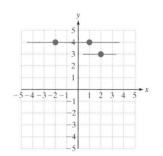

This function is one-to-one. No horizontal line intersects more than once.

This function is *not* one-to-one. A horizontal line intersects more than once.

Skill Practice

Use the horizontal line test to determine if the functions are one-to-one.

4.
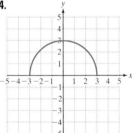

5.
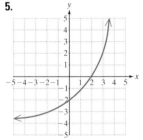

example 1   Identifying One-to-One Functions

Determine whether the function is one-to-one.

a.
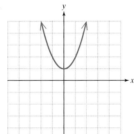

b.
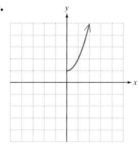

Solution:

a. Function is not one-to-one. A horizontal line intersects in more than one point.

b. Function is one-to-one. No horizontal line intersects more than once.

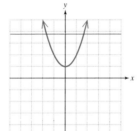 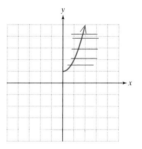

3. Finding an Equation of the Inverse of a Function

Another way to view the construction of the inverse of a function is to find a function that performs the inverse operations in the reverse order. For example,

Answers
4. Not one-to-one
5. One-to-one

the function defined by $f(x) = 2x + 1$ multiplies x by 2 and then adds 1. Therefore, the inverse function must *subtract* 1 from x and *divide* by 2. We have

$$f^{-1}(x) = \frac{x-1}{2}$$

To facilitate the process of finding an equation of the inverse of a one-to-one function, we offer the following steps.

Finding an Equation of an Inverse of a Function

For a one-to-one function defined by $y = f(x)$, the equation of the inverse can be found as follows:

1. Replace $f(x)$ by y.
2. Interchange x and y.
3. Solve for y.
4. Replace y by $f^{-1}(x)$.

---

example 2 — Finding an Equation of the Inverse of a Function

Find the inverse. $f(x) = 2x + 1$

Skill Practice

6. Find the inverse of $f(x) = 4x + 6$.

Solution:

Foremost, we know the graph of f is a nonvertical line. Therefore, $f(x) = 2x + 1$ defines a one-to-one function. To find the inverse we have

$y = 2x + 1$	**Step 1:** Replace $f(x)$ by y.
$x = 2y + 1$	**Step 2:** Interchange x and y.
$x - 1 = 2y$	**Step 3:** Solve for y. Subtract 1 from both sides.
$\frac{x-1}{2} = y$	Divide both sides by 2.
$f^{-1}(x) = \frac{x-1}{2}$	**Step 4:** Replace y by $f^{-1}(x)$.

The key step in determining the equation of the inverse of a function is to interchange x and y. By so doing, a point (a, b) on f corresponds to a point (b, a) on f^{-1}. For this reason, the graphs of f and f^{-1} are symmetric with respect to the line $y = x$ (Figure 10-3). Notice that the point $(-3, -5)$ of the function f corresponds to the point $(-5, -3)$ of f^{-1}. Likewise, $(1, 3)$ of f corresponds to $(3, 1)$ of f^{-1}.

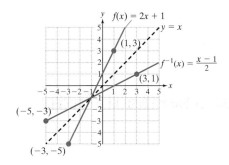

Figure 10-3

Answer

6. $f^{-1}(x) = \dfrac{x-6}{4}$

Skill Practice

7. Find the inverse of $h(x) = \sqrt[3]{2x - 1}$.

example 3 — Finding an Equation of the Inverse of a Function

Find the inverse of the one-to-one function. $g(x) = \sqrt[3]{5x} - 4$

Solution:

$y = \sqrt[3]{5x} - 4$	**Step 1:** Replace $g(x)$ by y.
$x = \sqrt[3]{5y} - 4$	**Step 2:** Interchange x and y.
$x + 4 = \sqrt[3]{5y}$	**Step 3:** Solve for y. Add 4 to both sides.
$(x + 4)^3 = (\sqrt[3]{5y})^3$	To eliminate the cube root, cube both sides.
$(x + 4)^3 = 5y$	Simplify the right side.
$\dfrac{(x + 4)^3}{5} = y$	Divide both sides by 5.
$g^{-1}(x) = \dfrac{(x + 4)^3}{5}$	**Step 4:** Replace y by $g^{-1}(x)$.

The graphs of g and g^{-1} are shown in Figure 10-4. Once again we see that the graphs of a function and its inverse are symmetric with respect to the line $y = x$.

Concept Connections

8. The graphs of a function and its inverse are symmetric with respect to what line?

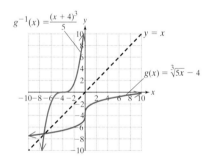

Figure 10-4

For a function that is not one-to-one, sometimes we can restrict its domain to create a new function that is one-to-one. This is demonstrated in Example 4.

Skill Practice

9. Find the inverse.
$g(x) = x^2 - 2 \quad x \geq 0$

example 4 — Finding the Equation of an Inverse of a Function with a Restricted Domain

Given the function defined by $m(x) = x^2 + 4$ for $x \geq 0$, find an equation defining m^{-1}.

Solution:

From Section 8.4, we know that $y = x^2 + 4$ is a parabola with vertex at $(0, 4)$ (Figure 10-5). The graph represents a function that is not one-to-one. However, with the restriction on the domain $x \geq 0$, the graph of $m(x) = x^2 + 4$, $x \geq 0$, consists of only the "right" branch of the parabola (Figure 10-6). This is a one-to-one function.

Answers

7. $h^{-1}(x) = \dfrac{x^3 + 1}{2}$ 8. $y = x$

9. $g^{-1}(x) = \sqrt{x + 2}$

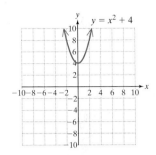

Figure 10-5   Figure 10-6

To find the inverse, we have

$y = x^2 + 4$	$x \geq 0$	**Step 1:** Replace $m(x)$ by y.
$x = y^2 + 4$	$y \geq 0$	**Step 2:** Interchange x and y. Notice that the restriction $x \geq 0$ becomes $y \geq 0$.
$x - 4 = y^2$	$y \geq 0$	**Step 3:** Solve for y. Subtract 4 from both sides.
$\sqrt{x - 4} = y$	$y \geq 0$	Apply the square root property. Notice that we obtain the *positive* square root of $x - 4$ because of the restriction $y \geq 0$.
$m^{-1}(x) = \sqrt{x - 4}$		**Step 4:** Replace y by $m^{-1}(x)$. Notice that the domain of m^{-1} has the same values as the range of m.

Figure 10-7 shows the graphs of m and m^{-1}.

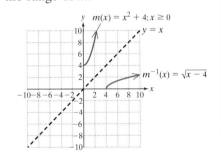

Figure 10-7

4. Definition of the Inverse of a Function

An important relationship between a function and its inverse is shown in Figure 10-8.

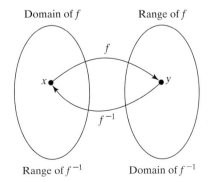

Figure 10-8

Recall that the domain of f is the range of f^{-1} and the range of f is the domain of f^{-1}. The operations performed by f are reversed by f^{-1}. This leads to a formal definition of the inverse of a function.

> **Definition of the Inverse of a Function**
>
> If f is a one-to-one function, then g is the inverse of f if and only if
>
> $$(f \circ g)(x) = x \quad \text{for all } x \text{ in the domain of } g$$
>
> and
>
> $$(g \circ f)(x) = x \quad \text{for all } x \text{ in the domain of } f$$

Skill Practice

10. Show that the functions are inverses.
    $f(x) = 3x - 2$ and
    $g(x) = \dfrac{x + 2}{3}$

example 5   Composing a Function with Its Inverse

Show that the functions are inverses.

$$h(x) = 2x + 1 \quad \text{and} \quad k(x) = \dfrac{x - 1}{2}$$

Solution:

To show that the functions h and k are inverses, we need to confirm that $(h \circ k)(x) = x$ and $(k \circ h)(x) = x$.

$$(h \circ k)(x) = h(k(x)) = h\left(\dfrac{x-1}{2}\right)$$

$$= 2\left(\dfrac{x-1}{2}\right) + 1$$

$$= x - 1 + 1$$

$$= x \checkmark \quad (h \circ k)(x) = x \text{ as desired.}$$

$$(k \circ h)(x) = k(h(x)) = k(2x + 1)$$

$$= \dfrac{(2x + 1) - 1}{2}$$

$$= \dfrac{2x + 1 - 1}{2}$$

$$= \dfrac{2x}{2}$$

$$= x \checkmark \quad (k \circ h)(x) = x \text{ as desired.}$$

The functions h and k are inverses because $(h \circ k)(x) = x$ and $(k \circ h)(x) = x$ for all real numbers x.

Answer

10. $(f \circ g)(x) = f(g(x))$
    $= 3\left(\dfrac{x+2}{3}\right) - 2 = x$
    $(g \circ f)(x) = g(f(x))$
    $= \dfrac{3x - 2 + 2}{3} = x$

Section 10.2 Practice Exercises

Boost your GRADE at mathzone.com!

- Practice Problems
- Self-Tests
- NetTutor
- e-Professors
- Videos

Study Skills Exercises

1. Try to not study math when you are tired. It is a subject that requires a clear mind. When you sit down to do your homework, you will be more productive if you do the more demanding assignments first. List the assignments you have due this week in the order that you want to complete them. Place the more difficult assignments on the top of the list.

2. Define the key terms.

    a. Inverse function   b. One-to-one function   c. Horizontal line test

Review Exercises

For Exercises 3–8, determine if the relation is a function by using the vertical line test. (See Section 4.2.)

3.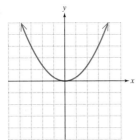

4.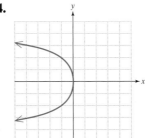

5.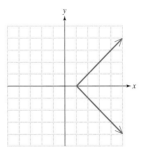

6.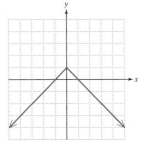

7.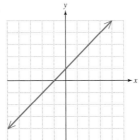

8.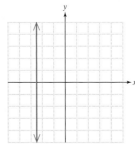

Objective 1: Introduction to Inverse Functions

For Exercises 9–12, write the inverse function for each function as defined.

9. $g = \{(3, 5), (8, 1), (-3, 9), (0, 2)\}$

10. $f = \{(-6, 2), (-9, 0), (-2, -1), (3, 4)\}$

11. $r = \{(a, 3), (b, 6), (c, 9)\}$

12. $s = \{(-1, x), (-2, y), (-3, z)\}$

Objective 2: Definition of a One-to-One Function

13. The table relates a state, x, to the number of representatives in the House of Representatives, y, in the year 2005. Does this relation define a one-to-one function? If so, write a function defining the inverse as a set of ordered pairs.

State x	Number of Representatives y
Colorado	7
California	53
Texas	32
Louisiana	7
Pennsylvania	19

14. The table relates a city x to its average January temperature y. Does this relation define a one-to-one function? If so, write a function defining the inverse as a set of ordered pairs.

City x	Temperature y (°C)
Gainesville, FL	13.6
Keene, NH	−6.0
Wooster, OH	−4.0
Rock Springs, WY	−6.0
Lafayette, LA	10.9

For Exercises 15–20, determine if the function is one-to-one by using the horizontal line test. **(See Example 1.)**

15.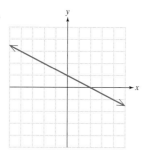

16.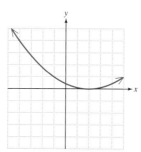

17.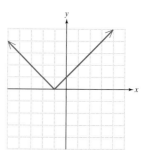

18.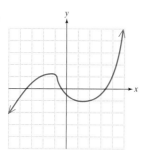

19.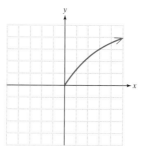

20.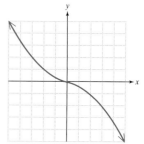

Objective 3: Finding an Equation of the Inverse of a Function

For Exercises 21–30, write an equation of the inverse for each one-to-one function as defined. **(See Examples 2–3.)**

21. $h(x) = x + 4$
22. $k(x) = x - 3$
23. $m(x) = \frac{1}{3}x - 2$
24. $n(x) = 4x + 2$

25. $p(x) = -x + 10$ **26.** $q(x) = -x - \dfrac{2}{3}$ **27.** $f(x) = x^3$ **28.** $g(x) = \sqrt[3]{x}$

29. $g(x) = \sqrt[3]{2x - 1}$ **30.** $f(x) = x^3 - 4$

31. The function defined by $f(x) = 0.3048x$ converts a length of x feet into $f(x)$ meters.

  a. Find the equivalent length in meters for a 4-ft board and a 50-ft wire.

  b. Find an equation defining $y = f^{-1}(x)$.

  c. Use the inverse function from part (b) to find the equivalent length in feet for a 1500-m race in track and field. Round to the nearest tenth of a foot.

32. The function defined by $s(x) = 1.47x$ converts a speed of x mph to $s(x)$ ft/sec.

  a. Find the equivalent speed in feet per second for a car traveling 60 mph.

  b. Find an equation defining $y = s^{-1}(x)$.

  c. Use the inverse function from part (b) to find the equivalent speed in miles per hour for a train traveling 132 ft/sec. Round to the nearest tenth.

For Exercises 33–39, answer true or false.

33. The function defined by $y = 2$ has an inverse function defined by $x = 2$.

34. The domain of any one-to-one function is the same as the domain of its inverse.

35. All linear functions have an inverse function.

36. The function defined by $g(x) = |x|$ is one-to-one.

37. The function defined by $k(x) = x^2$ is one-to-one.

38. The function defined by $h(x) = |x|$ for $x \geq 0$ is one-to-one.

39. The function defined by $L(x) = x^2$ for $x \geq 0$ is one-to-one.

40. Explain how the domain and range of a one-to-one function and its inverse are related.

41. If $(0, b)$ is the y-intercept of a one-to-one function, what is the x-intercept of its inverse?

42. If $(a, 0)$ is the x-intercept of a one-to-one function, what is the y-intercept of its inverse?

43. Can you think of any function that is its own inverse?

44. a. What are the domain and range of the function defined by $f(x) = \sqrt{x - 1}$? **(See Example 4.)**

b. What are the domain and range of the function defined by $f^{-1}(x) = x^2 + 1, x \geq 0$?

45. a. What are the domain and range of the function defined by $g(x) = x^2 - 4, x \leq 0$?

b. What are the domain and range of the function defined by $g^{-1}(x) = -\sqrt{x + 4}$?

For Exercises 46–49, the graph of $y = f(x)$ is given.

    a. State the domain of f.   **b.** State the range of f.

    c. State the domain of f^{-1}.   **d.** State the range of f^{-1}.

    e. Graph the function defined by $y = f^{-1}(x)$. The line $y = x$ is provided for your reference.

46.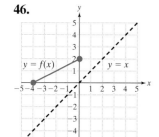

47.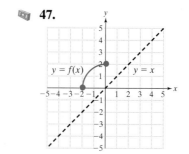

48.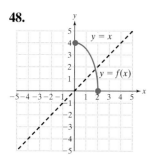

49.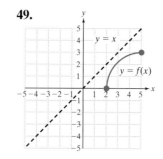

Objective 4: Definition of the Inverse of a Function

For Exercises 50–55, verify that f and g are inverse functions by showing that

a. $(f \circ g)(x) = x$ **b.** $(g \circ f)(x) = x$

(See Example 5.)

50. $f(x) = 6x + 1$ and $g(x) = \dfrac{x-1}{6}$

51. $f(x) = 5x - 2$ and $g(x) = \dfrac{x+2}{5}$

52. $f(x) = \dfrac{\sqrt[3]{x}}{2}$ and $g(x) = 8x^3$

53. $f(x) = \dfrac{\sqrt[3]{x}}{3}$ and $g(x) = 27x^3$

54. $f(x) = x^2 + 1$, $x \geq 0$, and $g(x) = \sqrt{x-1}$, $x \geq 1$

55. $f(x) = x^2 - 3$, $x \geq 0$, and $g(x) = \sqrt{x+3}$, $x \geq -3$

Expanding Your Skills

For Exercises 56–67, write an equation of the inverse of the one-to-one function.

56. $f(x) = \dfrac{x-1}{x+1}$

57. $p(x) = \dfrac{3-x}{x+3}$

58. $t(x) = \dfrac{2}{x-1}$

59. $w(x) = \dfrac{4}{x+2}$

60. $g(x) = x^2 + 9$ $x \geq 0$

61. $m(x) = x^2 - 1$ $x \geq 0$

62. $n(x) = x^2 + 9$ $x \leq 0$

63. $g(x) = x^2 - 1$ $x \leq 0$

64. $q(x) = \sqrt{x+4}$

65. $v(x) = \sqrt{x+16}$

66. $z(x) = -\sqrt{x+4}$

67. $u(x) = -\sqrt{x+16}$

Graphing Calculator Exercises

For Exercises 68–71, use a graphing calculator to graph each function on the standard viewing window defined by $-10 \leq x \leq 10$ and $-10 \leq y \leq 10$. Use the graph of the function to determine if the function is one-to-one on the interval $-10 \leq x \leq 10$. If the function is one-to-one, find its inverse and graph both functions on the standard viewing window.

68. $f(x) = \sqrt[3]{x+5}$

69. $k(x) = x^3 - 4$

70. $g(x) = 0.5x^3 - 2$

71. $m(x) = 3x - 4$

Objectives

1. Definition of an Exponential Function
2. Approximating Exponential Expressions with a Calculator
3. Graphs of Exponential Functions
4. Applications of Exponential Functions

section 10.3 Exponential Functions

1. Definition of an Exponential Function

The concept of a function was first introduced in Section 4.2. Since then we have learned to recognize several categories of functions, including constant, linear, rational, and quadratic functions. In this section and the next, we will define two new types of functions called exponential and logarithmic functions.

To introduce the concept of an exponential function, consider the following salary plans for a new job. Plan A pays $1 million for a month's work. Plan B starts with a 1¢ signing bonus, and every day thereafter the salary is doubled.

At first glance, the million-dollar plan appears to be more favorable. Look, however, at Table 10-1, which shows the daily payments for 30 days under plan B.

table 10-1

Day	Payment	Day	Payment	Day	Payment
1	2¢	11	$20.48	21	$20,971.52
2	4¢	12	$40.96	22	$41,943.04
3	8¢	13	$81.92	23	$83,886.08
4	16¢	14	$163.84	24	$167,772.16
5	32¢	15	$327.68	25	$335,544.32
6	64¢	16	$655.36	26	$671,088.64
7	$1.28	17	$1310.72	27	$1,342,177.28
8	$2.56	18	$2621.44	28	$2,684,354.56
9	$5.12	19	$5242.88	29	$5,368,709.12
10	$10.24	20	$10,485.76	30	$10,737,418.24

Notice that the salary on the 30th day for plan B is over $10 million. Taking the sum of the payments, we see the total salary for the 30-day period is $21,474,836.46.

The daily salary for plan B can be represented by the function $y = 2^x$, where x is the number of days on the job and y is the salary (in cents) for that day. An interesting feature of this function is that for every positive 1-unit change in x, the y-value doubles. The function $y = 2^x$ is called an exponential function.

Definition of an Exponential Function

Let b be any real number such that $b > 0$ and $b \neq 1$. Then for any real number x, a function of the form $y = b^x$ is called an **exponential function**.

An exponential function is easily recognized as a function with a constant base and variable exponent. Notice that the base of an exponential function must be a positive real number not equal to 1.

Concept Connections

Determine which are exponential functions.

1. $y = x^3$
2. $y = 3^x$
3. $y = \left(\dfrac{1}{2}\right)^x$

Answers

1. No   2. Yes   3. Yes

2. Approximating Exponential Expressions with a Calculator

Up to this point, we have evaluated exponential expressions with integer exponents and with rational exponents, for example, $4^3 = 64$ and $4^{1/2} = \sqrt{4} = 2$. However, how do we evaluate an exponential expression with an irrational exponent such as 4^π? In such a case, the exponent is a nonterminating and nonrepeating decimal. The value of 4^π can be thought of as the limiting value of a sequence of approximations using rational exponents:

$$4^{3.14} \approx 77.7084726$$
$$4^{3.141} \approx 77.81627412$$
$$4^{3.1415} \approx 77.87023095$$
$$\vdots$$
$$4^\pi \approx 77.88023365$$

Calculator Connections

On a graphing calculator, use the $\boxed{\wedge}$ key to approximate an expression with an irrational exponent.

```
4^π
        77.88023365
```

An exponential expression can be evaluated at all rational numbers and at all irrational numbers. Hence, the domain of an exponential function is all real numbers.

example 1 — Approximating Exponential Expressions with a Calculator

Approximate the expressions. Round the answers to 4 decimal places.

a. $8^{\sqrt{3}}$ b. $5^{-\sqrt{17}}$ c. $\sqrt{10}^{\sqrt{2}}$

Solution:

a. $8^{\sqrt{3}} \approx 36.6604$ b. $5^{-\sqrt{17}} \approx 0.0013$ c. $\sqrt{10}^{\sqrt{2}} \approx 5.0946$

Skill Practice

Approximate the value of the expressions. Round the answers to 4 decimal places.

4. $9^{1.2}$
5. $15^{\sqrt{5}}$
6. $\sqrt{7}^{\sqrt{3}}$

Calculator Connections

```
8^√(3)
        36.66044576
5^-√(17)
         .00131242
(√(10))^√(2)
         5.09456117
```

3. Graphs of Exponential Functions

The functions defined by $f(x) = 2^x$, $g(x) = 3^x$, $h(x) = 5^x$, and $k(x) = (\frac{1}{2})^x$ are all examples of exponential functions. Example 2 illustrates the two general shapes of exponential functions.

Answers
4. 13.9666
5. 426.4028
6. 5.3936

Skill Practice

Graph the functions f and g.

7. $f(x) = 3^x$

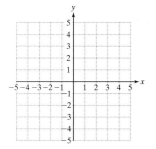

8. $g(x) = \left(\dfrac{1}{3}\right)^x$

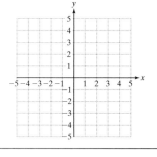

---

example 2   Graphing an Exponential Function

Graph the functions f and g.

a. $f(x) = 2^x$ **b.** $g(x) = \left(\dfrac{1}{2}\right)^x$

Solution:

Table 10-2 shows several function values $f(x)$ and $g(x)$ for both positive and negative values of x. The graph is shown in Figure 10-9.

table 10-2

x	$f(x) = 2^x$	$g(x) = \left(\dfrac{1}{2}\right)^x$
-4	$\dfrac{1}{16}$	16
-3	$\dfrac{1}{8}$	8
-2	$\dfrac{1}{4}$	4
-1	$\dfrac{1}{2}$	2
0	1	1
1	2	$\dfrac{1}{2}$
2	4	$\dfrac{1}{4}$
3	8	$\dfrac{1}{8}$
4	16	$\dfrac{1}{16}$

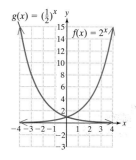

Figure 10-9

The graphs in Figure 10-9 illustrate several important features of exponential functions.

Graphs of $f(x) = b^x$

The graph of an exponential function defined by $f(x) = b^x$ ($b > 0$ and $b \neq 1$) has the following properties.

1. If $b > 1$, f is an *increasing* exponential function, sometimes called an **exponential growth function**.

   If $0 < b < 1$, f is a *decreasing* exponential function, sometimes called an **exponential decay function**.

2. The domain is the set of all real numbers, $(-\infty, \infty)$.
3. The range is $(0, \infty)$.
4. The x-axis is a horizontal asymptote.
5. The function passes through the point $(0, 1)$ because $f(0) = b^0 = 1$.

These properties indicate that the graph of an exponential function is an increasing function if the base is greater than 1. Furthermore, the base affects its "steepness." Consider the graphs of $f(x) = 2^x$, $h(x) = 3^x$, and $k(x) = 5^x$ (Figure 10-10). For every positive 1-unit change in x, $f(x) = 2^x$ increases by 2 times, $h(x) = 3^x$ increases by 3 times, and $k(x) = 5^x$ increases by 5 times (Table 10-3).

Answers

7.

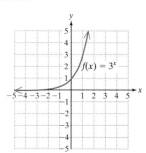

8.

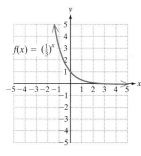

table 10-3

x	$f(x) = 2^x$	$h(x) = 3^x$	$k(x) = 5^x$
-3	$\frac{1}{8}$	$\frac{1}{27}$	$\frac{1}{125}$
-2	$\frac{1}{4}$	$\frac{1}{9}$	$\frac{1}{25}$
-1	$\frac{1}{2}$	$\frac{1}{3}$	$\frac{1}{5}$
0	1	1	1
1	2	3	5
2	4	9	25
3	8	27	125

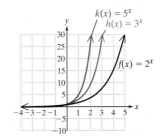

Figure 10-10

Concept Connections

9.

Given the graph of the exponential function $y = a^x$, which is true?

a. $a > 1$
b. $0 < a < 1$
c. It cannot be determined.

The graph of an exponential function is a *decreasing function* if the base is between 0 and 1. Consider the graphs of $g(x) = (\frac{1}{2})^x$, $m(x) = (\frac{1}{3})^x$, and $n(x) = (\frac{1}{5})^x$ (Table 10-4 and Figure 10-11).

table 10-4

x	$g(x) = (\frac{1}{2})^x$	$m(x) = (\frac{1}{3})^x$	$n(x) = (\frac{1}{5})^x$
-3	8	27	125
-2	4	9	25
-1	2	3	5
0	1	1	1
1	$\frac{1}{2}$	$\frac{1}{3}$	$\frac{1}{5}$
2	$\frac{1}{4}$	$\frac{1}{9}$	$\frac{1}{25}$
3	$\frac{1}{8}$	$\frac{1}{27}$	$\frac{1}{125}$

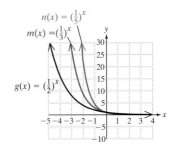

Figure 10-11

4. Applications of Exponential Functions

Exponential growth and decay can be found in a variety of real-world phenomena; for example,

- Population growth can often be modeled by an exponential function.
- The growth of an investment under compound interest increases exponentially.
- The mass of a radioactive substance decreases exponentially with time.
- The temperature of a cup of coffee decreases exponentially as it approaches room temperature.

A substance that undergoes radioactive decay is said to be radioactive. The **half-life** of a radioactive substance is the amount of time it takes for one-half of the original amount of the substance to change into something else. That is, after each half-life the amount of the original substance decreases by one-half.

In 1898, Marie Curie discovered the highly radioactive element radium. She shared the 1903 Nobel Prize in physics for her research on radioactivity and was awarded the 1911 Nobel Prize in chemistry for her discovery of radium and polonium. Radium 226 (an isotope of radium) has a half-life of 1620 years and decays into radon 222 (a radioactive gas).

Marie and Pierre Curie

Answer
9. b

Skill Practice

Cesium 137 is a radioactive metal with a short half-life of 30 years. In a sample originally having 1 g of cesium 137, the amount of cesium 137 present after t years is given by

$$A(t) = \left(\frac{1}{2}\right)^{t/30}$$

10. How much cesium 137 will be present after 30 years?

11. How much cesium 137 will be present after 90 years?

example 3 — Applying an Exponential Decay Function

In a sample originally having 1 g of radium 226, the amount of radium 226 present after t years is given by

$$A(t) = \left(\frac{1}{2}\right)^{t/1620}$$

where A is the amount of radium in grams and t is the time in years.

a. How much radium 226 will be present after 1620 years?

b. How much radium 226 will be present after 3240 years?

c. How much radium 226 will be present after 4860 years?

Solution:

$$A(t) = \left(\frac{1}{2}\right)^{t/1620}$$

a. $A(1620) = \left(\frac{1}{2}\right)^{1620/1620}$ Substitute $t = 1620$.

$\qquad = \left(\frac{1}{2}\right)^{1}$

$\qquad = 0.5$

After 1620 years (1 half-life), 0.5 g remains.

b. $A(3240) = \left(\frac{1}{2}\right)^{3240/1620}$ Substitute $t = 3240$.

$\qquad = \left(\frac{1}{2}\right)^{2}$

$\qquad = 0.25$

After 3240 years (2 half-lives), the amount of the original substance is reduced by one-half, 2 times: 0.25 g remains.

c. $A(4860) = \left(\frac{1}{2}\right)^{4860/1620}$ Substitute $t = 4860$.

$\qquad = \left(\frac{1}{2}\right)^{3}$

$\qquad = 0.125$

After 4860 years (3 half-lives), the amount of the original substance is reduced by one-half, 3 times: 0.125 g remains.

Exponential functions are often used to model population growth. Suppose the initial value of a population at some time $t = 0$ is P_0. If the rate of increase of a population is r, then after 1, 2, and 3 years, the new population can be found as follows:

Answers

10. 0.5 g   11. 0.125 g

After 1 year: $\begin{pmatrix} \text{Total} \\ \text{population} \end{pmatrix} = \begin{pmatrix} \text{initial} \\ \text{population} \end{pmatrix} + \begin{pmatrix} \text{increase in} \\ \text{population} \end{pmatrix}$

$= P_0 + P_0 r$

$= P_0(1 + r)$ Factor out P_0.

After 2 years: $\begin{pmatrix} \text{Total} \\ \text{population} \end{pmatrix} = \begin{pmatrix} \text{population} \\ \text{after 1 year} \end{pmatrix} + \begin{pmatrix} \text{increase in} \\ \text{population} \end{pmatrix}$

$= P_0(1 + r) + P_0(1 + r)r$

$= P_0(1 + r)1 + P_0(1 + r)r$

$= P_0(1 + r)(1 + r)$ Factor out $P_0(1 + r)$.

$= P_0(1 + r)^2$

After 3 years: $\begin{pmatrix} \text{Total} \\ \text{population} \end{pmatrix} = \begin{pmatrix} \text{population} \\ \text{after 2 years} \end{pmatrix} + \begin{pmatrix} \text{increase in} \\ \text{population} \end{pmatrix}$

$= P_0(1 + r)^2 + P_0(1 + r)^2 r$

$= P_0(1 + r)^2 1 + P_0(1 + r)^2 r$

$= P_0(1 + r)^2(1 + r)$ Factor out $P_0(1 + r)^2$.

$= P_0(1 + r)^3$

This pattern continues, and after t years, the population $P(t)$ is given by

$$P(t) = P_0(1 + r)^t$$

example 4   Applying an Exponential Growth Function

The population of the Bahamas in 1998 was estimated at 280,000 with an annual rate of increase of 1.39%.

a. Find a mathematical model that relates the population of the Bahamas as a function of the number of years after 1998.

b. If the annual rate of increase remains the same, use this model to predict the population of the Bahamas in the year 2010. Round to the nearest thousand.

Solution:

a. The initial population is $P_0 = 280,000$, and the rate of increase is $r = 1.39\%$.

$P(t) = P_0(1 + r)^t$ Substitute $P_0 = 280,000$ and $r = 0.0139$.

$= 280,000(1 + 0.0139)^t$

$= 280,000(1.0139)^t$ Here $t = 0$ corresponds to the year 1998.

b. Because the initial population ($t = 0$) corresponds to the year 1998, we use $t = 12$ to find the population in the year 2010.

$P(12) = 280,000(1.0139)^{12}$

$\approx 330,000$

Skill Practice

The population of Colorado in 1990 was approximately 3,000,000 with an annual increase of 2%.

12. Find a mathematical model that relates the population of Colorado as function of the number of years after 1990.

13. If the annual increase remains the same, use this model to predict the population of Colorado in 2010.

Answers

12. $P(t) = 3,000,000(1.02)^t$
13. Approximately 4,460,000

Section 10.3 Practice Exercises

Boost your GRADE at mathzone.com!

- Practice Problems
- Self-Tests
- NetTutor
- e-Professors
- Videos

Study Skills Exercises

1. Organize all the things that you consistently need in order to do your math homework. Keep them all together in the same place, such as a desk drawer. When you sit down to study, everything you need will be right there. Make a list of the things you use to do your homework.

2. Define the key terms.
   a. Exponential function
   b. Exponential growth function
   c. Exponential decay function

Review Exercises

For Exercises 3–8, find the functions, using f and g as given. $f(x) = 2x^2 + x + 2 \quad g(x) = 3x - 1$

3. $(f + g)(x)$
4. $(g - f)(x)$
5. $(f \cdot g)(x)$

6. $\left(\dfrac{g}{f}\right)(x)$
7. $(f \circ g)(x)$
8. $(g \circ f)(x)$

For Exercises 9–10, find the inverse function.

9. $\{(2, 3), (0, 0), (-8, 4), (10, 12)\}$

10. $\left\{(-13, 14), \left(\dfrac{1}{2}, -\dfrac{1}{2}\right), (6, 30), \left(-\dfrac{5}{3}, 0\right), (0, 1)\right\}$

For Exercises 11–18, evaluate the expression without the use of a calculator.

11. 5^2
12. 2^{-3}
13. 10^{-3}
14. 3^4

15. $36^{1/2}$
16. $27^{1/3}$
17. $16^{3/4}$
18. $8^{2/3}$

Objective 2: Approximating Exponential Expressions with a Calculator

For Exercises 19–26, evaluate the expression by using a calculator. Round to 4 decimal places. **(See Example 1.)**

19. $5^{1.1}$
20. $2^{\sqrt{3}}$
21. 10^{π}
22. $3^{4.8}$

23. $36^{-\sqrt{2}}$
24. $27^{-0.5126}$
25. $16^{-0.04}$
26. $8^{-0.61}$

27. Solve for x.
   a. $3^x = 9$
   b. $3^x = 27$
   c. Between what two consecutive integers must the solution to $3^x = 11$ lie?

28. Solve for x.
    a. $5^x = 125$
    b. $5^x = 625$
    c. Between what two consecutive integers must the solution to $5^x = 130$ lie?

29. Solve for x.
    a. $2^x = 16$
    b. $2^x = 32$
    c. Between what two consecutive integers must the solution to $2^x = 30$ lie?

30. Solve for x.
    a. $4^x = 16$
    b. $4^x = 64$
    c. Between what two consecutive integers must the solution to $4^x = 20$ lie?

31. For $f(x) = \left(\frac{1}{5}\right)^x$ find $f(0), f(1), f(2), f(-1)$, and $f(-2)$.

32. For $g(x) = \left(\frac{2}{3}\right)^x$ find $g(0), g(1), g(2), g(-1)$, and $g(-2)$.

33. For $h(x) = \pi^x$ use a calculator to find $h(0), h(1), h(-1), h(\sqrt{2})$, and $h(\pi)$. Round to 2 decimal places.

34. For $k(x) = (\sqrt{5})^x$ use a calculator to find $k(0), k(1), k(-1), k(\pi)$, and $k(\sqrt{2})$. Round to 2 decimal places.

35. For $r(x) = 3^{x+2}$ find $r(0), r(1), r(2), r(-1), r(-2)$, and $r(-3)$.

36. For $s(x) = 2^{2x-1}$ find $s(0), s(1), s(2), s(-1)$, and $s(-2)$.

Objective 3: Graphs of Exponential Functions

37. How do you determine whether the graph of $f(x) = b^x$ is increasing or decreasing?

38. For $f(x) = b^x$ $(b > 0, b \neq 1)$, find $f(0)$.

Graph the functions defined in Exercises 39–46. Plot at least three points for each function. **(See Example 2.)**

39. $f(x) = 4^x$

40. $g(x) = 6^x$

41. $m(x) = \left(\frac{1}{8}\right)^x$

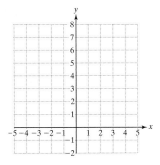
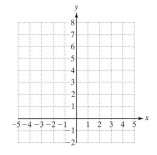
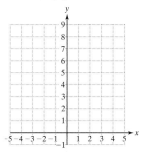

42. $n(x) = \left(\dfrac{1}{3}\right)^x$

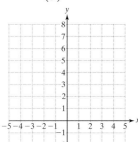

43. $h(x) = 2^{x+1}$

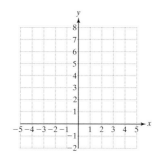

44. $k(x) = 5^{x-1}$

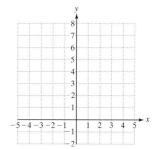

45. $g(x) = 5^{-x}$

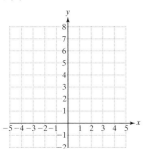

46. $f(x) = 2^{-x}$

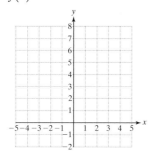

Objective 4: Applications of Exponential Functions

47. The half-life of the element radon (Rn 86) is 3.8 days. In a sample originally containing 1 g of radon, the amount left after t days is given by $A(t) = (0.5)^{t/3.8}$. (Round to 2 decimal places, if necessary.) **(See Example 3.)**

   a. How much radon will be present after 7.6 days?

   b. How much radon will be present after 10 days?

48. Nobelium, an element discovered in 1958, has a half-life of 10 min under certain conditions. In a sample containing 1 g of nobelium, the amount left after t min is given by $A(t) = (0.5)^{t/10}$. (Round to 3 decimal places.)

   a. How much nobelium is left after 5 min?

   b. How much nobelium is left after 1 hr?

49. Once an antibiotic is introduced to bacteria, the number of bacteria decreases exponentially. For example, beginning with 1 million bacteria, the amount present t days from the time penicillin is introduced is given by the function $A(t) = 1,000,000(2)^{-t/5}$. Rounding to the nearest thousand, determine how many bacteria are present after

   a. 2 days   b. 1 week   c. 2 weeks

50. Once an antibiotic is introduced to bacteria, the number of bacteria decreases exponentially. For example, beginning with 1 million bacteria, the amount present t days from the time streptomycin is introduced is given by the function $A(t) = 1,000,000(2)^{-t/10}$. Rounding to the nearest thousand, determine how many bacteria are present after

   a. 5 days   **b.** 1 week   **c.** 2 weeks

51. The population of Bangladesh was 141,340,000 in 2004 with an annual growth rate of 1.5%. **(See Example 4.)**

   a. Find a mathematical model that relates the population of Bangladesh as a function of the number of years after 2004.

   b. If the annual rate of increase remains the same, use this model to predict the population of Bangladesh in the year 2050. Round to the nearest million.

52. The population of Figi was 886,000 in 2004 with an annual growth rate of 1.07%.

   a. Find a mathematical model that relates the population of Figi as a function of the number of years after 2004.

   b. If the annual rate of increase remains the same, use this model to predict the population of Figi in the year 2050. Round to the nearest thousand.

53. Suppose $1000 is initially invested in an account and the value of the account grows exponentially. If the investment doubles in 7 years, then the amount in the account t years after the initial investment is given by

   $$A(t) = 1000(2)^{t/7}$$

   where t is expressed in years and $A(t)$ is the amount in the account.

   a. Find the amount in the account after 5 years.

   b. Find the amount in the account after 10 years.

   c. Find $A(0)$ and $A(7)$ and interpret the answers in the context of this problem.

54. Suppose $1500 is initially invested in an account and the value of the account grows exponentially. If the investment doubles in 8 years, then the amount in the account t years after the initial investment is given by

   $$A(t) = 1500(2)^{t/8}$$

   where t is expressed in years and $A(t)$ is the amount in the account.

   a. Find the amount in the account after 5 years.

   b. Find the amount in the account after 10 years.

   c. Find $A(0)$ and $A(8)$ and interpret the answers in the context of this problem.

Graphing Calculator Exercises

For Exercises 55–62, graph the functions on your calculator to support your solutions to the indicated exercises.

55. $f(x) = 4^x$ (see Exercise 39)

56. $g(x) = 6^x$ (see Exercise 40)

57. $m(x) = \left(\frac{1}{8}\right)^x$ (see Exercise 41)

58. $n(x) = \left(\frac{1}{3}\right)^x$ (see Exercise 42)

59. $h(x) = 2^{x+1}$ (see Exercise 43)

60. $k(x) = 5^{x-1}$ (see Exercise 44)

61. $g(x) = 5^{-x}$ (see Exercise 45)

62. $f(x) = 2^{-x}$ (see Exercise 46)

63. The function defined by $A(x) = 1000(2)^{x/7}$ represents the total amount A in an account x years after an initial investment of $1000.

    a. Graph $y = A(x)$ on the window where $0 \leq x \leq 25$ and $0 \leq y \leq 10{,}000$.

    b. Use *Zoom* and *Trace* to approximate the times required for the account to reach $2000, $4000, and $8000.

64. The function defined by $A(x) = 1500(2)^{x/8}$ represents the total amount A in an account x years after the initial investment of $1500.

    a. Graph $y = A(x)$ on the window where $0 \leq x \leq 40$ and $0 \leq y \leq 25{,}000$.

    b. Use *Zoom* and *Trace* to approximate the times required for the account to reach $3000, $6000, and $12,000.

Section 10.4 Logarithmic Functions

1. Definition of a Logarithmic Function

Consider the following equations in which the variable is located in the exponent of an expression. In some cases the solution can be found by inspection because the constant on the right-hand side of the equation is a perfect power of the base.

Equation	Solution
$5^x = 5$	$x = 1$
$5^x = 20$	$x = ?$
$5^x = 25$	$x = 2$
$5^x = 60$	$x = ?$
$5^x = 125$	$x = 3$

The equation $5^x = 20$ cannot be solved by inspection. However, we might suspect that x is between 1 and 2. Similarly, the solution to the equation $5^x = 60$ is between 2 and 3. To solve an exponential equation for an unknown exponent, we must use a new type of function called a logarithmic function.

Objectives
1. Definition of a Logarithmic Function
2. Evaluating Logarithmic Expressions
3. The Common Logarithmic Function
4. Graphs of Logarithmic Functions
5. Applications of the Common Logarithmic Function

Definition of a Logarithm Function

If x and b are positive real numbers such that $b \neq 1$, then $y = \log_b(x)$ is called the **logarithmic function** with base b and

$$y = \log_b x \quad \text{is equivalent to} \quad b^y = x$$

Note: In the expression $y = \log_b x$, y is called the **logarithm**, b is called the **base**, and x is called the **argument**.

The expression $y = \log_b x$ is equivalent to $b^y = x$ and indicates that *the logarithm y is the exponent to which b must be raised to obtain x*. The expression $y = \log_b x$ is called the logarithmic form of the equation, and the expression $b^y = x$ is called the exponential form of the equation.

The definition of a logarithmic function suggests a close relationship with an exponential function of the same base. In fact, a logarithmic function is the inverse of the corresponding exponential function. For example, the following steps find the inverse of the exponential function defined by $f(x) = b^x$.

$f(x) = b^x$

$y = b^x$ Replace $f(x)$ by y.

$x = b^y$ Interchange x and y.

$y = \log_b x$ Solve for y using the definition of a logarithmic function.

$f^{-1}(x) = \log_b x$ Replace y by $f^{-1}(x)$.

Concept Connections

1. Write the logarithmic equation in exponential form.
  $a = \log_r w$

Skill Practice

Rewrite the logarithmic equations in exponential form.

2. $\log_3 9 = 2$
3. $\log_{10}\left(\dfrac{1}{100}\right) = -2$
4. $\log_8 1 = 0$

example 1   Converting from Logarithmic Form to Exponential Form

Rewrite the logarithmic equations in exponential form.

a. $\log_2 32 = 5$ **b.** $\log_{10}\left(\dfrac{1}{1000}\right) = -3$ **c.** $\log_5 1 = 0$

Answers
1. $r^a = w$ 2. $3^2 = 9$
3. $10^{-2} = \dfrac{1}{100}$ 4. $8^0 = 1$

Solution:

Because the concept of a logarithm is new and unfamiliar, it may be advantageous to rewrite a logarithm in its equivalent exponential form.

Logarithmic Form		Exponential Form
a. $\log_2 32 = 5$	$\Leftrightarrow$	$2^5 = 32$
b. $\log_{10}\left(\dfrac{1}{1000}\right) = -3$	$\Leftrightarrow$	$10^{-3} = \dfrac{1}{1000}$
c. $\log_5 1 = 0$	$\Leftrightarrow$	$5^0 = 1$

2. Evaluating Logarithmic Expressions

Skill Practice

Evaluate the logarithmic expressions.

5. $\log_{10} 1000$
6. $\log_4\left(\dfrac{1}{16}\right)$
7. $\log_{1/3} 3$
8. $\log_x x$
9. $\log_b (b^{10})$
10. $\log_5 (\sqrt[3]{5})$

example 2 Evaluating Logarithmic Expressions

Evaluate the logarithmic expressions.

a. $\log_{10} 10{,}000$
b. $\log_5\left(\dfrac{1}{125}\right)$
c. $\log_{1/2}\left(\dfrac{1}{8}\right)$
d. $\log_b b$
e. $\log_c(c^7)$
f. $\log_3(\sqrt[4]{3})$

Solution:

a. $\log_{10} 10{,}000$ is the exponent to which 10 must be raised to obtain 10,000.

$y = \log_{10} 10{,}000$ Let y represent the value of the logarithm.

$10^y = 10{,}000$ Rewrite the expression in exponential form.

$y = 4$

Therefore, $\log_{10} 10{,}000 = 4$.

b. $\log_5\left(\dfrac{1}{125}\right)$ is the exponent to which 5 must be raised to obtain $\dfrac{1}{125}$.

$y = \log_5\left(\dfrac{1}{125}\right)$ Let y represent the value of the logarithm.

$5^y = \dfrac{1}{125}$ Rewrite the expression in exponential form.

$y = -3$

Therefore, $\log_5\left(\dfrac{1}{125}\right) = -3$.

c. $\log_{1/2}\left(\dfrac{1}{8}\right)$ is the exponent to which $\dfrac{1}{2}$ must be raised to obtain $\dfrac{1}{8}$.

$y = \log_{1/2}\left(\dfrac{1}{8}\right)$ Let y represent the value of the logarithm.

$\left(\dfrac{1}{2}\right)^y = \dfrac{1}{8}$ Rewrite the expression in exponential form.

$y = 3$

Therefore, $\log_{1/2}\left(\dfrac{1}{8}\right) = 3$.

Answers
5. 3   6. -2 7. -1
8. 1   9. 10   10. $\dfrac{1}{3}$

d. $\log_b b$ is the exponent to which b must be raised to obtain b.

$y = \log_b b$ Let y represent the value of the logarithm.

$b^y = b$ Rewrite the expression in exponential form.

$y = 1$

Therefore, $\log_b b = 1$.

e. $\log_c (c^7)$ is the exponent to which c must be raised to obtain c^7.

$y = \log_c (c^7)$ Let y represent the value of the logarithm.

$c^y = c^7$ Rewrite the expression in exponential form.

$y = 7$

Therefore, $\log_c (c^7) = 7$.

f. $\log_3 (\sqrt[4]{3}) = \log_3 (3^{1/4})$ is the exponent to which 3 must be raised to obtain $3^{1/4}$.

$y = \log_3 (3^{1/4})$ Let y represent the value of the logarithm.

$3^y = 3^{1/4}$ Rewrite the expression in exponential form.

$y = \dfrac{1}{4}$

Therefore, $\log_3 (\sqrt[4]{3}) = \dfrac{1}{4}$.

3. The Common Logarithmic Function

The logarithmic function with base 10 is called the **common logarithmic function** and is denoted by $y = \log (x)$. Notice that the base is not explicitly written but is understood to be 10. That is, $y = \log_{10} (x)$ is written simply as $y = \log (x)$.

Calculator Connections

On most calculators, the $\boxed{\log}$ key is used to compute logarithms with base 10. For example, we know the expression $\log (1{,}000{,}000) = 6$ because $10^6 = 1{,}000{,}000$. Use the $\boxed{\log}$ key to show this result on a calculator.

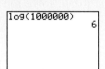

example 3   Evaluating Common Logarithms on a Calculator

Evaluate the common logarithms. Round the answers to 4 decimal places.

a. $\log 420$ **b.** $\log (8.2 \times 10^9)$ **c.** $\log (0.0002)$

Solution:

a. $\log 420 \approx 2.6232$

b. $\log (8.2 \times 10^9) \approx 9.9138$

c. $\log (0.0002) \approx -3.6990$

Skill Practice

Evaluate the common logarithms. Round answers to 4 decimal places.

11. $\log 1200$

12. $\log (6.3 \times 10^5)$

13. $\log (0.00025)$

Answers

11. 3.0792   **12.** 5.7993

13. -3.6021

4. Graphs of Logarithmic Functions

In Section 10.3 we studied the graphs of exponential functions. In this section, we will graph logarithmic functions. First, recall that $f(x) = \log_b x$ is the inverse of $g(x) = b^x$. Therefore, the graph of $y = f(x)$ is symmetric to the graph of $y = g(x)$ about the line $y = x$, as shown in Figures 10-12 and 10-13.

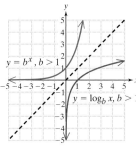
Figure 10-12

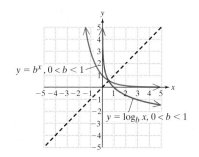
Figure 10-13

From Figures 10-12 and 10-13, we see that the range of $y = b^x$ is the set of positive real numbers. As expected, the domain of its inverse, the logarithmic function $y = \log_b x$, is also the set of positive real numbers. Therefore, the **domain of the logarithmic function** $y = \log_b x$ is the set of positive real numbers.

Skill Practice

Graph the functions

14. $y = \log_3 x$.

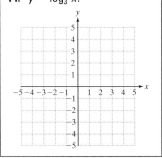

example 4 — Graphing Logarithmic Functions

Graph the functions and compare the graphs to examine the effect of the base on the shape of the graph.

a. $y = \log_2 x$ **b.** $y = \log x$

Solution:

We can write each equation in its equivalent exponential form and create a table of values (Table 10-5). To simplify the calculations, choose integer values of y and then solve for x.

$$y = \log_2 x \quad \text{or} \quad 2^y = x \qquad y = \log x \quad \text{or} \quad 10^y = x$$

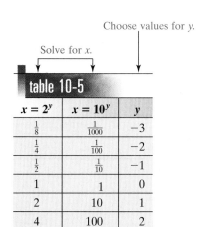

table 10-5

$x = 2^y$	$x = 10^y$	y
$\frac{1}{8}$	$\frac{1}{1000}$	-3
$\frac{1}{4}$	$\frac{1}{100}$	-2
$\frac{1}{2}$	$\frac{1}{10}$	-1
1	1	0
2	10	1
4	100	2
8	1000	3

Answer

14.

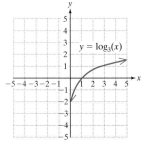

The graphs of $y = \log_2 x$ and $y = \log x$ are shown in Figure 10-14. Both graphs exhibit the same general behavior, and the steepness of the curve is affected by the base. The function $y = \log x$ requires a 10-fold increase in x to increase the y-value by 1 unit. The function $y = \log_2 x$ requires a 2-fold increase in x to increase the y-value by 1 unit. In addition, the following characteristics are true for both graphs.

- The domain is the set of real numbers x such that $x > 0$.
- The range is the set of real numbers.
- The y-axis is a vertical asymptote.
- Both graphs pass through the point $(1, 0)$.

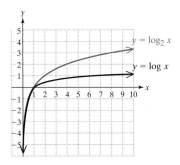

Figure 10-14

Example 4 illustrates that a logarithmic function with base $b > 1$ is an increasing function. In Example 5, we see that if the base b is between 0 and 1, the function decreases over its entire domain.

example 5 Graphing a Logarithmic Function

Graph $y = \log_{1/4}(x)$.

Solution:

The equation $y = \log_{1/4}(x)$ can be written in exponential form as $\left(\frac{1}{4}\right)^y = x$. By choosing several values for y, we can solve for x and plot the corresponding points (Table 10-6).

The expression $y = \log_{1/4}(x)$ defines a decreasing logarithmic function (Figure 10-15). Notice that the vertical asymptote, domain, and range are the same for both increasing and decreasing logarithmic functions.

Choose values for y.
Solve for x.

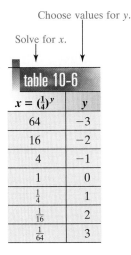

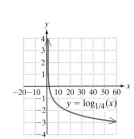

Figure 10-15

Skill Practice

15. Graph $y = \log_{1/2}(x)$.

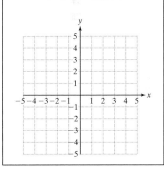

Answer

15.

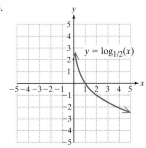

When graphing a logarithmic equation, it is helpful to know its domain.

Skill Practice

Find the domain of the functions.

16. $f(x) = \log_3 (x + 7)$
17. $g(x) = \log (4 - 8x)$

example 6 Identifying the Domain of a Logarithmic Function

Find the domain of the functions.

a. $f(x) = \log (4 - x)$ **b.** $g(x) = \log (2x + 6)$

Solution:

The domain of the function $y = \log_b (x)$ is the set of all positive real numbers. That is, the argument x must be greater than zero: $x > 0$.

a. $f(x) = \log (4 - x)$ The argument is $4 - x$.

$4 - x > 0$ The argument of the logarithm must be greater than zero.

$-x > -4$ Solve for x.

$x < 4$ Divide by -1 and reverse the inequality sign.

The domain is $(-\infty, 4)$.

b. $g(x) = \log (2x + 6)$ The argument is $2x + 6$.

$2x + 6 > 0$ The argument of the logarithm must be greater than zero.

$2x > -6$ Solve for x.

$x > -3$

The domain is $(-3, \infty)$.

Calculator Connections

The graphs of $Y_1 = \log (4 - x)$ and $Y_2 = \log (2x + 6)$ are shown here and can be used to confirm the solutions to Example 6. Notice that each function has a vertical asymptote at the value of x where the argument equals zero.

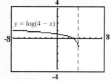 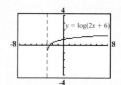

The general shape and important features of exponential and logarithmic functions are summarized as follows.

Answers

16. Domain: $(-7, \infty)$
17. Domain: $\left(-\infty, \dfrac{1}{2}\right)$

Graphs of Exponential and Logarithmic Functions—A Summary

Exponential Functions

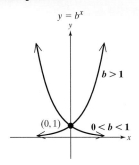

$y = b^x$

Logarithmic Functions

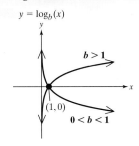

$y = \log_b(x)$

Domain: $(-\infty, \infty)$
Range: $(0, \infty)$
Horizontal asymptote: $y = 0$
Passes through $(0, 1)$
If $b > 1$, the function is increasing.
If $0 < b < 1$, the function is decreasing.

Domain: $(0, \infty)$
Range: $(-\infty, \infty)$
Vertical asymptote: $x = 0$
Passes through $(1, 0)$
If $b > 1$, the function is increasing.
If $0 < b < 1$, the function is decreasing.

Notice that the roles of x and y are interchanged for the functions $y = b^x$ and $b^y = x$. Therefore, it is not surprising that the domain and range are reversed between exponential and logarithmic functions. Moreover, an exponential function passes through $(0, 1)$, whereas a logarithmic function passes through $(1, 0)$. An exponential function has a horizontal asymptote at $y = 0$, whereas a logarithmic function has a vertical asymptote at $x = 0$.

5. Applications of the Common Logarithmic Function

example 7 — Applying a Common Logarithm to Compute pH

The pH (hydrogen potential) of a solution is defined as

$$pH = -\log[H^+]$$

where $[H^+]$ represents the concentration of hydrogen ions in moles per liter (mol/L). The pH scale ranges from 0 to 14. The midpoint of this range, 7, represents a neutral solution. Values below 7 are progressively more acidic, and values above 7 are progressively more alkaline. Based on the equation $pH = -\log[H^+]$, a 1-unit change in pH means a 10-fold change in hydrogen ion concentration.

a. Normal rain has a pH of 5.6. However, in some areas of the northeastern United States the rainwater is more acidic. What is the pH of a rain sample for which the concentration of hydrogen ions is 0.0002 mol/L?

b. Find the pH of household ammonia if the concentration of hydrogen ions is 1.0×10^{-11} mol/L.

Skill Practice

18. A new all-natural shampoo on the market claims its hydrogen ion concentration is 5.88×10^{-7} mol/L. Use the formula $pH = -\log[H^+]$ to calculate the pH level of the shampoo.

Answer

18. $pH \approx 6.23$

Solution:

a. $\text{pH} = -\log[\text{H}^+]$

   $= -\log(0.0002)$ Substitute $[\text{H}^+] = 0.0002$.

   ≈ 3.7 (To compare this value with a familiar substance, note that the pH of orange juice is roughly 3.5.)

b. $\text{pH} = -\log[\text{H}^+]$

   $= -\log(1.0 \times 10^{-11})$ Substitute $[\text{H}^+] = 1.0 \times 10^{-11}$.

   $= -\log(10^{-11})$

   $= -(-11)$

   $= 11$

   The pH of household ammonia is 11.

Skill Practice

The memory model for a student's score on a statistics test t months after completion of the course in statistics is approximated by

$S(t) = 92 - 28 \log(t + 1)$

19. What was the student's score at the time the course was completed ($t = 0$)?

20. What was her score after 1 month?

21. What was the score after 2 months?

---

example 8   Applying Logarithmic Functions to a Memory Model

One method of measuring a student's retention of material after taking a course is to retest the student at specified time intervals after the course has been completed. A student's score on a calculus test t months after completing a course in calculus is approximated by

$$S(t) = 85 - 25 \log(t + 1)$$

where t is the time in months after completing the course and $S(t)$ is the student's score as a percent.

a. What was the student's score at $t = 0$?

b. What was the student's score after 2 months?

c. What was the student's score after 1 year?

Solution:

a. $S(t) = 85 - 25 \log(t + 1)$

   $S(0) = 85 - 25 \log(0 + 1)$ Substitute $t = 0$.

   $= 85 - 25 \log 1$ $\log 1 = 0$ because $10^0 = 1$.

   $= 85 - 25(0)$

   $= 85 - 0$

   $= 85$ The student's score at the time the course was completed was 85%.

b. $S(t) = 85 - 25 \log(t + 1)$

   $S(2) = 85 - 25 \log(2 + 1)$

   $= 85 - 25 \log(3)$ Use a calculator to approximate $\log(3)$.

   ≈ 73.1 The student's score dropped to 73.1%.

c. $S(t) = 85 - 25 \log(t + 1)$

   $S(12) = 85 - 25 \log(12 + 1)$

   $= 85 - 25 \log(13)$ Use a calculator to approximate $\log(13)$.

   ≈ 57.2 The student's score dropped to 57.2%.

Answers

19. 92   20. 83.6   21. 78.6

section 10.4 Practice Exercises

Boost your GRADE at mathzone.com!

- Practice Problems
- Self-Tests
- NetTutor
- e-Professors
- Videos

Study Skills Exercises

1. Sometimes the problems on a test do not appear in the same order as the concepts appeared in the text. To better prepare for a test, try to practice problems taken from the book and placed in random order. Choose 10 problems from various sections of the chapter review, randomize the order, and use the problems to review for the test. Repeat the process several times for additional practice.

2. Define the key terms.

   a. Logarithmic function   b. Logarithm   c. Base of a logarithm

   d. Argument   e. Common logarithmic function   f. Domain of a logarithmic function

Review Exercises

3. For which graph of $y = b^x$ is $0 < b < 1$?

   i.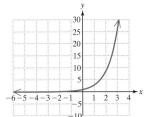

   ii.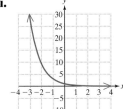

4. Let $f(x) = 6^x$.

   a. Find $f(-2), f(-1), f(0), f(1)$, and $f(2)$.

   b. Graph $y = f(x)$ by plotting the points found in part (a).

   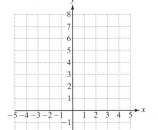

5. Let $g(x) = 3^x$.

   a. Find $g(-2), g(-1), g(0), g(1)$, and $g(2)$.

   b. Graph $y = g(x)$ by plotting the points found in part (a).

   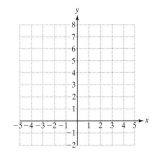

6. Let $r(x) = \left(\frac{3}{4}\right)^x$.

   a. Find $r(-2), r(-1), r(0), r(1)$, and $r(2)$.

   b. Graph $y = r(x)$ by plotting the points found in part (a).

   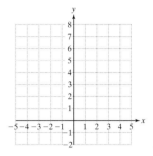

7. Let $s(x) = \left(\frac{2}{5}\right)^x$.

   a. Find $s(-2), s(-1), s(0), s(1)$, and $s(2)$.

   b. Graph $y = s(x)$ by plotting the points found in part (a).

   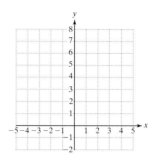

Objective 1: Definition of a Logarithmic Function

8. For the equation $y = \log_b(x)$, identify the base, the argument, and the logarithm.

9. Rewrite the equation in exponential form. $y = \log_b(x)$

For Exercises 10–21, write the equation in logarithmic form.

10. $3^x = 81$

11. $10^3 = 1000$

12. $5^2 = 25$

13. $8^{1/3} = 2$

14. $7^{-1} = \frac{1}{7}$

15. $8^{-2} = \frac{1}{64}$

16. $b^x = y$

17. $b^y = x$

18. $e^x = y$

19. $e^y = x$

20. $\left(\frac{1}{3}\right)^{-2} = 9$

21. $\left(\frac{5}{2}\right)^{-1} = \frac{2}{5}$

For Exercises 22–33, write the equation in exponential form. (See Example 1.)

22. $\log_5 625 = 4$

23. $\log_{125} 25 = \frac{2}{3}$

24. $\log_{10} 0.0001 = -4$

25. $\log_{25}\left(\frac{1}{5}\right) = -\frac{1}{2}$

26. $\log_6 36 = 2$

27. $\log_2 128 = 7$

28. $\log_b 15 = x$

29. $\log_b 82 = y$

30. $\log_3 5 = x$

31. $\log_2 7 = x$

32. $\log_{1/4} x = 10$

33. $\log_{1/2} x = 6$

Objective 2: Evaluating Logarithmic Expressions

For Exercises 34–49, find the logarithms without the use of a calculator. (See Example 2.)

34. $\log_7 49$ **35.** $\log_3 27$ **36.** $\log_{10} 0.1$ **37.** $\log_2 \left(\dfrac{1}{16}\right)$

38. $\log_{16} 4$ **39.** $\log_8 2$ **40.** $\log_{7/2} 1$ **41.** $\log_{1/2} 2$

42. $\log_3 3^5$ **43.** $\log_9 9^3$ **44.** $\log_{10} 10$ **45.** $\log_7 1$

46. $\log_a (a^3)$ **47.** $\log_r (r^4)$ **48.** $\log_x \sqrt{x}$ **49.** $\log_y \sqrt[3]{y}$

Objective 3: The Common Logarithmic Function

For Exercises 50–58, find the common logarithm without the use of a calculator.

50. $\log 10$ **51.** $\log 100$ **52.** $\log 1000$

53. $\log 10{,}000$ **54.** $\log (1.0 \times 10^6)$ **55.** $\log 0.1$

56. $\log 0.01$ **57.** $\log 0.001$ **58.** $\log (1.0 \times 10^{-6})$

For Exercises 59–70, use a calculator to approximate the logarithms. Round to 4 decimal places. (See Example 3.)

59. $\log 6$ **60.** $\log 18$ **61.** $\log \pi$ **62.** $\log \left(\dfrac{1}{8}\right)$

63. $\log \left(\dfrac{1}{32}\right)$ **64.** $\log \sqrt{5}$ **65.** $\log (0.0054)$ **66.** $\log (0.0000062)$

67. $\log (3.4 \times 10^5)$ **68.** $\log (4.78 \times 10^9)$ **69.** $\log (3.8 \times 10^{-8})$ **70.** $\log (2.77 \times 10^{-4})$

71. Given: $\log (10) = 1$ and $\log (100) = 2$
   a. Estimate $\log 93$.

   b. Estimate $\log 12$.

   c. Evaluate the logarithms in parts (a) and (b) on a calculator and compare to your estimates.

72. Given: $\log \left(\frac{1}{10}\right) = -1$ and $\log 1 = 0$
   a. Estimate $\log \left(\frac{9}{10}\right)$.

   b. Estimate $\log \left(\frac{1}{5}\right)$.

   c. Evaluate the logarithms in parts (a) and (b) on a calculator and compare to your estimates.

Objective 4: Graphs of Logarithmic Functions

73. Let $f(x) = \log_4(x)$.

    a. Find the values of $f(\frac{1}{64}), f(\frac{1}{16}), f(\frac{1}{4}), f(1), f(4),$ $f(16),$ and $f(64)$.

    b. Graph $y = f(x)$ by plotting the points found in part (a).

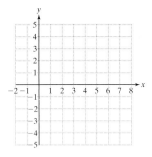

74. Let $g(x) = \log_2(x)$.

    a. Find the values of $g(\frac{1}{8}), g(\frac{1}{4}), g(\frac{1}{2}), g(1), g(2),$ $g(4),$ and $g(8)$.

    b. Graph $y = g(x)$ by plotting the points found in part (a).

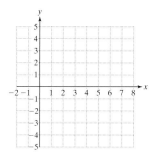

Graph the logarithmic functions in Exercises 75–78 by writing the function in exponential form and making a table of points. (See Examples 4–5.)

75. $y = \log_3 x$

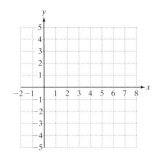

76. $y = \log_5 x$

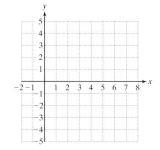

77. $y = \log_{1/2} x$

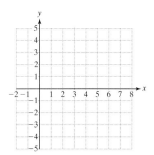

78. $y = \log_{1/3} x$

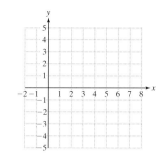

For Exercises 79–84, find the domain of the function and express the domain in interval notation. **(See Example 6.)**

79. $y = \log_7 (x - 5)$

80. $y = \log_3 (2x + 1)$

81. $y = \log_3 (x + 1.2)$

82. $y = \log \left(x - \dfrac{1}{2} \right)$

83. $y = \log (x^2)$

84. $y = \log (x^2 + 1)$

Objective 5: Applications of the Common Logarithmic Function

For Exercises 85–86, use the formula pH $= -\log[\text{H}^+]$, where $[\text{H}^+]$ represents the concentration of hydrogen ions in moles per liter. Round to the hundredths place. **(See Example 7.)**

85. Normally, the level of hydrogen ions in the blood is approximately 4.47×10^{-8} mol/L. Find the pH level of blood.

86. The normal pH level for streams and rivers is between 6.5 and 8.5. A high level of bacteria in a particular stream caused environmentalists to test the water. The level of hydrogen ions was found to be 0.006 mol/L. What is the pH level of the stream?

87. A graduate student in education is doing research to compare the effectiveness of two different teaching techniques designed to teach vocabulary to sixth-graders. The first group of students (group 1) was taught with method I, in which the students worked individually to complete the assignments in a workbook. The second group (group 2) was taught with method II, in which the students worked cooperatively in groups of four to complete the assignments in the same workbook.

None of the students knew any of the vocabulary words before the study began. After completing the assignments in the workbook, the students were then tested on the vocabulary at 1-month intervals to assess how much material they had retained over time. The students' average score t months after completing the assignments are given by the following functions:

Method I: $S_1(t) = 91 - 30 \log (t + 1)$, where t is the time in months and $S_1(t)$ is the average score of students in group 1.

Method II: $S_2(t) = 88 - 15 \log (t + 1)$, where t is the time in months and $S_2(t)$ is the average score of students in group 2.

a. Complete the table to find the average scores for each group of students after the indicated number of months. Round to 1 decimal place. **(See Example 8.)**

t (months)	0	1	2	6	12	24
$S_1(t)$						
$S_2(t)$						

b. Based on the table of values, what were the average scores for each group immediately after completion of the assigned material ($t = 0$)?

c. Based on the table of values, which teaching method helped students retain the material better for a long time?

88. Generally, the more money a company spends on advertising, the higher the sales. Let a represent the amount of money spent on advertising (in $100s). Then the amount of money in sales $S(a)$ (in $1000s) is given by

$$S(a) = 10 + 20 \log(a + 1)$$

where $a \geq 0$.

   a. The value of $S(1) \approx 16.0$ means that if $100 is spent on advertising, $16,000 is returned in sales. Find the values of $S(11)$, $S(21)$, and $S(31)$. Round to 1 decimal place. Interpret the meaning of each function value in the context of this problem.

   b. The graph of $y = S(a)$ is shown here. Use the graph and your answers from part (a) to explain why the money spent in advertising becomes less "efficient" as it is used in larger amounts.

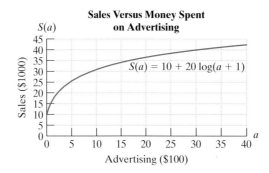

Graphing Calculator Exercises

For Exercises 89–94, graph the function on an appropriate viewing window. From the graph, identify the domain of the function and the location of the vertical asymptote.

89. $y = \log(x + 6)$

90. $y = \log(2x + 4)$

91. $y = \log(0.5x - 1)$

92. $y = \log(x + 8)$

93. $y = \log(2 - x)$

94. $y = \log(3 - x)$

chapter 10 | midchapter review

For Exercises 1–6, use $f(x) = 5 - x$ and $g(x) = x^2 + x$ to find the functions.

1. $(f + g)(x)$
2. $(f - g)(x)$
3. $(f \cdot g)(x)$
4. $\left(\dfrac{f}{g}\right)(x)$
5. $(f \circ g)(x)$
6. $(g \circ f)(x)$

7. Determine if the function shown is one-to-one. If it is not, explain why. $\{(3, 5), (9, -1), (0, 5), (6, 20)\}$

8. Given $f(x) = \dfrac{1}{2}x + 2$, find the inverse.

9. a. For $f(x) = 7^x$ find $f(2), f(-1), f(0),$ and $f(1)$.

   b. For $g(x) = \log_7 x$, evaluate without using a calculator.
   $g(7), g(49), g\left(\dfrac{1}{7}\right), g(1)$

10. a. For $h(x) = 10^x$ find $h(2), h(-1), h(0),$ and $h(1)$.

    b. For $k(x) = \log x$ use a calculator to find $k(0.5)$, $k(1)$, and $k(3)$. Round to 4 decimal places.

11. a. Graph $y = 3^x$

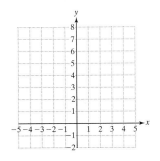

    b. Identify the domain and range.

12. a. Graph $y = \log_3 x$

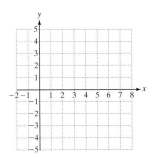

    b. Identify the domain and range.

13. Convert the exponential equation to logarithmic form. $x^y = z$

14. Convert the logarithmic equation to exponential form. $\log_p q = t$

For Exercises 15–19, evaluate the exponential and logarithmic expressions without the use of a calculator.

15. $\log 1000$
16. $\log_3 27$
17. $\log 10^{-6}$
18. $\log 1$
19. $\log_2 \left(\dfrac{1}{16}\right)$

Objectives

1. Properties of Logarithms
2. Expanded Logarithmic Expressions
3. Single Logarithmic Expressions

section 10.5 Properties of Logarithms

1. Properties of Logarithms

You have already been exposed to certain properties of logarithms that follow directly from the definition. Recall

$$y = \log_b x \quad \text{is equivalent to} \quad b^y = x \quad \text{for } x > 0, b > 0, \text{ and } b \neq 1$$

The following properties follow directly from the definition.

$$\log_b 1 = 0 \quad \text{Property 1}$$
$$\log_b b = 1 \quad \text{Property 2}$$
$$\log_b (b^p) = p \quad \text{Property 3}$$
$$b^{\log_b(x)} = x \quad \text{Property 4}$$

Skill Practice

Use the properties of logarithms to simplify the expressions.

1. $\log_5 1 + \log_5 5$
2. $15^{\log_{15} 7}$
3. $\log_{1/3}\left(\dfrac{1}{3}\right)^c$

example 1 Applying the Properties of Logarithms to Simplify Expressions

Use the properties of logarithms to simplify the expressions. Assume that all variable expressions within the logarithms represent positive real numbers.

a. $\log_8 8 + \log_8 1$ **b.** $10^{\log(x+2)}$ **c.** $\log_{1/2}\left(\dfrac{1}{2}\right)^x$

Solution:

a. $\log_8 8 + \log_8 1 = 1 + 0 = 1$ Properties 2 and 1

b. $10^{\log(x+2)} = x + 2$ Property 4

c. $\log_{1/2}\left(\dfrac{1}{2}\right)^x = x$ Property 3

Three additional properties are useful when simplifying logarithmic expressions. The first is the product property for logarithms.

Product Property for Logarithms

Let b, x, and y be positive real numbers where $b \neq 1$. Then

$$\log_b (xy) = \log_b x + \log_b y$$

The logarithm of a product equals the sum of the logarithms of the factors.

Proof:

Let $M = \log_b x$, which implies $b^M = x$.

Let $N = \log_b y$, which implies $b^N = y$.

Then $xy = b^M b^N = b^{M+N}$.

Writing the expression $xy = b^{M+N}$ in logarithmic form, we have,

$$\log_b (xy) = M + N$$
$$\log_b (xy) = \log_b x + \log_b y \checkmark$$

Answers

1. 1 2. 7 3. c

To demonstrate the product property for logarithms, simplify the following expressions by using the order of operations.

$$\log_3 (3 \cdot 9) \stackrel{?}{=} \log_3 3 + \log_3 9$$
$$\log_3 27 \stackrel{?}{=} 1 + 2$$
$$3 = 3 \checkmark$$

Quotient Property for Logarithms

Let b, x, and y be positive real numbers where $b \neq 1$. Then

$$\log_b \left(\frac{x}{y}\right) = \log_b x - \log_b y$$

The logarithm of a quotient equals the difference of the logarithms of the numerator and denominator.

The proof of the quotient property for logarithms is similar to the proof of the product property and is omitted here. To demonstrate the quotient property for logarithms, simplify the following expressions by using the order of operations.

$$\log \left(\frac{1,000,000}{100}\right) \stackrel{?}{=} \log (1,000,000) - \log (100)$$
$$\log (10,000) \stackrel{?}{=} 6 - 2$$
$$4 = 4 \checkmark$$

Power Property for Logarithms

Let b and x be positive real numbers where $b \neq 1$. Let p be any real number. Then

$$\log_b (x^p) = p \log_b x$$

Proof:

Let $M = \log_b (x)$, which implies $b^M = x$.

Raise both sides to the power p: $(b^M)^p = (x)^p$ or equivalently $b^{Mp} = (x^p)$.

Write the expression $b^{Mp} = x^p$ in logarithmic form: $\log_b (x^p) = Mp = pM$ or equivalently $\log_b (x^p) = p \log_b (x)$. ✓

To demonstrate the power property for logarithms, simplify the following expressions by using the order of operations.

$$\log_4 (4^2) \stackrel{?}{=} 2 \log_4 (4)$$
$$2 \stackrel{?}{=} 2 \cdot 1$$
$$2 = 2 \checkmark$$

The properties of logarithms are summarized in the box.

Concept Connections

4. Use the product property of logarithms to simplify $\log_6 3 + \log_6 12$.
5. Use the quotient property for logarithms to simplify $\log_4 24 - \log_4 6$.
6. Use the power property for logarithms to simplify $\log_7 49^{12}$.

Answers
4. 2   5. 1   6. 24

Properties of Logarithms

Let b, x, and y be positive real numbers where $b \neq 1$, and let p be a real number. Then the following **properties of logarithms** are true.

1. $\log_b 1 = 0$
2. $\log_b b = 1$
3. $\log_b (b^p) = p$
4. $b^{\log_b (x)} = x$
5. $\log_b (xy) = \log_b x + \log_b y$ Product property for logarithms
6. $\log_b \left(\dfrac{x}{y}\right) = \log_b x - \log_b y$ Quotient property for logarithms
7. $\log_b (x^p) = p \log_b x$ Power property for logarithms

2. Expanded Logarithmic Expressions

In many applications it is advantageous to expand a logarithm into a sum or difference of simpler logarithms.

Skill Practice

Write the expression as the sum or difference of logarithms of a, b, and c. Assume all variables represent positive real numbers.

7. $\log_5 \left(\dfrac{a^2 b^3}{c}\right)$

8. $\log \dfrac{(a-b)^2}{\sqrt{10}}$

9. $\log_3 \sqrt[4]{\dfrac{a}{bc^5}}$

example 2   Writing a Logarithmic Expression in Expanded Form

Write the expressions as the sum or difference of logarithms of x, y, and z. Assume all variables represent positive real numbers.

a. $\log_3 \left(\dfrac{xy^3}{z^2}\right)$ **b.** $\log \left(\dfrac{\sqrt{x+y}}{10}\right)$ **c.** $\log_b \sqrt[5]{\dfrac{x^4}{yz^3}}$

Solution:

a. $\log_3 \left(\dfrac{xy^3}{z^2}\right)$

$= \log_3 xy^3 - \log_3 z^2$ Quotient property for logarithms (property 6)

$= [\log_3 x + \log_3 y^3] - \log_3 z^2$ Product property for logarithms (property 5)

$= \log_3 x + 3 \log_3 y - 2 \log_3 z$ Power property for logarithms (property 7)

b. $\log \left(\dfrac{\sqrt{x+y}}{10}\right)$

$= \log (\sqrt{x+y}) - \log (10)$ Quotient property for logarithms (property 6)

$= \log (x+y)^{1/2} - 1$ Write $\sqrt{x+y}$ as $(x+y)^{1/2}$ and simplify $\log 10 = 1$.

$= \dfrac{1}{2} \log (x+y) - 1$ Power property for logarithms (property 7)

Answers

7. $2 \log_5 a + 3 \log_5 b - \log_5 c$
8. $2 \log (a-b) - \dfrac{1}{2}$
9. $\dfrac{1}{4} \log_3 a - \dfrac{1}{4} \log_3 b - \dfrac{5}{4} \log_3 c$

c. $\log_b \sqrt[5]{\dfrac{x^4}{yz^3}}$

$ = \log_b \left(\dfrac{x^4}{yz^3}\right)^{1/5}$ Write $\sqrt[5]{\dfrac{x^4}{yz^3}}$ as $\left(\dfrac{x^4}{yz^3}\right)^{1/5}$.

$ = \dfrac{1}{5}\log_b \left(\dfrac{x^4}{yz^3}\right)$ Power property for logarithms (property 7)

$ = \dfrac{1}{5}[\log_b x^4 - \log_b (yz^3)]$ Quotient property for logarithms (property 6)

$ = \dfrac{1}{5}[\log_b x^4 - (\log_b y + \log_b z^3)]$ Product property for logarithms (property 5)

$ = \dfrac{1}{5}[\log_b x^4 - \log_b y - \log_b z^3]$ Distributive property

$ = \dfrac{1}{5}[4\log_b x - \log_b y - 3\log_b z]$ Power property for logarithms (property 7)

or $ \dfrac{4}{5}\log_b x - \dfrac{1}{5}\log_b y - \dfrac{3}{5}\log_b z$

3. Single Logarithmic Expressions

In some applications it is necessary to write a sum or difference of logarithms as a single logarithm.

example 3 — Writing a Sum or Difference of Logarithms as a Single Logarithm

Rewrite the expressions as a single logarithm, and simplify the result, if possible. Assume all variable expressions within the logarithms represent positive real numbers.

a. $\log_2 560 - \log_2 7 - \log_2 5$
b. $2\log x - \dfrac{1}{2}\log y + 3\log z$
c. $\dfrac{1}{2}[\log_5 (x^2 - y^2) - \log_5 (x + y)]$

Skill Practice

Write the expression as a single logarithm, and simplify the result, if possible.

10. $\log_3 54 + \log_3 10 - \log_3 20$

11. $3\log x + \dfrac{1}{3}\log y - 2\log z$

12. $\dfrac{1}{2}[\log_2 (a^2 + ab) - \log_2 a]$

Solution:

a. $\log_2 560 - \log_2 7 - \log_2 5$

$ = \log_2 560 - (\log_2 7 + \log_2 5)$ Factor out -1 from the last two terms.

$ = \log_2 560 - \log_2 (7 \cdot 5)$ Product property for logarithms (property 5)

$ = \log_2 \left(\dfrac{560}{7 \cdot 5}\right)$ Quotient property for logarithms (property 6)

$ = \log_2 (16)$ Simplify inside parentheses.

$ = 4$

Answers

10. 3   **11.** $\log \left(\dfrac{x^3 \sqrt[3]{y}}{z^2}\right)$

12. $\log_2 \sqrt{a + b}$

b. $2 \log x - \dfrac{1}{2} \log y + 3 \log z$

$\quad = \log x^2 - \log y^{1/2} + \log z^3$ Power property for logarithms (property 7)

$\quad = \log x^2 + \log z^3 - \log y^{1/2}$ Group terms with positive coefficients.

$\quad = \log (x^2 z^3) - \log (y^{1/2})$ Product property for logarithms (property 5)

$\quad = \log \left(\dfrac{x^2 z^3}{y^{1/2}}\right) \text{ or } \log \left(\dfrac{x^2 z^3}{\sqrt{y}}\right)$ Quotient property for logarithms (property 6)

c. $\dfrac{1}{2} [\log_5 (x^2 - y^2) - \log_5 (x + y)]$

$\quad = \dfrac{1}{2} \log_5 \left(\dfrac{x^2 - y^2}{x + y}\right)$ Quotient property for logarithms (property 6)

$\quad = \dfrac{1}{2} \log_5 \left[\dfrac{(x + y)(x - y)}{x + y}\right]$ Factor and simplify within the parentheses.

$\quad = \dfrac{1}{2} \log_5 (x - y)$

$\quad = \log_5 (x - y)^{1/2} \text{ or } \log_5 \sqrt{x - y}$ Power property for logarithms (property 7)

It is important to note that the properties of logarithms may be used to write a single logarithm as a sum or difference of logarithms. Furthermore, the properties may be used to write a sum or difference of logarithms as a single logarithm. In either case, these operations may change the domain.

For example, consider the function $y = \log_b (x^2)$. Using the power property for logarithms, we have $y = 2 \log_b (x)$. Consider the domain of each function:

$y = \log_b (x^2)$ Domain: $(-\infty, 0) \cup (0, \infty)$

$y = 2 \log_b (x)$ Domain: $(0, \infty)$

These two functions are equivalent only for values of x in the intersection of the two domains, that is, for $(0, \infty)$.

section 10.5 Practice Exercises

Boost your GRADE at mathzone.com!

- Practice Problems
- Self-Tests
- NetTutor
- e-Professors
- Videos

Study Skills Exercises

1. On 3 × 5 cards, write a logarithm property on one side and give two examples of that property on the other side. Keep the cards with you so that when you have a free moment, you can look at the cards to memorize the properties and how they work.

Section 10.5 Properties of Logarithms 749

2. Define the key terms.
   a. Product property of logarithms
   b. Quotient property of logarithms
   c. Power property of logarithms

Review Exercises

For Exercises 3–6, find the values of the logarithmic and exponential expressions without using a calculator.

3. 8^{-2}
4. $\log 10{,}000$
5. $\log_2 32$
6. 6^{-1}

For Exercises 7–10, approximate the values of the logarithmic and exponential expressions by using a calculator.

7. $(\sqrt{2})^{\pi}$
8. $\log 8$
9. $\log 27$
10. $\pi^{\sqrt{2}}$

For Exercises 11–14, match the function with the appropriate graph.

11. $f(x) = 4^x$
12. $q(x) = \left(\dfrac{1}{5}\right)^x$
13. $h(x) = \log_5 x$
14. $k(x) = \log_{1/3} x$

a.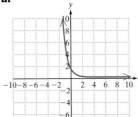
b.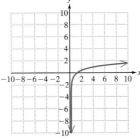
c.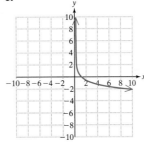
d.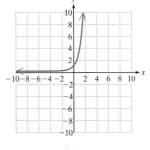

Objective 1: Properties of Logarithms

15. Property 1 of logarithms states that $\log_b 1 = 0$. Write an example of this property.

16. Property 2 of logarithms states that $\log_b b = 1$. Write an example of this property.

17. Property 3 of logarithms states that $\log_b (b^n) = n$. An example is $\log_6 (6^3) = 3$. Write another example of this property.

18. Property 4 of logarithms states that $b^{\log_b (x)} = x$. An example is $2^{\log_2 5} = 5$. Write another example of this property.

For Exercises 19–40, evaluate each expression. (See Example 1.)

19. $\log_3 3$
20. $\log 10$
21. $\log_5 (5^4)$
22. $\log_4 (4^5)$

23. $6^{\log_6 11}$
24. $7^{\log_7 2}$
25. $\log (10^3)$
26. $\log_6 (6^3)$

27. $\log_3 1$

28. $\log_8 1$

29. $10^{\log 9}$

30. $8^{\log_8 5}$

31. $\log_{1/2} 1$

32. $\log_{1/3}\left(\dfrac{1}{3}\right)$

33. $\log_2 1 + \log_2 (2^3)$

34. $\log 10^4 + \log 10$

35. $\log_4 4 + \log_2 1$

36. $\log_7 7 + \log_4 (4^2)$

37. $\log_{1/4}\left(\dfrac{1}{4}\right)^{2x}$

38. $\log_{2/3}\left(\dfrac{2}{3}\right)^p$

39. $\log_a (a^4)$

40. $\log_y (y^2)$

41. Compare the expressions by approximating their values on a calculator. Which two expressions appear to be equivalent?

   a. $\log(3 \cdot 5)$ b. $\log 3 \cdot \log 5$ c. $\log 3 + \log 5$

42. Compare the expressions by approximating their values on a calculator. Which two expressions appear to be equivalent?

   a. $\log\left(\dfrac{6}{5}\right)$ b. $\dfrac{\log 6}{\log 5}$ c. $\log 6 - \log 5$

43. Compare the expressions by approximating their values on a calculator. Which two expressions appear to be equivalent?

   a. $\log(20^2)$ b. $[\log 20]^2$ c. $2\log 20$

44. Compare the expressions by approximating their values on a calculator. Which two expressions appear to be equivalent?

   a. $\log \sqrt{4}$ b. $\dfrac{1}{2}\log 4$ c. $\sqrt{\log 4}$

Objective 2: Expanded Logarithmic Expressions

For Exercises 45–62, expand into sums and/or differences of logarithms. Assume all variables represent positive real numbers. **(See Example 2.)**

45. $\log_3\left(\dfrac{x}{5}\right)$

46. $\log_2\left(\dfrac{y}{z}\right)$

47. $\log(2x)$

48. $\log_6(xyz)$

49. $\log_5(x^4)$

50. $\log_7(z^{1/3})$

51. $\log_4\left(\dfrac{ab}{c}\right)$

52. $\log_2\left(\dfrac{x}{yz}\right)$

53. $\log_b\left(\dfrac{\sqrt{x}y}{z^3 w}\right)$

54. $\log\left(\dfrac{a\sqrt[3]{b}}{cd^2}\right)$

55. $\log_2\left(\dfrac{x+1}{y^2\sqrt{z}}\right)$

56. $\log\left(\dfrac{a+1}{b\sqrt[3]{c}}\right)$

57. $\log\left(\sqrt[3]{\dfrac{ab^2}{c}}\right)$

58. $\log_5\left(\sqrt[4]{\dfrac{w^3z}{x^2}}\right)$

59. $\log\left(\dfrac{1}{w^5}\right)$

60. $\log_3\left(\dfrac{1}{z^4}\right)$

61. $\log_b\left(\dfrac{\sqrt{a}}{b^3c}\right)$

62. $\log_x\left(\dfrac{x}{y\sqrt{z}}\right)$

Objective 3: Single Logarithmic Expressions

For Exercises 63–76, write the expressions as a single logarithm. Assume all variables represent positive real numbers. (See Example 3.)

63. $\log C + \log A + \log B + \log I + \log N$

64. $\log x + \log y - \log z$

65. $2\log_3 x - 3\log_3 y + \log_3 z$

66. $\dfrac{1}{3}\log_2 x - 5\log_2 y - 3\log_2 z$

67. $2\log_3 a - \dfrac{1}{4}\log_3 b + \log_3 c$

68. $\log_5 a - \dfrac{1}{2}\log_5 b - 3\log_5 c$

69. $\log_b x - 3\log_b x + 4\log_b x$

70. $2\log_3 z + \log_3 z - \dfrac{1}{2}\log_3 z$

71. $5\log_8 a - \log_8 1 + \log_8 8$

72. $\log_2 2 + 2\log_2 b - \log_2 1$

73. $2\log(x+6) + \dfrac{1}{3}\log y - 5\log z$

74. $\dfrac{1}{4}\log(a+1) - 2\log b - 4\log c$

75. $\log_b(x+1) - \log_b(x^2-1)$

76. $\log_x(p^2-4) - \log_x(p-2)$

Expanding Your Skills

For Exercises 77–86, find the values of the logarithms given that $\log_b 2 \approx 0.693$, $\log_b 3 \approx 1.099$, and $\log_b 5 \approx 1.609$.

77. $\log_b 6$

78. $\log_b 4$

79. $\log_b 12$

80. $\log_b 25$

81. $\log_b 81$

82. $\log_b 30$

83. $\log_b\left(\dfrac{5}{2}\right)$

84. $\log_b\left(\dfrac{25}{3}\right)$

85. $\log_b(10^6)$

86. $\log_b(2^{12})$

87. The intensity of sound waves is measured in decibels and is calculated by the formula

$$B = 10 \log \left(\frac{I}{I_0}\right)$$

where I_0 is the minimum detectable decibel level.

a. Expand this formula by using the properties of logarithms.

b. Let $I_0 = 10^{-16}$ W/cm^2 and simplify.

88. The Richter scale is used to measure the intensity of an earthquake and is calculated by the formula

$$R = \log \left(\frac{I}{I_0}\right)$$

where I_0 is the minimum level detectable by a seismograph.

a. Expand this formula by using the properties of logarithms.

b. Let $I_0 = 1$ and simplify.

Graphing Calculator Exercises

89. a. Graph $Y_1 = \log (x - 1)^2$ and state its domain.

b. Graph $Y_2 = 2 \log (x - 1)$ and state its domain.

c. For what values of x are the expressions $\log (x - 1)^2$ and $2 \log (x - 1)$ equivalent?

90. a. Graph $Y_1 = \log (x^2)$ and state its domain.

b. Graph $Y_2 = 2 \log(x)$ and state its domain.

c. For what values of x are the expressions $\log (x^2)$ and $2 \log (x)$ equivalent?

section 10.6 The Irrational Number e

Objectives

1. The Irrational Number e
2. Computing Compound Interest
3. The Natural Logarithmic Function
4. Change-of-Base Formula
5. Applications of the Natural Logarithmic Function

1. The Irrational Number e

The exponential function base 10 is particularly easy to work with because integral powers of 10 represent different place positions in the base-10 numbering system. In this section, we introduce another important exponential function whose base is an irrational number called e.

Consider the expression $(1 + \frac{1}{x})^x$. The value of the expression for increasingly large values of x approaches a constant (Table 10-7).

table 10-7

x	$\left(1 + \dfrac{1}{x}\right)^x$
100	2.70481382942
1000	2.71692393224
10,000	2.71814592683
100,000	2.71826823717
1,000,000	2.71828046932
1,000,000,000	2.71828182710

As x approaches infinity, the expression $(1 + \frac{1}{x})^x$ approaches a constant value that we call e. From Table 10-7, this value is approximately 2.718281828.

$$e \approx 2.718281828$$

The value of e is an irrational number (a nonterminating, nonrepeating decimal) and is a universal constant, as is π.

Concept Connections

1. Explain why the number e is an irrational number.

example 1 Graphing $f(x) = e^x$

Graph the function defined by $f(x) = e^x$.

Solution:

Because the base of the function is greater than 1 ($e \approx 2.718281828$), the graph is an increasing exponential function. We can use a calculator to evaluate $f(x) = e^x$ at several values of x.

Calculator Connections

Practice using your calculator by evaluating e^x for $x = 1$, $x = 2$, and $x = -1$.

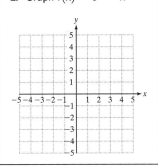

Skill Practice

2. Graph $f(x) = e^x + 1$.

Answers

1. The number e is irrational because it is a nonterminating, nonrepeating decimal. It cannot be written as the ratio of two integers.

2.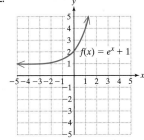

If you are using your calculator correctly, your answers should match those found in Table 10-8. Values are rounded to 3 decimal places. The corresponding graph of $f(x) = e^x$ is shown in Figure 10-16.

table 10-8

x	$f(x) = e^x$
−3	0.050
−2	0.135
−1	0.368
0	1.000
1	2.718
2	7.389
3	20.086

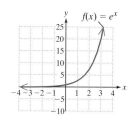

Figure 10-16

2. Computing Compound Interest

One particularly interesting application of exponential functions is in computing compound interest.

1. If the number of compounding periods per year is finite, then the amount in an account is given by

$$A(t) = P\left(1 + \frac{r}{n}\right)^{nt}$$

where P is the initial principal, r is the annual interest rate, n is the number of times compounded per year, and t is the time in years that the money is invested.

2. If the number of compound periods per year is infinite, then interest is said to be **compounded continuously**. In such a case, the amount in an account is given by

$$A(t) = Pe^{rt}$$

where P is the initial principal, r is the annual interest rate, and t is the time in years that the money is invested.

Skill Practice

3. Suppose $1000 is invested at 5%. Find the balance after 8 years under the following options.
   a. Compounded annually
   b. Compounded quarterly
   c. Compounded monthly
   d. Compounded daily
   e. Compounded continuously

example 2   Computing the Balance on an Account

Suppose $5000 is invested in an account earning 6.5% interest. Find the balance in the account after 10 years under the following compounding options.

a. Compounded annually
b. Compounded quarterly
c. Compounded monthly
d. Compounded daily
e. Compounded continuously

Answer

3. a. $1477.46   b. $1488.13
   c. $1490.59   d. $1491.78
   e. $1491.82

Solution:

Compounding Option	n Value	Formula	Result
Annually	$n = 1$	$A(10) = 5000\left(1 + \dfrac{0.065}{1}\right)^{(1)(10)}$	$9385.69
Quarterly	$n = 4$	$A(10) = 5000\left(1 + \dfrac{0.065}{4}\right)^{(4)(10)}$	$9527.79
Monthly	$n = 12$	$A(10) = 5000\left(1 + \dfrac{0.065}{12}\right)^{(12)(10)}$	$9560.92
Daily	$n = 365$	$A(10) = 5000\left(1 + \dfrac{0.065}{365}\right)^{(365)(10)}$	$9577.15
Continuously	Not applicable	$A(10) = 5000e^{(0.065)(10)}$	$9577.70

Notice that there is a $191.46 difference in the account balance between annual compounding and daily compounding. However, the difference between compounding daily and compounding continuously is small—$0.55. As n gets infinitely large, the function defined by

$$A(t) = P\left(1 + \frac{r}{n}\right)^{nt}$$

converges to $A(t) = Pe^{rt}$.

3. The Natural Logarithmic Function

Recall that the common logarithmic function $y = \log x$ has a base of 10. Another important logarithmic function is called the **natural logarithmic function**. The natural logarithmic function has a base of e and is written as $y = \ln x$. That is,

$$y = \ln x = \log_e x$$

example 3 Graphing $y = \ln(x)$

Graph $y = \ln(x)$.

Solution:

Because the base of the function $y = \ln x$ is e and $e > 1$, the graph is an increasing logarithmic function. We can use a calculator to find specific points on the graph of $y = \ln x$ by pressing the $\boxed{\ln}$ key.

Practice using your calculator by evaluating $\ln x$ for the following values of x. If you are using your calculator correctly, your answers should match those found in Table 10-9.

table 10-9

x	$\ln(x)$
1	0.000
2	0.693
3	1.099
4	1.386
5	1.609
6	1.792
7	1.946

Calculator Connections

```
ln(1)
           0
ln(2)
           .6931471806
ln(3)
           1.098612289
```

Skill Practice

4. Graph. $y = \ln x + 1$.

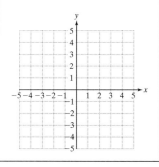

Values are rounded to 3 decimal places. The corresponding graph of $y = \ln x$ is shown in Figure 10-17.

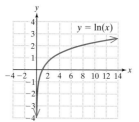

Figure 10-17

The properties of logarithms stated in Section 10.5 are also true for natural logarithms.

Properties of the Natural Logarithmic Function

Let x and y be positive real numbers, and let p be a real number. Then the following properties are true.

1. $\ln 1 = 0$
2. $\ln e = 1$
3. $\ln (e^p) = p$
4. $e^{\ln (x)} = x$
5. $\ln (xy) = \ln x + \ln y$ Product property for logarithms
6. $\ln \left(\dfrac{x}{y}\right) = \ln x - \ln y$ Quotient property for logarithms
7. $\ln (x^p) = p \ln x$ Power property for logarithms

Skill Practice

Simplify.

5. $\ln e^2$
6. $-3 \ln 1$
7. $\ln e^{(x+y)}$
8. $e^{\ln (3x)}$

example 4 — Simplifying Expressions with Natural Logarithms

Simplify the expressions. Assume that all variable expressions within the logarithms represent positive real numbers.

a. $\ln e$ **b.** $\ln 1$ **c.** $\ln (e^{x+1})$ **d.** $e^{\ln (x+1)}$

Solution:

a. $\ln e = 1$ Property 2
b. $\ln 1 = 0$ Property 1
c. $\ln (e^{x+1}) = x + 1$ Property 3
d. $e^{\ln (x+1)} = x + 1$ Property 4

example 5 — Writing a Sum or Difference of Natural Logarithms as a Single Logarithm

Write the expression as a single logarithm. Assume that all variable expressions within the logarithms represent positive real numbers.

$$\ln (x^2 - 9) - \ln (x - 3) - 2 \ln x$$

Answers

4.
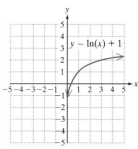

5. 2   6. 0   7. $x + y$
8. $3x$

Solution:

$\ln(x^2 - 9) - \ln(x - 3) - 2\ln x$

$= \ln(x^2 - 9) - \ln(x - 3) - \ln(x^2)$ Power property for logarithms (property 7)

$= \ln(x^2 - 9) - [\ln(x - 3) + \ln(x^2)]$ Factor out -1 from the last two terms.

$= \ln(x^2 - 9) - \ln[(x - 3)x^2]$ Product property for logarithms (property 5)

$= \ln\left[\dfrac{x^2 - 9}{(x - 3)x^2}\right]$ Quotient property for logarithms (property 6)

$= \ln\left[\dfrac{(x - 3)(x + 3)}{(x - 3)x^2}\right]$ Factor.

$= \ln\left(\dfrac{x + 3}{x^2}\right)$ provided $x \neq 3$ Simplify.

> **Skill Practice**
>
> 9. Write as a single logarithm.
> $\ln(x + 4) - \ln(x^2 - 16) + \ln x$

---

example 6   Writing a Logarithmic Expression in Expanded Form

Write the expression

$$\ln\left(\dfrac{e}{x^2\sqrt{y}}\right)$$

as a sum or difference of logarithms of x and y. Assume all variable expressions within the logarithm represent positive real numbers.

Solution:

$\ln\left(\dfrac{e}{x^2\sqrt{y}}\right)$

$= \ln e - \ln(x^2\sqrt{y})$ Quotient property for logarithms (property 6)

$= \ln e - [\ln x^2 + \ln \sqrt{y}]$ Product property for logarithms (property 5)

$= 1 - \ln x^2 - \ln(y^{1/2})$ Distributive property. Also simplify $\ln e = 1$.

$= 1 - 2\ln x - \dfrac{1}{2}\ln y$ Power property for logarithms (property 7)

> **Skill Practice**
>
> 10. Write as a sum or difference of logarithms of x and y. Assume all variables represent positive real numbers.
> $\ln\left(\dfrac{x^3\sqrt{y}}{e^2}\right)$

4. Change-of-Base Formula

A calculator can be used to approximate the value of a logarithm with a base of 10 or a base of e by using the $\boxed{\text{log}}$ key or the $\boxed{\text{ln}}$ key, respectively. However, to use a calculator to evaluate a logarithmic expression with a base other than 10 or e, we must use the **change-of-base formula**.

Answers

9. $\ln\left(\dfrac{x}{x-4}\right)$

10. $3\ln x + \dfrac{1}{2}\ln y - 2$

Concept Connections

11. Change to base 10.

$\log_7 19$

12. Change to base e.

$\log_7 19$

Change-of-Base Formula

Let a and b be positive real numbers such that $a \neq 1$ and $b \neq 1$. Then for any positive real number x,

$$\log_b x = \frac{\log_a x}{\log_a b}$$

Proof:

Let $M = \log_b x$, which implies that $b^M = x$.

Take the logarithm, base a, on both sides: $\quad \log_a(b^M) = \log_a x$

Apply the power property for logarithms: $\quad M \cdot \log_a b = \log_a x$

Divide both sides by $\log_a(b)$: $\quad \dfrac{M \cdot \cancel{\log_a b}}{\cancel{\log_a b}} = \dfrac{\log_a x}{\log_a b}$

$$M = \frac{\log_a x}{\log_a b}$$

Because $M = \log_b(x)$, we have $\quad \log_b(x) = \dfrac{\log_a x}{\log_a b}$ ✓

The change-of-base formula converts a logarithm of one base to a ratio of logarithms of a different base. For the sake of using a calculator, we often apply the change-of-base formula with base 10 or base e.

Skill Practice

13. Use the change-of-base formula to evaluate $\log_5 95$ by using base 10. Round to 3 decimal places.

14. Use the change-of-base formula to evaluate $\log_5 95$ by using base e. Round to 3 decimal places.

example 7 — Using the Change-of-Base Formula

a. Use the change-of-base formula to evaluate $\log_4 80$ by using base 10. (Round to 3 decimal places.)

b. Use the change-of-base formula to evaluate $\log_4 80$ by using base e. (Round to 3 decimal places.)

Solution:

a. $\log_4(80) = \dfrac{\log_{10} 80}{\log_{10} 4} = \dfrac{\log 80}{\log 4} \approx \dfrac{1.903089987}{0.6020599913} \approx 3.161$

b. $\log_4(80) = \dfrac{\log_e 80}{\log_e 4} = \dfrac{\ln 80}{\ln 4} \approx \dfrac{4.382026635}{1.386294361} \approx 3.161$

To check the result, we see that $4^{3.161} \approx 80$.

Calculator Connections

The change-of-base formula can be used to graph logarithmic functions with bases other than 10 or e. For example, to graph $Y_1 = \log_2 x$, we can enter the function as either

$$Y_1 = \frac{\log x}{\log 2} \quad \text{or} \quad Y_1 = \frac{\ln x}{\ln 2}$$

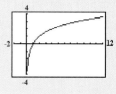

Answers

11. $\dfrac{\log 19}{\log 7}$ **12.** $\dfrac{\ln 19}{\ln 7}$

13. 2.829 **14.** 2.829

5. Applications of the Natural Logarithmic Function

Plant and animal tissue contains both carbon 12 and carbon 14. Carbon 12 is a stable form of carbon, whereas carbon 14 is a radioactive isotope with a half-life of approximately 5730 years. While a plant or animal is living, it takes in carbon from the atmosphere either through photosynthesis or through its food. The ratio of carbon 14 to carbon 12 in a living organism is constant and is the same as the ratio found in the atmosphere.

When a plant or animal dies, it no longer ingests carbon from the atmosphere. The amount of stable carbon 12 remains unchanged from the time of death, but the carbon 14 begins to decay. Because the rate of decay is constant, a tissue sample can be dated by comparing the percent of carbon 14 still present to the percentage of carbon 14 assumed to be in its original living state.

The age of a tissue sample is a function of the percent of carbon 14 still present in the organism according to the following model:

$$A(p) = \frac{\ln p}{-0.000121}$$

where $A(p)$ is the age in years and p is the percentage (in decimal form) of carbon 14 still present.

example 8   Applying the Natural Logarithmic Function to Radioactive Decay

Using the formula

$$A(p) = \frac{\ln p}{-0.000121}$$

a. Find the age of a bone that has 72% of its initial carbon 14.

b. Find the age of the Iceman, a body uncovered in the mountains of northern Italy in 1991. Samples of his hair revealed that 51.4% of the original carbon 14 was present after his death.

Solution:

a. $A(p) = \dfrac{\ln p}{-0.000121}$

     $A(0.72) = \dfrac{\ln(0.72)}{-0.000121}$ Substitute 0.72 for p.

     ≈ 2715 years

b. $A(p) = \dfrac{\ln p}{-0.000121}$

     $A(0.514) = \dfrac{\ln(0.514)}{-0.000121}$ Substitute 0.514 for p.

     ≈ 5500 years   The body of the Iceman is approximately 5500 years old.

Skill Practice

15. Use the formula
$$A(p) = \frac{\ln(p)}{-0.000121}$$
(where $A(p)$ is the age in years and p is the percent of carbon 14 still present) to determine the age of a human skull that has 90% of its initial carbon 14.

Answer

15. ≈ 871 years old

Section 10.6 Practice Exercises

Boost your GRADE at mathzone.com!

- Practice Problems
- Self-Tests
- NetTutor
- e-Professors
- Videos

Study Skills Exercises

1. If you can work the same problem in more than one way, it can help you to practice different methods as well as check your answer. Write down the change-of-base formula. Now find a decimal approximation for $\log_5 10$ in two ways, using two different bases.

2. Define the key terms.
   a. e
   b. Continuously compounded interest
   c. Natural logarithmic function
   d. Change-of-base formula

Review Exercises

For Exercises 3–6, fill out the tables and graph the functions. For Exercises 5 and 6, round to 2 decimal places.

3. $f(x) = \left(\dfrac{3}{2}\right)^x$

x	$f(x)$
-3	
-2	
-1	
0	
1	
2	
3	

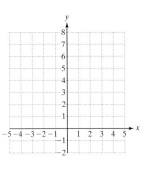

4. $g(x) = \left(\dfrac{1}{5}\right)^x$

x	$g(x)$
-3	
-2	
-1	
0	
1	
2	
3	

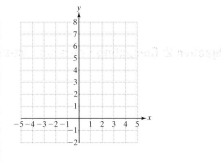

5. $q(x) = \log(x + 1)$

x	$q(x)$
-0.5	
0	
4	
9	

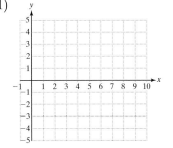

6. $r(x) = \log x$

x	$r(x)$
0.5	
1	
5	
10	

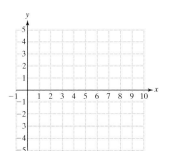

Objective 1: The Irrational Number e

For Exercises 7–10, graph the equation by completing the table and plotting points. Identify the domain. Round to 2 decimal places when necessary. (See Example 1.)

7. $y = e^{x+1}$

x	y
-4	
-3	
-2	
-1	
0	
1	

8. $y = e^{x+2}$

x	y
-5	
-4	
-3	
-2	
-1	
0	

9. $y = e^x + 2$

x	y
-2	
-1	
0	
1	
2	
3	

10. $y = e^x - 1$

x	y
-4	
-3	
-2	
-1	
0	
1	

Objective 2: Computing Compound Interest

In Exercises 11–16, use the model

$$A(t) = P\left(1 + \frac{r}{n}\right)^{nt}$$

for interest compounded n times per year. Use the model $A(t) = Pe^{rt}$ for interest compounded continuously.

11. Suppose an investor deposits $10,000 in a certificate of deposit for 5 years for which the interest is compounded monthly. Find the total amount of money in the account for the following interest rates. Compare your answers and comment on the effect of interest rate on an investment.

    a. $r = 4.0\%$ **b.** $r = 6.0\%$ **c.** $r = 8.0\%$ **d.** $r = 9.5\%$

12. Suppose an investor deposits $5000 in a certificate of deposit for 8 years for which the interest is compounded quarterly. Find the total amount of money in the account for the following interest rates. Compare your answers and comment on the effect of interest rate on an investment.

    a. $r = 4.5\%$ **b.** $r = 5.5\%$ **c.** $r = 7.0\%$ **d.** $r = 9.0\%$

13. Suppose an investor deposits $8000 in a savings account for 10 years at 4.5% interest. Find the total amount of money in the account for the following compounding options. Compare your answers. How does the number of compound periods per year affect the total investment? (See Example 2.)

   a. Compounded annually
   b. Compounded quarterly
   c. Compounded monthly
   d. Compounded daily
   e. Compounded continuously

14. Suppose an investor deposits $15,000 in a savings account for 8 years at 5.0% interest. Find the total amount of money in the account for the following compounding options. Compare your answers. How does the number of compound periods per year affect the total investment?

   a. Compounded annually
   b. Compounded quarterly
   c. Compounded monthly
   d. Compounded daily
   e. Compounded continuously

15. Suppose an investor deposits $5000 in an account bearing 6.5% interest compounded continuously. Find the total amount in the account for the following time periods. How does the length of time affect the amount of interest earned?

   a. 5 years   b. 10 years   c. 15 years   d. 20 years   e. 30 years

16. Suppose an investor deposits $10,000 in an account bearing 6.0% interest compounded continuously. Find the total amount in the account for the following time periods. How does the length of time affect the amount of interest earned?

   a. 5 years   b. 10 years   c. 15 years   d. 20 years   e. 30 years

Objective 3: The Natural Logarithmic Function

For Exercises 17–20, graph the equation by completing the table and plotting the points. Identify the domain. Round to 2 decimal places when necessary. (See Example 3.)

17. $y = \ln(x - 2)$

x	y
2.25	
2.50	
2.75	
3	
4	
5	
6	

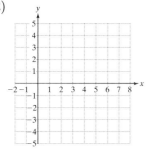

18. $y = \ln(x - 1)$

x	y
1.25	
1.50	
1.75	
2	
3	
4	
5	

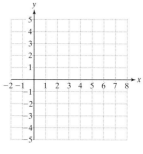

19. $y = \ln(x) - 1$

x	y
0.25	
0.5	
0.75	
1	
2	
3	
4	

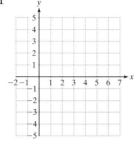

20. $y = \ln(x) + 2$

x.	y
0.25	
0.5	
0.75	
1	
2	
3	
4	

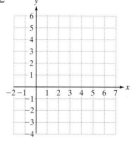

21. a. Graph $f(x) = e^x$.

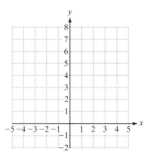

b. Identify the domain and range of f.

c. Graph $g(x) = \ln x$.

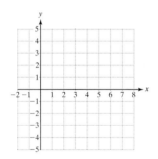

d. Identify the domain and range of g.

22. a. Graph $f(x) = 10^x$

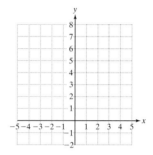

b. Identify the domain and range of f.

c. Graph $g(x) = \log x$.

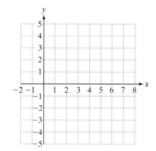

d. Identify the domain and range of g.

For Exercises 23–30, simplify the expressions. Assume all variables represent positive real numbers. **(See Example 4.)**

23. $\ln e$

24. $\ln e^2$

25. $\ln 1$

26. $e^{\ln x}$

27. $\ln e^{-6}$

28. $\ln e^{(5-3x)}$

29. $e^{\ln(2x+3)}$

30. $e^{\ln 4}$

For Exercises 31–38, write the expression as a single logarithm. Assume all variables represent positive real numbers. **(See Example 5.)**

31. $6 \ln p + \dfrac{1}{3} \ln q$

32. $2 \ln w + \ln z$

33. $\dfrac{1}{2}(\ln x - 3 \ln y)$

34. $\dfrac{1}{3}(4 \ln a - \ln b)$

35. $2 \ln a - \ln b - \dfrac{1}{3} \ln c$

36. $-\ln x + 3 \ln y - \ln z$

37. $4 \ln x - 3 \ln y - \ln z$

38. $\dfrac{1}{2} \ln c + \ln a - 2 \ln b$

For Exercises 39–46, write the expression as a sum and/or difference of $\ln a$, $\ln b$, and $\ln c$. Assume all variables represent positive real numbers. **(See Example 6.)**

39. $\ln \left(\dfrac{a}{b}\right)^2$

40. $\ln \sqrt[3]{\dfrac{a}{b}}$

41. $\ln (b^2 \cdot e)$

42. $\ln (\sqrt{c} \cdot e)$

43. $\ln \left(\dfrac{a^4 \sqrt{b}}{c}\right)$

44. $\ln \left(\dfrac{\sqrt{ab}}{c^3}\right)$

45. $\ln \left(\dfrac{ab}{c^2}\right)^{1/5}$

46. $\ln \sqrt{2ab}$

Objective 4: Change-of-Base Formula

47. a. Evaluate $\log_6 (200)$ by computing $\log (200)/\log (6)$ to 4 decimal places.

  b. Evaluate $\log_6 (200)$ by computing $\ln (200)/\ln (6)$ to 4 decimal places.

  c. How do your answers to parts (a) and (b) compare?

48. a. Evaluate $\log_8 (120)$ by computing $\log (120)/\log (8)$ to 4 decimal places.

  b. Evaluate $\log_8 (120)$ by computing $\ln (120)/\ln (8)$ to 4 decimal places.

  c. How do your answers to parts (a) and (b) compare?

For Exercises 49–60, use the change-of-base formula to approximate the logarithms to 4 decimal places. Check your answers by using the exponential key on your calculator. **(See Example 7.)**

49. $\log_2 7$

50. $\log_3 5$

51. $\log_8 24$

52. $\log_4 17$

53. $\log_8 (0.012)$

54. $\log_7 (0.251)$

55. $\log_9 1$

56. $\log_2 \left(\dfrac{1}{5}\right)$

57. $\log_4 \left(\dfrac{1}{100}\right)$

58. $\log_5 (0.0025)$

59. $\log_7 (0.0006)$

60. $\log_2 (0.24)$

Objective 5: Applications of the Natural Logarithmic Function

Under continuous compounding, the amount of time t in years required for an investment to double is a function of the interest rate r:

$$t = \frac{\ln 2}{r}$$

Use the formula for Exercises 61–62. (See Example 8.)

61. a. If you invest $5000, how long will it take the investment to reach $10,000 if the interest rate is 4.5%? Round to 1 decimal place.

b. If you invest $5000, how long will it take the investment to reach $10,000 if the interest rate is 10%? Round to 1 decimal place.

c. Using the doubling time found in part (b), how long would it take a $5000 investment to reach $20,000 if the interest rate is 10%?

62. a. If you invest $3000, how long will it take the investment to reach $6000 if the interest rate is 5.5%? Round to 1 decimal place.

b. If you invest $3000, how long will it take the investment to reach $6000 if the interest rate is 8%? Round to 1 decimal place.

c. Using the doubling time found in part (b), how long would it take a $3000 investment to reach $12,000 if the interest rate is 8%?

63. On August 31, 1854, an epidemic of cholera was discovered in London, England, resulting from a contaminated community water pump at Broad Street. By the end of September more than 600 citizens who drank water from the pump had died.

The cumulative number of deaths from cholera in the 1854 London epidemic can be approximated by

$$D(t) = 91 + 160 \ln (t + 1)$$

where t is the number of days after the start of the epidemic ($t = 0$ corresponds to September 1, 1854).

a. Approximate the total number of deaths as of September 1 ($t = 0$).

b. Approximate the total number of deaths as of September 5, September 10, and September 20.

Graphing Calculator Connections

64. a. Graph the function defined by $f(x) = \log_3 x$ by graphing $Y_1 = \dfrac{\log x}{\log 3}$.

b. Graph the function defined by $f(x) = \log_3 x$ by graphing $Y_2 = \dfrac{\ln x}{\ln 3}$.

c. Does it appear that $Y_1 = Y_2$ on the standard viewing window?

65. a. Graph the function defined by $f(x) = \log_7 x$ by graphing $Y_1 = \dfrac{\log x}{\log 7}$.

b. Graph the function defined by $f(x) = \log_7(x)$ by graphing $Y_2 = \dfrac{\ln x}{\ln 7}$.

c. Does it appear that $Y_1 = Y_2$ on the standard viewing window?

66. a. Graph the function defined by $f(x) = \log_{1/3} x$ by graphing $Y_1 = \dfrac{\log x}{\log\left(\frac{1}{3}\right)}$.

b. Graph the function defined by $f(x) = \log_{1/3} x$ by graphing $Y_2 = \dfrac{\ln x}{\ln\left(\frac{1}{3}\right)}$.

c. Does it appear that $Y_1 = Y_2$ on the standard viewing window?

67. a. Graph the function defined by $f(x) = \log_{1/7}(x)$ by graphing $Y_1 = \dfrac{\log x}{\log\left(\frac{1}{7}\right)}$.

b. Graph the function defined by $f(x) = \log_{1/7}(x)$ by graphing $Y_2 = \dfrac{\ln x}{\ln\left(\frac{1}{7}\right)}$.

c. Does it appear that $Y_1 = Y_2$ on the standard viewing window?

For Exercises 68–73, graph the functions on your calculator.

68. Graph $s(x) = \log_{1/2} x$

69. Graph $w(x) = \log_{1/3} x$

70. Graph $y = e^{x-2}$

71. Graph $y = e^{x-1}$

72. Graph $y = \ln(x + 1)$

73. Graph $y = \ln(x + 2)$

Section 10.7 Exponential and Logarithmic Equations

Objectives
1. Solving Logarithmic Equations
2. Applications of Logarithmic Equations
3. Solving Exponential Equations
4. Applications of Exponential Equations

1. Solving Logarithmic Equations

Equations containing one or more logarithms are called **logarithmic equations**. For example,

$$\log_4 x = 1 - \log_4(x-3) \quad \text{and} \quad \ln(x+2) + \ln(x-1) = \ln(9x - 17)$$

are logarithmic equations. To solve logarithmic equations of first degree, use the following guidelines.

Guidelines to Solve Logarithmic Equations

1. Isolate the logarithms on one side of the equation.
2. Write a sum or difference of logarithms as a single logarithm.
3. Rewrite the equation in step 2 in exponential form.
4. Solve the resulting equation from step 3.
5. Check all solutions to verify that they are within the domain of the logarithmic expressions in the original equation.

---

example 1   Solving a Logarithmic Equation

Solve the equation.

$$\log_4 x = 1 - \log_4(x-3)$$

Solution:

$\log_4 x = 1 - \log_4(x-3)$

$\log_4 x + \log_4(x-3) = 1$ Isolate the logarithms on one side of the equation.

$\log_4[x(x-3)] = 1$ Write as a single logarithm.

$\log_4(x^2 - 3x) = 1$ Simplify inside the parentheses.

$x^2 - 3x = 4^1$ Write the equation in exponential form.

$x^2 - 3x - 4 = 0$ The resulting equation is quadratic.

$(x-4)(x+1) = 0$ Factor.

$x = 4 \quad \text{or} \quad x = -1$ Apply the zero product rule.

Notice that $x = -1$ is *not* a solution because $\log_4 x$ is not defined at $x = -1$. However, $x = 4$ *is* defined in both expressions $\log_4 x$ and $\log_4(x-3)$. We can substitute $x = 4$ into the original equation to show that it checks.

Skill Practice

1. Solve the equation.
$$\log_3 x + \log_3(x-8) = 2$$

Answer

1. $x = 9$ ($x = -1$ does not check)

Check: $x = 4$

$\log_4 x = 1 - \log_4 (x - 3)$

$\log_4 4 \stackrel{?}{=} 1 - \log_4 (4 - 3)$

$1 \stackrel{?}{=} 1 - \log_4 (1)$

$1 \stackrel{?}{=} 1 - 0$ ✓ True

The solution is $x = -4$.

Tip: The equation from Example 1 involved the logarithmic functions $y = \log_4 (x)$ and $y = \log_4 (x - 3)$. The domains of these functions are $\{x \mid x > 0\}$ and $\{x \mid x > 3\}$, respectively. Therefore, the solutions to the equation are restricted to x-values in the intersection of these two sets, that is, $\{x \mid x > 3\}$. The solution $x = 4$ satisfies this requirement, whereas $x = -1$ does not.

Skill Practice

Solve the equations.

2. $\ln (t - 3) + \ln (t - 1) = \ln (2t - 5)$
3. $\log (2p + 6) = 1$
4. $\ln (x - 3) = 1$

example 2   **Solving Logarithmic Equations**

Solve the equations.

a. $\ln (x + 2) + \ln (x - 1) = \ln (9x - 17)$

b. $\log (x + 300) = 3.7$ c. $\ln (2x + 5) = 1$

Solution:

a. $\ln (x + 2) + \ln (x - 1) = \ln (9x - 17)$

$\ln (x + 2) + \ln (x - 1) - \ln (9x - 17) = 0$ Isolate the logarithms on one side.

$\ln \left[\dfrac{(x + 2)(x - 1)}{9x - 17} \right] = 0$ Write as a single logarithm.

$e^0 = \dfrac{(x + 2)(x - 1)}{9x - 17}$ Write the equation in exponential form.

$1 = \dfrac{(x + 2)(x - 1)}{9x - 17}$ Simplify.

$(1) \cdot (9x - 17) = \left[\dfrac{(x + 2)(x - 1)}{9x - 17} \right] \cdot (9x - 17)$ Multiply by the LCD.

$9x - 17 = (x + 2)(x - 1)$ The equation is quadratic.

$9x - 17 = x^2 + x - 2$

$0 = x^2 - 8x + 15$

$0 = (x - 5)(x - 3)$

$x = 5$ or $x = 3$

The solutions $x = 5$ and $x = 3$ are both within the domain of the logarithmic functions in the original equation.

Answers

2. $t = 4$ ($t = 2$ does not check)
3. $p = 2$
4. $x = e + 3 \approx 5.718$

Check: $x = 5$ | Check: $x = 3$
$\ln(x+2) + \ln(x-1) = \ln(9x-17)$ | $\ln(x+2) + \ln(x-1) = \ln(9x-17)$
$\ln(5+2) + \ln(5-1) \stackrel{?}{=} \ln[9(5)-17]$ | $\ln(3+2) + \ln(3-1) \stackrel{?}{=} \ln[9(3)-17]$
$\ln(7) + \ln(4) \stackrel{?}{=} \ln(45-17)$ | $\ln(5) + \ln(2) \stackrel{?}{=} \ln(27-17)$
$\ln(7 \cdot 4) \stackrel{?}{=} \ln(28)$ | $\ln(5 \cdot 2) \stackrel{?}{=} \ln(10)$
$\ln(28) \stackrel{?}{=} \ln(28)$ ✓ True | $\ln(10) \stackrel{?}{=} \ln(10)$ ✓ True

Both solutions check.

b. $\log(x+300) = 3.7$ — The equation has a single logarithm that is already isolated.

$10^{3.7} = x + 300$ — Write the equation in exponential form.

$10^{3.7} - 300 = x$ — Solve for x.

$x = 10^{3.7} - 300 \approx 4711.87$

Check: $x = 10^{3.7} - 300$ — Check the exact value of x in the original equation.

$\log(x + 300) = 3.7$
$\log[(10^{3.7} - 300) + 300] \stackrel{?}{=} 3.7$
$\log(10^{3.7} - 300 + 300) \stackrel{?}{=} 3.7$
$\log(10^{3.7}) \stackrel{?}{=} 3.7$ — Property 3 of logarithms: $\log_b(b^p) = p$
$3.7 \stackrel{?}{=} 3.7$ ✓ True

The solution $x = 10^{3.7} - 300$ checks.

c. $\ln(2x+5) = 1$ — The equation has a single logarithm that is already isolated.

$e^1 = 2x + 5$ — Write the equation in exponential form.

$2x + 5 = e$ — The resulting equation is linear.

$2x = e - 5$ — Solve for x.

$x = \dfrac{e-5}{2} \approx -1.14$

Check: $x = \dfrac{e-5}{2}$

$\ln(2x+5) = 1$

$\ln\left[2\left(\dfrac{e-5}{2}\right) + 5\right] \stackrel{?}{=} 1$

$\ln(e - 5 + 5) \stackrel{?}{=} 1$

$\ln e \stackrel{?}{=} 1$ ✓ True. — The solution checks.

Calculator Connections

To support the solution to Example 2(c), graph $Y_1 = \ln(2x + 5)$ and $Y_2 = 1$. Use the graph to approximate the x-coordinate where $Y_1 = Y_2$.

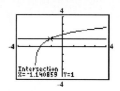

2. Applications of Logarithmic Equations

example 3 — Applying a Logarithmic Equation to Earthquake Intensity

The magnitude of an earthquake (the amount of seismic energy released at the hypocenter of the earthquake) is measured on the Richter scale. The Richter scale value R is determined by the formula

$$R = \log\left(\frac{I}{I_0}\right)$$

where I is the intensity of the earthquake and I_0 is the minimum measurable intensity of an earthquake. (I_0 is a "zero-level" quake—one that is barely detected by a seismograph.)

a. Compare the Richter scale values of earthquakes that are (i) 100,000 times (10^5 times) more intense than I_0 and (ii) 1,000,000 times (10^6 times) more intense than I_0.

b. On October 17, 1989, an earthquake measuring 7.1 on the Richter scale occurred in the Loma Prieta area in the Santa Cruz Mountains. The quake devastated parts of San Francisco and Oakland, California, bringing 63 deaths and over 3700 injuries. Determine how many times more intense this earthquake was than a zero-level quake.

Solution:

a. $R = \log\left(\dfrac{I}{I_0}\right)$

  i. Earthquake 100,000 times I_0

  $R = \log\left(\dfrac{10^5 \cdot I_0}{I_0}\right)$ Substitute $I = 10^5 I_0$

  $= \log(10^5)$

  $= 5$

  ii. Earthquake 1,000,000 times I_0

  $R = \log\left(\dfrac{10^6 \cdot I_0}{I_0}\right)$ Substitute $I = 10^6 I_0$.

  $= \log(10^6)$

  $= 6$

Skill Practice

5. In December 2004, an earthquake in the Indian Ocean measuring 9.0 on the Richter scale caused a tsunami that killed more than 300,000 people. Determine how many times more intense this earthquake was than a zero-level quake. Use the formula

   $R = \log\left(\dfrac{I}{I_0}\right)$

   where I is the intensity of the quake, I_0 is the measure of a zero-level quake, and R is the Richter scale value.

Answer

5. The earthquake was 10^9 times more intense than a zero-level quake.

Notice that the value on the Richter scale corresponds to the magnitude (power of 10) of the energy released. That is, a 1-unit increase on the Richter scale represents a 10-fold increase in the intensity of an earthquake.

b. $R = \log\left(\dfrac{I}{I_0}\right)$

$7.1 = \log\left(\dfrac{I}{I_0}\right)$ Substitute $R = 7.1$.

$\dfrac{I}{I_0} = 10^{7.1}$ Write the equation in exponential form.

$I = 10^{7.1} \cdot I_0$ Solve for I.

The Loma Prieta earthquake in 1989 was $10^{7.1}$ times ($\approx 12{,}590{,}000$ times) more intense than a zero-level earthquake.

3. Solving Exponential Equations

An equation with one or more exponential expressions is called an **exponential equation**. The following property is often useful in solving exponential equations.

> **Equivalence of Exponential Expressions**
>
> Let x, y, and b be real numbers such that $b > 0$ and $b \neq 1$. Then
>
> $$b^x = b^y \quad \text{implies} \quad x = y$$

The equivalence of exponential expressions indicates that if two exponential expressions of the same base are equal, their exponents must be equal.

example 4   Solving Exponential Equations

Solve the equations.

a. $4^{2x-9} = 64$ **b.** $(2^x)^{x+3} = \dfrac{1}{4}$

Solution:

a. $4^{2x-9} = 64$

$4^{2x-9} = 4^3$ Write both sides with a common base.

$2x - 9 = 3$ If $b^x = b^y$, then $x = y$.

$2x = 12$ Solve for x.

$x = 6$

To check, substitute $x = 6$ into the original equation.

$4^{2(6)-9} \stackrel{?}{=} 64$

$4^{12-9} \stackrel{?}{=} 64$

$4^3 = 64$ ✓ True

Skill Practice

Solve the equations.

6. $2^{3x+1} = 8$

7. $(3^x)^{x-3} = \dfrac{1}{9}$

Answers

6. $x = \dfrac{2}{3}$ **7.** $x = 1$ or $x = 2$

b. $(2^x)^{x+3} = \dfrac{1}{4}$

$2^{x^2+3x} = 2^{-2}$ Apply the multiplication property of exponents. Write both sides of the equation with a common base.

$x^2 + 3x = -2$ If $b^x = b^y$, then $x = y$.

$x^2 + 3x + 2 = 0$ The resulting equation is quadratic.

$(x + 2)(x + 1) = 0$ Solve for x.

$x = -2$ or $x = -1$

Check: $x = -2$ Check: $x = -1$

$(2^{-2})^{(-2)+3} \stackrel{?}{=} \dfrac{1}{4}$ $(2^{-1})^{(-1)+3} \stackrel{?}{=} \dfrac{1}{4}$

$(2^{-2})^1 \stackrel{?}{=} \dfrac{1}{4}$ $(2^{-1})^2 \stackrel{?}{=} \dfrac{1}{4}$

$2^{-2} \stackrel{?}{=} \dfrac{1}{4}$ ✓ True $2^{-2} \stackrel{?}{=} \dfrac{1}{4}$ ✓ True

Both solutions check.

Skill Practice

8. Solve the equation.

   $5^x = 32$

---

example 5   Solving an Exponential Equation

Solve the equation. $4^x = 25$

Solution:

Because 25 cannot be written as an integral power of 4, we cannot immediately use the property that if $b^x = b^y$, then $x = y$. Instead we can rewrite the equation in its corresponding logarithmic form to solve for x.

$4^x = 25$

$x = \log_4 25$ Write the equation in logarithmic form

$ = \dfrac{\ln 25}{\ln 4} \approx 2.322$ Change-of-base formula

Calculator Connections

Graph $Y_1 = 4^x$ and $Y_2 = 25$.

An *Intersect* feature can be used to find the x-coordinate where $Y_1 = Y_2$.

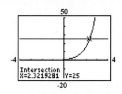

The same result can be reached by taking a logarithm of any base on both sides of the equation. Then by applying the power property of logarithms, the unknown exponent can be written as a factor.

Answer

8. $x = \dfrac{\ln 32}{\ln 5} \approx 2.153$

$$4^x = 25$$

$\log(4^x) = \log 25$ Take the common logarithm of both sides.

$x \log 4 = \log 25$ Apply the power property of logarithms to express the exponent as a factor. This is now a linear equation in x.

$\dfrac{x \cancel{\log 4}}{\cancel{\log 4}} = \dfrac{\log 25}{\log 4}$ Solve for x.

$x = \dfrac{\log 25}{\log 4} \approx 2.322$

Guidelines to Solve Exponential Equations

1. Isolate one of the exponential expressions in the equation.
2. Take a logarithm on both sides of the equation. (The natural logarithmic function or the common logarithmic function is often used so that the final answer can be approximated with a calculator.)
3. Use the power property of logarithms (property 7) to write exponents as factors. Recall: $\log_b(M^p) = p \log_b(M)$.
4. Solve the resulting equation from step 3.

example 6   Solving Exponential Equations by Taking a Logarithm on Both Sides

Solve the equations.

a. $2^{x+3} = 7^x$ **b.** $e^{-3.6x} = 9.74$

Skill Practice

Solve the equations.

9. $3^x = 8^{x+2}$
10. $e^{-0.2t} = 7.52$

Solution:

a.
$$2^{x+3} = 7^x$$

$\ln 2^{(x+3)} = \ln 7^x$ Take the natural logarithm of both sides.

$(x + 3) \ln 2 = x \ln 7$ Express the exponents as factors.

$x(\ln 2) + 3(\ln 2) = x \ln 7$ Apply the distributive property.

$x(\ln 2) - x(\ln 7) = -3 \ln 2$ Collect x-terms on one side.

$x(\ln 2 - \ln 7) = -3 \ln 2$ Factor out x.

$\dfrac{x\cancel{(\ln 2 - \ln 7)}}{\cancel{(\ln 2 - \ln 7)}} = \dfrac{-3 \ln 2}{(\ln 2 - \ln 7)}$ Solve for x.

$x = \dfrac{-3 \ln 2}{(\ln 2 - \ln 7)} \approx 1.660$

Answers

9. $x = \dfrac{2 \ln 8}{\ln 3 - \ln 8} \approx -4.240$

10. $t = \dfrac{\ln 7.52}{-0.2} \approx -10.088$

> **Tip:** The exponential equation $2^{x+3} = 7^x$ could have been solved by taking a logarithm of *any* base on both sides of the equation. For example, using base 10 yields
>
> $\log 2^{x+3} = \log 7^x$
>
> $(x + 3) \log 2 = x \log 7$ Apply the power property for logarithms.
>
> $x \log 2 + 3 \log 2 = x \log 7$ Apply the distributive property.
>
> $x \log 2 - x \log 7 = -3 \log 2$ Collect x-terms on one side of the equation.
>
> $x(\log 2 - \log 7) = -3 \log 2$ Factor out x.
>
> $x = \dfrac{-3 \log 2}{\log 2 - \log 7} \approx 1.660$

b. $e^{-3.6x} = 9.74$

$\ln e^{-3.6x} = \ln 9.74$ The exponential expression has a base of e, so it is convenient to take the natural logarithm of both sides.

$(-3.6x) \ln e = \ln 9.74$ Express the exponent as a factor.

$-3.6x = \ln 9.74$ Simplify (recall that $\ln e = 1$).

$x = \dfrac{\ln 9.74}{-3.6} \approx -0.632$

4. Applications of Exponential Equations

Skill Practice

Use the population function from Example 7.

11. Estimate the world population in October 2020.

12. Estimate the year in which the world population will reach 9 billion.

example 7   Applying an Exponential Function to World Population

The population of the world was estimated to have reached 6 billion in October 1999. The population growth rate for the world is estimated to be 1.4%. (Source: *Information Please Almanac.*)

$$P(t) = 6(1.014)^t$$

represents the world population in billions as a function of the number of years after October 1999 ($t = 0$ represents October 1999).

a. Use the function to estimate the world population in October 2005 and in October 2010.

b. Use the function to estimate the amount of time after October 1999 required for the world population to reach 12 billion.

Solution:

a. $P(t) = 6(1.014)^t$

$P(6) = 6(1.014)^6$ The year 2005 corresponds to $t = 6$.

≈ 6.5 In 2005, the world's population will be approximately 6.5 billion.

Answers

11. Approximately 8 billion
12. The year 2028

$$P(11) = 6(1.014)^{11}$$ The year 2010 corresponds to $t = 11$.

$$\approx 7.0$$ In 2010, the world's population will be approximately 7.0 billion.

b. $P(t) = 6(1.014)^t$

$12 = 6(1.014)^t$ Substitute $P(t) = 12$ and solve for t.

$$\dfrac{12}{6} = \dfrac{\cancel{6}(1.014)^t}{\cancel{6}}$$ Isolate the exponential expression on one side of the equation.

$2 = 1.014^t$

$\ln 2 = \ln 1.014^t$ Take the natural logarithm of both sides.

$\ln 2 = t \ln 1.014$ Express the exponent as a factor.

$$\dfrac{\ln 2}{\ln 1.014} = \dfrac{t \,\cancel{\ln 1.014}}{\cancel{\ln 1.014}}$$ Solve for t.

$$t = \dfrac{\ln 2}{\ln 1.014} \approx 50$$ The population will reach 12 billion (double the October 1999 value) approximately 50 years after 1999.

Note: It has taken thousands of years for the world's population to reach 6 billion. However, with a growth rate of 1.4%, it will take only 50 years to gain an additional 6 billion.

On Friday, April 25, 1986, a nuclear accident occurred at the Chernobyl nuclear reactor, resulting in radioactive contaminates being released into the atmosphere. The most hazardous isotopes released in this accident were ^{137}Cs (cesium 137), ^{131}I (iodine 131), and ^{90}Sr (strontium 90). People living close to Chernobyl (in Ukraine) were at risk of radiation exposure from inhalation, from absorption through the skin, and from food contamination. Years after the incident, scientists have seen an increase in the incidence of thyroid disease among children living in the contaminated areas. Because iodine is readily absorbed in the thyroid gland, scientists suspect that radiation from iodine 131 is the cause.

example 8   Applying an Exponential Equation to Radioactive Decay

The half-life of radioactive iodine ^{131}I is 8.04 days. If 10 g of iodine 131 is initially present, then the amount of radioactive iodine still present after t days is approximated by

$$A(t) = 10e^{-0.0862t}$$

where t is the time in days.

a. Use the model to approximate the amount of ^{131}I still present after 2 weeks. Round to the nearest 0.1 g.

b. How long will it take for the amount of ^{131}I to decay to 0.5 g? Round to the nearest 0.1 year.

Skill Practice

Radioactive strontium 90 (^{90}Sr) has a half-life of 28 years. If 100 g of strontium 90 is initially present, the amount left after t years is approximated by

$$A(t) = 100e^{-0.0248t}$$

13. Find the amount of ^{90}Sr present after 85 years.

14. How long will it take for the amount of ^{90}Sr to decay to 40 g?

Answers

13. 12.2 g   **14.** 36.9 years

Graphing Calculator Exercises

91. The amount of money a company receives from sales is related to the money spent on advertising according to

$$S(x) = 400 + 250 \log x \quad x \geq 1$$

where $S(x)$ is the amount in sales (in $1000s) and x is the amount spent on advertising (in $1000s).

a. The value of $S(1) = 400$ means that if $1000 is spent on advertising, the total sales will be $400,000.

  i. Find the total sales for this company if $11,000 is spent on advertising.

  ii. Find the total sales for this company if $21,000 is spent on advertising.

  iii. Find the total sales for this company if $31,000 is spent on advertising.

b. Graph the function $y = S(x)$ on a window where $0 \leq x \leq 40$ and $0 \leq y \leq 1000$. Using the graph and your answers from part (a), describe the relationship between total sales and the money spent on advertising. As the money spent on advertising is increased, what happens to the rate of increase of total sales?

c. How many advertising dollars are required for the total sales to reach $1,000,000? That is, for what value of x will $S(x) = 1000$?

92. Graph $Y_1 = 8\wedge x$ and $Y_2 = 21$ on a window where $0 \leq x \leq 5$ and $0 \leq y \leq 40$. Use the graph and an *Intersect* feature or *Zoom* and *Trace* to support your answer to Exercise 63.

93. Graph $Y_1 = 6\wedge x$ and $Y_2 = 39$ on a window where $0 \leq x \leq 5$ and $0 \leq y \leq 50$. Use the graph and an *Intersect* feature or *Zoom* and *Trace* to support your answer to Exercise 64.

94. Graph $Y_1 = \log_3 x$ (use the change-of-base formula) and $Y_2 = 2$ on a window where $0 \leq x \leq 40$ and $-4 \leq y \leq 4$. Use the graph and an *Intersect* feature or *Zoom* and *Trace* to support your answer to Exercise 11.

95. Graph $Y_1 = \log_4 x$ (use the change-of-base formula) and $Y_2 = 9$ on a window where $0 \leq x \leq 1,000,000$ and $-2 \leq y \leq 12$. Use the graph and an *Intersect* feature or *Zoom* and *Trace* to support your answer to Exercise 12.

chapter 10 | summary

section 10.1 Algebra and Composition of Functions

Key Concepts

The Algebra of Functions

Given two functions f and g, the functions $f + g, f - g, f \cdot g,$ and $\dfrac{f}{g}$ are defined as

$(f + g)(x) = f(x) + g(x)$

$(f - g)(x) = f(x) - g(x)$

$(f \cdot g)(x) = f(x) \cdot g(x)$

$\left(\dfrac{f}{g}\right)(x) = \dfrac{f(x)}{g(x)}$ provided $g(x) \neq 0$

Composition of Functions

The **composition** of f and g, denoted $f \circ g$, is defined by the rule

$(f \circ g)(x) = f(g(x))$ provided that $g(x)$ is in the domain of f

The **composition** of g and f, denoted $g \circ f$, is defined by the rule

$(g \circ f)(x) = g(f(x))$ provided that $f(x)$ is in the domain of g

Examples

Example 1

Let $g(x) = 5x + 1$ and $h(x) = x^3$. Find:

1. $(g + h)(3) = g(3) + h(3) = 16 + 27 = 43$
2. $(g \cdot h)(-1) = g(-1) \cdot h(-1) = (-4) \cdot (-1) = 4$
3. $(g - h)(x) = 5x + 1 - x^3$
4. $\left(\dfrac{g}{h}\right)(x) = \dfrac{5x + 1}{x^3}$

Example 2

Find $(f \circ g)(x)$ and $(g \circ f)(x)$ given the functions defined by $f(x) = 4x + 3$ and $g(x) = 7x$.

$(f \circ g)(x) = f(g(x))$

$\quad = f(7x) + 3$

$\quad = 4(7x) + 3$

$\quad = 28x + 3$

$(g \circ f)(x) = g(f(x))$

$\quad = g(4x + 3)$

$\quad = 7(4x + 3)$

$\quad = 28x + 21$

section 10.2 Inverse Functions

Key Concepts

Horizontal Line Test

Consider a function defined by a set of points (x, y) in a rectangular coordinate system. Then the graph defines y as a one-to-one function of x if no horizontal line intersects the graph in more than one point.

Examples

Example 1

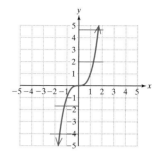

The function is one-to-one because it passes the horizontal line test.

Finding an Equation of the Inverse of a Function

For a one-to-one function defined by $y = f(x)$, the equation of the inverse can be found as follows:

1. Replace $f(x)$ by y.
2. Interchange x and y.
3. Solve for y.
4. Replace y by $f^{-1}(x)$.

Example 2

Find the inverse of the one-to-one function defined by $f(x) = 3 - x^3$.

1. $y = 3 - x^3$
2. $x = 3 - y^3$
3. $x - 3 = -y^3$
   $-x + 3 = y^3$
   $\sqrt[3]{-x + 3} = y$
4. $f^{-1}(x) = \sqrt[3]{-x + 3}$

The graphs defined by $y = f(x)$ and $y = f^{-1}(x)$ are symmetric with respect to the line $y = x$.

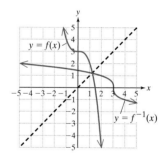

Definition of the Inverse of a Function

If f is a one-to-one function, then g is the inverse of f if and only if $(f \circ g)(x) = x$ for all x in the domain of g, and $(g \circ f)(x) = x$ for all x in the domain of f.

Example 3

Verify that the functions defined by $f(x) = x - 1$ and $g(x) = x + 1$ are inverses.

$(f \circ g)(x) = f(x + 1) = (x + 1) - 1 = x$

$(g \circ f)(x) = g(x - 1) = (x - 1) + 1 = x$

section 10.3 Exponential Functions

Key Concepts

A function $y = b^x$ $(b > 0, b \neq 1)$ is an **exponential function**.

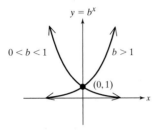

The domain is $(-\infty, \infty)$.

The range is $(0, \infty)$

The line $y = 0$ is a horizontal asymptote.

The y-intercept is $(0, 1)$.

Examples

Example 1

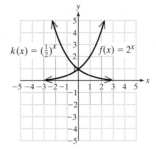

section 10.4 Logarithmic Functions

Key Concepts

The function $y = \log_b x$ is a **logarithmic function**.

$y = \log_b x \quad \Leftrightarrow \quad b^y = x \quad (x > 0, b > 0, b \neq 1)$

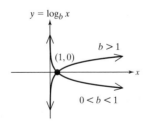

For $y = \log_b x$, the domain is $(0, \infty)$.

The range is $(-\infty, \infty)$.

The line $x = 0$ is a vertical asymptote.

The x-intercept is $(1, 0)$.

The function $y = \log x$ is the **common logarithmic function** (base 10).

Examples

Example 1

$\log_4 64 = 3 \quad$ because $4^3 = 64$

Example 2

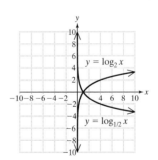

Example 3

$\log(10{,}000) = 4 \quad$ because $10^4 = 10{,}000$

section 10.5 Properties of Logarithms

Key Concepts

Let b, x, and y be positive real numbers where $b \neq 1$, and let p be a real number. Then the following properties are true.

1. $\log_b 1 = 0$
2. $\log_b b = 1$
3. $\log_b b^p = p$
4. $b^{\log_b(x)} = x$
5. $\log_b (xy) = \log_b x + \log_b y$
6. $\log_b \left(\dfrac{x}{y}\right) = \log_b x - \log_b y$
7. $\log_b x^p = p \log_b x$

The **properties of logarithms** can be used to write multiple logarithms as a single logarithm.

The **properties of logarithms** can be used to write a single logarithm as a sum or difference of logarithms.

Examples

Example 1

1. $\log_5 1 = 0$
2. $\log_6 6 = 1$
3. $\log_4 4^7 = 7$
4. $2^{\log_2 (5)} = 5$
5. $\log (5x) = \log 5 + \log x$
6. $\log_7 \left(\dfrac{z}{10}\right) = \log_7 z - \log_7 10$
7. $\log x^5 = 5 \log x$

Example 2

$\log x - \dfrac{1}{2} \log y - 3 \log z$

$= \log x - (\log y^{1/2} + \log z^3)$

$= \log x - \log (\sqrt{y} z^3)$

$= \log \left(\dfrac{x}{\sqrt{y} z^3}\right)$

Example 3

$\log \sqrt[3]{\dfrac{x}{y^2}}$

$= \dfrac{1}{3} \log \left(\dfrac{x}{y^2}\right)$

$= \dfrac{1}{3}(\log x - \log y^2)$

$= \dfrac{1}{3}(\log x - 2 \log y)$

$= \dfrac{1}{3} \log x - \dfrac{2}{3} \log y$

section 10.6 The Irrational Number e

Key Concepts

As x becomes infinitely large, the expression $\left(1 + \dfrac{1}{x}\right)^x$ approaches the irrational number e, where $e \approx 2.718281$.

The balance of an account earning compound interest n times per year is given by

$$A(t) = P\left(1 + \dfrac{r}{n}\right)^{nt}$$

where $P =$ principal, $r =$ interest rate, $t =$ time in years, and $n =$ number of compound periods per year.

The balance of an account earning interest continuously is given by

$$A(t) = Pe^{rt}$$

The function $y = e^x$ is the exponential function with base e.
The **natural logarithm function** $y = \ln(x)$ is the logarithm function with base e.

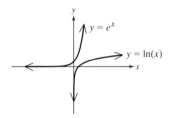

Change-of-Base Formula

$\log_b x = \dfrac{\log_a x}{\log_a b} \quad a > 0, a \neq 1, b > 0, b \neq 1$

Examples

Example 1

Find the account balance for $8000 invested for 10 years at 7% compounded quarterly.

$P = 8000 \quad t = 10 \quad r = 0.07 \quad n = 4$

$A(10) = 8000\left(1 + \dfrac{0.07}{4}\right)^{(4)(10)}$

$= \$16,012.78$

Example 2

Find the account balance for the same investment compounded continuously.

$P = 8000 \quad t = 10 \quad r = 0.07$

$A(t) = 8000e^{(0.07)(10)}$

$= \$16,110.02$

Example 3

Use a calculator to approximate the value of the expressions.

$e^{7.5} \approx 1808.04$

$e^{-\pi} \approx 0.0432$

$\ln(107) \approx 4.6728$

$\ln\left(\dfrac{1}{\sqrt{2}}\right) \approx -0.3466$

Example 4

$\log_3 59 = \dfrac{\log 59}{\log 3} \approx 3.7115$

section 10.7 Exponential and Logarithmic Equations

Key Concepts

Guidelines to Solve Logarithmic Equations

1. Isolate the logarithms on one side of the equation.
2. Write a sum or difference of logarithms as a single logarithm.
3. Rewrite the equation in step 2 in exponential form.
4. Solve the resulting equation from step 3.
5. Check all solutions to verify that they are within the domain of the logarithmic expressions in the equation.

The equivalence of exponential expressions can be used to solve **exponential equations**.

If $b^x = b^y$ then $x = y$

Guidelines to Solve Exponential Equations

1. Isolate one of the exponential expressions in the equation.
2. Take a logarithm of both sides of the equation.
3. Use the power property of logarithms to write exponents as factors.
4. Solve the resulting equation from step 3.

Examples

Example 1

$\log(3x - 1) + 1 = \log(2x + 1)$

Step 1: $\log(3x - 1) - \log(2x + 1) = -1$

Step 2: $\log\left(\dfrac{3x - 1}{2x + 1}\right) = -1$

Step 3: $10^{-1} = \dfrac{3x - 1}{2x + 1}$

Step 4: $\dfrac{1}{10} = \dfrac{3x - 1}{2x + 1}$

$2x + 1 = 10(3x - 1)$

$2x + 1 = 30x - 10$

$-28x = -11$

$x = \dfrac{11}{28}$

Step 5: $x = \dfrac{11}{28}$ Checks in original equation

Example 2

$5^{2x} = 125$

$5^{2x} = 5^3$ implies that $2x = 3$

$x = \dfrac{3}{2}$

Example 3

$4^{x+1} - 2 = 1055$

Step 1: $4^{x+1} = 1057$

Step 2: $\ln(4^{x+1}) = \ln(1057)$

Step 3: $(x + 1)\ln 4 = \ln 1057$

Step 4: $x + 1 = \dfrac{\ln 1057}{\ln 4}$

$x = \dfrac{\ln 1057}{\ln 4} - 1 \approx 4.023$

chapter 10 review exercises

Section 10.1

For Exercises 1–8, refer to the functions defined here.

$$f(x) = x - 7 \qquad g(x) = -2x^3 - 8x$$
$$m(x) = \sqrt{x} \qquad n(x) = \frac{1}{x-2}$$

Find the indicated function values. Write the domain in interval notation.

1. $(f - g)(x)$
2. $(f + g)(x)$
3. $(f \cdot n)(x)$
4. $(f \cdot m)(x)$
5. $\left(\dfrac{f}{g}\right)(x)$
6. $\left(\dfrac{g}{f}\right)(x)$
7. $(m \circ f)(x)$
8. $(n \circ f)(x)$

For Exercises 9–12, refer to the functions defined for Exercises 1–8. Find the function values, if possible.

9. $(m \circ g)(-2)$
10. $(n \circ g)(-1)$
11. $(f \circ g)(4)$
12. $(g \circ f)(8)$

13. Given $f(x) = 2x + 1$ and $g(x) = x^2$
   a. Find $(g \circ f)(x)$.
   b. Find $(f \circ g)(x)$.
   c. Based on your answers to part (a), is $f \circ g$ equal to $g \circ f$?

For Exercises 14–19, refer to the graph. Approximate the function values, if possible.

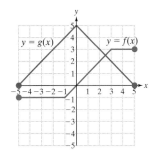

14. $\left(\dfrac{f}{g}\right)(1)$
15. $(f \cdot g)(-2)$
16. $(f + g)(-4)$
17. $(f - g)(2)$
18. $(g \circ f)(-3)$
19. $(f \circ g)(4)$

Section 10.2

For Exercises 20–21, determine if the function is one-to-one by using the horizontal line test.

20.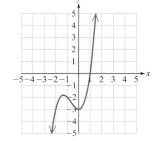
21.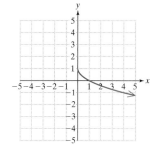

For Exercises 22–26, write the inverse for each one-to-one function.

22. $\{(3, 5), (2, 9), (0, -1), (4, 1)\}$
23. $q(x) = \dfrac{3}{4}x - 2$
24. $g(x) = \sqrt[5]{x} + 3$
25. $f(x) = (x - 1)^3$
26. $n(x) = \dfrac{4}{x - 2}$

27. Verify that the functions defined by $f(x) = 5x - 2$ and $g(x) = \frac{1}{5}x + \frac{2}{5}$ are inverses by showing that $(f \circ g)(x) = x$ and $(g \circ f)(x) = x$.

28. Graph the functions q and q^{-1} from Exercise 23 on the same coordinate axes. What can you say about the relationship between these two graphs?

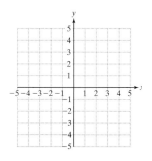

29. a. What are the domain and range of the function defined by $h(x) = \sqrt{x+1}$?

b. What are the domain and range of the function defined by $k(x) = x^2 - 1, x \geq 0$?

30. Determine the inverse of the function $p(x) = \sqrt{x} + 2$.

Section 10.3

For Exercises 31–38, evaluate the exponential expressions. Use a calculator and round to 3 decimal places, if necessary.

31. 4^5

32. 6^{-2}

33. $8^{1/3}$

34. $\left(\dfrac{1}{100}\right)^{-1/2}$

35. 2^π

36. $5^{\sqrt{3}}$

37. $(\sqrt{7})^{1/2}$

38. $\left(\dfrac{3}{4}\right)^{4/3}$

For Exercises 39–42, graph the functions.

39. $f(x) = 3^x$

40. $g(x) = \left(\dfrac{1}{4}\right)^x$

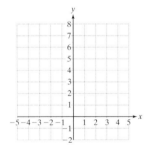
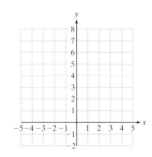

41. $h(x) = 5^{-x}$

42. $k(x) = \left(\dfrac{2}{5}\right)^{-x}$

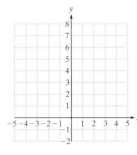
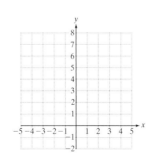

43. a. Does the graph of $y = b^x, b > 0, b \neq 1$, have a vertical or a horizontal asymptote?

b. Write an equation of the asymptote.

44. Background radiation is radiation that we are exposed to from naturally occurring sources including the soil, the foods we eat, and the sun. Background radiation varies depending on where we live. A typical background radiation level is 150 millirems (mrem) per year. (A rem is a measure of energy produced from radiation.) Suppose a substance emits 30,000 mrem per year and has a half-life of 5 years. The function defined by

$$A(t) = 30,000\left(\dfrac{1}{2}\right)^{t/5}$$

gives the radiation level (in millirems) of this substance after t years.

a. What is the radiation level after 5 years?

b. What is the radiation level after 15 years?

c. Will the radiation level of this substance be below the background level of 150 mrem after 50 years?

Section 10.4

For Exercises 45–52, evaluate the logarithms without using a calculator.

45. $\log_3\left(\dfrac{1}{27}\right)$

46. $\log_5 1$

47. $\log_7 7$

48. $\log_2 2^8$

49. $\log_2 16$

50. $\log_3 81$

51. $\log(100,000)$

52. $\log_8\left(\dfrac{1}{8}\right)$

For Exercises 53–54, graph the logarithmic functions.

53. $q(x) = \log_3 x$

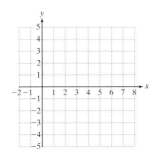

54. $r(x) = \log_{1/2} x$

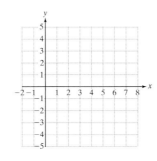

55. a. Does the graph of $y = \log_b x$ have a vertical or a horizontal asymptote?

b. Write an equation of the asymptote.

56. Acidity of a substance is measured by its pH. The pH can be calculated by the formula pH $= -\log [H^+]$, where $[H^+]$ is the hydrogen ion concentration.

a. What is the pH of a fruit with a hydrogen ion concentration of 0.00316 mol/L? Round to 1 decimal place.

b. What is the pH of an antacid tablet with $[H^+] = 3.16 \times 10^{-10}$? Round to 1 decimal place.

Section 10.5

For Exercises 57–60, evaluate the logarithms without using a calculator.

57. $\log_8 8$

58. $\log_{11} 11^6$

59. $\log_{1/2} 1$

60. $12^{\log_{12}(7)}$

61. Complete the properties. Assume x, y, and b are positive real numbers such that $b \neq 1$.

a. $\log_b(xy) =$

b. $\log_b x - \log_b y =$

c. $\log_b(x^p) =$

For Exercises 62–65, write the logarithmic expressions as a single logarithm.

62. $\dfrac{1}{4}(\log_b y - 4\log_b z + 3\log_b x)$

63. $\dfrac{1}{2}\log_3 a + \dfrac{1}{2}\log_3 b - 2\log_3 c - 4\log_3 d$

64. $\log 540 - 3\log 3 - 2\log 2$

65. $-\log_4 18 + \log_4 6 + \log_4 3 - \log_4 1$

66. Which of the following is equivalent to $\dfrac{2\log 7}{\log 7 + \log 6}$?

a. $\dfrac{\log 7}{\log 6}$ **b.** $\dfrac{\log 49}{\log 42}$ **c.** $\log\left(\dfrac{7}{6}\right)$

67. Which of the following is equivalent to $\dfrac{\log 8^{-3}}{\log 2 + \log 4}$?

a. -3 **b.** $-3\log\left(\dfrac{4}{3}\right)$ **c.** $\dfrac{-3\log 4}{\log 3}$

Section 10.6

For Exercises 68–75, use a calculator to approximate the expressions to 4 decimal places.

68. e^5

69. $e^{\sqrt{7}}$

70. $32e^{0.008}$

71. $58e^{-0.0125}$

72. $\ln(6)$

73. $\ln\left(\dfrac{1}{9}\right)$

74. $\log 22$

75. $\log e^3$

For Exercises 76–79, use the change-of-base formula to approximate the logarithms to 4 decimal places.

76. $\log_2 10$

77. $\log_9 80$

78. $\log_5(0.26)$

79. $\log_4(0.0062)$

80. An investor wants to deposit $20,000 in an account for 10 years at 5.25% interest. Compare the amount she would have if her money were invested with the following different compounding options. Use

$$A(t) = P\left(1 + \frac{r}{n}\right)^{nt}$$

for interest compounded n times per year and $A(t) = Pe^{rt}$ for interest compounded continuously.

a. Compounded annually

b. Compounded quarterly

c. Compounded monthly

d. Compounded continuously

81. To measure a student's retention of material at the end of a course, researchers give the student a test on the material every month for 24 months after the course is over. The student's average score t months after completing the course is given by

$$S(t) = 75e^{-0.5t} + 20$$

where t is the time in months and $S(t)$ is the test score.

a. Find $S(0)$ and interpret the result.

b. Find $S(6)$ and interpret the result.

c. Find $S(12)$ and interpret the result.

d. The graph of $y = S(t)$ is shown here. Does it appear that the student's average score is approaching a limiting value? Explain.

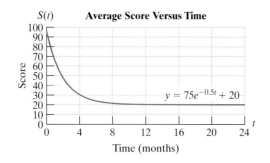

Section 10.7

For Exercises 82–89, identify the domain. Write the answer in interval notation.

82. $f(x) = e^x$

83. $g(x) = e^{x+6}$

84. $h(x) = e^{x-3}$

85. $k(x) = \ln x$

86. $q(x) = \ln(x + 5)$

87. $p(x) = \ln(x - 7)$

88. $r(x) = \ln(3x - 4)$

89. $w(x) = \ln(5 - x)$

Solve the logarithmic equations in Exercises 90–97. If necessary, round to 2 decimal places.

90. $\log_5 x = 3$

91. $\log_7 x = -2$

92. $\log_6 y = 3$

93. $\log_3 y = \dfrac{1}{12}$

94. $\log(2w - 1) = 3$

95. $\log_2(3w + 5) = 5$

96. $\log(p) - 1 = -\log(p - 3)$

97. $\log_4(2 + t) - 3 = \log_4(3 - 5t)$

Solve the exponential equations in Exercises 98–105. If necessary, round to 4 decimal places.

98. $4^{3x+5} = 16$

99. $5^{7x} = 625$

100. $4^a = 21$

101. $5^a = 18$

102. $e^{-x} = 0.1$

103. $e^{-2x} = 0.06$

104. $10^{2n} = 1512$

105. $10^{-3m} = \dfrac{1}{821}$

106. Radioactive iodine (^{131}I) is used to treat patients with a hyperactive (overactive) thyroid. Patients with this condition may have symptoms that include rapid weight loss, heart palpitations, and high blood pressure. The half-life of radioactive iodine is 8.04 days. If a patient is given an initial

dose of 2 µg, then the amount of iodine in the body after t days is approximated by

$$A(t) = 2e^{-0.0862t}$$

where t is the time in days and $A(t)$ is the amount (in micrograms) of ^{131}I remaining.

a. How much radioactive iodine is present after a week? Round to 2 decimal places.

b. How much radioactive iodine is present after 30 days? Round to 2 decimal places.

c. How long will it take for the level of radioactive iodine to reach 0.5 µg?

107. The growth of a certain bacterium in a culture is given by the model $A(t) = 150e^{0.007t}$, where $A(t)$ is the number of bacteria and t is time in minutes. Let $t = 0$ correspond to the initial number of bacteria.

a. What is the initial number of bacteria?

b. What is the population after $\frac{1}{2}$ hr?

c. How long will it take for the population to double?

108. The value of a car is depreciated with time according to

$$V(t) = 15{,}000e^{-0.15t}$$

where $V(t)$ is the value in dollars and t is the time in years.

a. Find $V(0)$ and interpret the result in the context of this problem.

b. Find $V(10)$ and interpret the result in the context of this problem. Round to the nearest dollar.

c. Find the time required for the value of the car to drop to $5000. Round to the nearest tenth of a year.

d. The graph of $y = V(t)$ is shown here. Does it appear that the value of the car is approaching a limiting value? Explain.

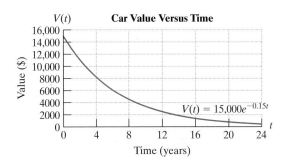

Car Value Versus Time

---

chapter 10 test

For Exercises 1–8, refer to these functions.

$$f(x) = x - 4 \qquad g(x) = \sqrt{x+2} \qquad h(x) = \frac{1}{x}$$

Find the function values if possible.

1. $\left(\dfrac{f}{g}\right)(x)$

2. $(h \cdot g)(x)$

3. $(g \circ f)(x)$

4. $(h \circ f)(x)$

5. $(f - g)(7)$

6. $(h + f)(2)$

7. $(h \circ g)(14)$

8. $(g \circ f)(0)$

9. If $f(x) = x - 4$ and $g(x) = \sqrt{x+2}$, write the domain of the function $\dfrac{g}{f}$.

10. Explain how to determine graphically if a function is one-to-one.

11. Which of the functions is one-to-one?

a.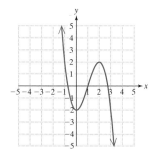
b.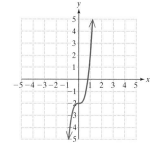

12. Write an equation of the inverse of the one-to-one function defined by $f(x) = \frac{1}{4}x + 3$.

13. Write an equation of the inverse of the function defined by $g(x) = (x-1)^2$, $x \geq 1$.

14. Given the graph of the function $y = p(x)$, graph its inverse $p^{-1}(x)$.

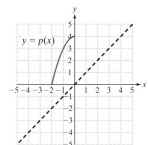 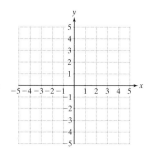

15. Use a calculator to approximate the expression to 4 decimal places.

a. $10^{2/3}$ b. $3^{\sqrt{10}}$ c. 8^{π}

16. Graph $f(x) = 4^{x-1}$.

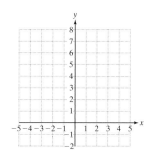

17. a. Write in logarithmic form. $16^{3/4} = 8$

b. Write in exponential form. $\log_x 31 = 5$

18. Graph $g(x) = \log_3 x$.

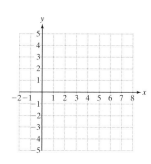

19. Complete the change-of-base formula: $\log_b n = $ _____

20. Use a calculator to approximate the expression to 4 decimal places:

a. $\log 21$ b. $\log_4 13$ c. $\log_{1/2} 6$

21. Using the properties of logarithms, expand and simplify. Assume all variables represent positive real numbers.

a. $-\log_3 \left(\dfrac{3}{9x}\right)$ b. $\log\left(\dfrac{1}{10^5}\right)$

22. Write as a single logarithm. Assume all variables represent positive real numbers.

a. $\dfrac{1}{2}\log_b x + 3\log_b y$ b. $\log a - 4\log a$

23. Use a calculator to approximate the expression to 4 decimal places, if necessary.

a. $e^{1/2}$ b. e^{-3}

c. $\ln\left(\dfrac{1}{3}\right)$ d. $\ln e$

24. Identify the graphs as $y = e^x$ or $y = \ln(x)$.

a.  b.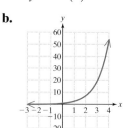

25. Researchers found that t months after taking a course, students remembered $p\%$ of the material according to

$$p(t) = 92 - 20 \ln(t + 1)$$

where $0 \leq t \leq 24$ is the time in months.

a. Find $p(4)$ and interpret the results.

b. Find $p(12)$ and interpret the results.

c. Find $p(0)$ and interpret the results.

26. The population of New York City has a 2% growth rate and can be modeled by the function $P(t) = 8008(1.02)^t$, where $P(t)$ is in thousands and t is in years ($t = 0$ corresponds to the year 2000).

a. Using this model, predict the population in the year 2010.

b. If this growth rate continues, in what year will the population reach 12 million (12 million is 12,000 thousand)?

27. A certain bacterial culture grows according to

$$P(t) = \frac{1{,}500{,}000}{1 + 5000e^{-0.8t}}$$

where P is the population of the bacteria and t is the time in hours.

a. Find $P(0)$ and interpret the result. Round to the nearest whole number.

b. How many bacteria will be present after 6 hr?

c. How many bacteria will be present after 12 hr?

d. How many bacteria will be present after 18 hr?

e. From the graph does it appear that the population of bacteria is reaching a limiting value? Explain.

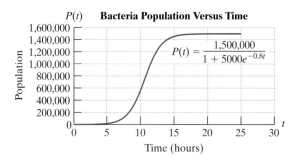

For Exercises 28–33, solve the exponential and logarithmic equations. If necessary, round to 3 decimal places.

28. $\log x + \log(x - 21) = 2$

29. $\log_{1/2} x = -5$

30. $\ln(x + 7) = 2.4$ **31.** $3^{x+4} = \dfrac{1}{27}$

32. $4^x = 50$ **33.** $e^{2.4x} = 250$

34. Atmospheric pressure P decreases exponentially with altitude x according to

$$P(x) = 760e^{-0.000122x}$$

where $P(x)$ is the pressure measured in millimeters of mercury (mm Hg) and x is the altitude measured in meters.

a. Find $P(2500)$ and interpret the result. Round to 1 decimal place.

b. Find the pressure at sea level.

c. Find the altitude at which the pressure is 633 mm Hg.

35. Use the formula $A(t) = Pe^{rt}$ to compute the value of an investment under continuous compounding.

a. If $2000 is invested at 7.5% compounded continuously, find the value of the investment after 5 years.

b. How long will it take the investment to double? Round to 2 decimal places.

chapters 1–10 | cumulative review exercises

1. Simplify completely.
$$\frac{8 - 4 \cdot 2^2 + 15 \div 5}{|-3 + 7|}$$

2. Divide.
$$\frac{-8p^2 + 4p^3 + 6p^5}{8p^2}$$

3. Divide $(t^4 - 13t^2 + 36) \div (t - 2)$. Identify the quotient and remainder.

4. Simplify. $\sqrt{x^2 - 6x + 9}$

5. Simplify. $\dfrac{4}{\sqrt[3]{40}}$

6. Find the length of the missing side.

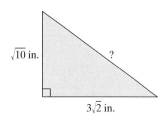

7. Simplify. Write the answer with positive exponents only.
$$\frac{2^{2/5} c^{-1/4} d^{1/5}}{2^{-8/5} c^{3/4} d^{1/10}}$$

8. Find the area of the rectangle.

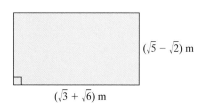

9. Perform the indicated operation.
$$\frac{4 - 3i}{2 + 5i}$$

10. Find the measure of each angle in the right triangle.

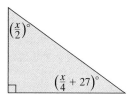

11. Find the positive slope of the sides of a pyramid with a square base 66 ft on a side and height of 22 ft.

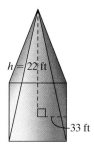

12. Solve for x. $\quad 2x(x - 7) = x - 18$

13. How many liters of pure alcohol must be mixed with 8 L of 20% alcohol to bring the concentration up to 50% alcohol?

14. Bank robbers leave the scene of a crime and head north through winding mountain roads to their hideaway. Their average rate of speed is 40 mph. The police leave 6 min later in pursuit. If the police car averages 50 mph traveling the same route, how long will it take the police to catch the bank robbers?

15. Solve the system by using the Gauss-Jordan method.
$$5x + 10y = 25$$
$$-2x + 6y = -20$$

16. Solve for w. $-2[w - 3(w + 1)] = 4 - 7(w + 3)$

17. Solve for x. $ax - c = bx + d$

18. Solve for t. $s = \frac{1}{2}gt^2$ $t \geq 0$

19. Solve for T. $\sqrt{1 - kT} = \dfrac{V_0}{V}$

20. Find the x-intercepts of the function defined by $f(x) = |x - 5| - 2$.

21. Let $f(t) = 6$, $g(t) = -5t$, and $h(t) = 2t^2$. Find
    a. $(f \cdot g)(t)$ b. $(g \circ h)(t)$ c. $(h - g)(t)$

22. Solve for q. $|2q - 5| = |2q + 5|$

23. a. Find an equation of the line parallel to the y-axis and passing through the point (2, 6).

    b. Find an equation of the line perpendicular to the y-axis and passing through the point (2, 6).

    c. Find an equation of the line perpendicular to the line $2x + y = 4$ and passing through the point (2, 6). Write the answer in slope-intercept form.

24. The number of inmates in U.S. state and federal prisons has increased with time between 1990 and 1995. See the following table.

Year	x	Number of Inmates y
1990	0	1,148,000
1991	1	1,129,000
1992	2	1,295,000
1993	3	1,369,000
1994	4	1,478,000
1995	5	1,585,000

Source: U.S. Department of Justice.

a. Let x represent the year, where $x = 0$ corresponds to 1990. Let y represent the number of inmates. Plot the ordered pairs.

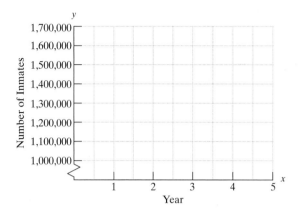

b. Use the ordered pairs (0, 1,148,000) and (4, 1,478,000) to find a linear equation to model the number of inmates as a function of the year after 1990. Write the equation in slope-intercept form.

c. What is the slope of the line from part (b)? What does the slope mean in the context of this problem?

d. Use the equation from part (b) to estimate the prison population in the year 2000.

25. The smallest angle in a triangle measures one-half the largest angle. The smallest angle measures 20° less than the middle angle. Find the measures of all three angles.

26. Solve the system.

$$\frac{1}{2}x - \frac{1}{4}y = 1$$
$$-2x + y = -4$$

27. Match the function with the appropriate graph.

i. $f(x) = \ln(x)$
ii. $g(x) = 3^x$
iii. $h(x) = x^2$
iv. $k(x) = -2x - 3$
v. $L(x) = |x|$
vi. $m(x) = \sqrt{x}$
vii. $n(x) = \sqrt[3]{x}$
viii. $p(x) = x^3$
ix. $q(x) = \dfrac{1}{x}$
x. $r(x) = x$

a.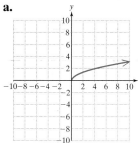
b.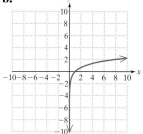
c.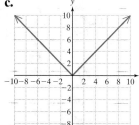
d.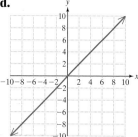
e.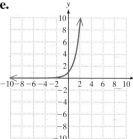
f.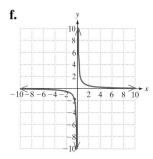
g.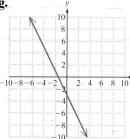
h.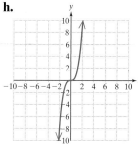
i.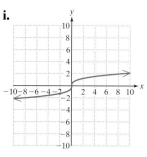
j.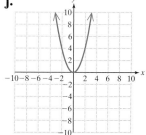

28. Find the domain. $f(x) = \sqrt{2x - 1}$.

29. Find $f^{-1}(x)$, given $f(x) = 5x - \frac{2}{3}$.

30. The volume of a gas varies directly as its temperature and inversely with pressure. At a temperature of 100 kelvins (K) and a pressure of 10 newtons per square meter (N/m²), the gas occupies a volume of 30 m³. Find the volume at a temperature of 200 K and pressure of 15 N/m².

31. Perform the indicated operations.
$$\frac{5x - 10}{x^2 - 4x + 4} \div \frac{5x^2 - 125}{25 - 5x} \cdot \frac{x^3 + 125}{10x + 5}$$

32. Perform the indicated operations.
$$\frac{x}{x - y} + \frac{y}{y - x} + x$$

33. Given the equation
$$\frac{2}{x - 4} = \frac{5}{x + 2}$$

a. Are there any restrictions on x for which the rational expressions are undefined?

b. Solve the equation.

c. Solve the related inequality.
$$\frac{2}{x - 4} \geq \frac{5}{x + 2}$$

Write the answer in interval notation.

34. Two more than 3 times the reciprocal of a number is $\frac{5}{4}$ less than the number. Find all such numbers.

35. Solve the equation. $\sqrt{-x} = x + 6$

36. Solve the inequality $2|x - 3| + 1 > -7$. Write the answer in interval notation.

37. Four million *Escherichia coli (E. coli)* bacteria are present in a laboratory culture. An antibacterial agent is introduced and the population of bacteria decreases by one-half every 6 hr according to

$$P(t) = 4{,}000{,}000 \left(\frac{1}{2}\right)^{t/6} \quad t \geq 0$$

where t is the time in hours.

a. Find the population of bacteria after 6, 12, 18, 24, and 30 hr.

b. Sketch a graph of $y = P(t)$ based on the function values found in part (a).

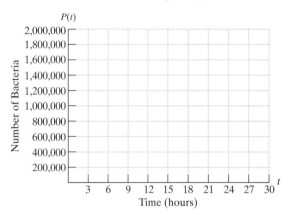

c. Predict the time required for the population to decrease to 15,625 bacteria.

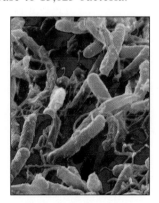

38. Evaluate the expressions without a calculator.

a. $\log_7 49$
b. $\log_4 \left(\frac{1}{64}\right)$
c. $\log(1{,}000{,}000)$
d. $\ln e^3$

39. Use a calculator to approximate the expressions to 4 decimal places.

a. $\pi^{4.7}$
b. e^{π}
c. $(\sqrt{2})^{-5}$
d. $\log 5362$
e. $\ln(0.67)$
f. $\log_4 37$

40. Solve the equation. $5^2 = 125^x$

41. Solve the equation. $e^x = 100$

42. Solve the equation. $\log_3(x + 6) - 3 = -\log_3 x$

43. Write the following expression as a single logarithm.

$$\frac{1}{2}\log z - 2\log x - 3\log y$$

44. Write the following expression as a sum or difference of logarithms

$$\ln\sqrt[3]{\frac{x^2}{y}}$$

Conic Sections

11

11.1 Distance Formula and Circles

11.2 More on the Parabola

11.3 The Ellipse and Hyperbola

11.4 Nonlinear Systems of Equations in Two Variables

11.5 Nonlinear Inequalities and Systems of Inequalities

In this chapter we will study several types of equations whose solution sets represent nonlinear graphs. These equations are called *conic sections* and consist of circles, parabolas, ellipses, and hyperbolas. These shapes are found in nature and are also used in engineering applications. For example, a reflecting telescope has a mirror whose cross section is in the shape of a parabola. Planetary orbits are often in the shape of an ellipse, as are many artificial structures such as the tunnel modeled in Exercise 41 in Section 11.3.

chapter 11 preview

The exercises in this preview contain concepts that have not yet been presented. These exercises are provided for students who want to compare their levels of understanding before and after studying the chapter. Alternatively, you may prefer to work these exercises when the chapter is completed and before taking the exam.

Section 11.1

1. Find the distance between the points $(-8, -2)$ and $(-6, 3)$.

2. Find the center and radius of the circle $(x + 2)^2 + (y + 1)^2 = 4$.

3. Find the center and the radius for the circle given by $x^2 + y^2 - 6x + 2y - 15 = 0$.

4. Write an equation for a circle with center at $(0, 0)$ and radius of 6.

Section 11.2

5. Given the equation of the parabola
$$x = -\frac{1}{2}(y + 3)^2 + 2,$$

   a. Determine the coordinates of the vertex.

   b. Write the equation of the axis of symmetry.

6. Find the vertex of $x = y^2 - 2y + 4$ by using the vertex formula.

Section 11.3

7. Graph the ellipse $\dfrac{x^2}{4} + \dfrac{y^2}{16} = 1$.

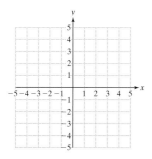

8. Graph the hyperbola $\dfrac{y^2}{16} - \dfrac{x^2}{16} = 1$.

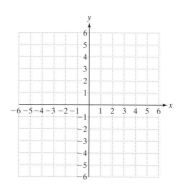

Section 11.4

For Exercises 9–10, solve the system of equations.

9. $x^2 + y^2 = 5$
$y = x + 1$

10. $2x^2 + 3y^2 = 30$
$x^2 - y^2 = 5$

Section 11.5

11. Graph the solution set to $y \geq -x^2 - 2$.

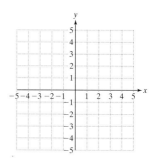

12. Graph the solution set to the system of inequalities. $y > x^2 - 1$, $\dfrac{x^2}{9} + y^2 < 1$

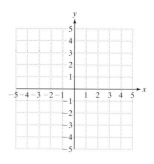

Section 11.1 Distance Formula and Circles

1. Distance Formula

Suppose we are given two points (x_1, y_1) and (x_2, y_2) in a rectangular coordinate system. The distance between the two points can be found by using the Pythagorean theorem (Figure 11-1).

First draw a right triangle with the distance d as the hypotenuse. The length of the horizontal leg a is $|x_2 - x_1|$, and the length of the vertical leg b is $|y_2 - y_1|$. From the Pythagorean theorem we have

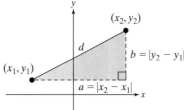

Figure 11-1

Objectives

1. Distance Formula
2. Circles
3. Writing an Equation of a Circle

$$d^2 = a^2 + b^2 \qquad \text{Pythagorean theorem}$$
$$= (x_2 - x_1)^2 + (y_2 - y_1)^2$$
$$d = \pm\sqrt{(x_2 - x_1)^2 + (y_2 - y_1)^2}$$
$$= \sqrt{(x_2 - x_1)^2 + (y_2 - y_1)^2} \qquad \text{Because distance is positive, reject the negative value.}$$

The Distance Formula

The distance d between the points (x_1, y_1) and (x_2, y_2) is
$$d = \sqrt{(x_2 - x_1)^2 + (y_2 - y_1)^2}$$

---

example 1 Finding the Distance Between Two Points

Find the distance between the points $(-2, 3)$ and $(4, -1)$ (Figure 11-2).

Solution:

$(-2, 3)$ and $(4, -1)$
(x_1, y_1) and (x_2, y_2) Label the points.

$d = \sqrt{(x_2 - x_1)^2 + (y_2 - y_1)^2}$
$= \sqrt{[(4) - (-2)]^2 + [(-1) - (3)]^2}$ Apply the distance formula.
$= \sqrt{(6)^2 + (-4)^2}$
$= \sqrt{36 + 16}$
$= \sqrt{52}$
$= \sqrt{2^2 \cdot 13}$
$= 2\sqrt{13}$

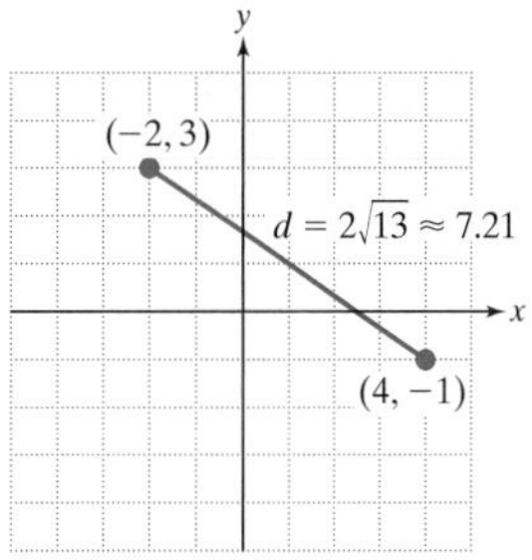

Figure 11-2

Skill Practice

1. Find the distance between the points $(-4, -2)$ and $(2, -5)$.

Answer

1. $3\sqrt{5}$

Tip: The order in which the points are labeled does not affect the result of the distance formula. For example, if the points in Example 1 had been labeled in reverse, the distance formula would still yield the same result:

$(-2, 3)$ and $(4, -1)$ $\quad d = \sqrt{(x_2 - x_1)^2 + (y_2 - y_1)^2}$

(x_2, y_2) and (x_1, y_1) $\quad\quad = \sqrt{[(-2) - (4)]^2 + [3 - (-1)]^2}$

$\quad\quad = \sqrt{(-6)^2 + (4)^2}$

$\quad\quad = \sqrt{36 + 16}$

$\quad\quad = \sqrt{52}$

$\quad\quad = 2\sqrt{13}$

2. Circles

A **circle** is defined as the set of all points in a plane that are equidistant from a fixed point called the **center**. The fixed distance from the center is called the **radius** and is denoted by r, where $r > 0$.

Suppose a circle is centered at the point (h, k) and has radius, r (Figure 11-3). The distance formula can be used to derive an equation of the circle.

Let (x, y) be any arbitrary point on the circle. Then, by definition, the distance between (h, k) and (x, y) must be r.

$$\sqrt{(x - h)^2 + (y - k)^2} = r$$

$$(x - h)^2 + (y - k)^2 = r^2 \quad \text{Square both sides.}$$

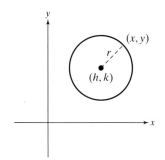

Figure 11-3

Standard Equation of a Circle

The **standard equation of a circle**, centered at (h, k) with radius r, is given by

$$(x - h)^2 + (y - k)^2 = r^2 \quad \text{where } r > 0$$

Note: If a circle is centered at the origin $(0, 0)$, then $h = 0$ and $k = 0$, and the equation simplifies to $x^2 + y^2 = r^2$.

Section 11.1 Distance Formula and Circles

example 2 Graphing a Circle

Find the center and radius of each circle. Then graph the circle.

a. $(x - 3)^2 + (y + 4)^2 = 36$
b. $x^2 + \left(y - \dfrac{10}{3}\right)^2 = \dfrac{25}{9}$

c. $x^2 + y^2 = 10$

Solution:

a. $(x - 3)^2 + (y + 4)^2 = 36$

$(x - 3)^2 + [y - (-4)]^2 = (6)^2$

$h = 3, k = -4,$ and $r = 6$

The center is at $(3, -4)$ and the radius is 6 (Figure 11-4).

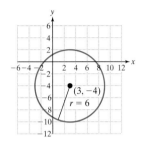

Figure 11-4

Skill Practice

Find the center and radius of each circle. Then graph the circle.

2. $(x + 1)^2 + (y - 2)^2 = 9$

3. $\left(x + \dfrac{7}{2}\right)^2 + y^2 = \dfrac{9}{4}$

Calculator Connections

Graphing calculators are designed to graph *functions*, in which y is written in terms of x. A circle is not a function. However, it can be graphed as the union of two functions—one representing the top semicircle and the other representing the bottom semicircle.

Solving for y in Example 2(a), we have

$(x - 3)^2 + (y + 4)^2 = 36$ Graph these functions as Y_1 and Y_2, using a square viewing window.

$(y + 4)^2 = 36 - (x - 3)^2$

$y + 4 = \pm\sqrt{36 - (x - 3)^2}$

$y = -4 \pm \sqrt{36 - (x - 3)^2}$

$Y_1 = -4 + \sqrt{36 - (x - 3)^2}$

$Y_2 = -4 - \sqrt{36 - (x - 3)^2}$

 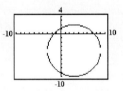

Notice that the image from the calculator does not show the upper and lower semicircles connecting at their endpoints, when in fact the semicircles should "hook up." This is due to the calculator's limited resolution.

b. $x^2 + \left(y - \dfrac{10}{3}\right)^2 = \dfrac{25}{9}$

$(x - 0)^2 + \left(y - \dfrac{10}{3}\right)^2 = \left(\dfrac{5}{3}\right)^2$

The center is at $(0, \dfrac{10}{3})$ and the radius is $\dfrac{5}{3}$ (Figure 11-5).

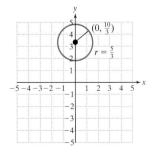

Figure 11-5

Answers

2. Center; $(-1, 2)$; radius: 3

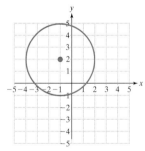

3. Center: $(-\dfrac{7}{2}, 0)$; radius: $\dfrac{3}{2}$

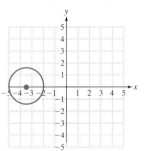

c. $x^2 + y^2 = 10$

$(x - 0)^2 + (y - 0)^2 = (\sqrt{10})^2$

The center is at $(0, 0)$ and the radius is $\sqrt{10} \approx 3.16$ (Figure 11-6).

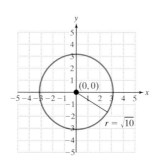

Figure 11-6

Sometimes it is necessary to complete the square to write an equation of a circle in standard form.

Skill Practice

4. Identify the center and radius of the circle given by the equation
$x^2 + y^2 - 10x + 4y - 7 = 0$

example 3   Writing the Equation of a Circle in the Form $(x - h)^2 + (y - k)^2 = r^2$

Identify the center and radius of the circle given by the equation $x^2 + y^2 + 2x - 16y + 61 = 0$.

Solution:

$x^2 + y^2 + 2x - 16y + 61 = 0$

To identify the center and radius, write the equation in the form $(x - h)^2 + (y - k)^2 = r^2$.

$(x^2 + 2x \quad) + (y^2 - 16y \quad) = -61$

Group the x-terms and group the y-terms. Move the constant to the right-hand side.

$(x^2 + 2x + 1) + (y^2 - 16y + 64) = -61 + 1 + 64$

- Complete the square on x. Add $[\frac{1}{2}(2)]^2 = 1$ to both sides of the equation.
- Complete the square on y. Add $[\frac{1}{2}(-16)]^2 = 64$ to both sides of the equation.

$(x + 1)^2 + (y - 8)^2 = 4$

$[x - (-1)]^2 + (y - 8)^2 = 2^2$

Factor and simplify.

The center is at $(-1, 8)$ and the radius is 2.

Answer

4. Center: $(5, -2)$; radius: 6

3. Writing an Equation of a Circle

example 4   Writing an Equation of a Circle

Write an equation of the circle shown in Figure 11-7.

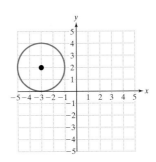

Figure 11-7

Solution:

The center is at $(-3, 2)$; therefore, $h = -3$ and $k = 2$.
From the graph, $r = 2$.

$$(x - h)^2 + (y - k)^2 = r^2$$
$$[x - (-3)]^2 + (y - 2)^2 = (2)^2$$
$$(x + 3)^2 + (y - 2)^2 = 4$$

Skill Practice

5. Write an equation for a circle whose center is the point $(6, -1)$ and whose radius is 8.

Answer

5. $(x - 6)^2 + (y + 1)^2 = 64$

---

section 11.1   Practice Exercises

Boost your GRADE at mathzone.com!

- Practice Problems
- Self-Tests
- NetTutor
- e-Professors
- Videos

Study Skills Exercises

1. What is the relationship between the formula for the distance between two points and the standard form of a circle? Understanding this relationship will help you memorize these formulas.

2. Define the key terms.
      a. Distance formula   **b.** Circle   **c.** Center   **d.** Radius   **e.** Standard equation of a circle

Objective 1: Distance Formula

For Exercises 3–16, use the distance formula to find the distance between the two points. **(See Example 1.)**

   3. $(-2, 7)$ and $(3, -9)$ **4.** $(1, 10)$ and $(-2, 4)$ **5.** $(0, 5)$ and $(-3, 8)$

6. (6, 7) and (3, 2)

7. $\left(\dfrac{2}{3}, \dfrac{1}{5}\right)$ and $\left(-\dfrac{5}{6}, \dfrac{3}{10}\right)$

8. $\left(-\dfrac{1}{2}, \dfrac{5}{8}\right)$ and $\left(-\dfrac{3}{2}, \dfrac{1}{4}\right)$

9. (4, 13) and (4, −6)

10. (−2, 5) and (−2, 9)

11. (8, −6) and (−2, −6)

12. (7, 2) and (15, 2)

13. (−6, −2) and (−1, −4)

14. (−1, −5) and (−3, −2)

15. $(3\sqrt{5}, 2\sqrt{7})$ and $(-\sqrt{5}, -3\sqrt{7})$

16. $(4\sqrt{6}, -2\sqrt{2})$ and $(2\sqrt{6}, \sqrt{2})$

17. Explain how to find the distance between 5 and −7 on the y-axis.

18. Explain how to find the distance between 15 and −37 on the x-axis.

19. Find the values of y such that the distance between the points (4, 7) and (−4, y) is 10 units.

20. Find the values of x such that the distance between the points (−4, −2) and (x, 3) is 13 units.

21. Find the values of x such that the distance between the points (x, 2) and (4, −1) is 5 units.

22. Find the values of y such that the distance between the points (−5, 2) and (3, y) is 10 units.

For Exercises 23–26, determine if the three points define the vertices of a right triangle.

23. (−3, 2), (−2, −4), and (3, 3)

24. (1, −2), (−2, 4), and (7, 1)

25. (−3, −2), (4, −3), and (1, 5)

26. (1, 4), (5, 3), and (2, 0)

Objective 2: Circles

For Exercises 27–48, identify the center and radius of the circle and then graph the circle. Complete the square, if necessary. **(See Examples 2–3.)**

27. $(x - 4)^2 + (y + 2)^2 = 9$

28. $(x - 3)^2 + (y + 1)^2 = 16$

29. $(x + 1)^2 + (y + 1)^2 = 1$

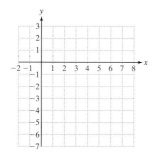

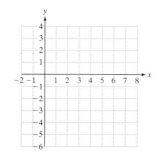

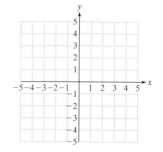

30. $(x-4)^2 + (y-4)^2 = 4$

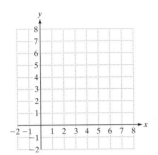

31. $x^2 + (y-5)^2 = 25$

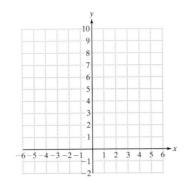

32. $(x+1)^2 + y^2 = 1$

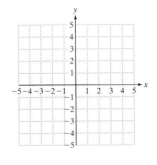

33. $(x-3)^2 + y^2 = 8$

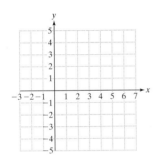

34. $x^2 + (y+2)^2 = 20$

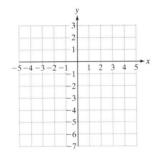

35. $x^2 + y^2 = 6$

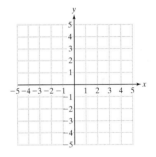

36. $x^2 + y^2 = 15$

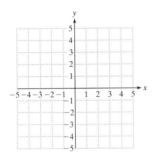

37. $\left(x + \dfrac{4}{5}\right)^2 + y^2 = \dfrac{64}{25}$

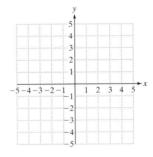

38. $x^2 + \left(y - \dfrac{5}{2}\right)^2 = \dfrac{9}{4}$

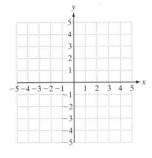

39. $x^2 + y^2 - 2x - 6y - 26 = 0$

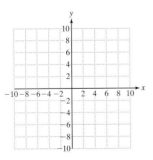

40. $x^2 + y^2 + 4x - 8y + 16 = 0$

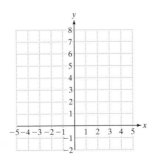

41. $x^2 + y^2 - 6y + 5 = 0$

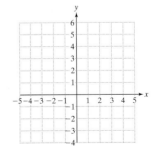

42. $x^2 + 2x + y^2 - 24 = 0$

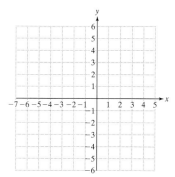

43. $x^2 + y^2 + 6y + \dfrac{65}{9} = 0$

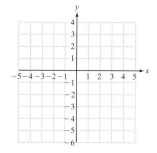

44. $x^2 + y^2 + 12x + \dfrac{143}{4} = 0$

45. $x^2 + y^2 - 12x + 12y + 71 = 0$

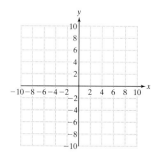

46. $x^2 + y^2 + 2x + 4y - 4 = 0$

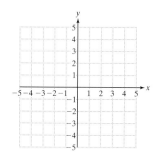

47. $2x^2 + 2y^2 = 32$

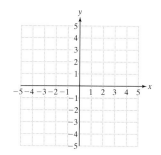

48. $3x^2 + 3y^2 = 3$

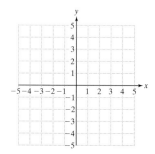

Objective 3: Writing an Equation of a Circle

For Exercises 49–54, write an equation that represents the graph of the circle. (**See Example 4.**)

49.

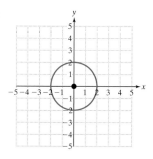

50.

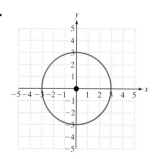

51.

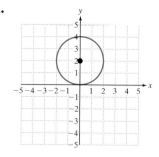

52.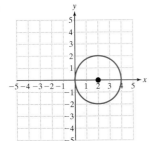
53.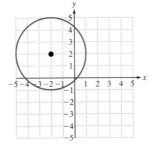
54.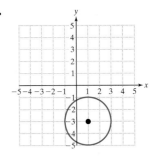

55. Write an equation of a circle centered at the origin with a radius of 7 m.

56. Write an equation of a circle centered at the origin with a radius of 12 m.

57. Write an equation of a circle centered at $(-3, -4)$ with a diameter of 12 ft.

58. Write an equation of a circle centered at $(5, -1)$ with a diameter of 8 ft.

Expanding Your Skills

59. Write an equation of a circle that has the points $(-2, 3)$ and $(2, 3)$ as endpoints of a diameter.

60. Write an equation of a circle that has the points $(-1, 3)$ and $(-1, -3)$ as endpoints of a diameter.

61. Write an equation of a circle whose center is at $(4, 4)$ and is tangent to the x- and y-axes. (*Hint:* Sketch the circle first.)

62. Write an equation of a circle whose center is at $(-3, 3)$ and is tangent to the x- and y-axes. (*Hint:* Sketch the circle first.)

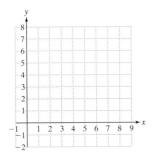
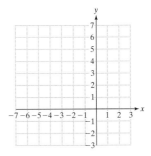

63. Write an equation of a circle whose center is at $(1, 1)$ and that passes through the point $(-4, 3)$.

64. Write an equation of a circle whose center is at $(-3, -1)$ and that passes through the point $(5, -2)$.

Graphing Calculator Exercises

For Exercises 65–70, graph the circles from the indicated exercise on a square viewing window, and approximate the center and the radius from the graph.

65. $(x - 4)^2 + (y + 2)^2 = 9$ (Exercise 27)

66. $(x - 3)^2 + (y + 1)^2 = 16$ (Exercise 28)

67. $x^2 + (y - 5)^2 = 25$ (Exercise 31)

68. $(x + 1)^2 + y^2 = 1$ (Exercise 32)

69. $x^2 + y^2 = 6$ (Exercise 35)

70. $x^2 + y^2 = 15$ (Exercise 36)

---

section 11.2 More on the Parabola

Objectives
1. Introduction to Conic Sections
2. Parabola—Vertical Axis of Symmetry
3. Parabola—Horizontal Axis of Symmetry
4. Vertex Formula

1. Introduction to Conic Sections

Recall that the graph of a second-degree equation of the form $y = ax^2 + bx + c$ ($a \neq 0$) is a parabola. In Section 11.1 we learned that the graph of $(x - h)^2 + (y - k)^2 = r^2$ is a circle. These and two other types of figures called ellipses and hyperbolas are called **conic sections**. Conic sections derive their names because each is the intersection of a plane and a double-napped cone (Figure 11-8).

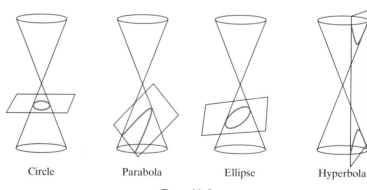

Figure 11-8

2. Parabola—Vertical Axis of Symmetry

A **parabola** is defined by a set of points in a plane that are equidistant from a fixed line (called the directrix) and a fixed point (called the focus) not on the directrix. Parabolas have numerous real-world applications. For example, a reflecting telescope has a mirror with the cross section in the shape of a parabola.

A parabolic mirror has the property that incoming rays of light are reflected from the surface of the mirror to the focus.

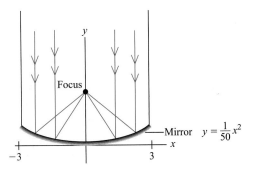

The graph of the solutions to the quadratic equation $y = ax^2 + by + c$ is a parabola. We graphed parabolas of this type in Section 8.4. Recall that the **vertex** is the highest or the lowest point of a parabola. The **axis of symmetry** of the parabola is a line that passes through the vertex and is perpendicular to the directrix (Figure 11-9).

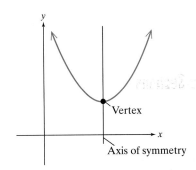

Figure 11-9

Standard Form of the Equation of a Parabola—Vertical Axis of Symmetry

The standard form of the equation of a parabola with vertex (h, k) and vertical axis of symmetry is

$$y = a(x - h)^2 + k \quad \text{where } a \neq 0$$

If $a > 0$, then the parabola opens upward; and if $a < 0$, the parabola opens downward.

The axis of symmetry is given by $x = h$.

In Section 8.5, we learned that it is sometimes necessary to complete the square to write the equation of a parabola in standard form.

example 1 Graphing a Parabola by First Completing the Square

Given: the equation of the parabola $y = -2x^2 + 4x + 1$

a. Write the equation in standard form $y = a(x - h)^2 + k$.

b. Identify the vertex and axis of symmetry. Determine if the parabola opens upward or downward.

c. Graph the parabola.

Skill Practice

Given: the equation of the parabola $y = 2x^2 - 4x + 5$

1. Write the equation in standard form.
2. Identify the vertex and axis of symmetry.
3. Graph the parabola.

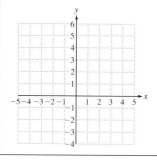

Solution:

a. Complete the square to write the equation in the form $y = a(x - h)^2 + k$.

$y = -2x^2 + 4x + 1$

$y = -2(x^2 - 2x) + 1$ Factor out -2 from the variable terms.

$y = -2(x^2 - 2x + 1 - 1) + 1$ Add and subtract the quantity $\left[\frac{1}{2}(-2)\right]^2 = 1$.

$y = -2(x^2 - 2x + 1) + (-2)(-1) + 1$ Remove the -1 term from within the parentheses by first applying the distributive property. When -1 is removed from the parentheses, it carries with it the factor of -2 from outside the parentheses.

$y = -2(x - 1)^2 + 2 + 1$

$y = -2(x - 1)^2 + 3$

The equation is in the form $y = a(x - h)^2 + k$ where $a = -2$, $h = 1$, and $k = 3$.

b. The vertex is $(1, 3)$. Because a is negative, $(a = -2 < 0)$, the parabola opens downward. The axis of symmetry is $x = 1$.

c. To graph the parabola, we know that its orientation is downward. Furthermore, we know the vertex is $(1, 3)$. To find other points on the parabola, select several values of x and solve for y. Recall that the y-intercept is found by substituting $x = 0$ and solving for y.

x	y	
1	3	← Vertex
−1	−5	
0	1	← y-intercept
2	1	
3	−5	

Choose x.   Solve for y.

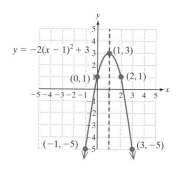

Answers

1. $y = 2(x - 1)^2 + 3$
2. Vertex: $(1, 3)$; axis of symmetry: $x = 1$
3.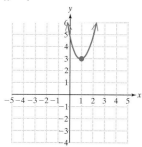

3. Parabola—Horizontal Axis of Symmetry

We have seen that the graph of a parabola $y = ax^2 + bx + c$ opens upward if $a > 0$ and downward if $a < 0$. A parabola can also open to the left or right. In such a case, the "roles" of x and y are essentially interchanged in the equation. Thus, the graph of $x = ay^2 + by + c$ opens to the right if $a > 0$ (Figure 11-10) and to the left if $a < 0$ (Figure 11-11).

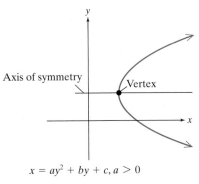

$x = ay^2 + by + c, a > 0$

Figure 11-10

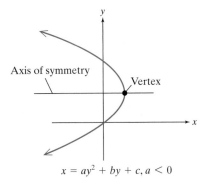

$x = ay^2 + by + c, a < 0$

Figure 11-11

Standard Form of the Equation of a Parabola—Horizontal Axis of Symmetry

The standard form of the equation of a parabola with vertex (h, k) and horizontal axis of symmetry is

$$x = a(y - k)^2 + h \qquad \text{where } a \neq 0$$

If $a > 0$, then the parabola opens to the right and if $a < 0$, the parabola opens to the left.

The axis of symmetry is given by $y = k$.

Skill Practice

Given: the equation $x = -y^2 + 1$

4. Identify the vertex and the axis of symmetry.
5. Determine if the parabola opens to the right or to the left.
6. Graph the parabola.

example 2   Graphing a Parabola with a Horizontal Axis of Symmetry

Given the equation of the parabola $x = 4y^2$,

a. Determine the coordinates of the vertex and the equation of the axis of symmetry.

b. Use the value of a to determine if the parabola opens to the right or left.

c. Plot several points and graph the parabola.

Solution:

a. The equation can be written in the form $x = a(y - k)^2 + h$:

$$x = 4(y - 0)^2 + 0$$

Therefore, $h = 0$ and $k = 0$.

The vertex is $(0, 0)$. The axis of symmetry is $y = 0$ (the x-axis).

b. Because a is positive ($a = 4 > 0$), the parabola opens to the right.

c. The vertex of the parabola is $(0, 0)$. To find other points on the graph, select values for y and solve for x.

x	y
0	0 ← Vertex
4	1
4	−1
16	2
16	−2

↑ Solve for x.   ↑ Choose y.

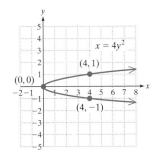

Answers

4. Vertex: $(1, 0)$; axis of symmetry: $y = 0$
5. Parabola opens left.
6.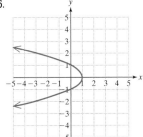

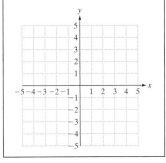

Calculator Connections

Graphing calculators are designed to graph functions, where y is written as a function of x. The parabola from Example 2 is not a function. It can be graphed, however, as the union of two functions, one representing the top branch and the other representing the bottom branch.

Solving for y in Example 2, we have: $x = 4y^2$

$$y^2 = \frac{x}{4}$$

$$y = \pm\sqrt{\frac{x}{4}}$$

$$y = \pm\frac{\sqrt{x}}{2}$$

Graph these functions as Y_1 and Y_2.

$$Y_1 = \frac{\sqrt{x}}{2} \quad \text{and} \quad Y_2 = -\frac{\sqrt{x}}{2}$$

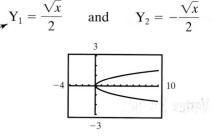

Skill Practice

Given the equation of the parabola $x = \frac{1}{8}(y + 2)^2 + 3$,

7. Determine the coordinates of the vertex.

8. Determine if the parabola opens to the right or to the left.

9. Graph the parabola.

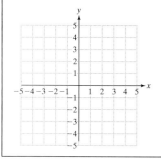

Answers
7. Vertex: $(3, -2)$
8. Right
9.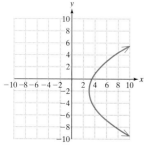

example 3 — Graphing a Parabola by First Completing the Square

Given the equation of the parabola $x = -y^2 + 8y - 14$,

a. Write the equation in standard form $x = a(y - k)^2 + h$.

b. Identify the vertex and axis of symmetry. Determine if the parabola opens to the right or left.

c. Graph the parabola.

Solution:

a. Complete the square to write the equation in the form $x = a(y - k)^2 + h$.

$x = -y^2 + 8y - 14$

$x = -1(y^2 - 8y) - 14$ Factor out -1 from the variable terms.

$x = -1(y^2 - 8y + 16 - 16) - 14$ Add and subtract the quantity $[\frac{1}{2}(-8)]^2 = 16$.

$x = -1(y^2 - 8y + 16) + (-1)(-16) - 14$ Remove the -16 term from within the parentheses by first applying the distributive property. When -16 is removed from the parentheses, it carries with it the factor of -1 from outside the parentheses.

$x = -1(y - 4)^2 + 16 - 14$

$x = -1(y - 4)^2 + 2$

The equation is in the form $x = a(y - k)^2 + h$, where $a = -1$, $h = 2$, and $k = 4$.

b. The vertex is at (2, 4). Because a is negative ($a = -1 < 0$), the parabola opens to the left. The axis of symmetry is $y = 4$.

c.

x	y	
2	4	←Vertex
1	5	
1	3	
-2	6	
-2	2	

↑ Solve for x. ↑ Choose y.

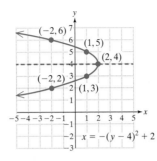

4. Vertex Formula

From Section 8.5, we learned that the vertex formula can also be used to find the vertex of a parabola.

For a parabola defined by $y = ax^2 + bx + c$,

- The x-coordinate of the vertex is given by $x = \frac{-b}{2a}$.
- The y-coordinate of the vertex is found by substituting this value for x into the original equation and solving for y.

For a parabola defined by $x = ay^2 + by + c$,

- The y-coordinate of the vertex is given by $y = \frac{-b}{2a}$.
- The x-coordinate of the vertex is found by substituting this value for y into the original equation and solving for x.

example 4 Finding the Vertex of a Parabola by Using the Vertex Formula

Find the vertex by using the vertex formula.

a. $x = y^2 + 4y + 5$ **b.** $y = \frac{1}{2}x^2 - 3x + \frac{5}{2}$

Solution:

a. $x = y^2 + 4y + 5$ The parabola is in the form $x = ay^2 + by + c$.

$a = 1 \quad b = 4 \quad c = 5$ Identify a, b, and c.

The y-coordinate of the vertex is given by $y = \frac{-b}{2a} = \frac{-(4)}{2(1)} = -2$.

The x-coordinate of the vertex is found by substitution:
$x = (-2)^2 + 4(-2) + 5 = 1$.

The vertex is at $(1, -2)$.

Skill Practice

10. Find the vertex by using the vertex formula.

$x = -4y^2 + 12y$

Answer

10. Vertex: $\left(9, \frac{3}{2}\right)$

b. $y = \dfrac{1}{2}x^2 - 3x + \dfrac{5}{2}$ The parabola is in the form $y = ax^2 + bx + c$.

$a = \dfrac{1}{2} \quad b = -3 \quad c = \dfrac{5}{2}$ Identify a, b, and c.

The x-coordinate of the vertex is given by $x = \dfrac{-b}{2a} = \dfrac{-(-3)}{2(\frac{1}{2})} = 3$.

The y-coordinate of the vertex is found by substitution.

$$y = \dfrac{1}{2}(3)^2 - 3(3) + \dfrac{5}{2}$$
$$= \dfrac{9}{2} - 9 + \dfrac{5}{2}$$
$$= \dfrac{14}{2} - 9$$
$$= 7 - 9$$
$$= -2$$

The vertex is at $(3, -2)$.

section 11.2 Practice Exercises

Boost your GRADE at mathzone.com!
MathZone

- Practice Problems
- Self-Tests
- NetTutor
- e-Professors
- Videos

Study Skills Exercises

1. You should not wait until the last minute to study for any test in math. It is particularly important that you begin your review early for the final exam. Check all the activities that will help you study for the final.

   ☐ Rework all old exams.

   ☐ Rework the chapter test for the chapters you covered in class.

   ☐ Do the cumulative review exercises at the end of Chapters 2–11.

2. Define the key terms.

   a. Conic sections   **b. Parabola**   **c. Axis of symmetry**   **d. Vertex**

Review Exercises

3. Determine the distance between the points $(1, 1)$ and $(2, -2)$.

4. Determine the distance from the origin to the point $(4, -3)$.

For Exercises 5–8, identify the center and radius of the circle and then graph the circle.

5. $x^2 + (y + 1)^2 = 16$

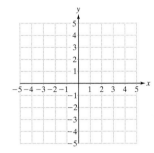

6. $(x - 3)^2 + y^2 = 4$

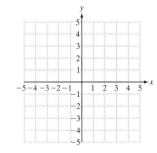

7. $(x - 3)^2 + (y + 3)^2 = 1$

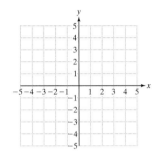

8. $(x + 2)^2 + (y + 4)^2 = 9$

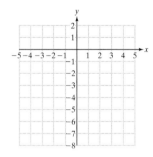

Objective 2: Parabola—Vertical Axis of Symmetry

For Exercises 9–16, use the equation of the parabola in standard form $y = a(x - h)^2 + k$ to determine the coordinates of the vertex and the equation of the axis of symmetry. Then graph the parabola. (See Example 1.)

9. $y = (x + 2)^2 + 1$

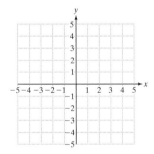

10. $y = (x - 1)^2 + 1$

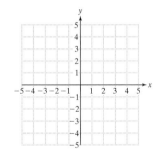

11. $y = x^2 - 4x + 3$

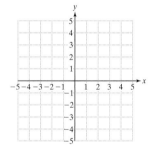

12. $y = x^2 + 6x + 5$

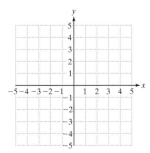

13. $y = -2x^2 + 8x$

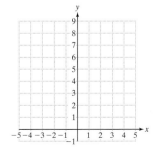

14. $y = -3x^2 - 6x$

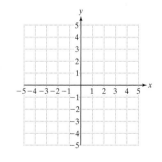

15. $y = -x^2 - 3x + 2$

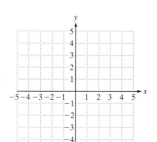

16. $y = -x^2 + x - 4$

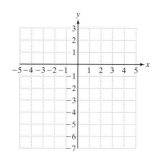

Objective 3: Parabola—Horizontal Axis of Symmetry

For Exercises 17–24, use the equation of the parabola in standard form $x = a(y - k)^2 + h$ to determine the coordinates of the vertex and the axis of symmetry. Then graph the parabola. **(See Examples 2–3.)**

17. $x = y^2 - 3$

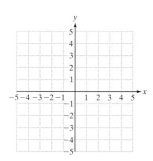

18. $x = y^2 + 1$

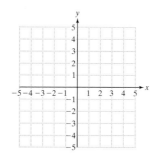

19. $x = -(y - 3)^2 - 3$

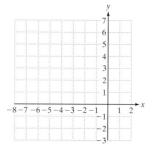

20. $x = -2(y + 2)^2 + 1$

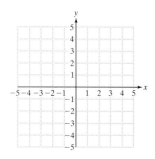

21. $x = -4y^2 - 4y - 2$

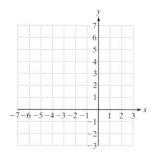

22. $x = -y^2 + 4y - 4$

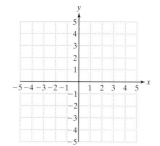

23. $x = y^2 - 2y + 2$

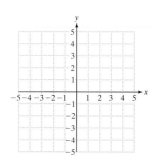

24. $x = y^2 + 4y + 1$

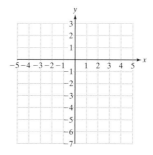

Objective 4: Vertex Formula

For Exercises 25–34, determine the vertex by using the vertex formula. **(See Example 4.)**

25. $y = x^2 - 4x + 3$

26. $y = x^2 + 6x - 2$

27. $x = y^2 + 2y + 6$

28. $x = y^2 - 8y + 3$

29. $y = -2x^2 + 8x$

30. $y = -3x^2 - 6x$

31. $y = x^2 - 3x + 2$

32. $y = x^2 + x - 4$

33. $x = -2y^2 + 16y + 1$

34. $x = -3y^2 - 6y + 7$

Mixed Exercises

35. Explain how to determine whether a parabola opens upward, downward, left, or right.

36. Explain how to determine whether a parabola has a vertical or horizontal axis of symmetry.

For Exercises 37–46, use the equation of the parabola to first determine whether the axis of symmetry is vertical or horizontal. Then determine if the parabola opens upward, downward, left, or right.

37. $y = (x - 2)^2 + 3$

38. $y = (x - 4)^2 + 2$

39. $y = -2(x + 1)^2 - 4$

40. $y = -3(x + 2)^2 - 1$

41. $x = y^2 + 4$

42. $x = y^2 - 2$

43. $x = -(y + 3)^2 + 2$

44. $x = -2(y - 1)^2 - 3$

45. $y = -2x^2 - 5$

46. $y = -x^2 + 3$

---

section 11.3 The Ellipse and Hyperbola

1. Standard Form of an Equation of an Ellipse

Objectives
1. Standard Form of an Equation of an Ellipse
2. Standard Forms of an Equation of a Hyperbola

In this section we will study the two remaining conic sections: the ellipse and the hyperbola. An **ellipse** is the set of all points (x, y) such that the sum of the distance between (x, y) and two distinct points is a constant. The fixed points are called the foci (plural of *focus*) of the ellipse.

To visualize an ellipse, consider the following application. Suppose Sonya wants to cut an elliptical rug from a rectangular rug to avoid a stain made by the family dog. She places two tacks along the center horizontal line. Then she ties the ends of a slack piece of rope to each tack. With the rope pulled tight, she traces out a curve. This curve is an ellipse, and the tacks are located at the foci of the ellipse (Figure 11-12).

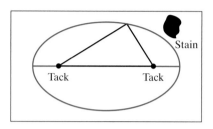

Figure 11-12

In this section, we will graph ellipses that are centered at the origin.

Standard Form of an Equation of an Ellipse Centered at the Origin

An ellipse with the center at the origin has the equation $\dfrac{x^2}{a^2} + \dfrac{y^2}{b^2} = 1$ where a and b are positive real numbers. In the standard form of the equation, the right side must equal 1.

To graph an ellipse centered at the origin, find the x- and y-intercepts.

To find the x-intercepts, let $y = 0$.

$$\frac{x^2}{a^2} + \frac{0}{b^2} = 1$$

$$\frac{x^2}{a^2} = 1$$

$$x^2 = a^2$$

$$x = \pm\sqrt{a^2}$$

$$x = \pm a$$

The x-intercepts are $(a, 0)$ and $(-a, 0)$.

To find the y-intercepts, let $x = 0$.

$$\frac{0}{a^2} + \frac{y^2}{b^2} = 1$$

$$\frac{y^2}{b^2} = 1$$

$$y^2 = b^2$$

$$y = \pm\sqrt{b^2}$$

$$y = \pm b$$

The y-intercepts are $(0, b)$ and $(0, -b)$.

Concept Connections

1. Refer to the graph of the ellipse to determine an equation of the ellipse.

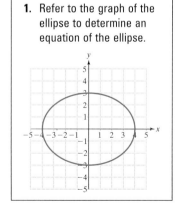

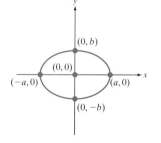

Answer

1. $\dfrac{x^2}{16} + \dfrac{y^2}{9} = 1$

example 1 — Graphing an Ellipse

Graph the ellipse given by the equation. $\dfrac{x^2}{9} + \dfrac{y^2}{4} = 1$

Solution:

The equation can be written as $\dfrac{x^2}{3^2} + \dfrac{y^2}{2^2} = 1$; therefore, $a = 3$ and $b = 2$.

Graph the intercepts $(3, 0)$, $(-3, 0)$, $(0, 2)$, $(0, -2)$ and sketch the ellipse.

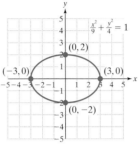

Skill Practice

2. Graph the ellipse given by the equation $x^2 + \dfrac{y^2}{16} = 1$.

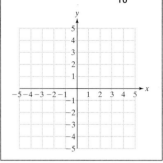

example 2 — Graphing an Ellipse

Graph the ellipse given by the equation. $25x^2 + y^2 = 25$

Solution:

First, to write the equation in standard form, divide both sides by 25:

$x^2 + \dfrac{y^2}{25} = 1$

The equation can then be written as $\dfrac{x^2}{1^2} + \dfrac{y^2}{5^2} = 1$; therefore, $a = 1$ and $b = 5$.

Graph the intercepts $(1, 0)$, $(-1, 0)$, $(0, 5)$, and $(0, -5)$ and sketch the ellipse.

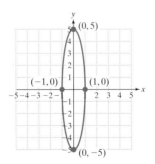

Skill Practice

3. Graph the ellipse given by the equation $25x^2 + 16y^2 = 400$.

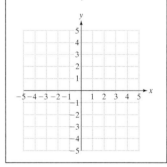

Answers

2.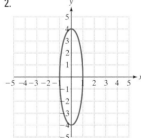

3.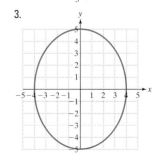

2. Standard Forms of an Equation of a Hyperbola

A **hyperbola** is the set of all points (x, y) such that the *difference* of the distances between (x, y) and two distinct points is a constant. The fixed points are called the foci of the hyperbola. The graph of a hyperbola has two parts, called branches. Each part resembles a parabola but is a slightly different shape. A hyperbola has two vertices that lie on an axis of symmetry called the **transverse axis**. For the hyperbolas studied here, the transverse axis is either horizontal or vertical.

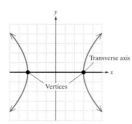
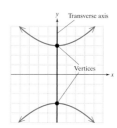

Standard Forms of an Equation of a Hyperbola with Center at the Origin

Let a and b represent positive real numbers.

Horizontal transverse axis:

The standard form of an equation of a hyperbola with a *horizontal transverse axis* and center at the origin is given by $\dfrac{x^2}{a^2} - \dfrac{y^2}{b^2} = 1$.

Note: The x-term is positive. The branches of the hyperbola open left and right.

Vertical transverse axis:

The standard form of an equation of a hyperbola with a *vertical transverse axis* and center at the origin is given by $\dfrac{y^2}{b^2} - \dfrac{x^2}{a^2} = 1$.

Note: The y-term is positive. The branches of the hyperbola open up and down.
In the standard forms of an equation of a hyperbola, the right side must equal 1.

To graph a hyperbola centered at the origin, first construct a reference rectangle. Draw this rectangle by using the points (a, b), $(-a, b)$, $(a, -b)$, and $(-a, -b)$. Asymptotes lie on the diagonals of the rectangle. The branches of the hyperbola are drawn to approach the asymptotes.

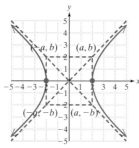
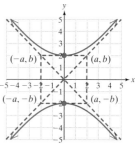

Skill Practice

4. Graph the hyperbola
$\dfrac{y^2}{4} - \dfrac{x^2}{9} = 1$.

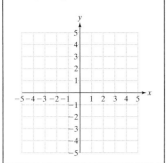

Answer

4.

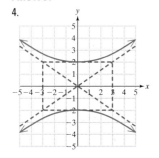

example 3   Graphing a Hyperbola

Graph the hyperbola given by the equation. $\quad \dfrac{x^2}{36} - \dfrac{y^2}{9} = 1$

a. Determine whether the transverse axis is horizontal or vertical.

b. Draw the reference rectangle and asymptotes.

c. Graph the hyperbola and label the vertices.

Solution:

a. Since the x-term is positive, the transverse axis is horizontal.

b. The equation can be written $\dfrac{x^2}{6^2} - \dfrac{y^2}{3^2} = 1$; therefore, $a = 6$ and $b = 3$. Graph the reference rectangle from the points $(6, 3), (6, -3), (-6, 3), (-6, -3)$.

Tip: In the equation $\dfrac{x^2}{36} - \dfrac{y^2}{9} = 1$, the x-term is positive. Therefore, the hyperbola opens in the x-direction (left/right).

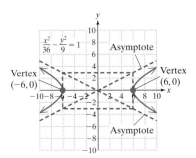

c. The coordinates of the vertices are $(-6, 0)$ and $(6, 0)$.

---

example 4   Graphing a Hyperbola

Graph the hyperbola given by the equation. $y^2 - 4x^2 - 16 = 0$

a. Write the equation in standard form to determine whether the transverse axis is horizontal or vertical.

b. Draw the reference rectangle and asymptotes.

c. Graph the hyperbola and label the vertices.

Solution:

a. Isolate the variable terms and divide by 16: $\dfrac{y^2}{16} - \dfrac{x^2}{4} = 1.$

Since the y-term is positive, the transverse axis is vertical.

b. The equation can be written $\dfrac{y^2}{4^2} - \dfrac{x^2}{2^2} = 1$; therefore, $a = 2$ and $b = 4$. Graph the reference rectangle from the points $(2, 4), (2, -4), (-2, 4), (-2, -4)$.

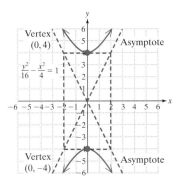

c. The coordinates of the vertices are $(0, 4)$ and $(0, -4)$.

Skill Practice

5. Graph the hyperbola $\dfrac{x^2}{1} - \dfrac{y^2}{9} = 1.$

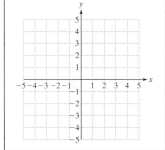

Answer

5.
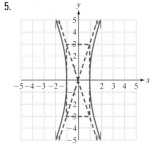

Section 11.3 Practice Exercises

Boost your GRADE at mathzone.com!

- Practice Problems
- Self-Tests
- NetTutor
- e-Professors
- Videos

Study Skills Exercises

1. To help remember formulas for circles, parabolas, ellipses, and hyperbolas, list them together and note their similarities and differences.

   a. Circle   b. Parabola   c. Ellipse   d. Hyperbola

2. Define the key terms.

   a. Ellipse   b. Hyperbola   c. Transverse axis of a hyperbola

Review Exercises

For Exercises 3–4, identify the center and radius of the circle.

3. $x^2 + y^2 - 16x + 12y = 0$

4. $x^2 + y^2 + 4x + 4y - 1 = 0$

For Exercises 5–6, identify the vertex and the axis of symmetry.

5. $y = 3(x + 3)^2 - 1$

6. $x = -\dfrac{1}{4}(y - 1)^2 - 6$

7. Write an equation for a circle whose center has coordinates $\left(\tfrac{1}{2}, \tfrac{5}{2}\right)$ with radius equal to $\tfrac{1}{2}$.

8. Write the equation for the circle centered at the origin and with radius equal to $\tfrac{1}{8}$.

Objective 1: Standard Form of an Equation of an Ellipse

For Exercises 9–16, find the x- and y-intercepts and graph the ellipse. (**See Examples 1–2.**)

9. $\dfrac{x^2}{4} + \dfrac{y^2}{9} = 1$

10. $\dfrac{x^2}{16} + \dfrac{y^2}{25} = 1$

11. $\dfrac{x^2}{16} + \dfrac{y^2}{9} = 1$

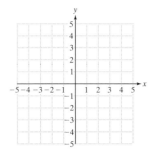
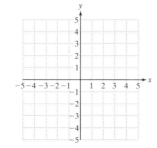
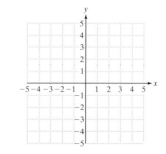

12. $\dfrac{x^2}{36} + \dfrac{y^2}{4} = 1$

13. $4x^2 + y^2 = 4$

14. $9x^2 + y^2 = 36$

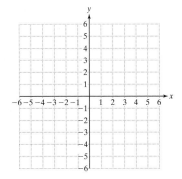
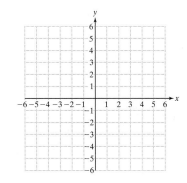
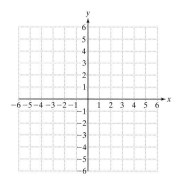

15. $x^2 + 25y^2 - 25 = 0$

16. $4x^2 + 9y^2 = 144$

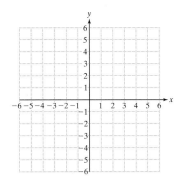
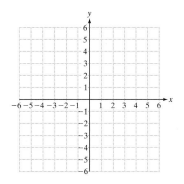

Objective 2: Standard Forms of an Equation of a Hyperbola

For Exercises 17–24, determine whether the transverse axis is horizontal or vertical.

17. $\dfrac{y^2}{6} - \dfrac{x^2}{18} = 1$

18. $\dfrac{y^2}{10} - \dfrac{x^2}{4} = 1$

19. $\dfrac{x^2}{20} - \dfrac{y^2}{15} = 1$

20. $\dfrac{x^2}{12} - \dfrac{y^2}{9} = 1$

21. $x^2 - y^2 = 12$

22. $x^2 - y^2 = 15$

23. $x^2 - 3y^2 = -9$

24. $2x^2 - y^2 = -10$

828 Chapter 11 Conic Sections

For Exercises 25–32, use the equation in standard form to graph the hyperbola. Label the vertices of the hyperbola. **(See Examples 3–4.)**

25. $\dfrac{x^2}{25} - \dfrac{y^2}{16} = 1$

26. $\dfrac{x^2}{9} - \dfrac{y^2}{36} = 1$

27. $\dfrac{y^2}{4} - \dfrac{x^2}{4} = 1$

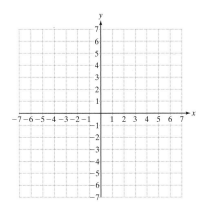
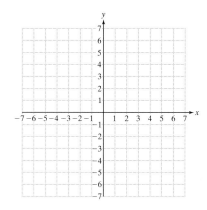

28. $\dfrac{y^2}{9} - \dfrac{x^2}{9} = 1$

29. $36x^2 - y^2 = 36$

30. $x^2 - 25y^2 = 25$

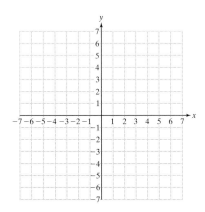
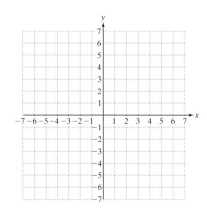
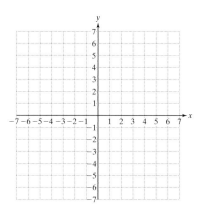

31. $y^2 - 4x^2 - 16 = 0$

32. $y^2 - 4x^2 - 36 = 0$

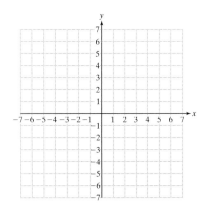
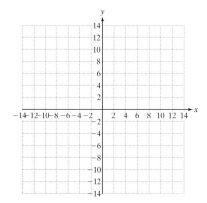

Mixed Exercises

For Exercises 33–40, determine if the equation represents an ellipse or a hyperbola.

33. $\dfrac{x^2}{6} - \dfrac{y^2}{10} = 1$

34. $\dfrac{x^2}{14} + \dfrac{y^2}{2} = 1$

35. $\dfrac{y^2}{4} + \dfrac{x^2}{16} = 1$

36. $\dfrac{x^2}{5} + \dfrac{y^2}{10} = 1$

37. $4x^2 + y^2 = 16$

38. $-3x^2 - 4y^2 = -36$

39. $-y^2 + 2x^2 = -10$

40. $x^2 - y^2 = -1$

Expanding Your Skills

41. An arch for a tunnel is in the shape of a semiellipse. The distance between vertices is 120 ft, and the height to the top of the arch is 50 ft. Find the height of the arch 10 ft from the center. Round to the nearest foot.

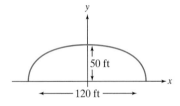

42. A bridge over a gorge is supported by an arch in the shape of a semiellipse. The length of the bridge is 400 ft, and the height is 100 ft. Find the height of the arch 50 ft from the center. Round to the nearest foot.

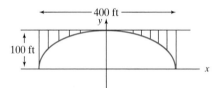

An ellipse (or hyperbola) does not have to be centered at the origin. Suppose that a and b are positive real numbers with $a > b$. Then

- The equation $\dfrac{(x - h)^2}{a^2} + \dfrac{(y - k)^2}{b^2} = 1$ represents an ellipse with center at (h, k) and elongated in the *horizontal* direction.

- The equation $\dfrac{(x - h)^2}{b^2} + \dfrac{(y - k)^2}{a^2} = 1$ represents an ellipse with center at (h, k) and elongated in the *vertical* direction.

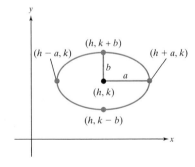
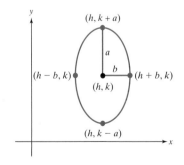

For Exercises 43–46, identify the center (h, k) of the ellipse and the values of a and b. Then graph the ellipse.

43. $\dfrac{(x-4)^2}{4} + \dfrac{(y-5)^2}{9} = 1$

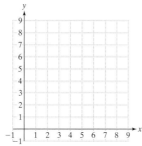

44. $\dfrac{(x+2)^2}{9} + \dfrac{(y-1)^2}{16} = 1$

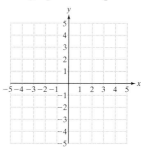

45. $\dfrac{(x-2)^2}{9} + (y+3)^2 = 1$

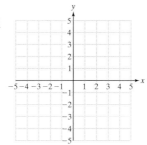

46. $\dfrac{(x-4)^2}{25} + \dfrac{(y+1)^2}{16} = 1$

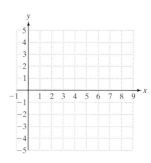

---

chapter 11 | midchapter review

1. State the distance formula from memory.

For Exercises 2–3, find the distance between the points.

2. $(-3, -4)$ and $(2, 1)$

3. $(4, -6)$ and $(-2, -4)$

4. Write an equation for the circle whose center has coordinates $(5, -10)$ and that has a radius of 9.

5. Identify the equation as representing a circle, parabola, ellipse, or hyperbola.

   a. $\dfrac{x^2}{4} - y^2 = 1$

   b. $x^2 + y^2 = 4$

   c. $x = y^2 - 3$

   d. $x^2 + 4y^2 = 4$

For Exercises 6–7, identify the center and radius of the circle, and then graph the circle.

6. $(x-3)^2 + (y-1)^2 = 4$

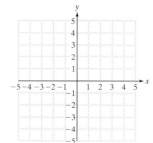

7. $x^2 + y^2 + 4x - 6y + 4 = 0$

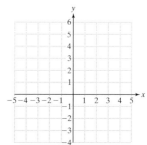

For Exercises 8–10, sketch the graph.

8. $\dfrac{x^2}{4} + \dfrac{y^2}{9} = 1$

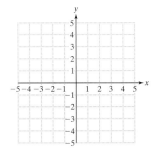

10. $\dfrac{x^2}{4} - \dfrac{y^2}{9} = 1$

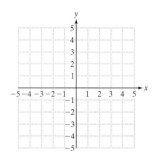

9. $x = 2(y - 3)^2 + 1$

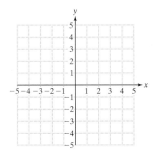

For Exercises 11–14, write the equation in standard form. Then identify it as an ellipse or hyperbola.

11. $5x^2 + y^2 - 10 = 0$

12. $5x^2 - 3y^2 = 15$

13. $y^2 - 6x^2 = 6$

14. $3x^2 + 5y^2 = 15$

section 11.4 Nonlinear Systems of Equations in Two Variables

Objectives
1. Solving Nonlinear Systems of Equations by the Substitution Method
2. Solving Nonlinear Systems of Equations by the Addition Method

1. Solving Nonlinear Systems of Equations by the Substitution Method

Recall that a linear equation in two variables x and y is an equation that can be written in the form $ax + by = c$, where a and b are not both zero. In Sections 3.1–3.3, we solved systems of linear equations in two variables by using the graphing method, the substitution method, and the addition method. In this section, we will solve *nonlinear* systems of equations by using the same methods. A **nonlinear system of equations** is a system in which at least one of the equations is nonlinear.

Graphing the equations in a nonlinear system helps to determine the number of solutions and to approximate the coordinates of the solutions. The substitution method is used most often to solve a nonlinear system of equations analytically.

example 1 Solving a Nonlinear System of Equations

Given the system

$$x - 7y = -25$$
$$x^2 + y^2 = 25$$

a. Solve the system by graphing.
b. Solve the system by the substitution method.

Skill Practice

Given the system

$2x + y = 5$
$x^2 + y^2 = 50$

1. Solve the system by graphing.

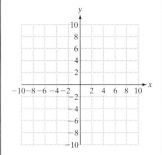

2. Solve the system by the substitution method.

Solution:

a. $x - 7y = -25$ is a line (the slope-intercept form is $y = \frac{1}{7}x + \frac{25}{7}$).

$x^2 + y^2 = 25$ is a circle centered at the origin with radius 5.

From Figure 11-13, we appear to have two solutions at $(-4, 3)$ and $(3, 4)$.

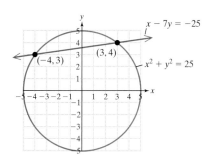

Figure 11-13

b. To use the substitution method, isolate one of the variables from one of the equations. We will solve for x in the first equation.

A $x - 7y = -25$ Solve for x. $x = 7y - 25$
B $x^2 + y^2 = 25$

B $(7y - 25)^2 + y^2 = 25$ Substitute $(7y - 25)$ for x in the second equation.

$49y^2 - 350y + 625 + y^2 = 25$ The resulting equation is quadratic in y.

$50y^2 - 350y + 600 = 0$ Set the equation equal to zero.

$50(y^2 - 7y + 12) = 0$ Factor.

$50(y - 3)(y - 4) = 0$

$y = 3$ or $y = 4$

For each value of y, find the corresponding x value from the equation $x = 7y - 25$.

$y = 3$: $x = 7(3) - 25 = -4$ The solution point is $(-4, 3)$.

$y = 4$: $x = 7(4) - 25 = 3$ The solution point is $(3, 4)$. (See Figure 11-13.)

example 2   Solving a Nonlinear System by the Substitution Method

Given the system

$$y = \sqrt{x}$$
$$x^2 + y^2 = 20$$

a. Sketch the graphs.

b. Solve the system by the substitution method.

Answers

1.

The solutions appear to be $(-1, 7)$ and $(5, -5)$.

2. $(-1, 7)$ and $(5, -5)$

Solution:

a. $y = \sqrt{x}$ is one of the six basic functions graphed in Section 4.3.

$x^2 + y^2 = 20$ is a circle centered at the origin with radius $\sqrt{20} \approx 4.5$.

From Figure 11-14, we see that there is one solution.

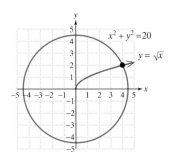

Figure 11-14

Skill Practice

Given the system

$y = \sqrt{2x}$

$x^2 + y^2 = 8$

3. Sketch the graphs.

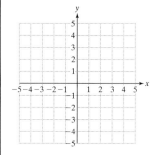

4. Solve the system by using the substitution method.

b. To use the substitution method, we will substitute $y = \sqrt{x}$ into equation $\boxed{B}$.

$\boxed{A}$ $\qquad y = \sqrt{x}$

$\boxed{B} \qquad x^2 + y^2 = 20$

$\boxed{B} \qquad x^2 + (\sqrt{x})^2 = 20$ Substitute $y = \sqrt{x}$ into the second equation.

$\qquad\qquad x^2 + x = 20$

$\qquad\qquad x^2 + x - 20 = 0$ Set the second equation equal to zero.

$\qquad\qquad (x + 5)(x - 4) = 0$ Factor.

$\qquad x \ne -5 \quad \text{or} \quad x = 4$ Reject $x = -5$ because it is not in the domain of $y = \sqrt{x}$.

Substitute $x = 4$ into the equation $y = \sqrt{x}$.

If $x = 4$, then $y = \sqrt{4} = 2$.   The solution point is $(4, 2)$.

Calculator Connections

Graph the equations from Example 2 to confirm your solution to the system of equations. Use an *Intersect* feature or *Zoom* and *Trace* to approximate the point of intersection. Recall that the circle must be entered into the calculator as two functions:

$Y_1 = \sqrt{20 - x^2}$

$Y_2 = -\sqrt{20 - x^2}$

$Y_3 = \sqrt{x}$

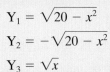

Answers

3.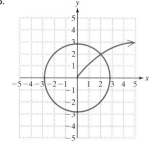

4. $(2, 2)$

Skill Practice

5. Solve the system by using the substitution method.
   $y = \sqrt[3]{9x}$
   $y = x$

example 3 — Solving a Nonlinear System by Using the Substitution Method

Solve the system by using the substitution method.

$$y = \sqrt[3]{x}$$
$$y = x$$

Solution:

A $y = \sqrt[3]{x}$
B $y = x$

$\sqrt[3]{x} = x$	Because y is isolated in both equations, we can equate the expressions for y.
$(\sqrt[3]{x})^3 = (x)^3$	To solve the radical equation, raise both sides to the third power.
$x = x^3$	This is a third-degree polynomial equation.
$0 = x^3 - x$	Set the equation equal to zero.
$0 = x(x^2 - 1)$	Factor out the GCF.
$0 = x(x + 1)(x - 1)$	Factor completely.

$x = 0$ or $x = -1$ or $x = 1$

For each value of x, find the corresponding y-value from either original equation. We will use equation B: $y = x$.

If $x = 0$, then $y = 0$.   The solution point is $(0, 0)$.
If $x = -1$, then $y = -1$.   The solution point is $(-1, -1)$.
If $x = 1$, then $y = 1$.   The solution point is $(1, 1)$.

Calculator Connections

Graph the equations $y = \sqrt[3]{x}$ and $y = x$ to support the solutions to Example 3.

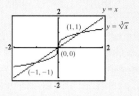

2. Solving Nonlinear Systems of Equations by the Addition Method

The substitution method is used most often to solve a system of nonlinear equations. In some situations, however, the addition method offers an efficient means of finding a solution. Example 4 demonstrates that we can eliminate a variable from both equations provided the terms containing that variable are *like* terms.

Answer

5. $(0, 0); (3, 3); (-3, -3)$

example 4 — Solving a Nonlinear System of Equations by the Addition Method

Solve the system. $2x^2 + y^2 = 17$
$x^2 + 2y^2 = 22$

Skill Practice

6. Solve the system by using the addition method.
$$x^2 - y^2 = 24$$
$$3x^2 + y^2 = 76$$

Solution:

[A] $2x^2 + y^2 = 17$ Notice that the y^2-terms are *like* in each equation.
[B] $x^2 + 2y^2 = 22$ To eliminate the y^2-terms, multiply the first equation by -2.

[A] $2x^2 + y^2 = 17$ $\xrightarrow{\text{Multiply by } -2.}$ $-4x^2 - 2y^2 = -34$
[B] $x^2 + 2y^2 = 22$ $\longrightarrow$ $\underline{\ x^2 + 2y^2 = 22}$
$-3x^2 = -12$ Eliminate the y^2 term.

$$\frac{-3x^2}{-3} = \frac{-12}{-3}$$

$$x^2 = 4$$

$$x = \pm 2$$

Substitute each value of x into one of the original equations to solve for y. We will use equation [A]: $2x^2 + y^2 = 17$.

$x = 2$:   [A] $2(2)^2 + y^2 = 17$
$8 + y^2 = 17$
$y^2 = 9$
$y = \pm 3$ The solution points are $(2, 3)$ and $(2, -3)$.

$x = -2$: [A] $2(-2)^2 + y^2 = 17$
$8 + y^2 = 17$
$y^2 = 9$
$y = \pm 3$ The solution points are $(-2, 3)$ and $(-2, -3)$.

Tip: In Example 4, the x^2-terms are also *like* in both equations. We could have eliminated the x^2-terms by multiplying equation [B] by -2.

Calculator Connections

The solutions to Example 4 can be checked from the graphs of the equations.

For the equation $2x^2 + y^2 = 17$, we have $y = \pm\sqrt{17 - 2x^2}$

For the equation $x^2 + 2y^2 = 22$, we have $y = \pm\sqrt{\dfrac{22 - x^2}{2}}$

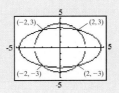

Answer

6. $(5, 1)$; $(5, -1)$; $(-5, 1)$; $(-5, -1)$

> **Tip:** It is important to note that the addition method can be used only if two equations share a pair of *like* terms. The substitution method is effective in solving a wider range of systems of equations. The system in Example 4 could also have been solved by using substitution.
>
> A $\quad 2x^2 + y^2 = 17 \xrightarrow{\text{Solve for } y^2} y^2 = 17 - 2x^2$
> B $\quad x^2 + 2y^2 = 22$
>
> B $\quad x^2 + 2(17 - 2x^2) = 22 \qquad\qquad x = 2: \quad y^2 = 17 - 2(2)^2$
> $\qquad x^2 + 34 - 4x^2 = 22 \qquad\qquad\qquad\qquad y^2 = 9$
> $\qquad\qquad\quad -3x^2 = -12 \qquad\qquad\qquad\qquad y = \pm 3 \quad$ The solutions are
> $\qquad\qquad\qquad\quad x^2 = 4 \qquad\qquad\qquad\qquad\qquad\qquad\qquad\quad (2, 3) \text{ and } (2, -3).$
> $\qquad\qquad\qquad\quad x = \pm 2 \qquad\quad x = -2: \quad y^2 = 17 - 2(-2)^2$
> $\qquad\qquad\qquad\qquad\qquad\qquad\qquad\qquad\qquad\quad y^2 = 9$
> $\qquad\qquad\qquad\qquad\qquad\qquad\qquad\qquad\qquad\quad y = \pm 3 \quad$ The solutions are
> $\qquad\qquad\qquad\qquad\qquad\qquad\qquad\qquad\qquad\qquad\qquad\quad (-2, 3) \text{ and } (-2, -3).$

section 11.4 Practice Exercises

Boost *your* GRADE at mathzone.com!
MathZone

- Practice Problems
- Self-Tests
- NetTutor
- e-Professors
- Videos

Study Skills Exercises

1. To study for the final exam, write the title of each chapter covered in this book at the top of a clean sheet of paper. Beneath each chapter title, write the names of the sections that were covered in that chapter. Try to make up exercises relating to each topic, then work the exercises.

2. Define the key term **nonlinear system of equations**.

Review Exercises

3. Write the distance formula between two points (x_1, y_1) and (x_2, y_2) from memory.

4. Find the distance between the two points $(8, -1)$ and $(1, -8)$.

5. Write an equation representing the set of all points 2 units from the point $(-1, 1)$.

6. Write an equation representing the set of all points 8 units from the point $(-5, 3)$.

For Exercises 7–14, determine if the equation represents a parabola, circle, ellipse, or hyperbola.

7. $x^2 + y^2 = 15$ **8.** $\dfrac{x^2}{4} - \dfrac{y^2}{2} = 1$ **9.** $y = (x-6)^2 + 4$ **10.** $\dfrac{(x+1)^2}{2} + \dfrac{(y+1)^2}{5} = 1$

11. $\dfrac{(y-1)^2}{3} - \dfrac{(x+2)^2}{3} = 1$ **12.** $3x^2 + 3y^2 = 1$ **13.** $\dfrac{x^2}{9} + \dfrac{y^2}{12} = 1$ **14.** $x = (y+2)^2 - 5$

Objective 1: Solving Nonlinear Systems of Equations by the Substitution Method

For Exercises 15–22, use sketches to explain.

15. How many points of intersection are possible between a line and a parabola?

16. How many points of intersection are possible between a line and a circle?

17. How many points of intersection are possible between two distinct circles?

18. How many points of intersection are possible between two distinct parabolas of the form $y = ax^2 + bx + c$, $a \neq 0$?

19. How many points of intersection are possible between a circle and a parabola?

20. How many points of intersection are possible between two distinct lines?

21. How many points of intersection are possible with an ellipse and a hyperbola?

22. How many points of intersection are possible with an ellipse and a parabola?

For Exercises 23–28, sketch each system of equations. Then solve the system by the substitution method. **(See Example 1.)**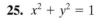

23. $y = x + 3$
       $x^2 + y = 9$

24. $y = x - 2$
       $x^2 + y = 4$

25. $x^2 + y^2 = 1$
       $y = x + 1$

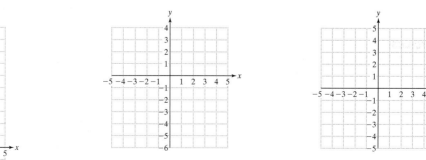

26. $x^2 + y^2 = 25$
$y = 2x$

27. $x^2 + y^2 = 6$
$y = x^2$

28. $x^2 + y^2 = 12$
$y = x^2$

For Exercises 29–36, solve the system by the substitution method. **(See Examples 2–3.)**

29. $x^2 + y^2 = 20$
$y = \sqrt{x}$

30. $x^2 + y^2 = 30$
$y = \sqrt{x}$

31. $y = x^2$
$y = -\sqrt{x}$

32. $y = -x^2$
$y = -\sqrt{x}$

33. $y = x^2$
$y = (x - 3)^2$

34. $y = (x + 4)^2$
$y = x^2$

35. $y = x^2 + 6x$
$y = 4x$

36. $y = 3x^2 - 6x$
$y = 3x$

Objective 2: Solving Nonlinear Systems of Equations by the Addition Method

For Exercises 37–50, solve the system of nonlinear equations by the addition method. **(See Example 4.)**

37. $x^2 + y^2 = 13$
$x^2 - y^2 = 5$

38. $4x^2 - y^2 = 4$
$4x^2 + y^2 = 4$

39. $9x^2 + 4y^2 = 36$
$x^2 + y^2 = 9$

40. $x^2 + y^2 = 4$
$2x^2 + y^2 = 8$

41. $3x^2 + 4y^2 = 16$
$2x^2 - 3y^2 = 5$

42. $2x^2 - 5y^2 = -2$
$3x^2 + 2y^2 = 35$

43. $y = x^2 - 2$
$y = -x^2 + 2$

44. $y = x^2$
$y = -x^2 + 8$

45. $\dfrac{x^2}{4} + \dfrac{y^2}{9} = 1$
$x^2 + y^2 = 4$

46. $\dfrac{x^2}{16} + \dfrac{y^2}{4} = 1$
$x^2 + y^2 = 4$

47. $x^2 + 6y^2 = 9$
$\dfrac{x^2}{9} + \dfrac{y^2}{12} = 1$

48. $\dfrac{x^2}{10} + \dfrac{y^2}{10} = 1$
$2x^2 + y^2 = 11$

49. $x^2 - xy = -4$
$2x^2 - xy = 12$

50. $x^2 - xy = 3$
$2x^2 + xy = 6$

Expanding Your Skills

51. The sum of two numbers is 7. The sum of the squares of the numbers is 25. Find the numbers.

52. The sum of the squares of two numbers is 100. The sum of the numbers is 2. Find the numbers.

53. The sum of the squares of two numbers is 32. The difference of the squares of the numbers is 18. Find the numbers.

54. The sum of the squares of two numbers is 24. The difference of the squares of the numbers is 8. Find the numbers.

Graphing Calculator Exercises

For Exercises 55–58, use the *Intersect* feature or *Zoom* and *Trace* to approximate the solutions to the system.

55. $y = x + 3$ (Exercise 23)
$x^2 + y = 9$

56. $y = x - 2$ (Exercise 24)
$x^2 + y = 4$

57. $y = x^2$ (Exercise 31)
$y = -\sqrt{x}$

58. $y = -x^2$ (Exercise 32)
$y = -\sqrt{x}$

For Exercises 59–60, graph the system on a square viewing window. What can be said about the solution to the system?

59. $x^2 + y^2 = 4$
$y = x^2 + 3$

60. $x^2 + y^2 = 16$
$y = -x^2 - 5$

Objectives

1. Nonlinear Inequalities in Two Variables
2. Systems of Nonlinear Inequalities in Two Variables

Section 11.5 Nonlinear Inequalities and Systems of Inequalities

1. Nonlinear Inequalities in Two Variables

In Section 9.5 we graphed the solution sets to linear inequalities in two variables, such as $y \leq 2x + 1$. This involved graphing the related equation (a line in the xy-plane) and then shading the appropriate region above or below the line. See Figure 11-15. In this section, we will employ the same technique to solve nonlinear inequalities in two variables.

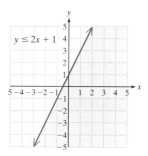

Figure 11-15

Skill Practice

1. Graph the solution set of the inequality $x^2 + y^2 \geq 9$.

example 1 Graphing a Nonlinear Inequality in Two Variables

Graph the solution set of the inequality $x^2 + y^2 < 16$.

Solution:

The related equation $x^2 + y^2 = 16$ is a circle of radius 4, centered at the origin. Graph the related equation by using a dashed curve because the points satisfying the equation $x^2 + y^2 = 16$ are not part of the solution to the strict inequality $x^2 + y^2 < 16$. See Figure 11-16.

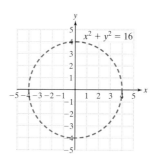

Figure 11-16

Notice that the dashed curve separates the xy-plane into two regions, one "inside" the circle, the other "outside" the circle. Select a test point from each region and test the point in the original inequality.

Answer

1.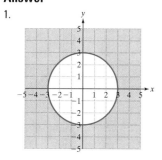

Test Point "Inside": (0, 0)	Test Point "Outside": (4, 4)
$x^2 + y^2 < 16$	$x^2 + y^2 < 16$
$(0)^2 + (0)^2 \stackrel{?}{<} 16$	$(4)^2 + (4)^2 \stackrel{?}{<} 16$
$0 \stackrel{?}{<} 16$ True	$32 \stackrel{?}{<} 16$ False

The inequality $x^2 + y^2 < 16$ is true at the test point $(0, 0)$. Therefore, the solution set is the region "inside" the circle. See Figure 11-17.

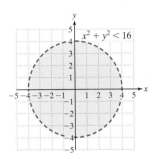

Figure 11-17

example 2 Graphing a Nonlinear Inequality in Two Variables

Graph the solution set of the inequality $9y^2 \geq 36 + 4x^2$.

Solution:

First graph the related equation $9y^2 = 36 + 4x^2$. Notice that the equation can be written in the standard form of a hyperbola.

$$9y^2 = 36 + 4x^2$$

$$9y^2 - 4x^2 = 36 \qquad \text{Subtract } 4x^2 \text{ from both sides.}$$

$$\frac{y^2}{4} - \frac{x^2}{9} = 1 \qquad \text{Divide both sides by 36.}$$

Graph the hyperbola as a solid curve, because the original inequality includes equality. See Figure 11-18.

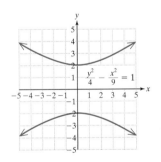

Figure 11-18

The hyperbola divides the xy-plane into three regions: a region above the upper branch, a region between the branches, and a region below the lower branch. Select a test point from each region.

$$9y^2 \geq 36 + 4x^2$$

Test: (0, 3)

$9(3)^2 \overset{?}{\geq} 36 + 4(0)^2$

$81 \overset{?}{\geq} 36$ True

Test: (0, 0)

$9(0)^2 \overset{?}{\geq} 36 + 4(0)^2$

$0 \overset{?}{\geq} 36$ False

Test: (0, −3)

$9(-3)^2 \overset{?}{\geq} 36 + 4(0)^2$

$81 \overset{?}{\geq} 36$ True

Skill Practice

2. Graph the solution set of the inequality $9x^2 < 144 - 16y^2$.

Answer

2.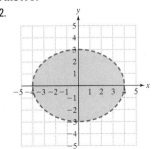

Shade the regions above the top branch and below the bottom branch of the hyperbola. See Figure 11-19.

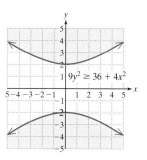

Figure 11-19

2. Systems of Nonlinear Inequalities in Two Variables

In Section 11.4 we solved systems of nonlinear equations in two variables. The solution set for such a system is the set of ordered pairs that satisfy both equations simultaneously. We will now solve systems of nonlinear inequalities in two variables. Similarly, the solution set is the set of all ordered pairs that simultaneously satisfy each inequality. To solve a system of inequalities, we will graph the solution to each individual inequality and then take the intersection of the solution sets.

Skill Practice

3. Graph the solution set.
$$y > \frac{1}{2}x^2$$
$$x > y^2$$

example 3   Graphing a System of Nonlinear Inequalities in Two Variables

Graph the solution set.
$$y > e^x$$
$$y < -x^2 + 4$$

Solution:

The solution to $y > e^x$ is the set of points above the curve $y = e^x$. See Figure 11-20. The solution to $y < -x^2 + 4$ is the set of points below the parabola $y = -x^2 + 4$. See Figure 11-21.

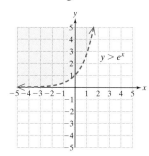 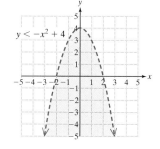

Figure 11-20   **Figure 11-21**

The solution to the system of inequalities is the intersection of the solution sets of the individual inequalities. See Figure 11-22.

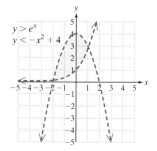

Figure 11-22

Answer

3.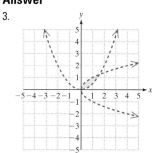

Section 11.5 Nonlinear Inequalities and Systems of Inequalities 843

section 11.5 Practice Exercises

Boost your GRADE at mathzone.com!

- Practice Problems
- Self-Tests
- NetTutor
- e-Professors
- Videos

Study Skills Exercise

1. Taking a pretest can help you prepare for a final exam. One way to make up a pretest is to take five problems from each chapter test that you covered in this text. Then write the questions in random order, but note the chapter and number of the problem so that you can check your answers.

Review Exercises

For Exercises 2–11, match the equation with its graph.

2. $y = 4x - 1$
3. $y = -4x^2$
4. $y = e^x$
5. $y = x^3$
6. $\dfrac{x^2}{4} + \dfrac{y^2}{9} = 1$
7. $\dfrac{x^2}{4} - \dfrac{y^2}{9} = 1$
8. $y = \dfrac{1}{x}$
9. $y = \log_2(x)$
10. $y = \sqrt{x}$
11. $(x + 2)^2 + (y - 1)^2 = 4$

a.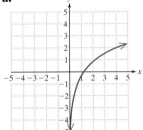
b.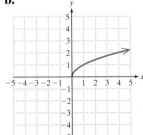
c.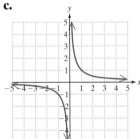
d.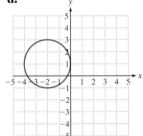
e.
f.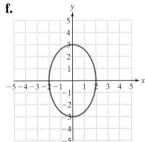
g.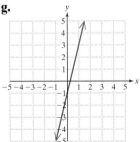
h.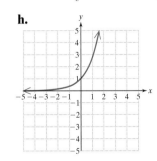
i.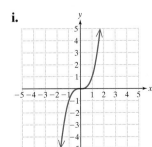
j.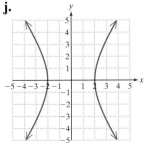

Objective 1: Nonlinear Inequalities in Two Variables

12. True or false? The point $(1, 1)$ satisfies the inequality $-x^2 + y^3 > 1$.

13. True or false? The point $(4, -2)$ satisfies the inequality $4x^2 - 2x + 1 + y^2 < 3$.

14. True or false? The point $(5, 4)$ satisfies the system of inequalities.
$$\frac{x^2}{36} + \frac{y^2}{25} < 1$$
$$x^2 + y^2 \geq 4$$

15. True or false? The point $(1, -2)$ satisfies the system of inequalities.
$$y < x^2$$
$$y > x^2 - 4$$

16. True or false? The point $(-3, 5)$ satisfies the system of inequalities.
$$(x + 3)^2 + (y - 5)^2 \leq 2$$
$$y > x^2$$

17. a. Graph the solution set of $x^2 + y^2 \leq 9$.

18. a. Graph the solution set of $\dfrac{x^2}{4} + \dfrac{y^2}{9} \geq 1$.

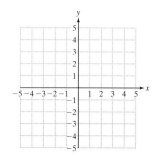

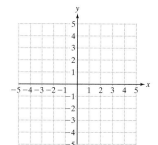

b. Describe the solution set for the inequality $x^2 + y^2 \geq 9$.

b. Describe the solution set for the inequality $\dfrac{x^2}{4} + \dfrac{y^2}{9} \leq 1$.

c. Describe the solution set of the equation $x^2 + y^2 = 9$.

c. Describe the solution set for the equation $\dfrac{x^2}{4} + \dfrac{y^2}{9} = 1$.

19. a. Graph the solution set of $y \geq x^2 + 1$.

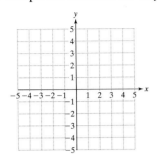

b. How would the solution change for the strict inequality $y > x^2 + 1$?

20. a. Graph the solution set of $\dfrac{x^2}{4} - \dfrac{y^2}{9} \leq 1$.

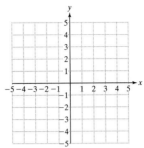

b. How would the solution change for the strict inequality $\dfrac{x^2}{4} - \dfrac{y^2}{9} < 1$?

For Exercises 21–36, graph the solution set. **(See Examples 1–2.)**

21. $2x + y \geq 1$

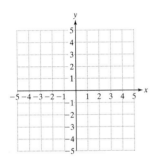

22. $3x + 2y \geq 6$

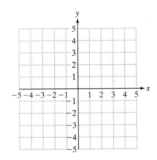

23. $x \leq y^2$

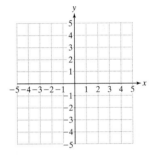

24. $y \leq -x^2$

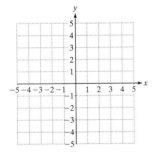

25. $(x - 1)^2 + (y + 2)^2 > 9$

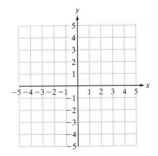

26. $(x + 1)^2 + (y - 4)^2 > 1$

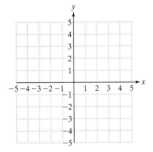

27. $x + y^2 \geq 4$

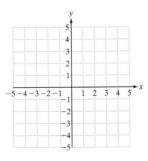

28. $x^2 + 2x + y - 1 \leq 0$

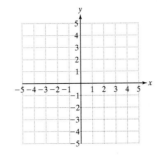

29. $9x^2 - y^2 > 9$

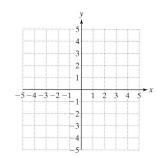

30. $y^2 - 4x^2 \leq 4$

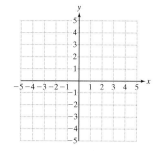

31. $x^2 + 16y^2 \leq 16$

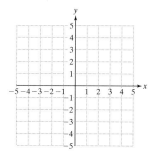

32. $4x^2 + y^2 \leq 4$

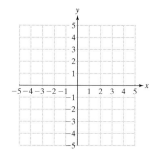

33. $y \leq \ln x$

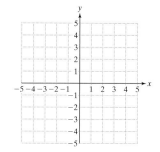

34. $y \leq \log x$

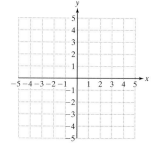

35. $y > 5^x$

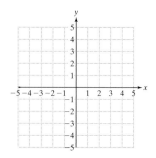

36. $y \geq \left(\dfrac{1}{3}\right)^x$

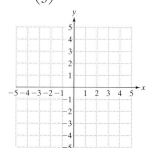

Objective 2: Systems of Nonlinear Inequalities in Two Variables

For Exercises 37–50, graph the solution set to the system of inequalities. (See Example 3.)

37. $y < \sqrt{x}$
$x > 1$

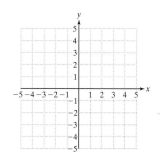

38. $y \geq \sqrt{x}$
$x \geq 0$

39. $\dfrac{x^2}{36} + \dfrac{y^2}{25} < 1$
$x^2 + y^2 \geq 4$

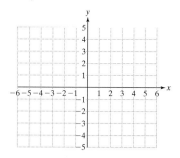

40. $x^2 - y^2 \geq 1$
$x \leq 0$

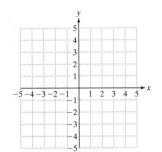

41. $y < x^2$
$y > x^2 - 4$

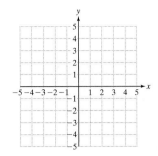

42. $y^2 - x^2 \geq 1$
$y \geq 0$

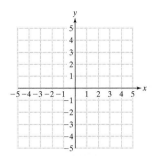

43. $y < \dfrac{1}{x}$
$y > 0$
$y < x$

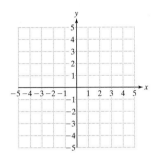

44. $y > x^3$
$y < 8$
$x > 0$

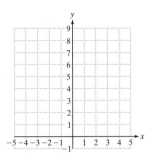

45. $x^2 + y^2 \geq 25$
$x^2 + y^2 \leq 9$

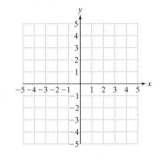

46. $\dfrac{x^2}{4} + \dfrac{y^2}{25} \geq 1$
$x^2 + \dfrac{y^2}{4} \leq 1$

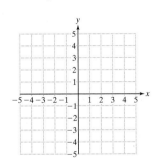

47. $x < -(y-1)^2 + 3$
$x + y > 2$

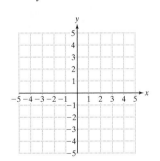

48. $x > (y-2)^2 + 1$
$x - y < 1$

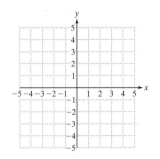

49. $x^2 + y^2 < 25$

$y < \dfrac{4}{3}x$

$y > -\dfrac{4}{3}x$

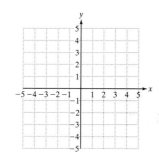

50. $y < e^x$

$y > 1$

$x < 2$

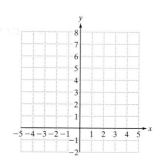

Expanding Your Skills

For Exercises 51–54, graph the compound inequalities.

51. $x^2 + y^2 \leq 36$ or $x + y \geq 0$

52. $y \leq -x^2 + 4$ or $y \geq x^2 - 4$

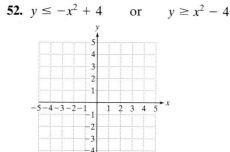

53. $y + 1 \geq x^2$ or $y + 1 \leq -x^2$

54. $(x + 2)^2 + (y + 3)^2 \leq 4$ or $x \geq y^2$

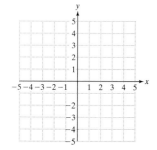

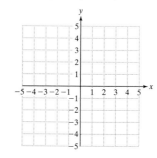

chapter 11 | summary

section 11.1 Distance Formula and Circles

Key Concepts

The **distance between two points** (x_1, y_1) and (x_2, y_2) is
$$d = \sqrt{(x_2 - x_1)^2 + (y_2 - y_1)^2}$$

The standard equation of a **circle** with center (h, k) and radius r is
$$(x - h)^2 + (y - k)^2 = r^2$$

Examples

Example 1

Find the distance between two points.

$(5, -2)$ and $(-1, -6)$

$$\begin{aligned} d &= \sqrt{(-1 - 5)^2 + [-6 - (-2)]^2} \\ &= \sqrt{(-6)^2 + (-4)^2} \\ &= \sqrt{36 + 16} \\ &= \sqrt{52} = 2\sqrt{13} \end{aligned}$$

Example 2

Find the center and radius of the circle.
$$x^2 + y^2 - 8x + 6y = 0$$
$$(x^2 - 8x + 16) + (y^2 + 6y + 9) = 16 + 9$$
$$(x - 4)^2 + (y + 3)^2 = 25$$

The center is $(4, -3)$ and the radius is 5.

section 11.2 More on the Parabola

Key Concepts

A **parabola** is the set of points in a plane that are equidistant from a fixed line (called the directrix) and a fixed point (called the focus) not on the directrix.

The standard form of an equation of a parabola with **vertex** (h, k) and vertical **axis of symmetry** is

$$y = a(x - h)^2 + k$$

- The equation of the axis of symmetry is $x = h$.
- If $a > 0$, then the parabola opens upward.
- If $a < 0$, the parabola opens downward.

The standard form of an equation of a parabola with vertex (h, k) and horizontal axis of symmetry is

$$x = a(y - k)^2 + h$$

- The equation of the axis of symmetry is $y = k$.
- If $a > 0$, then the parabola opens to the right.
- If $a < 0$, the parabola opens to the left.

Examples

Example 1

Given the parabola, $y = (x - 2)^2 + 1$

The vertex is $(2, 1)$.

The axis of symmetry is $x = 2$.

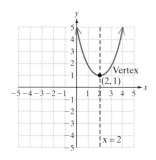

Example 2

Given the parabola $x = -\dfrac{1}{4}y^2 + 1$, determine the coordinates of the vertex and the equation of the axis of symmetry.

$$x = -\frac{1}{4}(y - 0)^2 + 1$$

The vertex is $(1, 0)$.

The axis of symmetry is $y = 0$.

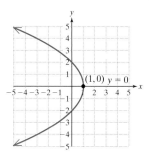

section 11.3 The Ellipse and Hyperbola

Key Concepts

An **ellipse** is the set of all points (x, y) such that the sum of the distances between (x, y) and two distinct points (called foci) is constant.

Standard Form of an Ellipse with Center at the Origin
An ellipse with the center at the origin has the equation

$$\frac{x^2}{a^2} + \frac{y^2}{b^2} = 1$$

where a and b are positive real numbers.

For an ellipse centered at the origin, the x-intercepts are given by $(a, 0)$ and $(-a, 0)$, and the y-intercepts are given by $(0, b)$ and $(0, -b)$.

A **hyperbola** is the set of all points (x, y) such that the difference of the distances between (x, y) and two distinct points is a constant. The fixed points are called the foci of the hyperbola.

Standard Forms of an Equation of a Hyperbola
Let a and b represent positive real numbers.

Horizontal Transverse Axis. The standard form of an equation of a hyperbola with a horizontal transverse axis and center at the origin is given by

$$\frac{x^2}{a^2} - \frac{y^2}{b^2} = 1$$

Vertical Transverse Axis. The standard form of an equation of a hyperbola with a vertical transverse axis and center at the origin is given by

$$\frac{y^2}{a^2} - \frac{x^2}{b^2} = 1$$

Examples

Example 1

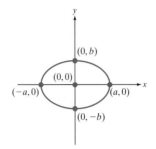

Example 2

$$\frac{x^2}{25} + \frac{y^2}{9} = 1$$

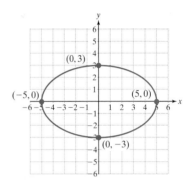

Example 3

$$\frac{x^2}{a^2} - \frac{y^2}{b^2} = 1$$

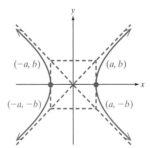

Example 4

$$\frac{y^2}{a^2} - \frac{x^2}{b^2} = 1$$

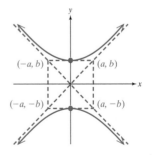

section 11.4 Nonlinear Systems of Equations in Two Variables

Key Concepts

A **nonlinear system of equations** can be solved by graphing or by the substitution method.

$2x^2 + y^2 = 15$
$x^2 - y = 0$

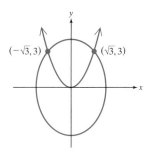

A nonlinear system may also be solved by using the **addition method** when the equations share *like* terms.

Examples

Example 1

$\boxed{A}$ $\quad 2x^2 + y^2 = 15$
$\boxed{B}$ $\quad x^2 - y = 0 \qquad$ Solve for y: $\quad y = x^2$

$\boxed{A}$ $\quad 2x^2 + (x^2)^2 = 15 \qquad$ Substitute in first equation.
$\qquad 2x^2 + x^4 = 15$
$\quad x^4 + 2x^2 - 15 = 0$
$(x^2 + 5)(x^2 - 3) = 0$

$x^2 + 5 = 0 \quad$ or $\quad x^2 - 3 = 0$
$\quad \cancel{x^2 = -5} \quad$ or $\qquad x^2 = 3$
$\qquad\qquad\qquad\qquad\qquad x = \pm\sqrt{3}$

If $x = \sqrt{3}$, then $y = (\sqrt{3})^2 = 3$.
If $x = -\sqrt{3}$, then $y = (-\sqrt{3})^2 = 3$.
Points of intersection are $(\sqrt{3}, 3)$ and $(-\sqrt{3}, 3)$.

Example 2

$2x^2 + y^2 = 4 \quad \xrightarrow{\text{Mult. by } -5.} \quad -10x^2 - 5y^2 = -20$
$3x^2 + 5y^2 = 13 \quad \longrightarrow \qquad\quad \underline{\;3x^2 + 5y^2 = 13\;}$
$\qquad\qquad\qquad\qquad\qquad\qquad\quad -7x^2 \qquad\quad = -7$

$\dfrac{-7x^2}{-7} = \dfrac{-7}{-7}$

$x^2 = 1 \longrightarrow x = \pm 1$

If $x = 1$, $\qquad 2(1)^2 + y^2 = 4$
$\qquad\qquad\qquad\qquad y^2 = 2$
$\qquad\qquad\qquad\qquad y = \pm\sqrt{2}$
If $x = -1$, $\qquad 2(-1)^2 + y^2 = 4$
$\qquad\qquad\qquad\qquad y^2 = 2$
$\qquad\qquad\qquad\qquad y = \pm\sqrt{2}$

The points of intersection are $(1, \sqrt{2})$, $(1, -\sqrt{2})$, $(-1, \sqrt{2})$, and $(-1, -\sqrt{2})$.

section 11.5 Nonlinear Inequalities and Systems of Inequalities

Key Concepts

Graph a nonlinear inequality by using the test point method. That is, graph the related equation. Then choose test points in each region to determine where the inequality is true.

Examples

Example 1

$y \geq x^2$

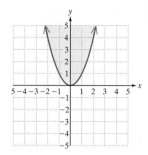

Example 2

$x^2 + y^2 < 4$

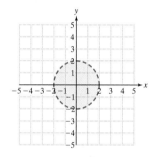

Graph a system of nonlinear inequalities by finding the intersection of the solution sets. That is, graph the solution set for each individual inequality, then take the intersection.

Example 3

$x \geq 0 \quad y > x^2 \quad$ and $\quad x^2 + y^2 < 4$

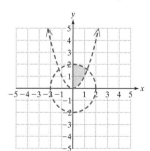

chapter 11 review exercises

Section 11.1

For Exercises 1–4, find the distance between the two points by using the distance formula.

1. $(-6, 3)$ and $(0, 1)$

2. $(4, 13)$ and $(-1, 5)$

3. Find x such that $(x, 5)$ is 5 units from $(2, 9)$.

4. Find x such that $(-3, 4)$ is 3 units from $(x, 1)$.

Points are said to be collinear if they all lie on the same line. If three points are collinear, then the distance between the outermost points will equal the sum of the distances between the middle point and each of the outermost points. For Exercises 5–6, determine if the three points are collinear.

5. $(-2, -3), (1, 3),$ and $(5, 11)$

6. $(-2, 11), (0, 5),$ and $(4, -7)$

For Exercises 7–10, find the center and the radius of the circle.

7. $(x - 12)^2 + (y - 3)^2 = 16$

8. $(x + 7)^2 + (y - 5)^2 = 81$

9. $(x + 3)^2 + (y + 8)^2 = 20$

10. $(x - 1)^2 + (y + 6)^2 = 32$

11. A stained glass window is in the shape of a circle with a 16-in. diameter. Find an equation of the circle relative to the origin for each of the following graphs.

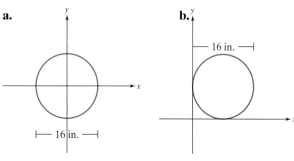

For Exercises 12–15, write the equation of the circle in standard form by completing the square.

12. $x^2 + y^2 + 12x - 10y + 51 = 0$

13. $x^2 + y^2 + 4x + 16y + 60 = 0$

14. $x^2 + y^2 - x - 4y + \dfrac{1}{4} = 0$

15. $x^2 + y^2 - 6x - \dfrac{2}{3}y + \dfrac{1}{9} = 0$

16. Write an equation of a circle with center at the origin and a diameter of 7 m.

17. Write an equation of a circle with center at $(0, 2)$ and a diameter of 6 m.

Section 11.2

For Exercises 18–21, determine whether the axis of symmetry is vertical or horizontal and if the parabola opens upward, downward, left, or right.

18. $y = -2(x - 3)^2 + 2$

19. $x = 3(y - 9)^2 + 1$

20. $x = -(y + 4)^2 - 8$

21. $y = (x + 3)^2 - 10$

For Exercises 22–25, determine the coordinates of the vertex and the equation of the axis of symmetry. Then use this information to graph the parabola.

22. $x = -(y - 1)^2$

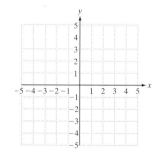

23. $y = (x + 2)^2$

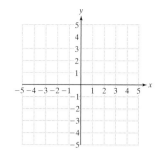

24. $y = -\dfrac{1}{4}x^2$ **25.** $x = 2y^2 - 1$

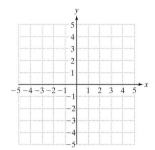 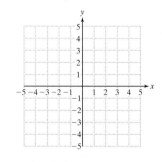

For Exercises 26–29, write the equation in standard form $y = a(x - h)^2 + k$ or $x = a(y - k)^2 + h$. Then identify the vertex, and axis of symmetry.

26. $y = x^2 - 6x + 5$

27. $x = y^2 + 4y + 2$

28. $x = -4y^2 + 4y$

29. $y = -2x^2 - 2x$

Section 11.3

For Exercises 30–31, identify the x- and y-intercepts. Then graph the ellipse.

30. $\dfrac{x^2}{9} + \dfrac{y^2}{25} = 1$

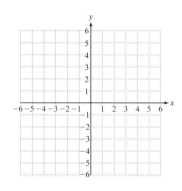

31. $x^2 + 4y^2 = 36$

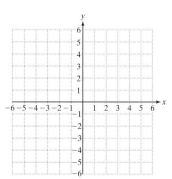

For Exercises 32–35, determine whether the transverse axis is horizontal or vertical.

32. $\dfrac{x^2}{12} - \dfrac{y^2}{16} = 1$ **33.** $\dfrac{y^2}{9} - \dfrac{x^2}{9} = 1$

34. $\dfrac{y^2}{24} - \dfrac{x^2}{10} = 1$

35. $\dfrac{x^2}{6} - \dfrac{y^2}{16} = 1$

For Exercises 36–37, graph the hyperbola by first drawing the reference rectangle and the asymptotes. Label the vertices.

36. $\dfrac{x^2}{4} - y^2 = 1$

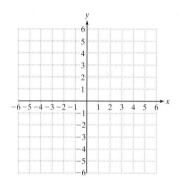

37. $y^2 - x^2 = 16$

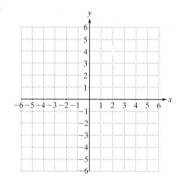

For Exercises 38–41, identify the equations as representing an ellipse or a hyperbola.

38. $\dfrac{x^2}{4} - \dfrac{y^2}{9} = 1$ **39.** $\dfrac{x^2}{16} + \dfrac{y^2}{9} = 1$

40. $\dfrac{x^2}{4} + \dfrac{y^2}{1} = 1$ **41.** $\dfrac{y^2}{1} - \dfrac{x^2}{16} = 1$

Section 11.4

For Exercises 42–45:

a. Identify each equation as representing a line, a parabola, a circle, an ellipse, or a hyperbola.

b. Graph both equations on the same coordinate system.

c. Solve the system analytically and verify the answers from the graph.

42. $3x + 2y = 10$
$y = x^2 - 5$

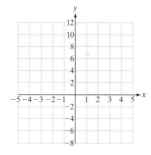

43. $4x + 2y = 10$
$y = x^2 - 10$

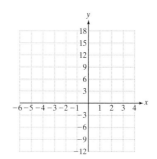

44. $x^2 + y^2 = 9$
$2x + y = 3$

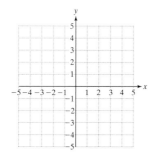

45. $x^2 + y^2 = 16$
$x - 2y = 8$

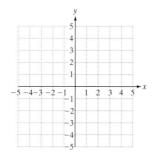

For Exercises 46–51, solve the system of nonlinear equations by using either the substitution method or the addition method.

46. $x^2 + 2y^2 = 8$
$2x - y = 2$

47. $x^2 + 4y^2 = 29$
$x - y = -4$

48. $x - y = 4$
$y^2 = 2x$

49. $y = x^2$
$6x^2 - y^2 = 8$

50. $x^2 + y^2 = 10$
$x^2 + 9y^2 = 18$

51. $x^2 + y^2 = 61$
$x^2 - y^2 = 11$

Section 11.5

For Exercises 52–59, graph the solution set to the inequality.

52. $3x + y \leq 4$

53. $x - 2y \geq -2$

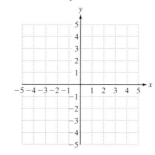
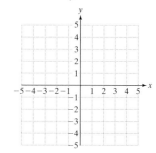

54. $\dfrac{x^2}{16} + \dfrac{y^2}{81} < 1$

55. $\dfrac{x^2}{25} + \dfrac{y^2}{4} > 1$

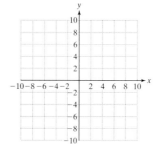
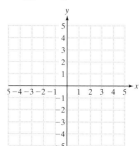

56. $(x - 3)^2 + (y + 1)^2 \geq 9$

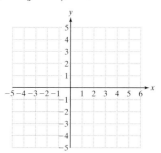

57. $(x + 2)^2 + (y + 1)^2 \leq 4$

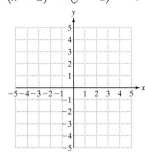

For Exercises 60–62, graph the solution set to the system of nonlinear inequalities.

60. $y > 4^x$
$x^2 + y^2 < 4$

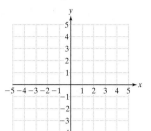

61. $y < 3^x$
$x^2 + y^2 < 9$

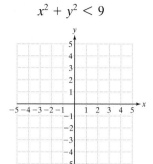

58. $y > (x - 1)^2$

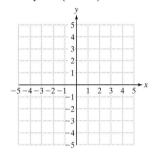

59. $y > x^2 - 1$

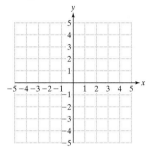

62. $\dfrac{x^2}{4} - y^2 < 1$
$x^2 + y^2 < 9$

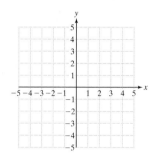

chapter 11 test

1. Determine the coordinates of the vertex and the equation of the axis of symmetry. Then graph the parabola.

$$x = -(y - 2)^2 + 3$$

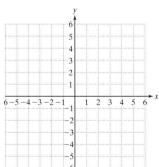

2. Determine the coordinates of the center and radius of the circle.

$$\left(x - \dfrac{5}{6}\right)^2 + \left(y + \dfrac{1}{3}\right)^2 = \dfrac{25}{49}$$

3. Write the equation in standard form $y = a(x - h)^2 + k$, and graph.

$$y = x^2 + 4x + 5$$

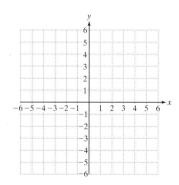

4. Use the distance formula to find the distance between the two points $(5, 19)$ and $(-2, 13)$.

5. Determine the coordinates of the center and radius of the circle.

$$x^2 + y^2 - 4y - 5 = 0$$

6. Let $(0, 4)$ be the center of a circle that passes through the point $(-2, 5)$.

   a. What is the radius of the circle?

   b. Write the equation of the circle in standard form.

7. Graph the ellipse.

$$\frac{x^2}{16} + \frac{y^2}{49} = 1$$

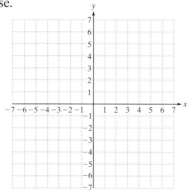

8. Graph the hyperbola.

$$\frac{y^2}{1} - \frac{x^2}{4} = 1$$

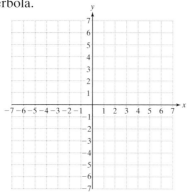

9. Solve the systems and identify the correct graph of the equations.

   a. $16x^2 + 9y^2 = 144$
      $4x - 3y = -12$

   b. $x^2 + 4y^2 = 4$
      $4x - 3y = -12$

   i.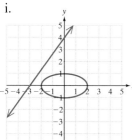

   ii.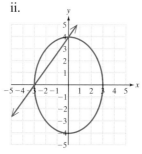

10. Describe the circumstances in which a nonlinear system of equations can be solved by using the addition method.

11. Solve the system by using either the substitution method or the addition method.

$$25x^2 + 4y^2 = 100$$
$$25x^2 - 4y^2 = 100$$

For Exercises 12–15, graph the solution set.

12. $x \leq y^2 + 1$

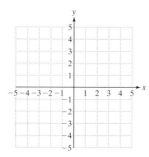

13. $y \geq -\frac{1}{3}x + 1$

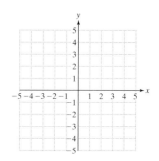

14. $x < y^2 + 1$
    $y > -\frac{1}{3}x + 1$

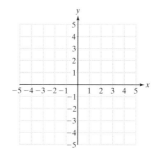

15. $y < \sqrt{x}$
    $y > x - 2$
    $x > 0$

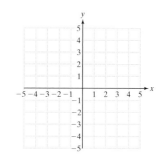

chapters 1–11 | cumulative review exercises

1. Solve the equation.
   $$5(2y - 1) = 2y - 4 + 8y - 1$$

2. Solve the inequality. Graph the solution and write the solution in interval notation.
   $$4(x - 1) + 2 > 3x + 8 - 2x$$

3. The product of two integers is 150. If one integer is 5 less than twice the other, find the integers.

4. For $5y - 3x - 15 = 0$:
   a. Find the x- and y-intercepts.
   b. Find the slope.
   c. Graph the line.

5. The amount of money spent on motor vehicles and parts each year since the year 2000 is shown in the graph (see figure). Let $x = 0$ correspond to the year 2000. Let y represent the amount of money spent on motor vehicles and parts (in billions of dollars).

   a. Use any two data points to find the slope of the line.

   b. Find an equation of the line. Write the answer in slope-intercept form.

   c. Use the linear model found in part (b) to approximate the amount spent on motor vehicles and parts in the year 2005.

   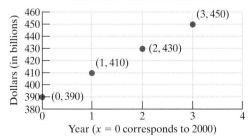

   Amount Spent (in $ billions) on Motor Vehicles and Parts in United States

   Source: Bureau of Economic Analysis, U.S. Department of Commerce

6. A collection of dimes and quarters has a total value of $2.45. If there are 17 coins, how many of each type are there?

7. Solve the system.
   $$\begin{aligned} x + y &= -1 \\ 2x \phantom{{}+y} - z &= 3 \\ y + 2z &= -1 \end{aligned}$$

8. a. Given the matrix $\mathbf{A} = \begin{bmatrix} 1 & -2 & | & -8 \\ 0 & 3 & | & 6 \end{bmatrix}$, write the matrix obtained by multiplying the elements in the second row by $\frac{1}{3}$.

   b. Using the matrix obtained from part (a), write the matrix obtained by multiplying the second row by 2 and adding the result to the first row.

9. Solve the following system.
   $$\begin{aligned} 4x - 2y &= 7 \\ -3x + 5y &= 0 \end{aligned}$$

10. For $f(x) = 3x - x^2 - 12$, find the function values $f(0), f(-1), f(2)$, and $f(4)$.

11. For $g = \{(2, 5), (8, -1), (3, 0), (-5, 5)\}$ find the function values $g(2), g(8), g(3)$, and $g(-5)$.

12. The quantity z varies jointly as y and as the square of x. If z is 80 when x is 5 and y is 2, find z when $x = 2$ and $y = 5$.

13. a. Find the value of the expression $x^3 + x^2 + x + 1$ for $x = -2$.

    b. Factor the expression $x^3 + x^2 + x + 1$ and find the value when x is -2.

    c. Compare the values for parts (a) and (b).

14. Solve the radical equations.
    a. $\sqrt{2x - 5} = -3$
    b. $\sqrt[3]{2x - 5} = -3$

15. Perform the indicated operations with complex numbers.

    a. $6i(4 + 5i)$
    b. $\dfrac{3}{4 - 5i}$

16. An automobile starts from rest and accelerates at a constant rate for 10 sec. The distance, $d(t)$, in feet traveled by the car is given by
    $$d(t) = 4.4t^2$$
    where $0 \leq t \leq 10$ is the time in seconds.

    a. How far has the car traveled after 2, 3, and 4 sec, respectively?

    b. How long will it take for the car to travel 281.6 ft?

17. Solve the equation $125w^3 + 1 = 0$ by factoring and using the quadratic formula. (*Hint:* You will find one real solution and two imaginary solutions.)

18. Solve the rational equation.
    $$\dfrac{x}{x+2} - \dfrac{3}{x-1} = \dfrac{1}{x^2+x-2}$$

19. Solve the inequality and write the answer in interval notation.
    $$|x - 9| - 3 < 7$$

20. Write the expression in logarithmic form.
    $8^{5/3} = 32$

21. Find the coordinates of the vertex of the parabola defined by $f(x) = x^2 + 10x - 11$ by completing the square.

22. Graph the quadratic function defined by $g(x) = -x^2 - 2x + 3$.

    a. Label the x-intercepts.
    b. Label the y-intercept.
    c. Label the vertex.

    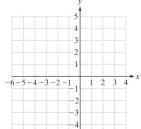

23. Write an equation representing the set of all points 4 units from the point (0, 5).

24. Can a circle and a parabola intersect in only one point? Explain.

25. Solve the system of nonlinear equations.
    $$x^2 + y^2 = 16$$
    $$y = -x^2 - 4$$

26. Graph the solution set.
    $$y^2 - x^2 < 1$$

    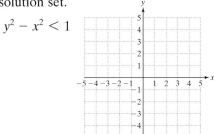

27. Graph the solution set to this system.
    $$y > \left(\dfrac{1}{2}\right)^x$$
    $$x < 0$$

    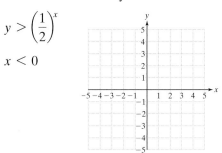

28. Factor completely.
    $$x^2 - y^2 - 6x - 6y$$

Additional Topics Appendix

section A.1 Binomial Expansions

Objectives

1. Binomial Expansions and Pascal's Triangle
2. Factorial Notation
3. The Binomial Theorem

1. Binomial Expansions and Pascal's Triangle

In Section 5.2 we learned how to square a binomial.

$$(a + b)^2 = a^2 + 2ab + b^2$$

The expression $a^2 + 2ab + b^2$ is called the **binomial expansion** of $(a + b)^2$. To expand $(a + b)^3$, we can find the product $(a + b)(a + b)^2$.

$$(a + b)(a + b)^2 = (a + b)(a^2 + 2ab + b^2)$$

$$= a^3 + 2a^2b + ab^2 + a^2b + 2ab^2 + b^3$$

$$= a^3 + 3a^2b + 3ab^2 + b^3$$

Similarly, to expand $(a + b)^4$, we can multiply $(a + b)$ by $(a + b)^3$. Using this method, we can expand several powers of $(a + b)$ to find the following pattern:

$$(a + b)^0 = 1$$

$$(a + b)^1 = a + b$$

$$(a + b)^2 = a^2 + 2ab + b^2$$

$$(a + b)^3 = a^3 + 3a^2b + 3ab^2 + b^3$$

$$(a + b)^4 = a^4 + 4a^3b + 6a^2b^2 + 4ab^3 + b^4$$

$$(a + b)^5 = a^5 + 5a^4b + 10a^3b^2 + 10a^2b^3 + 5ab^4 + b^5$$

Notice that the exponent on a decreases from left to right while the exponents on b increase from left to right. Also observe that for each term, the sum of the exponents is equal to the exponent to which $(a + b)$ is raised. Finally, notice that the number of terms in the expansion is exactly 1 more than the power to which $(a + b)$ is raised. For example, the expansion of $(a + b)^4$ has five terms, and the expansion of $(a + b)^5$ has six terms.

With these guidelines in mind, we know that $(a + b)^6$ will contain seven terms involving

$$a^6 \quad a^5b \quad a^4b^2 \quad a^3b^3 \quad a^2b^4 \quad ab^5 \quad b^6$$

We can complete the expansion of $(a + b)^6$ if we can determine the correct coefficients of each term.

If we write the coefficients for several expansions of $(a + b)^n$, where $n \geq 0$, we have a triangular array of numbers.

$(a + b)^0 = 1$

$(a + b)^1 = 1a + 1b$

$(a + b)^2 = 1a^2 + 2ab + 1b^2$

$(a + b)^3 = 1a^3 + 3a^2b + 3ab^2 + 1b^3$

$(a + b)^4 = 1a^4 + 4a^3b + 6a^2b^2 + 4ab^3 + 1b^4$

$(a + b)^5 = 1a^5 + 5a^4b + 10a^3b^2 + 10a^2b^3 + 5ab^4 + 1b^5$

```
            1
          1   1
        1   2   1
      1   3   3   1
    1   4   6   4   1
  1   5   10  10  5   1
```

Each row begins and ends with a 1, and each entry in between is the sum of the two entries from the line above. For example, in the last row, $5 = 1 + 4$, $10 = 4 + 6$, and so on. This triangular array of coefficients for a binomial expansion is called **Pascal's triangle**, named after French mathematician Blaise Pascal (1623–1662).

```
              1
            1   1
          1   2   1
        1   3   3   1
      1   4   6   4   1
    1   5   10  10  5   1
  1   6   15  20  15  6   1
```

By using the pattern shown in Pascal's triangle, the coefficients corresponding to $(a + b)^6$ would be 1, 6, 15, 20, 15, 6, 1. By inserting the coefficients, the sum becomes

$$(a + b)^6 = 1a^6 + 6a^5b + 15a^4b^2 + 20a^3b^3 + 15a^2b^4 + 6ab^5 + 1b^6$$

2. Factorial Notation

Although Pascal's triangle provides an easy method to find the coefficients of $(a + b)^n$, it is impractical for large values of n. A more efficient method to find the coefficients of a binomial expansion involves **factorial notation**.

Definition of $n!$

Let n be a positive integer. Then $n!$ (read as "n factorial") is defined as the product of integers from 1 through n. That is,

$$n! = n(n - 1)(n - 2) \cdots (2)(1)$$

Note: We define $0! = 1$.

Skill Practice

1. Evaluate.
   a. 3!   b. 8!   c. 1!

example 1 Evaluating Factorial Notation

Evaluate the expressions.

a. 4!   b. 10!   c. 0!

Answers

1. a. 6   b. 40,320   c. 1

Section A.1 Binomial Expansions A-3

Solution:

a. $4! = 4 \cdot 3 \cdot 2 \cdot 1 = 24$

b. $10! = 10 \cdot 9 \cdot 8 \cdot 7 \cdot 6 \cdot 5 \cdot 4 \cdot 3 \cdot 2 \cdot 1 = 3{,}628{,}800$

c. $0! = 1$ by definition

Calculator Connections

Most calculators have a $\boxed{!}$ function. Try evaluating the expressions from Example 1 on a calculator:

```
4!
           24
10!
      3628800
0!
            1
```

Sometimes factorial notation is used with other operations such as multiplication and division.

example 2 Operations with Factorials

Evaluate the expressions.

a. $\dfrac{4!}{4! \cdot 0!}$ b. $\dfrac{4!}{3! \cdot 1!}$ c. $\dfrac{4!}{2! \cdot 2!}$ d. $\dfrac{4!}{1! \cdot 3!}$ e. $\dfrac{4!}{0! \cdot 4!}$

Skill Practice

2. Evaluate.

a. $\dfrac{5!}{5!0!}$ b. $\dfrac{5!}{4!1!}$ c. $\dfrac{5!}{3!2!}$

Solution:

a. $\dfrac{4!}{4! \cdot 0!} = \dfrac{(4 \cdot 3 \cdot 2 \cdot 1)}{(4 \cdot 3 \cdot 2 \cdot 1) \cdot (1)} = \dfrac{4 \cdot 3 \cdot 2 \cdot 1}{4 \cdot 3 \cdot 2 \cdot 1 \cdot 1} = 1$

b. $\dfrac{4!}{3! \cdot 1!} = \dfrac{(4 \cdot 3 \cdot 2 \cdot 1)}{(3 \cdot 2 \cdot 1) \cdot (1)} = \dfrac{4 \cdot 3 \cdot 2 \cdot 1}{3 \cdot 2 \cdot 1 \cdot 1} = 4$

c. $\dfrac{4!}{2! \cdot 2!} = \dfrac{(4 \cdot 3 \cdot 2 \cdot 1)}{(2 \cdot 1) \cdot (2 \cdot 1)} = \dfrac{4 \cdot 3 \cdot 2 \cdot 1}{2 \cdot 1 \cdot 2 \cdot 1} = 6$

d. $\dfrac{4!}{1! \cdot 3!} = \dfrac{(4 \cdot 3 \cdot 2 \cdot 1)}{(1) \cdot (3 \cdot 2 \cdot 1)} = \dfrac{4 \cdot 3 \cdot 2 \cdot 1}{1 \cdot 3 \cdot 2 \cdot 1} = 4$

e. $\dfrac{4!}{0! \cdot 4!} = \dfrac{(4 \cdot 3 \cdot 2 \cdot 1)}{(1) \cdot (4 \cdot 3 \cdot 2 \cdot 1)} = \dfrac{4 \cdot 3 \cdot 2 \cdot 1}{1 \cdot 4 \cdot 3 \cdot 2 \cdot 1} = 1$

Calculator Connections

To evaluate the expressions from Example 2 on a calculator, use parentheses around the factors in the denominator:

```
4!/(4!*0!)
            1
4!/(3!*1!)
            4
4!/(2!*2!)
            6
```

Answers

2. a. 1   b. 5   c. 10

3. The Binomial Theorem

Notice from Example 2 that the values of

$$\frac{4!}{4! \cdot 0!} \quad \frac{4!}{3! \cdot 1!} \quad \frac{4!}{2! \cdot 2!} \quad \frac{4!}{1! \cdot 3!} \quad \text{and} \quad \frac{4!}{0! \cdot 4!}$$

correspond to the values 1, 4, 6, 4, 1, which are the coefficients for the expansion of $(a + b)^4$. Generalizing this pattern, we see the coefficients for the terms in the expansion of $(a + b)^n$ are given by

$$\frac{n!}{r! \cdot (n - r)!}$$

where r corresponds to the exponent on the factor of a and $(n - r)$ corresponds to the exponent on the factor of b. Using this formula for the coefficients in a binomial expansion results in the **binomial theorem**.

The Binomial Theorem

For any positive integer n,

$$(a + b)^n = \frac{n!}{n! \cdot 0!} a^n + \frac{n!}{(n - 1)! \cdot 1!} a^{(n-1)}b + \frac{n!}{(n - 2)! \cdot 2!} a^{(n-2)}b^2$$
$$+ \cdots + \frac{n!}{0! \cdot n!} b^n$$

---

example 3 Applying the Binomial Theorem

Write out the first three terms of the expansion of $(a + b)^{10}$.

Solution:

The first three terms of $(a + b)^{10}$ are

$$\frac{10!}{10! \cdot 0!} a^{10} + \frac{10!}{9! \cdot 1!} a^9 b + \frac{10!}{8! \cdot 2!} a^8 b^2$$

$$= \frac{\cancel{10!}}{\cancel{10!} \cdot 1} a^{10} + \frac{10 \cdot \cancel{9!}}{\cancel{9!} \cdot 1!} a^9 b + \frac{10 \cdot 9 \cdot \cancel{8!}}{\cancel{8!} \cdot 2 \cdot 1} a^8 b^2$$

$$= a^{10} + 10a^9 b + 45a^8 b^2$$

Skill Practice

3. Write out the first three terms of $(x - 2y)^5$.

---

example 4 Applying the Binomial Theorem

Use the binomial theorem to find the expansion of $(3x^2 - 5y)^4$.

Solution:

Write $(3x^2 - 5y)^4$ as $[(3x^2) + (-5y)]^4$. In this case, the expressions $3x^2$ and $-5y$ may be substituted for a and b in the expansion of $(a + b)^4$.

Answer

3. $x^5 - 10x^4 y + 40x^3 y^2$

$$(a+b)^4 = \frac{4!}{4! \cdot 0!}a^4 + \frac{4!}{3! \cdot 1!}a^3b + \frac{4!}{2! \cdot 2!}a^2b^2 + \frac{4!}{1! \cdot 3!}ab^3 + \frac{4!}{0! \cdot 4!}b^4$$

$$= \frac{4!}{4! \cdot 0!}(3x^2)^4 + \frac{4!}{3! \cdot 1!}(3x^2)^3(-5y) + \frac{4!}{2! \cdot 2!}(3x^2)^2(-5y)^2$$
$$+ \frac{4!}{1! \cdot 3!}(3x^2)(-5y)^3 + \frac{4!}{0! \cdot 4!}(-5y)^4$$

$$= 1 \cdot (81x^8) + 4 \cdot (27x^6)(-5y) + 6 \cdot (9x^4)(25y^2) + 4 \cdot (3x^2)(-125y^3)$$
$$+ 1 \cdot (625y^4)$$

$$= 81x^8 - 540x^6y + 1350x^4y^2 - 1500x^2y^3 + 625y^4$$

Tip: The values of $\frac{n!}{r!(n-r)!}$ can also be found by using Pascal's Triangle.

```
        1
       1 1
      1 2 1
     1 3 3 1
    1 4 6 4 1
```

The binomial theorem may also be used to find a specific term in a binomial expansion.

example 5 Finding a Specific Term in a Binomial Expansion

Find the fourth term of the expansion $(a+b)^{13}$.

Solution:

There are 14 terms in the expansion of $(a+b)^{13}$. The first term will have variable factors $a^{13}b^0$. The second term will have variable factors $a^{12}b^1$. The third term will have $a^{11}b^2$, and the fourth term will have $a^{10}b^3$. Hence the fourth term is

$$\frac{13!}{10! \cdot 3!}a^{10}b^3 = 286a^{10}b^3$$

Skill Practice

4. Find the 4th term of $(x+y)^8$.

From Example 5, we see that for the kth term in the expansion $(a+b)^n$, where k is an integer greater than zero, the variable factors are $a^{n-(k-1)}$ and b^{k-1}. Therefore, to find the kth term of $(a+b)^n$, we can make the following generalization.

Finding a Specific Term in a Binomial Expansion

Let n and k be positive integers such that $k \leq n$. Then the kth term in the expansion of $(a+b)^n$ is given by

$$\frac{n!}{[n-(k-1)]! \cdot (k-1)!} \cdot a^{n-(k-1)} \cdot b^{k-1}$$

example 6 Finding a Specific Term in a Binomial Expansion

Find the sixth term of $(p^3 + 2w)^8$.

Answer

4. $56x^5y^3$

Skill Practice

5. Find the 5th term of $(t - m^2)^8$.

Solution:

Apply the formula.

$$\frac{n!}{[n-(k-1)]! \cdot (k-1)!} \cdot a^{n-(k-1)} \cdot b^{k-1} \quad \text{with } n = 8, k = 6, a = p^3, \text{ and } b = 2w$$

$$\frac{8!}{[8-(6-1)]! \cdot (6-1)!} \cdot (p^3)^{8-(6-1)} \cdot (2w)^{6-1}$$

$$= \frac{8!}{[8-(5)]! \cdot (5)!} \cdot (p^3)^{8-(5)} \cdot (2w)^5$$

$$= \frac{8!}{(3)! \cdot (5)!} \cdot (p^3)^3 \cdot (2w)^5$$

$$= 56 \cdot (p^9) \cdot (32w^5)$$

$$= 1792 p^9 w^5$$

Answer

5. $70 t^4 m^8$

---

section A.1 Practice Exercises

Boost your GRADE at mathzone.com!
MathZone

- Practice Problems
- Self-Tests
- NetTutor
- e-Professors
- Videos

Objective 1: Binomial Expansions and Pascal's Triangle

For Exercises 1–6, expand the binomials. Use Pascal's triangle to find the coefficients.

1. $(x + y)^4$
2. $(a + b)^3$
3. $(4 + p)^3$
4. $(1 + g)^4$

5. $(a^2 + b)^6$
6. $(p + q^2)^7$

For Exercises 7–8, rewrite each binomial of the form $(a - b)^n$ as $[a + (-b)]^n$. Then expand the binomials. Use Pascal's triangle to find the coefficients.

7. $(p^2 - w)^3$
8. $(5 - u^3)^4$

9. For $a > 0$ and $b > 0$, what happens to the signs of the terms when expanding the binomial $(a - b)^n$ compared with $(a + b)^n$?

Objective 2: Factorial Notation

For Exercises 10–13, evaluate the expression. **(See Example 1.)**

10. $5!$
11. $3!$
12. $0!$
13. $1!$

14. True or false: $0! \ne 1!$

15. True or false: $n!$ is defined for negative integers.

16. True or false: $n! = n$ for $n = 1$ and 2. **17.** Show that $9! = 9 \cdot 8!$

18. Show that $6! = 6 \cdot 5!$

For Exercises 19–26, evaluate the expression. (See Example 2.)

19. $\dfrac{8!}{4!}$ **20.** $\dfrac{7!}{5!}$ **21.** $\dfrac{3!}{0!}$ **22.** $\dfrac{4!}{0!}$

23. $\dfrac{8!}{3!\,5!}$ **24.** $\dfrac{6!}{2!\,4!}$ **25.** $\dfrac{4!}{0!\,4!}$ **26.** $\dfrac{6!}{0!\,6!}$

Objective 3: Binomial Theorem

For Exercises 27–36, use the binomial theorem to expand the binomials. (See Example 4.)

27. $(s + t)^6$ **28.** $(h + k)^4$ **29.** $(b - 3)^3$ **30.** $(c - 2)^5$

31. $(2x + y)^4$ **32.** $(p + 3q)^3$ **33.** $(c^2 - d)^7$ **34.** $(u - v^3)^6$

35. $\left(\dfrac{a}{2} - b\right)^5$ **36.** $\left(\dfrac{s}{3} + t\right)^5$

For Exercises 37–40, find the first three terms of the expansion. (See Example 3.)

37. $(m - n)^{11}$ **38.** $(p - q)^9$ **39.** $(u^2 - v)^{12}$ **40.** $(r - s^2)^8$

41. How many terms are in the expansion of $(a + b)^8$?

42. How many terms are in the expansion of $(x + y)^{13}$?

For Exercises 43–48, find the indicated term of the binomial expansion. (See Examples 5–6.)

43. $(m - n)^{11}$; sixth term **44.** $(p - q)^9$; fourth term **45.** $(u^2 - v)^{12}$; fifth term

46. $(r - s^2)^8$; sixth term **47.** $(5f + g)^9$; 10th term **48.** $(4m + n)^{10}$; 11th term

Additional Topics Appendix

Objectives

1. Introduction to Determinants
2. Determinant of a 3 × 3 Matrix
3. Cramer's Rule

section A.2 Determinants and Cramer's Rule

1. Introduction to Determinants

Associated with every square matrix is a real number called the **determinant** of the matrix. A determinant of a square matrix **A**, denoted det **A**, is written by enclosing the elements of the matrix within two vertical bars. For example,

$$\text{If} \quad \mathbf{A} = \begin{bmatrix} 2 & -1 \\ 6 & 0 \end{bmatrix} \quad \text{then} \quad \det \mathbf{A} = \begin{vmatrix} 2 & -1 \\ 6 & 0 \end{vmatrix}$$

$$\text{If} \quad \mathbf{B} = \begin{bmatrix} 0 & -5 & 1 \\ 4 & 0 & \frac{1}{2} \\ -2 & 10 & 1 \end{bmatrix} \quad \text{then} \quad \det \mathbf{B} = \begin{vmatrix} 0 & -5 & 1 \\ 4 & 0 & \frac{1}{2} \\ -2 & 10 & 1 \end{vmatrix}$$

Determinants have many applications in mathematics, including solving systems of linear equations, finding the area of a triangle, determining whether three points are collinear, and finding an equation of a line between two points.

The determinant of a 2 × 2 matrix is defined as follows:

Determinant of a 2 × 2 Matrix

The determinant of the matrix $\begin{bmatrix} a & b \\ c & d \end{bmatrix}$ is the real number $ad - bc$. It is written as

$$\begin{vmatrix} a & b \\ c & d \end{vmatrix} = ad - bc$$

Skill Practice

1. Evaluate the determinants.

   a. $\begin{vmatrix} 2 & 8 \\ -1 & 5 \end{vmatrix}$

   b. $\begin{vmatrix} -6 & 0 \\ 4 & 0 \end{vmatrix}$

example 1 Evaluating a 2 × 2 Determinant

Evaluate the determinants.

a. $\begin{vmatrix} 6 & -2 \\ 5 & \frac{1}{3} \end{vmatrix}$ b. $\begin{vmatrix} 2 & -11 \\ 0 & 0 \end{vmatrix}$

Solution:

a. $\begin{vmatrix} 6 & -2 \\ 5 & \frac{1}{3} \end{vmatrix}$ For this determinant, $a = 6$, $b = -2$, $c = 5$, and $d = \frac{1}{3}$.

$$ad - bc = (6)\left(\frac{1}{3}\right) - (-2)(5)$$

$$= 2 + 10$$

$$= 12$$

Answers

1. a. 18   b. 0

b. $\begin{vmatrix} 2 & -11 \\ 0 & 0 \end{vmatrix}$ For this determinant, $a = 2, b = -11, c = 0, d = 0$.

$$ad - bc = (2)(0) - (-11)(0)$$
$$= 0 - 0$$
$$= 0$$

Tip: Example 1(b) illustrates that the value of a determinant having a row of all zeros is 0. The same is true for a determinant having a column of all zeros.

2. Determinant of a 3 × 3 Matrix

To find the determinant of a 3 × 3 matrix, we first need to define the **minor** of an element of the matrix. For any element of a 3 × 3 matrix, the minor of that element is the determinant of the 2 × 2 matrix obtained by deleting the row and column in which the element resides. For example, consider the matrix

$$\begin{bmatrix} 5 & -1 & 6 \\ 0 & -7 & 1 \\ 4 & 2 & 6 \end{bmatrix}$$

The minor of the element 5 is found by deleting the first row and first column and then evaluating the determinant of the remaining 2 × 2 matrix:

$\begin{bmatrix} 5 & -1 & 6 \\ 0 & -7 & 1 \\ 4 & 2 & 6 \end{bmatrix}$ Now evaluate the determinant: $\begin{vmatrix} -7 & 1 \\ 2 & 6 \end{vmatrix} = (-7)(6) - (1)(2)$
$$= -44$$

For this matrix, the minor for the element 5 is -44.

To find the minor of the element -7, delete the second row and second column, and then evaluate the determinant of the remaining 2 × 2 matrix.

$\begin{bmatrix} 5 & -1 & 6 \\ 0 & -7 & 1 \\ 4 & 2 & 6 \end{bmatrix}$ Now evaluate the determinant: $\begin{vmatrix} 5 & 6 \\ 4 & 6 \end{vmatrix} = (5)(6) - (6)(4) = 6$

For this matrix, the minor for the element -7 is 6.

example 2   Determining the Minor for Elements in a 3 × 3 matrix

Find the minor for each element in the first column of the matrix.

$$\begin{bmatrix} 3 & 4 & -1 \\ 2 & -4 & 5 \\ 0 & 1 & -6 \end{bmatrix}$$

Skill Practice

2. Find the minor for the element 3.
$$\begin{bmatrix} -1 & 8 & -6 \\ \frac{1}{2} & 3 & 2 \\ 5 & 7 & 4 \end{bmatrix}$$

Solution:

For 3: $\begin{bmatrix} 3 & 4 & -1 \\ 2 & -4 & 5 \\ 0 & 1 & -6 \end{bmatrix}$ The minor is: $\begin{vmatrix} -4 & 5 \\ 1 & -6 \end{vmatrix} = (-4)(-6) - (5)(1) = 19$

Answer

2. $\begin{vmatrix} -1 & -6 \\ 5 & 4 \end{vmatrix} = 26$

For 2: $\begin{bmatrix} 3 & 4 & -1 \\ 2 & -4 & 5 \\ 0 & 1 & -6 \end{bmatrix}$ The minor is: $\begin{vmatrix} 4 & -1 \\ 1 & -6 \end{vmatrix} = (4)(-6) - (-1)(1) = -23$

For 0: $\begin{bmatrix} 3 & 4 & -1 \\ 2 & -4 & 5 \\ 0 & 1 & -6 \end{bmatrix}$ The minor is $\begin{vmatrix} 4 & -1 \\ -4 & 5 \end{vmatrix} = (4)(5) - (-1)(-4) = 16$

The determinant of a 3 × 3 matrix is defined as follows.

Definition of a Determinant of a 3 × 3 Matrix

$$\begin{vmatrix} a_1 & b_1 & c_1 \\ a_2 & b_2 & c_2 \\ a_3 & b_3 & c_3 \end{vmatrix} = a_1 \cdot \begin{vmatrix} b_2 & c_2 \\ b_3 & c_3 \end{vmatrix} - a_2 \cdot \begin{vmatrix} b_1 & c_1 \\ b_3 & c_3 \end{vmatrix} + a_3 \cdot \begin{vmatrix} b_1 & c_1 \\ b_2 & c_2 \end{vmatrix}$$

From this definition, we see that the determinant of a 3 × 3 matrix can be written as

$$a_1 \cdot (\text{minor of } a_1) - a_2 \cdot (\text{minor of } a_2) + a_3 \cdot (\text{minor of } a_3)$$

Evaluating determinants in this way is called expanding minors.

Skill Practice

3. Evaluate the determinant.
$$\begin{vmatrix} -2 & 4 & 9 \\ 5 & -1 & 2 \\ 1 & 1 & 6 \end{vmatrix}$$

example 3 Evaluating a 3 × 3 Determinant

Evaluate the determinant. $\begin{vmatrix} 2 & 4 & 2 \\ 1 & -3 & 0 \\ -5 & 5 & -1 \end{vmatrix}$

Solution:

$$\begin{vmatrix} 2 & 4 & 2 \\ 1 & -3 & 0 \\ -5 & 5 & -1 \end{vmatrix} = 2 \cdot \begin{vmatrix} -3 & 0 \\ 5 & -1 \end{vmatrix} - (1) \cdot \begin{vmatrix} 4 & 2 \\ 5 & -1 \end{vmatrix} + (-5) \cdot \begin{vmatrix} 4 & 2 \\ -3 & 0 \end{vmatrix}$$

$= 2[(-3)(-1) - (0)(5)] - 1[(4)(-1) - (2)(5)] - 5[(4)(0) - (2)(-3)]$

$= 2(3) - 1(-14) - 5(6)$

$= 6 + 14 - 30$

$= -10$

Answer

3. -42

Although we defined the determinant of a matrix by expanding the minors of the elements in the first column, *any row or column may be used*. However, we must choose the correct sign to apply to each term in the expansion. The following array of signs is helpful.

$$\begin{matrix} + & - & + \\ - & + & - \\ + & - & + \end{matrix}$$

Tip: There is another method to determine the signs for each term of the expansion. For the a_{ij} element, multiply the term by $(-1)^{i+j}$.

The signs alternate for each row and column, beginning with $+$ in the first row, first column.

example 4   Evaluating a 3 × 3 Determinant

Evaluate the determinant, by expanding minors about the elements in the second row.

$$\begin{vmatrix} 2 & 4 & 2 \\ 1 & -3 & 0 \\ -5 & 5 & -1 \end{vmatrix}$$

Skill Practice

4. Evaluate the determinant.
$$\begin{vmatrix} 4 & -1 & 2 \\ 3 & 6 & -8 \\ 0 & \frac{1}{2} & 5 \end{vmatrix}$$

Solution:

signs obtained from the array of signs

$$\begin{vmatrix} 2 & 4 & 2 \\ 1 & -3 & 0 \\ -5 & 5 & -1 \end{vmatrix} = -(1) \cdot \begin{vmatrix} 4 & 2 \\ 5 & -1 \end{vmatrix} + (-3) \cdot \begin{vmatrix} 2 & 2 \\ -5 & -1 \end{vmatrix} - (0) \cdot \begin{vmatrix} 2 & 4 \\ -5 & 5 \end{vmatrix}$$

$$= -1[(4)(-1) - (2)(5)] - 3[(2)(-1) - (2)(-5)] - 0$$

$$= -1(-14) - 3(8)$$

$$= 14 - 24$$

$$= -10$$

Notice that the value of the determinant obtained in Examples 3 and 4 is the same.

Calculator Connections

The determinant of a matrix can be evaluated on a graphing calculator. First use the matrix editor to enter the elements of the matrix. Then use a *det* function to evaluate the determinant. The determinant from Examples 3 and 4 is evaluated below.

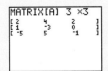 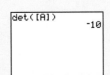

Answer

4. 154

In Example 4, the third term in the expansion of minors was zero because the element 0 when multiplied by its minor is zero. To simplify the arithmetic in evaluating a determinant of a 3×3 matrix, expand about the row or column that has the most 0 elements.

3. Cramer's Rule

In Sections 3.2, 3.3, and 3.6, we learned three methods to solve a system of linear equations: the substitution method, the addition method, and the Gauss-Jordan method. In this section, we will learn another method called **Cramer's rule** to solve a system of linear equations.

Cramer's Rule for a 2 × 2 System of Linear Equations

The solution to the system
$$a_1 x + b_1 y = c_1$$
$$a_2 x + b_2 y = c_2$$

is given by $\quad x = \dfrac{\mathbf{D}_x}{\mathbf{D}} \quad$ and $\quad y = \dfrac{\mathbf{D}_y}{\mathbf{D}}$

where $\mathbf{D} = \begin{vmatrix} a_1 & b_1 \\ a_2 & b_2 \end{vmatrix}$ (and $\mathbf{D} \neq 0$) $\quad \mathbf{D}_x = \begin{vmatrix} c_1 & b_1 \\ c_2 & b_2 \end{vmatrix} \quad \mathbf{D}_y = \begin{vmatrix} a_1 & c_1 \\ a_2 & c_2 \end{vmatrix}$

Skill Practice

5. Solve using Cramer's rule.
$$2x + y = 5$$
$$-x - 3y = 5$$

example 5 — Using Cramer's Rule to Solve a 2 × 2 System of Linear Equations

Solve the system by using Cramer's rule.
$$3x - 5y = 11$$
$$-x + 3y = -5$$

Solution:

For this system: $\quad a_1 = 3 \quad\quad b_1 = -5 \quad\quad c_1 = 11$
$\quad\quad\quad\quad\quad\quad\quad\quad a_2 = -1 \quad\quad b_2 = 3 \quad\quad\quad c_2 = -5$

$\mathbf{D} = \begin{vmatrix} 3 & -5 \\ -1 & 3 \end{vmatrix} = (3)(3) - (-5)(-1) = 9 - 5 = 4$

$\mathbf{D}_x = \begin{vmatrix} 11 & -5 \\ -5 & 3 \end{vmatrix} = (11)(3) - (-5)(-5) = 33 - 25 = 8$

$\mathbf{D}_y = \begin{vmatrix} 3 & 11 \\ -1 & -5 \end{vmatrix} = (3)(-5) - (11)(-1) = -15 + 11 = -4$

Therefore, $\quad x = \dfrac{\mathbf{D}_x}{\mathbf{D}} = \dfrac{8}{4} = 2 \quad\quad y = \dfrac{\mathbf{D}_y}{\mathbf{D}} = \dfrac{-4}{4} = -1$

The solution is $(2, -1)$. Check: $\quad 3x - 5y = 11 \longrightarrow 3(2) - 5(-1) \stackrel{?}{=} 11 \checkmark$
$\quad\quad\quad\quad\quad\quad\quad\quad\quad\quad\quad\quad\quad -x + 3y = -5 \longrightarrow -(2) + 3(-1) \stackrel{?}{=} -5 \checkmark$

Answer

5. $(4, -3)$

> **Tip:** Here are some memory tips to help you remember Cramer's rule.
>
> 1. The determinant D is the determinant of the coefficients of x and y.
>
> $D = \begin{vmatrix} a_1 & b_1 \\ a_2 & b_2 \end{vmatrix}$ ← coefficients of x-terms and y-terms
>
> 2. The determinant D_x has the column of x-term coefficients replaced by c_1 and c_2.
>
> $D_x = \begin{vmatrix} c_1 & b_1 \\ c_2 & b_2 \end{vmatrix}$ ← x-coefficients replaced by c_1 and c_2
>
> 3. The determinant D_y has the column of y-term coefficients replaced by c_1 and c_2.
>
> $D_y = \begin{vmatrix} a_1 & c_1 \\ a_2 & c_2 \end{vmatrix}$ ← y-coefficients replaced by c_1 and c_2

It is important to note that the linear equations must be written in standard form to apply Cramer's rule.

example 6   Using Cramer's Rule to Solve a 2 × 2 System of Linear Equations

Solve the system by using Cramer's rule.

$$-16y = -40x - 7$$
$$40y = 24x + 27$$

Solution:

$-16y = -40x - 7 \longrightarrow 40x - 16y = -7$ Rewrite each equation
$40y = 24x + 27 \longrightarrow -24x + 40y = 27$ in standard form.

For this system: $a_1 = 40$ $b_1 = -16$ $c_1 = -7$
                   $a_2 = -24$ $b_2 = 40$ $c_2 = 27$

$D = \begin{vmatrix} 40 & -16 \\ -24 & 40 \end{vmatrix} = (40)(40) - (-16)(-24) = 1216$

$D_x = \begin{vmatrix} -7 & -16 \\ 27 & 40 \end{vmatrix} = (-7)(40) - (-16)(27) = 152$

$D_y = \begin{vmatrix} 40 & -7 \\ -24 & 27 \end{vmatrix} = (40)(27) - (-7)(-24) = 912$

Therefore, $x = \dfrac{D_x}{D} = \dfrac{152}{1216} = \dfrac{1}{8}$ $y = \dfrac{D_y}{D} = \dfrac{912}{1216} = \dfrac{3}{4}$

The solution $\left(\dfrac{1}{8}, \dfrac{3}{4}\right)$ checks in the original equations.

Skill Practice

6. Solve using Cramer's rule.
   $$9x - 12y = -8$$
   $$18x + 30y = -7$$

Answer

6. $\left(-\dfrac{2}{3}, \dfrac{1}{6}\right)$

Cramer's rule can be used to solve a 3×3 system of linear equations by using a similar pattern of determinants.

Cramer's Rule for a 3×3 System of Linear Equations

The solution to the system

$$a_1 x + b_1 y + c_1 z = d_1$$
$$a_2 x + b_2 y + c_2 z = d_2$$
$$a_3 x + b_3 y + c_3 z = d_3$$

is given by

$$x = \frac{\mathbf{D}_x}{\mathbf{D}} \qquad y = \frac{\mathbf{D}_y}{\mathbf{D}} \quad \text{and} \quad z = \frac{\mathbf{D}_z}{\mathbf{D}}$$

where $\mathbf{D} = \begin{vmatrix} a_1 & b_1 & c_1 \\ a_2 & b_2 & c_2 \\ a_3 & b_3 & c_3 \end{vmatrix}$ (and $\mathbf{D} \neq 0$) $\qquad \mathbf{D}_x = \begin{vmatrix} d_1 & b_1 & c_1 \\ d_2 & b_2 & c_2 \\ d_3 & b_3 & c_3 \end{vmatrix}$

$\mathbf{D}_y = \begin{vmatrix} a_1 & d_1 & c_1 \\ a_2 & d_2 & c_2 \\ a_3 & d_3 & c_3 \end{vmatrix} \qquad \mathbf{D}_z = \begin{vmatrix} a_1 & b_1 & d_1 \\ a_2 & b_2 & d_2 \\ a_3 & b_3 & d_3 \end{vmatrix}$

Skill Practice

7. Solve using Cramer's rule.
   $x + 3y - 3z = -14$
   $x - 4y + z = 2$
   $x + y + 2z = 6$

example 7 Using Cramer's Rule to Solve a 3×3 System of Linear Equations

Solve the system by using Cramer's rule.

$$x - 2y + 4z = 3$$
$$x - 4y + 3z = -5$$
$$x + 3y - 2z = 6$$

Solution:

$$\mathbf{D} = \begin{vmatrix} 1 & -2 & 4 \\ 1 & -4 & 3 \\ 1 & 3 & -2 \end{vmatrix} = 1 \cdot \begin{vmatrix} -4 & 3 \\ 3 & -2 \end{vmatrix} - 1 \cdot \begin{vmatrix} -2 & 4 \\ 3 & -2 \end{vmatrix} + 1 \cdot \begin{vmatrix} -2 & 4 \\ -4 & 3 \end{vmatrix}$$

$$= 1(-1) - 1(-8) + 1(10)$$
$$= 17$$

Tip: In Example 7, we expanded the determinants about the first column.

$$\mathbf{D}_x = \begin{vmatrix} 3 & -2 & 4 \\ -5 & -4 & 3 \\ 6 & 3 & -2 \end{vmatrix} = 3 \cdot \begin{vmatrix} -4 & 3 \\ 3 & -2 \end{vmatrix} - (-5) \cdot \begin{vmatrix} -2 & 4 \\ 3 & -2 \end{vmatrix} + 6 \cdot \begin{vmatrix} -2 & 4 \\ -4 & 3 \end{vmatrix}$$

$$= 3(-1) + 5(-8) + 6(10)$$
$$= 17$$

Answer

7. $(-2, 0, 4)$

$$D_y = \begin{vmatrix} 1 & 3 & 4 \\ 1 & -5 & 3 \\ 1 & 6 & -2 \end{vmatrix} = 1 \cdot \begin{vmatrix} -5 & 3 \\ 6 & -2 \end{vmatrix} - 1 \cdot \begin{vmatrix} 3 & 4 \\ 6 & -2 \end{vmatrix} + 1 \cdot \begin{vmatrix} 3 & 4 \\ -5 & 3 \end{vmatrix}$$

$$= 1(-8) - 1(-30) + 1(29)$$

$$= 51$$

$$D_z = \begin{vmatrix} 1 & -2 & 3 \\ 1 & -4 & -5 \\ 1 & 3 & 6 \end{vmatrix} = 1 \cdot \begin{vmatrix} -4 & -5 \\ 3 & 6 \end{vmatrix} - 1 \cdot \begin{vmatrix} -2 & 3 \\ 3 & 6 \end{vmatrix} + 1 \cdot \begin{vmatrix} -2 & 3 \\ -4 & -5 \end{vmatrix}$$

$$= 1(-9) - 1(-21) + 1(22)$$

$$= 34$$

Hence

$$x = \frac{D_x}{D} = \frac{17}{17} = 1 \quad y = \frac{D_y}{D} = \frac{51}{17} = 3 \quad \text{and} \quad z = \frac{D_z}{D} = \frac{34}{17} = 2$$

The solution is $(1, 3, 2)$.

Check: $x - 2y + 4z = 3$ $(1) - 2(3) + 4(2) \stackrel{?}{=} 3$ ✓

$x - 4y + 3z = -5$ $(1) - 4(3) + 3(2) \stackrel{?}{=} -5$ ✓

$x + 3y - 2z = 6$ $(1) + 3(3) - 2(2) \stackrel{?}{=} 6$ ✓

Cramer's rule may seem cumbersome for solving a 3×3 system of linear equations. However, it provides convenient formulas that can be programmed into a computer or calculator to solve for x, y, and z. Cramer's rule can also be extended to solve a 4×4 system of linear equations, a 5×5 system of linear equations, and in general an $n \times n$ system of linear equations.

It is important to remember that Cramer's rule does not apply if $D = 0$. In such a case, the system of equations is either dependent or inconsistent, and another method must be used to analyze the system.

example 8 Analyzing a Dependent System of Equations

Solve the system. Use Cramer's rule if possible.

$$2x - 3y = 6$$
$$-6x + 9y = -18$$

Solution:

$$D = \begin{vmatrix} 2 & -3 \\ -6 & 9 \end{vmatrix} = (2)(9) - (-3)(-6) = 18 - 18 = 0$$

Skill Practice

8. Solve. Use Cramer's rule if possible.

   $x - 6y = 2$
   $2x - 12y = -2$

Answer

6. No solution (inconsistent system)

Because **D** = 0, Cramer's rule does not apply. Using the addition method to solve the system, we have

$$2x - 3y = 6 \xrightarrow{\text{Multiply by 3.}} 6x - 9y = 18$$
$$-6x + 9y = -18 \longrightarrow \underline{-6x + 9y = -18}$$
$$0 = 0 \quad \text{The system is dependent.}$$

The solution is $\{(x, y) \mid 2x - 3y = 6\}$.

section A.2 Practice Exercises

Boost your GRADE at mathzone.com! MathZone

- Practice Problems
- Self-Tests
- NetTutor
- e-Professors
- Videos

Objective 1: Introduction to Determinants

For Exercises 1–6, evaluate the determinant of the 2 × 2 matrix. **(See Example 1.)**

1. $\begin{vmatrix} -3 & 1 \\ 5 & 2 \end{vmatrix}$

2. $\begin{vmatrix} 5 & 6 \\ 4 & 8 \end{vmatrix}$

3. $\begin{vmatrix} -2 & 2 \\ -3 & -5 \end{vmatrix}$

4. $\begin{vmatrix} 5 & -1 \\ 1 & 0 \end{vmatrix}$

5. $\begin{vmatrix} \frac{1}{2} & 3 \\ -2 & 4 \end{vmatrix}$

6. $\begin{vmatrix} -3 & \frac{1}{4} \\ 8 & -2 \end{vmatrix}$

Objective 2: Determinant of a 3 × 3 Matrix

For Exercises 7–10, evaluate the minor corresponding to the given element from matrix **A**. **(See Example 2.)**

$$\mathbf{A} = \begin{bmatrix} 4 & -1 & 8 \\ 2 & 6 & 0 \\ -7 & 5 & 3 \end{bmatrix}$$

7. 4

8. −1

9. 2

10. 3

11. Construct the sign array for a 3 × 3 matrix.

12. Evaluate the determinant of matrix **B**, using expansion by minors.

    $$\mathbf{B} = \begin{bmatrix} 0 & 1 & 2 \\ 3 & -1 & 2 \\ 3 & 2 & -2 \end{bmatrix}$$

    a. About the first column
    b. About the second row

13. Evaluate the determinant of matrix **C**, using expansion by minors.

    $$\mathbf{C} = \begin{bmatrix} 4 & 1 & 3 \\ 2 & -2 & 1 \\ 3 & 1 & 2 \end{bmatrix}$$

    a. About the first row
    b. About the second column

14. When evaluating the determinant of a 3 × 3 matrix, explain the advantage of being able to choose any row or column about which to expand minors.

For Exercises 15–20, evaluate the determinants. (See Examples 3–4.)

15. $\begin{vmatrix} 8 & 2 & -4 \\ 4 & 0 & 2 \\ 3 & 0 & -1 \end{vmatrix}$

16. $\begin{vmatrix} 5 & 2 & 1 \\ 3 & -6 & 0 \\ -2 & 8 & 0 \end{vmatrix}$

17. $\begin{vmatrix} -2 & 1 & 3 \\ 1 & 4 & 4 \\ 1 & 0 & 2 \end{vmatrix}$

18. $\begin{vmatrix} 3 & 2 & 1 \\ 1 & -1 & 2 \\ 1 & 0 & 4 \end{vmatrix}$

19. $\begin{vmatrix} -5 & 4 & 2 \\ 0 & 0 & 0 \\ 3 & -1 & 5 \end{vmatrix}$

20. $\begin{vmatrix} 0 & 5 & -8 \\ 0 & -4 & 1 \\ 0 & 3 & 6 \end{vmatrix}$

For Exercises 21–26, evaluate the determinants.

21. $\begin{vmatrix} x & 3 \\ y & -2 \end{vmatrix}$

22. $\begin{vmatrix} a & 2 \\ b & 8 \end{vmatrix}$

23. $\begin{vmatrix} a & 5 & -1 \\ b & -3 & 0 \\ c & 3 & 4 \end{vmatrix}$

24. $\begin{vmatrix} x & 0 & 3 \\ y & -2 & 6 \\ z & -1 & 1 \end{vmatrix}$

25. $\begin{vmatrix} p & 0 & q \\ r & 0 & s \\ t & 0 & u \end{vmatrix}$

26. $\begin{vmatrix} f & e & 0 \\ d & c & 0 \\ b & a & 0 \end{vmatrix}$

Objective 3: Cramer's Rule

For Exercises 27–29, evaluate the determinants represented by D, D_x, and D_y.

27. $x - 4y = 2$
    $3x + 2y = 1$

28. $4x + 6y = 9$
    $-2x + y = 12$

29. $-3x + 8y = -10$
    $5x + 5y = -13$

For Exercises 30–35, solve the system by using Cramer's rule. (See Examples 5–6.)

30. $2x + y = 3$
    $x - 4y = 6$

31. $2x - y = -1$
    $3x + y = 6$

32. $x - 4y = 8$
    $3x + 7y = 5$

33. $7x + 3y = 4$
    $5x - 4y = 9$

34. $4x - 3y = 5$
    $2x + 5y = 7$

35. $2x + 3y = 4$
    $6x - 12y = -5$

36. When does Cramer's rule not apply in solving a system of equations?

37. How can a system be solved if Cramer's rule does not apply?

For Exercises 38–43, solve the system of equations by using Cramer's rule, if possible. If not possible, use another method. (See Example 8.)

38. $4x - 2y = 3$
    $-2x + y = 1$

39. $6x - 6y = 5$
    $x - y = 8$

40. $4x + y = 0$
    $x - 7y = 0$

41. $-3x - 2y = 0$
    $-x + 5y = 0$

42. $x + 5y = 3$
    $2x + 10y = 6$

43. $-2x - 10y = -4$
    $x + 5y = 2$

For Exercises 44–49, solve for the indicated variable by using Cramer's rule. (See Example 7.)

44. $2x - y + 3z = 9$
    $x + 4y + 4z = 5$ for x
    $3x + 2y + 2z = 5$

45. $x + 2y + 3z = 8$
    $2x - 3y + z = 5$ for y
    $3x - 4y + 2z = 9$

46. $3x - 2y + 2z = 5$
    $6x + 3y - 4z = -1$ for z
    $3x - y + 2z = 4$

47. $4x + 4y - 3z = 3$
    $8x + 2y + 3y = 0$ for x
    $4x - 4y + 6z = -3$

48. $5x + 6z = 5$
    $-2x + y = -6$ for y
    $3y - z = 3$

49. $8x + y = 1$
    $7y + z = 0$ for y
    $x - 3z = -2$

For Exercises 50–53, solve the system by using Cramer's rule, if possible.

50. $x = 3$
    $-x + 3y = 3$
    $y + 2z = 4$

51. $4x + z = 7$
    $y = 2$
    $x + z = 4$

52. $x + y + 8z = 3$
    $2x + y + 11z = 4$
    $x + 3z = 0$

53. $-8x + y + z = 6$
    $2x - y + z = 3$
    $3x - z = 0$

Expanding Your Skills

For Exercises 54–57, solve the equation.

54. $\begin{vmatrix} 6 & x \\ 2 & -4 \end{vmatrix} = 14$

55. $\begin{vmatrix} y & -2 \\ 8 & 7 \end{vmatrix} = 30$

56. $\begin{vmatrix} 3 & 1 & 0 \\ 0 & 4 & -2 \\ 1 & 0 & w \end{vmatrix} = 10$

57. $\begin{vmatrix} -1 & 0 & 2 \\ 4 & t & 0 \\ 0 & -5 & 3 \end{vmatrix} = -4$

For Exercises 58–59, evaluate the determinant by using expansion by minors about the first column.

58. $\begin{vmatrix} 1 & 0 & 3 & 0 \\ 0 & 1 & 2 & 4 \\ -2 & 0 & 0 & 1 \\ 4 & -1 & -2 & 0 \end{vmatrix}$

59. $\begin{vmatrix} 5 & 2 & 0 & 0 \\ 0 & 4 & -1 & 1 \\ -1 & 0 & 3 & 0 \\ 0 & -2 & 1 & 0 \end{vmatrix}$

For Exercises 60–61, refer to the following system of four variables.

$$x + y + z + w = 0$$
$$2x - z + w = 5$$
$$2x + y - w = 0$$
$$y + z = -1$$

60. a. Evaluate the determinant **D**.

b. Evaluate the determinant $\mathbf{D}_x$.

c. Solve for x by computing $\dfrac{\mathbf{D}_x}{\mathbf{D}}$.

61. a. Evaluate the determinant $\mathbf{D}_y$.

b. Solve for y by computing $\dfrac{\mathbf{D}_y}{\mathbf{D}}$.

Section A.3 Sequences and Series

Objectives
1. Finite and Infinite Sequences
2. Series

1. Finite and Infinite Sequences

In everyday life we think of a sequence as a set of events or items with some order or pattern. In mathematics, a sequence is a list of terms that correspond with the set of positive integers. For example, the sequence

$$1, 4, 9, 16, 25$$

represents the squares of the first five positive integers. This sequence has a finite number of terms and is called a *finite sequence*. The sequence

$$1, 4, 9, 16, 25, \ldots$$

represents the squares of *all* positive integers. This sequence continues indefinitely and is called an *infinite sequence*.

Because the terms in a sequence are related to the set of positive integers, we give a formal definition of finite and infinite sequences, using the language of functions.

Definition of Finite and Infinite Sequences

An **infinite sequence** is a function whose domain is the set of positive integers.
A **finite sequence** is a function whose domain is the set of the first n positive integers.

For any positive integer n, the value of the sequence is denoted by a_n (read as "a sub n"). The values $a_1, a_2, a_3, \ldots$ are called the **terms of the sequence**. The expression a_n defines the **nth term** (or general term) **of the sequence**.

Skill Practice

1. List the terms of the sequences.
   a. $a_n = n^3 - 2$; $1 \leq n \leq 3$
   b. $a_n = \left(\dfrac{2}{3}\right)^n$

example 1 — Listing the Terms of a Sequence

List the terms of the following sequences.

a. $a_n = 3n^2 - 4$; $1 \leq n \leq 4$
b. $a_n = 3 \cdot 2^n$

Solution:

a. The domain is restricted to the first four positive integers, indicating that the sequence is finite.

n	a_n
1	$3(1)^2 - 4 = -1$
2	$3(2)^2 - 4 = 8$
3	$3(3)^2 - 4 = 23$
4	$3(4)^2 - 4 = 44$

The sequence is $-1, 8, 23, 44$.

b. The sequence $a_n = 3 \cdot 2^n$ has no restrictions on its domain; therefore, it is an infinite sequence.

n	a_n
1	$3 \cdot 2^1 = 6$
2	$3 \cdot 2^2 = 12$
3	$3 \cdot 2^3 = 24$
4	$3 \cdot 2^4 = 48$
...	

The sequence is $6, 12, 24, 48, \ldots$

Calculator Connections

If the nth term of a sequence is known, a *Seq* function on a graphing calculator can quickly display a list of terms. It is necessary to enter the formula for the nth term, the independent variable, the starting value for n, the ending value for n, and the increment for n.

$$\text{Seq}(a_n, n, \text{begin}, \text{end}, \text{increment})$$

The first four terms of the sequence $a_n = 3 \cdot 2^n$ from Example 1(b) are shown.

```
seq(3*2^n,n,1,4,
1)
          {6 12 24 48}
```

Sometimes the terms of a sequence may have alternating signs. Such a sequence is called an **alternating sequence**.

example 2 — Listing the Terms of an Alternating Sequence

List the first four terms of each alternating sequence.

a. $a_n = (-1)^n \cdot \dfrac{1}{n}$
b. $a_n = (-1)^{n+1} \cdot \left(\dfrac{2}{3}\right)^n$

Answers

1. a. $-1, 6, 25$ b. $\dfrac{2}{3}, \dfrac{4}{9}, \dfrac{8}{27}, \ldots$

Solution:

a.

n	a_n	
1	$(-1)^1 \cdot \dfrac{1}{1} = -1$	
2	$(-1)^2 \cdot \dfrac{1}{2} = \dfrac{1}{2}$	
3	$(-1)^3 \cdot \dfrac{1}{3} = -\dfrac{1}{3}$	
4	$(-1)^4 \cdot \dfrac{1}{4} = \dfrac{1}{4}$	

The first four terms are $-1, \dfrac{1}{2}, -\dfrac{1}{3}, \dfrac{1}{4}$.

Tip: Notice that the factor $(-1)^n$ makes the even-numbered terms positive and the odd-numbered terms negative.

b.

n	a_n
1	$(-1)^{1+1} \cdot \left(\dfrac{2}{3}\right)^1 = (-1)^2 \cdot \left(\dfrac{2}{3}\right) = \dfrac{2}{3}$
2	$(-1)^{2+1} \cdot \left(\dfrac{2}{3}\right)^2 = (-1)^3 \cdot \left(\dfrac{4}{9}\right) = -\dfrac{4}{9}$
3	$(-1)^{3+1} \cdot \left(\dfrac{2}{3}\right)^3 = (-1)^4 \cdot \left(\dfrac{8}{27}\right) = \dfrac{8}{27}$
4	$(-1)^{4+1} \cdot \left(\dfrac{2}{3}\right)^4 = (-1)^5 \cdot \left(\dfrac{16}{81}\right) = -\dfrac{16}{81}$

Tip: Notice that the factor $(-1)^{n+1}$ makes the odd-numbered terms positive and the even-numbered terms negative.

The first four terms are $\dfrac{2}{3}, -\dfrac{4}{9}, \dfrac{8}{27}, -\dfrac{16}{81}$.

In Examples 1 and 2, we were given the formula for the nth term of a sequence and asked to list several terms of the sequence. We now consider the reverse process. Given several terms of the sequence, we will find a formula for the nth term. To do so, look for a pattern that establishes each term as a function of the term number.

example 3   Finding the nth Term of a Sequence

Find a formula for the nth term of the sequence.

a. $\dfrac{1}{2}, \dfrac{2}{3}, \dfrac{3}{4}, \dfrac{4}{5}, \ldots$

b. $-2, 4, -6, 8, -10, \ldots$

c. $\dfrac{1}{2}, \dfrac{1}{4}, \dfrac{1}{8}, \dfrac{1}{16}, \ldots$

Skill Practice

2. Find a formula for the nth term of the sequence.

    a. $\dfrac{1}{3}, \dfrac{1}{9}, \dfrac{1}{27}, \dfrac{1}{81}, \ldots$

    b. $-1, 4, -9, 16, -25, \ldots$

Solution:

a. For each term in the sequence, the numerator is equal to the term number, and the denominator is equal to 1 more than the term number. Therefore, the nth term may be given by

$$a_n = \dfrac{n}{n+1}$$

Answers

2. a. $a_n = \dfrac{1}{3^n}$ b. $a_n = (-1)^n n^2$

b. The odd-numbered terms are negative, and the even-numbered terms are positive. The factor $(-1)^n$ will produce the required alternation of signs. The numbers 2, 4, 6, 8, 10, ... are equal to 2(1), 2(2), 2(3), 2(4), 2(5), ... Therefore, the nth term may be given by

$$a_n = (-1)^n \cdot 2n$$

c. The denominators are consecutive powers of 2. The sequence can be written as

$$\frac{1}{2^1}, \frac{1}{2^2}, \frac{1}{2^3}, \frac{1}{2^4}, \ldots$$

Therefore, the nth term may be given by

$$a_n = \frac{1}{2^n}$$

Tip: The first few terms of a sequence are not sufficient to define the sequence uniquely. For example, consider the sequence

$$\frac{1}{2}, \frac{1}{4}, \frac{1}{8}, \ldots$$

The following formulas both produce the first three terms, but differ at the fourth term:

$$a_n = \frac{1}{2^n} \quad \text{and} \quad b_n = \frac{1}{n^2 - n + 2}$$

$$a_4 = \frac{1}{16} \quad \text{whereas} \quad b_4 = \frac{1}{14}$$

To define a sequence uniquely, the nth term must be provided.

Skill Practice

3. The first swing of a pendulum measures 30°. If each swing that follows is 25% less, list the first 3 terms of this sequence.

example 4 Using a Sequence in an Application

A child drops a ball from a height of 4 ft. With each bounce, the ball rebounds to 50% of its height. Write a sequence whose terms represent the heights from which the ball falls (begin with the initial height from which the ball was dropped).

Solution:

The ball first drops 4 ft and then rebounds to a new height of 0.50(4 ft) = 2 ft. Similarly, it falls from 2 ft and rebounds 0.50(2 ft) = 1 ft. Repeating this process, we have

$$4, 2, 1, \frac{1}{2}, \frac{1}{4}, \frac{1}{8}, \ldots$$

The nth term can be represented by $a_n = 4 \cdot (0.50)^{n-1}$.

Answer

3. 30°, 22.5°, 16.875°

2. Series

In many mathematical applications it is important to find the sum of the terms of a sequence. For example, suppose that the yearly interest earned in an account over a 4-year period is given by the sequence

$$\$250, \quad \$265, \quad \$278.25, \quad \$292.16$$

The sum of the terms gives the total interest earned

$$\$250 + \$265 + \$278.25 + \$292.16 = \$1085.41$$

By adding the terms of a sequence, we obtain a series.

> **Definition of a Series**
>
> The indicated sum of the terms of a sequence is called a **series**.

As with a sequence, a series may be a finite or an infinite sum of terms.

A convenient notation used to denote the sum of a set of terms is called **sigma notation**. The Greek letter Σ (sigma) is used to indicate sums. For example, the first four terms of the sequence defined by $a_n = n^3$ are denoted by

$$\sum_{n=1}^{4} n^3$$

This is read as "the sum from n equals 1 to 4 of n^3" and is simplified as

$$\sum_{n=1}^{4} n^3 = (1)^3 + (2)^3 + (3)^3 + (4)^3$$
$$= 1 + 8 + 27 + 64$$
$$= 100$$

In this example, the letter n is called the **index of summation**. Many times, the letters $i, j,$ and k are also used for the index of summation.

Calculator Connections

The sum of a finite series may be found on a graphing calculator if the nth term of the corresponding sequence is known. The calculator takes the sum of a sequence by using the *Sum* and *Seq* functions.

```
sum(seq(n^3,n,1,
4,1)
              100
```

example 5   Finding a Sum from Sigma Notation

Find the sum.

$$\sum_{i=1}^{3} (-1)^{i+1} \cdot (3i + 4)$$

Skill Practice

4. Find the sum.

$$\sum_{i=1}^{4} (-1)^i (4i - 3)$$

Answer

4. 8

Solution:

$$\sum_{i=1}^{3}(-1)^{i+1} \cdot (3i+4) = (-1)^{1+1} \cdot [3(1)+4] + (-1)^{2+1} \cdot [3(2)+4]$$
$$+ (-1)^{3+1} \cdot [3(3)+4]$$
$$= (-1)^2 \cdot (7) + (-1)^3 \cdot (10) + (-1)^4 \cdot (13)$$
$$= 7 - 10 + 13$$
$$= 10$$

Avoiding Mistakes: The letter i is used as a variable for the index of summation. In this context, it does not represent an imaginary number.

Skill Practice

4. Write the series in summation notation.
$$\frac{1}{2} - \frac{1}{4} + \frac{1}{6} - \frac{1}{8} + \frac{1}{10}$$

example 6 — Converting to Summation Notation

Write the series in summation notation.

a. $\frac{2}{3} + \frac{4}{9} + \frac{8}{27} + \frac{16}{81}$ Use n as the index of summation.

b. $1 - \frac{1}{2} + \frac{1}{3} - \frac{1}{4} + \frac{1}{5} - \frac{1}{6}$ Use j as the index of summation.

Solution:

a. The sum can be written as

$$\left(\frac{2}{3}\right)^1 + \left(\frac{2}{3}\right)^2 + \left(\frac{2}{3}\right)^3 + \left(\frac{2}{3}\right)^4$$

Taking n from 1 to 4, we have

$$\sum_{n=1}^{4}\left(\frac{2}{3}\right)^n$$

b. The even-numbered terms are negative. The factor $(-1)^{j+1}$ is negative for even values of j. Therefore, the series can be written as

$$\sum_{j=1}^{6}(-1)^{j+1} \cdot \frac{1}{j}$$

Answer

5. $\sum_{i=1}^{5}(-1)^{i+1}\frac{1}{2i}$

---

section A.3 Practice Exercises

Boost your GRADE at mathzone.com!

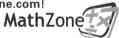

- Practice Problems
- Self-Tests
- NetTutor
- e-Professors
- Videos

Objective 1: Finite and Infinite Sequences

For Exercises 1–16, list the terms of each sequence. (See Examples 1–2.)

1. $a_n = 3n + 1, \quad 1 \leq n \leq 4$

2. $a_n = -2n + 3, \quad 1 \leq n \leq 4$

3. $a_n = \sqrt{n+2}, \quad 1 \leq n \leq 4$

4. $a_n = \sqrt{n-1}, \quad 1 \leq n \leq 4$

5. $a_n = \dfrac{3}{n}$, $1 \leq n \leq 5$

6. $a_n = \dfrac{n}{n+2}$, $1 \leq n \leq 5$

7. $a_n = (-1)^n \dfrac{n+1}{n+2}$, $1 \leq n \leq 4$

8. $a_n = (-1)^n \dfrac{n-1}{n+2}$, $1 \leq n \leq 4$

9. $a_n = (-1)^{n+1}(n^2 - 1)$, $1 \leq n \leq 4$

10. $a_n = (-1)^{n+1}(n^2)$, $1 \leq n \leq 4$

11. $a_n = 3 - \dfrac{1}{n}$, $2 \leq n \leq 5$

12. $a_n = 2 - \dfrac{1}{n+1}$, $2 \leq n \leq 5$

13. $a_n = n^2 - n$, $1 \leq n \leq 4$

14. $a_n = n(n^2 - 1)$, $1 \leq n \leq 4$

15. $a_n = (-1)^n 3^n$, $1 \leq n \leq 4$

16. $a_n = (-1)^n n$, $1 \leq n \leq 4$

17. If the nth term of a sequence is $a_n = (-1)^n \cdot n^2$, which terms are positive and which are negative?

18. If the nth term of a sequence is $a_n = (-1)^{n-1} \cdot \dfrac{1}{n}$, which terms are positive and which are negative?

19. Edmond borrowed $500. To pay off the loan, he agreed to pay 2% of the balance plus $50 each month. Write a sequence representing the amount Edmond will pay each month for the next 6 months. Round each term to the nearest cent. (See Example 4.)

20. Janice deposited $1000 in a savings account that pays 3% interest compounded annually. Write a sequence representing the amount Janice receives in interest each year for the first 4 years. Round each term to the nearest cent.

21. A certain bacteria culture doubles its size each day. If there are 25,000 bacteria on the first day, write a sequence representing the population each day for the first week (7 days).

22. A radioactive chemical decays by one-half of its amount each week. If there is 16 g of the chemical in week 1, write a sequence representing the amount present each week for 2 months (8 weeks).

For Exercises 23–34, find a formula for the nth term of the sequence. Answers may vary. (See Example 3.)

23. $2, 4, 6, 8, \ldots$

24. $3, 6, 9, 12, \ldots$

25. $1, 3, 5, 7, \ldots$

26. $3, 5, 7, 9, \ldots$

27. $1, \dfrac{1}{4}, \dfrac{1}{9}, \dfrac{1}{16}, \ldots$

28. $\dfrac{1}{2}, \dfrac{2}{3}, \dfrac{3}{4}, \dfrac{4}{5}, \ldots$

29. $1, -1, 1, -1, \ldots$

30. $-1, 1, -1, 1, \ldots$

31. $-2, 4, -8, 16, \ldots$

32. $3, -9, 27, -81, \ldots$

33. $\dfrac{3}{5}, \dfrac{3}{25}, \dfrac{3}{125}, \dfrac{3}{625}, \ldots$

34. $\dfrac{1}{4}, \dfrac{1}{16}, \dfrac{1}{64}, \dfrac{1}{256}, \ldots$

Objective 2: Series

35. What is the difference between a sequence and a series?

36. What is the index of a summation?

For Exercises 37–52, find the sums. **(See Example 5.)**

37. $\sum_{i=1}^{4} (3i^2)$

38. $\sum_{i=1}^{4} (2i^2)$

39. $\sum_{j=0}^{4} \left(\frac{1}{2}\right)^j$

40. $\sum_{j=0}^{4} \left(\frac{1}{3}\right)^j$

41. $\sum_{i=1}^{6} 5$

42. $\sum_{i=1}^{7} 3$

43. $\sum_{j=1}^{4} (-1)^j (5j)$

44. $\sum_{j=1}^{4} (-1)^j (4j)$

45. $\sum_{i=1}^{4} \frac{i+1}{i}$

46. $\sum_{i=2}^{5} \frac{i-1}{i}$

47. $\sum_{j=1}^{3} (j+1)(j+2)$

48. $\sum_{j=1}^{3} j(j+2)$

49. $\sum_{k=1}^{7} (-1)^k$

50. $\sum_{k=0}^{5} (-1)^{k+1}$

51. $\sum_{k=1}^{5} k^2$

52. $\sum_{k=1}^{5} 2^k$

For Exercises 53–62, write the series in summation notation. **(See Example 6.)**

53. $1 + 2 + 3 + 4 + 5 + 6$

54. $1 - 2 + 3 - 4 + 5 - 6$

55. $4 + 4 + 4 + 4 + 4$

56. $8 + 8 + 8 + 8 + 8$

57. $4 + 8 + 12 + 16 + 20$

58. $3 + 6 + 9 + 12 + 15$

59. $\frac{1}{3} - \frac{1}{9} + \frac{1}{27} - \frac{1}{81}$

60. $\frac{1}{2} - \frac{1}{4} + \frac{1}{8} - \frac{1}{16}$

61. $x + x^2 + x^3 + x^4 + x^5$

62. $y^2 + y^4 + y^6 + y^8 + y^{10}$

63. A certain plant will grow $1\frac{1}{2}$ in. each day for 1 week (7 days). If the plant begins with a height of 1 in., the height of the plant can be represented as a series. For example, after 1 day the plant will be $1 + 1\frac{1}{2}$ in. tall. Write out a series to represent the height of the plant each day for 1 week (7 days). Then write the series in summation notation. Finally, determine the height after 1 week.

64. A company produces a product and sells 5000 units. The company's research department found that for each year, 10% of the units cease to operate.

  a. Write out a sum representing how many of the original 5000 units will become inoperative in the first 3 years.

  b. How many of the original 5000 units will still be operational at the end of the 3 years?

Summation notation is used extensively in statistics. The sample mean (average), $\bar{x}$, of a set of n values $x_1, x_2, x_3, \ldots, x_n$ is given by

$$\bar{x} = \frac{1}{n}\sum_{i=1}^{n} x_i$$

Use this formula for Exercises 65–66.

65. Find the mean number of grams of protein for a sample of five energy bars: 10, 15, 12, 18, 22.

66. Find the mean age of a sample of viewers who regularly watch CNN on television: 29, 56, 62, 39, 58, 74.

The sample standard deviation, s, of a set of n values $x_1, x_2, x_3, \ldots, x_n$ is given by

$$s = \sqrt{\frac{n \cdot \sum_{i=1}^{n} x_i^2 - \left(\sum_{i=1}^{n} x_i\right)^2}{n(n-1)}}$$

Use this formula for Exercises 67–68.

67. Find the standard deviation for the number of grams of protein for a sample of five energy bars: 10, 15, 12, 18, 22. Round to 1 decimal place.

68. Find the standard deviation of a sample of viewers who regularly watch CNN on television: 29, 56, 62, 39, 58, 74. Round to 1 decimal place.

Expanding Your Skills

Some sequences are defined by a recursion formula, which defines each term of a sequence in terms of one or more of its preceding terms. For example, if $a_1 = 5$ and $a_n = 2a_{n-1} + 1$ for $n > 1$, then the terms of the sequence are 5, 11, 23, 47, ... In this case, each term after the first is one more than twice the term before it.

For Exercises 69–72, list the first five terms of the sequence.

69. $a_1 = -3$, $a_n = a_{n-1} + 5$ for $n > 1$

70. $a_1 = 10$, $a_n = a_{n-1} - 3$ for $n > 1$

71. $a_1 = 5$, $a_n = 4a_{n-1} + 1$ for $n > 1$

72. $a_1 = -2$, $a_n = -3a_{n-1} + 4$ for $n > 1$

73. A famous sequence in mathematics is called the Fibonacci sequence, named after the Italian mathematician Leonardo Fibonacci of the 13th century. The Fibonacci sequence is defined by

$$a_1 = 1$$
$$a_2 = 1$$
$$a_n = a_{n-1} + a_{n-2} \quad \text{for } n > 2$$

This definition implies that beginning with the third term, each term is the sum of the preceding two terms. Write out the first 10 terms of the Fibonacci sequence.

Graphing Calculator Exercises

For Exercises 74–77, use a graphing calculator to list the first five terms of the sequence.

74. $a_n = 1 + 3n$ **75.** $a_n = \dfrac{8}{n}$ **76.** $a_n = (-1)^n \left(\dfrac{7}{8}\right)^n$ **77.** $a_n = (-1)^n (n^2)$

For Exercises 78–81, use the *Sum* and *Seq* functions on a graphing calculator to find the sum.

78. $\displaystyle\sum_{n=1}^{20} (1 + 3n)$ **79.** $\displaystyle\sum_{n=1}^{4} \dfrac{8}{n}$ **80.** $\displaystyle\sum_{n=1}^{6} (-1)^n \left(\dfrac{1}{2}\right)^n$ **81.** $\displaystyle\sum_{n=1}^{100} (-1)^n (n^2)$

---

section A.4 Arithmetic and Geometric Sequences and Series

Objectives

1. Arithmetic Sequences
2. Arithmetic Series
3. Geometric Sequences
4. Geometric Series

1. Arithmetic Sequences

In this section we will study two special types of sequences and series. The first is called an arithmetic sequence, for example:

$$4, 7, 10, 13, 16, \ldots$$

This sequence is an arithmetic sequence. Note the characteristic that each successive term after the first is a fixed value more than the previous term (in this case the terms differ by 3).

> **Definition of an Arithmetic Sequence**
>
> An **arithmetic sequence** is a sequence in which the difference between consecutive terms is constant.

The fixed difference between a term and its predecessor is called the **common difference** and is denoted by the letter d. The common difference is the difference between a term and its predecessor. That is,

$$d = a_{n+1} - a_n$$

Furthermore, if a_1 is the first term, then

$a_2 = a_1 + d$ is the second term

$a_3 = a_1 + 2d$ is the third term

$a_4 = a_1 + 3d$ is the fourth term and so on

$\ldots$

In general, $a_n = a_1 + (n - 1)d$.

Section A.4 Arithmetic and Geometric Sequences and Series   A-29

nth Term of an Arithmetic Sequence

The nth term of an arithmetic sequence is given by

$$a_n = a_1 + (n-1)d$$

where a_1 is the first term of the sequence and d is the common difference.

---

example 1   **Writing the nth Term of an Arithmetic Sequence**

Write the nth term of the sequence $9, 2, -5, -12, \ldots$.

Solution:

$9, \quad 2, \quad -5, \quad -12, \ldots$
$\;\;-7\;\;-7\;\;-7$

The common difference can be found by subtracting a term from its predecessor: $2 - 9 = -7$

With $a_1 = 9$ and $d = -7$, we have

$$a_n = 9 + (n-1)(-7)$$
$$= 9 - 7n + 7$$
$$= -7n + 16$$

Skill Practice

1. Write the nth term of the sequence.
   $-3, -1, 1, 3, 5, \ldots$

In Example 1, the common difference between terms is -7. Accordingly, each term of the sequence *decreases* by 7.

The formula $a_n = a_1 + (n-1)d$ contains four variables: $a_n, a_1, n,$ and d. Consequently, if we know the value of three of the four variables, we can solve for the fourth.

---

example 2   **Finding a Specified Term of an Arithmetic Sequence**

Find the ninth term of the arithmetic sequence in which $a_1 = -4$ and $a_{22} = 164$.

Solution:

To find the value of the ninth term a_9, we need to determine the value of d. To find d, substitute $a_1 = -4$, $n = 22$, and $a_{22} = 164$ into the formula $a_n = a_1 + (n-1)d$.

$$a_n = a_1 + (n-1)d$$
$$164 = -4 + (22-1)d$$
$$164 = -4 + 21d$$
$$168 = 21d$$
$$d = 8$$

Therefore, $a_n = -4 + (n-1)(8)$

Skill Practice

2. Find the 10th term of the arithmetic sequence in which $a_1 = 14$ and $a_{15} = -28$.

Answers

1. $2n - 5$ 2. $d = -3$, $a_{10} = -13$

$$a_9 = -4 + (9-1)(8)$$
$$= -4 + (8)(8)$$
$$= 60$$

Skill Practice

3. Find the number of terms of the sequence.

   $-15, -11, -7, -3, \ldots 81$

example 3   Finding the Number of Terms in an Arithmetic Sequence

Find the number of terms of the sequence $7, 3, -1, -5, \ldots, -113$.

Solution:

To find the number of terms n, we can substitute $a_1 = 7$, $d = -4$, and $a_n = -113$ into the formula for the nth term.

$$a_n = a_1 + (n-1)d$$
$$-113 = 7 + (n-1)(-4)$$
$$-113 = 7 - 4n + 4$$
$$-113 = 11 - 4n$$
$$-124 = -4n$$
$$n = 31$$

2. Arithmetic Series

The indicated sum of an arithmetic sequence is called an **arithmetic series**. For example, the series

$$3 + 7 + 11 + 15 + 19 + 23$$

is an arithmetic series because the common difference between terms is constant (4). Adding the terms in a lengthy sum is cumbersome, so we offer the following "short cut," which is developed here. Let S represent the sum of the terms in the series.

$S = 3 + 7 + 11 + 15 + 19 + 23$	Add the terms in ascending order.
$S = 23 + 19 + 15 + 11 + 7 + 3$	Add the terms in descending order.
$2S = 26 + 26 + 26 + 26 + 26 + 26$	Adding the two series produces six terms of 26.

$$2S = 6 \cdot 26$$
$$S = \frac{6 \cdot 26}{2}$$
$$= 78$$

By adding the terms in ascending and descending order, we double the sum but create a pattern that is easily added. This is true in general. To find the sum, S_n, of the first n terms of the arithmetic series $a_1 + a_2 + a_3 + \cdots + a_n$, we have

$S_n = a_1 + (a_1 + d) + (a_1 + 2d) + \cdots + a_n$	Ascending order
$S_n = a_n + (a_n - d) + (a_n - 2d) + \cdots + a_1$	Descending order

$$2S_n = (a_1 + a_n) + (a_1 + a_n) + (a_1 + a_n) + \cdots + (a_1 + a_n)$$
$$2S_n = n(a_1 + a_n)$$
$$S_n = \frac{n}{2}(a_1 + a_n)$$

Answer

3. 25

Sum of an Arithmetic Series

The sum, S_n, of the first n terms of an arithmetic series is given by

$$S_n = \frac{n}{2}(a_1 + a_n)$$

where a_1 is the first term of the series and a_n is the nth term of the series.

example 4 — Finding the Sum of an Arithmetic Series

Find the sum of the series.

$$\sum_{i=1}^{25}(2i+3)$$

Solution:

In this series, $n = 25$. Furthermore, $a_1 = 2(1) + 3 = 5$ and $a_{25} = 2(25) + 3 = 53$. Therefore,

$S_{25} = \dfrac{25}{2}(5 + 53)$ Apply the formula $S_n = \dfrac{n}{2}(a_1 + a_n)$.

$= \dfrac{25}{2}(58)$ Simplify.

$= 725$

Skill Practice

4. Find the sum of the series.
$$\sum_{i=1}^{20}(3i - 2)$$

Avoiding Mistakes: When we applying the sum formula $S_n = \dfrac{n}{2}(a_1 + a_n)$ to find the sum $\sum_{i=1}^{n} a_n$, the index of summation, i, must begin at 1.

example 5 — Finding the Sum of an Arithmetic Series

Find the sum of the series.

$$-3 + (-5) + (-7) + \cdots + (-127)$$

Solution:

For this series, $a_1 = -3$ and $a_n = -127$. However, to determine the sum, we also need to find the value of n. The difference between the second term and its predecessor is $-5 - (-3) = -2$. Thus, $d = -2$. We have

$-127 = -3 + (n-1)(-2)$ Apply the formula $a_n = a_1 + (n-1)d$.

$-127 = -3 - 2n + 2$ Apply the distributive property.

$-127 = -1 - 2n$ Combine *like* terms.

$-126 = -2n$ Solve for n.

$n = 63$

Using $n = 63$, $a_1 = -3$, and $a_{63} = -127$, we have

$$S_n = \frac{n}{2}(a_1 + a_n) = \frac{63}{2}[-3 + (-127)] = \frac{63}{2}(-130) = -4095$$

Skill Practice

5. Find the sum of the series.
$$8 + 3 + (-2) + \cdots + (-137)$$

Answers

4. 590   5. −1935

3. Geometric Sequences

The sequence 2, 4, 8, 16, 32, ... is not an arithmetic sequence because the difference between terms is not constant. However, a different pattern exists. Notice that each term after the first one is 2 times the preceding term. This sequence is called a geometric sequence.

> **Definition of a Geometric Sequence**
>
> A **geometric sequence** is a sequence in which each term after the first term is a constant multiple of the preceding term.

The constant multiple between a term and its predecessor is called the **common ratio** and is denoted by r. The common ratio is found by dividing a term by the preceding term. That is,

$$r = \frac{a_{n+1}}{a_n}$$

For the sequence 2, 4, 8, 16, 32, ... we have $r = \frac{4}{2} = \frac{8}{4} = \frac{16}{8} = \frac{32}{16} = 2$.

If a_1 denotes the first term of a geometric sequence, then

$a_2 = a_1 r$ is the second term.

$a_3 = a_1 r^2$ is the third term.

$a_4 = a_1 r^3$ is the fourth term, and so on.

...

This pattern gives $a_n = a_1 r^{n-1}$.

> **nth Term of a Geometric Sequence**
>
> The nth term of a geometric sequence is given by
>
> $$a_n = a_1 r^{n-1}$$
>
> where a_1 is the first term and r is the common ratio.

Skill Practice

6. Find the nth term of the geometric sequence.

   $6, 2, \dfrac{2}{3}, \dfrac{2}{9}, \ldots$

example 6   Finding the nth Term of a Geometric Sequence

Find the nth term of the sequence.

a. $-1, 4, -16, 64, \ldots$ **b.** $12, 8, \dfrac{16}{3}, \dfrac{32}{9}, \ldots$

Solution:

a. The common ratio is found by dividing any term (after the first) by its predecessor.

$$r = \frac{4}{-1} = -4$$

With $r = -4$ and $a_1 = -1$, we have $a_n = -1(-4)^{n-1}$.

Answer

6. $a_n = 6\left(\dfrac{1}{3}\right)^{n-1}$

b. The common ratio is $r = \frac{8}{12} = \frac{2}{3}$. With $a_1 = 12$ and $r = \frac{2}{3}$, we have

$$a_n = 12\left(\frac{2}{3}\right)^{n-1}$$

The formula $a_n = a_1 r^{n-1}$ contains the variables $a_n, a_1, n,$ and r. If we know the value of three of the four variables, we can find the fourth.

example 7 — Finding a Specified Term of a Geometric Sequence

Given $a_n = 6\left(\frac{1}{2}\right)^{n-1}$, find a_5.

Solution:

$$a_n = 6\left(\frac{1}{2}\right)^{n-1}$$

$$a_5 = 6\left(\frac{1}{2}\right)^{5-1} = 6\left(\frac{1}{2}\right)^4 = 6\left(\frac{1}{16}\right) = \frac{3}{8}$$

Skill Practice

7. Given $a_n = 3(-2)^{n-1}$, find a_6.

example 8 — Finding a Specified Term of a Geometric Sequence

Find the first term of the geometric sequence where $a_5 = -162$ and $r = 3$.

Solution:

$-162 = a_1(3)^{5-1}$ Substitute $a_5 = -162$, $n = 5$, and $r = 3$ into the formula $a_n = a_1 r^{n-1}$.

$-162 = a_1(3)^4$ Simplify and solve for a_1.

$-162 = a_1(81)$

$a_1 = -2$

Skill Practice

8. Find the first term of the geometric sequence when $a_4 = \frac{25}{32}$ and $r = \frac{1}{4}$.

4. Geometric Series

The indicated sum of a geometric sequence is called a **geometric series**. For example,

$$1 + 3 + 9 + 27 + 81 + 243$$

is a geometric series. To find the sum, consider the following procedure. Let S represent the sum

$$S = 1 + 3 + 9 + 27 + 81 + 243$$

Now multiply S by the common ratio, which in this case is 3.

$$3S = 3 + 9 + 27 + 81 + 243 + 729$$

Then

$3S - S = (3 + 9 + 27 + 81 + 243 + 729) - (1 + 3 + 9 + 27 + 81 + 243)$

$2S = 3 + 9 + 27 + 81 + 243 + 729 - 1 - 3 - 9 - 27 - 81 - 243$

$2S = 729 - 1$ The terms in red form a sum of zero.

$2S = 728$

$S = 364$

Answers

7. -96 8. 50

A similar procedure can be used to find the sum S_n of the first n terms of any geometric series. Subtract rS_n from S_n.

$$S_n = a_1 + a_1r + a_1r^2 + \cdots + a_1r^{n-1}$$
$$rS_n = a_1r + a_1r^2 + a_1r^3 + \cdots + a_1r^n$$
$$S_n - rS_n = (a_1 - a_1r) + (a_1r - a_1r^2) + (a_1r^2 - a_1r^3) + \cdots + (a_1r^{n-1} - a_1r^n)$$
$$S_n - rS_n = a_1 - a_1r^n \quad \text{The terms in red form a sum of zero.}$$
$$S_n(1-r) = a_1(1-r^n) \quad \text{Factor each side of the equation.}$$
$$S_n = \frac{a_1(1-r^n)}{1-r} \quad \text{Divide by } (1-r).$$

Sum of a Geometric Series

The sum, S_n, of the first n terms of a geometric series $\sum_{i=1}^{n} (a_1 r^{i-1})$ is given by

$$S_n = \frac{a_1(1-r^n)}{1-r}$$

where a_1 is the first term of the series, r is the common ratio, and $r \neq 1$.

Skill Practice

9. Find the sum of the geometric series.

$$\frac{1}{9} + \frac{1}{3} + 1 + 3 + 9 + 27$$

example 9   Finding the Sum of a Geometric Series

Find the sum of the series.

$$4 + 2 + 1 + \frac{1}{2} + \frac{1}{4} + \frac{1}{8}$$

Solution:

$$a_1 = 4 \quad r = \frac{1}{2} \quad \text{and} \quad n = 6$$

$$S_n = \frac{a_1(1-r^n)}{1-r} = \frac{4[1-(\frac{1}{2})^6]}{1-\frac{1}{2}} = \frac{4(1-\frac{1}{64})}{\frac{1}{2}} = 8\left(\frac{63}{64}\right) = \frac{63}{8}$$

Skill Practice

10. Find the sum of the geometric series.

$$12 + 6 + 3 + \cdots + \frac{3}{32}$$

example 10   Finding the Sum of a Geometric Series

Find the sum of the series $5 + 10 + 20 + \cdots + 5120$.

Solution:

The common ratio is 2 and $a_1 = 5$. The nth term of the sequence can be written as $a_n = 5(2)^{n-1}$. To find the value of n, substitute 5120 for a_n.

$$5120 = 5(2)^{n-1}$$

$$\frac{5120}{5} = \frac{5(2)^{n-1}}{5} \quad \text{Divide both sides by 5.}$$

$$1024 = 2^{n-1} \quad \text{To solve the exponential equation, write each side as a power of 2.}$$

$$2^{10} = 2^{n-1}$$

$$10 = n - 1 \quad \text{From Section 10.7, we know that if } b^x = b^y, \text{ then } x = y.$$

$$n = 11$$

Answers

9. $\dfrac{364}{9}$ 10. $\dfrac{765}{32}$

With $a_1 = 5$, $r = 2$, and $n = 11$, we have

$$S_n = \frac{a_1(1 - r^n)}{1 - r} = \frac{5(1 - 2^{11})}{1 - 2} = \frac{5(1 - 2048)}{-1} = \frac{5(-2047)}{-1} = 10{,}235$$

Consider a geometric series where $|r| < 1$. For increasing values of n, r^n decreases. For example,

$$\left(\frac{1}{2}\right)^5 = 0.03125 \quad \left(\frac{1}{2}\right)^{10} \approx 0.00097656 \quad \left(\frac{1}{2}\right)^{15} \approx 0.00003052$$

For $|r| < 1$, r^n approaches 0 as n gets larger and larger. As n approaches infinity, the sum

$$S = \frac{a_1(1 - r^n)}{1 - r} \quad \text{approaches} \quad \frac{a_1(1 - 0)}{1 - r} = \frac{a_1}{1 - r}$$

Sum of an Infinite Geometric Series

Given an infinite geometric series $a_1 + a_1 r + a_1 r^2 + \cdots$, with $|r| < 1$, the sum S of all terms in the series is given by

$$S = \frac{a_1}{1 - r}$$

Note: If $|r| \geq 1$, then the sum does not exist.

---

example 11   Finding the Sum of an Infinite Geometric Series

Find the sum of the series.

$$1 + \frac{1}{3} + \frac{1}{9} + \frac{1}{27} + \frac{1}{81} + \cdots$$

Solution:

This is a geometric series with $a_1 = 1$ and $r = \frac{1}{3}$. Because $|r| = |\frac{1}{3}| < 1$, we have

$$S = \frac{a_1}{1 - r} = \frac{1}{1 - \frac{1}{3}} = \frac{1}{\frac{2}{3}} = \frac{3}{2}$$

The sum is $\frac{3}{2}$.

Skill Practice

11. Find the sum of the series.

$$4 + 3 + \frac{9}{4} + \frac{27}{16} + \cdots$$

---

example 12   Using Geometric Series in a Physics Application

A child drops a ball from a height of 4 ft. With each bounce, the ball rebounds to 50% of its original height. Determine the total distance traveled by the ball.

Solution:

The heights from which the ball drops are given by the sequence

$$4, 2, 1, \frac{1}{2}, \frac{1}{4}, \ldots$$

Answer

11. 16

After the ball falls from its initial height of 4 ft, the distance traveled for every bounce thereafter is doubled (the ball travels up and down). Therefore, the total distance traveled is given by the series

$$4 + 2 \cdot 2 + 2 \cdot 1 + 2 \cdot \frac{1}{2} + 2 \cdot \frac{1}{4} + \cdots$$

or equivalently

$$4 + 4 + 2 + 1 + \frac{1}{2} + \cdots$$

The series $4 + 2 + 1 + \frac{1}{2} + \cdots$ is an infinite geometric series with $a_1 = 4$ and $r = \frac{1}{2}$.

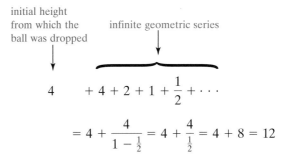

$$= 4 + \frac{4}{1 - \frac{1}{2}} = 4 + \frac{4}{\frac{1}{2}} = 4 + 8 = 12$$

The ball traveled a total of 12 ft.

section A.4 Practice Exercises

Boost *your* GRADE at mathzone.com!

- Practice Problems
- Self-Tests
- NetTutor
- e-Professors
- Videos

Objective 1: Arithmetic Sequences

1. Explain how to determine if a sequence is arithmetic.

For Exercises 2–7, find the common difference d for each arithmetic sequence.

2. $1, 3, 5, 7, 9, \ldots$

3. $2, 8, 14, 20, 26, \ldots$

4. $6, 3, 0, -3, -6, \ldots$

5. $8, 3, -2, -7, -12, \ldots$

6. $-7, -9, -11, -13, -15, \ldots$

7. $-15, -11, -7, -3, 1, \ldots$

For Exercises 8–13, write the first five terms of the arithmetic sequence.

8. $a_1 = 3, d = 5$

9. $a_1 = -3, d = 2$

10. $a_1 = 2, d = \frac{1}{2}$

11. $a_1 = 0, d = \frac{1}{3}$

12. $a_1 = 2, d = -4$

13. $a_1 = 10, d = -6$

For Exercises 14–23, write the *n*th term of the sequence. (See Example 1.)

14. $0, 5, 10, 15, 20, \ldots$

15. $7, 12, 17, 22, 27, \ldots$

16. $-2, -4, -6, -8, -10, \ldots$

17. $1, -3, -7, -11, -15, \ldots$

18. $2, \frac{5}{2}, 3, \frac{7}{2}, 4, \ldots$

19. $1, \frac{4}{3}, \frac{5}{3}, 2, \frac{7}{3}, \ldots$

20. $21, 17, 13, 9, 5, \ldots$

21. $9, 6, 3, 0, -3, \ldots$

22. $-8, -2, 4, 10, 16, \ldots$

23. $-9, -1, 7, 15, 23, \ldots$

For Exercises 24–31, find the indicated term of each arithmetic sequence. (See Example 2.)

24. Find the eighth term given $a_1 = -6$ and $d = -3$.

25. Find the sixth term given $a_1 = -3$ and $d = 4$.

26. Find the 12th term given $a_1 = -8$ and $d = -2$.

27. Find the ninth term given $a_1 = -1$ and $d = 6$.

28. Find the seventh term given $a_1 = 0$ and $d = -5$.

29. Find the 10th term given $a_1 = 1$ and $d = 5$.

30. Find the sixth term given $a_1 = 3$ and $d = 3$.

31. Find the 11th term given $a_1 = 12$ and $d = -6$.

For Exercises 32–39, find the number of terms, *n*, of each arithmetic sequence. (See Example 3.)

32. $2, 0, -2, \ldots, -56$

33. $8, 13, 18, \ldots, 98$

34. $1, -3, -7, \ldots, -67$

35. $1, 5, 9, \ldots, 85$

36. $1, \frac{3}{4}, \frac{1}{2}, \ldots, -4$

37. $2, \frac{5}{2}, 3, \ldots, 13$

38. $-\frac{5}{3}, -1, -\frac{1}{3}, \ldots, \frac{37}{3}$

39. $\frac{13}{3}, \frac{19}{3}, \frac{25}{3}, \ldots, \frac{73}{3}$

40. If the third and fourth terms of an arithmetic sequence are 18 and 21, what are the first and second terms?

41. If the third and fourth terms of an arithmetic sequence are -8 and -11, what are the first and second terms?

Objective 2: Arithmetic Series

42. Explain the difference between an arithmetic sequence and an arithmetic series.

For Exercises 43–56, find the sum of the arithmetic series. (See Examples 4–5.)

43. $\sum_{i=1}^{20} (3i + 2)$

44. $\sum_{i=1}^{15} (2i - 3)$

45. $\sum_{i=1}^{20} (i + 4)$

46. $\sum_{i=1}^{25} (i - 3)$

47. $\sum_{j=1}^{10} (4 - j)$

48. $\sum_{j=1}^{10} (6 - j)$

49. $\sum_{j=1}^{15} \left(\frac{2}{3}j + 1\right)$

50. $\sum_{j=1}^{15} \left(\frac{1}{2}j - 2\right)$

51. $4 + 8 + 12 + \cdots + 84$

52. $4 + 9 + 14 + \cdots + 49$

53. $6 + 8 + 10 + \cdots + 34$

54. $-4 + (-3) + (-2) + \cdots + 12$

55. $-3 + (-7) + (-11) + \cdots + (-39)$

56. $2 + 5 + 8 + \cdots + 53$

57. Find the sum of the first 100 positive integers.

58. Find the sum of the first 50 positive even integers.

59. Find the sum of the first 50 positive odd integers.

60. Find the sum of the first 20 positive multiples of 5.

61. The seating in a certain theater is arranged so that there are 30 seats in row 1, 32 in row 2, 34 in row 3, and so on. If there are 20 rows, how many total seats are there? What is the total revenue if the average ticket price is $15 per seat and the theater is sold out?

62. A triangular array of dominoes has one domino in the first row, two dominoes in the second row, three dominoes in the third row, and so on. If there are 15 rows, how many dominoes are in the array?

Objective 3: Geometric Sequences

63. Explain how to determine if a sequence is geometric.

For Exercises 64–69, determine the common ratio, r, for the geometric sequence.

64. $5, 10, 20, 40, \ldots$

65. $-2, -1, -\dfrac{1}{2}, -\dfrac{1}{4}, \ldots$

66. $8, -2, \dfrac{1}{2}, -\dfrac{1}{8}, \ldots$

67. $4, -12, 36, -108, \ldots$

68. $3, -6, 12, -24, \ldots$

69. $1, 4, 16, 64, \ldots$

For Exercises 70–75, write the first five terms of the geometric sequence.

70. $a_1 = -3, r = -2$

71. $a_1 = -4, r = -1$

72. $a_1 = 6, r = \dfrac{1}{2}$

73. $a_1 = 8, r = \dfrac{1}{4}$

74. $a_1 = -1, r = 6$

75. $a_1 = 2, r = -3$

For Exercises 76–81, find the nth term of each geometric sequence. **(See Example 6.)**

76. $3, 12, 48, 192, \ldots$

77. $2, 6, 18, 54, \ldots$

78. $-5, 15, -45, 135, \ldots$

79. $-6, 12, -24, 48, \ldots$

80. $\dfrac{1}{2}, 2, 8, 32, \ldots$

81. $\dfrac{16}{3}, 4, 3, \dfrac{9}{4}, \ldots$

For Exercises 82–91, find the indicated term of each sequence. (See Examples 7–8.)

82. Given $a_n = 2(\frac{1}{2})^{n-1}$, find a_8.

83. Given $a_n = -3(\frac{1}{2})^{n-1}$, find a_8.

84. Given $a_n = 4(-\frac{3}{2})^{n-1}$, find a_6.

85. Given $a_n = 6(-\frac{1}{3})^{n-1}$, find a_6.

86. Given $a_n = -3(2)^{n-1}$, find a_5.

87. Given $a_n = 5(3)^{n-1}$, find a_4.

88. Given $a_5 = -\frac{16}{9}$ and $r = -\frac{2}{3}$, find a_1.

89. Given $a_6 = \frac{5}{16}$ and $r = -\frac{1}{2}$, find a_1.

90. Given $a_7 = 8$ and $r = 2$, find a_1.

91. Given $a_6 = 27$ and $r = 3$, find a_1.

92. If the second and third terms of a geometric sequence are 16 and 64, what is the first term?

93. If the second and third terms of a geometric sequence are $\frac{1}{3}$ and $\frac{1}{9}$, what is the first term?

Objective 4: Geometric Series

94. Explain the difference between a geometric sequence and a geometric series.

For Exercises 95–104, find the sum of the geometric series. (See Examples 9–10.)

95. $10 + 2 + \frac{2}{5} + \frac{2}{25} + \frac{2}{125}$

96. $1 + 3 + 9 + 27 + 81 + 243$

97. $-2 + 1 + \left(-\frac{1}{2}\right) + \frac{1}{4} + \left(-\frac{1}{8}\right)$

98. $\frac{1}{4} + (-1) + 4 + (-16) + 64$

99. $12 + 16 + \frac{64}{3} + \frac{256}{9} + \frac{1024}{27}$

100. $9 + 6 + 4 + \frac{8}{3} + \frac{16}{9}$

101. $1 + \frac{2}{3} + \frac{4}{9} + \cdots + \frac{64}{729}$

102. $\frac{8}{3} + 2 + \frac{3}{2} + \cdots + \frac{243}{512}$

103. $-4 + 8 + (-16) + \cdots + (-256)$

104. $-\frac{7}{3} + 7 + (-21) + \cdots + (-1701)$

105. A deposit of $1000 is made in an account that earns 5% interest compounded annually. The balance in the account after n years is given by

$$a_n = 1000(1.05)^n \quad \text{for } n \geq 1$$

   a. List the first four terms of the sequence. Round to the nearest cent.

   b. Find the balance after 10 years, 20 years, and 40 years by computing a_{10}, a_{20}, and a_{40}. Round to the nearest cent.

106. A home purchased for $125,000 increases by 4% of its value each year. The value of the home after n years is given by
$$a_n = 125{,}000(1.04)^n \quad \text{for } n \geq 1$$

a. List the first four terms of the sequence. Round to the nearest dollar.

b. Find the value of the home after 5 years, 10 years, and 20 years by computing a_5, a_{10}, and a_{20}. Round to the nearest dollar.

For Exercises 107–112, first find the common ratio r. Then determine the sum of the infinite series, if it exists. (See Example 11.)

107. $1 + \dfrac{1}{6} + \dfrac{1}{36} + \dfrac{1}{216} + \cdots$

108. $-2 + \left(-\dfrac{1}{2}\right) + \left(-\dfrac{1}{8}\right) + \left(-\dfrac{1}{32}\right) + \cdots$

109. $-3 + 1 + \left(-\dfrac{1}{3}\right) + \dfrac{1}{9} + \cdots$

110. $\dfrac{1}{2} + \left(-\dfrac{1}{10}\right) + \dfrac{1}{50} + \left(-\dfrac{1}{250}\right) + \cdots$

111. $\dfrac{2}{3} + (-1) + \dfrac{3}{2} + \left(-\dfrac{9}{4}\right) + \cdots$

112. $3 + 5 + \dfrac{25}{3} + \dfrac{125}{9} + \cdots$

113. Suppose $200 million is spent by tourists at a certain resort town. Further suppose that 75% of the revenue is respent in the community and then respent over and over, each time at a rate of 75%. The series
$$200 + 200(0.75) + 200(0.75)^2 + 200(0.75)^3 + \cdots$$
gives the total amount spent (and respent) in the community. Find the sum of the infinite series.

114. A bungee jumper jumps off a platform and stretches the cord 80 ft before rebounding upward. Each successive bounce stretches the cord 60% of its previous length. The total vertical distance traveled is given by
$$80 + 2(0.60)(80) + 2(0.60)^2(80) + 2(0.60)^3(80) + \cdots$$
Ignoring the first term, the series is an infinite geometric series. Compute the total vertical distance traveled.

115. A ball drops from a height of 4 ft. With each bounce, the ball rebounds to $\frac{3}{4}$ of its height. The total vertical distance traveled is given by
$$4 + 2\left(\dfrac{3}{4}\right)(4) + 2\left(\dfrac{3}{4}\right)^2(4) + 2\left(\dfrac{3}{4}\right)^3(4) + \cdots$$
Ignoring the first term, the series is an infinite geometric series. Compute the total vertical distance traveled. (See Example 12.)

116. The repeating decimal number $0.\overline{2}$ can be written as an infinite geometric series by
$$\dfrac{2}{10} + \dfrac{2}{100} + \dfrac{2}{1000} + \cdots$$

a. What is a_1? **b.** What is r? **c.** Find the sum of the series.

117. The repeating decimal number $0.\overline{7}$ can be written as an infinite geometric series by

$$\frac{7}{10} + \frac{7}{100} + \frac{7}{1000} + \cdots$$

   a. What is a_1?   **b.** What is r?   **c.** Find the sum of the series.

Mixed Exercises

For Exercises 118–131, determine if the sequence is arithmetic, geometric, or neither. If the sequence is arithmetic, find d. If the sequence is geometric, find r.

118. $5, -\dfrac{5}{2}, \dfrac{5}{4}, -\dfrac{5}{8}, \ldots$ **119.** $1, -\dfrac{3}{2}, \dfrac{9}{4}, -\dfrac{27}{8}, \ldots$ **120.** $\dfrac{1}{2}, 1, \dfrac{3}{2}, 2, \dfrac{5}{2}, \ldots$

121. $-\dfrac{1}{3}, \dfrac{1}{3}, 1, \dfrac{5}{3}, \ldots$ **122.** $-2, -4, -8, -16, -32, \ldots$ **123.** $2, 6, 18, 54, 162, \ldots$

124. $-2, -4, -6, -8, -10, \ldots$ **125.** $2, 6, 10, 14, 18, 22, \ldots$ **126.** $-2, -4, -7, -11, -16, \ldots$

127. $2, 6, 11, 17, 24, \ldots$ **128.** $1, 3, 1, 3, 1, \ldots$ **129.** $0, 1, 0, 1, 0, 1, \ldots$

130. $2, -2, 2, -2, 2, \ldots$ **131.** $-1, 1, -1, 1, -1, \ldots$

Expanding Your Skills

132. The yearly salary for job A is $30,000 initially with an annual raise of $2000 per year. The yearly salary for job B is $29,000 initially with an annual raise of 6% per year.

   a. Find the total earnings for job A over 20 years (there will be 19 raises). Is this an arithmetic or geometric series?

   b. For job B, what is the amount of the raise after 1 year?

   c. Find the total earnings for job B over 20 years. Round to the nearest dollar. Is this an arithmetic or geometric series?

   d. What is the difference in total salary earned over 20 years between job A and job B?

133. **a.** Brook has a job that pays $40,000 the first year. She receives a 4% raise each year. Find the sum of her yearly salaries over a 20-year period. Round to the nearest dollar.

   b. Chamille has a job that pays $40,000 the first year. She receives a 4.5% raise each year. Find the sum of her yearly salaries over a 20-year period. Round to the nearest dollar.

   c. Chamille's raise each year was 0.5% higher than Brook's raise. How much more total income did Chamille receive than Brook over 20 years?

Graphing Calculator Exercises

134. Use the *Seq* function on a graphing calculator to list the first four terms of the sequence $a_n = 1000(1.05)^n$. Verify your answer to Exercise 105(a).

135. Use the *Seq* function on a graphing calculator to list the first four terms of the sequence $a_n = 125,000(1.04)^n$. Verify your answer to Exercise 106(a).

136. Use the *Sum* and *Seq* functions on a graphing calculator to confirm your answer to Exercise 61.

137. Use the *Sum* and *Seq* functions on a graphing calculator to confirm your answer to Exercise 62.

138. Use the *Sum* and *Seq* functions on a graphing calculator to find the sum of the first 10 terms and the first 20 terms of the series given below. Then compute the exact value of the infinite geometric series.

$$10 + 10(0.6) + 10(0.6)^2 + 10(0.6)^3 + \cdots$$

139. Use the *Sum* and *Seq* functions on a graphing calculator to find the sum of the first 10 terms and the first 20 terms of the series given below. Then compute the exact value of the infinite geometric series.

$$4 + 2 + 1 + \frac{1}{2} + \cdots$$

Student Answer Appendix

Chapter 1

Chapter 1 Preview, p. 2

1. a. Irrational **b.** Rational **c.** Rational **d.** Irrational **3.** -24
5. Undefined **7.** a **9.** $x = -4$ **11.** $x = \frac{29}{2}$ **13.** The integers are 9, 11, and 13. **15.** The length is 132 ft and the width is 40 ft.
17. $(-\infty, -6)$ **19.** $\frac{1}{a^2}$ **21. a.** 8.1×10^{-6} **b.** 2.52×10^8

Section 1.1 Practice Exercises, pp. 11–15

3.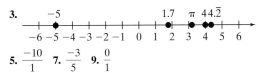

5. $\frac{-10}{1}$ **7.** $\frac{-3}{5}$ **9.** $\frac{0}{1}$

11.

	Real Numbers	Irrational Numbers	Rational Numbers	Integers	Whole Numbers	Natural Numbers
$\frac{6}{8}$	✓		✓			
$1\frac{1}{2}$	✓		✓			
π	✓	✓				
0	✓		✓	✓	✓	
$0.\overline{8}$	✓		✓			
$\frac{8}{2}$	✓		✓	✓	✓	✓
$4.\overline{2}$	✓		✓			

13. > **15.** < **17.** > **19.** $\left(\frac{5}{6}, \infty\right)$ **21.** $(-\infty, 9]$ **23.** $[-1, 15]$
25. $[12.8, \infty)$ **27.** $(-\infty, 3)$
29. $[-4, \infty)$
31. $\left[-\frac{7}{3}, \infty\right)$
33. $[-6, 0)$
35. $(2.34, \infty)$
37. $\left(-\infty, \frac{4}{7}\right)$
39. $(-\infty, 5]$
41. $(-7, -1)$
43. $[-1, 6]$

45. All real numbers greater than or equal to 2 **47.** All real numbers between -3.9 and 0 **49.** All real numbers **51.** All real numbers less than or equal to -1 **53. a.** $\{a, e, i\}$ **b.** $\{a, b, c, d, e, f, g, h, i, o, u\}$
55. $(-\infty, \infty)$ $\{x | x \text{ is a real number}\}$
57. $[-1, 0]$ $\{x | -1 \leq x \leq 0\}$
59. $(1, 3)$ $\{x | 1 < x < 3\}$
61. $(-3, \infty)$ $\{x | x > -3\}$
63. $(-\infty, \infty)$ **65.** $(-1, 1)$ **67.** $\{\ \}$ **69.** $p \geq 90$ **71.** $p \geq 140$
73. $2.2 \leq \text{pH} \leq 2.4$ acidic **75.** $3.0 \leq \text{pH} \leq 3.5$ acidic
77. $10.0 \leq \text{pH} \leq 11.0$ alkaline **79.** $(-2, 0)$ **81.** $[-2, 0)$
83. $(-\infty, 1)$ **85.** $[-6, \infty)$ **87.** $\{\ \}$ **89.** $(-\infty, \infty)$

Section 1.2 Practice Exercises, pp. 26–30

3. Integers **5.** Whole numbers **7.** Distance can never be negative. **9.** Negative
11.

Number	Opposite	Reciprocal	Absolute Value
6	-6	$\frac{1}{6}$	6
$\frac{1}{11}$	$-\frac{1}{11}$	11	$\frac{1}{11}$
-8	8	$-\frac{1}{8}$	8
$-\frac{13}{10}$	$\frac{13}{10}$	$-\frac{10}{13}$	$\frac{13}{10}$
0	0	Undefined	0
-3	3	$-0.\overline{3}$	3

13. < **15.** = **17.** < **19.** < **21.** -4 **23.** -19 **25.** -7
27. 14 **29.** -21 **31.** -8.1 **33.** $-\frac{5}{3}$ or $-1\frac{2}{3}$ **35.** $-\frac{67}{45}$ **37.** -32
39. $\frac{8}{21}$ **41.** $\frac{7}{18}$ **43.** Undefined **45.** 0 **47.** 3.72 **49.** 6.02
51. 64 **53.** -49 **55.** 49 **57.** $\frac{125}{27}$ **59.** 9 **61.** Not a real number
63. $\frac{1}{2}$ **65.** -7 **67.** $-\frac{9}{16}$ **69.** 32 **71.** 63 **73.** 40 **75.** 17
77. -11 **79.** -603 **81.** $\frac{109}{150}$ **83.** 5.4375 **85.** $\frac{2}{3}$ **87.** -1 **89.** 6
91. $\frac{9}{10}$ **93.** $-10.1°\text{C}$ **95. a.** $25°\text{C}$ **b.** $100°\text{C}$ **c.** $0°\text{C}$ **d.** $-40°\text{C}$
97. $9\frac{3}{4}$ in.^{2} **99.** 8.06 cm^2 **101.** 14.1 ft^3 **103.** 26.8 ft^3
105. 141.4 in.^{3}
107. $12/(6 - 2)$ **109.** $\frac{2}{3}$ or 0.6666667

Section 1.3 Practice Exercises, pp. 35–38

3. **a.** Rational number, integer, real number **b.** $-\frac{1}{4}$ **c.** 4 **d.** 4
5. $(3, \infty)$ 7. $\left(-\frac{5}{2}, 3\right]$ 9. **a.** 3 terms **b.** 6 **c.** 2, -5, 6 11. **a.** 5 terms **b.** -7 **c.** 1, -7, 1, -4, 1 13. a 15. e 17. b 19. d 21. a
23. $14y - 2x$ 25. $6p^2 + p - 6$ 27. $-p^2 - 3p$ 29. $n^3 + m - 6$
31. $7ab + 8a$ 33. $16xy^2 - 5y^2$ 35. $8x - 23$ 37. $-4c - 6$
39. $-9w + 10$ 41. $4z - 16$ 43. $7s - 26$ 45. $-12w + 13$
47. 0 49. $4c + 2$ 51. $1.4x + 10.2$ 53. $-2a^2 + 3a + 38$
55. $2y^2 - 3y - 5$ 57. 0, for example: $3 + 0 = 3$ 59. Reciprocal
61. No, for example: $6 - 5 \neq 5 - 6$; $1 \neq -1$ 63. **a.** $x(y + z)$ **b.** xy **c.** xz **d.** $xy + xz$ **e.** $x(y + z) = xy + xz$ the distributive property of multiplication over addition

Section 1.4 Practice Exercises, pp. 47–50

3. $3x - 3y + 14xy$ 5. $5z - 20$ 7. Linear 9. Nonlinear
11. Linear 13. b 15. $x = 12$ 17. $x = -\frac{1}{32}$ 19. $z = \frac{21}{20}$
21. $a = \frac{8}{5}$ 23. $t = -1.1$ 25. $p = 6.7$ 27. $q = 11$ 29. $y = 13$
31. $b = -7$ 33. $x = 13$ 35. $t = \frac{3}{2}$ 37. $a = 0$ 39. $z = 3$
41. $y = \frac{7}{4}$ 43. $x = 0$ 45. $x = \frac{4}{3}$ 47. $p = -4$ 49. $x = 3$
51. $q = -6$ 53. $w = 1$ 55. $m = 2$ 57. An equation that is true for some values of the variable but false for other values. 59. All real numbers; an identity 61. $x = 0$; A conditional equation
63. No solution; a contradiction 65. $b = 16$ 67. $x = 2$ 69. $c = 3$
71. $b = -1$ 73. No solution 75. $c = \frac{8}{5}$ 77. $x = -\frac{3}{2}$ 79. All real numbers 81. $b = \frac{33}{5}$ 83. $x = -8$ 85. $y = \frac{3}{7}$ 87. $y = \frac{3}{4}$
89. $x = 60$ 91. $x = 6$ 93. $b = 40$ 95. $y = -\frac{9}{2}$ 97. **a.** $y + 8$ **b.** $y = -8$ **c.** Simplifying an expression clears parentheses and combines *like* terms. Solving an equation isolates a variable to find a solution.

Midchapter Review, p. 51

1. **a.** $\left\{-7.1, -2, -\frac{1}{8}, 0, 0.\overline{3}, \frac{7}{8}, 6, \frac{9}{2}\right\}$ **b.** $\{6\}$
  c. $\left\{-7.1, -5\pi, -2, -\frac{1}{8}, 0, 0.\overline{3}, \sqrt{2}, \frac{7}{8}, 6, \frac{9}{2}\right\}$ **d.** $\{-5\pi, \sqrt{2}\}$
  e. $\{0, 6\}$ 2. **a.** $\{10, 20\}$ **b.** $\{\,\}$ **c.** $\{-10, -5, 0, 5, 10, 15, 20, 25\}$
  d. $\{-5, 0, 5, 10, 15, 20, 25, 30, 40, 50\}$ 3. 8 4. 15 5. $-\frac{6}{5}$ 6. $\frac{9}{10}$
7. 59 8. 29 9. -58 10. 2 11. 18 12. 64 13. Equation; $x = -34$ 14. Expression; $0.17a + 4.495$ 15. Expression; $-x - 4$
16. Equation; $b = 68$ 17. Equation; $n = -\frac{3}{32}$ 18. Equation; all real numbers 19. Equation; no solution 20. Expression; $0.39q + 500$ 21. Expression; $\frac{3}{5}c - \frac{1}{5}$ 22. Equation; $y = 4$
23. Equation; $x = 16$ 24. $(-3, \infty)$
25. $[6, \infty)$ 26. $\left(-\infty, 2\frac{1}{2}\right]$
27. $(-\infty, 4.8)$ 28. $(-\infty, 0) \cup (4, \infty)$
29. $[1, 13]$

Section 1.5 Practice Exercises, pp. 59–63

3. $a = \frac{13}{7}$ 5. $x = 6$ 7. $b = -1$ 9. $p = \frac{18}{5}$ 11. $d = 10,000$
13. 9, 6 15. -7 17. 7, 3 19. 404,227 and 404,228 21. $-22, -20$
23. 6, 8, 10 25. $49{,}200 27. $26.88 29. A 2-year loan at $8\frac{1}{2}\%$ is better. 31. Perot 8.5%, Dole 41.5%, Clinton 50.0% 33. $12,980
35. $45.00 37. $46 billion 39. $6800 at 9% $8200 at 10%
41. Sienna deposited $8000 in her money market account.
43. $8000 at 4.5% $13,000 at 6% 45. 60 oz 47. 12.5 L of 18% solution 7.5 L of 10% solution 49. 2.5 lb of black tea 1.5 lb of orange pekoe tea 51. 112 of $2 tickets 96 of $1 tickets
53. 9.12 hr 55. 2.4 mph 57. $1\frac{3}{4}$ hr

Section 1.6 Practice Exercises, pp. 68–72

3. $x = -\frac{14}{3}$ 5. $z = 1$ 7. 29.5 ft wide, 59 ft long 9. 6 m, 8 m, and 10 m 11. **a.** $11\frac{1}{2}$ yd by 8 yd **b.** 39 yd 13. $4\frac{1}{2}$ ft on a side 15. 30°, 30°, 120° 17. 15°, 75° 19. $x = 20$; 139°, 41° 21. $x = 27.5$; 60°, 30°
23. $x = 18$; 36°, 91°, 53° 25. $x = 11$; 20° 27. a, b, c 29. a, b
31. $l = \frac{A}{w}$ 33. $P = \frac{I}{rt}$ 35. $K_1 = K_2 - W$
37. $C = \frac{5}{9}(F - 32)$ or $C = \frac{5F - 160}{9}$ 39. $v^2 = \frac{2K}{m}$ 41. $a = \frac{v - v_0}{t}$
43. $v_2 = \frac{w}{p} + v_1$ or $v_2 = \frac{w + pv_1}{p}$ 45. $y = \frac{c - ax}{b}$ 47. $B = \frac{3V}{h}$
49. $y = -3x + 6$ 51. $y = \frac{5}{4}x - 5$ 53. $y = -3x - \frac{13}{2}$
55. $y = x - 2$ 57. $y = -\frac{27}{4}x + \frac{15}{4}$ 59. $y = \frac{3}{2}x$ 61. **a.** $r = \frac{d}{t}$
b. 145.2 mph 63. **a.** $m = \frac{F}{a}$ **b.** 2.5 kg 65. **a.** $x = z\sigma + \mu$
b. $x = 130$ 67. $t = \frac{12}{6 - r}$ 69. $x = \frac{-2}{a - 6}$ or $x = \frac{2}{6 - a}$
71. $P = \frac{A}{1 + rt}$ 73. $m = \frac{T}{g - f}$ 75. $x = \frac{z - by}{a - c}$ or $x = \frac{by - z}{c - a}$

Section 1.7 Practice Exercises, pp. 80–83

3. $v = \frac{d + 16t^2}{t}$ 5. -11 7. **a.** $h = \frac{2A}{b}$ **b.** $6\frac{2}{3}$ cm
9. $(-\infty, -1]$
11. $(-\infty, 5)$
13. $\left(-\infty, \frac{15}{8}\right]$
15. $[-20, \infty)$
17. $(13, \infty)$
19. $(-\infty, 1)$
21. $(-\infty, -5)$
23. $[-6, \infty)$

25. $(-\infty, \frac{7}{8})$
27. $(-\infty, -36]$
29. $(-\infty, \frac{1}{2})$
31. $(-\infty, 3.25]$
33. $5 < x$ and $x \leq 7$ 35. $(-\frac{1}{4}, \frac{9}{2})$
37. $[4, 5)$
39. $[16, 32)$
41. $[33, 51]$
43. $(-\frac{11}{3}, -3]$
45. $[-7, -\frac{1}{2}]$

47. $-1 > x > 3$ is describing a number that is both less than -1 and greater than 3. This is not possible.   49. **a.** It will take at least 15.4 years.   **b.** It will take at least 30.8 years.   51. More than 509 bikes must be sold.   53. $68° \leq F \leq 84.2°$ 55. **a.** 1960 to 1967   **b.** 1981 to 1985

57. $[3, \infty)$
59. $(-\infty, -20)$
61. $(8, \infty)$
63. $[1, 2)$
65. $(-\infty, -\frac{2}{7})$
67. $(-\infty, 6]$
69. $(-\frac{9}{7}, \frac{2}{7})$
71. $(-\infty, -\frac{5}{2}]$
73. $(0, \infty)$ 75. $>$ 77. $<$

Section 1.8 Practice Exercises, pp. 89–93

3. $a = -\frac{8}{5}$ 5. $[-3, \infty)$ 7. $x = \frac{9y+11}{5}$ or $x = \frac{9}{5}y + \frac{11}{5}$
9. $b^4 \cdot b^3 = (b \cdot b \cdot b \cdot b) \cdot (b \cdot b \cdot b) = b^7$ $(b^4)^3 = b^4 \cdot b^4 \cdot b^4 = (b \cdot b \cdot b \cdot b) \cdot (b \cdot b \cdot b \cdot b) \cdot (b \cdot b \cdot b \cdot b) = b^{12}$ 11. For example: $3^2 \cdot 3^4 = 3^6$; $x^8 \cdot x^2 = x^{10}$ 13. For example: $(x^2)^4 = x^8$; $(2^3)^5 = 2^{15}$ 15. For example: $\left(\frac{x}{y}\right)^3 = \frac{x^3}{y^3}$; $\left(\frac{2}{7}\right)^2 = \frac{2^2}{7^2}$ 17. $\frac{9}{4}$ 19. $\frac{1}{25}$
21. $-\frac{1}{25}$ 23. $\frac{1}{25}$ 25. 64   27. $\frac{16}{81}$ 29. $-\frac{125}{8}$ 31. 1   33. $10a$
35. y^8 37. 13^2 or 169   39. y^8 41. 3^4x^8 or $81x^8$ 43. $\frac{1}{p^3}$
45. $\frac{1}{7^3}$ or $\frac{1}{343}$ 47. $\frac{1}{w^3}$ 49. $\frac{1}{a^7}$ 51. r^2 53. $\frac{1}{z^4}$ 55. a^3b^2 57. 1

59. $\frac{65}{4}$ 61. $\frac{26}{25}$ 63. 3   65. $\frac{q^2}{p^3}$ 67. $-\frac{3b^7}{2a^3}$ 69. $\frac{x^{16}}{81y^{20}z^8}$
71. $-\frac{4m^4}{n^2}$ 73. $\frac{4q^{11}}{p^4}$ 75. $\frac{5x^8}{y^2}$ 77. $\frac{16a^2}{b^6}$ 79. $-\frac{27y^{27}}{8x^{24}}$ 81. $\frac{4x^{18}}{y^{10}}$
83. $27x^3y^9$ 85. **a.** 4.2×10^{-3} **b.** 6.022×10^{23} **c.** 4.6×10^{-4}
87. **a.** 5,282,200,000   **b.** 0.000018   **c.** 6,378,000,000   89. 3.5×10^5
91. Proper   93. Proper   95. 3.38×10^{-4} 97. 1.608×10^4
99. 3.4×10^{13} 101. 2.5×10^3 103. 1.204×10^{24} hydrogen atoms and 6.02×10^{23} oxygen atoms   105. There are 2×10^4 or 20,000 people per square mile.   107. x^{2a+6} 109. y^{a+2} 111. $x^{b-3}y^{-b-1}$
113. 11.6 days   115. Anyone over 52 years old.

Chapter 1 Review Exercises, pp. 101–105

1. 0   3. For example: $-2, -1, 0, 1, 2$ 5. All real numbers greater than 0 but less than or equal to 2.6   7. All real numbers greater than 8   9. All real numbers
11. $(-\infty, 2)$
13. $(-1, 5)$
15. $[0, 5)$ 17. True   19. $8, -\frac{1}{8}, 8$
21. 16, 2   23. -2 25. -21.6 27. $-\frac{25}{13}$ 29. 3   31. 11   33. 256 ft
35. $3x + 15y$ 37. $4x - 10y + z$ 39. $7q - 14$ 41. $-6y - 5$
43. For example: $3 + x = x + 3$ 45. The empty set; no solution
47. $x = -5$; A conditional equation   49. $x = -18.075$; A conditional equation   51. $m = \frac{31}{6}$; A conditional equation   53. $x = -\frac{21}{8}$; A conditional equation   55. No solution; contradiction
57. $x, x+1, x+2$ 59. Distance equals rate times time.
61. 2 ft and $\frac{2}{3}$ ft   63. 4.125 mph   65. $1500 in 6% account $3500 in 9% account   67. 16,600   69. Width is 9 ft, length is 11 ft   71. $x = 30$; 29°, 61°   73. $y = 6x + 12$ 75. $b = \frac{2A}{h}$
77. $(-\infty, -\frac{4}{3})$
79. $[\frac{7}{3}, 8]$
81. $[\frac{46}{7}, \infty)$
83. $(-\infty, \frac{29}{2}]$
85. $[-\frac{2}{5}, \frac{9}{5})$ 87. 3^5x^5 or $243x^5$
89. $-3xy^2$ 91. $-\frac{b^{15}}{8a^6}$ 93. $\frac{5^4y^{24}}{4^4x^{20}}$ 95. **a.** 3.362994×10^9
b. 4.247079×10^9 97. **a.** 0.001   **b.** 0.000000001   99. 6.25×10^9
101. 3.24×10^7

Chapter 1 Test, p. 105–106

1. **a.** $-5, -4, -3, -2, -1, 0, 1, 2$ **b.** For example: $\frac{3}{2}, \frac{5}{4}, \frac{8}{5}$
2. $(-3, 4)$ does not include the endpoints. $[-3, 4]$ does include the endpoints.   3. **a.** $(-\infty, 6)$
b. $[-3, \infty)$

4. [-5, -2) **5. a.** $\frac{1}{2}, -2, \frac{1}{2}$

b. $-4, \frac{1}{4}, 4$ **c.** 0, no reciprocal exists, 0 **6.** 6 **7.** $z = 1.1$

8. a. False **b.** True **c.** True **d.** True **9. a.** $-2b - 6$ **b.** $-2x + 1$

10. $x = 133$ **11.** $z = \frac{4}{5}$ **12.** $x = 142{,}500$ **13.** $x = \frac{14}{9}$

14. a. Identity **b.** Contradiction **c.** A conditional equation
15. 18 and 90 **16. a.** 0.4 hr (24 min) **b.** 1.8 mi **17.** Shawnna invested $1000 in the CD. **18.** 27 in. on each side **19.** $y = -2x + 3$
20. $z = \frac{x - \mu}{\sigma}$ **21.** $(34, \infty)$

22. $(-\infty, \frac{18}{5}]$

23. $(-\frac{1}{3}, 2]$

24. It can carry at most seven more passengers. **25.** $5a^{13}$ **26.** x^{11}
27. $\frac{9x^{12}}{25y^{14}}$ **28.** $\frac{8y^{12}}{x^7}$ **29.** 5.68 **30.** 2.3×10^9

Chapter 2

Chapter 2 Preview, p. 108

1. IV **3.** III **5.**

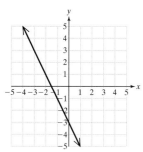

7. x-intercept: $(-5, 0)$; y-intercept: $\left(0, \frac{10}{3}\right)$ **9.** 0 **11. a.** $-\frac{4}{3}$

b. $\frac{3}{4}$ **13.** $y = 2x - 3$ or $2x - y = 3$ **15. a.** $y = \frac{3}{2}x$ **b.** 22.5 in. of rain **c.** 24 hr

Section 2.1 Practice Exercises, pp. 113–117

3. For (x, y), if $x > 0, y > 0$ the point is in quadrant I. If $x < 0$, $y > 0$ the point is in quadrant II. If $x < 0, y < 0$ the point is in quadrant III. If $x > 0, y < 0$ the point is in quadrant IV.

5. **7.**

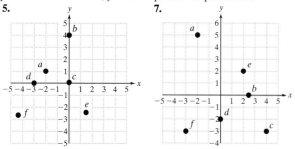

9. $A(-4, 5)$, II; $B(-2, 0)$, x-axis; $C(1, 1)$, I; $D(4, -2)$, IV; $E(-5, -3)$, III **11.** $A(-1, 3)$, II; $B(2, 1)$, I; $C(0, -3)$, y-axis; $D(4, -3)$, IV; $E(-2, -2)$, III **13. a.** 25 **b.** 178 lb **15.** $(1, 2)$
17. $(-1, 0)$ **19.** $(-1, 6)$ **21.** $(0, 3)$ **23.** $\left(-\frac{1}{2}, \frac{3}{2}\right)$ **25.** $(-0.4, -1)$

27. $(40, 7\frac{1}{2})$; they should meet 40 mi east, $7\frac{1}{2}$ mi north of the warehouse.

Section 2.2 Practice Exercises, pp. 125–132

3.

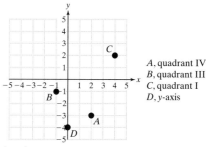

A, quadrant IV
B, quadrant III
C, quadrant I
D, y-axis

5. $\left(\frac{3}{2}, \frac{9}{2}\right)$ **7. a.** Yes **b.** No **c.** Yes **9. a.** No **b.** Yes **c.** No

11.

x	y
0	-2
4	4
-1	$-\frac{7}{2}$

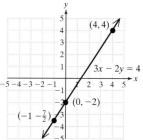

13.

x	y
0	0
5	-1
-5	1

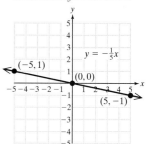

15. 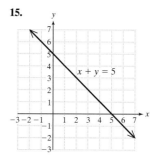 **17.**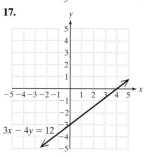

19. 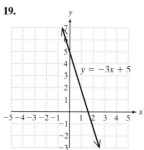 **21.**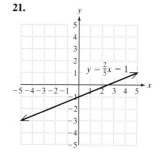

23.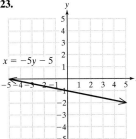

25.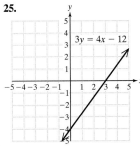

27.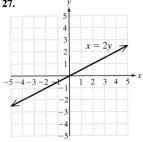

29. To find an x-intercept, substitute $y = 0$ and solve for x. To find a y-intercept, substitute $x = 0$ and solve for y.

31. a. $(9, 0)$ **b.** $(0, 6)$ **c.**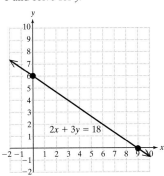

33. a. $(4, 0)$ **b.** $(0, -2)$ **c.**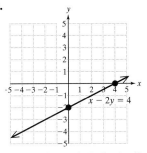

35. a. $(0, 0)$ **b.** $(0, 0)$ **c.**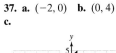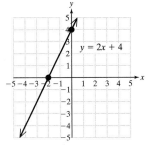

37. a. $(-2, 0)$ **b.** $(0, 4)$ **c.**

39. a. $\left(\frac{3}{2}, 0\right)$ **b.** $(0, 2)$ **c.**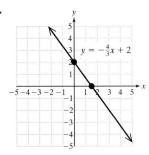

41. a. $(0, 0)$ **b.** $(0, 0)$ **c.**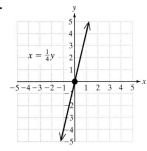

43. a. $35{,}000$ **b.** $25{,}000$ **c.** For \$0 sales the salary is \$10,000. **d.** Total sales cannot be negative.

45. Horizontal **47.** Vertical

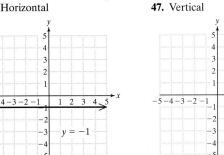

49. Vertical **51.** Horizontal

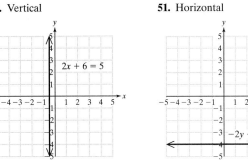

53. x-intercept $(2, 0)$ y-intercept $(0, 3)$
55. x-intercept $(a, 0)$ y-intercept $(0, b)$
57. Nonlinear **59.** Nonlinear

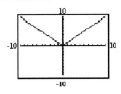

61. Nonlinear 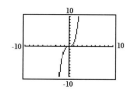   **63.** $y = -2x - 1$ 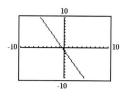   **57.** For example: (1, 2) 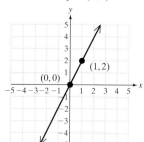   **59.** For example: (2, 0)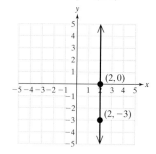

65. $y = -5$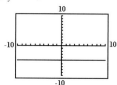

61. For example: (4, 0)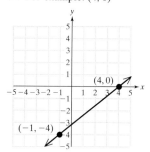

Section 2.3 Practice Exercises, pp. 139–144

3. a. 4 **b.** 8 **c.** 6
5. No x-intercept; (0, 2)

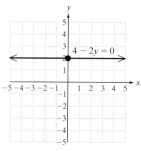

7. (6, 0), (0, 18)

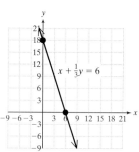

9. $m = \pm\dfrac{2}{5}$ **11.** $m = \dfrac{3}{10}$ **13.** $m = \dfrac{1}{12}$. This means that the plane gains 1 ft of elevation for every 12 ft it travels horizontally.
15. $m = -\dfrac{4}{5}$ **17.** $m = 1$ **19.** $m = -1$ **21.** $m = -1$ **23.** m is undefined. **25.** $m = 0$ **27.** $m = 1$ **29.** $m = \dfrac{6}{5}$ **31.** 8 units
33. $m = 1$ **35.** Undefined **37.** $m = -\dfrac{1}{2}$ **39. a.** $m = 5$
b. $m = -\dfrac{1}{5}$ **41. a.** $m = -\dfrac{4}{7}$ **b.** $m = \dfrac{7}{4}$ **43. a.** $m = 0$ **b.** m is undefined. **45.** $m = 2$; $m = -\dfrac{1}{2}$; perpendicular.
47. $m_1 = -1$; $m_2 = 3$; neither **49.** Undefined; $m = 0$; perpendicular. One line is horizontal and one is vertical. **51.** $m = 1$; $m = 1$; parallel.
53. a. $m = 18.75$ **b.** The number of cell phone subscriptions increased by 18.75 million per year during this period.
55. a. $m = 6$ **b.** The weight of boys tends to increase by 6 lb/yr during this period of growth.

Midchapter Review, pp. 144–145

1. a. $(-10, 8)$ **b.** $(1.7, 0.9)$
2. a. Quadrant III **b.** x-axis **c.** Quadrant II
3. 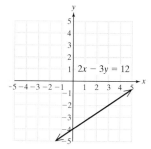   **4.**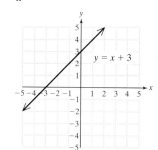

5. 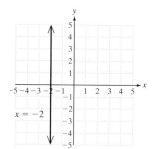   **6.**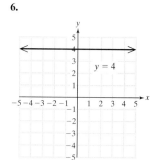

7. x-intercept: (3, 0); y-intercept: $\left(0, -\dfrac{4}{3}\right)$ **8. a.** y-intercept: (0, 12,000). The value of the machine when new (after 0 years) is $12,000. **b.** x-intercept: (8, 0). The value of the machine after 8 years is $0. **9.** $-\dfrac{12}{5}$ **10.** -2 **11.** Undefined **12. a.** $-\dfrac{2}{3}$ **b.** $\dfrac{3}{2}$

Section 2.4 Practice Exercises, pp. 152–157

3. a. x-intercept: $(2, 0)$ **b.** y-intercept: $(0, 3)$
c.
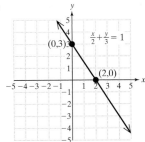

5. If the slope of one line is the opposite of the reciprocal of the other then the lines are perpendicular. **7. a.** 2 **b.** 2 **c.** $-\frac{1}{2}$
9. Slope: -7; y-intercept: $(0, 5)$ **11.** Slope: -1; y-intercept: $(0, 0)$
13. Slope: 0; y-intercept: $(0, -14)$ **15.** Slope: $\frac{9}{10}$; y-intercept: $\left(0, -\frac{2}{5}\right)$ **17.** Slope: -2.5; y-intercept: $(0, 1.8)$ **19.** c **21.** a **23.** e

25. 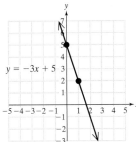   **27.**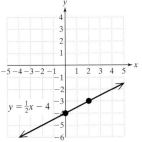

29. 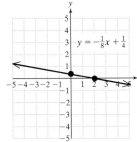   **31.**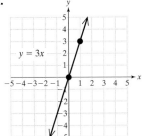

33. $m = -\frac{3}{7}$; y-intercept: $\left(0, \frac{9}{7}\right)$ **35.** Perpendicular
37. Parallel **39.** Neither **41.** $y = -\frac{1}{2}x + 5$ or $x + 2y = 10$
43. $y = -2x + 16$ or $2x + y = 16$ **45.** $y = 4x - 2$ or $4x - y = 2$
47. $y = \frac{7}{2}x - \frac{53}{2}$ or $7x - 2y = 53$ **49.** $y = 3x - 2$ or $3x - y = 2$
51. $y = -\frac{3}{5}x - \frac{11}{5}$ or $3x + 5y = -11$ **53.** $y = -1$ **55.** $y = \frac{1}{2}x + \frac{9}{2}$ or $x - 2y = -9$ **57.** $y = -2x + 1$ or $2x + y = 1$ **59.** $y = -\frac{2}{3}x - 5$ or $2x + 3y = -15$ **61.** $y = \frac{3}{4}x - 5$ or $3x - 4y = 20$
63. $y = -\frac{5}{6}x - \frac{5}{2}$ or $5x + 6y = -15$ **65.** $y = -7x - 90$ or $7x + y = -90$ **67.** $y = \frac{1}{4}x - \frac{19}{4}$ or $x - 4y = 19$ **69.** $x = \frac{5}{2}$
71. $y = 0$ **73.** $x = 4$ **75.** $y = 4$
77. $x = 1$ is not in slope-intercept form. No y-intercept, undefined slope. **79.** $y = -5$ is in slope-intercept form, $y = 0x - 5$. Slope is 0 and y-intercept is $(0, -5)$.
81. The lines have the same slope but different y-intercepts; they are parallel lines.

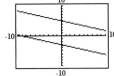

83. The lines have different slopes but the same y-intercept.

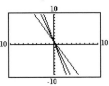

85. The lines are perpendicular.

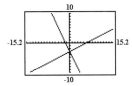

87.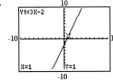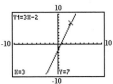

Section 2.5 Practice Exercises, pp. 162–171

3. $y = x - 2$

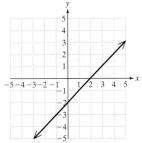

5. Cannot be written in slope-intercept form, $x = 5$

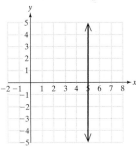

7. a. $m = -\frac{1}{3}$ **b.** $y = -\frac{1}{3}x - 5$

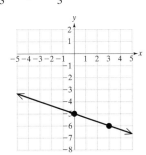

c. $y = -\frac{1}{3}x + c$ $(c \neq -5)$ **d.** $y = 3x + c$ (c = any real number)

9.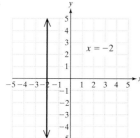

11. a. $y = 0.2x + 19.95$
b.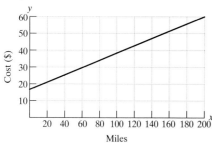

c. (0, 19.95); the cost is $19.95 for 0 mi. **d.** 50 mi costs $29.95; 100 mi costs $39.95; 200 mi costs $59.95 **e.** $42.35 **f.** No, one cannot drive a negative number of miles. **13. a.** $y = 52x + 2742$
b.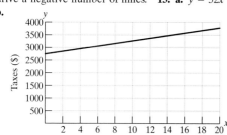

c. $m = 52$. Taxes increase $52 per year. **d.** (0, 2742). Initial year ($x = 0$) taxes were $2742. **e.** 10 years: $3262; 15 years: $3522 **15. a.** 0.8 mi, 2.4 mi, and 3.2 mi **b.** 21 sec
17. a. $195.9 thousand, or $195,900 **b.** $105,800
c. $m = 5.3$; there is a $5300 increase in median housing cost per year. **d.** (0, 63.4). In 1980 ($x = 0$) the median cost was $63,400.
19. a. $y = 0.027x + 3.3$ **b.** 3.84 m **c.** 5.352 m **d.** The linear model can only be used to approximate the winning heights.
e. $m = 0.027$. Winning heights have increased by approximately 0.027 m/yr.
21. a.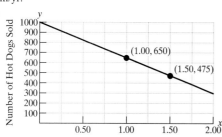

b. $y = -350x + 1000$ **c.** 405 hot dogs

23. a. Yes, there is a linear trend.
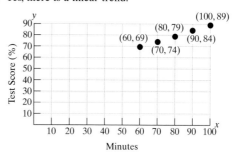

b. $y = \dfrac{1}{2}x + 39$

c. 102 min gives 90%. This model is not appropriate for other students because it is based on Loraine's scores. **d.** 54%
25. Collinear **27.** Not collinear **29.**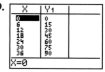

31.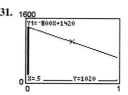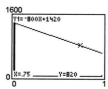

Chapter 2 Review Exercises, pp. 177–180

1. **3.** $(-1.45, -6)$

5.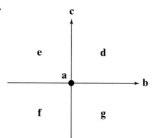

x	y
0	3
-2	0
1	$\dfrac{9}{2}$

7.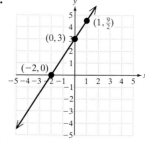

x	y
4	0
4	1
4	-2

9. x- and y-intercept (0, 0)   **11.** x-intercept (−2, 0)   **5.** x-intercept (4, 0) y-intercept (0, −3)

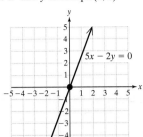 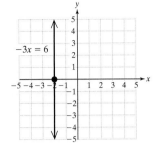 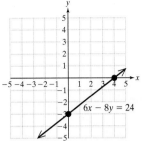

13. For example: $y = 2x$

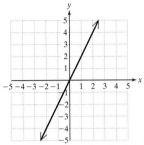

6. x-intercept (−4, 0) no y-intercept

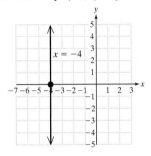

15. $m = 2$ **17.** $m = 0$ **19.** Perpendicular **21.** Neither
23. a. $m = 53$ **b.** There is an increase of 53 students per year.
25. a. $y = k$ **b.** $y - y_1 = m(x - x_1)$ **c.** $Ax + By = C$ **d.** $x = k$
e. $y = mx + b$ **27.** $y = -\frac{2}{3}x + 2$ or $2x + 3y = 6$ **29.** $y = 3x - 20$
or $3x - y = 20$ **31. a.** $y = -2$ **b.** $x = -3$ **c.** $x = -3$ **d.** $y = -2$
33. a. $y = 0.25x + 20$
b.

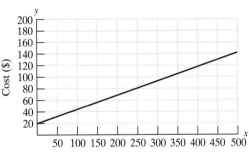

c. The daily cost is $20 if no ice cream is sold. **d.** $132.50 **e.** 0.25
f. The daily cost increases by $0.25 per ice cream product.

7. x- and y-intercept (0, 0)

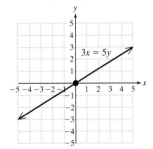

8. no x-intercept y-intercept (0, −3)

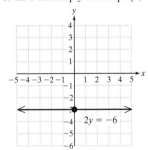

Chapter 2 Test, pp. 180–182

1. (0, −9) (6, 0) (4, −3)

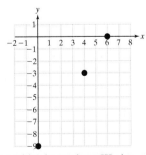

2. a. False, the product is positive in quadrant III also. **b.** False, the quotient is negative in quadrant II also. **c.** True **d.** True
3. $\left(13, \frac{17}{2}\right)$ **4.** To find the x-intercept, let $y = 0$ and solve for x. To find the y-intercept, let $x = 0$ and solve for y.

9. a. $m = \frac{5}{8}$ **b.** $m = \frac{6}{5}$ **10. a.** The slopes are the same. **b.** The slope of one line is the opposite of the reciprocal of the slope of the other. **11. a.** -7 **b.** $\frac{1}{7}$ **12.** Neither **13. a.** $y = \frac{3}{4}x + 1$
b. Slope: $\frac{3}{4}$; y-intercept: (0, 1) **c.**

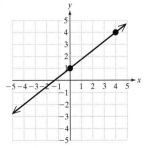

14. a. Perpendicular **b.** Parallel **c.** Perpendicular **d.** Neither
15. a. For example: $y = 3x + 2$ **b.** For example: $x = 2$ **c.** For example: $y = 3$ $m = 0$ **d.** For example: $y = -2x$
16. $y = -2x + \frac{31}{2}$ **17.** $y = \frac{3}{2}x - 6$ or $3x - 2y = 12$
18. $y = 2x - 11$ or $2x - y = 11$ **19.** $y = \frac{1}{3}x + \frac{1}{3}$
20. a. $y = 300x + 800$
b.

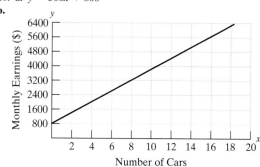

c. Jack earns a base salary of $800. **d.** $5900 **21. a.** $(0, 66)$. For a woman born in 1940, life expectancy was about 66 years.
b. $m = \frac{3}{10}$ Life expectancy rises by 3 years every 10 years.
c. $y = \frac{3}{10}x + 66$ **d.** 82.2 years old

Chapters 1–2 Cumulative Review Exercises, p. 183

1. -1 **2.** 4 **3.** $z = -7$ **4.** $b = -15$ **5.** 60 mi **6. a.** $h = \frac{V}{\pi r^2}$
b. 10.4 cm **7. a.** $[-2, \infty)$ **b.** $(-\infty, 3)$ **8.** $-\frac{1}{5}$ **9.** $(-1, 6)$
10. a. $\left(\frac{10}{3}, 0\right)(0, -2)$ **b.** $m = \frac{3}{5}$ **c.**

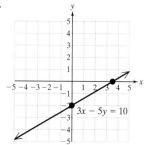

11. a. no x-intercept, y-intercept: $(0, 3)$ **b.** $m = 0$
c.

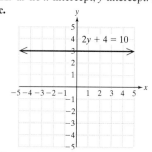

12. $x = 7$ **13.** $y = 20$ **14.** $y = -2x - 2$ **15.** $y = -4x$ **16.** She spent $5.98 on drinks and $11.96 on popcorn.

Chapter 3

Chapter 3 Preview, p. 186

1. Not a solution **3.** $(-4, 3)$ **5.** $(1, 5)$ **7.** No solution
9. The speed of the plane is 225 mph, and the speed of the wind is 25 mph. **11.** $(0, -3, 5)$ **13.** 22, 30, and 28 **15.** $(10, 4)$

Section 3.1 Practice Exercises, pp. 192–196

3. $(2, 11)$ is a solution. **5.** $(-1, 4)$ is a solution. **7.** $(2, -4)$ is a solution. **9. a.** Consistent **b.** Independent **c.** One solution
11. a. Inconsistent **b.** Independent **c.** Zero solutions
13. a. Consistent **b.** Dependent **c.** Infinitely many solutions
15. 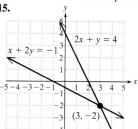 **17.**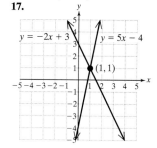

19. 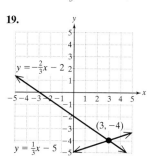 **21.**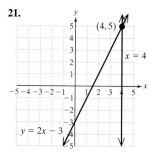

23. Inconsistent system, no solution. **25.** Dependent system: $\{(x, y) | 2x = 3y + 3\}$

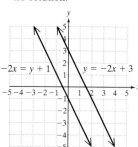 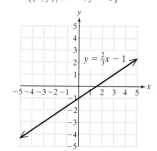

27. **29.** Dependent system: $\{(x, y) | -x + 3y = 6\}$

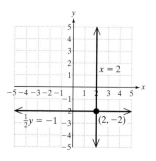 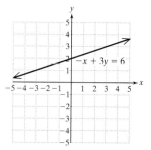

31. Inconsistent system, no solution. **33.** False **35.** True

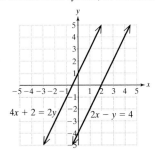

37. (2, 1)   **39.** (3, 1)

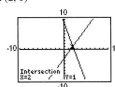 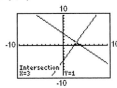

41. No solution

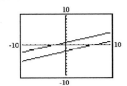

Section 3.2 Practice Exercises, pp. 201–202

3. a. Consistent **b.** Dependent **5. a.** Inconsistent **b.** Independent **7.** (0, 3) **9.** (−2, 1) **11.** (4, −3) **13.** (1, 6) **15.** $\left(-3, \frac{1}{3}\right)$ **17.** $\left(\frac{8}{3}, \frac{5}{3}\right)$ **19.** Dependent system: $\{(x, y) \,|\, x = 3y - 1\}$ **21.** Inconsistent system, no solution **23.** Inconsistent system, no solution **25.** Dependent system: $\{(x, y) \,|\, 3x - y = 7\}$ **27.** When solving a system, if you get an identity, such as 0 = 0 or 5 = 5, then the system is dependent. **29.** (8, 5) **31.** (5, 3) **33.** Inconsistent system, no solution **35.** (2, −3) **37.** (−7, 6) **39.** Dependent system: $\{(x, y) \,|\, 2x - y = 6\}$ **41.** (4, 480) **43.** (10, −32.1) **45.** $\left(\frac{1}{2}, \frac{3}{4}\right)$

Section 3.3 Practice Exercises, pp. 207–209

3. One solution **5.** (−2, −5) **7.** (3, −1) **9.** (12, −8) **11.** (1, −1) **13.** (2, −1) **15.** Dependent system: $\{(x, y) \,|\, 3x - 2y = 1\}$ **17.** Inconsistent system, no solution **19.** Dependent system: $\{(x, y) \,|\, 12x - 4y = 2\}$ **21.** Inconsistent system, no solution **23.** Use the substitution method if one equation has x or y already isolated. **25.** (−2, −3) **27.** $\left(-\frac{1}{2}, 2\right)$ **29.** (3, −3) **31.** Inconsistent system, no solution **33.** (12, 30) **35.** Dependent system: $\{(x, y) \,|\, x = \frac{3}{2}y\}$ **37.** (0, 4) **39.** $\left(-\frac{1}{2}, -\frac{5}{2}\right)$ **41.** (9, 9) **43.** $\left(\frac{4}{3}, -1\right)$ **45.** $\left(\frac{11}{28}, -\frac{37}{28}\right)$ **47.** $\left(-\frac{37}{27}, \frac{52}{27}\right)$

Midchapter Review, p. 210

1. (−1, 2) **2.** Dependent system: $\{(x, y) \,|\, 4x - 2y = 6\}$ **3.** No solution **4.** (3, 11)

Section 3.4 Practice Exercises, pp. 215–218

3. (5, −1) **5.** (12, −27) **7.** 102 nonstudent tickets, 84 student tickets **9.** Hamburger $1.39; fish sandwich $1.59 **11.** 4 dimes, 11 quarters **13.** Eleven 50¢ pieces, ten $1 coins **15.** 4 oz of 18% moisturizer; 8 oz of 24% moisturizer **17.** 1 oz of bleach; 11 oz of 4% solution **19.** $12,500 in the 2% account; $14,500 in the 3% account **21.** $2800 in CD; $2600 in savings **23.** Boat 6 mph; current 2 mph **25.** Plane 210 mph; wind 30 mph **27.** 69°; 21° **29.** 134.5°; 45.5° **31.** 29°; 61° **33.** $30 **35. a.** $y_c = 20 + 0.25x$ **b.** $y_m = 30 + 0.20x$ **c.** 200 mi

Section 3.5 Practice Exercises, pp. 225–227

3. (1, 1) **5.** The speeds are 65 mph and 58 mph. **7.** (4, 0, 2) is a solution. **9.** (1, 1, 1) is a solution. **11.** (1, 2, 3) **13.** (−2, −1, −3) **15.** (−1, 3, 4) **17.** (−6, 1, 7) **19.** $\left(\frac{1}{2}, \frac{2}{3}, -\frac{5}{6}\right)$ **21.** 67°, 82°, 31° **23.** 9 cm, 18 cm, 27 cm **25.** 148 adult tickets, 51 children's tickets, 23 senior tickets **27.** 24 oz peanuts, 8 oz pecans, 16 oz cashews **29.** Vanderbilt 6100, Baylor 12,200, Pace 8900 **31.** (−9, 5, 5) **33.** Inconsistent **35.** (1, 3, 1) **37.** Dependent **39.** (1, 0, 1) **41.** (0, 0, 0) **43.** Dependent

Section 3.6 Practice Exercises, pp. 233–237

3. $\left(12, \frac{1}{2}\right)$ **5.** (1, 3, −3) **7.** A coefficient matrix is one constructed from just the coefficients of the variable terms. **9.** A square matrix has the same number of rows and columns. **11.** 3 × 1, column matrix **13.** 2 × 2, square matrix **15.** 1 × 4, row matrix **17.** 2 × 3, none of these **19.** $\begin{bmatrix} 1 & -3 & | & 3 \\ 2 & -5 & | & 4 \end{bmatrix}$ **21.** $\begin{bmatrix} 5 & 0 & 2 & | & 17 \\ 8 & -1 & 6 & | & 26 \\ 8 & 3 & -12 & | & 24 \end{bmatrix}$ **23.** $\begin{matrix} -2x + 5y = -15 \\ -7x + 15y = -45 \end{matrix}$ **25.** $x = 0.5, y = 6.1, z = 3.9$ **27. a.** −13 **b.** 0 **29.** $\begin{bmatrix} 1 & 1 & | & 7 \\ 0 & 1 & | & -2 \end{bmatrix}$ **31.** $\begin{bmatrix} -7 & 2 & | & 19 \\ 9 & 6 & | & 13 \end{bmatrix}$ **33.** $\begin{bmatrix} 1 & 3 & | & -5 \\ 0 & 8 & | & 2 \end{bmatrix}$ **35. a.** $\begin{bmatrix} 1 & 2 & 0 & | & 10 \\ 0 & -9 & -4 & | & -47 \\ -3 & 4 & 5 & | & 2 \end{bmatrix}$ **b.** $\begin{bmatrix} 1 & 2 & 0 & | & 10 \\ 0 & -9 & -4 & | & -47 \\ 0 & 10 & 5 & | & 32 \end{bmatrix}$ **37.** True **39.** False **41.** Multiply row 3 by 2. Replace row 3 with the result. **43.** Multiply row 2 by 4 and add to row 3. Replace row 3 with the result. **45.** (−3, −2) **47.** (5, 4) **49.** Dependent: $\{(x, y) \,|\, 2x + 5y = 1\}$ **51.** (1, 2) **53.** (2, 2) **55.** Inconsistent, no solution **57.** $\left(7, 2, -\frac{3}{2}\right)$ **59.** (13, 20, 5) **61.** $\begin{bmatrix} 1 & 0 & | & -3 \\ 0 & 1 & | & -2 \end{bmatrix}$ **63.** $\begin{bmatrix} 1 & 0 & | & 5 \\ 0 & 1 & | & 4 \end{bmatrix}$ **65.** $\begin{bmatrix} 1 & 0 & 0 & | & 7 \\ 0 & 1 & 0 & | & 2 \\ 0 & 0 & 1 & | & -\frac{3}{2} \end{bmatrix}$

Chapter 3 Review Exercises, pp. 243–245

1. a. No **b.** Yes **3.** True **5.**

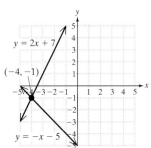

7. Inconsistent system, no solution

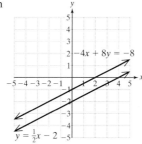

9. Inconsistent system, no solution   **11.** $(2, -7)$ **13.** $(2, -1)$
15. Dependent system: $\{(x, y) \mid 3x + y = 1\}$ **17.** Inconsistent system, no solution   **19.** $(-10, 25)$ **21.** $(-2, 7)$ **23.** 21 student tickets, 33 adult tickets   **25.** Plane 150 mph, wind 10 mph
27. $76°, 14°$ **29.** Inconsistent system, no solution   **31.** Dependent system   **33.** The pumps can drain 250, 300, and 400 gal/hr.
35. 3×2 **37.** 3×1 **39.** $\begin{bmatrix} 1 & -1 & 1 & | & 4 \\ 2 & -1 & 3 & | & 8 \\ -2 & 2 & -1 & | & -9 \end{bmatrix}$ **41.** $\begin{matrix} x = -5 \\ y = 2 \\ z = -8 \end{matrix}$
43. a. $\begin{bmatrix} 1 & 2 & 0 & | & -3 \\ 0 & -9 & 1 & | & 12 \\ -3 & 2 & 2 & | & 5 \end{bmatrix}$ **b.** $\begin{bmatrix} 1 & 2 & 0 & | & -3 \\ 0 & -9 & 1 & | & 12 \\ 0 & 8 & 2 & | & -4 \end{bmatrix}$
45. $(-3, 6)$ **47.** $(6, 1, -1)$

Chapter 3 Test, pp. 245–246

1. Yes   **2.** b   **3.** c   **4.** a   **5.**

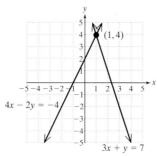

6. $(-4, 5)$ **7.** $\left(\frac{1}{2}, \frac{1}{4}\right)$ **8.** $(0, -3)$ **9.** Dependent: $\{(x, y) \mid 3x - 5y = -7\}$ **10.** $(2, -6)$ **11.** Inconsistent system, no solution   **12.** $(5, 0)$ **13.** $(16, -37, 9)$ **14.** 80 L of 20%, 120 L of 60% solution   **15.** The angles are 80° and 10°.   **16.** Joanne 142 orders; Kent 162 orders; Geoff 200 orders
17. For example: $\begin{bmatrix} 2 & 1 \\ 0 & -4 \\ 2.6 & 7 \end{bmatrix}$ **18. a.** $\begin{bmatrix} 1 & 2 & 1 & | & -3 \\ 0 & -8 & -3 & | & 10 \\ -5 & -6 & 3 & | & 0 \end{bmatrix}$
b. $\begin{bmatrix} 1 & 2 & 1 & | & -3 \\ 0 & -8 & -3 & | & 10 \\ 0 & 4 & 8 & | & -15 \end{bmatrix}$ **19.** $(6, -1)$ **20.** $(2, -4, 3)$

Chapters 1–3 Cumulative Review Exercises, p. 247

1. No solution   **2.** $-\frac{19}{12}$ **3.** $\left(-\frac{7}{5}, \infty\right)$ **4.** Slope $\frac{5}{2}$; x-intercept $(3, 0)$; y-intercept $(0, -\frac{15}{2})$
5.   **6.**

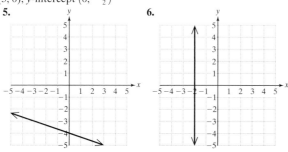

7. 0   **8.** $y = -4x + 4$ **9.** Inconsistent; no solution   **10.** $(1, 2)$
11. Four nickels, six dimes, nine quarters   **12. a.** $y = 2.50x + 25$
b. $y = 3x + 10$ **c.** 30 tapes   **13.** $(2, 0, -1)$ **14.** 2×3
15. For example: $\begin{bmatrix} 2 & 3 & 1 & 7 \\ -1 & 4 & 2 & 6 \end{bmatrix}$ **16.** Interchange two rows. Multiply a row by a nonzero constant. Add a multiple of one row to another row.   **17.** $(1, 1)$

Chapter 4

Chapter 4 Preview, p. 250

1. a. $\{(8, 88), (2, 70), (0, 45), (6, 95), (8, 90)\}$ **b.** Domain $\{8, 2, 0, 6\}$, range $\{88, 70, 45, 95, 90\}$ **3.** Function   **5.** -1
7. Domain: $(0, \infty)$; Range: $(-2, \infty)$ **9. a.** Neither   **b.** Constant
c. Linear   **11.** $v = ku^2$

Section 4.1 Practice Exercises, pp. 255–259

3. $\{(A, 1), (A, 2), (B, 2), (C, 3), (D, 5), (E, 4)\}$ **5.** $\{$(Pregnant women, 60), (Nursing mothers, 65), (Infants under 1 year, 14), (Children 1–4 years, 16), (Adults, 50)$\}$ **7.** Domain $\{A, B, C, D, E\}$; range $\{1, 2, 3, 4, 5\}$ **9.** Domain {pregnant women, nursing mothers, infants under 1, children 1–4, adults}; range $\{60, 65, 14, 16, 50\}$
11. Domain $[-5, 3]$; range $[-2.1, 2.8]$ **13.** Domain $[0, 4.2]$; range $[-2.1, 2.1]$ **15.** Domain $(-\infty, 0]$; range $(-\infty, \infty)$ **17.** Domain $[-4, \infty)$; range $[-4, -2] \cup (2, \infty)$ **19.** Domain $(-\infty, \infty)$; range $\{-2\}$ **21.** Domain $\{-3, -1, 1, 3\}$; range $\{0, 1, 2, 3\}$ **23.** Domain $[-4, 5)$; range $\{-2, 1, 3\}$ **25. a.** 2.85   **b.** 9.33   **c.** Dec.   **d.** Nov.
e. 7.63   **f.** {Jan., Feb., Mar., Apr., May, June, July, Aug., Sept., Oct., Nov., Dec.}   **27. a.** 31.876 million or 31,876,000   **b.** The year 2012
29. a. For example: {(Julie, New York), (Peggy, Florida), (Stephen, Kansas), (Pat, New York)}   **b.** Domain {Julie, Peggy, Stephen, Pat}; range {New York, Florida, Kansas}   **31.** $y = 2x - 1$

Section 4.2 Practice Exercises, pp. 266–272

3. a. {(Doris, Mike), (Richard, Nora), (Doris, Molly), (Richard, Mike)}   **b.** {Doris, Richard}   **c.** {Mike, Nora, Molly}   **5.** Function
7. Not a function   **9.** Function   **11.** Not a function   **13.** Function
15. Not a function   **17.** -11 **19.** 1   **21.** 2   **23.** $6t - 2$ **25.** 7
27. 4   **29.** 4   **31.** $6x + 4$ **33.** $-x^2 + 5$ **35.** $|x + h - 2|$ **37.** 7
39. $-6a - 2$ **41.** $|-c - 2|$ **43.** 1   **45.** 7   **47.** -18.8 **49.** -7
51. 2π **53.** -5 **55.** 4   **57. a.** 2   **b.** 1   **c.** 1   **d.** $x = -3$
e. $x = 1$ **f.** $[-3, 3]$ **g.** $[-3, 3]$ **59. a.** 3   **b.** $H(4)$ is not defined because 4 is not in the domain of H.   **c.** 4   **d.** $x = -3$ and $x = 2$
e. All x on the interval $[-2, 1]$ **f.** $(-4, 4)$ **g.** $[2, 5)$ **61.** The domain is the set of all real numbers for which the denominator is not zero. Set the denominator equal to zero, and solve the resulting equation. The solution(s) to the equation must be excluded from the domain. In this case, setting $x - 2 = 0$ indicates that $x = 2$ must be excluded from the domain. The domain is $\{x \mid x$ is a real number and $x \neq 2\}$.   **63.** $(-\infty, -6) \cup (-6, \infty)$ **65.** $(-\infty, 0) \cup (0, \infty)$
67. $(-\infty, \infty)$ **69.** $[-7, \infty)$ **71.** $[3, \infty)$ **73.** $(-\infty, \frac{1}{2}]$ **75.** $(-\infty, \infty)$
77. $(-\infty, \infty)$ **79. a.** 64, 44   **b.** After 1 sec, the height of the ball is 64 ft. After 1.5 sec, the height of the ball is 44 ft.   **81. a.** 11.5, 17.25
b. After 1 hr, the distance is 11.5 mi. After 1.5 hr, the distance is 17.25 mi.   **83. a.** $P(0) = 0\%$
$P(5) = \left(\frac{100}{3}\right) \approx 33.3\%$ $P(10) = \left(\frac{200}{3}\right) \approx 66.7\%$
$P(15) = \left(\frac{900}{11}\right) \approx 81.8\%$ $P(20) = \left(\frac{800}{9}\right) \approx 88.9\%$
$P(25) = \left(\frac{2500}{27}\right) \approx 92.6\%$. If Brian studies 25 hours he will get a score of 92.6%.

b. $A(0, 0)$ $B\left(5, \dfrac{100}{3}\right)$ $C\left(10, \dfrac{200}{3}\right)$ $D\left(15, \dfrac{900}{11}\right)$ $E\left(20, \dfrac{800}{9}\right)$ $F\left(25, \dfrac{2500}{27}\right)$ **85.** $(4, \infty)$ **87.** $\{5, -3, 4\}$ **89.** $\{0, 2, 6, 1\}$
91. -7 **93.** 0 and 2 **95.** 6
97. 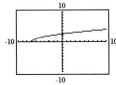 **99. a.**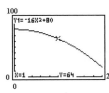
b. $h(1) = 64$

Section 4.3 Practice Exercises, pp. 278–283

3. a. Yes **b.** $\{6, 5, 4, 3\}$ **c.** $\{1, 2, 3, 4\}$
5. a. $f(0) = 2, f(-3) = 1, f(-4) = 0$, and $f(-5)$ cannot be evaluated because $x = -5$ is not in the domain of f. **b.** $[-4, \infty)$
7. $f(3) = 9$ means that 9 lb of force is required to stretch the spring 3 in. $f(10) = 30$ means that 30 lb of force is required to stretch the spring 10 in. **9.** Horizontal
11. Domain $(-\infty, \infty)$; range $\{2\}$ **13.**

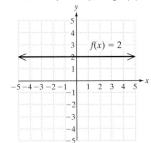 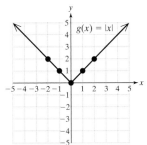

15. 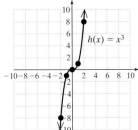 **17.**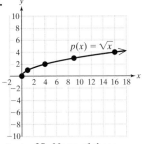

19. Quadratic **21.** Linear **23.** Constant **25.** None of these
27. Linear **29.** None of these

31. x-intercept: $(2, 0)$;
y-intercept: $(0, -10)$

33. x-intercept: $\left(\dfrac{5}{6}, 0\right)$;
y-intercept: $(0, 5)$

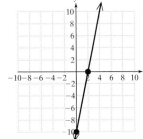 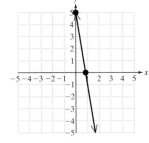

35. x-intercept: none;
y-intercept: $(0, 18)$

37. x-intercept: $\left(-\dfrac{3}{8}, 0\right)$;
y-intercept: $\left(0, \dfrac{1}{4}\right)$

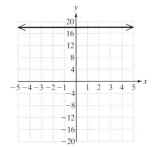 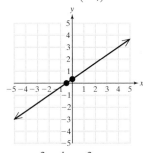

39. a. $x = -1$ **b.** $f(0) = 1$ **41. a.** $x = -2$ and $x = 2$
b. $f(0) = -2$ **43. a.** None **b.** $f(0) = 2$ **45. a.** $(-\infty, \infty)$
b. $(0, 0)$ **c.** vi **47. a.** $(-\infty, \infty)$ **b.** $(0, 1)$ **c.** viii
49. a. $[-1, \infty)$ **b.** $(0, 1)$ **c.** vii **51. a.** $(-\infty, 3) \cup (3, \infty)$
b. $\left(0, -\dfrac{1}{3}\right)$ **c.** ii **53. a.** $(-\infty, \infty)$ **b.** $(0, 2)$ **c.** iv

55. 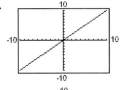 **57.**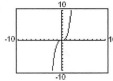

59.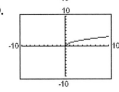

Section 4.4 Practice Exercises, pp. 288–292

3. 0 **5.** $x = -2$ **7.** $[-4, \infty)$ **9. a.** Increase **b.** Decrease
11. $T = kq$ **13.** $W = \dfrac{k}{p^2}$ **15.** $Q = \dfrac{kx}{y^3}$ **17.** $L = kw\sqrt{v}$
19. $k = \dfrac{9}{2}$ **21.** $k = 512$ **23.** $k = 1.75$ **25.** $Z = 56$ **27.** $L = 9$
29. $B = \dfrac{15}{2}$ **31.** 355,000 tons **33.** 42.6 ft **35.** 18.5 A **37.** 1.25 Ω
39. 20 lb **41.** 2224 lb **43. a.** $A = kl^2$ **b.** The area is increased by 4 times. **c.** The area is increased by 9 times.

Chapter 4 Review Exercises, pp. 296–299

1. For example: {(Peggy, Kent), (Charlie, Laura), (Tom, Matt), (Tom, Chris)} **3.** Domain $[-3, 9]$; range $[0, 60]$ **5.** Domain $\{-3, -1, 0, 2, 3\}$; range $\left\{-2, 0, 1, \dfrac{5}{2}\right\}$

7.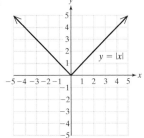

9. a. Function **b.** $(-\infty, \infty)$ **c.** $(-\infty, 0.35)$ **11. a.** Not a function
b. $\{0, 4\}$ **c.** $\{2, 3, 4, 5\}$ **13. a.** Function **b.** $\{6, 7, 8, 9\}$

c. {9, 10, 11, 12}   **15.** 2   **17.** $6t^2 - 4$ **19.** $6\pi^2 - 4$ **21.** 20
23. $(-\infty, 11) \cup (11, \infty)$ **25.** $[-2, \infty)$
27. 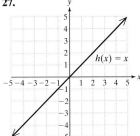   **29.**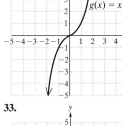

31. 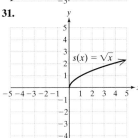   **33.**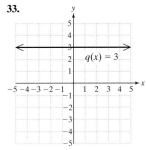

35. x-intercept: $\left(\dfrac{7}{4}, 0\right)$; y-intercept: $(0, -7)$ **37. a.** $b(0) = 4.5$. In 1985 consumption was 4.5 gal of bottled water per capita. $b(7) = 9.4$. In 1992 consumption was 9.4 gal of bottled water per capita.
b. $m = 0.7$. Consumption increased by 0.7 gal/year.   **39.** 0
41. There are no values of x for which $g(x) = -4$.   **43.** $(-\infty, 1]$
45. a. $r(4) = 0, r(5) = 2, r(8) = 4$ **b.** $[4, \infty)$
47. a. $k(-5) = -2, k(-4) = -1, k(-3) = 0, k(-2) = -1,$
$k(-1) = -2$ **b.** $(-\infty, \infty)$ **49.** $y = 256$ **51.** 48 km

Chapter 4 Test, pp. 299–300

1. a. Not a function   **b.** {−3, −1, 1, 3}   **c.** {−2, −1, 1, 3}
2. a. Function   **b.** $(-\infty, \infty)$ **c.** $(-\infty, 0]$ **3.** To find the x-intercept(s), solve for the real solutions of the equation $f(x) = 0$. To find the y-intercept, find $f(0)$.

4.   **5.**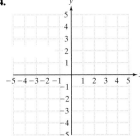

6. 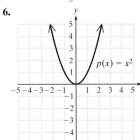   **7.**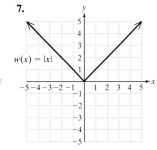

8. $(-\infty, -7) \cup (-7, \infty)$ **9.** $[-7, \infty)$ **10.** $(-\infty, \infty)$
11. a. $r(-2) = 9, r(-1) = 4, r(0) = 1, r(2) = 1, r(3) = 4$ **b.** $(-\infty, \infty)$
12. a. $s(0) = 36$. In 1985 the per capita consumption was 36 gal. $s(7) = 47.2$. In 1992 the per capita consumption was 47.2 gal.
b. $m = 1.6$. Consumption increases by 1.6 gal/year.   **13.** 1   **14.** 2

15. $[-1, 7]$ **16.** $[-1, 4]$ **17.** False   **18.** $(6, 0)$ **19.** $x = 6$
20. All x in the interval $[1, 3]$ and $x = 5$ **21.** Quadratic
22. Linear   **23.** Constant   **24.** None of these   **25.** x-intercept:
$(-12, 0)$; y-intercept:$(0, 9)$ **26.** $x = \dfrac{ky}{t^2}$ **27.** 3.3 sec

Chapters 1–4 Cumulative Review Exercises, pp. 301–302

1. $t = -\dfrac{12}{7}$ **2.** 14   **3. a.** $[6, \infty)$ **b.** $(-\infty, 17)$ **c.** $[-2, 3]$
4. $\left(-\infty, \dfrac{1}{6}\right]$ **5.** 1474 in.³   **6.** $y = -\dfrac{4}{3}x$ **7.** $m = \dfrac{2}{5}$ **8. a.** To find the x-intercepts, let $f(x) = 0$ and solve for x. To find the y-intercept, find $f(0)$.   **b.** $(0, 2)$ **c.** $\left(-\dfrac{2}{3}, 0\right)$ **9.** $(2, 1, 0)$ is not a solution because it is not a solution to all three equations.
10. $(5, 0)$ **11.** $(8, -7)$ **12.** $(4, 0)$ **13.** The numbers are 24 and 36.
14. Domain {3, 4}; range {−1, −5, −8}; not a function   **15. a.** 8620 students   **b.** The year 2011   **16.** Domain {12, 15, 18, 21}; range {a, c, e}; yes   **17. a.** 1   **b.** 33   **18.** $(-\infty, 15) \cup (15, \infty)$ **19.** $[6, \infty)$
20. $3500

Chapter 5

Chapter 5 Preview, p. 304

1. a. $4x^6 - 7x^5 - 2x^3 - 2$ **b.** 6   **c.** 4   **3.** $-7t^2 + 6t - 2$
5. $12x^4y^5 + 6x^3y^6 - 30x^5y^4$ **7.** $12z^3 - 14z^2 - 11z + 10$
9. $16d^2 - c^2$ **11.** $3x - 4 + \dfrac{10}{2x + 1}$ **13.** $2xy(7x^2y^3 - 10xy^2 + 9)$
15. $(a + 7)(a - 6)$ **17.** $2x(2x + 5)(x + 1)$ **19.** $(5y + 7)(5y - 7)$
21. $x = 9$ or $x = -2$ **23.** $c = 1$ or $c = -\dfrac{4}{3}$

Section 5.1 Practice Exercises, pp. 310–314

3. $-6a^3 + a^2 - a$; leading coefficient -6; degree 3
5. $3x^4 + 6x^2 - x - 1$; leading coefficient 3; degree 4
7. $-t^2 + 100$; leading coefficient -1; degree 2   **9.** For example: $3x^5$ **11.** For example: $x^2 + 2x + 1$ **13.** For example: $6x^4 - x^2$
15. $m^2 + 10m$ **17.** $3x^4 + 2x^3 - 8x^2 + 2x$ **19.** $2w^3 + \dfrac{1}{9}w^2 + 0.9w$
21. $17x^2 - 4x - 14$ **23.** $30y^3$ **25.** $-4p^3 - 2p + 12$
27. $11ab^2 - a^2b$ **29.** $6z^5 - 6z^2$ **31.** $-2x^3 + 4x^2 + 5$
33. $\dfrac{1}{2}a^2 - \dfrac{9}{10}ab + \dfrac{3}{5}b^2 + 8$ **35.** $-x^2 + 6x - 16$
37. $3x^5 - x^4 - 4x^3 + 11$ **39.** $4y^3 + 5y^2$ **41.** $-7r^4 - 11r$
43. $9x^2 + 5xy + 11y$ **45.** $18ab - 42b^2$ **47.** $3p - 9$ **49.** $2m^2 + 6$
51. Yes; degree 0   **53.** Yes; degree 2   **55.** No; the term $-\dfrac{3}{x} = 3x^{-1}$ and -1 is not a whole number.   **57.** No; the term $|x|$ is not of the form ax^n.   **59. a.** -17 **b.** -8 **c.** -5 **d.** -4 **61. a.** $\dfrac{1}{4}$ **b.** $\dfrac{9}{4}$
c. $-\dfrac{7}{4}$ **d.** $\dfrac{3}{4}$ **63.** $P(x) = 4x + 6$ **65. a.** $P(x) = 3.78x - 1$
b. $188   **67. a.** $D(0) = 2766$; in 1995 the yearly dormitory charges were $2766.   **b.** $D(2) = 2970$; in 1997 the yearly dormitory charges were $2970.   **c.** $D(4) = 3238$; in 1999 the yearly dormitory charges were $3238.   **d.** $D(6) = 3570$; in 2001 the yearly dormitory charges were $3570.   **69. a.** $W(0) = 5865; W(5) = 6580; W(10) = 7295$
b. $W(10) = 7295$ means that in the year 2005, 7295 thousand women (7,295,000) were due child support.   **71. a.** $(0, 0)$; at $t = 0$ sec, the position of the rocket is at the origin.   **b.** $(25, 27.3)$; after 1 sec the position of the rocket is $(25, 27.3)$.   **c.** $(50, 22.6)$

Section 5.2 Practice Exercises, pp. 321–324

3. a. 103 **b.** -5 **c.** -37 **5.** $2x^2 - 4x - 7$ **7.** They cannot be combined by addition because they are not *like* terms. They can be multiplied: $(2x^3)(3x^2) = 6x^5$ **9.** $3a^2b + 3ab^2$ **11.** $\frac{2}{5}a - \frac{3}{5}$
13. $2m^5n^5 - 6m^4n^4 + 8m^3n^3$ **15.** $x^2 - xy - 2y^2$
17. $12x^2 + 28x - 5$ **19.** $8a^2 - 22a + 9$ **21.** $2y^4 - 21y^2 - 36$
23. $25s^2 + 5st - 6t^2$ **25.** $5n^3 + 3n^2 + 50n + 30$
27. $3.25a^2 - 0.9ab - 28b^2$ **29.** $6x^3 + 7x^2y + 4xy^2 + y^3$
31. $x^3 - 343$ **33.** $4a^4 - 17a^3b + 8a^2b^2 - 5ab^3 + b^4$
35. $\frac{1}{2}a^2 + ab + \frac{1}{2}ac - 12b^2 + 8bc - c^2$ **37.** $-3x^3 + 11x^2 - 7x - 5$
39. $a^2 - 64$ **41.** $9p^2 - 1$ **43.** $x^2 - \frac{1}{9}$ **45.** $9h^2 - k^2$
47. $9h^2 - 6hk + k^2$ **49.** $t^2 - 14t + 49$ **51.** $u^2 + 6uv + 9v^2$
53. $h^2 + \frac{1}{3}hk + \frac{1}{36}k^2$ **55. a.** $A^2 - B^2$ **b.** $x^2 + 2xy + y^2 - B^2$
Both are examples of multiplying conjugates to get a difference of squares. **57.** $w^2 + 2wv + v^2 - 4$ **59.** $4 - x^2 - 2xy - y^2$
61. $9a^2 - 24a + 16 - b^2$ **63.** Write $(x + y)^3$ as $(x + y)^2(x + y)$. Square the binomial and then use the distributive property to multiply the resulting trinomial by the remaining factor of $x + y$.
65. $8x^3 + 12x^2y + 6xy^2 + y^3$ **67.** $64a^3 - 48a^2b + 12ab^2 - b^3$
69. Multiply and simplify the first two binomials. Then multiply the resulting trinomial to the third binomial, using the distributive property. **71.** $6a^4 + 32a^3 + 10a^2$ **73.** $x^3 + 5x^2 - 9x - 45$
75. $a^2 + b^3$ **77.** $x^2 - y^3$ **79.** The sum of p cubed and q squared
81. The product of x and the square of y
83. $f(x) = 4x^2 + 70x + 300$ **85.** $x^2 - 4x + 4$ **87.** $x^2 - 4$
89. $x^2 - 9$ **91.** $9x^3 + 30x^2$ **93.** $x^3 + 12x^2 + 48x + 64$
95. Multiply $(x + 2)^2(x + 2)^2$ by squaring the binomials. Then multiply the resulting trinomials, using the distributive property.
97. $(5x - 6)$ **99.** $(2y - 1)$

Section 5.3 Practice Exercises, pp. 332–335

3. a. $5x - 4$ **b.** $6x^2 - 13x - 5$ **5. a.** $y^2 + 5y$
b. $2y^4 - 10y^3 + 3y^2 - 5y + 1$
7. For example: $3x(x + 1) = 3x^2 + 3x$
9. For example: $(5w^5 + 1) + (3w^5 + 2w + 3) = 8w^5 + 2w + 4$
11. For example: $(6p - 1)(6p + 1) = 36p^2 - 1$ **13.** $p - 3p^3 + 5p^4$
15. $5m^2 - 2m + \frac{1}{5}$ **17.** $6y^2 - 4y^3 + 8y - 5$ **19.** $\frac{1}{3}q^2 - \frac{2}{3}q + \frac{1}{12}$
21. $2m^3 - 3m^2 + m - 1 + \frac{9}{m}$ **23.** $2rw + 1 - \frac{w^2}{2r^3}$
25. $20a^3b^4 - 20ab + 5 + \frac{6}{a^2b}$
27. a. Divisor $(x + 3)$; quotient $(x^2 + x + 4)$; remainder (-15)
b. Multiply the quotient and the divisor, then add the remainder. The result should equal the dividend. **29.** $x^2 - 5x + 3 + \frac{9}{x - 2}$
31. $z^2 + 2z + 10 + \frac{35}{z - 4}$ **33.** $7x - 2$ **35.** $3y + 1$
37. $27x^3 - 9x^2 + 3x - 1$ **39.** $a^3 - a^2 + 2a - 5$
41. $3y^2 - 16 - \frac{66}{y^2 - 3}$ **43.** $m^2 - 3m + 9$ **45.** No, the divisor must be of the form $x - r$. **47.** Yes, the divisor is of the form $x - r$. **49. a.** $x + 2$ **b.** $2x^3 - x^2 + 2x - 5$ **c.** 16 **51.** $x + 2$
53. $h + 4$ **55.** $3w - 5 + \frac{5}{w + 2}$ **57.** $z^2 - 5z + 17 + \frac{-56}{z + 3}$
59. $3y^3 + 9y^2 + 2y + 6$ **61.** $-12y^3 + 4y^2 - 4y + 4$
63. $\frac{8}{x} - \frac{9}{y} + 6$ **65.** $2m^2 + 2m + 11$ **67.** $10h^3 - 7 + \frac{5}{h} - \frac{4}{h^2}$
69. $y^3 - 5y^2 + 5y - 12 + \frac{29}{y + 2}$ **71.** $2m - 5 + \frac{-4m + 5}{2m^2 - 4m + 3}$
73. $5k^2 + k + 3 + \frac{k + 7}{3k^2 - 1}$

Midchapter review, p. 335

1. Leading coefficient: -5; degree: 6 **2.** 26 **3.** $2t^2 + t - 1$
4. $-15x^4 - 5x^3 + 10x^2$ **5.** $9x^2 + 6x + 1$ **6.** $3a^2 - a + 2$
7. $36z^2 - 25$ **8.** $9y^3 + 2y^2 - 3y + 1$ **9.** $2x + 4 + \frac{5}{2x - 1}$
10. $10a^3 + 19a^2 + 11a + 2$ **11.** $t^3 - 6t^2 + 8t + 3$ **12.** $2b^2 + 4b + 5$
13. $-p^2 - 28$ **14.** $k^2 + 4k + 25$

Section 5.4 Practice Exercises, pp. 341–344

3. $9t^4 + 5t^3 - 6t^2 - 6t$ **5.** $5y^4 + 5y^3 + 7y^2 - 3y - 6$
7. $-3v^2 + 6v - 1$ **9.** A common factor is an expression that divides evenly into each term of a polynomial. The greatest common factor is the *greatest* factor that divides evenly into each term. **11.** $3(x + 4)$ **13.** $2z(3z + 2)$ **15.** $4p(p^5 - 1)$
17. $12x^2(x^2 - 3)$ **19.** $9t(st + 3)$ **21.** $9(a^2 + 3a + 2)$
23. $5xy(2x + 3y - 7)$ **25.** $b(13b - 11a^2 - 12a)$
27. $-1(x^2 + 10x - 7)$ **29.** $-3xy(-4x^2 + 2x + 1)$
31. $-t(2t^2 - 11t + 3)$ **33.** $(3z - 2b)(2a - 5)$
35. $(2x - 3)(2x^2 + 1)$ **37.** $(2x + 1)^2(y - 3)$ **39.** $3(x - 2)^2(y + 2)$
41. $A = \dfrac{U}{v + cw}$ **43.** $y = \dfrac{bx}{c - a}$ or $y = \dfrac{-bx}{a - c}$
45. For example: $3x^3 + 6x^2 + 12x^4$ **47.** For example: $6(c + d) + y(c + d)$ **49. a.** $(2x - y)(a + 3b)$
b. $(2w - 1)(5w - 3b)$ **c.** In part (b), $-3b$ was factored out so that the signs in the last two terms were changed. The resulting binomial factor matches the binomial factor from the first two terms.
51. $(y + 4)(y^2 + 3)$ **53.** $(p - 7)(6 + q)$ **55.** $(m + n)(2x + 3y)$
57. $(2x - 3y)(5a - 4b)$ **59.** $(x^2 - 3)(x - 1)$ **61.** $6p(p + 3)(q - 5)$
63. $100(x - 3)(x^2 + 2)$ **65.** $(3a + b)(2x - y)$ **67.** $(4 - b)(a + 3)$
69. Cannot be factored **71.** It is not possible to get a common binomial factor regardless of the order of terms.
73. Length $= 2w + 1$ **75.** $(a + 3)^4(6a + 19)$
77. $18(3x + 5)^2(4x + 5)$ **79.** $(t + 4)(t + 3)$
81. $5w^2(2w - 1)^2(7w - 3)$

Section 5.5 Practice Exercises, pp. 356–359

3. $6c^2d^4e^7(6d^3e^4 + 2cde^8 - 1)$ **5.** $(3a - b)(2x - 1)$
7. $(z + 2)(wz - 33a)$ **9.** $(b - 8)(b - 4)$ **11.** $(y + 12)(y - 2)$
13. $(x + 10)(x + 3)$ **15.** $(c - 8)(c + 2)$ **17.** $(2x + 3)(x - 5)$
19. $(6a - 5)(a + 1)$ **21.** $(s + 3t)(s - 2t)$ **23.** $3(x - 18)(x - 2)$
25. $2(c - 4)(c + 3)$ **27.** $2(x - y)(x + 5y)$ **29.** Prime
31. $(3x + 5y)(x + 3y)$ **33.** $5uv(u - 3v)^2$ **35.** $x(x - 7)(x + 2)$
37. $(2z - 5)(5z + 1)$ **39.** Prime **41.** $-2(t + 4)(t - 10)$
43. $(7a - 4)(2a + 3)$ **45.** $2b(3a + 2)(a + 3)$ **47. a.** $x^2 + 10x + 25$
b. $(x + 5)^2$ **49. a.** $9x^2 - 12xy + 4y^2$ **b.** $(3x - 2y)^2$ **51.** $30x$
53. 36 **55.** 64 **57.** $(y - 4)^2$ **59.** Not a perfect square trinomial
61. $(3a - 5b)^2$ **63.** $4(4t^2 - 20tv + 5v^2)$ Not a perfect square trinomial **65. a.** $(u - 5)^2$ **b.** $(x^2 - 5)^2$ **c.** $(a - 4)^2$
67. a. $(u + 13)(u - 2)$ **b.** $(w^3 + 13)(w^3 - 2)$ **c.** $(y + 9)(y - 6)$
69. $(3y^3 + 2)(y^3 + 3)$ **71.** $(4p^2 + 1)(p^2 + 1)$ **73.** $(x^2 + 12)(x^2 + 3)$
75. $(3x - 4)(3x + 1)$ **77.** $(2x - 9)(x - 1)$ **79.** $(x - 3)(3x^2 + 5)$
81. $(a + 6)^2$ **83.** $(9w + 5)^2$ **85.** $3(a + b)(x - 2)$
87. $2abc^2(6a + 2b - 3c)$ **89.** $f(x) = (2x - 1)(x + 7)$
91. $m(t) = (t - 11)^2$ **93.** $P(x) = x(x + 1)(x + 3)$
95. $h(a) = (a + 5)(a^2 - 6)$ **97.** The factorization $(2y - 1)(2y - 4)$ is not factored completely because the factor $2y - 4$ has a GCF of 2.

Section 5.6 Practice Exercises, pp. 366–368

3. $(2x - 5)^2$ **5.** $(5 + 3y)(2x + 1)$ **7.** $4(8p + 1)(p - 1)$
9. $9a(5a - c)$ **11.** Look for a binomial of the form $a^2 - b^2$ which factors as $(a + b)(a - b)$. **13.** $(x - 3)(x + 3)$ **15.** $(4 - w)(4 + w)$
17. $2(2a - 9b)(2a + 9b)$ **19.** Prime **21.** $(2m - 11)(2m + 1)$
23. $(x - 4)(x + 4)(x - 1)$ **25.** $(2x + 1)(2x - 1)(x + 3)$
27. $(x + 6 - a)(x + 6 + a)$ **29.** $(p + q - 9)(p + q + 9)$

31. $(p - y + 3)(p + y - 3)$ **33.** Look for a binomial of the form $a^3 + b^3$ which factors as $(a + b)(a^2 - ab + b^2)$.
35. $(2x - 1)(4x^2 + 2x + 1)$ **37.** $(5c + 3)(25c^2 - 15c + 9)$
39. $(x - 10)(x^2 + 10x + 100)$ **41.** $(4t + 1)(16t^2 - 4t + 1)$
43. $\left(6y - \dfrac{1}{5}\right)\left(6y + \dfrac{1}{5}\right)$ **45.** $2(3d^6 - 4)(3d^6 + 4)$
47. $2(121v^2 + 16)$ **49.** $4(x - 2)(x + 2)$ **51.** $(5 - 7q)(5 + 7q)$
53. $(t + 2s - 6)(t + 2s + 6)$ **55.** $(3 - t)(9 + 3t + t^2)$
57. $\left(3a + \dfrac{1}{2}\right)\left(9a^2 - \dfrac{3}{2}a + \dfrac{1}{4}\right)$ **59.** $2(m + 2)(m^2 - 2m + 4)$
61. $(x - y)(x + y)(x^2 + y^2)$
63. $(a + b)(a^2 - ab + b^2)(a^6 - a^3b^3 + b^6)$
65. $(p - 7)(p^2 - 2p + 13)$
67. $(a + b)(a^2 - ab + b^2)(a - b)(a^2 + ab + b^2)$
69. $(2 + y)(4 - 2y + y^2)(2 - y)(4 + 2y + y^2)$
71. $(h^2 + k^2)(h^4 - h^2k^2 + k^4)$ **73.** $(2x^2 + 5)(4x^4 - 10x^2 + 25)$
75. $4x^2 - 9$ **77.** $8a^3 - 27$ **79.** $64x^6 + y^3$ **81. a.** $x^2 - y^2$
b. $(x + y)(x - y)$ **c.** 20 in.² **83.** $(x + y)(x - y + 1)$
85. $(x + y)(x^2 - xy + y^2)(5w - 2z)$

Section 5.6 A Factoring Strategy, p. 369

1. An expression whose only factors are 1 and itself
3. Difference of squares $a^2 - b^2$, difference of cubes $a^3 - b^3$, or sum of cubes $a^3 + b^3$ **5.** Look for a trinomial of the form $a^2 + 2ab + b^2$ or $a^2 - 2ab + b^2$. **7. a.** Trinomial
b. $3(2x + 3)(x - 5)$ **9. a.** Difference of squares
b. $2(2a - 5)(2a + 5)$ **11. a.** Trinomial **b.** $(2u - v)(7u - 2v)$
13. a. Difference of cubes **b.** $2(2x - 1)(4x^2 + 2x + 1)$
15. a. Sum of cubes **b.** $(3y + 5)(9y^2 - 15y + 25)$ **17. a.** Sum of cubes **b.** $2(4p^2 + 3q)(16p^4 - 12p^2q + 9q^2)$ **19. a.** Difference of squares **b.** $(2a - 1)(2a + 1)(4a^2 + 1)$ **21. a.** Grouping
b. $(p - 6 - c)(p - 6 + c)$ **23. a.** Grouping **b.** $2(2x - y)(3a + b)$
25. a. Trinomial **b.** $(5y - 1)(y + 3)$

Section 5.7 Practice Exercises, pp. 378–383

3. $x(10x + 3)$ **5.** $(p - 5)(2p + 1)$ **7.** $(t - 1)(t^2 + t + 1)$
9. The equation must be set equal to 0, and the expression must be factored. **11.** Correct form **13.** Incorrect form. Expression is not factored. **15.** Incorrect form. Equation is not set equal to 0.
17. $x = -3, x = -5$ **19.** $w = -\dfrac{9}{2}, w = \dfrac{1}{5}$
21. $x = 0, x = -4, x = \dfrac{3}{10}$ **23.** $y = 0.4, y = -2.1$
25. $x = -9, x = 3$ **27.** $x = -3, x = \dfrac{1}{2}$ **29.** $x = 0, x = \dfrac{3}{2}$
31. $y = \dfrac{23}{3}$ **33.** $y = -3$ **35.** $p = -\dfrac{1}{3}, p = 2$
37. $x = -1, x = \dfrac{1}{2}, x = 3$ **39.** $y = -5, y = 4$
41. $a = \dfrac{5}{2}, a = -1$ **43.** $p = -12, p = 5$ **45.** $t = -\dfrac{1}{11}$
47. $x = 0, x = 6, x = -2$ **49.** $w = 0, w = 4, w = -4$ **51.** 5, −5
53. 4, −3 **55.** −7, −6 or 6, 7 **57.** −9, −7 or 7, 9 **59.** The length is 7 ft, and the width is 5 ft. **61.** The length is 20 yd, and the width is 15 yd. **63. a.** Base 5 in.; height 6 in. **b.** 15 in.² **65.** The base is 10 ft, and the height is 5 ft. **67.** 4, 5 **69. a.** 14 mi **b.** The alternative route using superhighways **71.** The lengths are 6, 8, and 10 m. **73. a.** $x = 0, x = 3$ **b.** $f(0) = 0$ **75. a.** $x = 7$
b. $f(0) = -35$ **77.** x-intercepts: $(2, 0), (-1, 0), (0, 0)$; y-intercept: $(0, 0)$ **79.** x-intercept: $(1, 0)$; y-intercept: $(0, 1)$ **81.** $(3, 0)(-3, 0)$; d
83. $(-1, 0)$; a **85. a.** The function is in the form $s(t) = at^2 + bt + c$.
b. $(0, 0)$ and $(100, 0)$ **c.** At 0 sec and 100 sec, the rocket is at ground level (height = 0). **d.** At 1 sec and 99 sec
87. $f(x) = (x - 5)(x - 2)$; $x = 5$ and $x = 2$ represent x-intercepts.
89. $f(x) = (x + 1)^2$; $x = -1$ represents the x-intercept.

91. $(x + 3)(x - 1) = 0$ or $x^2 + 2x - 3 = 0$ **93.** $(x - 0)(x + 5) = 0$ or $x^2 + 5x = 0$
95. $(2, 0)(-1, 0)$

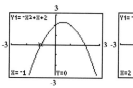
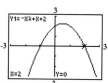

97. $(3, 0)$

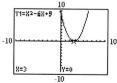

Chapter 5 Review Exercises, pp. 389–392

1. Trinomial; degree 4 **3. a.** −7 **b.** −23 **c.** 5
5. a. $S(5) = 276$; $S(13) = 1257$ **b.** $S(13) = 1257$ means that in 2003 there were approximately 1257 new Starbucks sites established.
7. $20xy - 18xz - 3yz$ **9.** $-2a^2 + 6a$ **11.** $x^4 + \dfrac{3}{4}x^2$ **13.** $-5x + 9y$
15. $4x^2 - 11x$ **17.** $-3x$ **19.** $-18x^3 + 15x^2 - 12x$
21. $x^2 - 11x + 18$ **23.** $2y^2 + \dfrac{1}{5}y - \dfrac{1}{25}$ **25.** $x^3 - y^3$
27. $\dfrac{1}{4}x^2 + 4x + 16$ **29.** $36w^2 - 1$ **31.** $z^2 - \dfrac{1}{16}$
33. $c^2 - w^2 - 6w - 9$ **35.** $y^6 - 9y^4 + 27y^2 - 27$
37. a. $P(x) = 8x + 4$ **b.** $A(x) = 3x^2 + 2x$ **39.** $2x^2 + 4x - 3$
41. a. **b.** Quotient $3y^3 - 2y^2 + 6y - 4$; no remainder **c.** Multiply the quotient and the divisor.
43. $x + 4 + \dfrac{-32}{x + 4}$ **45.** $2x^3 - 2x^2 + 5x - 4 + \dfrac{4x - 4}{x^2 + x}$
47. a. $x - 3$ **b.** $2x^3 + 11x^2 + 31x + 99$ **c.** 298
49. $x + 2 + \dfrac{4}{x + 5}$ **51.** $w^2 - 3w - 9 + \dfrac{-19}{w - 3}$
53. $x(-x^2 - 4x + 11)$ or $-x(x^2 + 4x - 11)$
55. $(x - 7)(5x - 2)$ **57.** $2x(x - 13)$ **59.** $12(2x - 3)(x^2 + 3)$
61. $(y - 6)(y^2 + 1)$ **63.** $(3x + 2y)(6x + 5y)$
65. $5a^2(3 + 4a)(4 - a)$ or $-5a^2(a - 4)(4a + 3)$
67. $(n + 5)^2$ **69.** $y(y + 5)(y - 2)$ **71.** $(3x - 2)^2$
73. $(5 - y)(5 + y)$ **75.** Prime **77.** $h(h^2 + 9)$
79. $y(3y - 2)(3y + 2)$ **81.** $(a + 6 - b)(a + 6 + b)$
83. It is a parabola. **85.** Quadratic **87.** Quadratic
89. $x = \dfrac{3}{8}, x = 7$ **91.** $x = 1, x = -5, x = \dfrac{9}{2}$
93. $(1, 0)(-1, 0)(0, -2)$; d **95.** $(2, 0)(-2, 0)\left(0, \dfrac{1}{2}\right)$; a
97. a.

Time t (sec)	Height $h(t)$ (ft)
0	−1280
1	−624
3	592
10	3840
20	5760
30	4480
40	0
42	−1280

b. The position of the missile is below sea level. **c.** The missile will be at sea level after 2 sec and again after 40 sec.

Chapter 5 Test, p. 393–394

1. $F(-1) = 1$ $F(2) = 40$ $F(0) = 8$
2. a. $C(2) = 3.8164$, $C(6) = 3.3524$, $C(12) = 1.7864$
b. $C(12) = 1.7864$ means that in 2002 there were 1.7864 million (1,786,400) violent crimes in the United States.
3. $8x^2 - 8x + 8$ **4.** $2a^3 - 13a^2 + 2a + 45$ **5.** $2x^2 - \frac{23}{3}x - 6$
6. $25x^2 - 16y^4$ **7.** The expression $25x^2 + 49$ does not account for the middle term $70x$.   **8.** $49x^2 - 56x + 16$
9. $x^2y^3 + \frac{5}{2}xy - 3y^2 - \frac{1}{2}$
10. $5p^2 - p + 1$ **11.** $y^3 + 2y^2 + 4y + 6 + \frac{17}{y-2}$ **12.** 1. Take out the GCF.   2. If there are more than three terms, try grouping.   3. If a trinomial, look for a perfect square trinomial. Otherwise, use the grouping method or trial-and-error method.   4. If a binomial, look for a difference of squares, a difference of cubes, or a sum of cubes.
13. 1. Set the equation equal to 0.   2. Factor the expression.
3. Set each factor equal to 0 and solve.
14. $x = 0, x = \frac{1}{4}, x = -\frac{1}{4}$ **15.** $(3y - 1)(y + 8)$
16. $(a - 6)(a + 1)(a - 1)$ **17.** $x = -\frac{1}{4}$ **18.** $3(x + 2)(x^2 - 2x + 4)$
19. $16(x^2 + 1)(x - 1)(x + 1)$ **20.** $(7x - 5y)^2$ **21.** $(4,0)(2,0)(0,8)$; c
22. $(-4,0)(3,0)(-3,0)(0,-36)$; b   **23.** $(-3,0)(-1,0)(0,-6)$; d
24. $(4,0)(-3,0)(0,0)$; a   **25.** 256 ft   **26. a.** 126.968 million
b. 124.52 million

Chapters 1–5 Cumulative Review Exercises, pp. 394–396

1. 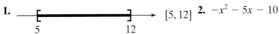 [5, 12]   **2.** $-x^2 - 5x - 10$

3. a. 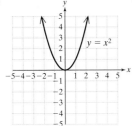   **b.**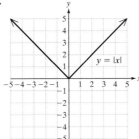

4. $\frac{71}{8}$ **5.** 1.97×10^8 people
6. Notre Dame scored 26 points; Florida State scored 31 points.
7. 37°, 74°, 69°   **8.** $x^2 - 4x + 16$
9. $m = \frac{4}{3}$; y-intercept (0, 3)

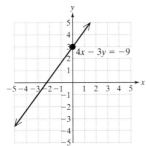

10. $\frac{1}{4}$ **11.** $\frac{a^6}{8b^{30}}$ **12.** (0, 3, 2)   **13. a.** Function   **b.** Not a function
14. $m = 7$ **15.** 7   **16.** $x = 8$ **17.** $y = \frac{3}{2}x - \frac{5}{2}$

18. a. The fourth test would have to be 107; therefore, it is not possible.   **b.** Student can get between 67 and 100, inclusive.
19. 15 L of 40% solution; 10 L of 15% solution
20. $8b^3 - 6b^2 + 4b - 3$ **21.** $3a^3 + 5a^2 - 2a + 5$
22. $3w - \frac{5}{2} - \frac{1}{w}$ **23.** $y = 7$ or $y = -2$
24. $x = -\frac{6}{5}$; or $x = \frac{6}{5}$ **25.** $a = 0, a = -4, a = -5$

Chapter 6

Chapter 6 Preview, p. 398

1. -3 **3.** $\{a \mid a \text{ is a real number and } a \neq -3, a \neq 2\}$
5. $\frac{9a^2c}{b^2}$ **7.** $\frac{-2(2x+5)}{x+1}$ **9.** $\frac{t+2}{5t}$ **11.** $\frac{7z-16}{z(z-3)(z-4)}$
13. $\frac{d+c}{2d-3c}$ **15.** $t = \frac{30}{13}$ **17.** $a = 3; a = -2$ **19.** $x = 60$
21. Kathy runs 6 mph and Dennis rides 14 mph.

Section 6.1 Practice Exercises, pp. 407–410

3. $5; \frac{5}{3}$; undefined; -1 **5.** 2; 1; 0; -1 **7.** $\{y \mid y \text{ is a real number and } y \neq 0\}$ **9.** $\{v \mid v \text{ is a real number and } v \neq 8\}$ **11.** $\{x \mid x \text{ is a real number and } x \neq \frac{5}{2}\}$ **13.** $\{q \mid q \text{ is a real number and } q \neq -9, q \neq 3\}$ **15.** $\{c \mid c \text{ is a real number}\}$ **17. a.** $\frac{(x+4)(x+2)}{(x+4)(x-1)}$
b. $\{x \mid x \text{ is a real number and } x \neq -4, x \neq 1\}$
c. $\frac{x+2}{x-1}$ provided $x \neq -4, x \neq 1$ **19. a.** $\frac{(x-9)(x-9)}{(x-9)(x+9)}$
b. $\{x \mid x \text{ is a real number and } x \neq 9, x \neq -9\}$
c. $\frac{x-9}{x+9}$ provided $x \neq 9, x \neq -9$ **21.** $\frac{25x^2}{9y^3}$ **23.** $\frac{w^8z^3}{2}$ **25.** $-\frac{1}{4m^2n^3}$
27. $\frac{2}{3}$ **29.** $\frac{1}{x+5}$ **31.** $-\frac{1}{3c-5}$ **33.** $\frac{t+4}{t+3}$ **35.** $\frac{(2p-1)^2}{p+1}$
37. $\frac{3-z}{2z-5}$ **39.** $-\frac{2(z^2-2z+4)}{z-5}$ **41.** $\frac{2x-5}{-2}$ **43.** 1   **45.** -1
47. -1 **49.** -2 **51.** $\frac{c+4}{c-4}$; cannot be simplified   **53.** $-\frac{1}{12(x+y)}$
55. $-\frac{2x}{5}$ **57.** $h(0) = 3$, $h(1)$ undefined, $h(-3) = \frac{3}{4}$, $h(-1) = \frac{3}{2}$,
$h\left(\frac{1}{2}\right) = 6$ **59.** b; $(-\infty, -4) \cup (-4, \infty)$ **61.** d; $(-\infty, 4) \cup (4, \infty)$
63. $\left(-\infty, \frac{1}{3}\right) \cup \left(\frac{1}{3}, \infty\right)$ **65.** $(-\infty, \infty)$ **67.** $(-\infty, 2) \cup (2, \infty)$
69. $\left(-\infty, -\frac{5}{6}\right) \cup \left(-\frac{5}{6}, 2\right) \cup (2, \infty)$
71. $\left(-\infty, -\frac{3}{2}\right) \cup \left(-\frac{3}{2}, 0\right) \cup \left(0, \frac{5}{3}\right) \cup \left(\frac{5}{3}, \infty\right)$

Section 6.2 Practice Exercises, pp. 413–415

3. For example: $\frac{1}{x-3}$ **5.** For example: $f(x) = \frac{1}{x+6}$
7. $\frac{1}{x-3}$ **9.** $\frac{bc^3}{2}$ **11.** $\frac{2}{z^3}$ **13.** $12r^3$ **15.** $\frac{x(2x-3)}{y^2(x+1)}$
17. $\frac{2(3w-7)}{5w+4}$ **19.** 15   **21.** $y^2(y-2)$ **23.** $\frac{r+3}{4r^2}$
25. $\frac{1}{(p+2)(6p-7)}$ **27.** $\frac{4}{(b+2)(b+3)}$ **29.** $\frac{2s+5t}{s+5t}$
31. $\frac{2a+b^2}{a^2+b^2}$ **33.** $\frac{-4x}{3(x+2)}$ **35.** $\frac{9}{4y^2}$ **37.** $\frac{24}{y^2}$ **39.** $\frac{2a}{5}$

41. $m - n$ **43.** $-\dfrac{x - 3y}{x + 3y}$ or $\dfrac{3y - x}{x + 3y}$ **45.** 1 **47.** $\dfrac{2k}{h^3}$ cm^2
49. $\dfrac{5x}{4}$ ft^2

Section 6.3 Practice Exercises, pp. 423–426

3. $-\dfrac{1}{x}$ **5.** $\dfrac{5a + 1}{2}$ **7.** $\dfrac{2}{x}$ **9.** $\dfrac{1}{x + 1}$ **11.** $\dfrac{2x + 5}{(2x + 9)(x - 6)}$
13. $\dfrac{3}{x - 5}$ or $\dfrac{-3}{5 - x}$ **15.** $\dfrac{2x}{x - 6}$ or $\dfrac{-2x}{6 - x}$ **17.** $40x$
19. $600x^2$ **21.** $30m^4n^7$ **23.** $(x - 4)(x + 2)(x - 6)$
25. $x^2(x - 1)(x + 7)^2$ **27.** $(x - 6)(x - 2)$ **29.** $15xy$
31. $2x^3 + 4x^2$ **33.** $y^2 - y$ **35.** $\dfrac{8p - 15}{6p^2}$ **37.** $\dfrac{x^2 + 7x + 6}{2x^2}$
39. $\dfrac{-t - s}{st}$ **41.** $\dfrac{1}{3}$ **43.** $\dfrac{3}{w - 2}$ **45.** $\dfrac{w^2 - 3}{w - 2}$ **47.** $\dfrac{6b^2 + 5b + 4}{(b - 4)(b + 1)}$
49. $\dfrac{t^2 + 30}{(t - 5)(t + 5)}$ **51.** $\dfrac{11x + 6}{(x - 6)(x + 6)(x + 3)}$ **53.** $\dfrac{-x^2 + 13x - 3}{x(x - 1)^2}$
55. $-\dfrac{2}{3}$ **57.** 1 **59.** $\dfrac{10 - x}{3(x - 5)(x + 5)}$ or $\dfrac{x - 10}{3(5 - x)(5 + x)}$
61. $\dfrac{5k^2 + 27k - 2}{(k + 7)(k - 1)}$ **63.** $\dfrac{-2y}{(x - y)(x + y)}$ or $\dfrac{2y}{(y - x)(y + x)}$
65. $\dfrac{5y - 18}{y(y - 6)}$ **67.** $\dfrac{3x^2 + 5x + 18}{3x^2}$ cm **69.** $\dfrac{4x^2 - 2x + 50}{(x - 3)(x + 5)}$ m

Midchapter Review, p. 426

1. Factor the numerator and denominator. Then simplify common factors whose ratio is 1 or -1. **2.** Factor the numerator and denominator. Multiply straight across and simplify. **3.** Factor the numerator and denominator. Then multiply the first expression by the reciprocal of the second expression and simplify. **4.** Factor the denominator to find the LCD. Rewrite all rational expressions to have the common denominator; then add or subtract the numerators and simplify.
5. $\dfrac{4y^2 - 8y + 9}{2y(2y - 3)}$ **6.** $\dfrac{x^2 + x - 13}{x - 4}$ **7.** $\dfrac{5x - 1}{4x - 3}$ **8.** $\dfrac{a - 5}{3a}$
9. $\dfrac{2y - 5}{(y - 1)(y + 1)}$ **10.** $\dfrac{4}{w + 4}$ **11.** $\dfrac{a + 4}{2}$ **12.** $\dfrac{(t - 3)(t + 2)}{t}$
13. $\dfrac{-x^2 + 4xy - y^2}{(x - y)(x + y)}$ or $\dfrac{x^2 - 4xy + y^2}{(y - x)(y + x)}$ **14.** $3(x - 4)$
15. $\dfrac{x - 3}{x + 3}$ **16.** $-(m + n)$ **17. a.** $\dfrac{4(x + 1)}{(x + 1)(x - 1)}$
b. $\{x \mid x$ is a real number and $x \neq 1, x \neq -1\}$ **c.** $\dfrac{4}{x - 1}$ provided $x \neq -1, x \neq 1$ **18. a.** $\dfrac{3(x + 2)}{(x + 2)(x - 5)}$ **b.** $\{x \mid x$ is a real number and $x \neq -2, x \neq 5\}$ **c.** $\dfrac{3}{x - 5}$ provided $x \neq -2, x \neq 5$

Section 6.4 Practice Exercises, pp. 433–435

3. $\dfrac{x^2 - xy + y^2}{5}$ **5.** $\dfrac{10 + 3x}{2x^2}$ **7.** $\dfrac{2a + 8}{(a - 5)(a + 1)}$ **9.** $\dfrac{5x^2}{27}$
11. $\dfrac{1}{x}$ **13.** $\dfrac{10}{3}$ **15.** $\dfrac{2y - 1}{4y + 1}$ **17.** $28y$ **19.** -8 **21.** $\dfrac{3 - p}{p - 1}$
23. $\dfrac{2a + 3}{4 - 9a}$ **25.** $-\dfrac{t}{t + 1}$ **27.** $-\dfrac{4(w - 1)}{w + 2}$ **29.** $\dfrac{1}{y + 4}$ **31.** $\dfrac{x + 2}{x - 1}$
33. $\dfrac{t^2}{(t + 1)^2}$ **35.** $\dfrac{-a + 2}{-a - 3}$ **37.** $-\dfrac{y + 1}{y - 5}$ **39.** $\dfrac{2}{x(x + h)}$
41. $\dfrac{1}{x(x^2 + 3)}$ **43.** $\dfrac{2b^2 + 3a}{b(b - a)}$ **45.** $m = \dfrac{y_2 - y_1}{x_2 - x_1}$ **47.** $\dfrac{63}{10}$ **49.** $\dfrac{4}{3}$

Section 6.5 Practice Exercises, pp. 440–443

3. $\dfrac{2x}{(x + 4)^2(x - 4)}$ **5.** $\dfrac{1}{m(m - 4)}$ **7.** $\dfrac{x}{x - 1}$ **9. a.** LCD = 12
b. $x = -14$ **11. a.** LCD = $5(x - 5)$ **b.** No solution ($x = 5$ does not check.) **13.** $y = 12$ **15.** $p = -\dfrac{15}{22}$ **17.** $x = 6$ **19.** $x = 5$
21. $w = 6$ **23.** $x = -25$ **25.** $a = 4, a = -4$ **27.** $a = 3, a = 1$
29. $t = -\dfrac{5}{2}$ **31.** $t = -2$ ($t = 2$ does not check.) **33.** $y = 60$
35. $x = \dfrac{31}{5}$ **37.** No solution ($k = 2$ does not check.)
39. No solution ($x = 4$ does not check.) **41.** $m = \dfrac{FK}{a}$ **43.** $E = \dfrac{IR}{K}$
45. $R = \dfrac{E - Ir}{I}$ or $R = \dfrac{E}{I} - r$ **47.** $B = \dfrac{2A - hb}{h}$ or
$B = \dfrac{2A}{h} - b$ **49.** $t = \dfrac{b}{x - a}$ or $t = \dfrac{-b}{a - x}$ **51.** $x = \dfrac{y}{1 - yz}$ or
$x = \dfrac{-y}{yz - 1}$ **53.** $h = \dfrac{2A}{a + b}$ **55.** $R = \dfrac{R_1 R_2}{R_2 + R_1}$
57. $t_2 = \dfrac{s_2 - s_1 + vt_1}{v}$ or $t_2 = \dfrac{s_2 - s_1}{v} + t_1$ **59. a.** $\dfrac{2w + 30}{(w - 5)(w + 5)}$
b. $w = -15$ **c.** The problem in part (a) is an expression, and the problem in part (b) is an equation. **61.** $\dfrac{1}{a + 1}$ **63.** $y = 1$
65. $\dfrac{x^2 - 12}{x(x - 1)}$ **67.** $w = -1$ **69.** $\dfrac{8p - 11}{4(2p - 3)}$ **71.** 3 **73.** $\dfrac{2}{5}$

Section 6.6 Practice Exercises, pp. 449–453

3. $x = -2, x = -3$ **5.** $\dfrac{11}{12(x - 2)}$ **7.** $y = -\dfrac{3}{5}, y = 3$
9. $\dfrac{8x - 3y}{x(x - y)(x + y)}$ **11.** $y = 8$ **13.** $m = 6$ **15.** $p = -15$
17. $x = \dfrac{1}{3}$ **19.** $x = \dfrac{20}{9}$ **21.** $y = 5, y = -1$ **23.** $w = \dfrac{3}{7}, w = -\dfrac{3}{7}$
25. Six adults **27.** 84 g of fat **29.** 1000 swordfish **31.** Pam needs 11.5 gal of gas. **33.** There are approximately 4000 bison in the park. **35.** 54 are men and 27 are women. **37.** 595 are men and 500 are women. **39.** $a = 8$ ft, $b = 8.4$ ft **41.** $x = 12$ in., $y = 13$ in., $z = 4.2$ in. **43.** 224 bikes **45.** The motorist drives 40 mph in the rain and 60 mph in sunny weather. **47.** 11 mph **49.** 12 mph
51. 8 days **53.** Gus would take 6 hr; Sid would take 12 hr.
55. $y = 5$ **57.** $x = 5$

Chapter 6 Review Exercises, pp. 458–460

1. a. $-\dfrac{2}{9}, -\dfrac{1}{10}, 0, -\dfrac{5}{6}$, undefined **b.** $\{t \mid t$ is a real number and $t \neq -9\}$ **3. a.** $k(2) = \dfrac{2}{3}, k(0) = 0, k(1)$ undefined, $k(-1)$ undefined,
$k\left(\dfrac{1}{2}\right) = -\dfrac{2}{3}$ **b.** $(-\infty, -1) \cup (-1, 1) \cup (1, \infty)$ **5.** $2a$ **7.** $x - 1$
9. $-\dfrac{x^2 + 3x + 9}{3 + x}$ **11.** $-\dfrac{2t + 5}{t + 7}$ **13.** c; $\{x \mid x$ is a real number and
$x \neq 3\}$ **15.** b; $\{x \mid x$ is a real number and $x \neq 0, x \neq 3\}$ **17.** $\dfrac{a}{2}$
19. $-\dfrac{x - y}{5x}$ or $\dfrac{y - x}{5x}$ **21.** $\dfrac{7(k - 4)}{2(k - 2)}$ **23.** $\dfrac{x - 5}{x - 4}$ **25.** $\dfrac{8}{9w^2}$
27. $\dfrac{5}{2}$ **29.** $\dfrac{x^2 + x - 1}{x^3}$ **31.** $\dfrac{y - 3}{2y - 1}$ or $\dfrac{3 - y}{1 - 2y}$

33. $\dfrac{4k^2 - k + 3}{(k+1)^2(k-1)}$ **35.** $\dfrac{2(a^2-5)}{(a-5)(a+3)}$ **37.** $\dfrac{6(7-4x)}{3x-5}$ or $\dfrac{-6(4x-7)}{3x-5}$ **39.** $\dfrac{9a^2+a+4}{(3a-1)(a-2)}$ **41.** $\dfrac{1}{x+1}$ **43.** $\dfrac{x(2y+1)}{4y}$
45. $\dfrac{1+a}{1-a}$ or $-\dfrac{a+1}{a-1}$ **47.** $\dfrac{y}{x-y}$ **49.** $m = \dfrac{1}{18}$
51. $x = 3$ ($x = 1$ does not check.) **53.** $x = 0, x = 17$
55. $y = 5, y = 1$ **57.** $x = \dfrac{b}{c-a}$ or $x = \dfrac{-b}{a-c}$ **59.** $x = \dfrac{15}{2}$
61. $x = -\dfrac{7}{11}$ **63.** Manning would gain 231 yd. **65.** 80 units
67. Larger pipe, 9 hr; smaller pipe, 18 hr

Chapter 6 Test, pp. 460–461

1. a. $\{x \mid x$ is a real number and $x \neq 4, x \neq -3\}$ **b.** $\dfrac{2}{x-4}$
2. a. $f(0) = \dfrac{2}{7}, f(5) = \dfrac{1}{6}, f(7)$ is undefined, $f(-7)$ is undefined
b. $\{x \mid x$ is a real number and $x \neq 7, x \neq -7\}$ **3.** $(-\infty, \infty)$ **4.** $\dfrac{2m^2}{3n}$
5. $\dfrac{9(x+1)}{3x+5}$ **6.** $m = -\dfrac{23}{9}$ **7.** $-(x-3)$ or $-x+3$ **8.** $x-4$
9. $\dfrac{x^2+5x+2}{x+1}$ **10.** $\dfrac{3}{4}$ **11.** $\dfrac{u^2v}{2(v^2-uv+u^2)}$ **12.** a
13. $\dfrac{1}{(x+5)(x+3)}$ **14.** $z = 3$ **15.** $y = 0, y = 4$ **16.** No solution
($x = 4$ does not check.) **17.** $T = \dfrac{1}{p-v}$ or $T = \dfrac{-1}{v-p}$
18. $m_1 = \dfrac{Fr^2}{Gm_2}$ **19.** $\dfrac{1}{6}$ or 2 **20.** $a = 14$ m, $b = 15$ m
21. 1960 mi **22.** 1000 units **23.** Lance rides 16 mph against the wind and 20 mph with the wind. **24.** $2\tfrac{6}{7}$ hr

Chapters 1–6 Cumulative Review Exercises, pp. 461–463

1.

Set \ Number	−22	π	6	−√2
Real numbers	✓	✓	✓	✓
Irrational numbers		✓		✓
Rational numbers	✓		✓	
Integers	✓		✓	
Whole numbers			✓	
Natural numbers			✓	

2. a. $21,839,777 **b.** $1,284,693 **3.** $x^2 - x - 13$
4. a. $b_1 = \dfrac{2A - hb_2}{h}$ or $b_1 = \dfrac{2A}{h} - b_2$ **b.** 10 cm
5. 50 m by 30 m **6.** No solution; inconsistent system
7. $y = -\dfrac{1}{3}x + 4$ **8. a.** Linear

b.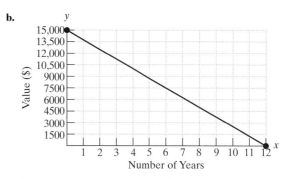

c. (0, 15,000). When the car is new (0 years old), the value is $15,000. **d.** (12, 0). When the car is 12 years old, the value is $0.
e. $m = -1250$. The car is depreciating at a rate of $1250 per year.
f. $V(5) = 8750$. After 5 years the value of the car is $8750.
g. $7\tfrac{1}{2}$ years **9.** 75 mph **10.** $(0, 0), \left(\dfrac{3}{4}, 0\right), \left(\dfrac{2}{3}, 0\right)$
11. $8(2y - z^2)(4y^2 + 2yz^2 + z^4)$ **12. a.** $\left(-\infty, \dfrac{3}{2}\right) \cup \left(\dfrac{3}{2}, \infty\right)$
b. $(-\infty, -3) \cup (-3, 4) \cup (4, \infty)$ **13.** $\dfrac{x+a}{4(x-7)}$
14. $\dfrac{-(x-2)}{5}$ or $\dfrac{2-x}{5}$ **15.** $\dfrac{x+3}{(x+10)(x-2)}$ **16.** $\dfrac{c-7}{c}$
17. $y = 3, y = -1$ **18.** 278 mi **19. a.** Vertical line; slope is undefined **b.** Horizontal line; $m = 0$ **20.** $6x^5y^6z^2$

Chapter 7

Chapter 7 Preview, p. 466

1. Not a real number **3.** $[-10, \infty)$ **5.** $\sqrt[3]{c^2}$
7. $\dfrac{1}{9}$ **9.** $2\sqrt[3]{10}$ **11.** 2 **13.** $10x\sqrt{3}$ **15.** $-18 + 2\sqrt{6}$
17. 1 **19.** $\dfrac{x\sqrt[3]{y^2}}{y}$ **21.** $x = 7$ **23.** $t = 1$ **25.** $-i$ **27.** $7 + 26i$

Section 7.1 Practice Exercises, pp. 475–479

3. a. $8, -8$ **b.** 8 **c.** There are two square roots for every positive number. $\sqrt{64}$ identifies the positive square root. **5. a.** 9 **b.** −9
7. There is no real number b such that $b^2 = -36$. **9.** 7 **11.** −7
13. Not a real number **15.** $\dfrac{8}{3}$ **17.** 0.9 **19.** −0.4 **21. a.** 8
b. 4 **c.** −8 **d.** −4 **e.** Not a real number **f.** −4 **23.** −3
25. $\dfrac{1}{2}$ **27.** 2 **29.** $-\dfrac{5}{4}$ **31.** Not a real number **33.** 10 **35.** −0.2
37. $|a|$ **39.** a **41.** $|a|$ **43.** x^2 **45.** $|x|y^2$ **47.** $-\dfrac{x}{y}$ **49.** $\dfrac{2}{|x|}$
51. −9 **53.** 2 **55.** xy^2 **57.** $\dfrac{a^3}{b}$ **59.** $-\dfrac{5}{q}$ **61.** $3xy^2z$ **63.** $\dfrac{hk^2}{4}$
65. $-\dfrac{t}{3}$ **67.** $2y^2$ **69.** $2p^2q^3$ **71.** 9 cm **73.** 13 ft **75.** 5 mi
77. 25 mi **79. a.** Not a real number **b.** Not a real number **c.** 0
d. 1 **e.** 2; Domain: $[2, \infty)$ **81. a.** −2 **b.** −1 **c.** 0 **d.** 1; Domain: $(-\infty, \infty)$ **83.** $[-5, \infty)$ **85.** $(-\infty, \infty)$ **87.** b **89.** d **91.** $q + p^2$
93. **95.** 8 in. **97.** 8.3066 **99.** 3.7100 **101.** 15.6525
103. −0.1235 **105.** 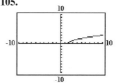 **107.**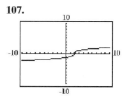

Section 7.2 Practice Exercises, pp. 484–487

3. a. 3 b. 27 5. 5 7. 3 9. 12 11. −12 13. Not a real number 15. −4 17. $\frac{1}{5}$ 19. $a^{m/n} = \sqrt[n]{a^m}$; The numerator of the exponent represents the power of the base. The denominator of the exponent represents the index of the radical. 21. a. 8 b. −8 c. Not a real number d. $\frac{1}{8}$ e. $-\frac{1}{8}$ f. Not a real number

23. a. 125 b. −125 c. Not a real number d. $\frac{1}{125}$ e. $-\frac{1}{125}$ f. Not a real number 25. $\frac{1}{512}$ 27. 27 29. $-\frac{1}{81}$ 31. $\frac{27}{1000}$

33. Not a real number 35. −2 37. −2 39. $\frac{1}{9}$ 41. 6 43. 10 45. $\frac{3}{4}$ 47. 1 49. $\frac{9}{2}$ 51. $\sqrt[3]{q^2}$ 53. $6\sqrt[6]{y^3}$ 55. $\sqrt[5]{x^2y}$ 57. $\frac{1}{\sqrt[5]{qr}}$ 59. $x^{1/3}$ 61. $10b^{1/2}$ 63. $y^{2/3}$ 65. $(a^2b^3)^{1/4}$ 67. $\frac{1}{x}$ 69. p 71. y^2 73. $6^{2/5}$ 75. $\frac{4}{t^{5/3}}$ 77. a^7 79. $\frac{25a^4d}{c}$ 81. $\frac{y^9}{x^8}$ 83. $\frac{2z^3}{w}$ 85. $5xy^2z^3$ 87. $x^{13}z^{4/3}$ 89. $\frac{x^3y^2}{z^5}$ 91. a. 10 in. b. 8.5 in. 93. a. 10.9% b. 8.8% c. The account in part (a). 95. $\sqrt[6]{x}$ 97. $\sqrt[8]{y}$ 99. $\sqrt[15]{w}$ 101. 3 103. 0.3761 105. 2.9240 107. 31.6228

Section 7.3 Practice Exercises, pp. 492–494

3. $\frac{p}{q^5}$ 5. $\sqrt[7]{x^4}$ 7. $y^{9/2}$ 9. $x^5\sqrt{x}$ 11. $q^2\sqrt[3]{q}$ 13. $a^2b^2\sqrt{a}$ 15. $-x^2y^3\sqrt[4]{y}$ 17. $2\sqrt{7}$ 19. $4\sqrt{5}$ 21. $3\sqrt{2}$ 23. $5b\sqrt{ab}$ 25. $3a^3b\sqrt{2b}$ 27. $-2x^2z\sqrt[3]{2y}$ 29. $2wz\sqrt[4]{5z^3}$ 31. x 33. p^2 35. 5 37. $\frac{1}{2}$ 39. $\frac{5\sqrt[3]{2}}{3}$ 41. $\frac{5\sqrt[3]{9}}{6}$ 43. $15\sqrt{2}$ 45. $-30\sqrt{3}$ 47. $5x^2y\sqrt{y}$ 49. $3yz\sqrt[3]{x^2z}$ 51. $\frac{2}{b}$ 53. $\frac{2\sqrt[5]{x}}{y^2}$ 55. $\frac{5x\sqrt{2xy}}{3y^2}$ 57. $2a^7b^4c^{15}d^{11}\sqrt{2c}$ 59. $\frac{1}{\sqrt[3]{w^6}}$ simplifies to $\frac{1}{w^2}$ 61. $\sqrt{k^3}$ simplifies to $k\sqrt{k}$ 63. $2\sqrt{41}$ ft 65. $6\sqrt{5}$ m 67. $90\sqrt{2}$ ft ≈ 127.3 ft 69. The path from A to B and B to C is faster.

Midchapter Review, p. 495

1. 4 2. 9 3. $\frac{1}{2}$ 4. $\frac{3}{5}$ 5. $3w\sqrt[3]{2x^2w}$ 6. $2tuv^3\sqrt{5uv}$ 7. $5xy^2\sqrt{2y}$ 8. $2wz^3\sqrt[3]{z}$ 9. $4r − 3$ 10. $\frac{ab}{a+b}$ 11. Write x^{51} as $x^{50} \cdot x^1$. Then apply the multiplication property of radicals. 12. 4 13. 8 14. 5 15. $\frac{5}{2}$ 16. $\frac{t^2}{sr^3}$ 17. $\frac{pr^{5/2}}{q^{3/2}}$ 18. $u^{2/3}v^{3/2}$ 19. $xy^{7/12}$ 20. First take the cube root of 125, then square the result. $125^{2/3} = (\sqrt[3]{125})^2 = (5)^2 = 25$ 21. a. $(7x^2)^{1/3}$ b. $(21a)^{1/2}$ 22. a. $\sqrt[3]{x^2}$ b. $(\sqrt[4]{c})^3$

Section 7.4 Practice Exercises, pp. 498–501

3. $-2st^3\sqrt[3]{2s}$ 5. $6ab\sqrt{b}$ 7. $\sqrt{(4x^2)^3}$ simplifies to $8x^3$ 9. $y^{11/12}$ 11. a. Not *like* radicals b. *Like* radicals c. Not *like* radicals 13. a. Both expressions can be simplified by using the distributive property. b. Neither expression can be simplified because they do not contain *like* terms or *like* radicals. 15. $9\sqrt{5}$ 17. $\sqrt[3]{t}$ 19. $5\sqrt{10}$ 21. $8\sqrt[4]{3} − \sqrt[4]{14}$ 23. $2\sqrt{x} + 2\sqrt{y}$ 25. Cannot be simplified further 27. Cannot be simplified further 29. $\frac{29}{18}z\sqrt[3]{6}$ 31. $0.70x\sqrt{y}$ 33. Simplify each radical: $3\sqrt{2} + 35\sqrt{2}$. Then add *like* radicals: $38\sqrt{2}$ 35. 15 37. $8\sqrt{3}$ 39. $3\sqrt{7}$ 41. $-\sqrt{2}$ 43. $\sqrt[3]{3}$ 45. $-5\sqrt{2a}$ 47. $8s^2t^2\sqrt[3]{s^2}$ 49. $6x\sqrt[3]{x}$ 51. $14p^2\sqrt{5}$ 53. $-\sqrt[3]{a^2b}$ 55. $33d\sqrt[3]{2c}$ 57. $2a^2b\sqrt{6a}$ 59. $5x\sqrt[3]{2} − 6\sqrt[3]{x}$ 61. False, $\sqrt{9} + \sqrt{16} \neq \sqrt{9+16}; 7 \neq 5$ 63. True 65. False, $\sqrt{y} + \sqrt{y} = 2\sqrt{y} \neq \sqrt{2y}$ 67. $\sqrt{48} + \sqrt{12}$ simplifies to $6\sqrt{3}$ 69. $5\sqrt[3]{x^6} − x^2$ simplifies to $4x^2$ 71. The difference of the principal square root of 18 and the square of 5 73. The sum of the principal fourth root of x and the cube of y 75. $9\sqrt{6}$ cm ≈ 22.0 cm 77. a. $10\sqrt{5}$ yd b. 22.36 yd c. $105.95

Section 7.5 Practice Exercises, pp. 507–510

3. a. 5 b. 4 5. $2 + h$ 7. $-2ab\sqrt{5bc}$ 9. 2 11. $\frac{q^2}{p^{1/8}}$ 13. $2\sqrt[3]{7}$ 15. $\sqrt[3]{21}$ 17. $2\sqrt{5}$ 19. $4\sqrt[4]{4}$ 21. $8\sqrt[3]{20}$ 23. $-24ab\sqrt{a}$ 25. $6\sqrt{10}$ 27. $6x\sqrt{2}$ 29. $10a^3b^2\sqrt{b}$ 31. $-24x^2y^2\sqrt{2y}$ 33. $6ab\sqrt[3]{2a^2b^2}$ 35. $12 − 6\sqrt{3}$ 37. $2\sqrt{3} − \sqrt{6}$ 39. $-3x − 21\sqrt{x}$ 41. $-8 + 7\sqrt{30}$ 43. $x − 5\sqrt{x} − 36$ 45. $\sqrt[3]{y^2} − \sqrt[3]{y} − 6$ 47. $9a − 28\sqrt{ab} + 3b$ 49. $8\sqrt{p} + 3p + 5\sqrt{pq} + 16\sqrt{q} − 2q$ 51. 15 53. $3y$ 55. 6 57. 709 59. a. $x^2 − y^2$ b. $x^2 − 25$ 61. $3 − x^2$ 63. 4 65. $64x − 4y$ 67. $29 + 8\sqrt{13}$ 69. $p − 2\sqrt{7p} + 7$ 71. $2a − 6\sqrt{2ab} + 9b$ 73. True 75. False; $(x − \sqrt{5})^2 = x^2 − 2x\sqrt{5} + 5$ 77. False; 5 is multiplied by 3 only. 79. True 81. 39 83. $6x$ 85. $3x + 1$ 87. $x + 19 − 8\sqrt{x+3}$ 89. $12\sqrt{5}$ ft² 91. $18\sqrt{15}$ in.² 93. $\sqrt[4]{x^3}$ 95. $\sqrt[15]{(2z)^8}$ 97. $p^2\sqrt[6]{p}$ 99. $u\sqrt[6]{u}$ 101. $\sqrt[6]{(a+b)}$ 103. $\sqrt[6]{x^2y}$ 105. $\sqrt[4]{2^3 \cdot 3^2}$ or $\sqrt[4]{72}$ 107. $a + b$

Section 7.6 Practice Exercises, pp. 515–517

3. $12y\sqrt{5}$ 5. $-18y + 3\sqrt{y} + 3$ 7. $64 − 16\sqrt{t} + t$ 9. -5 11. $\frac{\sqrt{5}}{\sqrt{5}}$ 13. $\frac{\sqrt[3]{x^2}}{\sqrt[3]{x^2}}$ 15. $\frac{\sqrt{3z}}{\sqrt{3z}}$ 17. $\frac{\sqrt[4]{2^3a^2}}{\sqrt[4]{2^3a^2}}$ 19. $\frac{\sqrt{3}}{3}$ 21. $\frac{\sqrt{x}}{x}$ 23. $\frac{3\sqrt{2y}}{y}$ 25. $-2\sqrt{a}$ 27. $\frac{3\sqrt{2}}{2}$ 29. $\frac{3\sqrt[3]{4}}{2}$ 31. $\frac{-6\sqrt[4]{x^3}}{x}$ 33. $\frac{7\sqrt[3]{2}}{2}$ 35. $\frac{\sqrt[3]{4w}}{w}$ 37. $\frac{2\sqrt[4]{27}}{3}$ 39. $\frac{\sqrt[3]{2x}}{x}$ 41. $\frac{2x\sqrt[3]{2y^2}}{y}$ 43. $\frac{xy^2\sqrt{10xy}}{10}$ 45. $\frac{2\sqrt{6}}{21}$ 47. $\frac{\sqrt{x}}{x^4}$ 49. $\frac{\sqrt{2x}}{2x^3}$ 51. $\sqrt{2} + \sqrt{6}$ 53. $\sqrt{x} − 23$ 55. $\frac{4\sqrt{2} − 12}{-7}$ or $\frac{-4\sqrt{2} + 12}{7}$ 57. $4\sqrt{6} + 8$ 59. $-\sqrt{21} + 2\sqrt{7}$ 61. $\frac{-\sqrt{p} + \sqrt{q}}{p − q}$ 63. $\sqrt{x} − \sqrt{5}$ 65. $\frac{-14\sqrt{a} − 35\sqrt{b}}{4a − 25b}$ 67. $9 − 2\sqrt{15}$ 69. $5 − \sqrt{10}$ 71. $\frac{5 + \sqrt{21}}{4}$ 73. $-6\sqrt{5} − 13$ 75. $\frac{16}{\sqrt[3]{4}}$ simplifies to $8\sqrt[3]{2}$ 77. $\frac{4}{x − \sqrt{2}}$ simplifies to $\frac{4x + 4\sqrt{2}}{x^2 − 2}$ 79. $\frac{\pi\sqrt{2}}{4}$ sec ≈ 1.11 sec 81. $\frac{2\sqrt{6}}{3}$ 83. $\frac{17\sqrt{15}}{15}$ 85. $\frac{8\sqrt[3]{25}}{5}$ 87. $\frac{-33}{2\sqrt{3} − 12}$ 89. $\frac{a − b}{a + 2\sqrt{ab} + b}$

Section 7.7 Practice Exercises, pp. 524–528

3. $4\sqrt{3}$ 5. $\frac{3w\sqrt{w}}{4}$ 7. Not a real number 9. $\frac{7\sqrt{5t}}{5t^2}$ 11. $4x − 6$ 13. $9p + 7$ 15. $2x$ 17. $x = 9$ 19. $y = 3$ 21. $z = 42$ 23. $x = 29$ 25. $w = 140$ 27. No solution 29. $a = 7$ ($a = -1$ does not check.) 31. No solution 33. $V = \frac{4\pi r^3}{3}$ 35. $h^2 = \frac{r^2 − \pi^2r^2}{\pi^2}$ or $h^2 = \frac{r^2}{\pi^2} − r^2$ 37. $a^2 + 10a + 25$ 39. $5a − 6\sqrt{5a} + 9$ 41. $r + 22 + 10\sqrt{r − 3}$ 43. $a = -3$ 45. No solution ($w = \frac{1}{2}$ does not check.)

47. $y = 0; y = 3$ **49.** $h = 9$ **51.** $a = -4$ **53.** $a = \frac{9}{5}$ **55.** $h = 2$
57. No solution ($t = 9$ does not check.) **59.** $x = -\frac{11}{4}$ **61.** No solution ($t = \frac{8}{3}$ does not check.) **63.** $z = -1$ ($z = 3$ does not check.)
65. $m = \frac{1}{3}, m = -1$ **67.** No solution ($t = 3$ and $t = 23$ do not check.) **69.** $y = \frac{49}{16}$ **71. a.** 30.25 ft **b.** 34.5 m **73. a.** 305 m **b.** 460 m **75. a.** 12 lb **b.** $t(18) = 5.1$. An 18-lb turkey will take about 5.1 hr to cook. **77.** $b = \sqrt{25 - h^2}$ **79.** $a = \sqrt{k^2 - 196}$
81. 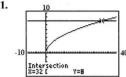 **83.**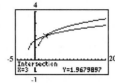

Section 7.8 Practice Exercises, pp. 536–539

3. $3\sqrt{5} - 15\sqrt{2}$ **5.** $9 - x$ **7.** $p = -8$ **9.** $c = \frac{49}{36}$
11. $\sqrt{-1} = i$ and $-\sqrt{1} = -1$ **13.** $12i$ **15.** $i\sqrt{3}$ **17.** $2i\sqrt{5}$
19. -60 **21.** $29i\sqrt{2}$ **23.** $13i\sqrt{7}$ **25.** -7 **27.** -12 **29.** $-3\sqrt{10}$
31. $i\sqrt{2}$ **33.** $3i$ **35.** $-i$ **37.** 1 **39.** i **41.** 1 **43.** $-i$ **45.** -1
47. $a - bi$ **49.** Real: -5; imaginary: 12 **51.** Real: 0; imaginary: -6
53. Real: 35; imaginary: 0 **55.** Real: $\frac{3}{5}$; imaginary: 1 **57.** $7 + 6i$
59. $\frac{3}{10} + \frac{3}{2}i$ **61.** $5 + 0i$ **63.** $-1 + 10i$ **65.** -24 **67.** $18 + 6i$
69. $26 - 26i$ **71.** $-29 + 0i$ **73.** $-9 + 40i$ **75.** $35 + 20i$
77. $-20 + 48i$ **79.** $\frac{13}{16}$ **81.** $\frac{1}{5} - \frac{3}{5}i$ **83.** $\frac{3}{25} - \frac{4}{25}i$
85. $\frac{21}{29} + \frac{20}{29}i$ **87.** $-\frac{17}{10} - \frac{1}{10}i$ **89.** $-\frac{1}{2} - \frac{5}{2}i$ **91.** $\frac{1 + 2i}{4}$
93. $-1 + i\sqrt{2}$ **95.** $-2 - i\sqrt{3}$

Chapter 7 Review Exercises, pp. 545–548

1. a. False; $\sqrt{0} = 0$ is not positive. **b.** False; $\sqrt[3]{-8} = -2$
3. a. False **b.** True **5.** 5 **7. a.** 3 **b.** 0 **c.** $\sqrt{7}$ **d.** $[1, \infty)$
9. $\frac{\sqrt[3]{2x}}{\sqrt[4]{2x}} + 4$ **11. a.** $2|y|$ **b.** $3y$ **c.** $|y|$ **d.** y **13.** Yes, provided the expressions are well defined. For example: $x^5 \cdot x^3 = x^8$ and $x^{1/5} \cdot x^{2/5} = x^{3/5}$ **15.** Take the reciprocal of the base and change the exponent to positive. **17.** $\frac{1}{2}$ **19.** b^{10} **21.** $x^{3/4}$ **23.** 2.1544
25. 54.1819 **27.** 1. Factors of the radicand must have powers less than the index. 2. No fractions in the radicand. 3. No radical in the denominator of a fraction. **29.** $xz\sqrt[4]{xy}$ **31.** $\frac{-2x^2y^2\sqrt[3]{2x}}{z^3}$
33. 31 ft **35.** Cannot be combined; the indices are different.
37. Can be combined: $3\sqrt[4]{3xy}$ **39.** $5\sqrt{7}$ **41.** $10\sqrt{2}$
43. False; 5 and $3\sqrt{x}$ are not *like* radicals. **45.** 6 **47.** 12
49. $4x - 9$ **51.** $7y - 2\sqrt{21xy} + 3x$ **53.** $-2z - 9\sqrt{6z} - 42$
55. $u^2\sqrt[6]{u^5}$ **57.** $\frac{\sqrt{14y}}{2y}$ **59.** $\frac{4\sqrt[3]{3p}}{3p}$ **61.** $-\sqrt{15} - \sqrt{10}$
63. $\sqrt{t} + \sqrt{3}$ **65.** The quotient of the principal square root of 2 and the square root of x **67.** $a = 31$ **69.** $p = 7$ **71.** No solution ($x = -2$ does not check.) **73.** $x = -2$ ($x = 2$ does not check.)
75. a. 25.3 ft/sec; when the water depth is 20 ft, a wave travels about 25.3 ft/sec. **b.** 8 ft **77.** $a + bi$, where $b \neq 0$ **79.** $4i$
81. -15 **83.** -1 **85.** $-i$ **87.** $-5 + 5i$ **89.** $25 + 0i$

91. $-\frac{17}{4} + i$; Real part: $-\frac{17}{4}$; Imaginary part: 1 **93.** $\frac{4}{13} - \frac{7}{13}i$
95. $\frac{-4 + i\sqrt{10}}{6}$

Chapter 7 Test, pp. 549–550

1. a. 6 **b.** -6 **2. a.** Real **b.** Not real **c.** Real **d.** Real
3. a. y **b.** $|y|$ **4.** 3 **5.** $\frac{4}{3}$ **6.** $2\sqrt[3]{4}$ **7.** $a^2bc^2\sqrt{bc}$ **8.** $3x^2\sqrt{2}$
9. $\frac{4w^2\sqrt{6w}}{3}$ **10.** $\sqrt[6]{7y^3}$ **11.** $\sqrt[12]{10}$ **12. a.** $f(-8) = 2\sqrt{3}$;
$f(-6) = 2\sqrt{2}; f(-4) = 2; f(-2) = 0$ **b.** $(-\infty, -2]$ **13.** -0.3080
14. -3 **15.** $\frac{1}{t^{3/4}}$ **16.** $3\sqrt{5}$ **17. a.** $3\sqrt{2x} - 3\sqrt{5x}$
b. $40 - 10\sqrt{5x} - 3x$ **18. a.** $\frac{-2\sqrt[3]{x^2}}{x}$ **b.** $\frac{x + 6 + 5\sqrt{x}}{9 - x}$
19. a. $2i\sqrt{2}$ **b.** $8i$ **c.** $\frac{1 + i\sqrt{2}}{2}$ **20.** $1 - 11i$ **21.** $30 + 16i$
22. -28 **23.** $-33 - 56i$ **24.** 104 **25.** $\frac{17}{25} + \frac{6}{25}i$ **26.** $36 - 11i$
27. $r(10) = 1.34$; the radius of a sphere of volume 10 cubic units is 1.34 units. **28.** 21 ft **29.** $x = -16$ **30.** $x = \frac{17}{5}$ **31.** $t = 2$; ($t = 42$ does not check)

Chapters 1–7 Cumulative Review Exercises, pp. 550–551

1. 54 **2.** $15x - 5y - 5$ **3.** $y = -3$ **4.** $(-\infty, -5)$
5. $y = -2x + 5$ **6.**

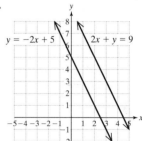

7. $\left(\frac{1}{2}, \frac{1}{3}\right)$ **8.** $\left(2, -2, \frac{1}{2}\right)$ is not a solution. **9.** $x = 6, y = 3, z = 8$
10. a. $f(-2) = -10; f(0) = -2; f(4) = 14; f\left(\frac{1}{2}\right) = 0$
b. $(-2, -10), (0, -2), (4, 14), \left(\frac{1}{2}, 0\right)$
c.

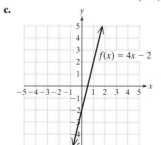

11. Not a function **12.** $a^4b^8c^2$ **13.** $a^6b^{12}c^4$ **14. a.** 1.4×10^{-4}
b. 3.14×10^9 **15.** $2x^2 - x - 15$; second degree
16. $\sqrt{15} + 3\sqrt{2} + 3$ **17.** $x - 4$ **18.** $-\frac{1}{6}$ **19.** $\frac{3c\sqrt[3]{2}}{d}$ **20.** $32b\sqrt{5b}$
21. $2 + 3i$ **22.** $y = 10$ **23.** $\frac{x - 15}{x(x + 5)(x - 5)}$ **24.** $\frac{-1}{(2a + 1)(a + 2)}$

Chapter 8

Chapter 8 Preview, p. 554

1. $k = 81; (y-9)^2$ 3. $a = -2 \pm 2\sqrt{3}$ 5. $r = \sqrt{\dfrac{V}{\pi h}}$ or
$r = \dfrac{\sqrt{V\pi h}}{\pi h}$ 7. $x = \dfrac{1 \pm i\sqrt{5}}{2}$ 9. The legs are $\sqrt{10}$ and $3\sqrt{10}$ in.
11. $a = 1$ ($a = -2$ does not check.) 13. $z = 1; z = -27$
15.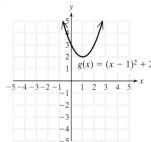
17. a. 192 ft b. 256 ft c. 5 sec

Section 8.1 Practice Exercises, pp. 560–563

3. $x = \pm 10$ 5. $a = \pm\sqrt{5}$ 7. $v = \pm i\sqrt{11}$ 9. $p = 8, p = 2$
11. $x = 2 \pm \sqrt{5}$ 13. $h = 4 \pm 2i\sqrt{2}$ 15. $p = \pm\dfrac{\sqrt{30}}{6}$
17. $x = \dfrac{3}{2} \pm \dfrac{i\sqrt{7}}{2}$ 19. 1. Factoring and applying the zero product rule. 2. Applying the square root property. $x = \pm 6$ 21. $k = 9;$ $(x-3)^2$ 23. $k = 16; (t+4)^2$ 25. $k = \dfrac{1}{4}; \left(c - \dfrac{1}{2}\right)^2$ 27. $k = \dfrac{25}{4};$ $\left(y + \dfrac{5}{2}\right)^2$ 29. $k = \dfrac{1}{25}; \left(b + \dfrac{1}{5}\right)^2$ 31. 1. Write equation in the form $ax^2 + bx + c = 0$. 2. Divide each term by a. 3. Isolate the variable terms. 4. Complete the square and factor. 5. Apply the square root property. 33. $t = -3, t = -5$ 35. $x = 2, x = -8$
37. $p = -2 \pm i\sqrt{2}$ 39. $y = 5, y = -2$ 41. $a = -1 \pm \dfrac{i\sqrt{6}}{2}$
43. $x = 2 \pm \dfrac{2}{3}i$ 45. $p = \dfrac{1}{5} \pm \dfrac{\sqrt{3}}{5}$ 47. $w = -\dfrac{3}{4} \pm \dfrac{\sqrt{65}}{4}$
49. $n = 2 \pm \sqrt{11}$ 51. $x = 1, x = -7$ 53. a. $t = \dfrac{\sqrt{d}}{4}$ b. 8 sec
55. $r = \sqrt{\dfrac{A}{\pi}}$ or $r = \dfrac{\sqrt{A\pi}}{\pi}$ 57. $a = \sqrt{d^2 - b^2 - c^2}$
59. $r = \sqrt{\dfrac{3V}{\pi h}}$ or $r = \dfrac{\sqrt{3V\pi h}}{\pi h}$ 61. 4.2 ft 63. a. 8% b. 11%
c. 14% 65. a. 4.5 thousand textbooks or 35.5 thousand textbooks
b. Profit increases to a point as more books are produced. Beyond that point, the market is "flooded," and profit decreases. Hence there are two points at which the profit is $20,000. Producing 4.5 thousand books makes the same profit using fewer resources as producing 35.5 thousand books.

Section 8.2 Practice Exercises, pp. 573–578

3. $x = 2, x = -12$ 5. $x = -3 \pm \sqrt{17}$ 7. $x = 1 \pm 3i$
9. $4 - 2\sqrt{10}$ 11. $2 - i\sqrt{3}$ 13. The form $ax^2 + bx + c = 0$
15. $x = \dfrac{-b \pm \sqrt{b^2 - 4ac}}{2a}$ 17. $b = -\dfrac{1}{5}, b = 3$ 19. $t = \dfrac{-3 \pm \sqrt{65}}{4}$
21. $n = \dfrac{2 \pm i\sqrt{26}}{5}$ 23. $x = \dfrac{2}{3}, x = -\dfrac{1}{4}$ 25. $k = 3 \pm \sqrt{5}$
27. $y = -\dfrac{2}{3}$ 29. $m = \dfrac{-3 \pm i\sqrt{7}}{4}$ 31. $y = -\dfrac{3}{4}, y = -2$
33. $m = -3, m = \dfrac{1}{2}$ 35. $p = \dfrac{1 \pm i\sqrt{47}}{8}$ 37. $w = \dfrac{-7 \pm \sqrt{33}}{2}$
39. $y = \dfrac{2}{5}, y = 1$ 41. $x = -2 \pm \sqrt{11}$
43. a. $(4x + 1)(16x^2 - 4x + 1)$ b. $x = -\dfrac{1}{4}, x = \dfrac{1 \pm i\sqrt{3}}{8}$
45. a. $5x(x^2 + x + 2)$ b. $x = 0, x = \dfrac{-1 \pm i\sqrt{7}}{2}$ 47. $1\dfrac{1}{3}$ ft by 4 ft
by 12 ft 49. 10.2 and 13.6 ft 51. $t = \dfrac{3 + \sqrt{5}}{2} \approx 2.62$ sec or
$t = \dfrac{3 - \sqrt{5}}{2} \approx 0.38$ sec 53. a. 46 mph b. 36 mph
55. a. $4y^2 - 12y + 9 = 0$ b. 0 c. 1 rational solution
57. a. $5n^2 - 2n + 0 = 0$ b. 4 c. 2 rational solutions
59. a. $3k^2 + 0k - 7 = 0$ b. 84 c. 2 irrational solutions
61. a. $2x^2 - 5x + 6 = 0$ b. -23 c. 2 imaginary solutions
63. x-intercepts: none; y-intercept: $(0, -1)$ 65. x-intercepts: $\left(\dfrac{1+\sqrt{7}}{3}, 0\right)$ and $\left(\dfrac{1-\sqrt{7}}{3}, 0\right)$; y-intercept: $(0, 2)$ 67. $z = 0, z = -\dfrac{7}{4}$
69. $b = \pm i\sqrt{7}$ 71. $k = \dfrac{1 \pm i\sqrt{31}}{2}$ 73. $y = 0, y = -3$
75. $w = \dfrac{5 \pm \sqrt{41}}{2}$ 77. $b = \dfrac{-1 \pm i\sqrt{11}}{2}$ 79. $x = 6, x = -5$
81. $k = \dfrac{3}{2}, k = \dfrac{1}{4}$ 83. $h = 5, h = -5$
85. 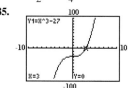 87.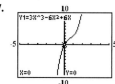
89. a. 12,769 thousand b. 15,507 thousand c. 1989
d.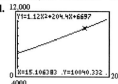

Section 8.3 Practice Exercises, pp. 582–585

3. $x = \dfrac{7}{2}, x = -\dfrac{1}{2}$ 5. $n = 3 \pm 2i$ 7. $k = \dfrac{-7 \pm \sqrt{193}}{12}$
9. $y = 4$ ($y = 64$ does not check.) 11. $x = \dfrac{1}{4}$ ($x = 4$ does not check.)
13. $b = 6, b = 2$ 15. $k = -1$ ($k = 15$ does not check.)
17. $t = 0, t = 2$ 19. $x = \pm 2, x = \pm 2i$ 21. $n = \pm 4, n = \pm 4i$
23. $a = -2, a = 1 \pm i\sqrt{3}$ 25. $p = 1, p = \dfrac{-1 \pm i\sqrt{3}}{2}$
27. $x = 16$ ($x = 9$ does not check.) 29. a. $u = -4, u = -6$
b. $y = -4, y = -1, y = -2, y = -3$ 31. a. $u = 6, u = -4$
b. $x = 6, x = -1, x = 4, x = 1$ 33. $x = -\dfrac{7}{4}, x = -\dfrac{3}{2}$ 35. $x = -7$
37. $x = 1 \pm i\sqrt{2}, x = 1 \pm \sqrt{2}$ 39. $x = 3, x = -3, x = 2, x = -2$
41. $m = \dfrac{1}{2}, m = -\dfrac{1}{2}, m = \sqrt{2}, m = -\sqrt{2}$
43. $x = 2, x = 1, x = -1 \pm i\sqrt{3}, x = \dfrac{-1 \pm i\sqrt{3}}{2}$
45. $m = 27, m = -8$ 47. $t = -\dfrac{1}{32}, t = -243$ 49. $x = \pm 2$
51. $a = \pm 4i, a = 1$ 53. $x = 4, x = \pm i\sqrt{5}$ 55. a. $x = \pm i\sqrt{2}$
b. Two imaginary solutions; zero real solutions c. No x-intercepts
d. 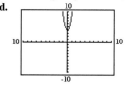 57. a. $x = 0, x = 3, x = -2$

b. Three real solutions; zero imaginary solutions
c. Three x-intercepts **d.**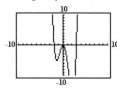

Midchapter Review, p. 585

1. $a = \pm i\sqrt{5}$ **2.** $y = -4 \pm 2\sqrt{3}$ **3.** $k = 100; (c - 10)^2$
4. $k = \dfrac{121}{4}; \left(v + \dfrac{11}{2}\right)^2$ **5.** $x = 5 \pm 2\sqrt{5}$ **6.** $t = -2 \pm i\sqrt{2}$
7. $x = \dfrac{-b \pm \sqrt{b^2 - 4ac}}{2a}$ **8.** $x = \dfrac{-1 \pm i\sqrt{47}}{12}$ **9.** $y = \dfrac{-3 \pm \sqrt{105}}{8}$
10. a. 64 **b.** 2 rational solutions **11.** $a = 3; a = -3; a = 3i;$
$a = -3i$ **12.** $x = 8; x = -125$

Section 8.4 Practice Exercises, pp. 593–599

3. $x = \dfrac{-1 \pm \sqrt{21}}{2}$ **5.** $a = \sqrt{6}$ **7.** $z = -\dfrac{3}{2}; z = 3$
9. The value of k shifts the graph of $y = x^2$ vertically.

11. 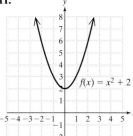 **13.**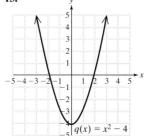

15. 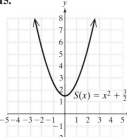 **17.**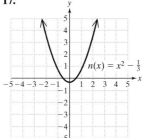

19. 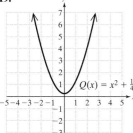 **21.**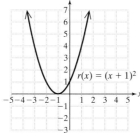

23. 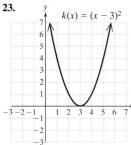 **25.**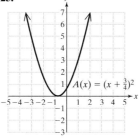

27. 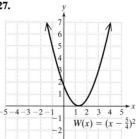 **29.**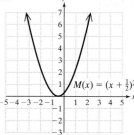

31. The value of a vertically stretches or shrinks the graph of $y = x^2$.

33. 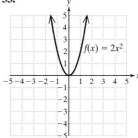 **35.**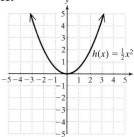

37. 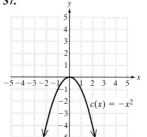 **39.**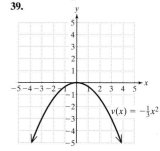

41. d **43.** g **45.** a **47.** e
49. 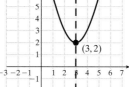 **51.**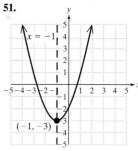

53. 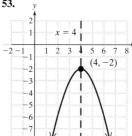 **55.**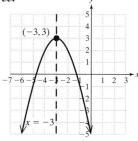

57. 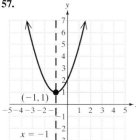 **59.**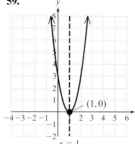

29. $P(x) = -2\left(x - \dfrac{1}{4}\right)^2 + \dfrac{1}{8}; \left(\dfrac{1}{4}, \dfrac{1}{8}\right)$ **31.** $(2, 3)$ **33.** $(-1, -2)$
35. $(-4, -15)$ **37.** $(-1, 2)$ **39.** $(1, 3)$ **41.** $(-1, 4)$
43. $\left(\dfrac{3}{2}, \dfrac{3}{4}\right)$ **45. a.** $\left(-\dfrac{9}{2}, -\dfrac{49}{4}\right)$ **b.** $(0, 8)$ **c.** $(-8, 0)(-1, 0)$
d.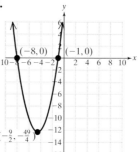

61. 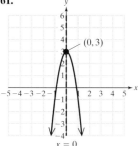 **63.**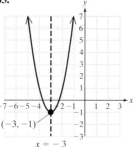

47. a. $\left(\dfrac{1}{2}, \dfrac{7}{2}\right)$ **b.** $(0, 4)$ **49. a.** $\left(\dfrac{3}{2}, 0\right)$ **b.** $\left(0, -\dfrac{9}{4}\right)$
c. No x-intercepts   **c.** $\left(\dfrac{3}{2}, 0\right)$
d. 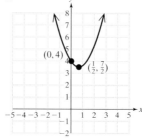   **d.**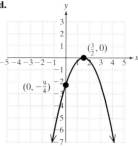

65.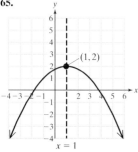

51. Mia must produce 10 MP3 players.   **53. a.** 33 psi
b. 38 thousand mi   **55. a.** $450   **b.** $(0, -550)$. If no cookbooks are produced, the organization will lose $550.   **c.** 50 books or 550 books
d. $(300, 1250)$
e.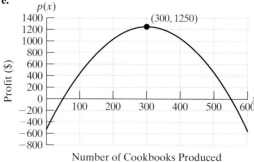
Number of Cookbooks Produced
f. 300 books produced will yield a maximum profit of $1250.
57. 45 mph   **59.** 48 hr   **61. a.** The sum of the three sides must equal the total amount of fencing.   **b.** $A = x(200 - 2x)$ **c.** 50 ft by 100 ft

67. Vertex $(6, -9)$ minimum point; minimum value: -9
69. Vertex $(2, 5)$; maximum point; maximum value: 5
71. Vertex $(-8, 0)$; minimum point; minimum value: 0
73. Vertex $\left(0, \dfrac{21}{4}\right)$; maximum point; maximum value: $\dfrac{21}{4}$
75. Vertex $\left(7, -\dfrac{3}{2}\right)$; minimum point; minimum value: $-\dfrac{3}{2}$
77. Vertex $(0, 0)$; minimum point; minimum value: 0
79. True   **81.** False   **83. a.** $(60, 30)$ **b.** 30 ft   **c.** 70 ft
85. 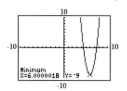 **87.**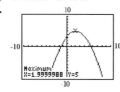

63. 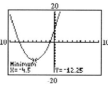 **65.**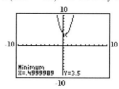

67.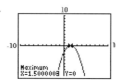

Section 8.5 Practice Exercises, pp. 607–612

3. The graph of f is the graph of $y = x^2$ but opening downward and stretched vertically by a factor of 2.   **5.** The graph of Q is the graph of $y = x^2$ shifted down $\dfrac{8}{3}$ units.   **7.** The graph of s is the graph of $y = x^2$ shifted to the right 4 units.   **9.** 16   **11.** $\dfrac{49}{4}$ **13.** $\dfrac{1}{81}$ **15.** $\dfrac{1}{36}$
17. $g(x) = (x - 4)^2 - 11; (4, -11)$
19. $n(x) = 2(x + 3)^2 - 5; (-3, -5)$
21. $p(x) = -3(x - 1)^2 - 2; (1, -2)$
23. $k(x) = \left(x + \dfrac{7}{2}\right)^2 - \dfrac{89}{4}; \left(-\dfrac{7}{2}, -\dfrac{89}{4}\right)$
25. $f(x) = (x + 4)^2 - 15; (-4, -15)$
27. $F(x) = 5(x + 1)^2 - 4; (-1, -4)$

Chapter 8 Review Exercises, pp. 617–620

1. $x = \pm\sqrt{5}$ 3. $a = \pm 9$ 5. $x = 2 \pm 6\sqrt{2}$ 7. $y = \dfrac{1 \pm \sqrt{3}}{3}$
9. $5\sqrt{3}$ in. ≈ 8.7 in. 11. $5\sqrt{6}$ in. ≈ 12.2 in. 13. $k = \dfrac{81}{4}; \left(x - \dfrac{9}{2}\right)^2$
15. $k = \dfrac{1}{25}; \left(z - \dfrac{1}{5}\right)^2$ 17. $y = \dfrac{3}{2} \pm i$ 19. $b = \dfrac{1}{2}, b = -4$
21. $t = 4 \pm 3i$ 23. Two rational solutions 25. Two irrational solutions 27. One rational solution 29. $y = 2 \pm \sqrt{3}$
31. $a = 2, a = -\dfrac{5}{6}$ 33. $b = \dfrac{4}{5}, b = -\dfrac{1}{5}$ 35. $x = 8, x = -4$
37. a. 1822 ft b. 115 ft/s 39. $x = 49$ ($x = 9$ does not check.)
41. $y = \pm 3, y = \pm\sqrt{2}$ 43. $t = -243, t = 32$ 45. $a = 3, a = 1$
47. $x = \pm 3i, x = \pm i\sqrt{3}$

49. 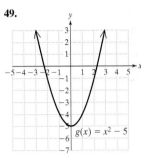 51.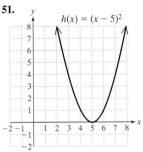

53. 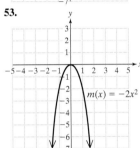 55.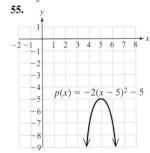

57. $\left(4, \dfrac{5}{3}\right)$ is the minimum point. The minimum value is $\dfrac{5}{3}$.
59. $x = -\dfrac{2}{11}$ 61. $z(x) = (x - 3)^2 - 2; (3, -2)$
63. $p(x) = -5(x + 1)^2 - 8; (-1, -8)$ 65. $(1, -15)$ 67. $\left(\dfrac{1}{2}, \dfrac{41}{4}\right)$
69. a. $(-2, 4)$ b. $(0, 0)$ $(-4, 0)$ c.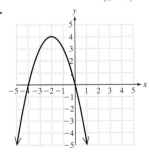

Chapter 8 Test, pp. 620–622

1. $x = 2, x = -8$ 2. $p = 2 \pm 2\sqrt{3}$ 3. $m = -1 \pm i$
4. $k = \dfrac{49}{4}; \left(d + \dfrac{7}{2}\right)^2$ 5. $x = -3 \pm 3\sqrt{3}$ 6. $x = \dfrac{3 \pm i\sqrt{47}}{4}$
7. a. $x^2 - 3x + 12 = 0$ b. $a = 1, b = -3, c = 12$ c. -39 d. Two imaginary solutions 8. a. $y^2 - 2y + 1 = 0$ b. $a = 1, b = -2, c = 1$
c. 0 d. One rational solution 9. $x = 1, x = \dfrac{1}{3}$ 10. $x = \dfrac{-7 \pm \sqrt{5}}{2}$
11. Height 4.6 ft; base 6.2 ft 12. Radius 12.0 ft 13. $x = 9$ ($x = 4$ does not check.) 14. $y = 8, y = -64$ 15. $y = \dfrac{11}{3}, y = 6$

16. $(4, 0)(2, 0)(0, 8)$; c 17. $(-4, 0)(3, 0)(-3, 0)$ $(0, -36)$; b
18. $(-3, 0)(-1, 0)(0, -6)$; d 19. $(4, 0)(-3, 0)(0, 0)$; a 20. 256 ft
21. a. 1320 million b. 1999 22. The graph of $y = x^2 - 2$ is the graph of $y = x^2$ shifted down 2 units. 23. The graph of $y = (x + 3)^2$ is the graph of $y = x^2$ shifted 3 units to the left. 24. The graph of $y = -4x^2$ is the graph of $y = 4x^2$ opening downward instead of upward. 25. a. $(4, 2)$ b. Down c. Maximum point d. The maximum value is 2. e. $x = 4$ 26. a. $g(x) = 2(x - 5)^2 + 1; (5, 1)$
b. $(5, 1)$ 27. 20,000 ft²

Chapter 8 Cumulative Review Exercises, pp. 622–624

1. a. $\{2, 4, 6, 8, 10, 12, 16\}$ b. $\{2, 8\}$ 2. $-3x^2 - 13x + 1$ 3. -16
4. 1.8×10^{10} 5. a. $(x + 2)(x + 3)(x - 3)$ b. Quotient: $x^2 + 5x + 6$; remainder: 0 6. $x - 2$ 7. $\dfrac{2\sqrt{2x}}{x}$ 8. $8000 in 12% account; $2000 in 3% account 9. $(8, 7)$ 10. a. 720 ft b. 720 ft
c. 12 sec 11. $x = 3 \pm 4i$ 12. $x = \dfrac{-5 \pm \sqrt{33}}{4}$ 13. 25
14. $2(x + 5)(x^2 - 5x + 25)$ 15.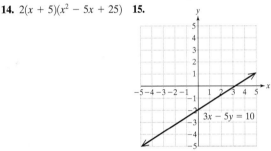

16. a. $\left(\dfrac{5}{2}, 0\right)(2, 0)$ b. $(0, 10)$
17. a.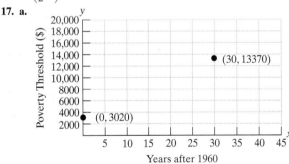

b. $y = 345x + 3020$ 18. Free throws: 565; 2-pt. shots: 851; 3-pt. shots: 30 19. The domain element 3 has more than one corresponding range element. 20.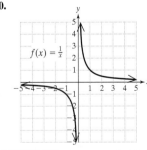

21. $y = 39$ 22. a. Linear b. $(0, 300,000)$. If there are no passengers, the airport runs 300,000 flights per year. c. $m = 0.008$ or $m = \dfrac{8}{1000}$. There are eight additional flights per 1000 passengers. 23. a. 3
b. $\sqrt{2}$ c. Not a real number 24. a. $[-4, \infty)$ b. $(-\infty, \infty)$
25. a. $(-\infty, 2]$ b. $(-\infty, 1) \cup \{2\} \cup [3, 4]$ c. 2 d. 3 e. 2
f. $x = -3$ 26. $x = \dfrac{1}{18}, x = \dfrac{1}{2}$ 27. $f = \dfrac{pq}{p + q}$ 28. $t = 7$
29. $\dfrac{y + 1}{y - 3}$ 30. a. $(3, 1)$ b. Upward c. $(0, 19)$ d. No x-intercept

e.

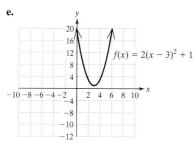

31. $x = 8 \pm \sqrt{62}$ **32.** Vertex: $(8, -62)$

Chapter 9

Chapter 9 Preview, p. 626

1. $\left(1, \dfrac{8}{5}\right]$ **3.** $200 < x < 239$ **5.** $\left(-\infty, -\dfrac{1}{2}\right) \cup (3, \infty)$

7. No solution **9.** $c = 3$ or $c = 5$ **11.** $x = -1$ or $x = \dfrac{2}{5}$

13. $\left[\dfrac{1}{3}, 3\right]$

15. 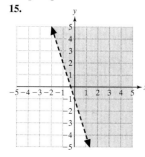 **17.**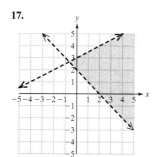

Section 9.1 Practice Exercises, pp. 633–636

3. $\left(-\dfrac{1}{2}, \infty\right)$ **5.** $[-16, \infty)$ **7.** $(66, \infty)$ **9. a.** $[-1, 5)$ **b.** $(-2, \infty)$

11. a. $(-1, 3)$ **b.** $\left(-\dfrac{5}{2}, \dfrac{9}{2}\right)$ **13. a.** $(0, 2]$ **b.** $(-4, 5]$

15. ⟵—[——]⟶ $[-2, 3]$
   $\quad\quad -2\ \ \ 3$

17. ⟵———(——)—⟶ $(3, 6)$
   $\quad\quad\quad 3\ \ \ 6$

19. ⟵————]⟶ $(-\infty, 2]$
   $\quad\quad\quad 2$

21. ⟵———[———⟶ $[8, \infty)$
   $\quad\quad\quad\ \ 8$

23. No solution **25.** $-4 \le t$ and $t < \dfrac{3}{4}$
27. The statement $6 < x < 2$ is equivalent to $6 < x$ and $x < 2$. However, no real number is greater than 6 and also less than 2.
29. The statement $-5 > y > -2$ is equivalent to $-5 > y$ and $y > -2$. However, no real number is less than -5 and also greater than -2. **31.** ⟵———[——)⟶ $\left[\dfrac{5}{2}, 7\right)$
   $\quad\quad\quad \frac{5}{2}\ \ \ 7$

33. ⟵—(———]—⟶ $(-6, 6]$
   $\quad -6\quad\quad 6$

35. ⟵———(——)⟶ $(2, 8)$
   $\quad\quad\quad 2\ \ 8$

37. ⟵—[————]⟶ $\left[-\dfrac{10}{3}, -\dfrac{7}{3}\right]$
   $\quad -\frac{10}{3}\quad -\frac{7}{3}$

39. ⟵—[————)⟶ $\left[-\dfrac{1}{2}, \dfrac{3}{2}\right)$
   $\quad -\frac{1}{2}\quad\ \ \frac{3}{2}$

41. ⟵——)——(——⟶ $(-\infty, -4) \cup (-2, \infty)$
   $\quad -4\ \ -2$

43. ⟵——)——[——⟶ $(-\infty, -2) \cup [2, \infty)$
   $\quad -2\ \ \ 2$

45. ⟵————]——⟶ $(-\infty, 2]$
   $\quad\quad\quad\ 2$

47. ⟵—[————⟶ $[-6, \infty)$
   $\quad -6$

49. ⟵————⟶ $(-\infty, \infty)$

51. ⟵—)——(——⟶ $\left(-\infty, -\dfrac{4}{3}\right) \cup (2, \infty)$
   $\ -\frac{4}{3}\ \ 2$

53. ⟵—]——(——⟶ $(-\infty, -4] \cup (4.5, \infty)$
   $\quad -4\quad \frac{9}{2}$

55. a. $(-10, 8)$ **b.** $(-\infty, \infty)$ **57. a.** $(-\infty, -8] \cup [1.3, \infty)$
b. No solution **59.** $\left(-\dfrac{11}{2}, \dfrac{7}{2}\right)$ **61.** $\left(-\dfrac{13}{2}, \infty\right)$ **63.** $(-\infty, -17)$
65. $(-\infty, 1) \cup (5, \infty)$ **67. a.** $4800 \le x \le 10{,}800$
b. $x < 4800$ or $x > 10{,}800$ **69. a.** $13 \le x \le 16$
b. $x < 13$ or $x > 16$ **71.** All real numbers between $-\dfrac{3}{2}$ and 6
73. All real numbers greater than 2 or less than -1

Section 9.2 Practice Exercises, pp. 644–648

3. $\left(-\infty, \dfrac{3}{8}\right) \cup (3, \infty)$ **5.** $(-3, 0)$ **7.** $[-1, 3]$ **9. a.** $(-2, 0) \cup (3, \infty)$
b. $(-\infty, -2) \cup (0, 3)$ **c.** $(-\infty, -2] \cup [0, 3]$ **d.** $[-2, 0] \cup [3, \infty)$
11. a. $(-1, 1]$ **b.** $(-\infty, -1) \cup [1, \infty)$ **c.** $(-\infty, -1) \cup (1, \infty)$
d. $(-1, 1)$ **13. a.** $x = 4, x = -\dfrac{1}{2}$ **b.** $\left(-\infty, -\dfrac{1}{2}\right) \cup (4, \infty)$ **c.** $\left(-\dfrac{1}{2}, 4\right)$
15. a. $x = -10, x = 3$ **b.** $(-10, 3)$ **c.** $(-\infty, -10) \cup (3, \infty)$
17. a. $p = -\dfrac{1}{2}, p = 3$ **b.** $\left[-\dfrac{1}{2}, 3\right]$ **c.** $\left(-\infty, -\dfrac{1}{2}\right] \cup [3, \infty)$
19. $(-1, 7)$ **21.** $(-\infty, -2) \cup (5, \infty)$ **23.** $[-5, 0]$ **25.** $[4, 8]$
27. $(-11, 11)$ **29.** $\left(-\infty, -\dfrac{1}{3}\right] \cup [3, \infty)$ **31.** $(-\infty, -3] \cup [0, 4]$
33. $(-2, -1) \cup (2, \infty)$ **35. a.** $x = 7$ **b.** $(-\infty, 5) \cup (7, \infty)$ **c.** $(5, 7)$
37. a. $z = 4$ **b.** $[4, 6)$ **c.** $(-\infty, 4] \cup (6, \infty)$ **39.** $(1, \infty)$
41. $(-\infty, -4) \cup (4, \infty)$ **43.** $\left(2, \dfrac{7}{2}\right)$ **45.** $(5, 7]$ **47.** $(-\infty, 0) \cup \left[\dfrac{1}{2}, \infty\right)$
49. $(0, \infty)$ **51.** All real numbers **53.** No solution **55.** No solution
57. $\{0\}$ **59.** $(-\infty, -12) \cup (-12, \infty)$ **61.** $\{-12\}$
63. Quadratic; $[-4, 4]$ **65.** Quadratic; $(-\infty, \infty)$
67. Linear; $\left(-\dfrac{5}{2}, \infty\right)$ **69.** Rational; $(-\infty, -1)$
71. Polynomial (degree > 2); $(0, 5) \cup (5, \infty)$ **73.** Polynomial
(degree > 2); $(2, \infty)$ **75.** Rational; no solution **77.** Quadratic;
$(-\sqrt{2}, \sqrt{2})$ **79.** Quadratic; $\left(-\infty, \dfrac{-5 - \sqrt{33}}{2}\right] \cup \left[\dfrac{-5 + \sqrt{33}}{2}, \infty\right)$
81. $(-\infty, 0) \cup (2, \infty)$ **83.** $(-1, 1)$

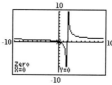   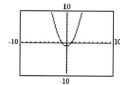

85. $\{-5\}$ **87.** No solution

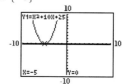   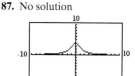

Midchapter Review, p. 648

1. $(-\infty, -9) \cup [10, \infty)$ **2.** $(-\infty, -1) \cup \left(\frac{7}{3}, \infty\right)$ **3.** $\left(\frac{4}{3}, 5\right]$
4. $(-\infty, 0) \cup (0, \infty)$ **5.** $(-3, 4)$ **6.** No solution
7. $[-2, -1] \cup [1, \infty)$ **8.** $[-1, 2)$ **9.** $\left(-\infty, -\frac{5}{4}\right) \cup (5, \infty)$
10. $(6.1, \infty)$ **11.** $(-\infty, -1) \cup (-1, \infty)$ **12.** $\left(-3, -\frac{3}{2}\right)$
13. No solution **14.** $(1, 9)$ **15.** No solution **16.** $(-\infty, -10)$
17. $(-\infty, -2] \cup (5, \infty)$ **18.** $(-\infty, 1]$

Section 9.3 Practice Exercises, pp. 654–656

3. $\left(\frac{2}{3}, \frac{23}{2}\right)$ **5.** $\left(4, \frac{16}{3}\right]$ **7.** All real numbers between -7 and 7 or equivalently $(-7, 7)$ **9.** $(-\infty, -4) \cup \left(\frac{1}{2}, 2\right)$ **11.** $p = 7, p = -7$
13. $x = 6, x = -6$ **15.** $y = \sqrt{2}, y = -\sqrt{2}$ **17.** No solution
19. $q = 0$ **21.** $x = \frac{1}{3}, x = 0$ **23.** $z = \frac{10}{7}, z = -\frac{8}{7}$
25. $y = 2, y = -\frac{14}{5}$ **27.** No solution **29.** $w = -12, w = 28$
31. $y = \frac{5}{2}, y = -\frac{7}{2}$ **33.** $b = \frac{7}{3}$ **35.** No solution **37.** $x = \frac{3}{2}$
39. $k = -4, k = \frac{16}{5}$ **41.** $x = -\frac{1}{2}, x = \frac{7}{6}$ **43.** $x = \frac{5}{2}, x = -\frac{3}{2}$
45. $w = -4, w = \frac{1}{3}$ **47.** $y = \frac{1}{2}$ **49.** $w = -\frac{1}{16}$ **51.** $h = \frac{1}{4}$
53. $m = -1.44, m = -0.4$ **55.** No solution **57.** $|x| = 6$
59. $|x| = \frac{4}{3}$ **61.** $y = \frac{4}{5}, y = \frac{2}{5}$ **63.** $x = 7 - \sqrt{3}, x = -7 - \sqrt{3}$
65. $w = -\sqrt{6}, w = 0$
67. $x = 2, x = -\frac{1}{2}$

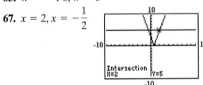

69. No solution **71.** $x = \frac{1}{2}$

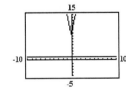

73. $x = \frac{4}{3}, x = -2$

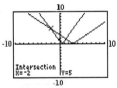
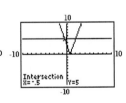
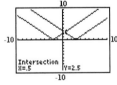
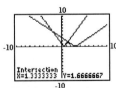

Section 9.4 Practice Exercises, pp. 664–667

3. No solution **5.** $x = -\frac{5}{3}, x = -\frac{1}{3}$
7. $(-3, -1]$
9. $(-\infty, \infty)$
11. a. $x = -5, x = 5$
  b. $(-\infty, -5) \cup (5, \infty)$
  c. $(-5, 5)$ **13. a.** $x = 10, x = -4$
  b. $(-\infty, -4) \cup (10, \infty)$
  c. $(-4, 10)$ **15. a.** No solution
  b. $(-\infty, \infty)$ **c.** No solution
17. a. No solution **b.** $(-\infty, \infty)$
  c. No solution **19.** b **21.** a
23. $(-\infty, -6) \cup (6, \infty)$
25. $[-3, 3]$
27. $(-\infty, \infty)$
29. $(-\infty, -2] \cup [3, \infty)$ **31.** No solution
33. $[-10, 14]$
35. $\left(-\infty, -\frac{5}{4}\right] \cup \left[\frac{23}{4}, \infty\right)$
37. $\left(-\frac{21}{2}, \frac{19}{2}\right)$
39. $(-\infty, \infty)$
41. No solution **43.** $\{-5\}$
45. $(-\infty, 6) \cup (6, \infty)$
47. $(-\infty, 2] \cup [4, \infty)$
49. $(4, 8)$
51. $(-10.5, 4.5)$
53. $|x| > 7$ **55.** $|x - 2| \leq 13$ **57.** $|x - 32| \leq 0.05$
59. $|x - 6\frac{3}{4}| \leq \frac{1}{8}$ **61.** $1.99 \leq w \leq 2.01$ or, equivalently in interval notation, $[1.99, 2.01]$. This means that the actual width of the bolt could be between 1.99 cm and 2.01 cm inclusive.
63. $(-\infty, -6) \cup (2, \infty)$

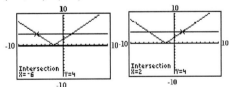

65. $(-7, 5)$

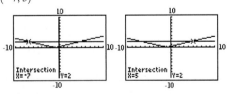

67. No solution **69.** $(-\infty, \infty)$

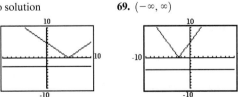

71. $x = -\dfrac{1}{6}$

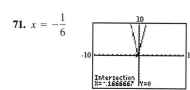

Section 9.5 Practice Exercises, pp. 676–682

3. $(1, 4)$ **5.** $\left[-\dfrac{5}{6}, \dfrac{7}{6}\right]$ **7.** $(-\infty, -2] \cup [1, \infty)$ **9. a.** Yes **b.** No **c.** Yes **d.** No **11. a.** Yes **b.** Yes **c.** No **d.** No **13.** $>$ **15.** $\geq$ **17.** $\leq$; $\geq$

19.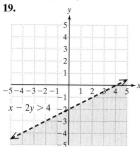

21.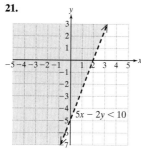

23.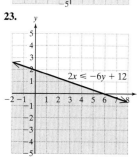

25.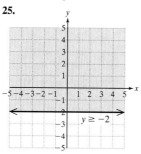

27.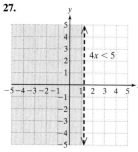

29.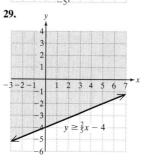

31.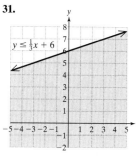

33.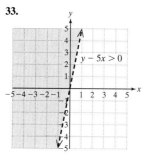

35.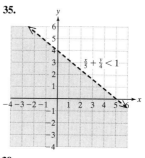

37.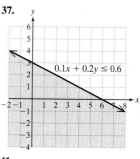

39.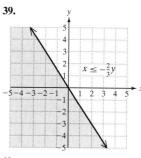

41.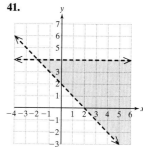

43.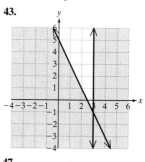

45.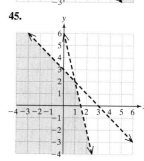

47.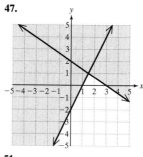

49.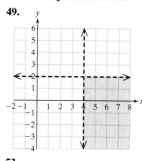

51.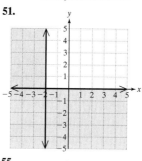

53.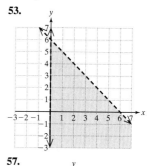

55.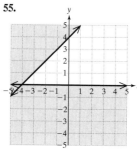

57.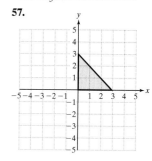

59. 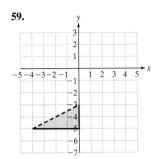 **61.** 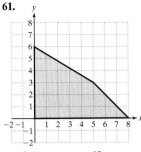 **65.** 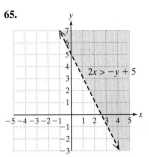 **67.**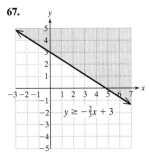

63. a. $x \geq 0, y \geq 0$ **b.** $x \leq 40, y \leq 40$ **c.** $x + y \geq 65$
d.

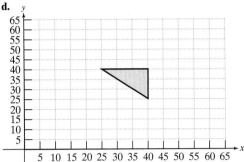

69. 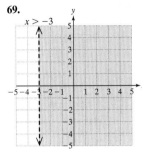 **71.**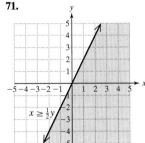

73. 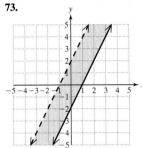 **75.**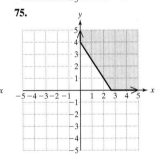

e. Yes; The point (35, 40) means that Karen works 35 hours and Todd works 40 hours. **f.** No; The point (20, 40) means that Karen works 20 hours and Todd works 40 hours. This does not satisfy the constraint that there must be at least 65 hours total.

Chapter 9 Review Exercises, pp. 688–692

1. $\left(-\dfrac{11}{4}, 4\right]$ **3.** No solution   **5.** $(-\infty, 6] \cup (12, \infty)$ **7.** $\left(-\infty, \dfrac{1}{2}\right)$
9. $\left[0, \dfrac{4}{3}\right]$ **11.** All real numbers between -6 and 12
13. a. $125 \leq x \leq 200$ **b.** $x < 125$ or $x > 200$ **15. a.** The solution is the intersection of the two inequalities. Answer: $\{x \mid -2 \leq x \leq 5\}$.   **b.** The solution is the union of the two inequalities. Answer: the set of all real numbers.   **17. a.** $x = 0$; $(0, 0)$ is the x-intercept.   **b.** $x = 2$ is the vertical asymptote.   **c.** On the intervals $(-\infty, 0]$ and $(2, \infty)$ the graph is on or above the x-axis.   **d.** On the interval $[0, 2)$ the graph is on or below the x-axis.   **19.** $(-\infty, \infty)$ **21.** $(-\infty, -1] \cup (1, \infty)$
23. $(-2, 0) \cup \left(\dfrac{5}{2}, \infty\right)$ **25.** $(-\infty, -5) \cup (1, \infty)$ **27.** $(-3, \infty)$
29. $(-\infty, -2) \cup (-2, \infty)$ **31.** $x = 17, x = -17$
33. $x = -0.44, x = 2.54$ **35.** $x = 3, x = 1$ **37.** No solution
39. $x = -\dfrac{5}{4}$ **41.** $x = -\dfrac{1}{2}$ **43.** $|x| > 5$ **45.** $|x| < 6$

47.  $(-\infty, -14] \cup [2, \infty)$

49. $\left(-\infty, \dfrac{1}{7}\right) \cup \left(\dfrac{1}{7}, \infty\right)$

51. $\left[-2, -\dfrac{2}{3}\right]$

53. $(2, 22)$

55. $(-\infty, \infty)$
57. $(-\infty, \infty)$
59. No solution   **61.** If an absolute value is less than a negative number, there will be no solution.   **63.** $0.17 \leq p \leq 0.23$ or, equivalently in interval notation, $[0.17, 0.23]$. This means that the actual percentage of viewers is estimated to be between 17% and 23% inclusive.

77.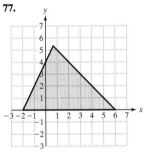

Chapter 9 Test, pp. 692–693

1. a. $\left[-\dfrac{1}{3}, 2\right]$ **b.** $(-\infty, -24] \cup [-15, \infty)$ **c.** $[-3, 0)$
2. a. $(-\infty, \infty)$ **b.** No solution   **3. a.** $9 \leq x \leq 33$
b. $x < 9$ or $x > 33$ **4.** $\left[\dfrac{1}{2}, 6\right)$ **5.** $(-5, 5)$
6. $(-\infty, -3) \cup (-2, 2)$ **7.** $\left(-3, -\dfrac{3}{2}\right)$
8. No solution   **9.** $\{-11\}$ **10. a.** $x = 10, x = -22$
b. $x = -8, x = 2$ **11. a.** $x = 7, x = -1$; $(7, 0)(-1, 0)$ are the x-intercepts.   **b.** $-1 < x < 7$. On the interval $(-1, 7)$ the graph is below the x-axis.   **c.** $x < -1$ or $x > 7$. On the intervals $(-\infty, -1)$ and $(7, \infty)$ the graph is above the x-axis.   **12.** No solution
13. $\left(-\infty, -\dfrac{1}{3}\right) \cup \left(\dfrac{17}{3}, \infty\right)$ **14.** $(-18.75, 17.25)$ **15.** $(-\infty, \infty)$
16. $|x - 15.41| \leq 0.01$

17. 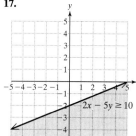 **18.**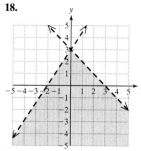

19. a. $x \geq 0, y \geq 0$ **b.** $300x + 400y \geq 1200$
c.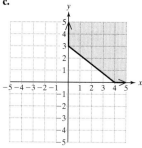

Chapters 1–9 Cumulative Review Exercises, pp. 693-695

1. $x^2 - x - 13$ **2.** $m = -\frac{3}{2}, m = 1$ **3. a.** $p = 0, p = 6$ **b.** $(0, 6)$
c. $(-\infty, 0) \cup (6, \infty)$ **4. a.** $y = 14, y = -10$ **b.** $(-10, 14)$
c. $(-\infty, -10) \cup (14, \infty)$ **5.**
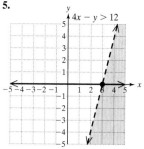

6. a. $t(1) = 9$. After one trial the rat requires 9 min to complete the maze. $t(50) = 3.24$. After 50 trials the rat requires 3.24 min to complete the maze. $t(500) = 3.02$. After 500 trials the rat requires 3.02 min to complete the maze. **b.** The limiting time is 3 min.
c. More than 5 trials **7. a.** $\left(-\infty, -\frac{5}{2}\right] \cup [2, \infty)$ **b.** On these intervals, the graph is on or above the x-axis.
8.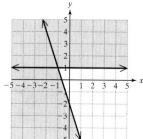

9. $-7x + 13$ **10.** Approximately $\$1.5 \times 10^6$ per restaurant
11. a. Quotient: $2x^2 - 5x + 12$; remainder: $-24x + 5$
12. a. $b_1 = \frac{2A - hb_2}{h}$ or $b_1 = \frac{2A}{h} - b_2$ **b.** 10 cm
13. 75 mph **14.** 57°, 123° **15.** $8000 in the 6.5% account; $5000 in the 5% account **16.** Neither

17. a. 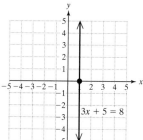 x-intercept $(1, 0)$; no y-intercept; undefined slope

b. 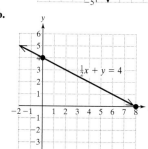 x-intercept $(8, 0)$; y-intercept $(0, 4)$; slope $-\frac{1}{2}$

18. $y = -\frac{2}{3}x - \frac{13}{3}$ **19.** $(-1, 3, -2)$ **20. a.** 4×3
b. 3×3 **21.** Plane: 500 mph; wind: 20 mph **22. a.** Quadratic
b. $P(0) = -2600$. The company will lose $2600 if no desks are produced. **c.** $P(x) = 0$ for $x = 20$ and $x = 650$. The company will break even ($0 profit) when 20 desks or 650 desks are produced.
23. $\{x | x \leq 50\}$ or $(-\infty, 50]$ **24.** $-\frac{xy}{x + y}$ **25.** $-\frac{a + 1}{a}$
26. $\frac{2x + 5}{(x - 5)(x - 2)(x + 10)}$

Chapter 10

Chapter 10 Preview, p. 698

1. $-2x^2 + x + 1$ **3.** $\frac{3x + 1}{2x^2 + 2x}$ **5.** $f^{-1}(x) = \frac{x - 2}{-4}$
7. 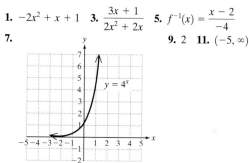 **9.** 2 **11.** $(-5, \infty)$

13. $\log_4 (xy)$ **15.** $\log_3 x + 2 \log_3 y - \log_3 z$ **17.** $\frac{\log 82}{\log 7}$
19. $y = 3$ ($y = -6$ does not check.) **21.** $t = \frac{\ln 30}{\ln 5} \approx 2.113$

Section 10.1 Practice Exercises, pp. 703–706

3. $(f + g)(x) = 2x^2 + 5x + 4; (-\infty, \infty)$
5. $(g - f)(x) = 2x^2 + 3x - 4; (-\infty, \infty)$
7. $(f \cdot h)(x) = (x + 4)\sqrt{x - 1}; [1, \infty)$
9. $(g \cdot f)(x) = 2x^3 + 12x^2 + 16x; (-\infty, \infty)$
11. $\left(\frac{h}{f}\right)(x) = \frac{\sqrt{x - 1}}{x + 4}; [1, \infty)$
13. $\left(\frac{f}{g}\right)(x) = \frac{x + 4}{2x^2 + 4x}; (-\infty, -2) \cup (-2, 0) \cup (0, \infty)$

15. $(f \circ g)(x) = 2x^2 + 4x + 4; (-\infty, \infty)$
17. $(f \circ k)(x) = \frac{1}{x} + 4; (-\infty, 0) \cup (0, \infty)$
19. $(k \circ h)(x) = \frac{1}{\sqrt{x-1}}; (1, \infty)$
21. $(k \circ g)(x) = \frac{1}{2x^2 + 4x}; (-\infty, -2) \cup (-2, 0) \cup (0, \infty)$
23. No   **25.** $(f \circ g)(x) = 25x^2 - 15x + 1; (g \circ f)(x) = 5x^2 - 15x + 5$
27. $(f \circ g)(x) = |x^3 - 1|; (g \circ f)(x) = |x|^3 - 1$
29. $(h \circ h)(x) = 25x - 24$ **31.** 0   **33.** -64 **35.** 2   **37.** 1   **39.** $\frac{1}{64}$
41. 0   **43.** Undefined   **45.** 0   **47.** 1   **49.** 0   **51.** Undefined
53. -1 **55.** 2   **57.** 2   **59.** 3   **61.** -6 **63.** -1 **65.** 4   **67.** 0
69. 2   **71. a.** $P(x) = 3.78x - 1$ **b.** $188   **73. a.** $F(t) = 0.2t + 6.4$; F represents the amount of child support (in billion dollars) not paid.   **b.** $F(0) = 6.4$ means that in 2000, $6.4 billion of child support was not paid. $F(2) = 6.8$ means that in 2002, $6.8 billion of child support was not paid. $F(4) = 7.2$ means that in 2004, $7.2 billion of child support was not paid.   **75. a.** $(D \circ r)(t) = 560t$; This function represents the total distance Joe travels as a function of time.
b. 5600 ft

Section 10.2 Practice Exercises, pp. 713–717

3. Yes   **5.** No   **7.** Yes   **9.** $g^{-1} = \{(5,3),(1,8),(9,-3),(2,0)\}$
11. $r^{-1} = \{(3,a),(6,b),(9,c)\}$ **13.** The function is not one-to-one.
15. Yes   **17.** No   **19.** Yes   **21.** $h^{-1}(x) = x - 4$
23. $m^{-1}(x) = 3(x+2)$ **25.** $p^{-1}(x) = -x + 10$ **27.** $f^{-1}(x) = \sqrt[3]{x}$
29. $g^{-1}(x) = \frac{x^3 + 1}{2}$ **31. a.** 1.2192 m, 15.24 m   **b.** $f^{-1}(x) = \frac{x}{0.3048}$
c. 4921.3 ft   **33.** False   **35.** True   **37.** False   **39.** True   **41.** $(b, 0)$
43. For example: $f(x) = x$ **45. a.** Domain $\{x \mid x \leq 0\}$, range $\{y \mid y \geq -4\}$ **b.** Domain $\{x \mid x \geq -4\}$, range $\{y \mid y \leq 0\}$
47. a. $\{x \mid -2 \leq x \leq 0\}$ **b.** $\{y \mid 0 \leq y \leq 2\}$ **c.** $\{x \mid 0 \leq x \leq 2\}$
d. $\{y \mid -2 \leq y \leq 0\}$
e.

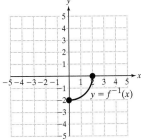

49. a. $\{x \mid 2 \leq x \leq 5\}$ **b.** $\{y \mid 0 \leq y \leq 3\}$ **c.** $\{x \mid 0 \leq x \leq 3\}$
d. $\{y \mid 2 \leq y \leq 5\}$
e.

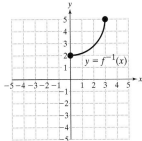

51. a. $(f \circ g)(x) = 5\left(\frac{x+2}{5}\right) - 2 = x$
b. $(g \circ f)(x) = \frac{(5x-2) + 2}{5} = x$
53. a. $(f \circ g)(x) = \frac{\sqrt[3]{27x^3}}{3} = x$ **b.** $(g \circ f)(x) = 27\left(\frac{\sqrt[3]{x}}{3}\right)^3 = x$

55. a. $(f \circ g)(x) = (\sqrt{x+3})^2 - 3 = x$
b. $(g \circ f)(x) = \sqrt{(x^2 - 3) + 3} = x$ **57.** $p^{-1}(x) = \frac{3 - 3x}{x + 1}$
59. $w^{-1}(x) = \frac{4 - 2x}{x}$ **61.** $m^{-1}(x) = \sqrt{x + 1}$
63. $g^{-1}(x) = -\sqrt{x+1}$ **65.** $v^{-1}(x) = x^2 - 16, x \geq 0$
67. $u^{-1}(x) = x^2 - 16, x \leq 0$
69. $k^{-1}(x) = \sqrt[3]{x+4}$ **71.** $m^{-1}(x) = \frac{x+4}{3}$

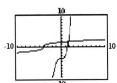   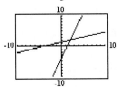

Section 10.3 Practice Exercises, pp. 724–728

3. $2x^2 + 4x + 1$ **5.** $6x^3 + x^2 + 5x - 2$ **7.** $18x^2 - 9x + 3$
9. $\{(3,2),(0,0),(4,-8),(12,10)\}$ **11.** 25   **13.** $\frac{1}{1000}$ **15.** 6   **17.** 8
19. 5.8731   **21.** 1385.4557   **23.** 0.0063   **25.** 0.8950   **27. a.** $x = 2$
b. $x = 3$ **c.** Between 2 and 3   **29. a.** $x = 4$ **b.** $x = 5$
c. Between 4 and 5   **31.** $f(0) = 1, f(1) = \frac{1}{5}, f(2) = \frac{1}{25}, f(-1) = 5$,
$f(-2) = 25$ **33.** $h(0) = 1, h(1) = \pi \approx 3.14, h(-1) = 0.32$,
$h(\sqrt{2}) = 5.05, h(\pi) = 36.46$ **35.** $r(0) = 9, r(1) = 27, r(2) = 81$,
$r(-1) = 3, r(-2) = 1, r(-3) = \frac{1}{3}$ **37.** If $b > 1$, the graph is increasing. If $0 < b < 1$, the graph is decreasing.

39.

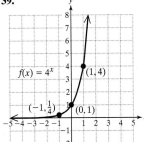

41.

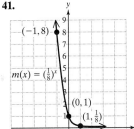

43.

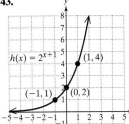

45.

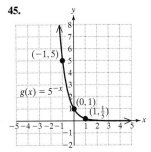

47. a. 0.25 g   **b.** ≈ 0.16 g   **49. a.** 758,000   **b.** 379,000   **c.** 144,000
51. a. $A(t) = 141,340,000(1.015)^t$ **b.** $A(46) \approx 280,000,000$
53. a. $1640.67   **b.** $2691.80   **c.** $A(0) = 1000$. The initial amount of the investment is $1000. $A(7) = 2000$. The amount of the investment doubles in 7 years.

55.

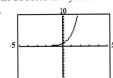

57.

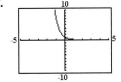

59. 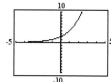 **61.**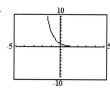

b.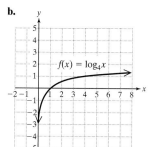

63. a.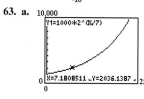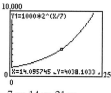

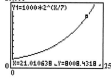 **b.** 7 yr, 14 yr, 21 yr

75. $3^y = x$

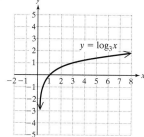

x	y
$\frac{1}{9}$	-2
$\frac{1}{3}$	-1
1	0
3	1
9	2

Section 10.4 Practice Exercises, pp. 737–742

3. ii **5. a.** $g(-2) = \frac{1}{9}, g(-1) = \frac{1}{3}, g(0) = 1, g(1) = 3, g(2) = 9$

b.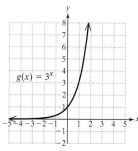

77. $\left(\frac{1}{2}\right)^y = x$

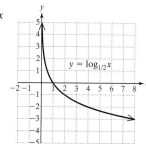

x	y
4	-2
2	-1
1	0
$\frac{1}{2}$	1
$\frac{1}{4}$	2

7. a. $s(-2) = \frac{25}{4}, s(-1) = \frac{5}{2}, s(0) = 1, s(1) = \frac{2}{5}, s(2) = \frac{4}{25}$

b.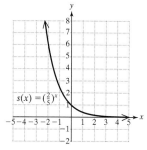

79. $(5, \infty)$ **81.** $(-1.2, \infty)$ **83.** $(-\infty, 0) \cup (0, \infty)$ **85.** ≈ 7.35

87. a.

t (months)	0	1	2	6	12	24
$S_1(t)$	91	82.0	76.7	65.6	57.6	49.1
$S_2(t)$	88	83.5	80.8	75.3	71.3	67.0

b. Group 1: 91; Group 2: 88 **c.** Method II

89. Domain: $(-6, \infty)$; asymptote: $x = -6$

91. Domain: $(2, \infty)$; asymptote: $x = 2$

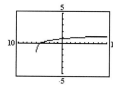 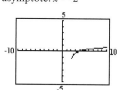

93. Domain: $(-\infty, 2)$; asymptote: $x = 2$

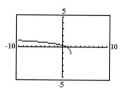

9. $b^y = x$ **11.** $\log_{10} 1000 = 3$ **13.** $\log_8 2 = \frac{1}{3}$

15. $\log_8\left(\frac{1}{64}\right) = -2$ **17.** $\log_b x = y$ **19.** $\log_e x = y$

21. $\log_{5/2}\left(\frac{2}{5}\right) = -1$ **23.** $125^{2/3} = 25$ **25.** $25^{-1/2} = \frac{1}{5}$ **27.** $2^7 = 128$

29. $b^y = 82$ **31.** $2^x = 7$ **33.** $\left(\frac{1}{2}\right)^6 = x$ **35.** 3 **37.** -4 **39.** $\frac{1}{3}$

41. -1 **43.** 3 **45.** 0 **47.** 4 **49.** $\frac{1}{3}$ **51.** 2 **53.** 4 **55.** -1

57. -3 **59.** 0.7782 **61.** 0.4971 **63.** -1.5051 **65.** -2.2676
67. 5.5315 **69.** -7.4202 **71. a.** Slightly less than 2 **b.** Slightly more than 1 **c.** $\log 93 \approx 1.9685$, $\log 12 \approx 1.0792$

73. a. $f\left(\frac{1}{64}\right) = -3, f\left(\frac{1}{16}\right) = -2, f\left(\frac{1}{4}\right) = -1, f(1) = 0, f(4) = 1,$ $f(16) = 2, f(64) = 3$

Midchapter Review, p. 743

1. $x^2 + 5$ **2.** $-x^2 - 2x + 5$ **3.** $-x^3 + 4x^2 + 5x$ **4.** $\frac{5 - x}{x^2 + x}$

5. $-x^2 - x + 5$ **6.** $x^2 - 11x + 30$ **7.** It is not one-to-one because the range element 5 corresponds to two different x-values.

8. $f^{-1}(x) = 2x - 4$ **9. a.** $f(2) = 49, f(-1) = \frac{1}{7}, f(0) = 1, f(1) = 7$

b. $1, 2, -1, 0$ **10. a.** $h(2) = 100, h(-1) = 0.1, h(0) = 1, h(1) = 10$
b. $k(0.5) = -0.3010, k(1) = 0, k(3) = 0.4771$
11. a. 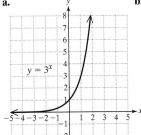 **b.** Domain: $(-\infty, \infty)$; range: $(0, \infty)$

12. a. 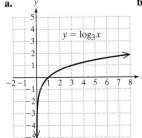 **b.** Domain: $(0, \infty)$; range: $(-\infty, \infty)$

13. $\log_x z = y$ **14.** $p^t = q$ **15.** 3 **16.** 3 **17.** -6 **18.** 0 **19.** -4

Section 10.5 Practice Exercises, pp. 748–752

3. $\dfrac{1}{64}$ **5.** 5 **7.** 2.9707 **9.** 1.4314 **11.** d **13.** b
15. For example: $\log_5 1 = 0$ **17.** For example: $\log_4(4^2) = 2$ **19.** 1
21. 4 **23.** 11 **25.** 3 **27.** 0 **29.** 9 **31.** 0 **33.** 3 **35.** 1 **37.** $2x$
39. 4 **41.** Expressions a and c are equivalent. **43.** Expressions a and c are equivalent. **45.** $\log_3 x - \log_3 5$ **47.** $\log 2 + \log x$
49. $4 \log_5 x$ **51.** $\log_4 a + \log_4 b - \log_4 c$
53. $\dfrac{1}{2} \log_b x + \log_b y - 3 \log_b z - \log_b w$
55. $\log_2(x+1) - 2\log_2 y - \dfrac{1}{2}\log_2 z$
57. $\dfrac{1}{3}\log a + \dfrac{2}{3}\log b - \dfrac{1}{3}\log c$ **59.** $-5 \log w$
61. $\dfrac{1}{2}\log_b a - 3 - \log_b c$ **63.** $\log(CABIN)$ **65.** $\log_3\left(\dfrac{x^2 z}{y^3}\right)$
67. $\log_3\left(\dfrac{a^2 c}{\sqrt[4]{b}}\right)$ **69.** $\log_b(x^2)$ **71.** $\log_8(8a^5)$ or $\log_8(a^5) + 1$
73. $\log\left[\dfrac{(x+6)^2 \sqrt[3]{y}}{z^5}\right]$ **75.** $\log_b\left(\dfrac{1}{x-1}\right)$ **77.** 1.792 **79.** 2.485
81. 4.396 **83.** 0.916 **85.** 13.812 **87. a.** $B = 10\log I - 10\log I_0$
b. $10\log I + 160$
89. a. Domain: $(-\infty, 1) \cup (1, \infty)$ **b.** Domain: $(1, \infty)$

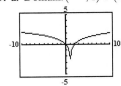 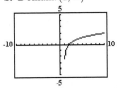

c. They are equivalent for all x in the intersection of their domains, $(1, \infty)$.

Section 10.6 Practice Exercises, pp. 760–766

3.

x	$f(x)$
-3	$\frac{8}{27}$
-2	$\frac{4}{9}$
-1	$\frac{2}{3}$
0	1
1	$\frac{3}{2}$
2	$\frac{9}{4}$
3	$\frac{27}{8}$

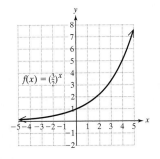

5.

x	$q(x)$
-0.5	-0.30
0	0
4	0.70
9	1.00

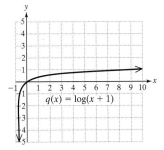

7.

x	y
-4	0.05
-3	0.14
-2	0.37
-1	1
0	2.72
1	7.39

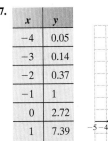

Domain: $(-\infty, \infty)$

9.

x	y
-2	2.14
-1	2.37
0	3
1	4.72
2	9.39
3	22.09

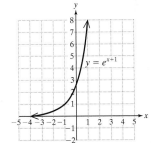

Domain: $(-\infty, \infty)$

11. a. \$12,209.97 **b.** \$13,488.50 **c.** \$14,898.46 **d.** \$16,050.09
An investment grows more rapidly at higher interest rates.
13. a. \$12,423.76 **b.** \$12,515.01 **c.** \$12,535.94 **d.** \$12,546.15
e. \$12,546.50 More money is earned at a greater number of compound periods per year. **15. a.** \$6920.15 **b.** \$9577.70
c. \$13,255.84 **d.** \$18,346.48 **e.** \$35,143.44 More money is earned over a longer period of time.

17.

x	y
2.25	-1.39
2.50	-0.69
2.75	-0.29
3	0
4	0.69
5	1.10
6	1.39

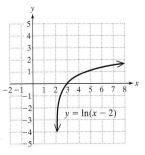

Domain: $(2, \infty)$

19.

x	y
0.25	−2.39
0.5	−1.69
0.75	−1.29
1	−1.00
2	−0.31
3	0.10
4	0.39

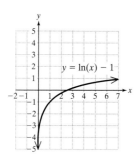

Domain: $(0, \infty)$

21. a. 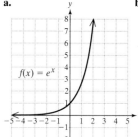 **b.** Domain: $(-\infty, \infty)$; range: $(0, \infty)$

c. 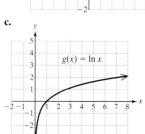 **d.** Domain: $(0, \infty)$; range: $(-\infty, \infty)$

23. 1 **25.** 0 **27.** −6 **29.** $2x + 3$ **31.** $\ln(p^6 \sqrt[3]{q})$ **33.** $\ln\sqrt{\dfrac{x}{y^3}}$

35. $\ln\left(\dfrac{a^2}{b\sqrt[3]{c}}\right)$ **37.** $\ln\left(\dfrac{x^4}{y^3 z}\right)$ **39.** $2\ln a - 2\ln b$ **41.** $2\ln b + 1$

43. $4\ln a + \dfrac{1}{2}\ln b - \ln c$ **45.** $\dfrac{1}{5}\ln a + \dfrac{1}{5}\ln b - \dfrac{2}{5}\ln c$

47. a. 2.9570 **b.** 2.9570 **c.** They are the same. **49.** 2.8074
51. 1.5283 **53.** −2.1269 **55.** 0 **57.** −3.3219 **59.** −3.8124
61. a. 15.4 years **b.** 6.9 years **c.** 13.8 years **63. a.** 91 deaths
 b. Sept. 5: 349 deaths; Sept. 10: 459 deaths; Sept. 20: 570 deaths
65. a., b. 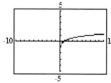 **c.** They appear to be the same.

67. a., b. 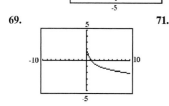 **c.** They appear to be the same.

69.   **71.**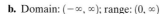

73.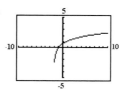

Section 10.7 Practice Exercises, pp. 776–782

3. a.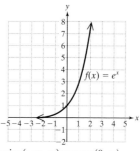

b. Domain: $(-\infty, \infty)$; range: $(0, \infty)$
5. a. 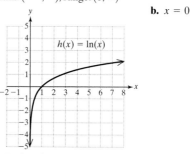 **b.** $x = 0$

c. Domain: $(0, \infty)$; range: $(-\infty, \infty)$ **7.** $\log_b[(x-1)(x+2)]$
9. $\log_b\left(\dfrac{x}{1-x}\right)$ **11.** $x = 9$ **13.** $p = 10^{42}$ **15.** $x = e^{0.08}$
17. $x = \dfrac{1}{16}$ **19.** $x = 5$ **21.** $b = 10$ **23.** $y = 25$ **25.** $c = 59$
27. $y = 1$ **29.** $x = 0$ **31.** $x = -\dfrac{37}{9}$ **33.** $k = \dfrac{3}{2}$ ($k = -3$ does not check.) **35.** $x = 4$ **37.** $x = 3$ **39.** $t = 2$ **41.** No solution ($m = -3$ does not check.) **43.** $10^{6.3}$ (or approximately 1,995,262) times more intense **45.** $10^{-3.07}$ W/m² **47.** $x = 4$ **49.** $x = -6$ **51.** $x = \dfrac{1}{2}$
53. $x = 2$ **55.** $x = \dfrac{11}{12}$ **57.** $x = 1$ **59.** $x = \dfrac{4}{19}$ **61.** $x = -2$, $x = 1$ **63.** $a = \dfrac{\ln 21}{\ln 8} \approx 1.464$ **65.** $x = \ln 8.1254 \approx 2.095$
67. $t = \log 0.0138 \approx -1.860$ **69.** $h = \dfrac{\ln 15}{0.07} \approx 38.686$
71. $m = \dfrac{\ln 4}{0.04} \approx 34.657$ **73.** $x = \dfrac{\ln 3}{\ln 5 - \ln 3} \approx 2.151$
75. a. ≈ 1285 million (or 1,285,000,000) people **b.** ≈ 1412 million (or 1,412,000,000) people **c.** The year 2049 ($t \approx 50.8$)
77. a. 500 bacteria **b.** ≈ 660 **c.** ≈ 25 min **79.** 9.9 years
81. a. 7.8 g **b.** 18.5 days **83.** 38.4 years **85. a.** 1.42 kg **b.** No
87. $x = 10^5$, $x = 10^{-3}$ **89.** $w = \dfrac{1}{27}$, $w = \dfrac{1}{9}$ **91. a. i.** $660,348
ii. $730,555 **iii.** $772,840 **b.** As the amount of money used for advertising increases, the amount of sales increases but at a slower rate.
c. $x \approx 251$ or $251,000 for advertising

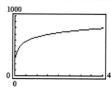

93. 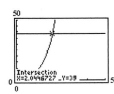 **95.**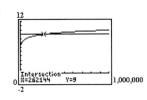

105. $m = \dfrac{\log\left(\dfrac{1}{821}\right)}{-3} \approx 0.9714$ **107. a.** 150 bacteria **b.** ≈ 185 **c.** ≈ 99 min

Chapter 10 Review Exercises, pp. 789–793

1. $(f - g)(x) = 2x^3 + 9x - 7; (-\infty, \infty)$
3. $(f \cdot n)(x) = \dfrac{x - 7}{x - 2}; (-\infty, 2) \cup (2, \infty)$
5. $\left(\dfrac{f}{g}\right)(x) = \dfrac{x - 7}{-2x^3 - 8x}; (-\infty, 0) \cup (0, \infty)$
7. $(m \circ f)(x) = \sqrt{x - 7}; [7, \infty)$ **9.** $\sqrt{32}$ or $4\sqrt{2}$
11. -167 **13. a.** $(2x + 1)^2$ or $4x^2 + 4x + 1$ **b.** $2x^2 + 1$
c. No, $f \circ g \neq g \circ f$ **15.** -3 **17.** -1 **19.** 1 **21.** Yes
23. $q^{-1}(x) = \dfrac{4}{3}(x + 2)$ **25.** $f^{-1}(x) = \sqrt[3]{x} + 1$
27. $(f \circ g)(x) = 5\left(\dfrac{1}{5}x + \dfrac{2}{5}\right) - 2 = x + 2 - 2 = x$
$(g \circ f)(x) = \dfrac{1}{5}(5x - 2) + \dfrac{2}{5} = x - \dfrac{2}{5} + \dfrac{2}{5} = x$
29. a. Domain: $\{x \mid x \geq -1\}$; range: $\{y \mid y \geq 0\}$
b. Domain: $\{x \mid x \geq 0\}$; range: $\{y \mid y \geq -1\}$ **31.** 1024
33. 2 **35.** 8.825 **37.** 1.627
39. 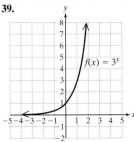 **41.**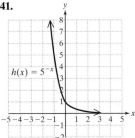

43. a. Horizontal **b.** $y = 0$ **45.** -3 **47.** 1 **49.** 4 **51.** 5
53.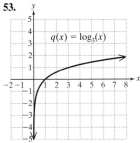

55. a. Vertical asymptote **b.** $x = 0$ **57.** 1 **59.** 0
61. a. $\log_b x + \log_b y$ **b.** $\log_b\left(\dfrac{x}{y}\right)$ **c.** $p \log_b x$ **63.** $\log_3\left(\dfrac{\sqrt{ab}}{c^2 d^4}\right)$
65. 0 **67.** a **69.** 14.0940 **71.** 57.2795 **73.** -2.1972 **75.** 1.3029
77. 1.9943 **79.** -3.6668 **81. a.** $S(0) = 95$; the student's score is 95 at the end of the course. **b.** $S(6) \approx 23.7$; the student's score is 23.7 after 6 months. **c.** $S(12) \approx 20.2$; the student's score is 20.2 after 1 year. **d.** The limiting value is 20. **83.** $(-\infty, \infty)$ **85.** $(0, \infty)$
87. $(7, \infty)$ **89.** $(-\infty, 5)$ **91.** $x = \dfrac{1}{49} \approx 0.02$ **93.** $y = 3^{1/12} \approx 1.10$
95. $w = 9$ **97.** $t = \dfrac{190}{321} \approx 0.59$ **99.** $x = \dfrac{4}{7}$
101. $a = \dfrac{\ln(18)}{\ln(5)} \approx 1.7959$ **103.** $x = -\dfrac{\ln(0.06)}{2} \approx 1.4067$

Chapter 10 Test, pp. 793–795

1. $\left(\dfrac{f}{g}\right)(x) = \dfrac{x - 4}{\sqrt{x + 2}}$ **2.** $(h \cdot g)(x) = \dfrac{\sqrt{x + 2}}{x}$
3. $(g \circ f)(x) = \sqrt{x - 2}$ **4.** $(h \circ f)(x) = \dfrac{1}{x - 4}$ **5.** 0 **6.** $-\dfrac{3}{2}$ **7.** $\dfrac{1}{4}$
8. Undefined **9.** $\left(\dfrac{g}{f}\right)(x) = \dfrac{\sqrt{x + 2}}{x - 4}$; domain: $[-2, 4) \cup (4, \infty)$
10. A function is one-to-one if it passes the horizontal line test.
11. b **12.** $f^{-1}(x) = 4x - 12$ **13.** $g^{-1}(x) = \sqrt{x} + 1$
14.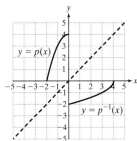

15. a. 4.6416 **b.** 32.2693 **c.** 687.2913
16.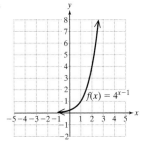

17. a. $\log_{16} 8 = \dfrac{3}{4}$ **b.** $x^5 = 31$ **18.**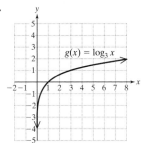

19. $\dfrac{\log_a n}{\log_a b}$ **20. a.** 1.3222 **b.** 1.8502 **c.** -2.5850 **21. a.** $1 + \log_3 x$
b. -5 **22. a.** $\log_b(\sqrt{x}\, y^3)$ **b.** $\log\left(\dfrac{1}{a^3}\right)$ or $-\log a^3$ **23. a.** 1.6487
b. 0.0498 **c.** -1.0986 **d.** 1 **24. a.** $y = \ln(x)$ **b.** $y = e^x$
25. a. $p(4) \approx 59.8$; 59.8% of the material is retained after 4 months.
b. $p(12) \approx 40.7$; 40.7% of the material is retained after 1 year.
c. $p(0) = 92$; 92% of the material is retained at the end of the course. **26. a.** 9762 thousand (or 9,762,000) people **b.** The year 2020 ($t \approx 20.4$) **27. a.** $P(0) = 300$; there are 300 bacteria initially.
b. 35,588 bacteria **c.** 1,120,537 bacteria **d.** 1,495,831 bacteria
e. The limiting value appears to be 1,500,000. **28.** $x = 25$ ($x = -4$ does not check.) **29.** $x = 32$ **30.** $x = e^{2.4} - 7 \approx 4.023$
31. $x = -7$ **32.** $x = \dfrac{\ln 50}{\ln 4} \approx 2.822$ **33.** $x = \dfrac{\ln 250}{2.4} \approx 2.301$
34. a. $P(2500) = 560.2$; at 2500 m the atmospheric pressure is 560.2 mm Hg. **b.** 760 mm Hg **c.** 1498.8 m **35. a.** $2909.98 **b.** 9.24 years to double

Chapters 1–10 Cumulative Review Exercises, pp. 796–799

1. $-\dfrac{5}{4}$ **2.** $-1 + \dfrac{p}{2} + \dfrac{3p^3}{4}$ **3.** Quotient: $t^3 + 2t^2 - 9t - 18$; remainder: 0 **4.** $|x - 3|$ **5.** $\dfrac{2\sqrt[3]{25}}{5}$ **6.** $2\sqrt{7}$ in. **7.** $\dfrac{4d^{1/10}}{c}$ **8.** $(\sqrt{15} - \sqrt{6} + \sqrt{30} - 2\sqrt{3})\,\text{m}^2$ **9.** $-\dfrac{7}{29} - \dfrac{26}{29}i$ **10.** $42°, 48°$ **11.** $m = \dfrac{2}{3}$ **12.** $x = 6, x = \dfrac{3}{2}$ **13.** 4.8 L **14.** 24 min **15.** $(7, -1)$ **16.** $w = -\dfrac{23}{11}$ **17.** $x = \dfrac{c + d}{a - b}$ **18.** $t = \dfrac{\sqrt{2sg}}{g}$ **19.** $T = \dfrac{1 - \left(\dfrac{V_0}{V}\right)^2}{k}$ or $\dfrac{V^2 - V_0^2}{kV^2}$ **20.** $(7, 0), (3, 0)$ **21. a.** $-30t$ **b.** $-10t^2$ **c.** $2t^2 + 5t$ **22.** $q = 0$ **23. a.** $x = 2$ **b.** $y = 6$ **c.** $y = \dfrac{1}{2}x + 5$

24. a.

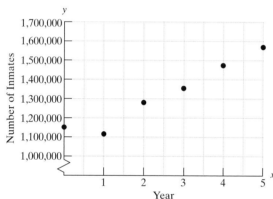

b. $y = 82{,}500x + 1{,}148{,}000$ **c.** $m = 82{,}500$; there is an increase of 82,500 inmates per year. **d.** 1,973,000 inmates **25.** $40°, 80°, 60°$ **26.** $\{(x, y)\,|\,-2x + y = -4\}$ **27. a.** vi **b.** i **c.** v **d.** x **e.** ii **f.** ix **g.** iv **h.** viii **i.** vii **j.** iii **28.** $\left[\dfrac{1}{2}, \infty\right)$ **29.** $f^{-1}(x) = \dfrac{1}{5}x + \dfrac{2}{15}$ **30.** $40\,\text{m}^3$ **31.** $\dfrac{-x^2 + 5x - 25}{(2x + 1)(x - 2)}$ **32.** $1 + x$ **33. a.** Yes; $x \neq 4, x \neq -2$ **b.** $x = 8$ **c.** $(-\infty, -2) \cup (4, 8]$ **34.** The numbers are $-\dfrac{3}{4}, 4$. **35.** $x = -4$ ($x = -9$ does not check.) **36.** $(-\infty, \infty)$ **37. a.** $P(6) = 2{,}000{,}000, P(12) = 1{,}000{,}000, P(18) = 500{,}000, P(24) = 250{,}000, P(30) = 125{,}000$

b.

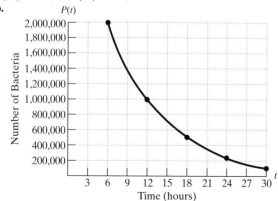

c. 48 hr **38. a.** 2 **b.** -3 **c.** 6 **d.** 3 **39. a.** 217.0723 **b.** 23.1407 **c.** 0.1768 **d.** 3.7293 **e.** -0.4005 **f.** 2.6047 **40.** $x = \dfrac{2}{3}$ **41.** $x = \ln(100) \approx 4.6052$ **42.** $x = 3$ ($x = -9$ does not check.) **43.** $\log\left(\dfrac{\sqrt{z}}{x^2 y^3}\right)$ **44.** $\dfrac{2}{3}\ln x - \dfrac{1}{3}\ln y$

Chapter 11

Chapter 11 Preview, p. 802

1. $\sqrt{29}$ **3.** Center: $(3, -1)$; radius: 5 **5. a.** Vertex: $(2, -3)$ **b.** $y = -3$ **7.**

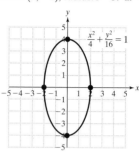

9. $(-2, -1)$ and $(1, 2)$ **11.**

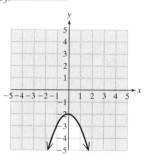

Section 11.1 Practice Exercises, pp. 807–812

3. $\sqrt{281}$ **5.** $3\sqrt{2}$ **7.** $\dfrac{\sqrt{226}}{10}$ **9.** 19 **11.** 10 **13.** $\sqrt{29}$ **15.** $\sqrt{255}$ **17.** Subtract 5 and -7. This becomes $5 - (-7) = 12$. **19.** $y = 13, y = 1$ **21.** $x = 0, x = 8$ **23.** Yes **25.** No

27.

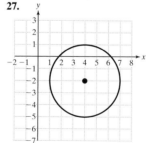

Center $(4, -2); r = 3$

29.

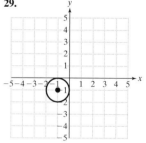

Center $(-1, -1); r = 1$

31.

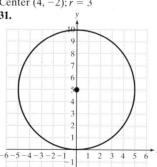

Center $(0, 5); r = 5$

33.

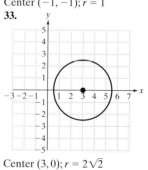

Center $(3, 0); r = 2\sqrt{2}$

35.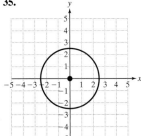
Center $(0,0); r = \sqrt{6}$

37.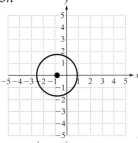
Center $\left(-\frac{4}{5}, 0\right); r = \frac{8}{5}$

39.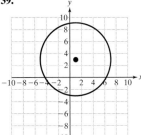
Center $(1, 3); r = 6$

41.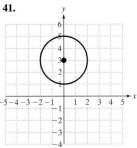
Center: $(0, 3); r = 2$

43.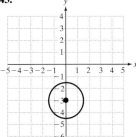
Center $(0, -3); r = \frac{4}{3}$

45.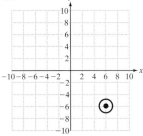
Center: $(6, -6); r = 1$

47.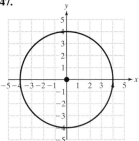
Center: $(0, 0); r = 4$

49. $x^2 + y^2 = 4$ **51.** $x^2 + (y - 2)^2 = 4$ **53.** $(x + 2)^2 + (y - 2)^2 = 9$
55. $x^2 + y^2 = 49$ **57.** $(x + 3)^2 + (y + 4)^2 = 36$
59. $x^2 + (y - 3)^2 = 4$
61. $(x - 4)^2 + (y - 4)^2 = 16$ **63.** $(x - 1)^2 + (y - 1)^2 = 29$

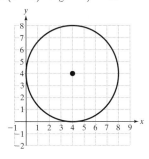

65.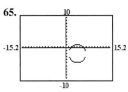

67.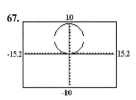

69.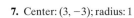

Section 11.2 Practice Exercises, pp. 818–821

3. $\sqrt{10}$
5. Center: $(0, -1)$; radius: 4
7. Center: $(3, -3)$; radius: 1

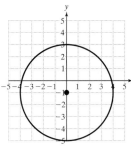
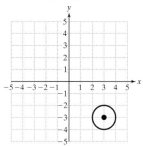

9.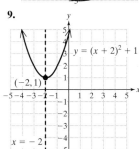

11.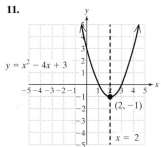

13.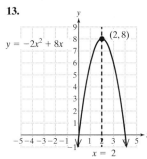

15.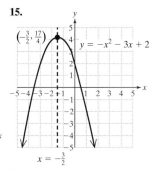

17.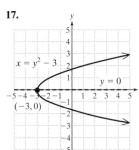

19.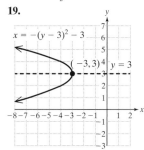

21. 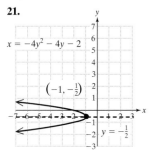 **23.**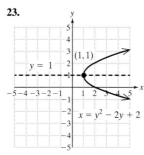

25. $(2, -1)$ **27.** $(5, -1)$ **29.** $(2, 8)$ **31.** $\left(\dfrac{3}{2}, -\dfrac{1}{4}\right)$ **33.** $(33, 4)$

35. For a parabola whose equation is written in the form $y = a(x - h)^2 + k$, if $a > 0$, the parabola opens upward; if $a < 0$, the parabola opens downward. For a parabola written in the form $x = a(y - k)^2 + h$, if $a > 0$, the parabola opens right; if $a < 0$, the parabola opens left. **37.** Vertical axis of symmetry; opens upward **39.** Vertical axis of symmetry; opens downward **41.** Horizontal axis of symmetry; opens right **43.** Horizontal axis of symmetry; opens left **45.** Vertical axis of symmetry; opens downward

Section 11.3 Practice Exercises, pp. 826–830

3. Center: $(8, -6)$; radius: 10 **5.** Vertex: $(-3, -1)$; $x = -3$
7. $\left(x - \dfrac{1}{2}\right)^2 + \left(y - \dfrac{5}{2}\right)^2 = \dfrac{1}{4}$

9. 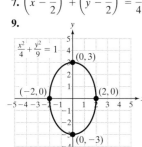 **11.**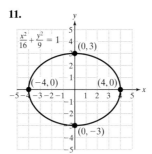

13.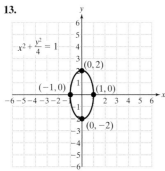

15.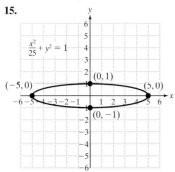

17. Vertical **19.** Horizontal **21.** Horizontal **23.** Vertical

25.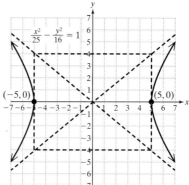

27.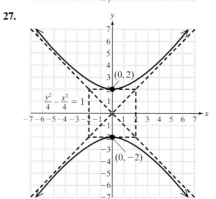

29.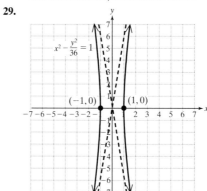

31.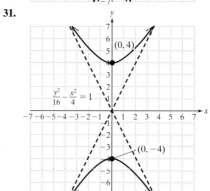

33. Hyperbola **35.** Ellipse **37.** Ellipse **39.** Hyperbola **41.** 49 ft

43. Center: $(4, 5)$; $a = 3, b = 2$ **45.** Center: $(2, -3)$; $a = 3, b = 1$

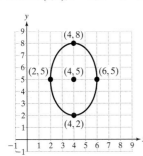
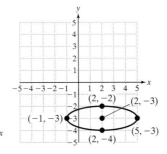

Midchapter Review, pp. 830–831

1. $d = \sqrt{(x_2 - x_1)^2 + (y_2 - y_1)^2}$ **2.** $5\sqrt{2}$ **3.** $2\sqrt{10}$
4. $(x - 5)^2 + (y + 10)^2 = 81$ **5. a.** Hyperbola   **b.** Circle
c. Parabola   **d.** Ellipse
6. 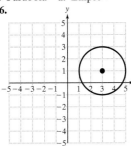   **7.**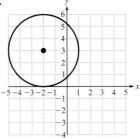

Center: $(3, 1)$; radius: 2   Center: $(-2, 3)$; radius: 3

8. 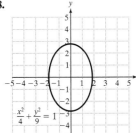   **9.**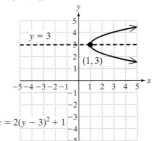

10.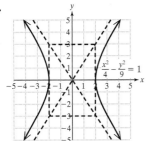

11. $\dfrac{x^2}{2} + \dfrac{y^2}{10} = 1$; ellipse   **12.** $\dfrac{x^2}{3} - \dfrac{y^2}{5} = 1$; hyperbola
13. $\dfrac{y^2}{6} - \dfrac{x^2}{1} = 1$; hyperbola   **14.** $\dfrac{x^2}{5} + \dfrac{y^2}{3} = 1$; ellipse

Section 11.4 Practice Exercises, pp. 836–839

3. $d = \sqrt{(x_2 - x_1)^2 + (y_2 - y_1)^2}$ **5.** $(x + 1)^2 + (y - 1)^2 = 4$
7. Circle   **9.** Parabola   **11.** Hyperbola   **13.** Ellipse   **15.** Zero, one, or two   **17.** Zero, one, or two   **19.** Zero, one, two, three, or four   **21.** Zero, one, two, three, or four

23. $(-3, 0)(2, 5)$

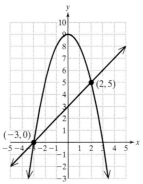

25. $(0, 1)(-1, 0)$ **27.** $(\sqrt{2}, 2)(-\sqrt{2}, 2)$

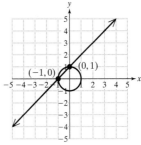
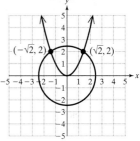

29. $(4, 2)$ **31.** $(0, 0)$ **33.** $\left(\dfrac{3}{2}, \dfrac{9}{4}\right)$ **35.** $(0, 0), (-2, -8)$
37. $(3, 2)(3, -2)(-3, 2)(-3, -2)$ **39.** $(0, 3)(0, -3)$
41. $(2, 1)(-2, 1)(2, -1)(-2, -1)$ **43.** $(\sqrt{2}, 0)(-\sqrt{2}, 0)$
45. $(2, 0)(-2, 0)$ **47.** $(3, 0)(-3, 0)$ **49.** $(4, 5)(-4, -5)$
51. 3 and 4   **53.** 5 and $\sqrt{7}$, -5 and $\sqrt{7}$, 5 and $-\sqrt{7}$, or -5 and $-\sqrt{7}$
55.

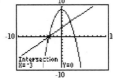
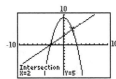

57.   **59.** No solution

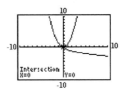
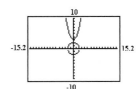

Section 11.5 Practice Exercises, pp. 843–848

3. e   **5.** i   **7.** j   **9.** a   **11.** d   **13.** False   **15.** True
17. a.

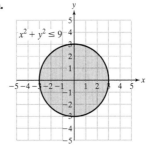

b. The set of points on and "outside" the circle $x^2 + y^2 = 9$
c. The set of points on the circle $x^2 + y^2 = 9$

19. a.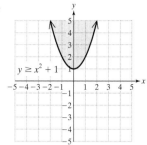

b. The parabola $y = x^2 + 1$ would be drawn as a dashed curve.

21. 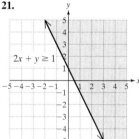   **23.**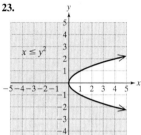

25.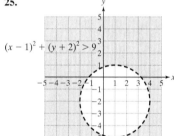

27. 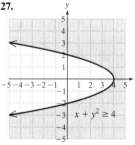   **29.**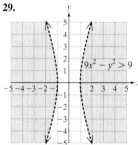

31. 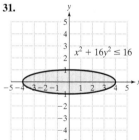   **33.**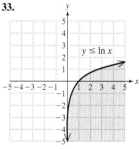

35. 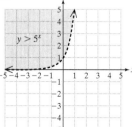   **37.**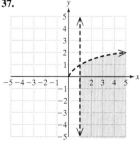

39. 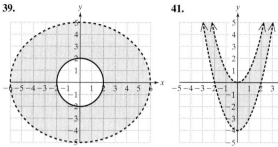   **41.**

43. 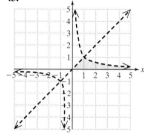   **45.** No Solution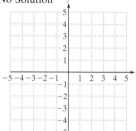

47. 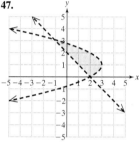   **49.**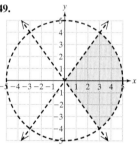

51. 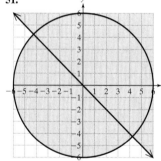   **53.**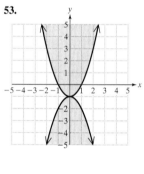

Chapter 11 Review Exercises, pp. 854–857

1. $2\sqrt{10}$ **3.** $x = 5$ or $x = -1$ **5.** Collinear   **7.** Center $(12, 3)$; $r = 4$ **9.** Center $(-3, -8)$; $r = 2\sqrt{5}$ **11. a.** $x^2 + y^2 = 64$ **b.** $(x - 8)^2 + (y - 8)^2 = 64$ **13.** $(x + 2)^2 + (y + 8)^2 = 8$ **15.** $(x - 3)^2 + \left(y - \dfrac{1}{3}\right)^2 = 9$ **17.** $x^2 + (y - 2)^2 = 9$

19. Horizontal axis of symmetry; parabola opens right
21. Vertical axis of symmetry; parabola opens upward

23.   **25.**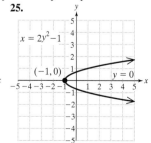

27. $x = (y + 2)^2 - 2$; vertex: $(-2, -2)$; axis of symmetry: $y = -2$
29. $y = -2\left(x + \frac{1}{2}\right)^2 + \frac{1}{2}$; vertex: $\left(-\frac{1}{2}, \frac{1}{2}\right)$; axis of symmetry: $x = -\frac{1}{2}$

31. 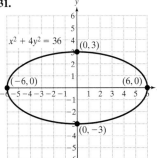 33. Vertical 35. Horizontal

37. 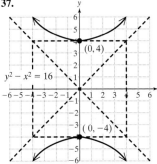 39. Ellipse 41. Hyperbola

43. a. Line and parabola
b. 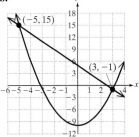 c. $(-5, 15), (3, -1)$

45. a. Circle and line
b. 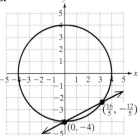 c. $(0, -4), \left(\frac{16}{5}, -\frac{12}{5}\right)$

47. $\left(-\frac{7}{5}, \frac{13}{5}\right)(-5, -1)$ 49. $(2, 4)(-2, 4)(\sqrt{2}, 2)(-\sqrt{2}, 2)$
51. $(6, 5)(-6, 5)$ $(6, -5)(-6, -5)$
53. 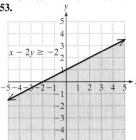 55.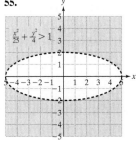

57.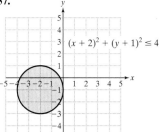

59. 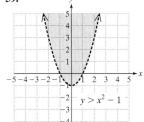 61.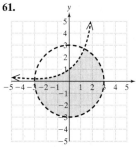

Chapter 11 Test, pp. 857–858

1. 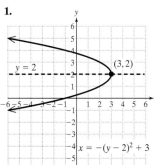 2. Center: $\left(\frac{5}{6}, -\frac{1}{3}\right); r = \frac{5}{7}$

3. 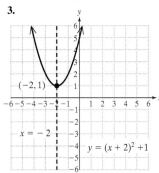 4. $\sqrt{85}$ 5. Center: $(0, 2); r = 3$

6. a. $\sqrt{5}$ b. $x^2 + (y - 4)^2 = 5$
7.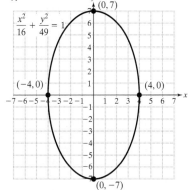

8.
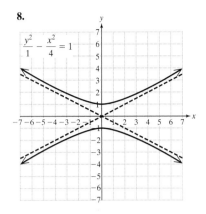
$\dfrac{y^2}{1} - \dfrac{x^2}{4} = 1$

9. a. $(-3, 0)(0, 4)$; ii **b.** No solution; i **10.** The addition method can be used if the equations have corresponding *like* terms.
11. $(2, 0)(-2, 0)$

12. 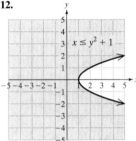 $x \le y^2 + 1$

13. 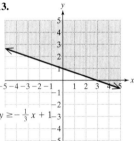 $y \ge -\dfrac{1}{3}x + 1$

14.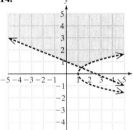

15.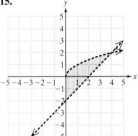

Chapters 11 Cumulative Review Exercises, p. 859

1. All real numbers
2. $\left(\dfrac{10}{3}, \infty\right)$ **3.** $10, 15$
4. a. $(-5, 0), (0, 3)$ **b.** $m = \dfrac{3}{5}$ **c.**
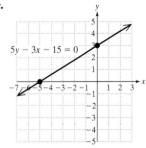
$5y - 3x - 15 = 0$

5. a. $m = 20$ **b.** $y = 20x + 390$ **c.** $490 billion **6.** 12 dimes, 5 quarters **7.** $(2, -3, 1)$ **8. a.** $\begin{bmatrix} 1 & -2 & | & -8 \\ 0 & 1 & | & 2 \end{bmatrix}$ **b.** $\begin{bmatrix} 1 & 0 & | & -4 \\ 0 & 1 & | & 2 \end{bmatrix}$
9. $\left(\dfrac{5}{2}, \dfrac{3}{2}\right)$ **10.** $f(0) = -12; f(-1) = -16; f(2) = -10; f(4) = -16$
11. $g(2) = 5; g(8) = -1; g(3) = 0; g(-5) = 5$ **12.** $z = 32$
13. a. -5 **b.** $(x+1)(x^2+1); -5$ **c.** They are the same.
14. a. No solution **b.** $x = -11$ **15. a.** $-30 + 24i$ **b.** $\dfrac{12}{41} + \dfrac{15}{41}i$

16. a. 17.6 ft; 39.6 ft; 70.4 ft **b.** 8 sec **17.** $w = -\dfrac{1}{5}; w = \dfrac{1}{10} \pm \dfrac{i\sqrt{3}}{10}$
18. $x = 2 \pm \sqrt{11}$ **19.** $(-1, 19)$ **20.** $\log_8 32 = \dfrac{5}{3}$ **21.** $(-5, -36)$
22. **a.** $(-3, 0), (1, 0)$ **b.** $(0, 3)$ **c.** $(-1, 4)$
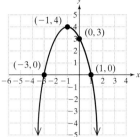

23. $x^2 + (y-5)^2 = 16$ **24.** Yes, the circle can be tangent to the parabola. **25.** $(0, -4)$
26. $y^2 - x^2 < 1$ 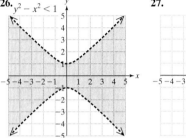 **27.**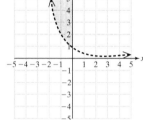

28. $(x+y)(x-y-6)$

Additional Topics Appendix

Section A.1 Practice Exercises, pp. 866–867

1. $x^4 + 4x^3y + 6x^2y^2 + 4xy^3 + y^4$ **3.** $64 + 48p + 12p^2 + p^3$
5. $a^{12} + 6a^{10}b + 15a^8b^2 + 20a^6b^3 + 15a^4b^4 + 6a^2b^5 + b^6$
7. $p^6 - 3p^4w + 3p^2w^2 - w^3$ **9.** The signs alternate on the terms of the expression $(a - b)^n$. The signs for the expression $(a + b)^n$ are all positive. **11.** 6 **13.** 1 **15.** False
17. $9! = 9 \cdot (8 \cdot 7 \cdot 6 \cdot 5 \cdot 4 \cdot 3 \cdot 2 \cdot 1) = 9 \cdot 8!$ **19.** 1680 **21.** 6
23. 56 **25.** 1 **27.** $s^6 + 6s^5t + 15s^4t^2 + 20s^3t^3 + 15s^2t^4 + 6st^5 + t^6$
29. $b^3 - 9b^2 + 27b - 27$ **31.** $16x^4 + 32x^3y + 24x^2y^2 + 8xy^3 + y^4$
33. $c^{14} - 7c^{12}d + 21c^{10}d^2 - 35c^8d^3 + 35c^6d^4 - 21c^4d^5 + 7c^2d^6 - d^7$
35. $\dfrac{1}{32}a^5 - \dfrac{5}{16}a^4b + \dfrac{5}{4}a^3b^2 - \dfrac{5}{2}a^2b^3 + \dfrac{5}{2}ab^4 - b^5$
37. $m^{11} - 11m^{10}n + 55m^9n^2$ **39.** $u^{24} - 12u^{22}v + 66u^{20}v^2$
41. 9 terms **43.** $-462m^6n^5$ **45.** $495u^{16}v^4$ **47.** g^9

Section A.2 Practice Exercises, pp. 876–879

1. -11 **3.** 16 **5.** 8 **7.** 18 **9.** -43
11. $\begin{bmatrix} + & - & + \\ - & + & - \\ + & - & + \end{bmatrix}$ **13. a.** 3 **b.** 3 **15.** 20 **17.** -26 **19.** 0
21. $-2x - 3y$ **23.** $-12a - 23b - 3c$ **25.** 0 **27.** $D = 14; D_x = 8; D_y = -5$ **29.** $D = -55; D_x = 54; D_y = 89$ **31.** $(1, 3)$ **33.** $(1, -1)$
35. $\left(\dfrac{11}{14}, \dfrac{17}{21}\right)$ **37.** The addition method, substitution method, or Gauss-Jordan method can be used to determine if a system is inconsistent or dependent. **39.** No solution (inconsistent system)
41. $(0, 0)$ **43.** $\{(x, y) | x + 5y = 2\}$ (dependent system)
45. $y = -2$ **47.** $x = \dfrac{1}{4}$ **49.** $y = -\dfrac{17}{167}$ **51.** $(1, 2, 3)$
53. Cramer's rule does not apply. **55.** $y = 2$ **57.** $t = -12$
59. 32 **61. a.** 8 **b.** $y = 4$

Section A.3 Practice Exercises, pp. 884–888

1. 4, 7, 10, 13   **3.** $\sqrt{3}, 2, \sqrt{5}, \sqrt{6}$ **5.** $3, \frac{3}{2}, 1, \frac{3}{4}, \frac{3}{5}$ **7.** $-\frac{2}{3}, \frac{3}{4}, -\frac{4}{5}, \frac{5}{6}$
9. $0, -3, 8, -15$ **11.** $\frac{5}{2}, \frac{8}{3}, \frac{11}{4}, \frac{14}{5}$ **13.** 0, 2, 6, 12
15. $-3, 9, -27, 81$ **17.** When n is odd, the term is negative. When n is even, the term is positive.   **19.** $60, $58.80, $57.62, $56.47, $55.34, $54.24   **21.** 25,000; 50,000; 100,000; 200,000; 400,000; 800,000; 1,600,000   **23.** $a_n = 2n$ **25.** $a_n = 2n - 1$ **27.** $a_n = \frac{1}{n^2}$
29. $a_n = (-1)^{n+1}$ **31.** $a_n = (-1)^n \, 2^n$ **33.** $a_n = \frac{3}{5^n}$
35. A sequence is an ordered list of terms. A series is the sum of the terms of a sequence.   **37.** 90   **39.** $\frac{31}{16}$ **41.** 30   **43.** 10   **45.** $\frac{73}{12}$
47. 38   **49.** -1 **51.** 55   **53.** $\sum_{n=1}^{6} n$ **55.** $\sum_{i=1}^{5} 4$ **57.** $\sum_{j=1}^{5} 4j$
59. $\sum_{k=1}^{4} (-1)^{k+1} \frac{1}{3^k}$ **61.** $\sum_{n=1}^{5} x^n$
63. $1 + 1\frac{1}{2} + 1\frac{1}{2} + 1\frac{1}{2} + 1\frac{1}{2} + 1\frac{1}{2} + 1\frac{1}{2} + 1\frac{1}{2}$; $1 + \sum_{i=1}^{7} 1\frac{1}{2}$; $11\frac{1}{2}$ in.
65. 15.4 g   **67.** 4.8 g   **69.** $-3, 2, 7, 12, 17$ **71.** 5, 21, 85, 341, 1365
73. 1, 1, 2, 3, 5, 8, 13, 21, 34, 55

75.
```
seq(8/n,n,1,5,1)
  {8 4 2.66666666…
Ans▶Frac
  {8 4 8/3 2 8/5}
```

77.
```
seq((-1)^n*n²,n,
1,5,1)
  {-1 4 -9 16 -25}
```

79.
```
sum(seq(8/n,n,1,
4,1)
       16.66666667
Ans▶Frac
              50/3
```

81.
```
sum(seq((-1)^n*n
²,n,1,100,1)
              5050
```

Section A.4 Practice Exercises, pp. 896–902

1. A sequence is arithmetic if the difference between any term and the preceding term is constant.   **3.** 6   **5.** -5 **7.** 4
9. $-3, -1, 1, 3, 5$ **11.** $0, \frac{1}{3}, \frac{2}{3}, 1, \frac{4}{3}$ **13.** $10, 4, -2, -8, -14$
15. $a_n = 2 + 5n$ **17.** $a_n = 5 - 4n$ **19.** $a_n = \frac{2}{3} + \frac{1}{3}n$
21. $a_n = 12 - 3n$ **23.** $a_n = -17 + 8n$ **25.** 17   **27.** 47   **29.** 46
31. -48 **33.** 19   **35.** 22   **37.** 23   **39.** 11   **41.** $a_1 = -2, a_2 = -5$
43. 670   **45.** 290   **47.** -15 **49.** 95   **51.** 924   **53.** 300   **55.** -210
57. 5050   **59.** 2500   **61.** 980 seats; $14,700   **63.** A sequence is geometric if the ratio between a term and the preceding term is constant.   **65.** $\frac{1}{2}$ **67.** -3 **69.** 4   **71.** $-4, 4, -4, 4, -4$
73. $8, 2, \frac{1}{2}, \frac{1}{8}, \frac{1}{32}$ **75.** $2, -6, 18, -54, 162$ **77.** $a_n = 2(3)^{n-1}$
79. $a_n = -6(-2)^{n-1}$ **81.** $a_n = \frac{16}{3}\left(\frac{3}{4}\right)^{n-1}$ **83.** $-\frac{3}{128}$ **85.** $-\frac{2}{81}$
87. 135   **89.** -10 **91.** $\frac{1}{9}$ **93.** 1   **95.** $\frac{1562}{125}$ **97.** $-\frac{11}{8}$ **99.** $\frac{3124}{27}$
101. $\frac{2059}{729}$ **103.** -172 **105. a.** $1050.00, $1102.50, $1157.63, $1215.51   **b.** $a_{10} = 1628.89; $a_{20} = 2653.30; $a_{40} = 7039.99
107. $r = \frac{1}{6}; \frac{6}{5}$ **109.** $r = -\frac{1}{3}; -\frac{9}{4}$ **111.** $r = -\frac{3}{2}$; sum does not exist
113. $800 million   **115.** 28 ft   **117. a.** $\frac{7}{10}$ **b.** $\frac{1}{10}$ **c.** $\frac{7}{9}$
119. Geometric, $r = -\frac{3}{2}$ **121.** Arithmetic, $d = \frac{2}{3}$
123. Geometric, $r = 3$ **125.** Arithmetic, $d = 4$ **127.** Neither
129. Neither   **131.** Geometric, $r = -1$
133. a. $1,191,123   **b.** $1,254,857   **c.** $63,734

135.
```
seq(125000*1.04^
n,n,1,4,1)
  {130000 135200 …
```

137.
```
sum(seq(n,n,1,15
,1)
              120
```

139. 8
```
sum(seq(4*(1/2)^
(n-1),n,1,10,1)
         7.9921875
sum(seq(4*(1/2)^
(n-1),n,1,20,1)
         7.999992371
```

Applications Index

BIOLOGY/HEALTH/LIFE SCIENCES

Adenosine deaminase levels in humans, 692
Animal population estimates, by sampling, 451
Bacteria population as function of time, 611, 726, 727, 780, 793, 795, 799, A25
Blood pressure, normal range of, 14
Body mass index vs. weight, 115–116
Calcium intake program, 693
Calorie totals for food, 450
Carbon dating, 759
Cholesterol level ranges, 626, 688
Disease epidemiology, 116
Doctor's visits as function of age, 606
Epidemic deaths as function of time, 765
Fat amount in foods, 450, 457
Feasible cage configurations at animal shelter, 674
Femur length vs. height in women, 251, 254
Hemoglobin range in humans, 636
Length of pregnancy in humans, 632
Life expectancy vs. year of birth, 182
Longevity of animals, by species, 252
Maximum heart rate vs. age, 104, 258
Mean protein content of foods, A27
Memory models, 736, 741, 792, 795
Plant growth calculation, A26
Platelets range in humans, 636
Precipitation per month, 257
Ratio of cats to dogs in shelter, 444
Reference daily intake of proteins, for various human subgroups, 255
Smoking in women, trends in, 310
Sodium content of foods, from total sodium intakes, 224
Thyroid-stimulating hormone ranges, 632
Weight vs. age for children, 143
White blood cell range in humans, 635

BUSINESS AND ECONOMICS

Airline flights as function of passenger volume, 624
Amount of each nut in nut mixture, from relative amounts, 226
Average cost as function of units produced, 609
Average sales per restaurant in chain, 694
Car rental cost vs. mileage, 161–162
Cell phone subscription trends, 142
Commission, 54, 61, 82, 98, 129–130, 165
Compensation package comparison, A41
Cost of airline operation, 527
Cost of cab ride per mile, 160
Cost of running business, 179–180
Cost per item as function of number produced, 446–447, 460
Crop planning, 691–692
Currency exchange rates, 460
Depreciation, 123, 145
Discounts and markdowns, 55, 62
Earnings needed to meet average salary level, 77
Employment trends, 138–139
Estimating ratio of male to female accountants, 451
Height of soda can, 183
Home value increase as function of time, A40
Income per capita, trends in, 176
Interest
  compound, 721, 727, 761, 762, 765, 780, 781, 792, 795, A25, A39
  simple, 54, 55, 56–57, 62, 82, 98, 103, 106, 287, 302, 344, 484, 487, 546, 562–563
Markups, 55–56, 62
Minimum wage trends, 159–160
Mixing foods for resale, 63
Motor vehicle and parts spending trends, 859
National debt per capita, 88–89
Number living below poverty line, 62
Number of fish, from weight of total catch, 450
Number of sales needed for profit, 82
Pay vs. hours worked, 284
Poverty threshold calculation, 82
Poverty threshold trends, 623
Product failure as function of time, A26
Production scheduling, 682
Productivity, from relative productivity, 246
Profit as function of number of items produced, 313, 382, 452, 605–606, 610, 695, 705
Profit as function of number of items sold, 461, 563
Profit of airline, 527
Radius of can, from volume, 562
Restaurant chain opening trend analysis, 166
Salary compounded daily, 718
Salary with commissions, vs. sales, 129–130, 165, 182
Salary with tips, vs. number of tables served, 297
Sales as function of advertising expenditures, 742, 782
Sales by each salesperson, from relative totals, 224
Sales tax, 2, 54, 61, 98
Sales vs. sale price, analysis of, 169–170
Starbucks sites added per year, 389
Supply and demand equilibrium, 218
SUV sales trends, 142
Tickets sold, by type, 63, 215, 226, 244
Time required for two workers to finish job, 461
Total economic impact of tourist spending, A40
Total revenue from sold-out theater, A38
Total salary over period, A41
Wholesale cost, 55–56
Work schedule for employees, 682

CONSTRUCTION AND DESIGN

Area of photograph as function of matting size, 320, 323
Area of washer, 368
Area of window, 25
Bridge support design, 599
Bulge height in heat-expansion of bridge, 547
Cost of carpeting room, 706
Cost of fencing area, 494
Cutting an elliptical rug, 821–822
Cutting proportional lengths of rope, 103
Diameter of fountain, 67
Dimensional variation in identical screws and bolts, 667, 690
Dimensions of fenced area, 69
Dimensions of octagonal stained glass window, 577
Dimensions of photograph from area, 320
Dimensions of pool, from perimeter, 462
Dimensions of space/object, from area, 380
Distance from ladder base to wall, 381
Elevator passenger capacity, 106
Fencing needed to enclose area, 69
Height of arch, 829
Length of front side of corner shelf, 562
Length of roof, from structure dimensions, 550
Length of tower guy wires, 548
Margin of error in measurement, 664, 666
Maximum enclosable area with given amount of fence, 611, 612, 622
Pump performance, from relative performance, 244
Radius of fountain, 67
Slope of ladder, 133, 140
Slope of leaning pole, 395
Slope of roof, 133, 140, 301
Slope of steps or ramp, 132–133, 179
Stained glass window design, 854
Strength of beam as function of dimensions, 291
Suspension bridge design, 599
Time required for two workers to finish job, 448–449, 453, 461
Time required for worker to finish job, 452, 453
Time required to fill tank, from pipe size, 460
Unit conversions, 715
Volume of box as function of height, 320
Window film needed for project, 25

CONSUMER APPLICATIONS

Car rental cost vs. mileage, 161–162
Car value vs. age, 462
Coins by type, from relative number, 216, 247, 859
Cook time vs. weight for turkey, 527
Cost of car rental, 218
Cost of items, from money spent, 210–211, 215, 216, 227
Cost of larger quantity of item, 450
Cost of long distance calls, 160
Depreciation of car value, 793
Discounts and markdowns, 55, 62
Dormitory fees, trends in, 313
Gas mileage, 10
Gas mileage as function of speed, 611

I-1

I-2 Applications Index

Gas mileage trends, analysis of, 83
Gas required to complete trip, 450
Home cost trends, 167, A40
Income after taxes, 103
Income tax owed, 103
Interest rate, from return, 562–563
Loan amount borrowed, from interest paid, 56
Loan cost comparisons, 61
Loan payment amounts, A25
Measurement error in filling product packages, 666
Mixing products, 62, 63, 216
Monthly budget ratios, 451
Moving van dimensions, 392
Price before sales tax, 2, 61
Property tax trends, 165–166
Sales tax, 2, 54, 61, 98
Spending on snacks at movie, 183
Telephone company charges compared, 218, 244
Telephone company costs per month, 250
Textbook cost *vs.* size, 161
Tire wear *vs.* tire pressure, 609
Tuition costs, trends in, 310
Video rental costs compared, 247
Wedding spending per year, 62
Weight of car *vs.* gas mileage, 254

DISTANCE/RATE/TIME

Airplane landing distance *vs.* speed, 618
Average speed of racecar, 71
Average speed of walker, 103
Car braking distance *vs.* speed, 575
Depth of water, from wave velocity, 548
Distance as function of acceleration, 860
Distance as function of time, 270
Distance as function of wheel revolutions, 706
Distance between cities, 63
Distance between cities, from map, 450, 461, 463
Distance between cities, from relative distance, 567
Distance between travelers moving at right angles, 473, 478
Distance from earth to celestial objects, 88–89, 93
Distance of free fall, from velocity, 526
Distance of lightning strike, calculating, 166
Distance to location, 106
Distance traveled by bouncing ball, A35–A36, A40
Distance traveled by bungee jumper, A40
Distance traveled by hiker, 58–59
Distance traveled, from final separation, 567
Distance traveled, from time and speeds, 183
Distance *vs.* time for vehicle, 284
Gas remaining *vs.* time, 123
Halfway point between travelers, 113, 117
Height *vs.* time in free fall, 270, 377, 527
Length of pendulum, from speed of swing, 526
Route travel time comparisons, 381, 494
Snowfall rate and totals, 158–159
Sound intensity *vs.* distance, 286–287
Speed, from relative speeds, 694
Speed of airplane discounting wind speed, 186, 213–214, 217, 244, 695
Speed of boat as function of length, 286
Speed of boat discounting water speed, 217, 452

Speed of car based on skid mark length, 523–524
Speed of travelers, from distance of separation, 63, 217, 225
Speed of travelers, from relative distances traveled, 452
Speed of travelers, from relative speeds, 398, 447–448, 460, 461
Speed of travelers, from relative time of trip, 452, 462
Speed of travelers, from time of meeting, 63, 452
Speed *vs.* time for vehicle, 284
Speed *vs.* time in free fall, 279
Stopping distance *vs.* speed, 290
Time required for object to fall, 517, 523, 559–560, 562, 563
Time required for pendulum swing, 517, 526
Time required for travelers to meet, 63
Time required for trip, 58–59, 63, 71, 106
Time required for two workers to finish job, 398, 448–449, 453, 461
Time required for worker to finish job, 452, 453
Time required to overtake traveler, 796
Time required to reach investment goal, 82, 158
Time spent driving at each speed, 58
Unit conversions, 715
Wave velocity in water, 548

EDUCATION AND SCHOOL

College attendance by males, trends in, 103
College enrollment, from relative enrollments, 227
College enrollment trends, 179
Dormitory fee trends, 313
Enrollment trends, 301
Grades needed to make desired average, 77–78, 396
Ratio of passing to failing students, 445
Student-teacher ratios, 450
Study time availability analysis, 674–676
Study time *vs.* grade, analysis of, 170, 271
Textbook cost *vs.* size, 161
Tuition costs, trends in, 310

ENVIRONMENT/EARTH SCIENCE/GEOGRAPHY

Atmospheric pressure as function of altitude, 795
Average low/high temperature, 25
Average temperature by city, 714
Buried treasure search area, 691
Conversion of degrees Celsius to/from Fahrenheit, 29, 82
Hurricane intensity, 10
Pollution as function of population, 290
Rainfall amounts per hour in storm, 108
Richter scale for earthquakes, 752, 770–771, 778
Slope of hill, 140
Snowfall rate and totals, 158–159
Snowstorm duration, 158–159
Total rainfall in storm, 108
Visibility *vs.* altitude, 299

GARDENING AND LANDSCAPING

Area of garden, from side length, 390
Area of walkway and garden as function of walkway width, 323, 390
Cost of edging garden, 501
Dimensions of area enclosable with available fencing/edging, 69
Dimensions of garden, from perimeter, 68, 99
Height of tree, from shadow cast, 446
Perimeter of garden, 501
Perimeter of garden as function of side length, 313
Radius of garden, from area, 620

GEOMETRY

Angle measure of supplementary angles, 65, 69, 214, 217, 694
Angle measures of complementary angles, 65, 69, 103, 217, 244, 246
Angle measures of triangle, 69–70, 103, 214–215, 217, 222–223, 226, 796, 797
Area as function of side length, 324, 368, 380, 390, 415
Area of region, 29, 38, 102, 509, 522, 796
Center of circle, finding, 113
Dimensions of region, from area, 2, 66, 304, 344, 374, 380, 381, 462, 479, 487, 562, 617, 620, 694
Dimensions of region, from perimeter, 2, 64, 65, 68, 69, 99, 103, 106, 222, 226, 244, 375, 381, 462, 796
Dimensions of region, from volume, 66, 183, 563, 574, 575
Height of triangle, 380, 617, 620
Length of triangle side, 2, 68, 222, 226, 244, 375, 380, 381, 472–473, 494, 554, 575, 620, 796
Perimeter as function of side length, 309, 313, 390, 425–426
Perimeter of figure, 501
Pythagorean Theorem, 375, 472–473, 477, 478, 494, 501, 528
Radius of circle, from area, 620
Radius of sphere, from volume, 484, 487, 549
Similar triangles, 445–446, 451, 461
Slope of figure sides, 796
Surface area as function of dimensions, 291
Volume as function of side length, 320, 324
Volume of figure, 30, 301, 522–523

INVESTMENT

Amount invested, from return, 55, 56–57, 62, 98, 103, 106, 186, 212–213, 216, 217, 240, 244, 344, 622, 694
Bond yield *vs.* price, 286
Interest
  compound, 721, 727, 761, 762, 765, 780, 781, 792, 795, A25, A39
  simple, 54, 55, 56–57, 62, 82, 98, 103, 106, 287, 302, 344, 484, 487, 546, 562–563
Savings deposits per month, 111–112
Savings totals with monthly deposits, 158
Stock price by month, 111
Stock price trends, 606
Time required to reach investment goal, 82, 158

POLITICS AND PUBLIC POLICY

Child support due *vs.* child support paid, 705–706
Margin of error in polls, 667
Minimum wage trends, 159–160
National debt per capita, 88–89
Number living below poverty line, 62
Percentage of votes received by candidates, 61
Poverty threshold calculation, 82
Poverty threshold trends, 623
Ratio of Democrats to Republicans in Senate, 444–445
Ratio of male to female police officers, 445
Representatives per state, 252, 714
States *vs.* year of statehood, 255

SCIENCE AND TECHNOLOGY

Amount of substances in mixture, 451
Atoms in substance, number of, 92
Conversion of degrees Celsius to/from Fahrenheit, 29, 82
Depth of water, from wave velocity, 548
Distance from earth to celestial objects, 88–89, 93
Distance of free fall, from velocity, 526
Distance of lightning strike, calculating, 166
Distance traveled by bouncing ball, A35–A36
Electrical current *vs.* voltage and resistance, 290–291
Electrical resistance *vs.* wire size, 291
Force on spring *vs.* distance stretched, 299
Force required to stretch spring, 166, 279
Frequency *vs.* length in vibrating string, 291
Height *vs.* time in free fall, 270, 377, 527
Impurities in soap, 61
Intensity of light as function of distance, 250, 290
Kinetic energy *vs.* speed, 287–288
Length of pendulum, from speed of swing, 526
Margin of error in measurement, 664, 693
Mixing solutions, 57–58, 62, 63, 103, 211–212, 216, 244, 246, 396, 796
Parabolic mirrors, 812–813
Pendulum period *vs.* length, 291, 300
pH of substances, 15, 735–736, 741, 791
Radioactive decay, 721, 722, 726, 759, 775–776, 780, 781, 790, 792–793, A25
Rat-in-maze performance data, analysis of, 694
Sequence of heights in bouncing ball, A22
Slope of airplane climb, 140
Sound intensity in decibels, 752, 778
Sound intensity *vs.* distance, 286–287
Speed of car based on skid mark length, 523–524
Speed *vs.* time in free fall, 279
Swing angle sequence of pendulum, A22
Temperature decay in liquids, 721
Time required for object to fall, 517, 523, 559–560, 562, 563
Time required for pendulum swing, 517, 526
Trajectory of missile/projectile, 314, 377–378, 382, 392, 394, 554, 568, 575, 604, 609, 620, 621, 622
Unit conversions, 715
Visibility *vs.* altitude, 299
Volume of gas as function of temperature and pressure, 798
Wave velocity in water, 548
Weight of ball as function of radius, 291

SPORTS/EXERCISE/ENTERTAINMENT

Buried treasure search area, 691
Coordinates of camping area, 113
Dimensions of court, 68, 374
Distance from home plate to second base, 494
Earnings per year of athlete, 461
Golf hole pars, from relative number of pars, 226
Halfway point between hikers, 113
Height of kite, 494
Men's pole vaulting heights, trends in, 168
Men's swimming times, trends in, 168–169
Neilsen ratings of TV shows, 690
Number of dominoes in an array, A38
Points scored, by type of shot, 623
Points scored, from relative scores, 395
Projected yardage gain from number of passes, 460
Projector image size *vs.* distance, 290
Slope of treadmill, 140
Sports participation by age group, 109
Water needed per time hiked, 286
Women's swimming times, trends in, 161–162
Women's track and field records, trends in, 258
Yards rushed *vs.* margin of victory, 180

STATISTICS AND DEMOGRAPHICS

Alcohol-related deaths, trends in, 103
Animal population estimates, by sampling, 451
Child support due *vs.* child support paid, 705–706
Child support paid, trends in, 314
Crime trends, 393
Days required to count to one million, 93
Employment trends, 138–139
Estimating ratio of men to women, 451
Growth in passenger cars, 78–79
Height of men, deviations in, 688
Height of stacked pennies, 89
Margin of error in polls, 667, 690
Mean age of population, A27
Mean protein content of foods, A27
Miles driven per year, trends in, 167
Number of lawyers in U.S., trends in, 575
Numbers, from relative size, 2, 52–54, 60–61, 103, 106, 186, 301, 373–374, 380, 381, 443, 632, 636, 688, 798, 859
Outliers in data analysis, 688
Population density, 93
Population density trends, 138
Population growth trends, 78, 258, 314, 394, 395, 578, 618, 621, 706, 721, 722–723, 727, 774–775, 779
Prison population as function of time, 797
Soft drink consumption per capita, trends in, 300
Sports participation by age group, 109
Standard deviation, A27
Water consumption per capita, trends in, 298
Women owed child support, trends in, 314

Index

A

Abscissa, 110
Absolute value
  applications, 663–664
  on calculator, 16
  on number line, 16, 95
  in order of operations, 23, 95
  of real number, 16, 95
  translating to/from words, 663–664
Absolute value equations
  definition of, 649
  solving
    on calculator, 651, 653
    one absolute value, 649–652
    steps in, 650, 685
    two absolute values, 652–653
Absolute value inequalities
  definition of, 656
  solving
    on calculator, 658, 659, 663
    by converting to compound inequality, 656–660
    by test point method, 660–663
    steps in, 660, 686
ac method. *See* Grouping method, factoring trinomials by
Addition
  associative property of, 32
  commutative property of, 32
  of complex numbers, 533–534, 545
  distributive property of multiplication over. *See* Distributive property of multiplication over addition
  of functions, 699–700
  identity property of, 32
  inverse property of, 32
  of like terms, 33–34, 96
  in order of operations, 23–24, 95
  of polynomials, 306, 384
  of radical expressions, 496–498, 542
  of rational expressions
    identifying least common denominator, 417–418
    like denominators, 416–417, 455
    steps in, 420, 455
    unlike denominators, 417–422, 455
    writing equivalent expressions, 418–419
  of real numbers, 17–18, 95
Addition method
  for solving nonlinear systems of equations, 834–836, 852
  for solving system of linear equations, 203–207
  steps in, 204, 239
Addition property of equality, 40
Addition property of inequality, 73
Algebra
  of functions, 699–700, 702–703, 783
  translating words into/out of. *See* Translating to/from words

Algebraic expressions
  definition of, 31
  evaluation of, 25
  simplification of, 34–35
  terms of. *See* Terms
Alternating sequences, A20–A21
"And." *See* Intersection of sets
Angles
  complementary
    definition of, 99
    finding measure of, 65
  supplementary
    definition of, 99
    finding measure of, 65
  of triangle, sum of, 99
Applications. *See also* Applications Index
  absolute value, 663–664
  complex fractions, 432
  compound inequalities in one variable, 632, 635–636
  compound inequalities in two variables, 674–676
  consecutive integers, 53–54, 60–61
  distance, rate, and time, 58–59, 63, 213–214, 217
  exponential equations, 774–776, 779–781
  exponential functions, 721–723, 726–727
  geometry, 64–65, 68–70, 214–215, 217
  graphs as, 109
  interest and principal, 56–57, 62, 212–213, 216–217
  linear equations in one variable, 52–59
  linear equations in three variables, 224, 226–227
  linear equations in two variables, 158–162, 164–170, 176
  linear inequalities in one variable, 77–79, 82–83
  linear models, 158–162, 164–170, 176
  logarithmic equations, 770–771, 778
  logarithmic functions
    common, 735–736, 741–742
    natural, 759
  mixtures, 56, 57–58, 62–63, 211–213, 216
  multiplication of polynomials, 320, 323
  percents and rates, 54–56, 61–62
  polynomial functions, 309–310, 313–314
  proportions, 444–445, 450–451
  Pythagorean theorem, 375, 472–473
  quadratic equations, 373–375, 380–381, 567–568, 575
  quadratic formula, 567–568, 575
  quadratic functions, 377–378, 382–383, 604–606, 609–611
  radical equations and functions, 522–524, 526–527
  rational equations, 446–449, 452–453
  rational exponents, 484, 487
  relations in x and y, 254, 257–258
  scientific notation, 88–89, 92–93
  series, A23
    geometric, A35–A36
  similar triangles, 446

  slope, 132–133, 138–139, 140, 142–143
  system of linear equations in two variables, 210–215, 216–218, 240
  variation, 285–288, 290–291
    steps for solving, 285, 295
  vertex of parabola, 604–606, 609–611
Approximately equal to, 6
Area, formulas for, 99
Argument, of logarithmic expressions, 729
Arithmetic sequence, A28–A30
  common difference of, A28
  definition of, A28
  nth term of, A29–A30
Arithmetic series, A30–A31
Associative property of addition, 32
Associative property of multiplication, 32
Asymptotes
  of exponential function, 274, 735, 785
  horizontal, 274
  of hyperbola, 824–825
  of logarithmic function, 274, 785
  vertical, 274
"At least," 6, 10
"At most," 6, 10
Augmented matrices, 228–229, 242
  equivalent, 229–230
Axis (axes), of rectangular coordinate system, 109, 172
Axis of symmetry, of parabola, 586, 813, 815, 850
  finding, 591–593, 600–602, 616
  horizontal, 814–817, 850
  vertical, 812–814, 850

B

Base
  of exponential expression, 21, 95
    division of like, 84, 483
    multiplication of like, 84, 483
  of logarithmic expressions, 729
"Between," 10, 94
Binomial(s)
  definition of, 305, 384
  factoring, 360–365, 388
    difference of cubes, 362–363, 388
    difference of squares, 360–362, 388
    form $x^6 - y^6$, 365
    summary of, 363–364, 369
    sum of cubes, 362–363, 388
  multiplication of, FOIL acronym for, 317
  square of. *See* Perfect square trinomials
Binomial expansion, A1–A2
  binomial theorem for, A4–A6
  finding specific term of, A5–A6
Binomial theorem, A4–A6
Boundary points, of inequality, 638
Braces
  in order of operations, 23
  in set notation, 3
Brackets
  in interval notation, 7–8
  order of operations and, 23

C

Calculator
  absolute value on, 16
  absolute value equations on, 651, 653
  absolute value inequalities on, 658, 659, 663
  change of base formula on, 758
  determinants on, A11
  domain of radical functions on, 474–475
  e^x, evaluating on, 753–754
  exponential expressions, approximating on, 719
  factorial notation on, A3
  functions on, evaluating, 263
  graphs and graphing on, 125
    circles, 805
    Dot Mode, 407
    Intersect feature, 191–192, 521, 833
    maximums and minimums of parabola, 605
    nonlinear systems of equations, 833, 835
    parabolas, 816
    rational functions, 407
    square window option, 150
    standard viewing window, 125
    systems of linear equations, 191–192
    Trace feature, 148, 191–192, 521, 605, 833
    Value (Eval) function, 148, 263
    Zoom feature, 521, 605, 833
  grouping symbols on, 24
  inequalities on, 641, 642, 643
  logarithmic expressions on
    common, 731, 757
    natural, 757
  matrices on, 233
  n^{th} roots on, 470
  quadratic formula on, 566
  radical equations, solving, 521
  rref. function, 233
  scientific notation on, 88
  sequences on, A20
  series on, A23
  square root on, 23, 468
  Table feature, 160, 263, 407
Center, of circle, 804
Change of base formula, 757–758, 787
Circle (geometric figure), 804–806
  center of, 804
  as conic section, 812
  definition of, 804
  equation of, standard form, 804, 806–807, 849
  graphing of, 805–806
  radius of, 804
Circle (in interval notation), 7
Clearing parentheses, 34–35
Closed circle, in interval notation, 7
Closed interval, 7
Coefficient
  leading, of polynomial, 305, 384
  of polynomial term, 305, 384
  of term, 31, 96
Coefficient matrices, 228
Column matrices, 228
Column method, for multiplication of polynomials, 316
Commission, formula for, 54, 98
Common difference, of arithmetic sequences, A28
Common logarithmic functions, 731, 785
  applications of, 735–736, 741–742
Common ratio, of geometric sequence, A32
Commutative property of addition, 32
Commutative property of multiplication, 32
Complementary angles
  definition of, 99
  finding measure of, 65
Completing the square
  in solving quadratic equations, 556–559, 613
  steps in, 557, 613
  to write quadratic function in form $a(x - h)^2 + k$, 600–602, 617
Complex fractions
  applications, 432
  definition of, 427
  simplifying
    method I, 427–429, 456
    method II, 429–432, 456
Complex numbers, 532–536, 545
  addition of, 533–534, 545
  conjugates, 532, 535, 545
  definition of, 532, 545
  division of, 535, 545
  imaginary part of, 532, 545
  multiplication of, 533–534, 545
  real part of, 532, 545
  simplifying, 535–536
  standard form of, 532
  subtraction of, 533–534, 545
Composition of functions, 700–701, 783
  inverse functions, 712
Compound inequalities in one variable
  applications of, 632, 635–636
  solving, 76–77, 627–632, 683
Compound inequalities in two variables
  applications of, 674–676
  solving, 670–674
Compound interest, 754–755, 787
Conditional equations, 46–47
Conic sections, 812
Conjugates
  complex numbers, 532, 535, 545
  multiplication of, 317–319, 384, 505–506
Consecutive integers, applications involving, 53–54, 60–61
Consistent system of linear equations
  in three variables, definition of, 219–220
  in two variables, definition of, 188, 237
Constant function, 272, 276, 294
Constant of variation *(k)*
  definition of, 283–284
  solving for, 285–288
Constant term
  of algebraic expression, 31, 96
  of trinomial, 556
Continuous compounding, 754–755, 787
Contradictions, 46–47, 97
Coordinate, first (abscissa), 110
Coordinate, second (ordinate), 110
Cost, applications, 210–211, 215–216. *See also Applications Index*
Cramer's rule, solving systems of linear equations with, A12–A16
Cube(s)
  difference of, factoring, 362–363, 388
  perfect
    of numbers, 470
    of variable expressions, 472
  sum of, factoring, 362–363, 388
Cube roots, 469
  of variable expressions
    definition of, 470–471
    simplifying, 471–472
Curie, Marie, 721

D

Decay functions, exponential, 720–721
  applications, 721, 722
Decimals
  clearing of, in solving linear equations, 43, 45–46
  non-terminating, non-repeating, 4
Degree
  of polynomial, 305, 384
  of polynomial term, 305, 384
Dependent system of linear equations
  in three variables
    definition of, 220, 241
    solving, 223
  in two variables
    definition of, 188, 237
    solving
      addition method, 206, 239
      with Cramer's rule, A15–A16
      Gauss-Jordan method, 232
      graphing method, 189, 191
      substitution method, 200
Determinants
  of 2×2 matrix, A8
  of 3×3 matrix, A9–A10
  definition of, A8
  evaluating
    2×2, A8–A9
    3×3, A10–A12
  sign of terms, determining, A11
Difference of cubes, factoring, 362–363, 388
Difference of squares, 317–319, 384
  factoring, 360–362, 388
  of radicals, 505–506
Directrix, of parabola, 812, 850
Direct variation
  applications, 286
  definition of, 283–284, 295
Discriminant, of quadratic equation, 568–571, 614
Distance
  expressing with absolute value, 663–664
  rate and time problems, 58–59, 63, 213–214, 217 (*See also Applications Index*)
Distance formula, 803–804, 849
Distributive property of multiplication over addition, 32–34, 96
Division
  of complex numbers, 535, 545
  of factorials, A3
  of functions, 699–700
  of like bases, 84, 483
  in order of operations, 23–24, 95
  of polynomial, 325–331, 385
    long division, 325–329, 331, 385
    by monomial, 325
    synthetic division, 329–331, 385
  of rational expressions, 412–413
  of real numbers, 20–21, 95
  by zero, 20, 95
  of zero, 20, 95
Division property of equality, 40
Division property of inequality, 74

Division property of radicals
    definition of, 488, 541
    negative radicands and, 529–530
    simplifying radicals with, 491–492
Domain
    of exponential function, 719, 720, 735, 785
    of function, 264–265, 293
    of inverse function, 707
    of logarithmic functions, 732, 734, 735, 785
    of rational expression
        definition of, 399, 453
        finding, 399–400, 453
    of relation, 251–254, 292

E

e (irrational number)
    definition of, 753, 787
    graphing of e^x, 753–754
Elementary row operations, for matrices, 229–230, 242
Elimination method. See Addition method, for solving system of linear equations
Ellipse
    as conic section, 812
    definition of, 821, 851
    equation of, standard form, 821–823, 851
    foci of, 821
    graphing of, 822–823, 829–830
    intercepts of, 822, 851
Equality
    addition property of, 40
    division property of, 40
    multiplication property of, 40
    subtraction property of, 40
Equal to, notation for, 6
Equation(s). See also Absolute value equations; Exponential equations; Linear equations; Logarithmic equations; Quadratic equations; Rational equations
    conditional, 46–47
    contradictions, 46–47, 97
    definition of, 39
    equivalent, 40
    identities, 46–47, 97
    in quadratic form
        definition of, 579
        solving, 578–582, 615
    solution to, defined, 39
Equivalence property of exponential expressions, 771, 788
Equivalent equations, 40
Expanded form
    of common logarithm, 746–747, 786
    of natural logarithm, 757
Expanding minors, A10–A12
    sign of terms in, A11
Exponent (power)
    of exponential expression, 21, 95
    negative, 84–85, 541
    zero as, 84, 541
Exponential equations
    applications, 774–776, 779–781
    definition of, 771
    solving, 771–774
        steps in, 773, 788
Exponential expressions, 21–22, 95. See also Rational exponents
    approximating, on calculator, 719
    base of. See Base

division of like bases, 84, 483
equivalence property of, 771, 788
exponent (power) of. See Exponent
multiplication of like bases, 84, 483
order of operations and, 23–24, 95
power of a product, 84, 483
power of a quotient, 84, 483
power rule, 84, 483
powers of i, 531–532
properties of, 84–85, 101, 482–483
simplifying, 85–86
Exponential functions, 718–723, 785
    applications of, 721–723, 726–727
    converting logarithmic functions to, 729–730
    decay functions, 720–721
        applications, 721, 722
    definition of, 718
    domain of, 719, 720, 735, 785
    graphing of, 719–721, 732, 735
    growth functions, 720–721
        applications, 721, 722–723
    inverse of, 729, 732
    range of, 720, 735, 785
Expressions. See Algebraic expressions
Extraneous solution, to radical equation, 518

F

Factor
    greatest common (GCF), of polynomial
        definition of, 336, 386
        factoring out, 336–337, 386
            binomial factors, 337
            negative factors, 337
    of term, 31
Factorial notation, A2–A3
Factoring
    binomials, 360–365, 388
        difference of cubes, 362–363, 388
        difference of squares, 360–362, 388
        of form $x^6 - y^6$, 365
        summary of, 363–364, 369
        sum of cubes, 362–363, 388
    integers, 336
    polynomials
        binomial factors, 337
        checking factorization, 336
        definition of, 336
        greatest common factor, 336–337, 386
        grouping method, 338–340, 361–362, 386
        negative factors, 337
        steps, 369
        substitution method, 354
    trinomials, 345–356, 387
        grouping method (ac method), 345–348
            steps in, 345, 387
        with leading coefficient of 1, 352–353
        perfect square trinomials, 353–354, 387
        steps in, 355–356, 369
        substitution method, 354
        summary of, 355–356, 369
        trial-and-error method, 348–352
            steps in, 349, 387
Feasible regions, graphing, 674–676
Fibonacci sequence, A27
Fifth powers, perfect, 470
Finite sequences, A19–A20
First coordinate (abscissa), 110
First-degree polynomial equations. See Linear equations in one variable

Focus (foci)
    of ellipse, 821
    of hyperbola, 823, 851
    of parabola, 812, 850
FOIL acronym, 317
Force, formula for, 72
Form fitting, in solving literal equations, 67
Formulas. See Literal equations
Fourth powers, perfect, 470
Fractions
    clearing of, in solving linear equations, 43–45
    complex. See Complex fractions
    in order of operations, 23
Function(s). See also Exponential functions; Inverse functions; Logarithmic functions; Polynomial functions; Quadratic functions; Radical functions; Rational functions
    addition of, 699–700
    algebra of, 699–700, 702–703, 783
    composition of, 700–701, 783
    constant, 272, 276, 294
    definition of, 258–259, 293
    division of, 699–700
    domain of, 264–265, 293
    evaluating
        algebraically, 262–264
        on calculator, 263
        from graph, 264
    graphs of basic functions, 273–275, 294
    linear, 272, 276, 294
    multiplication of, 699–700
    notation, 261–262, 293
    one-to-one
        definition of, 707
        horizontal line test for, 707–708, 784
    range of, 264
    subtraction of, 699–700
    vertical line test for, 260–261, 293
    x-intercept of, 276–278, 294
    y-intercept of, 276–278, 294
Fundamental principle of rational expressions, 401, 403

G

Gauss-Jordan method, 229–233, 242
    steps in, 231
GCF. See Greatest common factor (GCF)
Geometric sequences, A32–A33
    common ratio of, A32
    definition of, A32
    nth term of, A32–A33
Geometric series, A33–A36
    applications of, A35–A36
    infinite, sum of, A35
Geometry
    applications, 64–65, 68–70, 214–215, 217
        (See also Applications Index)
    dimensions of figures, solving for, 65–67
    formulas, 99
        evaluation of, 25
Graphs and graphing. See also Rectangular coordinate system
    circles, 805–806
    ellipses, 822–823, 829–830
    exponential functions, 719–721, 732, 735
    feasible regions, 674–676
    of functions, basic, 273–275, 294
    functions, evaluating by, 264
    horizontal lines, 124–125

hyperbolas, 823–825, 851
inequalities, solving by, 636–638
interpretation of, 111–112
linear equations in three variables, 219, 241
linear equations in two variables, 119–125, 173
linear inequalities in two variables, 668–670
logarithmic functions, 732–735
nonlinear inequalities in two variables, 840–842, 853
nonlinear systems of equations, 831–834
parabola, 276, 294, 375–376, 586–593, 616
  of form $(x - h)^2$, 588–589, 616
  of form $a(x - h)^2 + k$, 591–593, 616
  of form ax^2, 589–591, 616
  of form $x^2 + k$, 586–588, 616
  horizontal axis of symmetry, 814–817
  horizontal shifts, 588–589, 616
  vertical axis of symmetry, 812–814, 815–817
  vertical shifts, 586–588, 616
  vertical stretching/shrinking, 589–591, 616
plotting points, 110–111
quadratic functions. *See* Parabola, *above*
slope-intercept form of linear equation, 146
systems of linear equations, solving by, 188–192
systems of nonlinear inequalities in two variables, 842, 853
vertical lines, 124
of *x*- and *y*-intercepts. *See* *x*-intercept; *y*-intercept
Greater than
  definition of, 6
  notation for, 6, 94
Greater than or equal to, 6, 94
Greatest common factor (GCF), of polynomial
  definition of, 336, 386
  factoring out, 336–337, 386
    binomial factors, 337
    negative factors, 337
Grouping method
  factoring polynomials by, 338–340, 361–362, 386
  factoring trinomials by (*ac* method), 345–348
  steps in, 345, 387
Grouping symbols
  on calculator, 24
  order of operations and, 23–24, 95
Growth functions
  applications, 721, 722–723
  exponential, 720–721

H

Half-life, 721, 722
Horizontal asymptotes, 274, 735, 785
Horizontal line(s)
  equation of, 123–124, 151, 173, 175
  in function notation, 272
  graphing of, 124–125
  slope of, 135
Horizontal line test, 707–708, 784
Hyperbola
  as conic section, 812
  definition of, 823, 851
  focus (foci) of, 823, 851
  graph of, 823–825, 851
  transverse axis of, 824–825
  vertex of, 823–825

I

i (imaginary number)
  definition of, 529
  powers of, 531–532
  simplifying expressions in terms of, 529–530
Identities, 46–47, 97
Identity property of addition, 32
Identity property of multiplication, 32
Imaginary numbers, 529, 532
Imaginary part, of complex number, 532, 545
Inconsistent system of linear equations
  in three variables
    definition of, 219, 241
    solving, 223–224
  in two variables
    definition of, 188, 237
    solving
      addition method, 206–207, 239
      Gauss-Jordan method, 232–233
      graphing method, 189, 190
      substitution method, 199–200
Independent system of linear equations
  in three variables, definition of, 219
  in two variables, definition of, 188, 237
Index, of radical, 469
Index of summation, A23
Inequality(ies), 5–6. *See also* Compound inequalities; Linear inequalities; Nonlinear inequalities in two variables; Polynomial inequalities; Quadratic inequalities
  addition property of, 73
  boundary points of, 638
  division property of, 74
  multiplication property of, 74
  solving
    on calculator, 641, 642, 643
    graphically, 636–638
    special case solution sets, 642–644
    with test point method, 638–642
    steps, 639, 684
  subtraction property of, 73
  translating to/from words, 10
Inequality signs, 6
Infinite sequences, A19–A20
Infinity, symbol for, 7
Integers
  as complex numbers, 532
  consecutive, applications involving, 53–54, 60–61
  definition of, 4, 94
  factoring of, 336
  as subset of real numbers, 3, 5
Interest
  compound, 754–755, 787
  continuous compounding, 754–755, 787
  simple
    applications, 56–57, 62, 212–213, 216–217
    formula for, 54, 98
Intersect feature, on graphing calculator, 191–192, 521, 833
Intersection of sets, 8–10, 94
  definition of, 8, 627
  notation for, 627
  solving
    one variable, 627–630, 683
    two variables, 670–674
Interval
  closed, 7
  open, 7
Interval notation, 7–8, 94
Inverse functions, 707–712
  definition of, 707, 711–712, 784
  domain and range of, 707
  finding equation of, 708–711
    steps in, 709, 784
  graphs of, 709, 710, 784
  as one-to-one function, 707–708, 784
    restricting domain to achieve, 710–711
Inverse property of addition, 32
Inverse property of multiplication, 32
Inverse variation
  applications, 286–287
  definition of, 283–284, 295
Irrational numbers
  as complex numbers, 532
  decimal form of, 4, 468
  definition of, 4, 94, 468
  square roots and, 468
  as subset of real numbers, 3, 5

J

Joint variation
  applications, 287–288
  definition of, 284–285, 295

L

LCD. *See* Least common denominator
Leading coefficient, of polynomials, 305, 384
Leading term, of polynomial, 305, 384
Least common denominator (LCD), of rational expressions, 417–418
Less than
  definition of, 5–6, 94
  notation for, 5–6, 94
Less than or equal to, 6, 94
Like terms
  addition and subtraction of, 33–34, 96
  combining, 33–36, 96
Line(s). *See also* Linear equations in two variables
  horizontal
    equation of, 123–124, 151, 173, 175
    in function notation, 272
    graphing of, 124–125
    slope of, 135
  parallel
    in slope-intercept form, 146–147, 149–150
    slope of, 136–137, 174
  perpendicular
    in slope-intercept form, 146–147, 150
    slope of, 136–137, 174
  slope of. *See* Slope
  vertical
    equation of, 123–124, 151, 173, 175
    graphing of, 124
Linear equations, forms of, 151
Linear equations in one variable, 39–47, 97
  applications, 52–59
  definition of, 39, 97
  solving, 40–46
    decimals, clearing of, 43, 45–46
    fractions, clearing of, 43–45
    steps for, 40, 97

Linear equations in three variables. *See also* System of linear equations in three variables
  applications, 224, 226–227
  definition of, 219, 241
  graph of, 219, 241
  solution to, defined, 219, 241
Linear equations in two variables, 118–125. *See also* Line(s); System of linear equations in two variables
  applications, 158–162, 164–170, 176
  definition of, 118, 173
  graphing of, 119–125, 173
  slope-intercept form. *See* Slope-intercept form
  solution to, defined, 118
  solving for y, 67
  standard form, 118, 145, 151, 175
  x- and y-intercepts of, 121–123, 173
Linear function, 272, 276, 294
Linear inequalities in one variable
  applications, 77–79, 82–83
  definition of, 73, 100
  solving, 73–77
    compound inequalities, 76–77, 628–632
    test point method, 76
Linear inequalities in two variables
  applications, 674–676
  definition of, 668, 687
  graphing, 668–670
  solution to, defined, 668
  solving, 668–670
    compound, 670–674
Linear models, 158–162
  applications, 158–162, 164–170, 176
  interpretation of, 159–160
  writing, 158–159, 176
    from observed data, 161–162
Linear term, of trinomial, 556
Literal equations (formulas). *See also* Quadratic formula
  change of base formula, 757–758, 787
  commission, 54, 98
  definition of, 65, 99
  distance formula, 803–804, 849
  force, 72
  geometric, 99
    evaluation of, 25
  interest
    compound, 754–755, 787
    simple, 54, 98
  midpoint formula, 112–113, 172
  perimeter, 99
  point-slope formula, 147–150, 151, 175
  rational expressions in, 439–440
  sales tax, 54, 98
  slope, 134–135
  solving
    with square root property, 559–560
    for variable, 65–68
      form fitting in, 67
  vertex formula, 602–604, 617, 817–818
  z-score formula, 72
Logarithm, of logarithmic expression, 729
Logarithmic equations
  applications, 770–771, 778
  definition of, 767
  solving, 767–770
    steps in, 767, 788
Logarithmic expressions
  argument of, 729
  base of, 729
  change of base formula for, 757–758, 787
  evaluating, 730–731
  expanding into sum or difference, 746–747, 757, 786
  logarithm of, 729
  properties of, 744–746, 786
  reducing sum or difference into single, 747–748, 756–757, 786
  simplifying, 744, 756
Logarithmic functions, 729–736, 785
  common, 731, 785
    applications of, 735–736, 741–742
  converting to exponential form, 729–730
  definition of, 729–730
  domain of, 732, 734, 735, 785
  graphing of, 732–735
  inverse of, 729, 732
  natural, 755, 787
    applications, 759
    graphing of, 755–756
    properties of, 756
  range of, 732, 735, 785
Long division, of polynomials, 325–329, 331, 385

M

Magnitude, definition of, 87
Matrix (matrices)
  augmented, 228–229, 242
    equivalent, 229–230
  on calculator, 233
  coefficient, 228
  column, 228
  definition of, 228, 242
  determinant of. *See* Determinants
  elementary row operations, 229–230, 242
  minor of element of, A9–A10
  order of, 228, 242
  reduced row echelon form of, 230, 242
  row, 228
  solving systems of linear equations with, 229–233, 242
    steps in, 231
  square, 228
  systems of linear equations as, 228–229, 242
Maximum value, of parabola, 276, 586, 592–593, 600–602, 616
Measurement error, expressing with absolute value, 664
Midpoint, definition of, 112
Midpoint formula, 112–113, 172
Minimum value, of parabola, 276, 586, 592–593, 600–602, 616
Minors
  definition of, A9–A10
  expanding, A10–A12
  sign of terms in, A11
Mixture problems, 56, 57–58, 62–63, 211–213, 216. *See also* Applications Index
Monomials
  definition of, 305, 384
  division of polynomial by, 325
  multiplication of
    by monomial, 315
    by polynomial, 315
Multiplication
  associative property of, 32
  commutative property of, 32
  of complex numbers, 533–534, 545
  distributive property over addition. *See* Distributive property of multiplication over addition
  of factorials, A3
  of functions, 699–700
  identity property of, 32
  inverse property of, 32
  of like bases, 84, 483
  of monomial
    by monomial, 315
    by polynomial, 315
  in order of operations, 23–24, 95
  of polynomials, 315–317, 384
    applications, 320, 323
    column method, 316
    difference of squares, 317–319, 384, 505–506
    perfect square trinomials, 317–319, 384
    special products, 317–319, 384
  of radical expressions
    difference of squares, 505–506
    different indices, 506, 542
    nth root of nth power, 504
    same indices, 502–506, 542
    square of binomial, 504–505
  of rational expressions, 411–412, 454
  of real numbers, 19, 95
  reciprocals, 19
  by zero, 19, 95
Multiplication property of equality, 40
Multiplication property of inequality, 74
Multiplication property of radicals
  definition of, 488, 541
  multiplication of radicals with, 502–506
  negative radicands and, 529–530
  simplifying radicals with, 489–490

N

Natural logarithmic expressions
  expanding into sum or difference, 757
  reducing sum or difference into single, 756–757
  simplifying, 756
Natural logarithmic functions, 755, 787
  applications, 759
  graphing of, 755–756
  properties of, 756
Natural numbers
  as complex numbers, 532
  definition of, 4, 94
  as subset of real numbers, 3, 5
Negative exponents, 84–85, 541
Negative numbers
  addition of, 17–18, 95
  division of, 20–21, 95
  multiplication of, 19, 95
  square root of
    definition of, 529
    as non-real number, 23, 467, 468
  subtraction of, 18, 95
Negative square root, 467
"No less than," 6, 10
"No more than," 6, 10
Nonlinear inequalities in two variables
  graphing, 840–842, 853
  system of, solving, 842, 853
Nonlinear systems of equations
  definition of, 831
  graphing, 831–834

solving
    addition method, 834–836, 852
    substitution method, 831–834, 836, 852
Notation
    absolute value of real number, 16
    approximately equal to, 6
    closed interval, 7
    composition of functions, 700
    cube root, 469
    dashed line, 668, 687
    determinants, A8
    equal to, 6
    exponents, 21
    factorial notation, A2–A3
    function notation, 261–262, 293
    for graphs of linear inequalities in two variables, 668, 687
    greater than, 6, 94
    greater than or equal to, 6, 94
    inequalities, 5–6
    intersection of sets, 627
    interval notation, 7–8, 94
    inverse functions, 707
    less than, 5–6, 94
    less than or equal to, 6, 94
    not equal to, 6
    open interval, 7
    opposite of real number, 15
    roots, 469, 482
    scientific notation, 87–89, 101
    set-builder notation, 3, 94
    sigma notation, A23–A24
    solid line, 668, 687
    square root, 22, 95, 467
    union of sets, 627
Not equal to, 6
nth roots, 469–470, 540
    of variable expressions
        definition of, 470–471
        simplifying, 471–472
nth term, of sequence
    definition of, A19
    finding formula for, A21–A22
    geometric, A32–A33
Number(s). *See also* Complex numbers; Integers; Irrational numbers; Real numbers
    imaginary, 529, 532
    natural
        as complex numbers, 532
        definition of, 4, 94
        as subset of real numbers, 3, 5
    rational
        as complex numbers, 532
        definition of, 4, 94
        as subset of real numbers, 3, 5
    whole
        as complex numbers, 532
        definition of, 4, 94
        as subset of real numbers, 3, 5
Number line. *See* Real number line

O

Observed data, linear models from, 161–162
One-to-one functions
    definition of, 707
    horizontal line test for, 707–708, 784
    restricting domain to achieve, 710–711
Open circle, in interval notation, 7
Open interval, 7

Opposite
    of polynomial, 307
    of real number, 15, 95
"Or." *See* Union of sets
Order, of matrix, 228, 242
Ordered pair(s)
    plotting, in rectangular coordinate system, 110–112
    as solution to linear equation in two variables, 118
    as solution to linear inequalities in two variables, 668
    as solution to system of linear equations, 187, 237
Ordered triple
    as solution to linear equation in three variables, 219, 241
    as solution to system of linear equation in three variables, 219, 241
Order of magnitude, definition of, 87
Order of operations, 23–24, 95
Ordinate, 110
Origin, of rectangular coordinate system, 109, 172
Outliers, 688

P

Parabola, 586–593, 616
    applications, 812–813
    axis of symmetry, 586, 813, 815, 850
        finding, 591–593, 600–602, 616
        horizontal, 814–817, 850
        vertical, 812–814, 850
    as conic section, 812
    definition of, 812, 850
    directrix of, 812, 850
    equation of, 276, 294, 375–376
        form $(x - h)^2$, 588–589, 616
        form $a(x - h)^2 + k$, 591–593, 616
            writing equation in, 600–602
        form ax^2, 589–591, 616
        form $x^2 + k$, 586–588, 616
        standard form
            horizontal axis, 815–817, 850
            vertical axis, 813–814, 850
    focus of, 812, 850
    graphing of. *See* Graphs and graphing
    horizontal shifts, 588–589, 616
    maximum/minimum value of, 276, 586, 592–593, 600–602, 616
    up-opening and down-opening, 276, 586, 591, 616
    vertex of, 276, 586, 616, 813, 850
        applications of, 604–606, 609–611
        finding, 591–593, 600–604, 616, 617, 817–818
    vertical shifts, 586–588, 616
    vertical stretching/shrinking, 589–591, 616
Parallel lines
    in slope-intercept form, 146–147, 149–150
    slope of, 136–137, 174
Parentheses
    on calculator, 24
    clearing, 34–35
    in interval notation, 7–8
    order of operations and, 23–24, 95
Pascal, Blaise, A2
Pascal's triangle, A2
Percents, applications, 54–56, 61–62

Perfect cubes
    of numbers, 470
    of variable expressions, 472
Perfect fifth powers, 470
Perfect fourth powers, 470
Perfect square(s)
    of numbers, 468
    of variable expressions, 471
Perfect square trinomials, 317–319, 384
    as binomial expansion, A1
    completing the square, 556–557
    factoring of, 353–354, 387
    radical expressions, 504–505
Perimeter, formula for, 99
Perpendicular lines
    in slope-intercept form, 146–147, 150
    slope of, 136–137, 174
π(pi), decimal form of, 4
Plane, as solution to linear equation in three variables, 219, 241
Plotting points, in rectangular coordinate system, 110–112
Point-slope formula, 147–150, 151, 175
Polynomial(s). *See also* Binomial(s); Monomials; Trinomials
    addition of, 306, 384
    definition of, 305, 384
    degree of, 305, 384
    division of, 325–331, 385
        long division, 325–329, 331, 385
        by monomial, 325
        synthetic division, 329–331, 385
    factoring
        binomial factors, 337
        checking factorization, 336
        definition of, 336
        greatest common factor, 336–337, 386
        grouping method, 338–340, 361–362, 386
        negative factors, 337
        steps in, 369
        by substitution, 354
    greatest common factor of
        definition of, 336, 386
        factoring out, 336–337, 386
    leading coefficient of, 305, 384
    leading term of, 305, 384
    multiplication of, 315–317, 384
        applications, 320, 323
        column method, 316
        difference of squares, 317–319, 384, 505–506
        perfect square trinomials, 317–319, 384
        special products, 317–319, 384
    notation for, 305
    opposite of, 307
    prime, 345, 349
    subtraction of, 306–308, 384
    term of
        coefficient of, 305, 384
        degree of, 305, 384
    translating to/from words, 319
Polynomial equations
    first-degree. *See* Linear equations in one variable
    higher-degree, solving, with zero-product rule, 373
    second-degree. *See* Quadratic equations
Polynomial functions
    applications, 309–310, 313–314
    definition of, 308
    evaluation of, 309

Polynomial inequalities, solving
  graphical method, 636–638
  test point method, 639–641, 684
  steps in, 639, 684
Power. *See* Exponent
Power of a product, 84, 483
Power of a quotient, 84, 483
Power property of logarithms, 745–746, 756
Power rule, 84, 483
Prime polynomials, 345, 349
Principal square root, 22, 95, 467, 540
Product property of logarithms, 744, 746, 756
Proportions
  applications, 444–445, 450–451
  definition of, 443, 457
  similar triangles and, 445–446
  solving, 443–444
Pythagorean theorem, 472, 540
  applications, 375, 472–473

Q

Quadrants, of rectangular coordinate system, 109, 172
Quadratic equations
  applications of, 373–375, 380–381, 567–568, 575
  definition of, 370, 388
  discriminant of, 568–571, 614
  solutions, number and type of, 568–571, 614
  solving
    with quadratic formula, 565–568, 571–572
    with square root property, 555–559, 571–572
      completing the square, 556–559, 613
      steps in, 557, 613
    summary of, 571–572
    with zero product rule, 370–373, 388, 571–572
      steps in, 371
Quadratic formula, 564, 614
  applications, 567–568, 575
  on calculator, 566
  derivation of, 564
  solving quadratic equations with, 565–568, 571–572
Quadratic functions
  applications, 377–378, 382–383, 604–606, 609–611
  definition of, 275–276, 294, 375–376, 388
  form $(x - h)^2$, 588–589, 616
  form $a(x - h)^2 + k$, 591–593, 616
    writing equation in, 600–602
  form ax^2, 589–591, 616
  form $x^2 + k$, 586–588, 616
  graph of. *See* Parabola
  vertex of, 276, 586, 616
    applications of, 604–606, 609–611
    finding, 591–593, 600–604, 616, 617
  x- and y-intercepts of, 376–377, 388, 570–571, 601–602, 603–604
Quadratic inequalities
  definition of, 636
  solving
    graphical method, 636–638
    test point method, 639–641, 684
Quadratic term, of trinomial, 556
Quotient property of logarithms, 745, 746, 756

R

Radical equations
  applications, 522–524, 526–527
  definition of, 518
  solving, 518–524, 544
    on calculator, 521
    extraneous solutions, 518
    one radical, 519–520
    steps in, 518, 544
    two or more radicals, 520–522
Radical expressions
  addition of, 496–498, 542
  division property of, 488, 541
    negative radicands and, 529–530
    simplifying radicals with, 491–492
  like, definition of, 495–496, 542
  multiplication of
    difference of squares, 505–506
    different indices, 506, 542
    nth root of nth power, 504
    same indices, 502–506, 542
    square of binomial, 504–505
  multiplication property of, 488, 541
    multiplication of radicals with, 502–506
    negative radicands and, 529–530
    simplifying radicals with, 489–490
  rationalizing the denominator
    one term, 510–513, 543
    two terms, 513–514, 543
  simplified form of, defined, 488, 510, 541
  simplifying
    with division property of radicals, 491–492
    with multiplication property of radicals, 489–490
  subtraction of, 496–498, 542
Radical functions
  applications, 522–524, 526–527
  definition of, 473, 540
  domain of, determining, 474–475
Radical sign, 22, 95, 467
  index of, 469
  radicand of, 469
Radicand
  definition of, 95, 469
  negative, 529–530
Radius, of circle, 804
Range
  of exponential function, 720, 735, 785
  of function, 264
  of inverse function, 707
  of logarithmic functions, 732, 735, 785
  of relation, 251–254, 292
Rate(s), applications, 54–56, 61–62
Rate (speed), distance, and time problems, 58–59, 63, 213–214, 217. *See also* Applications Index
Ratio, definition of, 443
Rational equations
  applications, 446–449, 452–453
  definition of, 435
  solving, 435–440
    steps in, 437, 457
Rational exponents, 480–484
  applications of, 484, 487
  converting to radical notation, 482
  definition of, 480–481, 541
  evaluation of, 480–482
  properties of, 482–483

Rational expressions
  addition of
    identifying least common denominator, 417–418
    like denominators, 416–417, 455
    steps in, 420, 455
    unlike denominators, 417–422, 455
    writing equivalent expressions, 418–419
  definition of, 399, 453
  division of, 412–413
  domain of
    definition of, 399, 453
    finding, 399–400, 453
  fundamental principle of, 401, 403
  multiplication of, 411–412, 454
  simplification of, 400–405
  subtraction of
    identifying least common denominator, 417–418
    like denominators, 416–417, 455
    steps in, 420, 455
    unlike denominators, 417–422, 455
    writing equivalent expressions, 418–419
Rational functions
  definition of, 405, 453
  evaluating, 405–407
  graphing of, on calculator, 407
Rational inequalities
  definition of, 641
  solving
    graphical method, 638
    test point method, 641–642, 684
Rationalizing the denominator
  one term, 510–513, 543
  two terms, 513–514, 543
Rational numbers
  as complex numbers, 532
  definition of, 4, 94
  as subset of real numbers, 3, 5
Real number line, 3
  absolute value on, 16, 95
  absolute value equations on, 649
  absolute value inequalities on, 656–658, 661–663
  comparing number size on, 5–6
  compound inequalities on, 627–632
  intervals on, 7–8
  linear inequalities in one variable on, 73–77
Real numbers, 3–5, 94
  absolute value of, 16, 95
  addition of, 17–18, 95
  as complex numbers, 532
  division of, 20–21, 95
  multiplication of, 19, 95
  opposite of, 15, 95
  properties of, 31–34
  subsets of, 3–5
  subtraction of, 18, 95
Real part, of complex number, 532, 545
Reciprocals, 19, 95
Rectangular coordinate system, 109, 172. *See also* Graphs and graphing
  plotting points in, 110–112
Reduced row echelon form, 230, 242
Relation in x and y
  applications, 254, 257–258
  definition of, 251, 252–253, 292
  domain of, 251–254, 292
  vs. function, 258–259
  range of, 251–254, 292